P9-CPW-237

# HUMAN PHYSIOLOGY

# Human Physiology

## Third Edition

**Stuart Ira Fox, Ph.D.**

Pierce College

WCB Wm. C. Brown Publishers

**Book Team**

Editor *Edward G. Jaffe*
Developmental Editor *Colin H. Wheatley*
Production Editor *Ann Fuerste*
Designer *David C. Lansdon*
Art Editor *Gayle A. Salow*
Photo Editor *Mary Roussel*
Permissions Editor *Vicki Krug*
Visuals Processor *Joyce E. Watters*


**WCB** **Wm. C. Brown Publishers**

President *G. Franklin Lewis*
Vice President, Publisher *George Wm. Bergquist*
Vice President, Publisher *Thomas E. Doran*
Vice President, Operations and Production *Beverly Kolz*
National Sales Manager *Virginia S. Moffat*
Advertising Manager *Ann M. Knepper*
Marketing Manager *Craig S. Marty*
Editor in Chief *Edward G. Jaffe*
Production Editorial Manager *Colleen A. Yonda*
Production Editorial Manager *Julie A. Kennedy*
Publishing Services Manager *Karen J. Slaght*
Manager of Visuals and Design *Faye M. Schilling*





Library of Congress Catalog Card Number: 89-42502

ISBN 0-697-05740-2





Printed in the United States of America by Wm. C. Brown Publishers, 2460 Kerper Boulevard, Dubuque, IA 52001

10 9 8 7 6 5 4 3 2


**Cover Concept**

Artist's representation of the fibers of a sympathetic nerve making synapses with smooth muscle cells in an arteriole and in the intestine. The foreground illustrates some of the intracellular events that are stimulated by activity of these synapses. Such events include the opening of calcium channels, allowing the inward diffusion of $Ca^{++}$, and the conversion of ATP to cyclic AMP (cAMP) and two phosphate radicals.

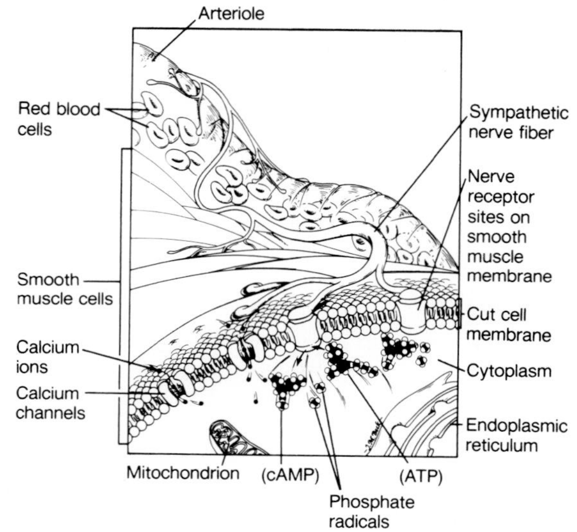

To my parents, Bess and Sam, and to the memory of my mother-in-law, Eileen

# Brief Contents

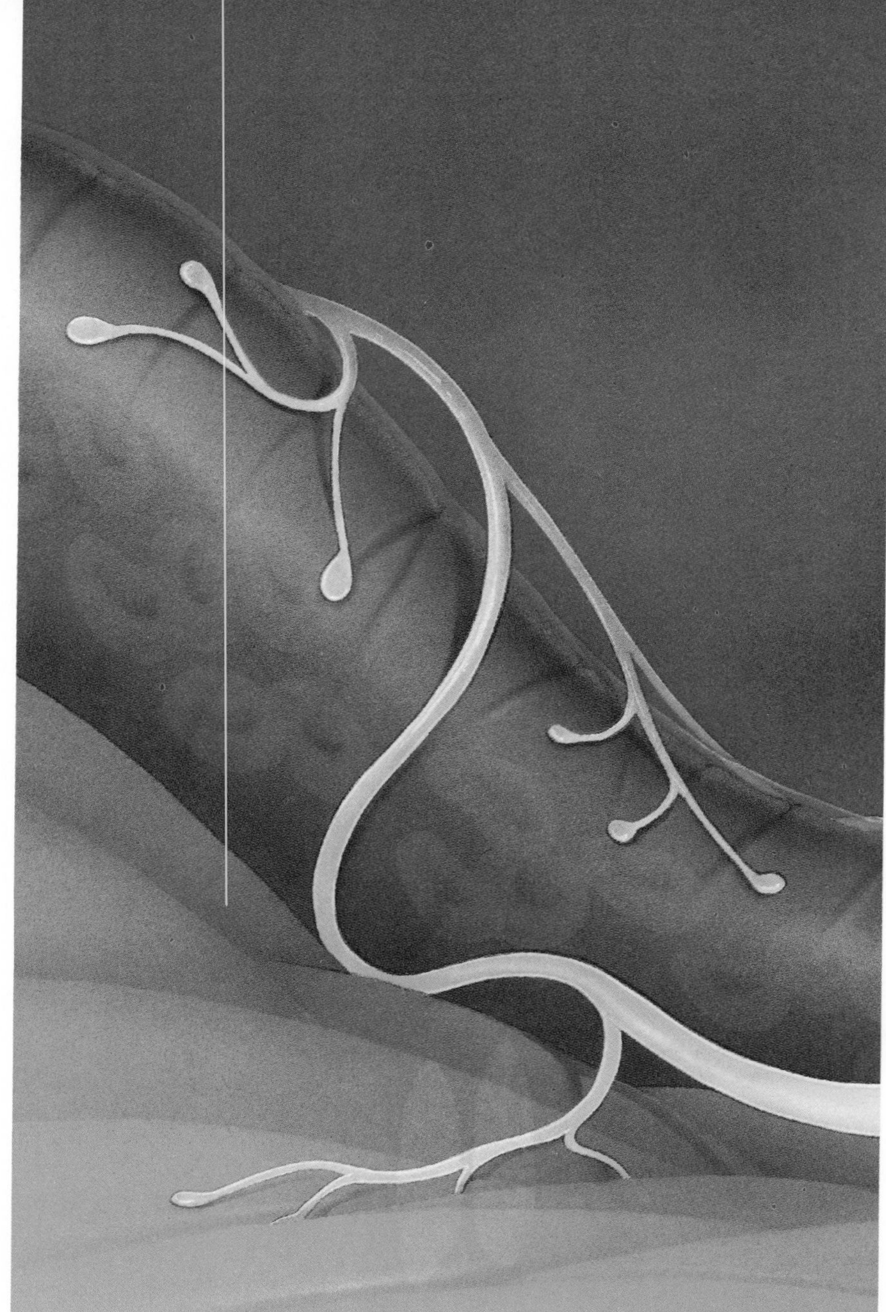

# EXPANDED CONTENTS

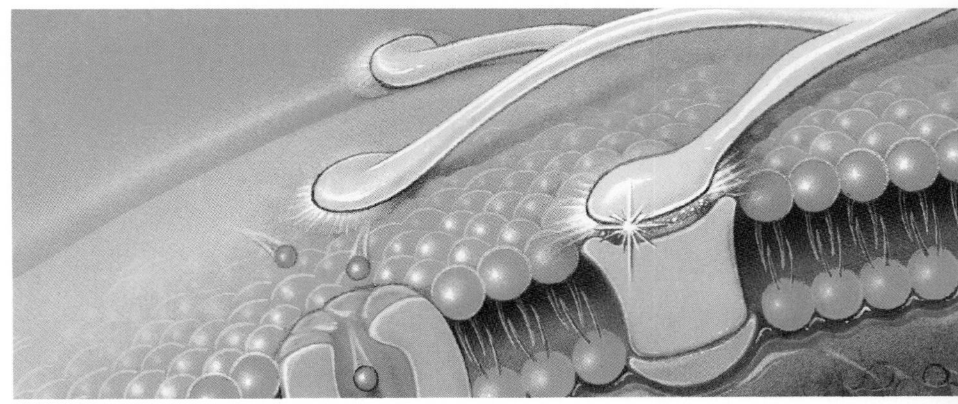

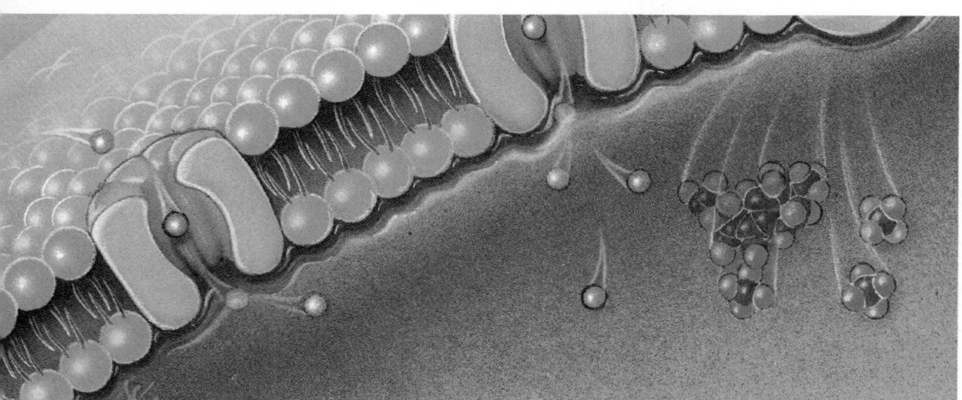

## 13 Heart and Circulation 354



## 14 Cardiac Output, Blood Flow, and Blood Pressure 396



## 15 Respiratory Physiology 436



## 16 Physiology of the Kidneys 486



## 17 The Digestive System 524



## 18 Regulation of Metabolism 562

# PREFACE

Human physiology is a subject that is fascinating and useful to almost everyone. For many students, human physiology is also a prerequisite course for their majors and intended careers. This is so for a very good reason: physiology is the major scientific foundation of medicine and all other technologies related to human health and physical performance. The scope of topics that can be covered in a human physiology course is therefore large, yet each topic must be covered in sufficient detail to provide a firm basis for future expansion and application. The scientific study of body function need not diminish the student's initial fascination with "how the body works." On the contrary, the understanding of physiological mechanisms can instill a deeper appreciation for the awesome complexity and beauty of the body.

This text is designed to serve the needs of students taking an undergraduate physiology course. Since most of these students do not have extensive science backgrounds, the beginning chapters introduce and review basic chemical and biological concepts that are needed to understand physiology. Physiological mechanisms are explained fully and in a straightforward style in order to promote understanding, rather than simple memorization, of the concepts. Summary flowcharts and tables are frequently provided to promote ease of review. Liberal use of beautifully rendered figures—in full color to enhance their value—help students to visualize the mechanisms presented. Every effort has been made to help students integrate related concepts and to understand the relationship between anatomical structures and their functions. Health applications are discussed often to heighten interest, to improve understanding of physiological concepts, and to aid the transfer of knowledge to the areas of the student's career interests. The ability of the text to serve all of these functions is aided by extensive, but not intrusive, use of a variety of pedagogical devices.

## Pedagogical Aids

The organization of this text maximizes students' ability to locate information, to associate facts in a conceptual framework, and to test their mastery of this material. The wide variety of pedagogical devices within the text further assist students to achieve these goals. These devices are described in detail under the heading Student Aids.

## Third Edition

All chapters in this text have been rewritten and revised to enhance the clarity of explanations, more fully integrate related concepts, and to maintain the currency of physiological information. In addition to the rewriting and updating of the text, the third edition has a number of new organizational features:

1. Each major heading within a chapter begins with a short paragraph that is set in a different color than the main body of the text. These are called "perspectives," and are a succinct summary of the major concepts of the sections they introduce. The perspectives help students to organize the details they will encounter around these major concepts, so that they can maintain a proper perspective on the subject covered in the sections.
2. Cross-referencing of physiological concepts has been greatly augmented within the third edition to improve the ability of students to integrate concepts presented in different sections and in different chapters.
3. Mathematical treatments of physiological concepts has been expanded in the third edition. This text now includes the Nernst equation (chapter 6), the equation for calculating the Starling forces (chapter 14), and the Henderson-Hasselbalch equation (chapter 16). These mathematical treatments are presented in such a way as to provide instructors the freedom to give these equations as much or as little attention as they deem appropriate.
4. The use of boxed information on the applications of basic physiological concepts has been greatly expanded in the third edition.
5. All of the chapters covering the nervous system (chapters 7, 8, 9, and 10) have been grouped together in the third edition. These are followed by the introductory chapter on the endocrine system (chapter 11) to allow integration of the control systems.
6. The treatment of cranial nerves has been expanded in chapter 8, and coverage of higher neural control of motor activity has been moved to this chapter.
7. The topic of the common aspects of neural and endocrine regulation has been expanded and moved to the endocrine system chapter.
8. Explanation of the physiology of cardiac and smooth muscles has been expanded and moved to the chapter on muscle physiology (chapter 12).
9. The sections on blood and blood clotting have been moved to the beginning of the chapter on the heart and circulation (chapter 13).
10. The description of blood typing has been moved from the immune system chapter to the section on blood in chapter 13.
11. The topic of the effects of exercise on calorie expenditure and the topic of thermoregulation has been added to the chapter on the regulation of metabolism (chapter 18).
12. The physiology of the lungs, kidneys, and digestive system, and the regulation of metabolism, is treated in successive chapters (chapters 15, 16, 17, and 18). The chapter covering the immune system is now chapter 19.
13. Many figures have been rerendered, new figures have been added, and most figures are now in full-color in the third edition. The judicious use of full-color figures greatly enhances their instructional value and increases the pleasure to be derived from studying physiological concepts.

## Supplementary Materials

Supplementary materials accompany the text and are designed to aid students in their learning activities and to help instructors plan coursework and presentations. These supplementary materials include:

1. An *Instructor's Manual*, including a test item file, by Stuart I. Fox. This manual can be used as an aid to planning lessons and as a bank of questions for constructing examinations.
2. *A Laboratory Guide to Human Physiology: Concepts and Clinical Applications*, fifth edition, by Stuart I. Fox. This laboratory manual is self-contained, so that students can prepare for the laboratory exercises and quizzes without having to bring the textbook to the laboratory. The manual provides exercises and instructional support for most of the topics covered in a human physiology course. These exercises have been classroom tested for a number of years.
3. An *Instructor's Manual* for the laboratory guide is available, which helps instructors set up the laboratory and which provides answers to the questions asked in the laboratory reports.
4. A *Student Study Guide to Accompany Human Physiology,* by Dr. Lawrence G. Thouin, Jr., is now available to accompany the third edition. This student study guide provides excellent sample objective questions, guides for answering essay questions, flowchart activities, crossword puzzles, and other devices to aid students in their efforts to understand the concepts of physiology.
5. A set of *100 acetate transparencies* is available to instructors who adopt this text. The transparencies are made from selected illustrations in the text and have been chosen for their value in reinforcing lecture presentations.
6. **wcb** *QuizPak*, a student self-testing program that operates on Apple® IIe or IIc, IBM® PC, or Macintosh computers, is available to instructors who adopt this text. The questions in QuizPak have been extensively revised and expanded by Dr. L. G. Thouin, Jr.
7. **wcb** *TestPak* is a computerized system that enables the instructor to make up customized exams quickly and easily. Test questions can be found in the test item file, which is printed in the Instructor's Manual.

## Student Aids

The organization and writing style of the text have been designed for optimum clarity so that you can understand, and not simply memorize, the concepts presented. Think of the sections in the text as analogous to the directions that accompany a device or appliance to be assembled. Take each concept step-by-step, reread when necessary, scrutinize the figures, and actually write out the study activities provided at the end of each major section. Do not be intimidated by a long explanation; after all, a detailed set of directions is easier to follow than one that is too brief. The more actively you interact with the information presented in the text, the better will be your understanding of physiology and the more enjoyable the study will become.

The following information about the organization and pedagogical devices in the text will help you to use this book to the best advantage.

## Chapter Outline and Objectives

Look over the chapter outline before reading the chapter to get a feel for the topics to be covered, and use the outline later to help you look up topics and integrate concepts with those that are covered later in the text. Check off the objectives as you complete each major heading to see if you are getting what is required from the text.

### Perspectives

These are the paragraphs at the beginning of each major heading that are set off in different color from the main body of the text. They are summaries of the major concepts to be presented in that section, and provide a bird's-eye view of the section. Read these carefully, because they will help you to identify the organizing concepts of the section and prevent you from becoming distracted by the details. The details provided later breathe life into these concepts, but should not be allowed to obscure the major themes covered in the sections.

### Boxed Information

Following a discussion of a basic concept in the text, you may find a colored box of text. These contain short discussions of the clinical or practical applications of the information preceding the boxes. You will find it enjoyable, as well as instructive, to see how your newly acquired basic knowledge can be applied to practical problems.

### Cross-References

Within an explanation of a particular physiological mechanism, you may find a reference to a concept that was discussed in a previous section or chapter. If you do not clearly remember this concept, look it up in the referenced chapter. This is a good way to review, and it helps you to more completely integrate related physiological concepts. You may also see a reference to a related topic that is covered in more detail in a later chapter. Go ahead—take a peek at this later chapter reference. You may not be responsible for the detail at this time, but you will be better prepared to integrate this information with previous knowledge when you reach this chapter later in the course.

### Study Activities

Each major heading in the chapters ends with a box containing study activities: pictures and flowcharts to draw, essays to answer, and other activities. The more you actually perform these activities, the better you will understand the material presented in the sections. Write these out, rather than just think about them.

### Illustrations and Tables

The text contains abundant tables and illustrations to support the concepts presented. Careful study of the table will help you to understand the text more completely, and the summary tables will be useful when you review for examinations. Though many of the figures are visually beautiful, they are constructed with one primary purpose—to illustrate concepts presented in the text. Therefore, analyze and try to understand each figure as it is referenced in the text. This will probably require rereading of sentences, but it is the only way to derive maximum benefit from the information presented.

### Chapter Summaries

At the end of each chapter, the material is summarized for you in outline form. This outline summary is organized by major headings followed by the major points of textual information. Read the summary after studying the chapter to be sure that you have not missed any points, and use the chapter summaries to help you review for examinations.

### Review Activities

Following each chapter summary is a section called Review Activities. These sections contain objective questions (with the answers in the Appendix) and essay questions. The first essay question in each chapter is answered in the Student Study Guide. Be sure to take these self-quizzes in a "closed book" fashion after studying the chapters, and then correct your answers using the Appendix. Review the information that relates to any missed questions. These practice exams will help you to anticipate the types of questions that could be on real exams, and they provide a "reality check" so you can be sure you learned the required information.

### Appendix

The Appendix contains the answers to the objective questions in the Review Activities at the end of each chapter. Be sure to look at these answers only after attempting the questions.

### Glossary

The Glossary provides definitions of the more important terms used in the text. Whenever you encounter an unfamiliar term or would like additional information about a term, look it up in the Glossary.

### Index

The Index at the end of the book allows you to locate the pages on which specific terms and concepts are discussed. Use the Index as frequently as needed; it is particularly useful when you are studying for exams and as a reference source in your later coursework.

### Student Study Guide

Written by Dr. Lawrence G. Thouin, Jr., this is an optional book that can help you to derive more benefit from the text. The answer to the first essay question in the Review Activities at the end of each chapter is provided here, along with helpful hints on how to answer essay questions on physiology. In addition, the study guide provides additional objective questions (with answers), fill-in-the-blank questions, crossword puzzles, and other devices to enable you to more effectively utilize the text to understand physiological concepts and take examinations in the subject.

### Acknowledgements

The third edition of *Human Physiology* has been greatly improved by comments I have received from users of the second edition. Though these are too many to acknowledge individually, I am grateful to each one. As in the past, my colleagues at Pierce College have been very supportive and helpful. In particular, I would like to thank Dr. Lawrence G. Thouin, Jr., Dr. James Rikel, and Mr. Edmont Katz.

Quality illustrations for this text were provided by a number of talented artists. I am grateful for their tremendous contributions.

The editorial and production staffs at Wm. C. Brown have guided and supported this enormous project at every stage in development. I owe a large debt of gratitude to Executive Editor Ed Jaffe, Project Editor Colin Wheatley, Art Editor Gayle Salow, and Designer David Lansdon, and many other talented people at Wm. C. Brown Publishers. Special thanks to Bea Sussman for her fine copy editing.

# HUMAN PHYSIOLOGY

# 1

# Tissues, Organs, and Control Systems

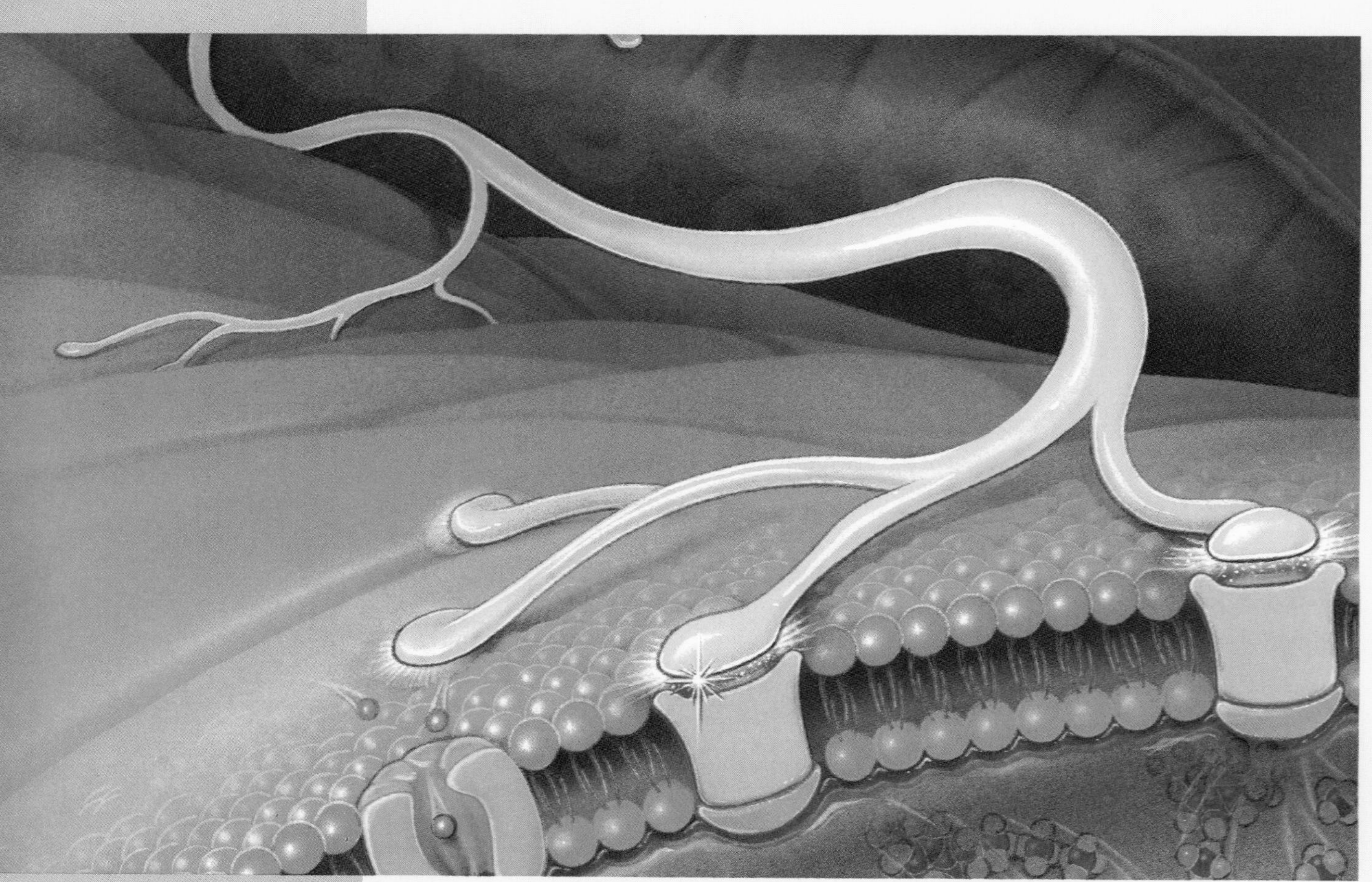

## Objectives

By studying this chapter, you should be able to

1. describe, in a general way, the topics studied in physiology and the importance of physiology in modern medicine
2. describe the characteristics and explain the importance of the scientific method
3. list the four primary tissues and their subtypes and describe the distinguishing features of each primary tissue
4. relate the structure of each primary tissue to its functions
5. describe how the primary tissues are organized into organs, using the skin as an example
6. define homeostasis and describe how this concept is used in physiology and medicine
7. explain the nature of negative feedback loops and how these mechanisms act to maintain homeostasis
8. explain how antagonistic effectors help to maintain homeostasis
9. explain the nature of positive feedback loops and how these function in the body
10. distinguish between intrinsic and extrinsic regulation, and explain, in a general way, the roles of the nervous and endocrine systems in body regulation
11. explain how negative feedback inhibition helps to regulate the secretion of hormones

## Outline

## *Introduction*

Human physiology is the study of how the human body functions, with emphasis upon specific cause-and-effect mechanisms. Knowledge of these mechanisms has been obtained experimentally through the application of the scientific method.

**Physiology** is the study of biological function—of how the body works, from cell to tissue, tissue to organ, organ to system, and of how the organism as a whole accomplishes particular tasks essential for life. In the study of physiology, the emphasis is on *mechanisms*—with questions that begin with the word *how* and answers that involve cause-and-effect sequences. These sequences can be woven into larger and larger stories that include descriptions of the structures involved (anatomy) and that overlap with the sciences of chemistry and physics.

The separate facts and relationships of these cause-and-effect sequences are derived empirically from experimental evidence. Explanations that seem logical are not necessarily true; they are only as valid as the data on which they are based, and they can change as new techniques are developed and further experiments are performed.

One standard technique for investigating the function of an organ is to observe what happens when it is surgically removed from an experimental animal or when its function is altered in a specific way. This study is often aided by "experiments of nature"—diseases—that involve specific damage to the function of an organ. The study of disease processes has thus aided our understanding of normal function, and the study of normal physiology has provided much of the scientific basis of modern medicine. This relationship is recognized by the Nobel Prize committee who award prizes in the category "Physiology or Medicine."

### Scientific Method

All of the information in this text has been gained by the application of the scientific method. Although many different techniques are involved in the scientific method, all share three attributes: (1) confidence that the natural world, including ourselves, is ultimately explainable in terms we can understand; (2) descriptions and explanations of the natural world that are honestly based on observations and that could be modified or refuted by other observations; and (3) humility, that is, the willingness to accept the fact that we could be wrong. If further study should yield conclusions that refute all or part of an idea, the idea must be accordingly modified. In short, the scientific method is based on a confidence in our rational ability, honesty, and humility. Practicing scientists may not always display these attributes, but the validity of the large body of scientific knowledge that has been accumulated—as shown by the technological applications and the predictive value of scientific hypotheses—are ample testimony to the fact that the scientific method works.

Suppose you measured the resting pulse rate of one athlete and one sedentary person and found that the athlete had a lower resting pulse rate. If you were following the scientific method, you could state that the first person had a lower resting pulse than the second at the time you made the measurements, but no generalization would be scientifically justified by this data. If you repeated these measurements at different times and found that the differences remained, you could construct a scientific hypothesis that athletes have lower resting pulse rates than sedentary people. This hypothesis is scientific because it is *testable;* you could measure the pulse rates of 100 athletes and 100 sedentary people and see if statistically significant differences were obtained. If they were, you would be justified in saying that athletes, on the average, have lower resting pulse rates than sedentary people *based on your data.* You must still be open to the fact that you could be wrong (your measurement techniques may have been biased, or your sample of people may not have been representative of the general population) or that there could be alternative explanations for your results. Before your discovery would become generally accepted and be included in textbooks, other scientists would consistently have to replicate your results—for scientific theories are based on *reproducible* data.

It is quite possible that when others attempt to replicate your experiment their results will be slightly different from your own. They may then construct scientific hypotheses that the differences in resting pulse rate also depend on factors such as the nature of the exercise performed by the athlete or on nutrition or on genetic influences. When other scientists attempt to test these hypotheses, they will likely encounter new problems, requiring new explanatory hypotheses, which must be tested by additional experiments.

In this way, a large body of highly specialized information is gradually accumulated and a more generalized explanation can be formulated. This explanation will almost always be different from preconceived notions. People who follow the scientific method will then appropriately modify their concepts, realizing that their new ideas will probably have to be changed again in the future as additional experiments are performed.

New concepts that emerge from the scientific method are not always welcome; people often prefer comfortable, older ideas, particularly when these are part of a large, cherished belief structure. In this situation, the knowledge gained by the scientific method may be rejected, and the scientists involved villified, persecuted, or (in previous ages) even murdered. Because the open communication of knowledge is a scientific tradition, however, those ideas that are closest to the truth have survived and eventually supplanted their competitors in the minds of most educated people.

**Figure 1.1.** Three skeletal muscle fibers showing the characteristic cross-striations.

1. *Define the science of physiology, and describe the relationship of this science to medicine.*
2. *Describe the characteristics of the scientific method. Explain how this differs from nonscientific thinking.*

## *The Primary Tissues*

The organs of the body are composed of four different primary tissues. Each of these tissues has its own characteristic structure and function, and the activities and interactions of these tissues determine the physiology of the organs.

Although physiology is the study of function, it is difficult to properly understand the function of the body without some knowledge of its anatomy, particularly at a microscopic level. Microscopic anatomy comprises a field of study known as *histology.* The anatomy and histology of specific organs will be discussed together with their functions in later chapters. In this section, the common "fabric" of all organs is described.

Cells that have similar functions are grouped into categories called *tissues.* The entire body is composed of only four types of tissues. These **primary tissues** include (1) muscle, (2) nervous, (3) epithelial, and (4) connective tissues. Groupings of these four primary tissues into anatomical and functional units are called *organs.* Organs, in turn, may be grouped together by common functions into *systems.*

### Muscle

Muscle tissue is specialized for contraction. There are three types of muscles: (1) **skeletal muscle,** (2) **cardiac muscle,** and (3) **smooth muscle.** Skeletal muscle is often called *voluntary muscle,* because we have conscious control of its contraction without special training. Both skeletal and cardiac muscles are **striated;** they have striations, or stripes, that extend across the width of the muscle cell (figs. 1.1 and 1.2), and for this reason they have similar mechanisms of contraction. Smooth muscle (fig. 1.3) lacks these cross-striations and has a different mechanism of contraction.

**Figure 1.2.** Human cardiac muscle. Notice the striated appearance and dark-staining intercalated discs.

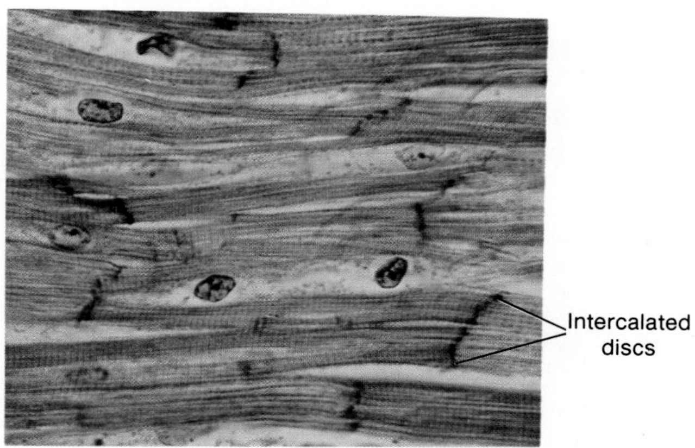


**Figure 1.3.** Photomicrograph of smooth muscle cells. Notice that these cells lack striations and contain single, centrally located nuclei.

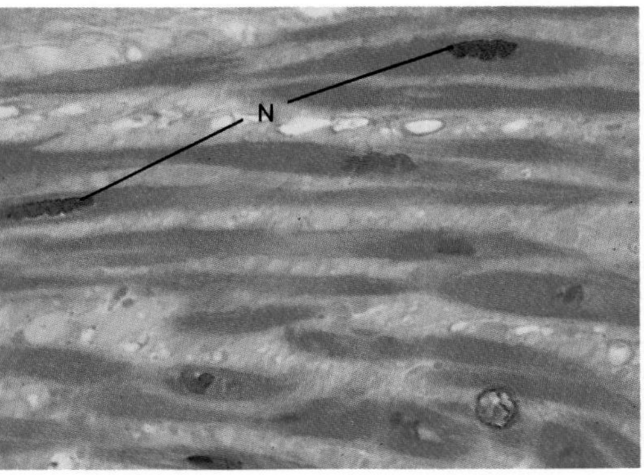


***Skeletal Muscle.*** Skeletal muscles are generally attached by means of tendons to bones at both ends, so that contraction produces movements of the skeleton. There are, however, exceptions to this pattern; the tongue, superior portion of the esophagus, anal sphincter, and diaphragm are also composed of skeletal muscle.

Since skeletal muscle cells are long and thin they are called **fibers,** or **myofibers** (*myo* = muscle). Despite their specialized structure and function, each myofiber contains structures common to all cells (nuclei, mitochondria, and other organelles, described in chapter 3).

**Figure 1.4.** The neuron. (*a*) A photomicrograph of nerve tissue. (*b*) A simplified diagram of a neuron and its principal parts. The arrows indicate the direction of the nerve impulse.

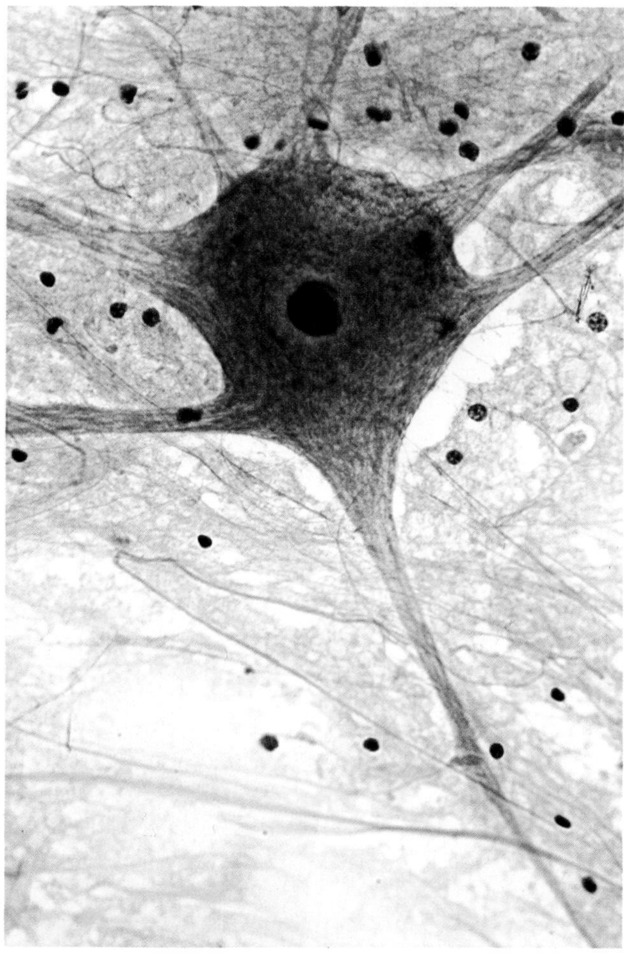

(a)

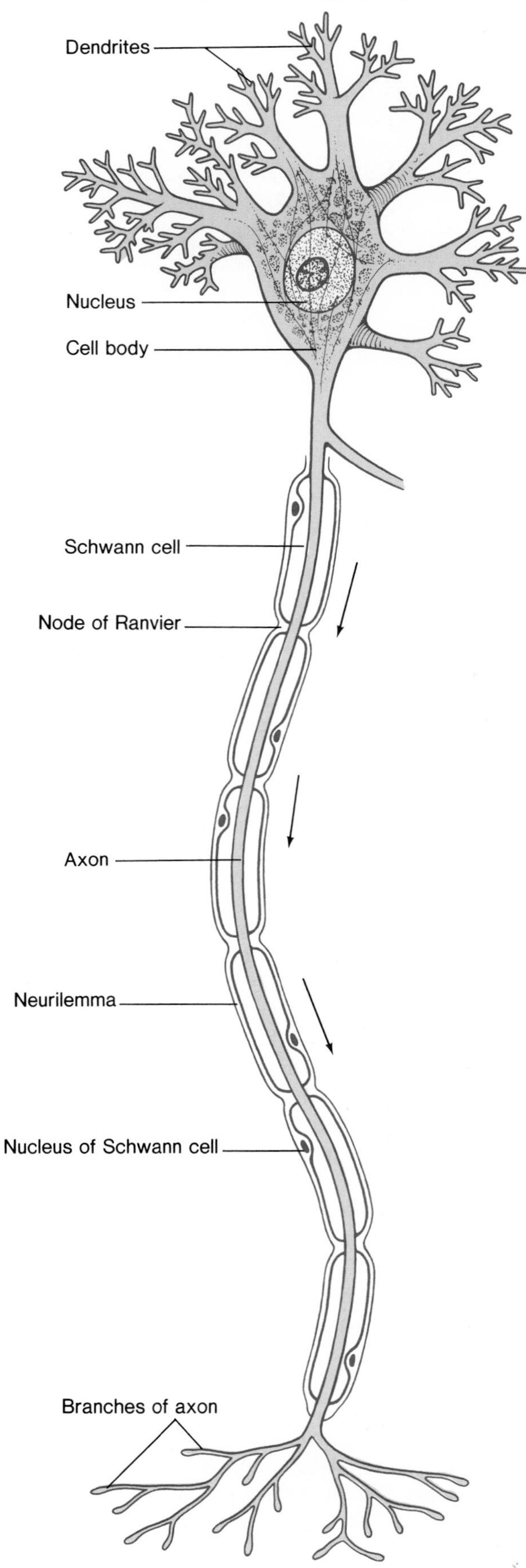


(b)

Within a skeletal muscle the muscle fibers are arranged in bundles, and within these bundles the fibers extend in parallel from one end to the other of the bundle. The parallel arrangement of muscle fibers (seen in fig. 1.1) allows each fiber to be controlled individually: one can thus contract fewer or more muscle fibers and, in this way, vary the strength of the whole muscle's contraction. The ability to vary, or "grade," the strength of skeletal muscle contraction is obviously needed for proper control of skeletal movements.

***Cardiac Muscle.*** Although cardiac muscle is striated, it has a very different appearance from skeletal muscle. Cardiac muscle is found only in the heart, where the **myocardial cells** are short, branched, and intimately interconnected to form a continuous fabric. Special areas of contact between adjacent cells stain darkly to show *intercalated discs* (fig. 1.2), which are characteristic of heart muscle.

The intercalated discs couple myocardial cells together mechanically and electrically. Unlike skeletal muscles, therefore, the heart cannot produce a graded contraction by varying the number of cells stimulated to contract. Because of the way it is constructed, the stimulation of one myocardial cell results in the stimulation of all other cells in the mass and a whole-hearted contraction.

***Smooth Muscle.*** As implied by the name, smooth muscle cells (fig. 1.3) do not have the cross-striations characteristic of skeletal and cardiac muscle. Smooth muscle is found in the digestive tract, blood vessels, bronchioles (small air passages in the lungs), and in the urinary and reproductive systems. Circular arrangements of smooth muscle in these organs produce constriction of the *lumen* (cavity) when the muscle cells contract. The digestive tract also contains longitudinally arranged layers of smooth muscle. Rhythmic contractions of circular and longitudinal layers of muscle produce *peristalsis,* a process that pushes food from one end of the digestive tract to the other.

## Nervous Tissue

Nervous tissue consists of nerve cells, or **neurons,** which are specialized for the generation and conduction of electrical events, and of **neuroglia,** which provide anatomical and functional support to the neurons.

Each neuron consists of three parts (fig. 1.4): (1) a *cell body,* which contains the nucleus and serves as the metabolic center of the cell; (2) *dendrites* (literally, "branches"), which are highly branched cytoplasmic extensions of the cell body that receive input from other neurons or from receptor cells; and (3) an *axon,* which is a single cytoplasmic extension of the cell body, which can be quite long (up to a few feet in length) and is specialized for conducting nerve impulses from the cell body to another neuron or an effector (muscle or gland) cell.

The neuroglia, composed of *neuroglial cells,* do not conduct impulses but instead serve to bind neurons together, modify the extracellular environment of the nervous system, and influence the nourishment and electrical activity of neurons. Neuroglial cells are about five times more abundant than neurons in the nervous system and, unlike neurons, maintain a limited ability to divide by mitosis throughout life.

## Epithelial Tissue

Epithelial tissue consists of cells that form **membranes,** which cover and line the body surfaces, and of **glands** that are derived from these membranes. There are two categories of glands. *Exocrine glands* (*exo* = outside) secrete chemicals through a duct that leads to the outside of the membrane and thus to the outside of the body. *Endocrine glands* (*endo* = within) secrete chemicals called *hormones* into the blood.

***Epithelial Membranes.*** Epithelial membranes are classified according to the number of their layers and the shape of the cells in the upper layer (table 1.1). Epithelial cells that are flattened in shape are *squamous,* those that are taller than they are wide are *columnar,* and those that are as wide as they are tall are *cuboidal* (fig. 1.5). Those epithelial membranes that are only one cell layer in thickness are known as *simple* membranes; those that are composed of a number of layers are *stratified* membranes.

A simple squamous membrane is adapted for diffusion and filtration; such a membrane lines all blood vessels, where it is known as an *endothelium.* A simple cuboidal epithelium lines the ducts of exocrine glands and part of the tubules of the kidney. A simple columnar epithelium lines the lumen of the stomach and intestine; this epithelium contains specialized unicellular glands, called *goblet cells,* which secrete mucus and are dispersed among the columnar epithelial cells. The columnar epithelial cells in the uterine (fallopian) tubes of females and in the respiratory passages contain numerous *cilia* (hairlike structures described in chapter 3), which can move in a coordinated fashion and aid the functions of these organs.

The epithelial covering of the esophagus and vagina that provides protection for these organs is a stratified squamous epithelium (fig. 1.6). This is a *nonkeratinized* membrane, and all layers of this epithelium consist of living cells. The *epidermis* of the skin, in contrast, is *keratinized,* or *cornified* (fig. 1.7). Since the epidermis is dry and exposed to the potentially desiccating effects of the air, the surface is covered with dead cells that are filled with a water-resistant protein known as *keratin.* This protective layer is constantly flaked off from the surface of the skin and therefore must be constantly replaced by the division of cells in the deeper layers of the epidermis.

**Table 1.1** Summary of epithelial tissues

| Type | Structure and Function | Location |
|---|---|---|
| **Simple Epithelia** | Single layer of cells; diffusion and filtration | Covering visceral organs, linings of lumina and body cavities |
| Simple squamous epithelium | Single layer of flattened, tightly bound cells; diffusion and filtration | Capillary walls, air sacs of lungs, covering visceral organs, linings of body cavities |
| Simple cuboidal epithelium | Single layer of cube-shaped cells; excretion, secretion, or absorption | Surface of ovaries; linings of kidney tubules, salivary ducts, and pancreatic ducts |
| Simple columnar epithelium | Single, nonciliated layer of tall, columnar-shaped cells; protection, secretion, and absorption | Lining of digestive tract |
| Simple ciliated columnar epithelium | Single, ciliated layer of columnar-shaped cells; transportive role through ciliary motion | Lining the lumen of the uterine tubes |

**Figure 1.5.** (*a*) Simple squamous, (*b*) simple cuboidal, and (*c*) simple columnar epithelial membranes. The tissue beneath each membrane is connective tissue.

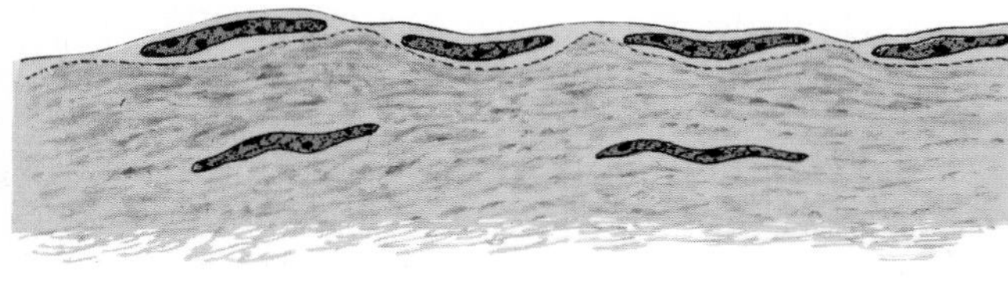

(a)

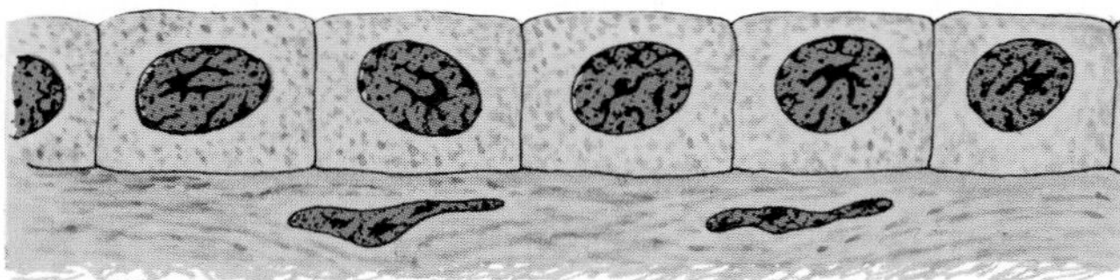

(b)

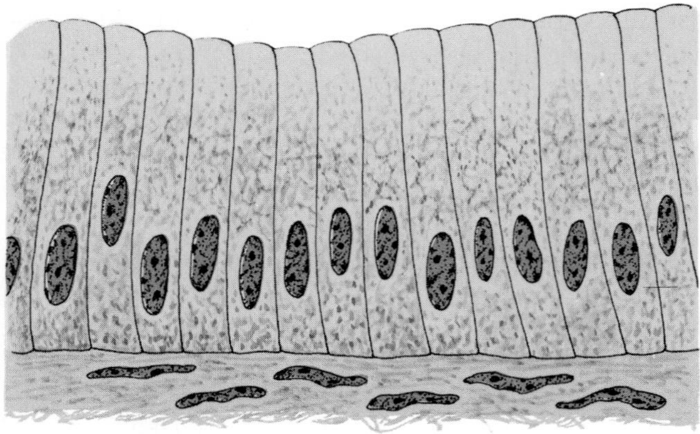

(c)

**Figure 1.6.** The stratified squamous nonkeratinized epithelial membrane of the vagina.

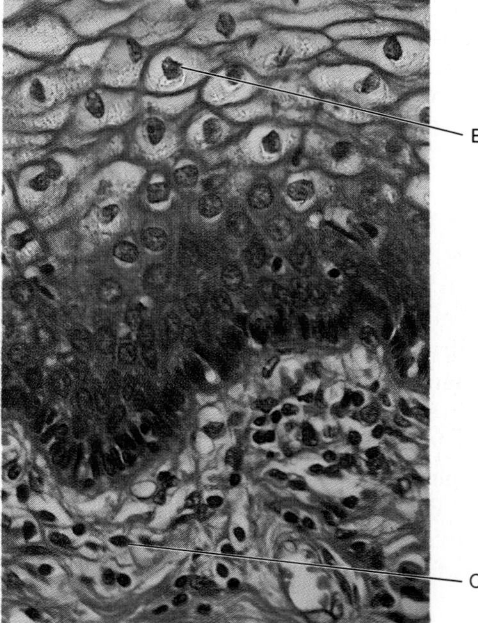

**Table 1.1** (Cont.)

| Type | Structure and Function | Location |
|---|---|---|
| Pseudostratified ciliated columnar epithelium | Single layer of ciliated, irregularly shaped cells, many goblet cells; protection, secretion, ciliary movement | Lining of respiratory passageways |
| **Stratified Epithelia** | Two or more layers of cells; protection, strengthening, or distension | Epidermal layer of skin; linings of body openings, ducts, urinary bladder |
| Stratified squamous epithelium (keratinized) | Numerous layers, contains keratin, outer layers flattened and dead; protection | Epidermis of skin |
| Stratified squamous epithelium (nonkeratinized) | Numerous layers, lacks keratin, outer layers moistened and alive; protection and pliability | Linings of oral and nasal cavities, vagina, and anal canal |
| Stratified cuboidal epithelium | Usually two layers of cube-shaped cells; strengthen luminal walls | Larger ducts of sweat glands, salivary glands, and pancreas |
| Transitional epithelium | Numerous layers of rounded, nonkeratinized cells; distension | Luminal walls of ureters and urinary bladder |

From Kent M. Van De Graaff, *Human Anatomy,* 2d ed. 

**Figure 1.7.** A section of skin showing the loose connective tissue dermis beneath the cornified epidermis. Loose connective tissue contains scattered collagen fibers in a matrix of protein-rich fluid. The intercellular spaces also contain cells and blood vessels.

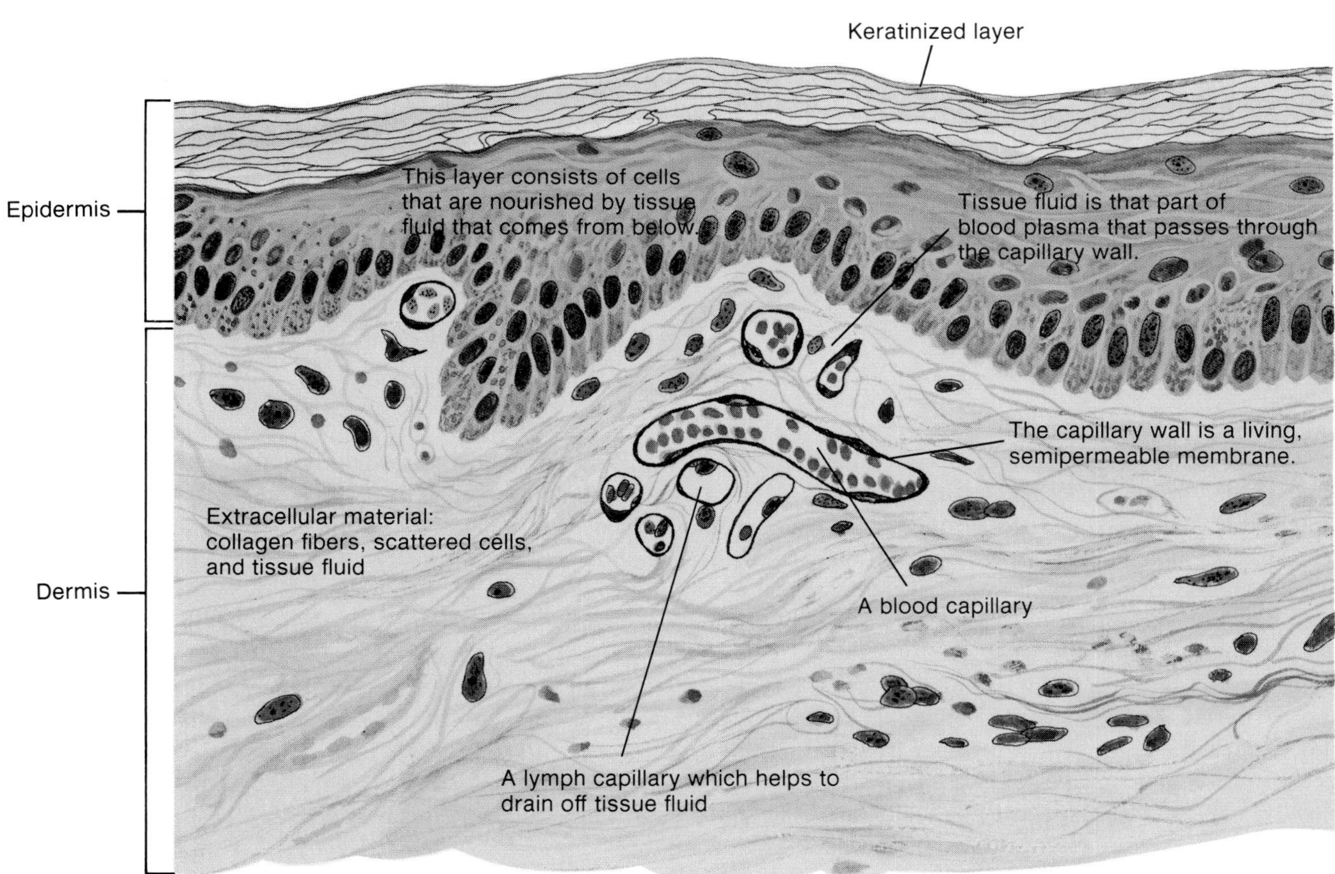

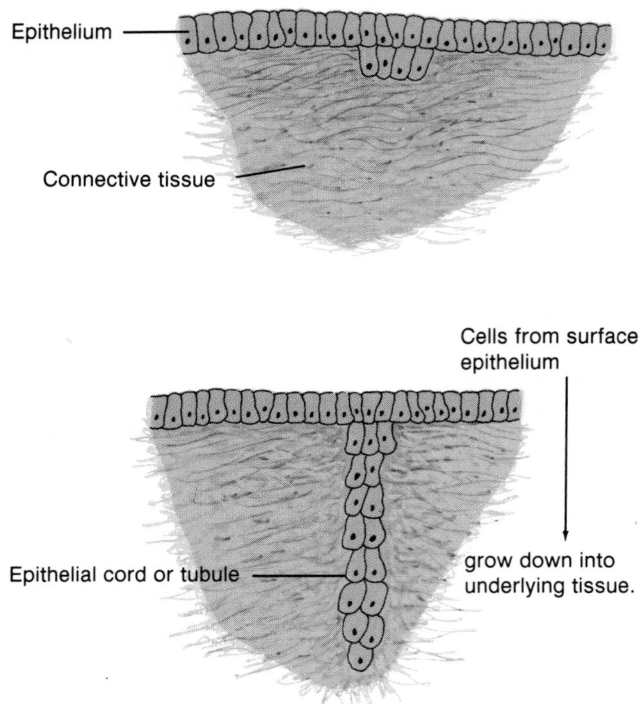

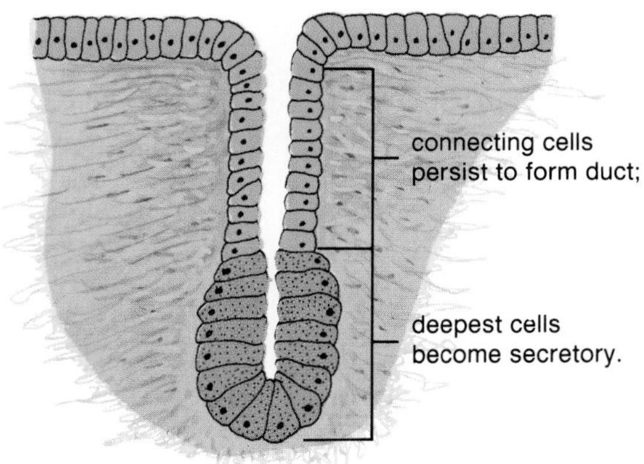


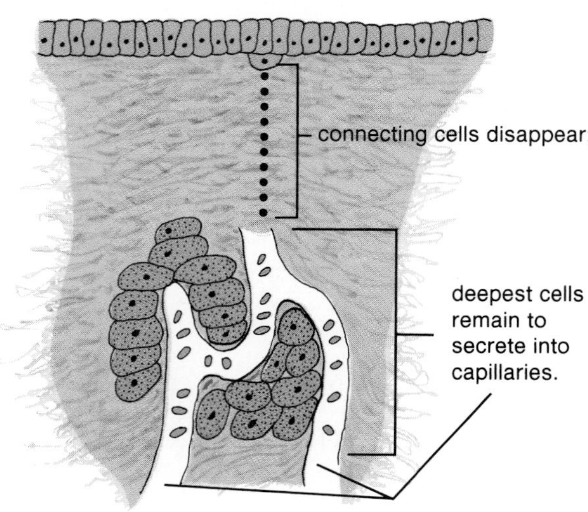


**Figure 1.8.** The formation of exocrine and endocrine glands from epithelial membranes.

The constant loss and renewal of cells is characteristic of epithelial membranes. The entire epidermis is completely replaced every two weeks; the stomach lining is renewed every two to three days. Examination of the cells that are lost, or "exfoliated," from the surface of the female genital tract is a common procedure in gynecology (as in the Pap smear).

In order to form a strong membrane that is effective as a barrier at the body surfaces, epithelial cells are very closely packed and are joined together by structures collectively called **junctional complexes.** There is no room for blood vessels between adjacent epithelial cells. The epithelium must therefore receive nourishment from the tissue beneath, which has large intercellular spaces that can accommodate blood vessels and nerves. This underlying tissue is called *connective tissue.* Epithelial membranes are attached to the underlying connective tissue by a layer of proteins and polysaccharides known as the **basement membrane,** which can only be observed under the microscope using specialized staining techniques.

***Exocrine Glands.*** Exocrine glands are derived from cells of epithelial membranes that cover and line the body surfaces. The secretions of these cells are expressed to the outside of the epithelial membranes (and hence to the outside of the body) through *ducts.* This is in contrast to endocrine glands, which lack ducts and which therefore secrete into capillaries within the body (fig. 1.8). The structure of endocrine glands will be described in chapter 11.

The secretory units of exocrine glands may be simple tubes, or they may be modified to form clusters of units around branched ducts (fig. 1.9). These clusters, or **acini,** are often surrounded by tentaclelike extensions of *myoepithelial cells,* which contract and squeeze the secretions through the ducts. The rate of secretion and the action of

**Figure 1.9.** Exocrine glands.

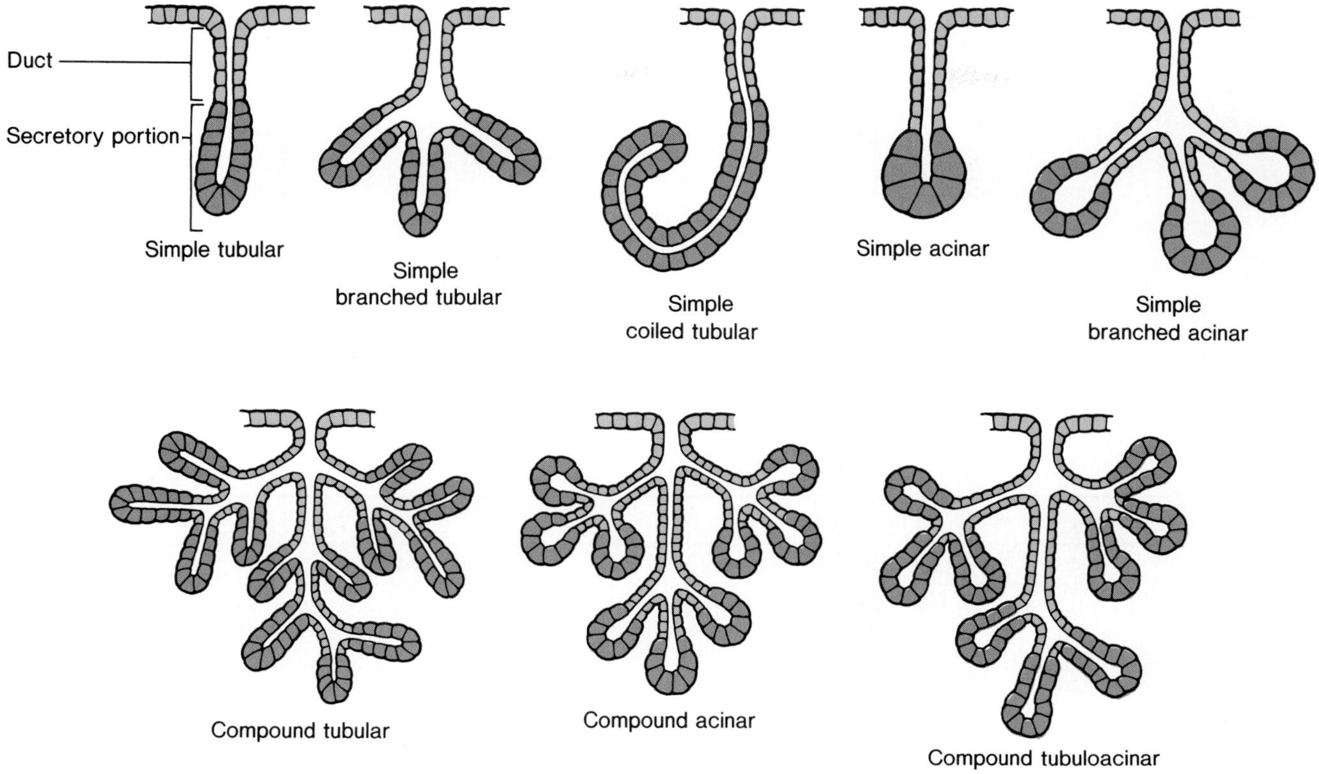

myoepithelial cells are influenced by hormones and by the autonomic nervous system, as will be described in later chapters.

Examples of exocrine glands in the skin include the lacrimal (tear) glands, sebaceous glands (which secrete oily sebum into hair follicles), and sweat glands. There are two types of sweat glands. The most numerous, the *eccrine sweat glands,* secrete a dilute salt solution that serves in thermoregulation (evaporation cools the skin). The *apocrine sweat glands,* located in the axilla (underarms) and pubic regions, secrete a protein-rich fluid. This provides nourishment for bacteria that produce the characteristic odor of this type of sweat.

All of the glands that secrete into the digestive tract are also exocrine. This is because the lumen of the digestive tract is a part of the external environment and secretions of these glands go to the outside of the membrane that lines this tract. Mucous glands are located throughout the length of the digestive tract. Other relatively simple glands of the tract include salivary glands, gastric glands, and simple tubular glands in the intestine.

The *liver* and *pancreas* are exocrine (as well as endocrine) glands, derived embryologically from the digestive tract. The exocrine secretion of the pancreas is pancreatic juice, containing digestive enzymes and bicarbonate, which is secreted into the small intestine via the pancreatic duct. The liver produces and secretes bile (an emulsifier of fat) into the small intestine via the gallbladder and bile duct.

Exocrine glands are also prominent in the reproductive system. The female reproductive tract contains numerous mucus-secreting exocrine glands. The male accessory sexual organs—the *prostate* and *seminal vesicles*—are exocrine glands that contribute to the semen. The testes and ovaries (the gonads) are both endocrine and exocrine glands. They are endocrine because they secrete sex steroid hormones into the blood; they are exocrine because they release gametes (ova and sperm) into the reproductive tracts.

## Connective Tissue

Connective tissue is characterized by large amounts of extracellular material in the spaces between the connective tissue cells. This extracellular material may be of various types and arrangements and, on this basis, several types of connective tissues are recognized: (1) connective tissue proper; (2) cartilage; (3) bone; and (4) blood. Blood is usually classified as connective tissue because about half its volume is composed of an extracellular fluid known as *plasma.*

**Connective tissue proper** includes a variety of subtypes. An example of *loose connective tissue* (or *areolar tissue*) is the *dermis* of the skin (fig. 1.8). This connective tissue consists of scattered fibrous proteins called *collagen* and tissue fluid, which provides abundant space for the entry of blood and lymphatic vessels and nerve fibers. Another type of connective tissue proper is *dense fibrous connective tissue,* which contains densely packed fibers of

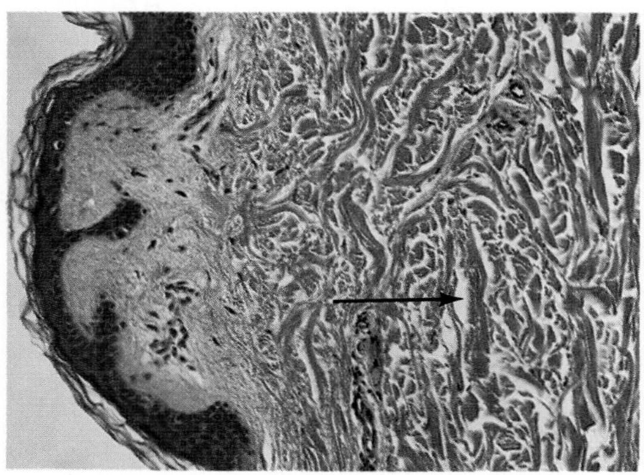

**Figure 1.10.** A photomicrograph of dense irregular connective tissue. Note the tightly packed, irregularly arranged collagen proteins.

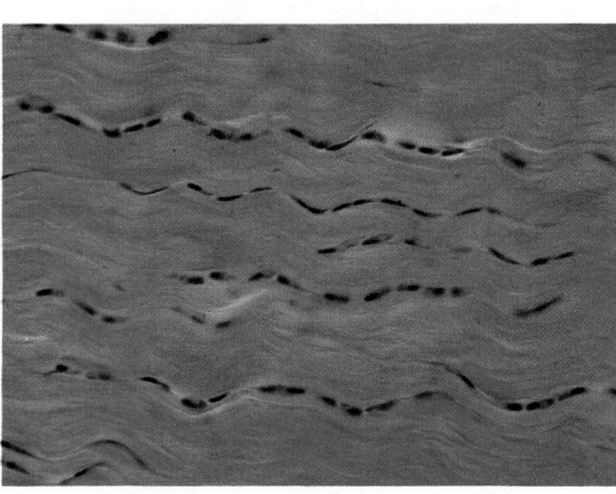

**Figure 1.11.** A photomicrograph of a tendon showing a dense, regular arrangement of collagen fibers.

**Table 1.2** Summary of connective tissue proper

| Type | Structure and Function | Location |
|---|---|---|
| Loose connective (areolar) tissue | Predominantly fibroblast cells with lesser amounts of collagen and elastin cells; binds organs, holds tissue fluids, diffusion | Surrounding nerves and vessels, between muscles, beneath the skin |
| Dense fibrous connective tissue | Densely packed collagen fibers; provides strong, flexible support | Tendons, ligaments, sclera of eye, deep skin layers |
| Elastic connective tissue | Predominantly irregularly arranged elastic fibers; supports, provides framework | Large arteries, lower respiratory tract, between vertebrae |
| Reticular connective tissue | Reticular fibers forming supportive network; stores, phagocytic | Lymph nodes, liver, spleen, thymus, bone marrow |
| Adipose connective tissue | Adipose cells; protects, stores fat, insulates | Hypodermis of skin, surface of heart, omentum, around kidneys, back of eyeball, surrounding joints |



collagen that may be in an irregular or a regular arrangement. Dense irregular connective tissue contains a meshwork of collagen fibers and forms the tough capsules and sheaths around organs (fig. 1.10). Tendons, which connect muscle to bone, and ligaments, which connect bones together at joints, are examples of dense regular connective tissue. This tissue contains a dense arrangement of collagen fibers that are parallel to each other (fig. 1.11). The characteristics of these and other types of connective tissue proper are summarized in table 1.2.

**Cartilage** consists of cells, called *chondrocytes,* surrounded by a semisolid ground substance that imparts elastic properties to the tissue. Cartilage is a type of supportive and protective tissue commonly called "gristle." It forms the precursor to many bones in the body and persists at the articular surfaces on the bones at all movable joints.

**Bone** is produced as concentric layers, or *lamellae,* of calcified material laid around blood vessels. The bone-forming cells, or *osteoblasts,* surrounded by their calcified products, become trapped within cavities (called *lacunae*). The trapped cells, which are now called *osteocytes,* remain alive because they are nourished by "lifelines" of cytoplasm that extend from the cells to the blood vessels in *canaliculi* (little canals). The blood vessels lie within central canals, surrounded by concentric rings of bone lamellae with their trapped osteocytes; these units are called *haversian systems* (fig. 1.12).

The *dentin* of a tooth (fig. 1.13) is similar in composition to bone, but the cells that form this calcified tissue are located in the pulp (composed of loose connective tissue). These cells send cytoplasmic extensions, called *dentinal tubules,* into the dentin. Dentin, like bone, is thus a living tissue which can be remodeled in response to stresses. The cells that form the outer *enamel* of a tooth, in contrast, are lost as the tooth erupts. Enamel is a highly calcified material, harder than bone or dentin, that cannot be regenerated; artificial "fillings" are therefore required to patch holes in the enamel.

**Figure 1.12.** A diagram illustrating how a bone grows in width. Cells within the outer connective tissue covering of the bone (the periosteum) add new bone lamellae around blood vessels within the periosteum. This produces new haversian systems with a central canal containing the blood vessel and lined by the same connective tissue (which is now an endosteum).

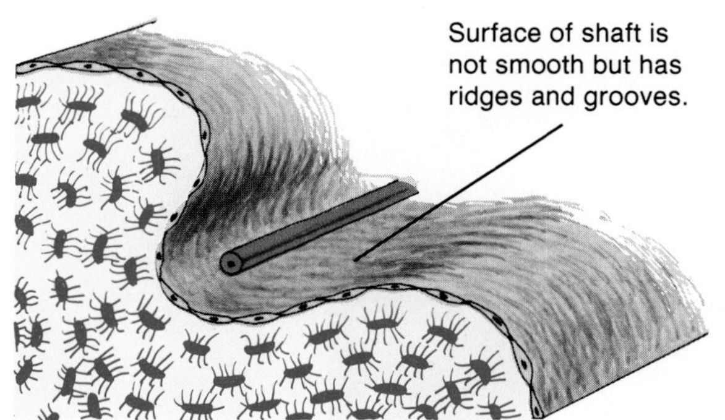


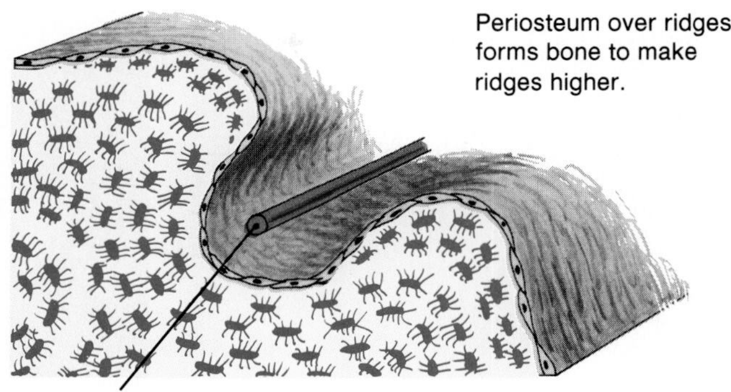


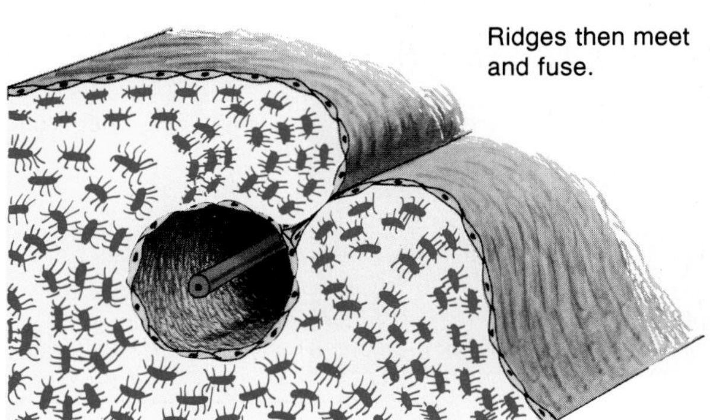


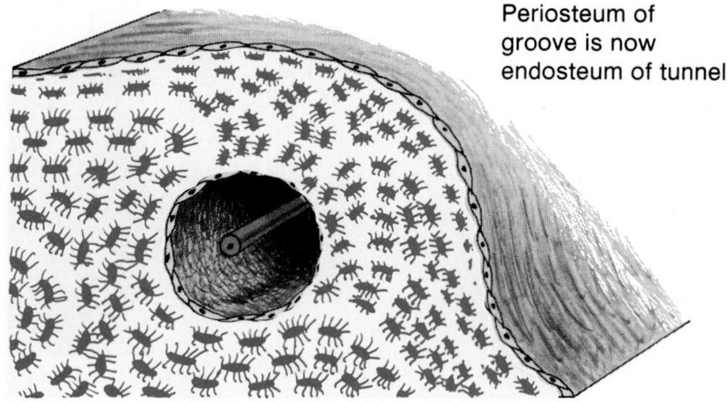


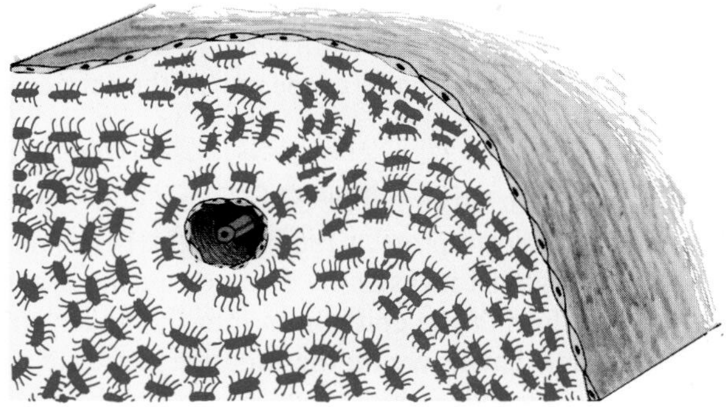


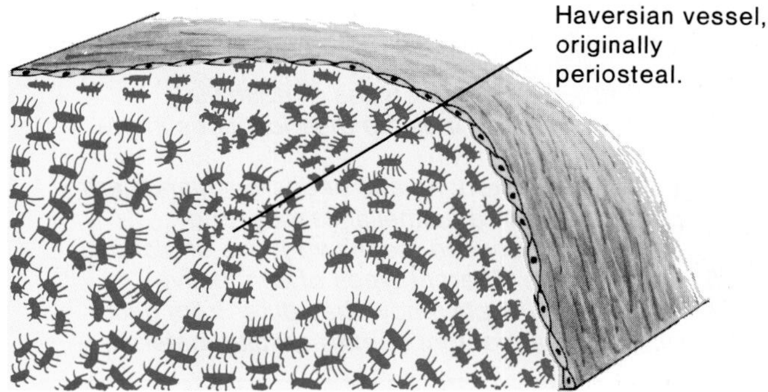

**Figure 1.13.** A cross section of a tooth showing pulp, dentin, and enamel. The cells that form dentin (odontoblasts) are located in the pulp, and their processes extend into the dentin-forming tubules. When these processes die, the tubules fill with air and appear dark. The odontoblasts in the pulp can form new reparative dentin to seal off the pulp from the dead tracts.

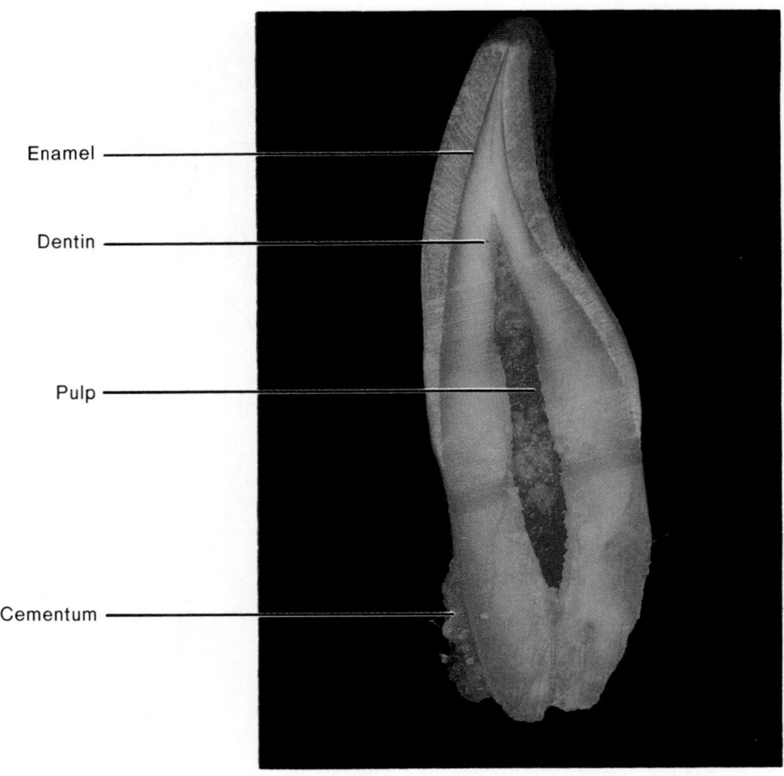


1. *List the four primary tissues, and describe the distinguishing features of each type.*
2. *Name the three types of muscle tissue, and distinguish between them.*
3. *Describe the different types of epithelial membranes, and indicate their locations in the body.*
4. *Explain why exocrine and endocrine glands are considered to be epithelial tissues, and distinguish between these two types of glands.*
5. *Describe the different types of connective tissues, and explain how they differ from each other in their content of extracellular material.*

## Organs and Systems

Organs are composed of all four primary tissues which serve the different functions of the organ. The skin is an organ that has numerous functions provided by its constituent tissues.

An organ is a structure composed of at least two, and usually all four, types of primary tissues. The largest organ in the body, in terms of its surface area, is the skin. The numerous functions of the skin will serve in this section to illustrate how primary tissues cooperate in the service of organ physiology.

### An Example of an Organ: The Skin

The cornified epidermis protects the skin (fig. 1.14) against water loss and against invasion by disease-causing organisms. Invaginations of the epithelium into the underlying connective tissue dermis create the exocrine glands of the skin. These include hair follicles (which produce the hair), sweat glands, and sebaceous glands. The secretion of sweat glands cools the body by evaporation and produces odors that, at least in lower animals, serve as sexual attractants. Sebaceous glands secrete oily sebum into hair follicles, where it is transported to the surface of the skin (unless these ducts are blocked, in which case blackheads are produced). Sebum lubricates the cornified surface of the skin, helping to prevent it from drying and cracking (as in

**Figure 1.14.** A diagram of the skin. Notice that all four types of primary tissues are present.

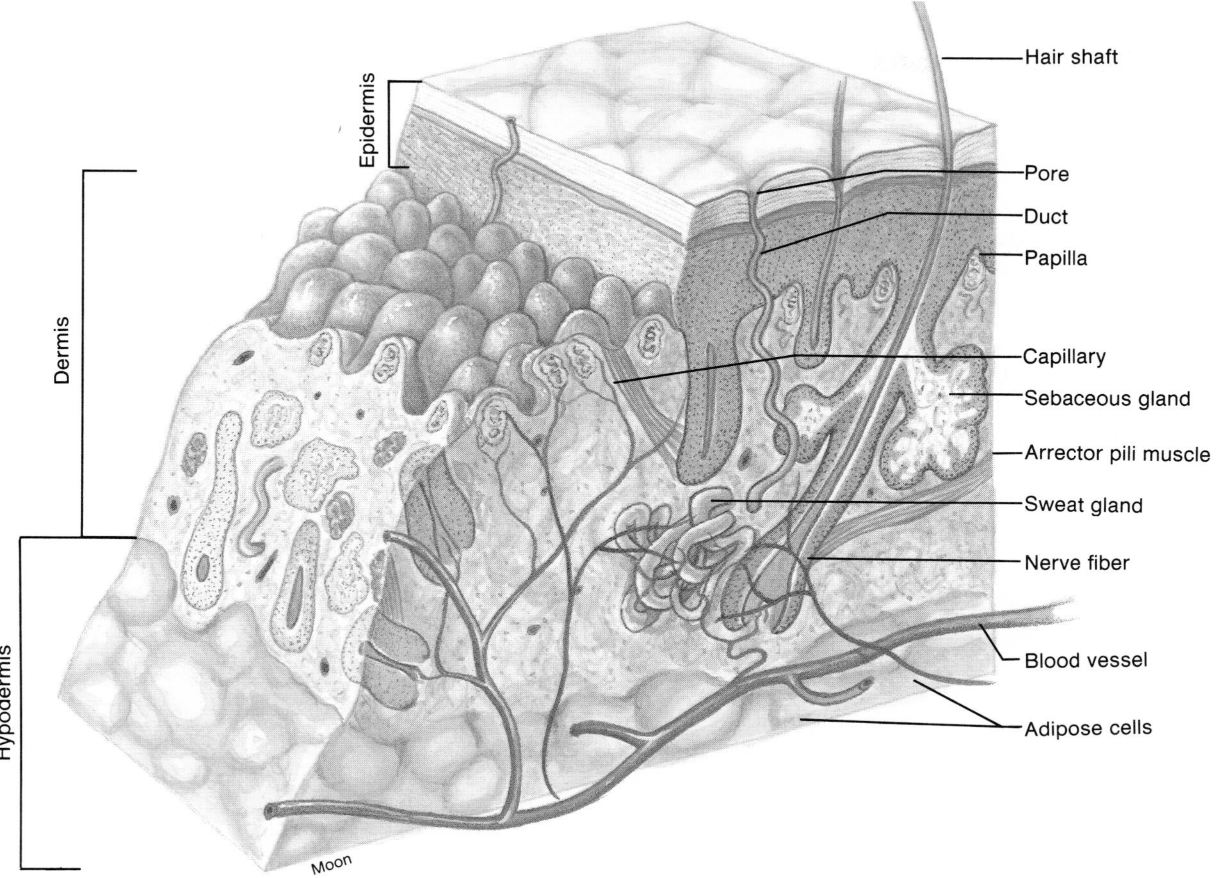


chapped lips—there are no sebaceous glands in the lips; one must, therefore, moisten the lips periodically with the tongue).

The skin is nourished by blood vessels within the dermis. In addition to blood vessels, the dermis contains wandering white blood cells and other types of cells that protect against invading disease-causing organisms, as well as nerve fibers and fat cells. Most of the fat cells, however, are grouped together to form the *hypodermis* (a layer beneath the dermis). Although fat cells are a type of connective tissue, masses of fat deposits throughout the body—such as subcutaneous fat—are referred to as **adipose tissue.**

Sensory nerve endings within the dermis mediate the cutaneous sensations of touch, pressure, heat, cold, and pain. Some of these sensory stimuli directly affect the sensory nerve endings. Others act via sensory structures derived from nonneural primary tissues. The Pacinian corpuscles in the dermis of the skin (fig. 1.15), for example, monitor sensations of pressure. Motor nerve fibers in the skin stimulate effector organs, resulting in, for example, the secretions of exocrine glands and contractions of the arrector pili muscles, which attach to hair follicles

**Figure 1.15.** The Pacinian corpuscle is a receptor for deep pressure. It consists of epithelial cells and connective tissue proteins that form concentric layers around the ending of a sensory neuron.

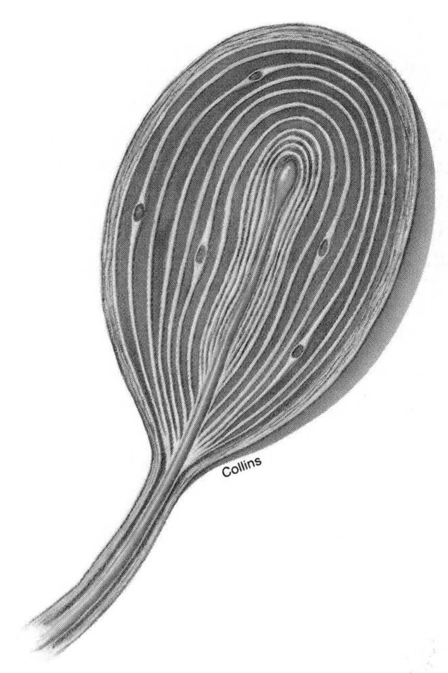

**Table 1.3** Approximate normal ranges for measurements of some blood values

| Measurement | Normal Range | Measurement | Normal Range |
|---|---|---|---|
| Arterial pH | 7.35–7.43 | Urea | 12–35 mg/100 ml |
| Bicarbonate | 21.3–28.5 mEq/L | Amino acids | 3.3–5.1 mg/100 ml |
| Sodium | 136–151 mEq/L | Protein | 6.5–8.0 g/100 ml |
| Calcium | 4.6–5.2 mEq/L | Total lipids | 350–850 mg/100 ml |
| Oxygen content | 17.2–22.0 ml/100 ml | Glucose | 75–110 mg/100 ml |

and surrounding connective tissue (producing goose bumps). The degree of constriction or dilation of cutaneous blood vessels—and therefore the rate of blood flow—is also regulated by motor nerve fibers.

The epidermis itself is a dynamic structure that can respond to environmental stimuli. The rate of its cell division—and consequently the thickness of the cornified layer—increases under the stimulus of constant abrasion. This produces calluses. The skin also protects itself against the dangers of ultraviolet light by increasing its production of *melanin* pigment, which absorbs ultraviolet light while producing a tan. In addition, the skin is an endocrine gland that produces and secretes vitamin D (derived from cholesterol under the influence of ultraviolet light), which functions as a hormone.

The architecture of most organs is similar to that of the skin. Most are covered by an epithelium immediately over a connective tissue layer. The connective tissue contains blood vessels, nerve endings, scattered cells for fighting infection, and possibly glandular tissue as well. If the organ is hollow—as in the digestive tract or in blood vessels—the lumen is also lined with an epithelium immediately over a connective tissue layer. The presence, type, and distribution of muscular and nervous tissue vary in different organs.

### Systems

Organs that are located in different regions of the body and that perform related functions are grouped into **systems.** These include the nervous system, endocrine system, cardiovascular system, respiratory system, excretory system, musculoskeletal system, integumentary system, reproductive system, digestive system, and immune system. By means of numerous regulatory mechanisms, these systems work together to maintain the life and health of the entire organism.

1. *Describe the location of each type of primary tissue in the skin.*
2. *Describe the functions of nerve, muscle, and connective tissue in the skin.*
3. *Describe the functions of the epidermis, and explain why this tissue is called "dynamic."*

## *Homeostasis and Feedback Control*

The regulatory mechanisms of the body can be understood in terms of a single, shared function: that of maintaining homeostasis, which is defined as the dynamic constancy of the internal environment. Homeostasis is maintained by effectors which are regulated by sensory information from the internal environment. The activity of effectors is thus controlled by the very changes that these organs help to produce—that is, by feedback control mechanisms.

Over a century ago the French physiologist Claude Bernard observed that the *milieu interieur* ("internal environment") remains remarkably constant despite changing conditions in the external environment. In a book entitled *The Wisdom of the Body* (published in 1932), Walter Cannon coined the term **homeostasis** to describe this internal constancy. Cannon further suggested that mechanisms of physiological regulation exist for one purpose—the maintenance of internal constancy.

The concept of homeostasis has been of inestimable value in the study of anatomy and physiology, because it allows diverse regulatory mechanisms to be understood in terms of their "why" as well as their "how." The concept of homeostasis also provides a major foundation for medical diagnostic procedures. When a particular measurement of the internal environment, such as blood measurements (table 1.3), deviates significantly from the normal range of values, it can be concluded that homeostasis is not maintained and the person is sick. A number of such measurements, combined with clinical observations, may allow the particular defective mechanism to be identified.

### Negative Feedback Loops

In order for internal constancy to be maintained the body must have sensors that are able to detect deviations from a *set point*. The set point is analogous to the temperature set on a house thermostat. In a similar manner, there is a set point for body temperature, blood glucose concentration, the tension on a tendon, and so on. When a sensor detects a deviation from a particular set point, it must relay this information to an **integrating center**, which usually receives information from many different sensors. The integrating center is often a particular region of the brain

**Figure 1.16.** A rise in some factor of the internal environment (↑X) is detected by a sensor. Acting through an integrating center, this caused an effector to produce a change in the opposite direction (↓X). The initial deviation is thus reversed, completing a negative feedback loop (shown by dashed arrow and negative sign). The numbers indicate the sequence of changes.

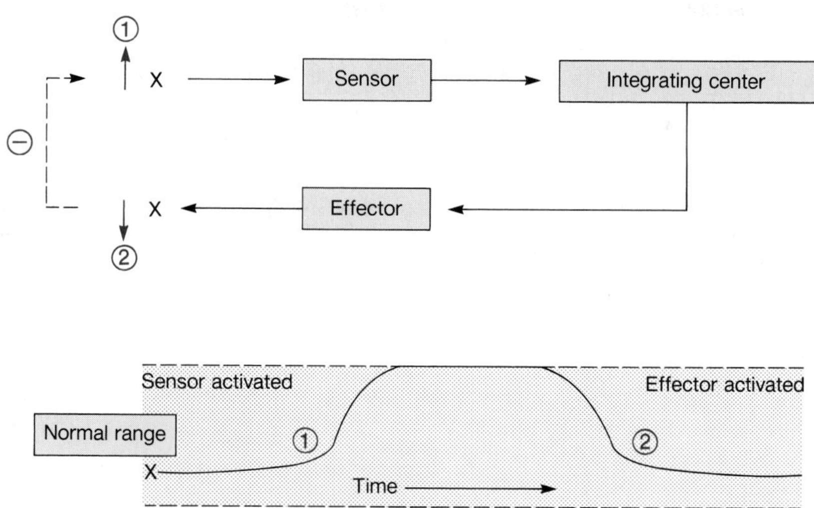


**Figure 1.17.** A negative feedback loop that compensates for a fall in some factor of the internal environment (↓X). Compare this figure with figure 1.16.

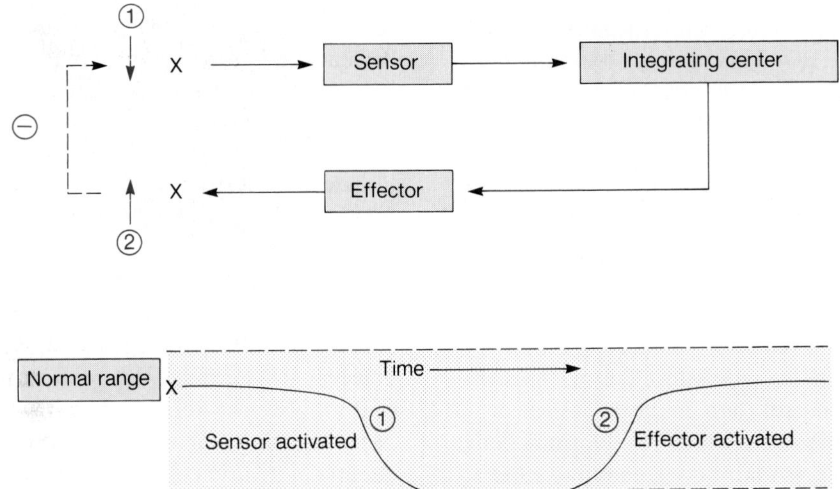


or spinal cord, but in some cases it can also be cells of endocrine glands. The relative strengths of different sensory inputs are weighed in the integrating center, and, in response, the integrating center either increases or decreases the activity of particular **effectors,** which are generally muscles or glands.

In response to sensory information about a deviation from a set point, therefore, effectors act to promote a reverse change in the internal environment. If the body temperature exceeds the set point of 37° C, for example, effectors act to lower the temperature. If, as another example, the blood glucose concentration falls below normal, the effectors act to increase the blood glucose. Since the activity of the effectors is influenced by the effects they produce, and since this regulation is in a negative, or reverse, direction, this type of control system is known as a **negative feedback loop** (fig. 1.16).

It is important to realize that these negative feedback loops are continuous, ongoing processes. Thus, a particular nerve fiber which is part of an effector mechanism may always display some activity, and a particular hormone, which is part of another effector mechanism, may always be present in the blood. The nerve activity and hormone concentration may decrease in response to deviations of the internal environment in one direction (fig. 1.16), or they may increase in response to deviations in the opposite direction (fig. 1.17). Changes from the normal range in either direction are thus compensated by reverse changes in effector activity.

**Figure 1.18.** Negative feedback loops (indicated by negative signs) maintain a state of dynamic constancy within the internal environment.

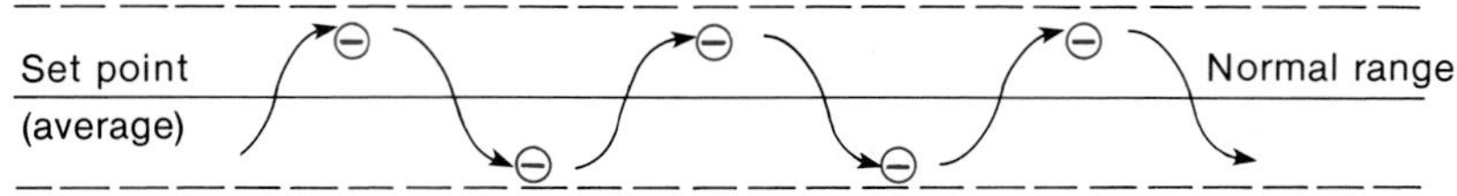


**Figure 1.19.** A simplified scheme by which body temperature is maintained within the normal range (with a set point of 37° C) by two antagonistic mechanisms—shivering and sweating. Shivering is induced when the body temperature falls too low and gradually subsides as the temperature rises. Sweating occurs when the body temperature is too high and diminishes as the temperature falls. Most aspects of the internal environment are regulated by the antagonistic actions of different effector mechanisms.

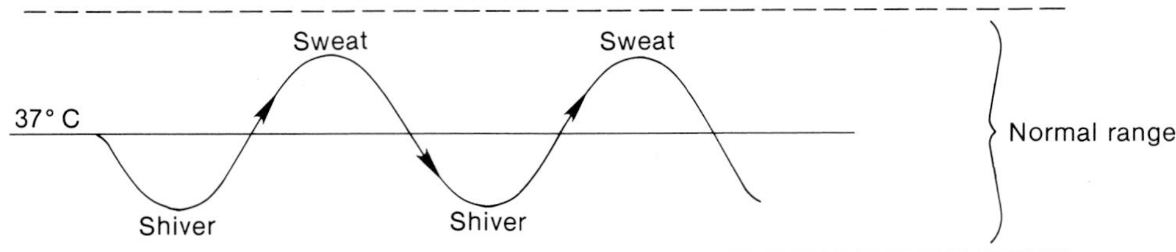


Homeostasis is best conceived as a state of **dynamic constancy**, rather than a state of absolute constancy. The values of particular measurements of the internal environment fluctuate above and below the set point, which can be taken as the average value within the normal range of measurements (fig. 1.18). This state of dynamic constancy results from greater or lesser activation of effectors in response to sensory feedback, and from the competing actions of antagonistic effectors.

***Antagonistic Effectors.*** Most factors in the internal environment are controlled by several effectors, which often have antagonistic actions. Control by antagonistic effectors is sometimes described as "push-pull," where the increasing activity of one effector is accompanied by decreasing activity of an antagonistic effector. This affords a finer degree of control than could be achieved by simply switching one effector on and off. Normal body temperature, for example, is maintained about a set point of 37° C by the antagonistic effects of sweating, shivering, and other mechanisms (fig. 1.19).

The blood concentrations of glucose, calcium, and other substances are regulated by negative feedback loops that involve hormones which promote opposite effects. While insulin, for example, lowers blood glucose, other hormones raise the blood glucose concentration. The heart rate, similarly, is controlled by nerve fibers that produce opposite effects: stimulation of one group of nerve fibers increases heart rate, and stimulation of another group slows the heart rate.

## Positive Feedback

Constancy of the internal environment is maintained by effectors which act to compensate the change that served as the stimulus for their activation; in short, by negative feedback loops. A thermostat, for example, maintains a constant temperature by increasing heat production when it is cold and decreasing heat production when it is warm. The opposite occurs during **positive feedback**—in this case, the action of effectors *amplifies* those changes that stimulated the effectors. A thermostat that works by positive feedback, for example, would increase heat production in response to a rise in temperature.

It is clear that homeostasis must ultimately be maintained by negative rather than by positive feedback mechanisms. The effectiveness of some negative feedback loops, however, is increased by positive feedback mechanisms that amplify the actions of a negative feedback response. Blood clotting, for example, occurs as a result of a sequential activation of clotting factors; the activation of one clotting factor results in activation of many in a positive feedback, avalanchelike, manner. In this way, a single change is amplified to produce a blood clot. Formation of the clot, however, can prevent further loss of blood and thus represents the completion of a negative feedback loop.

## Neural and Endocrine Regulation

The effectors of most negative feedback loops include the actions of nerves and hormones. In both neural and endocrine regulation, particular chemical regulators released by nerve fibers or endocrine glands stimulate target cells by interacting with specific receptor proteins in these cells. The mechanisms by which this regulation is achieved will be described in later chapters.

Homeostasis is maintained by two general categories of regulatory mechanisms: (1) those that are *intrinsic*, or "built-in," to the organs that produce them; and (2) those that are *extrinsic*; that is, regulation of an organ by the nervous and endocrine systems.

The endocrine system functions closely with the nervous system in regulating and integrating body processes and maintaining homeostasis. The nervous system controls the secretion of many endocrine glands, and some hormones in turn affect the function of the nervous system. Together, the nervous and endocrine system regulate the activities of most of the other systems of the body.

Regulation by the endocrine system is achieved by the secretion of chemical regulators called **hormones** into the blood. Since hormones are secreted into the blood, they are carried by the blood to all organs in the body. Only specific organs can respond to a particular hormone, however; these are known as the *target organs* of that hormone.

Nerve fibers, or axons (previously described in this chapter), are said to *innervate* the organs that they regulate. When stimulated, these axons produce electrochemical nerve impulses which are conducted from the origin of the axon at the cell body to the end of the axon in the target organ innervated by the axon. These target organs can be muscles or glands, which may function as effectors in the maintenance of homeostasis.

**Figure 1.20.** The regulation of hormone secretion by the hypothalamus, anterior pituitary, and target glands (thyroid, adrenal cortex, and gonads). Hormones secreted by the target glands exert negative feedback inhibition on the release of hormones from the hypothalamus and anterior pituitary.

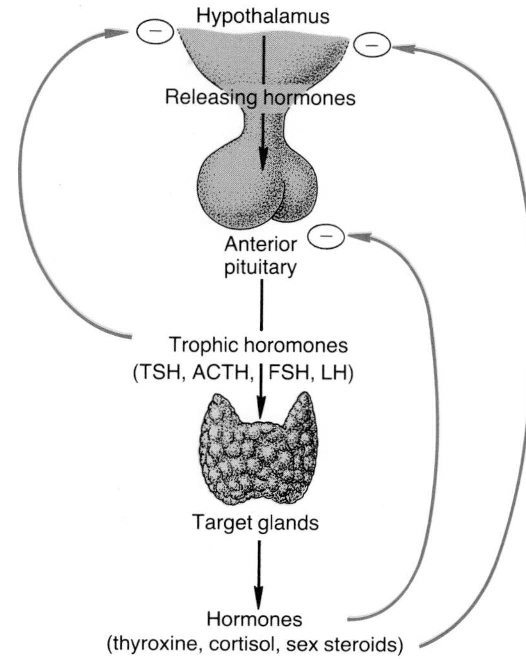


## Feedback Control of Hormone Secretion

The details of the nature of the endocrine glands, the interaction of the nervous and endocrine system, and the actions of hormones, will be explained in later chapters. The purpose of this section is to describe the regulation of hormone secretion in a brief, general manner, because this subject so superbly illustrates the principles of homeostasis and negative feedback regulation.

Hormones are secreted in response to specific chemical stimuli. The presence of proteins in the stomach, for example, stimulates the secretion of the hormone gastrin, and a rise in plasma glucose concentration stimulates insulin secretion from the islets of Langerhans in the pancreas. Hormones are also secreted in response to neurotransmitter molecules released from nerve endings and to stimulation by other hormones.

The nervous system helps to regulate hormone secretion in a variety of ways. Autonomic nerves (chapter 10), for example, stimulate the secretion of epinephrine from the adrenal medulla and also participate in the regulation of insulin and glucagon secretion from the islets of Langerhans in the pancreas. Hormonal secretion by the adrenal cortex, thyroid, and gonads is stimulated by hormones secreted by the anterior pituitary gland. Secretions of the anterior pituitary, in turn, are regulated by hormones released by neurons in the hypothalamus of the brain. This control is not one-way; many neural centers in the brain are affected by the actions of hormones (fig. 1.20).

The secretion of hormones is controlled by inhibitory as well as by stimulatory influences. The effect of a given hormone's action can inhibit its own secretion. The secretion of insulin—which acts to lower the plasma glucose concentration—is stimulated by a rise in glucose concentration, for example, and is inhibited by a fall in blood glucose. The lowering of blood glucose levels by insulin thus has an inhibitory feedback effect on further insulin secretion. This closed loop control system is called **negative feedback inhibition** (see fig. 1.21).

1. *Define homeostasis, and describe how this concept can be used to understand physiological control mechanisms.*
2. *Describe the meaning of the term negative feedback, and explain how it contributes to homeostasis. Illustrate this concept by drawing a negative feedback loop.*
3. *Describe positive feedback, and explain how this process functions in the body.*
4. *Explain how the secretion of a hormone is controlled by negative feedback inhibition. Illustrate this process with an example.*

**Figure 1.21.** The negative feedback control of insulin secretion by changes in the blood glucose concentration. Dashed arrows and negative signs indicate that negative feedback loops compensate the initial changes in blood glucose concentrations produced by eating or fasting.

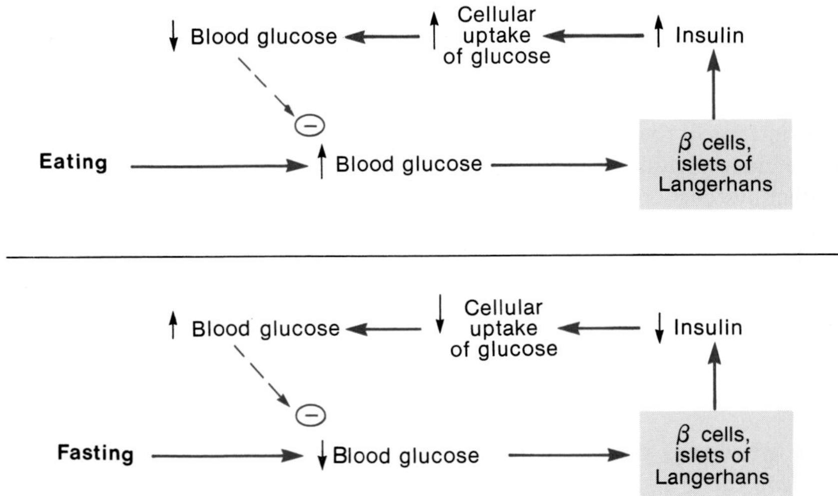


## Summary

### Introduction p. 4

I. Physiology is the study of how cells, tissues, and organs function.
   A. In the study of physiology, cause-and-effect sequences are emphasized.
   B. Knowledge of physiological mechanisms is deduced from data obtained experimentally.

II. All of the information in this book has been gained by applications of the scientific method. This method has three essential characteristics.
   A. It is assumed that the subject under study can ultimately be explained in terms we can understand.
   B. Descriptions and explanations are honestly based on observations of the natural world and can be changed when warranted by new observations.
   C. Humility is an important characteristic of the scientific method; the scientist must be willing to change his or her theories when warranted by the weight of the evidence.

### The Primary Tissues, Organs, and Systems p. 5

I. The body is composed of four primary tissues: muscular, nervous, epithelial, and connective tissues.
   A. There are three types of muscle tissue: skeletal, cardiac, and smooth muscle.
      1. Skeletal and cardiac muscle are striated.
      2. Smooth muscle is found in the walls of the internal organs.
   B. Nervous tissue is composed of neurons and neuroglial cells.
      1. Neurons are specialized for the generation and conduction of electrical impulses.
      2. Neuroglial cells provide anatomical and functional support to the neurons.
   C. Epithelial tissue includes membranes and glands.
      1. Epithelial membranes cover and line the body surfaces, and their cells are tightly joined by junctional complexes.
      2. Epithelial membranes may be simple or stratified, and their cells may be squamous, cuboidal, or columnar.
      3. Exocrine glands, which secrete into ducts, and endocrine glands, which lack ducts and secrete hormones into the blood, are derived from epithelial membranes.
   D. Connective tissue is characterized by large intercellular spaces that contain extracellular material.
      1. Connective tissue proper includes subtypes such as loose, dense fibrous, adipose, and others.
      2. Cartilage, bone, and blood are classified as connective tissues because their cells are widely spaced with abundant extracellular material between them.

II. Organs are units of structure and function that are composed of at least two, and usually all four, primary tissues.
   A. The skin is a good example of an organ.
      1. The epidermis is a stratified squamous keratinized epithelium, which serves to protect underlying structures and also produces vitamin D.
      2. The dermis is an example of loose connective tissue.
      3. Hair follicles, sweat glands, and sebaceous glands are exocrine glands found in the dermis.
      4. Sensory and motor nerve fibers enter the spaces within the dermis to innervate sensory organs and smooth muscles.

5. The arrector pili muscles that attach to the hair follicles are composed of smooth muscle.

B. Organs that are located in different regions of the body and perform related functions are grouped into systems; these include the cardiovascular system, digestive system, endocrine system, and others.

### Homeostasis and Feedback Control p. 16

I. Homeostasis refers to the dynamic constancy of the internal environment.

A. Homeostasis is maintained by mechanisms that act through negative feedback loops.

1. A negative feedback loop requires a sensor that can detect a change in the internal environment, and an effector that can be activated by the sensor.
2. In a negative feedback loop the effector acts to cause changes in the internal environment that compensate the initial deviations that are detected by the sensor.

B. Positive feedback loops serve to amplify changes and may be part of the action of an overall negative feedback mechanism.

C. The nervous and endocrine systems provide extrinsic regulation of other body systems and act to maintain homeostasis.

1. Many nerve fibers release chemicals called neurotransmitters; endocrine glands secrete chemicals called hormones into the blood.
2. Both neurotransmitters and hormones combine with specific receptor proteins and regulate specific target cells.

D. The secretion of hormones is stimulated by specific chemicals and is inhibited by negative feedback mechanisms.

---

## *Review Activities*

### Objective Questions

Match the following:

1. Glands are derived from
2. Cells are joined closely together in
3. Cells are separated by large extracellular spaces in
4. Blood vessels and nerves are usually located within

(a) nervous tissue
(b) connective tissue
(c) muscular tissue
(d) epithelial tissue

### Multiple Choice

5. The wall of a blood vessel is composed of
   (a) epithelial tissue
   (b) muscle tissue
   (c) connective tissue
   (d) all of these
6. Sweat is secreted by exocrine glands. This means that
   (a) it is produced by epithelial cells
   (b) it is a hormone
   (c) it is secreted into a duct
   (d) it is produced outside the body
7. Which of the following statements about homeostasis is *true?*
   (a) The internal environment is maintained absolutely constant.
   (b) Negative feedback mechanisms act to correct deviations from a normal range within the internal environment.
   (c) Homeostasis is maintained by switching effector actions on and off.
   (d) All of these are true.
8. In a negative feedback loop the effector organ produces changes that are
   (a) similar in direction to that of the initial stimulus
   (b) opposite in direction to that of the initial stimulus
   (c) unrelated to the initial stimulus
9. A hormone called *parathyroid hormone* acts to help raise the blood-calcium concentration. According to the principles of negative feedback, an effective stimulus for parathyroid hormone secretion would be
   (a) a fall in blood calcium
   (b) a rise in blood calcium
10. Which of the following consists of dense, parallel arrangements of collagen fibers?
    (a) skeletal muscle tissue
    (b) nervous tissue
    (c) tendons
    (d) dermis
11. The act of breathing raises the blood oxygen level, lowers the blood carbon dioxide concentration, and raises the blood pH. According to the principles of negative feedback, sensors that regulate breathing should respond to
    (a) a rise in blood oxygen
    (b) a rise in blood pH
    (c) a rise in blood carbon dioxide concentration
    (d) all of the above

### Essay Questions

1. Describe the structure of different epithelial membranes, and explain how their structures relate to their functions.
2. What are the similarities between the dermis of the skin, bone, and blood? What are the major structural differences between these tissues?
3. Explain the role of antagonistic negative feedback processes in the maintenance of homeostasis.
4. Explain, using examples, how the secretion of a hormone is controlled by the effects of that hormone's actions.
5. Describe the marriage of neural and endocrine regulation in the control of the anterior pituitary gland.

## *Selected Readings*


Adolph, E. F. 1968. *Origins of physiological regulations.* New York: Academic Press.

Benison, S., A. C. Barger, and E. L. Wolfe. 1987. *Walter B. Cannon: The Life and Times of a Young Scientist.* Cambridge, Mass.: Harvard University Press.

Bloom, W. B., and D. W. Fawcett. 1975. *A textbook of histology.* 10th ed. Philadelphia: W. B. Saunders Co.

Cannon, W. B. 1932. *The wisdom of the body.* New York: Norton and Company.

Di Fiore, M. S. H. 1974. *An atlas of histology.* 4th ed. Philadelphia: Lea and Febiger.

Fye, B. 1987. *The Development of American Physiology: Scientific Medicine in the Nineteenth Century.* Baltimore: Johns Hopkins University Press.

Jacobs, S., and P. Cuatrecasas. 1977. Cell receptors in disease. *New England Journal of Medicine* 297:1383.

Jones, R. W. 1973. *Principles of biological regulation: an introduction to feedback systems.* New York: Academic Press.

Kessel, R. G., and R. H. Kardon. 1979. *Tissues and organs: a text-atlas of scanning electron microscopy.* San Francisco: W. H. Freeman and Co.

Snyder, S. H. October 1985. The molecular basis for communication between cells. *Scientific American.*

Soderberg, U. 1964. Neurophysiological aspects of homeostasis. *Annual Review of Physiology* 26:271.

# 2

# Chemical Composition of the Body

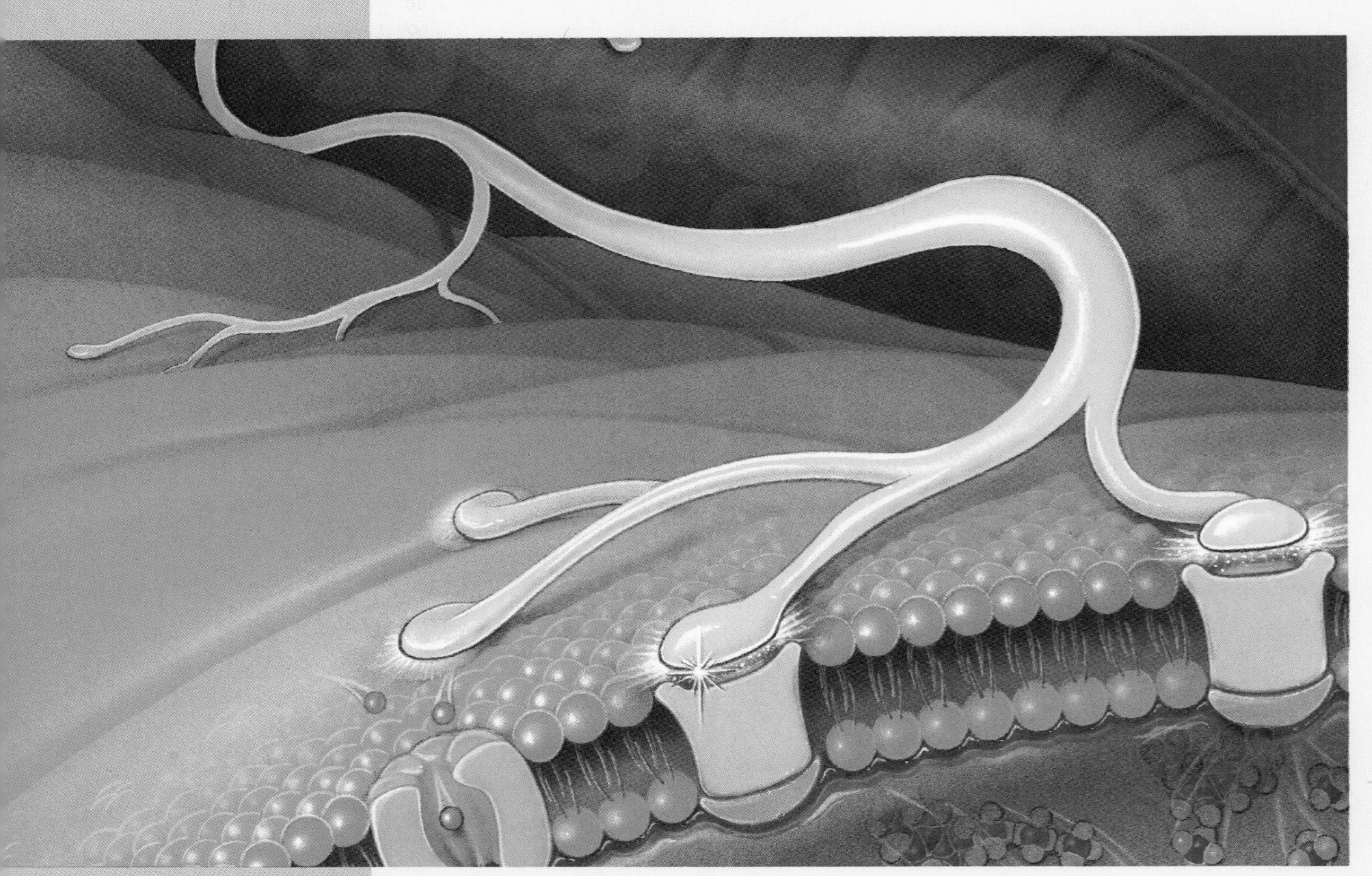

## Objectives

By studying this chapter, you should be able to

1. describe the structure of an atom and explain the meaning of the terms *atomic mass* and *atomic number*
2. explain how covalent bonds are formed and distinguish between nonpolar and polar covalent bonds
3. explain how ions and ionic bonds are formed and describe the nature of hydrogen bonds
4. describe the structure of a water molecule and explain why some compounds are hydrophilic and others are hydrophobic
5. define the terms *acid* and *base* and explain the meaning of the pH scale
6. describe the different types of carbohydrates and give examples of each type
7. explain the mechanisms and significance of dehydration synthesis and hydrolysis reactions
8. explain the common characteristic of lipids and describe the different categories of lipids
9. describe how peptide bonds are formed and the different orders of protein structure
10. list some of the functions of proteins and explain why proteins provide specificity in their functions

## Outline

## Atoms, Ions, and Chemical Bonds

The study of physiology requires some familiarity with the concepts and terminology of basic chemistry. A knowledge of atomic and molecular structure, the nature of chemical bonds, and the nature of pH and associated concepts provide the foundation for much of physiology.

The anatomical structures and physiological processes of the body are based, to a large degree, on the properties and interactions of atoms, ions, and molecules. Water is the major solvent in the body and contributes 65% to 75% of the total weight of an average adult. Of this amount, two-thirds is within the body cells (in the *intracellular compartment*); the remainder is in the *extracellular compartment,* including the blood and tissue fluids. Dissolved in this water are many organic molecules (carbon-containing molecules such as carbohydrates, lipids, proteins, and nucleic acids) and inorganic molecules and ions (atoms with a net charge). Before describing the structure and function of organic molecules within the body, some basic chemical concepts, terminology, and symbols will be introduced.

### Atoms

Atoms are much too small to be seen individually, even with the most powerful electron microscope. Through the efforts of many generations of scientists, however, the structure of an atom is now well understood. At the center of an atom is its *nucleus*. The nucleus contains two types of particles—*protons,* which have a positive charge, and *neutrons,* which are noncharged. The mass of a proton is approximately equal to the mass of a neutron, and the sum of the protons and neutrons in an atom is equal to the **atomic mass** of the atom. For example, an atom of carbon, which contains six protons and six neutrons, has an atomic mass of 12 (table 2.1).

The number of protons in an atom is given as its **atomic number.** Carbon has six protons and thus has an atomic number of six. Outside the positively charged nucleus are negatively charged *electrons*. Since the number of electrons in an atom is equal to the number of protons, atoms have a net charge of zero.

Although it is often convenient to think of electrons as orbiting the nucleus like planets orbiting the sun, this older view of atomic structure is no longer believed to be correct. A given electron can occupy any position in a certain volume of space called the *orbital* of the electron. The orbital is like an energy "shell," or barrier, beyond which the electron usually does not pass.

There are potentially several such orbitals around a nucleus, with each successive orbital being farther from the nucleus. The first orbital, closest to the nucleus, can contain only two electrons. If an atom has more than two electrons (as do all atoms except hydrogen and helium), the additional electrons must occupy orbitals that are farther removed from the nucleus. The second orbital can contain a maximum of eight electrons; the third can also contain a maximum of eight, and the fourth can contain a maximum of eighteen. Thus, the orbitals are filled from the innermost outward. Carbon, with six electrons, has two electrons in its first orbital and four electrons in its second orbital (fig. 2.1).

It is always the electrons in the outermost orbital, if this orbital is incomplete, that participate in chemical reactions and form chemical bonds. These outermost electrons are known as the *valence electrons* of the atom.

***Isotopes.*** A particular atom with a given number of protons in its nucleus may exist in several forms that differ from each other in their number of neutrons. The atomic number of these forms is thus the same, but their atomic mass is different. These different forms are called **isotopes.** All of the isotopic forms of a given atom are included in the term **chemical element.** The element hydrogen, for example, has three isotopes. The most common of these has a nucleus consisting of only one proton. Another isotope of hydrogen (called *deuterium*) has one proton and one neutron in the nucleus, whereas the third isotope (*tritium*) has one proton and two neutrons. Tritium is a radioactive isotope that is commonly used in physiological research and in many clinical laboratory procedures.

**Table 2.1** Atoms commonly present in organic molecules

| Atom | Symbol | Atomic Number | Atomic Mass | Orbital 1 | Orbital 2 | Orbital 3 | Number of Chemical Bonds |
|---|---|---|---|---|---|---|---|
| Hydrogen | H | 1 | 1 | 1 | 0 | 0 | 1 |
| Carbon | C | 6 | 12 | 2 | 4 | 0 | 4 |
| Nitrogen | N | 7 | 14 | 2 | 5 | 0 | 3 |
| Oxygen | O | 8 | 16 | 2 | 6 | 0 | 2 |
| Phosphorus | P | 15 | 31 | 2 | 8 | 5 | 5 |
| Sulfur | S | 16 | 32 | 2 | 8 | 6 | 2 |

**Figure 2.1.** Diagrams of the hydrogen and carbon atoms. The electron orbitals on the left are represented by dots indicating probable positions of the electrons. The orbitals on the right are represented by concentric circles.

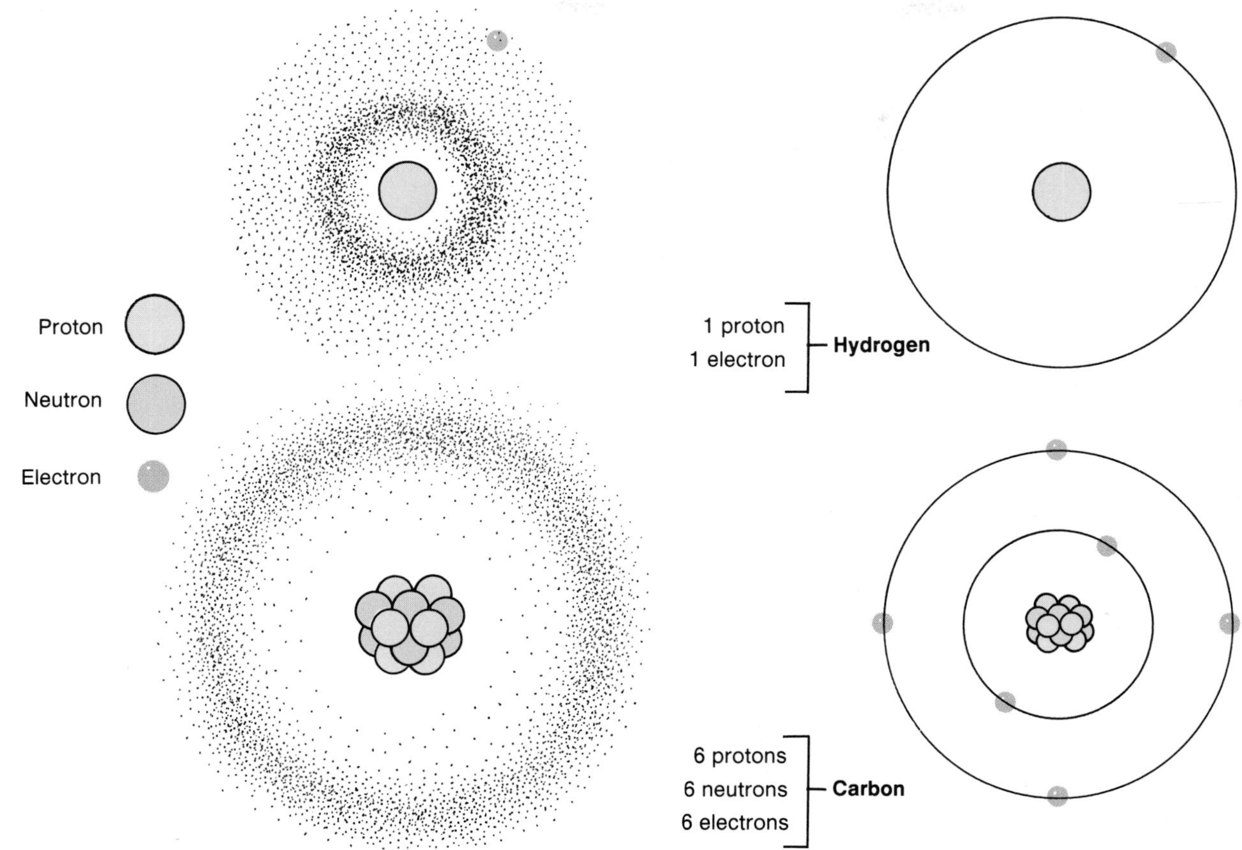


## Chemical Bonds, Molecules, and Ionic Compounds

Molecules are formed when two or more atoms are joined together by sharing or some other interaction of the electrons in their outer orbitals. Such sharing of electrons or other attractive forces produces *chemical bonds* (fig. 2.2). The number of bonds that each atom can have is determined by the number of electrons in its outer orbital. Hydrogen, for example, must obtain only one more electron—and can thus form one chemical bond—to complete the first orbital of two electrons. Carbon, in contrast, must obtain four more electrons—and can thus form four chemical bonds—to complete the second orbital of eight electrons (fig. 2.3).

***Covalent Bonds.*** In **covalent bonds,** atoms are joined together by the sharing of electrons—equally in nonpolar covalent bonds, unequally in polar covalent bonds. Covalent bonds that are formed between identical atoms—as in oxygen gas ($O_2$) and hydrogen gas ($H_2$)—are the strongest because their electrons are equally shared. Since the electrons are equally distributed between the two atoms, these molecules are said to be *nonpolar.* When covalent bonds are formed between two different atoms, however, the electrons may be pulled more toward one atom than the other. The side of the molecule toward which the electrons are pulled is electrically negative in comparison to the other end. Such a molecule is said to be *polar* (has a positive and negative "pole"). Atoms of oxygen, nitrogen, and phosphorus have a particularly strong tendency to pull electrons toward themselves when they bond with other atoms.

Water is the most abundant molecule in the body and serves as the solvent of body fluids. Water is a good solvent because it is polar; the oxygen atom pulls electrons from the two hydrogens toward its side of the water molecule, so that the oxygen side is negatively charged compared to the hydrogen side of the molecule (fig. 2.4). The significance of the polar nature of water in its function as a solvent is discussed in the next section.

**Figure 2.2.** The hydrogen molecule, showing the covalent bonds between hydrogen atoms formed by the equal sharing of electrons.

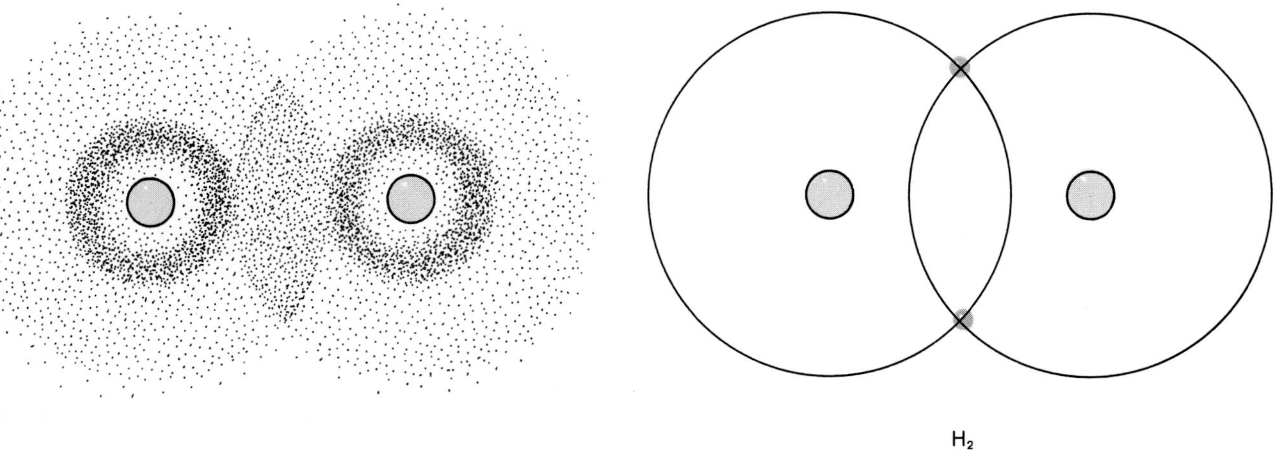


**Figure 2.3.** The molecules methane and ammonia represented in three different ways. Notice that a bond between two atoms consists of a pair of shared electrons (the electrons from the outer orbital of each atom).

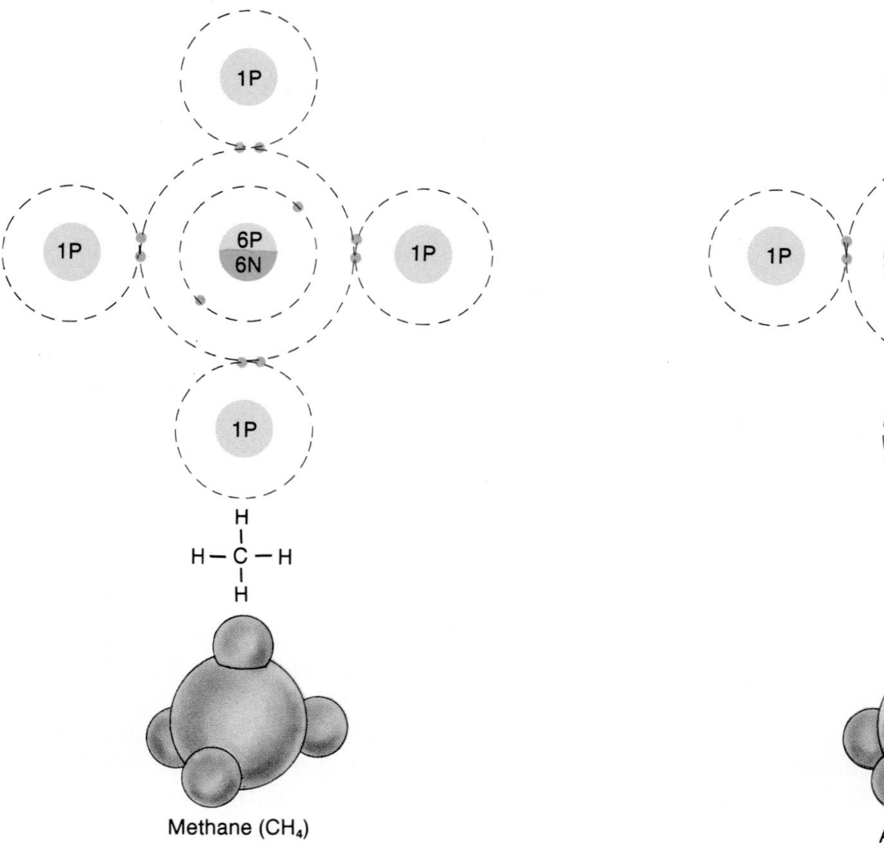

**Figure 2.4.** A model of a water molecule showing its polar nature. Notice that the oxygen side of the molecule is negative whereas the hydrogen side is positive. Polar covalent bonds are weaker than nonpolar covalent bonds. As a result, some water molecules ionize to form $OH^-$ (hydroxyl ion) and $H^+$ (hydrogen ion).

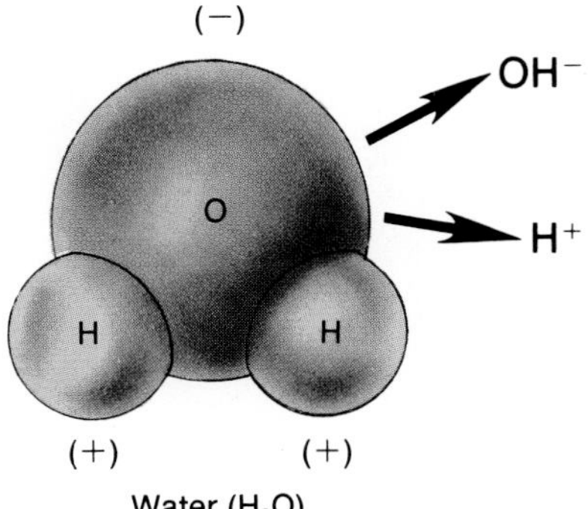


***Ionic Bonds.*** In **ionic bonds** the electrons are not shared at all. Instead, one or more valence electrons from one atom are completely transferred to a second atom. The first atom thus loses electrons, so that its number of electrons becomes less than its number of protons; it becomes a positively charged **ion.** Positively charged ions are called *cations* because they move toward the negative pole, or cathode, in an electric field. The second atom gains more electrons than it has protons and becomes a negatively charged ion, or *anion* (so-called because it moves toward the positive pole, or anode, in an electric field). The cation and anion attract each other to form an **ionic compound.**

Common table salt, sodium chloride (NaCl), is an example of an ionic compound. Sodium, with a total of eleven electrons, has two in its first orbital, eight in its second orbital, and only one in its third orbital. Chlorine, conversely, is one electron short of completing its outer orbital of eight electrons. The lone electron in sodium's outer orbital is attacted to chlorine's outer orbital. This creates a chloride ion (represented as $Cl^-$) and a sodium ion ($Na^+$). Although table salt is shown as NaCl, it is actually composed of $Na^+Cl^-$ (fig. 2.5).

**Figure 2.5.** The ionization of sodium chloride to produce sodium and chloride ions. The positive sodium and the negative chloride ions attract each other to produce the ionic compound sodium chloride (NaCl).

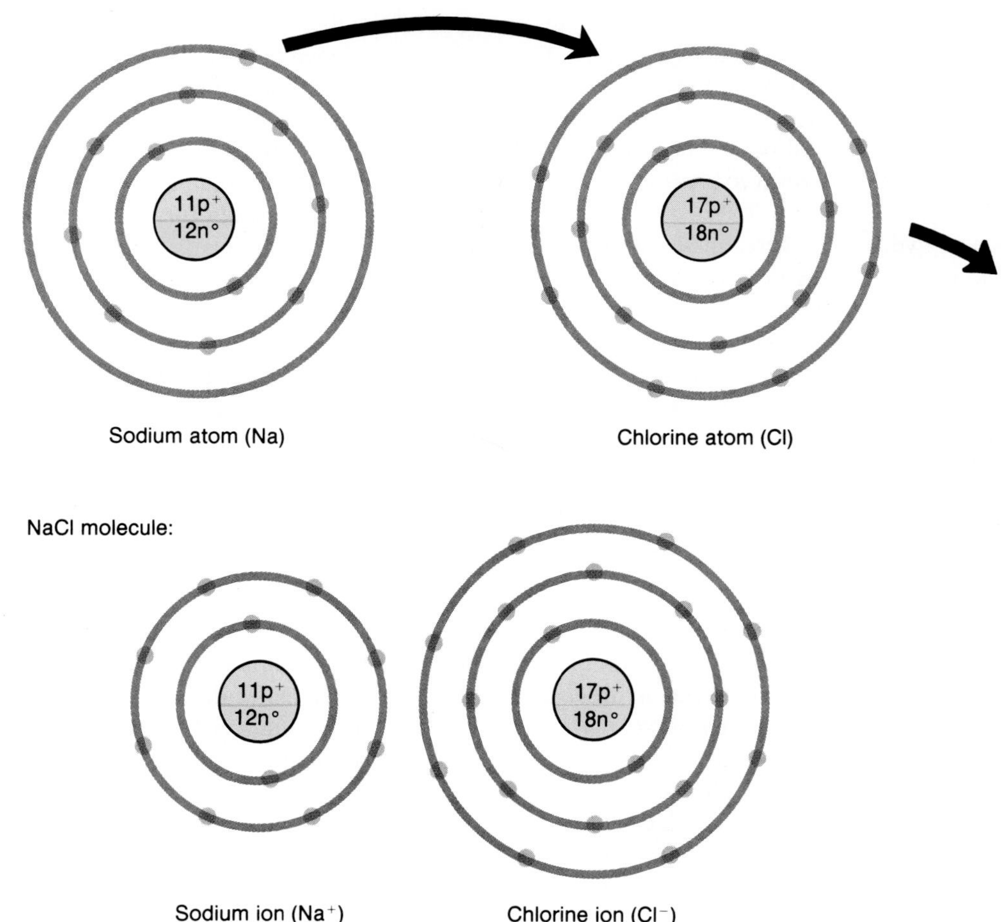

**Figure 2.6.** The negatively charged oxygen-ends of water molecules are attracted to the positively charged $Na^+$, whereas the positively charged hydrogen-ends of water molecules are attracted to the negatively charged $Cl^-$. Other water molecules are attracted to this first concentric layer of water, forming hydration spheres around the sodium and chloride ions.

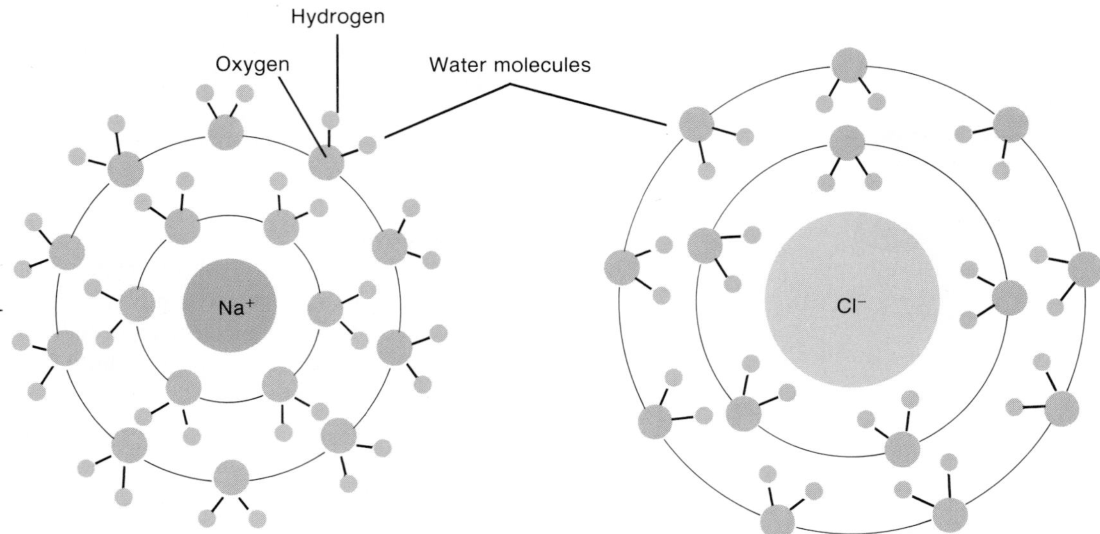


Ionic bonds are weaker than polar covalent bonds, and therefore ionic compounds easily dissociate when dissolved in water to yield their separate ions. Dissociation of NaCl, for example, yields $Na^+$ and $Cl^-$. Each of these ions attracts polar water molecules; the negative ends of water molecules are attracted to the $Na^+$, and the positive ends of water molecules are attracted to the $Cl^-$ (fig. 2.6). The water molecules that surround these ions in turn attract other molecules of water to form *hydration spheres* around each ion.

The formation of hydration spheres makes an ion or a molecule soluble in water. Glucose, amino acids, and many other organic molecules are water-soluble because hydration spheres can form around atoms of oxygen, nitrogen, and phosphorus, which are joined by polar covalent bonds to other atoms. Such molecules are said to be **hydrophilic.** In contrast, molecules composed primarily of nonpolar covalent bonds, such as the hydrocarbon chains of fat molecules, have few charges and, thus, cannot form hydration spheres. They are insoluble in water and, in fact, actually avoid water—they are **hydrophobic.**

***Hydrogen Bonds.*** Hydrogen bonds are very weak bonds that help to stabilize the delicate folding and bending of long organic molecules such as proteins. When hydrogen forms a polar covalent bond with an atom of oxygen or nitrogen, the hydrogen gains a slight positive charge as its electron is pulled toward the other atom. This also imparts a slight negative charge to the other atom, which is said to be electronegative. Since the hydrogen has a slight positive charge, it will have a weak attraction for a second electronegative atom (oxygen or nitrogen) that may be located nearby. This weak attraction is called a **hydrogen bond.** Hydrogen bonds are usually shown with dotted lines (fig. 2.7) to distinguish them from strong covalent bonds, which are shown with solid lines.

Hydrogen bonds can be formed within the folds of a large molecule, such as a protein (discussed in a later section). They can also be formed between adjacent water molecules (fig. 2.7). The hydrogen bonding between water molecules is responsible for many of the physical properties of water, including its *surface tension* and its ability to be pulled as a column through narrow channels in a process called *capillary action.*

## Acids, Bases, and the pH Scale

The bonds in water molecules joining hydrogen and oxygen atoms together are, as previously discussed, polar covalent bonds. Although these bonds are strong, a proportion of them break as the electron from the hydrogen atom is completely transferred to oxygen. When this occurs, the water molecule ionizes to form a *hydroxyl* ion ($OH^-$) and a hydrogen ion ($H^+$), which is simply a free proton. A proton released in this way does not remain free for long, however, because it is attracted to the electrons of oxygen atoms in water molecules. This forms a *hydronium ion,* shown by the formula $H_3O^+$. For the sake of clarity in the following discussion, however, $H^+$ will be used to represent the ion resulting from the dissociation of water.

**Ionization of water molecules produces equal amounts of $OH^-$ and $H^+$. Since only a small proportion of water molecules ionize, the concentration of $H^+$ ions are each**

**Figure 2.7.** The oxygen atoms of water molecules are weakly joined together by the attraction of the electronegative oxygen for the positively charged hydrogen. These weak bonds are called *hydrogen bonds.*

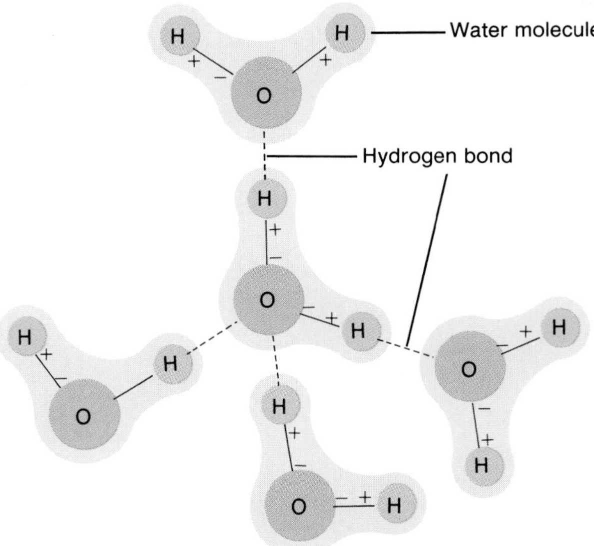


**Table 2.2** Common acids and bases

| Acid | Symbol | Base | Symbol |
|---|---|---|---|
| Hydrochloric acid | $HCl$ | Sodium hydroxide | $NaOH$ |
| Phosphoric acid | $H_3PO_4$ | Potassium hydroxide | $KOH$ |
| Nitric acid | $HNO_3$ | Calcium hydroxide | $Ca(OH)_2$ |
| Sulfuric acid | $H_2SO_4$ | Ammonium hydroxide | $NH_4OH$ |
| Carbonic acid | $H_2CO_3$ | | |

**Table 2.3** The pH scale

| | $H^+$ Concentration (molar) | pH | $OH^-$ Concentration (molar) |
|---|---|---|---|
| | 1.0 | 0 | $10^{-14}$ |
| | 0.1 | 1 | $10^{-13}$ |
| | 0.01 | 2 | $10^{-12}$ |
| Acids | 0.001 | 3 | $10^{-11}$ |
| | 0.0001 | 4 | $10^{-10}$ |
| | $10^{-5}$ | 5 | $10^{-9}$ |
| | $10^{-6}$ | 6 | $10^{-8}$ |
| Neutral | $10^{-7}$ | 7 | $10^{-7}$ |
| | $10^{-8}$ | 8 | $10^{-6}$ |
| | $10^{-9}$ | 9 | $10^{-5}$ |
| | $10^{-10}$ | 10 | 0.0001 |
| Bases | $10^{-11}$ | 11 | 0.001 |
| | $10^{-12}$ | 12 | 0.01 |
| | $10^{-13}$ | 13 | 0.1 |
| | $10^{-14}$ | 14 | 1.0 |

only equal to $10^{-7}$ molar (the term *molar* is a unit of concentration described in chapter 6; for hydrogen, one molar equals one gram per liter). A solution with $10^{-7}$ molar hydrogen ion concentration, which is produced by the ionization of water molecules in which the $H^+$ and $OH^-$ concentrations are equal, is said to be **neutral.**

A solution that contains a higher $H^+$ concentration than that of water is called *acidic,* and a solution with a lower $H^+$ concentration is called *basic.* An **acid** is defined as a molecule that can release protons ($H^+$) to a solution; it is a "proton donor." A **base** is a negatively charged ion (anion), or a molecule that ionizes to produce the anion, which can combine with $H^+$ and thus remove the $H^+$ from solution; it is a "proton acceptor." Most strong bases release $OH^-$ into a solution, which combines with $H^+$ to form water and which thus lowers the $H^+$ concentration. Examples of common acids and bases are shown in table 2.2.

***pH.*** The $H^+$ concentration of a solution is usually indicated in pH units on a pH scale that runs from 0 to 14. The pH number is equal to the logarithm of one over the $H^+$ concentration:

$$pH = \log \frac{1}{[H^+]}$$

where $[H^+]$ = molar $H^+$ concentration.

Pure water has a $H^+$ concentration of $10^{-7}$ molar and, thus, has a pH of 7 (neutral). Because of the logarithmic relationship, a solution with ten times the hydrogen ion concentration ($10^{-6}$ M) has a pH of 6, whereas a solution with one-tenth the $H^+$ concentration ($10^{-8}$ M) has a pH of 8. The pH number is easier to write than the molar $H^+$ concentration, but it is admittedly confusing because it is *inversely related* to the $H^+$ concentration: a solution with a higher $H^+$ concentration has a lower pH number; one with a lower $H^+$ concentration has a higher pH number. A strong acid with a high $H^+$ concentration of $10^{-2}$ molar, for example, has a pH of 2, whereas a solution with only $10^{-10}$ molar $H^+$ has a pH of 10. **Acidic solutions,** therefore, have a pH of less than 7 (that of pure water), whereas **basic solutions** have a pH of between 7 and 14 (table 2.3).

***Buffers.*** A *buffer* is a system of molecules and ions that acts to prevent changes in $H^+$ concentration and thus that serves to stabilize the pH of a solution. In blood plasma, for example, the pH is stabilized by the following reversible reaction involving the bicarbonate ion ($HCO_3^-$) and carbonic acid ($H_2CO_3$):

$$HCO_3^- + H^+ \rightleftharpoons H_2CO_3$$

The double arrows indicate that the reaction could go either to the right or to the left; the net direction depends on the concentration of molecules and ions on each side. If an acid (such as lactic acid) should release $H^+$ into the

**Figure 2.8.** Two carbon atoms joined by a single covalent bond (*above*) or by a double covalent bond (*below*). In both cases each carbon atom shares four pairs of electrons (has four bonds) to complete the eight electrons required to fill its outer orbital.

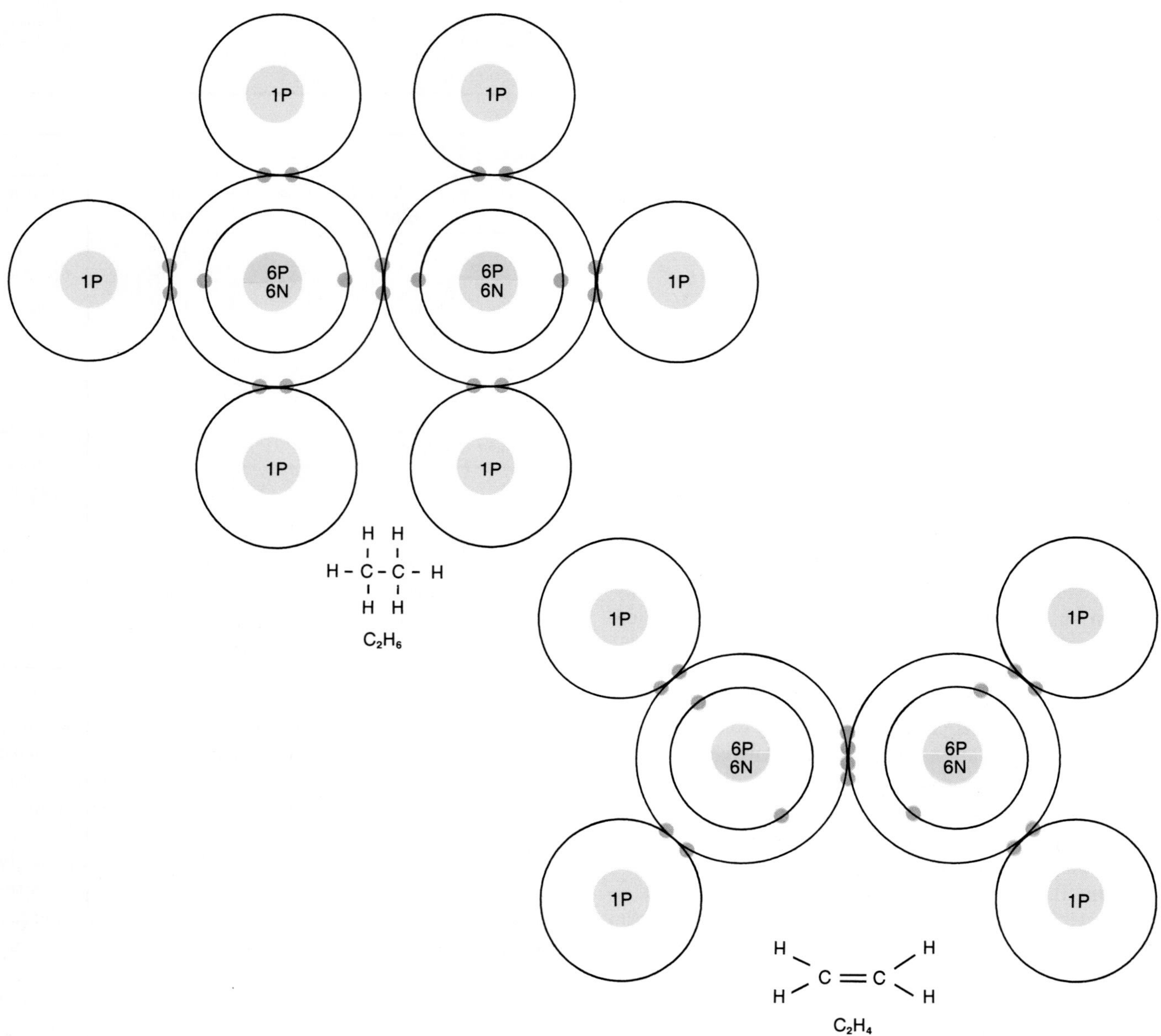


solution, for example, the increased concentration of $H^+$ would drive the equilibrium to the right and the following reaction would be promoted:

$$HCO_3^- + H^+ \longrightarrow H_2CO_3$$

The above reaction serves to decrease the effect of added $H^+$ on the pH of the blood. Bicarbonate, in fact, is the major buffer of the blood. Under the opposite condition, when the concentration of free $H^+$ in the blood is falling, the reaction previously described can be reversed:

$$H_2CO_3 \longrightarrow H^+ + HCO_3^-$$

The dissociation of carbonic acid yields free $H^+$, which helps to prevent an increase in pH. Bicarbonate ions and carbonic acid thus act as a *buffer pair* to prevent either decreases or increases in pH, respectively. This buffering action normally maintains the blood pH at a very stable $7.40 \pm 0.05$.

## Organic Molecules

Organic molecules are those that contain the atom *carbon.* Since the carbon atom has four electrons in its outer orbital, it must share four additional electrons by covalent bonding with other atoms to fill its outer orbital with eight

**Figure 2.9.** Hydrocarbons that are (*a*) linear, (*b*) cyclic, and (*c*) aromatic rings.

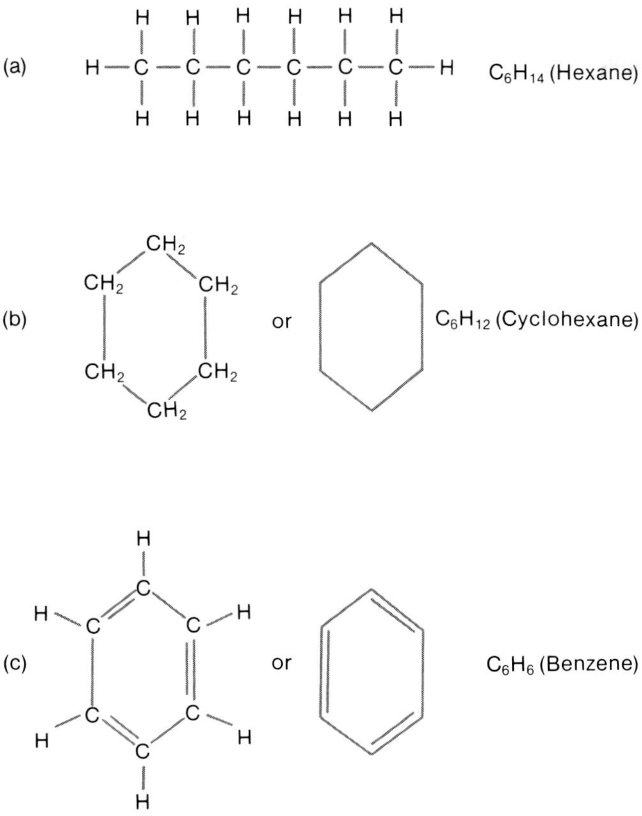

(a) $C_6H_{14}$ (Hexane)

(b) $C_6H_{12}$ (Cyclohexane)

(c) $C_6H_6$ (Benzene)

**Figure 2.10.** Various functional groups of organic molecules.

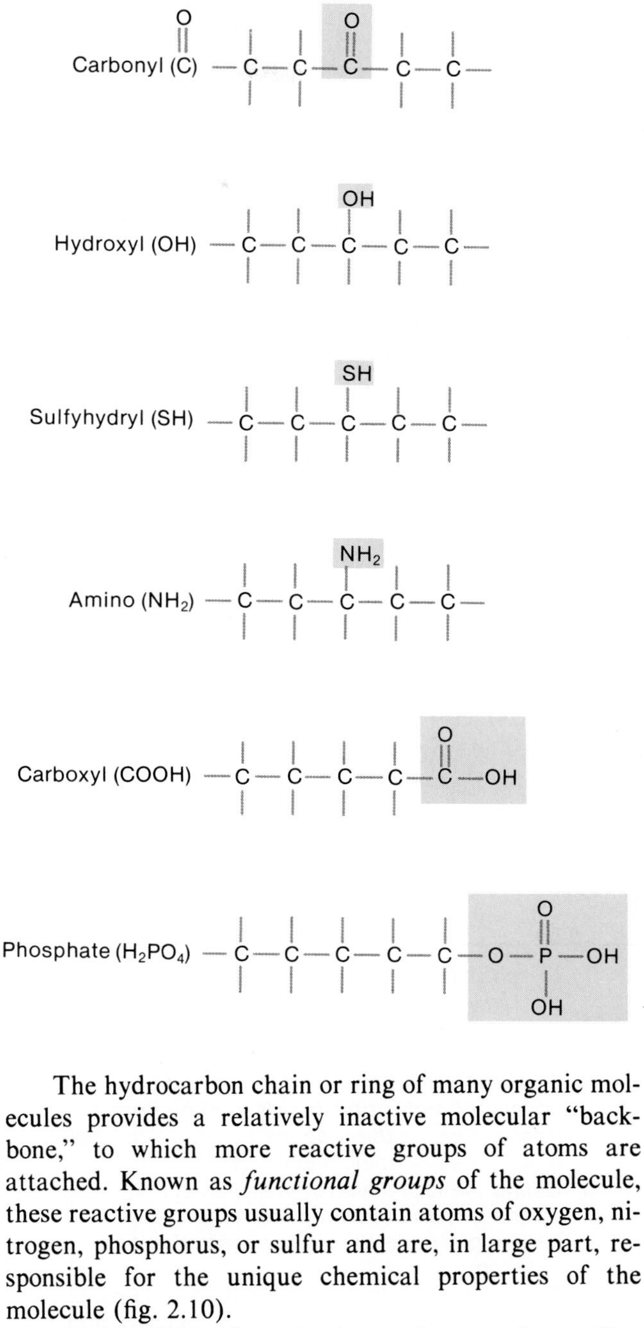

Carbonyl (C)

Hydroxyl (OH)

Sulfyhydryl (SH)

Amino ($NH_2$)

Carboxyl (COOH)

Phosphate ($H_2PO_4$)

electrons. The unique bonding requirements of carbon enable it to join with other carbon atoms to form chains and rings, while still allowing the carbon atoms to bond with hydrogen and other atoms.

Most organic molecules in the body contain hydrocarbon chains and rings as well as other atoms bonded to carbon. Two adjacent carbon atoms in a chain or ring may share one or two pairs of electrons. If the two carbon atoms share one pair of electrons, they are said to have a *single covalent bond;* this leaves each carbon atom free to bond to as many as three other atoms. If the two carbon atoms share two pairs of electrons, they have a *double covalent bond,* and each carbon atom can only bond to a maximum of two additional atoms (fig. 2.8).

The ends of some hydrocarbons are joined together to form rings. In the shorthand structural formulas of these molecules, the carbon atoms are not shown but are understood to be located at the corners of the ring. Some of these cyclic molecules have a double bond between two adjacent carbon atoms. Benzene and related molecules are shown as a six-sided ring with alternating double bonds. Such compounds are called *aromatic.* Since all of the carbons in an aromatic ring are equivalent, double bonds can be shown between any two adjacent carbons in the ring (fig. 2.9).

The hydrocarbon chain or ring of many organic molecules provides a relatively inactive molecular "backbone," to which more reactive groups of atoms are attached. Known as *functional groups* of the molecule, these reactive groups usually contain atoms of oxygen, nitrogen, phosphorus, or sulfur and are, in large part, responsible for the unique chemical properties of the molecule (fig. 2.10).

Classes of organic molecules can be named according to their functional groups. *Ketones,* for example, have a carbonyl group within the carbon chain. An organic molecule is an *alcohol* if it has a hydroxyl group at one end of the chain. All *organic acids* (such as acetic acid, citric acids, and others) have a *carboxyl* group (fig. 2.11).

***Stereoisomers.*** Two molecules may have exactly the same atoms arranged in exactly the same sequence, yet may differ with respect to the spatial orientation of key functional groups. Such molecules are called *stereoisomers* of each other. The isomers that have a key functional

**Figure 2.11.** Categories of organic molecules based on functional groups.

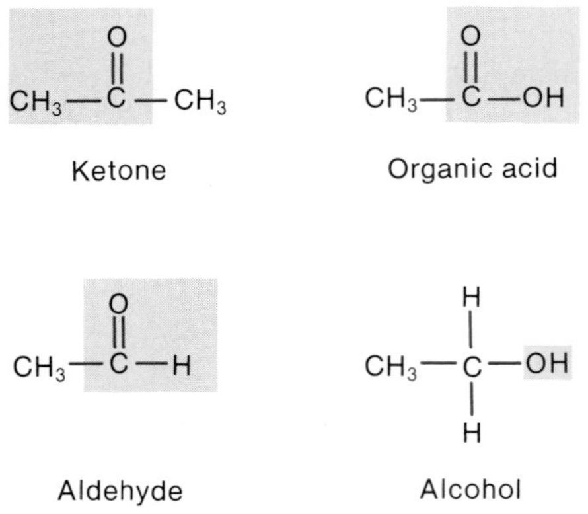

$CH_3—C(=O)—CH_3$

Ketone

$CH_3—C(=O)—OH$

Organic acid

$CH_3—C(=O)—H$

Aldehyde

$CH_3—CH_2—OH$

Alcohol

group represented on the right side of the molecule are called **D-isomers** (for *dextro,* or right-handed). Molecules that are represented by structures showing functional groups on the left side are called **L-isomers** (for *levo,* or left-handed).

The two stereoisomers are mirror images of each other—they cannot be superimposed. These subtle differences in structure are extremely important biologically, because enzymes—which interact with such molecules in a stereospecific way in chemical reactions—cannot combine with the "wrong" stereoisomer. The enzymes of all cells (human and others) can only combine with L-amino acids and D-sugars, for example. The opposite stereoisomers (D-amino acids and L-sugars) cannot be used by the body.

1. *Describe the structure of an atom, and define the terms* atomic weight *and* atomic number. *Explain why different atoms are able to form characteristic numbers of chemical bonds.*
2. *Describe the nature of nonpolar and polar covalent bonds, ionic bonds, and hydrogen bonds. Explain the reasons why ions and polar molecules are soluble in water.*
3. *Define the terms* acidic, basic, acid, and base. *Define* pH, *and describe the relationship between pH and the $H^+$ concentration of a solution.*
4. *Explain how carbon atoms can bond to each other and to atoms of hydrogen, oxygen, and nitrogen. Describe some of the functional groups to be found in organic molecules, and explain the significance of these compounds.*

## Carbohydrates and Lipids

Carbohydrates are a class of organic molecules which consist of monosaccharides, disaccharides, and polysaccharides. All of these molecules are based upon a characteristic ratio of carbon, hydrogen, and oxygen atoms. Lipids are a diverse category of organic molecules which share the physical property of being nonpolar and thus insoluble in water.

Carbohydrates and lipids are similar in many ways. Both groups of molecules consist primarily of the atoms carbon, hydrogen, and oxygen, and both serve as major sources of energy in the body (comprising most of the calories consumed in food). Carbohydrates and lipids differ, however, in some important aspects of their chemical structures and physical properties. Such differences significantly affect the functions of these molecules in the body.

### Carbohydrates

Carbohydrates are organic molecules that contain carbon, hydrogen, and oxygen in the ratio described by their name—*carbo* (carbon) and *hydrate* (water, $H_2O$). The general formula of a carbohydrate molecule is thus $CH_2O$; the molecule contains twice the number of hydrogen atoms as it contains carbon or oxygen atoms.

***Monosaccharides, Disaccharides, and Polysaccharides.*** Carbohydrates include simple sugars, or **monosaccharides,** and longer molecules that contain a number of monosaccharides joined together. The suffix *-ose* denotes a sugar molecule; the term *hexose,* for example, refers to a six-carbon monosaccharide with the formula $C_6H_{12}O_6$. This formula is adequate for some purposes, but it does not distinguish between related hexose sugars, which are *structural isomers* of each other. The structural isomers glucose, fructose, and galactose, for example, are monosaccharides that have the same ratio of atoms arranged in slightly different ways (fig. 2.12).

Two monosaccharides can be joined covalently to form a **disaccharide,** or double sugar. Common disaccharides include table sugar, or *sucrose* (composed of glucose and fructose), milk sugar, or *lactose* (composed of glucose and galactose), and malt sugar, or *maltose* (composed of two glucose molecules). When many monosaccharides are joined together, the resulting molecule is called a **polysaccharide.** *Starch,* for example, is a polysaccharide found in many plants which is formed by the bonding together of thousands of glucose subunits. Animal starch (**glycogen**), found in the liver and muscles, likewise consists of repeating glucose molecules but differs from plant starch in that it is more highly branched (fig. 2.13).

**Figure 2.12.** The structural formulas of three hexose sugars—(*a*) glucose, (*b*) galactose, and (*c*) fructose. All three have the same ratio of atoms—$C_6H_{12}O_6$.

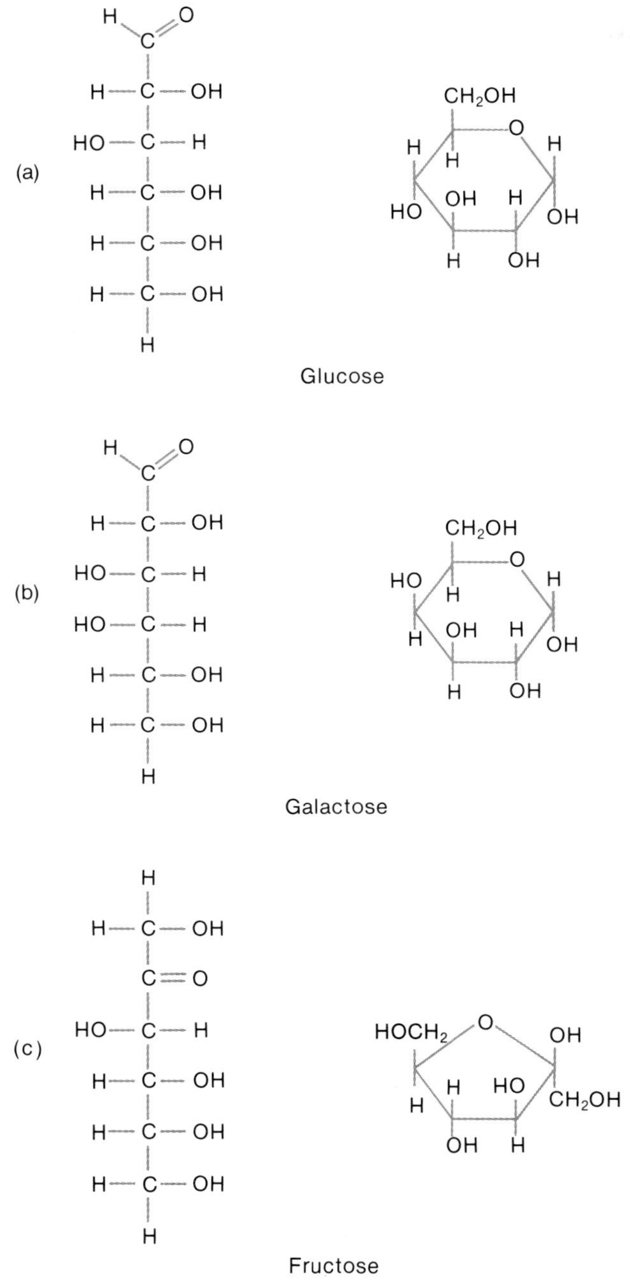

(a) Glucose

(b) Galactose

(c) Fructose

***Dehydration Synthesis and Hydrolysis.*** In the formation of disaccharides and polysaccharides, the separate subunits (monosaccharides) are bonded together covalently by a type of reaction called **dehydration synthesis,** or **condensation.** In this reaction, which requires the participation of specific enzymes (chapter 4), a hydrogen atom is removed from one monosaccharide and a hydroxyl group (OH) is removed from another. As a covalent bond is formed between the two monosaccharides, water ($H_2O$) is produced. Dehydration synthesis reactions are illustrated in figure 2.14.

When a person eats disaccharides and polysaccharides or when the stored glycogen in the liver and muscles is to be used by tissue cells, the covalent bonds that join monosaccharides into disaccharides and polysaccharides must be broken. These *digestion* reactions occur by means of **hydrolysis.** Hydrolysis is the reverse of dehydration synthesis. A water molecule is split, as implied by the word *hydrolysis,* and the resulting hydrogen atom is added to one of the free glucose molecules as the hydroxyl group is added to the other (fig. 2.15).

When a potato is eaten, the starch within it is hydrolyzed into separate glucose molecules within the intestine. This glucose is absorbed into the blood and carried to the tissues. Some tissue cells may use this glucose for energy. Liver and muscles, however, can store excess glucose in the form of glycogen by dehydration synthesis reactions in these cells. During fasting or prolonged exercise, the liver can add glucose to the blood through hydrolysis of its stored glycogen.

Dehydration synthesis, or condensation, and hydrolysis reactions do not occur spontaneously; they require the action of specific enzymes. Similar reactions, in the presence of other enzymes, build and break down lipids, proteins, and nucleic acids. In general, therefore, hydrolysis reactions digest molecules into their subunits, and dehydration synthesis reactions build larger molecules by the bonding together of their subunits.

## Lipids

The category of molecules known as lipids includes several types of molecules that differ greatly in chemical structure. These diverse molecules are all in the lipid category by virtue of a common physical property—they are all *insoluble in polar solvents* such as water. This is because lipids consist primarily of hydrocarbon chains and rings, which are nonpolar and, thus, hydrophobic. Although lipids are insoluble in water, they can be dissolved in nonpolar solvents such as ether, benzene, and related compounds.

***Triglycerides.*** Triglycerides are a subcategory of lipids that includes fat and oil. These molecules are formed by the condensation of one molecule of *glycerol* (a three-carbon alcohol) with three molecules of *fatty acids.* Each fatty acid molecule consists of a nonpolar hydrocarbon chain with a carboxylic acid group (abbreviated COOH) on one end. If the carbon atoms within the hydrocarbon chain are joined by single covalent bonds, so that each carbon atom can also bond to two hydrogen atoms, the fatty acid is said to be *saturated.* If there are a number

**Figure 2.13.** Glycogen is a polysaccharide composed of glucose subunits joined together to form a large, highly branched molecule.

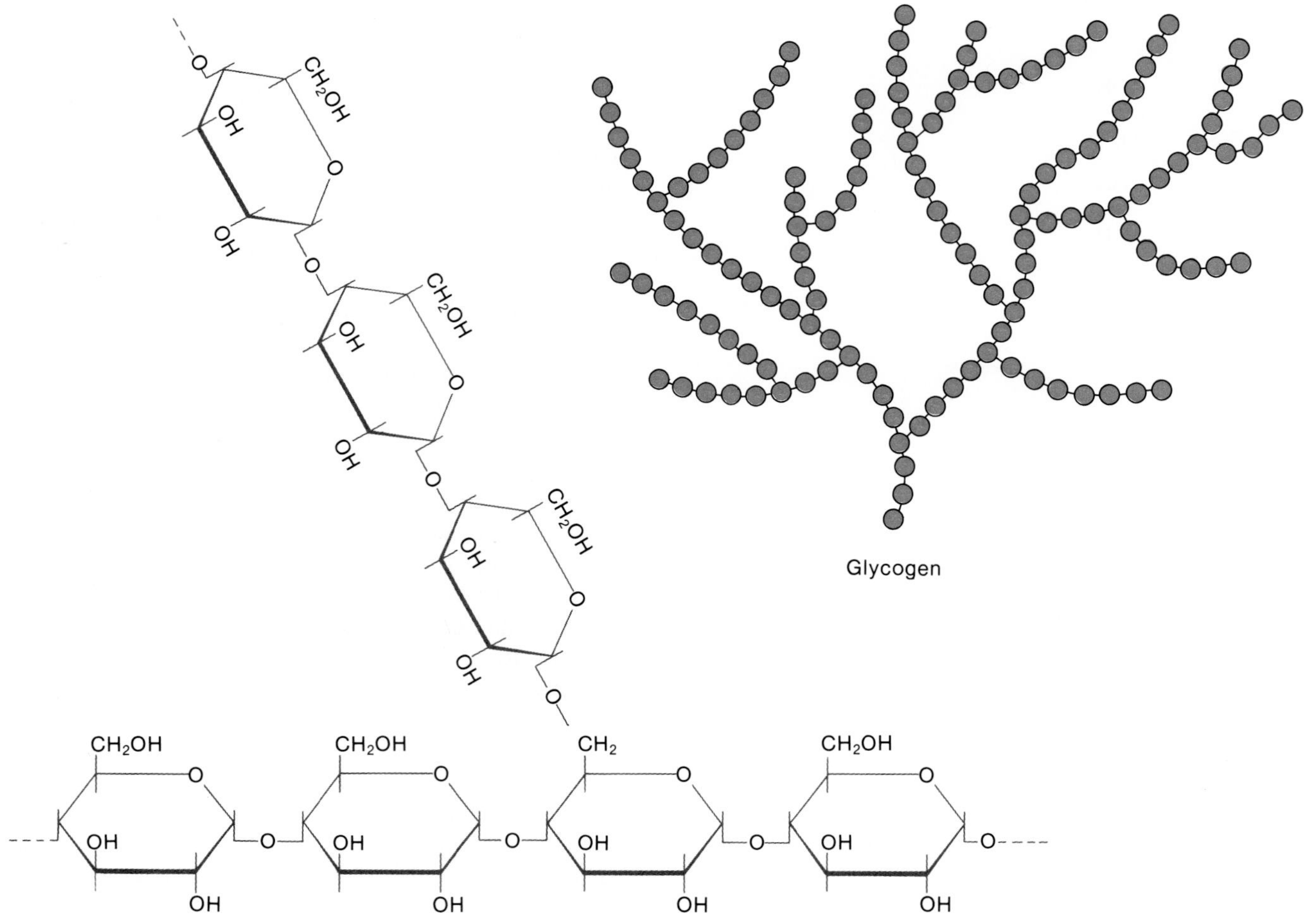


of double covalent bonds within the hydrocarbon chain, so that each carbon atom can only bond to one hydrogen atom, the fatty acid is said to be *unsaturated.* Triglycerides that contain saturated fatty acids are called **saturated fats;** those that contain unsaturated fatty acids are **unsaturated fats** (fig. 2.16).

Within the adipose cells of the body, triglycerides are formed as the carboxylic acid ends of fatty acid molecules condense with the hydroxyl groups of a glycerol molecule (fig. 2.17). Since the hydrogen atoms from the carboxyl ends of fatty acid molecules form water molecules during dehydration synthesis, fatty acids that are combined with glycerol can no longer release $H^+$ and function as acids. For this reason, triglycerides are described as *neutral fats.*

***Ketone Bodies.*** Hydrolysis of triglycerides within adipose tissue releases *free fatty acids* into the blood. Free fatty acids can be used as an immediate source of energy by many organs; they can also be converted by the liver into derivatives called *ketone bodies*. These include four-carbon-long acidic molecules (acetoacetic acid and $\beta$-hydroxybutyric acid) and acetone (the active ingredient in nail polish remover). A rapid breakdown of fat, such as occurs during dieting and in uncontrolled diabetes mellitus, results in elevated levels of ketone bodies in the blood, a condition called **ketosis.** If there are sufficient amounts of ketone bodies in the blood to lower the blood pH, the condition is called **ketoacidosis.**

***Phospholipids.*** The class of lipids known as *phospholipids* contains a number of different categories, which have in common the fact that they are lipids that contain a phosphate group. The most common type of phospholipid molecule has this structure: the three-carbon alcohol molecule glycerol is attached to two fatty acid molecules; the third carbon atom of the glycerol molecule is attached to a phosphate group, and the phosphate group in turn is bonded to other molecules. If the phosphate group is attached to a nitrogen-containing choline molecule, the phospholipid molecule thus formed is known as *lecithin*

**Figure 2.14.** Dehydration synthesis of two disaccharides, (*a*) maltose and (*b*) sucrose. Notice that as the disaccharides are formed a molecule of water is produced.

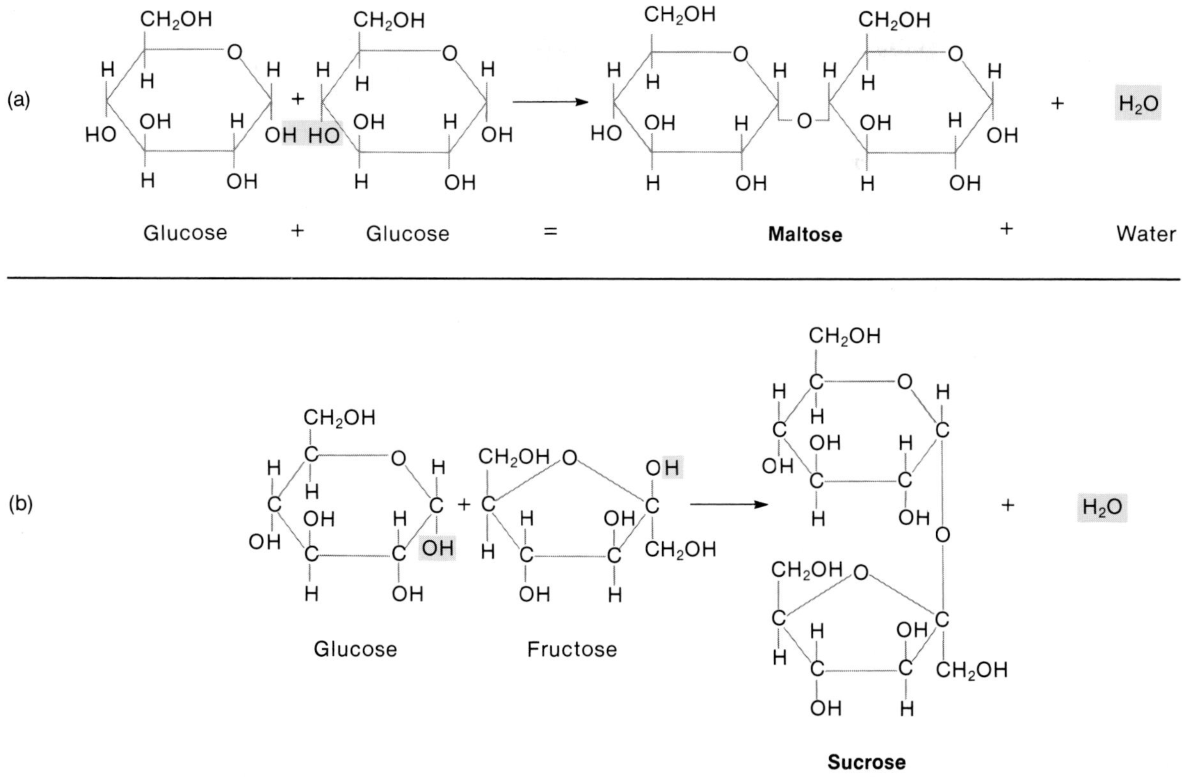

(a)

Glucose + Glucose = **Maltose** + Water ($H_2O$)

(b)

Glucose + Fructose → **Sucrose** + $H_2O$

**Figure 2.15.** The hydrolysis of starch (*a*) into disaccharides (maltose) and (*b*) into monosaccharides (glucose). Notice that as the covalent bond between the subunits breaks, a molecule of water is split. In this way the hydrogen and hydroxyl from water is added to the ends of the released subunits.

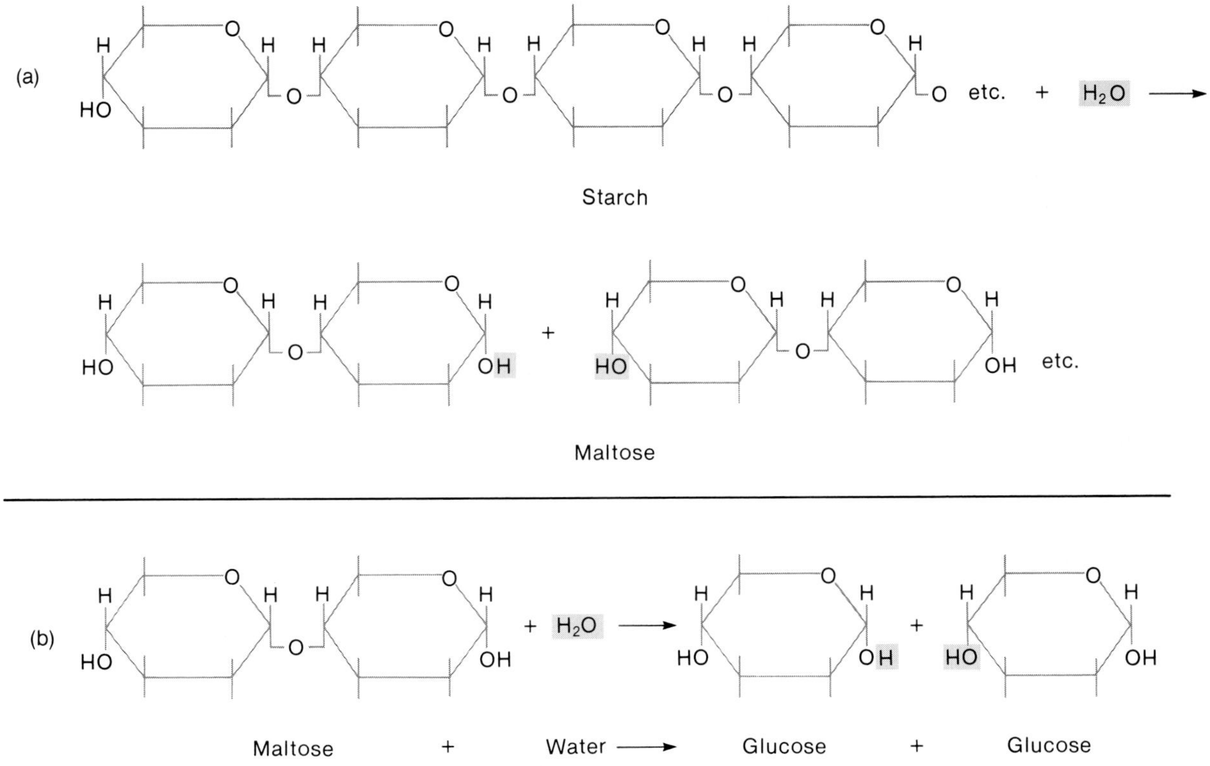

(a)

Starch etc. + $H_2O$ →

Maltose + Maltose etc.

(b)

Maltose + Water → Glucose + Glucose

**Figure 2.16.** Structural formulas for (*a*) saturated and (*b*) unsaturated fatty acids.

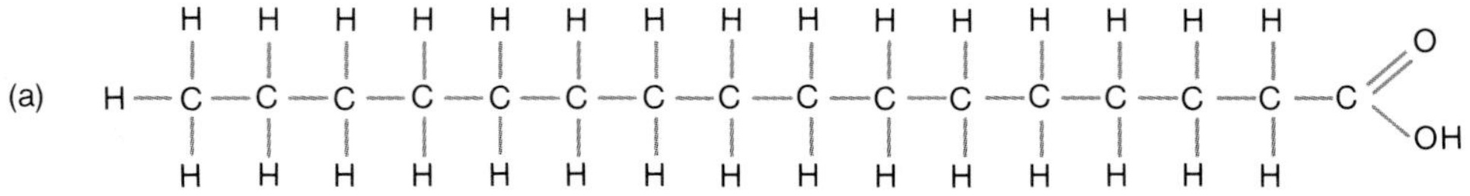

(a)

Palmitic acid, a saturated fatty acid

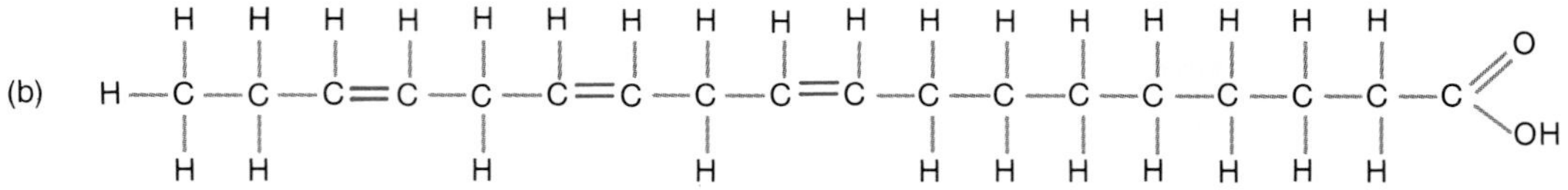

(b)

Linolenic acid, an unsaturated fatty acid

**Figure 2.17.** Dehydration synthesis of a triglyceride molecule from a glycerol and three fatty acids. A molecule of water is produced as an ester bond forms between each fatty acid and the glycerol. Sawtooth lines represent carbon chains, which are symbolized by an *R*.

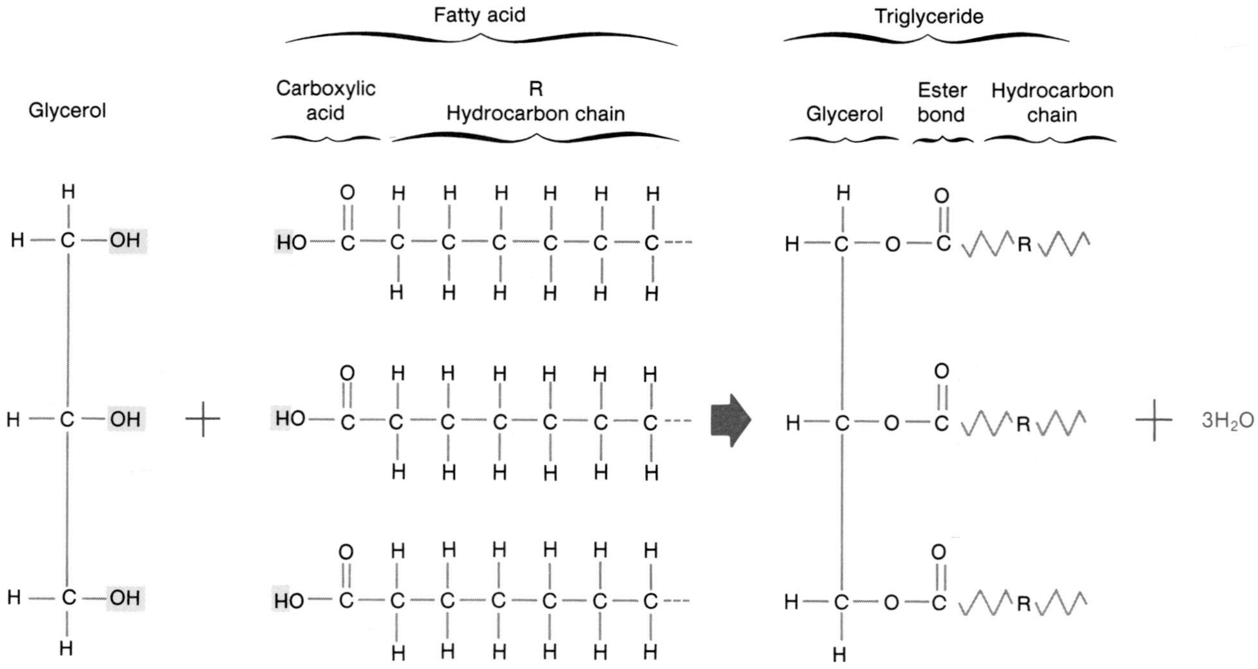

Glycerol

Fatty acid

Carboxylic acid

R Hydrocarbon chain

Triglyceride

Glycerol

Ester bond

Hydrocarbon chain

+ $3H_2O$

**Figure 2.18.** The structure of lecithin, a typical phospholipid (*above*), and its more simplified representation (*below*).

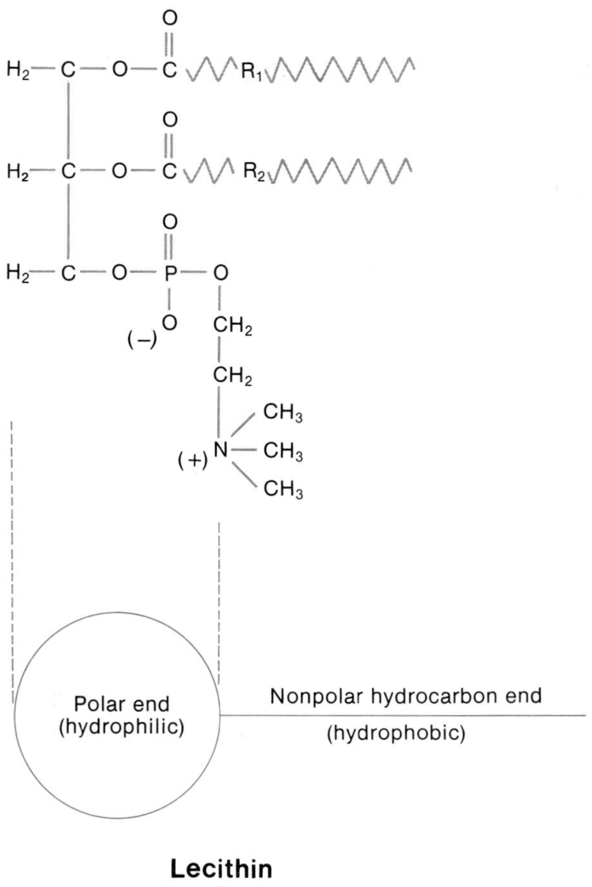

Polar end (hydrophilic)

Nonpolar hydrocarbon end (hydrophobic)

**Lecithin**

(fig. 2.18). Figure 2.18 shows a simple way of illustrating the structure of a phospholipid—the parts of the molecule capable of ionizing and thus becoming charged are shown as a circle, whereas the nonpolar parts of the molecule are represented by lines.

Since the nonpolar ends of phospholipids are hydrophobic, they tend to group together when mixed in water; this allows the hydrophilic parts (which are polar) to face the surrounding water molecules (fig. 2.19). Such aggregates of molecules are called **micelles.** The dual nature of phospholipid molecules (part polar, part nonpolar) allows them to form the major component of the cell membrane as well as to alter the interaction of water molecules and thus decrease the surface tension of water. This latter function of phospholipids, which makes them *surfactants* (surface-active-agents), prevents collapse of the lungs.

***Steroids.*** The structure of steroid molecules is quite different from that of triglycerides or phospholipids, and yet steroids are still included in the lipid category of molecules because they are nonpolar and insoluble in water. All steroid molecules have the same basic structure; three six-carbon rings are joined to one five-carbon ring (fig. 2.20). However, different kinds of steroids have different functional groups attached to this basic structure, and they vary in the number and position of the double covalent bonds between the carbon atoms in the rings.

*Cholesterol* is an important molecule in the body because it serves as the precursor (parent molecule) for the steroid hormones produced by the gonads and adrenal cortex. The testes and ovaries (collectively called the *gonads*) secrete **sex steroids,** which include estradiol and progesterone from the ovaries and testosterone from the testes. The adrenal cortex secretes the **corticosteroids,** including hydrocortisone and aldosterone, among others.

***Prostaglandins.*** Prostaglandins are a type of fatty acid (with a cyclic hydrocarbon group) which have a variety of regulatory functions. Although their name is derived from the fact that they were originally found in the semen as a secretion of the prostate gland, it has since been shown that they are produced by and active in almost all tissues. Prostaglandins are implicated in the regulation of blood vessel diameter, ovulation, uterine contraction during labor,

**Figure 2.19.** The formation of a micelle structure by phospholipids such as lecithin. The straight lines represent the hydrophobic fatty acid parts of the molecule, and the circles represent the polar phosphate part of the molecule. The detailed structure of lecithin is shown in one part of the micelle.

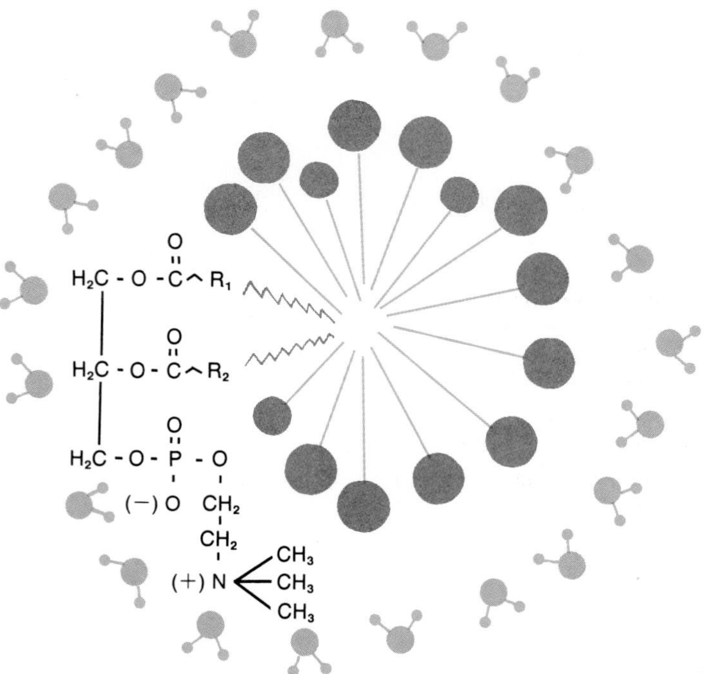


inflammation reactions, blood clotting, and many other functions. Some of the different types of prostaglandins are shown in figure 2.21.

1. *Describe the common structure of all carbohydrates, and distinguish between monosaccharides, disaccharides, and polysaccharides.*
2. *Using dehydration synthesis and hydrolysis reactions, explain how disaccharides and monosaccharides can be interconverted and how triglycerides can be formed and broken down.*
3. *Explain the characteristics of a lipid, and describe the different subcategories of lipids.*
4. *Relate the functions of phospholipids to their structure, and describe the significance of the prostaglandins.*

## PROTEINS

Proteins are large molecules composed of amino acid subunits. Since there are twenty different types of amino acids that can be used in constructing a given protein, the variety of protein structures is immense. This variety allows each type of protein to be constructed so that it can perform very specific functions.

The enormous diversity of protein structure results from the fact that there are twenty different building blocks—the amino acids—that can be used to form a protein. These amino acids, as will be described in the next section, are joined together to form a chain which can twist and fold in a specific manner due to chemical interactions between the amino acids. The specific sequence of amino acids in a protein, and thus the specific functional structure of the protein, is determined by genetic information. This genetic information for protein synthesis is contained in another category of organic molecules, the nucleic acids. The structure of nucleic acids, and the mechanisms by which the genetic information they encode directs protein synthesis, is described in chapter 3.

**Figure 2.20.** Cholesterol and some steroid hormones derived from cholesterol.

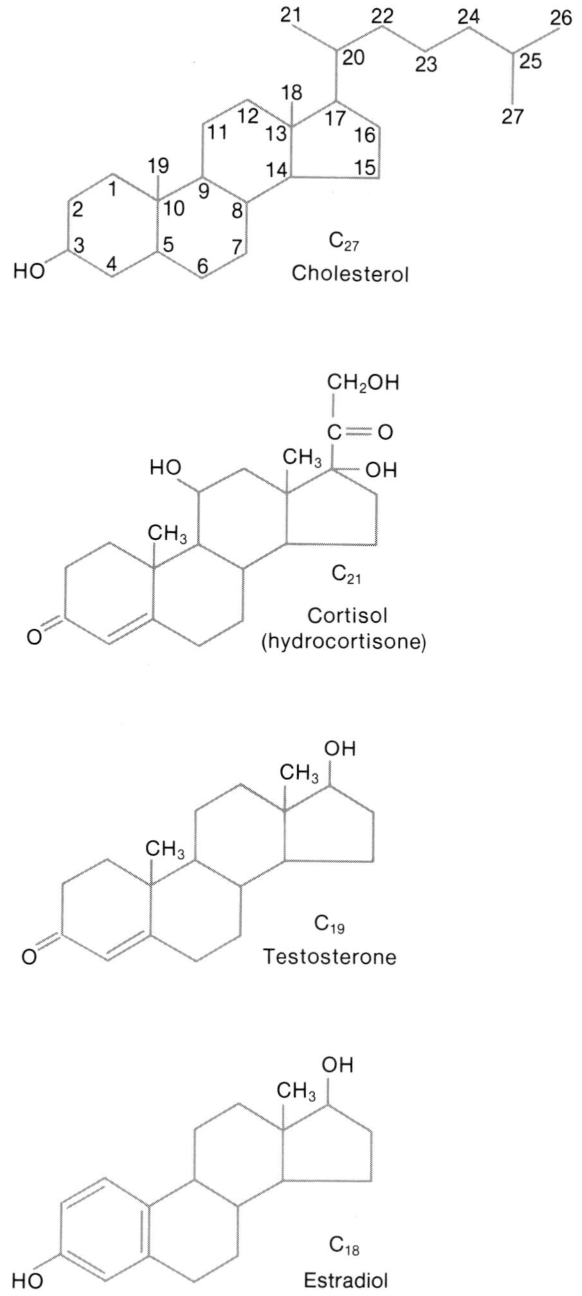

**Figure 2.21.** Structural formulas of some prostaglandins.

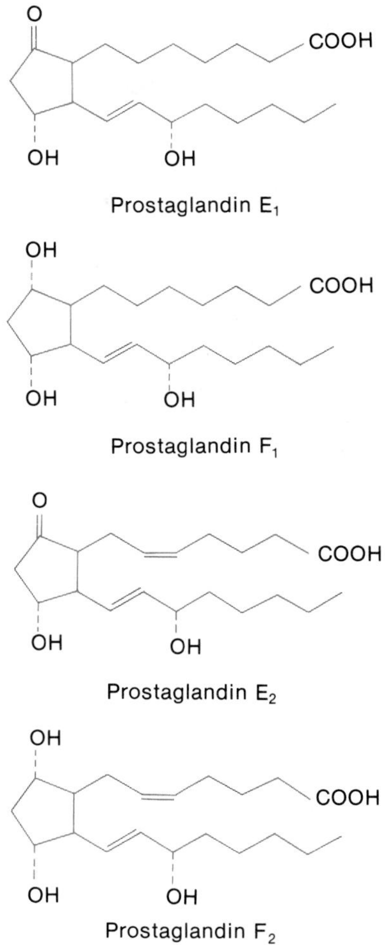

## Structure of Proteins

Proteins consist of long chains of subunits called **amino acids.** As the name implies, each amino acid contains an *amino group* ($NH_2$) on one end of the molecule and a *carboxylic acid group* (COOH) on another end. There are approximately twenty different amino acids, with different structures and chemical properties that are used to build proteins. These differences are due to differences in the *functional groups* of these amino acids. *R* is the abbreviation for *functional group* in the general formula for an amino acid (fig. 2.22). The *R* symbol actually stands for the word *residue,* but it can be thought of as indicating the "rest of the molecule."

When amino acids are joined together by dehydration synthesis, the hydrogen from the amino end of one amino acid combines with the hydroxyl group of the carboxylic acid end of another amino acid. As a covalent bond is formed between the two amino acids, water is produced (fig. 2.23). The bond between adjacent amino acids is called a **peptide bond,** and the compound formed is called a *peptide.* When many amino acids are joined in this way, a chain of amino acids, or **polypeptide,** is produced.

**Figure 2.22.** Representative amino acids, showing different types of functional (*R*) groups.

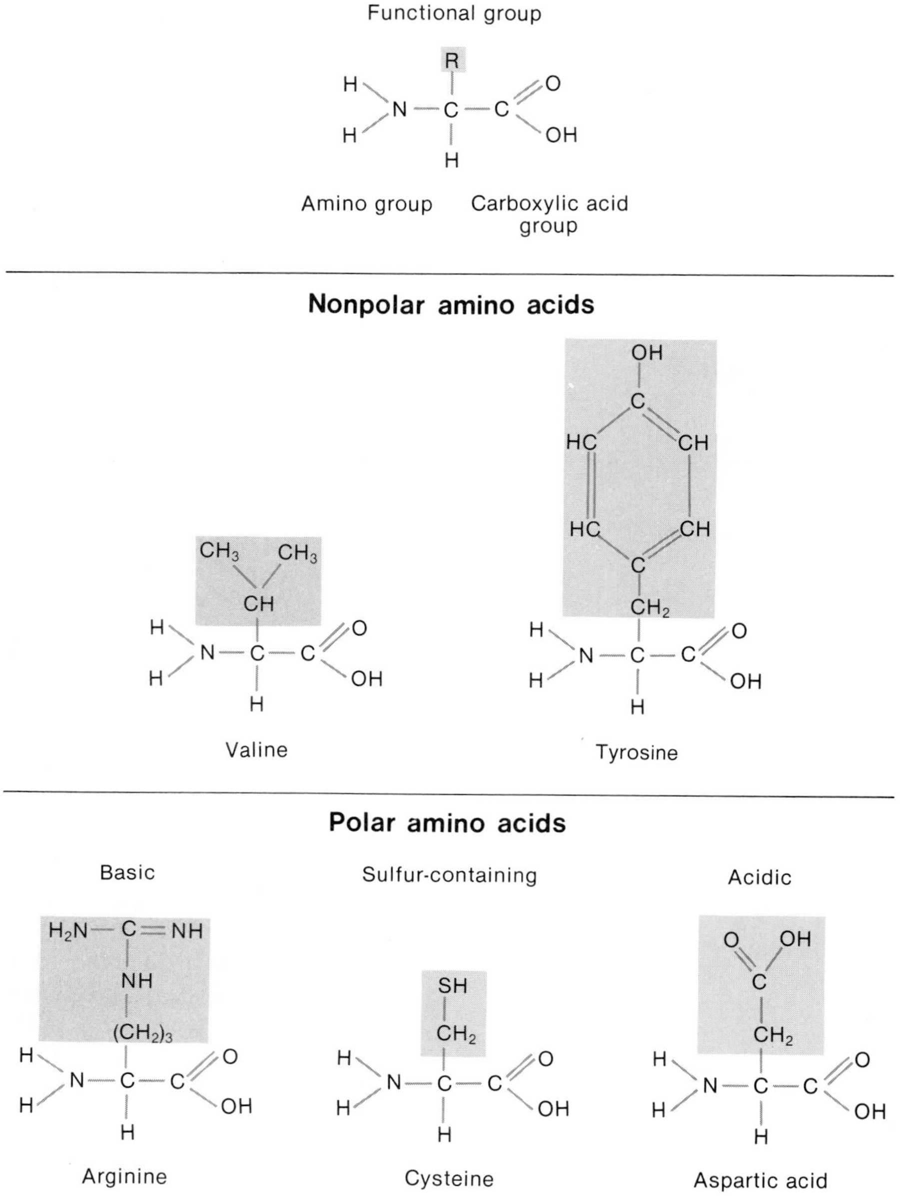

Functional group

Amino group

Carboxylic acid group

**Nonpolar amino acids**

Valine

Tyrosine

**Polar amino acids**

Basic

Sulfur-containing

Acidic

Arginine

Cysteine

Aspartic acid

The lengths of polypeptide chains vary greatly. A hormone called *thyrotropin-releasing hormone,* for example, is only three amino acids long, whereas *myosin,* a muscle protein, contains about forty-five hundred amino acids. When the length of a polypeptide chain becomes very long (greater than about a hundred amino acids), the molecule is called a **protein.**

The structure of a protein can be described at four different levels. At the first level, the sequence of amino acids in the protein, called the **primary structure** of the protein, is described. Each type of protein has a different primary structure. All of the billions of *copies* of a given type of protein in the body, however, have the same structure, because the structure of a given protein is coded by the genes. The primary structure of a protein is illustrated in figure 2.24.

Weak interactions (such as hydrogen bonds) between functional (*R*) groups of amino acids in nearby positions in the polypeptide chain cause this chain to twist into a *helix.* The extent and location of the helical structure is different for each protein because of differences in amino acid composition. A description of the helical structure of a protein is termed its **secondary structure** (fig. 2.24).

Most polypeptide chains bend and fold on themselves to produce complex three-dimensional shapes, called the **tertiary structure** of the proteins. Each type of protein has its own characteristic tertiary structure. This is because

**Figure 2.23.** The formation of peptide bonds by a dehydration synthesis reaction between amino acids.

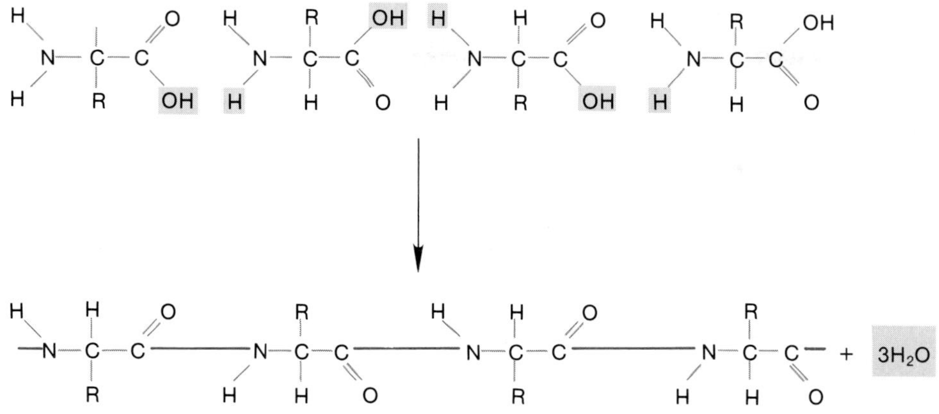

**Figure 2.24.** A polypeptide chain, showing (*a*) its primary structure and (*b*) secondary structure.

(a)

(b) Secondary structure

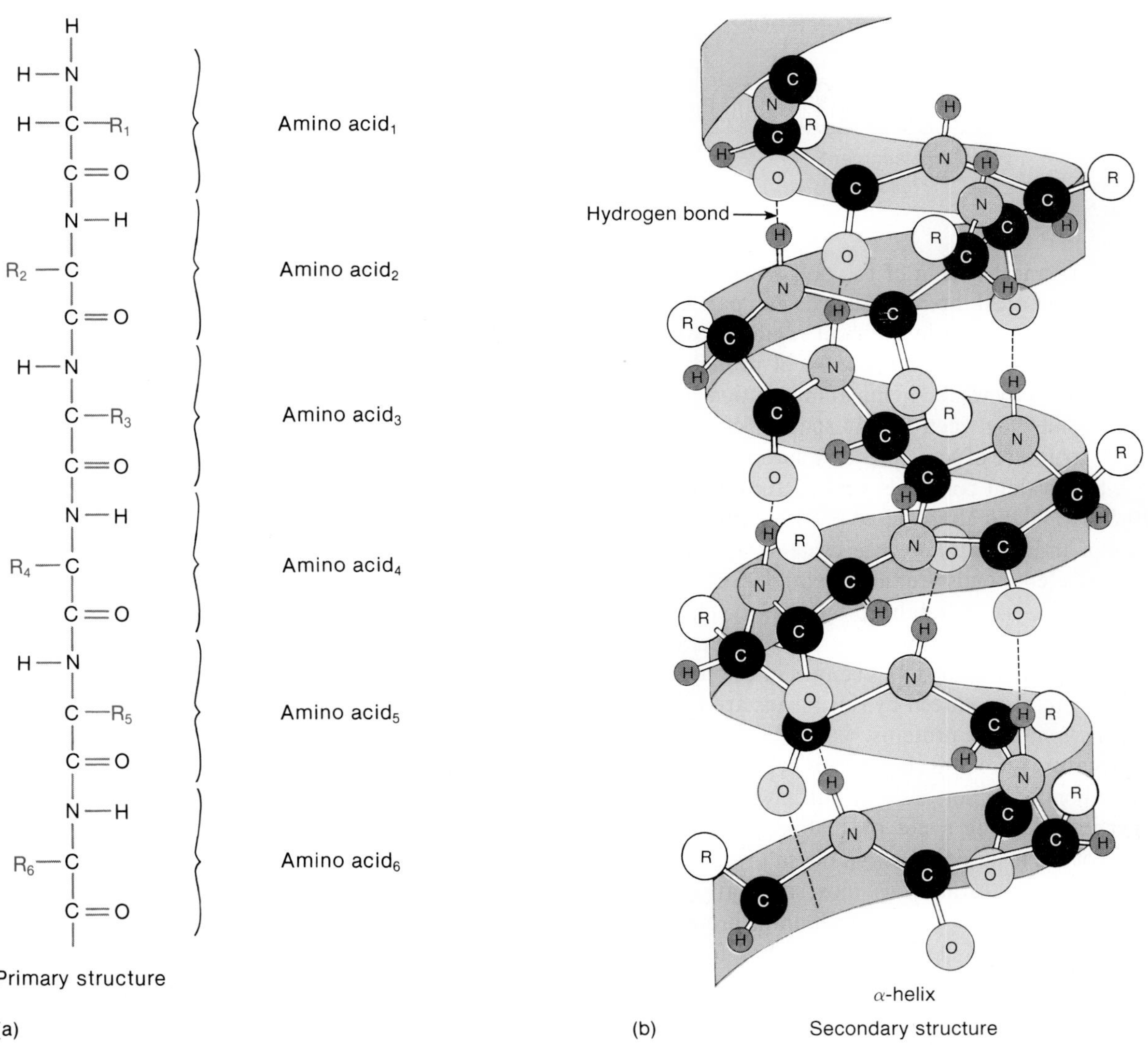

**Figure 2.25.** The tertiary structure of a protein. (*a*) interactions between functional (*R*) groups of amino acids result in (*b*) the formation of complex three-dimensional shapes of proteins.

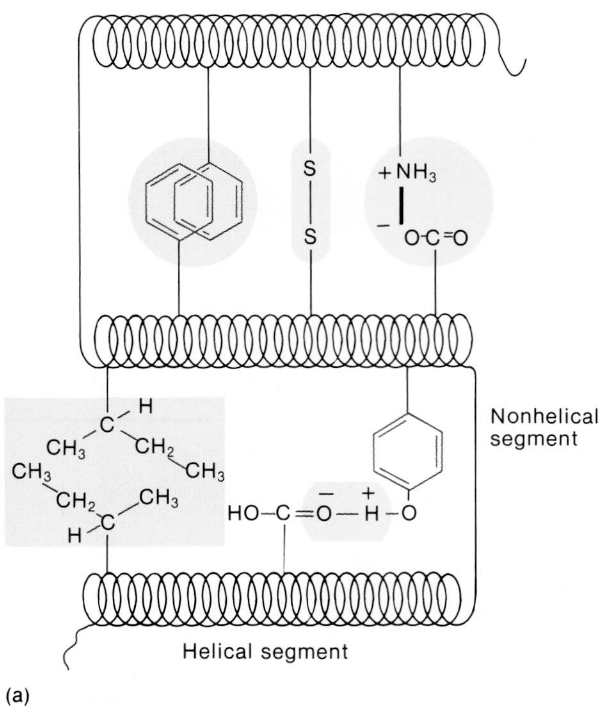


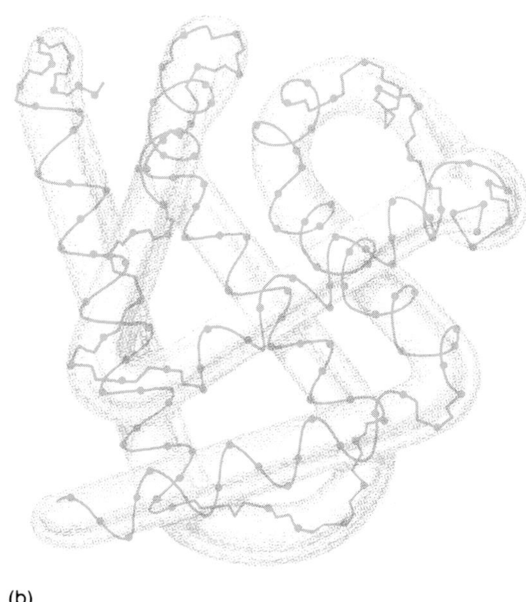


the folding and bending of the polypeptide chain is produced by chemical interactions between particular amino acids that are located in different regions of the chain.

Most of the tertiary structure of proteins is formed and stabilized by weak chemical interactions (such as hydrogen bonds) between widely spaced amino acids. The tertiary structure of some proteins, however, is made more stable by covalent bonds between sulfur atoms (called *disulfide bonds* and abbreviated S-S) in the functional (*R*) groups of amino acids known as cysteines (fig. 2.25). These strong covalent bonds are the exception. Since most of the tertiary structure is stabilized by weak bonds, this structure can easily be disrupted by high temperature or by changes in pH. Irreversible changes in the tertiary structure of proteins produced by this means are referred to as *denaturation* of the proteins.

Denatured proteins retain their primary structure (the peptide bonds are not broken) but have altered chemical properties. Cooking a pot roast, for example, alters the texture of the meat proteins—it doesn't result in an amino acid soup. Denaturation is most dramatically demonstrated by frying an egg. Egg-albumin proteins are soluble in their native state, in which they form the clear, viscous fluid of a raw egg. When denatured by cooking, these proteins change shape, cross-bond with each other, and by this means form an insoluble white precipitate—the egg white.

Some proteins (such as hemoglobin and insulin) are composed of a number of polypeptide chains covalently bonded together. This is the **quaternary structure** of these proteins. Insulin, for example, is composed of two polypeptide chains, one that is twenty-one amino acids long, the other that is thirty amino acids long. Hemoglobin (the protein in red blood cells that carries oxygen) is composed of four separate polypeptide chains. The composition of various body proteins is shown in table 2.4.

Many proteins in the body are normally found combined, or *conjugated,* with other types of molecules. *Glycoproteins* are proteins conjugated with carbohydrates. Examples of such molecules include certain hormones and some proteins found in the cell membrane. *Lipoproteins* are proteins conjugated with lipids. These are found in cell membranes and in the plasma (the fluid portion of the blood). Proteins conjugated with pigment molecules are *chromoproteins.* These include hemoglobin, which transports oxygen in red blood cells, and the cytochromes, which are needed for oxygen utilization and energy production within cells.

## Functions of Proteins

Because of their tremendous structural diversity, proteins can serve a wider variety of functions than any other type of molecule in the body. Many proteins, for example, contribute significantly to the structure of different tissues and in this way play a passive role in the functions of these

**Table 2.4** Composition of selected proteins in the body

| Protein | Number of Polypeptide Chains | Nonprotein Component | Function |
|---|---|---|---|
| Hemoglobin | 4 | Heme pigment | Carries oxygen in the blood |
| Myoglobin | 1 | Heme pigment | Stores oxygen in muscle |
| Insulin | 2 | None | Hormone-regulating metabolism |
| Luteinizing hormone | 1 | Carbohydrate | Hormone that stimulates gonads |
| Fibrinogen | 1 | Carbohydrate | Involved in blood clotting |
| Mucin | 1 | Carbohydrate | Forms mucus |
| Blood group proteins | 1 | Carbohydrate | Produces blood types |
| Lipoprpteins | 1 | Lipids | Transports lipids in blood |

**Figure 2.26.** A photomicrograph of collagen fibers.

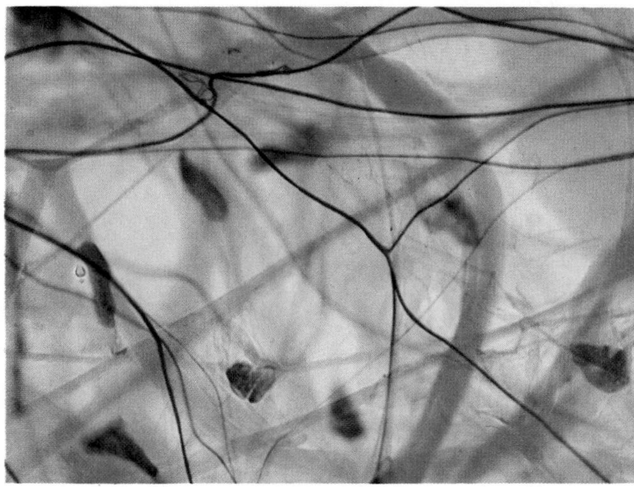

tissues. Examples of such *structural proteins* include collagen (fig. 2.26) and keratin. Collagen is a fibrous protein that provides tensile strength to connective tissues, such as tendons and ligaments. Keratin is found in the outer layer of dead cells in the epidermis, where it serves to prevent water loss through the skin.

Many proteins serve a more active role in the body where specialized structure and function are required. *Enzymes* and *antibodies,* for example, are proteins—no other type of molecule could provide the vast array of different structures needed for these functions. Proteins in cell membranes serve as *receptors* for specific regulator molecules (such as hormones) and as *carriers* that transport specific molecules across the membrane. Proteins provide the diversity of shape and chemical properties for the specificity required by these functions.

1. *Write the general formula for an amino acid, and describe how amino acids differ from each other.*
2. *Describe the different levels of protein structure and explain how these different levels of structure are produced.*
3. *Describe the different categories of protein function in the body, and explain why proteins can have such diverse functions.*

## SUMMARY

### Atoms, Ions, and Chemical Bonds p. 26

I. Covalent bonds are formed by atoms that share electrons; these are the strongest type of chemical bonds.
   A. Electrons are equally shared in nonpolar covalent bonds and unequally shared in polar covalent bonds.
   B. Atoms of oxygen, nitrogen, and phosphorus strongly attract electrons and become electrically negative compared to the other atoms sharing electrons with them.

II. Ionic bonds are formed by atoms that transfer electrons; these weak bonds join atoms together in an ionic compound.
   A. If one atom in this compound takes the electron from another atom, it gains a net negative charge and the other becomes positively charged.
   B. Ionic bonds easily break when the ionic compound is dissolved in water; dissociation of the ionic compound yields charged atoms called ions.

III. When hydrogen is bonded to an electronegative atom, it gains a slight positive charge and is weakly attracted to another electronegative atom; this weak attraction is a hydrogen bond.

IV. Acids donate hydrogen ions to solution, whereas bases lower the hydrogen ion concentration of a solution.
   A. The pH scale is a negative function of the logarithm of the hydrogen ion concentration.

B. In a neutral solution the concentration of $H^+$ is equal to the concentration of $OH^-$, and the pH is 7.
C. Acids raise the $H^+$ concentration and thus lower the pH below 7; bases lower the $H^+$ concentration and thus raise the pH above 7.

V. Organic molecules contain atoms of carbon joined together by covalent bonds; atoms of nitrogen, oxygen, phosphorus, or sulfur may be present as specific functional groups in the organic molecule.

## Carbohydrates and Lipids p. 34

I. Carbohydrates contain carbon, hydrogen, and oxygen, usually in a ratio of 1:2:1.
A. Carbohydrates consist of simple sugars (monosaccharides), disaccharides, and polysaccharides (such as glycogen).
B. Covalent bonds between monosaccharides are formed by dehydration synthesis, or condensation; bonds are broken by hydrolysis reactions.

II. Lipids are organic molecules that are insoluble in polar solvents such as water.
A. Triglycerides (fat and oil) consist of three fatty acid molecules joined to a molecule of glycerol.
B. Ketone bodies are smaller derivatives of fatty acids.
C. Phospholipids (such as lecithin) are phosphate-containing lipids that have a polar group, which is hydrophilic; the rest of the molecule is hydrophobic.
D. Steroids (including the hormones of the adrenal cortex and gonads) are lipids with a characteristic five-ring structure.
E. Prostaglandins are a family of cyclic fatty acids, which serve a variety of regulatory functions.

## Proteins p. 40

I. Proteins are composed of long chains of amino acids bonded together by covalent peptide bonds.
A. Each amino acid contains an amino group, a carboxyl group, and a functional group that is different for each of the more than twenty different amino acids.
B. The polypeptide chain may be twisted into a helix (secondary structure) and bent and folded to form the tertiary structure of the protein.
C. Proteins that are composed of two or more polypeptide chains are said to have a quaternary structure.
D. Proteins may be combined with carbohydrates, lipids, or other molecules.
E. Because of their great variety of possible structures, proteins serve a wider variety of specific functions than any other type of

# Review Activities

## Objective Questions

1. Which of the following statements about atoms is *true?*
   (a) They have more protons than electrons.
   (b) They have more electrons than protons.
   (c) They are electrically neutral.
   (d) They have as many neutrons as they have electrons.
2. The bond between oxygen and hydrogen in a water molecule is a(n)
   (a) hydrogen bond
   (b) polar covalent bond
   (c) nonpolar covalent bond
   (d) ionic bond
3. Which of the following is a nonpolar covalent bond?
   (a) The bond between two carbons.
   (b) The bond between sodium and chloride.
   (c) The bond between two water molecules.
   (d) The bond between nitrogen and hydrogen.
4. Solution A has a pH of 2, and solution B has a pH of 10. Which of the following statements about these solutions is *true?*
   (a) Solution A has a higher $H^+$ concentration than solution B.
   (b) Solution B is basic.
   (c) Solution A is acidic.
   (d) All of these are true.
5. Glucose is a
   (a) disaccharide
   (b) polysaccharide
   (c) monosaccharide
   (d) phospholipid
6. Digestion reactions occur by means of
   (a) dehydration synthesis
   (b) hydrolysis
7. Carbohydrates are stored in the liver and muscles in the form of
   (a) glucose
   (b) triglycerides
   (c) glycogen
   (d) cholesterol
8. Lecithin is a
   (a) carbohydrate
   (b) protein
   (c) steroid
   (d) phospholipid
9. Which of the following lipids have regulatory roles in the body?
   (a) steroids
   (b) prostaglandins
   (c) triglycerides
   (d) both *a* and *b*
   (e) both *b* and *c*
10. The tertiary structure of a protein is *directly* determined by
   (a) the genes
   (b) the primary structure of the protein
   (c) enzymes that "mold" the shape of the protein
   (d) the position of peptide bonds
11. The type of bond formed between two molecules of water is a
   (a) hydrolytic bond
   (b) polar covalent bond
   (c) nonpolar covalent bond
   (d) hydrogen bond
12. The carbon-to-nitrogen bond that joins amino acids together is called a
   (a) glycosidic bond
   (b) peptide bond
   (c) hydrogen bond
   (d) double bond

## Essay Questions

1. Compare and contrast nonpolar covalent bonds, polar covalent bonds, and ionic bonds.
2. Give the definition of an acid and base, and explain how these influence the pH of a solution.
3. Using dehydration synthesis and hydrolysis reactions, explain the relationships between starch in an ingested potato, liver glycogen, and blood glucose.
4. "All fats are lipids, but not all lipids are fats." Explain why this statement is true.
5. Explain the relationship between the primary structure of a protein and its secondary and tertiary structures.

## Selected Readings


Demers, L. M. September 1984. The effects of prostaglandins. *Diagnostic Medicine,* p. 37.

Doolittle, R. F. October 1985. Proteins. *Scientific American.*

Hakomori, Sen-itiroh. May 1986. Glycosphingolipids. *Scientific American.*

Jackson, R. W., and A. M. Gotto. 1974. Phospholipids in biology and medicine. *New England Journal of Medicine* 290:24.

Kuehl, F. A., Jr., and R. W. Egan. 1980. Prostaglandins, arachidonic acid, and inflammation. *Science* 210:978.

Prockop, D. J., and N. A. Guzman. 1977. Collagen disease and the biosynthesis of collagen. *Hospital Practice* 12:61.

Sharon, N. May 1974. Glycoproteins. *Scientific American.*

Sharon, N. November 1980. Carbohydrates. *Scientific American.*

Weinberg, R. A. October 1985. The molecules of life. *Scientific American.*

# Cell Structure and Genetic Control

# 3

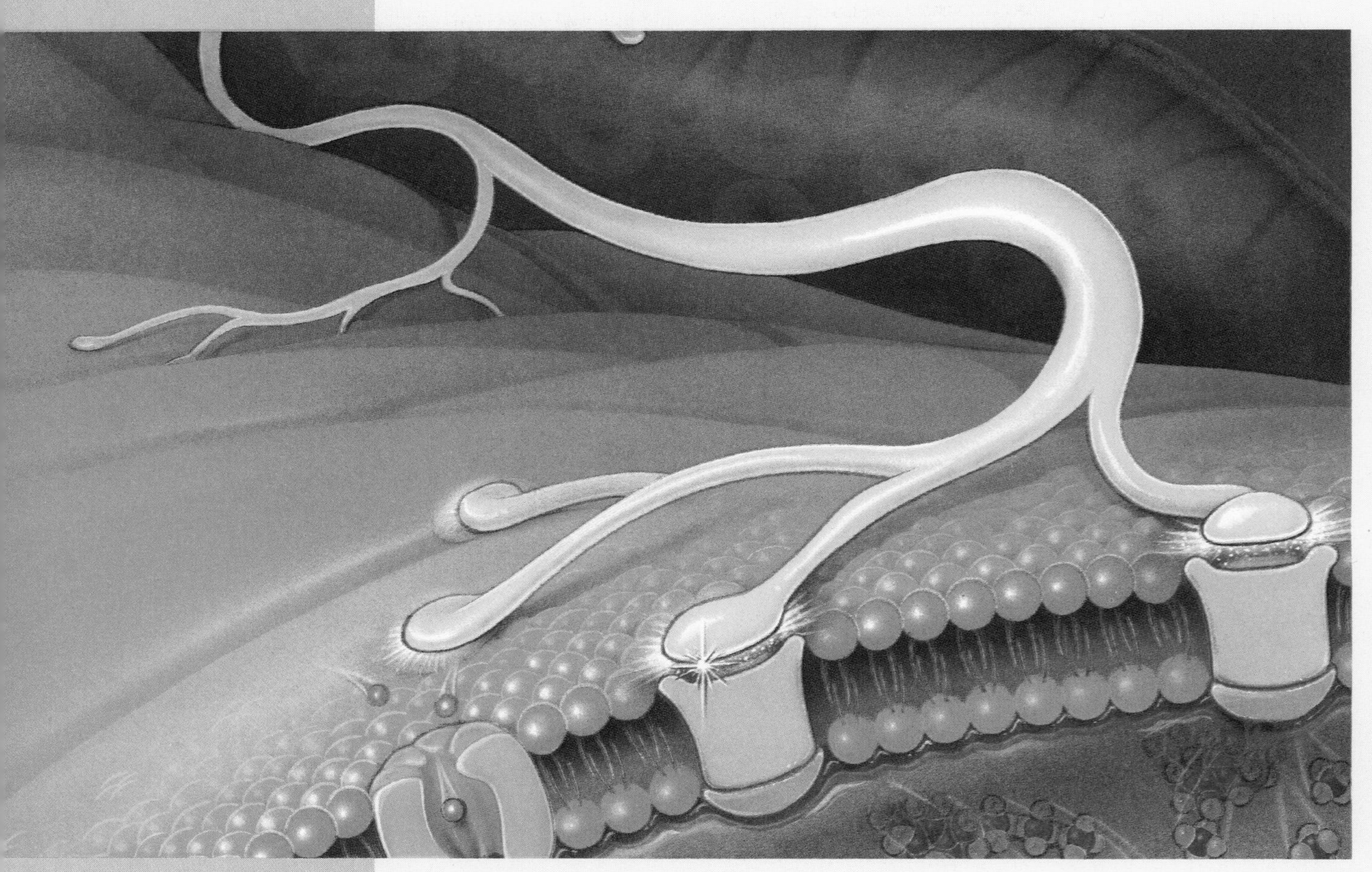

## Objectives

By studying this chapter, you should be able to

1. describe the structure of the cell membrane and explain its functional importance
2. describe how cells move by amoeboid motion and the structure and significance of cilia and flagella
3. explain the processes of phagocytosis, pinocytosis, receptor-mediated endocytosis, and exocytosis
4. explain the functions of the cytoskeleton, lysosomes, mitochondria, and the endoplasmic reticulum
5. describe the structure and significance of the cell nucleus
6. describe the structure of nucleotides and distinguish between the nucleotides of DNA and RNA
7. explain how DNA is constructed and the law of complementary base pairing
8. explain how RNA is produced according to the genetic information in DNA and distinguish between the different types of RNA
9. describe how proteins are produced according to the information contained in messenger RNA
10. describe the structure and function of the rough endoplasmic reticulum and Golgi apparatus in the secretion of proteins
11. explain the semiconservative mechanism of DNA replication
12. describe the different stages of the cell cycle and the events that occur in the different phases of mitosis
13. define the terms *hypertrophy* and *hyperplasia* and explain their physiological importance
14. describe the events that occur in meiosis; compare these to those that occur in mitosis and explain the function of meiotic cell division in human physiology

## Outline

## Cell Membrane and Associated Structures

The cell is the basic unit of structure and function in the body. Many of the functions of cells are performed by particular subcellular structures known as organelles. The cell membrane is an extremely important organelle which allows selective communication between the intracellular and extracellular compartments and which participates in cellular movements.

The cell is so small and so simple in appearance when viewed with the ordinary (light) microscope that it is difficult to conceive that each cell is a living entity unto itself. Equally amazing is the fact that the physiology of our organs and systems derive from the complex functions of the cells of which they are composed. Complexity of function demands complexity of structure even at the subcellular level.

As the basic functional unit of the body, each cell is a highly organized molecular factory. Cells come in a great variety of shapes and sizes. This great variation, which is also apparent in the subcellular structures within different cells, reflects the variation of functions of different cells in the body. All cells, however, share certain characteristics—such as the fact that they are surrounded by a cell membrane—and most cells possess the structures listed in table 3.1. Thus, although no single cell can be considered "typical," the general structure of cells can be shown with a single illustration (fig. 3.1).

For descriptive purposes, a cell can be divided into three principal parts:

1. **Cell (plasma) membrane.** The selectively permeable cell membrane surrounds the cell, gives form to it, and separates the cell's internal structures from the extracellular environment.
2. **Cytoplasm and organelles.** The cytoplasm is the aqueous content of a cell between the nucleus and the cell membrane. Organelles are subcellular structures within the cytoplasm of a cell that perform specific functions.
3. **Nucleus.** The nucleus is a large, generally spheroid body within a cell that contains the DNA, or genetic material, of a cell and which thus directs the activities of the cell.

### Cell Membrane

Because both the intracellular and extracellular environments (or "compartments") are aqueous, a barrier must be present to prevent the loss of cellular molecules such

**Table 3.1** Structure and function of cellular components

| Component | Structure | Function |
|---|---|---|
| Cell (plasma) membrane | Membrane composed of phospholipid and protein molecules | Gives form to cell and controls passage of materials in and out of cell |
| Cytoplasm | Fluid, jellylike substance in which organelles are suspended | Serves as matrix substance in which chemical reactions occur |
| Endoplasmic reticulum | System of interconnected membrane-forming canals and tubules | Smooth endoplasmic reticulum metabolizes nonpolar compounds and stores $Ca^{++}$ in striated muscle cells; rough endoplasmic reticulum assists in protein synthesis |
| Ribosomes | Granular particles composed of protein and RNA | Synthesize proteins |
| Golgi apparatus | Cluster of flattened, membranous sacs | Synthesizes carbohydrates and packages molecules for secretion; secretes lipids and glycoproteins |
| Mitochondria | Membranous sacs with folded inner partitions | Release energy from food molecules and transform energy into usable ATP |
| Lysosomes | Membranous sacs | Digest foreign molecules and worn and damaged cells |
| Peroxisomes | Spherical membranous vesicles | Contain certain enzymes; form hydrogen peroxide |
| Centrosome | Nonmembranous mass of two rodlike centrioles | Helps organize spindle fibers and distribute chromosomes during mitosis |
| Vacuoles | Membranous sacs | Store and excrete various substances within the cytoplasm |
| Fibrils and microtubules | Thin, hollow tubes | Support cytoplasm and transport materials within the cytoplasm |
| Cilia and flagella | Minute cytoplasmic extensions from cell | Move particles along surface of cell or move cell |
| Nuclear membrane | Membrane surrounding nucleus, composed of protein and lipid molecules | Supports nucleus and controls passage of materials between nucleus and cytoplasm |
| Nucleolus | Dense, nonmembranous mass composed of protein and RNA molecules | Forms ribosomes |
| Chromatin | Fibrous strands composed of protein and DNA molecules | Controls cellular activity for carrying on life processes |

**Figure 3.1.** A generalized cell and the principal organelles.

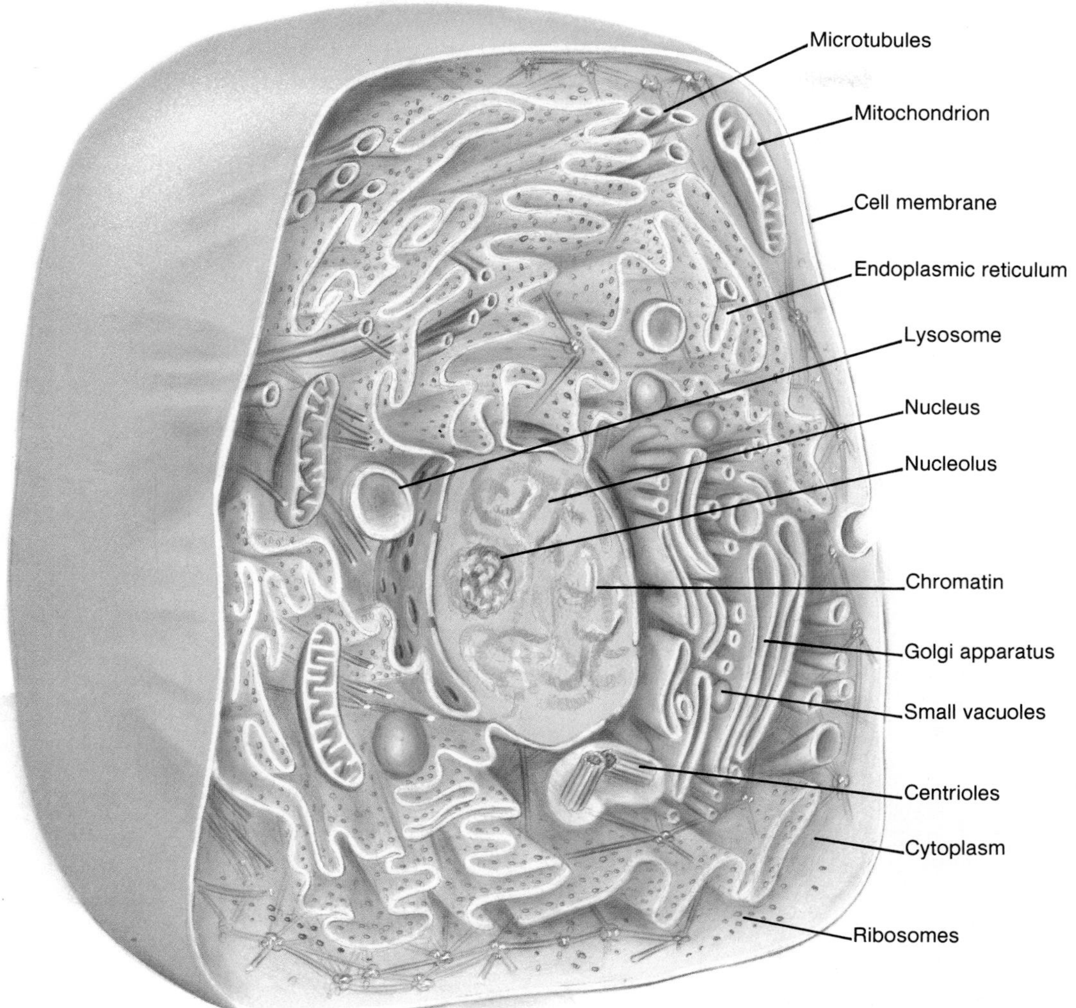


as enzymes, nucleotides, and others that are water soluble. Since this barrier cannot itself be composed of water-soluble molecules, it makes sense that the cell membrane is composed of lipids.

The **cell membrane** (also called the **plasma membrane,** or **plasmalemma**), and indeed all of the membranes surrounding organelles within the cell, are composed primarily of phospholipids and proteins. Phospholipids, as described in chapter 2, are polar on the end that contains the phosphate group and nonpolar (and hydrophobic) throughout the rest of the molecule. Since there is an aqueous environment on each side of the membrane, the hydrophobic parts of the molecules "huddle together" in the center of the membrane, leaving the polar ends exposed to water on both surfaces. This results in the formation of a double layer of phospholipids in the cell membrane.

The hydrophobic core of the membrane restricts the passage of water and water-soluble molecules and ions. Certain of these polar compounds, however, do pass through the membrane. The specialized functions and selective transport properties of the membrane are believed to be due to its protein content. Some proteins are found partially submerged on each side of the membrane; other proteins span the membrane completely from one side to the other. Since the membrane is not solid—phospholipids and proteins are free to move laterally—the proteins within the phospholipid "sea" are not uniformly distributed, but rather present a mosaic pattern. This structure is known as the **fluid-mosaic model** of membrane structure (fig. 3.2).

The proteins found in the cell membrane serve a variety of functions, including (1) structural support; (2) transport of molecules across the membrane; (3) enzymatic control of chemical reactions at the cell surface; (4) receptors for hormones and other regulatory molecules that arrive at the outer surface of the membrane; and (5) cellular "markers" (antigens), which identify the blood and tissue type.

**Figure 3.2.** The fluid-mosaic model of the cell membrane. The membrane consists of a double layer of phospholipids, with the phosphates (*open circles*) oriented outward and the hydrophobic hydrocarbons (*wavy lines*) oriented toward the center. Proteins may completely or partially span the membrane. Carbohydrates are attached to the outer surface.

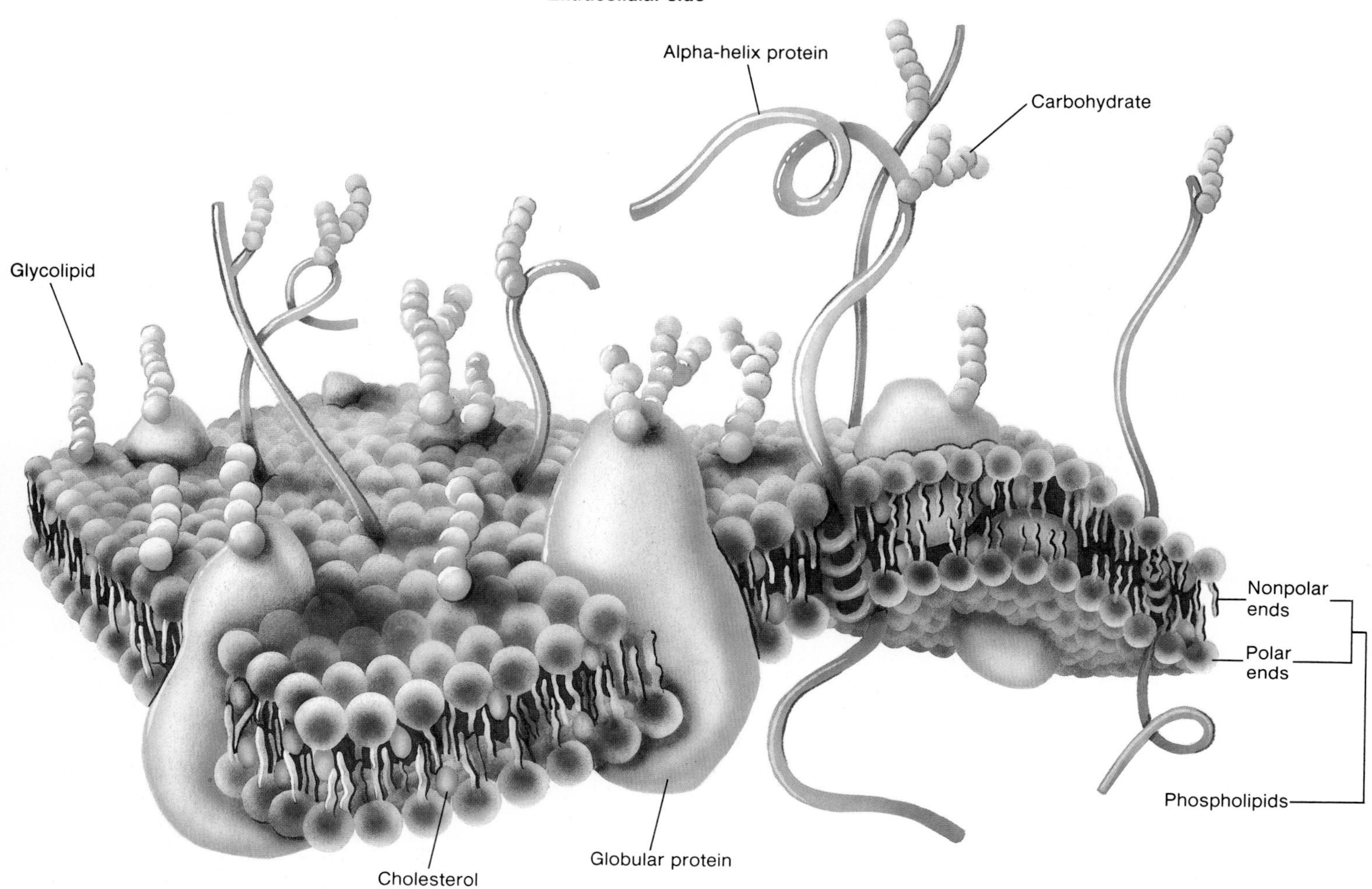


The cell membranes of all higher organisms contain cholesterol. The ratio of cholesterol to phospholipids, as well as the ratio of different types of phospholipids, determines the flexibility of the cell membrane. When there is an inherited defect in these ratios, the flexibility of the cell may be reduced. This could result, for example, in the inability of red blood cells to flex at the middle when passing through narrow blood channels (thereby causing the occlusion of these small vessels).

In addition to lipids and proteins, the cell membrane also contains carbohydrates, which are primarily attached to the outer surface of the membrane as glycoproteins and glycolipids. These surface carbohydrates have many negative charges and, as a result, affect the interaction of regulatory molecules with the membrane. The negative charges at the surface also affect interactions between cells—they help keep red blood cells apart, for example. Stripping the carbohydrates from the outer red blood cell surface results in their more rapid destruction by the liver, spleen, and bone marrow.

**Figure 3.3.** Electron micrographs of cilia, showing (*a*) longitudinal and (*b*) cross sections. Notice the characteristic "9 + 2" arrangement of microtubules in the cross sections.

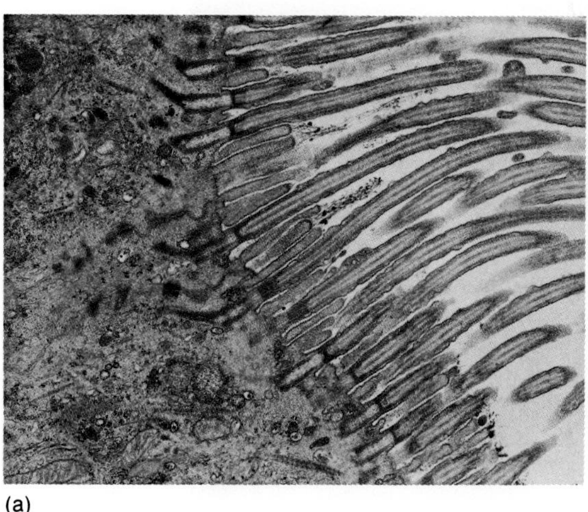

(a)

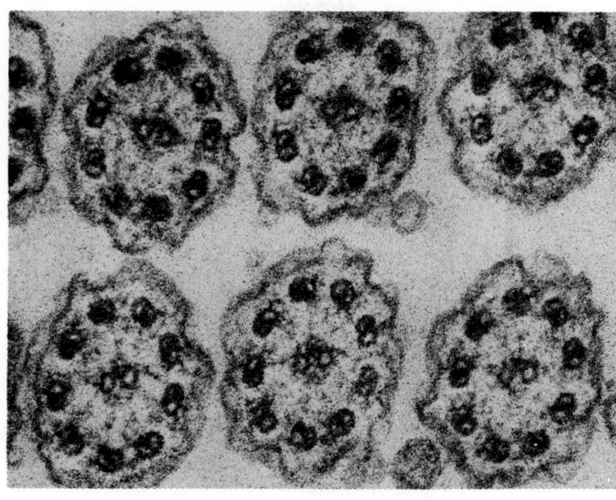

(b)

## Cellular Movements

Some body cells—including certain white blood cells and macrophages in connective tissues and microglial cells in the brain—are able to move like an amoeba (a single-celled animal). This "amoeboid" movement is performed by the extension of parts of the cytoplasm to form *pseudopods*, which attach to a substrate and pull the cell along.

***Cilia and Flagella.*** **Cilia** are tiny hairlike structures that protrude from the cell and, like the coordinated action of oarsmen in a boat, stroke in unison. Cilia in the human body are found on the apical surface (the surface facing the lumen, or cavity) of stationary epithelial cells in the respiratory and female genital tracts. In the respiratory system, the cilia transport strands of mucus, which are then conveyed by the cilia to a region (the pharynx) where the mucus can either be swallowed or expectorated. In the female genital tract, ciliary movements in the epithelial lining draw the egg (ovum) into the uterine tube and move it toward the uterus.

Sperm are the only cells in the human body that have **flagella.** The flagellum is a single, whiplike structure that propels the sperm through its environment. Both cilia and flagella are composed of *microtubules* (formed from proteins) arranged in a characteristic way. One pair of microtubules in the center of a cilum or flagellum is surrounded by nine other pairs of microtubules, to produce what is often described as a "9 + 2" arrangement (fig. 3.3).

## Endocytosis and Exocytosis

Regions of the cell membrane can invaginate (move inward to form a pouch) and pinch off to produce a membrane-enclosed body within the cytoplasm. This process removes regions of cell membrane as it brings part of the extracellular environment into the cell. The process by which part of the extracellular environment is brought into a cell by invagination of the cell membrane is called **endocytosis.** There are three types of endocytosis: phagocytosis, pinocytosis, and receptor-mediated endocytosis.

***Phagocytosis and Pinocytosis.*** Cells that move by amoeboid motion (such as white blood cells)—as well as liver cells, which are not mobile—use pseudopods to surround and engulf particles of organic matter (such as bacteria). This process is a type of cellular "eating" called **phagocytosis,** which serves to protect the body from invading microorganisms and to remove extracellular debris.

Phagocytic cells surround their victim with pseudopods, which join together and fuse (fig. 3.4). After the inner membrane of the pseudopods becomes a continuous membrane around the ingested particle, it pinches off from the cell membrane. The ingested particle is now contained in an organelle called a *food vacuole* within the cell. The particle will subsequently be digested by enzymes contained in a different organelle (the lysosome, described in a later section).

**Pinocytosis** is a related process performed by many cells. Instead of forming pseudopods, the cell membrane invaginates to produce a deep, narrow furrow. The membrane near the surface of this furrow then fuses, and a small vacuole containing the extracellular fluid is pinched off and enters the cell. In this way a cell can take in large molecules such as proteins which may be present in the extracellular fluid.

**Figure 3.4.** Scanning electron micrographs of phagocytosis, showing the formation of pseudopods and the entrapment of the prey within a food vacuole.

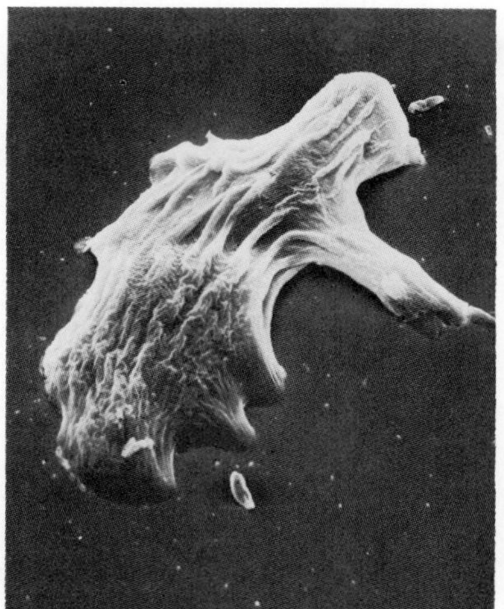

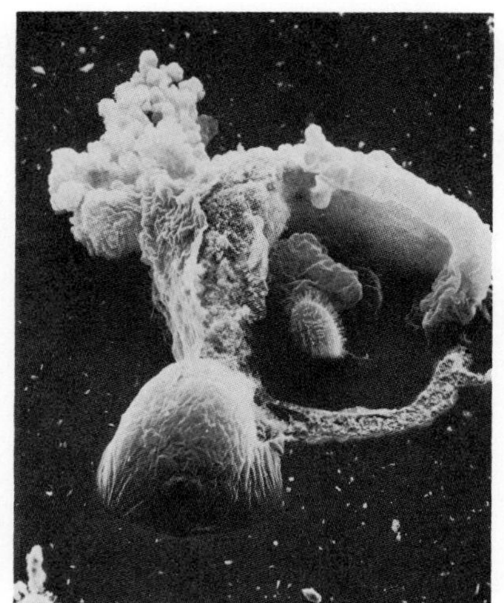

***Receptor-Mediated Endocytosis.*** This type of endocytosis involves the smallest area of the cell membrane, and it occurs only in response to specific molecules in the extracellular environment. In receptor-mediated endocytosis, the interaction of very specific molecules in the extracellular environment with specific membrane receptor proteins causes the membrane to invaginate, fuse, and pinch off to form a *vesicle*—a small vacuole (fig. 3.5). Vesicles formed in this way contain extracellular fluid and molecules that could not have passed by other means into the cell. Cholesterol attached to specific proteins, for example, is taken into cells that line arteries by receptor-mediated endocytosis. (This is in part responsible for atherosclerosis, as will be described in chapter 13.)

***Exocytosis.*** Proteins and other molecules produced within the cell that are destined for export (secretion) are packaged within vesicles by an organelle known as the Golgi apparatus (described in a later section). In the process of **exocytosis**, these secretory vesicles fuse with the cell membrane and release their contents into the extracellular environment (see fig. 3.25). This process adds new membrane material, which replaces that which was lost from the cell membrane during endocytosis.

Endocytosis and exocytosis account for only part of the two-way traffic between the intracellular and extracellular compartments. Most of this traffic is due to membrane transport processes involving the movement of molecules and ions through the cell membrane (described in chapter 6).

1. *Draw the fluid-mosaic model of the cell membrane, and describe the structure of the membrane.*
2. *Describe the structure of cilia and flagella and some of their functions.*
3. *Draw a figure showing phagocytosis and pinocytosis, and explain the significance of these processes.*
4. *Describe the events that occur in receptor-mediated endocytosis, and explain the significance of this process.*

## Cytoplasm and Its Organelles

Many of the functions of a cell that are performed in the cytoplasmic compartment result from the activity of specific structures called organelles. Among these are the microtubules and microfilaments of the cytoskeleton; the lysosomes, which contain digestive enzymes; and the mitochondria, where most of the cellular energy is produced. Other organelles include the endoplasmic reticulum and Golgi apparatus, which have specialized functions described in later sections.

### Cytoplasm and Cytoskeleton

The jellylike matrix within a cell (exclusive of that within the nucleus) is known as **cytoplasm.** When viewed in a microscope without special techniques, the cytoplasm appears to be uniform and unstructured. According to recent evidence, however, the cytoplasm is not a homogenous solution; it is, rather, a highly organized structure in which protein fibers—in the form of *microtubules* and *microfilaments*—are arranged in a complex latticework. These can be seen by fluorescence microscopy with the aid of

**Figure 3.5.** Stages (1–4) of endocytosis, where specific bonding of extracellular particles to membrane receptor proteins is believed to occur.

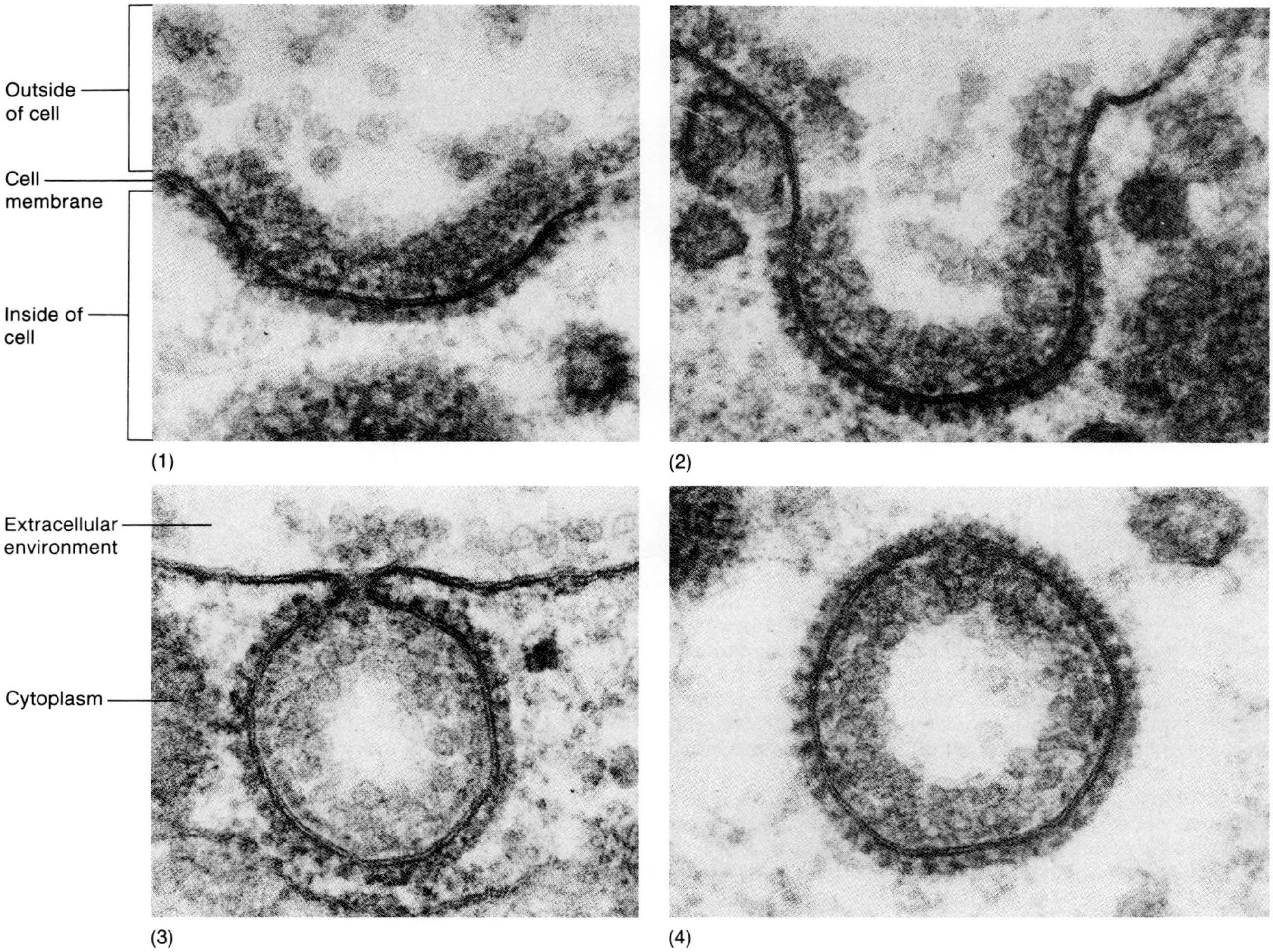

Outside of cell

Cell membrane

Inside of cell

(1)

(2)

Extracellular environment

Cytoplasm

(3)

(4)

antibodies against the proteins that compose these structures (fig. 3.6). The interconnected microfilaments and microtubules are believed to provide structural organization for cytoplasmic enzymes and support for various organelles.

The latticework of microfilaments and microtubules is thus said to function as a **cytoskeleton** (fig. 3.7). The structure of this "skeleton" is not rigid; it has been shown to be capable of quite rapid reorganization. Contractile proteins—including actin and myosin, which are responsible for muscle contraction—may be able to shorten the length of some microfilaments. The cytoskeleton may thus represent the cellular "musculature." Microtubules, for example, form the *spindle apparatus* that pulls chromosomes away from each other in cell division; they also form the central parts of cilia and flagella. Recent evidence suggests that the rapid movement of organelles within nerve fibers is also dependant upon microtubules, which can move different organelles in opposite directions at the same time.

**Figure 3.6.** An immunofluorescence photograph of microtubules forming the cytoskeleton of a cell. Microtubules are visualized with the aid of antibodies against tubulin, the major protein component of the microtubules.

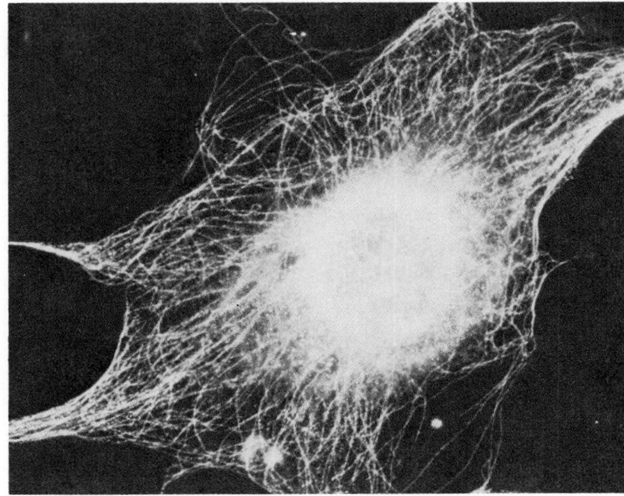

**Figure 3.7.** A diagram of a proposed structure of the cytoskeleton. (From "The Ground Substance of the Living Cell," by Keith Porter and Jonathon Tucker. Copyright © 1981 by Scientific American, Inc. All rights reserved.)

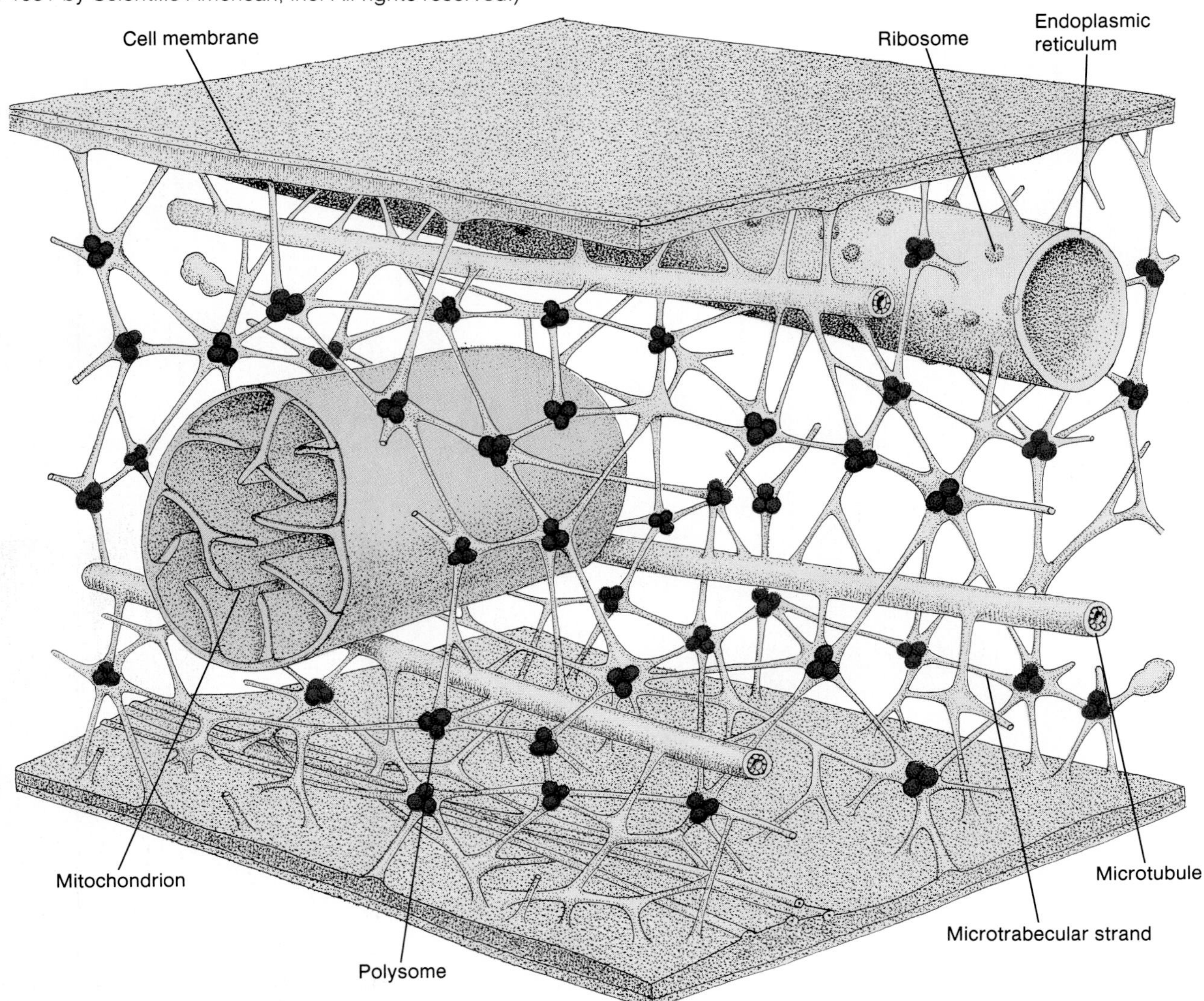


## Lysosomes

After a phagocytic cell has engulfed the proteins, polysaccharides, and lipids present in a particle of "food" (such as a bacterium), these molecules are still kept isolated from the cytoplasm by the membranes surrounding the food vacuole. The large molecules of proteins, polysaccharides, and lipids must first be digested into their smaller subunits (amino acids, monosaccharides, and so on) before they can cross the vacuole membrane and enter the cytoplasm.

The digestive enzymes of a cell are isolated from the cytoplasm and concentrated within membrane-bound organelles called **lysosomes** (fig. 3.8). A *primary lysosome* is one which contains only digestive enzymes (about forty different species) within an environment that is considerably more acidic than the surrounding cytoplasm. A primary lysosome may fuse with a food vacuole (or with another cellular organelle) to form a *secondary lysosome* in which worn-out organelles and the products of phagocytosis can be digested. Thus, a secondary lysosome contains partially digested remnants of other organelles and ingested organic material. A lysosome that contains undigested wastes is called a *residual body*. Residual bodies may eliminate their wastes by exocytosis, or the wastes may accumulate within the cell as the cell ages.

Partly digested membranes of various organelles and other cellular debris are often observed within secondary lysosomes. This is a result of **autophagy,** a process that destroys worn-out organelles so that they can be continuously replaced. Lysosomes are thus aptly referred to as the "digestive system" of the cell.

Lysosomes have also been called "suicide bags," because a break in their membranes would release their digestive enzymes and thus destroy the cell. This happens normally as part of *programmed cell death,* in which the

**Figure 3.8.** Electron micrograph showing primary lysosomes ($Lys_1$) and secondary lysosomes ($Lys_2$). Mitochondria (Mi), Golgi apparatus (GA), and the nuclear envelope (NE) are also seen.

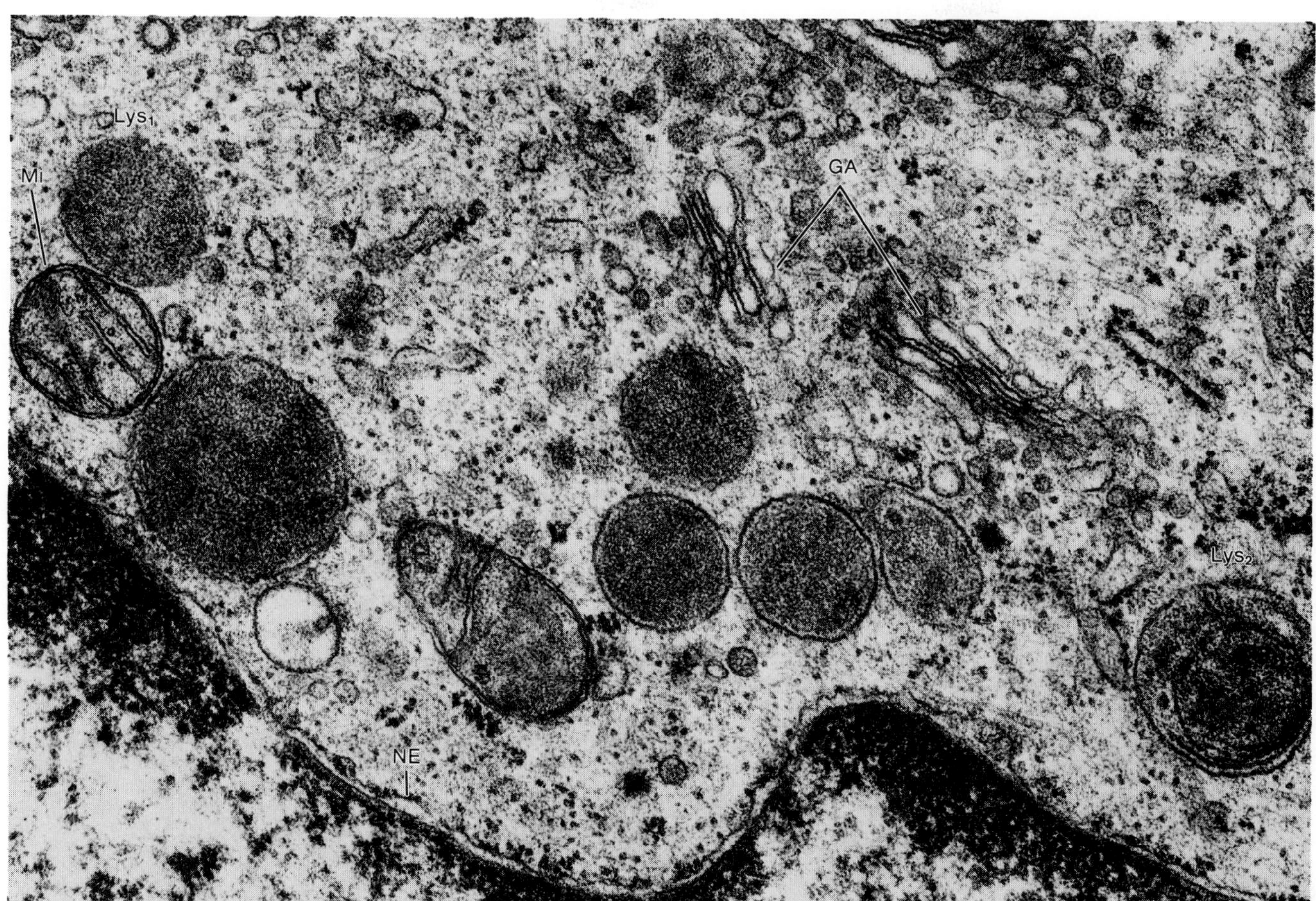


destruction of tissues is part of embryological development. It also occurs in white blood cells during an inflammation reaction.

Most, if not all, molecules in the cell have a limited life span. They are continuously destroyed and must be continuously replaced. Glycogen and some complex lipids in the brain, for example, are digested normally at a particular rate by lysosomes. If a person, because of some genetic defect, does not have the proper amount of these lysosomal enzymes, the resulting abnormal accumulation of glycogen and lipids could destroy the tissues. Examples of such diseases include Tay-Sach's disease and Gaucher's disease.

### Mitochondria

All cells in the body, with the exception of mature red blood cells, have a hundred to a few thousand organelles called **mitochondria.** Mitochondria serve as sites for the production of most of the cellular energy (chapter 5). For this reason, mitochondria are sometimes called the "powerhouses" of the cell.

Mitochondria vary in size and shape, but all have the same basic structure (fig. 3.9). Each is surrounded by an *outer membrane* that is separated by a narrow space from an *inner membrane.* The inner membrane has many folds, called *cristae,* which extend into the central area (or *matrix*) of the mitochondrion. The cristae and the matrix provide different compartments in the mitochondrion and have different roles in the generation of cellular energy. The detailed structure and function of mitochondria will be described in the context of cellular metabolism in chapter 5.

Mitochondria are able to migrate through the cytoplasm of a cell, and it is believed that they are able to reproduce themselves. Indeed, mitochondria contain their own DNA! This is a more primitive form of DNA than that found within the cell nucleus. For these and other reasons, many scientists believe that mitochondria evolved from separate organisms, related to bacteria, which entered animal cells and remained in a state of symbiosis.

**Figure 3.9.** A mitochondrion (*a*). The outer membrane (OM) and the infoldings of the inner membrane—the cristae—are clearly seen. The fluid in the center is the matrix. The structure of a mitochondrion is illustrated in (*b*).

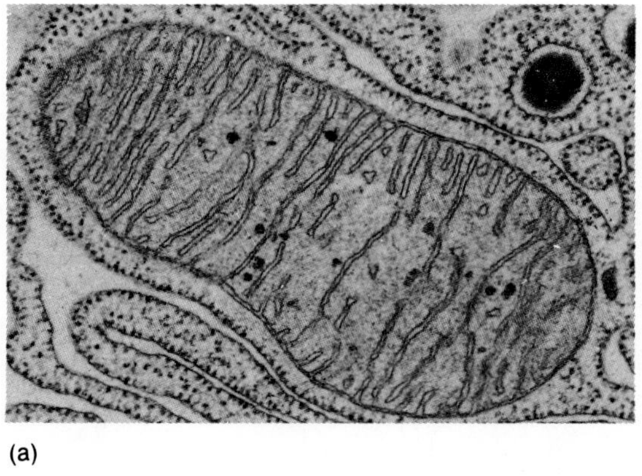

(a)

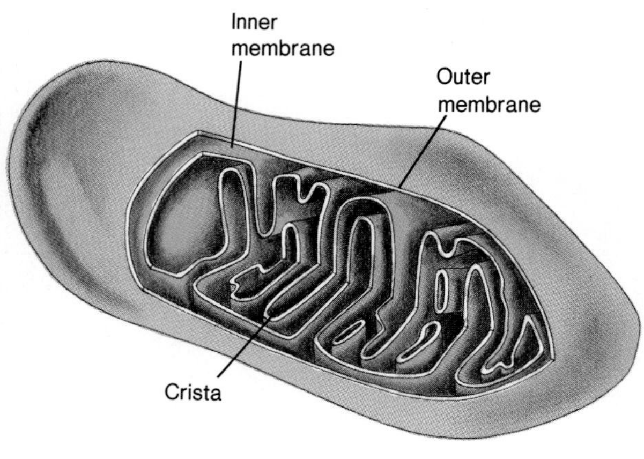


(b)

**Figure 3.10.** (*a*) An electron micrograph of endoplasmic reticulum magnified about 100,000 times. (*b*) Rough endoplasmic reticulum has ribosomes attached to its surface, whereas (*c*) smooth endoplasmic reticulum lacks ribosomes.

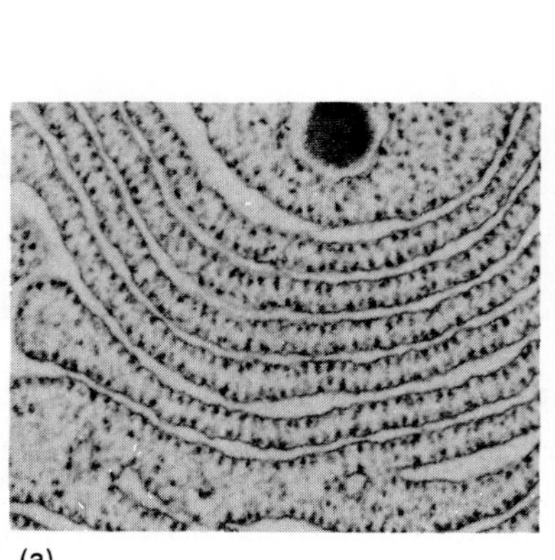

(a)

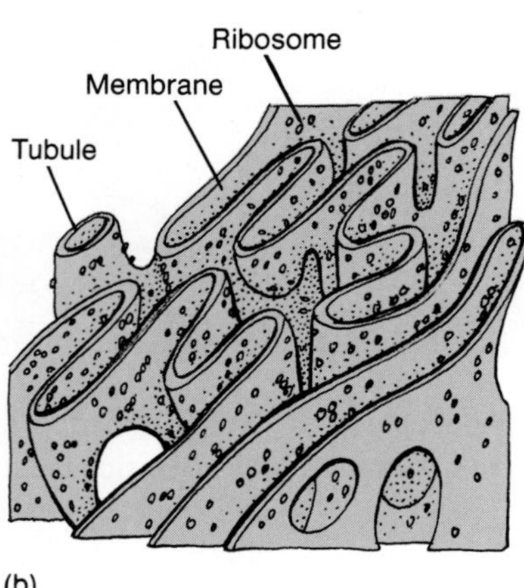


(b)

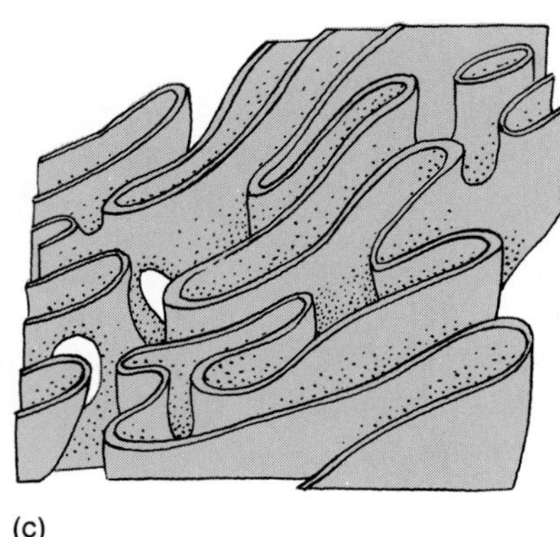

(c)

An ovum (egg cell) contains mitochondria; the head of a sperm contains none. Therefore, all of the mitochondria in a fertilized egg are derived from the mother. The mitochondrial DNA replicates itself and the mitochondria divide, so that all of the mitochondria in the fertilized ovum and the cells derived from it during embryonic and fetal development are genetically identical to those in the original ovum. This provides a unique form of inheritance that is passed only from mother to child. A rare cause of blindness—*Leber's hereditary optic neuropathy*—and perhaps genetically based neuromuscular disorders, are believed to be inherited in this manner.

## Endoplasmic Reticulum

Most cells contain a system of membranes known as the endoplasmic reticulum, of which there are two types: (1) a **rough,** or **granular, endoplasmic reticulum;** and (2) a **smooth endoplasmic reticulum** (fig. 3.10). A rough endoplasmic reticulum contains ribosomes on its surface, and a smooth endoplasmic reticulum does not contain ribosomes. The smooth endoplasmic reticulum is used for a variety of purposes in different cells; it serves as a site for enzyme reactions in steroid hormone production and inactivation, for example, and as a site for the storage of $Ca^{++}$ in striated muscle cells. The rough endoplasmic reticulum is found in cells, such as those of exocrine and endocrine glands, that are active secretors of proteins.

**Figure 3.11.** An electron micrograph of a freeze-fractured nuclear envelope, showing the nuclear pores.

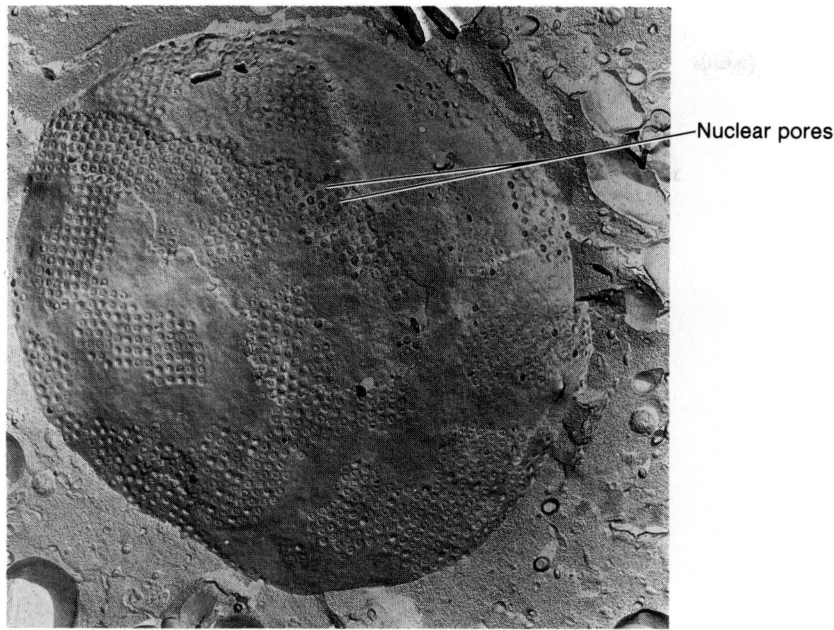


The smooth endoplasmic reticulum in liver cells contains enzymes used for the inactivation of steroid hormones and many drugs. This inactivation is generally achieved by reactions that convert these compounds to more water-soluble and less active forms, which can be more easily excreted by the kidneys. When people take certain drugs such as alcohol and phenobarbital for a long period of time, a larger dose of these compounds is required to produce a given effect. This phenomenon, called *tolerance,* is accompanied by an increase in the smooth endoplasmic reticulum and thus an increase in the enzymes charged with inactivation of these drugs.

Details of the structure and function of the rough endoplasmic reticulum and its associated ribosomes, and of another organelle called the Golgi apparatus, will be described in a later section in protein synthesis. The structure of centrioles and the spindle apparatus, which are involved in DNA replication and cell division, will also be described in a separate section.

1. *Explain why microtubules and microfilaments can be thought of as the skeleton and musculature of a cell.*
2. *Describe the contents of lysosomes and explain the significance of autophagy.*
3. *Describe the structure and function of mitochondria.*
4. *Explain how mitochondria can provide a genetic inheritance that is derived only from the mother.*
5. *Distinguish between the structure and function of a rough versus a smooth endoplasmic reticulum.*

## Cell Nucleus and Nucleic Acids

The nucleus contains threads of chromatin, which is a combination of DNA and protein. DNA and RNA are composed of subunits called nucleotides, and together these molecules are known as nucleic acids. The base sequence of DNA comprises the genetic code and serves to direct the synthesis of RNA molecules. It is through the synthesis of RNA, and eventually of protein, that the genetic code is expressed.

Most cells in the body have a single nucleus, although some—such as skeletal muscle cells—are multinucleate. The nucleus is surrounded by a *nuclear envelope* composed of an inner and an outer membrane; these two membranes fuse together to form thin sacs with openings called *nuclear pores* (figs. 3.11 and 3.12). These pores allow RNA to exit the nucleus (where it is formed) and enter the cytoplasm but prevent DNA from leaving the nucleus.

### Nucleic Acids

Nucleic acids include the macromolecules of **DNA** and **RNA,** which are critically important in genetic regulation, and the subunits from which these molecules are formed. These subunits are known as *nucleotides.*

Nucleotides are used as subunits in the formation of long polynucleotide chains. The nucleotides themselves, however, are composed of subunits. Each nucleotide is composed of three parts—a five-carbon sugar, a phosphate group bonded to one end of the sugar, and a *nitrogenous base* bonded to the other end of the sugar (fig. 3.13). The nucleotide bases are cyclic nitrogen-containing molecules with either one ring of carbons (the *pyrimidines*) or two rings (the *purines*).

**Figure 3.12.** The nucleus of a liver cell showing the nuclear envelope, heterochromatin, and nucleolus.

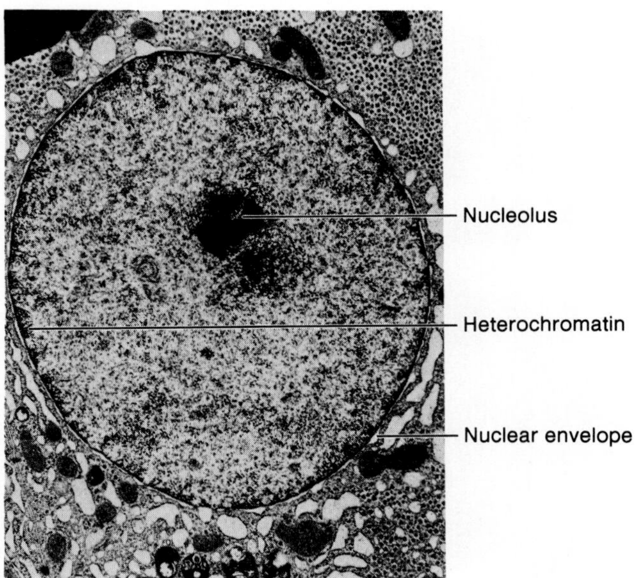


***Deoxyribonucleic Acid.*** The structure of *DNA* (*deoxyribonucleic acid*) serves as the basis for the genetic code. One might, therefore, expect DNA to have an extremely complex structure. Actually, although DNA is the largest molecule in the cell, it has a simpler structure than that of most proteins. This simplicity of structure deceived some of the early scientists into believing that the protein content of chromosomes, rather than their DNA content, provided the basis for the genetic code.

Sugar molecules in the nucleotides of DNA are a type of pentose (five-carbon) sugar called **deoxyribose** (hence the name for this nucleic acid). Each deoxyribose sugar can be covalently bonded to one of four possible bases. These bases include the two purines (adenine and guanine) and the two pyrimidines (cytosine and thymine). There are thus four different types of nucleotides that can be used to produce the long DNA chains.

When nucleotides combine to form a chain, the phosphate group of one condenses with the deoxyribose sugar of another nucleotide. This forms a sugar-phosphate chain as water is removed in dehydration synthesis. Since the nitrogenous bases are attached to the sugar molecules, the sugar-phosphate chain looks like a "backbone" from which the bases project. Each of these bases can form hydrogen bonds with other bases, which are in turn joined to a different chain of nucleotides. Such hydrogen bonding between bases thus produces a *double-stranded* DNA molecule; the two strands are like a staircase, with the paired bases as steps (fig. 3.14).

Actually, the two chains of DNA twist about each other to form a **double helix**—the molecule is like a spiral staircase (fig. 3.15). It has been shown that the number

**Figure 3.13.** The general structure of a nucleotide and the formation of sugar-phosphate bonds between nucleotides to form a polymer.

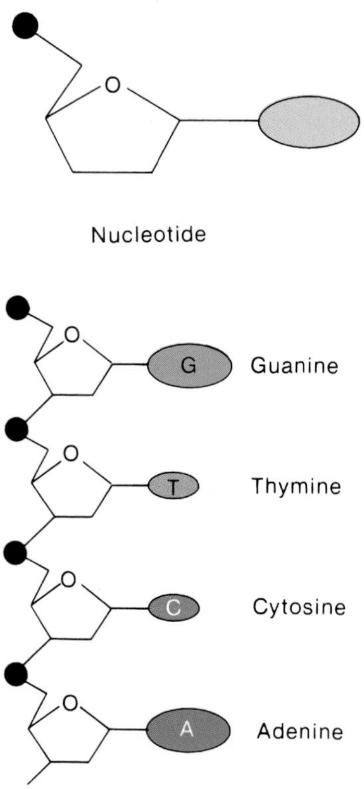

Phosphate Deoxyribose sugar Base

O

Nucleotide

O G Guanine

O T Thymine

O C Cytosine

O A Adenine

of purine bases in DNA is equal to the number of pyrimidine bases. The reason for this is explained by the **law of complementary base pairing;** adenine can only pair with thymine (through two hydrogen bonds), whereas guanine can only pair with cytosine (through three hydrogen bonds). Knowing this rule, we could predict the base sequence of one DNA strand if we knew the sequence of bases in the complementary strand.

Although we can predict which base is opposite a given base in DNA, we cannot predict which bases are above or below that position within a single polynucleotide chain. Although there are only four bases, the number of possible base sequences along a stretch of several thousand nucleotides (the length of a gene) is almost infinite. Despite the almost infinite possible variety of sequences, almost all of the billions of copies of a particular gene in a person are identical. The mechanisms by which this is achieved will be described in a later section.

***Ribonucleic Acid.*** The genetic information contained in DNA functions to direct the activities of the cell through its production of another type of nucleic acid—*RNA (ribonucleic acid).* Like DNA, RNA consists of long chains of nucleotides joined together by sugar-phosphate bonds. Nucleotides in RNA, however, differ from those in DNA

**Figure 3.14.** The four nitrogenous bases in deoxyribonucleic acid (DNA). Notice that hydrogen bonds can form between guanine and cytosine and between thymine and adenine.

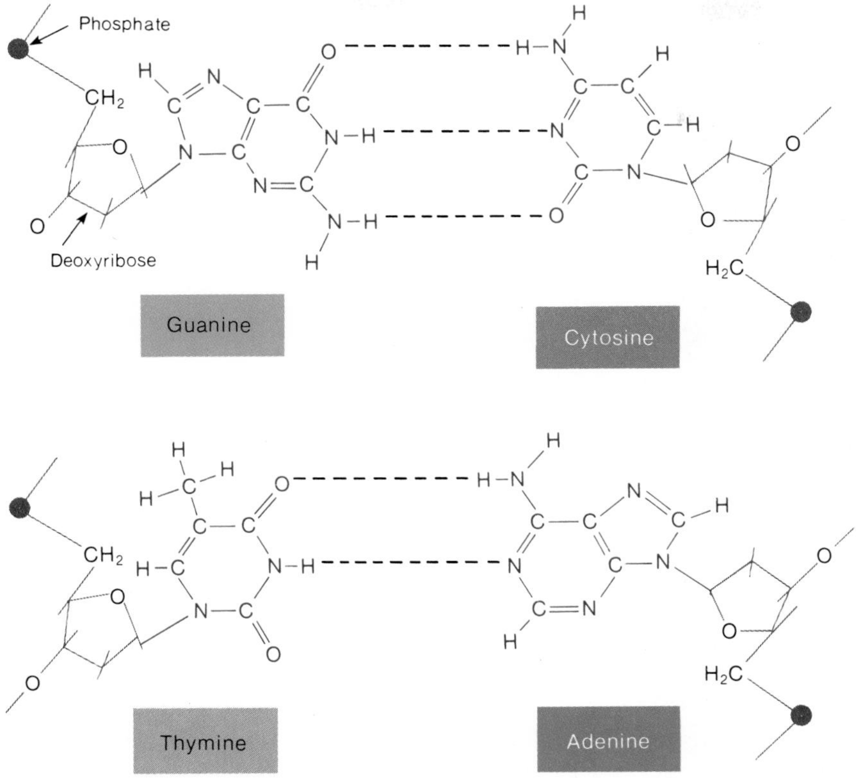

Phosphate

Deoxyribose

Guanine

Cytosine

Thymine

Adenine

(fig. 3.16) in three ways: (1) a **ribonucleotide** contains the sugar *ribose* (instead of deoxyribose); (2) the base *uracil* is found in place of thymine; and (3) RNA is composed of a single polynucleotide strand (it is not double-stranded like DNA).

There are three types of RNA molecules that function in the cytoplasm of cells: *messenger RNA (mRNA), transfer RNA (tRNA),* and *ribosomal RNA (rRNA).* All three types are made within the cell nucleus by using information contained in DNA as a guide.

## Chromatin

Many granulated threads in the nuclear fluid can be seen with an electron microscope. These threads are called *chromatin* and consist of a combination of DNA and protein. There are two forms of chromatin. Thin, extended chromatin—or *euchromatin*—appears to be the active form of DNA in a nondividing cell. Regions of condensed, "blotchy"-appearing chromatin, known as *heterochromatin,* are believed to contain inactive DNA.

One or more dark areas within each nucleus can be seen. These regions, which are not surrounded by membranes, are called *nucleoli.* The DNA within nucleoli contains genes that code for the production of ribosomal RNA (rRNA), an essential component of ribosomes.

## Genetic Transcription—RNA Synthesis

The thin, extended euchromatin is the "working" form of DNA; the more familiar short, stubby form of chromosomes seen during cell division are inactive packages of DNA. The genes do not become active until the chromosomes unravel. Active DNA directs the metabolism of the cell indirectly through its regulation of RNA and protein synthesis.

*One gene codes for one polypeptide chain.* Each gene is a stretch of DNA that is several thousand nucleotide pairs long. The DNA in a human cell contains three to four billion base pairs—enough to code for at least three million proteins. Since the average human cell contains less than this amount (30,000 to 150,000 different proteins), it follows that only a fraction of the DNA in each cell is used to code for proteins. The remainder of the DNA may be inactive, may be redundant, and may serve to regulate those regions that do code for proteins.

In order for the genetic code to be translated into the synthesis of specific proteins, the DNA code must first be transcribed into an RNA code (fig. 3.17). This is accomplished by DNA-directed RNA synthesis, or **genetic transcription.**

**Figure 3.15.** The double helix structure of DNA.

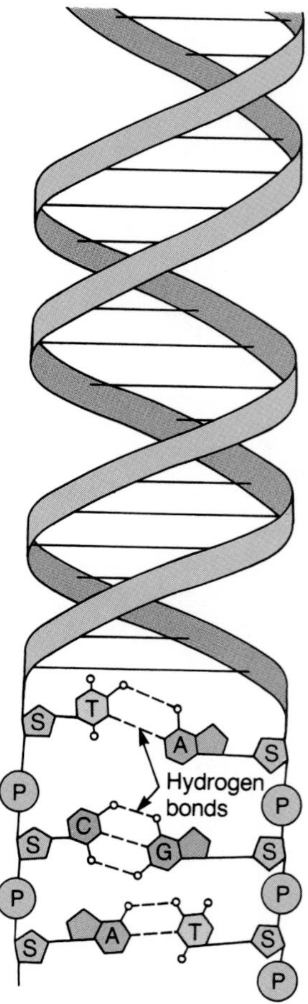


**Figure 3.16.** Differences between the nucleotides and sugars in DNA and RNA.

| DNA nucleotides contain | RNA nucleotides contain |
| --- | --- |
| Deoxyribose | Ribose |
| Thymine | Uracil |

**Figure 3.17.** The synthesis of ribosomal RNA on DNA in the nucleolus.

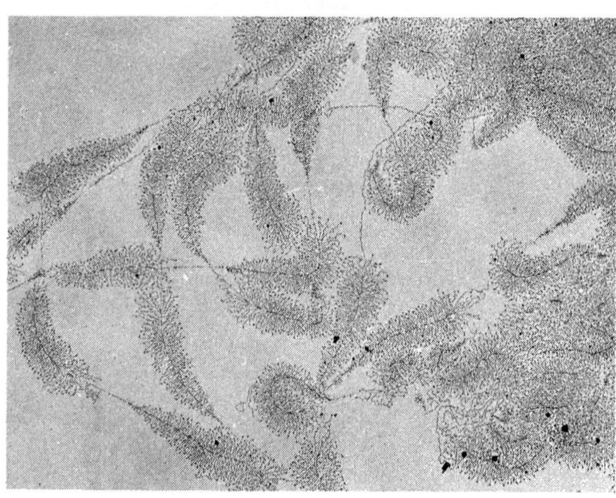

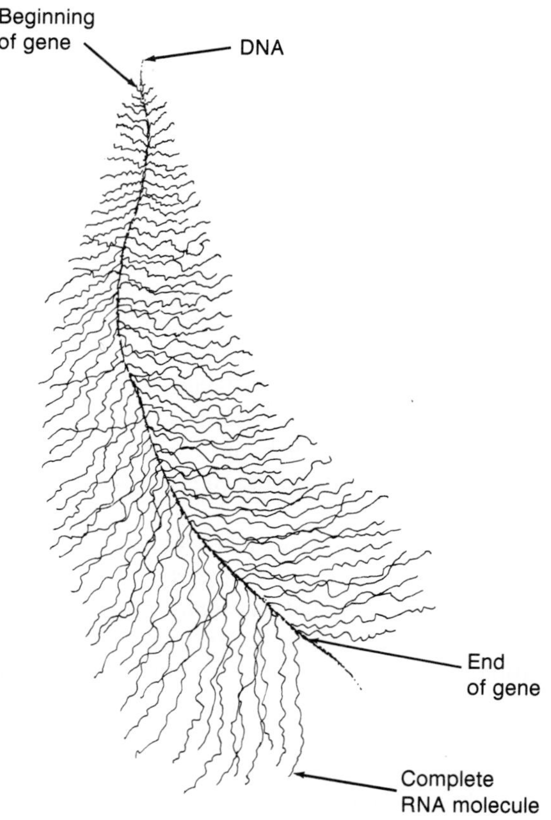


In RNA synthesis, the enzyme *RNA polymerase* breaks the weak hydrogen bonds between paired DNA bases. This does not occur throughout the length of DNA, but only in the regions that are to be transcribed (there are base sequences that code for "start" and "stop"). Double-stranded DNA, therefore, separates in these regions so that the freed bases can pair with the complementary RNA nucleotide bases, which are freely available in the nucleoplasm.

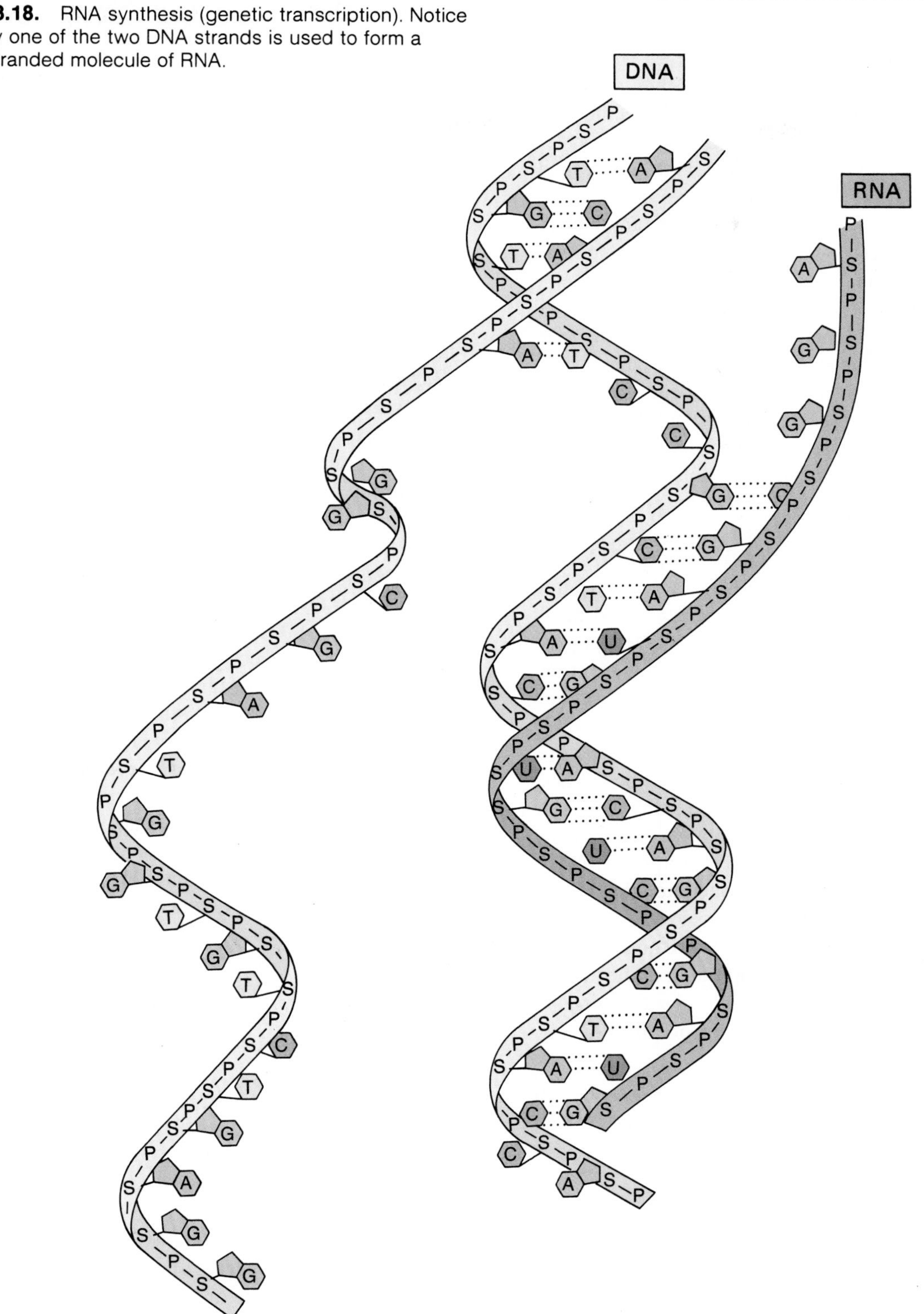


**Figure 3.18.** RNA synthesis (genetic transcription). Notice that only one of the two DNA strands is used to form a single-stranded molecule of RNA.

This pairing of bases, like that which occurs in DNA replication (described in a later section), follows the law of complementary base pairing: guanine bonds with cytosine (and vice versa), and adenine bonds with uracil (because uracil in RNA is equivalent to thymine in DNA). Unlike DNA replication, however, only *one* of the two freed strands of DNA serves as a guide for RNA synthesis (fig. 3.18). Once an RNA molecule has been produced it detaches from the DNA strand on which it was formed. This process can continue indefinitely, producing many thousands of RNA copies of the DNA strand which is being transcribed. When the gene is no longer to be transcribed, the separated DNA strands can then go back together again.

***Types of RNA.*** There are four types of RNA produced within the nucleus by genetic transcription: (1) **precursor messenger RNA (pre-mRNA),** which is altered within the nucleus to form mRNA; (2) **messenger RNA (mRNA),** which contains the code for the synthesis of specific proteins; (3) **transfer RNA (tRNA),** which is needed for decoding the genetic message contained in mRNA; and (4) **ribosomal RNA (rRNA),** which forms part of the structure of ribosomes. The DNA that code for rRNA synthesis are located in the part of the nucleus called the nucleolus. The DNA that code for pre-mRNA and tRNA synthesis are located elsewhere in the nucleus.

In bacteria, where the molecular biology of the gene is best understood, a gene that codes for one type of protein produces an mRNA molecule that begins to direct protein synthesis as soon as it is transcribed. This is not the case in higher organisms, including humans. In higher cells a pre-mRNA is produced that must be modified within the nucleus before it can enter the cytoplasm as mRNA and direct protein synthesis.

Precursor mRNA is much larger than the mRNA that it forms. This large size of pre-mRNA is, surprisingly, not due to excess bases at the ends of the molecule that must be trimmed. Rather, the excess bases are *within* the pre-mRNA. The genetic code for a particular protein, in other words, is split up by stretches of base pairs that do not contribute to the code. These regions of noncoding DNA within a gene are called *introns*; the coding regions are known as *exons*. As a result, pre-mRNA must be cut and spliced to make mRNA. This cutting and splicing can be quite extensive; a single gene may contain up to fifty introns which must be removed from the pre-mRNA. After cutting and splicing, the mRNA is further modified by the addition of characteristic bases at its ends; about two hundred adenine-containing nucleotides are added to the tail end, and a modified guanine-containing nucleotide is added to the other end of mRNA. These *post-transcriptional modifications* produce the form of mRNA that enters the cytoplasm (fig. 3.19).

1. *Explain the difference between euchromatin and heterochromatin.*
2. *Describe the structure of DNA, and explain the law of complementary base pairing.*
3. *Describe the structure of RNA, and list the different types of RNA.*
4. *Explain how RNA is produced within the nucleus according to the information contained in DNA.*
5. *Explain how precursor mRNA is modified to produce mRNA.*

## *Protein Synthesis and Secretion*

In order for a gene to be expressed, it first must be used as a guide, or template, in the production of a complementary strand of messenger RNA. This process is called genetic transcription. The mRNA is then used as a guide to produce a particular type of protein, which has a sequence of amino acids determined by the sequence of base triplets (codons) in the mRNA. This process is called genetic translation. The secretion of proteins by the cell requires the participation of a number of organelles within the cytoplasm of the cell.

When mRNA enters the cytoplasm it attaches to ribosomes, which are seen in the electron microscope as numerous small particles. A **ribosome** is composed of three molecules of ribosomal RNA and fifty-two proteins, arranged to form two subunits of unequal size. The mRNA passes through a number of ribosomes to form a "string-of-pearls" structure called a *polyribosome* (or *polysome,* for short), shown in figure 3.20. The association of mRNA with ribosomes is needed for **genetic translation**—the production of specific proteins according to the code contained in the mRNA base sequence.

Each mRNA molecule contains several hundred or more nucleotides, arranged in the sequence determined by complementary base pairing with DNA during genetic transcription (RNA synthesis). Every three bases, or *base triplet,* is a code word—called a **codon**—for a specific amino acid. Sample codons and their amino acid "translations" are shown in table 3.2 and in figure 3.21. As mRNA moves through the ribosome, the sequence of codons is translated into a sequence of specific amino acids within a growing polypeptide chain.

### Transfer RNA

Translation of the codons is accomplished by tRNA and particular enzymes. Each tRNA molecule, like mRNA and rRNA, is single stranded. Although tRNA is single stranded, it bends on itself to form three loop regions (fig. 3.22). One of these loops contains the **anticodon**—three nucleotides that are complementary to a specific codon in mRNA.

Enzymes in the cell cytoplasm called *aminoacyl-tRNA synthetase* enzymes join specific amino acids to the ends of tRNA, so that a tRNA with a given anticodon is always bonded to one specific amino acid. These enzymes don't actually "read" the anticodon. Rather, each tRNA with a different anticodon has a different "recognition loop" (fig. 3.22) that is detected by these enzymes. Recent evidence suggests that there are twenty aminoacyl-tRNA synthetase enzymes which each recognize only one pair of bases in tRNA, and join specific amino acids to specific tRNA molecules on this basis. The cytoplasm of a cell, therefore, contains tRNA molecules that are each bonded

**Figure 3.19.** Processing of a pre-mRNA into mRNA. Noncoding regions of the genes, called introns, produce excess bases within the pre-mRNA. These excess bases are removed (as lariatlike strands), and the coding regions of mRNA are spliced together.

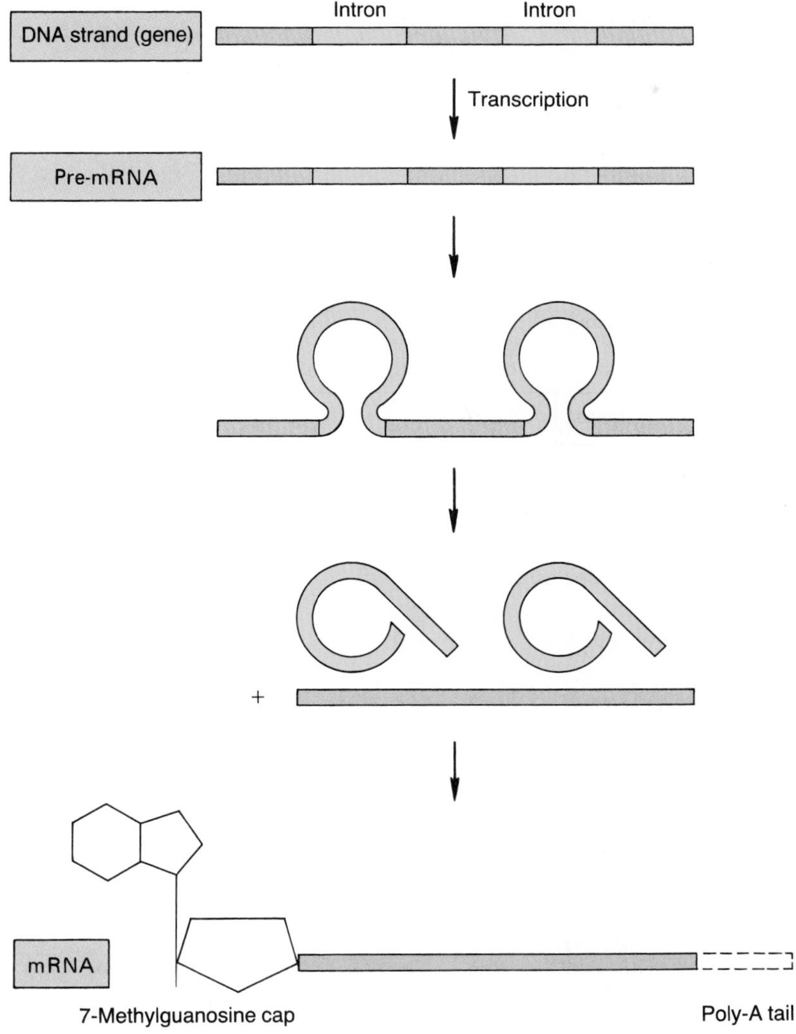


**Figure 3.20.** An electron micrograph of polyribosomes. An RNA strand (*arrow*) joins the ribosomes together.

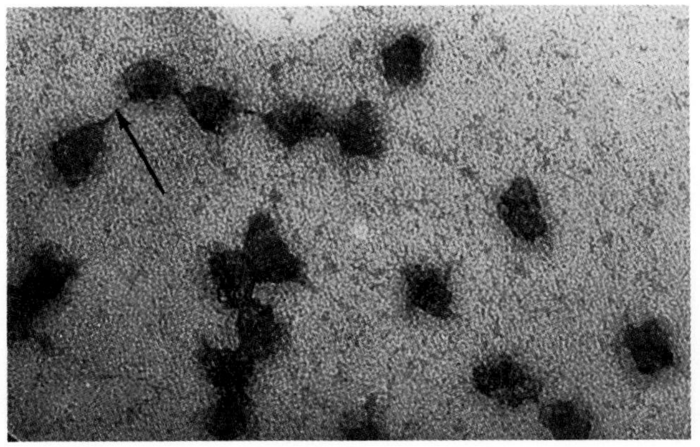

**Table 3.2** Selected DNA base triplets and mRNA codons

| DNA Triplet | RNA Codon | Amino Acid |
|---|---|---|
| TAC | AUG | "Start" |
| ATC | UAG | "Stop" |
| AAA | UUU | Phenylalanine |
| AGG | UCC | Serine |
| ACA | UGU | Cysteine |
| GGG | CCC | Proline |
| GAA | CUU | Leucine |
| GCT | CGA | Arginine |
| TTT | AAA | Lysine |
| TGC | ACG | Tyrosine |
| CCG | GGC | Glycine |
| CTC | GAG | Aspartic acid |

**Figure 3.21.** The genetic code is first transcribed into base triplets (codons) in mRNA and then translated into a specific sequence of amino acids in a protein.

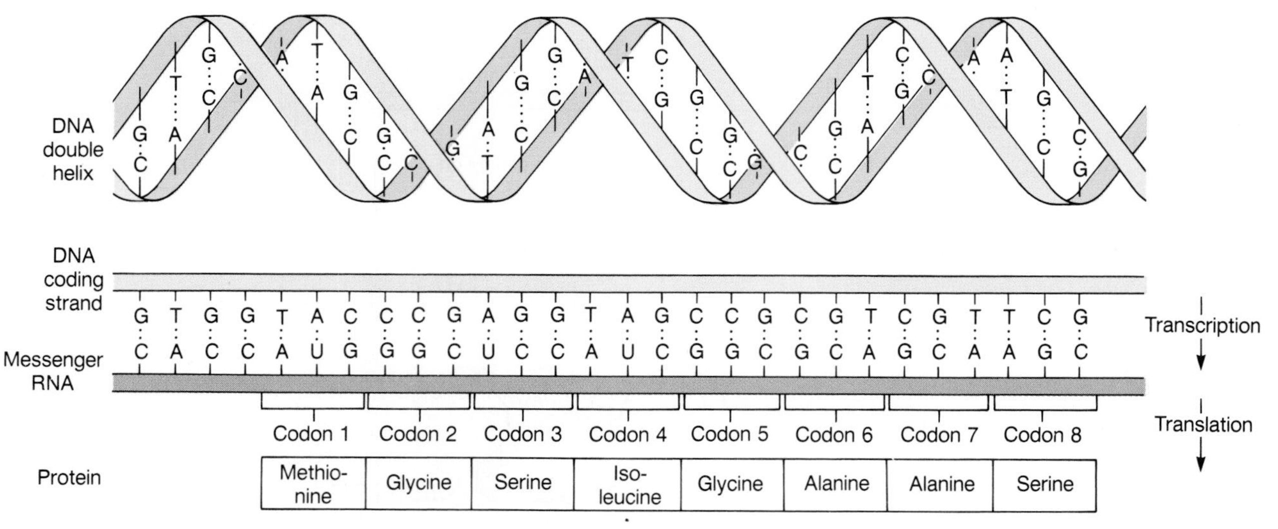


**Figure 3.22.** The structure of transfer RNA (tRNA).

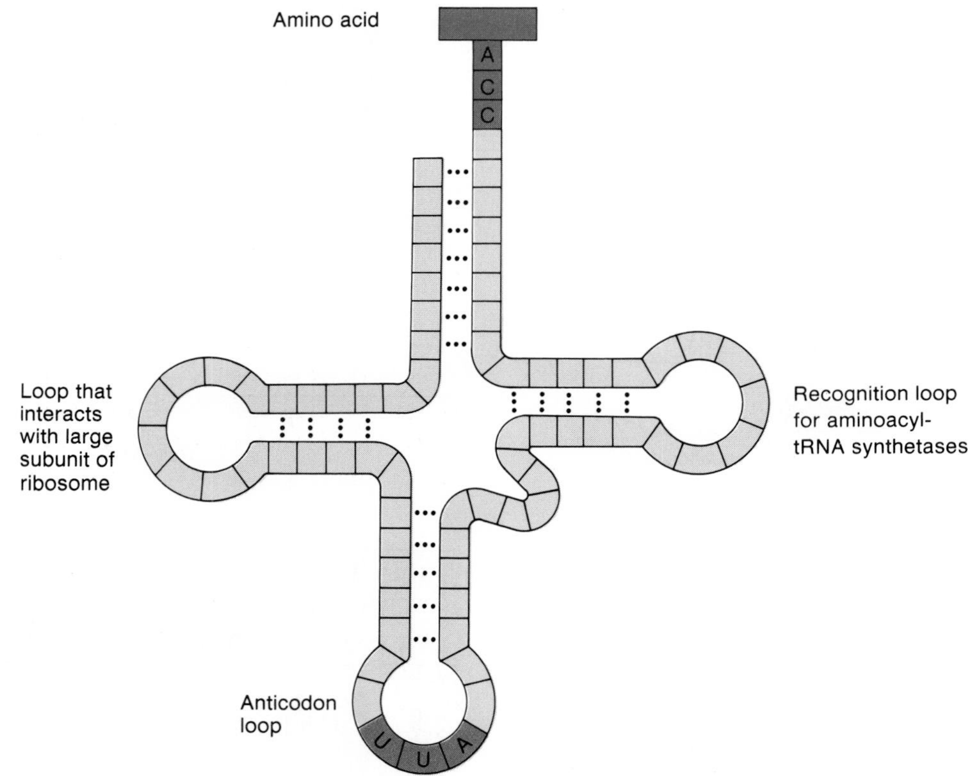

**Figure 3.23.** The translation of messenger RNA (mRNA). As the anticodon of each new aminoacyl-tRNA bonds to a codon of the mRNA, new amino acids are joined to the growing tip of the polypeptide chain.

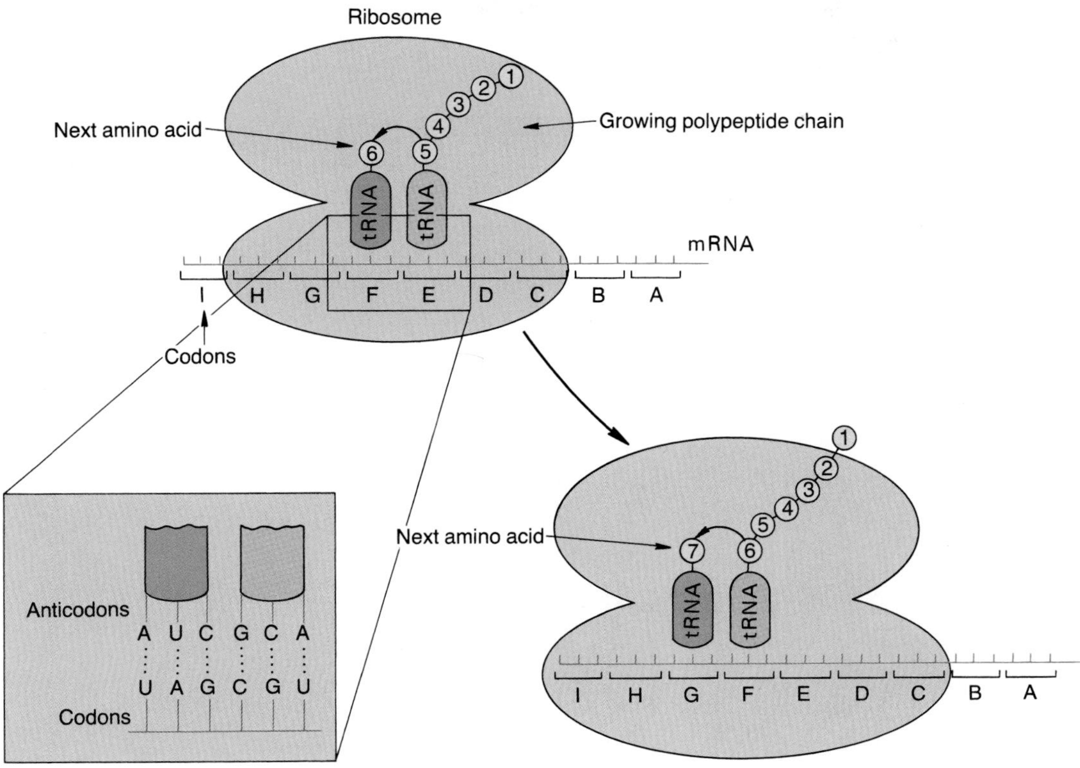


to a specific amino acid, and which are each capable of bonding with their anticodon base triplet to a specific codon in mRNA.

## Formation of a Polypeptide

The anticodons of tRNA bond to the codons of mRNA as the mRNA moves through the ribosome. Since each tRNA molecule carries a specific amino acid, the joining together of these amino acids by peptide bonds creates a polypeptide whose amino acid sequence has been determined by the sequence of codons in mRNA.

When the first and second tRNA bring the first and second amino acids together and a peptide bond forms between them, the first amino acid detaches from its tRNA so that a dipeptide is linked by the second amino acid to the second tRNA. When the third tRNA bonds to the third codon, the third amino acid forms a peptide bond with the second amino acid (which detaches from its tRNA); a tripeptide is then attached by the third amino acid to the third tRNA. The polypeptide chain thus grows as new amino acids are added to its growing tip (fig. 3.23). This growing polypeptide chain is always attached by means of only one tRNA to the strand of mRNA, and this tRNA molecule is always the one which has added the latest amino acid to the growing polypeptide.

As the polypeptide chain grows in length, interactions between its amino acids cause the chain to twist into a helix (secondary structure) and to fold and bend on itself (tertiary structure). At the end of this process, the new protein detaches from the tRNA as the last amino acid is added. Many proteins are further modified after they are formed; these modifications occur in the rough endoplasmic reticulum and Golgi apparatus.

## Function of the Rough Endoplasmic Reticulum

Proteins that are to be used within the cell are produced in polyribosomes that are free in the cytoplasm. If the protein is a secretory product of the cell, however, it is made by mRNA-ribosome complexes located in the rough endoplasmic reticulum. The membranes of this system enclose fluid-filled spaces (cisternae), which the newly formed proteins may enter.

When proteins that are destined for secretion are produced, the first thirty or so amino acids are primarily hydrophobic. This *leader sequence* is attracted to the lipid component of the membranes of the endoplasmic reticulum. As the polypeptide chain elongates, it is "injected" into the cisterna within the endoplasmic reticulum. The leader sequence is, in a sense, an "address" that directs secretory proteins into the endoplasmic reticulum. Once

**Figure 3.24.** A protein destined for secretion begins with a *leader sequence* that enables it to be inserted into the endoplasmic reticulum. Once it has been inserted, the leader sequence is removed and carbohydrate is added to the protein.

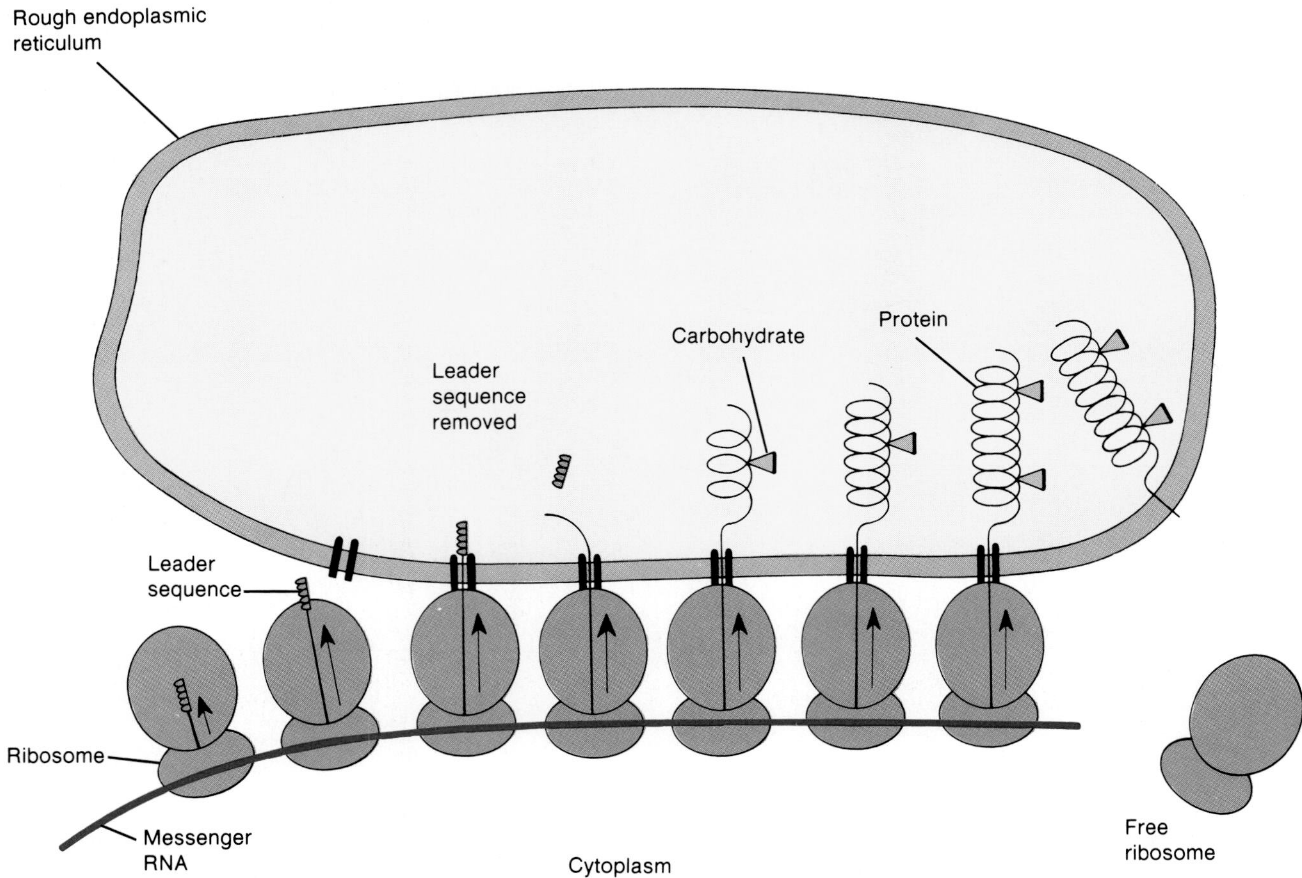


the proteins are in the cisterna, the leader sequence is removed so the protein cannot reenter the cytoplasm (fig. 3.24).

The processing of the hormone insulin can serve as an example of the post-translational changes that occur within the endoplasmic reticulum. The original molecule enters the cisterna as a single polypeptide composed of 109 amino acids. This molecule is called preproinsulin. The first twenty-three amino acids serve as a leader sequence and are quickly removed (to produce proinsulin). The remaining chain folds within the cisterna so that the first and last amino acids in the polypeptide are brought close together. The central region is then enzymatically removed, producing two chains—one is twenty-one amino acids long; the other is thirty amino acids long—which are subsequently joined together by disulfide bonds. This is the form of insulin that is normally secreted from the cell.

### Function of the Golgi Apparatus

Secretory proteins do not remain trapped within the rough endoplasmic reticulum; they are transported to another organelle within the cell—the **Golgi apparatus.** This organelle serves two interrelated functions: (1) further modifications of proteins (such as addition of carbohydrates to form *glycoproteins*) occur in the Golgi apparatus; (2) different types of proteins are separated according to their function and destination; and (3) the final products are packaged into secretory vesicles. For example, proteins that are to be secreted are separated from those that will be incorporated into the cell membrane and from those that will be introduced into lysosomes, and these are packaged into separate membrane-enclosed vesicles.

The Golgi apparatus consists of several flattened sacs that enclose cisternae (spaces). Proteins produced by the rough endoplasmic reticulum are believed to travel in membrane-enclosed vesicles to the sac on one end of the Golgi apparatus. After specialized modifications of the proteins are made within one sac, the modified proteins are passed by means of vesicles to the next sac until the finished products finally leave the Golgi apparatus in vesicles that fuse with the cell membrane or with the membranes of lysosomes (fig. 3.25).

The Golgi apparatus and the rough endoplasmic reticulum contain enzymes that modify the structure of proteins. These changes, combined with the events that occur

**Figure 3.25.** (*a*) An electron micrograph of a Golgi apparatus. Notice the formation of vesicles at the ends of some of the flattened sacs. (*b*) An illustration of the processing of proteins by the rough endoplasmic reticulum and Golgi apparatus. ([b] From "The Compartmental Organization of the Golgi Apparatus," by J. E. Rothman. Copyright © 1985 by Scientific American, Inc. All rights reserved.)

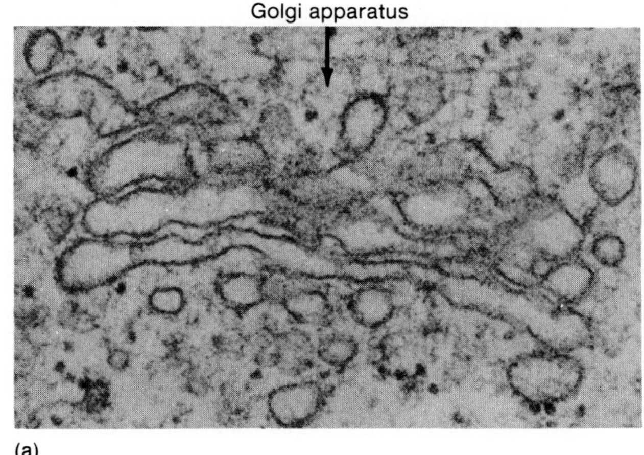

(a)

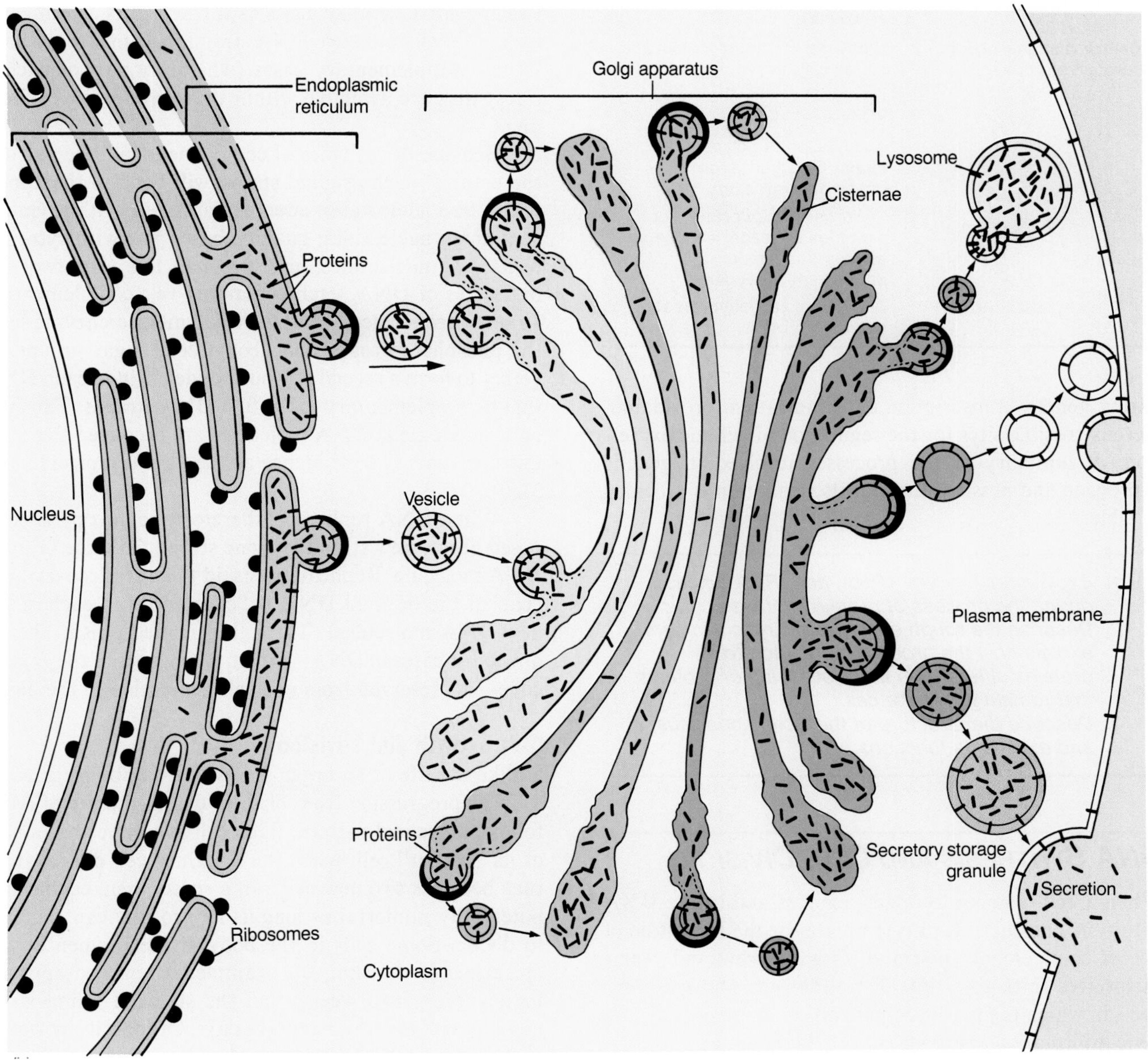

(b)

**Table 3.3** Stages in genetic expression and possible sites for its regulation

| Stage of gene expression | Possible sites for regulation |
|---|---|
| RNA synthesis | Duplication of genes; cutting and splicing of genes to different positions on a chromosome |
| | Association of histone and nonhistone proteins with chromatin |
| Nuclear processing of RNA | Cutting and splicing of nuclear RNA |
| | Capping of messenger RNA with 7-methyl guanosine and addition of about two hundred adenylate nucleotides |
| Translation of mRNA | Availability of specific tRNA molecules |
| Post-translational changes in protein structure | Cleavage of parent protein into biologically active fragments |
| | Chemical modification of amino acids |
| | Association of polypeptide chains together to form quaternary structure |
| | "Addressing" of protein for secretion by leader sequence and insertion into rough endoplasmic reticulum |
| | Addition of carbohydrate to secreted protein |

during genetic transcription and translation, provide numerous possible sites for the regulation of genetic expression. A summary of the processes involved in genetic expression and possible sites for its regulation is provided in table 3.3.

1. *Explain how mRNA, rRNA, and tRNA function during the process of protein synthesis.*
2. *Describe the rough endoplasmic reticulum, and explain how the processing of secretory proteins differs from the processing of proteins that remain within the cell.*
3. *Describe the structure of the Golgi apparatus, and explain its functions.*

## DNA Synthesis and Cell Division

When a cell is going to divide, each strand of the DNA within its nucleus acts as a template for the formation of a new complementary strand. Organs grow and repair themselves through a type of cell division known as mitosis, in which the two daughter cells receive the same genetic information as the parent cell. Gametes contain only half the number of chromosomes as their parent cell and are formed by a type of cell division called meiosis.

Genetic information is required for the life of the cell and for the ability of the cell to perform its functions in the body. Each cell obtains this genetic information from its parent cell through the process of DNA replication and cell division. DNA is the only type of molecule in the body capable of replicating itself, and mechanisms exist within the dividing cell to ensure that the duplicate copies of DNA are properly distributed to the daughter cells.

### DNA Replication

When a cell is going to divide, each DNA molecule replicates itself, and each of the identical DNA copies thus produced is distributed to the two daughter cells. Replication of DNA requires the action of a specific enzyme known as *DNA polymerase.* This enzyme moves along the DNA molecule, breaking the weak hydrogen bonds between complementary bases as it travels. As a result, the bases of each of the two DNA strands become free to bond to new complementary bases (which are part of nucleotides) that are available within the surrounding environment.

Because of the rules of complementary base pairing, the bases of each original strand will bond to the appropriate free nucleotides: adenine bases pair with thymine-containing nucleotides; guanine bases pair with cytosine-containing nucleotides, and so on. In this way, two new molecules of DNA, each containing two complementary strands, are formed. The DNA polymerase enzyme links the phosphate groups and deoxyribose sugar groups together to form a second polynucleotide chain in each DNA that is complementary to the first DNA strands. Thus two new double-helix DNA molecules are produced that contain the same base sequence as the parent molecule (fig. 3.26).

When DNA replicates, therefore, each copy is composed of one new strand and one strand from the original DNA molecule. Replication is said to be **semiconservative** (half of the original DNA is "conserved" in each of the new DNA molecules). Through this mechanism, the sequence of bases in DNA—which is the basis of the genetic code—is preserved from one cell generation to the next.

### Cell Growth and Division

Unlike the life of an organism, which can be pictured as a linear progression from birth to death, the life of a cell follows a cyclical pattern. Each cell is produced as a part of its "parent" cell; when the daughter cell divides, it in turn becomes two new cells. In a sense, then, each cell is potentially immortal as long as its progeny can continue to divide. Some cells in the body divide frequently; the epidermis of the skin, for example, is renewed approximately every two weeks, and the stomach lining is renewed about every two or three days. Other cells, however, such as nerve and striated muscle cells in the adult, do not divide at all. All cells in the body, of course, live only as long as the person lives (some cells live longer than others, but eventually all cells die when vital functions cease).

**Figure 3.26.** The replication of DNA. Each new double helix is composed of one old and one new strand. The base sequences of each of the new molecules is identical to that of the parent DNA because of complementary base pairing.

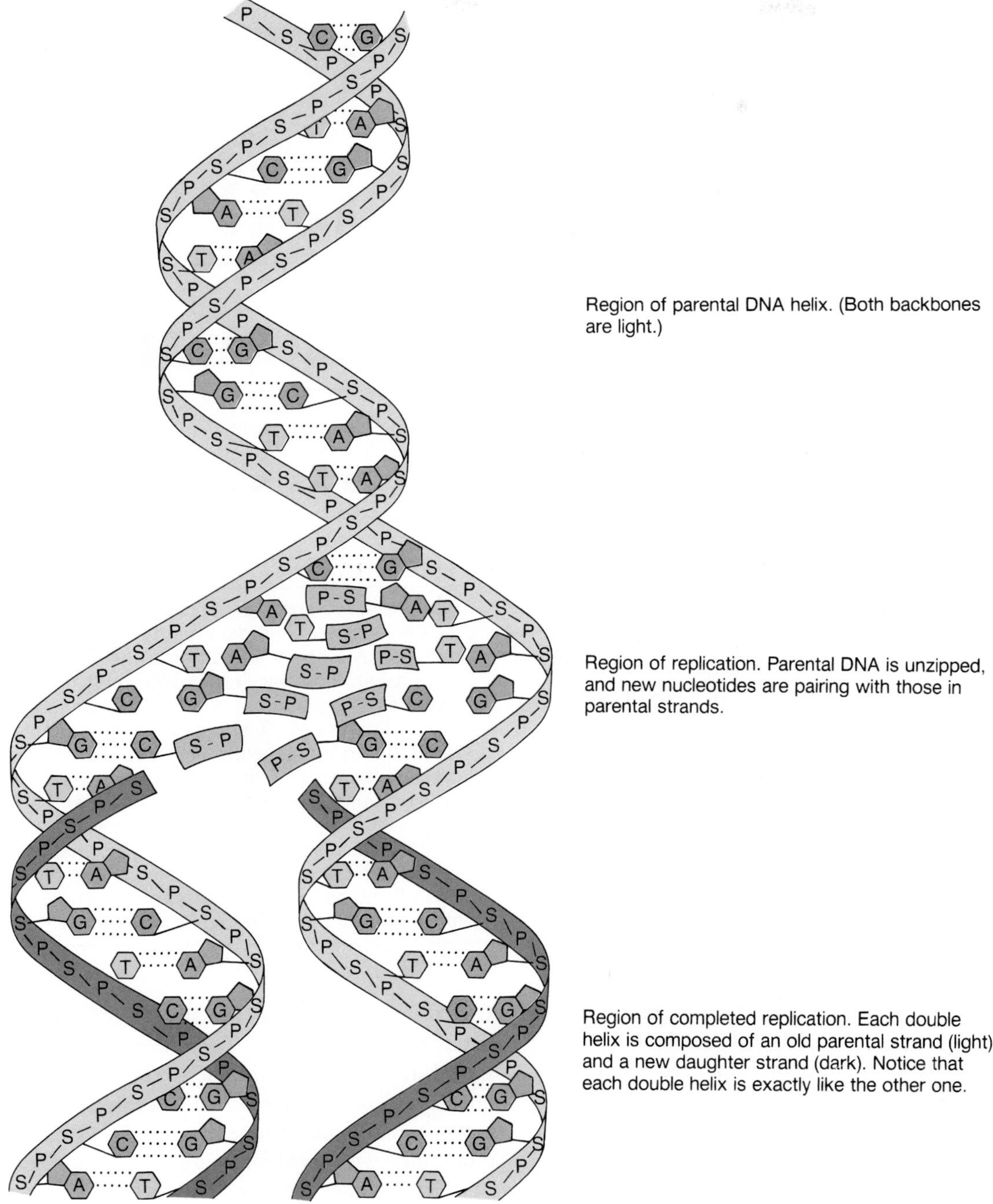


***The Cell Cycle.*** The nondividing cell is in a part of its life cycle known as **interphase** (fig. 3.27), which is subdivided into $G_1$, S, and $G_2$ phases, as will be described. The chromosomes are in their extended form (as euchromatin), and their genes actively direct the synthesis of RNA. Through their direction of RNA synthesis, genes control the metabolism of the cell. During this time the cell may be growing, so this part of interphase is known as the $G_1$ *phase* (*G* stands for *growth*). Although sometimes described as "resting," cells in the $G_1$ phase perform the physiological functions characteristic of the tissue in which they are found. The DNA of resting cells in the $G_1$ phase thus produces mRNA and proteins as previously described.

**Figure 3.27.** The life cycle of a cell.

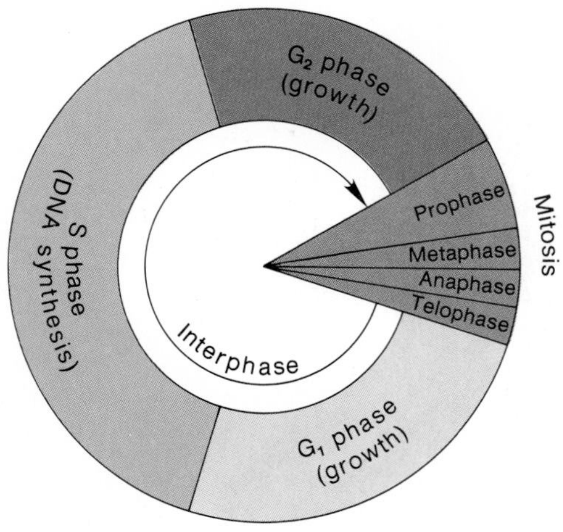


If a cell is going to divide, it replicates its DNA in a part of interphase known as the *S phase* (*S* stands for *synthesis*). The mechanisms that cause transformation of a cell from the $G_1$ to the S phase are not known. Some experiments suggest that two nucleotides—cyclic adenosine monophosphate (cAMP) and cyclic guanosine monophosphate (cGMP)—may regulate this transformation. An increase in the intracellular concentration of cAMP and a decrease in cGMP can stimulate cell division. Reverse changes in the concentration of the nucleotides inhibit cell division. Besides helping to regulate cell division, these cyclic nucleotides have a wide variety of other regulatory functions in the cell (as described in later chapters).

Once DNA has replicated in the S phase, the chromatin condenses in the *$G_2$* phase to form short, thick, rodlike structures by the end of $G_2$. This is the more familiar form of chromosomes because they are easily seen in the ordinary (light) microscope. It should be remembered that this form of the chromatin represents a "packaged" state of DNA—not the extended, threadlike form that is active in directing the metabolism of the cell during the $G_1$ phase.

The matched pairs of chromosomes are called **homologous chromosomes.** One member of each homologous pair is derived from a chromosome inherited from the father, and the other member is a copy of one of the chromosomes inherited from the mother. Homologous chromosomes do not have identical DNA base sequences; one member of the pair may code for blue eyes, for example, and the other for brown eyes. There are twenty-two homologous pairs of *autosomal chromosomes* and one pair of *sex chromosomes,* described as X and Y. Females have two X chromosomes, whereas males have one X and one Y chromosome (fig. 3.28).

**Figure 3.28.** A photograph of homologous pairs of chromosomes from a human male cell at metaphase of mitosis (homologous chromosomes have been paired and numbered according to convention).

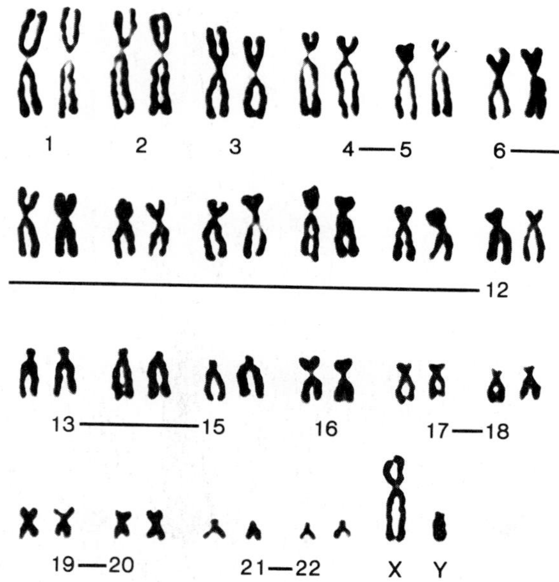


Mature nerve and muscle cells do not replicate at all; neurons are thus particularly susceptible to damage by alcohol and other drugs. Epithelial cells, in contrast, have very rapid cell cycles that help to replace the continuous loss of cells. *Cancers* have rapid rates of cell division but not necessarily more rapid than normal tissue. The fast growth of some cancers is thus not due simply to a rapid rate of cell division but is rather due to the fact that the rate of cell division is much greater than the rate of cell death. The observation that many normal tissues are at least as rapid in their rates of cell divisions as cancer makes the chemotherapy of cancer much more difficult than it would be otherwise.

***Centrioles.*** **Centrioles** are small, rodlike structures located near the nucleus. There are two centrioles, positioned at right angles to each other, within a spherical nonmembranous mass called a **centrosome.** Each centriole is composed of nine evenly spaced bundles of microtubules, with three microtubules per bundle (fig. 3.29).

Centrosomes are found only in those cells that are capable of division. During the process of cell division, the centrioles take up positions on opposite sides of the nucleus. The centrioles are then involved in the production of, and are attached to, **spindle fibers,** which are also composed of microtubules. The spindle fibers and centrioles help to pull the duplicated chromosomes to opposite poles of the cell during cell division. Cells that lack centrioles, such as mature muscle and nerve cells, cannot divide.

**Figure 3.29.** The centrioles. (*a*) A micrograph of the two centrioles in a centrosome. (*b*) A diagram showing that the centrioles are positioned at right angles to one another.

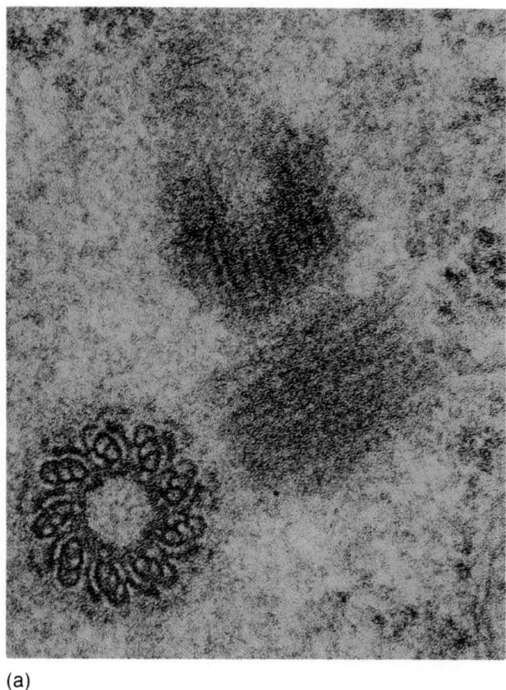

(a)

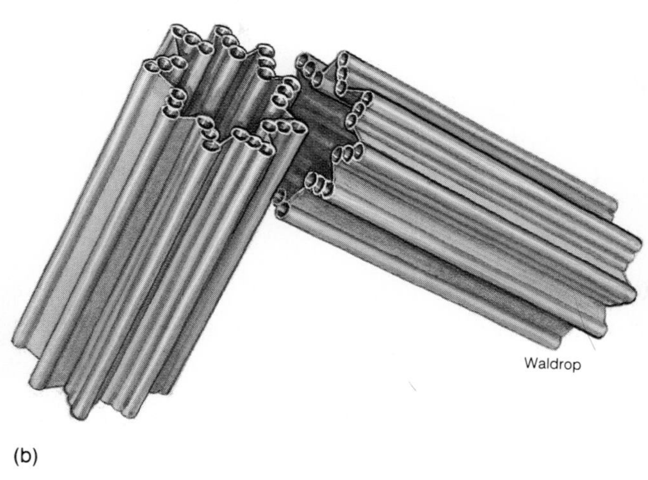


(b)

***Mitosis.*** At the end of the $G_2$ phase of the cell cycle, which is generally shorter than $G_1$, each chromosome consists of two strands called **chromatids,** which are joined together by a *centromere.* The two chromatids within a chromosome contain identical DNA base sequences because each is produced by the semiconservative replication of DNA. Each chromatid, therefore, contains a complete double helix DNA molecule that is a copy of the single DNA molecule existing prior to replication. Each chromatid will become a separate chromosome once cell division has been completed.

The $G_2$ phase completes interphase. The cell next proceeds through the various stages of cell division, or **mitosis** (the *M phase* of the cell cycle). Mitosis is subdivided into four phases: prophase, metaphase, anaphase, and telophase (fig. 3.30). In mitosis, the chromosomes line up single file along the equator of the cell. Spindle fibers from the centrioles form and attach to the centromere of each chromosome (fig. 3.30).

When the spindle fibers shorten, the centromeres split apart and the two chromatids in each chromosome are pulled to opposite poles. Each pole therefore gets one copy of each of the forty-six chromosomes. Division of the cytoplasm (cytokinesis) results in the production of two daughter cells, which are genetically identical to each other and to the original parent cell.

Certain types of cells can be removed from the body and grown in nutrient solutions (outside the body, or *in vitro*). Under these artificial conditions the potential longevity of different cell lines can be studied. For unknown reasons, normal connective tissue cells (called fibroblasts) stop dividing *in vitro* after about forty to seventy population doublings. Cells that become transformed into cancer, however, apparently do not age and continue dividing indefinitely in culture. It is ironic that these potentially immortal cells may commit suicide by killing their host.

***Hypertrophy and Hyperplasia.*** The growth of an individual from a fertilized egg into an adult involves an increase in cell number and an increase in cell size. Growth due to an increase in cell number results from mitotic cell division and is termed **hyperplasia.** Growth of a tissue or organ due to an increase in cell size is termed **hypertrophy.**

Most growth is due to hyperplasia. A callus on the palm of the hand, for example, involves thickening of the skin by hyperplasia due to frequent abrasion. An increase in skeletal muscle size as a result of exercise, in contrast, is generally believed to be produced by hypertrophy.

**Figure 3.30.** The stages of mitosis.

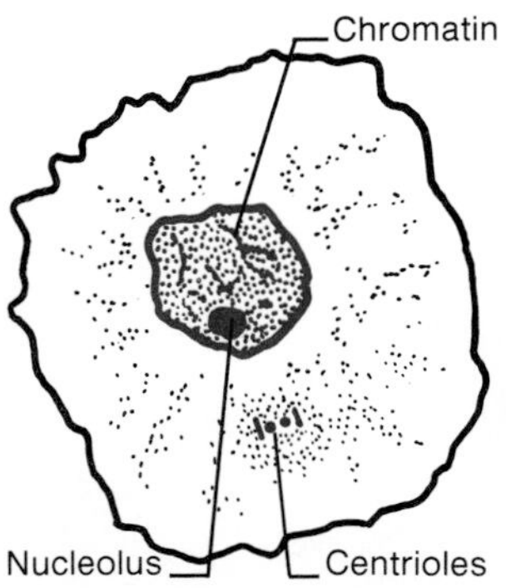


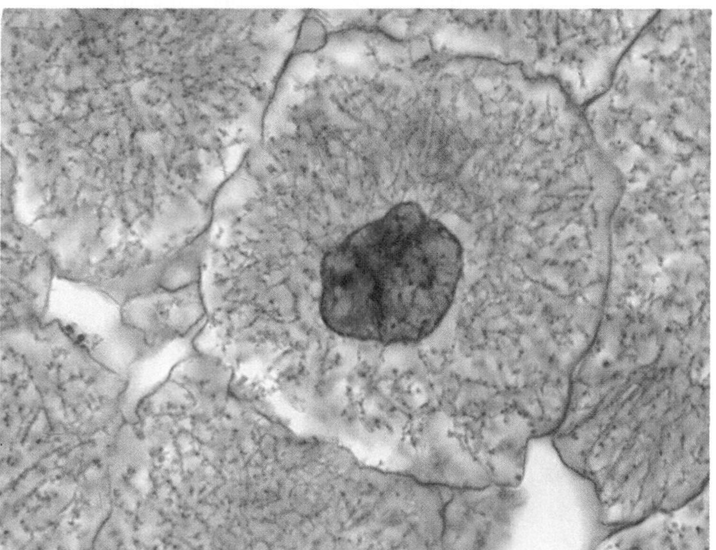

**(a) Interphase**

The chromosomes are in an extended form and seen as chromatin in the electron microscope.
The nucleus is visible.

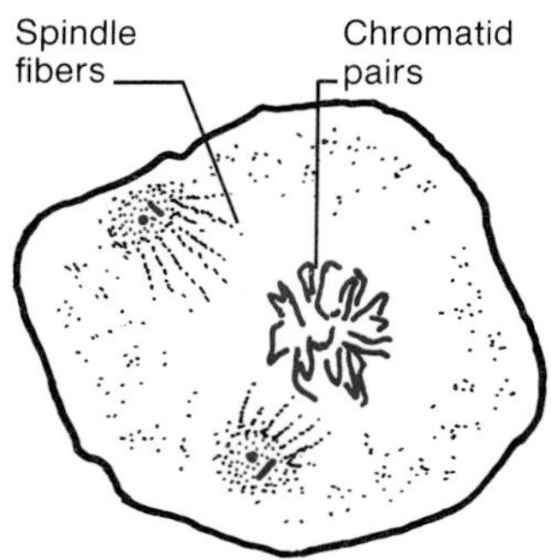


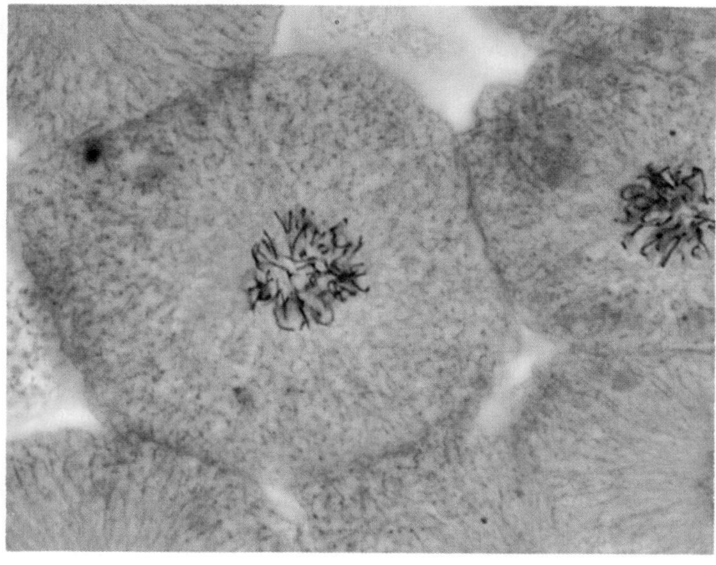

**(b) Prophase**

The chromosomes are seen and observed to consist of two chromatids joined together by a centromere.
The centrioles move apart towards opposite poles of the cell.
Spindle fibers are produced and extend from each centriole.
The nuclear membrane starts to disappear.
The nucleolus is no longer visible.

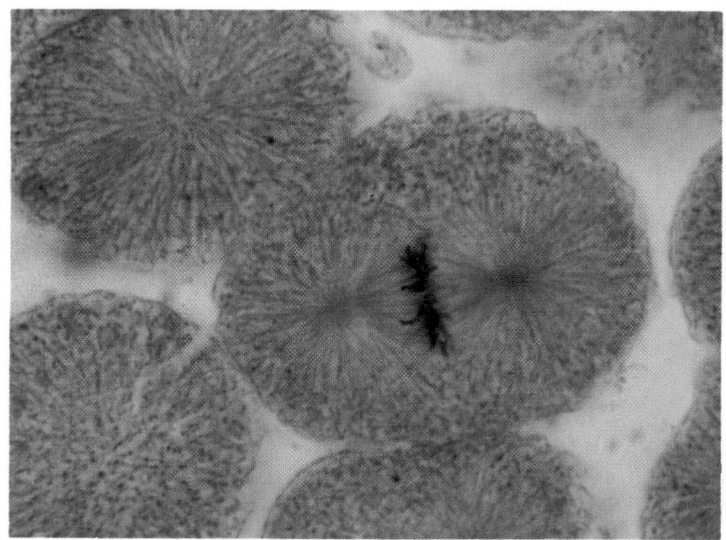

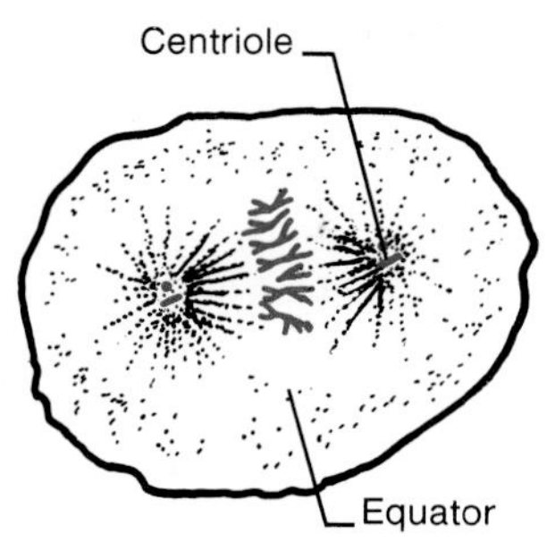


**(c) Metaphase**

The chromosomes are lined up at the equator of the cell.
The spindle fibers from each centriole are attached to the centromeres of the chromosomes.
The nuclear membrane has disappeared.

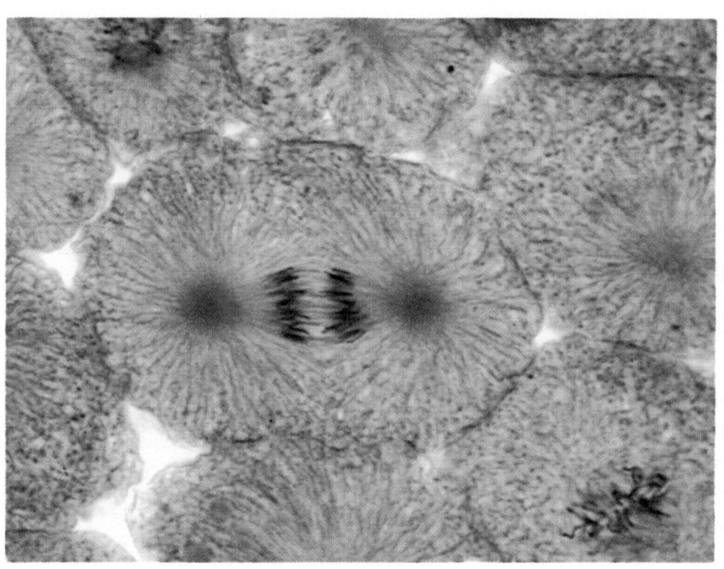

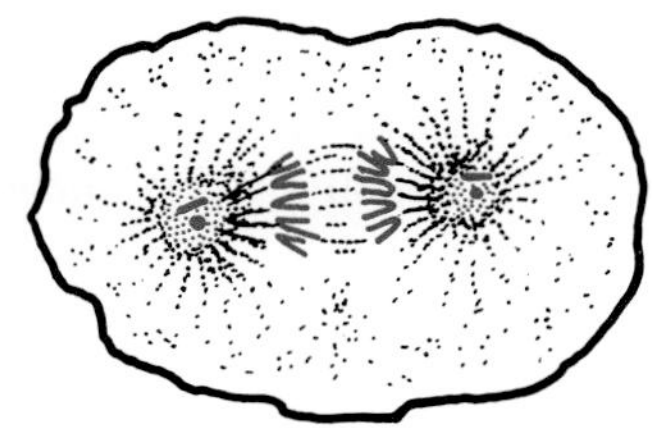

**(d) Anaphase**

The centromeres split, and the sister chromatids separate as each is pulled to an opposite pole.

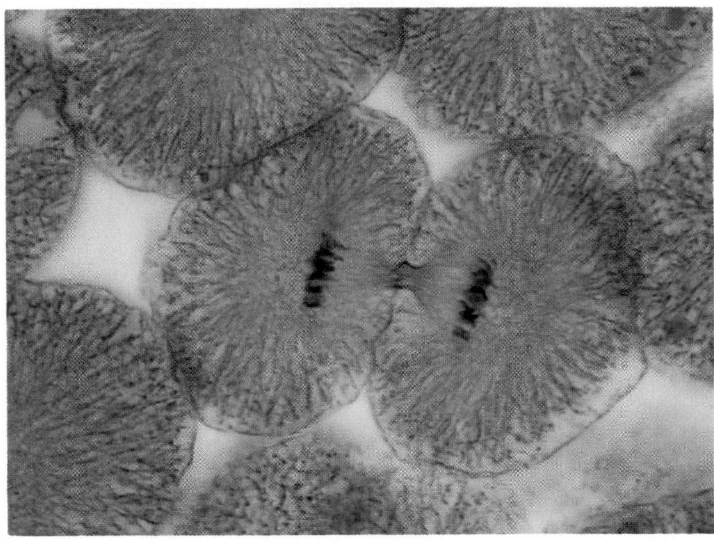

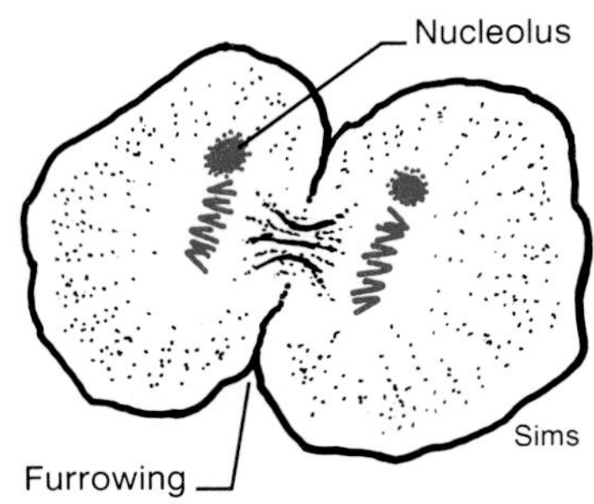


**(e) Telophase**

The chromosomes become longer, thinner, and less distinct.
New nuclear membranes form.
The nucleolus reappears.
Cell division (cytokinesis) is nearly complete.

Skeletal muscle and cardiac (heart) muscle can only grow by hypertrophy. When this occurs in skeletal muscles in response to an increased workload—during weight training, for example—it is called *compensatory hypertrophy.* The heart muscle may also demonstrate compensatory hypertrophy when its workload increases in response, for example, to hypertension (high blood pressure). The opposite of hypertrophy is *atrophy,* where the cells become smaller than normal. This may result from the disuse of skeletal muscles (during prolonged bed rest, for example), various diseases, and advanced age.

## Meiosis

When a cell is going to divide, either by mitosis or meiosis, the DNA is replicated (forming chromatids) and the chromosomes become shorter and thicker, as previously described. At this point, the cell has forty-six chromosomes (or twenty-three pairs of homologous chromosomes), which each consist of two duplicate chromatids.

When mitosis occurs, the homologous chromosomes line up in single file along the equator of a cell. One chromatid from each chromosome is then pulled by the spindle fibers into each daughter cell, so that each daughter cell receives one chromatid from each of the forty-six chromosomes of the parent cell. In mitosis, therefore, each daughter cell is genetically identical to the parent cell. Such is not the case with a type of cell division known as meiosis.

**Meiosis** is a special type of cell division that occurs only in the gonads (testes and ovaries) and is used only in the production of gametes (sperm and ova). In meiosis, the homologous chromosomes line up side-by-side, rather than single file, along the equator of the cell. The spindle fibers then pull one member of a homologous pair to one pole of the cell, and the other member of the pair to the other pole of the cell. Each of the two daughter cells thus acquires only one chromosome from each of the twenty-three homologous pairs contained in the parent. The daughter cells, in other words, contain twenty-three rather than forty-six chromosomes. For this reason, meiosis is also known as **reduction division.**

Meiosis, however, consists of two cell divisions. The necessity for this is obvious when you realize that, at the end of the cell division previously described, each daughter cell contains twenty-three chromosomes—but *each of these consists of two chromatids.* (Since the two chromatids per chromosome are identical, this does not make forty-six chromosomes; there are still only twenty-three *different* chromosomes per cell at this point.) Each of the daughter cells from the first cell division then itself divides, with the duplicate chromatids going to each of two new daughter cells. A grand total of four daughter cells can thus be produced from the meiotic cell division of one parent cell. One parent cell thus produces four sperm (which each contains twenty-three chromosomes). One parent cell also produces four egg cells, but the development of ova is complicated by the fact that three of the four potential egg cells die (this will be described in chapter 20).

The stages of meiosis are subdivided according to whether they occur in the first or the second meiotic cell division. These stages are labeled prophase I, metaphase I, anaphase I, telophase I; and then prophase II, metaphase II, anaphase II, and telophase II (table 3.4 and fig. 3.31).

The reduction of the chromosome number from forty-six to twenty-three is of obvious necessity for sexual reproduction, where the sex cells join and add their content of chromosomes together to produce a new individual. The significance of meiosis, however, goes deeper than simply a reduction in chromosome number. At metaphase I, the pairs of homologous chromosomes can line up with either member facing a given pole of the cell. (Recall that each member of a homologous pair came from a different parent.) Maternal and paternal members of homologous pairs are thus randomly shuffled. When the first meiotic division occurs, each daughter cell will thus obtain a complement of twenty-three chromosomes that are randomly derived from the maternal or paternal contribution to the homologous pairs of chromosomes of the parent.

**Table 3.4** Stages of meiosis

| Stage | Events |
|---|---|
| **First Meiotic Division** | |
| Prophase I | Chromosomes appear double-stranded. Each strand, called a chromatid, contains duplicate DNA joined together by a structure known as a centromere. Homologous chromosomes pair up side by side. |
| Metaphase I | Homologous chromosome pairs line up at equator. Spindle apparatus complete. |
| Anaphase I | Homologous chromosomes are separated; each member of a homologous pair moves to opposite poles. |
| Telophase I | Cytoplasm divides to produce two haploid cells. |
| **Second Meiotic Division** | |
| Prophase II | Chromosomes appear, each containing two chromatids. |
| Metaphase II | Chromosomes line up single file along equator as spindle formation is completed. |
| Anaphase II | Centromeres split and chromatids move to opposite poles. |
| Telophase II | Cytoplasm divides to produce two haploid cells from each of the haploid cells formed at telophase I. |

**Figure 3.31.** Meiosis, or reduction division. In the first meiotic division the homologous chromosomes of a diploid parent cell are separated into two haploid daughter cells. Each of these chromosomes contain duplicate strands, or chromatids. In the second meiotic division these chromosomes are distributed to two new haploid daughter cells.

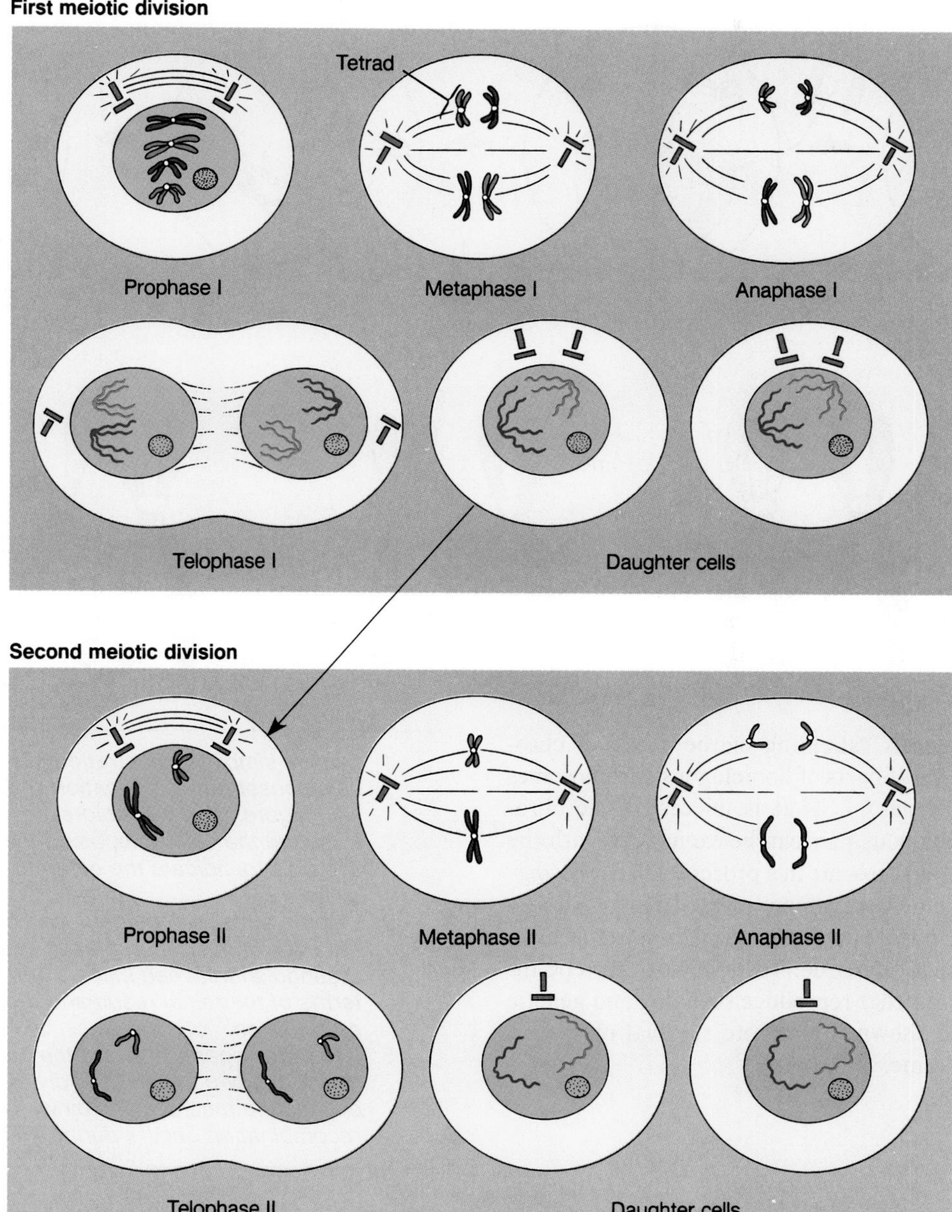

**Figure 3.32.** (*a*) Genetic variation results from the crossing-over of tetrads, which occurs during the first meiotic prophase. (*b*) A diagram of the pairing of homologous chromosomes of a single tetrad: (*1*) before the crossing-over; (*2*) chromatids crossing over; and (*3*) results of the crossing-over.

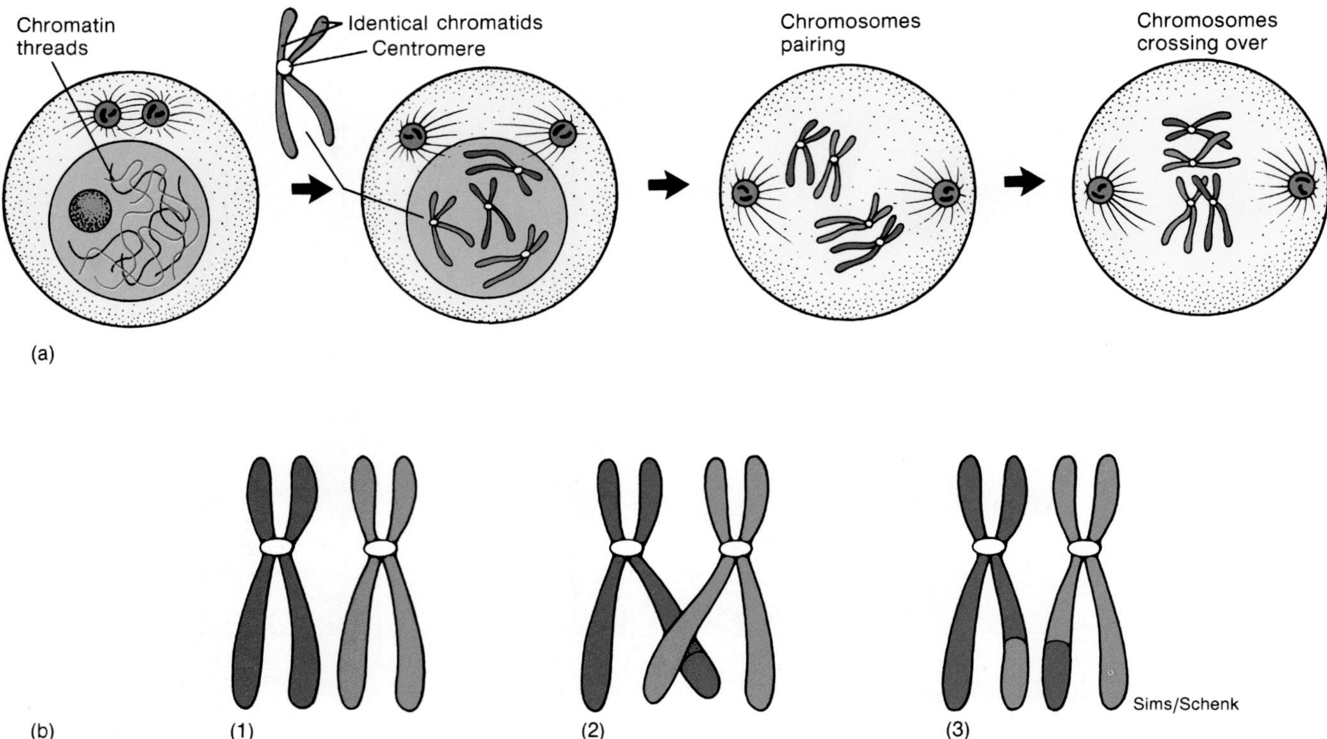


In addition to this "shuffling of the deck" of chromosomes, exchanges of parts of homologous chromosomes can occur at metaphase I. That is, pieces of one chromosome of a homologous pair can be exchanged with the other homologous chromosome in a process called *crossing-over* (fig. 3.32). These events together result in **genetic recombination,** and ensure that the gametes produced by meiosis are genetically unique. This provides genetic diversity for organisms that reproduce sexually, and genetic diversity has been shown to promote survival of species over evolutionary time.

1. *Draw a simplified illustration of the semiconservative replication of DNA, using stick figures and two colors.*
2. *Describe the cell cycle, using the proper symbols to indicate the different stages of the cycle.*
3. *List the phases of mitosis, and briefly describe the events that occur in each phase.*
4. *Distinguish between mitosis and meiosis in terms of their final result and their functional significance.*
5. *Describe in general terms the events that occur during the two meiotic cell divisions, and explain the mechanisms by which genetic recombination occurs during meiosis.*

## SUMMARY

### Cell Membrane and Associated Structures p. 50

I. The structure of the cell, or plasma, membrane is described by a fluid-mosaic model.
   A. The membrane is composed predominately of a double layer of phospholipids.
   B. The membrane also contains proteins in a mosaic pattern.

II. Some cells move by extending pseudopods; cilia and flagella protrude from the cell membrane of some specialized cells.

III. Invaginations of the cell membrane, in a process of endocytosis, allows the cells to take up molecules from the external environment.
   A. Phagocytosis is a process in which the cell extends pseudopods, which eventually fuse together to create a food vacuole; pinocytosis involves the formation of a narrow furrow in the membrane which eventually fuses.

B. Receptor-mediated endocytosis requires the interaction of a specific molecule in the extracellular environment with a specific receptor protein in the cell membrane.

C. Exocytosis is the reverse of endocytosis and is a process that allows the cell to secrete its products.

## Cytoplasm and Its Organelles p. 54

I. Microfilaments and microtubules produce a cytoskeleton, which aids movements of organelles within a cell.

II. Lysosomes contain digestive enzymes and are responsible for the elimination of structures and molecules within the cell and for digestion of the contents of phagocytic food vacuoles.

III. Mitochondria contain an outer and an inner membrane, which is folded into cristae, and serve as the major sites for energy production in the cell.

IV. The endoplasmic reticulum is a system of membranous tubules in the cell.

A. The rough endoplasmic reticulum is covered with ribosomes and is involved in protein synthesis.

B. The smooth endoplasmic reticulum is the site for many enzymatic reactions and, in skeletal muscles, serves to store $Ca^{++}$.

## Cell Nucleus and Nucleic Acids p. 59

I. The cell nucleus is surrounded by a nuclear membrane and contains chromatin, which consists of DNA and protein.

II. Nucleic acids include DNA, RNA, and their nucleotide subunits.

A. The DNA nucleotides contain the sugar deoxyribose, whereas the RNA nucleotides contain the sugar ribose.

B. There are four different types of DNA nucleotides, which contain one of four possible bases: adenine, guanine, cytosine, and thymine; in RNA, the base uracil substitutes for the base thymine.

C. DNA consists of two long polynucleotide strands twisted into a double helix; the two strands are held together by hydrogen bonds between specific bases; adenine pairs with thymine, and guanine pairs with cytosine.

D. RNA is single stranded; there are four types of RNA molecules: ribosomal RNA, transfer RNA, precursor messenger RNA, and messenger RNA.

III. Active euchromatin directs the synthesis of RNA; this process is called genetic transcription.

A. The enzyme RNA polymerase causes separation of the two strands of DNA along the region of the DNA that constitutes a gene.

B. One of the two separated strands of DNA serves as a template for the production of RNA; this occurs by complementary base pairing between the DNA bases and ribonucleotide bases.

## Protein Synthesis and Secretion p. 64

I. Messenger RNA leaves the nucleus and attaches to the ribosomes.

II. Each transfer RNA, with a specific base triplet in its anticodon, bonds to a specific amino acid.

A. As the mRNA moves through the ribosomes, complementary base pairing between tRNA anticodons and mRNA codons occurs.

B. As each successive tRNA molecule bonds to its complementary codon, the amino acid it carries is added to the end of a growing polypeptide chain.

III. Proteins destined for secretion are produced in ribosomes located in the rough endoplasmic reticulum and enter the cisternae of this organelle.

IV. Secretory proteins move from the rough endoplasmic reticulum to the Golgi apparatus, which consists of a stack of membranous sacs.

A. The Golgi apparatus modifies the proteins it contains, separates different proteins, and packages them in vesicles.

B. Secretory vesicles from the Golgi apparatus fuse with the cell membrane and release their products by exocytosis.

## DNA Synthesis and Cell Division p. 70

I. Replication of DNA is semiconservative; each DNA strand serves as a template for the production of a new strand.

A. The strands of the original DNA molecule gradually separate along their entire length and, through complementary base pairing, form a new complementary strand.

B. In this way, each DNA molecule consists of one old and one new strand.

II. During the $G_1$ phase of the cell cycle, the DNA directs the synthesis of RNA and hence of proteins.

III. During the *S* phase of the cycle, DNA directs the synthesis of new DNA and replicates itself.

IV. After a brief rest ($G_2$), the cell begins mitosis (*M* stage of the cycle).

A. Mitosis consists of the following phases: interphase, prophase, metaphase, anaphase, and telophase.

B. In mitosis, the homologous chromosomes line up single-file and are pulled by spindle fibers to opposite poles.

C. This results in the production of two daughter cells which each contain forty-six chromosomes, just like the parent cell.

V. Meiosis is a special type of cell division that results in the production of gametes in the gonads.

A. The homologous chromosomes line up side-by-side, so that only one of each pair is pulled to each pole.

B. This results in the production of two daughter cells that each contain only twenty-three chromosomes, which are duplicated.

C. The duplicate chromatids are separated into two new daughter cells during the second meiotic cell division.

## Review Activities

### Objective Questions

1. According to the fluid-mosaic model of the cell membrane
   (a) protein and phospholipids form a regular, repeating structure
   (b) the membrane is a rigid structure
   (c) phospholipids form a double layer, with the polar parts facing each other
   (d) proteins are free to move within a double layer of phospholipids
2. After the DNA molecule has replicated itself, the duplicate strands are called
   (a) homologous chromosomes
   (b) chromatids
   (c) centromeres
   (d) spindle fibers
3. Nerve and skeletal muscle cells in the adult, which do not divide, remain in the
   (a) $G_1$ phase
   (b) S phase
   (c) $G_2$ phase
   (d) M phase
4. The phase of mitosis in which the chromosomes line up at the equator of the cell is called
   (a) interphase
   (b) prophase
   (c) metaphase
   (d) anaphase
   (e) telophase
5. The phase of mitosis in which the chromatids separate is called
   (a) interphase
   (b) prophase
   (c) metaphase
   (d) anaphase
   (e) telophase
6. The RNA nucleotide base that pairs with adenine in DNA is
   (a) thymine
   (b) uracil
   (c) guanine
   (d) cytosine
7. Which of the following statements about RNA is *true?*
   (a) It is made in the nucleus.
   (b) It is double stranded.
   (c) It contains the sugar deoxyribose.
   (d) It is a complementary copy of the entire DNA molecule.
8. Which of the following statements about mRNA is *false?*
   (a) It is produced as a larger pre-mRNA.
   (b) It forms associations with ribosomes.
   (c) Its base triplets are called anticodons.
   (d) It codes for the synthesis of specific proteins.
9. The organelle that combines proteins with carbohydrates and packages them within vesicles for secretion is the
   (a) Golgi apparatus
   (b) rough endoplasmic reticulum
   (c) smooth endoplasmic reticulum
   (d) ribosome
10. The organelles that contain digestive enzymes are the
    (a) mitochondria
    (b) lysosomes
    (c) endoplasmic reticulum
    (d) Golgi apparatus
11. If four bases in one DNA strand are A (adenine), G (guanine), C (cytosine), and T (thymine), the complementary bases in the RNA strand made from this region are:
    (a) T,C,G,A
    (b) C,G,A,U
    (c) A,G,C,U
    (d) U,C,G,A
12. Which of the following statements about tRNA is *true?*
    (a) It is made in the nucleus.
    (b) It is looped back on itself.
    (c) It contains the anticodon.
    (d) There are over 20 different types of tRNA.
    (e) All of these.
13. The step in protein synthesis during which tRNA, rRNA, and mRNA are all active is known as:
    (a) transcription
    (b) translation
    (c) replication
    (d) RNA polymerization
14. The anticodons are located in:
    (a) tRNA
    (b) rRNA
    (c) mRNA
    (d) ribosomes
    (e) endoplasmic reticulum

### Essay Questions

1. The cell membrane is an extremely dynamic structure. Using examples, explain why this statement is true.
2. Explain how one DNA molecule serves as a template for the formation of another DNA and why DNA synthesis is said to be semiconservative.
3. What is the genetic code, and how does it affect the structure and function of the body?
4. Why may tRNA be considered the "interpreter" of the genetic code?
5. Compare the processing of cellular proteins with that of proteins that are secreted by a cell.

## Selected Readings


Afzelius, Björn. 1986. Disorders of ciliary motility. *Hospital Practice* 21:73.

Allen, R. D. February 1987. The microtubule as an intracellular engine. *Scientific American.*

Berlin, R. D., T. E. Oliver, and H. H. Yin. 1975. The cell surface. *New England Journal of Medicine* 292:515.

Bretscher, M. S. October 1985. The molecules of the cell membrane. *Scientific American.*

Brown, D. D. 1981. Gene expression in eukaryotes. *Science* 211:667.

Capaldi, R. A. March 1974. A dynamic model of cell membranes. *Scientific American.*

Chambon, P. May 1981. Split genes. *Scientific American.*

Crick, F. October 1962. The genetic code. *Scientific American.*

Crick, F. 1979. Split genes and RNA splicing. *Science* 204:264.

Danielli, J. F. 1973. The bilayer hypothesis of membrane structure. *Hospital Practice* 8:63.

Darnell, J. E., Jr. October 1985. RNA. *Scientific American.*

Dautry-Varsat, A., and H. F. Lodish. May 1984. How receptors bring proteins and particles into cells. *Scientific American.*

Davidson, E. H., and R. J. Britten. 1979. Regulation of gene expression: possible role of repetitive sequences. *Science* 204:1052.

DeDuve, C. May 1963. The lysosome. *Scientific American.*

DeDuve, C. May 1983. Microbodies in the living cell. *Scientific American.*

Dustin, P. August 1980. Microtubules. *Scientific American.*

Felsenfeld, G. October 1985. DNA. *Scientific American.*

Fox, C. F. February 1972. The structure of cell membranes. *Scientific American.*

Grivell, L. A. March 1983. Mitochondria DNA. *Scientific American.*

Hayflick, H. January 1980. The cell biology of human aging. *Scientific American.*

Kornfeld, S., and W. S. Sly. 1985. Lysosomal storage defects. *Hospital Practice* 20:71.

Lake, J. A. August 1981. The ribosome. *Scientific American.*

Lazarides, E., and J. P. Revel. May 1979. The molecular basis of cell movement. *Scientific American.*

Lodish, H. F., and J. E. Rothman. January 1979. The assembly of cell membranes. *Scientific American.*

McKusick, V. A. 1981. The anatomy of the human genome. *Hospital Practice* 16:82.

Mazia, D. January 1974. The cell cycle. *Scientific American.*

Miller, O. L., Jr. March 1973. The visualization of genes in action. *Scientific American.*

Nomura, M. October 1969. Ribosomes. *Scientific American.*

Palade, G. 1975. Intracellular aspects of the process of protein synthesis. *Science* 189:347.

Porter, K. R., and J. B. Tucker. March 1981. The ground substance of the living cell. *Scientific American.*

Ptashne, M. January 1989. How gene activators work. *Scientific American.*

Racker, E. February 1968. The membrane of the mitochondrion. *Scientific American.*

Rich, A., and S. H. Kim. January 1978. The three-dimensional structure of transfer RNA. *Scientific American.*

Rothman, J. E. September 1985. The compartmental organization of the Golgi apparatus. *Scientific American.*

Singer, S. J., and G. L. Nicolson. 1972. The fluid mosaic model of the structure of cell membranes. *Science* 175:720.

Singer, S. J. 1973. Biological membranes. *Hospital Practice* 8:81.

Sloboda, R. D. 1980. The role of microtubules in cell structure and cell division. *American Scientist* 68:290.

Stein, G., J. S. Stein, and L. J. Kleinsmith. February 1975. Chromosomal proteins and gene regulation. *Scientific American.*

Steitz, J. A. June 1988. "Snurps." *Scientific American.*

Wallace, D. C. 1986. Mitochondria genes and disease. *Hospital Practice* 21:77.

Weber, K., and M. Osborn. October 1985. The molecules of the cell matrix. *Scientific American.*

White, R., and J. M. Lalouel. February 1988. Chromosome mapping with DNA markers. *Scientific American.*

# ENZYMES AND ENERGY

# 4

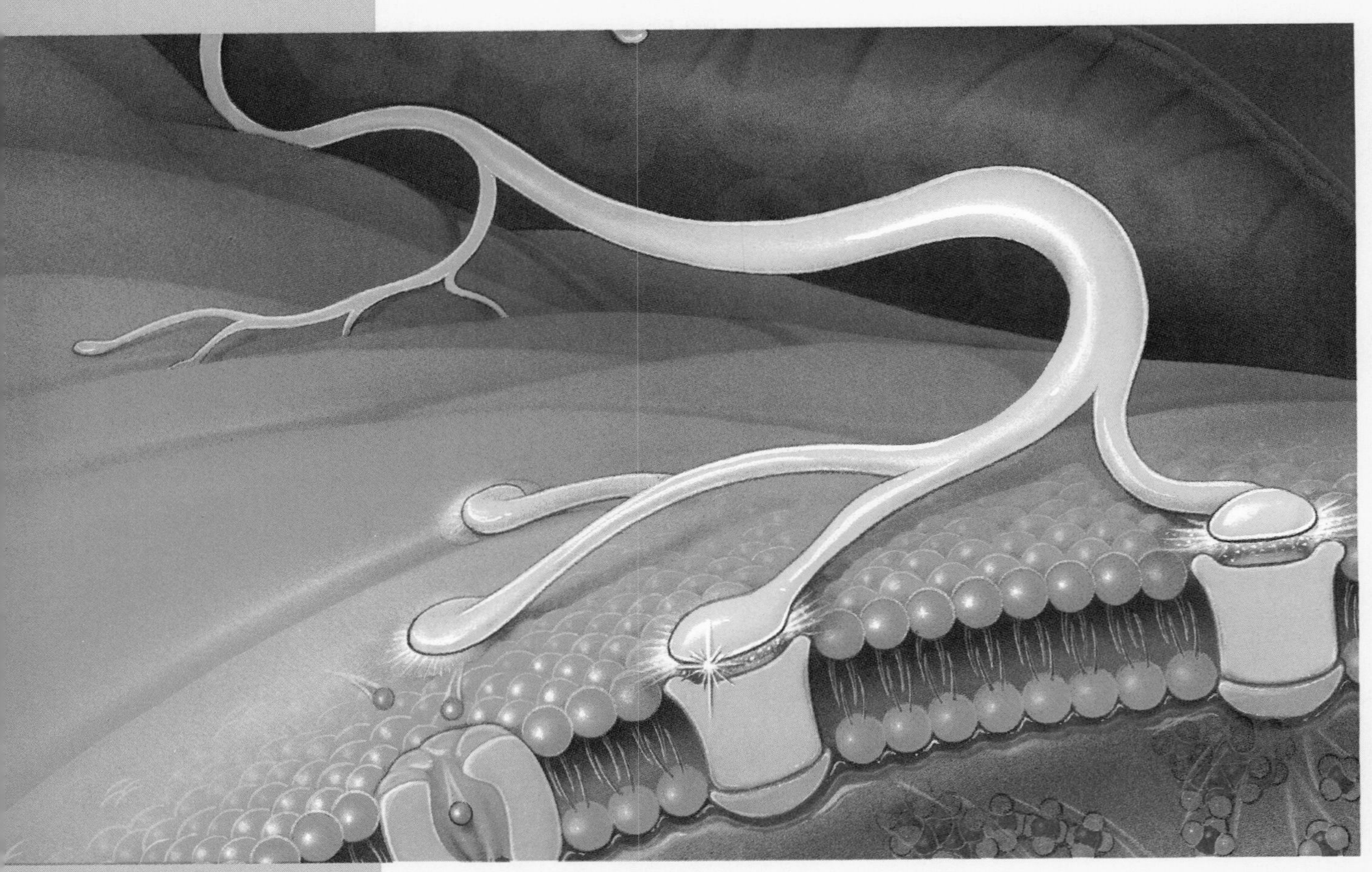

## Objectives

By studying this chapter, you should be able to

1. describe the principles of catalysis and explain how enzymes function as catalysts
2. describe how enzymes are named and explain the significance of isoenzymes
3. describe the effects of pH and temperature on the rate of enzyme-catalyzed reactions and explain how these effects are produced
4. describe the roles of cofactors and coenzymes in enzymatic reactions
5. describe how the law of mass action helps to explain the direction of reversible reactions
6. explain how enzymes cooperate to produce a metabolic pathway and how this pathway may be affected by end-product inhibition and inborn errors of metabolism
7. explain how the first and second laws of thermodynamics can be used to predict if metabolic reactions will be endergonic or exergonic
8. describe how ATP is produced and explain its significance as the universal energy carrier
9. define the terms *oxidation, reduction, oxidizing agent,* and *reducing agent*
10. describe the derivation and functional significance of NAD and FAD in oxidation-reduction reactions

## Outline

## *Enzymes as Catalysts*

Enzymes are proteins that function as biological catalysts. The catalytic action of enzymes results from the complex structure of these proteins, and the great diversity in the structure of proteins allows different enzymes to be specialized in their actions.

The ability of yeast cells to make alcohol from glucose (a process called *fermentation*) had been known since antiquity, yet no chemist by the mid-nineteenth century could duplicate the trick in the absence of living yeast. Also, yeast and other living cells could perform a vast array of chemical reactions at body temperature that could not be duplicated in the chemical laboratory without adding a substantial amount of heat energy. These observations led many mid-nineteenth-century scientists to believe that chemical reactions in living cells were aided by a "vital force" that operated beyond the laws of the physical world. This "vitalist" concept was squashed along with the yeast cells when a pioneering biochemist, Eduard Buchner, demonstrated that juice obtained from yeast could ferment glucose to alcohol. The yeast juice was not alive—evidently some chemicals in the cells were responsible for fermentation. Buchner didn't know what these chemicals were, so he simply named them **enzymes** (Greek for "in yeast").

In more recent times, biochemists have demonstrated that *all enzymes are proteins* and that enzymes act as *biological catalysts.* (This description must be somewhat modified by recent evidence that RNA may also have limited, but important, catalytic ability.) A catalyst is a chemical that (1) increases the rate of a reaction, (2) is not itself changed at the end of the reaction, and (3) does not change the nature of the reaction or its final result. The same reaction would have occurred to the same degree in the absence of the catalyst, but it would have progressed at a much slower rate.

In order for a given reaction to occur, the reactants must have sufficient energy. The amount of energy required for a reaction to proceed is called the **energy of activation.** In a large population of molecules, only a small fraction will possess sufficient energy for a reaction. Adding heat will raise the energy level of all the reactant molecules, thus increasing the fraction of the population that has the activation energy. Heat would make reactions go faster, but it would produce undesirable side effects in cells. Catalysts make the reaction go faster at lower temperatures by *lowering the activation energy* required so that a larger fraction of the population of reactant molecules has sufficient energy to participate in the reaction (fig. 4.1).

Since a small fraction of the reactants have the activation energy required for the reaction even in the absence of a catalyst, the reaction could theoretically occur spontaneously at a slow rate. This rate, however, would be much too slow for the needs of a cell. So, from a biological standpoint, the presence or absence of a specific enzyme catalyst acts as a switch—the reaction will occur if the enzyme is present and will not occur if the enzyme is absent.

### Mechanism of Enzyme Action

The ability of enzymes to lower the activation energy of a reaction is a result of their structure. Enzymes are proteins—they are thus very large molecules with complex, highly ordered, three-dimensional shapes produced by chemical interactions between their amino acids. Each type of enzyme protein has characteristic ridges, grooves, and pockets that are lined with specific amino acids. The particular pockets that are active in catalyzing a reaction are called the *active sites* of the enzyme.

The reactant molecules, which are the *substrates* of the enzyme, have shapes that allow them to fit into the active sites. The fit may not be perfect at first, but a perfect fit may be induced as the substrate gradually slips into the active site. This induced fit, together with temporary bonds that form between the substrate and the amino acids lining the active sites of the enzyme, weaken the existing bonds within the substrate molecules, which allows them to be more easily broken. New bonds are more easily formed as substrates are brought close together in the proper orientation. The *enzyme-substrate complex,* formed temporarily in the course of the reaction, then dissociates to yield *products* and the free unaltered enzyme. This model of how enzymes work is known as the **lock-and-key** model of enzyme activity (fig. 4.2).

By this mechanism, enzymes lower the activation energy needed and greatly increase the rate of reactions. Hydrogen peroxide ($H_2O_2$), for example, will slowly decompose by itself if left uncapped for a few days. The addition of *catalase,* an enzyme found in many tissues such as blood and the liver, will make hydrogen peroxide break down into water and oxygen gas a trillion times faster. The ability of enzymes to increase the rate of reactions can be measured by their **turnover number,** which is the number of substrate molecules that the enzyme can convert into products per minute (see table 4.1 for examples).

**Figure 4.1.** A comparison of a noncatalyzed reaction with a catalyzed reaction. Upper figures compare the proportion of reactant molecules that have sufficient activation energy to participate in the reaction (color). This proportion is increased in the enzyme-catalyzed reaction, because enzymes lower the activation energy required for the reaction (shown as a barrier on top of an energy "hill" in the bottom figures). Reactants that can overcome this barrier are able to participate in the reaction, as shown by arrows pointing to the bottom of the energy hill.

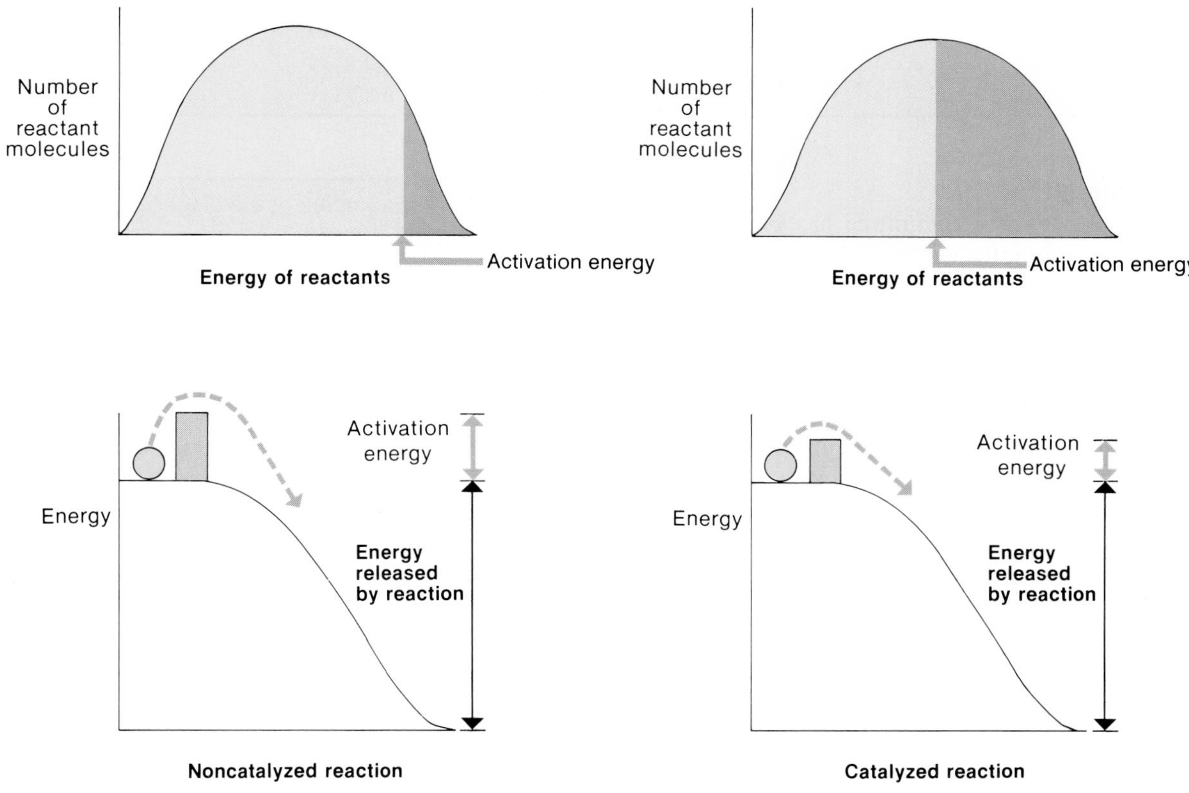


**Figure 4.2.** The lock-and-key model of enzyme action.

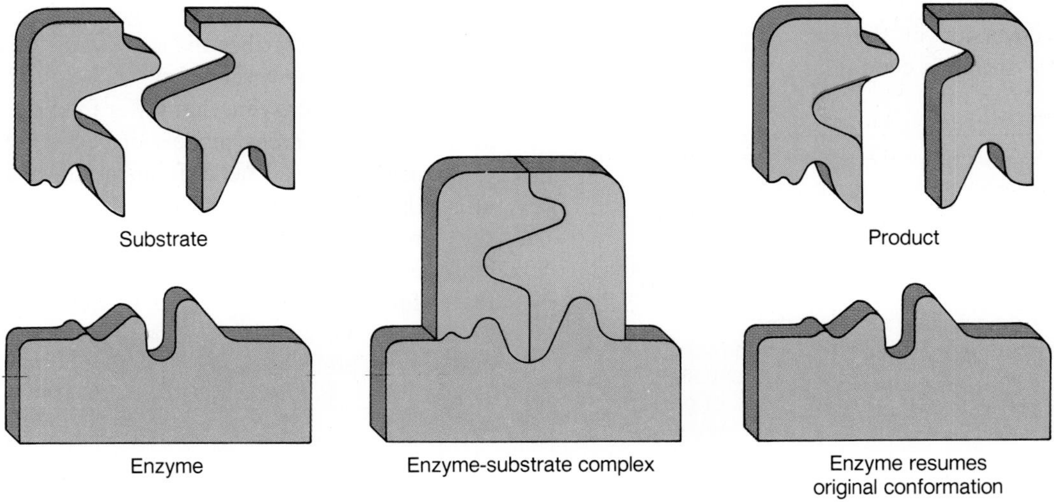

**Table 4.1** Some examples of the names of enzymes, the reactions they catalyze, and their turnover numbers (molecules of substrate converted to product per minute under optimum conditions)

| Enzyme | Reaction Catalyzed | Turnover Number |
|---|---|---|
| Catalase | $2\,H_2O_2 \rightarrow 2\,H_2O + O_2$ | $600 \times 10^6$ |
| Carbonic anhydrase | $H_2CO_3 \rightarrow H_2O + CO_2$ | $3.6 \times 10^6$ |
| Amylase | starch + $H_2O \rightarrow$ maltose | $1.0 \times 10^5$ |
| Lactate dehydrogenase | lactic acid $\rightarrow$ pyruvic acid + $H_2$ | $6.0 \times 10^4$ |
| Ribonuclease | RNA + $H_2O \rightarrow$ ribonucleotides | 600 |

## Naming of Enzymes

Although an international committee has established a uniform naming system for enzymes, the names that are in common use do not follow a completely consistent pattern. With the exception of some older enzyme names (such as pepsin, trypsin, and renin), all enzyme names end with the suffix *-ase.* Classes of enzymes are named according to their job category. *Hydrolases,* for example, promote hydrolysis reactions. Other enzyme categories include *phosphatases,* which catalyze the removal of phosphate groups; *synthetases,* which catalyze dehydration synthesis reactions; and *dehydrogenases,* which remove hydrogen atoms from their substrates. Enzymes called *isomerases* rearrange atoms within their substrate molecules to form structural isomers (examples of such structural isomers include glucose and fructose).

The names of many enzymes specify both the substrate of the enzyme and the activity ("job category") of the enzyme. Lactic acid dehydrogenase, for example, removes hydrogens from lactic acid. Since enzymes are very specific as to their substrates and activity, the concentration of a specific enzyme in a sample of fluid can be measured relatively easily. This is usually done by measuring the rate of conversion of the enzyme's substrates into products under specified conditions.

When tissues become damaged due to diseases, some of the dead cells disintegrate and release their enzymes into the blood. Most of these enzymes are not normally active in the blood because of the absence of their specific substrates, but their enzymatic activity can be measured in a test tube by the addition of the appropriate substrates to samples of plasma. Such measurements are clinically useful, because abnormally high plasma concentrations of particular enzymes are characteristic of certain diseases (table 4.2).

**Table 4.2** Examples of the diagnostic value of some enzymes found in plasma

| Enzyme | Diseases Associated with Abnormal Plasma Enzyme Concentrations |
|---|---|
| Alkaline phosphatase | Obstructive jaundice, Paget's disease (osteitis deformans), carcinoma of bone |
| Acid phosphatase | Benign hypertrophy of prostate, cancer of prostate |
| Amylase | Pancreatitus, perforated peptic ulcer |
| Aldolase | Muscular dystrophy |
| Creatine kinase (or creatine phosphokinase-CPK) | Muscular dystrophy, myocardial infarction |
| Lactate dehydrogenase (LDH) | Myocardial infarction, liver disease, renal disease, pernicious anemia |
| Transaminases (GOT and GPT) | Myocardial infarction, hepatitis, muscular dystrophy |

Enzymes that do exactly the same job (that catalyze the same reaction) in different organs have the same name, since the name describes the activity of the enzyme. Different organs, however, may make slightly different "models" of the enzyme that differ in one or a few amino acids. These different models of the same enzyme are called **isoenzymes.** The differences in structure do not affect the active sites (otherwise they would not catalyze the same reaction), but they do alter the structure of the enzymes at other locations, so that the different isoenzymatic forms can be separated by standard biochemical procedures. These techniques are useful in the diagnosis of diseases (table 4.3).

1. *Define the "energy of activation" of a reaction; explain how it is affected by a catalyst and how the structure of an enzyme makes it function as a catalyst.*
2. *Draw a labeled picture of the lock-and-key model of enzyme action, and write a formula for this reaction.*
3. *Explain how enzymes are named.*

**Table 4.3** Some clinical uses of isoenzyme measurements

| Enzyme | Isoenzyme Form Number | Disease Associated with Abnormal Elevation of Isoenzyme |
|---|---|---|
| Creatine phosphokinase (CPK) | 1<br>2 | Muscular dystrophy<br>Myocardial infarction |
| Lactic acid dehydrogenase (LDH) | 1<br>5 | Myocardial infarction<br>Liver disease (such as hepatitis) |

## Control of Enzyme Activity

The rate of an enzyme-catalyzed reaction depends on the concentration of the enzyme and of the substrates, the temperature and pH of the reaction, the availability of cofactors and coenzymes, and the concentration of products formed. Variations in some of these factors control the rate of progress along particular metabolic pathways and thus help to regulate cellular metabolism.

The activity of an enzyme, as measured by the rate at which its substrates are converted to products, is influenced by a variety of factors, including (1) the temperature and pH of the solution; (2) the concentration of cofactors and coenzymes, which are needed by many enzymes as "helpers" for their catalytic activity; (3) the concentration of enzyme and substrate molecules in the solution; and (4) the stimulatory and inhibitory effects of some products of enzyme action on the activity of the enzymes that helped form these molecules.

### Effects of Temperature and pH

An increase in temperature, as previously described, will increase the rate of non-enzyme-catalyzed reactions because a larger number of reactant molecules will have the activation energy required. A similar relationship between temperature and reaction rate occurs in enzyme-catalyzed reactions. At a temperature of 0° C the reaction rate is unmeasurably slow. As the temperature is raised above 0° C the reaction rate increases but only up to a point. At a few degrees above body temperature (which is 37° C) the reaction rate reaches a plateau; further increases in temperature actually *decrease* the rate of the reaction (fig. 4.3). This decrease is due to the fact that the tertiary structure of enzymes becomes altered at higher temperatures.

A similar relationship is observed when the rate of an enzymatic reaction is measured at different pH values. Each enzyme characteristically has its peak activity in a very narrow pH range, which is the **pH optimum** for the enzyme. If the pH is changed from this optimum, the reaction rate decreases (fig. 4.4). This decreased enzyme activity is due to changes in the conformation of the enzyme and in the charges of the R groups of the amino acids lining the active sites.

The pH optimum of an enzyme usually reflects the pH of the body fluid in which the enzyme is found. The acidic pH optimum of the protein-digesting enzyme

**Figure 4.3.** The effect of temperature on enzyme activity, as measured by the rate of the enzyme-catalyzed reaction under standardized conditions.

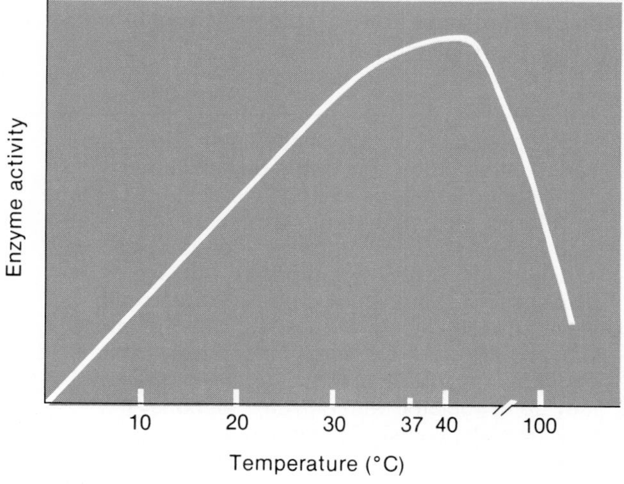


**Figure 4.4.** The effect of pH on the activity of three digestive enzymes.

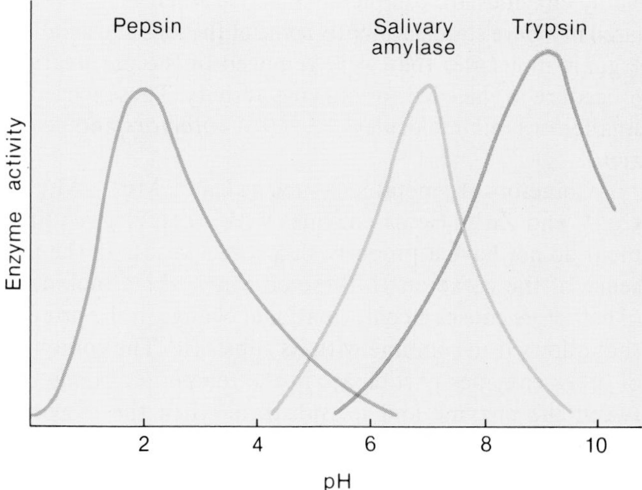

*pepsin,* for example, allows it to be active in the strong hydrochloric acid of gastric juice. Similarly, the neutral pH optimum of *salivary amylase* and the alkaline pH optimum of *trypsin* in pancreatic juice allow these enzymes to digest starch and protein, respectively, in other parts of the digestive tract.

Although the pH of other body fluids shows less variation than the fluids of the digestive tract, significant differences exist between the pH optima of different enzymes found throughout the body (see table 4.4). Some of these differences can be exploited for diagnostic purposes. Disease of the prostate, for example, may be associated with elevated blood levels of a prostatic phosphatase with an acidic pH optimum (descriptively called acid phosphatase). Bone disease, on the other hand, may be associated with elevated blood levels of alkaline phosphatase, which has a higher pH optimum than the similar enzyme released from the diseased prostate.

### Cofactors and Coenzymes

Many enzymes are completely inactive when they are isolated in a pure state. Evidently some of the ions and smaller organic molecules that were removed in the purification procedure are needed for enzyme activity. These ions and smaller organic molecules are called *cofactors* and *coenzymes.*

Cofactors are metal ions such as $Ca^{++}$, $Mg^{++}$, $Mn^{++}$, $Cu^{++}$, and $Zn^{++}$. Some enzymes with a cofactor requirement do not have a properly shaped active site in the absence of the cofactor. In these enzymes, the attachment of cofactors causes a conformational change in the protein that allows it to combine with its substrate. The cofactors of other enzymes participate in the temporary bonds between the enzyme and its substrate when the enzyme-substrate complex is formed (fig. 4.5).

Coenzymes are organic molecules that are derived from water-soluble vitamins, such as niacin and riboflavin. Coenzymes participate in enzyme-catalyzed reactions by transporting hydrogen atoms and small molecules from one enzyme to another. Examples of the actions of cofactors and coenzymes in specific reactions will be given in the context of their roles in cellular metabolism later in this chapter.

**Table 4.4** The pH optima of selected enzymes

| Enzyme | Reaction Catalyzed | pH Optimum |
|---|---|---|
| Pepsin (stomach) | Digestion of protein | 2.0 |
| Acid phosphatase (prostate) | Removal of phosphate group | 5.5 |
| Salivary amylase (saliva) | Digestion of starch | 6.8 |
| Lipase (pancreatic juice) | Digestion of fat | 7.0 |
| Alkaline phosphatase (bone) | Removal of phosphate group | 9.0 |
| Trypsin (pancreatic juice) | Digestion of protein | 9.5 |
| Monoamine oxidase (nerve endings) | Removal of amine group from norepinephrine | 9.8 |

### Substrate Concentration and Reversible Reactions

The rate at which an enzymatic reaction converts substrates into products depends on the enzyme concentration and on the concentration of substrates. When the enzyme concentration is at a given level, the rate of product formation will increase as the substrate concentration increases. Eventually, however, a point will be reached where additional increases in substrate concentration do not result in comparable increases in reaction rate. When the relationship between substrate concentration and reaction rate reaches a plateau, the enzyme is said to be *saturated.* If one thinks of enzymes as workers and substrates as jobs, there is 100% employment when the enzyme is saturated; further availability of jobs (substrate) cannot further increase employment (conversion of substrate to product). This is illustrated in figure 4.6.

Some enzymatic reactions within a cell are reversible, with both the forward and backward reactions catalyzed by the same enzyme. The enzyme *carbonic anhydrase,* for example, is named because it can catalyze the following reaction:

$$H_2CO_3 \rightarrow H_2O + CO_2$$

The same enzyme, however, can also catalyze the reverse reaction:

$$H_2O + CO_2 \rightarrow H_2CO_3$$

The two reactions can be more conveniently illustrated by a single equation:

$$H_2O + CO_2 \rightleftharpoons H_2CO_3$$

**Figure 4.5.** The roles of cofactors in enzyme function. In (*a*) the cofactor changes the conformation of the active site, allowing a better fit between the enzyme and its substrates. In (*b*) the cofactor participates in the temporary bonding between the active site and the substrates.

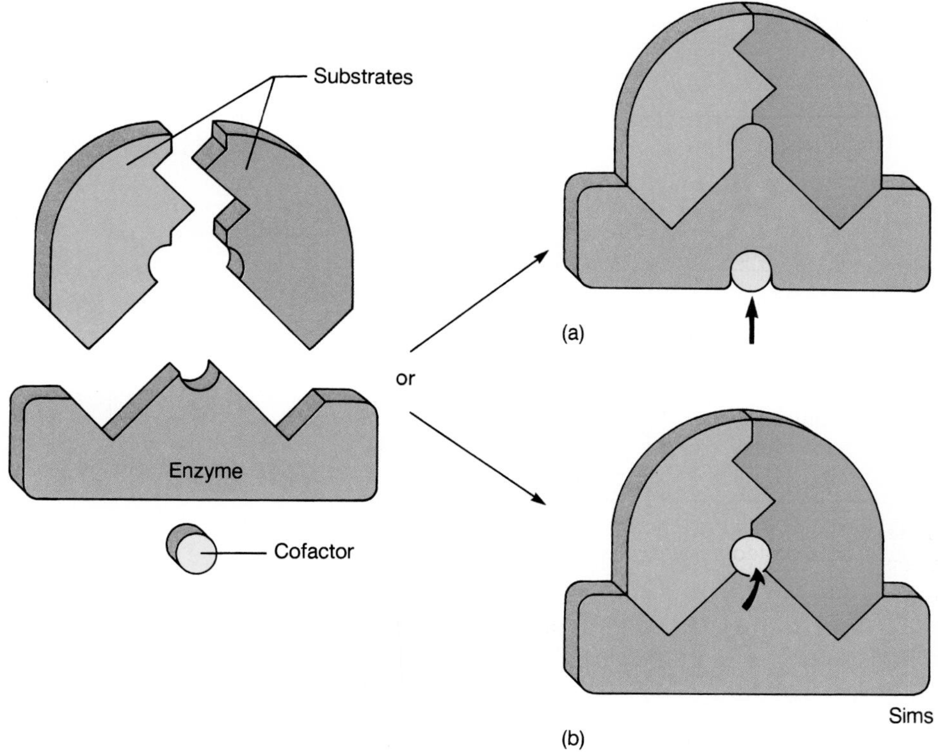


**Figure 4.6.** The effect of substrate concentration on the reaction rate of an enzyme-catalyzed reaction. When the reaction rate is maximum, the enzyme is said to be *saturated.*

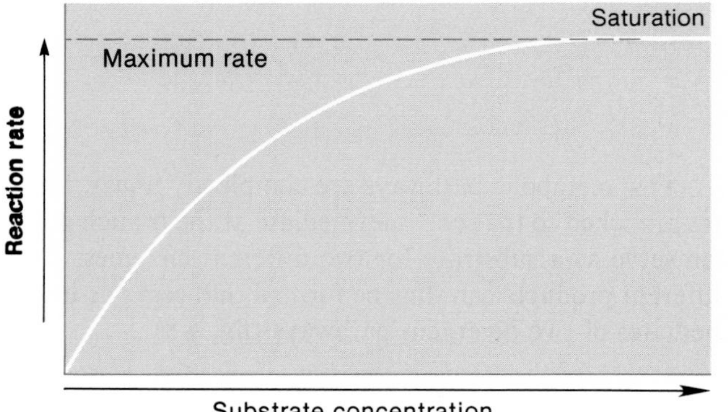


The direction of the reversible reaction depends, in part, on the relative concentrations of the molecules to the left and right of the arrows. If the concentration of $CO_2$ is very high (as it is in the tissues), the reaction will be driven to the right. If the concentration of $CO_2$ is low and that of $H_2CO_3$ is high (as it is in the lungs), the reaction will be driven to the left. The ability of reversible reactions to be driven from the side of the equation where the concentration is higher to the side where the concentration is lower is known as the **law of mass action.**

Although some enzymatic reactions are not directly reversible, the net effects of the reactions can be reversed by the action of different enzymes. The enzymes that convert glucose to pyruvic acid, for example, are different from those that reverse the pathway and produce glucose from pyruvic acid. Likewise, the formation and breakdown of glycogen (a polymer of glucose) are catalyzed by different enzymes.

**Figure 4.7.** A metabolic pathway, where the product of one enzyme becomes the substrate of the next in a multi-enzyme system.

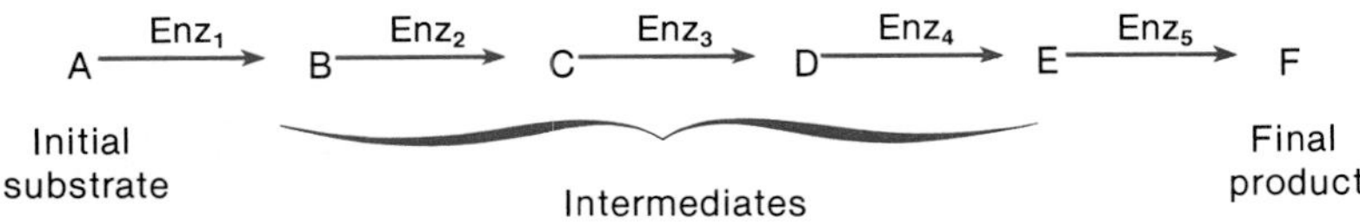


**Figure 4.8.** A branched metabolic pathway.

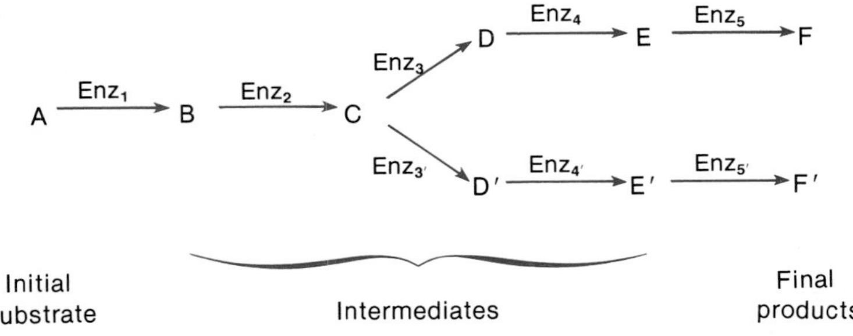


**Figure 4.9.** End-product inhibition in a branched metabolic pathway. Inhibition is shown by a dotted arrow and a negative sign.

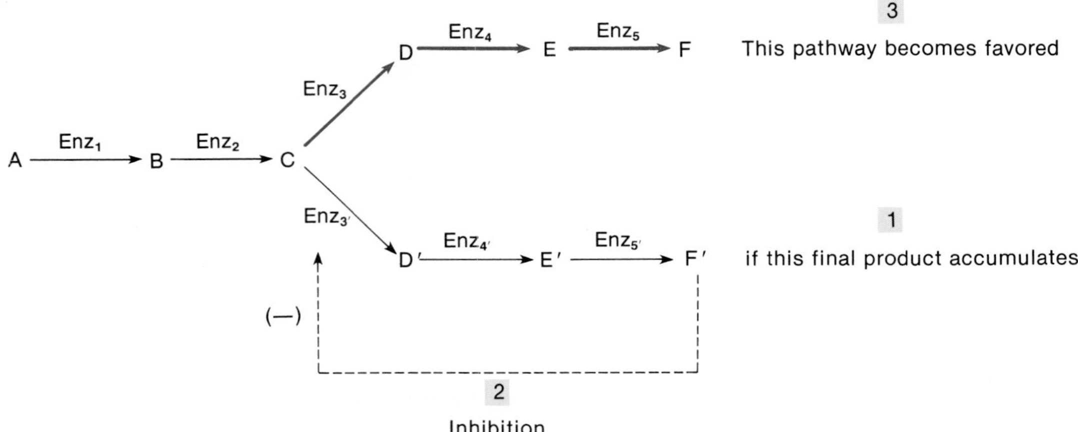


## Metabolic Pathways

The many thousands of different types of enzymatic reactions within a cell do not occur independently of each other. They are, rather, all linked together by intricate webs of interrelationships, the total pattern of which constitutes cellular metabolism. A part of this web that begins with an *initial substrate,* progresses through a number of *intermediates,* and ends with a *final product* is known as a **metabolic pathway.**

The enzymes in a metabolic pathway cooperate in a manner analogous to workers on an assembly line where each contributes a small part to the final product. In this process, the product of one enzyme in the line becomes the substrate of the next enzyme, and so on (fig. 4.7).

Few metabolic pathways are completely linear. Most are branched so that one intermediate at the branch point can serve as a substrate for two different enzymes. Two different products can thus be formed that serve as intermediates of two divergent pathways (fig. 4.8).

***End-Product Inhibition.*** The activities of enzymes at the branch points of metabolic pathways are often regulated by a process called **end-product inhibition.** In this process, one of the final products of a divergent pathway inhibits the branch point enzyme that began the path toward the production of this inhibitor. This inhibition prevents that final product from accumulating excessively and results in a shift toward the final product of the alternate divergent pathway (fig. 4.9).

**Figure 4.10.** The effects of an inborn error of metabolism on a branched metabolic pathway.

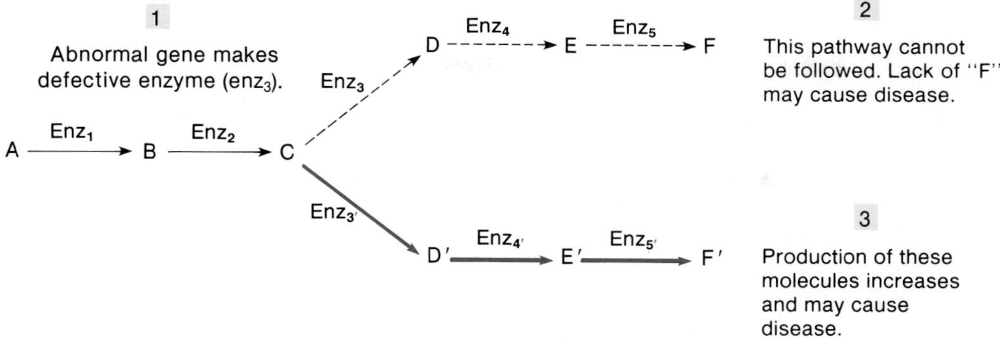


**Figure 4.11.** Metabolic pathways for the degradation of the amino acid phenylalanine. Defective enzyme$_1$ produces phenylketonuria (PKU), defective enzyme$_2$ produces alcaptonuria (not a clinically significant condition), and defective enzyme$_3$ produces albinism.

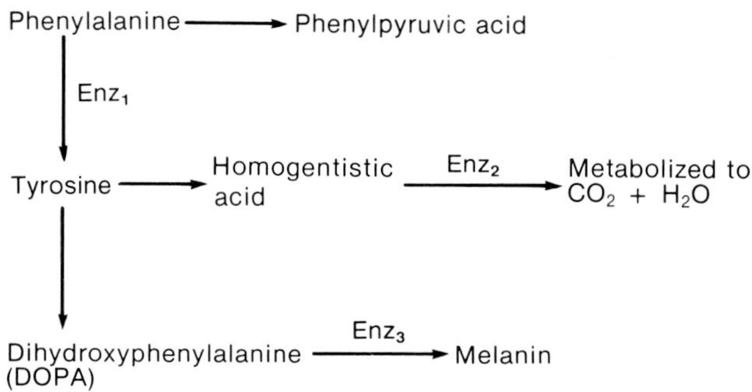


The mechanism by which a final product inhibits an earlier enzymatic step in its pathway is known as **allosteric inhibition.** The allosteric inhibitor combines with a part of the enzyme located away from the active site, causing the active site to change shape so that it can no longer combine properly with its substrate.

***Inborn Errors of Metabolism.*** Each enzyme in a metabolic pathway is coded by a different gene. An inherited defect in one of these genes may result in a disease known as an "inborn error of metabolism." In these diseases there is an *increased* amount of intermediates formed *prior* to the defective step and a *decrease* in the intermediates and final products formed *after* the defective enzymatic step. Diseases may result from lack of the normal end product, or from accumulation to toxic levels of intermediates formed prior to the defective step. If the defective enzyme is active at a step after a branch point in a pathway, the intermediates and final products of the alternate pathway will increase (fig. 4.10). Abnormally increased production of these alternate products can be the cause of some metabolic diseases.

The branched metabolic pathway that begins with phenylalanine as the initial substrate is subject to a number of inborn errors of metabolism (fig. 4.11). When the enzyme that converts this amino acid to the amino acid *tyrosine* is defective, the final products of a divergent pathway accumulate and can be detected in the blood and urine. This disease—*phenylketonuria (PKU)*—can result in severe mental retardation and a shortened life span. Although no inborn error of metabolism is common, PKU occurs so frequently and is so easy to detect that all newborn babies are tested for this defect. If this disease is detected early, brain damage can be prevented by placing the child on an artificial diet low in the amino acid phenylalanine.

One of the conversion products of phenylalanine is a molecule known as *DOPA,* which is an acronym for dihydroxyphenylalanine. DOPA is a precursor of the pigment molecule *melanin.* An inherited defect in the enzyme that catalyzes the formation of melanin from DOPA results in an albino. Besides PKU and albinism, there are a

**Table 4.5** Selected examples of inborn errors in the metabolism of amino acids, carbohydrates, and lipids

| Metabolic Defect | Disease | Abnormality | Clinical Result |
|---|---|---|---|
| Amino acid metabolism | Phenylketonuria (PKU) | Increase in phenylalanine | Mental retardation, epilepsy |
| | Albinism | Lack of melanin | Susceptibility to skin cancer |
| | Maple-syrup disease | Increase in leucine, isoleucine, and valine | Degeneration of brain, early death |
| | Homocystinuria | Accumulation of homocystine | Mental retardation, eye problems |
| Carbohydrate metabolism | Lactose intolerance | Lactose not utilized | Diarrhea |
| | Glucose-6-phosphatase deficiency (Gierke's disease) | Accumulation of glycogen in liver | Liver enlargement, hypoglycemia |
| | Glycogen phosphorylase deficiency (McArdle syndrome) | Accumulation of glycogen in muscle | Muscle fatigue and pain |
| Lipid metabolism | Gaucher's disease | Lipid accumulation (glucocerebroside) | Liver and spleen enlargement, brain degeneration |
| | Tay-Sachs disease | Lipid accumulation (ganglioside $G_{M_2}$) | Brain degeneration, death by age 5 |
| | Hypercholestremia | High blood cholesterol | Atherosclerosis of coronary and large arteries |

large number of other inborn errors of amino acid metabolism, as well as errors in carbohydrate and lipid metabolism (table 4.5).

1. *Draw graphs to represent the effects of changes in temperature, pH, enzyme and substrate concentration, cofactors, and coenzymes on the rate of enzymatic reactions. Explain the mechanisms responsible for the appearance of these graphs.*
2. *Draw a flowchart of a metabolic pathway (using arrows and letters such as* A, B, C, *etc.) with one branch point.*
3. *Describe a reversible reaction, and explain how the law of mass action affects this reaction.*
4. *Define end-product inhibition, and use your diagram of a branched metabolic pathway to explain how this process will affect the concentration of different intermediates.*
5. *Suppose, due to an inborn error of metabolism, that the enzyme that catalyzes the third reaction in your pathway (question no. 2) is defective. Describe the effects this would have on the concentrations of the intermediates in your pathway. Give two real-life examples of inborn errors of metabolism.*

## Bioenergetics

Living organisms require the constant expenditure of energy to maintain their complex structures and processes. Central to life processes are chemical reactions that are coupled, so that the energy released by one reaction is incorporated into the products of another reaction. The transformation of energy in living systems is largely based on reactions that produce and destroy molecules of ATP and on oxidation-reduction reactions.

*Bioenergetics* refers to the flow of energy in living systems. Organisms maintain their highly ordered structure and life-sustaining activities through the constant expenditure of energy obtained ultimately from the environment. The energy flow in living systems obeys the first and second laws of a branch of physics known as *thermodynamics*.

According to the **first law of thermodynamics,** energy can be transformed (changed from one form to another), but it can neither be created nor destroyed. This is sometimes called the law of *conservation of energy*. As a result of energy transformations, according to the **second law of thermodynamics,** the universe and its parts (including living systems) become increasingly disorganized. The term *entropy* is used to describe the degree of disorganization of a system. Energy transformations thus increase the amount of entropy of a system. Only energy that is in an organized state—called *free energy*—can be used to do work. Thus, since entropy increases in every energy transformation, the amount of free energy available to do work decreases. As a result of the increased entropy described by the second law, systems tend to go from states of higher to states of lower free energy.

The chemical bonding of atoms into molecules obeys the laws of thermodynamics. Atoms that are organized into complex organic molecules, such as glucose, have more free energy (less entropy) than six separate molecules each of carbon dioxide and water. Therefore, in order to convert carbon dioxide and water to glucose, energy must be added. Plants perform this feat using energy from the sun in the process of *photosynthesis* (fig. 4.12).

**Figure 4.12.** A simplified diagram of photosynthesis. Some of the sun's radiant energy is captured by plants and used to produce glucose from carbon dioxide and water. As the product of this endergonic reaction, glucose has a higher free energy content than the initial reactants.

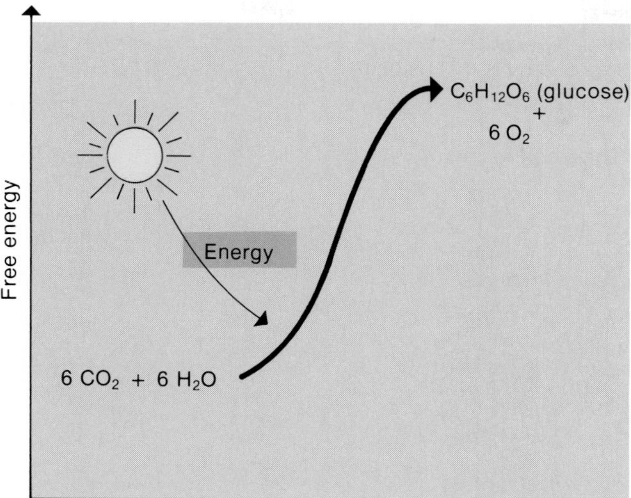


**Figure 4.13.** Since glucose contains more energy than carbon dioxide and water, the combustion of glucose is an exergonic reaction. The same amount of energy is released if glucose is broken down stepwise within the cell.

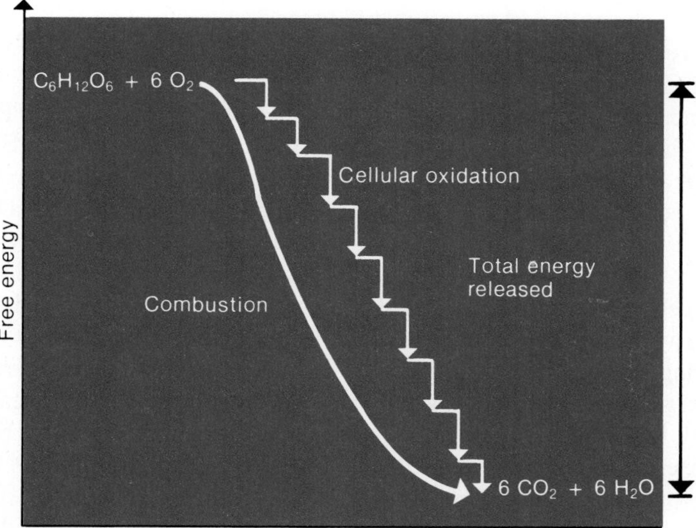


## Endergonic and Exergonic Reactions

Chemical reactions that require an input of energy are known as **endergonic reactions.** Since energy is added to make these reactions "go," the products of endergonic reactions must contain more free energy than the reactants. A portion of the energy added, in other words, is contained within the product molecules. This follows from the fact that energy cannot be created or destroyed (first law of thermodynamics) and from the fact that a more organized state of matter contains more free energy (less entropy) than a less organized state (as described by the second law).

The fact that glucose contains more free energy than carbon dioxide and water can be easily proven by the combustion of glucose to $CO_2$ and $H_2O$. This reaction releases energy in the form of heat. Reactions that convert molecules with more free energy to molecules with less—and, therefore, release energy as they proceed—are called **exergonic reactions.**

As illustrated in figure 4.13, the amount of energy released by an exergonic reaction is the same whether the energy is released in a single combustion reaction or in the many small, enzymatically controlled steps that occur in tissue cells. The energy that the body obtains from the

**Figure 4.14.** A model of the coupling of exergonic and endergonic reactions. The drive shaft (representing the energy of activation) turns the exergonic gear, which turns the endergonic gear. The reactants of the exergonic reaction (represented by the larger gear) have more free energy than the products of the endergonic reaction because the coupling is not 100% efficient—some energy is lost as heat.

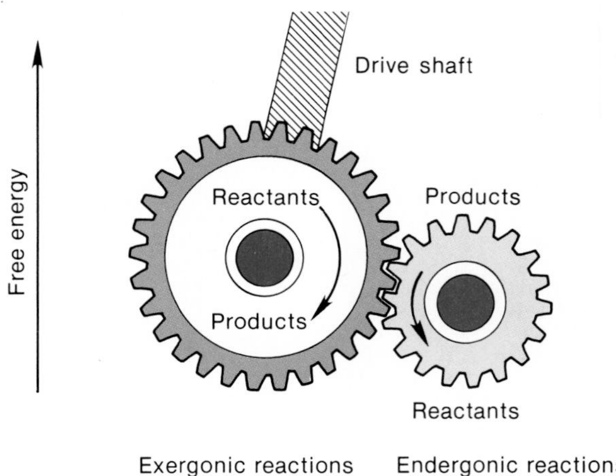


consumption of particular foods can, therefore, be measured as the amount of heat energy released when these foods are combusted.

Heat is measured in units called *calories.* One calorie is defined as the amount of heat required to raise the temperature of one cubic centimeter of water one degree Celsius. The caloric value of food is usually indicated in kilocalories (one kilocalorie = 1000 calories), which is also often called a Calorie with a capital letter *C.*

## Coupled Reactions: ATP

In order to remain alive a cell must maintain its highly organized, low entropy state at the expense of free energy in its environment. Accordingly, the cell contains many enzymes that catalyze exergonic reactions, using substrates that come ultimately from the environment. The energy released by these exergonic reactions is used to drive the energy-requiring processes (endergonic reactions) in the cell. Since the cell cannot use heat energy to drive energy-requiring processes, chemical bond energy that is released in the exergonic reactions must be directly transferred to chemical bond energy in the products of endergonic reactions. Energy-liberating reactions are thus *coupled* to energy-requiring reactions. This relationship is like two meshed gears; the turning of one (the energy-releasing, exergonic gear) causes turning of the other (the energy-requiring, endergonic gear—fig. 4.14).

The energy released by most exergonic reactions in the cell is used, either directly or indirectly, to drive *one* endergonic reaction (fig. 4.15): the formation of **adenosine triphosphate (ATP)** from adenosine diphosphate (ADP) and inorganic phosphate (abbreviated $P_i$).

**Figure 4.15.** The formation and structure of adenosine triphosphate (ATP).

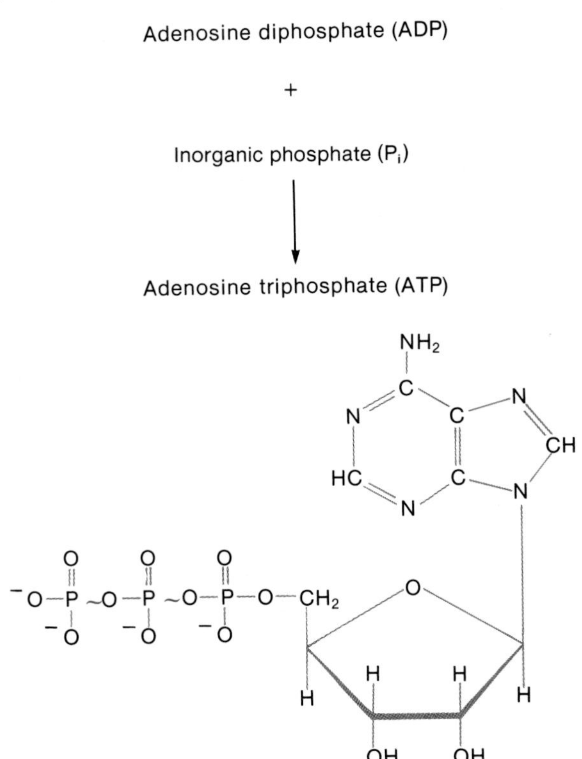


The formation of ATP requires the input of a fairly large amount of energy. Since this energy must be conserved (first law of thermodynamics), the bond that is produced by joining $P_i$ to ADP must contain a part of this energy. Thus, when enzymes reverse this reaction and convert ATP to ADP and $P_i$, a large amount of energy is released. Energy released from the breakdown of ATP is used to power the energy-requiring processes in all cells. As the **universal energy carrier,** ATP serves to couple more efficiently the energy released by the breakdown of food molecules to the energy required by the diverse endergonic processes in the cell (fig. 4.16).

## Coupled Reactions: Oxidation-Reduction

When an atom or a molecule gains electrons, it is said to become **reduced;** when it loses electrons, it is said to become **oxidized.** Reduction and oxidation are always coupled reactions: an atom or a molecule cannot become oxidized unless it donates electrons to another, which therefore becomes reduced. The atom or molecule that donates electrons *to* another is a **reducing agent**, and the one that accepts electrons *from* another is an **oxidizing agent.** It should be noted that an atom or a molecule may function as an oxidizing agent in one reaction and as a reducing agent in another reaction; it may gain electrons from one atom or molecule and pass them on to another in a series of coupled oxidation-reduction reactions—like a bucket brigade.

**Figure 4.16.** A model of ATP as the universal energy carrier of the cell. Exergonic reactions are shown as gears with arrows going down (reactions produce decrease in free energy); endergonic reactions are shown as gears with arrows going up (reactions produce increase in free energy).

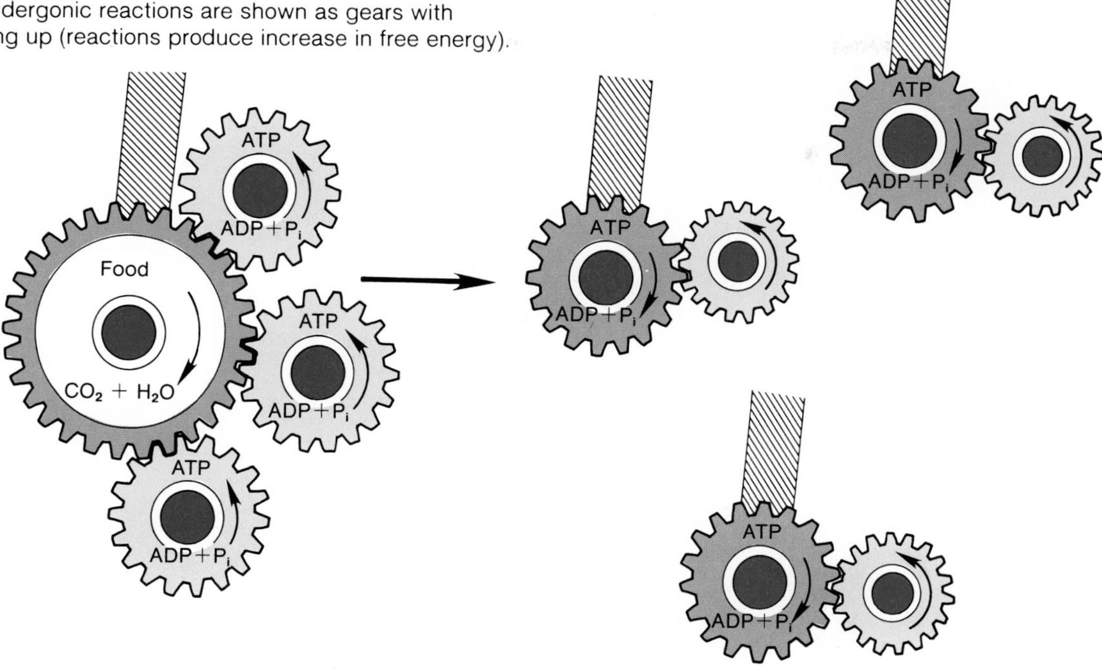


Notice that the term *oxidation* does not imply that oxygen participates in the reaction. This term is derived from the fact that oxygen has a great tendency to accept electrons, that is, to act as a strong oxidizing agent. This property of oxygen is exploited by cells; oxygen acts as the final electron acceptor in a chain of oxidation-reduction reactions that provides energy for ATP production.

Oxidation-reduction reactions in cells often involve the transfer of hydrogen atoms rather than of free electrons. Since a hydrogen atom contains one electron (and one proton in the nucleus) a molecule that loses hydrogen becomes oxidized, and one that gains hydrogen becomes reduced. In many oxidation-reduction reactions, pairs of electrons—either as free electrons or as a pair of hydrogen atoms—are transferred from the reducing agent to the oxidizing agent.

Two molecules that serve important roles in the transfer of hydrogens are **nicotinamide adenine dinucleotide (NAD),** which is derived from the vitamin niacin (vitamin $B_3$), and **flavin adenine dinucleotide (FAD),** which is derived from the vitamin riboflavin (vitamin $B_2$). These molecules (fig. 4.17) are coenzymes that function as *hydrogen carriers,* because they accept hydrogens (becoming reduced) in one enzyme reaction and donate hydrogens (becoming oxidized in) in a different enzyme reaction (fig. 4.18). The oxidized form of these molecules is written simply as NAD and FAD.

Each FAD can accept two electrons and can bind two protons. The reduced form of FAD is, therefore, combined with the equivalent of two hydrogen atoms and can be written as $FADH_2$. Each NAD can also accept two electrons but can only bind one proton. The reduced form of NAD, therefore, is shown as NADH + $H^+$ (the $H^+$ represents a free proton). When the reduced forms of these two coenzymes participate in an oxidation-reduction reaction they transfer two hydrogen atoms to the oxidizing agent (fig. 4.18).

1. *Describe the first and second laws of thermodynamics, and use these laws to explain why the chemical bonds in glucose represent a source of potential energy and how cells can obtain this energy.*
2. *Define the terms* exergonic reaction *and* endergonic reaction. *Use these terms to explain the function of ATP in cells.*
3. *Using the symbols X-$H_2$ and Y, draw a coupled oxidation-reduction reaction. Identify the molecule that is reduced and the one that is oxidized, and tell which one is the reducing agent and which is the oxidizing agent.*
4. *Describe the functions of NAD, FAD, and oxygen (in terms of oxidation-reduction reactions), and explain the meaning of the symbols NAD, NADH + $H^+$, FAD, and $FADH_2$.*

**Figure 4.17.** Structures of (*a*) the oxidized form of NAD (nicotinamide adenine dinucleotide) and (*b*) the reduced form of FAD (flavin adenine dinucleotide). Note the two additional hydrogen atoms (shown in color) that reduce FAD.

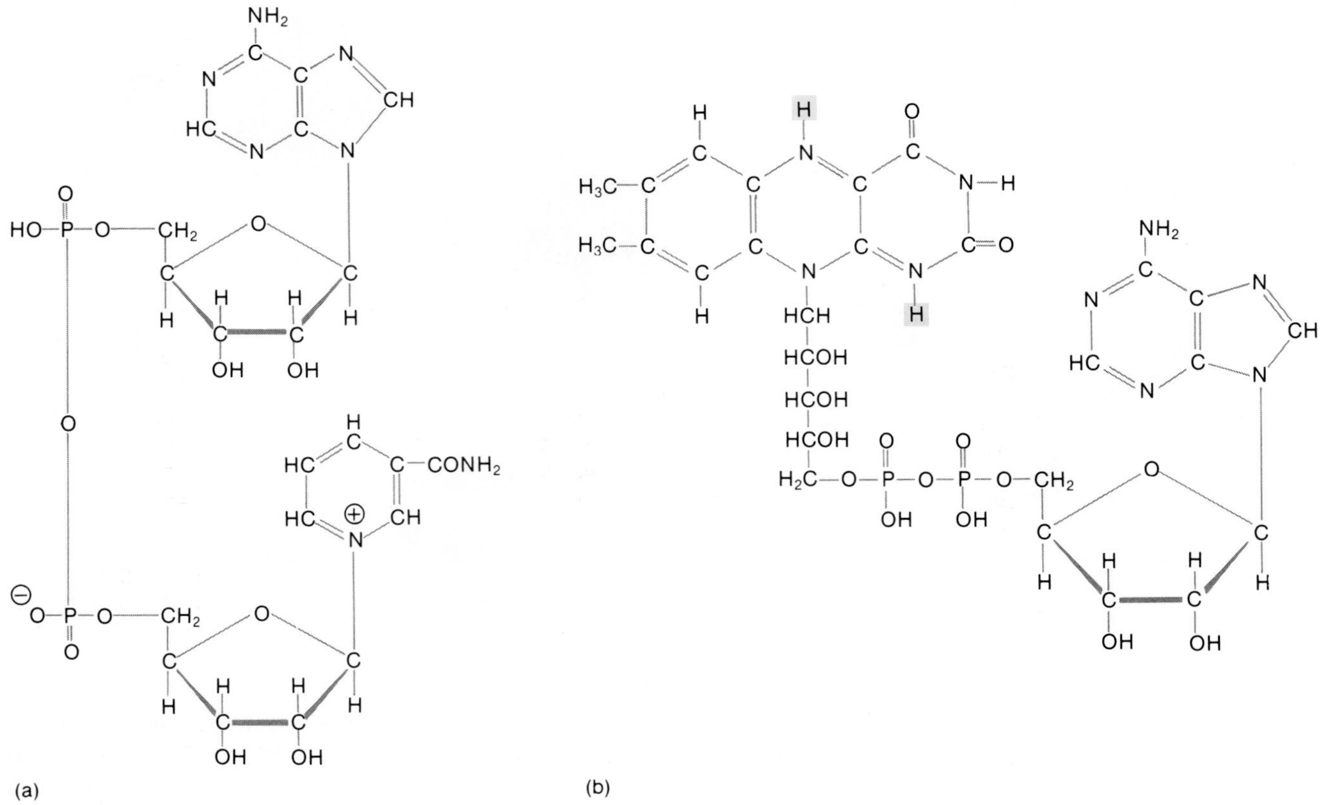

(a)

(b)

**Figure 4.18.** NAD is a coenzyme that functions to transfer pairs of hydrogen atoms from one molecule to another. In the first reaction, NAD is reduced (acts as an oxidizing agent); in the second reaction, NAD is oxidized (acts as a reducing agent).

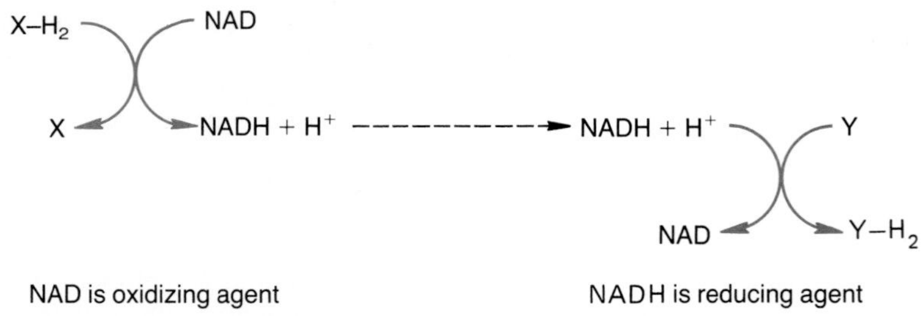

$X\text{–}H_2 + NAD \rightarrow X + NADH + H^+$ - - - - - - - - → $NADH + H^+ + Y \rightarrow NAD + Y\text{–}H_2$

NAD is oxidizing agent

NADH is reducing agent

## Summary

### Enzymes as Catalysts p. 84

I. Enzymes are biological catalysts.
  A. Catalysts increase the rate of chemical reactions.
    1. A catalyst is not altered by the reaction.
    2. Catalysts do not change the final result of a reaction.
  B. Catalysts lower the activation energy of chemical reactions.
    1. The activation energy is the amount of energy needed by the reactant molecules to participate in a reaction.
    2. In the absence of a catalyst, only a small proportion of the reactants have the activation energy.
    3. By lowering the activation energy, enzymes allow a larger proportion of the reactants to participate in the reaction, thus increasing the reaction rate.

II. All enzymes are proteins.
  A. Protein enzymes have specific three-dimensional shapes, which are determined by the amino acid sequence and, ultimately, by the genes.
  B. The reactants in an enzyme-catalyzed reaction—called the substrates of the enzyme—fit into a specific pocket in the enzyme called the active site.
  C. By forming an enzyme-substrate complex, substrate molecules are brought into proper orientation and existing bonds are weakened; this allows new bonds to be more easily formed.

### Control of Enzyme Activity p. 87

I. The activity of an enzyme is affected by a variety of factors.
  A. The rate of enzyme-catalyzed reactions increases with increasing temperature, up to a maximum.
    1. This is because increasing the temperature increases the energy in the total population of reactant molecules, thus increasing the proportion of reactants that have the activation energy.
    2. At a few centigrade degrees above body temperature, however, most enzymes start to denature, and the rate of the reactions which they catalyze therefore decreases.
  B. Each enzyme has optimal activity at a characteristic pH—called the pH optimum for that enzyme.
    1. This is because the pH affects the shape and charges within the active site.
    2. The pH optima of different enzymes can be quite different—pepsin has a pH optimum of 2, for example, while trypsin is most active at a pH of 9.
  C. Many enzymes require metal ions in order to be active—these ions are therefore said to be cofactors for the enzymes.
  D. Many enzymes require smaller organic molecules for activity. These smaller organic molecules are called coenzymes.
    1. Many coenzymes are derived from water-soluble vitamins.
    2. Coenzymes transport hydrogen atoms and small substrate molecules from one enzyme to another.
  E. The rate of enzymatic reactions increases when either the substrate concentration or the enzyme concentration is increased.
    1. If the enzyme concentration is constant, the rate of the reaction increases as the substrate concentration is raised, up to a maximum rate.
    2. When the rate of the reaction does not increase upon further addition of substrate, the enzyme is said to be saturated.

II. Metabolic pathways involve a number of enzyme-catalyzed reactions.
  A. A number of enzymes usually cooperate to convert an initial substrate to a final product by way of several intermediates.
  B. Metabolic pathways are produced by multi-enzyme systems in which the product of one enzyme becomes the substrate of the next.
  C. If an enzyme is defective due to an abnormal gene, the intermediates formed after the step catalyzed by the defective enzymes decrease, and the intermediates formed prior to the defective step accumulate.
    1. Diseases that result from defective enzymes are called inborn errors of metabolism.
    2. Accumulation of intermediates often results in damage to the organ that contains the defective enzyme.
  D. Many metabolic pathways are branched so that one intermediate can serve as the substrate for two different enzymes.
  E. The activity of a particular pathway can be regulated by end-product inhibition.
    1. In end-product inhibition, one of the products of the pathway inhibits the activity of a key enzyme.
    2. This is an example of allosteric inhibition, in which the product combines with its specific site on the enzyme, changing the conformation of the active site.

### Bioenergetics p. 92

I. The flow of energy in the cell is called bioenergetics.
  A. According to the first law of thermodynamics, energy can neither be created nor destroyed but only transformed from one form to another.
  B. According to the second law of thermodynamics, all energy transformation reactions result in an increase in entropy (disorder).
    1. As a result of the increase in entropy, there is a decrease in free (usable) energy.
    2. Atoms that are organized into large organic molecules thus contain more free energy than more disorganized, smaller molecules.
  C. In order to produce glucose from carbon dioxide and water, energy must be added as sunlight.
    1. This process is called photosynthesis.
    2. Reactions that require the input of energy to produce molecules with higher free energy than the reactants are called endergonic reactions.

D. The combustion of glucose to carbon dioxide and water releases energy in the form of heat.
   1. A reaction that releases energy and thus forms products that contain less free energy than the reactants is called an exergonic reaction.
   2. The same total amount of energy is released when glucose is converted into carbon dioxide and water within cells, even though this process occurs in many small steps.
E. The exergonic reactions that convert food molecules into carbon dioxide and water in cells are coupled to endergonic reactions that form adenosine triphosphate (ATP).
   1. Some of the chemical-bond energy in glucose is, therefore, transferred to the "high-energy" bonds of ATP.
   2. The breakdown of ATP into adenosine diphosphate (ADP) and inorganic phosphate results in the liberation of energy.
   3. The energy liberated by the breakdown of ATP is used to power all of the energy-requiring processes of the cell—ATP is thus the "universal energy carrier" of the cell.

II. Oxidation-reduction reactions have several characteristics.
A. A molecule is said to be oxidized when it loses electrons and to be reduced when it gains electrons.
B. A reducing agent is thus an electron donor, and an oxidizing agent is an electron acceptor.
C. Although oxygen is the final electron acceptor in the cell, other molecules can act as oxidizing agents.
D. A single molecule can be an electron acceptor in one reaction and an electron donor in another.
   1. NAD and FAD can become reduced by accepting electrons from hydrogen atoms removed from other molecules.
   2. NADH + $H^+$, and $FADH_2$, in turn, donate these electrons to other molecules in other locations within the cells.
   3. Oxygen is the final electron acceptor (oxidizing agent) in a chain of oxidation-reduction reactions that provide energy for ATP production.

---

## REVIEW ACTIVITIES

### Objective Questions

1. Which of the following statements about enzymes is *true?*
   (a) All proteins are enzymes.
   (b) All enzymes are proteins.
   (c) Enzymes are changed by the reactions they catalyze.
   (d) The active sites of enzymes have little specificity for substrates.
2. Which of the following statements about enzyme-catalyzed reactions is *true?*
   (a) The rate of reaction is independent of temperature.
   (b) The rate of all enzyme-catalyzed reactions is decreased when the pH is lowered from 7 to 2.
   (c) The rate of reaction is independent of substrate concentration.
   (d) Under given conditions of substrate concentration, pH, and temperature, the rate of product formation varies directly with enzyme concentration, up to a maximum, at which point the rate cannot be further increased.
3. Which of the following statements about lactate dehydrogenase is *true?*
   (a) It is a protein.
   (b) It oxidizes lactic acid.
   (c) It reduces another molecule (pyruvic acid).
   (d) All of these are true.
4. In a metabolic pathway
   (a) the product of one enzyme becomes the substrate of the next
   (b) the substrate of one enzyme becomes the product of the next
5. In an inborn error of metabolism
   (a) a genetic change results in the production of a defective enzyme
   (b) intermediates produced before the defective step accumulate
   (c) alternate pathways are taken by intermediates at branch points located before the defective step
   (d) All of these are true.
6. Which of the following represents an *endergonic* reaction?
   (a) ADP + $P_i \rightarrow$ ATP
   (b) ATP $\rightarrow$ ADP + $P_i$
   (c) glucose + $O_2 \rightarrow CO_2 + H_2O$
   (d) $CO_2 + H_2O \rightarrow$ glucose
   (e) both *a* and *d*
   (f) both *b* and *c*
7. Which of the following statements about ATP is *true?*
   (a) The bond joining ADP and the third phosphate is a high-energy bond.
   (b) The formation of ATP is coupled to energy-liberating reactions.
   (c) The conversion of ATP to ADP and $P_i$ provides energy for biosynthesis, cell movement, and other cellular processes that require energy.
   (d) ATP is the "universal energy carrier" of cells.
   (e) All of these are true.
8. When oxygen is combined with two hydrogens to make water,
   (a) oxygen is reduced
   (b) the molecule that donated the hydrogens becomes oxidized
   (c) oxygen acts as a reducing agent
   (d) both *a* and *b* apply
   (e) both *a* and *c* apply
9. Enzymes increase the rate of chemical reactions by
   (a) increasing the body temperature
   (b) decreasing the blood pH
   (c) increasing the affinity of reactant molecules for each other
   (d) decreasing the activation energy of the reactants

10. According the the law of mass action, the reaction is driven to the right in the reaction: $A + B \rightarrow C$ when
    (a) the concentration of A and B are increased
    (b) the concentration of C is decreased
    (c) the concentration of enzyme is increased
    (d) both a and b
    (e) both b and c

## Essay Questions

1. Explain the relationship between the chemical structure and the function of an enzyme, and describe how various conditions may alter both the structure and the function of an enzyme.
2. Explain how the rate of enzymatic reactions may be regulated by the relative concentrations of substrates and products.
3. Explain how end-product inhibition represents a form of negative feedback regulation.
4. Using the first and second laws of thermodynamics, explain how ATP is formed and how it serves as the universal energy carrier.
5. The coenzymes NAD and FAD can "shuttle" hydrogens from one reaction to another. Explain how this process serves to couple oxidation and reduction reactions.

## Selected Readings


Baker, J. J. W., and G. E. Allen. 1982. *Matter, energy, and life.* 4th ed. Mass.: Addison-Wesley.

Berry, H. K. March 1984. The spectrum of metabolic disorders. *Diagnostic Medicine,* p. 39.

Dyson, R. D. 1978. *Essentials of cell biology.* 2d ed. Boston: Allyn & Bacon.

Hinkle, P., and R. E. McCarty. March 1979. How cells make ATP. *Scientific American.*

Lehninger, A. L. September 1961. How cells transform energy. *Scientific American.*

Lehninger, A. L. 1982. *Principles of biochemistry.* New York: Worth.

Sheeler, P., and D. E. Bianchi. 1980. *Cell biology: structure, biochemistry, and function.* New York: John Wiley & Sons.

Stryker, L. 1981. *Biochemistry,* 2d ed. New York: Freeman.

# 5

# Cell Respiration and Metabolism

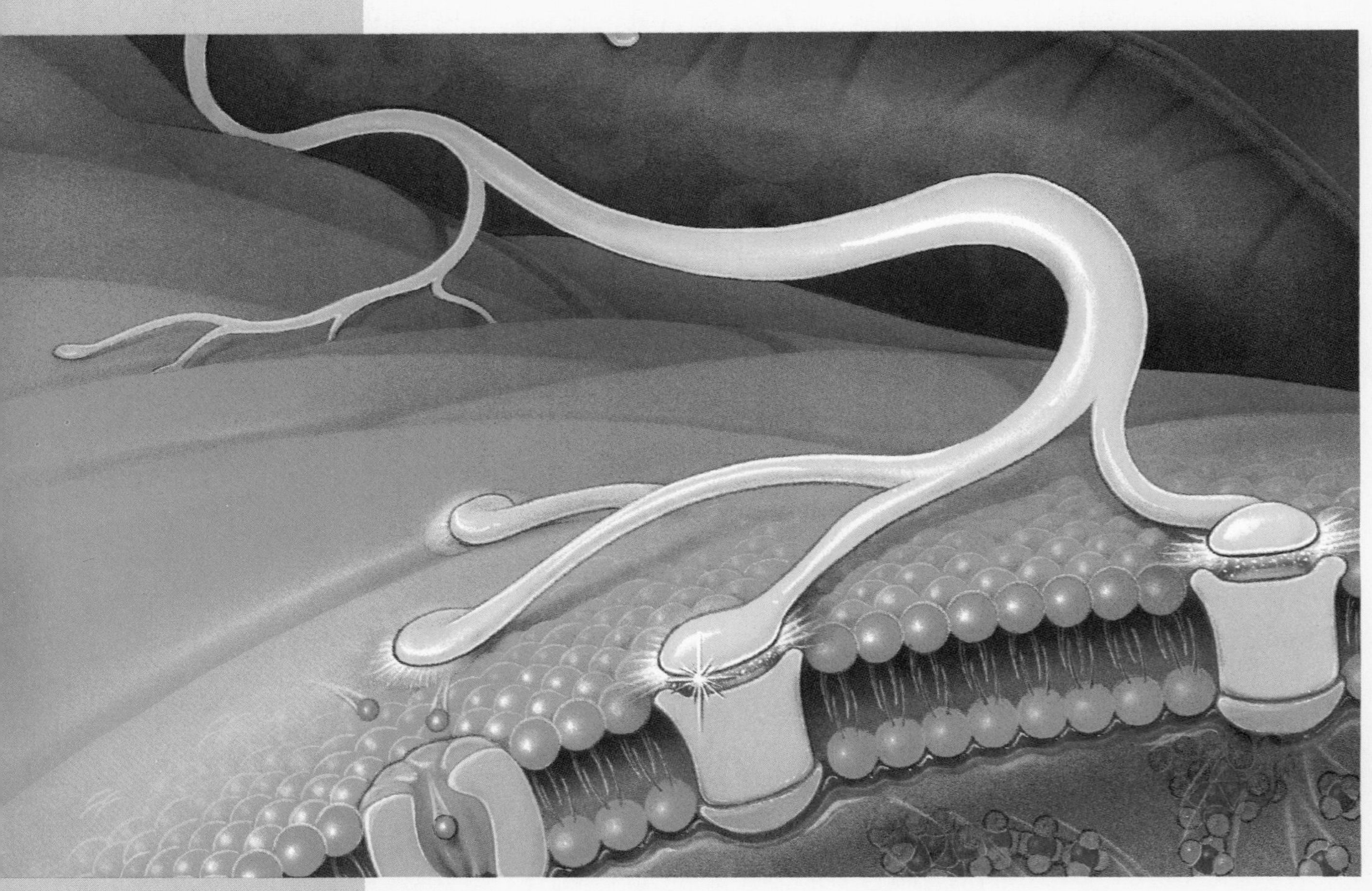

## OBJECTIVES

By studying this chapter, you should be able to

1. describe the steps of glycolysis and explain its significance
2. describe how lactic acid is formed in anaerobic respiration and explain the physiological significance of the anaerobic pathway
3. define the term *gluconeogenesis* and describe the Cori cycle
4. describe the pathway for the aerobic respiration of glucose through the steps of the Krebs cycle
5. explain the functional significance of the Krebs cycle in relation to the electron transport system
6. describe the electron transport system and oxidative phosphorylation
7. describe the role of oxygen in aerobic respiration
8. compare anaerobic and aerobic respiration in terms of initial substrates, final products, cellular locations, and the total number of ATP molecules produced per glucose respired
9. define the terms *lipolysis* and *β-oxidation* and describe how these processes function in cellular energy production
10. explain how ketone bodies are formed
11. describe the processes of oxidative deamination and transamination of amino acids and explain how these processes can contribute to energy production
12. explain, in terms of the metabolic pathways involved, how excess calories in the form of carbohydrates or protein can be converted to fat
13. describe the changes in skeletal muscle metabolism during exercise
14. explain the roles of anaerobic and aerobic respiration in the metabolism of skeletal muscles, and explain the significance of the oxygen debt
15. describe the changing sources of energy for skeletal muscle metabolism during exercise

## OUTLINE

## Glycolysis and Anaerobic Respiration

In cellular respiration, chemical reactions that transform food molecules into simpler molecules liberate energy, some of which is used to produce ATP. The complete combustion of a molecule requires the presence of oxygen; some energy is liberated, however, during incomplete combustion in the absence of oxygen. Anaerobic respiration provides a net gain of two ATP per glucose through the process of glycolysis and the conversion of pyruvic acid into lactic acid.

All of the reactions in the body that involve energy transformation are collectively termed **metabolism.** Metabolism may be divided into two categories: *anabolism* and *catabolism.* Catabolic reactions release energy, usually by the breakdown of larger organic molecules into smaller molecules. Anabolic reactions require the input of energy and include the synthesis of large, energy-storage molecules such as glycogen, fat, and protein.

The catabolic reactions that break down glucose, fatty acids, and amino acids serve as the primary sources of energy for the cellular synthesis of ATP. These metabolic pathways are known collectively as *cellular respiration.* When oxygen is utilized, these processes are called **aerobic cell respiration.** The final products of aerobic respiration are carbon dioxide, water, and energy (a part of which is trapped in the chemical bonds of ATP). The overall equation for aerobic respiration, therefore, is identical to the equation that describes combustion (fuel + $O_2 \rightarrow CO_2 + H_2O$ + energy).

Notice that the term *respiration* refers to chemical reactions that liberate energy for the production of ATP. The oxygen used in aerobic respiration by tissue cells is obtained from the blood; the blood, in turn, becomes oxygenated in the lungs by the process of breathing (described in chapter 16). Breathing (also called *ventilation* or *external respiration*) is thus needed for, but is different from, aerobic respiration.

Unlike combustion, the conversion of glucose to carbon dioxide and water within the cells occurs in small, enzymatically catalyzed steps. Oxygen is only used at the last step (this will be described in a later section). Since a small amount of the chemical bond energy of glucose is released at early steps in the metabolic pathway, some cells in the body can obtain energy for ATP production in the temporary absence of oxygen. This process is called **anaerobic respiration.**

### Glycolysis

Both the anaerobic and the aerobic respiration of glucose begin with a metabolic pathway known as **glycolysis.** Glycolysis (*glyco* = sugar; *lysis* = break) is the metabolic pathway by which glucose—a six-carbon (hexose) sugar—is converted into two molecules of *pyruvic acid.* Even though each pyruvic acid molecule is roughly half the size of a glucose, one must not think of glycolysis as simply the breaking in half of glucose. Glycolysis is a metabolic pathway involving many enzymatically controlled steps, as will be described in this section.

Each pyruvic acid molecule contains three carbons, three oxygens, and four hydrogens. The number of carbon and oxygen atoms in one molecule of glucose—$C_6H_{12}O_6$—can thus be accounted for in the two pyruvic acid molecules. Since the two pyruvic acids together account for only eight hydrogens, however, it is clear that four hydrogen atoms are removed from the intermediates in glycolysis. Each pair of these hydrogen atoms are used to reduce a molecule of NAD. In this process, each pair of hydrogen atoms donate two electrons to NAD, thus reducing it. The reduced NAD binds one proton from the hydrogen atoms, leaving one proton unbound as $H^+$ (described in chapter 4). Starting from one glucose molecule, therefore, glycolysis results in the production of two molecules of NADH and two $H^+$. The $H^+$ will follow the NADH in subsequent reactions, so for simplicity we can refer to reduced NAD simply as NADH.

Glycolysis is exergonic, and a portion of the energy that is released is used to drive the endergonic reaction ADP + $P_i \rightarrow$ ATP. At the end of the glycolytic pathway there is a net gain of two ATP per glucose molecule, as indicated in the overall equation for glycolysis:

$$\text{Glucose} + 2\text{NAD} + 2\text{ADP} + 2\text{P}_i \rightarrow$$
$$2 \text{ pyruvic acid} + 2\text{NADH} + 2\text{ATP}$$

Although the overall equation for glycolysis is exergonic, glucose must be "activated" at the beginning of the pathway before energy can be obtained. This activation requires the addition of two phosphate groups derived from two molecules of ATP. Energy from the reaction ATP $\rightarrow$ ADP + $P_i$ is therefore consumed at the beginning of glycolysis. This is shown as an "up-staircase" in figure 5.1. Notice that the $P_i$ is not shown in these reactions in figure 5.1; this is because the phosphate is not released but instead is added to the intermediate molecules of glycolysis. At later steps in glycolysis, four molecules of ATP are produced (and two molecules of NAD are reduced) as energy is liberated (the "down-staircase" in fig. 5.1). The two molecules of ATP used in the beginning, therefore, represent an energy investment; the net gain of two ATP and two NADH by the end of the pathway represent an energy profit.

**Figure 5.1.** The energy expenditure and gain in glycolysis. Notice that there is a "net profit" of 2 ATP and 2 NADH per glucose molecule in glycolysis. Molecules listed by number are (*1*) fructose-1, 6-diphosphate; (*2*) 1, 3-diphosphoglyceric acid; and (*3*) 3-phosphoglyceric acid (see fig. 5.2).

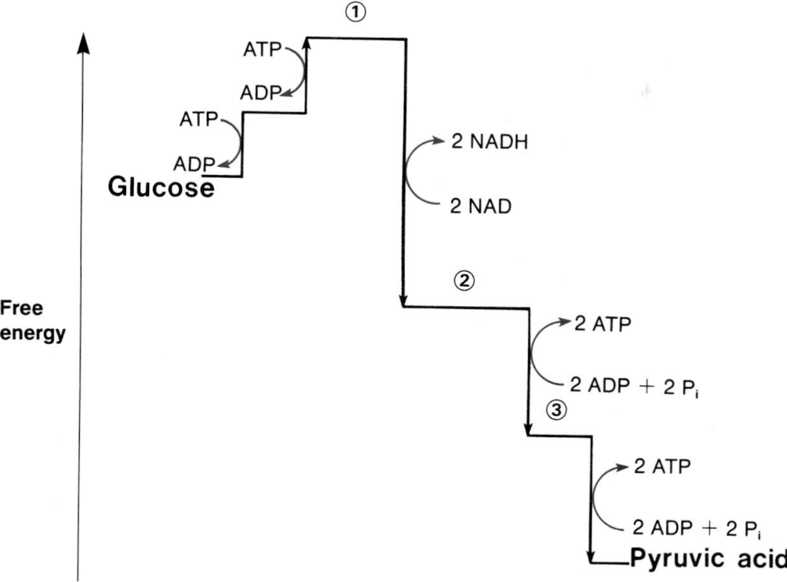


The overall equation for glycolysis obscures the fact that this is a metabolic pathway consisting of nine separate steps. The individual steps in this pathway are shown in figure 5.2, and the enzymes that catalyze these steps are listed in table 5.1.

## Anaerobic Respiration

In order for glycolysis to continue, there must be adequate amounts of NAD available to accept hydrogen atoms. The NADH that is produced in glycolysis, therefore, must become oxidized by donating its electrons to another molecule. (In aerobic respiration this other molecule is located in the mitochondria and ultimately passes its electrons to oxygen.)

When oxygen is *not* available in sufficient amounts, the NADH (+ $H^+$) produced in glycolysis is oxidized in the cytoplasm by donating its electrons to pyruvic acid. This results in the re-formation of NAD and the addition of two hydrogen atoms to pyruvic acid, which is thus reduced. This addition of two hydrogen atoms to pyruvic acid produces *lactic acid* (fig. 5.3). This metabolic pathway, by which glucose is converted through pyruvic acid to lactic acid, is called anaerobic respiration.

Anaerobic respiration yields a net gain of two ATP (produced by glycolysis) per glucose molecule. A cell can survive anaerobically as long as it can produce sufficient energy for its needs in this way and as long as lactic acid concentrations do not become excessive. Some tissues are better adapted to anaerobic respiration than others—skeletal muscles survive longer than cardiac muscle, which in turn can survive under anaerobic conditions longer than can the brain.

Except for red blood cells, which can only respire anaerobically (thus sparing the oxygen they carry), anaerobic respiration provides only a temporary sustenance for tissues that have energy requirements in excess of their aerobic ability. Anaerobic respiration can only occur for a limited period of time (longer for skeletal muscles, shorter for the heart, and shortest for the brain) when the *ratio of oxygen supply to oxygen need* (related to the concentration of NADH) falls below a critical level. Anaerobic respiration is, in a sense, an emergency procedure that provides some ATP until the emergency (oxygen deficiency) has passed.

The term "emergency" is actually overstated in the case of skeletal muscles, in which anaerobic respiration is a normal, daily occurrence that does not harm the tissue or the individual. The metabolism of skeletal muscles will be discussed in a separate summary section at the end of this chapter. Anaerobic respiration does not normally occur in the heart, however, and when it does occur a potentially dangerous condition may be present.

**Figure 5.2.** In glycolysis, one glucose molecule is converted into two pyruvic acid molecules in nine separate steps. Since two pyruvic acids are produced from one glucose, the products are multiplied by two in the pathway shown. In addition to two pyruvic acids, these products include two molecules of NADH and four molecules of ATP. Since two ATP molecules were used at the beginning of glycolysis, however, the net gain is 2 ATP per glucose.

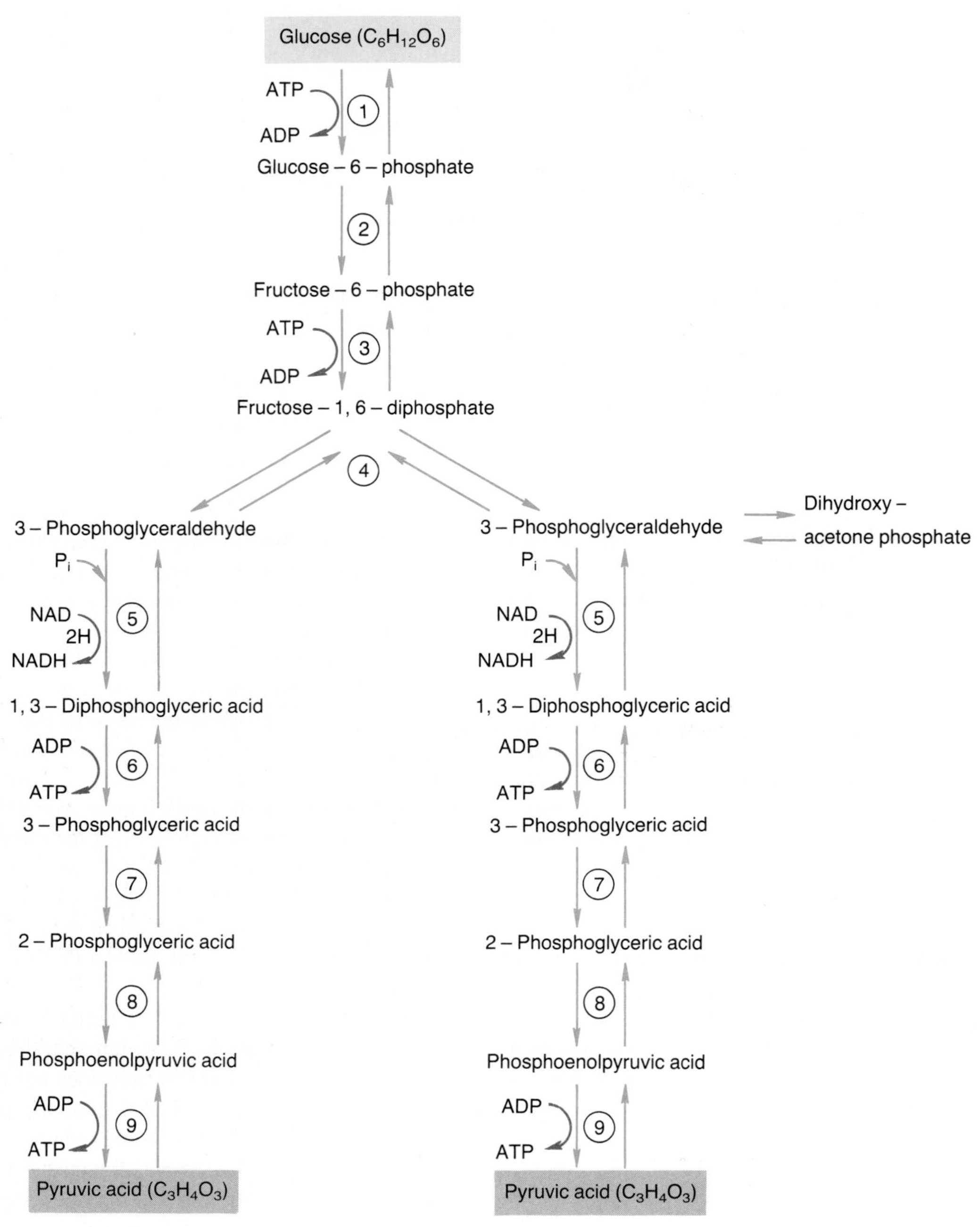

**Table 5.1** Enzymes, cofactors, and coenzymes required for glycolysis

| Step | Enzyme | Coenzyme or Cofactor | Comments |
|---|---|---|---|
| 1 | Hexokinase | $Mg^{++}$ | Liver enzyme catalyzes the phosphorylation of glucose, fructose, or mannose |
| 2 | Hexose phosphate isomerase | — | Interconverts glucose and fructose |
| 3 | Phosphofructokinase | $Mg^{++}$, ATP | Allosteric enzyme that is inhibited by high ATP |
| 4 | Aldolase | — | Splits hexose sugar into two three-carbon compounds |
| 5 | Phosphoglyceraldehyde dehydrogenase | NAD | Adds inorganic phosphate and oxidizes aldehyde to acid as NAD is reduced by removal of two hydrogens |
| 6 | Phosphoglycerate kinase | $Mg^{++}$ | Two molecules of ATP formed at this step |
| 7 | Phosphoglyceromutase | — | Phosphate group transferred to different carbon |
| 8 | Enolase | $Mg^{++}$, $Mn^{++}$ | Catalyzes molecular rearrangement |
| 9 | Pyruvate kinase | $Mg^{++}$, $K^{+}$ | Two molecules of ATP formed at this step |

**Figure 5.3.** The addition of two hydrogen atoms (colored boxes) from reduced NAD to pyruvic acid produces lactic acid and oxidized NAD. This reaction is catalyzed by lactate dehydrogenase (LDH).

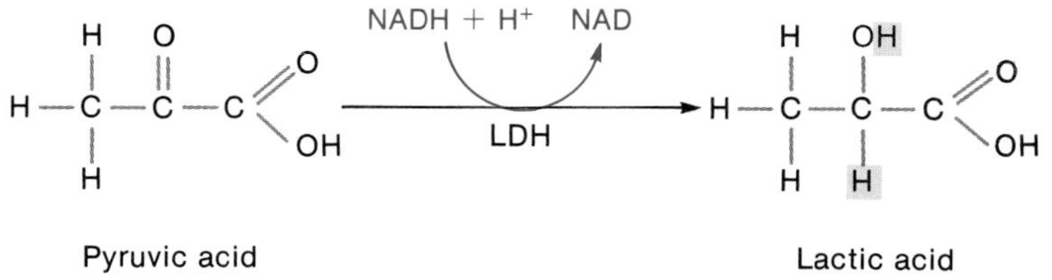

$$\text{H}-\underset{\text{H}}{\overset{\text{H}}{\text{C}}}-\overset{\text{O}}{\overset{\|}{\text{C}}}-\text{C}\begin{smallmatrix}\text{O}\\ \text{OH}\end{smallmatrix} \xrightarrow[\text{LDH}]{\text{NADH} + \text{H}^+ \;\; \text{NAD}} \text{H}-\underset{\text{H}}{\overset{\text{H}}{\text{C}}}-\underset{\text{H}}{\overset{\text{OH}}{\text{C}}}-\text{C}\begin{smallmatrix}\text{O}\\ \text{OH}\end{smallmatrix}$$

Pyruvic acid Lactic acid

**I**schemia refers to inadequate blood flow to an organ, such that the rate of oxygen delivery is insufficient to maintain aerobic respiration. Inadequate blood flow to the heart, or *myocardial ischemia,* may occur if the coronary blood flow is occluded by atherosclerosis, a blood clot, or by an artery spasm. People with myocardial ischemia often experience *angina pectoris,* severe pain in the chest and left (and sometimes the right) arm area. This pain is associated with increased blood levels of lactic acid, which are produced by anaerobic respiration by the heart muscle. The degree of ischemia and angina can be decreased by vasodilator drugs such as nitroglycerin and amyl nitrite, which improve blood flow to the heart and also decrease the work of the heart by dilating peripheral blood vessels.

***Gluconeogenesis and the Cori Cycle.*** Some of the lactic acid produced by exercising skeletal muscles is delivered by the blood to the liver. The enzyme lactic acid dehydrogenase within liver cells is then able to convert lactic acid to pyruvic acid. In the process, NAD is reduced to NADH + $H^+$. Unlike most other organs, the liver contains the enzymes needed to take pyruvic acid molecules and convert them to glucose-6-phosphate, a process that is essentially the reverse of glycolysis. Glucose-6-phosphate in liver cells can then be used as an intermediate for glycogen synthesis, or it can be converted to free glucose, which is secreted into the blood.

The conversion of noncarbohydrate molecules (lactic acid, amino acids, and glycerol) through pyruvic acid into glucose is an extremely important process called **gluconeogenesis.** In starvation and in prolonged exercise, when glycogen stores are depleted, the formation of new glucose in this way becomes the only means for maintaining constant blood sugar levels. Under these conditions, gluconeogenesis in the liver is the only way that adequate blood glucose levels can be maintained to prevent brain death.

During exercise some of the lactic acid produced by skeletal muscles may be transformed through gluconeogenesis in the liver to blood glucose. This new glucose can serve as an energy source during exercise and can be used after exercise in order to replenish the depleted muscle glycogen. This two-way traffic between skeletal muscles and the liver is called the **Cori cycle** (fig. 5.4). Recent evidence also suggests that some muscle fibers may be able to convert lactic acid to glucose. Through the Cori cycle, gluconeogenesis in the liver (together perhaps with gluconeogenesis in some muscle fibers) allows depleted skeletal muscle glycogen to be restored within forty-eight hours after exercise.

**Figure 5.4.** The Cori cycle. The direction of reversible reactions that occurs in the Cori cycle is shown by solid arrows; the sequence of steps is indicated by numbers 1 through 9.

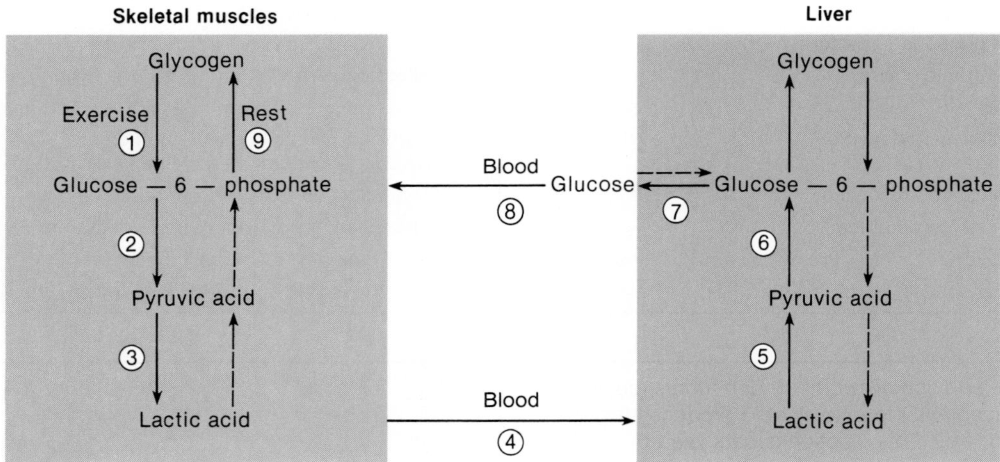


1. *Define the term* glycolysis *in terms of its initial substrates and products. Explain why there is a net gain of two ATP in this process.*
2. *Define the term* anaerobic respiration *in terms of its initial substrates and final products. Explain the significance of lactic acid formation at the end of this process.*
3. *Explain why anaerobic respiration occurs. In which tissue(s) is anaerobic respiration normal? In which tissue is it abnormal?*
4. *Define the term* gluconeogenesis, *and explain how this process functions to replenish the glycogen stores of skeletal muscles following exercise.*

## Aerobic Respiration

In the aerobic respiration of glucose, pyruvic acid is formed by glycolysis and then converted into acetyl coenzyme A. This begins a cyclic metabolic pathway called the Krebs cycle. As a result of these pathways, a large number of reduced NAD and FAD are generated. These reduced coenzymes provide electrons for an energy-generating process that drives the formation of many ATP. A total of thirty-eight ATP can be produced by the aerobic respiration of a single molecule of glucose.

Aerobic respiration is equivalent to combustion in terms of its final products ($CO_2$ and $H_2O$) and in terms of the total amount of energy liberated. In aerobic respiration, however, the energy is released in small, enzymatically controlled oxidation reactions, and a portion (38%–40%) of the energy released in this process is captured in the high-energy bonds of ATP.

The aerobic respiration of glucose begins with glycolysis. Glycolysis in both anaerobic and aerobic respiration results in the production of two molecules of pyruvic acid, two molecules of ATP, and two molecules of NADH + $H^+$ per glucose. In aerobic respiration, however, the electrons in NADH are *not* donated to pyruvic acid and lactic acid is not formed, as happens in anaerobic respiration. Instead, the pyruvic acids will move to a different cellular location and undergo a different reaction; the NADH produced by glycolysis will eventually be oxidized, but that occurs later in the story.

The enzymes that catalyze glycolysis are located in the cell cytoplasm. In aerobic respiration, pyruvic acid leaves the cell cytoplasm and enters the interior (the matrix) of mitochondria. Once pyruvic acid is inside a mitochondrion, carbon dioxide is enzymatically removed from each three-carbon-long pyruvic acid to form a two-carbon-long organic acid—acetic acid. The enzyme that catalyzes this reaction combines the acetic acid with a coenzyme (derived from the vitamin pantothenic acid) called *coenzyme A.* The combination thus produced is called **acetyl coenzyme A,** abbreviated **acetyl CoA** (fig. 5.5).

Glycolysis converts one glucose molecule to two molecules of pyruvic acid. Since each pyruvic acid molecule is converted into one molecule of acetyl CoA and one $CO_2$, two molecules of acetyl CoA and two molecules of $CO_2$ are derived from each glucose. The acetyl CoA molecules serve as substrates for mitochondrial enzymes in the aerobic pathway; the carbon dioxide is a waste product in this process, which is carried by the blood to the lungs for elimination. It should be noted that the oxygen in $CO_2$ is derived from pyruvic acid, not from oxygen gas.

**Figure 5.5.** The formation of acetyl coenzyme A in aerobic respiration.

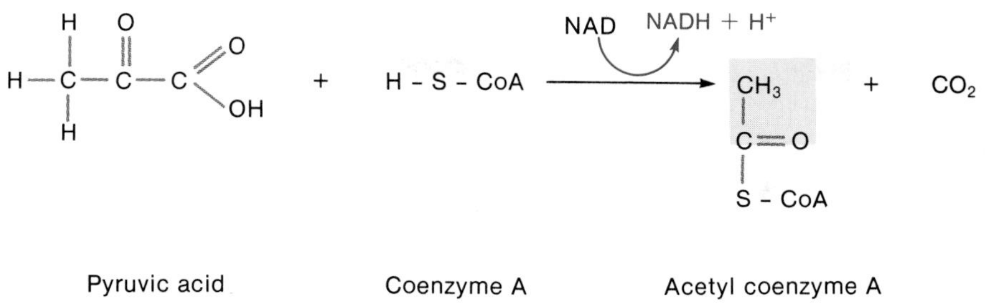

$$H_3C{-}CO{-}COOH + H{-}S{-}CoA \xrightarrow{NAD \;\to\; NADH + H^+} CH_3{-}C({=}O){-}S{-}CoA + CO_2$$

Pyruvic acid    Coenzyme A    Acetyl coenzyme A

**Figure 5.6.** A simplified diagram of the Krebs cycle, showing how the original four-carbon-long oxaloacetic acid is regenerated at the end of the cyclic pathway. Only the numbers of carbon atoms in the Krebs cycle intermediates are shown; the numbers of hydrogens and oxygens are not accounted for in this simplified scheme.

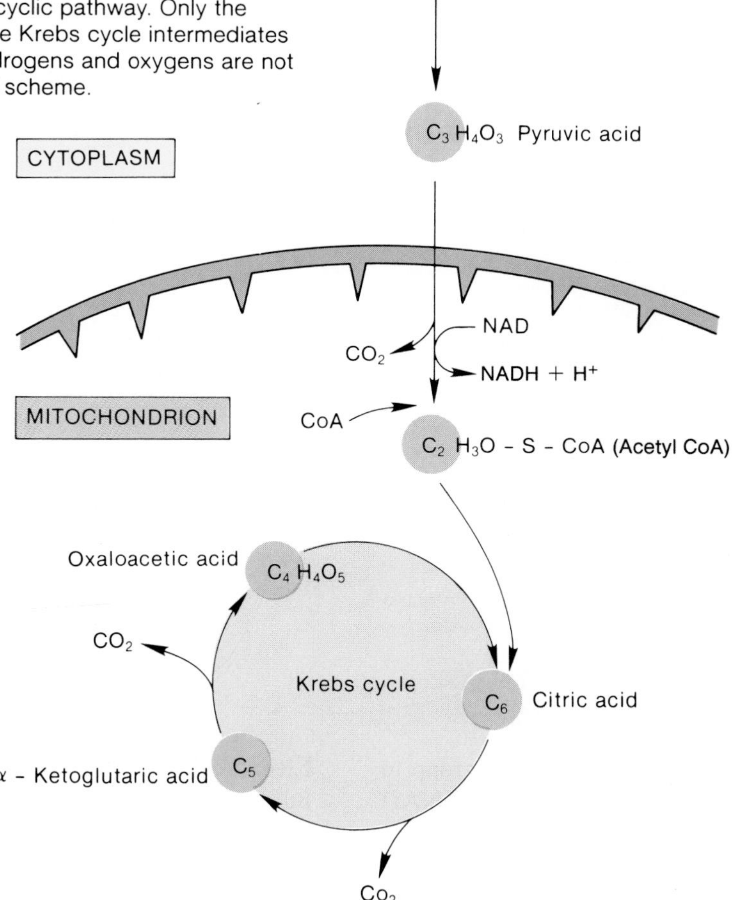


## The Krebs Cycle

Once acetyl CoA is formed, the acetic acid subunit (two carbons long) is combined with oxaloacetic acid (four carbons long) to form a molecule of citric acid (six carbons long). Coenzyme A acts only as a transporter of acetic acid from one enzyme to another (similar to the transport of hydrogen by NAD). The formation of citric acid begins a cyclic metabolic pathway known as the **citric acid cycle,** or **TCA cycle** (for tricarboxylic acid; citric acid has three carboxylic acid groups). Most commonly, however, this cyclic pathway is named after its principal discoverer, Sir Hans A. Krebs, and is called the **Krebs cycle.** A simplified illustration of this pathway is shown in figure 5.6.

Through a series of reactions involving the elimination of two carbons and four oxygens (as two $CO_2$ molecules) and the removal of hydrogens, citric acid is eventually converted to oxaloacetic acid, which completes the cyclic metabolic pathway. In this process the following occur: (1) one guanosine triphosphate (GTP) is produced

**Figure 5.7.** The Krebs cycle.

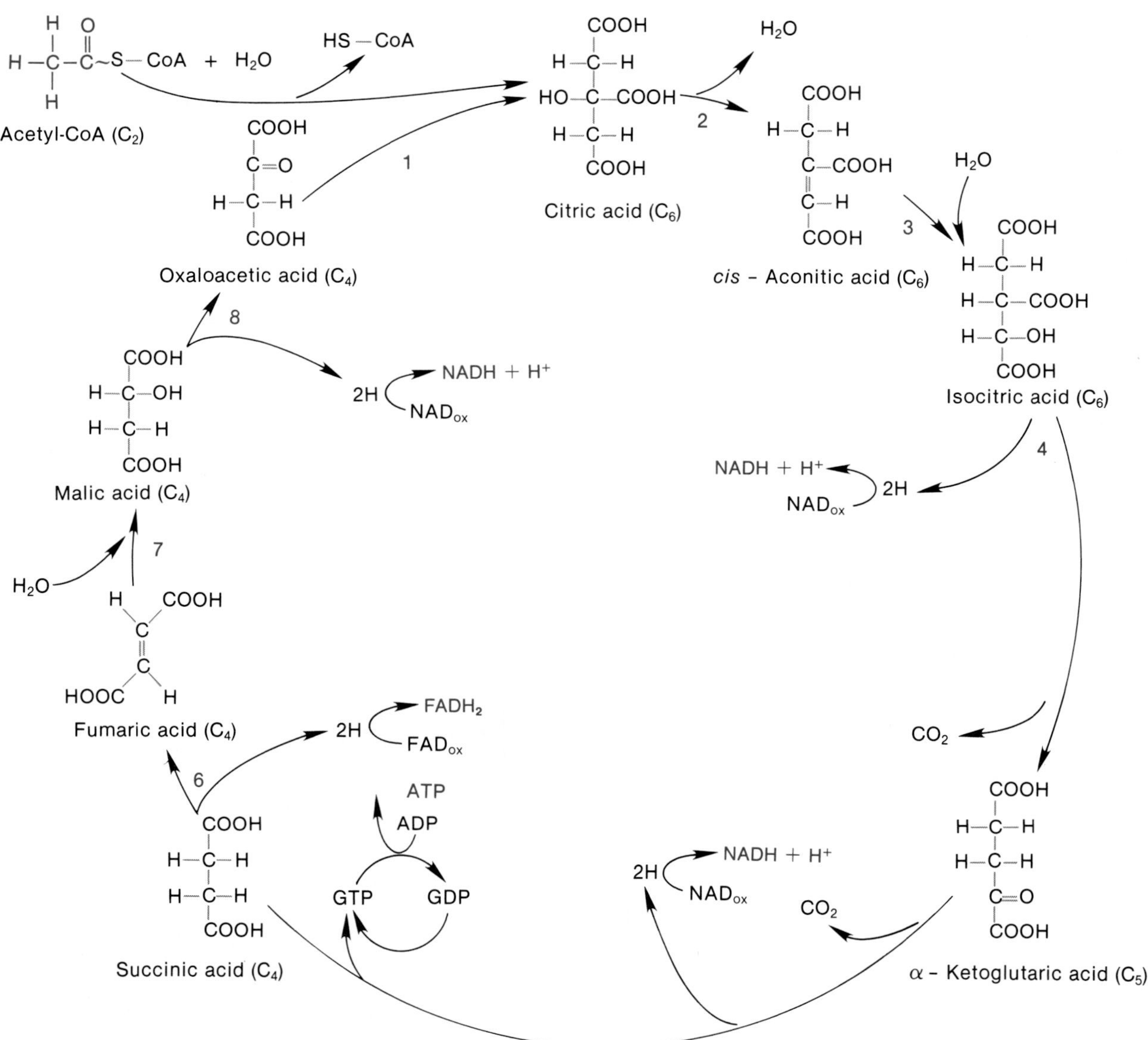


(step 5 of fig. 5.7), which donates a phosphate group to ADP to produce one ATP; (2) three molecules of NAD are reduced (steps 4, 5, and 8 of fig. 5.7); and (3) one molecule of FAD is reduced (step 6).

The production of reduced NAD and FAD by each "turn" of the Krebs cycle is far more significant, in terms of energy production, than the single GTP (converted to ATP) produced directly by the cycle. This is because NADH and $FADH_2$ eventually donate their electrons to an energy-generating process that results in the formation of many molecules of ATP.

## Electron Transport and Oxidative Phosphorylation

Built into the foldings, or cristae, of the inner mitochondrial membrane are a series of molecules that serve in **electron transport** during aerobic respiration. This electron transport chain of molecules consists of a flavoprotein (derived from the vitamin riboflavin), coenzyme Q (derived from vitamin E), and a group of iron-containing pigments called *cytochromes*. The last of these cytochromes is cytochrome $a_3$, which donates electrons to oxygen in the final oxidation-reduction reaction (as will be described). These molecules of the electron transport system are fixed in position within the inner mitochondrial membrane in such a way that they can pick up electrons from NADH and $FADH_2$ and transport them in a definite sequence and direction.

**Figure 5.8.** Electron transport and oxidative phosphorylation. Each element in the electron transport chain alternately becomes reduced and then oxidized as it transports electrons to the next member of the chain. This process provides energy for the formation of ATP. At the end of the electron transport chain the electrons are donated to oxygen, which becomes reduced (by the addition of two hydrogen atoms) to water.

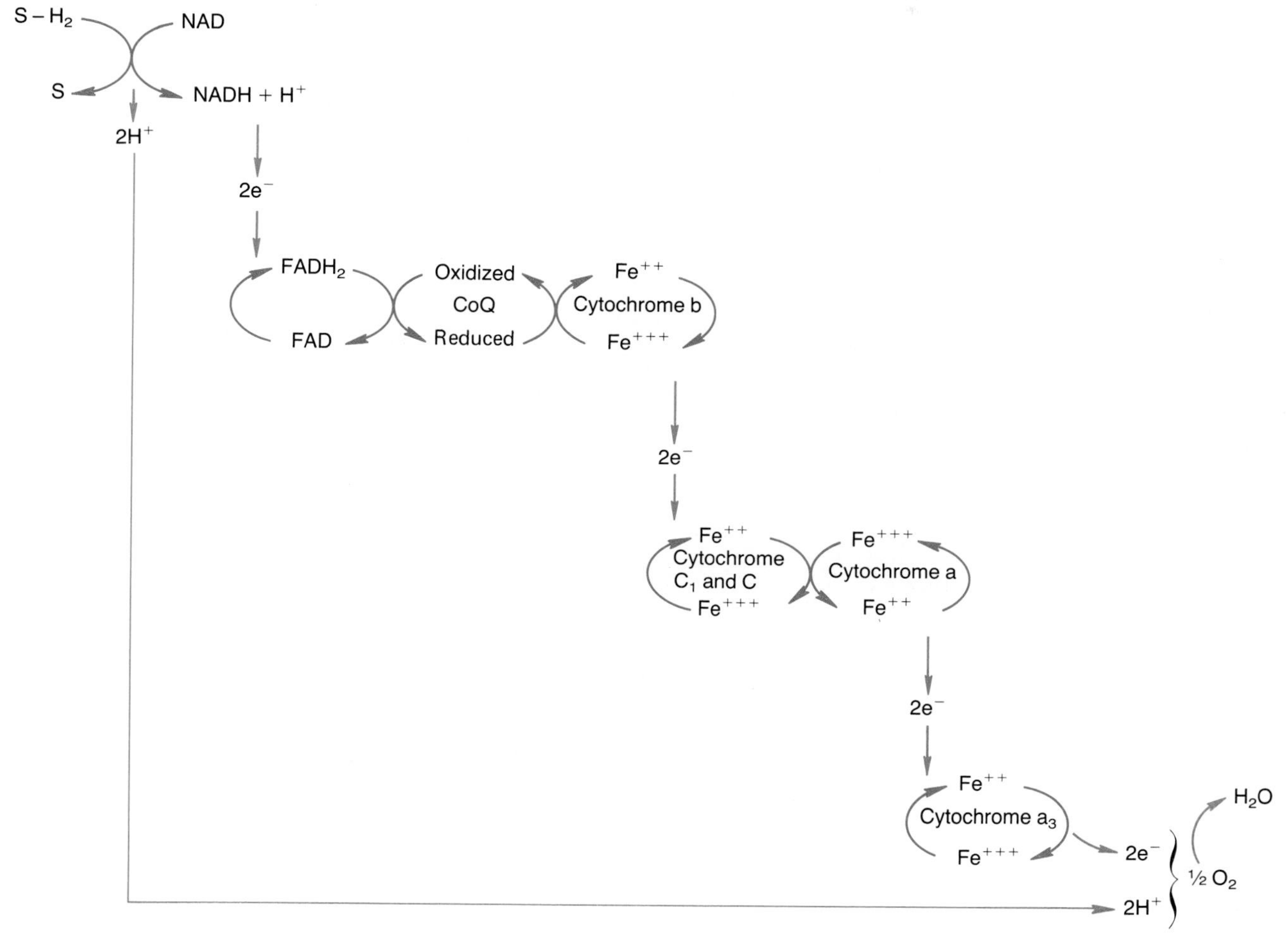

In aerobic respiration, NADH and $FADH_2$ become oxidized by transferring their pairs of electrons to the electron transport system of the cristae. It should be noted that the protons ($H^+$) are not transported together with the electrons; their fate will be described a little later. The oxidized forms of NAD and FAD are thus regenerated and able to continue to "shuttle" electrons from the Krebs cycle to the electron transport chain. The first molecule of the electron transport chain, in turn, becomes reduced when it accepts the electron pair from NADH. When the cytochromes receive a pair of electrons, two ferric ions ($Fe^{+++}$) become reduced to two ferrous ions ($Fe^{++}$). Notice that the gain of an electron is indicated by the reduction of the number of positive charges.

The electron transport chain thus acts as an oxidizing agent for NAD and FAD. Each element in the chain, however, also functions as a reducing agent; one reduced cytochrome transfers its electron pair to the next cytochrome in the chain. In this way, the iron ions in each cytochrome alternately become reduced (to ferrous ions) and oxidized (to ferric ions)—like a "ferrous wheel" (fig. 5.8). This is an exergonic process, and the energy derived is used to phosphorylate ADP to ATP. The production of ATP in this manner is thus appropriately termed **oxidative phosphorylation.** A more detailed explanation of how ATP is produced by the electron transport system is provided by the *chemiosmotic theory.*

**Figure 5.9.** The ATPase at the end of the electron transport system is attached to the cristae by a stalk. This unit produces ATP, using energy from electron transport (a process called oxidative phosphorylation).

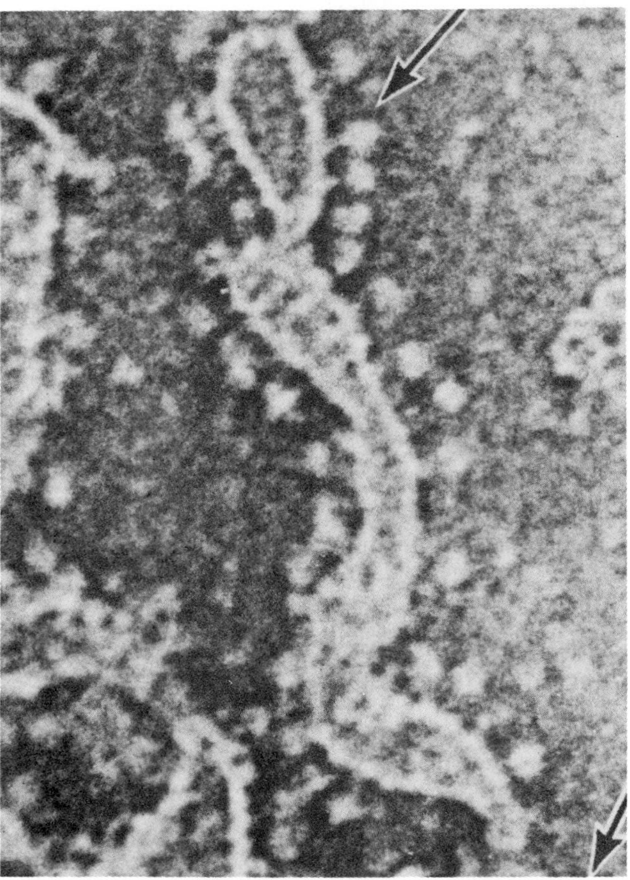

***Chemiosmotic Theory.*** One key to the understanding of how oxidative phosphorylation occurs is the observation that, under very high magnification with the electron microscope, lollipoplike structures protrude from the inner mitochondrial membrane into the matrix; indeed, the cristae are covered with these structures (fig. 5.9). The lollipops—sometimes called *respiratory assemblies*—consist of a group of proteins that form a stem and a globular subunit. The stem contains a proton channel through the inner mitochondrial membrane. The globular subunit, which protrudes into the matrix, contains an ATP synthetase enzyme that is capable of catalyzing the reaction ADP + $P_i \rightarrow$ ATP.

According to the chemiosmotic theory, three members of the electron transport system become active **proton pumps** when they accept a pair of electrons. When these members become reduced, they move a pair of protons ($H^+$) from the matrix to the space between the inner and outer mitochondrial membranes. The elements of the electron transport system that act as proton pumps are NADH dehydrogenase (which accepts electrons from NADH), cytochrome b, and cytochrome a. Note that for every two transported electrons that originate from NADH, six protons will be pumped. Since $FADH_2$ donates electrons to coenzyme Q and then to cytochrome b, electrons from $FADH_2$ only activate two of the three proton pumps. As a result of electron transport and activation of the proton pumps, the $H^+$ concentration in the intramembranous space is much higher than in the matrix (fig. 5.10). A steep electrochemical gradient is thus established across the inner mitochondrial membrane.

The inner mitochondrial membrane is not permeable to $H^+$; however, the protons can diffuse through the channels provided by the "stalks" of the respiratory assemblies. As each pair of protons moves into and through the bulbous portion of the respiratory assemblies the ATPase enzyme in this portion is activated, and a molecule of ATP is produced. Since each NADH results in the pumping of three pairs of protons by the electron transport system, three ATP are produced. Since each $FADH_2$ results in the pumping of two pairs of protons, two ATP are produced as the $H^+$ diffuse back into the matrix through the respiratory assemblies.

**Figure 5.10.** Schematic diagram of the chemiosmotic theory. (*a*) A mitochondrion. (*b*) Matrix and the compartment between the inner and outer mitochondria membrane, showing how the electron transport system results in the production of a $H^+$ gradient and of $H_2O$ at the end of the electron transport system. (*c*) Enlarged view of ATP synthesis, showing how the diffusion of $H^+$ results in the production of ATP.

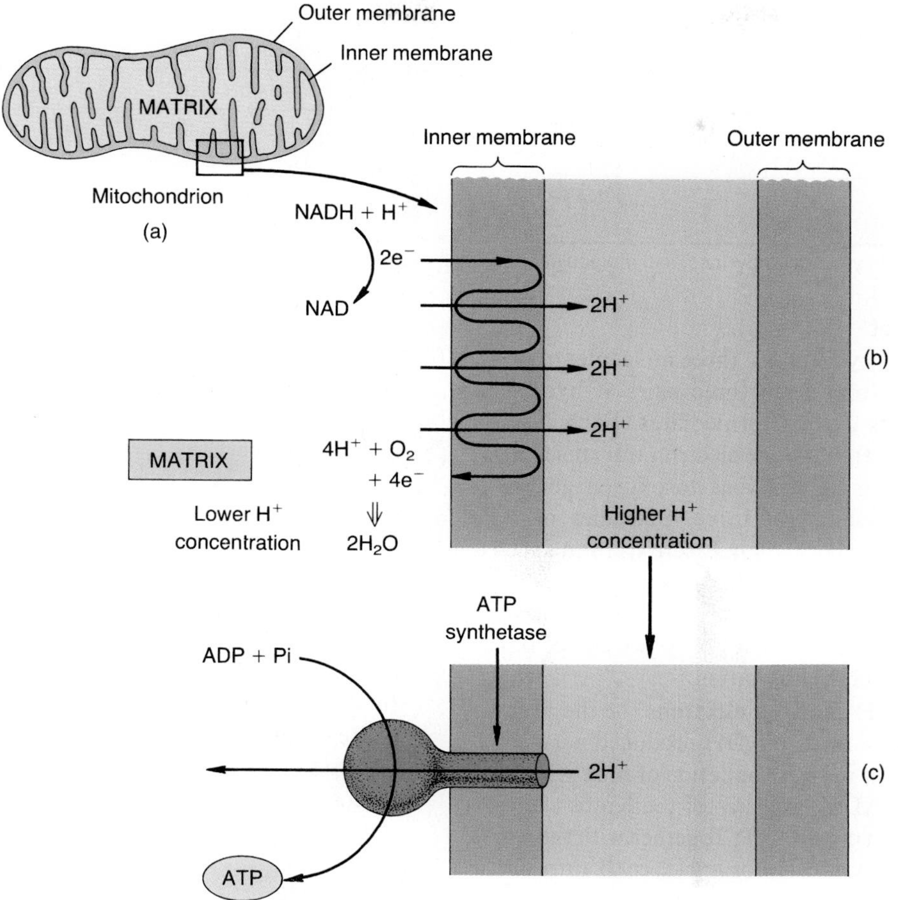


***Function of Oxygen.*** If the last cytochrome remained in a reduced state, it would be unable to accept more electrons. Electron transport would then progress only to the next-to-last cytochrome. This process would continue until all of the elements of the electron transport chain remained in the reduced state. At this point, the electron transport system would stop functioning and no ATP could be produced within the mitochondria. With the electron transport system incapacitated, NADH and $FADH_2$ could not become oxidized by donating their electrons to the cytochrome chain, and through inhibition of Krebs cycle enzymes, no more NADH and $FADH_2$ could be produced in the mitochondria (the Krebs cycle would stop). Respiration would then become anaerobic.

Oxygen, from the air we breathe, allows electron transport to continue by functioning as the **final electron acceptor** of the electron transport chain. This oxidizes cytochrome $a_3$ so that electron transport and oxidative phosphorylation can continue. At the very last step of aerobic respiration, therefore, oxygen becomes reduced by the two electrons that were passed to the chain from NADH and $FADH_2$. This reduced oxygen binds two protons, and a molecule of water is formed. Since the oxygen atom is part of a molecule of oxygen gas ($O_2$), this last reaction can be shown as follows:

$$O_2 + 4\ e^- + 4\ H^+ \rightarrow 2\ H_2O$$

*Cyanide* is a very rapidly lethal poison (1–15 minutes), which produces symptoms of rapid heart rate, hypotension, coma, and ultimately death if not quickly treated. Cyanide produces these effects because it has one very specific action: it blocks the transfer of electrons from cytochrome $a_3$ to oxygen. The effects of this deadly poison are thus the same as would occur if oxygen were completely removed; aerobic cell respiration and the production of ATP by oxidative phosphorylation comes to a halt.

**Table 5.2** Maximum ATP yield per glucose in aerobic respiration*

| Phase of Respiration | ATP Produced Directly | Reduced Coeyzymes | ATP from Oxidative Phosphorylation |
|---|---|---|---|
| Glycolysis (glucose to pyruvic acid) | 2 ATP | 2 NADH + $H^+$ | 6 ATP |
| Pyruvic acid to acetyl CoA | —— | 1 NADH + $H^+$ | 3 ATP |
| Pyruvic acid to acetyl CoA | —— | 1 NADH + $H^+$ | 3 ATP |
| Krebs cycle | 1 ATP | 3 NADH + $H^+$<br>1 $FADH_2$ | 9 ATP<br>2 ATP |
| Krebs cycle | 1 ATP | 3 NADH + $H^+$<br>1 $FADH_2$ | 9 ATP<br>2 ATP |

*One glucose molecule yields two pyruvic acid molecules, and each of these results in two turns of the Krebs cycle.

## ATP Balance Sheet

Each time the Krebs cycle turns, three molecules of NAD are reduced by electrons from three pairs of hydrogens removed from Krebs cycle intermediates. Each NADH donates a pair of electrons to the electron transport chain. Transport of this pair of electrons to oxygen generates energy for the production of three molecules of ATP through oxidative phosphorylation. Electrons from $FADH_2$ enter the electron transport chain "down the line" from where the first ATP is produced. Each pair of electrons from $FADH_2$, therefore, produces only two molecules of ATP from oxidative phosphorylation.

Since one NADH provides electrons for the production of three ATP, the three NADH produced per turn of the Krebs cycle result in the production of nine ATP molecules. The single $FADH_2$ per turn of the Krebs cycle results in the production of two ATP. Together with the single ATP made directly by the Krebs cycle, each turn of the Krebs cycle, therefore, yields a total of twelve ATP molecules. Since one molecule of glucose produces two pyruvic acids and thus two turns of the Krebs cycle, a total of twenty-four ATP molecules are produced by a single molecule of glucose, if only the ATP made by the Krebs cycle and its NADH and $FADH_2$ products are considered.

The conversion of pyruvic acid to acetyl CoA, however, also involves the reduction of one NAD (see fig. 5.5). Since two pyruvic acids are produced per glucose, two NADH are formed. And, since each NADH yields three ATP by oxidative phosphorylation, a total of twenty-four plus six, or thirty, ATP molecules are made in the mitochondrion from the steps that occur beyond pyruvic acid.

Now recall that two molecules of NADH are produced in the cytoplasm during glycolysis (conversion of glucose to pyruvic acid). These NADH cannot directly enter the mitochondria; instead, they donate their electrons to other molecules that "shuttle" these electrons into the mitochondria. Depending on which shuttle is used, either two or three ATP can be produced from each of these cytoplasmic NADH electrons through oxidative phosphorylation. A total of four or six ATP molecules are thus produced. Added to the thirty ATP previously mentioned, this brings the total to thirty-four or thirty-six (depending on which shuttle is used for the cytoplasmic electrons). Now, when we add the two molecules of ATP produced directly by glycolysis, a grand total of thirty-six to thirty-eight ATP are produced by the aerobic respiration of glucose (table 5.2).

## Glycogenesis and Glycogenolysis

Many tissues, particularly the liver, skeletal muscles, and heart, store carbohydrates in the form of glycogen. The formation of glycogen from glucose is called **glycogenesis.** In this process, glucose is converted to glucose-6-phosphate by utilizing the terminal phosphate group of ATP. Glucose-6-phosphate is then converted into its isomer, glucose-1-phosphate. Finally, the enzyme *glycogen synthetase* removes these phosphate groups as it polymerizes glucose to form glycogen.

The reverse reactions are similar. The enzyme *glycogen phosphorylase* catalyzes the breakdown of glycogen to glucose-1-phosphate. (The phosphates are derived from inorganic phosphate, not from ATP, so glycogen breakdown does not cost metabolic energy). Glucose-1-phosphate is then converted to glucose-6-phosphate. The conversion of glycogen to glucose-6-phosphate is called **glycogenolysis.** In most tissues, glucose-6-phosphate can then be respired for energy (through glycolysis), or used to resynthesize glycogen. Only in the liver, for reasons that will now be explained, can the glucose-6-phosphate also be used to produce free glucose for secretion into the blood.

Organic molecules with phosphate groups cannot cross cell membranes. This has important consequences; the glucose derived from glycogen is in the form of glucose-1-phosphate and then glucose-6-phosphate, so it cannot leak out of the cell. Similarly, glucose that enters the cell from the blood is "trapped" within the cell by conversion to glucose-6-phosphate. Skeletal muscles, which have large amounts of glycogen, can generate glucose-6-phosphate for their own glycolytic needs but cannot secrete glucose into the blood because they lack the ability to remove the phosphate group.

**Figure 5.11.** Blood glucose that enters tissue cells is rapidly converted to glucose-6-phosphate. This intermediate can be metabolized for energy in glycolysis or can be converted to glycogen (*1*), a process called *glycogenesis.* Glycogen represents a storage form of carbohydrates, which can be used as a source for new glucose-6-phosphate (*2*), in a process called *glycogenolysis.* The liver contains an enzyme that can remove the phosphate from glucose-6-phosphate; liver glycogen thus serves as a source for new blood glucose.

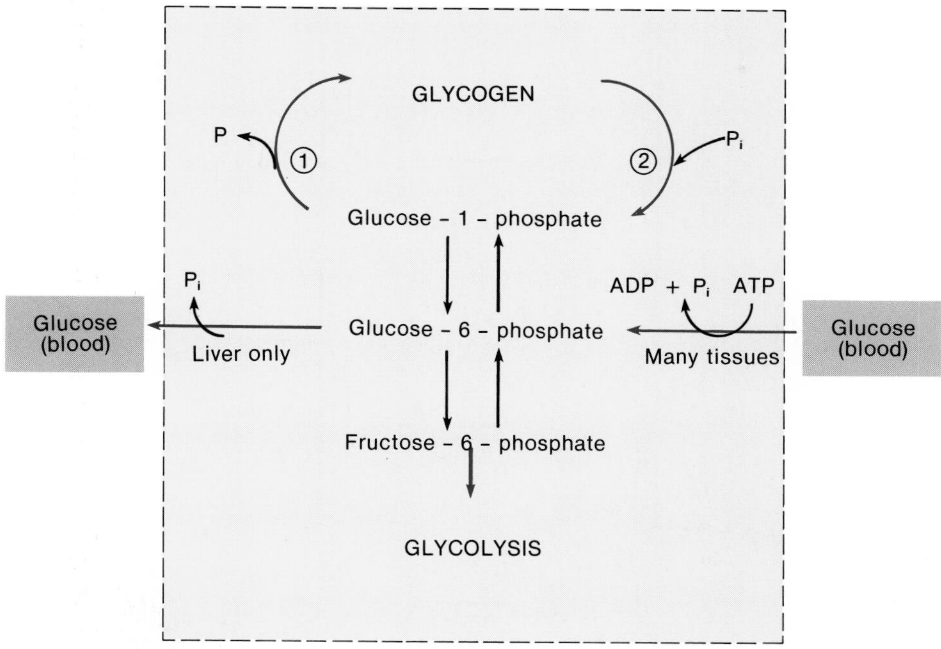


Unlike skeletal muscles, the liver contains an enzyme—known as *glucose-6-phosphatase*—which can remove the phosphate groups and produce free glucose (fig. 5.11). This free glucose can then be transported through the cell membrane; the liver, therefore, can secrete glucose into the blood, whereas skeletal muscles cannot. Liver glycogen can thus supply blood glucose for use by other organs, including exercising skeletal muscles that may have depleted much of their own stored glycogen during exercise.

## Control of Carbohydrate Metabolism

The rate of glucose utilization by the tissues is affected by the action of hormones such as insulin (described in chapter 11) and by the availability of oxygen. When oxygen is not available in sufficient amounts, the activity of the Krebs cycle enzymes decreases, and glycolysis (and anaerobic respiration) increases. This switching from aerobic to anaerobic respiration is accomplished largely by the allosteric inhibition of enzymes (review "Metabolic Pathways" in chapter 4). Key glycolytic enzymes are inhibited by high ATP levels. When the Krebs cycle is inhibited (by lack of sufficient NAD during oxygen deficiency), the decreased ATP concentrations that result stimulate glycolysis. This increased rate of glycolysis under anaerobic conditions—called the *Pasteur effect*—makes good sense, since only two ATP are produced per glucose anaerobically (vs. thirty-eight produced aerobically). More glucose molecules per unit time would have to be respired anaerobically than aerobically to produce a sufficient amount of ATP to sustain the tissue.

The fate of glucose molecules within tissues is also affected in another way by the cellular ATP concentrations. When food intake by the body occurs at a faster rate than energy consumption, the cellular concentration of ATP rises. Cells, however, do not store extra energy in the form of extra ATP. When cellular ATP concentrations rise, because more energy (from food) is available than can be immediately used, high ATP concentrations inhibit glycolysis. Under conditions of high cellular ATP concentrations, when glycolysis is inhibited, glucose is instead converted into glycogen and fat (fig. 5.12).

Conversion of glucose to fat is favored by high ATP concentrations, which inhibit Krebs cycle enzymes. This prevents acetyl CoA from joining oxaloacetic acid and thereby forming citric acid. The inhibition of Krebs cycle enzymes thus results in the increased availability of acetyl CoA, as well as of the glycolytic intermediates phosphoglyceraldehyde and dihydroxyacetone phosphate, for alternate pathways. These intermediates are instead channeled into pathways leading to triglyceride (fat) production. The interconversion of glucose, glycogen, and fat is depicted in figure 5.12, with the heavy arrows indicating the pathways that are favored when a person's intake of carbohydrates exceeds the energy requirements.

**Figure 5.12.** The conversion of glucose into glycogen and fat due to the allosteric inhibition of respiratory enzymes when the cell has adequate amounts of ATP. Favored pathways are indicated by heavy arrows.

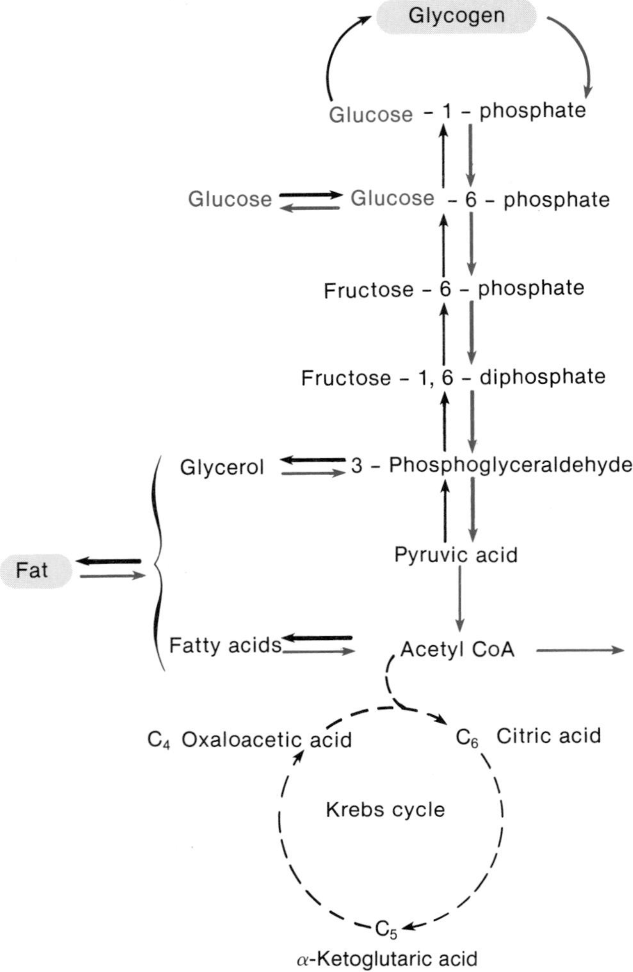


1. *Describe the fate of pyruvic acid in aerobic respiration, and explain how this differs from its fate in anaerobic respiration.*
2. *Draw a simplified Krebs cycle, using $C_2$ for acetic acid, $C_4$ for oxaloacetic acid, $C_5$ for alpha-ketoglutaric acid, and $C_6$ for citric acid. List the high-energy products that are produced at each turn of the Krebs cycle.*
3. *Using a diagram, show how electrons from NADH and $FADH_2$ are passed by the cytochromes. Represent the oxidized and reduced forms of the cytochromes with $Fe^{+++}$ and $Fe^{++}$, respectively.*
4. *Explain the function of oxygen in the body and how a lack of sufficient oxygen in an organ can make respiration become anaerobic.*
5. *Explain how end-product inhibition by ATP controls the rate of cell respiration and the rate of the formation of glycogen and fat.*

## Metabolism of Lipids and Proteins

Triglycerides can be hydrolyzed into glycerol and fatty acids. The latter are of particular importance, because they can be converted into numerous molecules of acetyl CoA, which can enter Krebs cycles and generate a large amount of ATP. Amino acids derived from proteins may also be used for energy. This involves deamination (removal of the amine group) and the conversion of the remaining molecule into either pyruvic acid or one of the Krebs cycle molecules.

Energy can be derived by the cellular respiration of lipids and proteins, using the same aerobic pathway previously described for the metabolism of pyruvic acid. Indeed, some organs preferentially use molecules other than glucose as an energy source. Pyruvic acid and the Krebs cycle acids also serve as common intermediates in the interconversion of glucose, lipids, and amino acids.

**Figure 5.13.** Divergent metabolic pathways for acetyl coenzyme A.

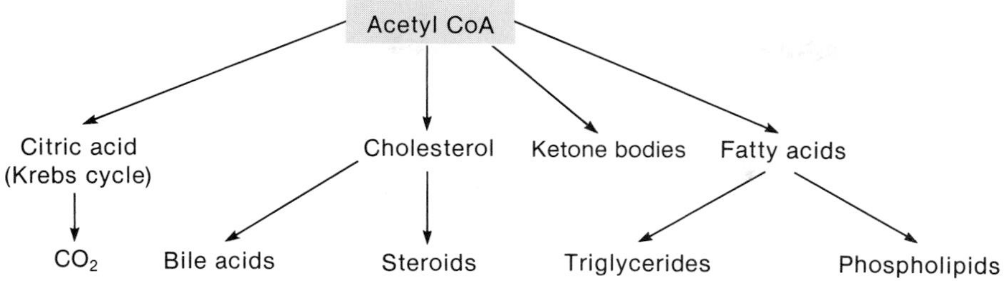


It is common experience that the ingestion of excessive calories in the form of carbohydrates (cakes, ice cream, candy, and so on) increases fat production. The rise in blood glucose that follows carbohydrate-rich meals stimulates insulin secretion, and this hormone, in turn, promotes the entry of blood glucose into adipose cells. Increased availability of glucose within adipose cells, under conditions of high-insulin secretion, promotes the conversion of glucose to fat (fig. 5.12). The hormonal control of carbohydrate and fat metabolism is discussed in chapter 18.

## Lipid Metabolism

When glucose is going to be converted into fat, glycolysis occurs and pyruvic acid is converted into acetyl CoA. Some of the glycolytic intermediates phosphoglyceraldehyde and dihydroxyacetone phosphate, however, do not complete their conversion to pyruvic acid, and acetyl CoA does not enter a Krebs cycle. The two-carbon acetic acid subunits of these acetyl CoA molecules can instead be used to produce a variety of lipids, including steroids such as cholesterol as well as ketone bodies and fatty acids (fig. 5.13). Acetyl CoA may thus be considered as a branch point at which a number of different possible metabolic pathways may progress.

In the formation of fatty acids, a number of acetic acid (two-carbon) subunits are joined together to form the fatty acid chain. Six acetyl CoA molecules, for example, will produce a fatty acid that is twelve carbons long. When three of these fatty acids condense with one glycerol (derived from phosphoglyceraldehyde), a triglyceride molecule is formed. The formation of fat, or **lipogenesis,** occurs primarily in adipose tissue and in the liver when the concentration of blood glucose is elevated following a meal.

Fat represents the major form of energy storage in the body. One gram of fat contains 9 kilocalories of energy, compared to 4 kilocalories for a gram of carbohydrates or protein. In a nonobese 70-kilogram man, 80%–85% of the body's energy is stored as fat, which amounts to about 140,000 kilocalories. Stored glycogen, in contrast, accounts for less than 2,000 kilocalories, most of which (about 350 g) is stored in skeletal muscles and is available only for the muscle's use. The liver contains about 80–90 grams of glycogen, which can be converted to glucose and used by other organs. Protein accounts for about 15%–20% of the stored calories in the body, but this is usually not used extensively as an energy source because that would involve the loss of muscle mass.

***Breakdown of Fat (Lipolysis).*** When fat stored in adipose tissue is going to be used as an energy source, lipase enzymes hydrolyze triglycerides into *glycerol* and *free fatty acids* in a process called **lipolysis.** These molecules (primarily the free fatty acids) serve as *blood-borne energy carriers* that can be used by the liver, skeletal muscles, and other organs for aerobic respiration.

A few organs can utilize glycerol for energy, by virtue of an enzyme that converts glycerol to phosphoglyceraldehyde. Free fatty acids, however, serve as the major energy source derived from triglycerides. Most fatty acids consist of a long hydrocarbon chain with a carboxylic acid group (COOH) at one end. In a process known as **$\beta$-oxidation** ($\beta$ is the Greek letter *beta*), enzymes remove two-carbon acetic acid molecules from the acid end of a fatty acid. This results in the formation of acetyl CoA, as the third carbon from the end becomes oxidized to produce a new carboxylic acid group. The fatty acid chain is thus decreased in length by two carbons. The process of $\beta$-oxidation continues until the entire fatty acid molecule is converted to acetyl CoA (fig. 5.14).

A sixteen-carbon-long fatty acid, for example, yields eight acetyl CoA molecules. Each of these can enter a Krebs cycle and produce twelve ATP per turn of the cycle, so that eight times twelve, or ninety-six, ATP are produced. In addition, each time an acetyl CoA is formed and the end-carbon of the fatty acid chain is oxidized, one NADH and one $FADH_2$ are produced. Oxidative phosphorylation produces three ATP per NADH and two ATP per $FADH_2$. For a sixteen-carbon-long fatty acid, these five ATP molecules would be formed seven times (producing five times seven, or thirty-five, ATP). Not counting the single ATP used to start $\beta$-oxidation (fig. 5.14), this fatty acid could yield a grand total of 35 + 96, or 131,

**Figure 5.14.** $\beta$-oxidation of a fatty acid. After the attachment of coenzyme A to the carboxylic acid group (step *1*), a pair of hydrogens is removed from the fatty acid and used to reduce one molecule of FAD (step *2*). When this electron pair is donated to the cytochrome chain, two ATP are produced. The addition of a hydroxyl group from water (step *3*), followed by the oxidation of the $\beta$-carbon (step *4*), results in the production of three ATP from the electron pair donated by reduced NAD. The bond between the $\alpha$ and $\beta$ carbons in the fatty acid is broken (step *5*), releasing acetyl coenzyme A and a fatty acid chain that is two carbons shorter than the original. With the addition of a new coenzyme A to the shorter fatty acid the process begins again (step *2*), as acetyl CoA enters the Krebs cycle and generates twelve ATP.

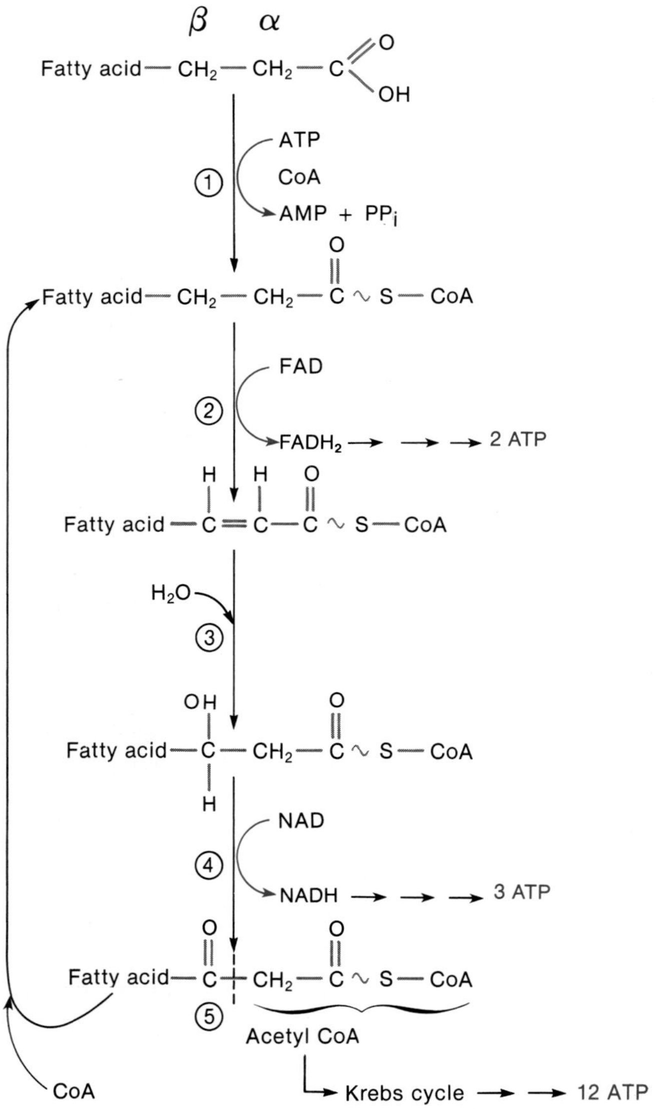

$\beta$ $\alpha$
Fatty acid—$CH_2$—$CH_2$—C(=O)OH
① ATP, CoA → AMP + $PP_i$
Fatty acid—$CH_2$—$CH_2$—C(=O) ∿ S—CoA
② FAD → $FADH_2$ → → → 2 ATP
Fatty acid—CH=CH—C(=O) ∿ S—CoA
③ $H_2O$
Fatty acid—CH(OH)—$CH_2$—C(=O) ∿ S—CoA
④ NAD → NADH → → → 3 ATP
Fatty acid—C(=O) ⑤ $CH_2$—C(=O) ∿ S—CoA
Acetyl CoA → Krebs cycle → → 12 ATP
CoA

ATP molecules! Since one triglyceride molecule consists of three fatty acids and one glycerol, the aerobic respiration of one triglyceride molecule yields more than 400 ATP.

***Ketone Bodies.*** There is a continuous turnover of triglycerides in adipose tissue, even when a person is not losing weight. New triglycerides are produced, while others are hydrolyzed into glycerol and fatty acids. This turnover ensures that the blood normally contains a sufficient amount of fatty acids for aerobic respiration by skeletal muscles, the liver, and other organs. When the rate of lipolysis exceeds the rate of fatty acid utilization—as it may do in starvation, dieting, and in diabetes mellitus—the blood concentrations of fatty acids increase.

If the liver cells contain sufficient amounts of ATP, so that further production of ATP is not needed, some of the acetyl CoA derived from fatty acids is channeled into an alternate pathway. This pathway involves the conversion of two molecules of acetyl CoA into four-carbon-long acidic derivatives, *acetoacetic acid* and *$\beta$-hydroxybutyric acid.* Together with *acetone,* which is a three-carbon-long derivative of acetoacetic acid, these products are known as **ketone bodies.**

Ketone bodies, which can be used for energy by many organs, are found in the blood under normal conditions. Under conditions of fasting or of diabetes mellitus, however, the increased liberation of free fatty acids from adipose tissue results in an elevated production of ketone bodies by the liver. The secretion of abnormally high amounts of ketone bodies into the blood produces **ketosis,** which is one of the signs of fasting or an uncontrolled diabetic state. A person in this condition may also have a sweet-smelling breath due to the presence of acetone, which is volatile and leaves the blood in the exhaled air.

## Amino Acid Metabolism

Nitrogen is ingested primarily as proteins, enters the body as amino acids, and is excreted mainly as urea in the urine. In childhood, the amount of nitrogen excreted may be less than the amount ingested because amino acids are incorporated into proteins during growth. Growing children are thus said to be in a state of *positive nitrogen balance.* People who are starving or suffering from prolonged wasting diseases, in contrast, are in a state of *negative nitrogen balance;* they excrete more nitrogen than they ingest because they are breaking down their tissue proteins.

Healthy adults maintain a state of nitrogen balance, in which the amount of nitrogen excreted is equal to the amount ingested. This does not imply that the amino acids ingested are unnecessary; on the contrary, they are needed to replace the approximately 400 grams of protein that is "turned over" each day. When more amino acids are ingested than are needed to replace proteins, the excess amino acids are not stored as additional protein (one cannot build muscles simply by eating large amounts of protein). Rather, the amine groups can be removed, and the "carbon skeletons" of the organic acids that are left can be used for energy or converted to carbohydrate and fat.

***Transamination.*** An adequate amount of all twenty amino acids is required to build proteins for growth and for replacement of the proteins that are turned over. Fortunately, only eight (in adults) or nine (in children) amino acids cannot be produced by the body and so must be obtained in the diet; these are the *essential amino acids* (table 5.3). Arginine is a "semiessential" amino acid because it can be produced by transamination, but not in sufficient quantities to classify it as a nonessential amino acid. The remaining amino acids are "nonessential" only in the sense that the body can produce them if it is given a sufficient amount of the essential amino acids and of carbohydrates.

Pyruvic acid and the Krebs cycle acids are collectively termed *keto acids* because they have a ketone group; these should not be confused with the ketone bodies (derived from acetyl CoA) discussed in the previous section.

**Table 5.3** The essential, semiessential, and nonessential amino acids

| Essential Amino Acids | Semiessential Amino Acid | Nonessential Amino Acids |
|---|---|---|
| Lysine | Arginine | Aspartic acid |
| Tryptophan | | Glutamic acid |
| Phenylalanine | | Proline |
| Threonine | | Glycine |
| Valine | | Serine |
| Methionine | | Alanine |
| Leucine | | Cysteine |
| Isoleucine | | |
| Histidine (children) | | |

Keto acids can be converted to amino acids by the addition of an amine ($NH_2$) group. This amine group is usually obtained by "cannibalizing" another amino acid; in this process, a new amino acid is formed as the one that was cannibalized is converted to a new keto acid. This type of reaction, in which the amine group is moved across from one amino acid to form another, is called **transamination** (fig. 5.15).

Each transamination reaction is catalyzed by a specific enzyme (a transaminase) that requires vitamin $B_6$ (pyridoxine) as a coenzyme. The amine group from glutamic acid, for example, may be transferred to either pyruvic acid or oxaloacetic acid. The former reaction is catalyzed by the enzyme *glutamate pyruvate transaminase* (GPT), and the latter reaction is catalyzed by *glutamate oxaloacetate transaminase* (GOT). The addition of an amine group to pyruvic acid produces the amino acid alanine; the addition of an amine group to oxaloacetic acid produces the amino acid known as aspartic acid (fig. 5.15).

***Oxidative Deamination.*** As shown in figure 5.16, glutamic acid can be formed through transamination by the combination of an amine group with $\alpha$-ketoglutaric acid. Glutamic acid is also produced in the liver from the ammonia that is generated by intestinal bacteria and carried to the liver in the hepatic portal vein. Since free ammonia is very toxic, its removal from the blood and incorporation into glutamic acid is an important function of the healthy liver.

If there are more amino acids than are needed for protein synthesis, the amine group from glutamic acid may be removed and excreted as *urea* in the urine (fig. 5.16). The metabolic pathway that removes amine groups from amino acids—leaving a keto acid and ammonia (which is converted to urea)—is known as **oxidative deamination.**

A number of amino acids can be converted into glutamic acid by transamination. Since glutamic acid can donate amine groups to urea (through deamination), it

**Figure 5.15.** The formation of the amino acids aspartic acid and alanine using glutamic acid as the amine donor in transamination (GOT = glutamate oxaloacetate transaminase; GPT = glutamate pyruvate transaminase). The shaded areas show the parts of the molecules that are changed by transamination reactions.

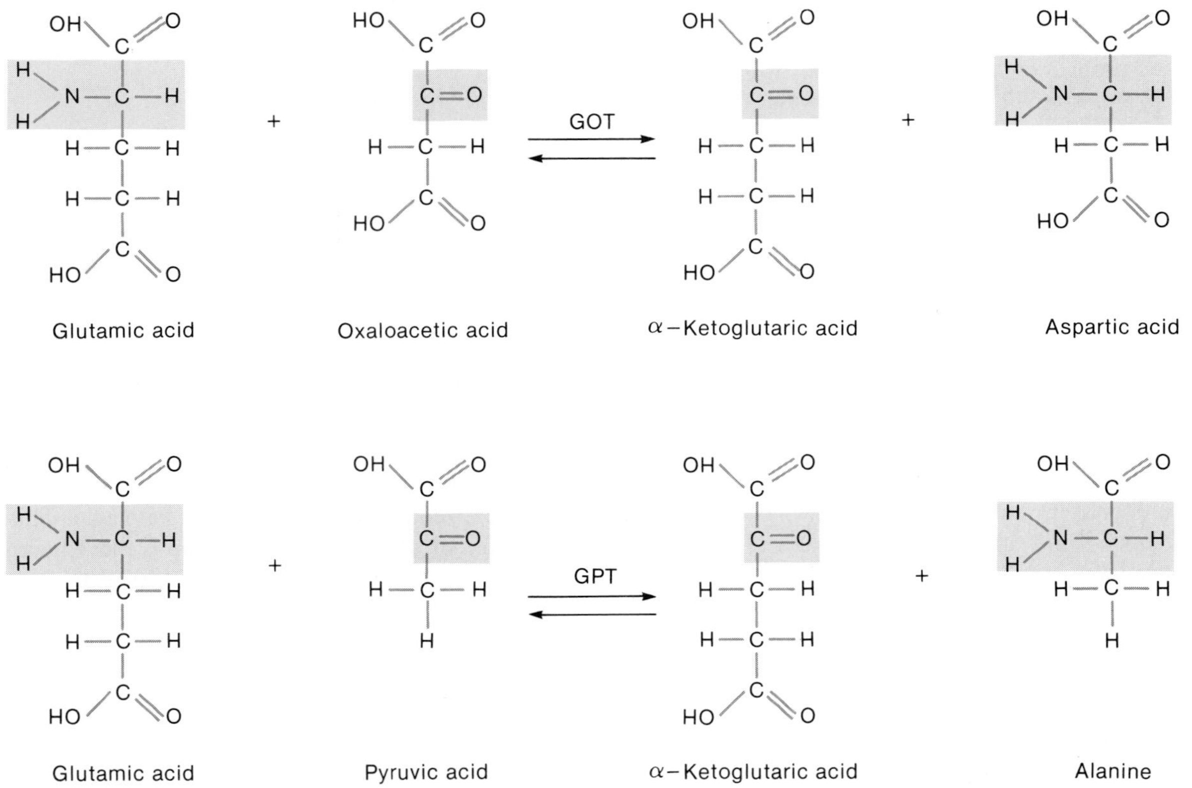

**Figure 5.16.** Oxidative deamination. Glutamic acid is converted to α-ketoglutaric acid as it donates its amine group to the metabolic pathway that results in the formation of urea.

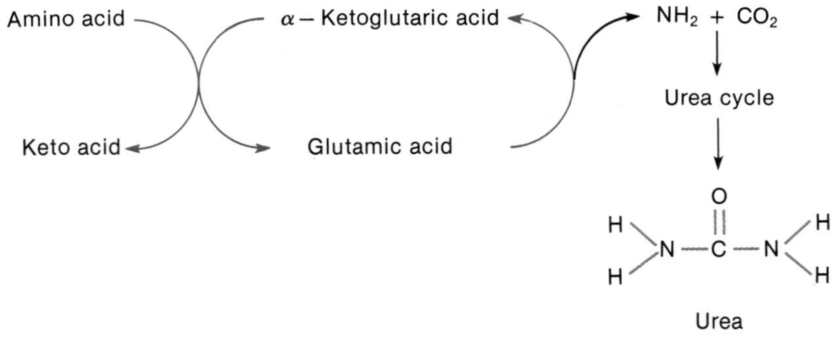

serves as a "funnel" through which other amino acids can be used to produce keto acids (pyruvic acid and Krebs cycle acids). These keto acids may then be used in the Krebs cycle as a source of energy (fig. 5.17).

Depending on which amino acid is deaminated, the keto acid left over may be either pyruvic acid or one of the Krebs cycle acids. These can be respired for energy, converted to fat, or converted to glucose. In the latter case, the amino acids are eventually changed to pyruvic acid, which is used to form glucose in the process of gluconeogenesis (described earlier in this chapter). Gluconeogenesis from amino acids occurs primarily in the liver, and is extremely important in maintaining blood glucose levels during prolonged fasting and exercise. The possible interrelationships between amino acids, carbohydrates, and fat are illustrated in figure 5.18.

Figure 5.17. Pathways by which amino acids can be catabolized for energy. These pathways are indirect for some amino acids, which must be transaminated into other amino acids before being converted into keto acids by deamination.

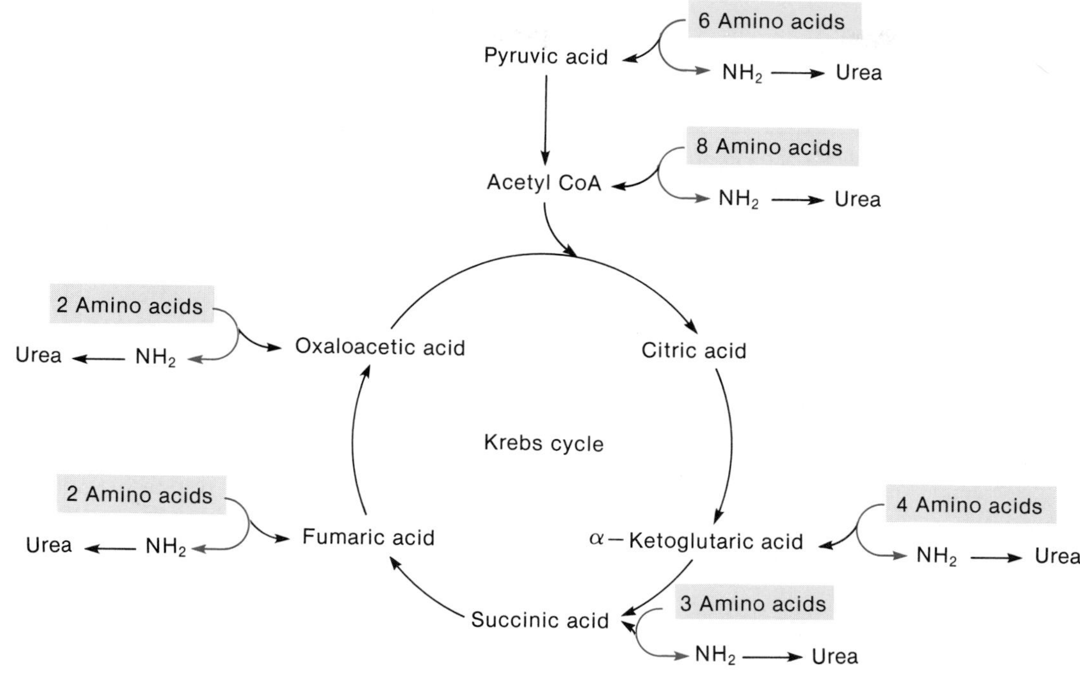


Figure 5.18. Simplified metabolic pathways showing how glycogen, fat, and protein can be interconverted.

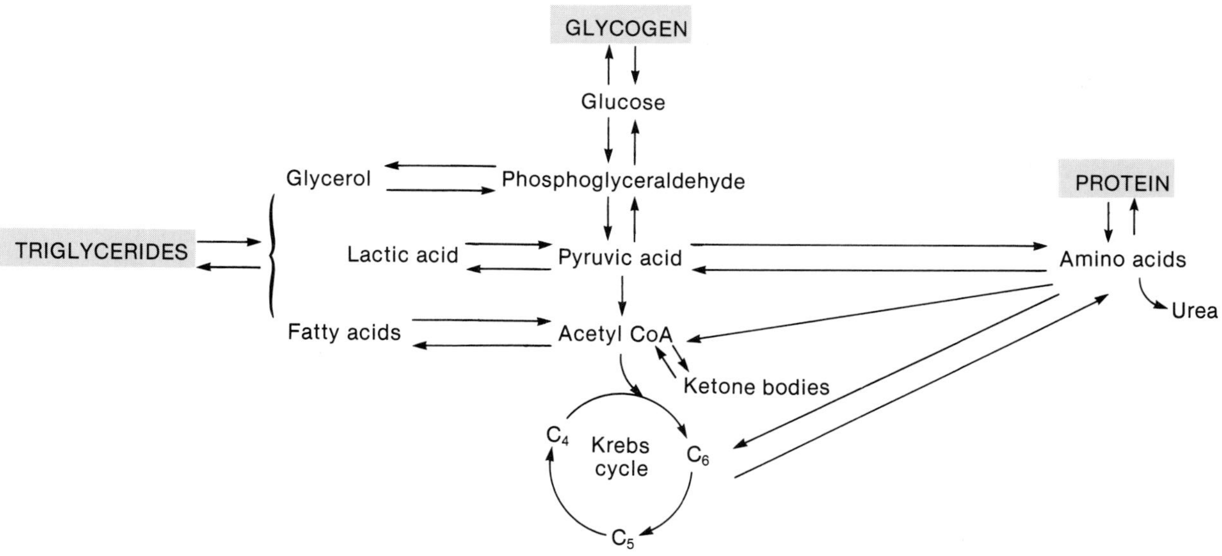


## Uses of Different Energy Sources

The blood serves as a common trough from which all the cells in the body are fed. If all cells used the same energy source, such as glucose, this source would quickly be depleted and cellular starvation would occur. This is normally prevented by the fact that the blood contains different energy sources: glucose and ketone bodies from the liver, fatty acids from adipose tissue, and lactic acid and amino acids from muscles. Some organs preferentially use one energy source more than the others, so that each energy source is "spared" for organs with strict energy needs.

Under normal conditions, for example, the brain uses blood glucose as its major energy source. This glucose is supplied primarily by the liver, which secretes about 150 milligrams of glucose into the blood per minute. About

**Table 5.4** Relative importance of different molecules in blood to the energy yield in various tissues

| Organ | Glucose | Fatty Acids | Ketone Bodies | Lactic Acid |
|---|---|---|---|---|
| Brain | + + + | + | − | − |
| Skeletal muscles (resting) | + | + + + | + | − |
| Liver | + | + + + | + + | + |
| Heart | + | + | + | + + + |

**Figure 5.19.** The relative contributions of anaerobic and aerobic respiration to the total energy in a well-trained person performing at maximal effort.

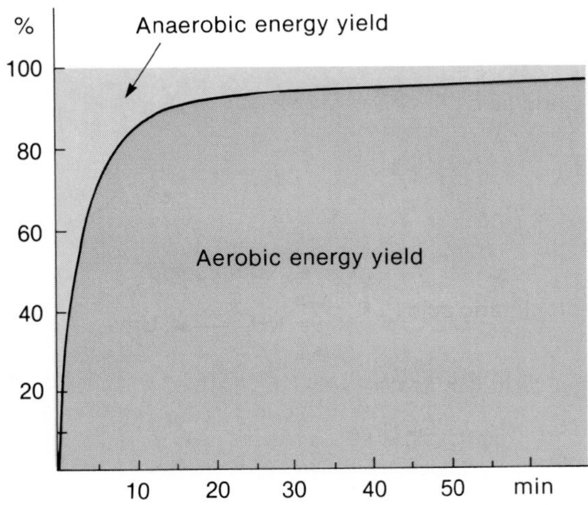


75% of this glucose is supplied by glycogenolysis and 25% is produced in the liver by gluconeogenesis, using amino acids, lactic acids, and glycerol as initial substrates. Many organs spare glucose by using fatty acids, ketone bodies, and lactic acid as energy sources (table 5.4).

## Metabolism of Skeletal Muscles

The metabolism of skeletal muscles, during rest and during exercise, can provide a summary and a practical example of some of the general principles described in this chapter. Skeletal muscle metabolism is of great significance because of the large contribution of skeletal muscles to total body oxygen requirements, particularly during exercise, and because knowledge of muscle metabolism is helpful to people interested in exercise and physical fitness.

***Anaerobic and Aerobic Respiration.*** Skeletal muscles respire anaerobically for the first forty-five to ninety seconds of moderate-to-heavy exercise, because the energy requirement (need for ATP) increases faster than the rate of oxygen supply by the cardiopulmonary system. If exercise is moderate and the person is in good physical condition, aerobic respiration contributes the major portion of the skeletal muscle energy requirements following the first two minutes of exercise (fig. 5.19).

The maximum rate of oxygen consumption (by aerobic respiration) in the body is called the **maximal oxygen uptake.** The maximal oxygen uptake in a given person is determined primarily by the person's age, size, and sex. It is about 15%–20% higher for males than for females and highest at age twenty for both sexes. Some world-class athletes have maximal oxygen uptakes that are twice the average for their age and sex—this appears to be due largely to genetic factors, but training can increase this value by about 20%.

When the oxygen uptake during exercise is about 60% of the maximum, anaerobic respiration contributes a small share of the energy required and some lactic acid is produced. The extent of anaerobic respiration and lactic acid production subsequently increases proportionately as more strenuous exercise results in the attainment of higher percentages of the maximal oxygen uptake. Since lactic acid contributes to muscle fatigue (described in chapter 12), people with higher maximal oxygen uptakes can achieve higher levels of exercise before fatigue than people with lower maximal oxygen uptakes.

When a person stops exercising, the rate of oxygen uptake does not immediately go back to pre-exercise levels; it returns slowly (the person continues to breathe heavily for some time afterwards). This extra oxygen is used to repay the **oxygen debt** incurred during exercise. The oxygen debt includes oxygen that was withdrawn from savings deposits—hemoglobin in blood and myoglobin in muscle (chapter 12), the extra oxygen required for metabolism by tissues warmed during exercise, and the oxygen needed for the metabolism of lactic acid, which was produced during anaerobic respiration. Under aerobic conditions lactic acid can be converted to pyruvic acid, which can serve as a substrate for aerobic respiration (as in the heart) or a source of new glucose (in the liver).

***Sources of Energy During Rest and Exercise.*** Skeletal muscles at rest use fatty acids as their major energy source. Metabolism changes at the beginning of exercise, when glucose-1-phosphate derived from stored muscle glycogen becomes the prime source of energy. At this time the energy contribution from blood glucose is still low.

After exercise has continued for ten to forty minutes, however, the metabolic pattern changes as muscle glycogen becomes depleted. The contribution of blood glucose to the energy requirements of exercising muscles

**Table 5.5** Relative energy yield from different sources for skeletal muscles during different phases of exercise

| Phase | Free Fatty Acids | Glucose from Muscle Glycogen | Blood Glucose | Liver Glucose Output (%) | |
|---|---|---|---|---|---|
| | | | | From Glycogenolysis | From Gluconeogenesis |
| Rest | + + + | − | + | 75% | 25% |
| 5–10 min | − | + + + | + | Decreases | Increases |
| 10–40 min | + | + + | + + | Decreases | Increases |
| 40–90 min | + + | + | + + + | Decreases | Increases |
| By 4 hours | + + + | − | + + | 55% | 45% |

increases to about 40%. This increased demand for blood glucose is met by the liver, which increases its rate of glycogenolysis two to five times over resting levels.

By forty minutes of exercise, the liver has added about 18 grams of glucose to the blood—roughly 25% of its stored glycogen. Liver glycogenolysis continues beyond this time, but much of the burden is removed by the increased use of fatty acids as an energy source for the exercising muscles. Since some consumption of blood glucose continues, however, the blood glucose concentration is slightly decreased by ninety minutes of exercise. Despite this slight decrease, true hypoglycemia (low blood glucose) is rare, which is fortunate because the glucose requirements of the brain do not change with exercise.

As glycogen stored in the liver becomes depleted during prolonged exercise, the contribution of gluconeogenesis to liver glucose output increases from 25% (at rest) to 45%. By four hours of exercise, the liver contains only about 25% of its original glycogen stores, and the muscles rely increasingly on free fatty acids as an energy source (table 5.5).

During and immediately following exercise, the lactic acid produced by skeletal muscles can be metabolized aerobically by the heart for energy and used by the liver as a substrate for gluconeogenesis (this is part of the Cori cycle). A recovery period of about forty-eight hours is needed before the muscle glycogen that was depleted during prolonged exercise is completely restored.

1. *Construct a flowchart to show the metabolic pathway by which glucose can be converted to fat. Indicate only the major intermediates involved (not all of the steps of glycolysis).*
2. *Define the terms* lipolysis *and* $\beta$-oxidation, *and explain in general terms how fat can be used for energy.*
3. *Describe transamination and deamination, and explain their functional significance.*
4. *List five blood-borne energy carriers, and explain in general terms how these are used as sources of energy.*
5. *Define the terms* glycogenolysis *and* gluconeogenesis, *and explain how these processes help to maintain constant blood glucose concentrations during exercise.*

## Summary

### Glycolysis and Anaerobic Respiration p. 102

I. Glycolysis refers to the conversion of glucose to two molecules of pyruvic acid.
   A. In the process, two molecules of ATP are consumed and four molecules of ATP are formed; there is thus a net gain of two ATP.
   B. In the steps of glycolysis two pairs of hydrogens are released; electrons from these reduce two molecules of NAD.
II. When respiration is anaerobic, reduced NAD is oxidized by pyruvic acid, which accepts two hydrogen atoms and is thereby reduced to lactic acid.
   A. Skeletal muscles use anaerobic respiration and thus produce lactic acid during the exercise; heart muscle only respires anaerobically for a short time when there is ischemia.
   B. Lactic acid can be converted to glucose in the liver by a process called gluconeogenesis.

### Aerobic Respiration p. 106

I. The Krebs cycle begins when coenzyme A donates acetic acid to an enzyme reaction that adds it to oxaloacetic acid to form citric acid.
   A. Acetyl CoA is formed from pyruvic acid by the removal of carbon dioxide and two hydrogens.
   B. The formation of citric acid begins a cyclic pathway that ultimately forms a new molecule of oxaloacetic acid.
   C. As the Krebs cycle progresses, one ATP is formed, and three molecules of NAD and one of FAD are reduced by hydrogens from the Krebs cycle.
II. Reduced NAD and FAD donate their electrons to an electron transport chain of molecules located in the cristae.
   A. The electrons from NAD and FAD are passed from one cytochrome to the next in the electron transport chain in a series of coupled oxidation-reduction reactions.

B. As each cytochrome iron gains an electron it becomes reduced, and as it passes the electron to the next cytochrome it becomes oxidized.
C. The last cytochrome becomes oxidized by donating its electron to oxygen, which functions as the final electron acceptor.
D. When one oxygen atom accepts two electrons and two protons, it becomes reduced to form water.
E. The energy provided by electron transport is used to form ATP from ADP and $P_i$, which is a process called oxidative phosphorylation.

III. Thirty-six to thirty-eight ATP are produced by the aerobic respiration of one glucose molecule; of these, two are produced in the cytoplasm by glycolysis and the remainder are produced in the mitochondria.

IV. The formation of glycogen from glucose is called glycogenesis, and the breakdown of glycogen is called glycogenolysis.
A. Glycogenolysis yields glucose-6-phosphate, which can enter the pathway of glycolysis.
B. The liver, but not skeletal muscles, contains an enzyme that can produce free glucose from glucose-6-phosphate; the liver can thus secrete glucose derived from glycogen.

V. Carbohydrate metabolism is influenced by the availability of oxygen and by a negative feedback effect of ATP on glycolysis and the Krebs cycle.

### Metabolism of Lipids and Proteins p. 114

I. In lipolysis, triglycerides yield glycerol and fatty acids.
A. Glycerol can be converted to phosphoglyceraldehyde and used for energy.
B. In the process of the $\beta$-oxidation of fatty acid, a number of acetyl CoA molecules are produced.
C. Processes that operate in the reverse direction can convert glucose to triglycerides.

II. Amino acids can serve as sources of energy.
A. Through transamination, a particular amino acid and a particular keto acid (pyruvic acid or one of the Krebs cycle acids) can serve as substrates to form a new amino acid and a new keto acid.
B. In oxidative deamination, amino acids are converted into keto acids as their amino group is incorporated into urea.

III. Each organ uses certain blood-borne energy carriers as their preferred energy source.
A. The brain has an almost absolute requirement for blood glucose as its energy source.
B. During exercise, the needs of skeletal muscles for blood glucose can be met by glycogenolysis and by gluconeogenesis in the liver.

IV. Skeletal muscles respire anaerobically during the early stages of exercise and as the maximal oxygen uptake is approached during heavy exercise.
A. Resting skeletal muscles respire aerobically using fatty acids as their preferred energy source.
B. During exercise, the skeletal muscles derive glucose-1-phosphate from their stored glycogen and from the blood.
C. The liver secretes glucose into the blood during exercise; this glucose is first derived primarily from glycogenolysis and then primarily from gluconeogenesis.

## *Review Activities*

### Objective Questions

1. The net gain of ATP per glucose molecule in anaerobic respiration is _____ ; the net gain in aerobic respiration is _____ .
   (a) 2;4
   (b) 2;38
   (c) 38;2
   (d) 24;30
2. In anaerobic respiration, the oxidizing agent for NADH (that is, the molecule that removes electrons from NADH) is
   (a) pyruvic acid
   (b) lactic acid
   (c) citric acid
   (d) oxygen
3. When organs respire anaerobically, there is an increased blood concentration of
   (a) oxygen
   (b) glucose
   (c) lactic acid
   (d) ATP
4. The conversion of lactic acid to pyruvic acid occurs
   (a) in anaerobic respiration
   (b) in the heart, where lactic acid is aerobically respired
   (c) in the liver, where lactic acid can be converted to glucose
   (d) in both *a* and *b*
   (e) in both *b* and *c*
5. The oxygen in the air we breathe
   (a) functions as the final electron acceptor of the electron transport chain
   (b) combines with hydrogen to form water
   (c) combines with carbon to form $CO_2$
   (d) both *a* and *b*
   (e) both *a* and *c*
6. In terms of the number of ATP molecules directly produced, the major energy-yielding process in the cell is
   (a) glycolysis
   (b) the Krebs cycle
   (c) oxidative phosphorylation
   (d) gluconeogenesis
7. Ketone bodies are derived from
   (a) fatty acids
   (b) glycerol
   (c) glucose
   (d) amino acids
8. The conversion of glycogen to glucose-6-phosphate occurs in the
   (a) liver
   (b) skeletal muscles
   (c) both of these organs
9. The conversion of glucose-6-phosphate to free glucose, which can be secreted into the blood, occurs in
   (a) the liver
   (b) the skeletal muscles
   (c) both of these organs
10. The formation of glucose from pyruvic acid that is derived from lactic acid, amino acids, or glycerol is called
   (a) glycogenesis
   (b) glycogenolysis
   (c) glycolysis
   (d) gluconeogenesis

11. Which of the following organs has an almost absolute requirement for blood glucose as its energy source? The
    (a) liver
    (b) brain
    (c) skeletal muscles
    (d) heart
12. When amino acids are used as an energy source
    (a) oxidative deamination occurs
    (b) pyruvic acid or one of the Krebs cycle acids (keto acids) are formed
    (c) urea is produced
    (d) all of the above occur
13. Intermediates formed during fatty acid metabolism can enter the Krebs cycle as
    (a) keto acids
    (b) acetyl CoA
    (c) Kreb's cycle molecules
    (d) pyruvic acid

## Essay Questions

1. Explain the advantages and disadvantages of anaerobic respiration.
2. What purpose is served by the formation of lactic acid during anaerobic respiration? How is this purpose achieved during aerobic respiration?
3. The poison cyanide blocks the transfer of electrons from the last cytochrome to oxygen. Describe the effect of this poison on oxidative phosphorylation and on the Krebs cycle, and explain why this poison is deadly.
4. Describe the metabolic pathway by which glucose can be converted into fat, and explain how end-product inhibition by ATP can favor this pathway.
5. Describe the metabolic pathway by which fat can be used as a source of energy, and explain why the metabolism of fatty acids can yield more ATP than the metabolism of glucose.
6. Explain how energy is obtained from the metabolism of amino acids. Why does a starving person have high concentrations of urea in the blood?
7. Explain why the liver is the only organ able to secrete glucose into the blood, and describe the possible sources of this hepatic glucose.
8. Describe the metabolism of glucose and fatty acids by resting and exercising skeletal muscles, and explain the functional significance of fatty acid metabolism by muscles.

## ***Selected Readings***


Baker, J. J. W., and G. E. Allen. 1982. *Matter, energy, and life.* 4th ed. Mass.: Addison-Wesley.

Berry, H. K. March 1984. The spectrum of metabolic disorders. *Diagnostic Medicine,* p. 39.

Dyson, R. D. 1978. *Essentials of cell biology.* 2d ed. Boston: Allyn & Bacon.

Hinkle, P., and R. E. McCarty. March 1979. How cells make ATP. *Scientific American.*

Lehninger, A. L. September 1961. How cells transform energy. *Scientific American.*

Lehninger, A. L. 1982. *Principles of biochemistry.* New York: Worth.

Nadel, E. R. 1985. Physiological adaptations to aerobic training. *American Scientist* 73:334.

Sheeler, P., and D. E. Bianchi. 1980. *Cell biology: structure, biochemistry, and function.* New York: John Wiley & Sons.

Stryker, L., 1981. *Biochemistry,* 2d ed. New York: Freeman.

# 6 Membrane Transport and the Membrane Potential

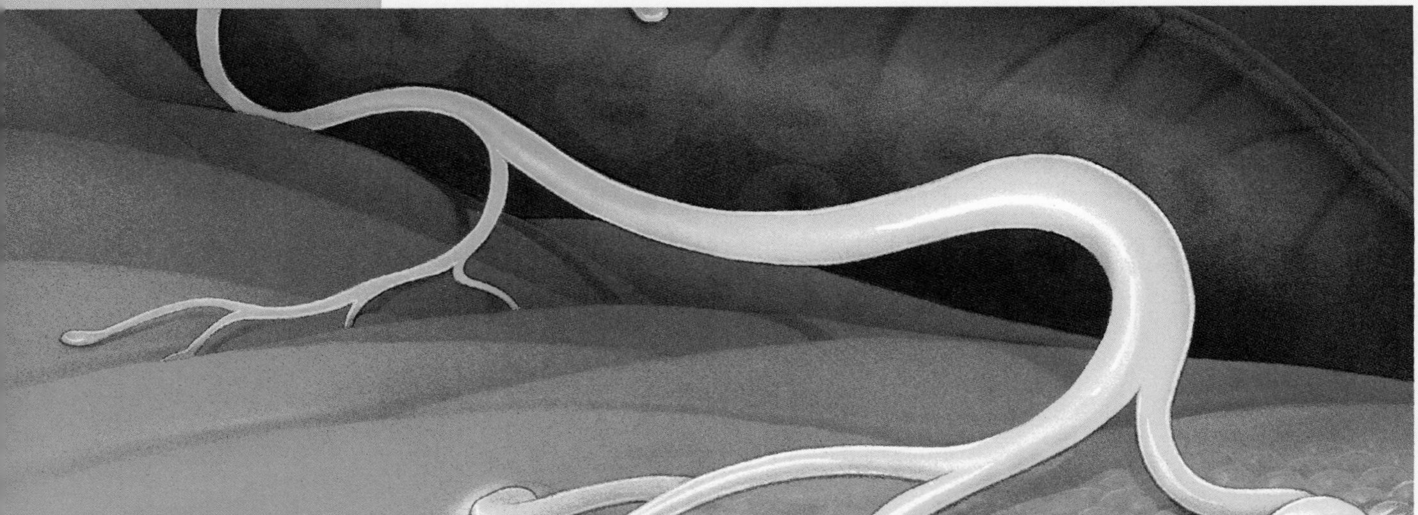

## Objectives

By studying this chapter, you should be able to

1. describe diffusion and explain its physical basis
2. explain how nonpolar molecules, inorganic ions, and water can diffuse through a cell membrane
3. describe the factors that influence the rate of diffusion through cell membranes
4. define osmosis and describe the conditions required for osmosis to occur
5. explain the meanings of the terms *osmolality* and *osmotic pressure* and how these factors relate to osmosis
6. define the terms *isotonic, hypertonic,* and *hypotonic* and explain their physiological significance
7. describe the characteristics of carrier-mediated transport
8. describe the facilitated diffusion of glucose through cell membranes, and give examples of where this occurs in the body
9. define active transport and explain how active transport may be accomplished
10. explain the active transport of $Na^+$ and $K^+$ by the $Na^+/K^+$ pump and the physiological significance of this activity
11. explain how an equilibrium potential is produced when only one ion is able to diffuse through a cell membrane
12. explain why the resting membrane potential is slightly different than the potassium equilibrium potential, and how the resting membrane potential is affected by the extracellular potassium concentration
13. explain the role of the $Na^+/K^+$ pump in the maintenance of the resting membrane potential

## Outline

## Diffusion and Osmosis

Net diffusion of a molecule or ion through a cell membrane always occurs in the direction of its lower concentration. Nonpolar molecules can penetrate the phospholipid barrier, and small inorganic ions can pass through channels in the membrane. The net diffusion of water through a membrane is known as osmosis.

The cell (plasma) membrane separates the intracellular environment from the extracellular environment. Proteins, nucleotides, and other molecules needed for the structure and function of the cell cannot penetrate, or "permeate," the membrane. The cell membrane is, however, **selectively permeable** to certain molecules and many ions; this allows two-way traffic in nutrients and wastes needed to sustain metabolism and provides electrical currents created by the movements of ions through the membrane.

The mechanisms involved in the transport of molecules and ions through the cell membrane may be divided into two categories: (1) transport that requires the action of specific *carrier proteins* in the membrane (*carrier-mediated transport*); and (2) transport through the membrane that is not carrier-mediated. Carrier-mediated transport includes *facilitated diffusion* and *active transport;* non-carrier-mediated transport consists of the *simple diffusion* of ions, lipid-soluble molecules, and water through the membrane. The diffusion of water (solvent) through a membrane is called *osmosis.*

### Diffusion

Molecules in a gas and molecules and ions dissolved in a solution are in a constant state of random motion as a result of their thermal (heat) energy. This random motion, called **diffusion,** tends to make the gas or solution evenly mixed, or diffusely spread out, within a given volume. Whenever a *concentration difference,* or *concentration gradient,* exists between two parts of a solution, therefore, random molecular motion tends to abolish the gradient and to make the molecules uniformly distributed (fig. 6.1). In terms of the second law of thermodynamics, the concentration difference represents an unstable state of high organization (low entropy), which changes to produce a uniformly distributed solution with maximum disorganization (high entropy).

As a result of random molecular motion, molecules in the part of the solution with a higher concentration will enter the area of lower concentration. Molecules will also move in the opposite direction, but not as frequently. As a result, there will be a *net movement* from the region of higher to the region of lower concentration until the concentration difference is abolished. This net movement is called **net diffusion.** Net diffusion is a physical process that occurs whenever there is a concentration difference; when the concentration difference exists across a membrane, diffusion becomes a type of membrane transport.

**Figure 6.1.** Net diffusion occurs when there is a concentration difference (or concentration gradient) between two regions of a solution (*a*) provided that the membrane separating these regions is permeable to the diffusing substance. Diffusion tends to equalize the concentration of these solutions (*b*) and thus to abolish the concentration differences.

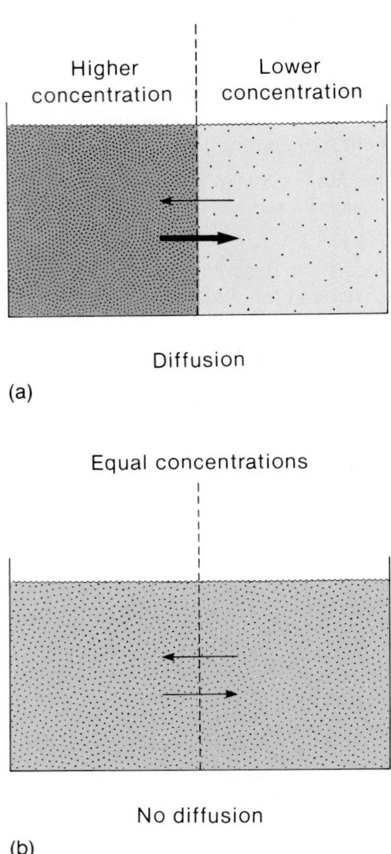


Normal functioning kidneys remove waste products from the blood. After the blood is filtered through pores in capillary walls—which are large enough to permit the passage of wastes and other molecules—the molecules needed by the body are reabsorbed back into the blood. The wastes generally remain in the filtrate and are excreted in the urine. When the kidneys do not function properly, waste molecules can be removed from the blood artificially by a process called **dialysis.** Dialysis refers to the process of removing particular molecules from a solution by having them pass, by means of diffusion, through an artificial porous membrane. Since the pores in this dialysis membrane are large enough to permit the passage of some molecules but too small to permit the passage of others (the plasma proteins), small waste molecules can be removed from the blood by this technique.

**Figure 6.2.** Gas exchange between the intracellular and extracellular compartments occurs by diffusion. The regions of higher concentration are represented by the larger symbols.

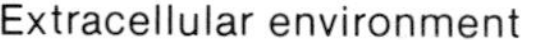


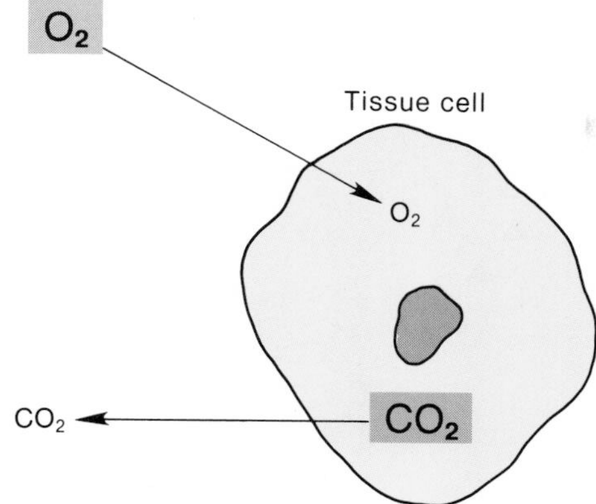


## Diffusion through the Cell Membrane

Since the cell membrane consists primarily of a double layer of phospholipids, molecules that are nonpolar and thus lipid-soluble can easily pass from one side of the membrane to the other. The cell membrane, in other words, does not present a barrier to the diffusion of nonpolar molecules such as oxygen gas ($O_2$) or steroid hormones. Small organic molecules that have polar covalent bonds but are uncharged, such as $CO_2$ (as well as ethanol and urea), are also able to penetrate the double-lipid layers. Net diffusion of these molecules can thus easily occur between the intracellular and extracellular compartments when concentration gradients are present.

The oxygen concentration is relatively high, for example, in the extracellular fluid because oxygen is carried from the lungs to the body tissues by the blood. Since oxygen is converted to water in aerobic cell respiration, the oxygen concentration within the cells is lower than in the extracellular fluid. The concentration gradient for carbon dioxide is in the opposite direction because cells produce $CO_2$. *Gas exchange* thus occurs by diffusion between the tissue cells and their extracellular environments (fig. 6.2).

Although water is not lipid-soluble, water molecules can diffuse through the cell membrane because of their small size and lack of net charge. The passage of water through the membrane may also be aided by the dipolar nature of water molecules (chapter 2), which permits interactions between the water molecules and the negative charges of the membrane phospholipids. In certain membranes the passage of water is restricted to specific channels that can open or close in response to physiological regulation. The net diffusion of water molecules across the membrane (called *osmosis*) occurs when the membrane is permeable to water and when the solution on one side of the membrane is more dilute (has a higher water concentration) than on the other side of the membrane.

Larger polar molecules, such as glucose, cannot pass through the double phospholipid layers of the membrane and thus require special *carrier proteins* in the membrane for transport (described later). The phospholipid portion of the membrane is similarly impermeable to charged inorganic ions, such as $Na^+$ and $K^+$. Passage of these ions through the cell membrane may be permitted by tiny **ion channels** through the membrane that are too small to be seen even with an electron microscope. Many scientists believe that these channels are provided by some of the *integral proteins* that span the thickness of the membrane (fig. 6.3).

## Rate of Diffusion

The rate of diffusion, measured by the number of diffusing molecules passing through the membrane per unit time, depends on (1) the magnitude of the concentration difference across the membrane (the "steepness" of the concentration gradient); (2) the permeability of the membrane to the diffusing substances; and (3) the surface area of the membrane through which the substances are diffusing.

The magnitude of the concentration difference across the membrane serves as the driving force for diffusion. Regardless of this concentration difference, however, the diffusion of a substance across a membrane will not occur if the membrane is not permeable to that substance. With a given concentration difference, the rate of diffusion through a membrane will vary directly with the degree of permeability. In a resting neuron, for example, the membrane is about twenty times more permeable to potassium ($K^+$) than to sodium ($Na^+$), and as a consequence, $K^+$ diffuses much more rapidly than does $Na^+$. Changes in the protein structure of the membrane channels, however, can change the permeability of the membrane. This occurs

**Figure 6.3.** Inorganic ions (such as $Na^+$ and $K^+$) may penetrate the membrane through pores within integral proteins that span the thickness of the double phospholipid layers.

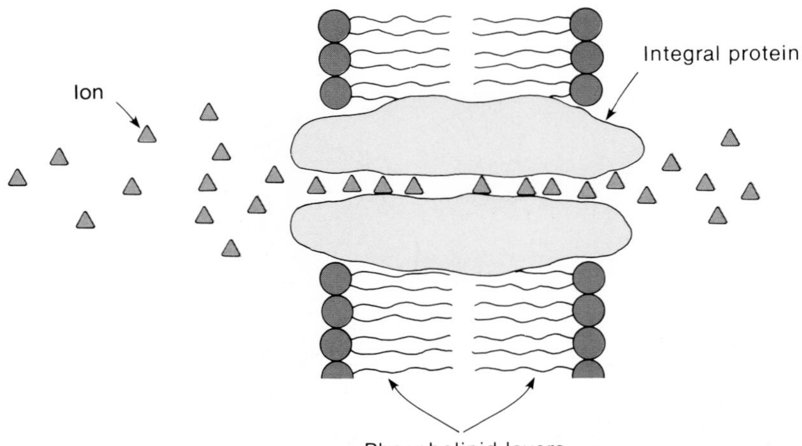


during the production of a nerve impulse, when specific stimulation opens $Na^+$ channels temporarily and allows a faster diffusion rate for $Na^+$ than for $K^+$.

In areas of the body that are specialized for rapid diffusion the surface area of the cell membranes may be increased by numerous folds. The rapid passage of the products of digestion across the epithelial membranes in the intestine, for example, is aided by such structural adaptations. The surface area of the apical membranes (the part facing the lumen) in the intestine is increased by many tiny folds that form fingerlike projections called **microvilli** (fig. 6.4). Similar microvilli are also found in the kidney tubule epithelium, which must reabsorb various molecules that are filtered out of the blood.

## Osmosis

Suppose that a cylinder is divided into two equal compartments by a membrane partition that can freely move and that one compartment initially contains 180 g/L (grams per liter) of glucose and the other compartment contains 360 g/L of glucose. If the membrane is permeable to glucose, glucose will diffuse from the 360 g/L compartment to the 180 g/L compartment until both compartments contain 270 g/L of glucose. If the membrane is not permeable to glucose but is permeable to water, the same result (270 g/L solutions on both sides of the membrane) will be achieved by the diffusion of water. As water diffuses from the 180 g/L compartment to the 360 g/L compartment, the former solution would become more concentrated while the latter becomes more dilute. This is accompanied by volume changes, as illustrated in figure 6.5.

**Osmosis** is the net diffusion of water (the solvent) across the membrane. In order for this to occur, the membrane must be *semipermeable;* that is, it must be more permeable to water molecules than to solutes. Like the diffusion of solute molecules, the diffusion of water occurs when the water is more concentrated on one side of the membrane than on the other side; that is, when one solution is more dilute than the other (fig. 6.6). The more dilute solution has a higher concentration of water molecules and a lower concentration of solute. The principles of osmosis apply to the diffusion of any molecule, but the terminology is backwards because the term *concentration* is usually used to refer to the density of solute rather than solvent molecules.

In order for osmosis to occur between two solutions, the two solutions must have different concentrations, and the membrane must be relatively impermeable to the solutes that produce the differences in concentration. Such impermeable solutes are said to be **osmotically active.** Water, for example, returns from tissue fluid to blood capillaries as a result of the fact that the protein concentration of blood plasma is higher than the protein concentration of tissue fluid. This occurs because the plasma proteins, in contrast to other plasma solutes, cannot pass from the capillaries into the tissue fluid. The plasma proteins, in this case, are osmotically active. When clinicians want to expand a patient's blood volume (to raise the blood pressure), therefore, they give intravenous infusions of an albumin solution or of plasma (which contains albumin and other proteins). If a person has an abnormally low concentration of plasma proteins, as may occur in liver disease (cirrhosis, for example), fluid may accumulate in the tissues and produce edema.

**Figure 6.4.** Microvilli (*MV*) in the small intestine, as seen with the transmission (*a*) and scanning (*b*) electron microscope.

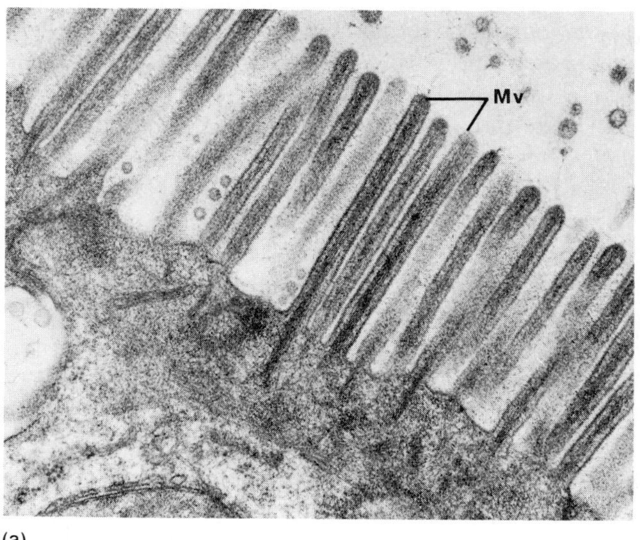


(a)

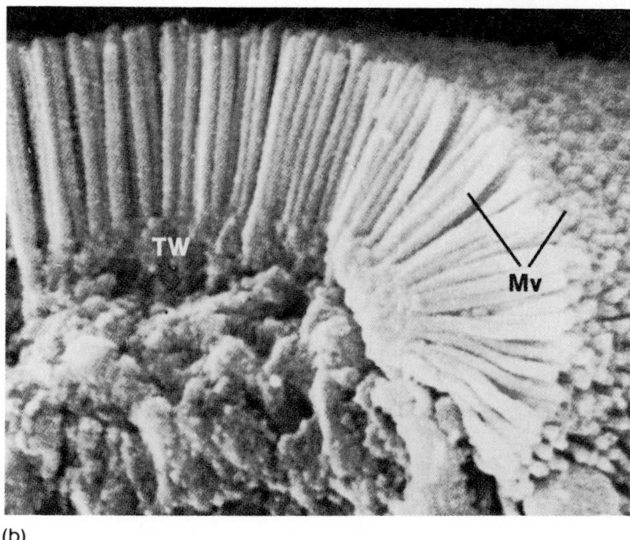


(b)

**Figure 6.5.** A model of osmosis, or the net movement of water from the solution of lesser solute concentration to the solution of greater solute concentration.

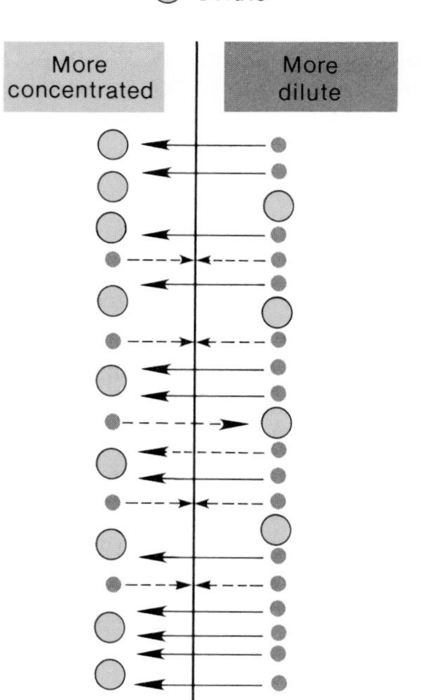


**Figure 6.6.** A movable semipermeable membrane (permeable to water but not glucose) separates two solutions of different glucose concentration (*a*). As a result, water moves by osmosis into the solution of greater concentration until (*b*) the volume changes equalize the concentrations on both sides of the membrane.

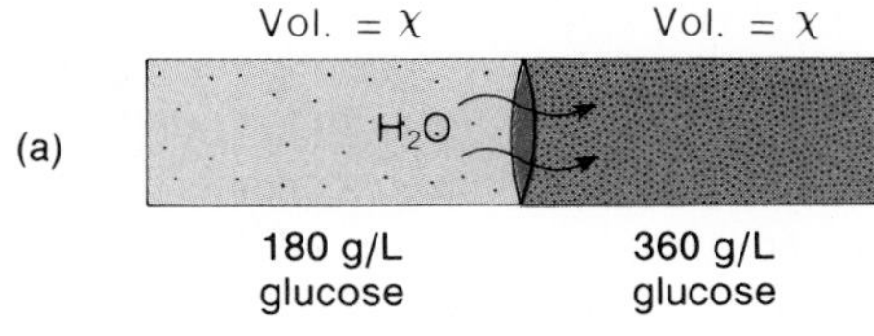


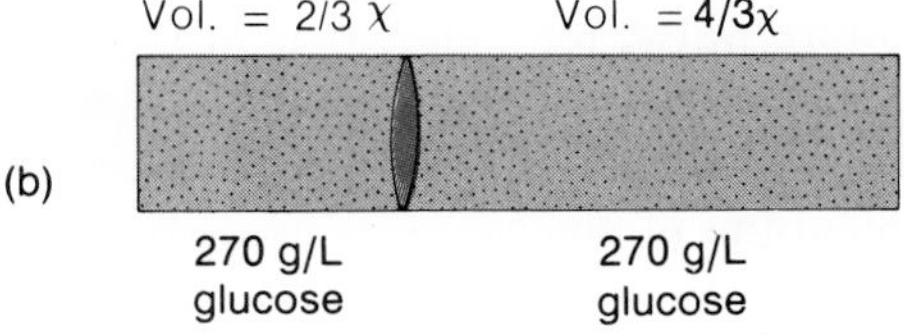


***Osmotic Pressure.*** Osmosis and the movement of the membrane partition could be prevented by an opposing force. If one compartment contained 180 g/L of glucose and the other compartment contained pure water, the osmosis of water into the glucose solution could be prevented by pushing against the membrane with a certain force (equal to 22.4 atmospheres pressure, in this example). This is illustrated in figure 6.7.

The force that would have to be exerted to prevent osmosis in this situation is the **osmotic pressure** of the solution. This backwards measurement indicates how strongly the solution "draws" water into it by osmosis. The greater the solute concentration of a solution, the greater its osmotic pressure. Pure water, thus, has an osmotic pressure of zero, and a 360 g/L glucose solution has twice the osmotic pressure of a 180 g/L glucose solution.

**Figure 6.7.** If a semipermeable membrane separates pure water from a 180 g/L glucose solution, water tends to move by osmosis into the glucose solution, thus creating a hydrostatic pressure that pushes the membrane to the left and expands the volume of the glucose solution. The amount of pressure that must be applied to just counteract this volume change is equal to the osmotic pressure of the glucose solution.

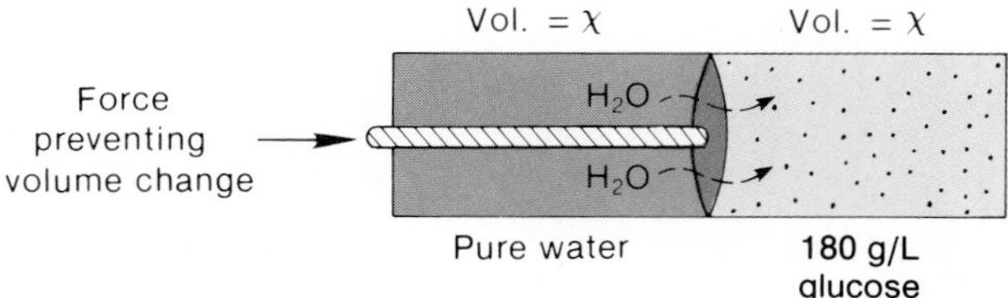


***Molarity and Molality.*** Glucose is a monosaccharide with a molecular weight of 180 (the sum of its atomic weights). Sucrose is a disaccharide of glucose and fructose, which have molecular weights of 180 each. When glucose and fructose join together by dehydration synthesis to form sucrose, a molecule of water (molecular weight = 18) is split off. Therefore, sucrose has a molecular weight of 342 (the sum of 180 + 180 − 18); each sucrose molecule weighs 342/180 times as much as each glucose molecule. It follows that 342 grams of sucrose must contain the same number of molecules as 180 grams of glucose.

Notice that the molecular weight in grams of any compound must contain the same number of molecules as the gram molecular weight of any other compound. This unit of weight is called a *mole,* and it always contains $6.02 \times 10^{23}$ molecules (**Avogadro's number**). One mole of solute dissolved in water to make one liter of solution is described as a one-**molar** solution (abbreviated 1.0 M). Although this unit of measurement is commonly used in chemistry, it is not completely desirable in discussions of osmosis because the exact ratio of solute to water is not specified. More water, for example, is needed to make a 1.0 M NaCl solution (where a mole of NaCl weighs 58.5 grams) than is needed to make a 1.0 M glucose solution, because 180 grams of glucose takes up more volume than 58.5 grams of salt.

Since the ratio of solute to water molecules is of critical importance in osmosis, a more desirable measurement of concentration is **molality.** In a one-molal solution (abbreviated 1.0 m), one mole of solute (180 grams of glucose, for example) is dissolved in one kilogram of water (equal to one liter at 4° C). A 1.0 m NaCl solution and a 1.0 m glucose solution both contain a mole of solute dissolved in exactly the same amount of water (fig. 6.8).

***Osmolality.*** If 180 grams of glucose and 180 grams of fructose were dissolved in the same kilogram of water, the osmotic pressure of the solution would be the same as that of a 360 g/L glucose solution. Osmotic pressure depends on the ratio of solute to solvent, *not* on the chemical nature of the solute molecules. The expression for the total molality of a solution is **osmolality (Osm).** Thus, the solution of 1.0 m glucose plus 1.0 m fructose has a total molality, or *osmolality,* of 2.0 osmol/L (abbreviated 2.0 Osm). This is the same as the 360 g/L glucose solution, which is 2.0 m and 2.0 Osm (fig. 6.9).

Unlike glucose, fructose, and sucrose, electrolytes such as NaCl ionize when they dissolve in water. One molecule of NaCl dissolved in water yields two ions ($Na^+$ and $Cl^-$); one mole of NaCl ionizes to form one mole of $Na^+$ and one mole of $Cl^-$. Thus a 1.0 m NaCl solution has a total concentration of 2.0 Osm. The effect of this on osmosis is illustrated in figure 6.10.

***Measurement of Osmolality.*** Plasma and other biological fluids contain many organic molecules and electrolytes. The osmolality of such complex solutions can only be estimated by calculations. Fortunately, however, there is a relatively simple method for measuring osmolality. This method is based on the fact that the freezing point of a solution, like its osmotic pressure, is affected by the total concentration of the solution and not by the chemical nature of the solute.

One mole of solute depresses the freezing point of water by −1.86° C. Accordingly, a 1.0 m glucose solution freezes at a temperature of −1.86° C, and a 1.0 m NaCl solution freezes at a temperature of $2 \times -1.86 = -3.72°$ C, because of ionization. The *freezing point depression* is, thus, a measure of the osmolality. Since plasma freezes at about −0.56° C, its osmolality is equal to 0.56/1.86, or 0.3 Osm, which is more commonly indicated as 300 milliosmolal (or 300 mOsm).

***Tonicity.*** A 0.3 m glucose solution has the same osmolality and osmotic pressure as plasma. The same is true of a 0.15 m NaCl solution, which ionizes to produce a total concentration of 300 mOsm. Both of these solutions are used clinically as intravenous infusions, labeled *5% dextrose* (5 g of glucose per 100 ml, which is 0.3 m) and *normal saline* (0.9 g of NaCl per 100 ml, which is 0.15 m). Since 5% dextrose and normal saline have the same osmolality as plasma, they are said to be *isosmotic* to plasma. If these solutions are separated from plasma by a membrane that is permeable to water but not to glucose or NaCl, osmosis will not occur. In this case the solutions are said to be **isotonic** (*iso* = same; *tonic* = strength) to plasma.

Red blood cells placed in an isotonic solution will neither gain nor lose water. It should be noted that a solution may be isosmotic but not isotonic; such is the case whenever the solute in the isosmotic solution can freely penetrate the membrane. A 0.3 m urea solution, for example, is isosmotic but not isotonic because the cell membrane is permeable to urea. When red blood cells are placed in a

**Figure 6.8.** Diagram illustrating the difference between a one molar (1.0 M) and a one molal (1.0 m) glucose solution.

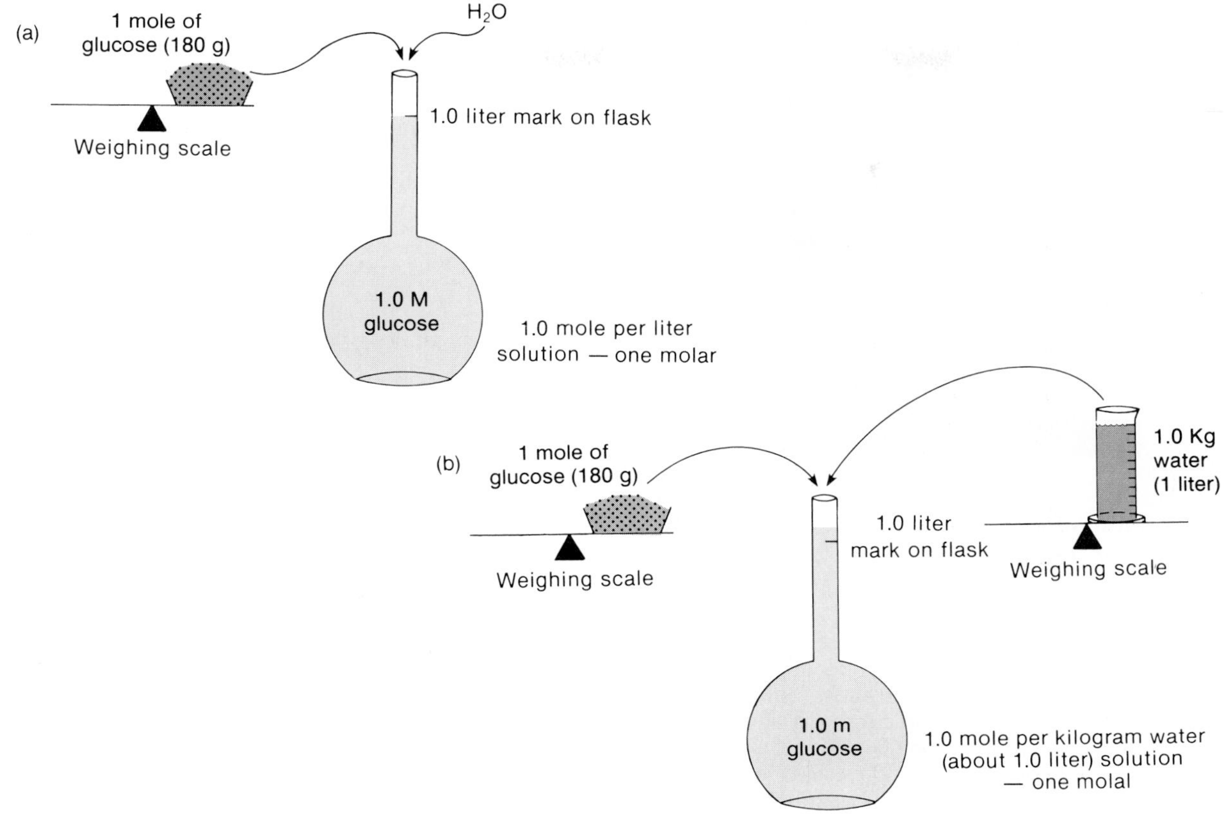


**Figure 6.9.** The osmolality (Osm) of a solution is equal to the sum of the molalities of each solute in the solution. If a semipermeable membrane separates two solutions with equal osmolalities, no osmosis will occur.

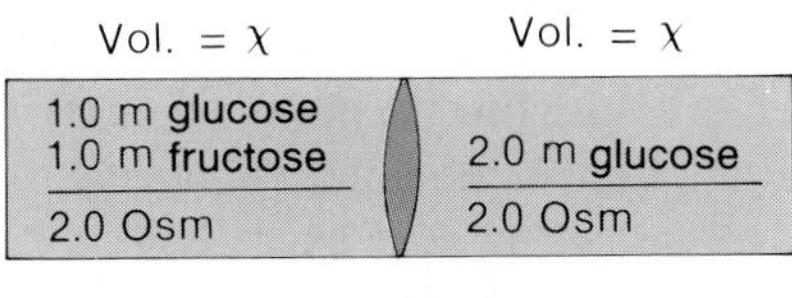


0.3 m urea solution, the urea diffuses into the cells until its concentration on both sides of the cell membranes becomes equal. Meanwhile, the solutes within the cells that cannot exit, and which are, therefore, osmotically active, cause osmosis of water into the cells. Red blood cells placed in 0.3 m urea will thus eventually burst.

Solutions that have a lower total concentration of osmotically active solutes and a lower osmotic pressure than plasma are said to be **hypotonic** to plasma. Red blood cells placed in hypotonic solutions gain water and may burst (*hemolysis*). When red blood cells are placed in a **hypertonic** solution (such as sea water), which has a higher osmolality and osmotic pressure than plasma, they shrink

**Figure 6.10.** If a semipermeable membrane (permeable to water but not to glucose, $Na^+$, or $Cl^-$) separates a 1.0 m glucose solution from a 1.0 m NaCl solution (*a*), water will move by osmosis into the NaCl solution. This is because NaCl can ionize to yield one molal $Na^+$ plus one molal $Cl^-$. After osmosis (*b*), the total concentration, or osmolality, of the two solutions are equal.

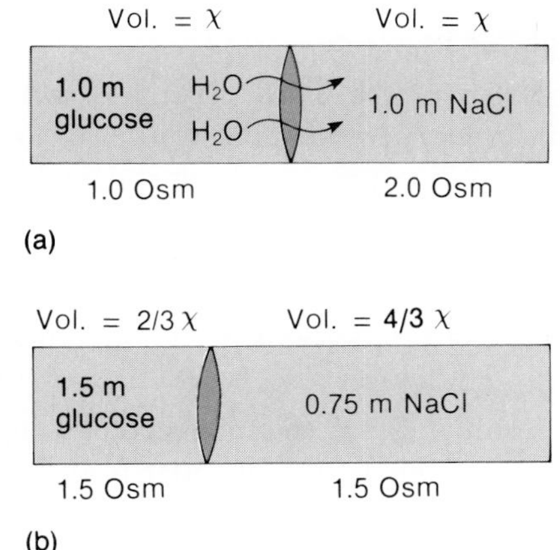

**Figure 6.11.** A scanning electron micrograph of normal and crenated red blood cells.

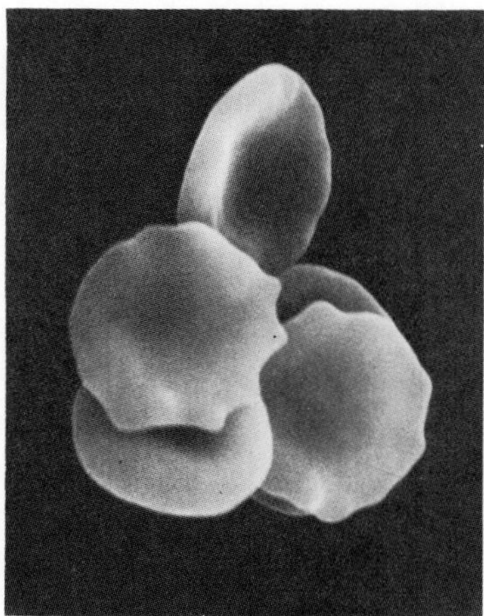

due to the osmosis of water out of the cells. In this process, called *crenation* (*crena* = notch), the cell surface becomes scalloped in appearance (fig. 6.11).

Fluids delivered intravenously must be isotonic to blood in order to maintain the correct osmotic pressure and prevent cells from either expanding or shrinking due to the gain or loss of water. Common fluids used for this purpose are normal saline and 5% dextrose, which, as previously described, have about the same osmolality as normal plasma (approximately 300 mOsm). Another isotonic solution frequently used in hospitals is *Ringer's Lactate,* which contains glucose and lactic acid in addition to a number of different salts. Isotonic solutions are also used in heart-lung machines, which take the place of the heart and lungs during open-heart surgery.

## Regulation of Blood Osmolality

The osmolality of the blood plasma is normally maintained within very narrow limits by a variety of regulatory mechanisms. When a person becomes dehydrated, for example, the blood becomes more concentrated as the total blood volume is reduced. The increased blood osmolality and osmotic pressure stimulates *osmoreceptors,* which are neurons located in a part of the brain called the hypothalamus.

As a result of increased osmoreceptor stimulation, the person becomes thirsty and, if water is available, drinks. Along with increased water intake, a person who is dehydrated excretes a lower volume of urine. This occurs as a result of the following sequence of events: (1) increased plasma osmolality stimulates osmoreceptors in the hypothalamus of the brain; (2) the osmoreceptors stimulate the

**Figure 6.12.** An increase in plasma osmolality (increased concentration and osmotic pressure) due to dehydration stimulates thirst and increased ADH secretion. These effects cause the person to drink more and urinate less. The blood volume, as a result, is increased while the plasma osmolality is decreased. These effects help bring the blood volume back to the normal range and complete the negative feedback loop (indicated by a negative sign).

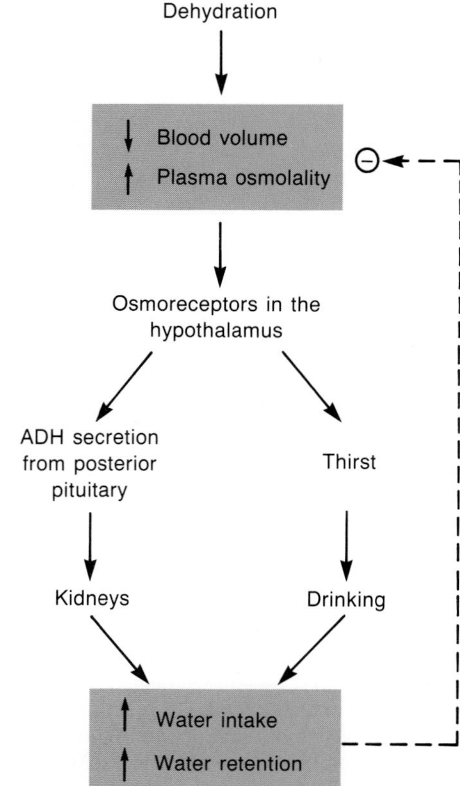


posterior pituitary gland, by means of a tract of nerve fibers, to secrete **antidiuretic hormone (ADH);** (3) ADH acts on the kidneys to promote water retention; so that (4) a lower volume of urine is excreted.

A person who is dehydrated, therefore, drinks more and urinates less. This represents a negative feedback loop (fig. 6.12), which acts to maintain homeostasis of the plasma concentration (osmolality) and, in the process, helps to maintain a proper blood volume.

1. *Define simple diffusion, and describe three factors that influence the diffusion rate.*
2. *Define the term* osmosis, *and describe the conditions required for it to occur; define the terms* osmolality *and* osmotic pressure.
3. *Define the terms* isotonic, hypotonic, *and* hypertonic, *and explain why hospitals use 5% dextrose and normal saline as intravenous infusions.*
4. *Explain how changes in the osmolality of plasma are detected and corrected by the body.*

## Carrier-Mediated Transport

Molecules such as glucose and amino acids are transported across the cell membranes by special protein carriers that are specific and that can be saturated, much like enzymes. Carrier-mediated transport in which the net movement is down a concentration gradient, and which is therefore passive, is called facilitated diffusion. Carrier-mediated transport that occurs against a concentration gradient and that requires metabolic energy is called active transport.

In order to sustain metabolism, cells must be able to take up glucose, amino acids, and other organic molecules from the extracellular environment. Molecules such as these, however, are too large and polar to pass through the lipid barrier by a process of simple diffusion.

The transport of glucose, amino acids, and some other molecules is mediated by protein **carriers** within the membrane. Although such carriers cannot be directly observed, their presence has been inferred by the observation that this transport has characteristics in common with enzyme activity. These characteristics include (1) *specificity,* (2) *competition,* and (3) *saturation.* Because of their similarity to enzymes, the transport carriers are sometimes called *permeases.*

Like enzyme proteins, carrier proteins only interact with specific molecules. Glucose carriers, for example, can only interact with glucose and not with closely related monosaccharides. As a further example of specificity, particular carriers for amino acids transport some types of amino acids but not others. Two amino acids that are transported by the same carrier compete with each other, so that the rate of transport of each when together is lower than it would be if each amino acid were present alone (fig. 6.13).

As the concentration of a transported molecule is increased, its rate of transport will also be increased—but only up to a maximum. Beyond this rate, called the *transport maximum* (or $T_m$), further increases in concentration do not further increase the transport rate. This indicates that the carriers have become saturated (fig. 6.13).

As an example of saturation, imagine a bus stop that is serviced once per hour by a bus that can hold a maximum of forty people (its "transport maximum"). If ten people wait at the bus stop, ten will be transported per hour. If twenty people wait at the bus stop, twenty will be transported per hour. This linear relationship will hold up to a maximum of forty people; if eighty people are at the bus stop, the transport rate will still be forty per hour.

**Figure 6.13.** Carrier-mediated transport displays the characteristics of saturation (illustrated by the *transport maximum*) and competition. Molecules *X* and *Y* compete for the same carrier, so that when they are present together the rate of transport of each is less than when either is present separately.

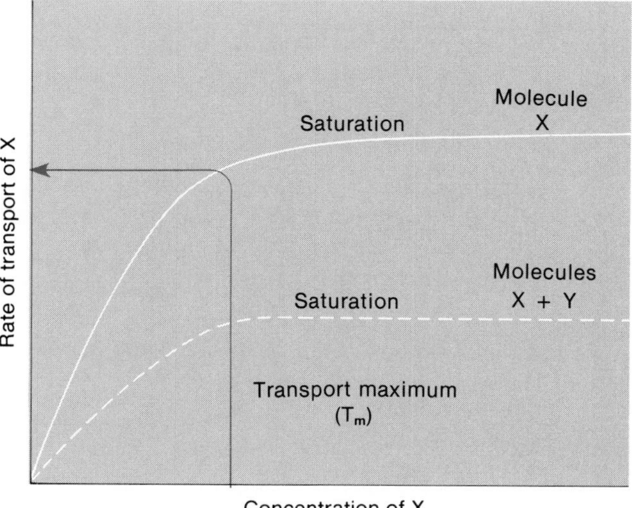


The kidneys transport a number of molecules from the blood filtrate (which will become urine) back into the blood. Glucose, for example, is normally completely reabsorbed so that urine normally is free of glucose. If the glucose concentration of the blood and filtrate is too high (a condition called *hyperglycemia*), however, the transport maximum will be exceeded. In this case, glucose will be found in the urine (a condition called *glycosuria*). This may result from eating too many sweets or from the inadequate action of the hormone *insulin* (in the disease **diabetes mellitus**).

### Facilitated Diffusion

The transport of glucose from the blood across the cell membranes of tissue cells occurs by **facilitated diffusion**. Facilitated diffusion, like simple diffusion, is powered by the thermal energy of the diffusing molecules and involves the net transport of substances through a cell membrane from the side of higher to the side of lower concentration. Active cellular metabolism is not required for either facilitated or simple diffusion.

Unlike simple diffusion of nonpolar molecules, water, and inorganic ions through a membrane, the diffusion of glucose through the cell membrane displays the properties of carrier-mediated transport: specificity, competition, and saturation. The diffusion of glucose through a cell membrane must therefore be mediated by carrier proteins. One conceptual model of the transport carriers is that they may

**Figure 6.14.** A model of facilitated diffusion, where a molecule is transported across the cell membrane by a carrier protein.

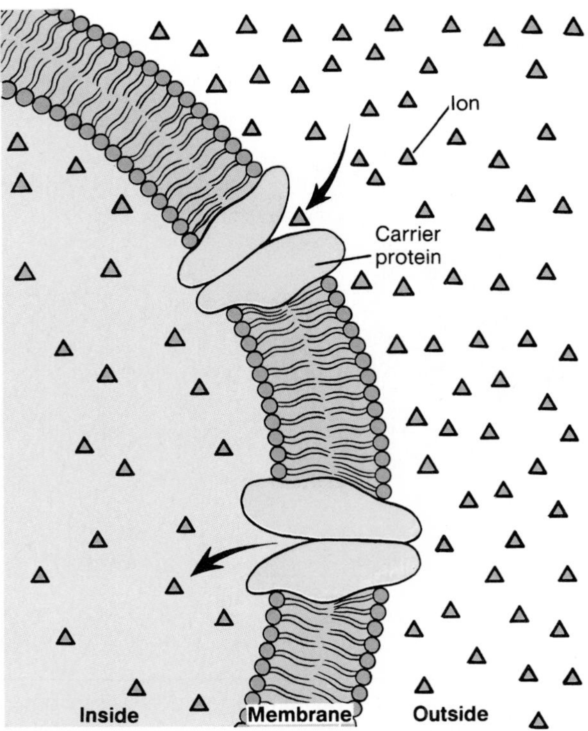


each be composed of two protein subunits that interact with glucose in a specific way that creates a channel through the membrane (fig. 6.14), so that the glucose can move from the side of higher to the side of lower concentration.

The rate of the facilitated diffusion of glucose into tissue cells depends directly on the plasma glucose concentration. When the plasma glucose concentration is abnormally low—a condition called *hypoglycemia*—the rate of transport of glucose into brain cells may be inadequate for the metabolic needs of the brain. Severe hypoglycemia, as may be produced in a diabetic person by an overdose of insulin, can thus result in the loss of consciousness and even death.

## Active Transport

There are some aspects of cell transport that cannot be explained by simple or facilitated diffusion. The epithelial lining of the intestine and of the kidney tubules, for example, moves glucose from the side of lower to the side of higher concentration (from the lumen to the blood). Similarly, all cells extrude $Ca^{++}$ into the extracellular environment and, by this means, maintain an intracellular $Ca^{++}$ concentration that is one thousand to ten thousand times lower than the extracellular $Ca^{++}$ concentration.

Since membrane transport carriers are proteins that are coded by specific genes, inherited defects in these carriers can result when there is an alteration in the genetic code. Defective carrier proteins in the cell membranes of epithelial cells that line the intestine may produce diseases that result from an inadequate absorption of ingested molecules, and defects in transport carriers within the epithelial cells of kidney tubules may result in the abnormal excretion of particular molecules in the urine (table 6.1).

The movement of molecules and ions against their concentration gradients, from lower to higher concentrations, requires the expenditure of cellular energy that is obtained from ATP. This type of transport is termed **active transport.** If a cell is poisoned with cyanide (which inhibits oxidative phosphorylation), active transport is inhibited. This contrasts with passive transport, which can continue even when metabolic poisons kill the cell by preventing the formation of ATP.

Active transport, like facilitated diffusion, is carrier-mediated. These carriers appear to be integral proteins that span the thickness of the membrane. According to one theory of active transport, the following events may occur: (1) the molecule or ion to be transported bonds to a specific "recognition site" on one side of the carrier protein; (2) this bonding stimulates the breakdown of ATP, which in turn results in phosphorylation of the carrier protein; (3) as a result of phosphorylation, the carrier protein undergoes a conformational change, like the change in enzyme proteins as a result of allosteric effects; and (4) the carrier protein undergoes a hingelike motion, which releases the transported molecule or ion on the other side of the membrane. This model of active transport is illustrated in figure 6.15.

***The Sodium-Potassium Pump.*** Active transport carriers are often referred to as "pumps." Although some of these carriers transport only one molecule or ion at a time, other carriers exchange one molecule or ion for another. The most important of the latter type of carriers is the **$Na^+/K^+$ pump.** This carrier protein, which is also an ATPase enzyme that converts ATP to ADP and $P_i$, actively extrudes three $Na^+$ ions from the cell as it transports two $K^+$ into the cell. This transport is energy dependent because $Na^+$ is more highly concentrated outside the cell and $K^+$ is more concentrated within the cell. Both ions, in other words, are moved against their concentration gradients (fig. 6.16).

All cells have numerous $Na^+/K^+$ pumps that are constantly active. This represents an enormous expenditure of energy used to maintain a steep gradient of $Na^+$ and $K^+$ across the cell membrane. This steep gradient serves three known functions: (1) the steep $Na^+$ gradient is used

**Table 6.1** Inherited defects of transport carriers in the kidney and intestine

| Disease | Defect | Clinical Significance |
|---|---|---|
| Cystinuria | Excessive urinary excretion of cystine, lysine, arginine, and ornithine | Calculi (stones) in urinary tract |
| Phosphaturia | Excessive urinary excretion of phosphate | Rickets: treated with large doses of vitamin D |
| Renal glycosuria | Kidney tubules have lower than normal $T_m$ for glucose | None known |
| Glucose malabsorption | Dietary glucose not absorbed from intestine | Sometimes fatal; must use fructose as only dietary carbohydrate |
| Hartnup disease | Delayed intestinal absorption of tryptophan and related molecules | Cerebellum dysfunction; photosensitive dermatitis |

**Figure 6.15.** A model of active transport, showing the hingelike motion of the integral protein subunits.

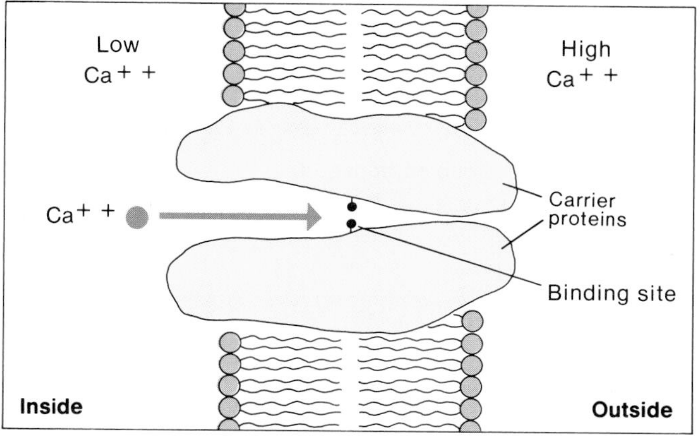


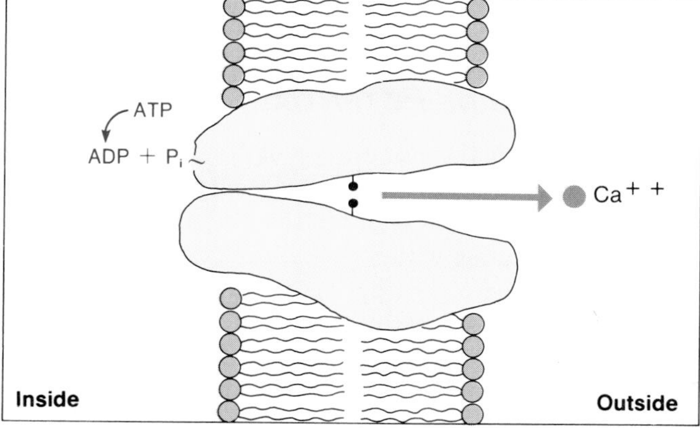


**Figure 6.16.** The $Na^+/K^+$ pump actively exchanges intracellular $Na^+$ for $K^+$. The carrier itself is an ATPase that breaks down ATP for energy. *Dotted lines* indicate the direction of passive transport (diffusion); *solid arrows* indicate the direction of active transport.

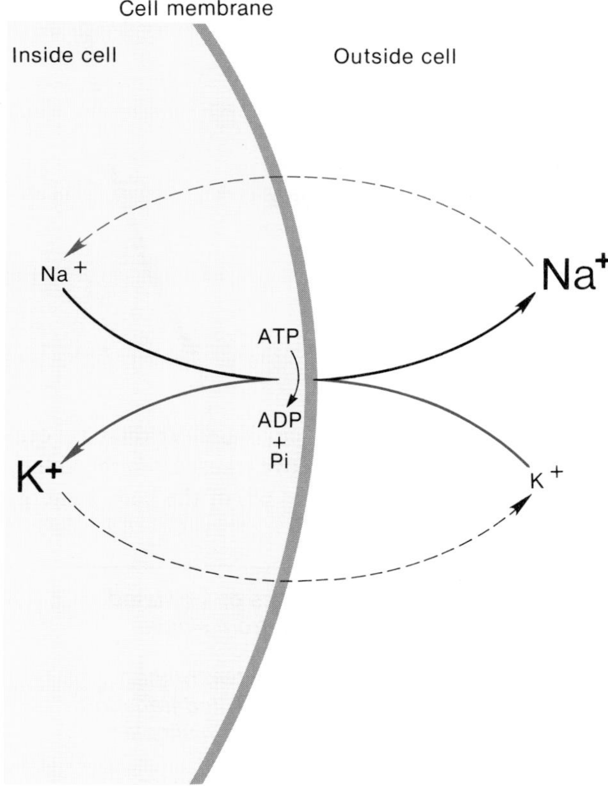


to provide energy for the "co-transport" of other molecules; (2) the activity of the $Na^+/K^+$ pumps can be adjusted (primarily by thyroid hormones) to regulate the resting calorie expenditure and basal metabolic rate of the body; and (3) the $Na^+$ and $K^+$ gradients across the cell membranes of nerve and muscle cells are used to produce electrical impulses.

***Coupled Transport (Co-transport).*** In co-transport, the energy needed for the "uphill" movement of a molecule or ion is obtained from the "downhill" transport of $Na^+$ into the cell. The active extrusion of $Ca^{++}$ from some cells, for example, is coupled to the passive diffusion of $Na^+$ into the cell. Cellular energy, obtained from ATP, is not used to move $Ca^{++}$ directly out of the cell, but energy is constantly required to maintain the steep $Na^+$ gradient. The inward diffusion of $Na^+$ due to this concentration gradient, in turn, is used to power the active extrusion of $Ca^{++}$.

A similar mechanism is used to actively transport glucose across the membranes of epithelial cells that line the digestive tract and kidney tubules. In this case, the inward diffusion of $Na^+$ is coupled to the cellular uptake

**Figure 6.17.** Co-transport of glucose. Glucose accumulates inside the cell against its concentration gradient, using energy derived from the passive transport (diffusion) of $Na^+$ into the cell. Metabolic energy is indirectly required because the steep electrochemical gradient for $Na^+$ is maintained by the $Na^+/K^+$ pump. *Dotted arrows* indicate direction of diffusion; *solid arrows* show direction of active transport.

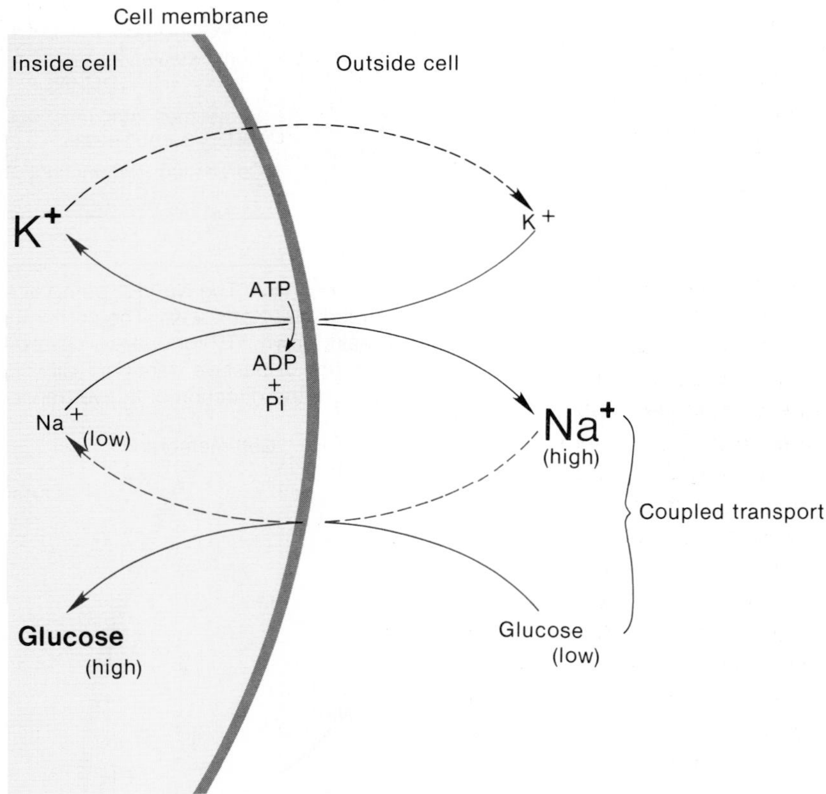


of glucose (as well as amino acids) against their concentration gradients (fig. 6.17). Active transport in the kidney tubules accounts for as much as 6% of the body's energy expenditure at rest.

---

1. *List the three characteristics of facilitated diffusion that distinguish it from simple diffusion.*
2. *Draw a figure that illustrates two of the characteristics of carrier-mediated transport, and explain how this differs from simple diffusion.*
3. *Describe active transport, including co-transport in your description. Explain how active transport differs from facilitated diffusion.*
4. *Explain the functional significance of the $Na^+/K^+$ pump.*

---

## THE MEMBRANE POTENTIAL

There is an unequal concentration of $Na^+$ and $K^+$ across the cell membrane; $Na^+$ is more concentrated in the extracellular fluid than it is inside the cell, and $K^+$ is at a higher concentration within the cell. There is also an unequal distribution of charges across the membrane; the inside of the cell is negatively charged in comparison to the outside of the cell. The diffusion gradient and the gradient of electrical attraction are in opposite directions for $K^+$ but do not balance each other. There is a constant net movement of $K^+$ out of the cell and of $Na^+$ into the cell, to the degree permitted by their permeabilities, and this movement requires constant correction by the $Na^+/K^+$ pumps.

In the preceeding section, the action of the $Na^+/K^+$ pumps was discussed in conjunction with the topic of active transport, and it was noted that these pumps move $Na^+$ and $K^+$ against their concentration gradients. This action would, by itself, create and amplify the difference in concentration of these ions across the cell membrane. There is another reason, in addition to the actions of the $Na^+/K^+$ pumps, why the concentration of these ions would be unequal across the membrane. This reason is related to the basis for the electrical properties of the membrane.

**Figure 6.18.** Proteins, organic phosphates, and other organic anions that cannot leave the cell create a fixed negative charge on the inside of the membrane. This attracts positively charged inorganic ions (cations), which therefore accumulate within the cell at a higher concentration than in the extracellular fluid. The amount of cations that accumulate within the cell is limited by the fact that a concentration gradient builds up, which favors the diffusion of the cations out of the cell.

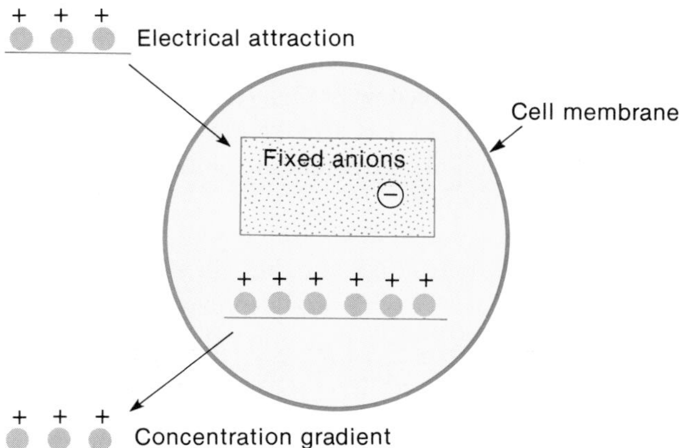


Cellular proteins and the phosphate groups of ATP and other organic molecules are negatively charged within the cell cytoplasm. These negative ions (anions) are "fixed" within the cell by the fact that they cannot penetrate the cell membrane. Since these negatively charged organic molecules cannot leave the cell, they attract positively charged inorganic ions (cations) from the extracellular fluid that are small enough to diffuse through the membrane pores. The distribution of small, inorganic cations (mainly $K^+$, $Na^+$, and $Ca^{++}$) between the intracellular and extracellular compartments, in other words, is influenced by the negatively charged fixed ions within the cell.

Since the cell membrane is much more permeable to $K^+$ than to any other cation, $K^+$ accumulates within the cell more than the others as a result of its electrical attraction for the fixed anions (fig. 6.18). Instead of being evenly distributed between the intracellular and extracellular compartments, therefore, $K^+$ becomes more highly concentrated within the cell. In the human body the intracellular $K^+$ concentration is 155 mEq/L compared to an extracellular concentration of 4 mEq/L (mEq = milliequivalents, which is the millimolar concentration multiplied by the valence of the ion—in this case, by one).

Even without the action of the $Na^+/K^+$ pumps, therefore, $K^+$ would be more highly concentrated within the cell than in the extracellular fluid. The magnitude of the intracellular $K^+$ concentration depends on the number of fixed anions and on the $K^+$ concentrations that are normally found in the extracellular compartment (including plasma and tissue fluid). An increase in the extracellular $K^+$ concentration above normal, for example, would cause a corresponding increase in the intracellular $K^+$ concentration.

**Figure 6.19.** If $K^+$ were the only ion able to diffuse through the cell membrane, it would distribute itself between the intracellular and extracellular compartment until an equilibrium would be established. At equilibrium the $K^+$ concentration within the cell would be higher than outside the cell due to the attraction of $K^+$ to the fixed anions. Not enough $K^+$ would accumulate within the cell to neutralize these anions, however, so the inside of the cell would be 90 millivolts negative compared to the outside of the cell. This membrane voltage is the equilibrium potential ($E_K$) for potassium.

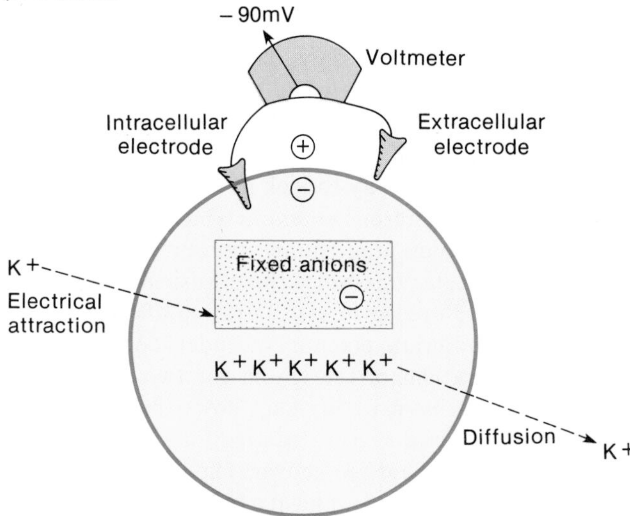


## Equilibrium Potentials

An equilibrium potential is a theoretical voltage that would be produced across a cell membrane if only one ion were able to diffuse through the membrane. Since the membrane is most permeable to $K^+$, we can construct a theoretical approximation to the true situation by considering what would happen if $K^+$ were the *only* ion able to cross the membrane. If this were the case, $K^+$ would diffuse until its concentration inside and outside of a cell became stable. An *equilibrium* would be established: if a certain amount of $K^+$ were to move inside the cell (by electrical attraction for the fixed anions), an identical amount of $K^+$ would diffuse out of the cell (down its concentration gradient). At equilibrium, in other words, the forces of electrical attraction and of the diffusion gradient are equal and opposite.

At this equilibrium, the concentration of $K^+$ would be higher inside the cell than outside the cell; a concentration difference would exist across the cell membrane, which was stabilized by the attraction of $K^+$ to the fixed anions. At this point, we could ask the question: are the fixed anions neutralized; are the charges balanced? The answer to that question depends on how much $K^+$ gets into the cell, which in turn depends on the $K^+$ concentration in the extracellular fluid. At the $K^+$ concentrations that are, in fact, found in the body, the answer to the question is *no.* Not enough $K^+$ is present in the cell to neutralize the fixed anions (fig. 6.19).

At equilibrium, therefore, the inside of the cell membrane would have a higher concentration of negative charges than the outside of the membrane. There is a difference in charge, as well as a difference in concentration, across the membrane. The magnitude of the difference in charge, or **potential difference,** on the two sides of the membrane under these conditions is 90 millivolts (mV). This is shown with a negative sign (as −90 mV) to indicate that the inside of the cell is the negative pole. (If one were to write +90 mV, the magnitude of the potential difference would be unchanged, but the inside of the cell would be shown as the positive pole.) The potential difference of −90 mV, which would be developed if $K^+$ were the only diffusible ion, is called the $K^+$ **equilibrium potential** (abbreviated $E_K$).

There is another way to look at the equilibrium potential: it is the membrane potential which would *exactly balance* the diffusion gradient and prevent the net movement of a particular ion. Since the diffusion gradient depends on the difference in concentration of the ion, the value of the equilibrium potential must depend on the ratio of the concentrations of the ion on the two sides of the membrane. The **Nernst equation** allows this theoretical equilibrium potential to be calculated for a particular ion when its concentrations are known. The following simplified form of the equation is valid at a temperature of 37° C:

$$E_x = \frac{61}{z} \log \frac{[X_0]}{[X_i]}$$

Where:

$E_x$ = Equilibrium potential in millivolts (mV) for ion $x$
$X_0$ = Concentration of the ion outside the cell
$X_i$ = Concentration of the ion inside the cell
z = Valence of the ion (+1 for $Na^+$ or $K^+$)

Note that, using this equation, the equilibrium potential for a cation will have a negative value when $[X_i]$ is greater than $[X_0]$. If we substitute $K^+$ for X, this is indeed the case (because of the attraction provided by the fixed anions, as previously discussed). As a hypothetical example, if the concentration of $K^+$ were ten times higher inside compared to outside the cell, the equilibrium potential would be 61 mV (log 1/10) = 61 X (−1) = −61 mV. In reality, the concentration of $K^+$ inside the cell is actually thirty times greater than outside (150 mM inside compared to 5 mM outside). Since the log of 1/30 is −1.477, the equilibrium potential for $K^+$ given realistic concentration values is −90 mV.

If we calculate the equilibrium potential for $Na^+$, different values must be used. The concentration of $Na^+$ in the extracellular fluid is 145 mEq/L, whereas its concentration inside cells is only 12 mEq/L. The diffusion gradient thus promotes the movement of $Na^+$ into the cell, and, in order to oppose this diffusion, the membrane potential would have to have a positive polarity on the inside of the cell. This is indeed what the Nernst equation would provide, since $[X_0]/[X_i]$ is greater than one. Using this equation, the equilibrium potential for $Na^+$ can be calculated to be +60 mV.

A membrane potential of +60 mV, therefore, would prevent the diffusion of $Na^+$ into the cell, whereas a membrane potential of −90 mV would prevent the diffusion of $K^+$ out of the cell. It is clear that the membrane potential cannot be both values at the same time; indeed, it is seldom either value but instead is somewhere between these two extremes. We will call this the **resting membrane potential**, to distinguish it from the theoretical equilibrium potential. The actual value of the resting membrane potential is dependent upon the permeability of the membrane to each ion and the equilibrium potential of each diffusible ion.

## Resting Membrane Potential

The resting membrane potential of most cells in the body is about −65 mV to −85 mV (in neurons it averages −70 mV). This value is very close to the one that we had predicted if $K^+$ were the only ion able to move through the cell membrane and establish an equilibrium. The resting membrane potential, however, is quite different from the $Na^+$ equilibrium potential. The reason that the resting membrane potential is closer to the $K^+$ than to the $Na^+$ equilibrium potential is because, as previously discussed, the membrane is much more permeable to $K^+$ than to $Na^+$. The actual value of the membrane potential thus depends on the permeability of the membrane to different ions. When neurons produce nerve impulses, the permeability to $Na^+$ increases dramatically, sending the membrane potential toward the $Na^+$ equilibrium potential (nerve impulses are discussed in chapter 7). This is the reason that the term "resting" is used to describe the membrane potential when it is not producing impulses.

**Figure 6.20.** Because some $Na^+$ leaks into the cell by diffusion, the actual resting membrane potential is less than the $K^+$ equilibrium potential. As a result, some $K^+$ diffuses out of the cell (*dotted lines*).

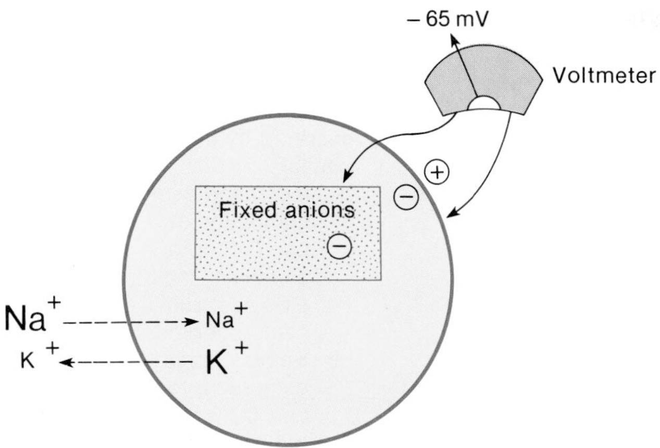


**Figure 6.21.** The concentrations of $Na^+$ and $K^+$ both inside and outside the cell do not change as a result of diffusion (*dotted arrows*) because of active transport (*solid arrows*) by the $Na^+/K^+$ pump. Since the pump transports three $Na^+$ for every two $K^+$, the pump itself helps to create a charge separation (a potential difference, or voltage) across the membrane.

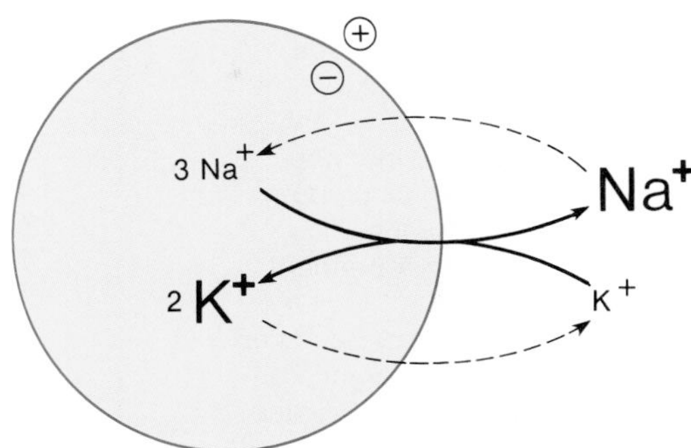


Although changes in the extracellular concentration of many ions can affect the membrane potential, this potential is particularly sensitive to changes in plasma potassium. Since the maintenance of a particular membrane potential is critical for the generation of electrical events in nerves and muscles (including the heart), the body has a variety of mechanisms that serve to maintain plasma $K^+$ concentrations within very narrow limits. These mechanisms act primarily through the kidneys, which can excrete $K^+$ in the urine or reabsorb it into the blood. The excretion of $K^+$ is stimulated by hormones of the adrenal cortex (particularly by aldosterone); if the adrenal glands of an experimental animal are removed, the animal may die as a result of an accumulation of $K^+$ in the blood. An abnormal increase in the blood concentration of $K^+$ is called **hyperkalemia.**

When hyperkalemia occurs, the diffusion gradient that favors the extrusion of $K^+$ from the cell is reduced and, as a result, more $K^+$ can enter the cell. In terms of the Nernst equation, the ratio $[K^+_o]/[K^+_i]$ is decreased. This reduces the membrane potential (brings it closer to zero) and thus alters the function of many organs, particularly the heart. For these reasons, the blood electrolyte concentrations are monitored very carefully in patients with heart or kidney disease.

***The Na⁺/K⁺ Pump.*** Since the membrane potential is less negative than $E_K$ as a result of some $Na^+$ entry, some $K^+$ leaks out of the cell (fig. 6.20). The cell is *not* at equilibrium with respect to $K^+$ and $Na^+$ concentrations. Despite this, the concentrations of $K^+$ and $Na^+$ are maintained constant; this is due to the constant expenditure of energy in active transport by the $Na^+/K^+$ pump. The $Na^+/K^+$ pump acts to counter the leaks and thus maintains the membrane potential.

Actually, the $Na^+/K^+$ pump does more than simply work against the ion leaks, since it transports *three* $Na^+$ ions out of the cell for every *two* $K^+$ ions that it moves in, its action helps generate a potential difference across the membrane (fig. 6.21). As a result of all of these activities, a real cell has (1) a relatively constant intracellular concentration of $Na^+$ and $K^+$; and (2) a constant membrane potential (in the absence of stimulation) in nerves and muscles of $-65$ mV to $-85$ mV.

1. *Define the term* membrane potential, *and describe how it is measured.*
2. *Describe how an equilibrium potential is produced when potassium is the only diffusible cation, and describe how the value of the equilibrium potential is affected by the potassium concentrations outside and inside the cell.*
3. *Explain why the resting membrane potential is close to but different from the potassium equilibrium potential.*
4. *Suppose a person has hyperkalemia such that the extracellular $K^+$ concentration increases from 5 mM to 10 mM (a potentially fatal condition). Use the Nernst equation to calculate the new $E_K$, and describe verbally how the resting membrane potential would be changed.*
5. *Explain the role of the $Na^+/K^+$ pump in the generation and maintenance of the resting membrane potential.*

## Summary

### Diffusion and Osmosis p. 126

I. Diffusion is the net movement of molecules or ions from regions of high to regions of low concentration.
   A. This is a type of passive transport—energy is provided by the thermal energy of the molecules, not by cellular metabolism.
   B. Net diffusion stops when the concentration is equal on both sides of the membrane.

II. The rate of diffusion is dependent on a variety of factors.
   A. The rate of diffusion depends on the concentration difference on the two sides of the membrane.
   B. The rate depends on the permeability of the cell membrane to the diffusing substance.
   C. The rate of diffusion through a membrane is also directly proportional to the surface area of the membrane, which can be increased by such adaptations as microvilli.

III. Simple diffusion is the type of passive transport in which small molecules and inorganic ions, such as $Na^+$ and $K^+$, move through the cell membrane.
   A. Inorganic ions pass through specific channels in the membrane.
   B. Lipids such as steroid hormones, and dipolar water molecules, can pass by simple diffusion directly through the phospholipid layers of the membrane.

IV. Osmosis is the simple diffusion of solvent (water) through a membrane that is more permeable to the solvent than it is to the solute.
   A. Water moves from the solution that is more dilute to the solution that has a higher solute concentration.
   B. Osmosis depends on a difference in total solute concentration, not on the chemical nature of the solute.
      1. The concentration of total solute (in moles) per kilogram (liter) of water is measured in osmolality units.
      2. The solution with the higher osmolality has the higher osmotic pressure.
      3. Water moves by osmosis from the solution of lower osmolality and osmotic pressure to the solution of higher osmolality and osmotic pressure.
   C. Solutions that have the same osmotic pressure as plasma (such as 0.9% NaCl and 5% glucose) are said to be isotonic.
      1. Solutions with a lower osmotic pressure are hypotonic; those with a higher osmotic pressure are hypertonic.
      2. Cells in a hypotonic solution gain water and swell; those in a hypertonic solution lose water and shrink (crenate).
   D. The osmolality and osmotic pressure of the plasma is detected by osmoreceptors in the hypothalamus of the brain and maintained within a normal range by the action of antidiuretic hormone (ADH) secreted by the posterior pituitary.
      1. Increased osmolality stimulates the osmoreceptors; decreased osmolality inhibits the electrical activity of the osmoreceptors.
      2. Stimulation of the osmoreceptors causes thirst and the secretion of antidiuretic hormone (ADH) from the pituitary.
      3. ADH stimulates water retention by the kidneys, which serves to maintain a normal blood volume and osmolality.

### Carrier-Mediated Transport p. 133

I. The passage of glucose, amino acids, and other polar molecules through the cell membrane is mediated by carrier proteins in the cell membrane.
   A. Carrier-mediated transport exhibits the properties of specificity, competition, and saturation.
   B. The transport rate of molecules such as glucose reaches a maximum when the carriers are saturated—this maximum rate is called the transport maximum, or $T_m$.

II. The transport of molecules such as glucose from the side of higher to the side of lower concentration by means of membrane carriers is called facilitated diffusion.
   A. Like simple diffusion, this is passive transport—cellular energy is not required.
   B. Unlike simple diffusion, facilitated diffusion is specific, exhibits competition, and can be saturated.

III. The active transport of molecules and ions across a membrane requires the expenditure of cellular energy (ATP).
   A. In active transport, carriers move molecules or ions from the side of lower to the side of higher concentration.
   B. One example of active transport is the action of the $Na^+/K^+$ pump.
      1. Sodium is more concentrated on the outside of the cell, whereas potassium is more concentrated on the inside of the cell.
      2. The $Na^+/K^+$ pump helps to maintain these concentration differences by transporting $Na^+$ out of the cell and $K^+$ into the cell.
      3. Three $Na^+$ are transported out for every two $K^+$ ions that are transported into the cell.

## The Membrane Potential p. 136

I. The cytoplasm of the cell contains negatively charged organic ions (anions) that cannot leave the cell—they are "fixed anions."
   A. These fixed anions attract $K^+$, which is the inorganic ion that can most easily pass through the cell membrane.
   B. As a result of this electrical attraction, the concentration of $K^+$ within the cell is greater than the concentration of $K^+$ in the extracellular fluid.
   C. If $K^+$ were the only diffusible ion, the concentrations of $K^+$ on the inside and outside of the cell would reach an equilibrium.
      1. At this point, the rate of $K^+$ entry (due to electrical attraction) would equal the rate of $K^+$ exit (due to diffusion).
      2. At this equilibrium, there would still be a higher concentration of negative charges within the cell (due to the fixed anions) than outside the cell.
      3. At this equilibrium, the inside of the cell would be ninety millivolts negative ($-90$ mV) compared to the outside of the cell—this is called the $K^+$ equilibrium potential ($E_K$).
   D. The resting membrane potential is less than $E_K$—it is usually $-65$ mV to $-85$ mV.
      1. This is because some $Na^+$ can also enter the cell.
      2. $Na^+$ is more highly concentrated outside than inside the cell, and the inside of the cell is negative—these forces attract $Na^+$ into the cell.
      3. The rate of $Na^+$ entry is generally slow because the membrane is usually not very permeable to $Na^+$.

II. The slow rate of $Na^+$ entry is accompanied by a slow rate of $K^+$ exit from the cell.
   A. These changes are corrected by the active transport $Na^+/K^+$ pump, which maintains constant concentrations and a constant resting membrane potential.
   B. There are numerous $Na^+/K^+$ pumps in all cells of the body that require a constant expenditure of energy.
   C. The $Na^+/K^+$ pump itself contributes to the membrane potential because it pumps more $Na^+$ out than it pumps $K^+$ in (by a ratio of three to two).

## Review Activities

### Objective Questions

1. The movement of water across a cell membrane occurs by
   (a) active transport
   (b) facilitated diffusion
   (c) simple diffusion (osmosis)
   (d) all of the above
2. Which of the following statements about the facilitated diffusion of glucose is *true*?
   (a) There is a net movement from the region of low to the region of high concentration.
   (b) Carrier proteins in the cell membrane are required for this transport.
   (c) This transport requires energy obtained from ATP.
   (d) This is an example of co-transport.
3. If a poison such as cyanide stops the production of ATP, which of the following transport processes would cease?
   (a) the movement of $Na^+$ out of a cell
   (b) osmosis
   (c) the movement of $K^+$ out of a cell
   (d) all of the above
4. Red blood cells crenate in
   (a) a hypotonic solution
   (b) an isotonic solution
   (c) a hypertonic solution
5. Plasma has an osmolality of about 300 mOsm. Isotonic saline has an osmolality of
   (a) 150 mOsm
   (b) 300 mOsm
   (c) 600 mOsm
   (d) none of the above
6. A 0.5 m NaCl solution and a 1.0 m glucose solution
   (a) have the same osmolality
   (b) have the same osmotic pressure
   (c) are isotonic to each other
   (d) are all of the above
7. The diffusible ion that is most important in the establishment of the membrane potential is
   (a) $K^+$
   (b) $Na^+$
   (c) $Ca^{++}$
   (d) $Cl^-$
8. An increase in blood osmolality
   (a) can occur as a result of dehydration
   (b) causes a decrease in blood osmotic pressure
   (c) is accompanied by a decrease in ADH secretion
   (d) all of the above
9. In hyperkalemia, the membrane potential
   (a) increases
   (b) decreases
   (c) is not changed
10. Which of the following statements about the $Na^+/K^+$ pump is *true*?
    (a) $Na^+$ is actively transported into the cell.
    (b) $K^+$ is actively transported out of the cell.
    (c) An equal number of $Na^+$ and $K^+$ ions are transported with each cycle of the pump.
    (d) The pumps are constantly active in all cells.
11. Carrier-mediated facilitated diffusion
    (a) uses cellular ATP
    (b) is used for cellular uptake of blood glucose
    (c) is a form of active transport
    (d) none of these

### Essay Questions

1. Describe the conditions required to produce osmosis, and explain why osmosis occurs under these conditions.
2. Explain how simple diffusion can be distinguished from facilitated diffusion and how active transport can be distinguished from passive transport.
3. Compare the theoretical membrane potential that occurs at $K^+$ equilibrium with the true resting membrane potential. Explain the reasons for the differences between these.
4. Explain how the $Na^+/K^+$ pump contributes to the resting membrane potential.

## Selected Readings


Benos, D. J. 1989. The biology of amiloride-sensitive sodium channels. *Hospital Practice* 24:149.

Christensen, H. N. 1975. *Biological transport,* 2nd ed. Menlo Park, CA: W. A. Benjamen, Inc.

Dubinsky, W. P., Jr. 1989. The physiology of epithelial chloride channels. *Hospital Practice* 24:69.

Dyson, R. D. 1978. *Essentials of cell biology,* 2d ed. Boston: Allyn & Bacon.

Finean, J. B., Coleman, R., and R. H. Mitchell. 1978. *Membranes and their cellular functions,* 2d ed. New York: John Wiley & Sons, Inc.

Johnson, L. G. 1983. *Biology.* Dubuque: Wm. C. Brown Company Publishers.

Kaplan, J. H. 1985. Ion movements through the sodium pump. *Annual Review of Physiology* 47:535.

Keeton, W. T. 1980. *Biological science,* 3rd ed. New York: W. W. Norton & Co.

Kotyk, A., and K. Janacek. 1975. *Cell membrane transport: principles and techniques,* 2nd ed. New York: Plenum Publishing Corp.

Mader, S. S. 1982. *Inquiry into life,* 3rd ed. Dubuque: Wm. C. Brown Company Publishers.

Sheeler, P., and D. E. Bianchi. 1980. *Cell biology: structure, biochemistry, and function.* New York: John Wiley & Sons.

Silverman, M. 1989. Molecular biology of the $Na^+$-D-glucose cotransporter. *Hospital Practice* 24:180.

Wheeler, T. J., and P. C. Hinkle. 1985. The glucose transporter of mammalian cells. *Annual Review of Physiology* 47:503.

# 7 The Nervous System: Organization, Electrical Activity, and Synaptic Transmission

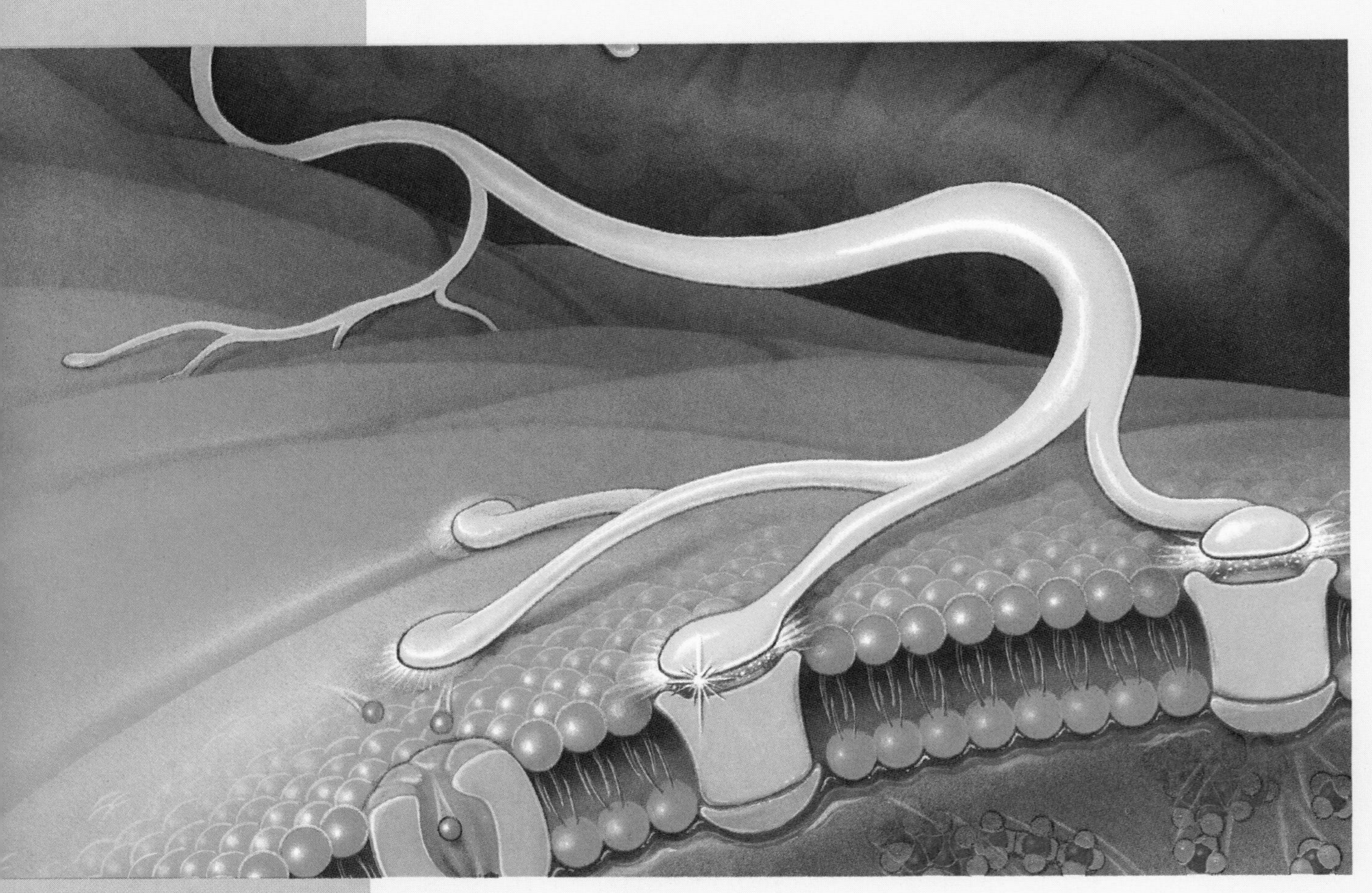

## Objectives

By studying this chapter, you should be able to

1. describe the parts of a neuron and their functional significance
2. describe the anatomical and functional divisions of the nervous system
3. describe the locations and functions of the different types of neuroglial cells
4. describe the blood-brain barrier and explain its significance
5. describe the sheath of Schwann and explain how this sheath helps the regeneration of cut peripheral nerve fibers
6. explain how a myelin sheath is formed
7. define depolarization, repolarization, and hyperpolarization
8. explain the actions of voltage-regulated $Na^+$ and $K^+$ gates and the events that occur during the production of an action potential
9. describe the properties of action potentials and explain the significance of the all-or-none law and the refractory periods
10. explain how action potentials are regenerated along a myelinated and a nonmyelinated axon and explain why saltatory conduction improves the conduction speed of nerve impulses
11. describe the events that occur between the electrical excitation of an axon and the release of neurotransmitter
12. describe how ACh stimulates a postsynaptic cell
13. describe the characteristics of an EPSP and compare these characteristics to those of an action potential
14. compare the mechanisms that inactivate ACh and catecholamine neurotransmitters
15. explain the role of cyclic AMP in the action of catecholamine neurotransmitters
16. explain the nature and significance of the inhibitory effects of glycine and GABA in the central nervous system
17. explain how EPSPs and IPSPs can interact and explain the significance of spatial and temporal summation
18. describe the processes of presynaptic inhibition and post-tetanic potentiation

## Outline

**Figure 7.1.** The structure of two kinds of neurons. (*a*) A motor neuron; (*b*) a sensory neuron.

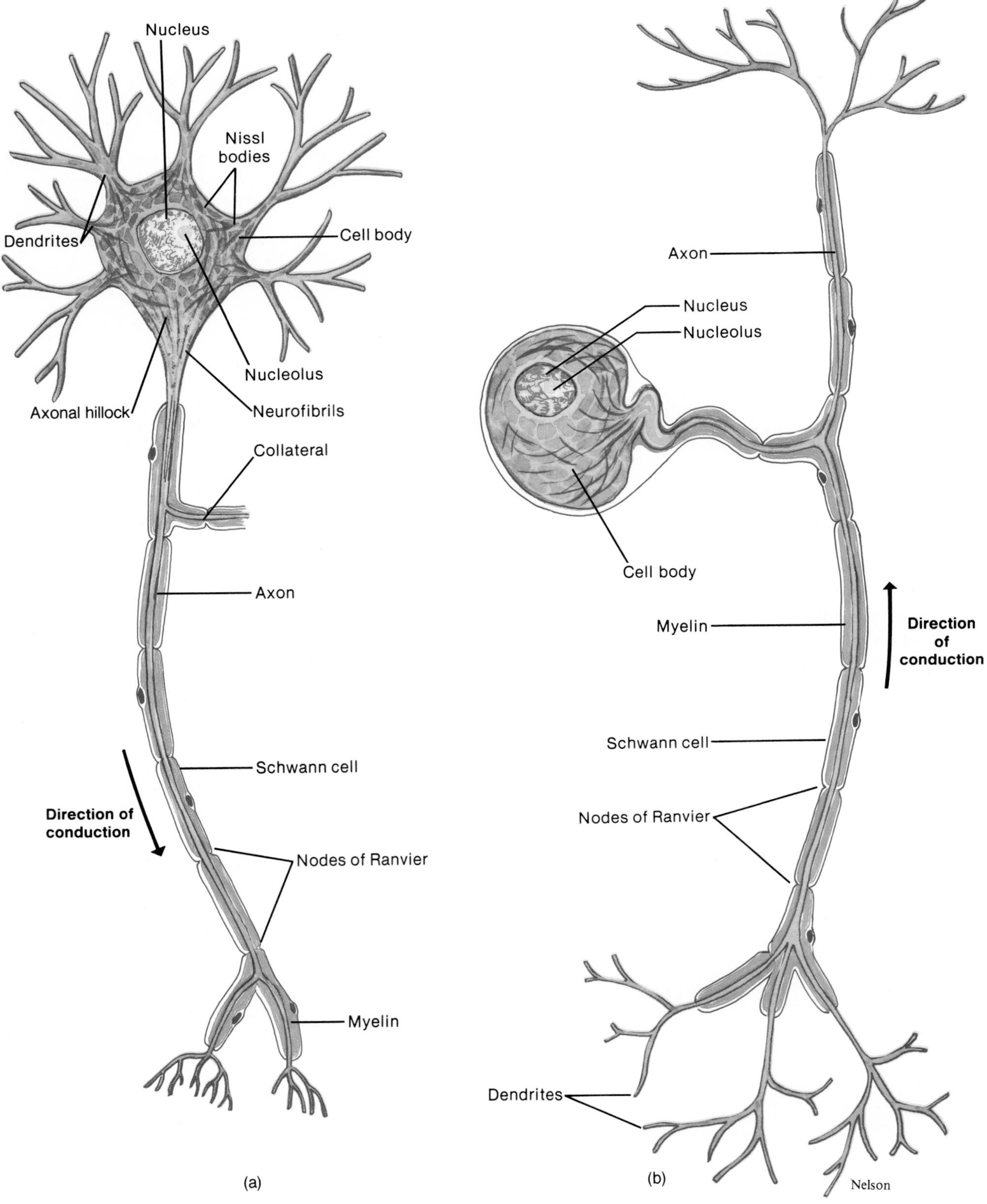

## Organization and Histology of the Nervous System

The nervous system is composed of neurons, which produce and conduct electrochemical impulses, and neuroglial cells, which support the functions of neurons. The nervous system is subdivided into the central nervous system and peripheral nervous system, and terms such as "nerve," "ganglion," "tract," and "nucleus" relate to specific structures in the two subdivisions.

The nervous system is divided into the **central nervous system** (**CNS**), which includes the brain and spinal cord, and the **peripheral nervous system** (**PNS**), which includes the *cranial nerves* arising from the brain and the *spinal nerves* arising from the spinal cord.

The nervous system is composed of only two principal types of cells, neurons and neuroglia. **Neurons** are the basic structural and functional units of the nervous system. They are specialized to respond to physical and chemical stimuli, conduct electrochemical impulses, and release specific chemical regulators. Through these activities, neurons perform such functions as the perception of sensory stimuli, learning, memory, and the control of muscles and glands. Neurons cannot divide by mitosis, although some neurons can regenerate a severed portion or sprout small new branches under some conditions.

**Neuroglia,** or **glial** (*glia* = glue) **cells,** are supportive cells in the nervous system that aid the function of neurons. Glial cells are about five times more abundant than neurons and have limited mitotic abilities (brain tumors that occur in adults are usually composed of glial cells rather than neurons).

### Neurons

Although neurons vary considerably in size and shape, they generally have three principal regions: (1) a cell body; (2) dendrites; and (3) an axon (figs. 7.1 and 7.2). Dendrites and axons can be referred to generically as *processes*, or extensions from the cell body.

The **cell body,** or **perikaryon** (*peri* = around; *karyon* = nucleus), is the enlarged portion of the neuron, which contains the nucleus and serves as the "nutritional center" of the neuron where macromolecules are produced. The perikaryon also contains granular, densely staining material known as *Nissl bodies,* which are not found in the dendrites or axon. The Nissl bodies are composed of granular (rough) endoplasmic reticulum, an organelle involved in protein synthesis. The cell bodies within the CNS are frequently clustered into groups called *nuclei* (not to be confused with the nucleus of a cell). Cell bodies in the PNS usually occur in clusters called *ganglia* (table 7.1).

**Figure 7.2.** A photomicrograph of neurons from the anterior column of gray matter of the spinal cord (120X).

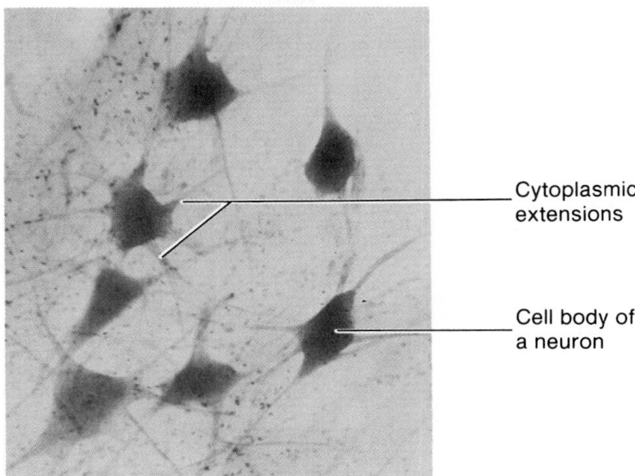


**Dendrites** (*dendron* = tree branch) are thin, branched processes that extend from the cytoplasm of the cell body. Dendrites serve as a receptive area that transmits electrical impulses to the cell body. The **axon,** or **nerve fiber,** is a longer process that conducts impulses away from the cell body. Axons vary in length from only a millimeter to as long as a meter or more (for axons that extend from the CNS to the foot) in length. The origin of the axon near the cell body is called the *axon hillock,* and side branches that may extend from the axon are called *axon collaterals.*

Proteins and other molecules are transported through the axon at faster rates than could be achieved by simple diffusion. This rapid movement is produced by two different mechanisms: axoplasmic flow and axonal transport (table 7.2). **Axoplasmic flow,** the slower of the two, results from rhythmic waves of contraction that push the cytoplasm from the axon hillock to the nerve endings. **Axonal transport,** which is more rapid and more selective, may occur in a reverse (retrograde) as well as a forward (orthograde) direction. Indeed, retrograde transport may be responsible for the movement of herpes virus, rabies virus, and tetanus toxin from the nerve terminals into cell bodies.

**Table 7.1** Anatomical terms used in describing the nervous system

| Term | Definition |
|---|---|
| Central nervous system (CNS) | Brain and spinal cord |
| Peripheral nervous system (PNS) | Nerves and ganglia |
| Interneuron | Multipolar neuron located entirely within CNS |
| Sensory neuron | Neuron that transmits impulses from sensory receptor into CNS (afferent fiber) |
| Motor neuron | Neuron that transmits impulses from CNS to an effector organ (e.g., muscle efferent fiber) |
| Nerve | Cablelike collection of nerve fibers; may be "mixed" (contain both sensory and motor fibers) |
| Somatic motor nerve | Nerve that stimulates contraction of skeletal muscles |
| Autonomic motor nerve | Nerve that stimulates contraction (or inhibits contraction) of smooth muscle and cardiac muscle and secretion of glands |
| Ganglion | Collection of neuron cell bodies located outside CNS |
| Nucleus | Groupings of neuron cell bodies within CNS |
| Tract | Collections of nerve fibers that interconnect regions of CNS |

**Table 7.2** Comparison of axoplasmic flow with axonal transport

| Axoplasmic flow | Axonal transport |
|---|---|
| Transport rate comparatively slow (1–2mm/day) | Transport rate comparatively fast (200–400mm/day) |
| Molecules transported only from cell body | Molecules transported from cell body to axon endings and in reverse direction |
| Bulk movement of proteins in axoplasm, including microfilaments and tubules | Transport of specific proteins, mainly of membrane proteins and acetylcholinesterase |
| Transport accompanied by peristaltic waves of axon membrane | Transport dependent on cagelike microtubule structure within axon and on actin and $Ca^{++}$ |

***Classification of Neurons and Nerves.*** Neurons may be classified according to their structure or function. The functional classification is based on the direction that they conduct impulses. **Sensory,** or **afferent, neurons** conduct impulses from sensory receptors into the CNS. **Motor,** or **efferent, neurons** (fig. 7.3) conduct impulses out of the CNS to effector organs (muscles and glands). **Association neurons,** or **interneurons,** are located entirely within the CNS and serve the associative, or integrative, functions of the nervous system.

There are two types of motor neurons: somatic and autonomic. **Somatic motor neurons** provide both reflex and voluntary control of skeletal muscles. **Autonomic motor neurons** innervate the involuntary effectors—smooth muscle, cardiac muscle, and glands. The cell bodies of the autonomic neurons that innervate these organs are located outside the CNS in autonomic ganglia (fig. 7.3). There are two subdivisions of autonomic neurons: *sympathetic* and *parasympathetic.* Autonomic motor neurons, together with their central control centers, comprise the *autonomic nervous system,* which will be discussed in chapter 10.

The structural classification of neurons is based on the number of processes that extend from the cell body of the neuron (fig. 7.4). **Bipolar neurons** have two processes, one at either end; this type is found in the retina of the eye. **Multipolar neurons** have several dendrites and one axon extending from the cell body; this is the most common type of neuron (motor neurons are good examples of this type). A **pseudounipolar neuron** has a single short process that divides like a T to form a longer process. Sensory neurons are pseudounipolar—one end of the process formed by the T receives sensory stimuli and produces nerve impulses; the other end of the T delivers these impulses to synapses within the brain or spinal cord. The long process that extends from the sensory receptor to the stalk of the T (toward the cell body) is a dendrite, while the process that extends from the stalk to the CNS is the axon. The cell bodies of these sensory neurons are located outside the CNS in the dorsal root ganglia of spinal and cranial nerves.

A **nerve** is a collection of axons outside the CNS. Most nerves are composed of both motor and sensory fibers and are thus called *mixed nerves.* Some of the cranial nerves, however, contain only sensory processes. These are the nerves that serve the special senses of sight, hearing, taste, and smell.

**Figure 7.3.** The relationship between sensory and motor fibers of the peripheral nervous system (*PNS*) and the central nervous system (*CNS*).

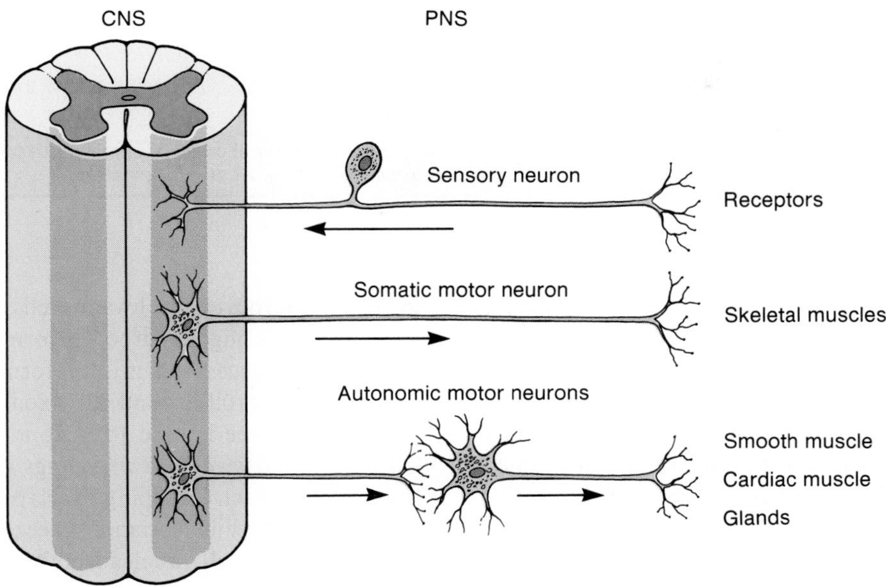


**Figure 7.4.** Three different types of neurons.

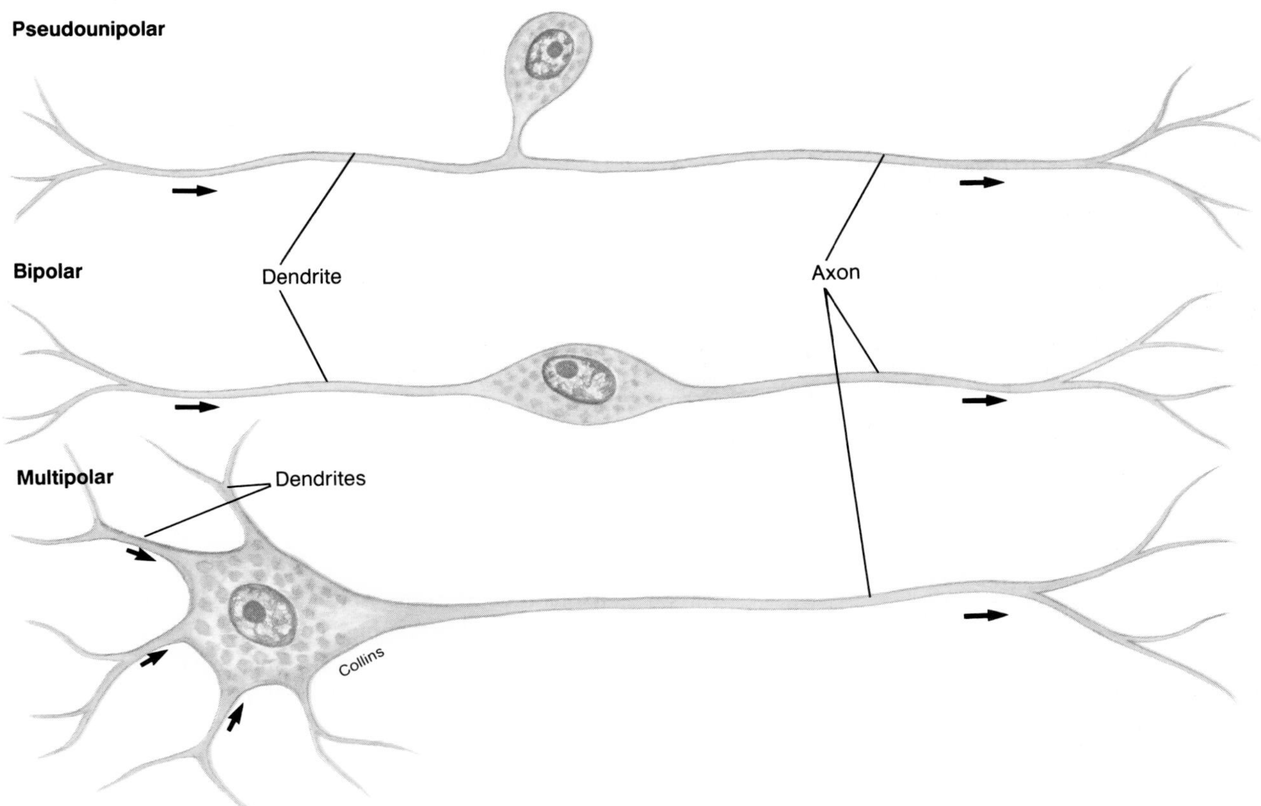

**Table 7.3** Some types of neuroglial cells and their functions

| Neuroglia | Functions |
|---|---|
| Schwann cells | Surround axons of all peripheral nerve fibers, forming neurilemmal sheath, or sheath of Schwann; wrap around many peripheral fibers to form myelin sheaths |
| Oligodendrocytes | Form myelin sheaths around central axons, producing "white matter" of CNS |
| Astrocytes | Perivascular foot processes, covering capillaries within brain and contributing to the blood-brain barrier |
| Microglia | Amoeboid cells within CNS that are phagocytic |
| Ependyma | Form epithelial lining of brain cavities (ventricles) and central canal of spinal cord; cover tufts of capillaries to form choroid plexus—structures that produce cerebrospinal fluid |

## Neuroglia

Unlike other organs that are "packaged" in connective tissue derived from mesoderm (the middle layer of embryonic tissue), the supporting neuroglial cells of the nervous system are derived from the same embryonic tissue layer (ectoderm) that produces neurons. There are six categories of neuroglial cells: (1) **Schwann cells,** which form myelin sheaths around peripheral axons; (2) **oligodendrocytes,** which form myelin sheaths around axons of the CNS; (3) **microglia,** which are phagocytic cells that migrate through the CNS and remove foreign and degenerated material; (4) **astrocytes,** which may help regulate the external environment of neurons in the CNS; (5) **ependyma,** which line the ventricles of the brain and the central canal of the spinal cord; and (6) **satellite cells,** which support neuron cell bodies within the ganglia of the PNS (table 7.3).

***Sheath of Schwann and Myelin Sheath.*** Some axons in the CNS and PNS are surrounded by a myelin sheath and are known as *myelinated* axons. Other axons do not have a myelin sheath and are *unmyelinated.*

All axons in the PNS, but not in the CNS, are surrounded by a living sheath of Schwann cells, known as the **sheath of Schwann.** The outer surface of this layer of Schwann cells is encased in a glycoprotein *basement membrane,* often called the *neurilemma,* which is analogous to the basement membrane that underlies epithelial membranes. The axons of the CNS, in contrast, lack a sheath of Schwann (because Schwann cells are only found in the PNS) and also lack a continuous basement membrane. This is significant in terms of nerve regeneration, as will be described in a later section.

Axons that are less than two micrometers in diameter are usually unmyelinated. Larger axons are generally surrounded by a **myelin sheath,** which is composed of successive wrappings of the cell membrane of Schwann cells or oligodendrocytes. Schwann cells form myelin in the PNS, whereas oligodendrocytes form myelin in the CNS.

In the process of myelin formation in the PNS, Schwann cells roll around the axon, much like a roll of electrician's tape is used to wrap a wire. Unlike electrician's tape, however, the wrappings are made in the same spot, so that each wrapping overlaps the others. The cytoplasm, meanwhile, becomes squeezed to the outer region of the Schwann cell, much as toothpaste is squeezed to the top of the tube as the bottom is rolled up (fig. 7.5). Each Schwann cell only wraps about 1 mm of axon, leaving gaps of exposed axon between the adjacent Schwann cells. These gaps in the myelin sheath are known as the **nodes of Ranvier.** The successive wrappings of Schwann cell membrane provide insulation around the axon, leaving only the nodes of Ranvier exposed to produce nerve impulses.

The Schwann cells remain alive as their cytoplasm is squeezed to the outside of the myelin sheath. As a result, myelinated axons of the PNS, like their unmyelinated counterparts, are surrounded by a living sheath of Schwann, to the outside of which is a continuous basement membrane (fig. 7.6).

The myelin sheaths of the CNS are formed by oligodendrocytes. Unlike a Schwann cell, which forms a myelin sheath around only one axon, each oligodendrocyte has extensions, like the tentacles of an octopus, that form myelin sheaths around several axons (fig. 7.7). Myelinated axons of the CNS, as a result, are not surrounded by a continuous basement membrane. The myelin sheaths around axons of the CNS give this tissue a white color; areas of the CNS that contain a high concentration of axons thus form the **white matter**. The **gray matter** of the CNS is composed of high concentrations of cell bodies and dendrites, which lack myelin sheaths. (The only dendrites that have myelin sheaths are those of the sensory neurons of the PNS.)

**Figure 7.5.** The formation of a myelin sheath in a peripheral axon. The myelin sheath is formed by successive wrappings of the Schwann cell membranes, leaving most of the Schwann cell cytoplasm outside the myelin. The neurilemmal sheath of Schwann cells is thus located outside the myelin sheath.

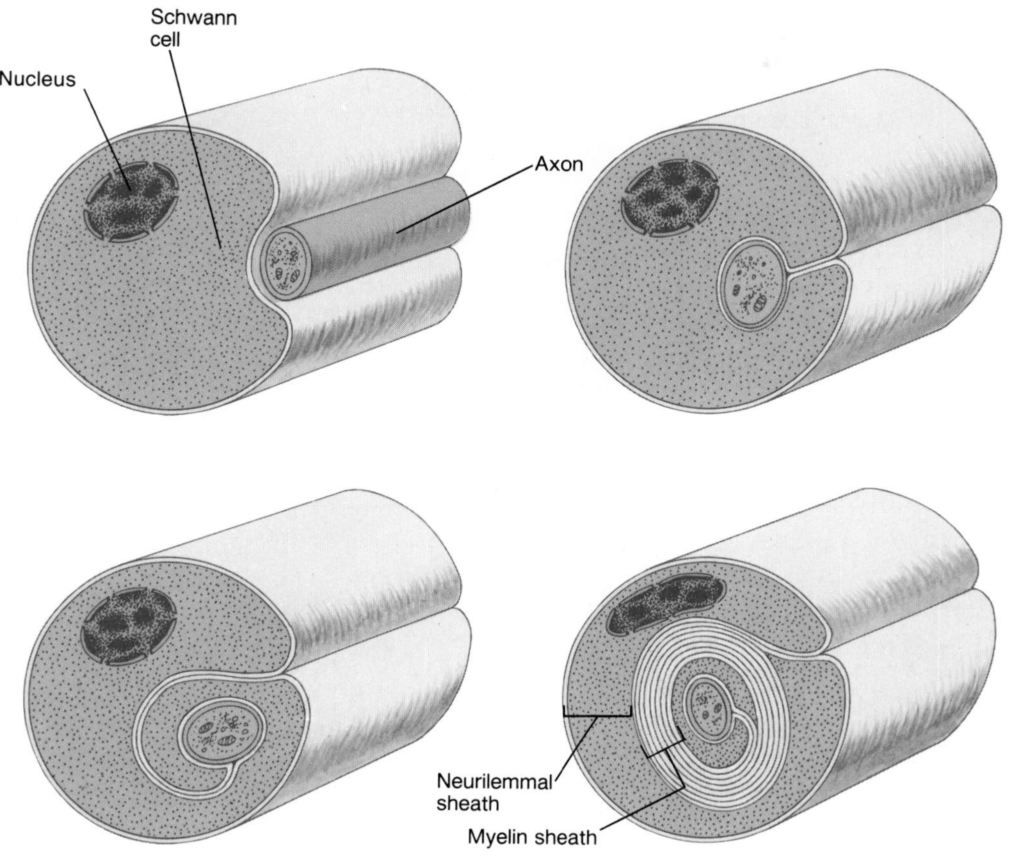


**Multiple sclerosis (MS)** is a relatively common neurological disease in persons between the ages of twenty and forty. MS is a chronic, degenerating, remitting, and relapsing disease that progressively destroys the myelin sheaths of neurons in multiple areas of the CNS. Initially, lesions form on the myelin sheaths and soon develop into hardened *scleroses* (*skleros* = hardened). Destruction of the myelin sheaths prohibits the normal conduction of impulses, resulting in a progressive loss of functions. Because myelin degeneration is widespread, MS has a greater variety of symptoms than any other neurological disease. This characteristic, coupled with remissions, frequently causes misdiagnosis of this disease.

***Regeneration of a Cut Axon.*** When an axon in a peripheral nerve is cut, the distal portion of the axon that was severed from the cell body degenerates and is phagocytosed by Schwann cells. The Schwann cells, surrounded by the basement membrane, then form a *regeneration tube* (fig. 7.8), as the part of the axon that is connected to the cell body begins to grow and exhibit amoeboid movement. The Schwann cells of the regeneration tube are believed to secrete chemicals that attract the growing axon tip, and the regeneration tube helps to guide the regenerating axon to its proper destination. Even a severed major nerve may be surgically reconnected and the function of the nerve largely reestablished if the surgery is performed before tissue death.

**Figure 7.6.** An electron micrograph of unmyelinated and myelinated axons.

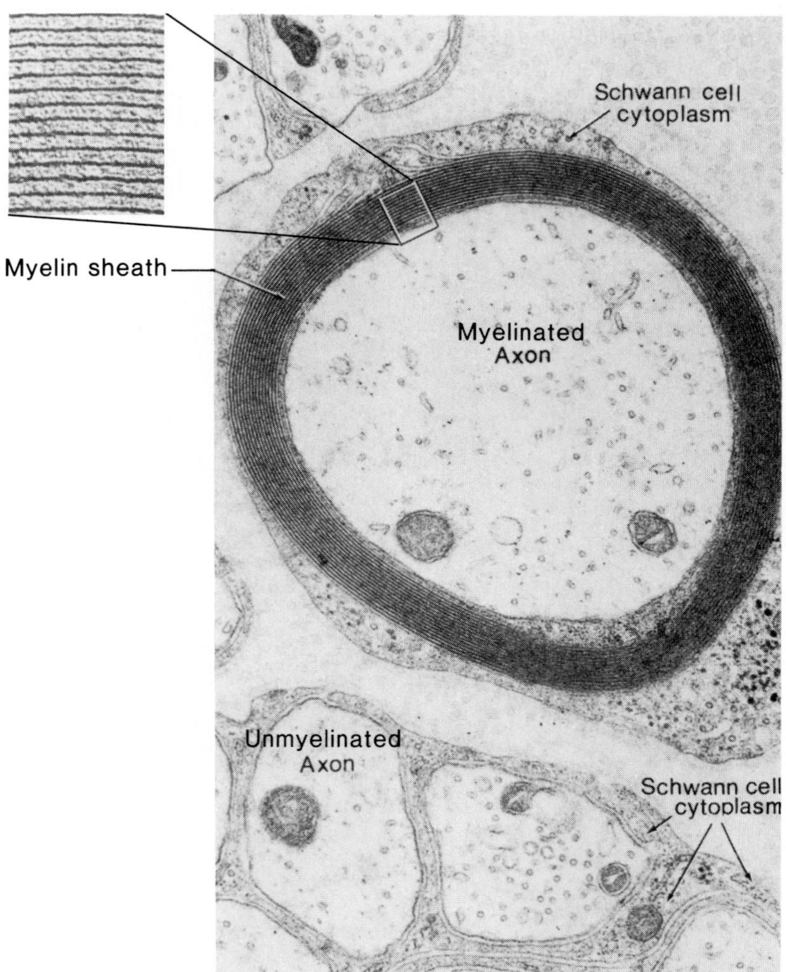


Injury in the CNS stimulates growth of axon collaterals, but central axons have a much more limited ability to regenerate than peripheral axons. This is believed to be primarily due to the absence of a continuous basement membrane, so that a regeneration tube cannot be formed.

Experiments *in vitro* suggest that central axons can regenerate if they are provided with the appropriate environment. In a developing fetal brain, chemicals that include *nerve growth factor* promote axon growth, and such chemicals have been shown in some experimental animals to promote neuron regeneration in adult brains. Grafts of fetal brain tissue into adult brains have similar effects. In another approach to the problem, researchers have shown that providing a basement membrane derived from a human amnion (the membrane surrounding a fetus) can promote axon regeneration in the adult brains of experimental animals. With continued experimentation, it may someday be possible to apply these findings clinically in the regeneration of brain tissue that has been damaged by disease or injury.

***The Functions of Astrocytes and the Blood-Brain Barrier.*** Astrocytes (*aster* = star) are large, stellate cells with numerous cytoplasmic processes that radiate outwards. They are the most abundant of the neuroglial cells in the CNS, constituting up to 90% of the nervous tissue in some areas of the brain.

Astrocytes (fig. 7.9) are known to interact with neurons in two different ways. First, they have been shown to take up potassium ions from the extracellular fluid. Since $K^+$ is released from active neurons during the production of nerve impulses (discussed in a later section), this action of astrocytes may be very important in maintaining a proper ionic environment for the neurons. Second, astrocytes have been shown to take up specific neurotransmitter chemicals (which are released from the axon endings, as described in a later section). These neurotransmitters, known as glutamic acid and gamma-aminobutyric acid (GABA), are broken down within the astrocytes. The molecule produced from this breakdown—glutamine—is released from the astrocytes and made available to the neurons in order for them to resynthesize these particular neurotransmitters.

**Figure 7.7.** Formation of myelin sheaths in the central nervous system by an oligodendrocyte. One oligodendrocyte forms myelin around several axons.

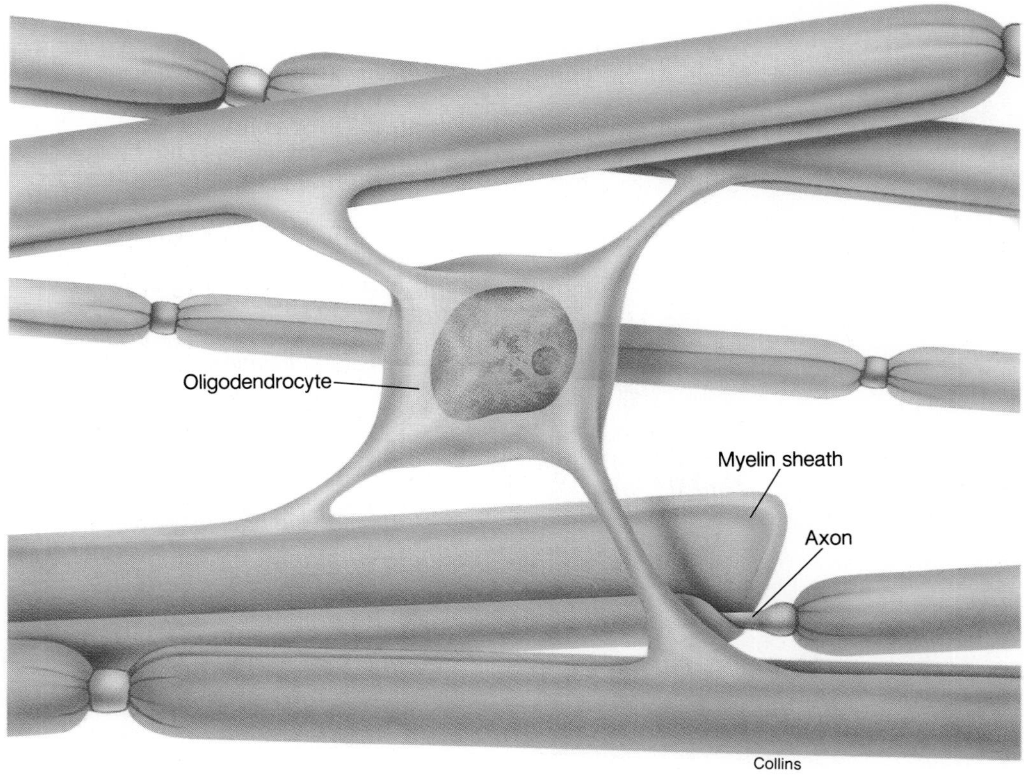


**Figure 7.8.** The process of neuron regeneration. (*a*) If a neuron is severed through a myelinated axon, the proximal portion may survive, but (*b*) the distal portion degenerates through phagocytosis. The myelin sheath provides a pathway for (*c* and *d*) the regeneration of an axon, and (*e*) innervation is restored.

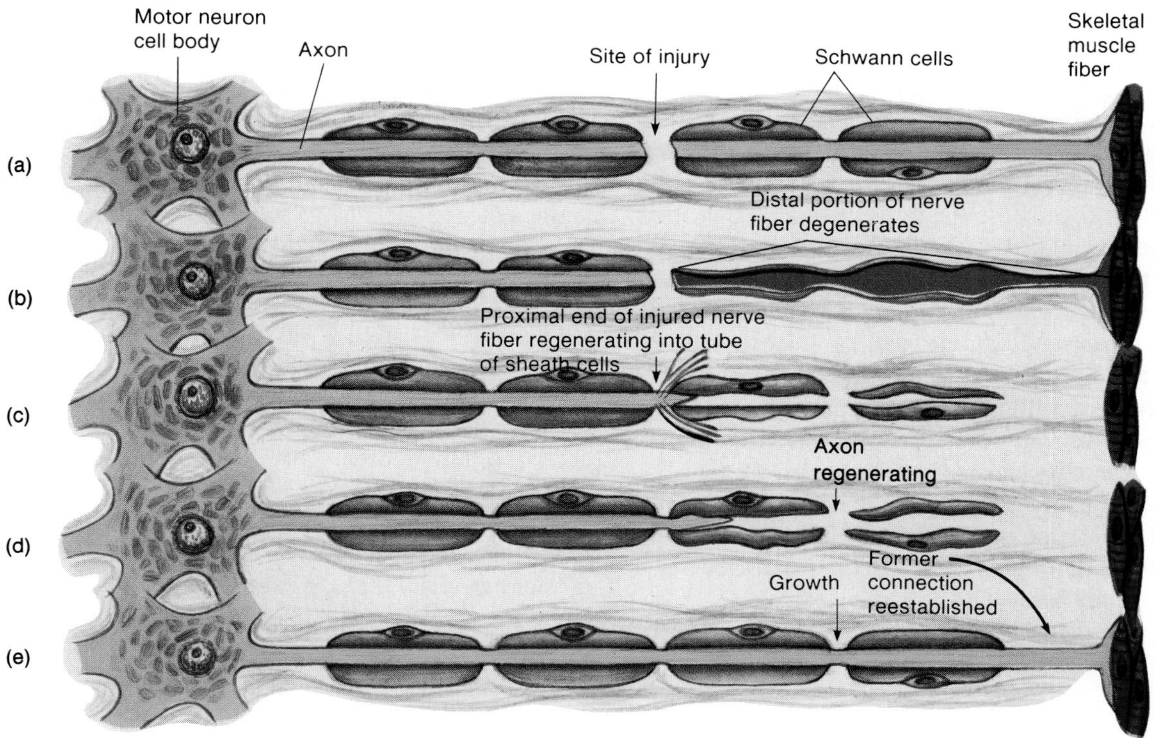

**Figure 7.9.** A photomicrograph showing the perivascular feet of astrocytes (a type of neuroglial cell), which cover most of the surface area of brain capillaries.

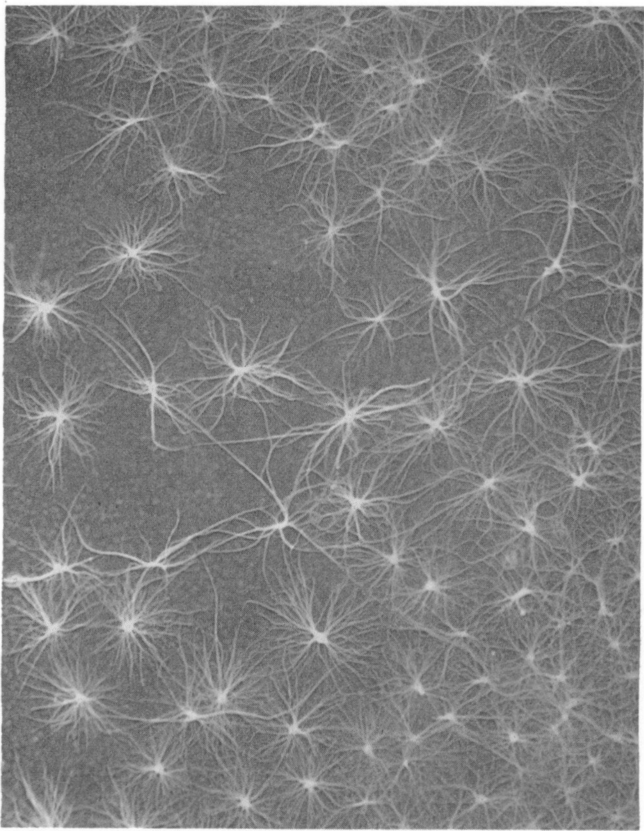

Astrocytes have also been shown to interact with blood capillaries within the brain. Indeed, the brain capillaries are almost entirely surrounded by extensions of the astrocytes known as *perivascular feet.* This association between astrocytes and brain capillaries has important physiological consequences.

Capillaries in the brain, unlike those of most other organs, do not have pores between adjacent endothelial cells (the cells that compose the walls of capillaries). Instead, the endothelial cells of brain capillaries are joined together by tight junctions. Unlike other organs, therefore, the brain cannot obtain molecules from the blood plasma by a nonspecific filtering process. Instead, molecules within brain capillaries must be moved through the endothelial cells by diffusion, active transport, endocytosis, and exocytosis. This imposes a very selective **blood-brain barrier.** There is evidence to suggest that the development of tight junctions between adjacent endothelial cells in brain capillaries, and thus the development of the blood-brain barrier, results from the effects of astrocytes on the brain capillaries.

The blood-brain barrier presents difficulties in the chemotherapy of brain diseases because drugs that could enter other organs may not be able to enter the brain. In the treatment of Parkinson's disease, for example, patients who need a chemical called dopamine in the brain must be given a precursor molecule called levodopa (L-dopa). This is because dopamine cannot cross the blood-brain barrier, whereas L-dopa can enter the neurons and be changed to dopamine in the brain.

1. *Draw a neuron, label its parts, and describe the functions of these parts.*
2. *Distinguish between sensory neurons, motor neurons, and interneurons in terms of their structure, location, and function.*
3. *Describe the structure of the sheath of Schwann and how it functions to promote nerve regeneration. Explain how a myelin sheath in the PNS is formed.*
4. *Explain how myelin sheaths in the CNS are formed, and how they compare with myelin sheaths in the PNS. Describe how the presence or absence of myelin sheaths in the CNS determines the color differences of this tissue.*
5. *Explain the nature of the blood-brain barrier, and describe its structure and clinical significance.*

## *Electrical Activity in Nerve Fibers*

The permeability of the axon membrane to $Na^+$ and $K^+$ is regulated by gates, which open in response to a reduction in the membrane potential. This permits the passive movement of these ions down their electrochemical gradients; first $Na^+$ moves into the axon, then $K^+$ moves out. This flow of ions occurs in a very limited region of membrane, and momentarily changes the potential across that membrane in a characteristic way. This flow of ions, and the changes in potential difference that result, is called an action potential or a nerve impulse.

All cells in the body maintain a potential difference (voltage) across the membrane, or *resting membrane potential,* in which the inside of the cell is negatively charged in comparison to the outside of the cell (for example, $-65$ mV). As explained in chapter 6, this potential difference is largely the result of the permeability properties of the cell membrane, which traps large, negatively charged organic molecules within the cell and which permits only limited diffusion of positively charged inorganic ions, which thus have an unequal distribution across the membrane. Also, the action of the $Na^+/K^+$ pumps help to maintain a potential difference because of the fact that they pump out three $Na^+$ ions for every two $K^+$ ions that they transport into the cell. $Na^+$ is thus more highly concentrated in the extracellular fluid than in the cell, whereas $K^+$ is at a higher concentration within the cell.

**Figure 7.10.** The difference in potential (in millivolts) between an intracellular and extracellular recording electrode is displayed on an oscilloscope screen. The resting membrane potential (*rmp*) of the axon may be reduced (depolarization) or increased (hyperpolarization).

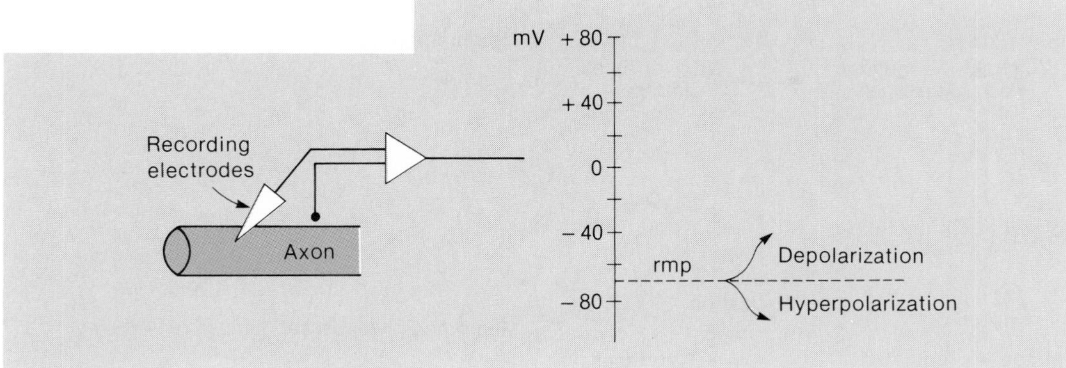


Although all cells have a membrane potential, only a few types of cells have been shown to alter their membrane potential in response to stimulation. Such alterations in membrane potential are achieved by varying the membrane permeability to specific ions in response to stimulation. A central aspect of the physiology of neurons and muscle cells is their ability to produce and conduct these changes in membrane potential. Such an ability is termed *excitability* or *irritability*.

An increase in membrane permeability to a specific ion results in the diffusion of that ion down its concentration gradient, either into or out of the cell. These *ion currents* occur only across limited patches of membrane (located fractions of a millimeter apart), where specific ion channels are located. Changes in the potential difference across the membrane at these points can be measured by the voltage developed between two electrodes—one placed inside the cell, the other placed outside the cell membrane at the region being recorded. The voltage between these two *recording electrodes* can be visualized by connecting these electrodes to an oscilloscope (fig. 7.10).

In an oscilloscope, electrons from a cathode ray "gun" are sprayed across a fluorescent screen, producing a line of light. Changes in the potential difference between the two recording electrodes produce a deflection of this line. The oscilloscope can be calibrated in such a way that an upward deflection of this line indicates that the inside of the membrane has become less negative (or more positive) compared to the outside of the membrane. A downwards deflection of the line, conversely, indicates that the inside of the cell has become more negative. The oscilloscope can thus function as a fast-responding voltmeter with an ability to display voltage changes as a function of time.

If both recording electrodes were placed outside of the cell, the potential difference between the two would be zero (because there is no charge separation). When one of the two electrodes penetrates the cell membrane, the oscilloscope shows that the intracellular electrode is electrically negative with respect to the extracellular electrode; a membrane potential is recorded. If appropriate stimulation causes positive charges to flow into the cell, the line would deflect upward; this change is called **depolarization,** or hypopolarization, because the potential difference between the two recording electrodes is reduced. If the inside of the membrane becomes more negative as a result of stimulation, the line on the oscilloscope would deflect downwards; this is called **hyperpolarization.**

## Ion Gating in Axons

The permeability of the membrane to $Na^+$, $K^+$, and other ions is regulated by parts of the ion channels through the membrane called **gates.** Gates are believed to be composed of polypeptide chains that can open or close a membrane channel according to specific conditions. When the gates of specific ion channels are closed, the membrane is not very permeable to that ion, and when the gates are opened, the permeability to that ion can be greatly increased.

It is believed that there are two types of channels for $K^+$; one type lacks gates and is always open, whereas the other type has gates that are closed in the resting cell. Channels for $Na^+$, in contrast, always have gates, and these gates are closed in the resting cell. The resting cell is thus more permeable to $K^+$ than to $Na^+$ (some $Na^+$ does leak into the cell, as described in chapter 6; this leakage may occur in a nonspecific manner through open $K^+$ channels). The resting membrane potential is thus close to, but slightly less than, the equilibrium potential for $K^+$ (described in chapter 6).

**Figure 7.11.** Depolarization of an axon has two effects: (*1*) $Na^+$ gates open and $Na^+$ diffuses into the cell and (*2*) after a brief period, $K^+$ gates open and $K^+$ diffuses out of the cell. An inward diffusion of $Na^+$ causes further depolarization—this causes further opening of $Na^+$ gates in a positive feedback (+) fashion. The opening of $K^+$ gates and outward diffusion of $K^+$ make the inside of the cell more negative and thus have a negative feedback effect (−) on the initial depolarization.

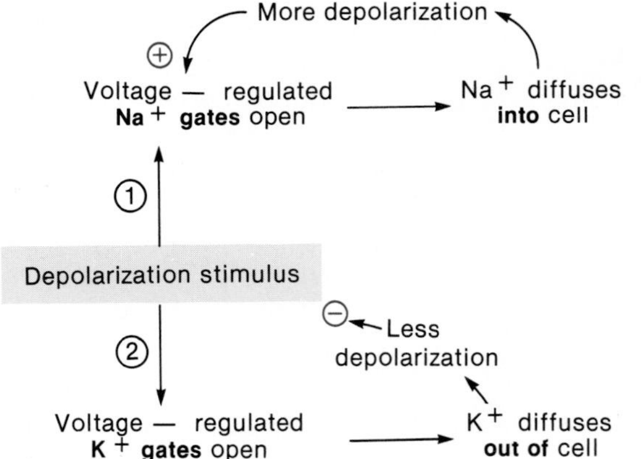
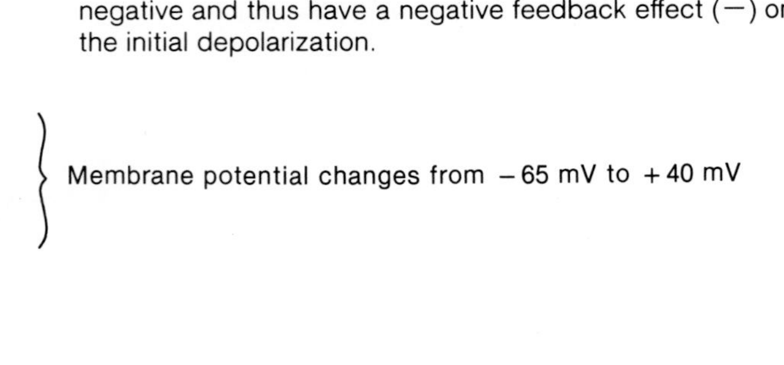


Whether the gates for the $Na^+$ and $K^+$ channels are open or closed depends on the membrane potential. The gated channels are closed at the resting membrane potential of −65 mV, but they open when the membrane is depolarized to a certain threshold level. Since the opening and closing of these gates is regulated by the membrane voltage, the gates are said to be **voltage regulated.**

Depolarization of a small region of an axon can be experimentally induced by a pair of stimulating electrodes, which act as if they inject positive charges into the axon. If a pair of recording electrodes are placed in the same region (one electrode within the axon and one outside), an upward deflection of the oscilloscope line will be observed as a result of this depolarization. If a certain level of depolarization is achieved (from −65 mV to −55 mV, for example) by this artificial stimulation, a sudden and very rapid change in the membrane potential will be observed. This is due to the fact that *depolarization to a threshold level causes the $Na^+$gates to open.* Now the permeability properties of the membrane are changed, and $Na^+$ diffuses down its concentration gradient into the cell.

A fraction of a second after the $Na^+$ gates open, they close again. At this time, *the depolarization stimulus then causes the $K^+$ gates to open.* This makes the membrane more permeable to $K^+$ than it is at rest, and $K^+$ diffuses down its concentration gradient out of the cell. The $K^+$ gates will then close and the permeability properties of the membrane will return to what they were at rest.

## Action Potentials

The information in this section relates to events that occur at one point in an axon, when a small region of axon membrane is stimulated artificially and responds with changes in ion permeabilities. The resulting changes in membrane potential at this point are detected by recording electrodes placed in this region of the axon. The nature of the stimulus *in vivo* (in the body), and the manner by which electrical events are conducted to different points along the axon, are discussed in later sections.

When the axon membrane has been depolarized to a threshold level—by stimulating electrodes, in the previous example—the $Na^+$ gates open and the membrane becomes permeable to $Na^+$. This permits $Na^+$ to enter the axon by diffusion, which further depolarizes the membrane (makes the inside less negative, or more positive). Since the $Na^+$ gates of the axon are voltage regulated, this further depolarization makes the membrane even more permeable to $Na^+$, so that even more $Na^+$ can enter the cell and open even more voltage-regulated $Na^+$ gates. A *positive feedback loop* (fig. 7.11) is thus created, which causes the rate of $Na^+$ entry and depolarization to accelerate in an explosive fashion.

After a slight time delay, depolarization of the axon membrane also causes the opening of voltage-regulated $K^+$ gates, and the diffusion of $K^+$ out of the cell. Since $K^+$ is positively charged, the diffusion of $K^+$ out of the cell makes the inside of the cell less positive, or more negative, and acts to restore the original resting membrane potential. This process is called **repolarization** and represents the completion of a *negative feedback loop* (fig. 7.11).

**Figure 7.12.** An action potential (*upper figure*) is produced by an increase in sodium diffusion followed, with a short time delay, by an increase in potassium diffusion (*lower figure*). This drives the membrane potential first toward the sodium equilibrium potential and then toward the potassium equilibrium potential.

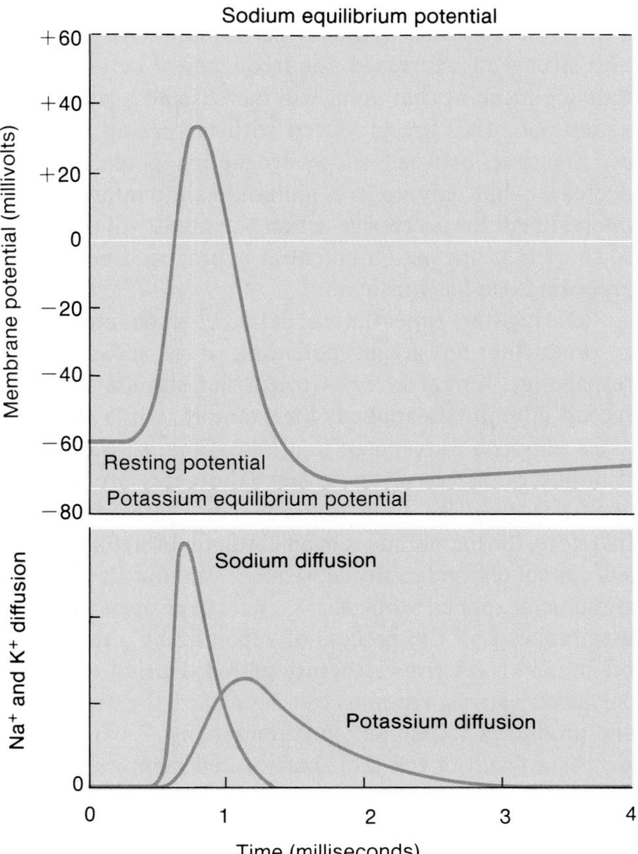


Figure 7.12 (bottom) illustrates the movement of $Na^+$ and $K^+$ through the axon membrane in response to a depolarization stimulus. Notice that the explosive increase in $Na^+$ diffusion causes rapid depolarization to 0 mV and then *overshoot* of the membrane potential so that the inside of the membrane actually becomes positively charged (almost +40 mV) compared to the outside (fig. 7.12, top). The greatly increased permeability to $Na^+$ thus drives the membrane potential toward the equilibrium potential for $Na^+$ (chapter 6). The $Na^+$ permeability then rapidly decreases as the diffusion of $K^+$ increases, resulting in repolarization to the resting membrane potential. These changes in $Na^+$ and $K^+$ diffusion and the resulting changes in the membrane potential that they produce constitute an event called the **action potential,** or **nerve impulse.**

Once an action potential is completed, the $Na^+/K^+$ pumps will, by active transport, extrude the extra $Na^+$ that has entered the axon and recover the $K^+$ that has diffused out of the axon. This occurs very quickly because the events described occur across only a very small area of membrane, and so only a relatively small amount of $Na^+$ and $K^+$ actually diffuse through the membrane during the production of an action potential. The total concentrations of $Na^+$ and $K^+$ in the axon and in the extracellular fluid are not significantly changed during an action potential. Even during the overshoot phase, for example, the concentration of $Na^+$ remains higher outside the axon; repolarization thus requires the outward diffusion of $K^+$, which has a concentration gradient in a direction opposite to the $Na^+$ gradient.

Notice that active transport processes are not directly involved in the production of an action potential; both depolarization and repolarization are produced by the diffusion of ions down their concentration gradients. A neuron poisoned with cyanide, so that it cannot produce ATP, can still produce action potentials for a period of time. After a while, however, the lack of ATP for active transport by the $Na^+/K^+$ pumps will result in a decline in the ability of the axon to produce action potentials. This shows that the $Na^+/K^+$ pumps are not directly involved, but are instead required to maintain the concentration gradients needed for the diffusion of $Na^+$ and $K^+$ during action potentials.

***All-or-None Law.*** Once a region of axon membrane has been depolarized to a threshold value, the positive feedback effect of depolarization on $Na^+$ permeability and of $Na^+$ permeability on depolarization causes the membrane potential to shoot toward about +40 mV. It does not normally become more positive because of the fact that the $Na^+$ gates quickly close and the $K^+$ gates open. The length of time that the $Na^+$ and $K^+$ gates stay open is independent of the strength of the depolarization stimulus.

The amplitude of action potentials is therefore **all-or-none.** When depolarization is below a threshold value, the voltage-regulated gates are closed; when depolarization reaches threshold, a maximum potential change (the action potential) is produced. Since the change from −65 mV to +40 mV and back to −65 mV lasts only about three msec (milliseconds), the image of an action potential on an oscilloscope screen looks like a spike. Action potentials are therefore sometimes called *spike potentials.*

**Figure 7.13.** Recordings from a single sensory fiber of a sciatic nerve of a frog stimulated by varying degrees of stretch of the gastrocnemius muscle. Note that increasing amounts of stretch (indicated by increasing weights attached to the muscle) result in an increased frequency of action potentials.

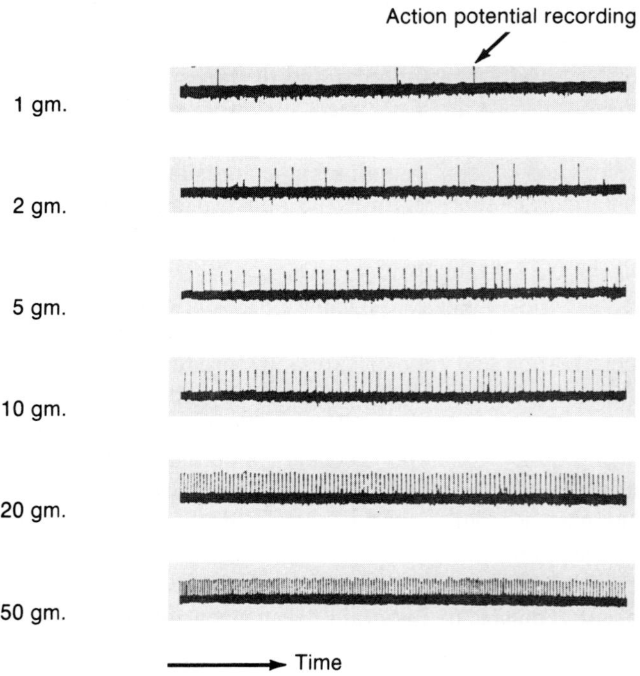


Both the rising (depolarization) and falling (repolarization) phases of an action potential are, in summary, produced by gating the *diffusion* of ions down their opposing concentration gradients. Since the gates are open for a fixed period of time, the duration of each action potential is about the same. Likewise, since the concentration gradient for $Na^+$ is relatively constant, the amplitude of each action potential is about the same in all axons at all times (from $-65$ mV to $+40$ mV, or about 100 mV in total amplitude).

***Coding for Stimulus Intensity.*** If one depolarization stimulus is greater than another, the greater stimulus strength is not coded by a greater amplitude of action potentials (because action potentials are all-or-none). The code for stimulus strength in the nervous system is not **AM** (amplitude modulated). When a greater stimulus strength is applied to a neuron, identical action potentials are produced more frequently (more are produced per minute). Therefore, the code for stimulus strength in the nervous system is frequency modulated (**FM**). This is shown in figure 7.13.

When an entire collection of axons (in a nerve) is stimulated, different axons will be stimulated at different stimulus intensities. A low-intensity stimulus will only activate those few fibers with low thresholds, whereas high-intensity stimuli can activate fibers with higher thresholds. As the intensity of stimulation increases, more and more fibers will become activated. This process, called **recruitment,** represents another mechanism by which the nervous system can code for stimulus strength.

***Refractory Periods.*** If a stimulus of a given intensity is maintained at one point of an axon and depolarizes it to threshold, action potentials will be produced at that point at a given frequency (number per minute). As the stimulus strength is increased, the frequency of action potentials produced at that point will increase accordingly. As action potentials are produced with increasing frequency, the time between successive action potentials will decrease—but only up to a minimum time interval. The interval between successive action potentials will never get so short that one action potential is produced before the preceding one has finished.

During the time that a patch of axon membrane is producing an action potential, it is incapable of responding—or *refractory*—to further stimulation. If a second stimulus is applied, for example, while the $Na^+$ gates are open in response to a first stimulus, the second stimulus cannot have any effect (the gates are already open). During the time that the $Na^+$ gates are open, therefore, the membrane is in an **absolute refractory period** and cannot respond to any subsequent stimulus. If a second stimulus is applied while the $K^+$ gates are open (and the membrane is in the process of repolarizing), the membrane is in a **relative refractory period.** During this time only a very strong stimulus can depolarize the membrane and produce a second action potential (fig. 7.14).

As a result of the fact that the cell membrane is refractory during the time it is producing an action potential, each action potential remains a separate, all-or-none event. In this way, as a continuously applied stimulus increases in intensity, its strength can be coded strictly by the frequency of the action potentials it produces at each point of the axon membrane.

One might think that as an axon produces a large number of action potentials, the relative concentrations of $Na^+$ and $K^+$ would be changed in the extracellular and intracellular compartments. This is not true. In a typical mammalian axon that is one micrometer in diameter, for example, only one intracellular $K^+$ ion in three thousand would be exchanged for a $Na^+$. Since a typical neuron has about one million $Na^+/K^+$ pumps, able to transport nearly 200 million ions per second, these small changes can be quickly corrected.

***Cable Properties of Neurons.*** If a pair of stimulating electrodes produces a depolarization that is too weak to cause the opening of voltage-regulated $Na^+$ gates—that is, if the depolarization is below threshold (about $-55$ mV)—the change in membrane potential will be *localized* to within one to two millimeters of the point of stimulation. For example, if the stimulus causes depolarization from $-65$ mV to $-60$ mV at one point, and the recording

**Figure 7.14.** The absolute and relative refractory periods. While a segment of axon is producing an action potential, the membrane is absolutely or relatively resistant (refractory) to further stimulation.

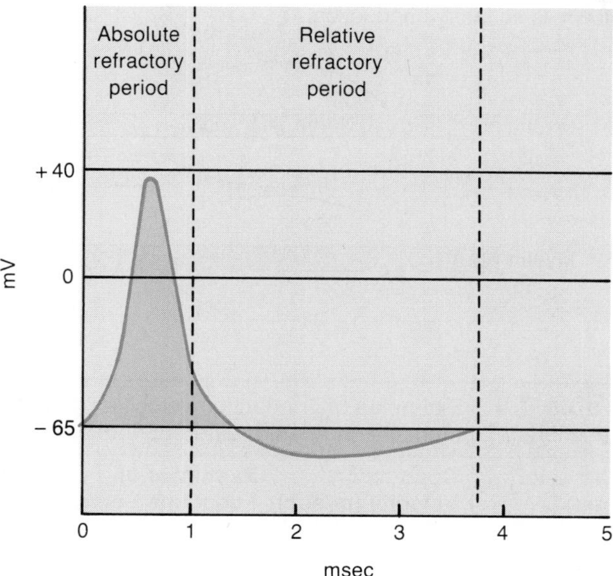


**Figure 7.15.** The conduction of a nerve impulse (action potential) in an unmyelinated nerve fiber (axon). Each action potential "injects" positive charges that spread to adjacent regions. The region that has previously produced an action potential is refractory. The previously unstimulated region is partially depolarized. As a result, its voltage-regulated $Na^+$ gates open, repeating the process.

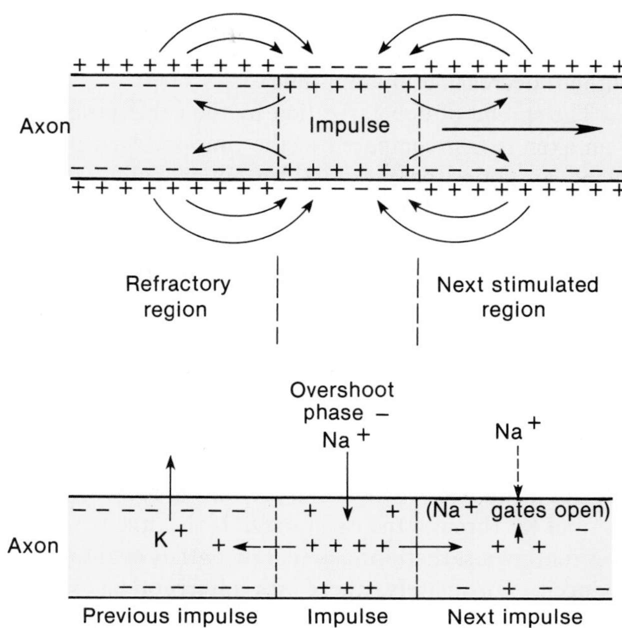


electrodes are placed only three millimeters away from the stimulus, the membrane potential recorded will remain at $-65$ mV (the resting potential). The axon is thus a very poor conductor compared to metal wires.

The ability of a neuron to transmit charges through its cytoplasm is known as its *cable properties.* These cable properties are quite poor because there is a high internal resistance to the spread of charges and because many charges leak out of the axon through its membrane. If an axon had to conduct only through its cable properties, therefore, no axon could be more than a millimeter in length. The fact that some axons are a meter or more in length suggests that the conduction of nerve impulses does not rely on the cable properties of the axon.

## Conduction of Nerve Impulses

When stimulating electrodes artificially depolarize one point of an axon membrane to a threshold level, voltage-regulated gates open and an action potential is produced at that small region of axon membrane containing those gates. For about the first millisecond of the action potential, when the membrane voltage changes from $-65$ mV to $+40$ mV, a current of $Na^+$ enters the cell by diffusion due to the opening of the $Na^+$ gates. Each action potential thus "injects" positive charges ($Na^+$ ions) into the axon.

These positively charged $Na^+$ ions are conducted, by the cable properties of the axon, to an adjacent region that still has a membrane potential of $-65$ mV. Within the limits of the cable properties of the axon (one to two millimeters), this helps to depolarize the adjacent region of axon membrane. When this adjacent region of membrane reaches a threshold level of depolarization, it too produces an action potential as its voltage-regulated gates open.

Each action potential thus acts as a stimulus for the production of another action potential at the next region of membrane that contains voltage-regulated gates. In the previous description of action potentials, the stimulus for their production was artificial—depolarization produced by a pair of stimulating electrodes. Now it can be seen that an action potential at one point along an axon is produced by depolarization that results from the production of a preceding action potential. This line of reasoning explains how all action potentials along an axon are produced after the first action potential is generated.

***Conduction in an Unmyelinated Axon.*** In an unmyelinated axon, every patch of membrane that contains $Na^+$ and $K^+$ gates can produce an action potential. Action potentials are thus produced at locations only a fraction of a micrometer apart all along the length of the axon.

The cablelike spread of depolarization induced by $Na^+$ influx during one action potential helps to depolarize the adjacent regions of membrane. This process is also aided by movements of ions on the outer surface of the axon membrane (fig. 7.15). This process would depolarize the adjacent membranes on each side of the region producing an action potential, but the area which had previously produced an action potential cannot produce a new one at this time because it is still in its refractory period. In this way, action potentials are passed in one direction only along the axon.

Notice that action potentials are not really conducted, although it is convenient to use that word. Each action potential is a separate, complete event that is repeated, or *regenerated,* along the axon's length. The action potential produced at the end of the axon is thus a completely new event that was produced in response to depolarization from the previous action potential. The last action potential has the same amplitude as the first. Action potentials are thus said to be **conducted without decrement** (without decreasing in amplitude).

The spread of depolarization by the cable properties of an axon is fast compared to the time involved in producing an action potential. Since action potentials are produced at every fraction of a micrometer in an unmyelinated axon, the conduction rate is relatively slow. This conduction rate is somewhat faster if the unmyelinated axon is thicker, because the ability of fibers to conduct charges by cable properties improves with increasing diameter. The conduction rate is substantially faster if the axon is myelinated.

***Conduction in a Myelinated Axon.*** The myelin sheath provides insulation for the axon, preventing movements of $Na^+$ and $K^+$ through the membrane. If the myelin sheath were continuous, therefore, action potentials could not be produced. Fortunately, there are interruptions in the myelin known as the *nodes of Ranvier.*

Since the cable properties of axons can only conduct depolarizations a very short distance (1 mm–2 mm), it follows that the nodes of Ranvier must be very close together (usually they are about 1 mm apart). Studies have shown that $Na^+$ channels are highly concentrated at the nodes (estimated at 10,000 per square micrometer) and almost absent in the regions of axon membrane between the nodes. Action potentials, therefore, occur only at the nodes of Ranvier (fig. 7.16) and seem to "leap" from node to node; this is called **saltatory conduction** (*saltario* = leap).

Since the cablelike spread of depolarization between the nodes is very fast and fewer action potentials need to be produced per given length of axon, saltatory conduction allows a *faster rate of conduction* than is possible in an unmyelinated fiber. Conduction rates in the human nervous system vary from 1.0 m/sec—in thin, unmyelinated fibers that mediate slow, visceral responses—to greater than 100 m/sec (225 miles per hour)—in thick myelinated fibers involved in quick stretch reflexes in skeletal muscles (table 7.4).

**Figure 7.16.** The conduction of the nerve impulse in a myelinated nerve fiber. Since the myelin sheath prevents inward $Na^+$ current, action potentials can only be produced at the interruptions in the myelin sheath, or nodes of Ranvier. This "leaping" of the action potential from node to node is known as saltatory conduction.

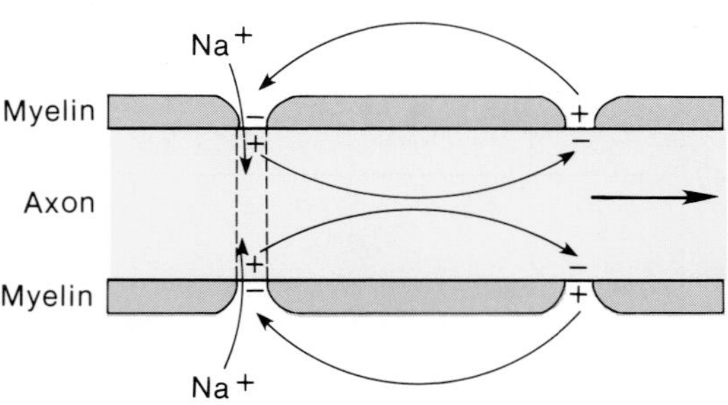


**Table 7.4** Examples of conduction velocities and functions of mammalian nerves of different diameters

| Diameter (μm) | Conduction Velocity (m/sec) | Examples of Functions Served |
|---|---|---|
| 12–22 | 70–120 | Sensory: muscle position |
| 5–13 | 30–90 | Somatic motor fibers |
| 3–8 | 15–40 | Sensory: touch, pressure |
| 1–5 | 12–30 | Sensory: pain, temperature |
| 1–3 | 3–15 | Autonomic fibers to ganglia |
| 0.3–1.3 | 0.7–2.2 | Autonomic fibers to muscles |

1. *Define the meaning of the terms* depolarization *and* repolarization, *and illustrate these processes with a figure.*
2. *Describe how the permeability of the axon membrane to $Na^+$ and $K^+$ is regulated and how changes in permeability to these ions affects the membrane potential.*
3. *Describe how gating of $Na^+$ and $K^+$ in the axon membrane results in the production of an action potential.*
4. *Define the all-or-none law of action potentials, and describe the effect of increased stimulus strength on action potential production. Explain how the refractory periods affect the maximum frequency of action potential production.*
5. *Describe how action potentials are conducted by unmyelinated nerve fibers, and explain why saltatory conduction in myelinated fibers is more rapid.*

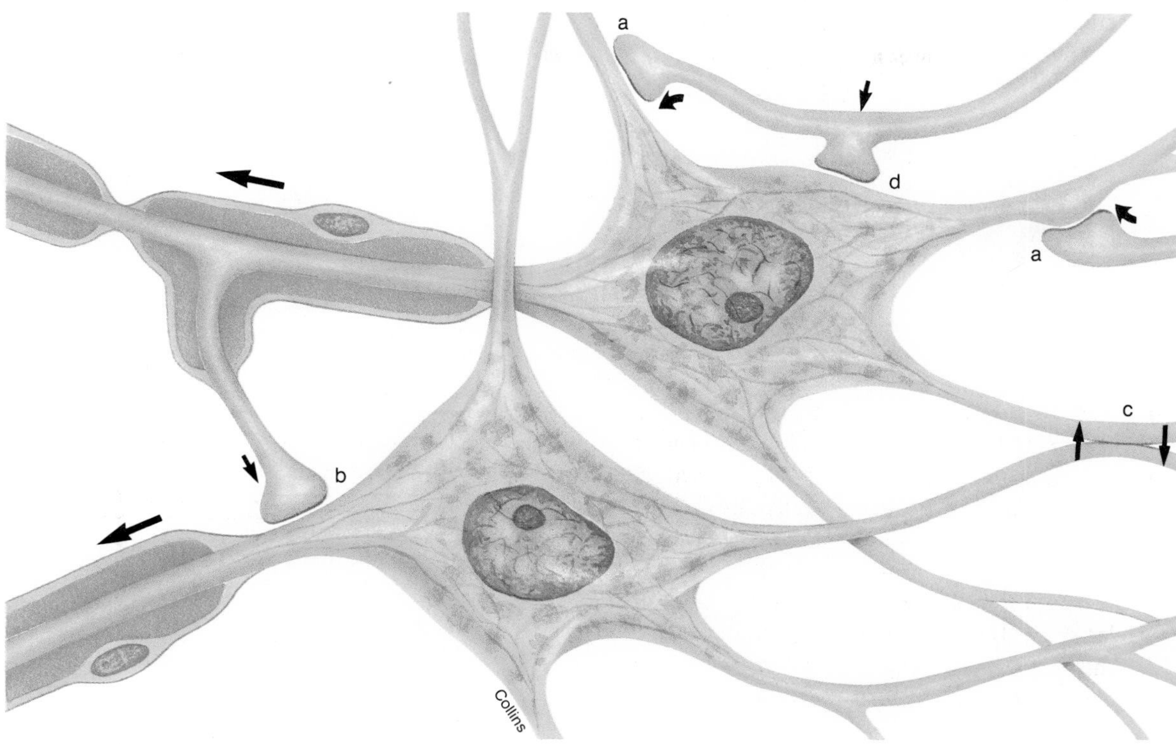


**Figure 7.17.** Different types of synapses: (*a*) axodendritic, (*b*) axoaxonic, (*c*) dendrodendritic, and (*d*) axosomatic.

## *Synaptic Transmission*

Axons end close to, or in some cases in contact with, another cell. Once action potentials reach the end of an axon, they directly or indirectly stimulate the other cell. In specialized cases, action potentials can directly pass from one cell to another. In most cases, however, the action potentials stop at the axon ending, where they stimulate the release of a chemical neurotransmitter that stimulates the next cell.

A **synapse** is the functional connection between a neuron and a second cell. In the CNS this other cell is also a neuron. In the PNS the other cell may be a neuron or an *effector cell* within either a muscle or a gland. Although the physiology of neuron-neuron synapses and neuron-muscle synapses is similar, the latter synapses are often distinguished by the names **myoneural** or **neuromuscular junctions.**

Neuron-neuron synapses usually involve a connection between the axon of one neuron and the dendrites, cell body, or axon of a second neuron. These are called, respectively, *axodendritic, axosomatic,* and *axoaxonic synapses* (fig. 7.17). In almost all synapses, transmission is in one direction only—from the axon of the first (or presynaptic) cell to the second (or postsynaptic) cell. *Dendrodendritic synapses* do not fit this classic pattern; in these synapses, two dendrites from different neurons make reciprocal innervations—some of these synapses conduct in one direction, whereas others conduct in the opposite direction.

In the early part of the twentieth century, most physiologists believed that synaptic transmission was *electrical*—that is, that action potentials were conducted directly from one cell to the next. This belief was reasonable in view of the facts that nerve endings appeared to touch the postsynaptic cells and that the delay in synaptic conduction was extremely short (about 0.5 msec). Improved histological techniques, however, revealed tiny gaps in the synapses, and experiments demonstrated that the actions of autonomic nerves could be duplicated by certain chemicals. This led to the suspicion that synaptic transmission might be *chemical*—that the presynaptic nerve endings might release chemical **neurotransmitters** that stimulated action potentials in the postsynaptic cells.

In 1921, a physiologist named Otto Loewi published the results of an experiment suggesting that synaptic transmission was chemical, at least at the junction between a branch of the vagus nerve (chapter 10) and the heart. He had isolated the heart of a frog and—while stimulating the branch of the vagus that innervates the heart—perfused the heart with an isotonic salt solution. Stimulation of this nerve slowed the heart rate, as expected. More importantly, application of this salt solution to the heart of a second frog caused the second heart to slow its rate of beat.

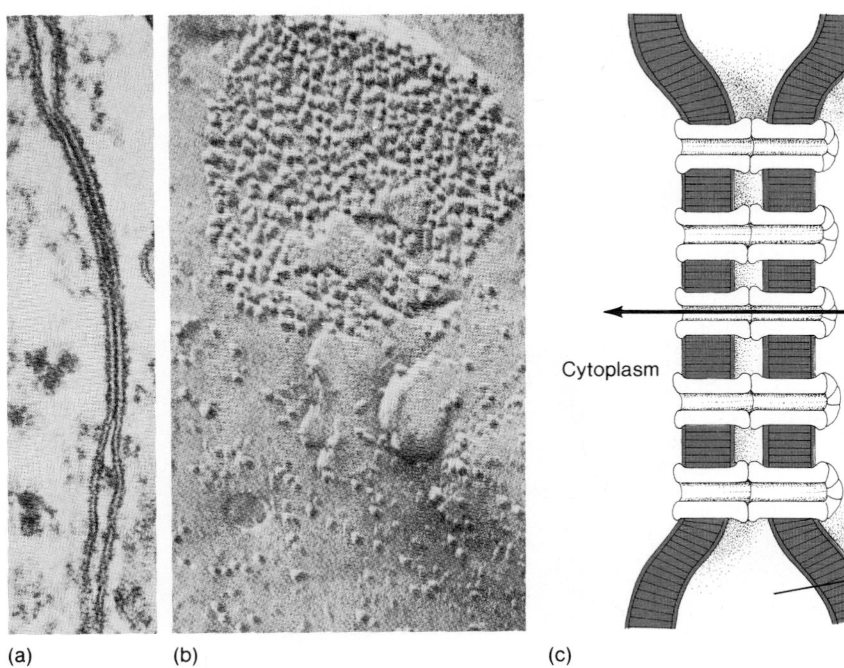


**Figure 7.18.** Gap junctions are shown in electron micrographs, (*a*) and (*b*). (*a*) The photograph shows that cell membranes of two cells are fused together in the gap junction; (*b*) a surface view of a gap junction is seen. (*c*) The information presented in these and other electron micrographs is interpreted by the illustration.

Loewi concluded that the nerve endings of the vagus must have released a chemical—which he called *vagusstoff*—that inhibited the heart rate. This chemical was subsequently identified as **acetylcholine,** or **ACh.** In the decades following Loewi's discovery, many other examples of chemical synapses were discovered, and the theory of electrical synaptic transmission fell into disrepute. More recent evidence, ironically, has shown that electrical synapses do exist in the nervous system (though they are the exception), within smooth muscles, and between cardiac cells in the heart.

## Electrical Synapses: Gap Junctions

In order for two cells to be electrically coupled they must be approximately equal in size and must be joined by areas of contact with low electrical resistance. In this way, impulses can be regenerated from one cell to the next without interruption—and without Frankenstein-like sparks between cells.

Adjacent cells that are electrically coupled are joined together by **gap junctions.** In gap junctions, the membranes of the two cells are separated by only two nanometers (1 nanometer equals $10^{-9}$ meter). When gap junctions are observed in the electron microscope from a surface view, hexagonal arrays of particles are seen that are believed to be channels through which ions and molecules may pass from one cell to the next (fig. 7.18).

Gap junctions are present in smooth and cardiac muscle, where they allow excitation and rhythmic contraction of large masses of muscle cells. Gap junctions have also been observed in various regions of the brain. Although their functional significance here is unknown, it has been speculated that they may allow a two-way transmission of impulses (in contrast to chemical synapses, which are always one-way). Gap junctions have also been observed between neuroglial cells, which do not produce electrical impulses; perhaps these act as channels for the passage of informational molecules between cells. It is interesting in this regard that many embryonic tissues have gap junctions and that these gap junctions disappear as the tissue becomes more specialized.

**Figure 7.19.** An electron micrograph of a chemical synapse, showing synaptic vesicles at the end of an axon.

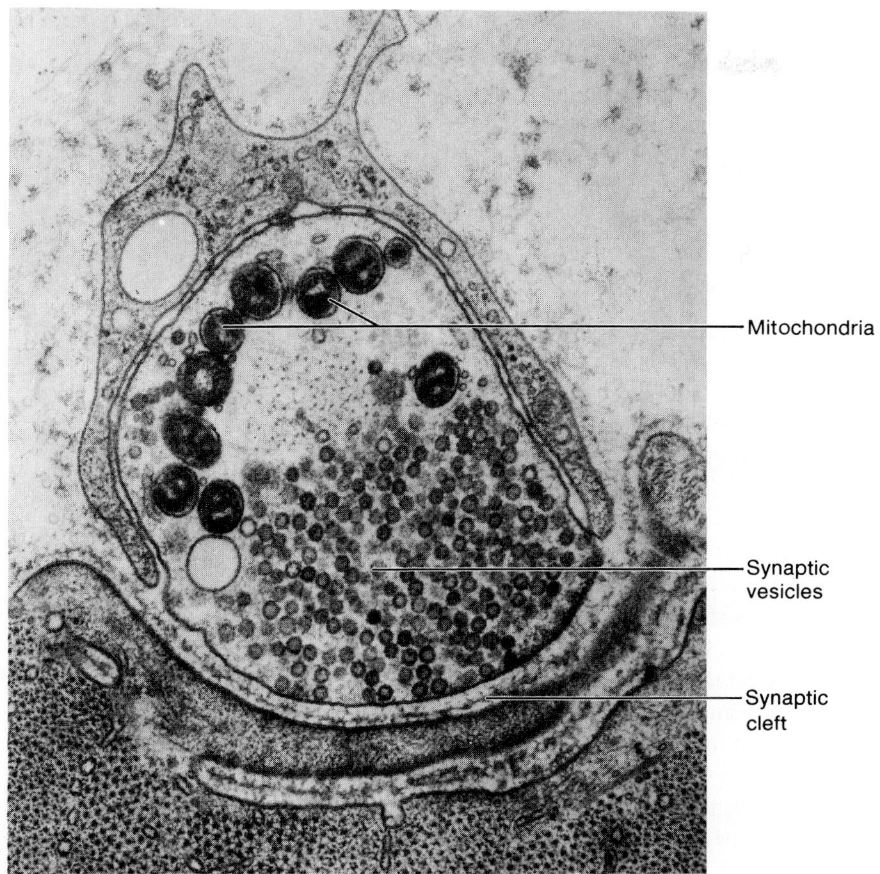


## Chemical Synapses

Transmission across the majority of synapses in the nervous system is one-way and occurs through the release of chemical neurotransmitters from presynaptic axon endings. These presynaptic endings, which are called **terminal boutons** (*bouton* = button) because of their swollen appearance, are separated from the postsynaptic cell by a **synaptic cleft** so narrow that it can only be seen clearly with an electron microscope (fig. 7.19).

Neurotransmitter molecules within the presynaptic neuron endings are contained within many small, membrane-enclosed **synaptic vesicles.** In order for the neurotransmitter within these vesicles to be released into the synaptic cleft, the vesicle membrane must fuse with the axon membrane in the process of *exocytosis.* The neurotransmitter is released in *quanta*—that is, in multiples of the amount contained in one vesicle. The number of vesicles that undergo exocytosis is directly related to the frequency of action potentials produced at the presynaptic axon ending.

The release of neurotransmitters following electrical excitation of the presynaptic endings accounts for most of the 0.5 msec time delay in synaptic transmission. During this time interval there is a sudden, transient inflow of $Ca^{++}$ into the presynaptic endings. This inflow of $Ca^{++}$ is apparently due to opening of $Ca^{++}$ gates in response to electrical excitation and is required for the release of neurotransmitters. The $Ca^{++}$ is believed to activate previously inactive enzymes within the axon terminals. This process may produce a change in the ratio of phospholipids in the axon membrane, which may change the fluidity of the membrane and thus allow exocytosis to occur. Recent evidence has suggested, alternatively, that the $Ca^{++}$ influx may stimulate the production of specific transporters of ACh in the axon membrane. By either mechanism, $Ca^{++}$ serves to couple electrical excitation of the axon ending to the release of neurotransmitter.

1. *Define a synapse, and describe the structures that form synapses.*
2. *Describe the structure, locations, and functions of gap junctions.*
3. *Explain how neurotransmitter chemicals are stored and released by presynaptic neuron endings.*

## *Acetylcholine*

The postsynaptic membrane contains chemically regulated gates for $Na^+$ and $K^+$, which open when a neurotransmitter, such as acetylcholine, binds to its receptor in the membrane. Opening of the chemically regulated gates can cause a depolarization to be produced. This depolarization is graded and capable of summation, and in fact differs in several respects from an action potential. The depolarization produced by opening of chemically regulated gates serves as the stimulus for the production of the first action potentials in the postsynaptic cell.

Acetylcholine (ACh) is used as a neurotransmitter by some neurons in the CNS, by somatic motor neurons at the neuromuscular junction, and by certain autonomic nerve endings. The effects of this chemical are excitatory in the first two synapses and either excitatory or inhibitory in the third. The excitatory effects of ACh will be discussed in this section; its inhibitory effects are considered in a later section on synaptic integration.

### Acetylcholine at the Neuromuscular Junction

The neuromuscular junction—the synapse between a somatic motor neuron and a skeletal muscle fiber—is the most accessible synapse to study and thus the best understood. The presynaptic neuron endings have synaptic vesicles, which each contain about ten thousand molecules of ACh. Once these molecules are released by exocytosis, they quickly diffuse across the narrow synaptic cleft to the membrane of the postsynaptic cell. Here they chemically bond to **receptor proteins** that are built into the postsynaptic membrane. These receptor proteins combine with ACh in a specific manner, analogous to the specific interaction between transport proteins and their substrates.

Acetylcholine is not, however, transported into the postsynaptic cell. Instead, the bonding of ACh to its receptor protein causes changes in membrane structure that result in the opening of ion channel gates for $Na^+$ and $K^+$. These gates, located only in the postsynaptic membrane, are called **chemically regulated gates** because they open in response to bonding by a chemical (ACh). Unlike the voltage-regulated gates previously described in the axon, the chemically regulated gates do not respond to changes in membrane potential.

Also in contrast to voltage-regulated gates, where the outward flow of $K^+$ occurs after the inward flow of $Na^+$, chemically regulated gates allow the simultaneous diffusion of $Na^+$ and $K^+$ (fig. 7.20). The depolarizing effect of $Na^+$ diffusion predominates because the electrochemical

**Figure 7.20.** (*a* and *b*) The binding of acetylcholine to receptor proteins causes the opening of chemically regulated gates in the postsynaptic membrane. (*c*) This results in the increased diffusion of $Na^+$ and $K^+$ through the membrane.

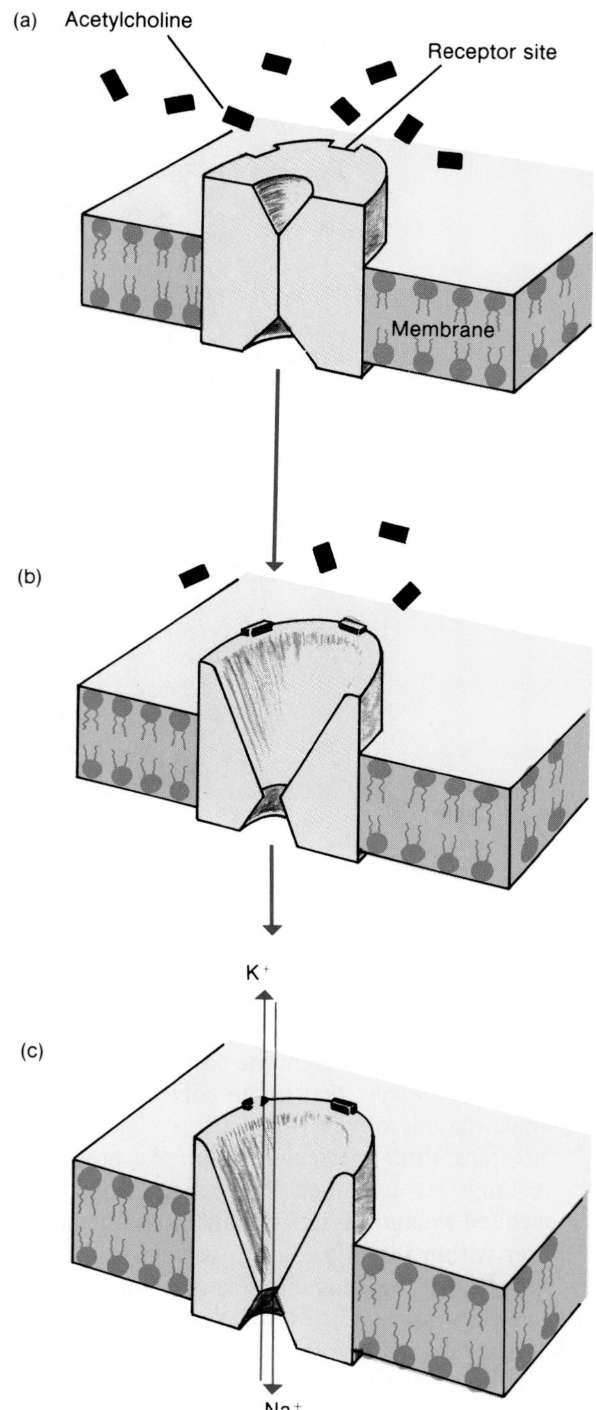

**Figure 7.21.** Mechanisms of the release of acetylcholine (*ACh*) from presynaptic nerve endings and the binding of ACh to receptor proteins (*R*) in the postsynaptic membrane. Acetylcholine that combines with acetylcholinesterase (*AChE*) is hydrolyzed and thus inactivated.

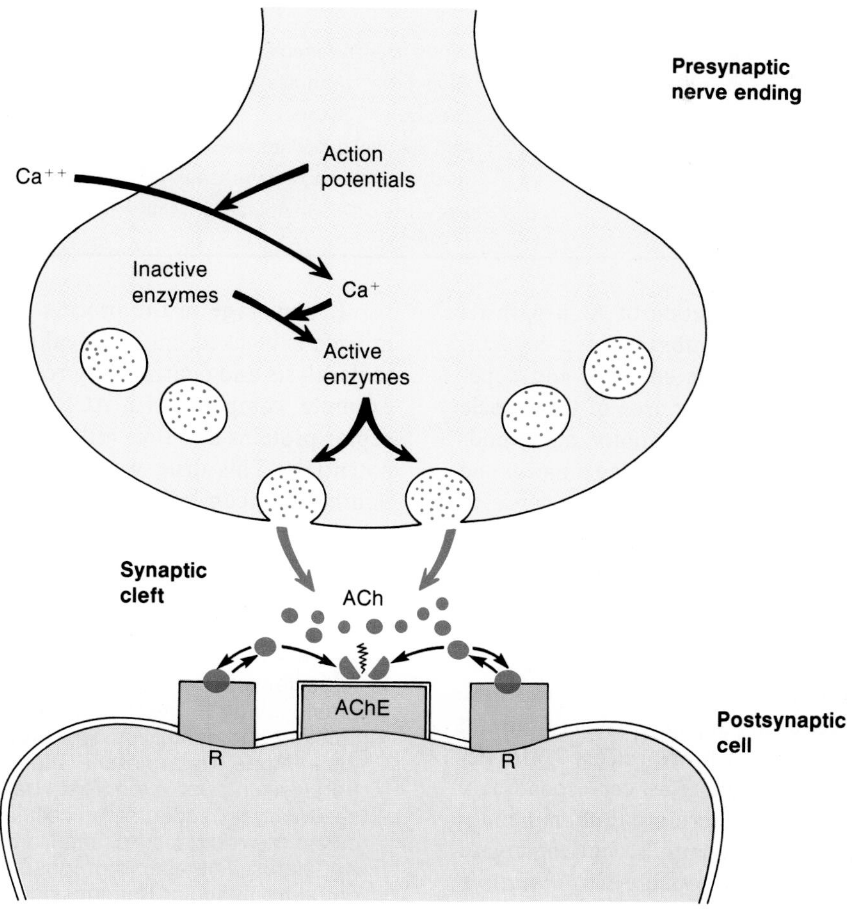


gradient for $Na^+$ is greater than for $K^+$. The outflow of $K^+$, however, does prevent the "overshoot" characteristic of action potentials—the membrane potential can reach 0 mV but cannot reverse polarity (as occurs in action potentials). Because of the characteristics of chemically regulated gates, neurotransmitters do not directly produce action potentials. They can only produce depolarization, which may stimulate the opening of voltage-regulated gates and production of action potentials a short distance away from the site of the synapse.

***Acetylcholinesterase.*** The bond between ACh and its receptor protein exists for only a brief instant. The ACh-receptor complex quickly dissociates but can be quickly reformed as long as free ACh is in the vicinity. In order for activity in the postsynaptic cell to be controlled (in this case, for control of skeletal muscle contraction), free ACh must be inactivated very soon after it is released. The inactivation of ACh is achieved by means of an enzyme called **acetylcholinesterase,** or **AChE,** which is on the postsynaptic membrane or immediately outside the membrane, with its active site facing the synaptic cleft (fig. 7.21).

Nerve gas exerts its odious effects by inhibiting AChE in skeletal muscles. Since ACh is not degraded it can continue to combine with receptor proteins and can continue to stimulate the postsynaptic cell, leading to spastic paralysis. Clinically, cholinesterase inhibitors (such as *neostigmine*) are used to enhance the effects of ACh on muscle contraction when neuromuscular transmission is weak, as in the disease myasthenia gravis.

**Table 7.5** Derivation and effect of various drugs that affect neural control of skeletal muscles

| Drug | Origin | Effects |
|---|---|---|
| Botulinus toxin | Produced by *Clostridium botulinum* (bacteria) | Inhibits release of acetylcholine (ACh) |
| Curare | Resin from a South American tree | Prevents interaction of ACh with the postsynaptic receptor protein |
| α-bungarotoxin | Venom of *Bungarus* snakes | Binds to ACh receptor proteins |
| Saxitoxin | Red tide *(Gonyaulax)* | Blocks $Na^+$ channels |
| Tetrodotoxin | Puffer fish | Blocks $Na^+$ channels |
| Nerve gas | Artificial | Inhibits acetylcholinesterase in postsynaptic cell |
| Neostigmine | Nigerian bean | Inhibits acetylcholinesterase in postsynaptic cell |
| Strychnine | Seeds of an Asian tree | Prevents IPSPs in spinal cord that inhibit contraction of antagonistic muscles |

***End Plate Potential.*** The interaction of ACh with its receptors in the postsynaptic membrane of a skeletal muscle cell opens chemically regulated gates and depolarizes that region of membrane. The area of the muscle cell membrane innervated by somatic motor nerve endings and affected in this way by ACh is called a **motor end plate.** Depolarizations of this membrane in response to ACh are thus known as **end plate potentials.** The depolarizations of the end plate potentials do not overshoot 0 mV, for reasons previously discussed, and differ from action potentials in a number of other respects.

Unlike action potentials, end plate potentials have *no threshold;* a single quantum of ACh (released from a single synaptic vesicle) produces a tiny depolarization of the end plate. When more quanta of ACh are released, the depolarization of the motor end plate is correspondingly greater. End plate potentials are therefore *graded* in magnitude, unlike all-or-none action potentials. Since end plate potentials can be graded, they are capable of *summation.* This is quite different from action potentials, which are prevented from summating by their all-or-none nature and by the presence of refractory periods.

When the end plate becomes sufficiently depolarized, the depolarization serves as a stimulus for the opening of voltage-regulated gates in adjacent areas of the skeletal muscle cell membrane. The opening of voltage-regulated gates causes action potentials to be produced. These action potentials in turn serve as stimuli for the production of other action potentials in the next area of membrane, similar to the way that action potentials are conducted in unmyelinated axons. This is significant because electrical excitation of a muscle fiber (through mechanisms that will be discussed in chapter 12), stimulates muscle contraction.

If any stage in the process of neuromuscular transmission is blocked, muscle weakness—sometimes leading to paralysis and death—may result. The drug *curare,* for example, competes with ACh for attachment to the receptor proteins and thus reduces the size of the end plate potentials. This drug was first used on poison darts by South American Indians because it produced flaccid paralysis of their victims. Clinically, curare is used as a muscle relaxant during anesthesia and to prevent muscle damage during electroconvulsive shock therapy.

Muscle weakness in the disease **myasthenia gravis** is due to the fact that ACh receptors are blocked and destroyed by antibodies secreted by the immune system of the affected person. Paralysis in people who eat shellfish poisoned with saxitoxin, produced by unicellular organisms that cause the red tides, results from the blockage of $Na^+$ gates. The effects of these and other poisons on neuromuscular transmission are summarized in table 7.5.

## Acetylcholine in Neuron-Neuron Synapses

Within the central nervous system, the axon terminals of one neuron typically synapse with the dendrites or cell body of another. The dendrites and cell body thus serve as the receptive area of the neuron, and it is in these regions that receptor proteins for neurotransmitters and chemically regulated gates are located. The first voltage-regulated gates are located at the beginning of the axon, at the axon hillock. It is therefore here that action potentials are first produced (fig. 7.22).

**Figure 7.22.** A diagram illustrating the functional specialization of different regions in a "typical" neuron.

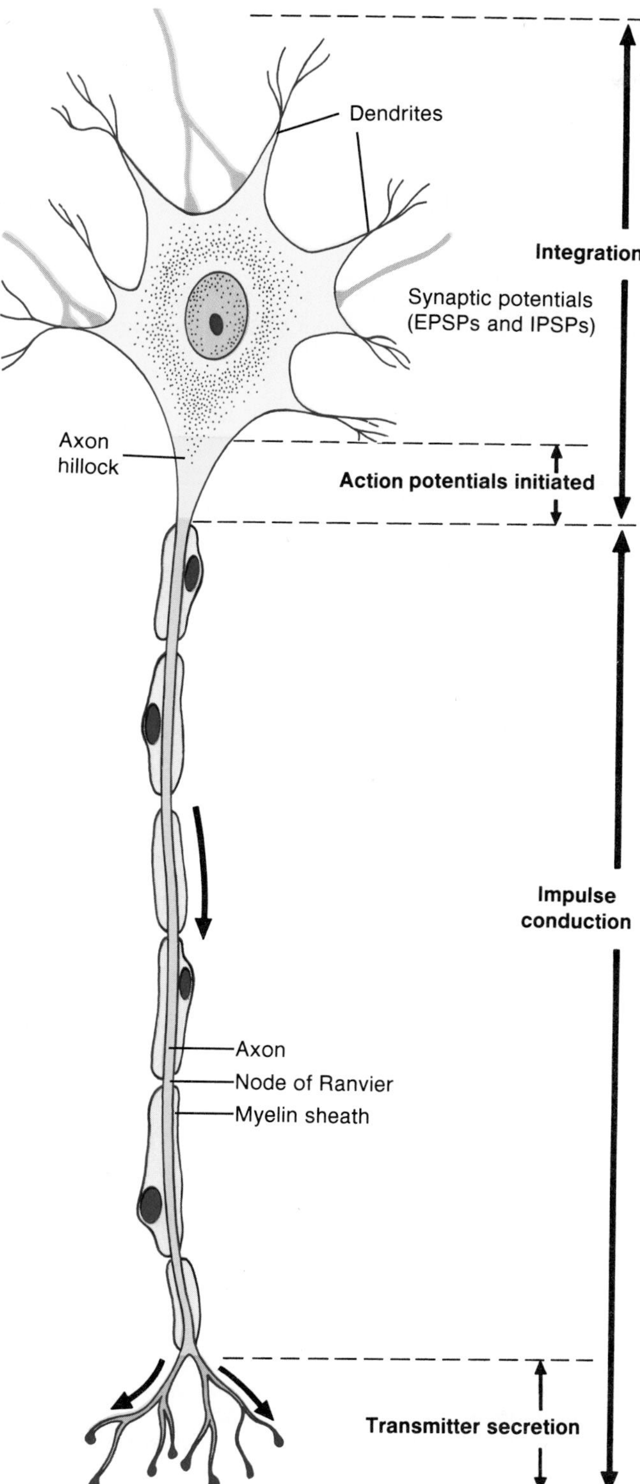


Release of ACh at these synapses results, through the opening of chemically regulated gates, in the production of graded depolarizations in the dendrites and cell body. These graded depolarizations in neurons are known as **excitatory postsynaptic potentials (EPSPs)**, and are exactly the same in their characteristics as the end plate potentials of skeletal muscle cells. As previously mentioned, these characteristics are quite different from those of action potentials. A comparison of EPSPs and action potentials is provided in table 7.6.

Voltage-regulated gates are absent in the dendrites of central neurons; although some may be present in the cell body, they are normally too few in number to produce an action potential. The *initial segment* of the axon, which is the unmyelinated region of the axon around the axon hillock, on the other hand, has a high concentration of voltage-regulated gates and a low threshold for the production of action potentials. Since the dendrites and cell body cannot regenerate a depolarization (the way action potentials are conducted in axons) due to the absence of voltage-regulated gates, depolarization in these regions must spread by cable properties to the initial segment of the axon.

If the depolarization is at or above threshold by the time it reaches the initial segment of the axon, the EPSP will stimulate the production of action potentials, which can then regenerate themselves along the axon. If, however, the EPSP is below threshold at the initial segment, no action potentials will be produced in the postsynaptic cell (fig. 7.23). Gradations in the strength of the EPSP above threshold determine the frequency with which action potentials will be produced at the axon hillock, and at each point in the axon where the impulse is conducted.

When the concept of action potential was introduced earlier in this chapter, the story began in the axon with a depolarization stimulus delivered by stimulating electrodes. It was pointed out that this stimulation was, of course, artificial, but that the true stimulus *in vivo* could not yet be described. Now you know that EPSPs, conducted from the dendrites and cell body, serve as the stimulus for the production of action potentials in the axon hillock, and that the action potentials at this point serve as the depolarization stimulus for the next region, and so on. This ends at the terminal boutons of the axon, where neurotransmitter is released. New EPSPs may then be produced in the postsynaptic cell, to continue the process if their depolarization is above threshold.

**Table 7.6** Comparison of action potentials with excitatory postsynaptic potentials (EPSPs)

| Characteristic | Action Potential | Excitatory Postsynaptic Potential |
|---|---|---|
| Stimulus for opening of ionic gates | Depolarization | Acetylcholine (ACh) |
| Initial effect of stimulus | $Na^+$ gates open | $Na^+$ and $K^+$ gates open |
| Production of repolarization | Opening of $K^+$ gates | Loss of intracellular positive charges with time and distance |
| Conduction distance | Not conducted—regenerated over length of axon | 1–2mm; a localized potential |
| Positive feedback between depolarization and opening of $Na^+$ gates | Present | Absent |
| Maximum depolarization | +40 mV | Close to zero |
| Summation | No summation—is all-or-none | Summation of EPSPs, producing graded depolarizations |
| Refractory period | Present | Absent |
| Effect of drugs | Inhibited by tetrodotoxin, not by curare | Inhibited by curare, not by tetrodotoxin |

**Figure 7.23.** The graded nature of excitatory postsynaptic potentials is shown, in which stimuli of increasing strength produce increasing amounts of depolarization. When a threshold level of depolarization is produced, action potentials are generated in the axon.

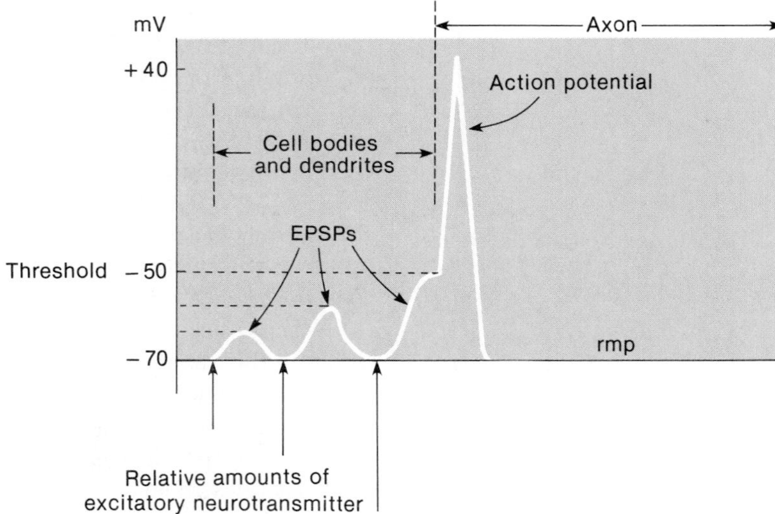


What caused the *very first* EPSP? If this question relates to a reflex, the first EPSP is produced in a sensory organ in response to some environmental stimulus (this is described in chapter 9). If this question relates to mental activity or voluntary movements, the answer is—nobody knows.

A skeletal muscle fiber receives synaptic input from only one neuron. A neuron, in contrast, may receive synaptic inputs from as many as one thousand other neurons. So, whereas the synaptic input from a single axon must be sufficient to excite a muscle cell, the excitation of a neuron requires the summation of many EPSPs before a threshold depolarization is produced at the initial segment of the axon. Summation is thus an extremely important process which can only occur between graded synaptic potentials. Summation of EPSPs allows the postsynaptic cell to "weigh" different combinations of synaptic inputs and respond appropriately.

1. *Describe the events that occur between the release of ACh and the production of end plate potentials in a skeletal muscle fiber.*
2. *Describe the function of acetylcholinesterase and its physiological significance.*
3. *Compare the properties of EPSPs and action potentials, and describe where these events occur in a postsynaptic neuron.*
4. *Explain how EPSPs produce action potentials in the postsynaptic neuron.*

## Neurotransmitters of the Central Nervous System

A variety of chemicals in the CNS are believed to function as neurotransmitters. These include ACh, the catecholamines (epinephrine and norepinephrine), certain amino acids, and a number of polypeptides. Some of these neurotransmitters stimulate the production of a molecule called cyclic AMP in the postsynaptic cell, which acts as a "second messenger" to cause the opening of ion gates and other effects. A knowledge of the neurotransmitters of the brain is required for understanding of brain function and disorders.

Acetylcholine is not only the neurotransmitter of somatic motor neurons and some autonomic neurons of the PNS, it is also an important neurotransmitter within the CNS. Since the excitatory effects of ACh were discussed in a previous section (and the inhibitory effects will be discussed later), this section is devoted primarily to CNS neurotransmitters other than ACh.

**Alzheimer's disease** is the most common cause of senile dementia, which often begins in middle age and produces progressive mental deterioration. The cause of Alzheimer's disease is not known, but there is evidence that it is associated with a loss of cholinergic neurons (those that use acetylcholine as a neurotransmitter), which terminate in the hippocampus and cerebral cortex of the brain (areas concerned with memory storage). Since acetylcholine is produced from acetyl coenzyme A and choline, attempts have been made to increase ACh in the brain by the ingestion of large amounts of lecithin (which contains choline). Thus far such nutritional treatments and possible drug treatments to increase ACh by inhibiting acetylcholinesterase have not proved successful.

In addition to acetylcholine, there is a great variety of other molecules that serve as neurotransmitters in the CNS. **Catecholamines** are a group of regulatory molecules, derived from the amino acid tyrosine, that include *dopamine, norepinephrine,* and *epinephrine.* Dopamine and norepinephrine function as neurotransmitters; epinephrine and norepinephrine additionally serve as hormones secreted by the adrenal medulla gland.

The catecholamines, together with a related molecule called *serotonin,* are included in a larger category of molecules called *monoamines.* There is good evidence that serotonin and particular amino acids and polypeptides function as important neurotransmitters in the CNS. It is extremely difficult, however, to prove conclusively that a particular molecule functions as a neurotransmitter at a central synapse. Because of such lack of conclusive proof of their transmitter function, dopamine, serotonin, histamine, glycine, glutamic acid, aspartic acid, and many polypeptides are often referred to as **putative neurotransmitters** (table 7.7). These molecules, together with the well-established neurotransmitters (acetylcholine, norepinephrine, and GABA), provide the diversity needed for the complex functions of the central nervous system.

**Table 7.7** Examples of chemicals that are either proven or putative neurotransmitters

| Category | Chemicals |
|---|---|
| Amines | Acetylcholine<br>Histamine<br>Serotonin |
| Catecholamines | Dopamine<br>Epinephrine<br>Norepinephrine |
| Amino acids | Aspartic acid<br>GABA (gamma-aminobutyric acid)<br>Glutamic acid<br>Glycine |
| Polypeptides | Glucagon<br>Insulin<br>Somatostatin<br>Substance P<br>ACTH (adrenocorticotrophic hormone)<br>Angiotensin II<br>Endorphins<br>LHRH (luteinizing hormone-releasing hormone)<br>TRH (thyrotrophin-releasing hormone)<br>Vasopressin (antidiuretic hormone) |

### Catecholamines as Neurotransmitters

Like ACh, catecholamine neurotransmitters are released by exocytosis from presynaptic vesicles and diffuse across the synaptic cleft to interact with specific receptor proteins in the membrane of the postsynaptic cell. The stimulatory effects of these catecholamines, like those of ACh, are quickly inhibited. The inhibition of catecholamine action is due to (1) *reuptake* of catecholamines into the presynaptic neuron endings; (2) enzymatic degradation of catecholamines in the presynaptic neuron endings by *monoamine oxidase* (*MAO*); and (3) their enzymatic degradation in the postsynaptic neuron by *catecholamine-O-methyltransferase* (*COMT*). This is shown in figure 7.24. Drugs that inhibit MAO and COMT thus promote the effects of catecholamine action.

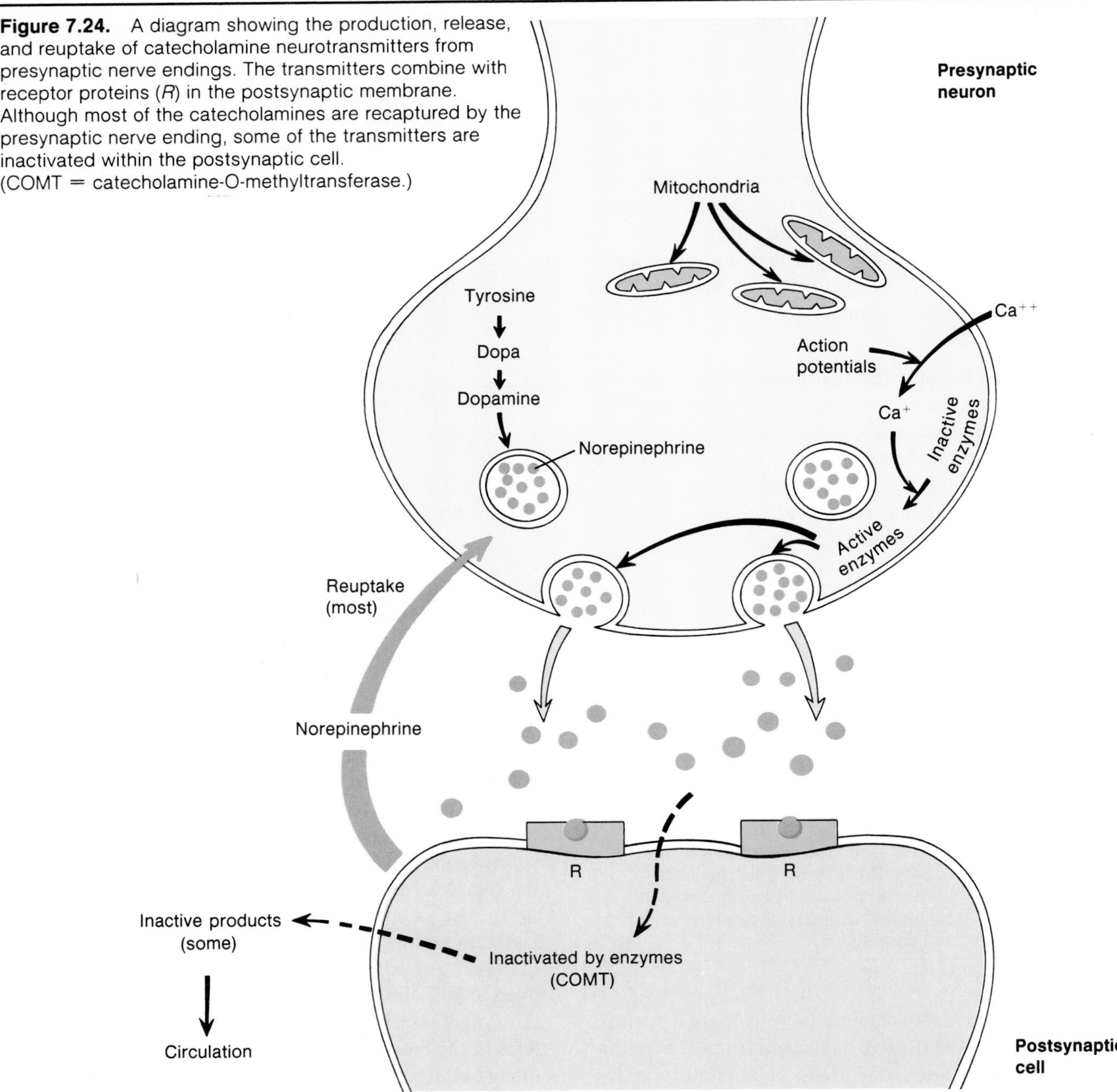


**Figure 7.24.** A diagram showing the production, release, and reuptake of catecholamine neurotransmitters from presynaptic nerve endings. The transmitters combine with receptor proteins (*R*) in the postsynaptic membrane. Although most of the catecholamines are recaptured by the presynaptic nerve ending, some of the transmitters are inactivated within the postsynaptic cell. (COMT = catecholamine-O-methyltransferase.)

Monoamine oxidase (MAO) is an enzyme found in the mitochondria of the endings of presynaptic axons. This enzyme breaks down catecholamine neurotransmitters that have been taken up from the synaptic cleft after prior release. Drugs that inhibit this enzyme—known as **MAO inhibitors**—thus increase the effectiveness of catecholamine neurotransmitters. Some of these drugs have been found to be effective in the treatment of people with psychotic depression. This and other sources of evidence suggest that catecholamines, specifically norepinephrine, may be involved in neural pathways that normally prevent severe emotional depression.

Whereas ACh opens chemically regulated gates immediately upon bonding to its receptor, the interaction of catecholamines with their receptors does not directly open chemically regulated $Na^+$ and $K^+$ gates. Instead, these neurotransmitters activate an enzyme in the postsynaptic cell membrane known as *adenylate cyclase.* This enzyme converts ATP to **cyclic AMP (cAMP)** and pyrophosphate (two inorganic phosphates) within the postsynaptic cell cytoplasm. Cyclic AMP in turn activates another enzyme, called *protein kinase*, which phosphorylates (adds a phosphate group to) the chemically regulated gates in the postsynaptic membrane. This action causes the gates to open. Cyclic AMP is thus a **second messenger** in the action of these neurotransmitters (fig. 7.25).

**Figure 7.25.** Catecholamine neurotransmitters induce the intracellular production of cyclic AMP (cAMP) in the postsynaptic cell. The cAMP, in turn, stimulates the opening of ionic gates (short-term effects) and genetic expression (long-term effects) through the activation of previously inactive enzymes.

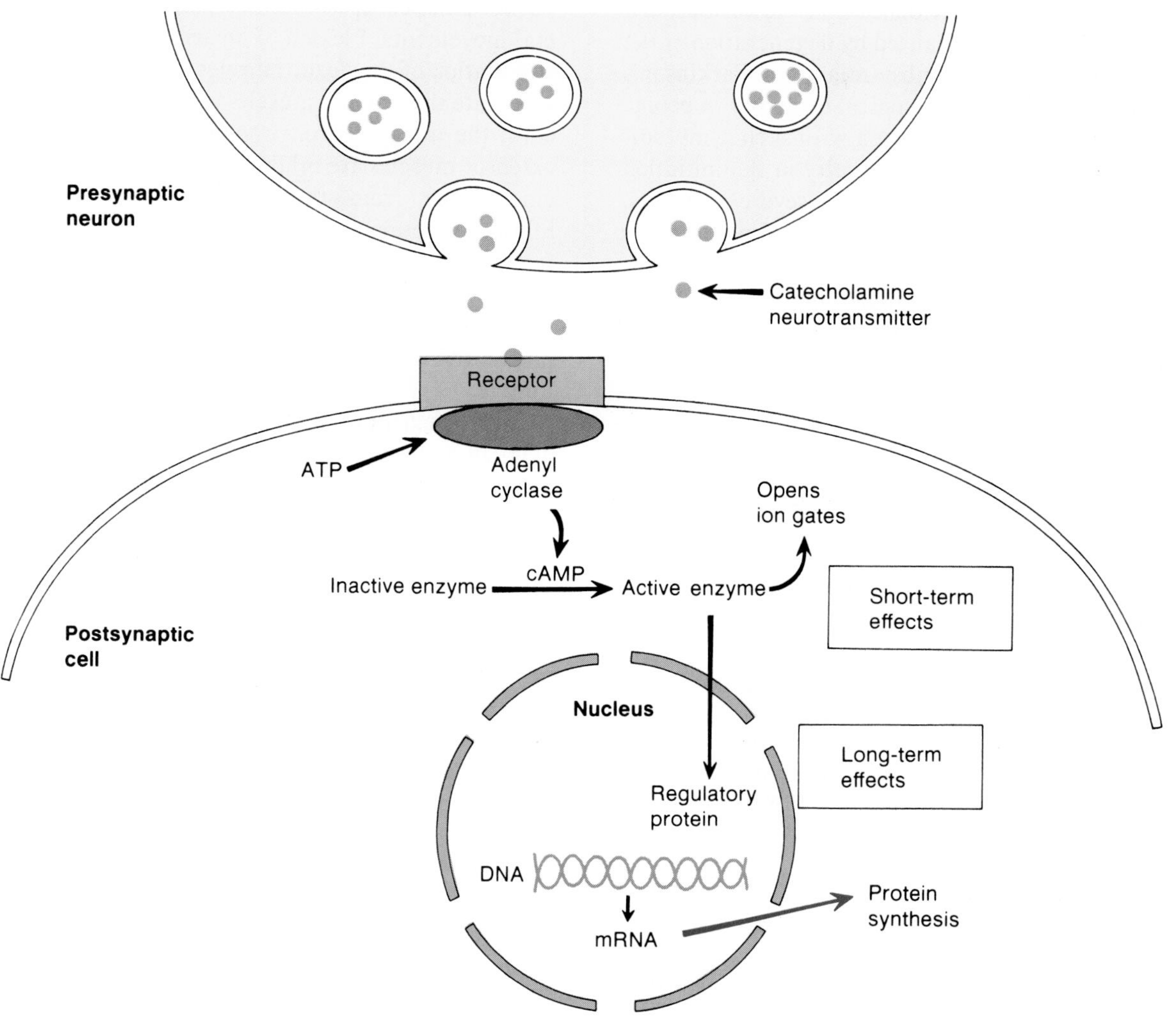


In addition to promoting the opening of ionic gates and the production of electrical impulses, cyclic AMP may also promote more long-term modifications of the postsynaptic neuron. These long-term modifications may involve cAMP-mediated changes in genetic expression.

***Dopamine and Norepinephrine.*** Neurons that use **dopamine** as a neurotransmitter and postsynaptic neurons with dopamine receptor proteins in their membranes can be identified in postmortem brain tissue. More recently, the position of dopamine receptor proteins has been observed in the living brain using the technique of *positron emission tomography* (*PET*). These investigations have been spurred by the great clinical interest in the effects of **dopaminergic neurons** (those that use dopamine as a neurotransmitter).

**Cocaine** is a stimulant, related to the amphetamines in its action, which is currently widely abused in the United States. Although early use of this drug produces feelings of euphoria and social adroitness, continued use leads to social withdrawal, depression, dependence upon ever-higher dosages, and serious organic disease that often results in death. It is surprising that the many effects of cocaine on the central nervous system appear to be mediated by one primary mechanism: cocaine blocks the reuptake of the neurotransmitter dopamine into the presynaptic axon endings. This results in overstimulation of those neural pathways that use dopamine as a neurotransmitter.

Dopaminergic neurons are highly concentrated in the *substantia nigra* (literally, the "dark substance," so-called because it contains melanin pigment) of the brain. Many neurons in the substantia nigra send fibers to the *basal ganglia,* which are large masses of cell bodies deep in the cerebrum that are involved in the coordination of skeletal movements. There is much evidence that **Parkinson's disease,** or **paralysis agitans,** is caused by degeneration of the dopaminergic neurons in the substantia nigra. Parkinson's disease, a major cause of neurological disability in people over sixty years of age, is associated with such symptoms as muscle tremors and rigidity, difficulty in the initiation of movements and in speech, and other severe problems. These patients are treated with L-dopa to increase the production of dopamine in the brain, as described previously in this chapter, and with anticholinergic drugs to decrease the production of acetylcholine (ACh is believed to antagonize the actions of dopaminergic neurons in the basal ganglia).

A side effect of L-dopa treatment in some patients with Parkinson's disease is the appearance of symptoms characteristic of **schizophrenia.** This effect is not surprising, in view of the fact that the drugs used to treat schizophrenic patients (chlorpromazine and related compounds) act as specific antagonists of dopamine receptors. As might be predicted from these observations, schizophrenic patients treated with these drugs often develop symptoms of Parkinson's disease. It seems reasonable to suppose, from this evidence, that schizophrenia may be caused, at least in part, by overactivity of the dopaminergic pathways. This hypothesis is strengthened by the additional evidence that the brains of schizophrenic patients appear to have an abnormally high content of dopamine receptors.

**Norepinephrine,** like ACh, is used as a neurotransmitter in both the PNS and the CNS. Sympathetic neurons of the PNS use norepinephrine as a neurotransmitter at their synapse with smooth muscles, cardiac muscle, and glands. Some neurons in the CNS also appear to use norepinephrine as a neurotransmitter; these neurons seem to be involved in general behavioral arousal. This would help to explain the effects of such drugs as *amphetamines,* which specifically stimulate pathways that use norepinephrine as a neurotransmitter.

### Amino Acids as Neurotransmitters

The amino acids **glutamic acid** and **aspartic acid** function as excitatory neurotransmitters in some neurons of the CNS. The amino acid **glycine,** in contrast, is inhibitory; instead of depolarizing the postsynaptic membrane and producing an EPSP, it hyperpolarizes the postsynaptic membrane. This hyperpolarization makes the membrane potential even more negative than it is at rest (changing the membrane potential from −65 mV to, for example −85 mV). Such hyperpolarization is known as an **inhibitory postsynaptic potential (IPSP)**, and will be discussed in more detail in the next section.

The inhibitory effects of glycine are very important in the spinal cord, where they help in the control of skeletal movements. Flexion of an arm, for example, involves stimulation of the flexor muscles. The motor neurons that innervate these flexor muscles are stimulated in the spinal cord; the motor neurons that innervate the antagonistic extensor muscles are inhibited by IPSPs produced by glycine released from other neurons. The importance of the inhibitory actions of glycine is revealed by the deadly effects of the poison *strychnine,* which causes spastic paralysis by specifically blocking the glycine receptor proteins. Animals poisoned with strychnine die from asphyxiation due to their inability to relax the diaphragm muscle.

The neurotransmitter **GABA (gamma-aminobutyric acid)** is a derivative of the amino acid glutamic acid. GABA is the most prevalent neurotransmitter in the brain; in fact, as many as one-third of all the neurons in the brain use GABA as a neurotransmitter. Like glycine, GABA is inhibitory—it hyperpolarizes the postsynaptic membrane. Also like glycine, some neurons that use GABA are involved in motor control (such as the large *Purkinje cells* of the cerebellum, which mediate the motor functions of the cerebellum by producing IPSPs in their postsynaptic neurons). A deficiency in those neurons that release GABA as a neurotransmitter produces the uncontrolled movements seen in people with *Huntington's chorea.*

In addition to its involvement in motor control, GABA also appears to function as a neurotransmitter involved in mood and emotion. Drugs given to treat extreme anxiety—including the *benzodiazepines* (e.g., Valium)—act by increasing the effectiveness of GABA in the brain. Drugs that antagonize the actions of GABA, conversely, can produce extreme feelings of anxiety.

### Polypeptide Neurotransmitters

Many polypeptides of various sizes are found in the brain and are believed to function as neurotransmitters. Interestingly, some of the polypeptides that function as hormones secreted by the intestine and other endocrine glands are also produced in the brain and may function there as neurotransmitters (table 7.7).

***Synaptic Plasticity.*** Although some of the polypeptides released from neurons may function as neurotransmitters in the traditional sense—by stimulating the opening of ionic gates and causing changes in the membrane potential—others may have more subtle and poorly understood effects. The name *neuromodulators* has been proposed to identify compounds with such alternative effects. An exciting recent discovery is that some neurons in both the PNS and CNS produce both a classical neurotransmitter (ACh or a catecholamine) and a putative polypeptide neurotransmitter. Additionally, there is evidence that these neurons may secrete one or the other of their transmitters depending upon the environmental conditions.

Recent evidence suggests that synapses are more changeable than was previously believed. This has been called **synaptic plasticity,** and it exists at both the molecular and cellular levels. At the molecular level, it has been shown that some neurons may secrete either a classical neurotransmitter or a putative neurotransmitter, as previously discussed. At the cellular level, there is evidence that axonal sprouting over short distances produces a turnover of synapses even in the mature CNS. This breakdown and reforming of synapses may occur within a time span of only a few hours. The physiological significance of these interesting discoveries is not yet fully understood.

***Endorphins.*** The ability of opium and its analogues—that is, the opioids—to relieve pain (promote analgesia) has been known for centuries. Morphine, for example, has long been used for this purpose. The discovery in 1973 of opioid receptor proteins in the brain suggested that the effects of these drugs might be due to the stimulation of specific neuron pathways. This implied that opioids—like LSD, mescaline, and other mind-altering drugs—might resemble neurotransmitters produced by the brain.

The analgesic effects of morphine are blocked in a specific manner by a drug called *naloxone.* In the same year that opioid receptor proteins were discovered, it was found that naloxone also blocked the analgesic effect of electrical brain stimulation. Subsequently, evidence suggested that the analgesic effects of hypnosis and acupuncture could also be blocked by naloxone. These experiments indicated that the brain might be producing its own morphinelike analgesic compounds.

These compounds have been identified as a family of chemicals called **endorphins** (for "endogenously produced morphinelike compounds") produced by the brain and pituitary gland. The endorphins include a group of five-amino-acid peptides called *enkephalins,* which may function as neurotransmitters, and a thirty-one-amino-acid polypeptide, produced by the pituitary gland, called *β-endorphin.*

Endorphins have been shown to block the transmission of pain. Current evidence for this includes results obtained both from neurophysiological studies—in which endorphins blocked the release of substance P (the chemical transmitter believed to be released by nerve fibers that mediate painful sensations)—and from behavioral studies. The pain threshold of pregnant rats, for example, has been found to decrease when they are treated with naloxone.

Endorphins—like opium and morphine—may also provide pleasant sensations and thus mediate reward or positive reinforcement pathways. Overeating in genetically obese mice, for example, appears to be blocked by naloxone. It has been found that blood levels of β-endorphin are increased in exercise. Some people have suggested that the "jogger's high" may thus be due to endorphins. (Naloxone, however, does not appear to block the jogger's high.) Although evidence for this particular effect is poor, it does appear that endorphins may promote some type of psychic reward system as well as analgesia.

1. *Describe three ways that catecholamine neurotransmitters are inactivated at the synapse.*
2. *Explain how catecholamines stimulate the production of cyclic AMP, and why cAMP is called a second messenger in the action of these neurotransmitters.*
3. *Describe the known significance of dopaminergic neurons, and explain how their function relates to the etiology of Parkinson's disease and schizophrenia.*
4. *Describe the nature and explain the significance of inhibitory neurotransmitters in the CNS.*
5. *Explain the meaning of the term* synaptic plasticity *at the molecular and cellular level.*
6. *Describe the nature and explain the proposed functions of the endorphins.*

## SYNAPTIC INTEGRATION

A number of EPSPs must generally summate to produce a depolarization of sufficient magnitude to stimulate the postsynaptic cell. The number of EPSPs required to do this, and their net effect on the frequency of action potential generated by the postsynaptic neuron, is affected by the hyperpolarizing IPSPs produced by inhibitory neurotransmitters. Such hyperpolarization of the postsynaptic membrane is called postsynaptic inhibition. Axoaxonic synapses may cause presynaptic inhibition, which reduces the amount of excitatory neurotransmitter released.

Since voltage-regulated $Na^+$ and $K^+$ gates are absent in the dendrites and cell bodies, changes in membrane potential induced by neurotransmitters in these areas do not have the all-or-none characteristics of action potentials. Synaptic potentials are graded and can add together, or summate. **Spatial summation** occurs because many presynaptic nerve fibers converge on a single postsynaptic neuron (fig. 7.26). In spatial summation, synaptic depolarizations (EPSPs) produced at different synapses may summate in the postsynaptic dendrites and cell body. In **temporal summation,** the successive activity of presynaptic axon terminals, causing successive waves of transmitter release, may result in the summation of EPSPs in

**Figure 7.26.** A diagram illustrating the convergence of large numbers of presynaptic fibers on the cell body of a spinal motor neuron.

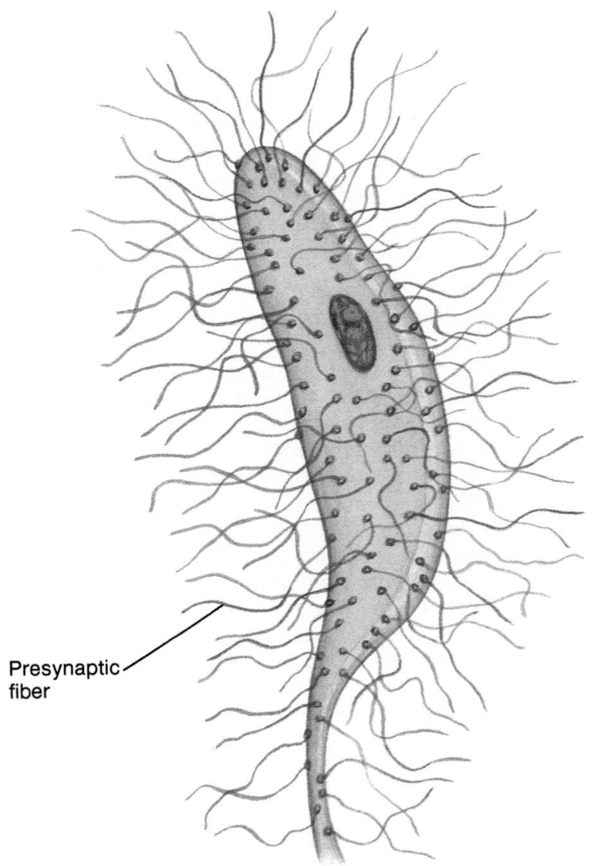


the postsynaptic neuron. The summation of EPSPs helps to ensure that the depolarization that reaches the axon hillock will be sufficient to generate new action potentials in the postsynaptic nerve fiber (fig. 7.27).

When a presynaptic neuron is stimulated continuously, even for as short a time as a few seconds, its ability to excite a postsynaptic neuron is enhanced—or potentiated—when this neuron pathway is subsequently stimulated. This potentiation of synaptic transmission may last for hours or even weeks following the tetanic (continuous) stimulation of the presynaptic neuron. This phenomenon is called **post-tetanic potentiation.**

Post-tetanic potentiation is due to the increased release of neurotransmitter as a result of previous tetanic stimulation of the presynaptic neuron. Experiments suggest that there is increased $Ca^{++}$ concentration in the axon terminals following tetanic stimulation. Since $Ca^{++}$ mediates the release of neurotransmitter in response to action potentials, this observation may explain the improved synaptic efficiency that results. Post-tetanic potentiation may favor transmission along frequently used neural pathways and thus may represent a mechanism of neural "learning."

## Synaptic Inhibition

Although most neurotransmitters depolarize the postsynaptic membrane (produce EPSPs), some transmitters have the opposite effect. These inhibitory neurotransmitters cause *hyperpolarization* of the postsynaptic membrane: they make the inside of the membrane more negative than it is at rest. Since hyperpolarization (from $-65$ mV to, for example, $-85$ mV) takes the membrane potential farther away from the threshold depolarization required to stimulate action potentials, such hyperpolarization inhibits the activity of the postsynaptic neuron. Hyperpolarizations produced by neurotransmitters are therefore called *inhibitory postsynaptic potentials (IPSPs)*. The inhibition produced in this way is called **postsynaptic inhibition.**

Excitatory and inhibitory inputs (EPSPs and IPSPs) to a postsynaptic neuron can summate in an algebraic fashion (fig. 7.28). The effects of IPSPs in this way reduce, or may even abolish, the ability of EPSPs to generate action potentials in the postsynaptic cell. Considering that the nervous system contains approximately $10^{12}$ neurons and that a given neuron may receive as many as 1,000 presynaptic inputs, one can see that the possibilities for synaptic integration in this way are awesome.

In addition to serving an integrative function in the CNS, IPSPs also have more specialized roles. Although ACh produces depolarization at the neuromuscular junction and stimulates skeletal muscle contraction, for example, ACh causes hyperpolarization in the heart. This is due to the fact that the combination of ACh with its receptor protein in the myocardial cells causes opening of only $K^+$ gates, and the outward diffusion of $K^+$ causes hyperpolarization. The baseline membrane potential is thus lowered, so that a longer time is required to reach threshold. The parasympathetic fibers that innervate the heart, which release ACh as a neurotransmitter, in this way cause a slowing of the heart rate (discussed in chapter 13).

The neurotransmitter called *gamma-aminobutyric acid (GABA)* is exclusively inhibitory. This neurotransmitter is believed to hyperpolarize central neurons by opening chloride gates. Chloride ions ($Cl^-$), as a result, diffuse into the neuron and make the membrane potential more negative on the inside. The amino acid *glycine* is also believed to function as a neurotransmitter. This transmitter may help muscular coordination by inhibiting those neurons in the spinal cord that activate antagonistic muscles. Glycine, like GABA, hyperpolarizes the postsynaptic membrane by stimulating the chloride gates to open.

**Figure 7.27.** Excitatory postsynaptic potentials (EPSPs) can summate over distance (spatial summation) and time (temporal summation). When summation results in a threshold level of depolarization at the axon hillock, voltage-regulated $Na^+$ gates are opened and an action potential is produced.

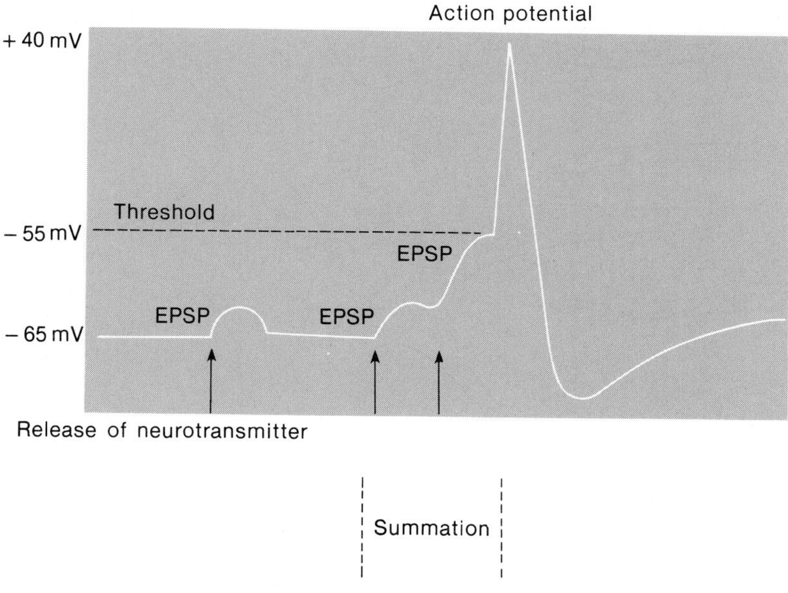


**Figure 7.28.** An inhibitory postsynaptic potential (IPSP) makes the inside of the postsynaptic membrane more negative than the resting potential—it hyperpolarizes the membrane. Subsequent or simultaneous excitatory postsynaptic potentials (EPSPs), which are depolarizations, must thus be stronger to reach the threshold required to generate action potentials at the axon hillock.

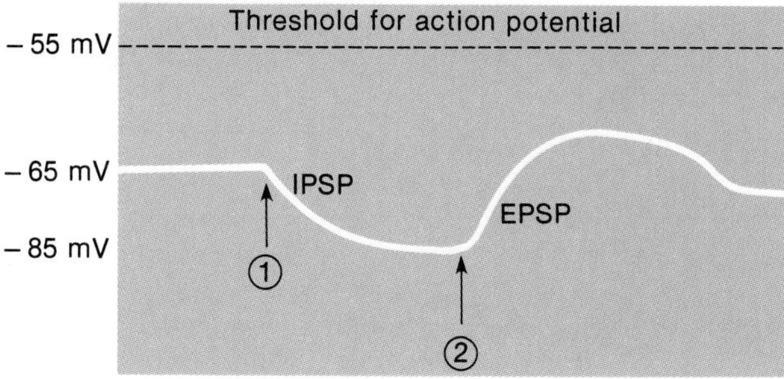

**Figure 7.29.** A diagram illustrating postsynaptic and presynaptic inhibition.

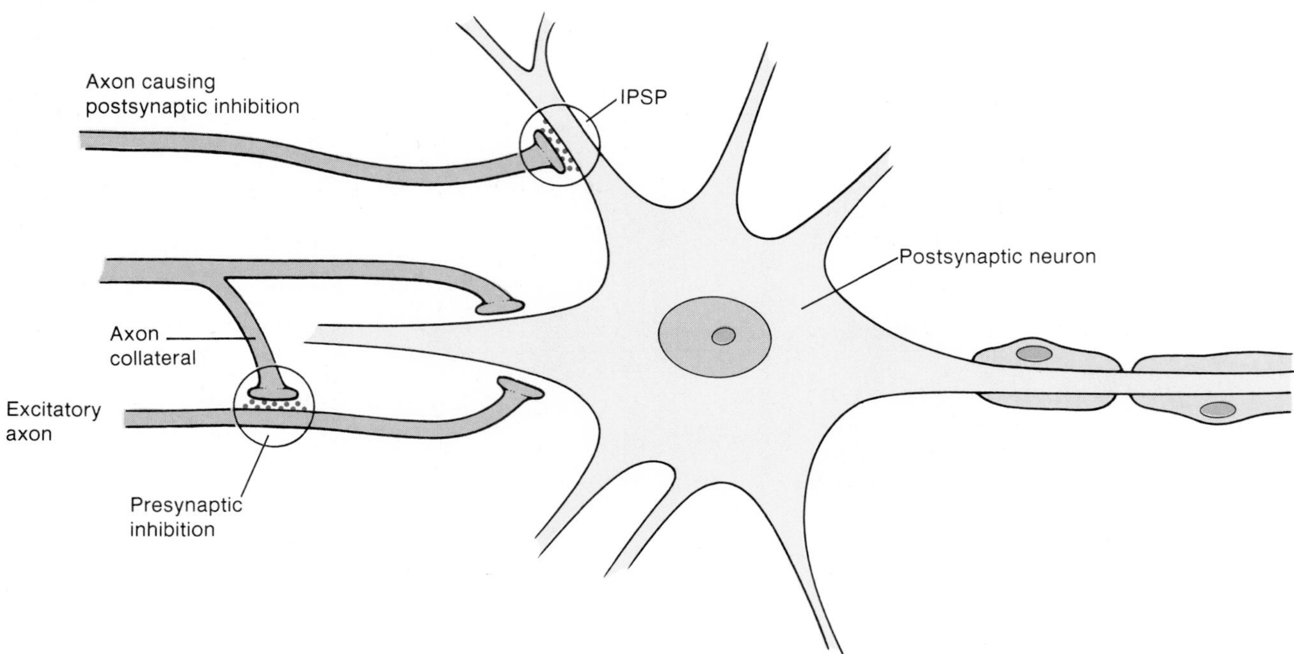


In **presynaptic inhibition** (fig. 7.29), the amount of excitatory neurotransmitter released at the end of an axon is decreased by a second neuron, whose axon makes a synapse with the axon of the first neuron (axoaxonic synapse). The neurotransmitter released at this axoaxonic synapse partially depolarizes the axon of the first neuron, bringing it closer to threshold and making it easier to "fire" action potentials. The amplitude of these action potentials (subtracting the new lower potential from the +40 mV "top" of the action potential), however, is less than normal. The smaller action potentials that result cause the release of lesser amounts of excitatory neurotransmitter by the first neuron, resulting in inhibition of its effect on its postsynaptic cell.

1. *Define spatial summation, temporal summation, and post-tetanic potentiation. Explain their functional importance.*
2. *Explain how postsynaptic inhibition is produced and how IPSPs and EPSPs can interact.*
3. *Explain the mechanism of presynaptic inhibition.*

## Summary

### Organization and Histology of the Nervous System p. 147

I. The nervous system is divided into the central nervous system (CNS) and the peripheral nervous system (PNS).
   A. The central nervous system includes the brain and spinal cord, which contain nuclei and tracts.
   B. The peripheral nervous system consists of nerves and ganglia.

II. A neuron consists of dendrites, a cell body, and an axon.
   A. The cell body contains the nucleus, Nissl bodies, neurofibrils, and other organelles.
   B. Dendrites receive stimuli, and the axon conducts nerve impulses away from the cell body.

III. A nerve is a collection of axons in the PNS.
   A. A sensory or afferent neuron is pseudounipolar and conducts impulses from sensory receptors into the CNS.
   B. A motor or efferent neuron is multipolar and conducts impulses from the CNS to effector organs.
   C. Interneurons, or association neurons, are located entirely within the CNS.
   D. Somatic motor nerves innervate skeletal muscle; visceral motor, or autonomic, nerves innervate smooth muscle, cardiac muscle, and glands.

IV. There are six different categories of neuroglial cells.
   A. Schwann cells form a *sheath of Schwann* around axons of the PNS.
   B. Some neurons are surrounded by successive wrappings of glial cell membranes called a myelin sheath; this is produced by Schwann cells in the PNS and oligodendrocytes in the CNS.
   C. Astrocytes in the CNS may contribute to the blood-brain barrier.

### Electrical Activity in Nerve Fibers p. 154

I. The permeability of the axon membrane to $Na^+$ and $K^+$ is regulated by gates.
   A. At the resting membrane potential of −65 mV, the membrane is relatively impermeable to $Na^+$ and only slightly permeable to $K^+$.
   B. These gates are voltage regulated; the $Na^+$ and $K^+$ gates open in response to the stimulus of depolarization.
   C. When the membrane is depolarized to a threshold level, the $Na^+$ gates open first, followed quickly by opening of the $K^+$ gates.

II. The opening of voltage-regulated gates produces an action potential.
   A. The opening of $Na^+$ gates in response to depolarization allows $Na^+$ to diffuse into the axon, thus further depolarizing the membrane in a positive feedback fashion.
   B. The inward diffusion of $Na^+$ causes a reversal of the membrane potential from −65 mV to +40 mV.
   C. The opening of $K^+$ gates and outward diffusion of $K^+$ causes the reestablishment of the resting membrane potential; this is called repolarization.
   D. Action potentials are all-or-none events.
   E. The refractory periods of an axon membrane prevent action potentials from running together.
   F. Stronger stimuli produce action potentials at greater frequency.

III. One action potential serves as the depolarization stimulus for production of the next action potential in the axon.
   A. In unmyelinated axons, action potentials are produced fractions of a micrometer apart.
   B. In myelinated actions, action potentials are only produced at the nodes of Ranvier; this saltatory conduction is faster than conduction in an unmyelinated nerve fiber.

### Synaptic Transmission p. 161

I. Gap junctions are electrical synapses, found in cardiac muscle, smooth muscle, and some synapses in the CNS.

II. In chemical synapses, neurotransmitters are packaged in synaptic vesicles and released by exocytosis into the synaptic cleft.

### Acetylcholine p. 164

I. The combination of ACh with its receptor protein in the postsynaptic membrane causes the chemically regulated gates to open and produces depolarizations called end plate potentials.
   A. ACh is inactivated by acetylcholinesterase.
   B. Depolarizations called end plate potentials are produced by ACh in skeletal muscle cells.
   C. End plate potentials are graded depolarizations that stimulate the production of action potentials in skeletal muscle fibers.

II. Depolarization produced by neurotransmitters in neurons is called an excitatory postsynaptic potential (EPSP).
   A. EPSPs are graded and capable of summation, and they decrease in amplitude as they are conducted.
   B. EPSPs produced at synapses in the dendrites or cell body travel to the axon hillock, stimulate opening of voltage-regulated gates, and generate action potentials in the axon.

### Neurotransmitters of the Central Nervous System p. 169

I. Catecholamines include dopamine, norepinephrine, and epinephrine.
   A. These neurotransmitters are inactivated after being released primarily by reuptake into the presynaptic nerve endings.
   B. Catecholamines activate adenylate cyclase in the postsynaptic cell, which catalyzes the formation of cyclic AMP.
   C. Cyclic AMP, formed in the postsynaptic cell, produces the effects of the neurotransmitter.

II. In addition to the classical neurotransmitters, neurons produce a large number of other chemicals that are believed to have a neurotransmitter function.
   A. These chemicals include glycine and GABA, which have inhibitory effects.
   B. The putative neurotransmitters include the endorphins and many other polypeptides.

### Synaptic Integration p. 173

I. Spatial and temporal summation of EPSPs allows a sufficient depolarization to be produced to cause the stimulation of action potentials in the postsynaptic neuron.

II. Neurotransmitters that cause hyperpolarization of the postsynaptic membrane produce inhibitory postsynaptic potentials (IPSPs).
   A. IPSPs and EPSPs from different synaptic inputs can summate.
   B. The production of IPSPs is called postsynaptic inhibition.
   C. Presynaptic inhibition occurs in an axoaxonic synapse and reduces the amount of neurotransmitter released by the inhibited neuron.

## Review Activities

### Objective Questions

1. The neuroglial cells that form myelin sheaths in the peripheral nervous system are
   (a) oligodendrocytes
   (b) satellite cells
   (c) Schwann cells
   (d) astrocytes
   (e) microglia
2. A collection of neuron cell bodies located outside the CNS is called a
   (a) tract
   (b) nerve
   (c) nucleus
   (d) ganglion
3. Which of the following neurons are pseudounipolar?
   (a) sensory neurons
   (b) somatic motor neurons
   (c) neurons in the retina
   (d) autonomic motor neurons
4. Depolarization of an axon is produced by the
   (a) inward diffusion of $Na^+$
   (b) active extrusion of $K^+$
   (c) outward diffusion of $K^+$
   (d) inward active transport of $Na^+$
5. Repolarization of an axon during an action potential is produced by the
   (a) inward diffusion of $Na^+$
   (b) active extrusion of $K^+$
   (c) outward diffusion of $K^+$
   (d) inward active transport of $Na^+$
6. As the strength of a depolarizing stimulus to an axon is increased,
   (a) the amplitude of action potentials increases
   (b) the duration of action potentials increases
   (c) the speed with which action potentials are conducted increases
   (d) the frequency with which action potentials are produced increases
7. The conduction of action potentials in a myelinated nerve fiber is
   (a) saltatory
   (b) without decrement
   (c) faster than in an unmyelinated fiber
   (d) all of the above
8. Which of the following is *not* a characteristic of synaptic potentials?
   (a) They are all-or-none in amplitude.
   (b) They decrease in amplitude with distance.
   (c) They are produced in dendrites and cell bodies.
   (d) They are graded in amplitude.
   (e) They are produced by chemically regulated gates.
9. Which of the following is *not* a characteristic of action potentials?
   (a) They are produced by voltage-regulated gates.
   (b) They are conducted without decrement.
   (c) $Na^+$ and $K^+$ gates open at the same time.
   (d) The membrane potential reverses polarity during depolarization.
10. A drug that inactivates acetylcholinesterase
    (a) inhibits the release of ACh from presynaptic endings
    (b) inhibits the attachment of ACh to its receptor protein
    (c) increases the ability of ACh to stimulate muscle contraction
    (d) all of the above
11. Postsynaptic inhibition is produced by
    (a) depolarization of the postsynaptic membrane
    (b) hyperpolarization of the postsynaptic membrane
    (c) axoaxonic synapses
    (d) post-tetanic potentiation
12. Hyperpolarization of the postsynaptic membrane in response to glycine or GABA is produced by the opening of
    (a) $Na^+$ gates
    (b) $K^+$ gates
    (c) $Ca^{++}$ gates
    (d) $Cl^-$ gates
13. The absolute refractory period of a neuron
    (a) is due to the high negative polarity of the inside of the neuron
    (b) occurs only during the repolarization phase
    (c) occurs only during the depolarization phase
    (d) occurs during depolarization and the first part of the repolarization phase
14. Which of the following statements about catecholamines is *false?*
    (a) They include norepinephrine, epinephrine, and dopamine.
    (b) Their effects are increased by action of the enzyme catechol-O-methyltransferase.
    (c) They are inactivated by monoamine oxidase.
    (d) They are inactivated by re-uptake into the presynaptic axon.
    (e) They stimulate the production of cyclic AMP in the postsynaptic axon.
15. The summation of EPSPs from many presynaptic nerve fibers converging onto one postsynaptic neuron is called
    (a) spatial summation
    (b) post-tetanic potentiation
    (c) temporal summation
    (d) synaptic plasticity

### Essay Questions

1. Compare the characteristics of action potentials with those of synaptic potentials.
2. Explain how voltage-regulated gates produce an all-or-none action potential.
3. Explain how action potentials are regenerated along an axon.
4. Explain why conduction in a myelinated axon is faster than in an unmyelinated axon.
5. Trace the course of events between the production of an EPSP and the generation of action potentials at the axon hillock. Explain the effect of spatial and temporal summation on this process.
6. Explain how an IPSP is produced and how IPSPs can inhibit activity of the postsynaptic neuron.

## SELECTED READINGS


Axelrod, J. 1974. Neurotransmitters. *Scientific American.*

Barchas, J. D. et al. 1978. Behavioral neurochemistry: Neuroregulators and behavioral states. *Science* 200:964.

Bartus, R. T. et al. 1982. The cholinergic hypothesis of geriatric memory dysfunction. *Science* 217:408.

Block, I. B. et al. 1984. Neurotransmitter plasticity at the molecular level. *Science* 225:1266.

Blusztajn, J. K., and R. J. Wurtman. 1983. Choline and cholinergic neurons. *Science* 221:614.

Cottman, C. W., and M. Nieto-Sampedro. 1984. Cell biology of synaptic plasticity. *Science* 225:1287.

Coyle, J. T., Prince, D. L., and M. R. DeLong. 1983. Alzheimer's disease: A disorder of cortical cholinergic innervation. *Science* 219:1184.

Dunant, Y., and M. Israel. April 1985. The release of acetylcholine. *Scientific American.*

Freed, W. J., de Medinaceli, L., and R. J. Wyatt. 1985. Promoting functional plasticity in the damaged nervous system. *Science* 227:1544.

Goldstein, G. W., and A. L. Betz. September 1986. The blood-brain barrier. *Scientific American.*

Gottlieb, D. I. February 1988. GABAergic Neurons. *Scientific American.*

Jacobs, B. L., and M. E. Trulson. 1979. Mechanism of action of LSD. *American Scientist* 67:396.

Kandel, E. R. 1979. Psychotherapy and the single synapse. *New England Journal of Medicine* 301:1028.

Katz, B. November 1952. The nerve impulse. *Scientific American.*

Keynes, R. D. March 1979. Ion channels in the nerve cell membrane. *Scientific American.*

Kimelberg, H. K., and M. D. Norenberg. April 1989. Astrocytes. *Scientific American.*

Krieger, D. T. 1983. Brain peptides: what, where, and why? *Science* 222:975.

Kuffler, S. W., and J. G. Nicholls. 1976. *From neuron to brain: a cellular approach to the function of the nervous system.* Sunderland, Mass.: Sinauer Associates, Inc., Publishers.

Lester, H. A. February 1977. The response of acetylcholine. *Scientific American.*

Morrel, P., and W. Norton. May 1980. Myelin. *Scientific American.*

Nathanson, J. A., and P. Greegard. August 1977. "Second messengers" in the brain. *Scientific American.*

Ritz, M. C. et al. 1987. Cocaine receptors on dopamine transporters are related to self-administration of cocaine. *Science* 237:1219.

Schwartz, J. H. April 1980. The transport of substances in nerve cells. *Scientific American.*

Snyder, S. H. 1980. Brain peptides as neurotransmitters. *Science* 209:976.

Snyder, S. H. 1984. Drug and neurotransmitter receptors in the brain. *Science* 224:22.

Stevens, C. F. September 1979. The neuron. *Scientific American.*

Wagner, H. N. 1984. Imaging CNS receptors: the dopaminergic system. *Hospital Practice* 19:187.

Waxman, S. G., and J. M. Ritchie. 1985. Organization of ion channels in the myelinated nerve fiber. *Science* 228:1502.

Wurtman, R. J. April 1982. Nutrients that modify brain function. *Scientific American.*

Wurtman, R. J. January 1985. Alzheimer's disease. *Scientific American.*

# 8

# Central Nervous System

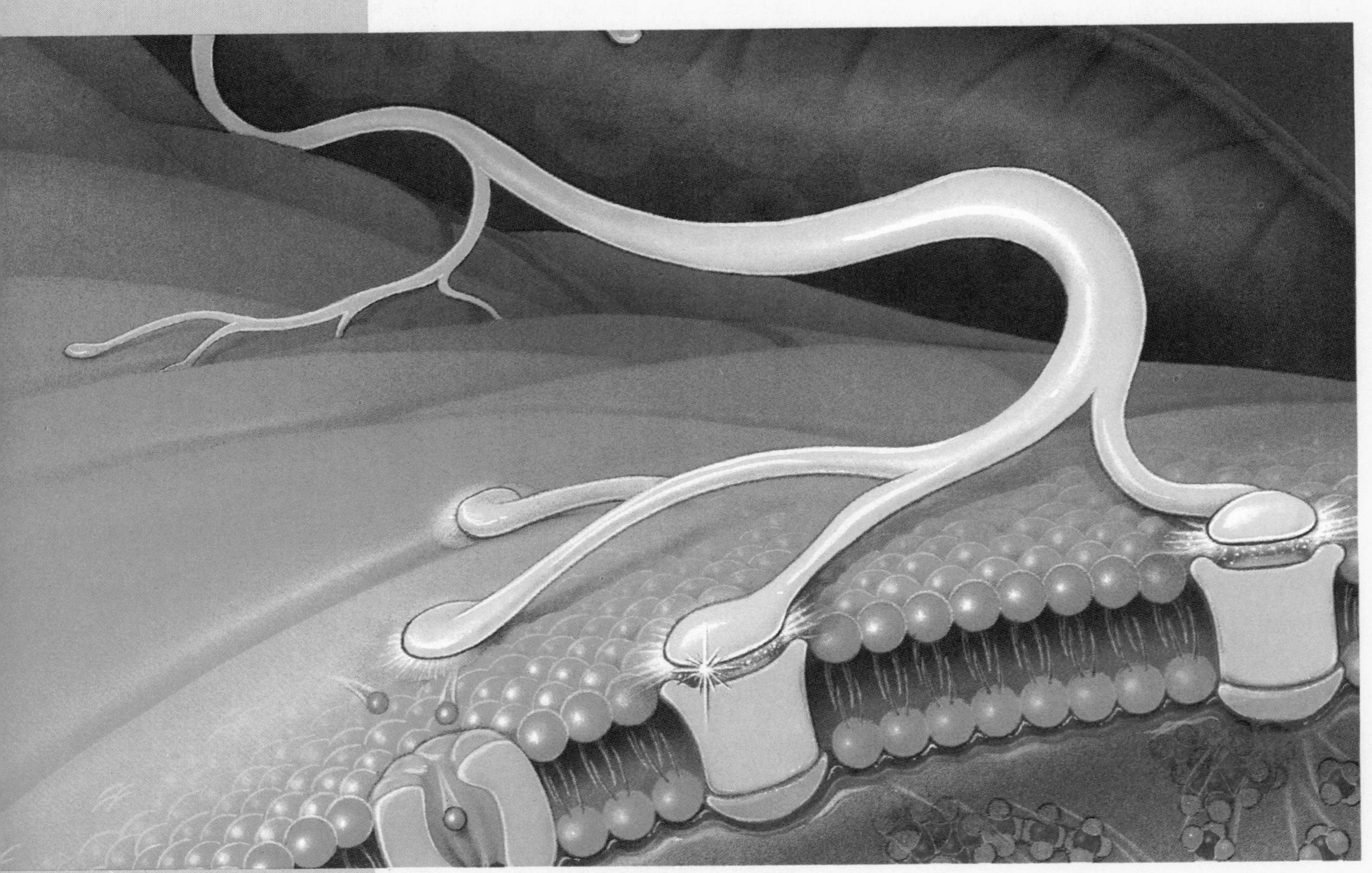

## OBJECTIVES

By studying this chapter you should be able to

1. describe the location of the major brain regions and the structures contained in each of these regions
2. describe the organization of the cerebrum and the primary roles of its lobes
3. describe the location and function of the sensory cortex and motor cortex
4. explain the lateralization of functions in the right and left cerebral hemispheres
5. identify the structures involved in the control of speech and explain their interrelationships
6. describe the different types of aphasias that result from damage to specific regions of the brain
7. describe the structures involved in the limbic system and explain the possible role of this system in emotion
8. distinguish between different types of memory, identify the brain regions known to function in memory, and explain the process of long-term potentiation
9. identify the location of the thalamus and explain its significance
10. identify the location of the hypothalamus and explain its significance
11. describe the locations and structures contained in the midbrain and hindbrain, and the particular importance of the medulla oblongata in the control of visceral functions
12. explain the organization of the spinal cord and how ascending and descending tracts are named
13. describe the origin, pathway, and significance of the pyramidal motor tracts
14. explain the role of the basal ganglia and cerebellum in motor control by means of the extrapyramidal system, and describe the pathways of this system
15. explain the structures and pathways involved in a reflex arc

## OUTLINE

**Figure 8.1.** The CNS consists of the brain and the spinal cord, both of which are covered with meninges and bathed in cerebrospinal fluid.

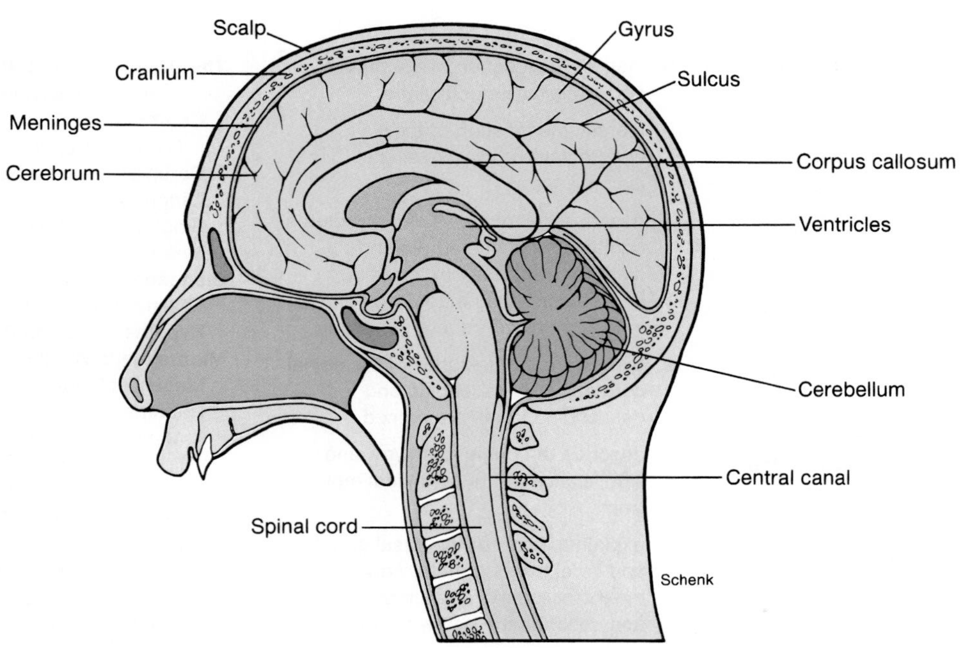


## The Brain: Structural Plan and Functions of the Cerebrum

The brain is composed of an enormous number of association neurons, with accompanying neuroglia, arranged into brain regions and divisions, lobes, nuclei, and fiber tracts. These neurons receive sensory information, direct the activity of motor neurons, and perform such higher brain functions as learning and memory. Analysis of sensory information, association of that information with past experience and emotional responses, and even self-awareness and consciousness may derive from complex interactions of different brain regions.

The **central nervous system (CNS)**, consisting of the brain and spinal cord (fig. 8.1), receives input from *sensory neurons* and directs the activity of *motor neurons,* which innervate muscles and glands. The *association neurons* within the brain and spinal cord are in a position, as their name implies, to associate appropriate motor responses with sensory stimuli and, thus, to maintain homeostasis in the internal environment and the continued existence of the organism in a changing external environment. Further, the central nervous systems of all vertebrates (and most invertebrates) are capable of at least rudimentary forms of learning and retaining the memories of past experiences. This ability, which is most highly developed in the human brain, permits behavior to be modified by experience and is thus of obvious benefit to survival. Perceptions, learning, memory, emotions, and the self-awareness that forms the basis of consciousness are creations of the brain. Whimsical though it seems, the study of brain physiology is the process of the brain studying itself.

The study of the structure and function of the central nervous system requires a knowledge of its basic "plan," which is established during the course of embryonic development. The early embryo is surrounded by an embryonic tissue layer known as surface *ectoderm.* This will eventually form the epidermis of the skin, among other structures. A groove then appears in this ectoderm along the dorsal midline of the embryo's body. This groove deepens, and by the twentieth day after conception, fuses to form a *neural tube.* The part of the ectoderm where the fusion occurs becomes a separate structure from the neural tube, called the *neural crest,* which is located between the neural tube and the surface ectoderm (fig. 8.2). Eventually, the neural tube will become the central nervous system and the neural crest will become the ganglia of the peripheral nervous system, among other structures.

By the middle of the fourth week after conception, the anterior end of the neural tube, which is going to form the brain, exhibits three swellings: the *prosencephalon (forebrain), mesencephalon (midbrain),* and *rhombencephalon (hindbrain)* (fig. 8.3). During the fifth week, these areas

**Figure 8.2.** A dorsal view of a twenty-two-day-old embryo, showing transverse sections at three levels of the developing central nervous system.

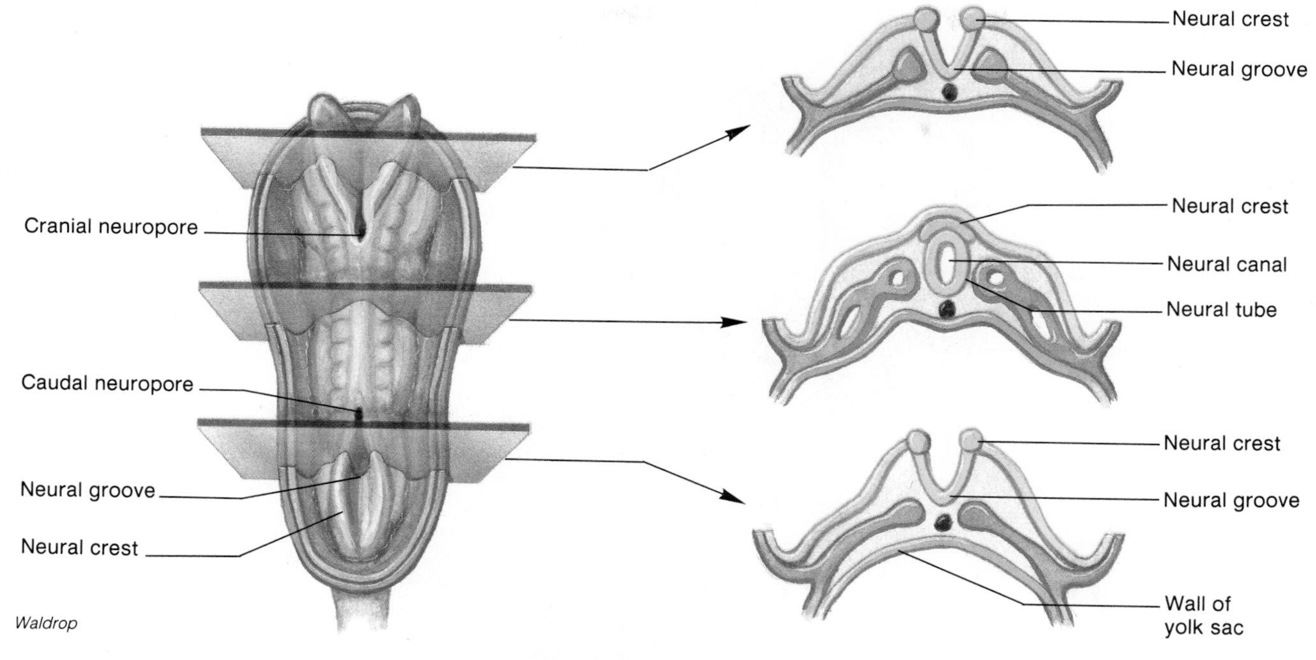


**Figure 8.3.** The developmental sequence of the brain. During the fourth week, three principal regions of the brain are formed. During the fifth week, a five-regioned brain develops, and specific structures begin to form.

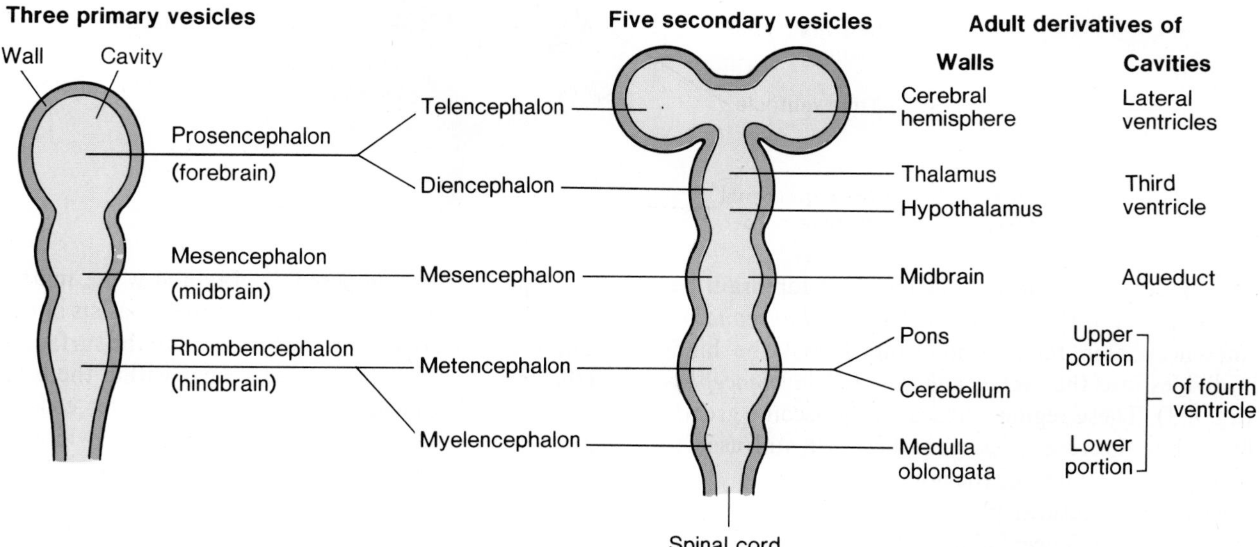

**Figure 8.4.** The ventricles of the brain. (*a*) An anterior view; (*b*) a lateral view.

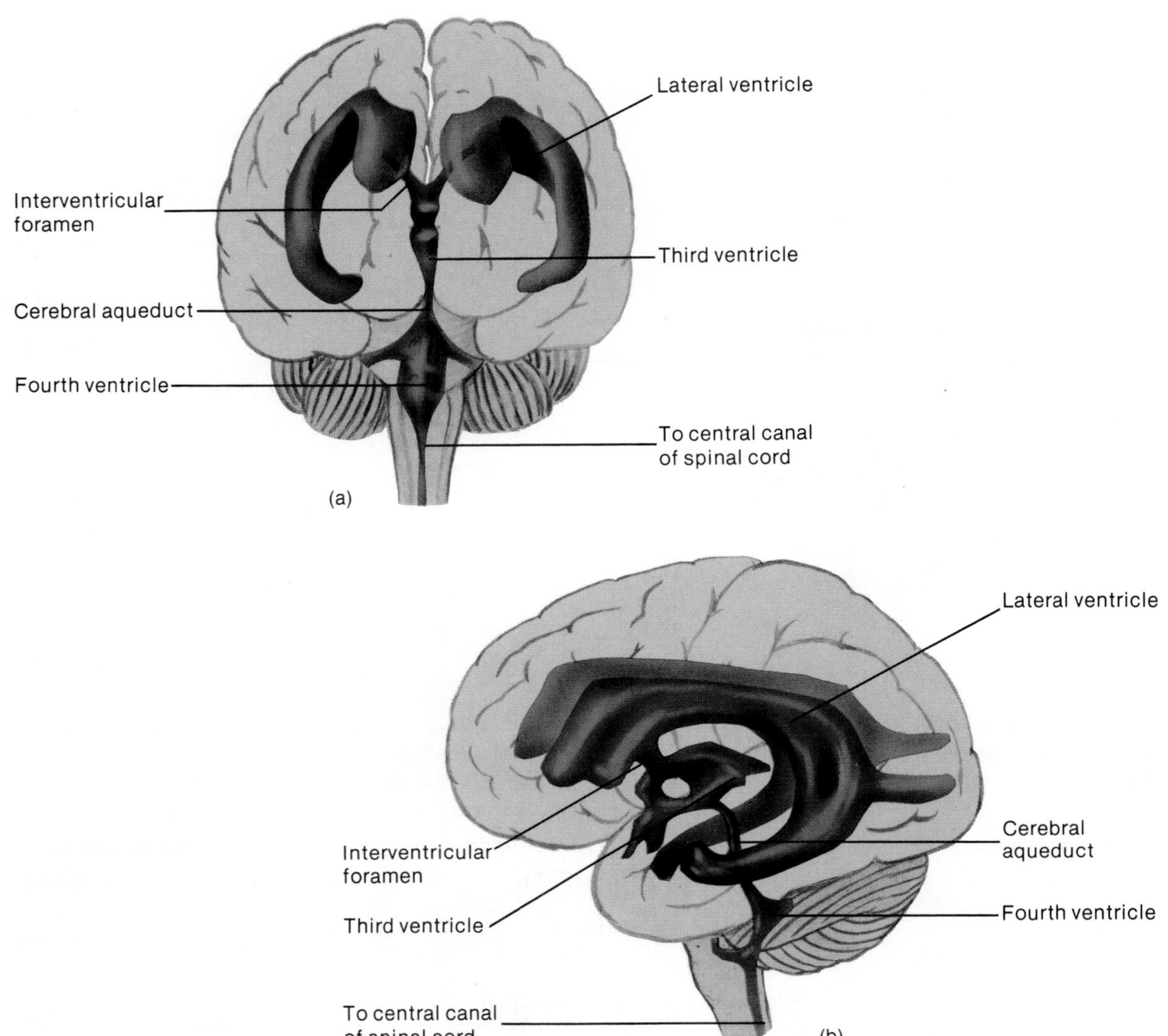


become modified to form five regions. The forebrain becomes divided into the *telencephalon* and *diencephalon*; the mesencephalon remains unchanged; and the hindbrain divides into the *metencephalon* and *myelencephalon* (fig. 8.3). These regions subsequently become greatly modified, but the terms described here are still used to indicate general regions of the adult brain.

The basic structural plan of the CNS can now be understood. The telencephalon (refer to fig. 8.3) grows disproportionately in humans, forming the two enormous hemispheres of the *cerebrum* which cover the diencephalon, the midbrain, and a portion of the hindbrain. Also, notice that the CNS begins as a hollow tube, and indeed remains hollow as the brain regions are formed. The cavities of the brain are known as *ventricles* and become filled with *cerebrospinal fluid (CSF)*. The cavity of the spinal cord is the *central canal*, which is also filled with CSF (fig. 8.4).

The CNS is composed of gray and white matter, as described in chapter 7. The gray matter consists of neuron cell bodies and dendrites, and is found in the surface layer (the *cortex*) of the brain, and deeper within the brain in aggregations known as *nuclei*. White matter consists of axon tracts (the myelin sheaths produce the white color) that underlie the cortex and surround the nuclei. The adult brain contains an estimated 100 billion ($10^{11}$) neurons, weighs 1.5 kg (3–3.5 lbs), and receives about 20% of the total blood flow to the body per minute. This high rate of blood flow is a consequence of the high metabolic requirements of the brain; it is not, as Aristotle believed, because the brain's function is to cool the blood. (This fanciful notion—completely incorrect—is a striking example of prescientific thought, where hypotheses and conclusions are not based on a foundation of experimentation and careful observation).

**Figure 8.5.** The cerebrum. (*a*) A lateral view; (*b*) a superior view.

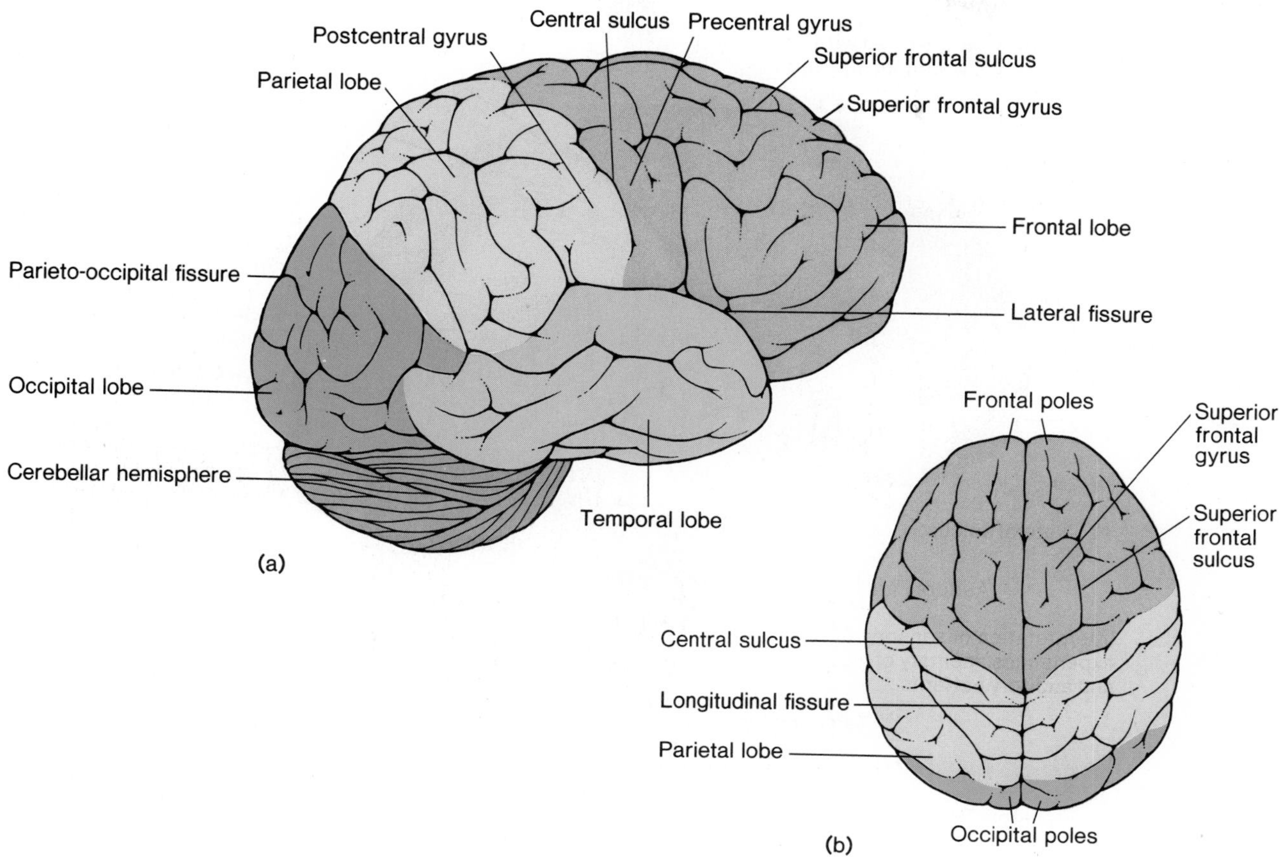


## Cerebrum

The **cerebrum** (fig. 8.5), which is the only structure of the telencephalon, is the largest portion of the brain (accounting for about 80% of its mass) and is believed to be the brain region primarily responsible for higher mental functions. The cerebrum consists of *right* and *left hemispheres,* which are connected internally by a large fiber tract called the *corpus callosum* (see fig. 8.1).

The cerebrum consists of an outer **cerebral cortex,** which is 2–4 mm of gray matter, and underlying white matter. The surface of the cerebrum is folded into convolutions. The elevated folds of the convolutions are called *gyri,* and the depressed grooves are the *sulci.* Each cerebral hemisphere is subdivided by deep sulci, or *fissures,* into five lobes, four of which are visible from the surface (fig. 8.6). These lobes are the *frontal, parietal, temporal,* and *occipital,* which are visible from the surface, and the deep *insula* (table 8.1).

The **frontal lobe** is the anterior portion of each cerebral hemisphere. A deep fissure, called the *central sulcus,* separates the frontal lobe from the **parietal lobe.** The *precentral gyrus,* (fig. 8.5) involved in motor control, is located in the frontal lobe just in front of the central sulcus. The *postcentral gyrus,* which is located just behind the central sulcus in the parietal lobe, is the primary area of the cortex responsible for the perception of *somatesthetic sensation*—sensation arising from cutaneous, muscle, tendon, and joint receptors.

The precentral (motor) and postcentral (sensory) gyri have been mapped in conscious patients undergoing brain surgery. Electrical stimulation of specific areas of the precentral gyrus was found to cause specific movements, and stimulation of different areas of the postcentral gyrus evoked sensations in specific parts of the body. Typical maps of these regions (fig. 8.7) show an upside-down picture of the body, with the superior regions of cortex devoted to the toes and the inferior regions devoted to the head.

A striking feature of these maps is that the areas of cortex responsible for different parts of the body do not correspond to the size of the body parts being served. Instead, the body regions with the highest densities of receptors are represented with the largest areas of the sensory cortex, and the body regions with the greatest number of motor innervations are represented with the largest area of motor cortex. The hands and face, therefore, which have a high density of sensory receptors and motor innervation, are served by larger areas of the precentral and postcentral gyri than is the rest of the body.

**Figure 8.6.** The lobes of the left cerebral hemisphere showing the principal motor and sensory areas of the cerebral cortex. Dotted lines show the division of the lobes.

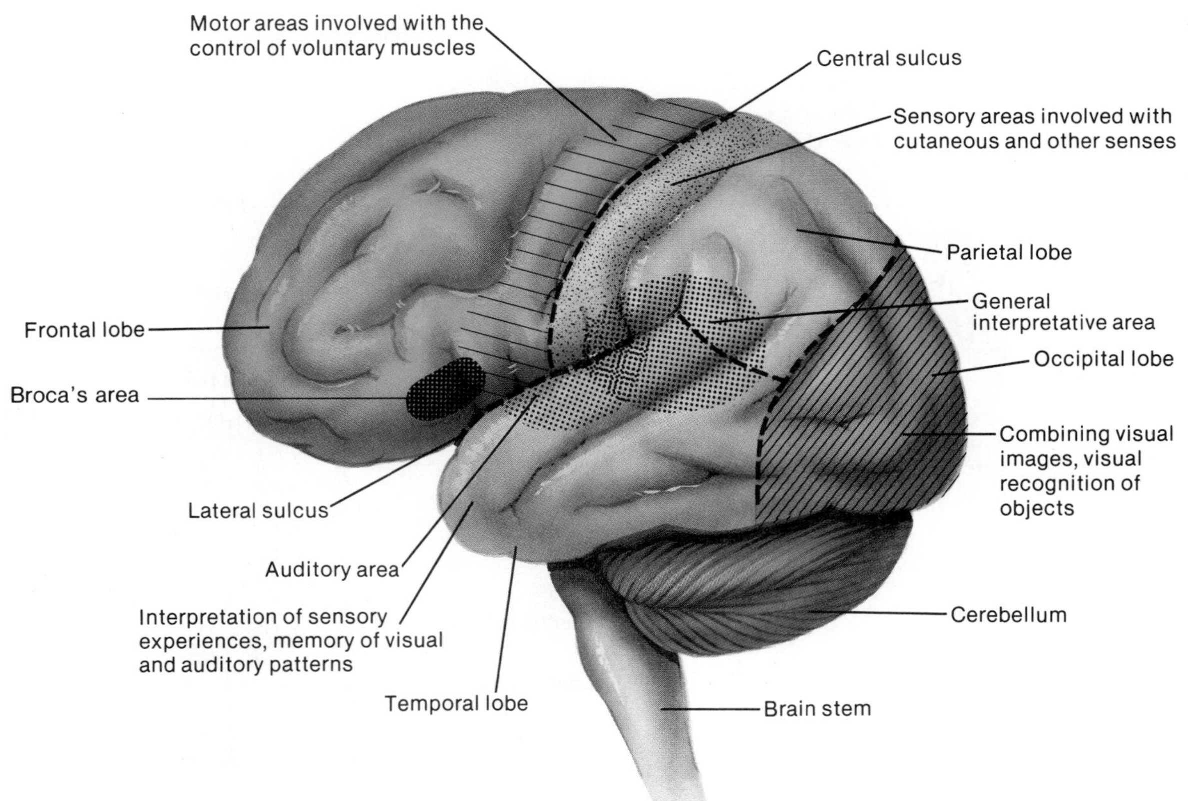


**Table 8.1** Functions of the cerebral lobes

| Lobe | Functions |
|---|---|
| Frontal | Voluntary motor control of skeletal muscles; personality; higher intellectual processes (e.g., concentration, planning, and decision making); verbal communication |
| Parietal | Somatesthetic interpretation (e.g., cutaneous and muscular sensations); understanding speech and formulating words to express thoughts and emotions; interpretation of textures and shapes |
| Temporal | Interpretation of auditory sensations; storage (memory) of auditory and visual experiences |
| Occipital | Integrates movements in focusing the eye; correlating visual images with previous visual experiences and other sensory stimuli; conscious perception of vision |
| Insula | Memory; integrates other cerebral activities |



The **temporal lobe** contains auditory centers that receive sensory fibers from the cochlea of each ear. This lobe is also involved in the interpretation and association of auditory and visual information. The **occipital lobe** is the primary area responsible for vision and for the coordination of eye movements.

Surgical removal of the temporal lobes in monkeys produces a condition called the *Kluver-Bucy syndrome,* and some aspects of this syndrome are observed in humans with temporal lobe damage. One interesting feature of this syndrome is the apparent lack of recognition of objects based on visual cues. Vision is not damaged, but the ability to associate visual information with past experience is impaired.

**Figure 8.7.** Motor and sensory areas of the cerebral cortex. (*a*) Motor areas that control skeletal muscles. (*b*) Sensory areas that receive somatesthetic sensations.

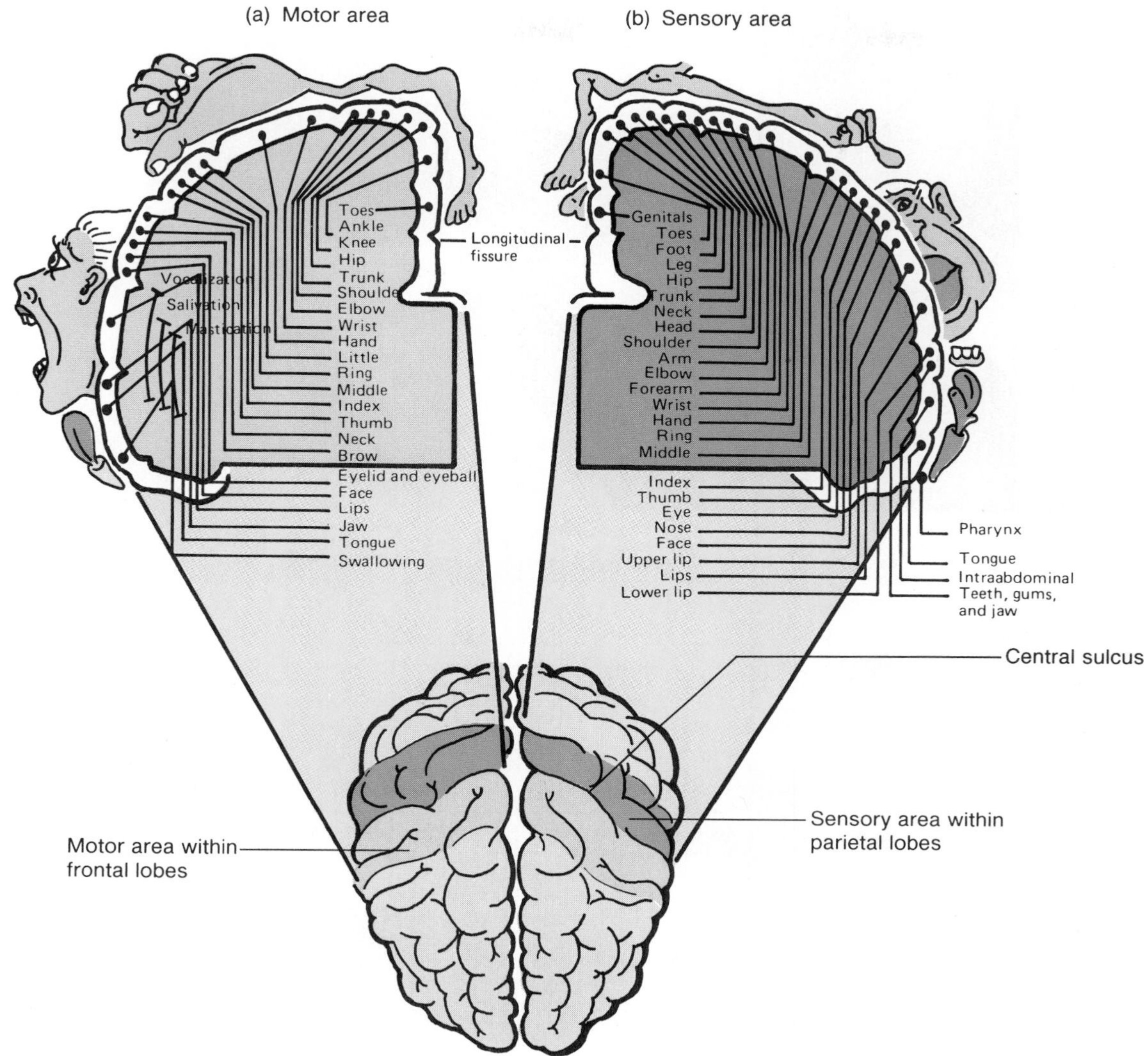


***Electroencephalogram.*** The synaptic potentials (discussed in chapter 7) produced at the cell bodies and dendrites of the cerebral cortex create electrical currents, which can be measured by electrodes placed on the scalp. A record of these electrical currents is called an *electroencephalogram,* or *EEG.* Deviations from normal EEG patterns can be used clinically to diagnose epilepsy and other abnormal states, and the absence of an EEG can be used to detect brain death.

There are normally four types of EEG patterns (fig. 8.8): **Alpha waves** are best recorded from the parietal and occipital regions while a person is awake and relaxed but with the eyes closed. These waves are rhythmic oscillations of about 10–12 cycles/second. The alpha rhythm of a child younger than eight years old occurs at a slightly lower frequency of 4–7 cycles/second.

**Beta waves** are strongest from the frontal lobes, especially the area near the precentral gyrus. These waves are produced by visual stimuli and mental activity. Because they respond to stimuli from receptors and are superimposed on the continuous activity patterns, they constitute *evoked activity.* The frequency of beta waves is 13–25 cycles/second.

**Theta waves** are emitted from the temporal and occipital lobes. They have a frequency of 5–8 cycles/second and are common in newborn infants. The recording of theta waves in adults generally indicates severe emotional stress and can be a forewarning of a nervous breakdown.

**Delta waves** are seemingly emitted in a general pattern from the cerebral cortex. These waves have a frequency of 1–5 cycles/second and are common during sleep and in an awake infant.The presence of delta waves in an awake adult indicates brain damage.

**Figure 8.8.** Brain waves. (*a*) Technician using an electroencephalograph to take the EEG of a teenaged boy. (*b*) Types of electroencephalograms.

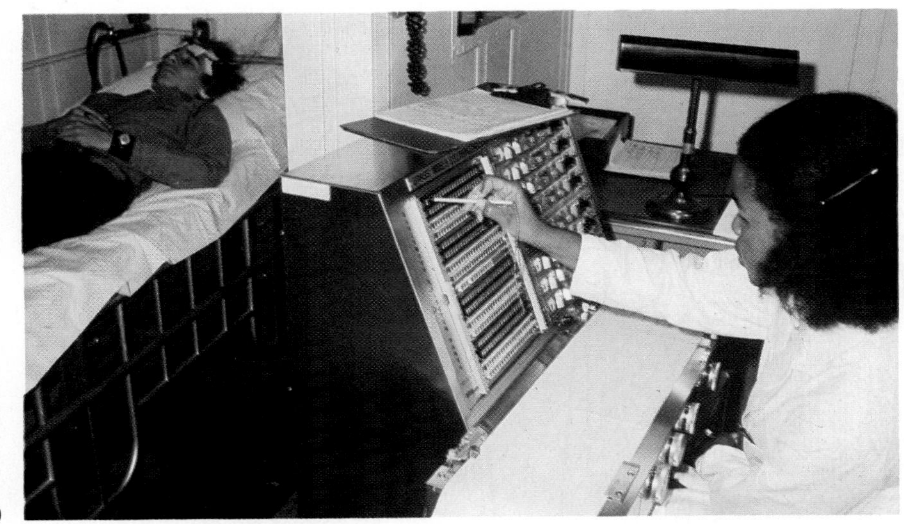
(a)

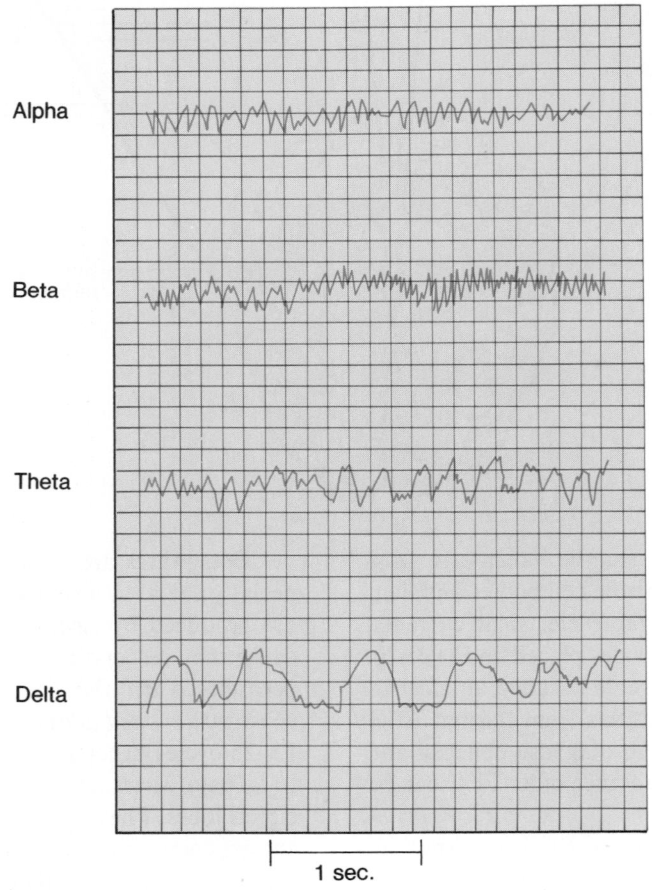

(b)

***Basal Ganglia.*** The *basal ganglia* are masses of gray matter, composed of neuron cell bodies, located deep within the white matter of the cerebrum (fig. 8.9). The most prominent of the basal ganglia is the **corpus striatum,** which consists of several masses of nuclei (a *nucleus* is a collection of cell bodies in the CNS). The upper mass is the *caudate nucleus,* which is separated from two lower masses that are collectively called the *lentiform nucleus.* The lentiform nucleus consists of a lateral portion, called the *putamen,* and a medial portion, called the *globus pallidus.* The basal ganglia function in the control of voluntary movements.

**Figure 8.9.** Sections through the cerebrum and diencephalon. (*a*) A coronal section. (*b*) A transverse section.

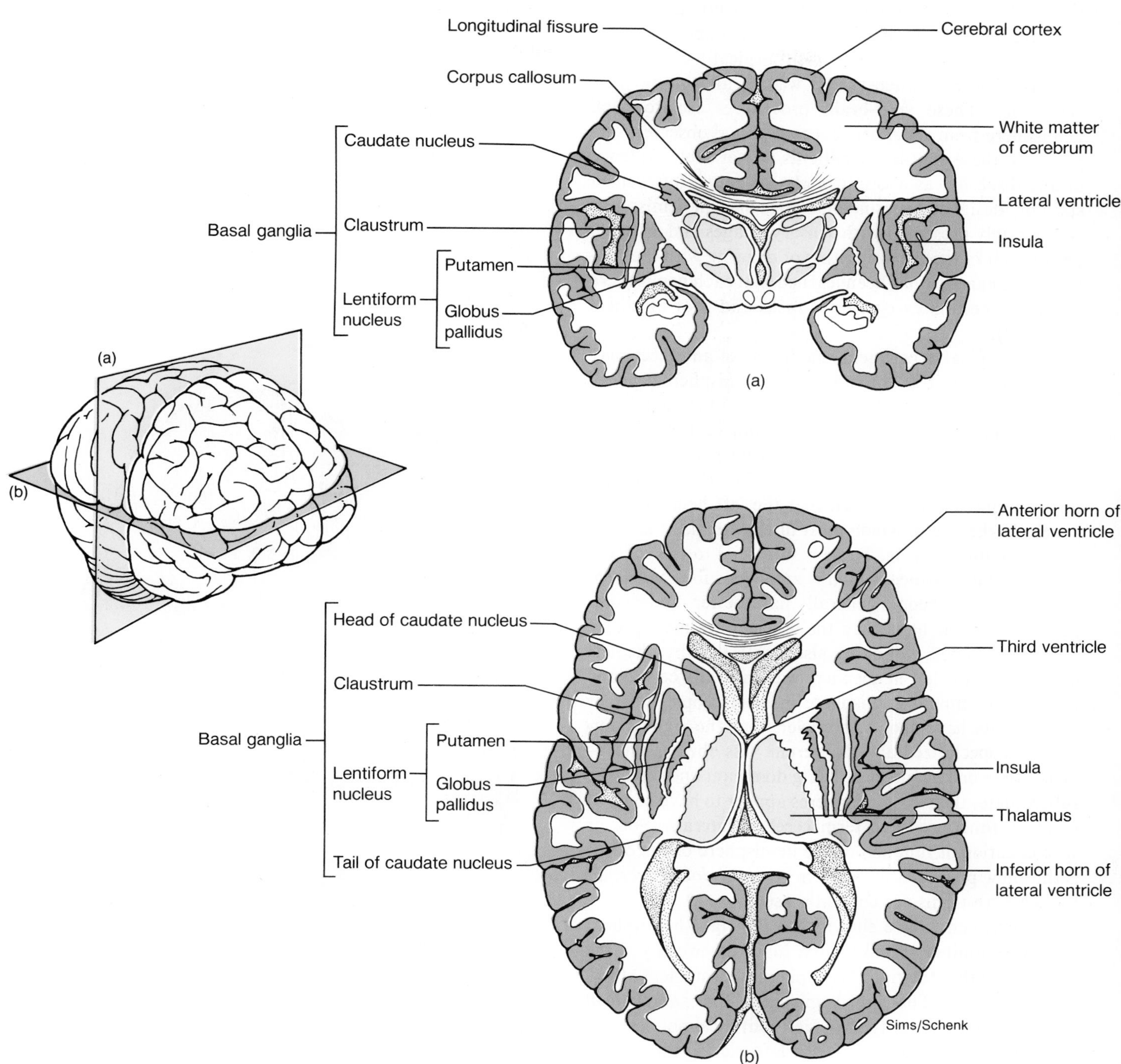


Degeneration of the caudate nucleus produces **chorea**—a hyperkinetic disorder characterized by rapid, uncontrolled, jerky movements. Degeneration of dopaminergic neurons from the substantia nigra, a small nucleus considered a part of the basal ganglia, to the caudate nucleus produces **Parkinson's disease.** As discussed in chapter 7, this disease is associated with rigidity, resting tremor, and difficulty in the initiation of voluntary movements.

## Cerebral Lateralization

The two hemispheres of the cerebral cortex control, via motor fibers originating in the precentral gyrus, skilled movements of the contralateral side of the body. At the same time, somatesthetic sensation from the right side of the body projects to the left postcentral gyrus, and vice versa, because of *decussation* (crossing-over) of fibers. In a similar manner, images falling in the left half of each retina project to the right occipital lobe, and images in the right half of each retina project to the left occipital lobe.

Each cerebral hemisphere, however, receives information from both sides of the body because the two hemispheres communicate with each other via the *corpus callosum,* a large tract composed of about 200 million fibers.

The corpus callosum has been surgically cut in some victims of severe epilepsy as a way of alleviating their symptoms. These **split-brain** procedures isolate each hemisphere from the other, but to a casual observer, surprisingly, these split-brain patients do not show evidence of any disability as a result of the surgery. However, in specially designed experiments in which each hemisphere is separately presented with sensory images and asked to perform tasks (speech or writing or drawing with the contralateral hand), it has been learned that each hemisphere is good at certain categories of tasks and poor at others (fig. 8.10).

In a typical experiment, the image of an object may be presented to either the right or left hemisphere (by presenting it to either the left or right visual field only) and the person may be asked to name the object. It was found that, in most people, the task could be performed successfully by the left hemisphere but not by the right. Similar experiments have shown that the left hemisphere is generally the one in which most of the language and analytical abilities reside. These results lead to the concept of **cerebral dominance,** which is analogous to the concept of handedness—people generally have better motor competence with one hand than the other. Since most people are right-handed, which is also controlled by the left hemisphere, the left hemisphere was naturally considered to be the dominant hemisphere in most people. Further experiments have shown, however, that the right hemisphere is specialized along different, less obvious lines—rather than one hemisphere being dominant and the other subordinate, the two hemispheres appear to have complementary functions. The term **cerebral lateralization,** or specialization of function to one hemisphere or the other, is thus now preferred to the term *cerebral dominance,* although both terms are currently used.

Experiments have shown that the right hemisphere does have limited verbal ability; more noteworthy is the observation that the right hemisphere is most adept at *visuospatial tasks.* The right hemisphere, for example, can recognize faces better than the left, but it cannot describe facial appearances as well as the left. Acting through its control of the left hand, the right hemisphere is better than the left (controlling the right hand) at arranging blocks or drawing cubes. Patients with damage to the right hemisphere, as might be predicted from the results of split-brain research, have difficulty finding their way around a house and reading maps.

Perhaps as a result of the role of the right hemisphere in the comprehension of patterns and part-whole relationships, the ability to compose music, but not to critically understand it, appears to depend on the right hemisphere.

**Figure 8.10.** Different functions of the right and left cerebral hemispheres, as revealed by experiments with people who have had the tract connecting the two hemispheres (the corpus callosum) surgically split.

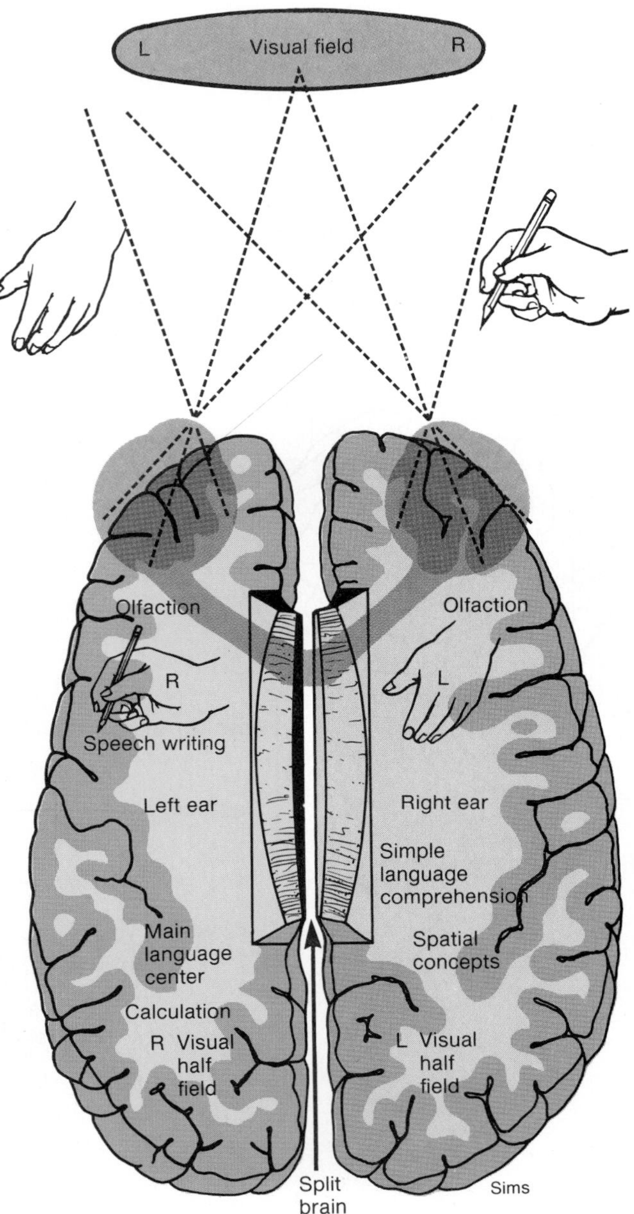


Interestingly, damage to the left hemisphere may cause severe speech problems while leaving the ability to sing unaffected.

The lateralization of functions just described—with the left hemisphere specialized for language and analytical ability, and the right hemisphere specialized for visuospatial ability—is true for 97% of all people. It is true for all right-handers (who comprise 90% of all people) and for 70% of all left-handers. The remaining left-handers are split roughly equally into those who have language-analytical ability in the right hemisphere and those in whom this ability is present in both hemispheres.

**Figure 8.11.** Brain areas involved in the control of speech.

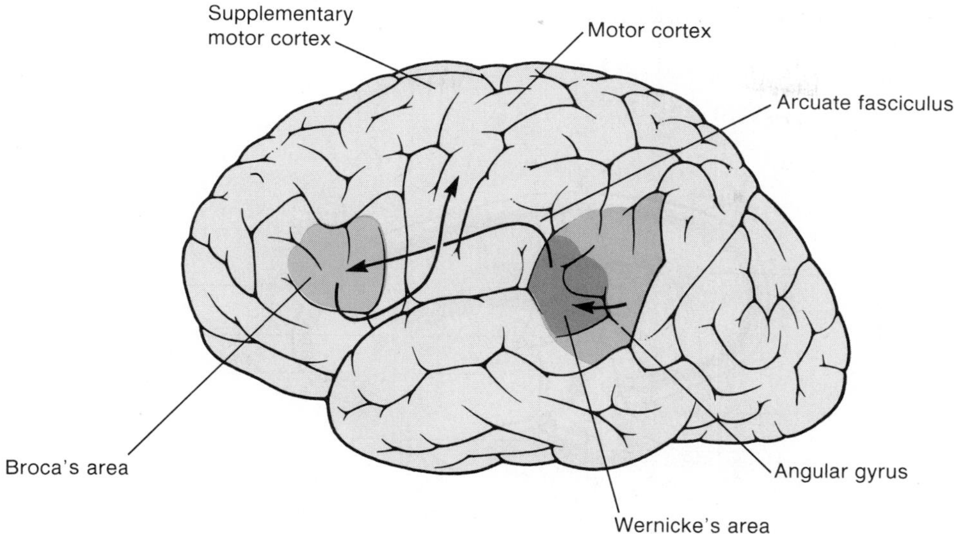


It is interesting to speculate that the creative ability of a person may be related to the interaction of information between the right and left hemispheres. This interaction may be greater in left-handed people; a study found that the number of left-handers among college art students was disproportionately higher than in the general population. The observation that Leonardo da Vinci and Michelangelo were left-handed is interesting in this regard, but clearly is not scientific proof of any hypothesis. Further research on the lateralization of function of the cerebral hemispheres may reveal much more about both brain function and the creative process.

## Language

Knowledge of the brain regions involved in language has been gained primarily by the study of *aphasias*—speech and language disorders caused by damage to the brain. The language areas of the brain are primarily located in the left hemisphere of the cerebral cortex in most people, as previously described. As long ago as the nineteenth century, two areas of the cortex—Broca's area and Wernicke's area (fig. 8.11)—were found to be of particular importance in the production of aphasias.

Damage to **Broca's area,** located in the left inferior frontal gyrus, produces an aphasia in which the person is reluctant to speak, and when speech is attempted, it is slow and poorly articulated. People with *Broca's aphasia,* however, have unimpaired comprehension of speech. It should be noted that this speech difficulty is not simply due to a problem in motor control, since the neural control over the musculature of the tongue, lips, larynx, and so on is unaffected. Damage to **Wernicke's area,** located in the superior temporal gyrus, results in speech that is rapid and fluid but does not convey any information. People with *Wernicke's aphasia* produce speech that has been described as a "word salad." The words used may be real words that are chaotically mixed together, or they may be made-up words. Language comprehension has been destroyed; people with Wernicke's aphasia cannot understand either spoken or written language.

It appears that the concept of words to be spoken originates in Wernicke's area and is communicated to Broca's area in a fiber tract called the **arcuate fasciculus.** Broca's area, in turn, sends fibers to the motor cortex (precentral gyrus), which directly controls the musculature of speech. Damage to the arcuate fasciculus produces *conduction aphasia,* which is fluent but nonsensical speech as in Wernicke's aphasia, even though both Broca's and Wernicke's areas are intact.

The **angular gyrus,** located at the junction of the parietal, temporal, and occipital lobes, is believed to be a center for the integration of auditory, visual, and somatesthetic information. Damage to the angular gyrus produces aphasias, which suggests that this area projects to Wernicke's area. Some patients with damage to the left angular gyrus can speak and understand spoken language but cannot read or write. Other patients can write a sentence but cannot read it, presumably due to damage to the projections from the occipital lobe (involved in vision) to the angular gyrus.

Recovery of language ability, by transfer to the right hemisphere after damage to the left hemisphere, is very good in children but decreases after adolescence. Recovery is reported to be faster in left-handed people, possibly because language ability is more evenly divided between the two hemispheres in left-handed people. Some recovery usually occurs after damage to Broca's area, but damage to Wernicke's area produces more severe and permanent aphasias.

**Figure 8.12.** The limbic system and the pathways that interconnect the structures of the limbic system (note: the left temporal lobe of the cerebral cortex has been removed).

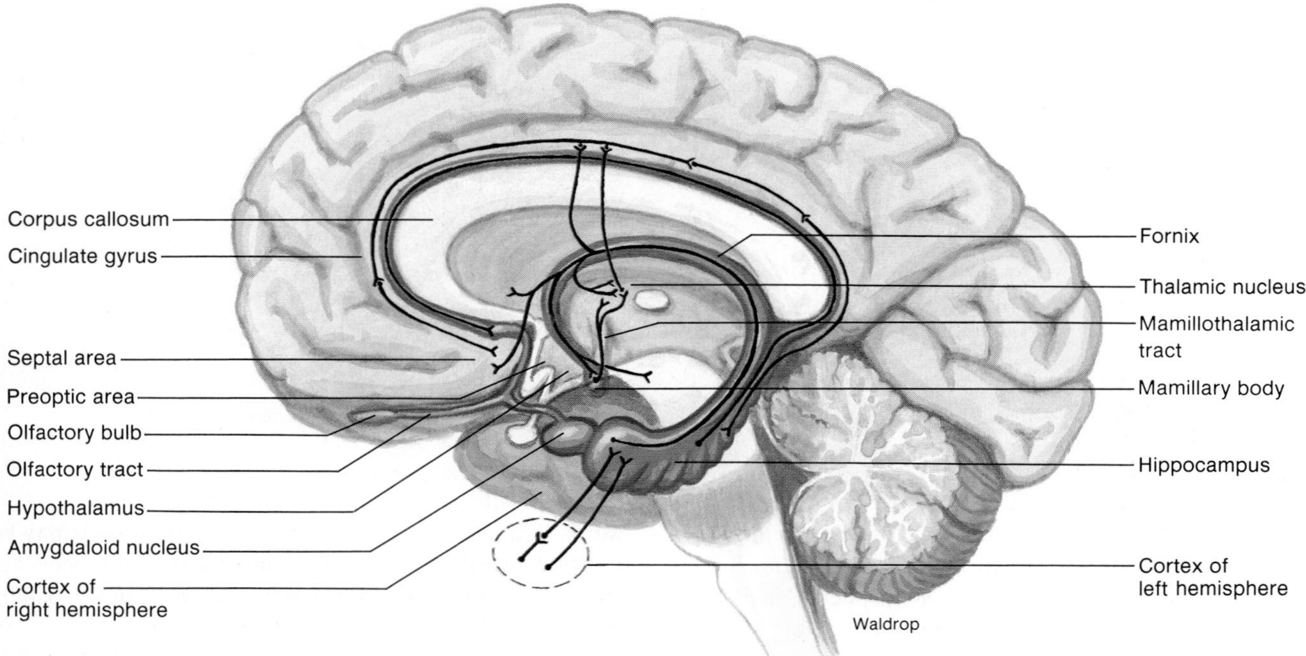


## Emotion and Motivation

The parts of the brain that appear to be of paramount importance in the neural basis of emotional states are the hypothalamus (in the diencephalon) and the **limbic system.** The limbic system is a group of nuclei and fiber tracts that form a ring around the brain stem. The structures of the limbic system include the *cingulate gyrus* (part of the cerebral cortex), *amygdaloid nucleus* (or *amygdala*), *hippocampus,* and the *septal nuclei* (fig. 8.12).

The limbic system was formerly called the *rhinencephalon,* or "smell brain," because it is involved in the central processing of olfactory information. This function may be the primary one of lower vertebrates, in which the limbic system may constitute the entire forebrain, but it is now known that the limbic system of humans is a center for basic emotional drives. The limbic system was derived early in the course of vertebrate evolution, and its tissue is thus phylogenetically older than the *neocortex* of the cerebral hemispheres. There are few synaptic connections between the neocortex and the structures of the limbic system, which perhaps helps to explain why we have so little conscious control over our emotions.

There is a closed circuit of information flow between the limbic system and the thalamus and hypothalamus (fig. 8.12; these areas are part of the diencephalon, and are described in a later section), called the *Papez circuit.* In this circuit, a fiber tract, the *fornix,* connects the hippocampus to the mammillary bodies, which in turn project to the anterior nuclei of the thalamus. The nuclei of the thalamus, in turn, send fibers to the cingulate gyrus, which then completes the circuit by sending fibers to the hippocampus. Through these interconnections, the limbic system and the hypothalamus appear to cooperate in the neural basis of emotional states.

Studies of the functions of these regions include electrical stimulation of specific locations, destruction of tissue (producing *lesions*) in particular sites, and surgical removal, or *ablation,* of specific structures. These studies suggest that the hypothalamus and limbic system are involved in the following processes:

**Aggression.** Stimulation of certain areas of the amygdala produce rage and aggression, and lesions of the amygdala can produce docility in experimental animals. Stimulation of particular areas of the hypothalamus can produce similar effects.

**Fear.** Fear can be produced by electrical stimulation of the amygdala and hypothalamus, and surgical removal of the limbic system can produce an absence of fear. Monkeys are normally terrified of snakes, for example, but if they have had their limbic system removed they will handle snakes without fear.

**Feeding.** The hypothalamus contains both a *feeding center* and a *satiety center.* Electrical stimulation of the former produces overeating, and stimulation of the latter will stop feeding behavior in experimental animals.

**Sex.** The hypothalamus and limbic system are involved in the regulation of the sexual drive and sexual behavior, as shown by stimulation and ablation studies in experimental animals. The cerebral cortex, however, is also critically important for the sex drive in lower animals, and the role of the cerebrum is believed to be even more important for the sex drive in humans.

**Reward and punishment system.** Electrodes placed in particular sites from the frontal cortex to the hypothalamus can deliver shocks that function as a reward. In rats, this reward is more powerful than food or sex in motivating behavior. Similar studies have been done in some humans, who report a feeling of relaxation and relief from tension, but not ecstasy. Electrodes placed in slightly different positions apparently stimulate a punishment system in experimental animals, who stop their behavior when stimulated in these regions.

## Memory

Clinical studies of amnesia (loss of memory) suggest that several different brain regions are involved in memory storage and retrieval. Amnesia has been found to result from damage to the temporal lobe of the cerebral cortex, hippocampus, head of the caudate (in Huntington's disease), or the dorsomedial thalamus (in alcoholics suffering from Korsakoff's syndrome with thiamine deficiency). A number of researchers now believe that there are two different systems of information storage in the brain. One system relates to the simple learning of stimulus-response that even invertebrates can do to a lesser degree; this may correspond in humans to unconscious memory. Even people with severe amnesia retain this ability. The other system is unique to humans and other higher organisms, and may correspond to conscious memory. Clinical studies also suggest that this latter system of memory can be divided into two major categories: **short-term memory** and **long-term memory.** People with head trauma, for example, and patients with suicidal depression who are treated by *electroconvulsive shock (ECS)* therapy, may lose their memory of recent events but retain their older memories.

Surgical removal of the right and left medial temporal lobes was performed in one patient, designated "H.M.," in an effort to treat his epilepsy. After the surgery he was unable to consolidate any short-term memory. He could repeat a phone number and carry out a normal conversation; he could not remember the phone number if momentarily distracted, however, and if the person to whom he was talking left the room and came back a few minutes later, H.M. would have no recollection of seeing that person or of having had a conversation with that person before. Although his memory of events that occurred before the operation was intact, all subsequent events in his life seemed as if they were happening for the first time.

The effects of bilateral removal of the medial temporal lobes on H.M. are believed to be due, at least in part, to the fact that the hippocampus (fig. 8.12) was also removed in the process. Surgical removal of the left hippocampus impairs the consolidation of short-term verbal memories into long-term memory, and removal of the right hippocampus impairs the consolidation of nonverbal memories. From these clinical observations, it appears that the hippocampus is an important structure involved in the change from short-term to long-term memory. Experiments in monkeys, however, have failed to duplicate the complete symptoms when only the hippocampus is removed. In these experiments, the amygdaloid nucleus (fig. 8.12), which, like the hippocampus, is also on the inner surface of the temporal lobes, must be destroyed as well to produce these effects. The amygdaloid nucleus has been proposed to play an important role in the association of memories with sensory information from other areas of the cortex, and in the association of emotions with both sensory perceptions and memory.

The cerebral cortex is thought to store factual information, with verbal memories lateralized to the left hemisphere and visuospatial information in the right hemisphere. The neurosurgeon Wilder Penfield has electrically stimulated various regions in the brain of awake patients, often evoking visual or auditory memories that were extremely vivid. Electrical stimulation of specific points in the temporal lobe evoked specific memories that were so detailed the patient felt that he was reliving the experience. The medial regions of the temporal lobes, however, cannot be the site where long-term memory is stored, because destruction of these areas in patients being treated for epilepsy did not destroy the memory of events prior to the surgery. The inferior region of the temporal lobe cortex, on the other hand, does appear to be a site of storage of visual memories.

The amount of memory destroyed by ablation (removal) of brain tissue appears to depend more on the amount of brain tissue removed than on the location of the surgery. On the basis of these observations, it was formerly believed that the memory may be diffusely located in the brain; stimulation of the correct location of the cortex then retrieves the memory. According to current thinking, however, particular aspects of the memory—visual, auditory, olfactory, spatial, and so on—are stored in particular areas, and the cooperation of all of these areas are required to produce the complete memory.

Since long-term memory is not abolished by electroconvulsive shock, it seems reasonable to conclude that the consolidation of memory depends on relatively permanent changes in the chemical structure of neurons and their synapses. Experiments suggest that protein synthesis is required for the consolidation of the "memory trace." According to one theory, these proteins may be secreted into the extracellular environment, where they influence synaptic connections. According to another theory, new receptor proteins in the membrane of the postsynaptic neuron are made available as a result of high-frequency stimulation of the presynaptic neuron. This would help to account for the increased sensitivity of postsynaptic neurons to neurotransmitter, as seen in post-tetanic potentiation (discussed in chapter 7). A related process observed in vertebrate brains is *long-term potentiation.* When a high-frequency electrical stimulus is applied for a few seconds to presynaptic axons along a particular pathway, the postsynaptic cells exhibit an increased excitability that can persist for days to weeks. Much more research is obviously needed in this exciting area of physiology before memory can be fully explained at a cellular and molecular level.

---

1. *List the regions of the brain, and explain why the brain and spinal cord contain cavities.*
2. *Identify the location of the sensory and motor cortex, and explain how these areas are organized.*
3. *Identify the location and composition of the basal ganglia, and describe their functions.*
4. *Describe the structures of the limbic system, the relationship between the limbic system and the hypothalamus and cerebral cortex, and explain the functional significance of the limbic system.*
5. *Explain the differences in function of the right and left cerebral hemispheres.*
6. *Describe the areas of the brain believed to be involved in the production of speech, the relationships between these areas, and the different types of aphasias produced by damage to these areas.*
7. *Describe the different forms of memory, the brain structures shown to be involved in memory, and some of the experimental evidence on which this information is based.*

---

## DIENCEPHALON

The diencephalon is the part of the forebrain which contains such important structures as the thalamus, hypothalamus, and pituitary gland. The hypothalamus, though small in area, has a large number of extremely important functions related to the control of visceral body functions by way of other brain regions and the autonomic nervous system.

The **diencephalon**, together with the telencephalon (cerebrum) previously discussed, constitutes the forebrain. The diencephalon is almost completely surrounded by the cerebral hemispheres and contains a cavity known as the third ventricle.

### Thalamus and Epithalamus

The **thalamus,** is a mass of gray matter that comprises about four-fifths of the diencephalon. It is located below the lateral ventricles and forms most of the walls of the third ventricle (see fig. 8.4). The thalamus acts primarily as a relay center through which all sensory information (except smell) passes on the way to the cerebrum (fig. 8.13). The **epithalamus** is the dorsal portion of the diencephalon that contains a *choroid plexus* over the third ventricle, where cerebrospinal fluid is formed, and the *pineal gland* (*epiphysis*). The pineal gland secretes the hormone *melatonin,* which may play a role in the endocrine control of reproduction (discussed in chapter 20).

### Hypothalamus and Pituitary Gland

The **hypothalamus** is a small portion of the diencephalon located below the thalamus, where it forms the floor and part of the lateral walls of the third ventricle (fig. 8.14). This extremely important brain region contains neural centers for hunger, thirst, and the regulation of body temperature and pituitary gland secretion. In addition, centers in the hypothalamus contribute to the regulation of sleep, wakefulness, sexual arousal and performance, and emotions such as anger, fear, pain, and pleasure. Acting through its connections with the medulla oblongata of the brain stem, the hypothalamus helps to evoke the visceral responses to various emotional states. In its regulation of emotion, the hypothalamus works together with the limbic system, as was discussed in the previous section.

Experimental stimulation of different areas of the hypothalamus can evoke the autonomic responses characteristic of aggression, sexual behavior, eating, or satiety. Chronic stimulation of the lateral hypothalamus, for example, can make an animal eat and become obese, whereas stimulation of the medial hypothalamus inhibits eating. Other areas contain osmoreceptors that stimulate thirst and the secretion of antidiuretic hormone (ADH) from the posterior pituitary.

The hypothalamus is also where the body's "thermostat" is located. Experimental cooling of the preoptic-anterior hypothalamus causes shivering (a somatic response) and nonshivering thermogenesis (a sympathetic response). Experimental heating of this hypothalamic area results in hyperventilation (stimulated by somatic motor nerves), vasodilation, salivation, and sweat gland secretion (stimulated by sympathetic nerves). These responses serve to correct the temperature deviations in a negative feedback fashion.

**Figure 8.13.** A midsagittal section through the brain.
(*a*) A diagram. (*b*) A photograph.

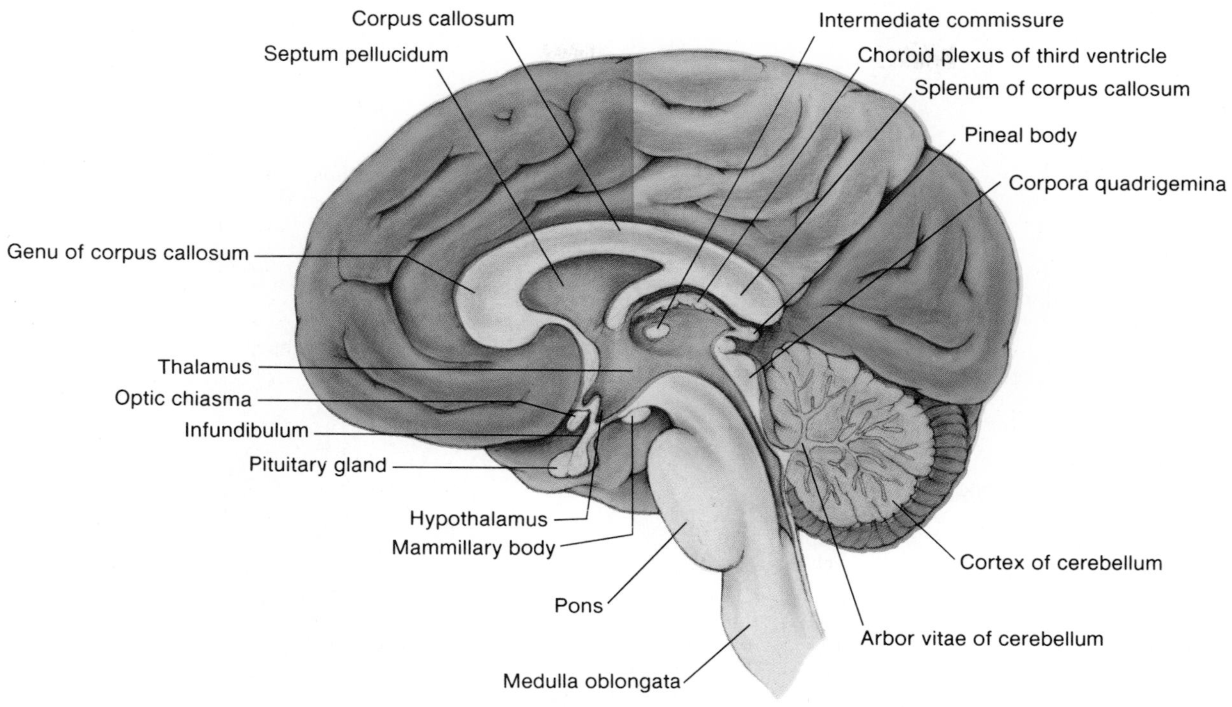


(a)

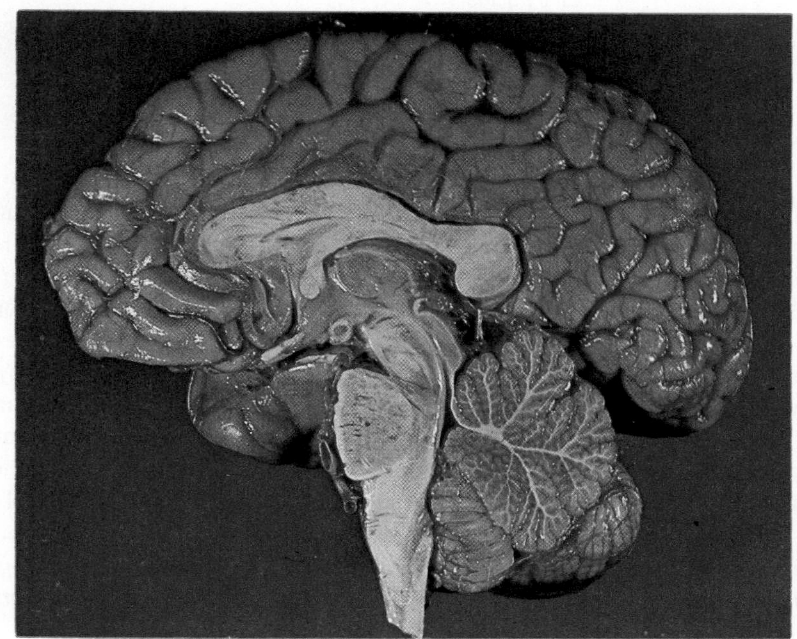

(b)

**Figure 8.14.** Diagram of the nuclei within the hypothalamus.

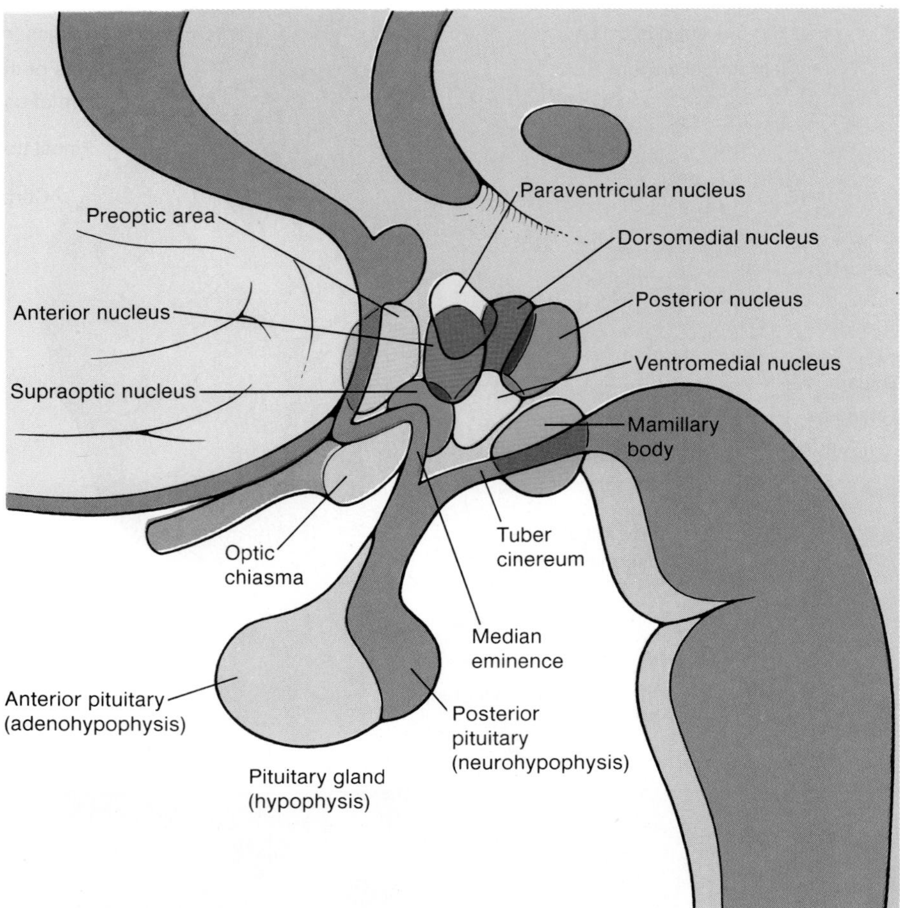


The coordination of sympathetic and parasympathetic reflexes is thus integrated with the control of somatic and endocrine responses by the hypothalamus. The activities of the hypothalamus are in turn influenced by higher brain centers.

The **pituitary gland** is located immediately inferior to the hypothalamus. Indeed, the posterior pituitary was derived embryonically from a downgrowth of the diencephalon, and the entire pituitary remains connected to the diencephalon by means of a stalk (this relationship will be described in more detail in chapter 11). Neurons within the *supraoptic* and *paraventricular nuclei* of the hypothalamus (fig. 8.14) produce two hormones—**antidiuretic hormone (ADH),** which is also known as *vasopressin,* and **oxytocin.** These two hormones are transported in axons of the *hypothalamo-hypophyseal tract* to the **neurohypophysis** (posterior pituitary), where they are stored and secreted in response to hypothalamic stimulation. Oxytocin stimulates contractions of the uterus during labor, and ADH stimulates the kidneys to reabsorb water and thus to excrete a smaller volume of urine. Neurons in the hypothalamus also produce hormones known as **releasing hormones** and **inhibiting hormones,** which are transported by the blood to the **adenohypophysis** (anterior pituitary). These hypothalamic releasing and inhibiting hormones regulate the secretions of the anterior pituitary and, by this means, regulate the secretions of other endocrine glands (as described in chapter 11).

1. *Describe the location of the diencephalon with respect to the cerebrum and the brain ventricles.*
2. *Make a list of the functions of the hypothalamus, and indicate the other brain regions that cooperate with the hypothalamus in the performance of these functions.*
3. *Explain the structural and functional relationships between the hypothalamus and the pituitary gland.*

**Figure 8.15.** Nuclei within the pons and medulla oblongata that constitute the respiratory center.

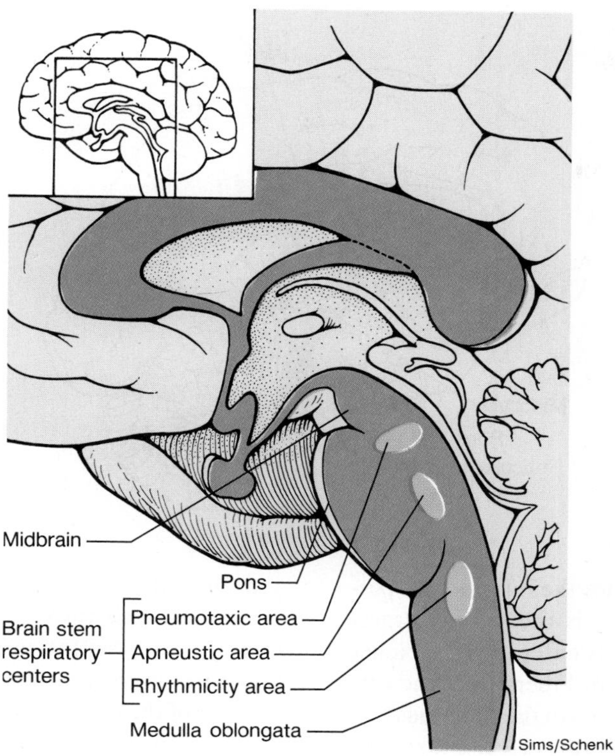


## Midbrain and Hindbrain

The midbrain and hindbrain contain many important relay centers for sensory and motor pathways, and are particularly important in the control of skeletal movements by the brain. The medulla oblongata is a vital region of the hindbrain which contains centers for the control of breathing and cardiovascular function.

### Mesencephalon

The *mesencephalon,* or **midbrain,** is located between the diencephalon and the pons. The **corpora quadrigemina** are four rounded elevations on the dorsal portion of the midbrain (see fig. 8.13). The upper two of these, called the *superior colliculi,* are involved in visual reflexes. The posterior two, called the *inferior colliculi,* are relay centers for auditory information.

The mesencephalon also contains the cerebral peduncles, red nucleus, and the substantia nigra (not illustrated). The **cerebral peduncles** are a pair of structures composed of ascending and descending fiber tracts. The **red nucleus** is an area of gray matter deep in the midbrain that maintains connections with the cerebrum and the cerebellum and is involved in motor coordination. Another nucleus, the **substantia nigra,** is ventral to the red nucleus and is also involved in motor coordination through its synaptic connections with the basal ganglia, as previously discussed.

### Rhombencephalon (Hindbrain)

The hindbrain is composed of two regions: the metencephalon and the myelencephalon. Each of these regions will be discussed separately.

***Metencephalon.*** The metencephalon is composed of the pons and the cerebellum. The **pons** can be seen as a rounded bulge on the underside of the brain, between the midbrain and the medulla oblongata (fig. 8.15). Surface fibers in the pons connect to the cerebellum, and deeper fibers are part of motor and sensory tracts that pass from the medulla oblongata, through the pons, and on to the midbrain. Within the pons are several nuclei associated with specific cranial nerves—the trigeminal (V), abducens (VI), facial (VII), and vestibulocochlear (VIII). Other nuclei of the pons cooperate with nuclei in the medulla oblongata to regulate breathing. The two respiratory control centers in the pons are known as the *apneustic* and the *pneumotaxic centers.*

**Figure 8.16.** The reticular activating system (RAS). The arrows indicate the direction of impulses along nerve pathways that connect with the RAS.

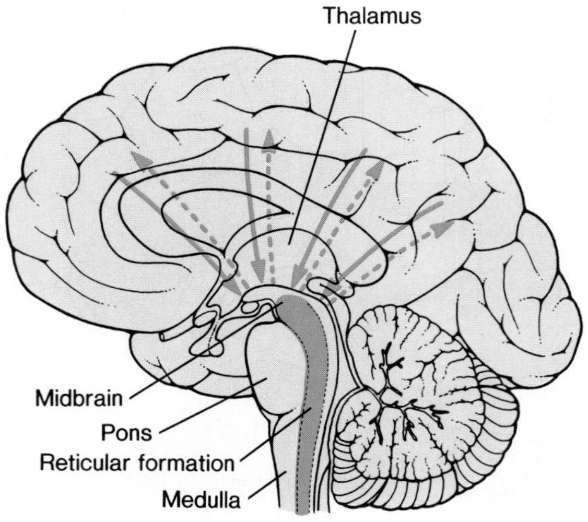


The **cerebellum,** which occupies the inferior and posterior aspect of the cranial cavity, is the second largest structure of the brain, and contains outer gray and inner white matter as does the cerebrum. Fibers from the cerebellum pass through the red nucleus to the thalamus and then to the motor areas of the cerebral cortex. Other fiber tracts connect the cerebellum with the pons, medulla oblongata, and spinal cord. The cerebellum receives input from *proprioceptors* (joint, tendon, and muscle receptors) and, working together with the basal ganglia and motor areas of the cerebral cortex, participates in the coordination of movement.

Damage to the cerebellum produces **ataxia,** which is lack of coordination due to errors in the speed, force, and direction of movement. The movements and speech of those afflicted by ataxia may resemble those of a drunken person. This condition is also characterized by *intention tremor,* which differs from the resting tremor of Parkinson's disease in that it only occurs when intentional movements are made. The person suffering cerebellar damage may reach for an object and miss it, by placing the hand too far to the left or right, and then attempt to compensate by moving the hand in the opposite direction; this can continue and produce oscillations of the limb.

***Myelencephalon.*** The only structure within the myelencephalon is the **medulla oblongata,** often simply called the *medulla.* The medulla is about 3 cm long and is continuous with the pons anteriorly and the spinal cord inferiorly. All the descending and ascending fiber tracts that provide communication between the spinal cord and the brain must pass through the medulla. These fiber tracts cross to the contralateral side in elevated, triangular structures in the medulla called the **pyramids,** so that the left side of the brain receives sensory information from the right side of the body and vice versa. Similarly, because of the decussation of fibers, the right side of the brain controls motor activity in the left side of the body and vice versa.

There are several important nuclei within the medulla. The *nucleus ambiguus* and *hypoglossal nucleus* give rise to several cranial nerves (VIII, IX, XI, and XII). The *vagus nuclei* (there is one on each lateral side of the medulla) give rise to the vagus (X) nerves. The *nucleus gracilis* and *nucleus cuneatus* relay sensory information to the thalamus and then to the cerebral cortex. The *inferior olivary nuclei* and the *accessory olivary nuclei* are along the synaptic pathway from the forebrain and midbrain to the cerebellum.

The medulla contains groupings of neurons known as the *vital centers,* because these neurons are required for the regulation of breathing and of cardiovascular responses. The **vasomotor center** controls the autonomic innervation of blood vessels; the **cardioinhibitory center** controls the parasympathetic innervation (via the vagus nerve) of the heart (there does not appear to be a separate cardioaccelerator center); and the **respiratory center** of the medulla acts together with centers in the pons to control breathing.

***Reticular Formation.*** The reticular formation is a complex network of nuclei and nerve fibers within the medulla, pons, midbrain, thalamus, and hypothalamus that functions as the **reticular activating system,** or **RAS** (fig. 8.16). Because of its many interconnections, the RAS is

**Figure 8.17.** A cross section showing the principal ascending and descending tracts within the spinal cord.

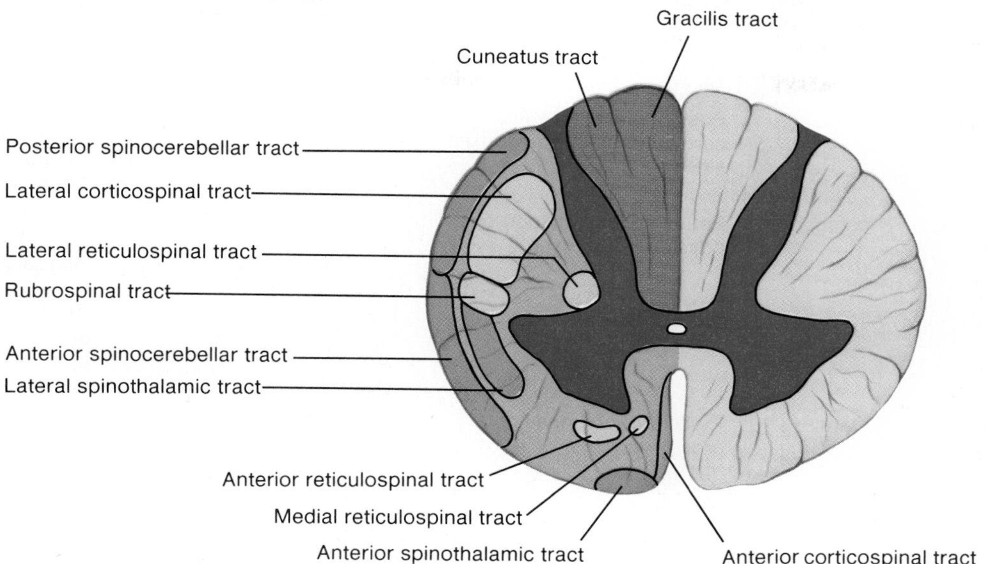


activated in a nonspecific fashion by any modality of sensory information. Nerve fibers from the RAS, in turn, project diffusely to the cerebral cortex; this results in *nonspecific arousal* of the cerebral cortex to incoming sensory information.

The RAS, through its nonspecific arousal of the cortex, helps to maintain a state of alert consciousness. Not surprisingly, there is evidence that general anesthetics may produce unconsciousness by depressing the RAS. Similarly, the ability to fall asleep may be due to the action of specific neurotransmitters that inhibit activity of the RAS.

1. *Identify the structures of the mesencephalon, and describe their functions.*
2. *Describe the functions of the medulla oblongata and pons.*
3. *Describe the composition and functions of the reticular activating system*

## Spinal Cord Tracts

Sensory information from receptors throughout most of the body are relayed to the brain by means of ascending tracts of fibers that conduct impulses up the spinal cord. When the brain directs motor activities, these directions are in the form of nerve impulses that travel down the spinal cord in descending tracts of fibers.

The spinal cord extends from the level of the foramen magnum of the skull to the first lumbar vertebra. Unlike the brain, in which the gray matter forms a cortex over white matter, the gray matter of the spinal cord is located centrally, surrounded by white matter. The central gray matter of the spinal cord is arranged in the form of an "H," with two *dorsal horns* and two *ventral horns* (also called posterior and anterior horns, respectively). The white matter of the spinal cord is composed of ascending and descending fiber tracts. These are arranged into six columns of white matter called **funiculi**. The arrangement of gray and white matter within the spinal cord is illustrated in figure 8.17.

The names of the fiber tracts within the white matter of the spinal cord tell whether they are ascending sensory or descending motor tracts. The names of all of the ascending tracts start with the prefix "spino-" and end with the name of the brain region where the spinal cord fibers first synapse. The anterior spinothalamic tract, for example, carries impulses conveying the sense of touch and pressure, and synapses in the thalamus. From there it is relayed to the cerebral cortex. The names of descending motor tracts, conversely, begin with a prefix denoting the brain region that give rise to the fibers and end with the suffix "-spinal." The lateral corticospinal tracts, for example, begin in the cerebral cortex and descend the spinal cord.

**Table 8.2** Principal ascending tracts of spinal cord

| Tract | Funiculus | Origin | Termination | Function |
|---|---|---|---|---|
| Anterior spinothalamic | Anterior | Posterior horn on one side of cord but crosses to opposite side | Thalamus, then cerebral cortex | Conducts sensory impulses for crude touch and pressure |
| Lateral spinothalamic | Lateral | Posterior horn on one side of cord but crosses to opposite side | Thalamus, then cerebral cortex | Conducts pain and temperature impulses that are interpreted within cerebral cortex |
| Fasciculus gracilis and fasciculus cuneatus | Posterior | Peripheral afferent neurons; does not cross over | Nucleus gracilis and nucleus cuneatus of medulla; eventually thalamus, then cerebral cortex | Conducts sensory impulses from skin, muscles, tendons, and joints, which are interpreted as sensations of fine touch, precise pressures, and body movements |
| Posterior spinocerebellar | Lateral | Posterior horn; does not cross over | Cerebellum | Conducts sensory impulses from one side of body to same side of cerebellum for subconscious proprioception necessary for coordinated muscular contractions |
| Anterior spinocerebellar | Lateral | Posterior horn; some fibers cross, others do not | Cerebellum | Conducts sensory impulses from both sides of body to cerebellum for subconscious proprioception necessary for coordinated muscular contractions |



## Ascending Tracts

The ascending fiber tracts convey sensory information from cutaneous receptors, proprioceptors (muscle and joint senses), and visceral receptors (table 8.2). Most of the sensory information that originates in the right side of the body crosses over to eventually reach the region on the left side of the brain, which analyzes this information. Similarly, the information arising in the left side of the body is ultimately analyzed by the right side of the brain. This decussation occurs in the medulla oblongata (fig. 8.18) for some sensory modalities, or in the spinal cord for other modalities of sensation. This is discussed in more detail in chapter 9.

## Descending Tracts

There are two major groups of descending tracts from the brain: the **corticospinal,** or **pyramidal, tracts,** and the **extrapyramidal tracts** (table 8.3). The pyramidal tracts descend directly, without synaptic interruption, from the cerebral cortex to the spinal cord. The cell bodies that contribute fibers to these pyramidal tracts are located primarily in the *precentral gyrus* (also called the *motor cortex*). Other areas of the cerebral cortex, however, also contribute to these tracts.

About 80%–90% of the corticospinal fibers decussate in the pyramids of the medulla oblongata (hence the name "pyramidal tracts") and descend as the *lateral corticospinal tracts.* The remaining uncrossed fibers form the *ventral corticospinal tracts,* which decussate in the spinal cord. Because of the crossing of fibers, the right cerebral hemisphere controls the musculature on the left side of the body (fig. 8.19), whereas the left hemisphere controls the right musculature. The corticospinal tracts are primarily concerned with the control of fine, skilled movements.

The corticospinal tracts appear to be particularly important in voluntary movements that require complex interactions between sensory input and the motor cortex. Speech, for example, is impaired when the corticospinal tracts are damaged in the thoracic region of the spinal cord, whereas involuntary breathing continues. Damage to the pyramidal motor system can be detected clinically by a positive **Babinski reflex,** in which stimulation of the sole of the foot causes extension (upward movement) of the toes. The positive Babinski reflex is normal in infants because neural control is not yet fully developed.

**Figure 8.18.** Ascending tracts composed of sensory fibers that cross over within the medulla.

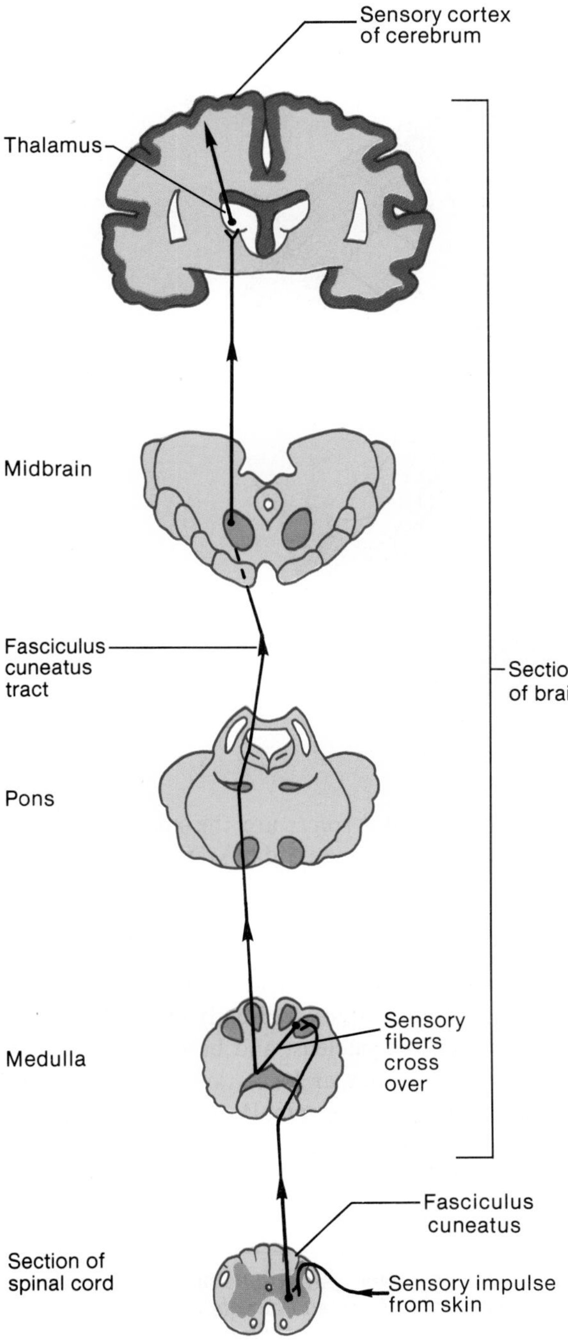

**Figure 8.19.** Descending tracts composed of motor fibers that cross over within the medulla oblongata.

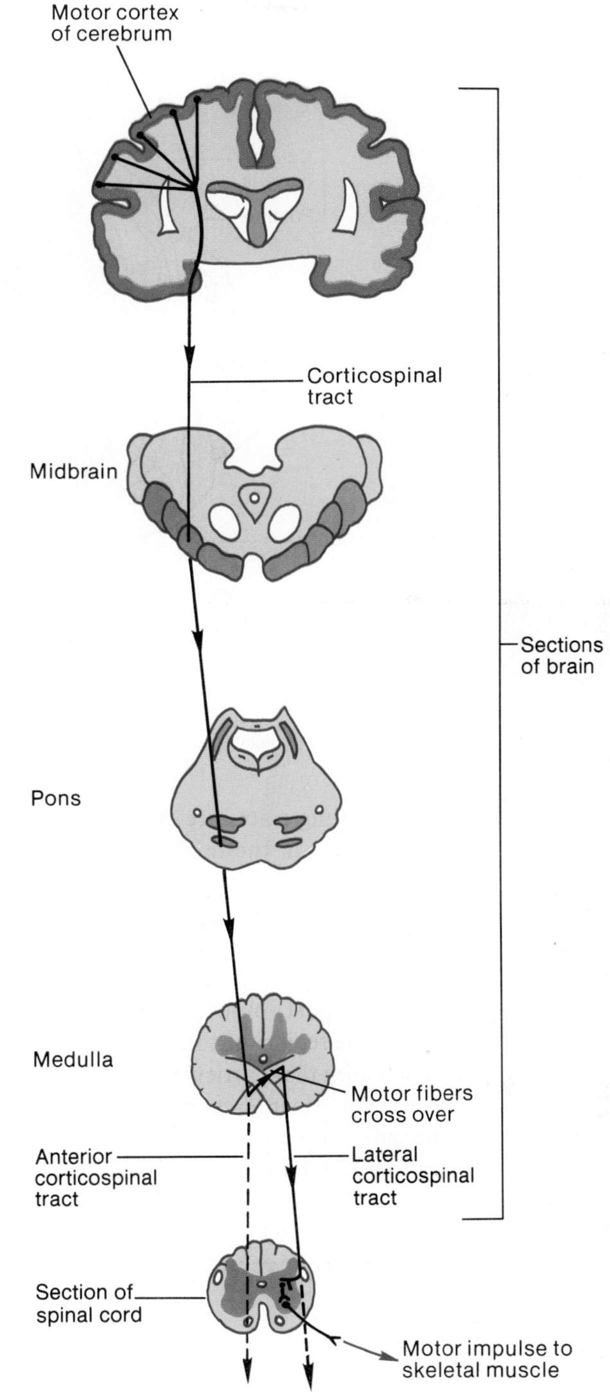


**Table 8.3** Descending motor tracts to spinal interneurons and motor neurons

| Tract | Category | Origin | Crossed/Uncrossed |
|---|---|---|---|
| Lateral corticospinal | Pyramidal | Cerebral cortex | Crossed |
| Anterior corticospinal | Pyramidal | Cerebral cortex | Uncrossed |
| Rubrospinal | Extrapyramidal | Red nucleus (midbrain) | Crossed |
| Tectospinal | Extrapyramidal | Superior colliculus (midbrain) | Crossed |
| Vestibulospinal | Extrapyramidal | Vestibular nuclei (medulla oblongata) | Uncrossed |
| Reticulospinal | Extrapyramidal | Brain stem reticular formation (medulla and pons) | Crossed |

**Figure 8.20.** Areas of the brain containing neurons involved in the control of skeletal muscles (higher motor neurons). The thalamus is a relay center between the motor cortex and other brain areas.

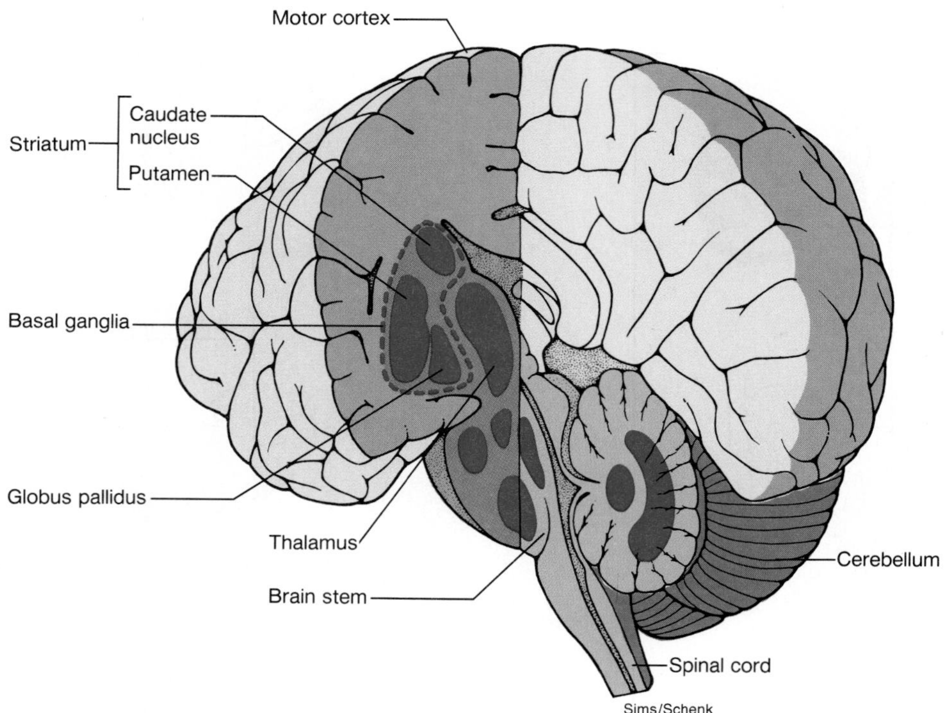


The remaining descending tracts are extrapyramidal motor tracts, which originate in the midbrain and brain stem regions (table 8.3). If the pyramidal tracts of an experimental animal are cut, electrical stimulation of the cerebral cortex, cerebellum, and basal ganglia (fig. 8.20) can still produce movements. The descending fibers that produce these movements must, by definition, be extrapyramidal motor tracts. The regions of the cerebral cortex, basal ganglia, and cerebellum that participate in this motor control have numerous synaptic interconnections, and can influence movement only indirectly by means of stimulation or inhibition of the nuclei that give rise to the extrapyramidal tracts. Notice that this differs from the neurons of the precentral gyrus, which send fibers directly down to the spinal cord in the pyramidal tracts.

The *reticulospinal tracts* are the major descending pathways of the extrapyramidal system. These tracts originate in the reticular formation of the brain stem, which receives either stimulatory or inhibitory input from the cerebrum and the cerebellum. There are no descending tracts from the cerebellum; the cerebellum can only influence motor activity indirectly by its effect on the vestibular nuclei, red nucleus, and basal ganglia (which sends axons to the reticular formation). These nuclei, in turn, send axons down the spinal cord via the *vestibulospinal tracts, rubrospinal tracts,* and reticulospinal tracts, respectively (fig. 8.21). Neural control of skeletal muscle is explained in more detail in chapter 12.

1. *Explain why each cerebral hemisphere receives sensory input from and directs motor output to the contralateral side of the body.*
2. *List the tracts of the pyramidal motor system, and explain the function of the pyramidal system.*
3. *List the tracts of the extrapyramidal system and explain how this system differs from the pyramidal motor system.*

**Figure 8.21.** Pathways involved in the higher motor neuron control of skeletal muscles.

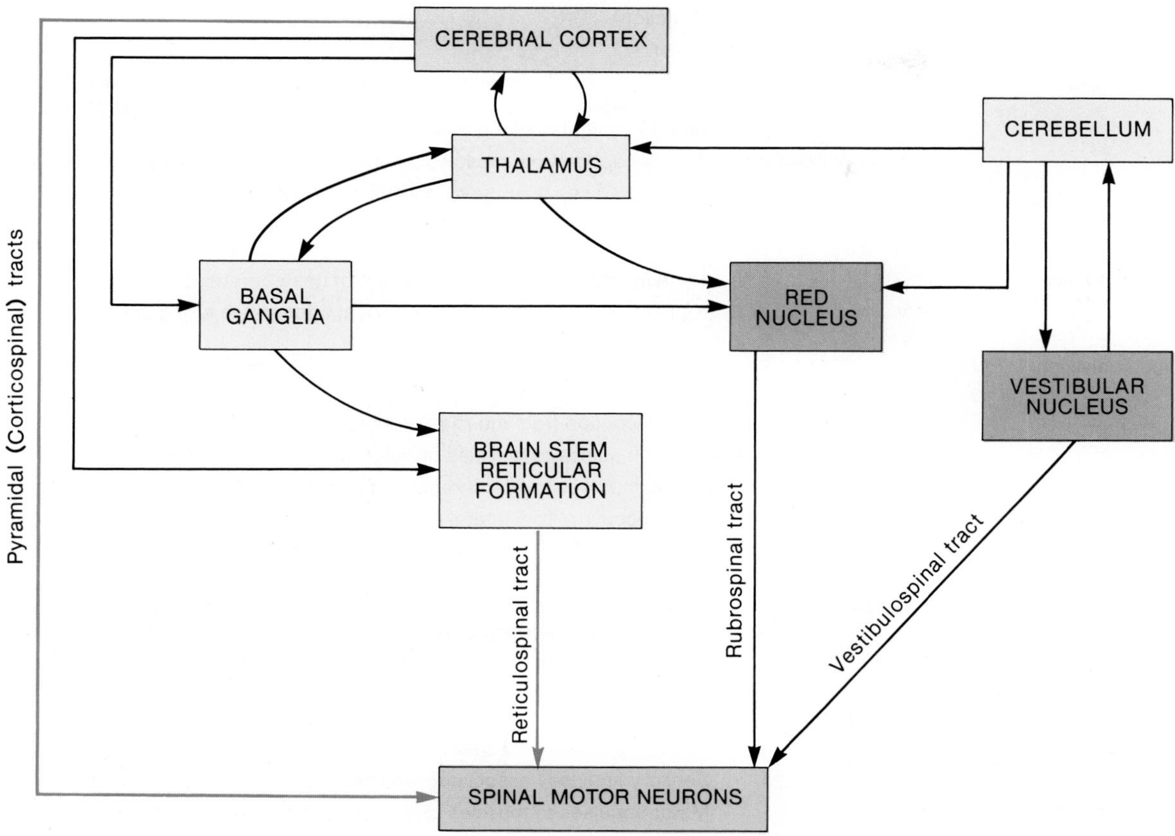


## *Cranial and Spinal Nerves*

The central nervous system communicates with the body by means of nerves that exit the CNS from the brain (cranial nerves) and spinal cord (spinal nerves). These nerves, together with aggregations of cell bodies located outside the CNS, constitute the peripheral nervous system.

As mentioned in chapter 7, the peripheral nervous system (PNS) consists of nerves (collections of axons and sensory dendrites) and their associated ganglia (collections of cell bodies). Although this chapter is devoted to the CNS, the CNS cannot function independently from the PNS. This section will thus serve to complete the explanation of the CNS and introduce concepts of the PNS, which will be explored more thoroughly in later chapters (particularly chapters 9, 10, and 12).

### Cranial Nerves

There are twelve pairs of cranial nerves. Two of these pairs arise from neuron cell bodies located in the forebrain and ten pairs arise from the midbrain and brain stem. The cranial nerves are designated by Roman numerals and by names. The Roman numerals refer to the order in which the nerves are positioned from the front of the brain to the back. The names indicate structures innervated by these nerves (e.g., facial) or the principal function of the nerves (e.g., occulomotor). A summary of the cranial nerves is presented in table 8.4.

**Table 8.4** Summary of cranial nerves

| Number and Name | Composition | Function |
|---|---|---|
| I Olfactory | Sensory | Olfaction |
| II Optic | Sensory | Vision |
| III Oculomotor | Motor | Motor impulses to levator palpebrae superioris and extrinsic eye muscles except superior oblique and lateral rectus; innervation to muscles that regulate amount of light entering eye and that focus the lens |
| | Sensory: proprioception | Proprioception from muscles innervated with motor fibers |
| IV Trochlear | Motor | Motor impulses to superior oblique muscle of eyeball |
| | Sensory: proprioception | Proprioception from superior oblique muscle of eyeball |
| V Trigeminal | | |
| Ophthalmic division | Sensory | Sensory impulses from cornea, skin of nose, forehead, and scalp |
| Maxillary division | Sensory | Sensory impulses from nasal mucosa, upper teeth and gums, palate, upper lip, and skin of cheek |
| Mandibular division | Sensory | Sensory impulses from temporal region, tongue, lower teeth and gums, and skin of chin and lower jaw |
| | Sensory: proprioception | Proprioception from muscles of mastication |
| | Motor | Motor impulses to muscles of mastication and muscle that tenses tympanum |
| VI Abducens | Motor | Motor impulses to lateral rectus muscle of eyeball |
| | Sensory: proprioception | Proprioception from lateral rectus muscle of eyeball |
| VII Facial | Motor | Motor impulses to muscles of facial expression and muscle that tenses the stapes |
| | Motor: parasympathetic | Secretion of tears from lacrimal gland and salivation from sublingual and submandibular salivary glands |
| | Sensory | Sensory impulses from taste buds on anterior two-thirds of tongue; nasal and palatal sensation |
| | Sensory: proprioception | Proprioception from muscles of facial expression |
| VIII Vestibulocochlear | Sensory | Sensory impulses associated with equilibrium |
| | | Sensory impulses associated with hearing |
| IX Glossopharyngeal | Motor | Motor impulses to muscles of pharynx used in swallowing |
| | Sensory: proprioception | Proprioception from muscles of pharynx |
| | Sensory | Sensory impulses from taste buds on posterior one-third of tongue; pharynx, middle-ear cavity, carotid sinus |
| | Parasympathetic | Salivation from parotid salivary gland |
| X Vagus | Motor | Contraction of muscles of pharynx (swallowing) and larynx (phonation) |
| | Sensory: proprioception | Proprioception from visceral muscles |
| | Sensory | Sensory impulses from taste buds on rear of tongue; sensations from auricle of ear; general visceral sensations |
| | Motor: parasympathetic | Regulate many visceral functions |
| XI Accessory | Motor | Laryngeal movement; soft palate |
| | | Motor impulses to trapezius and sternocleidomastoid muscles for movement of head, neck, and shoulders |
| | Sensory: proprioception | Proprioception from muscles that move head, neck, and shoulders |
| XII Hypoglossal | Motor | Motor impulses to intrinsic and extrinsic muscles of tongue and infrahyoid muscles |
| | Sensory: proprioception | Proprioception from muscles of tongue |

**Figure 8.22.** The distribution of the spinal nerves.

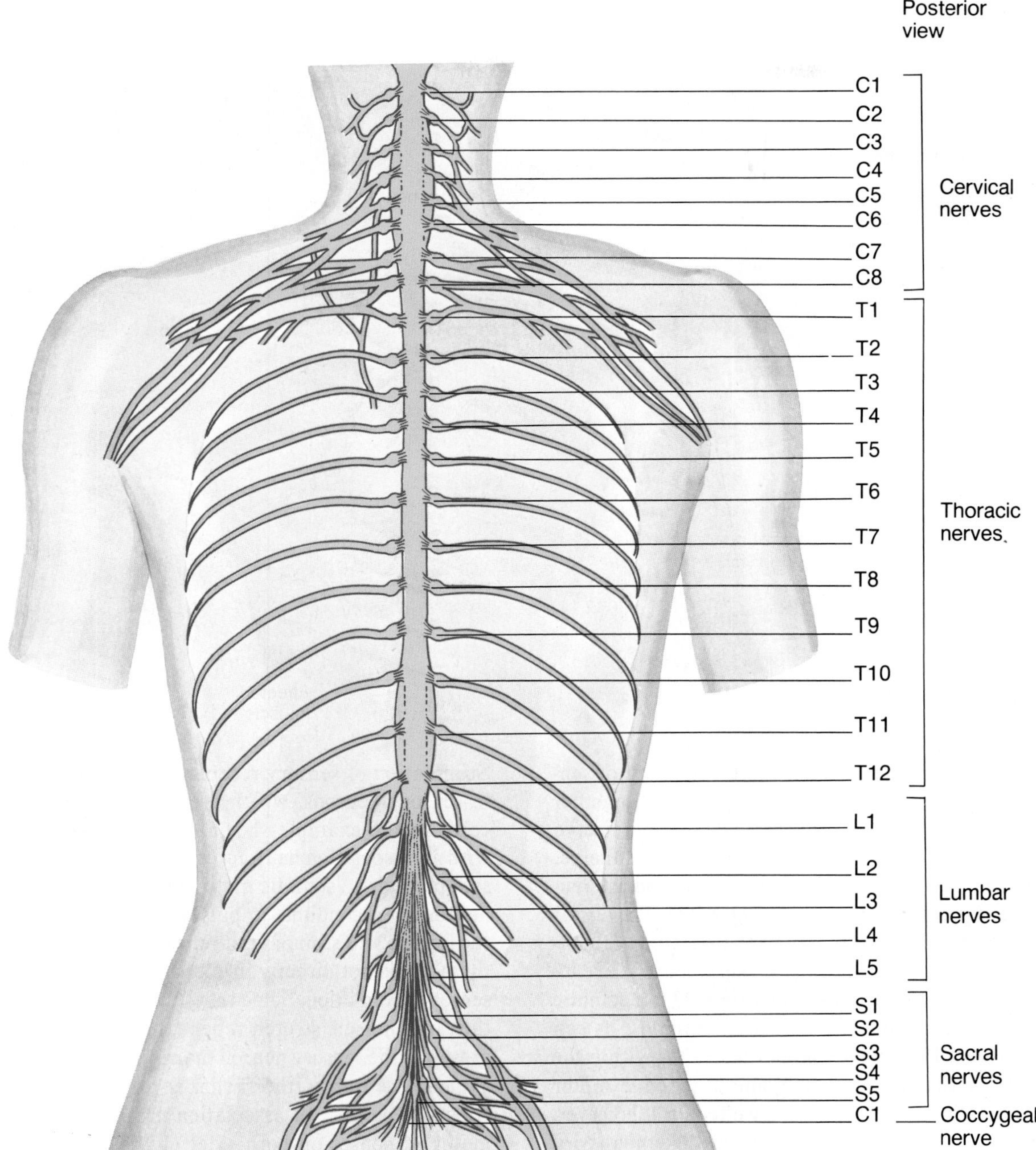


Most cranial nerves are classified as *mixed nerves*. This term indicates that the nerve contains both sensory and motor fibers. Those cranial nerves associated with the special senses (e.g., olfactory, optic), however, consist of sensory fibers only. The cell bodies of these sensory neurons are not located in the brain, but instead are found in ganglia near the sensory organ.

## Spinal Nerves

There are thirty-one pairs of spinal nerves. These nerves are grouped into eight cervical, twelve thoracic, five lumbar, five sacral, and one coccygeal according to the region of the vertebral column from which they arise (fig. 8.22).

**Figure 8.23.** Sensory neuron, association neuron (interneuron), and somatic motor neuron at the spinal cord level.

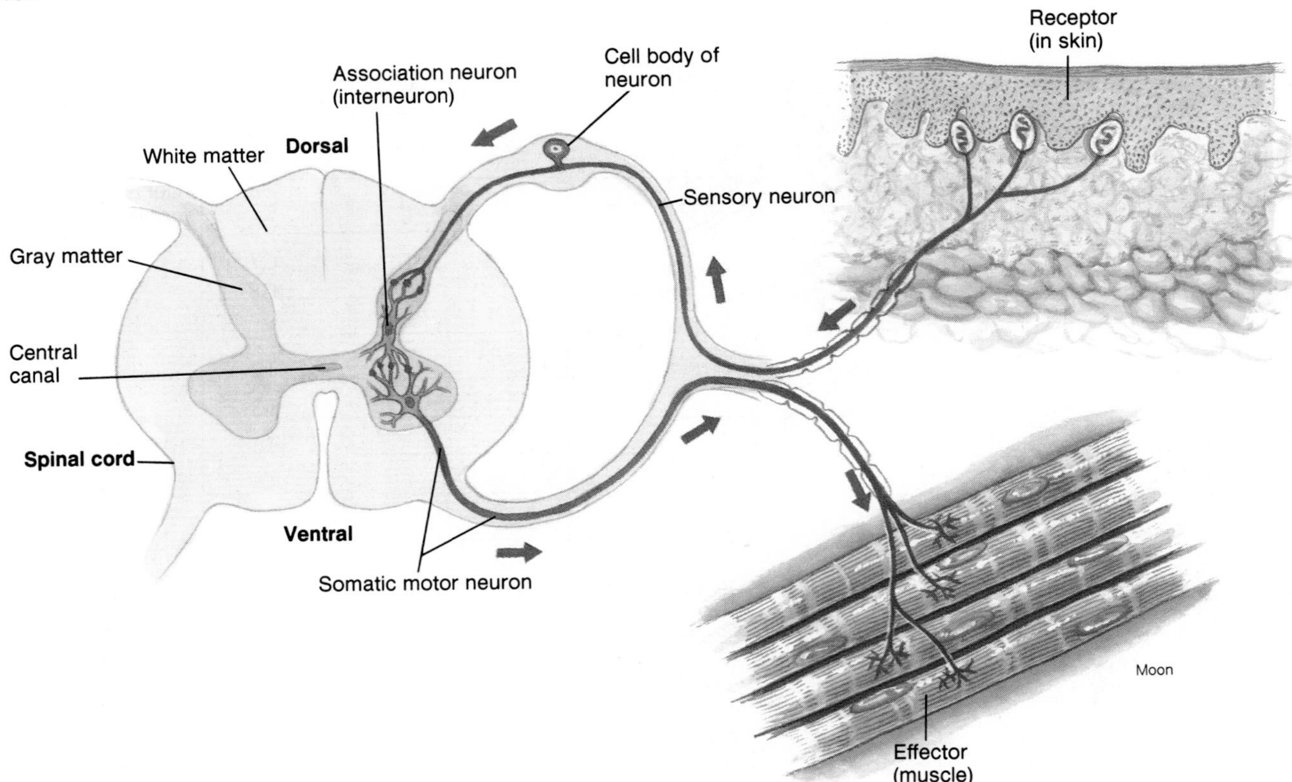


Each spinal nerve is a mixed nerve composed of sensory and motor fibers. These fibers are packaged together in the nerve, but separate near the attachment of the nerve to the spinal cord. This produces two "roots" to each nerve. The **dorsal root** is composed of sensory fibers, and the **ventral root** is composed of motor fibers (fig. 8.23). The dorsal root contains an enlargement called the *dorsal root ganglion,* where the cell bodies of the sensory neurons are located. The motor neuron shown in figure 8.23 is a somatic motor neuron, which innervates skeletal muscles; its cell body is not located in a ganglion, but instead is within the gray matter of the spinal cord. Some autonomic motor neurons (which innervate involuntary effectors), however, have their cell bodies in ganglia outside the spinal cord (the autonomic system is discussed separately in chapter 10).

***Reflex Arc.*** The functions of the sensory and motor components of a spinal nerve can most easily be understood by examination of a simple reflex; that is, an unconscious motor response to a sensory stimulus. Figure 8.23 demonstrates the neural pathway involved in a **reflex arc.** Stimulation of sensory receptors evokes action potentials in sensory neurons, which are conducted into the spinal cord. In the example shown, a sensory neuron synapses with an association neuron (or interneuron), which in turn synapses with a somatic motor neuron. The somatic motor neuron then conducts impulses out of the spinal cord to the muscle and stimulates a reflex contraction. Notice that the brain is not directly involved in this reflex response to sensory stimulation. Some reflex arcs are even simpler than this; in a muscle stretch reflex (the knee-jerk reflex), for example, the sensory neuron makes a synapse directly with a motor neuron. Other reflexes are more complex, involving a number of association neurons and resulting in motor responses on both sides of the spinal cord at different levels. (These skeletal reflexes are described together with muscle control in chapter 12.)

1. *Describe the meaning of the terms* dorsal root, dorsal root ganglion, ventral root, *and* mixed nerve.
2. *Describe the neural pathways and structures involved in a reflex arc.*

## SUMMARY

### The Brain: Structural Plan and Functions of the Cerebrum p. 182

I. During embryonic development, the brain develops five regions: telencephalon, diencephalon, mesencephalon, metencephalon, and myelencephalon.
   A. The telencephalon and diencephalon constitute the forebrain; the mesencephalon is the midbrain, and the hindbrain is composed of the metencephalon and the myelencephalon.
   B. The CNS begins as a hollow tube, and the brain and spinal cord are thus hollow; the cavities of the brain are known as ventricles.

II. The cerebrum consists of two hemispheres connected by a large fiber tract called the corpus callosum.
   A. The outer part of the cerebrum consists of gray matter and is the cerebral cortex.
   B. White matter is under the gray matter, but there are nuclei of grey matter within the cerebrum known as the basal ganglia.
   C. Synaptic potentials within the cerebral cortex produce the electrical activity seen in an electroencephalogram (EEG).

III. The two cerebral hemispheres have a degree of specialization of function, which is termed cerebral lateralization.
   A. In most people, the left hemisphere is dominant in language and analytical ability, whereas the right hemisphere is important in pattern recognition, musical creation, singing, and the recognition of faces.
   B. The two hemispheres cooperate in their functions; this is aided by communication between the two via the corpus callosum.

IV. Particular regions of the left cerebral cortex appear to be important in language ability; when these areas are damaged, the person develops characteristic types of aphasias.
   A. Wernicke's area is involved in speech comprehension, whereas Broca's area is required for the mechanical performance of speech.
   B. Wernicke's area is believed to control Broca's area by means of the arcuate fasciculus.
   C. The angular gyrus is believed to integrate different sources of sensory information and project to Wernicke's area.

V. The limbic system and hypothalamus are regions of the brain that have been implicated as centers for various emotions.

VI. Memory can be divided into a short-term and a long-term form.
   A. The medial temporal lobes, and in particular the hippocampus and perhaps the amygdaloid nucleus, appear to be required for the consolidation of short-term memory into long-term memory.
   B. Many brain regions may store particular aspects of a memory.
   C. Long-term potentiation is a process that may explain in part how memory is produced.

### Diencephalon p. 194

I. The diencephalon is the region of the forebrain which includes the thalamus, epithalamus, hypothalamus, and pituitary gland.
   A. The thalamus is an important relay center for sensory information and has other functions as well.
   B. The epithalamus contains a choroid plexus for the formation of cerebrospinal fluid and the pineal gland.
   C. The hypothalamus forms the floor of the third ventricle, and the pituitary gland is located immediately inferior to the hypothalamus.

II. The hypothalamus has a large number of important functions related to the control of visceral functions.
   A. The hypothalamus contains centers for the control of thirst, eating, body temperature, and (together with the limbic system) various emotions.
   B. The hypothalamus controls the pituitary gland: it controls the posterior pituitary by means of a fiber tract, and it controls the anterior pituitary by means of hormones.

### Midbrain and Hindbrain p. 197

I. The midbrain contains the superior and inferior colliculi, which are involved in visual and auditory reflexes, respectively.

II. The hindbrain consists of two regions: the metencephalon and the myelencephalon.
   A. The metencephalon contains the pons and cerebellum; the pons is the site of origination of some cranial nerves, and the cerebellum is an important organ involved in the control of skeletal movements.
   B. The myelencephalon consists of only one region, the medulla oblongata, which contains centers for the regulation of such vital functions as breathing and the control of the cardiovascular system.

### Spinal Cord Tracts p. 199

I. The ascending tracts carry sensory information from sensory organs up the spinal cord to the brain.

II. Descending tracts are motor tracts, and are divisible into the pyramidal and extrapyramidal system.
   A. Pyramidal tracts are the corticospinal tracts; these begin in the precentral gyrus and descend, without making any synapses, into the spinal cord.
   B. Most of the corticospinal fibers decussate in the pyramids of the medulla oblongata.
   C. The basal ganglia of the cerebrum, and the cerebellum, control movements indirectly by making synapses with other regions that produce descending extrapyramidal fiber tracts.
   D. The major extrapyramidal motor tract is the reticulospinal tract, which originates in the reticular formation of the midbrain.

### Cranial and Spinal Nerves p. 203

I. There are twelve pairs of cranial nerves; most of these are mixed, but some are exclusively sensory in function.

II. There are thirty-one pairs of spinal nerves; each of these contains both sensory and motor fibers.
   A. The dorsal root of a spinal nerve contains sensory fibers, and the cell bodies of these neurons is contained in the dorsal root ganglion.
   B. The ventral root of a spinal nerve contains motor fibers.

III. A reflex arc is the pathway that involves a sensory neuron and a motor neuron; one or more association neurons may also be involved in some reflexes.

## Review Activities

### Objective Questions

1. The precentral gyrus is
   (a) involved in motor control
   (b) involved in sensory perception
   (c) located in the frontal lobe
   (d) both *a* and *c*
   (e) both *b* and *c*
2. In most people, the right hemisphere controls movement
   (a) of the right side of the body primarily
   (b) of the left side of the body primarily
   (c) of both the right and left sides of the body equally
   (d) of the head and neck only
3. Which of the following statements about the basal ganglia is (are) *true?*
   (a) They are located in the cerebrum.
   (b) They contain the caudate nucleus.
   (c) They are involved in motor control.
   (d) They are part of the extrapyramidal system.
   (e) All of the above are true.
4. Which of the following acts as a relay center for somatesthetic sensation?
   (a) the thalamus
   (b) the hypothalamus
   (c) the red nucleus
   (d) the cerebellum
5. Which of the following statements about the medulla oblongata is *false?*
   (a) It contains nuclei for some cranial nerves.
   (b) It contains the apneustic center.
   (c) It contains the vasomotor center.
   (d) It contains ascending and descending fiber tracts.
6. The reticular activating system
   (a) is composed of neurons that are part of the reticular formation
   (b) is a loose arrangement of neurons with many interconnecting synapses
   (c) is located in the brain stem and midbrain
   (d) functions to arouse the cerebral cortex to incoming sensory information
   (e) all of the above
7. In the control of emotion and motivation, the limbic system works together with the
   (a) pons
   (b) thalamus
   (c) hypothalamus
   (d) cerebellum
   (e) basal ganglia
8. Verbal ability predominates in the
   (a) left hemisphere of right-handed people
   (b) left hemisphere of most left-handed people
   (c) right hemisphere of 97% of all people
   (d) both *a* and *b*
   (e) both *b* and *c*
9. The consolidation of short-term memory into long-term memory appears to be a function of the
   (a) substantia nigra
   (b) hippocampus
   (c) cerebral peduncles
   (d) arcuate fasciculus
   (e) precentral gyrus

Match the nature of the aphasia with its cause:

10. comprehension good, can speak and write, but cannot read (though can see)
11. comprehension good, but speech is slow and difficult (but motor ability not damaged)
12. comprehension poor, speech is fluent but meaningless
   (a) damage to Broca's area
   (b) damage to Wernicke's area
   (c) damage to angular gyrus
   (d) damage to precentral gyrus
13. Antidiuretic hormone (ADH) and oxytocin are synthesized by supraoptic and paraventricular nuclei, which are located in the
   (a) thalamus
   (b) pineal gland
   (c) pituitary gland
   (d) hypothalamus
   (e) pons
14. The superior colliculi are twin bodies within the corpora quadrigemina of the midbrain and are involved in
   (a) visual reflexes
   (b) auditory reflexes
   (c) relaying cutaneous information
   (d) release of pituitary hormones

### Essay Questions

1. Define the term *decussation,* and explain its significance in terms of the pyramidal motor system.
2. Electrical stimulation of the basal ganglia or cerebellum can produce skeletal movements. Describe the pathways by which these brain regions control motor activity.
3. Define the term *ablation,* and give two examples of how this experimental technique has been used to learn about the function of particular brain regions.
4. Explain how "split-brain" patients have been utilized in research on the function of the cerebral hemispheres. Propose experiments that would reveal the lateralization of function in the two hemispheres.
5. What is the evidence that Wernicke's area may control Broca's area? What is the evidence that the angular gyrus has input to Wernicke's area?
6. Provide two reasons why it is believed that there is a difference between short-term and long-term memory, and describe why it is believed that the hippocampus is involved in the consolidation of short-term memory.

## SELECTED READINGS


Andreasen, N. C. 1988. Brain imaging: Applications in psychiatry. *Science* 239:1381

Aoki, C., and P. Siekevitz. December 1988. Plasticity in brain development. *Scientific American.*

Benson, D. F., and N. Geschwind. 1972. Aphasia and related disturbances. In A. B. Baker, ed. *Clinical neurology.* New York: Harper and Row.

Brown, T. H. *et al.* 1988. Long-term potentiation. *Science* 242:724.

Cote, L. 1981. Basal ganglia, the extrapyramidal motor system, and disease of transmitter metabolism. In E. R. Kandel and J. H. Schwartz, eds. *Principles of neural science.* New York: Elsevier North Holland.

de Wied, D. 1989. Neuroendocrine aspects of learning and memory processes. *News in Physiological Sciences* 4:32.

Fine, A. August 1986. Transplantation in the central nervous system. *Scientific American.*

Ganong, W. F. 1985. *Review of medical physiology.* 12th ed. Los Altos, CA: Lange Medical Publishers.

Geschwind, N. April 1972. Language and the brain. *Scientific American.*

Ghez, C. 1981. Cortical control of voluntary movement. In E. R. Kandel and J. H. Schwartz, eds. *Principles of neural science.* New York: Elsevier North Holland.

Hubel, D. H. September 1979. The brain. *Scientific American.*

Kupferman, I. 1981. Learning. In E. R. Kandel and J. H. Schwartz, eds. *Principles of neural science.* New York: Elsevier North Holland.

Lynch, G., and M. Baudry. 1984. The biochemistry of memory: a new and specific hypothesis. *Science* 224:1057.

Lemay, M., and N. Geschwind. 1978. Asymmetries of the human cerebral hemispheres. In A. Caramazza and E. Zurif, eds. *Language acquisition and language breakdown.* Baltimore: Johns Hopkins University Press.

Mishkin, M., and T. Appenzeller. June 1987. The anatomy of memory. *Scientific American.*

Routtenberg, A. November 1978. The reward system of the brain. *Scientific American.*

Shashoua, V. E. 1985. The role of extracellular proteins in learning and memory. *American Scientist* 73:364.

Springer, S. P., and G. Deutch. 1985. *Left brain, right brain.* Rev. ed. New York: W. H. Freeman and Company.

Squire, L. R. 1986. Mechanisms of memory. *Science* 232:1612.

Thompson, R. F. 1985. *The brain.* New York: W. H. Freeman and Company.

Thompson, R. F. 1986. The neurobiology of learning and memory. *Science* 233:941.

Witelson, S. F. 1976. Sex and the single hemisphere: specialization of the right hemisphere for spatial processing. *Science* 193:425.

# SENSORY PHYSIOLOGY

# 9

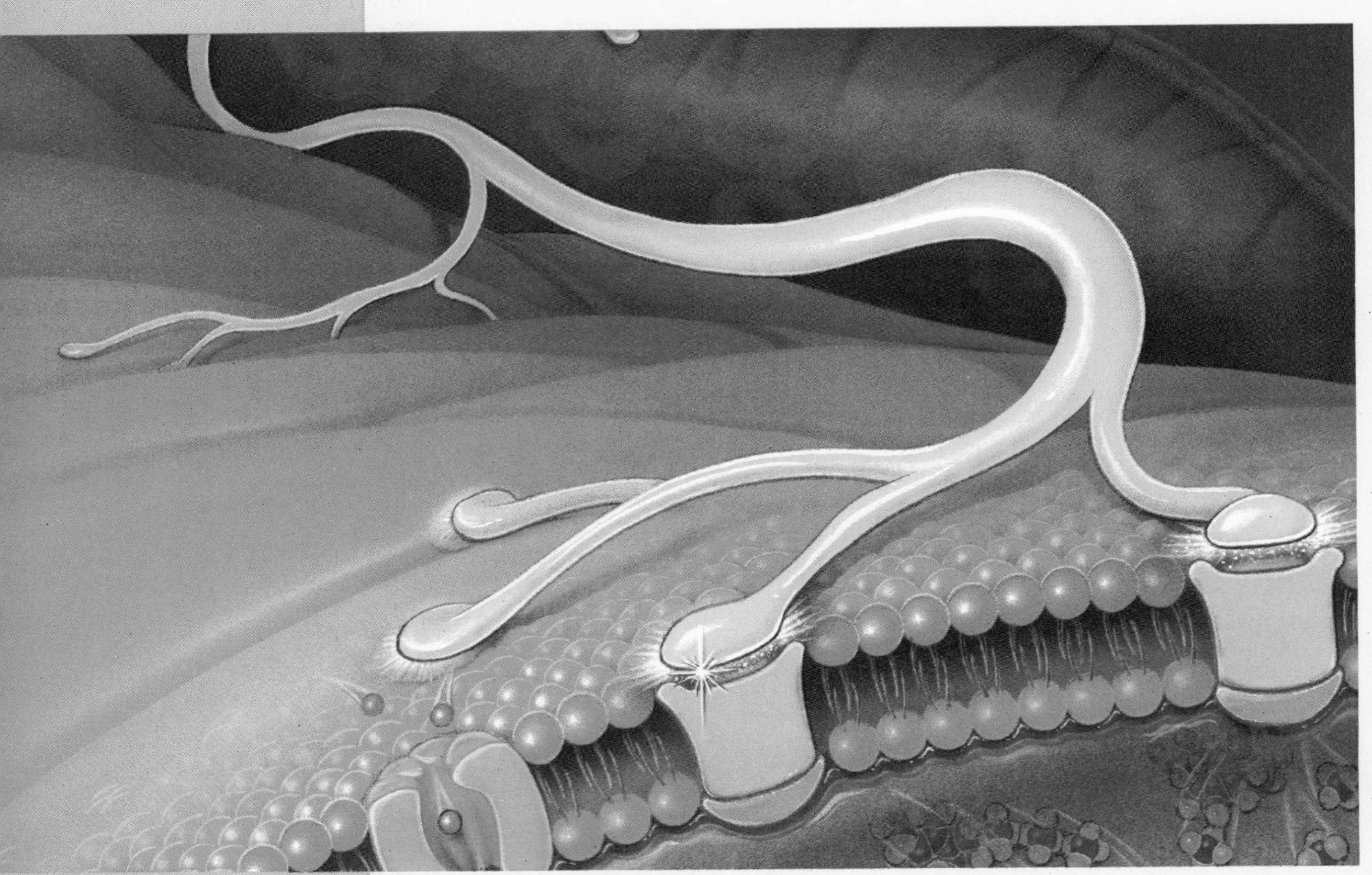

## Objectives

By studying this chapter, you should be able to

1. describe the different categories of sensory receptors and explain the differences between tonic and phasic receptors
2. explain the law of specific nerve energies
3. describe the characteristics of the generator potential
4. describe the different types of cutaneous receptors and the neural pathways for the cutaneous senses
5. explain the concepts of receptive fields and lateral inhibition in relation to the cutaneous senses
6. describe the sensory physiology of taste and olfaction
7. describe the structure of the vestibular apparatus and explain how it provides information about acceleration of the body in different directions
8. describe the functions of the outer and middle ear
9. describe the structure of the cochlea and explain how movements of the stapes against the oval window result in vibrations of the basilar membrane
10. explain how the organ of Corti converts mechanical energy into electrical nerve impulses and explain how pitch perception is accomplished
11. describe the structure of the eye and the manner in which images are brought to a focus on the retina
12. explain how visual accommodation is achieved and explain the defects involved in myopia, hyperopia, and astigmatism
13. describe the architecture of the retina and the pathways of light and nerve activity through the retina
14. explain how rhodopsin functions as the visual pigment in rods and explain dark adaptation
15. explain how light affects the electrical activity of rods and their synaptic input to bipolar cells
16. explain the trichromatic theory of color vision
17. describe the differences in synaptic connections and locations of rods and cones and explain why rods provide black-and-white vision under low illumination and cones provide color vision and high acuity under greater light intensities
18. describe the neural pathways from the retina, explaining the differences in pathways taken by input from different regions of the visual field
19. describe the receptive fields of ganglion cells and explain the significance of the arrangement of these receptive fields
20. explain the stimulus requirements of simple, complex, and hypercomplex cortical neurons

## Outline

## *Characteristics of Sensory Receptors*

Each type of sensory receptor is most sensitive to a particular modality of environmental stimulus, to which it responds by causing the production of action potentials in a sensory neuron. These impulses are conducted to parts of the brain that provide the proper interpretation of sensory perception when that particular neural pathway is activated.

Our perceptions of the world—its textures, colors, and sounds; its warmth, smells, and tastes—are created by the brain from electrochemical nerve impulses delivered to it from sensory receptors. These receptors **transduce** (change) different forms of energy in the "real world" into the energy of nerve impulses, which are conducted into the central nervous system by sensory neurons. Different sensory *modalities*—or qualities of sensation, such as sound, light, pressure, and so on—result from differences in neural pathways and synaptic connections. The brain thus interprets impulses in the auditory nerve as sound and in the optic nerve as sight, even though the impulses themselves are identical in the two nerves.

We know, through the use of scientific instruments, that our senses act as energy filters that allow us to perceive only a narrow range of energy. Vision, for example, is limited to light in the visible spectrum; ultraviolet and infrared light, X rays and radio waves, which are the same type of energy as visible light, cannot normally excite the photoreceptors in the eyes. The perception of cold is entirely a product of the nervous system—there is no such thing as cold in the physical world, only varying degrees of heat. The perception of cold, however, has obvious survival value. Although filtered and distorted by the limitations of sensory function, our perceptions of the world allow us to interact effectively with the environment.

### Categories of Sensory Receptors

Sensory receptors can be categorized by their structure and on the basis of different functional criteria. Structurally, the sensory receptors may be the dendritic endings of sensory neurons, which are either free (such as those in the skin, which mediate pain and temperature) or are encapsulated within nonneural structures, such as pressure receptors in the skin (fig. 9.1). The photoreceptors in the retina of the eyes (rods and cones) are highly specialized neurons, which synapse with other neurons in the retina. In the case of taste buds and of hair cells in the inner ears, modified epithelial cells respond to an environmental stimulus and activate sensory neurons.

***Functional Categories.*** Sensory receptors can be grouped according to the type of stimulus energy they transduce. These categories include (1) *chemoreceptors,* such as the taste buds, olfactory epithelium, and the aortic and carotid bodies, which sense chemical stimuli in the environment or the blood; (2) *photoreceptors*—the rods and cones in the retina of the eye; (3) *thermoreceptors,* which respond to heat and cold; and (4) *mechanoreceptors,* which are stimulated by mechanical deformation of the receptor cell membrane—these include touch and pressure receptors in the skin and hair cells within the inner ear. *Nocioreceptors*—or pain receptors—are stimulated by tissue damage.

Receptors can also be grouped according to the type of sensory information they deliver to the brain. *Proprioceptors* include the muscle spindles, Golgi tendon organs, and joint receptors. These provide a sense of body position and allow fine control of skeletal movements (as discussed in chapter 12). *Cutaneous receptors* include (1) touch and pressure receptors; (2) warmth and cold receptors; and (3) pain receptors. The receptors that mediate sight, hearing, and equilibrium are grouped together as the *special senses.*

***Tonic and Phasic Receptors: Sensory Adaptation.*** Some receptors respond with a burst of activity when a stimulus is first applied, but then quickly decrease their firing rate—adapt to the stimulus—when the stimulus is maintained. Receptors with this response pattern are called *phasic receptors.* Receptors that produce a relatively constant rate of firing as long as the stimulus is maintained are known as *tonic receptors* (see fig. 9.2).

Phasic receptors alert us to changes in sensory stimuli and are in part responsible for the fact that we can cease paying attention to constant stimuli. This ability is called **sensory adaptation.** Odor, touch, and temperature, for example, adapt rapidly; bathwater feels hotter when we first enter it. Sensations of pain, in contrast, adapt little if at all.

**Figure 9.1.** Different types of sensory receptors. Free nerve endings (*a*) mediate many cutaneous sensations. Some nerve endings are encapsulated within associated structures: e.g., (*b*) a Pacinian corpuscle and a Meissner's corpuscle. Some receptors, such as the taste bud (*d*), are modified epithelial cells that are innervated by sensory neurons.

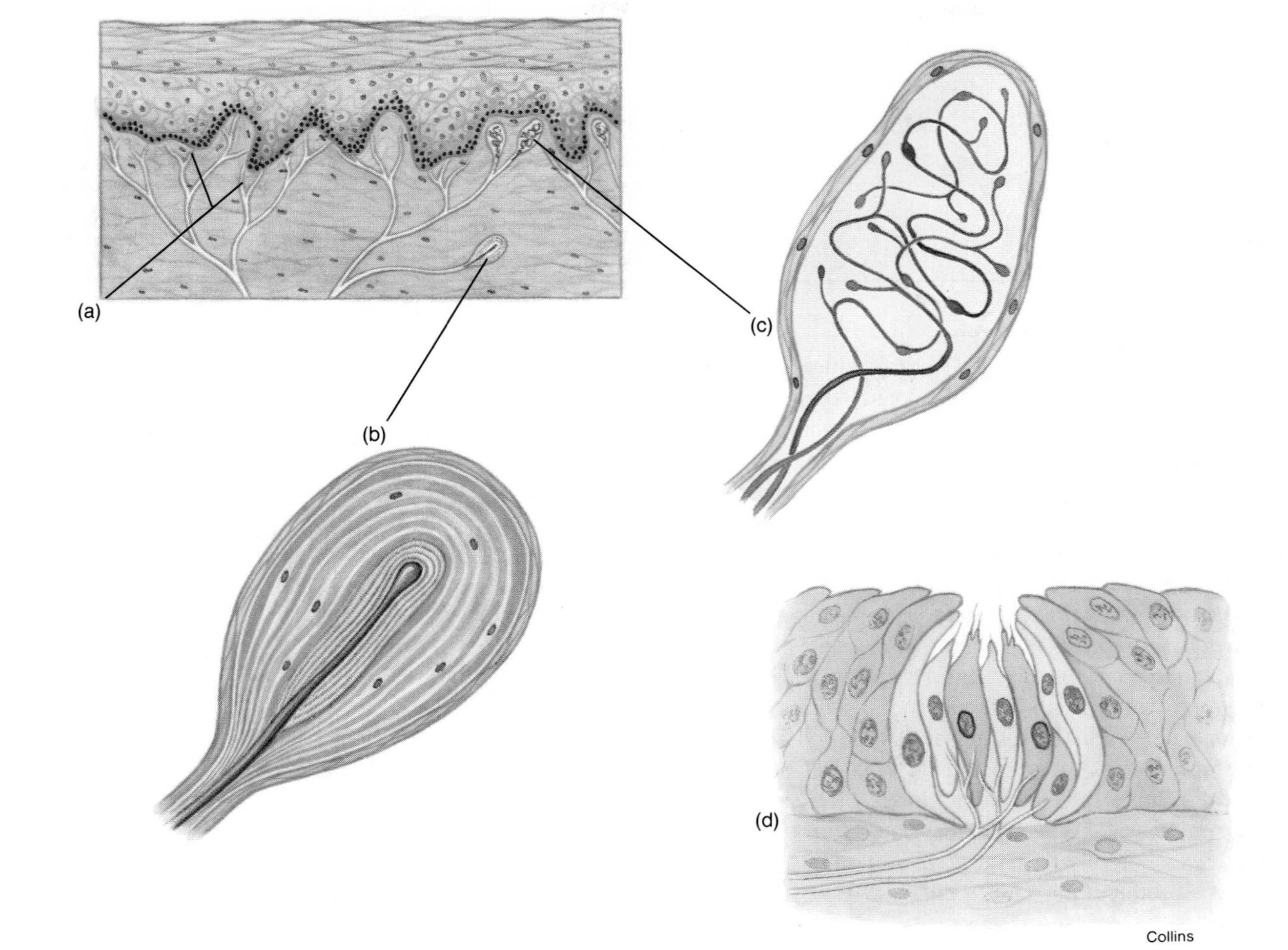

(a)
(b)
(c)
(d)
Collins

**Figure 9.2.** Tonic receptors (*a*) continue to fire at a relatively constant rate as long as the stimulus is maintained. These produce slowly adapting sensations. Phasic receptors (*b*) respond with a burst of action potentials when the stimulus is first applied, but then quickly reduce their rate of firing while the stimulus is maintained. This produces rapidly adapting sensations.

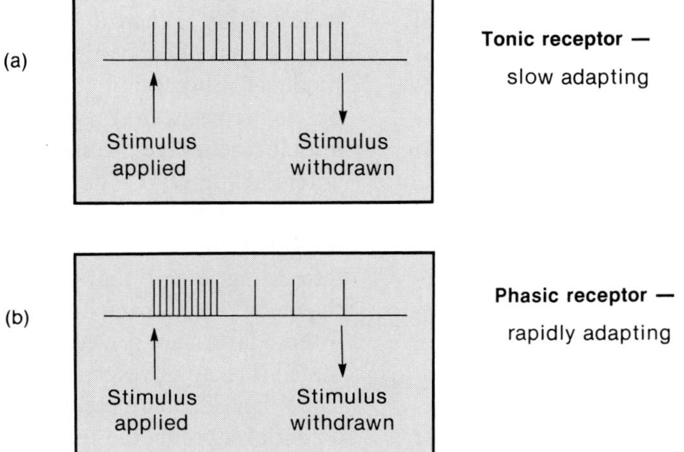

**Table 9.1** Classification of receptors based on their normal (or "adequate") stimulus

| Receptor | Normal Stimulus | Mechanisms | Examples |
|---|---|---|---|
| Mechanoreceptors | Mechanical force | Deforms cell membrane of sensory dendrites; or deforms hair cells that activate sensory nerve endings | Cutaneous touch and pressure receptors; vestibular apparatus and cochlea |
| Pain receptors | Tissue damage | Damaged tissues release chemicals that excite sensory endings | Cutaneous pain receptors |
| Chemoreceptors | Dissolved chemicals | Chemical interaction affects ionic permeability of sensory cells | Smell and taste (exteroreceptors); osmoreceptors and carotid body chemoreceptors (interoreceptors) |
| Photoreceptors | Light | Photochemical reaction affects ionic permeability of receptor cell | Rods and cones in retina of eyes |

## Law of Specific Nerve Energies

Stimulation of a sensory nerve fiber produces only one sensation—touch, cold, pain, and so on. According to the **law of specific nerve energies,** the sensation characteristic of each sensory neuron is that produced by its normal, or *adequate stimulus* (table 9.1). The adequate stimulus for the photoreceptors of the eye, for example, is light. If these receptors are stimulated by some other means—such as by pressure produced by a punch to the eye—a flash of light (the adequate stimulus) may be perceived.

*Paradoxical cold* provides another example of the law of specific nerve energies. First, a receptor for cold is located by touching the tip of a cold metal rod to the skin. Sensation then gradually disappears as the rod warms to body temperature. Applying the tip of a rod heated to 45°C to the same spot, however, causes the sensation of cold to reappear. This paradoxical cold is produced because the heat slightly damages receptor endings, and by this means produces an "injury current" that stimulates the receptor.

Regardless of how a sensory neuron is stimulated, therefore, only one sensory modality will be perceived. This specificity is due to the synaptic pathways within the brain that are activated by the sensory neuron. The ability of receptors to function as sensory filters and be stimulated by only one type of stimulus (the adequate stimulus) allows the brain to perceive the stimulus accurately under normal conditions.

**Figure 9.3.** Sensory stimuli result in the production of local, graded potential changes known as the receptor, or generator, potential (number 1 through 4). If the receptor potential reaches a threshold value of depolarization, it generates action potentials (number 5) in the sensory neuron.

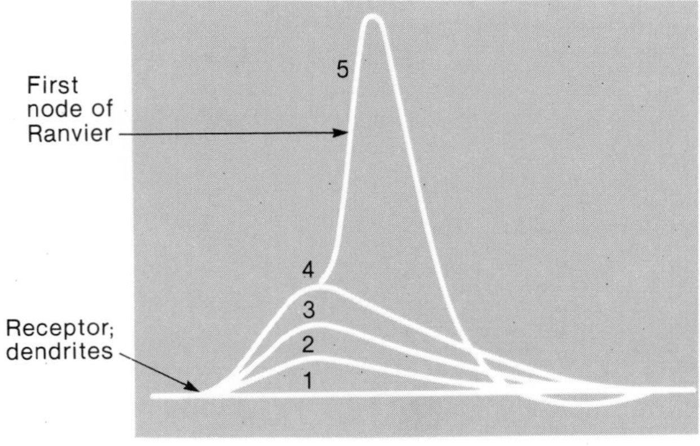


## Generator (Receptor) Potential

The electrical behavior of sensory nerve endings is similar to that of the dendrites of other neurons. In response to an environmental stimulus, the sensory endings produce local, graded changes in the membrane potential. In most cases these potential changes are depolarizations, analogous to excitatory postsynaptic potentials (EPSPs, as described in chapter 7). In the sensory endings, however, these potential changes in response to environmental stimulation are called **receptor,** or **generator, potentials,** because they serve to generate action potentials in response to the sensory stimulation. Since sensory neurons are pseudounipolar (chapter 7), the action potentials produced in response to the generator potential are conducted continuously from the periphery into the CNS.

The *pacinian corpuscle,* a cutaneous receptor for pressure (see fig. 9.1), can serve as an example of sensory transduction. When a light touch is applied to the receptor, a small depolarization (the generator potential) is produced. Increasing the pressure on the pacinian corpuscle increases the magnitude of the generator potential until it reaches the threshold required to produce an action potential (fig. 9.3). The pacinian corpuscle, however, is a phasic receptor; if the pressure is maintained the size of the generator potential produced quickly diminishes. It is interesting to note that this phasic response is a result of the onionlike covering around the dendritic nerve ending; if these layers are peeled off and the nerve ending is stimulated directly it responds in a tonic fashion.

When a tonic receptor is stimulated, the generator potential it produces is proportional to the intensity of the stimulus. After a threshold depolarization is produced, in-

**Figure 9.4.** The response of a tonic receptor to stimuli. As the strength of stimulation is increased, the generator potential increases (*a, b, c*). The amplitude of the generator potential and the length of time it remains above threshold determine the frequency and duration of action potentials (*b, c*).

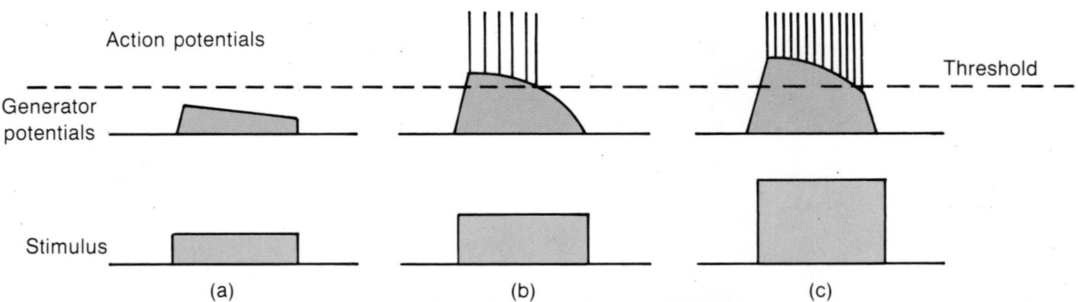


**Table 9.2** Types of cutaneous receptors

| Encapsulated or Free Nerve Endings | Type | Sensation |
|---|---|---|
| Encapsulated (dendrites within associated structures) | Pacinian corpuscles; Meissner's corpuscles; Krause's end bulbs; Ruffini's end organs | All serve touch and pressure |
| Free nerve endings | ——— | Touch, pressure, heat, cold, pain |

creases in the amplitude of the generator potential result in increases in the *frequency* with which action potentials are produced (fig. 9.4). In this way, the frequency of action potentials that are conducted into the central nervous system serves to code for the strength of the stimulus. As described in chapter 7, this frequency code is needed since the amplitude of action potentials is constant (all-or-none). Acting through changes in action potential frequency, tonic receptors thus provide information about the relative intensity of a stimulus.

1. *Our perceptions are products of our brains; they are incompletely and inconstantly related to physical reality. Explain this statement, using examples of vision and the perception of cold.*
2. *Define the law of specific nerve energies and the adequate stimulus, and relate these definitions to your answer for question 1.*
3. *Describe sensory adaptation in olfactory and pain receptors. Using a line drawing, relate sensory adaptation to the responses of phasic and tonic receptors.*
4. *Describe how the magnitude of a sensory stimulus is transduced into a receptor potential and how the magnitude of the receptor potential is coded in the sensory nerve fiber.*

## Cutaneous Sensations

There are several different types of sensory receptors in the skin, each of which is specialized to be maximally sensitive to one modality of sensation. Stimulation of a particular area of the skin that excites a specific receptor defines the receptive field of that receptor. A process known as lateral inhibition helps to sharpen the perceived location of the stimulus on the skin.

The cutaneous sensations of touch, pressure, hot and cold, and pain are mediated by the dendritic nerve endings of different sensory neurons. The receptors for hot, cold, and pain are the naked endings of sensory neurons. Sensations of touch and pressure are mediated by both naked dendritic endings and dendrites that are encapsulated within various structures (table 9.2). In pacinian corpuscles, for example, the dendritic endings are encased within thirty to fifty onionlike layers of connective tissue (fig. 9.5). These layers absorb some of the pressure when a stimulus is maintained, and thus help to accentuate the phasic response of this receptor.

**Figure 9.5.** A diagrammatic section of the skin showing the general location and magnified structure of cutaneous receptors.

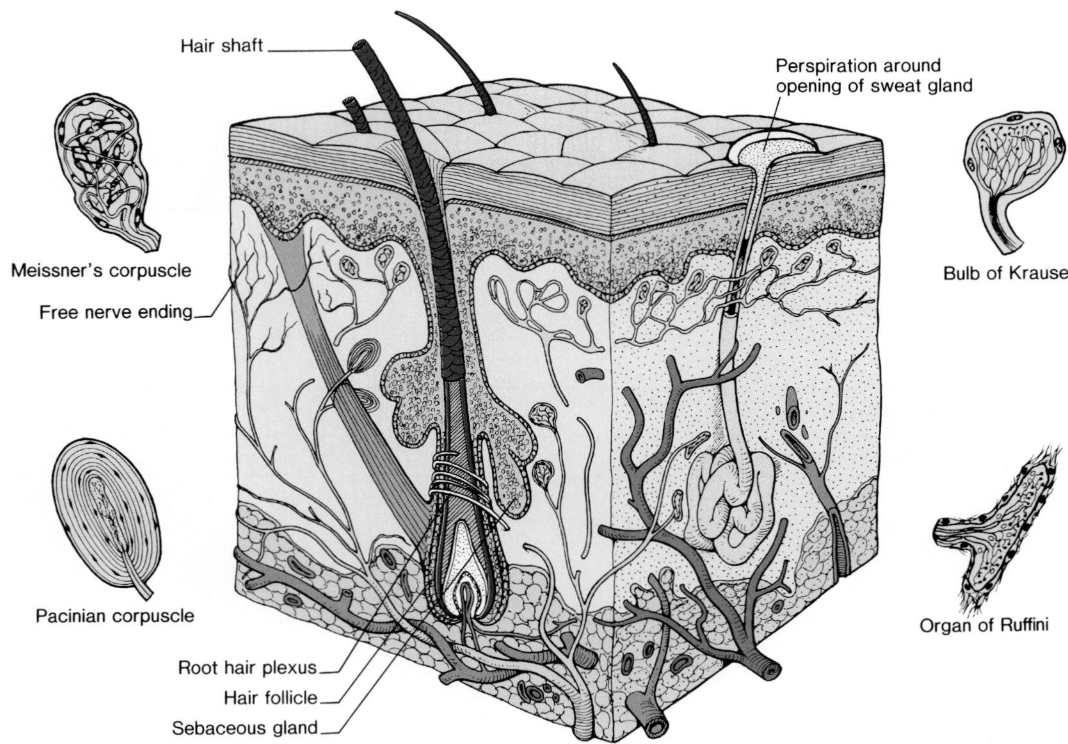


## Neural Pathways for Somatesthetic Sensations

The conduction pathways for the *somatesthetic senses*—a term that includes sensations from cutaneous receptors and proprioceptors—are shown in figure 9.6. *Proprioception* and *pressure* are carried by large, myelinated nerve fibers that ascend in the *dorsal columns* of the spinal cord on the same (ipsilateral) side. These fibers do not synapse until they reach the *medulla oblongata* of the brain stem; fibers that carry these sensations from the feet are thus incredibly long. After synapsing in the medulla with other, second-order sensory neurons, information in the latter neurons crosses over to the contralateral side as it ascends via a fiber tract, called the medial lemniscus, to the *thalamus*. Third-order sensory neurons in the thalamus that receive this input in turn project to the postcentral gyrus (the sensory cortex, as described in chapter 8).

Sensations of *hot, cold,* and *pain* are carried by thin, unmyelinated sensory neurons into the spinal cord. These synapse with second-order interneurons within the spinal cord, which cross over to the contralateral side and ascend to the brain in the *lateral spinothalamic tract*. Fibers that mediate *touch* and *pressure* ascend in the *ventral spinothalamic tract*. Fibers of both spinothalamic tracts synapse with third-order neurons in the thalamus, which in turn project to the postcentral gyrus. Notice that, in all cases, somatesthetic information is carried to the postcentral gyrus in third-order neurons. Also, because of crossing-over, somatesthetic information from each side of the body is projected to the postcentral gyrus of the contralateral cerebral hemisphere.

All somatesthetic information from the same area of the body projects to the same area of the postcentral gyrus, so that a "map" of the body can be drawn on the postcentral gyrus to represent sensory projection points (chapter 8, fig. 8.7). This map is very distorted, however, because it shows larger areas of cortex devoted to sensation in the face and hands than in other areas in the body. This disproportionately larger area of the cortex devoted to the face and hands reflects the fact that there is a higher density of sensory receptors in these regions.

## Receptive Fields and Sensory Acuity

The **receptive field** of a neuron serving cutaneous sensation is the area of skin whose stimulation results in changes in the firing rate of the neuron. Changes in the firing rate of primary sensory neurons affect the firing of second- and third-order neurons, which in turn affects the firing of those neurons in the postcentral gyrus that receive input from the third-order neurons. Indirectly, therefore, neurons in the postcentral gyrus can be said to have receptive fields in the skin.

**Figure 9.6.** Pathways that lead from the cutaneous receptors and proprioreceptors into the postcentral gyrus in the cerebral cortex.

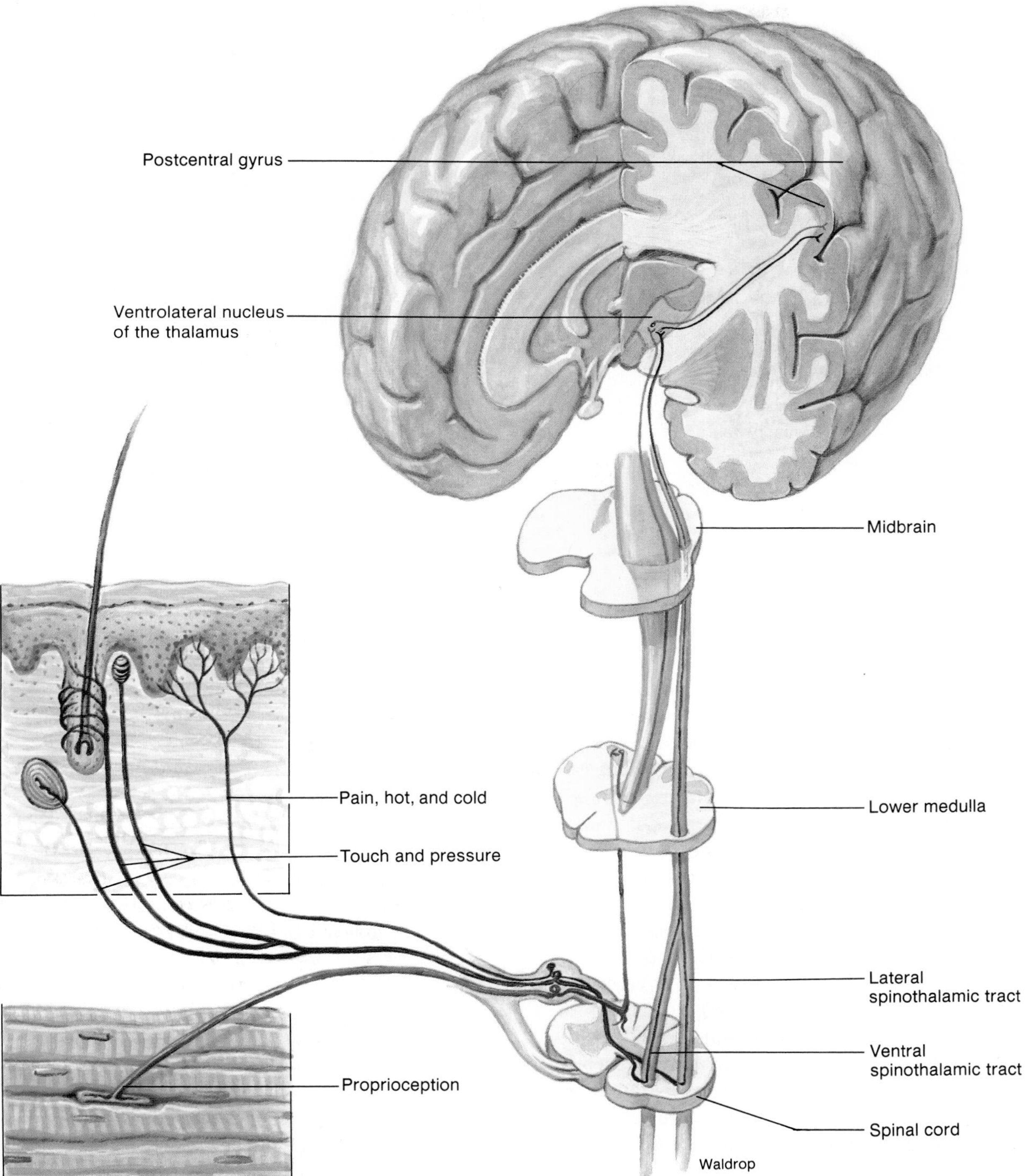

**Figure 9.7.** The two-point touch threshold test. If each point touches the receptive fields of different sensory neurons, two separate points of touch will be felt. If both caliper points touch the receptive field of one sensory neuron, only one point of touch will be felt.

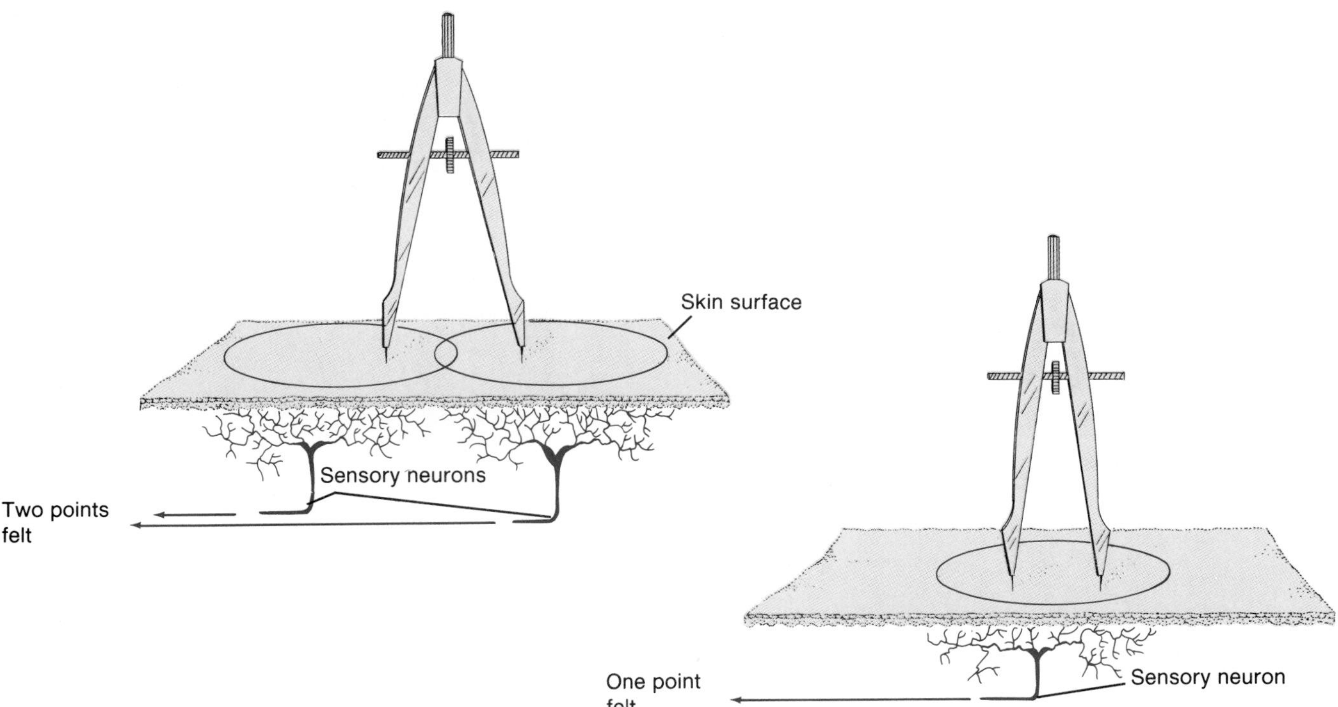


The area of each receptive field in the skin varies inversely with the density of receptors in the region. In the back and legs, where a large area of skin is served by relatively few sensory endings, the receptive field of each neuron is correspondingly large. In the fingertips—where a large number of cutaneous receptors serve a small area of skin—the receptive field of each sensory neuron is correspondingly small.

***Two-Point Touch Threshold.*** The approximate size of the receptive fields serving light touch can be measured by the *two-point touch threshold* test. In this procedure, two points of a pair of calipers are lightly touched to the skin at the same time. If the calipers are set sufficiently wide apart, each point will stimulate a different receptive field and a different sensory neuron—two separate points of touch will thus be felt. If the calipers are sufficiently closed, both points will touch the receptive field of only one sensory neuron, and only one point of touch will be felt (fig. 9.7).

The two-point touch threshold, which is the minimum distance that can be distinguished between two points of touch, is a measure of the distance between receptive fields. If the two points of the calipers are closer than this distance, only one "blurred" point of touch can be felt. The two-point touch threshold is thus an indication of tactile *acuity,* or the sharpness (acu = needle) of touch perception.

The high tactile acuity of the fingertips is exploited in the reading of *Braille.* Braille symbols consist of dots that are raised 1 mm up from the page and separated from each other by 2.5 mm, which is slightly above the two-point touch threshold in the fingertips (table 9.3). Experienced Braille readers can scan words at about the same speed that a sighted person can read aloud—a rate of about 100 words per minute.

## Lateral Inhibition

When a blunt object touches the skin a number of receptive fields may be stimulated. Those receptive fields in the center areas where the touch is strongest will be stimulated more than in neighboring fields where the touch is lighter. We do not usually feel a "halo" of light touch surrounding a center of stronger touch, however. Instead, only a single touch is felt, which is somewhat sharper than the actual shape of the blunt object. This sharpening of sensation is due to a process called *lateral inhibition* (fig. 9.8).

**Table 9.3** The two-point touch threshold for different regions of the body

| Body Region | Two-Point Touch Threshold (mm) |
|---|---|
| Big toe | 10 |
| Sole of foot | 22 |
| Calf | 48 |
| Thigh | 46 |
| Back | 42 |
| Abdomen | 36 |
| Upper arm | 47 |
| Forehead | 18 |
| Palm of hand | 13 |
| Thumb | 3 |
| First finger | 2 |

"From S. Weinstein and D. R. Kenshalo (editor), *The Skin Senses,* 1968. Courtesy of Charles C Thomas, Publisher, Springfield, Illinois."

Lateral inhibition and the sharpening of sensation that results occur within the central nervous system. Those sensory neurons whose receptive fields are stimulated most strongly inhibit—via interneurons that pass "laterally" within the CNS—sensory neurons that serve neighboring receptive fields. Lateral inhibition similarly plays a prominent role in the ability of the ears and brain to discriminate sounds of different pitch, as described in a later section.

1. *Using a flow diagram, describe the neural pathways leading from cutaneous pain and pressure receptors to the postcentral gyrus. Indicate where cross-over occurs.*
2. *Describe the meaning of the term* sensory acuity, *and explain how acuity is related to the density of receptive fields in different parts of the body.*
3. *Explain the significance of lateral inhibition in the acuity of cutaneous sensory perception.*

**Figure 9.8.** When an object touches the skin (*a*), receptors in the center of the touched skin are stimulated more than neighboring receptors (*b*). As a result of lateral inhibition within the central nervous system (*c*), input from these neighboring sensory neurons is reduced. Sensation, as a result, is more sharply localized to the area of skin that was stimulated the most (*d*).

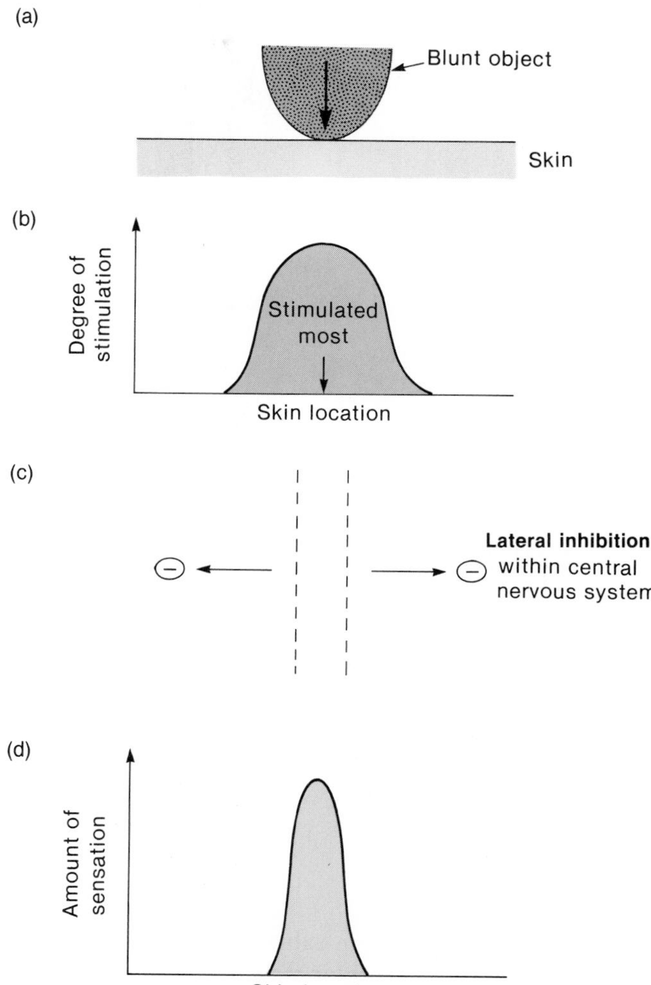


## Taste and Olfaction

The receptors for taste and olfaction respond to molecules that are dissolved in fluid, and are thus classified as chemoreceptors. Although there are only four basic modalities of taste, the combination of these, and of the taste sensations with those of olfaction, provide a wide variety of different sensory experiences.

Chemoreceptors that respond to chemical changes in the internal environment are called **interoceptors;** those that respond to chemical changes in the external environment are **exteroceptors.** Included in the latter category are *taste receptors,* which respond to chemicals dissolved in food or drink, and *olfactory receptors,* which respond to gaseous molecules in the air. This distinction is somewhat arbitrary, however, because odorant molecules in air must first dissolve in fluid within the olfactory mucosa before the sense of smell can be stimulated. Also, the sense of olfaction has a great effect on the sense of taste, as can easily be verified by eating an onion with the nostrils pinched together.

### Taste

Taste receptors are specialized epithelial cells that are grouped together into barrel-shaped arrangements called *taste buds,* located in the epithelium of the tongue (fig. 9.9). The cells of the taste buds have microvilli at their apical (top) surface, which is exposed to the external environment through a pore in the surface of the taste bud.

**Figure 9.9.** A taste bud.

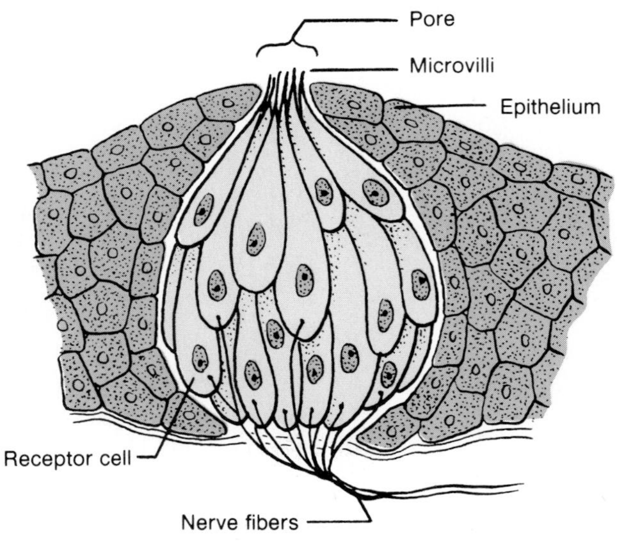


**Figure 9.10.** Patterns of taste receptor distribution on the dorsum of the tongue. (*a*) Sweet receptors; (*b*) sour receptors; (*c*) salt receptors; (*d*) bitter receptors.

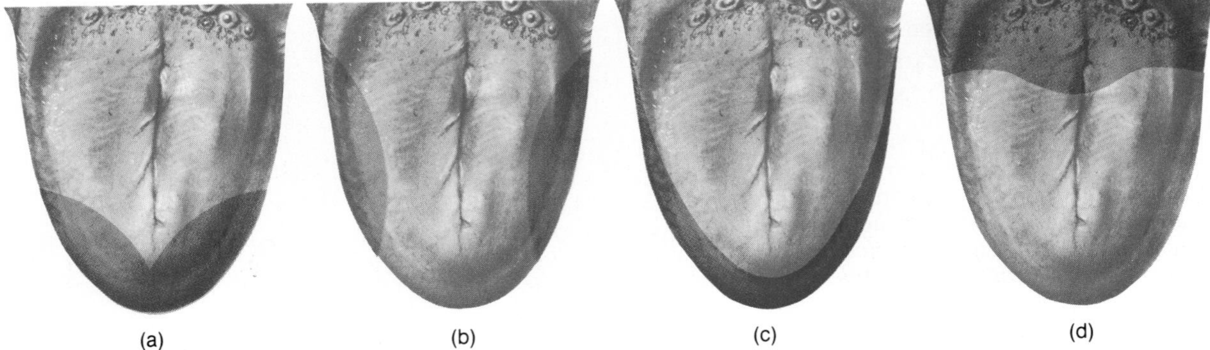


Molecules dissolved in saliva at the surface of the tongue interact with receptor molecules in the microvilli of the taste buds. This interaction stimulates the release of a neurotransmitter from the receptor cells, which in turn stimulates sensory nerve endings that innervate the taste buds. Taste buds in the posterior third of the tongue are innervated by the *glossopharyngeal (ninth cranial) nerve;* those in the anterior two-thirds of the tongue are innervated by the *facial (seventh cranial) nerve.*

There are only four basic modalities of taste, which are sensed most acutely in particular regions of the tongue. These are *sweet* (tip of the tongue), *sour* (sides of the tongue), *bitter* (back of the tongue), and *salty* (over most of the tongue). This distribution is illustrated in figure 9.10.

Sour taste is produced by hydrogen ions ($H^+$); all acids therefore taste sour. Most organic molecules, particularly sugars, taste sweet to varying degrees. Only pure table salt (NaCl) has a pure salty taste—other salts, such as KCl (commonly used in place of NaCl by people with hypertension) taste salty but have bitter overtones. Bitter taste is evoked by quinine and seemingly unrelated molecules.

Since there are only four basic modalities of taste, it might be assumed that the physiology of taste is well understood. This is not the case. Research has shown that particular types of molecules that evoke a characteristic taste cause changes in the permeability of sensory dendrites to various ions, including $Na^+$, $K^+$, and $Ca^{++}$. Some of the molecules tested have a direct effect on membrane permeability, but others appear to require the action of second messengers such as cyclic AMP (chapter 7). Thus far, however, no consistent pattern has been discovered that would allow a clear explanation of how a molecule such as sugar results in the perception of a sweet taste.

## Olfaction

The olfactory receptors are the dendritic endings of the *olfactory (first cranial) nerve,* in association with epithelial supporting cells. Unlike other sensory modalities, which are relayed to the cerebrum from the thalamus, the sense of olfaction is transmitted directly to the olfactory

**Figure 9.11.** The olfactory epithelium contains receptor neurons that synapse with neurons in the olfactory bulb of the brain.

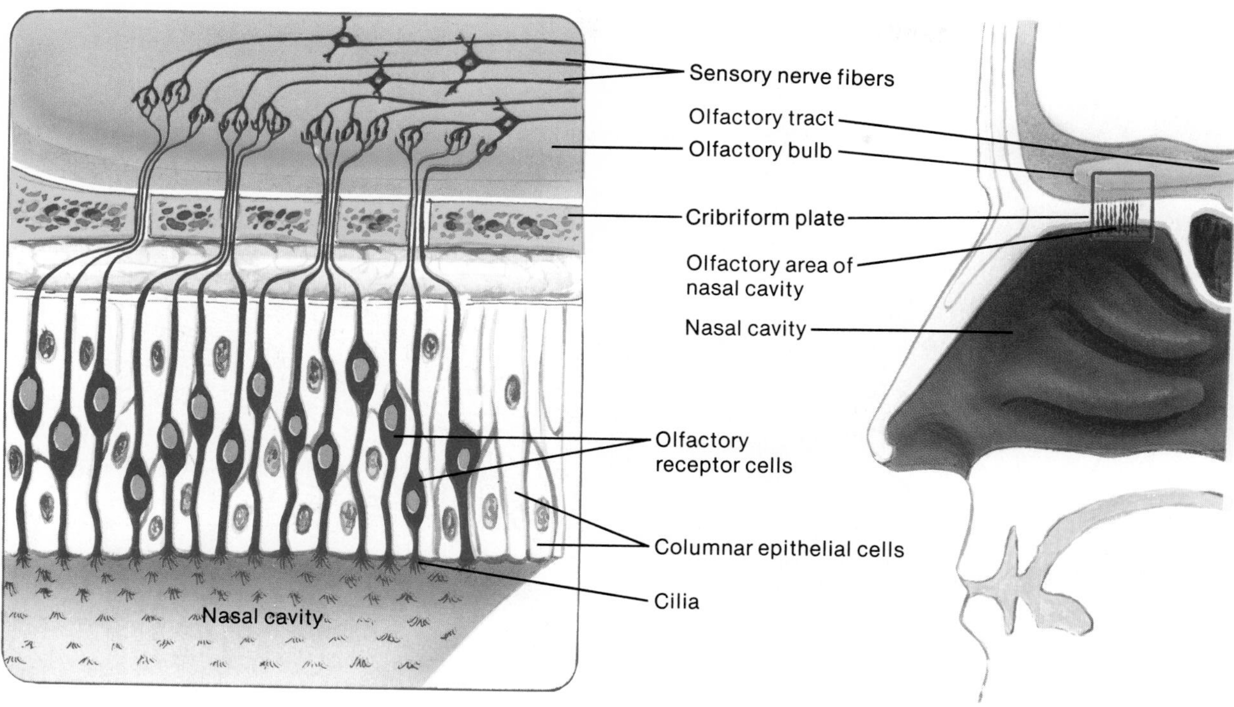


bulb of the cerebral cortex (fig. 9.11). This area of the brain is part of the limbic system, which was discussed in chapter 8 as having an important role in the generation of emotions and in memory. Perhaps this explains why the smell of a particular odor, more powerfully than other sensations, can evoke emotionally charged memories.

Unlike taste, which is divisible into only four modalities, many thousands of different odors can be distinguished by people who are trained in this capacity (as in the perfume and wine industries). The molecular basis of olfaction is not understood; although various theories have attempted to explain families of odors on the basis of similarities in molecular shape and/or charges, such attempts have been only partially successful. The extreme sensitivity of olfaction is possibly as amazing as its diversity—at maximum sensitivity, only one odorant molecule is needed to excite an olfactory receptor.

1. *Describe the distribution of taste receptors in the tongue, and explain what effect damage to the facial nerve might have on taste sensation.*
2. *To what types of stimuli do taste buds and olfactory receptors respond?*

## *Vestibular Apparatus and Equilibrium*

The sense of equilibrium is provided by structures in the inner ear collectively known as the vestibular apparatus. Movements of the head cause fluid within these structures to bend extensions of sensory cells, and this mechanical bending results in the production of action potentials.

The sense of equilibrium, which provides orientation with respect to gravity, is due to the function of an organ called the **vestibular apparatus.** The vestibular apparatus and a snail-like structure called the *cochlea,* which is involved in hearing, form the *inner ear* within the temporal bones of the skull. The vestibular apparatus consists of two parts: (1) the *otolith organs,* which include the *utricle* and *saccule;* and (2) the *semicircular canals* (fig. 9.12).

The sensory structures of the vestibular apparatus and cochlea are located within a tubular structure called the **membranous labyrinth,** which is filled with a fluid that is similar in composition to intracellular fluid. This fluid is

**Figure 9.12.** Structures within the inner ear include the cochlea and vestibular apparatus. The vestibular apparatus consists of the utricle and saccule (together called the otolith organs) and the three semicircular canals. Each semicircular canal contains a widened area (the ampulla, which is labeled for only one canal) with sensory hair cells.

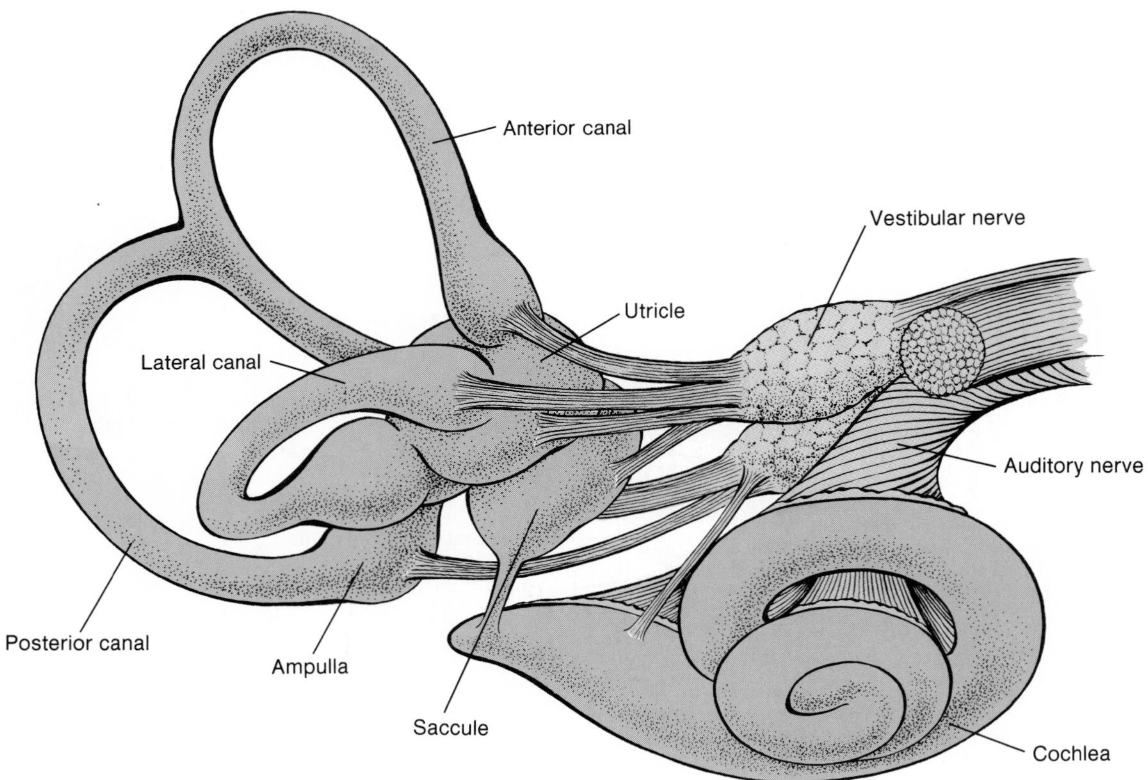


called *endolymph.* The bony structures surrounding the membranous labyrinth in the inner ear contain a fluid called *perilymph,* which is similar in composition to cerebrospinal fluid. The membranous labyrinth (fig. 9.13), in other words, is filled with endolymph and is surrounded by perilymph and bone.

## Sensory Hair Cells of the Vestibular Apparatus

The utricle and saccule provide information about *linear acceleration*—changes in velocity when traveling horizontally or vertically. We therefore have a sense of acceleration and deceleration when riding in a car or when skipping rope. A sense of *rotational* or *angular acceleration* is provided by the semicircular canals, which are oriented in three planes like the faces of a cube. This helps us maintain balance when turning the head, spinning, or tumbling.

The receptors for equilibrium are modified epithelial cells called *hair cells,* because they contain twenty to fifty hairlike extensions. All but one of these extensions contain filaments of protein and are known as *stereocilia.* There is one larger extension that has the structure of a true cilium (chapter 3) and is known as a *kinocilium* (fig. 9.14). When the stereocilia are bent in the direction of the kinocilium, the cell membrane is depressed and becomes depolarized. This causes the hair cell to release a synaptic transmitter which stimulates the dendrites of sensory neurons that are part of the eighth cranial nerve (chapter 8). When the stereocilia are bent in the opposite direction, the membrane of the hair cell becomes hyperpolarized (fig. 9.15) and, as a result, releases less synaptic transmitter. In this way, the frequency of action potentials in the sensory neurons that innervate the hair cells carries information about movements that cause the hair cell processes to bend.

**Figure 9.13.** The labyrinths of the inner ear. The membranous labyrinth (darker color) is contained within the bony labyrinth.

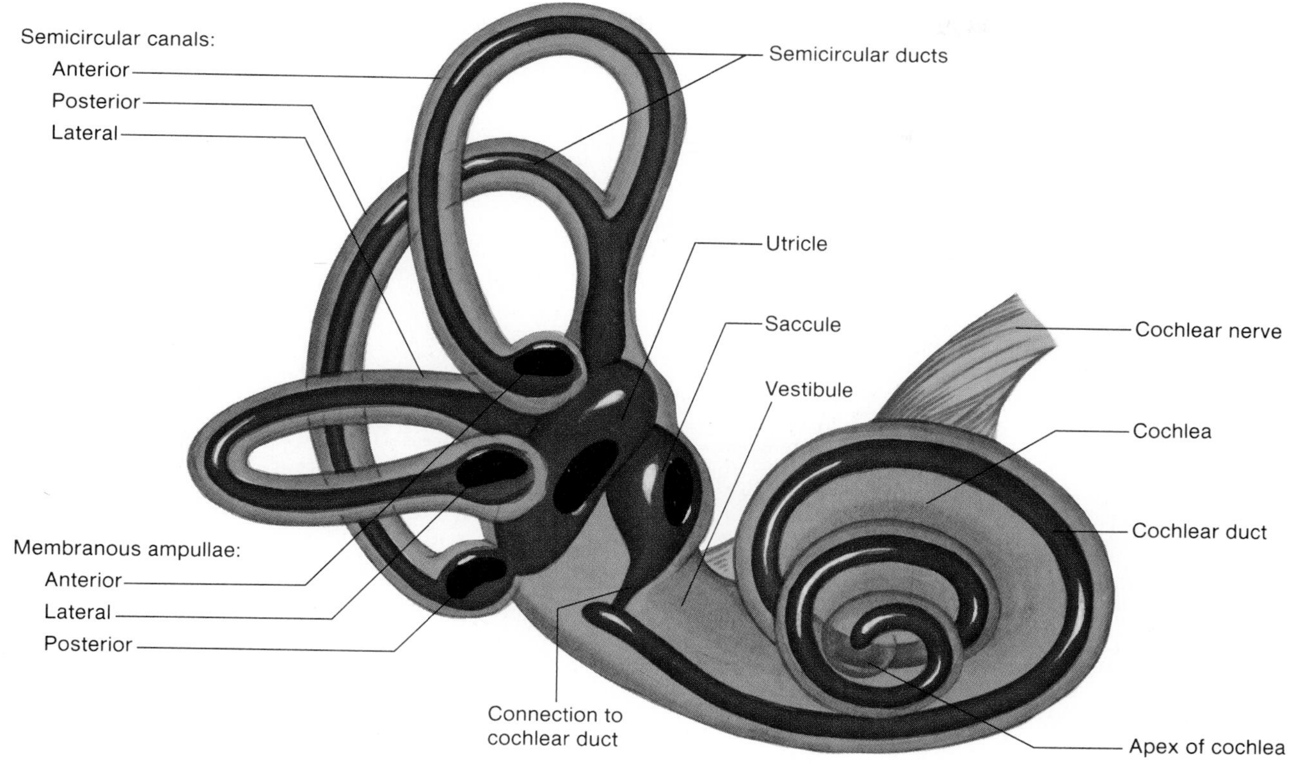


**Figure 9.14.** Scanning electron micrograph of hairs and kinocilium within the vestibular apparatus.

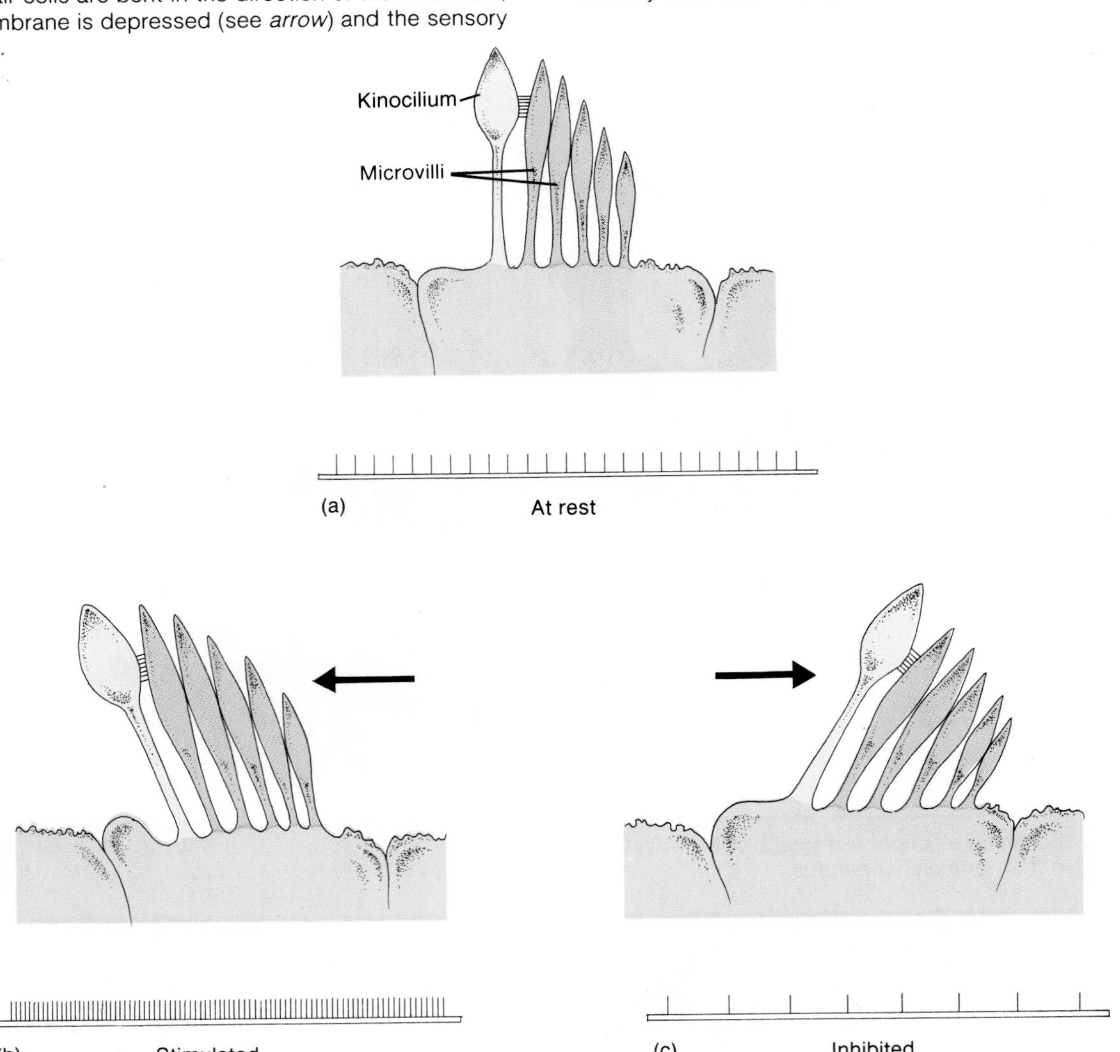


**Figure 9.15.** (*a*) Sensory hair cells in the vestibular apparatus contain hairs (microvilli) and one kinocilium. (*b*) When hair cells are bent in the direction of the kinocilium, the cell membrane is depressed (see *arrow*) and the sensory neuron innervating the hair cell is stimulated. (*c*) When the hairs are bent in a direction opposite to the kinocilium, the sensory neuron is inhibited.

## Utricle and Saccule

The hair cells of the utricle and saccule protrude into the endolymph-filled membranous labyrinth, with their hairs embedded within a gelatinous **otolith membrane.** The otolith membrane contains many crystals of calcium carbonate from which it derives its name (oto = ear; lith = stone). These stones increase the mass of the membrane, which results in a higher *inertia* (resistance to change in movement).

Because of the orientation of their hair cell processes into the otolith membrane, the utricle is most sensitive to horizontal acceleration whereas the saccule is most sensitive to vertical acceleration. During forward acceleration, the otolith membrane lags behind the hair cells, so the hairs of the utricle are pushed backwards (fig. 9.16). This is similar to the backwards thrust of the body when a car accelerates rapidly forward. The inertia of the otolith membrane similarly causes the hairs of the saccule to be pushed upwards when a person descends rapidly in an elevator. These effects, and the opposite ones that occur when a person accelerates backwards or upwards, produce a changed pattern of action potentials in sensory nerve fibers which allow us to maintain our equilibrium with respect to gravity during linear acceleration.

## Semicircular Canals

The three semicircular canals are oriented at right angles to each other. Each canal contains a bulge, called the *ampulla,* where the sensory hair cells are located. The processes of these sensory cells are embedded within a gelatinous membrane called the *cupula* (fig. 9.17), which projects into the endolymph of the membranous canals. Like a sail in the wind, the cupula can be pushed in one direction or the other by movements of the endolymph.

**Figure 9.16.** The otolith organ. (*a*) When the head is in an upright position, the weight of the otoliths applies direct pressure to the sensitive cytoplasmic extensions of the hair cells. (*b*) As the head is tilted forward, the extensions of the hair cells bend in response to gravitational force and cause the sensory nerve fibers to be stimulated.

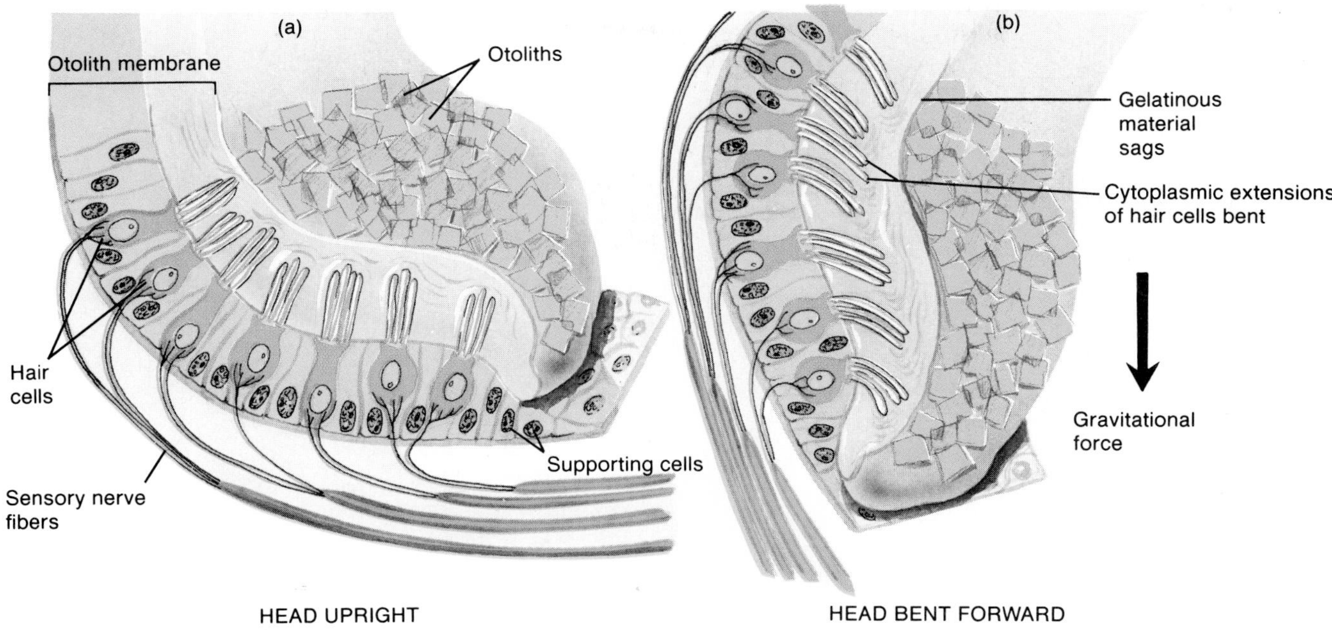


**Figure 9.17.** (*a*) The cupula and hair cells within the semicircular canals. (*b*) Movement of the endolymph during rotation causes the cupula to displace, thus stimulating the hair cells.

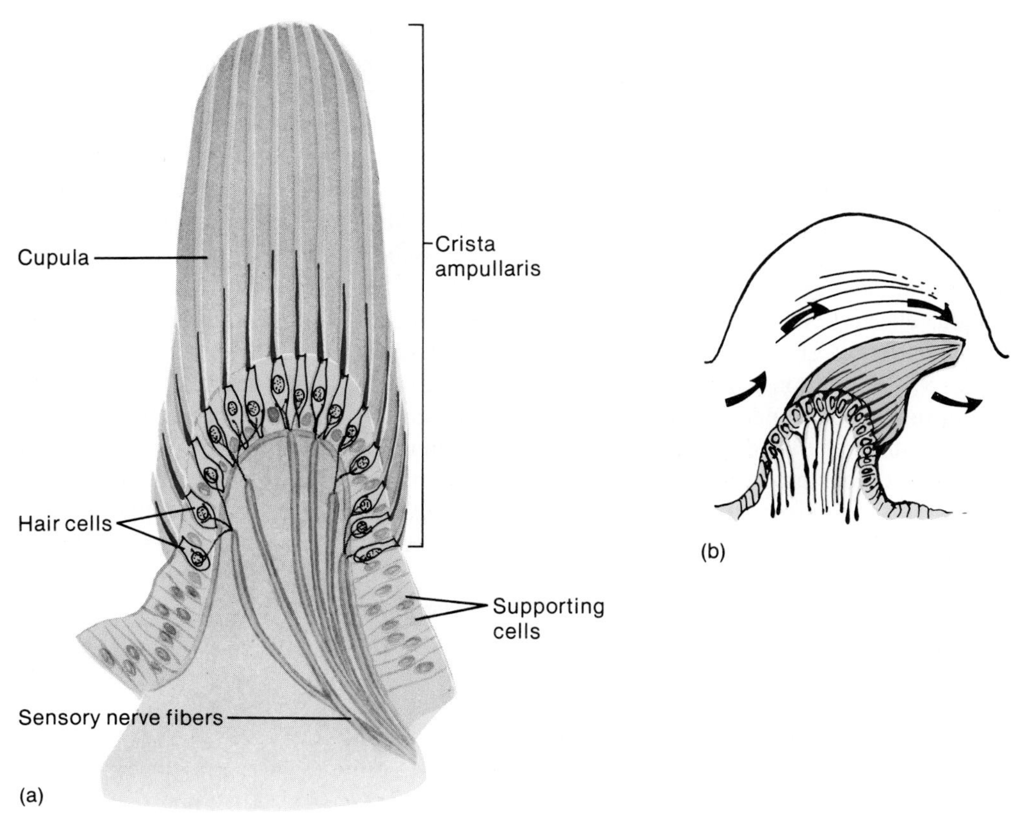

**Figure 9.18.** Neural processing involved in the maintenance of equilibrium and balance.

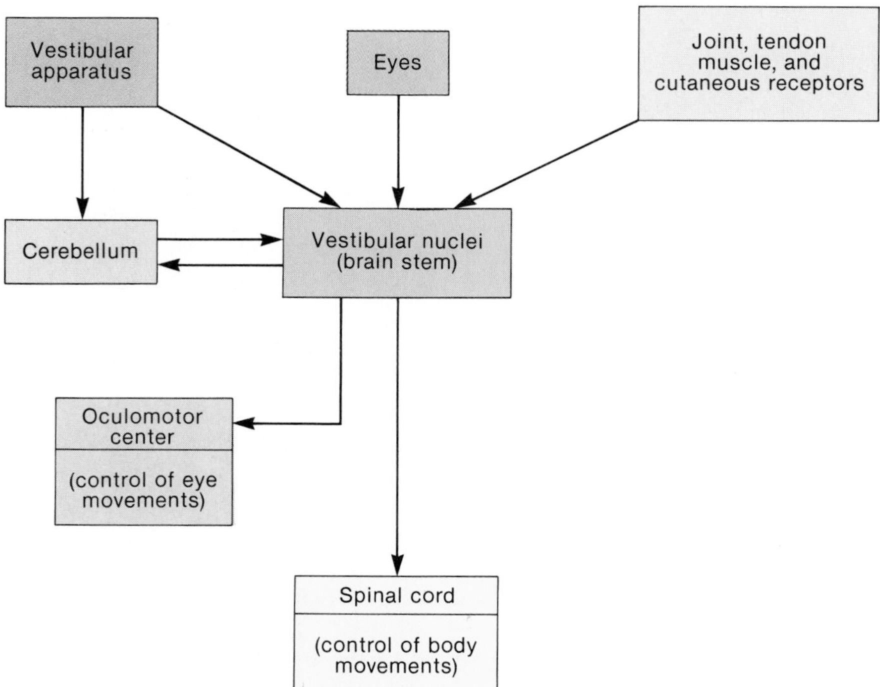


The endolymph of the semicircular canals serves a function analogous to that of the otolith membrane—it provides inertia so that the sensory processes will be bent in a direction opposite to that of the angular acceleration. As the head rotates to the right, for example, the endolymph causes the cupula to be bent toward the left, thereby stimulating the hair cells.

***Neural Pathways.*** Stimulation of hair cells in the vestibular apparatus activates sensory neurons of the *vestibulocochlear (eighth cranial) nerve.* These fibers transmit impulses to the cerebellum and to the vestibular nuclei of the medulla oblongata. The vestibular nuclei, in turn, send fibers to the oculomotor center of the brain stem and to the spinal cord (fig. 9.18). Neurons in the oculomotor center control eye movements, and neurons in the spinal cord stimulate movements of the head, neck, and limbs. Movements of the eyes and body produced by these pathways serve to maintain balance and "track" the visual field during rotation.

***Nystagmus and Vertigo.*** When a person first begins to spin, the inertia of endolymph within the semicircular canals causes the cupula to bend in the opposite direction. As the spin continues, however, the inertia of the endolymph is overcome and the cupula straightens. At this time the endolymph and the cupula are moving in the same direction and at the same speed. If movement is suddenly stopped, the greater inertia of the endolymph causes it to continue moving in the previous direction of spin and to bend the cupula in that direction.

Bending of the cupula after movement has stopped affects muscular control of the eyes and body through the neural pathways previously discussed. The eyes slowly drift in the direction of the previous spin, and then are rapidly jerked back to the midline position, producing involuntary oscillations. These movements are called **vestibular nystagmus.** People at this time may feel that they, or the room, are spinning. The loss of equilibrium that results is called *vertigo.* If the vertigo is sufficiently severe, or the person particularly susceptible, the autonomic system may become involved. This can produce dizziness, pallor, sweating, and nausea.

Vestibular nystagmus is one of the symptoms of an inner-ear disease called **Meniere's disease.** The early symptom of this disease is often "ringing in the ears," or *tinnitus.* Since the endolymph of the cochlea and the endolymph of the vestibular apparatus are continuous, through a tiny canal called the duct of Hensen, vestibular symptoms of vertigo and nystagmus often accompany hearing problems in this disease.

1. *Describe the structure of the utricle and saccule, and explain how linear acceleration results in stimulation of the hair cells within these organs.*
2. *Describe the structure of the semicircular canals, and explain how they provide a sense of angular acceleration.*

## The Ears and Hearing

Sound energy causes vibrations of the tympanic membrane. This, in turn, causes movements of the middle ear ossicles, which press against a membrane called the oval window in the cochlea. The movements of the oval window produce pressure waves within the endolymph of the cochlea, which cause movements of a membrane called the basilar membrane. Sensory hair cells are located on the basilar membrane, and the movements of this membrane in response to sound cause the hair cell processes to be bent. As a result, action potentials are produced in sensory fibers and transmitted to the brain where they are interpreted as sound.

Sound waves travel in all directions from their source, like ripples in a pond after a stone is dropped. These waves are characterized by their frequency and their intensity. The **frequency,** or distances between crests of the sound waves, is measured in *hertz (Hz),* which is the modern designation for *cycles per second (cps).* The *pitch* of a sound is directly related to its frequency—the greater the frequency of a sound, the higher its pitch.

The **intensity,** or loudness of a sound, is directly related to the amplitude of the sound waves. This is measured in units known as *decibels (db).* A sound that is barely audible—at the threshold of hearing—has an intensity of zero decibels. Every ten decibels indicates a tenfold increase in sound intensity; a sound is ten times louder than threshold at 10 db, 100 times louder at 20 db, a million times louder at 60 db, and ten billion times louder at 100 db.

The ear of a trained, young individual can hear sound over a frequency range of 20–20,000 Hz, yet can distinguish between two pitches that have only a 0.3% difference in frequency. The human ear can detect differences in sound intensities of only 0.1 to 0.5 db, while the range of audible intensities covers twelve order of magnitude ($10^{12}$), from the barely audible to the limits of painful loudness.

### The Outer Ear

Sound waves are funneled by the *pinna,* or *auricle* (flap), into the *external auditory meatus* (fig. 9.19). These two structures comprise the *outer ear.* The external auditory meatus channels the sound waves (while increasing their intensity) to the eardrum, or *tympanic membrane.* Sound waves in the external auditory meatus produce extremely small vibrations of the tympanic membrane; movements of the eardrum during speech (with an average sound intensity of 60 db) are estimated to be about equal to the diameter of a molecule of hydrogen!

### The Middle Ear

The middle ear is the cavity between the tympanic membrane on the outer side and the cochlea on the inner side (fig. 9.20). Within this cavity are three **middle ear ossicles**—the *malleus* (hammer), *incus* (anvil), and *stapes* (stirrup). The malleus is attached to the tympanic membrane, so that vibrations of this membrane are transmitted, via the malleus and incus, to the stapes. The stapes, in turn, is attached to a membrane in the cochlea called the *oval window,* which thus vibrates in response to vibrations of the tympanic membrane.

The *auditory (eustachian) tube* is a passageway leading from the middle ear to the nasopharynx (a cavity located posterior to the palate of the oral cavity). The auditory tube is usually collapsed, so that debris and infectious agents are prevented from traveling from the oral cavity to the middle ear. In order to open the auditory canal, the tensor veli tympani muscle must contract. This occurs when you swallow, yawn, or sneeze. People thus sense a "popping" sensation in the ears as they swallow when driving up a mountain, since the opening of the auditory canal permits air to move from the region of higher pressure in the middle ear to the lower pressure in the nasopharynx.

**Figure 9.19.** The ear. Note the external, middle, and inner regions of the ear.

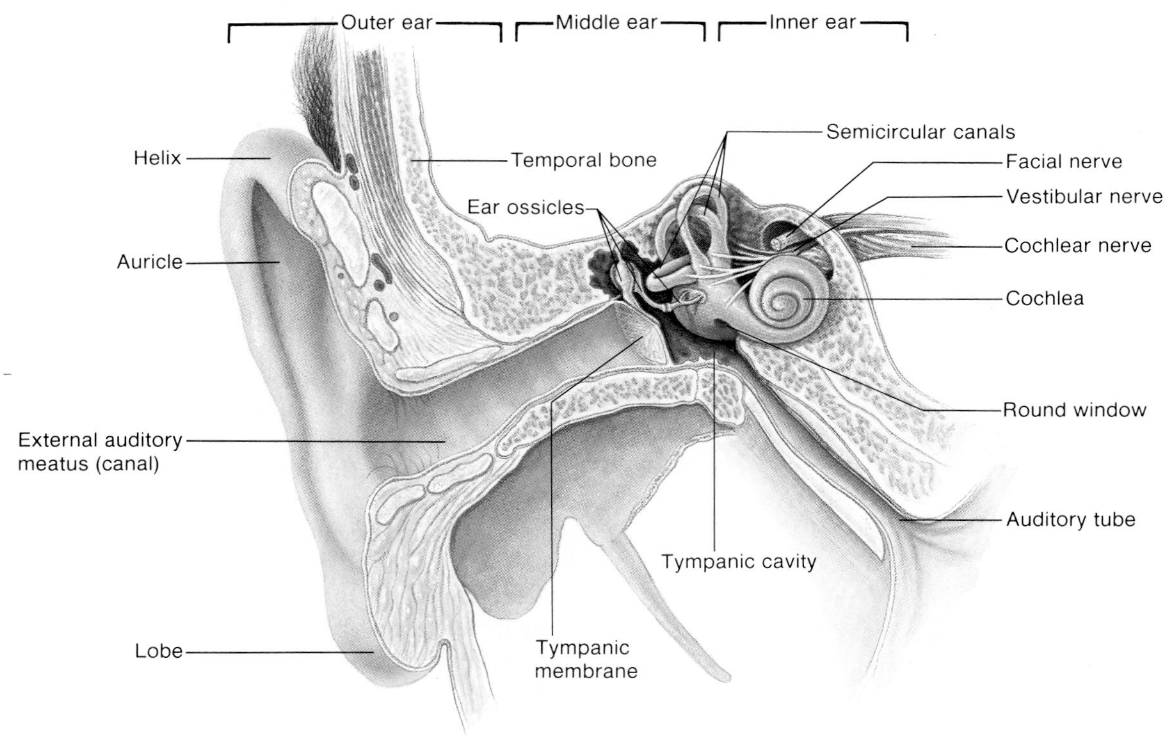


**Figure 9.20.** The position of the ear ossicles within the tympanic cavity.

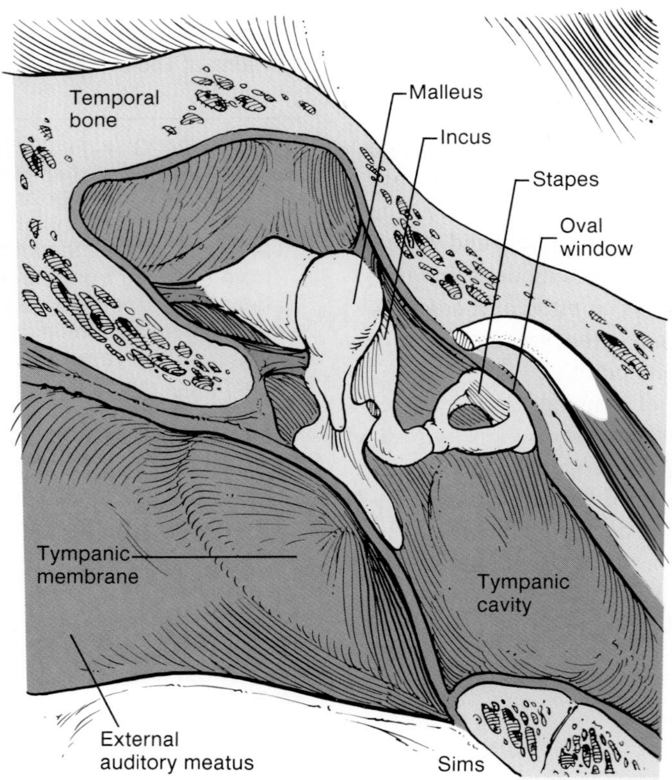

**Figure 9.21.** A medial view of the middle ear showing the position of the auditory muscles.

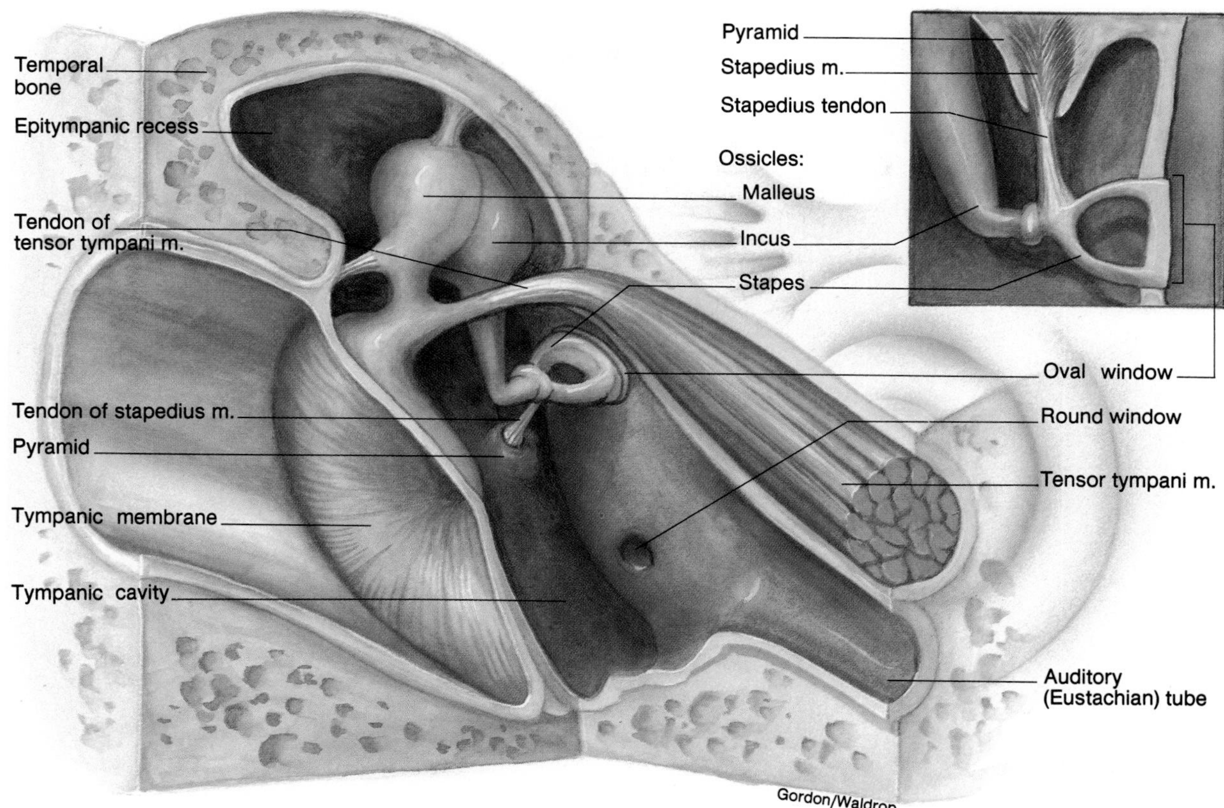


The fact that vibrations of the tympanic membrane are transferred through three bones instead of just one affords protection. If the sound is too intense, the ossicles can buckle. This protection is increased by the action of the *stapedius muscle,* which attaches to the neck of the stapes (fig. 9.21). When sound becomes too loud, this muscle contracts and moves the footplate of the stapes away from the oval window. This action helps to protect against nerve damage within the cochlea. If sounds reach loud amplitudes extremely rapidly—as in gunshots—however, the stapedius muscle may not respond rapidly enough to prevent nerve damage.

Damage to the tympanic membrane or middle ear ossicles produces **conduction deafness.** This can result from a variety of causes, including *otitis media* and *otosclerosis.* In otitis media, inflammation produces excessive fluid accumulation within the middle ear, which can in turn result in the excessive growth of epithelial tissue and damage to the eardrum. This can occur following allergic reactions or respiratory disease. In otosclerosis, bone is resorbed and replaced by "sclerotic bone" that grows over the oval window and immobilizes the footplate of the stapes (fig. 9.22). In conduction deafness these pathological changes hinder the transmission of sound waves from the air to the cochlea of the inner ear.

**Figure 9.22.** The development of otosclerosis in the middle ear immobilizes the stapes, leading to conduction deafness.

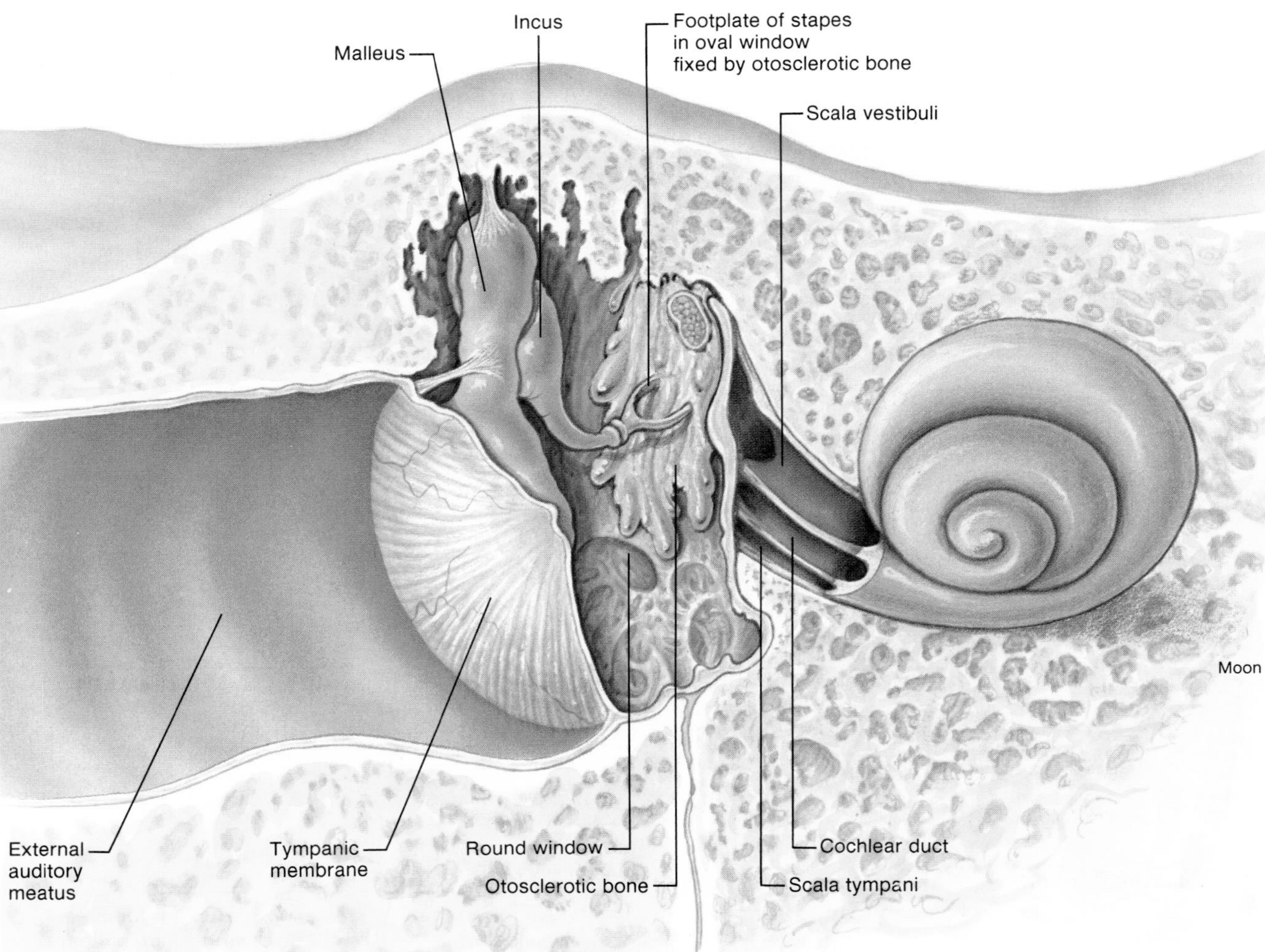


## The Cochlea

Vibrations of the stapes and oval window indirectly displace endolymph within a part of the membranous labyrinth known as the **cochlear duct,** or **scala media.** The latter name indicates that the cochlear duct is the middle part of the snail-shaped cochlea within the inner ear. Like the cochlea as a whole, the cochlear duct coils to form three levels (fig. 9.23), similar to the basal, middle, and apical portions of a snail shell. The part of the cochlea above the cochlear duct is called the *scala vestibuli,* and the part below is called the *scala tympani.* Unlike the central cochlear duct, which contains endolymph, the scala vestibuli and scala tympani are filled with perilymph.

The perilymph of the scala vestibuli and scala tympani is continuous at the apex of the cochlea because the cochlear duct ends blindly, leaving a small space called the *helicotrema* between the end of the cochlear duct and the wall of the cochlea. Vibrations of the oval window produced by movements of the stapes cause pressure waves within the scala vestibuli, which pass around the helicotrema to the scala tympani. Movements of perilymph within the scala tympani, in turn, travel to the base of the cochlea where they cause displacement of a membrane called the *round window* into the middle ear cavity (fig. 9.24). This occurs because fluid, such as perilymph, cannot be compressed; an inward movement of the oval window is thus compensated by an outward movement of the round window.

**Figure 9.23.** A cross section of the cochlea showing its three turns and its three compartments—scala vestibuli, cochlear duct (scala media), and scala tympani.

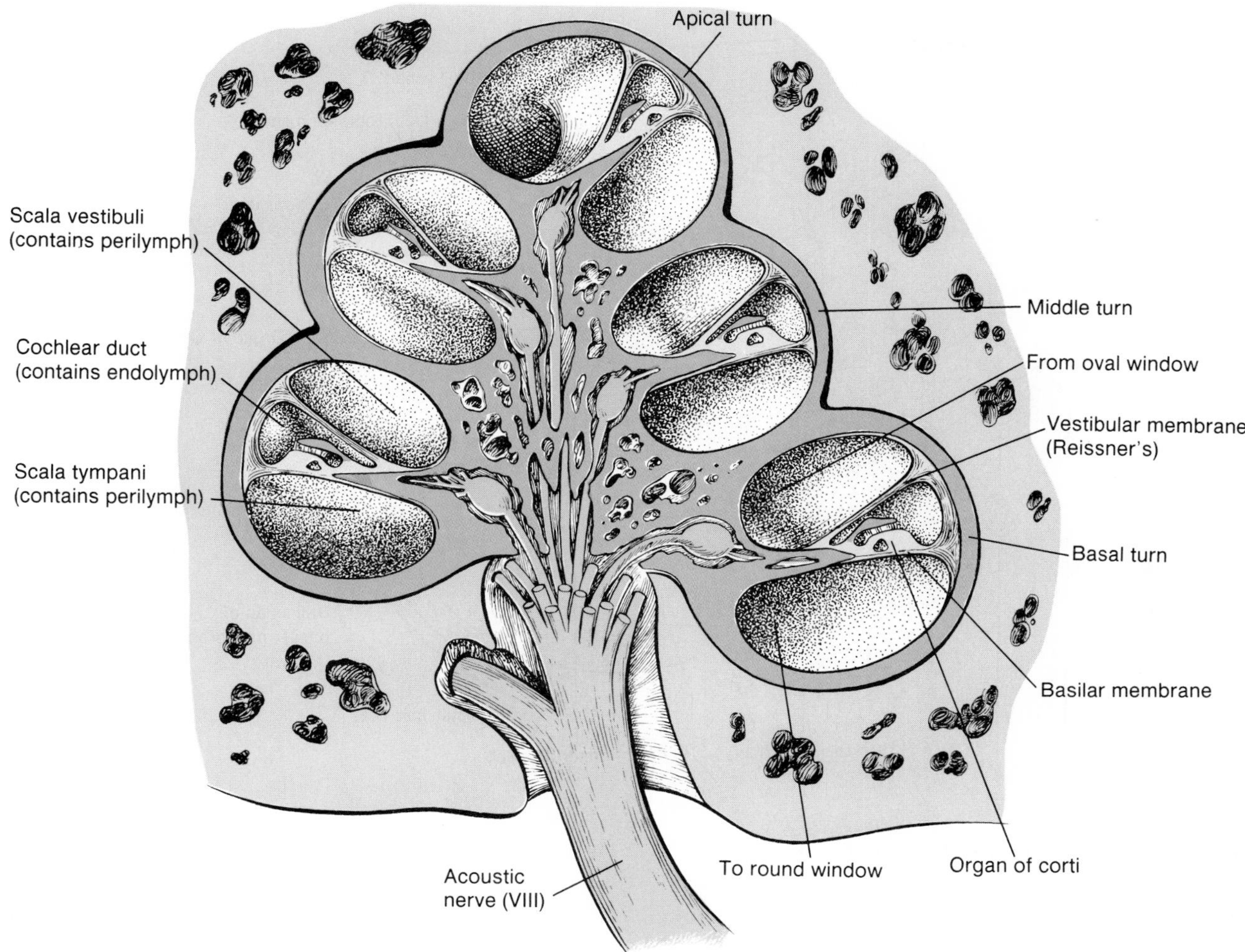


When the sound frequency (pitch) is sufficiently low, there is adequate time for the pressure waves of perilymph within the upper scala vestibuli to travel through the helicotrema to the scala tympani. As the sound frequency increases, however, pressure waves of perilymph within the scala vestibuli do not have time to travel all the way to the apex of the cochlea. Instead, they are transmitted through the *vestibular* (or *Reissner's*) *membrane,* which separates the scala vestibuli from the cochlear duct, and through the *basilar membrane,* which separates the cochlear duct from the scala tympani, to the perilymph of the scala tympani (fig. 9.25). The distance that these pressure waves travel, therefore, decreases as the sound frequency increases.

## The Organ of Corti

Movements of perilymph from the scala vestibuli to the scala tympani thus produce displacement of Reissner's membrane and the basilar membrane. Although the movement of Reissner's membrane does not directly contribute to hearing, displacement of the basilar membrane is central to pitch discrimination. The basilar membrane is fixed on the inner side of the cochlear wall to a bony ridge and is supported at its free end by a ligament.

Vibrations of the stapes against the oval window set up moving waves of perilymph in the scala vestibuli, which cause displacement of the basilar membrane into the scala

**Figure 9.24.** Pressure waves of perilymph in the scala vestibuli cause deflections of the vestibular and basilar membranes. In this way the pressure is transmitted to the scala tympani.

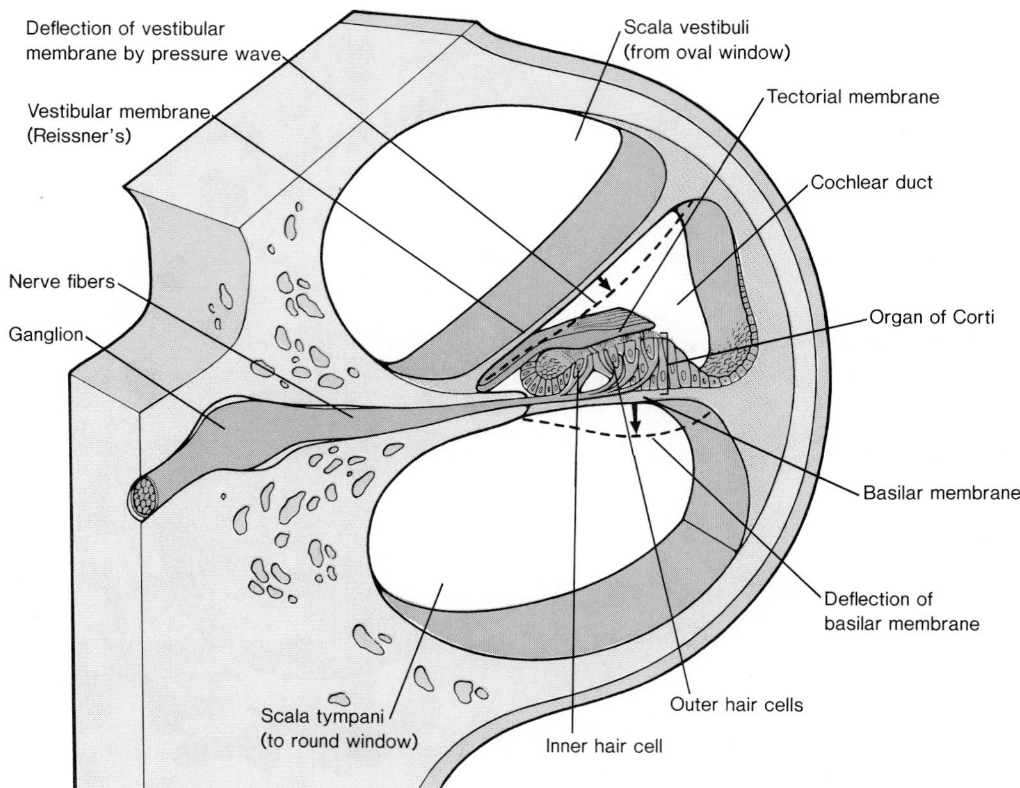


tympani. This produces vibrations of the basilar membrane; each region of the basilar membrane vibrates with maximum amplitude to a different sound frequency. Sounds of higher frequency (pitch) cause maximum vibrations of the basilar membrane closer to the stapes, as illustrated in figure 9.25.

The sensory *hair cells* are located on the basilar membrane, with their "hairs" (actually cilia) projecting into the endolymph of the cochlear duct. These hair cells are arranged to form one row of inner cells, which extend the length of the basilar membrane, and multiple rows of outer hair cells: three rows in the basal turn, four in the middle turn, and five in the apical turn of the cochlea (fig. 9.26).

The cilia of the outer hair cells are embedded within a gelatinous **tectorial membrane,** which overhangs the hair cells within the cochlear duct (fig. 9.27). The association of the basilar membrane hair cells with sensory fibers, and tectorial membrane form a functional unit called the **organ of Corti.** When the cochlear duct is displaced by pressure waves of perilymph, a shearing force is created between the basilar membrane and the tectorial membrane. This causes the cilia to bend, which in turn produces a generator potential in the sensory nerve endings that synapse with the hair cells.

The greater the displacement of the basilar membrane and the bending of the cilia, the greater is the generator potential that is produced, and the greater is the frequency of action potentials in the fibers of the cochlear (eighth cranial) nerve that synapse with the hair cells. Since a sound of a particular frequency causes a specific region of the basilar membrane to be maximally displaced, those neurons that originate in this region will be stimulated more than neurons which originate in other regions of the basilar membrane. This mechanism provides a neural code for *pitch discrimination.* The ability to discriminate different pitches of sounds with closely similar frequencies is aided by *lateral inhibition,* in which neural activity originating from the region of the basilar membrane with peak displacement is enhanced by inhibition of the neural activity from surrounding regions.

**Figure 9.25.** Sounds of low frequency cause pressure waves of perilymph to pass through the helicotrema. Sounds of higher frequency cause pressure waves to "short cut" through the cochlear duct. This causes displacement of the basilar membrane, which is central to the transduction of sound waves into nerve impulses. (cps = cycles per second.)

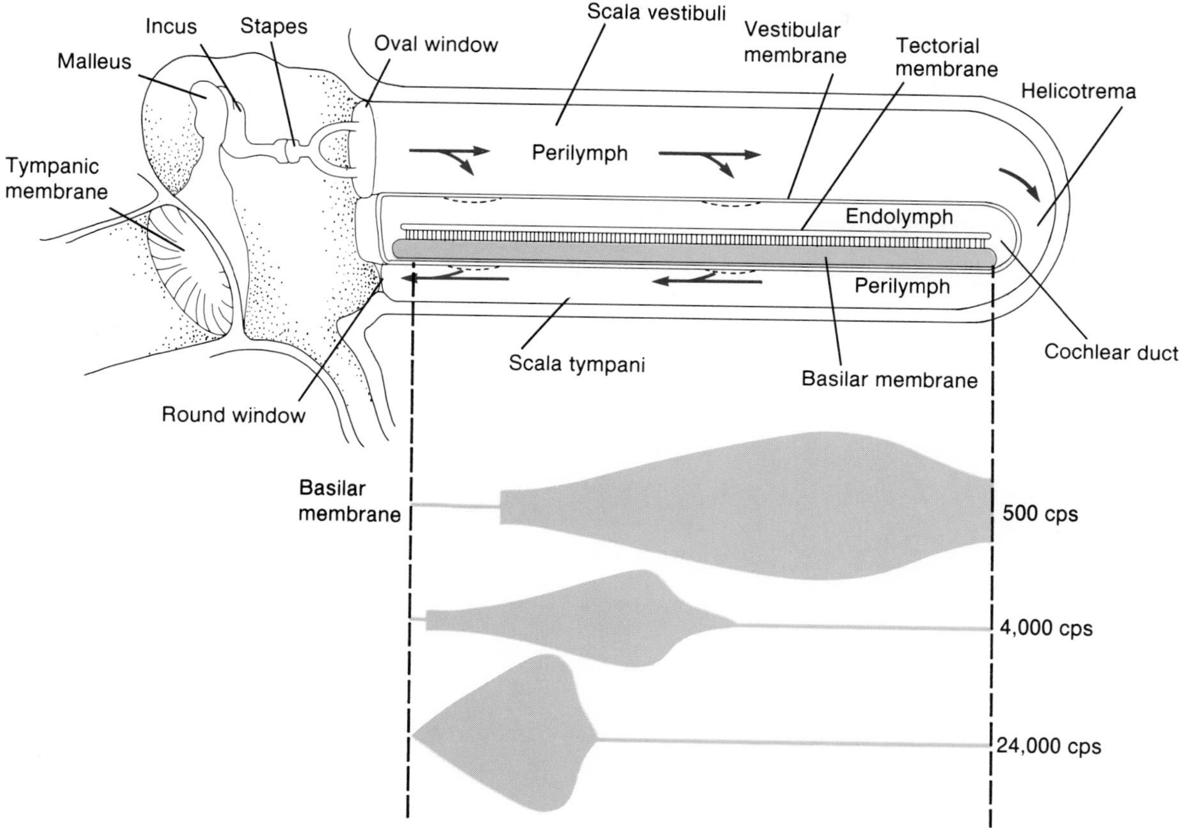


***Neural Pathways for Hearing.*** Sensory neurons in the eighth cranial nerve synapse with neurons in the medulla (fig. 9.28), which project to the inferior colliculus of the midbrain. Neurons in this area in turn project to the thalamus, which sends axons to the auditory cortex of the temporal lobe. By means of this pathway, neurons in different regions of the basilar membrane stimulate neurons in corresponding areas of the auditory cortex. Each area of this cortex thus represents a different part of the basilar membrane and a different pitch (fig. 9.29).

## Hearing Impairments

There are two major causes of hearing loss: (1) **conductive deafness,** in which the transmission of sound waves from air through the middle ear to the oval window is impaired; and (2) **nerve** or **sensory deafness,** in which the transmission of nerve impulses anywhere from the cochlea to the auditory cortex is impaired. Conductive deafness can be caused by middle ear damage from otitis media or otosclerosis, as previously discussed. Nerve deafness may result from a wide variety of pathological processes.

**Figure 9.26.** A scanning electron micrograph of the hair cells of the organ of Corti. From Kessel, R. G., and Kardon, R. H.: *Tissues and Organs: A Text-Atlas of Scanning Electron Microscopy.* © 1979 W. H. Freeman and Company.

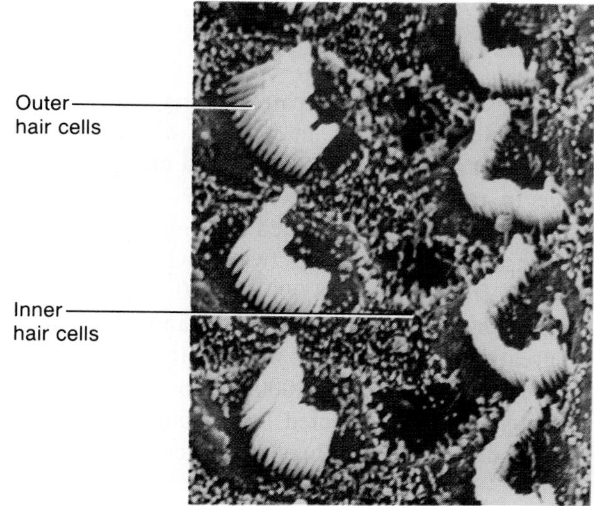

**Figure 9.27.** The organ of Corti (*a*) is located within the cochlear duct of the cochlea (*b*).

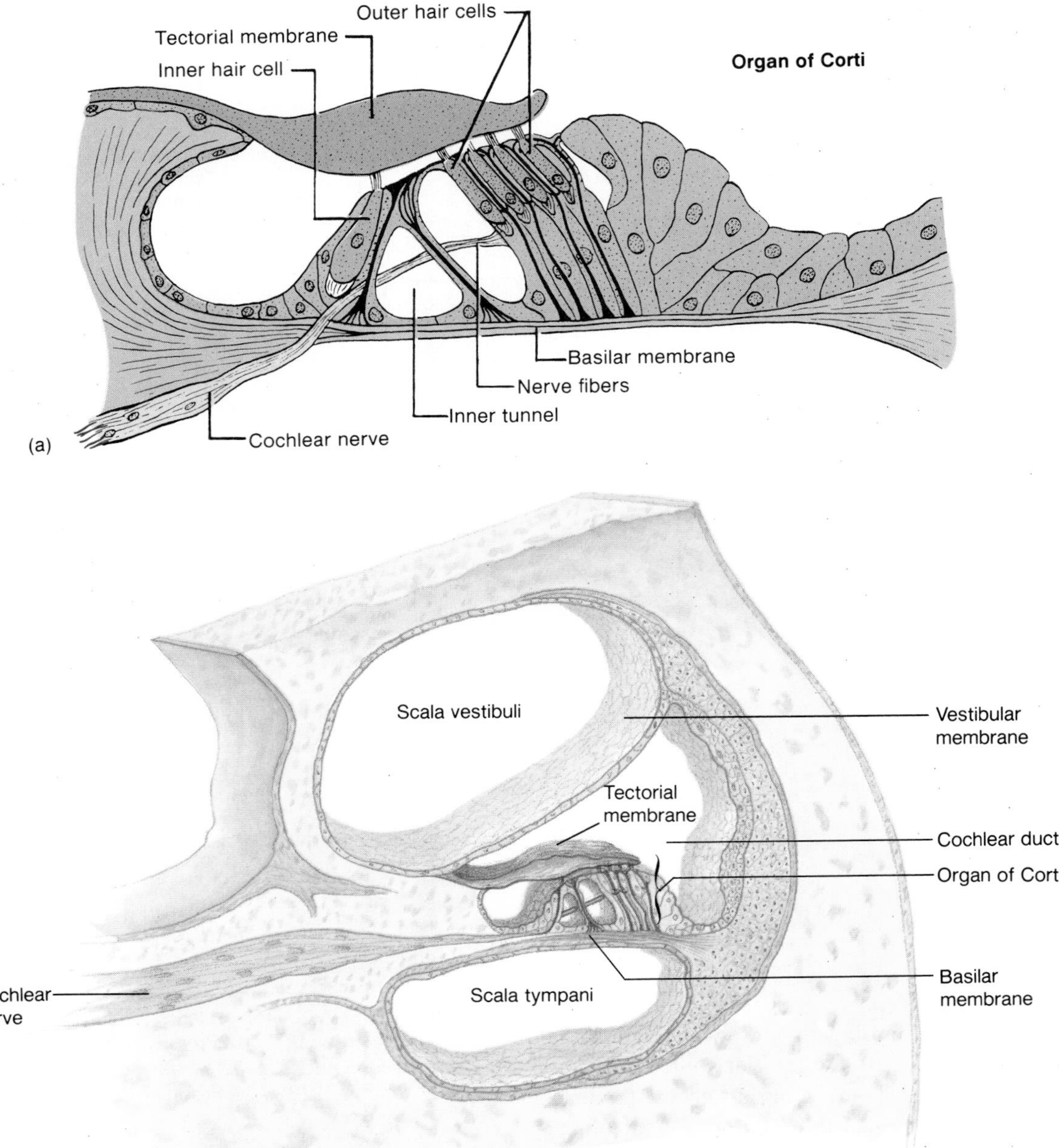


Conduction deafness impairs hearing at all sound frequencies. Nerve deafness, in contrast, often impairs the ability to hear some pitches more than others. This may be due to pathological procesesses or to changes that occur during aging. Age-related hearing deficits—called *presbycusis*—begin after age twenty when the ability to hear high frequencies (18,000–20,000 Hz) diminishes. Although the progression is variable, and occurs in men to a greater degree than in women, these deficits may gradually extend into the 4,000–8,000 Hz range. These impairments can be detected by a technique called *audiometry,* in which the threshold intensity of different pitches is determined. The ability to hear speech is particularly affected by hearing loss in the higher frequencies. People with these types of impairments can be helped by *hearing aids,* which amplify sounds and conduct the sound waves through bone to the inner ear.

1. *Use a flowchart to describe how sound waves in air within the external auditory meatus are transduced into movements of the basilar membrane.*
2. *Explain how movements of the basilar membrane of the organ of Corti can code for different sound frequencies (pitches).*

**Figure 9.28.** Neural pathways from the spiral ganglia of the cochlea to the auditory cortex.

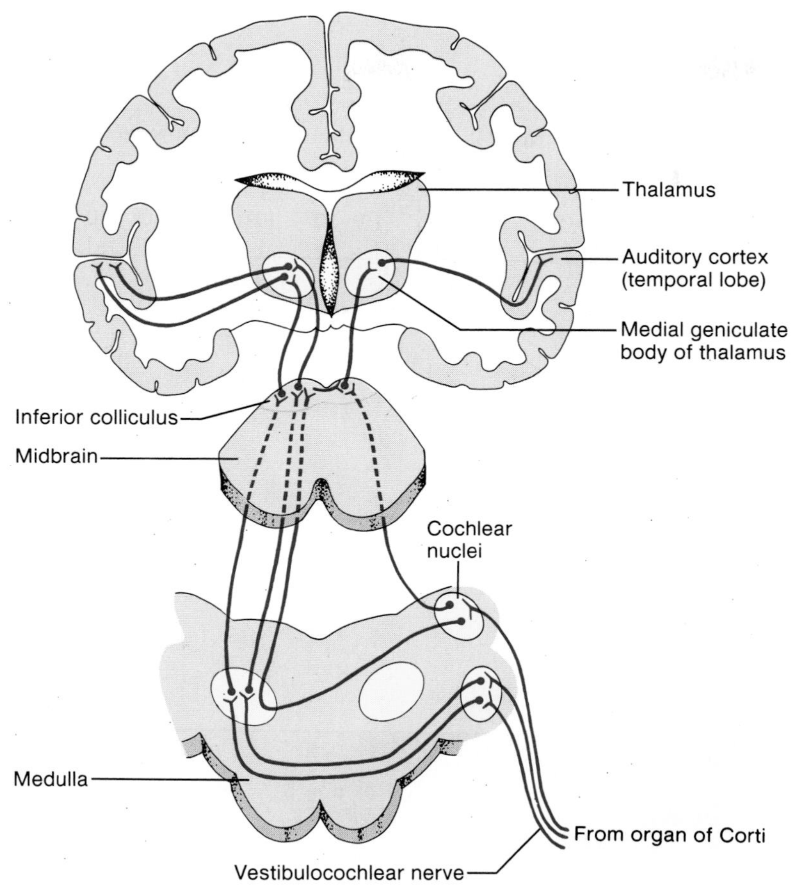


**Figure 9.29.** Sounds of different frequencies (pitches) excite different sensory neurons in the cochlea; these in turn send their input to different regions of the auditory cortex.

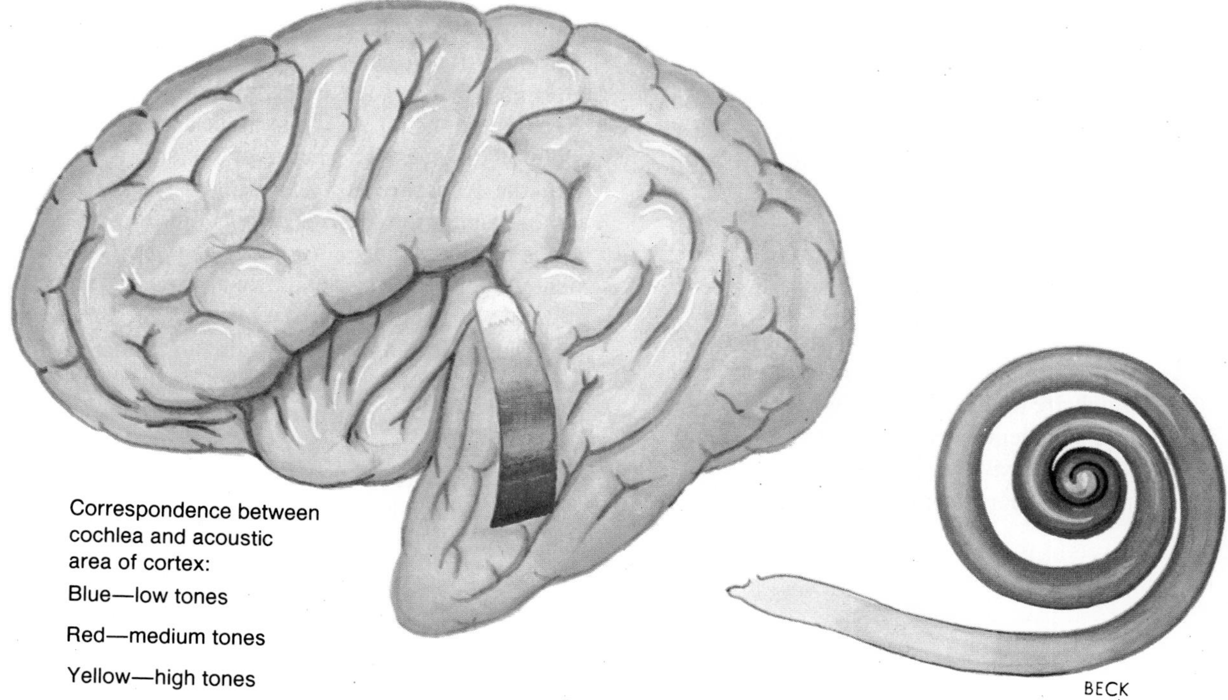

**Figure 9.30.** The electromagnetic spectrum (*top*) is shown in Angstrom units (1Å = $10^{-10}$ meter). The visible spectrum comprises only a small range of this spectrum (*bottom*), shown in nanometer units (1 nm = $10^{-9}$ meter).

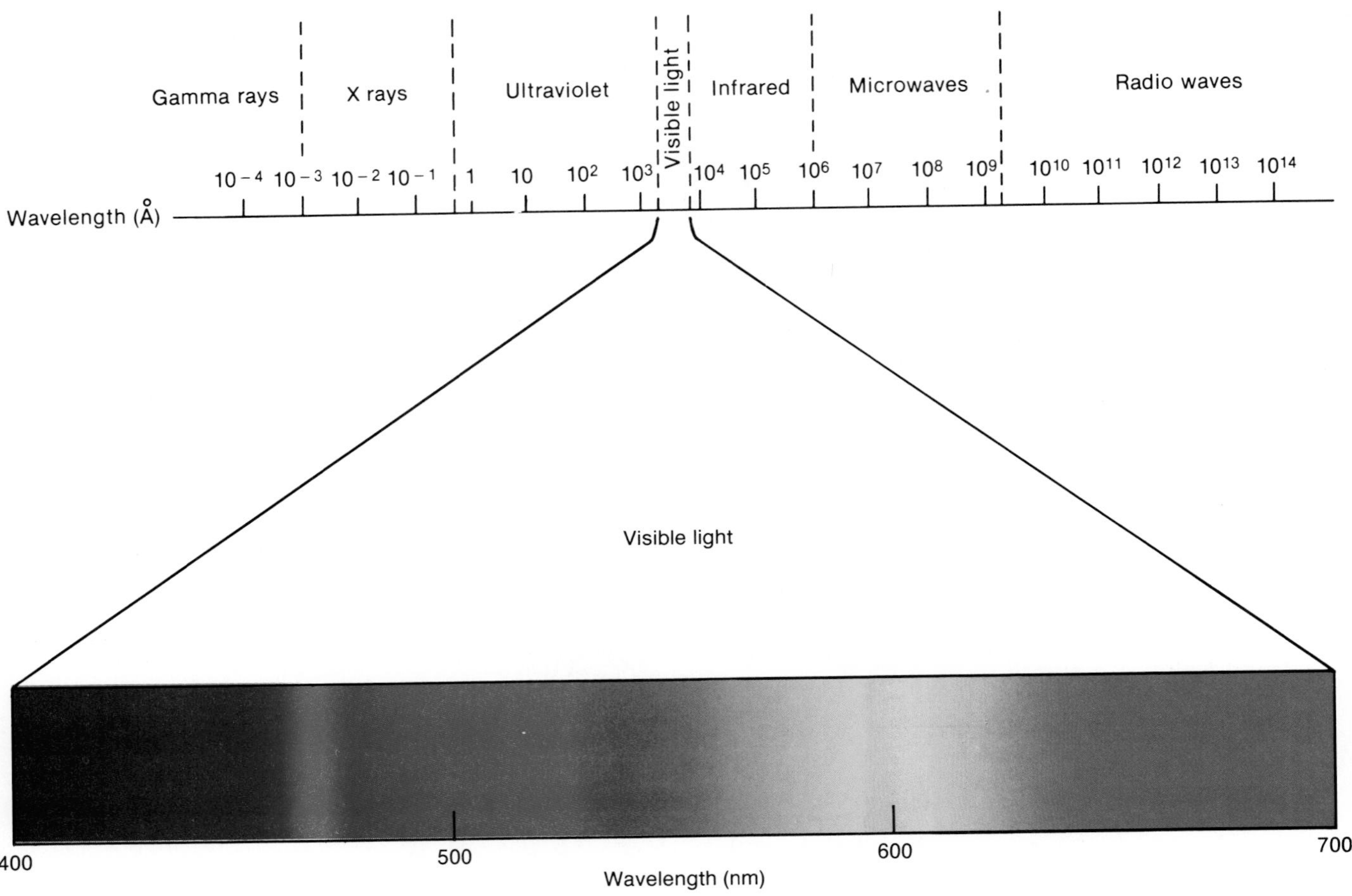


## The Eyes and Vision

Light from an observed object is focused by the cornea and lens onto the photoreceptive layer called the retina at the back of the eye. The focus is maintained on the retina at different distances between the eyes and the object by variations in the thickness and degree of curvature of the lens.

The eyes transduce energy in the *electromagnetic spectrum* (fig. 9.30) into nerve impulses. Only a limited part of this spectrum can excite the photoreceptors—electromagnetic energy with wavelengths between 400 and 700 nanometers (nm) comprise *visible light.* Light of longer wavelengths, which are in the infrared regions of the spectrum, do not have sufficient energy to excite the receptors but are felt as heat. Ultraviolet light, which has shorter wavelengths and more energy than visible light, is filtered out by the yellow color of the eye's lens. Honeybees—and people who have had their lens removed—can see light in the ultraviolet range.

The parts of the eye are summarized in table 9.4. The outermost layer of the eye is a tough coat of connective tissue called the *sclera.* This can be seen externally as the white of the eyes. The tissue of the sclera is continuous with the transparent *cornea.* Light passes through the cornea to enter the *anterior chamber* of the eye. Light then passes through an opening, called the *pupil,* within a pigmented (colored) muscle known as the *iris.* After passing through the pupil, light enters the *lens* (fig. 9.31).

The iris is like the diaphragm of a camera, which can increase or decrease the diameter of its aperture (the pupil) to admit more or less light. Constriction of the pupils is produced by contraction of circular muscles within the iris; dilation is produced by contraction of radial muscles. Constriction of the pupils results from parasympathetic stimulation, whereas dilation results from sympathetic

**Table 9.4** Location and functions of the structures of the eyeball

| Tunic and Structure | Location | Composition | Function |
|---|---|---|---|
| Fibrous tunic | Outer layer of eyeball | Avascular connective tissue | Provides shape of eyeball |
| Sclera | Posterior, outer layer; "white of the eye" | Tightly bound elastic and collagen fibers | Supports and protects eyeball |
| Cornea | Anterior surface of eyeball | Tightly packed dense connective tissue—transparent and convex | Transmits and refracts light |
| Vascular tunic | Middle layer of eyeball | Highly vascular pigmented tissue | Supplies blood; prevents reflection |
| Choroid | Middle layer in posterior portion of eyeball | Vascular layer | Supplies blood to eyeball |
| Ciliary body | Anterior portion of vascular tunic | Smooth muscle fibers and glandular epithelium | Supports the lens through suspensory ligaments and determines its shape; secretes aqueous humor |
| Iris | Anterior portion of vascular tunic continuous with ciliary body | Pigment cells and smooth muscle fibers | Regulates the diameter of the pupil and hence the amount of light entering the vitreous chamber |
| Retina | Inner layer in posterior portion of eyeball | Photoreceptor neurons (rods and cones), bipolar neurons, and ganglion neurons | Photoreception; transmits impulses |
| Lens | Between posterior and vitreous chambers; supported by suspensory ligaments of ciliary body | Tightly arranged protein fibers; transparent | Refracts light and focuses onto fovea centralis |

**Figure 9.31.** The internal anatomy of the eyeball.

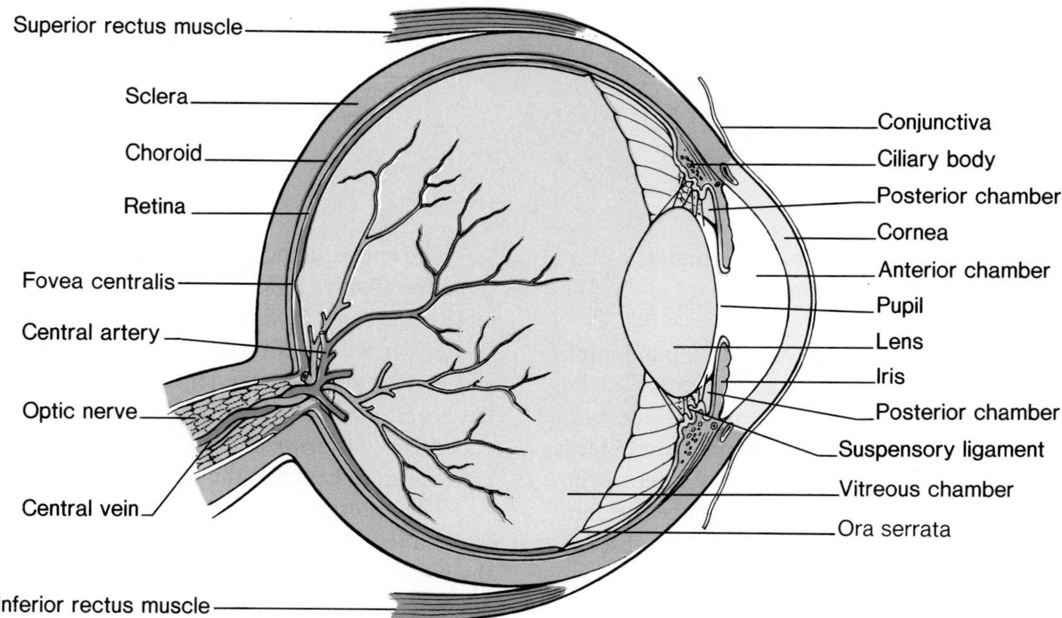

**Figure 9.32.** Dilation and constriction of the pupil. In dim light, the radially arranged smooth muscle fibers are stimulated to contract by sympathetic stimulation, dilating the pupil. In bright light, the circularly arranged smooth muscle fibers are stimulated to contract by parasympathetic stimulation, constricting the pupil.

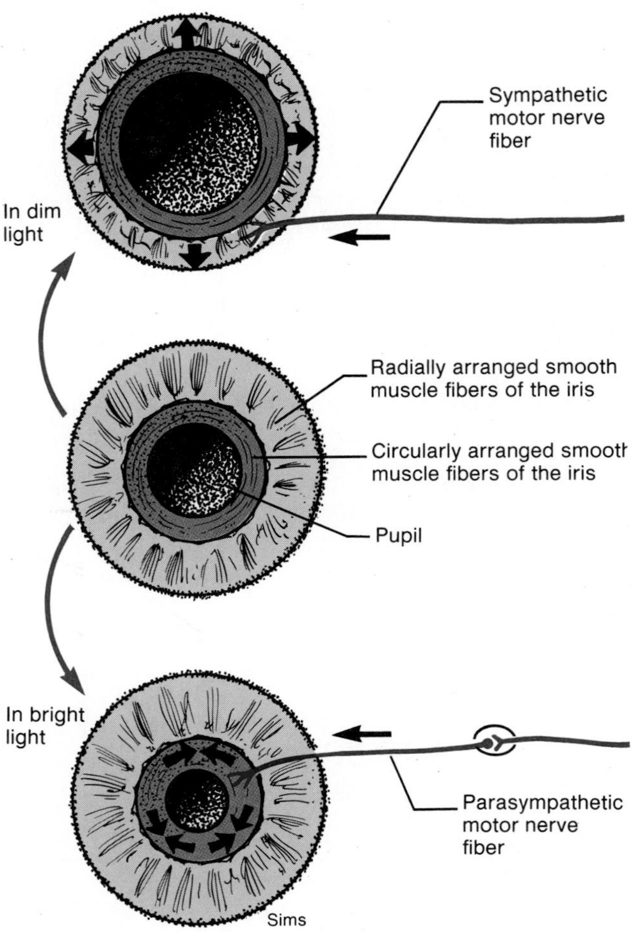


stimulation (fig. 9.32). Variations in the diameter of the pupil are similar in effect to variations in the "f-stop" of a camera.

The posterior part of the iris contains a pigmented epithelium that gives the eye its color. The color of the eye is determined by the amount of pigment—blue eyes have the least pigment, brown eyes have more, and black eyes have the greatest amount of pigment. Albinos, who have a congenital defect in the ability to produce melanin pigment, have eyes that appear pink because the absence of pigment allows blood vessels to be seen.

The lens is suspended from a muscular process called the **ciliary body,** which is connected to the sclera and encircles the lens. *Zonular fibers* (zon = girdle) suspend the lens from the ciliary body, forming a *suspensory ligament* that supports the lens. The space bounded by the iris anteriorly, and the ciliary body and lens posteriorly, is called the *posterior chamber* (as distinct from the anterior chamber between the iris and the cornea—fig. 9.33).

The anterior and posterior chambers are filled with a fluid called the **aqueous humor.** This fluid is produced by the ciliary body, secreted into the posterior chamber, and enters the anterior chamber through the pupil. Aqueous humor is drained from the anterior chamber into the *canal of Schlemm,* which returns this fluid to the venous blood (fig. 9.34). Inadequate drainage of aqueous humor can lead to excessive accumulation of fluid, which in turn results in increased intraocular pressure (a condition called *glaucoma*). This may cause serious damage to the retina.

**Figure 9.33.** The detailed structure of the anterior portion of the eyeball.

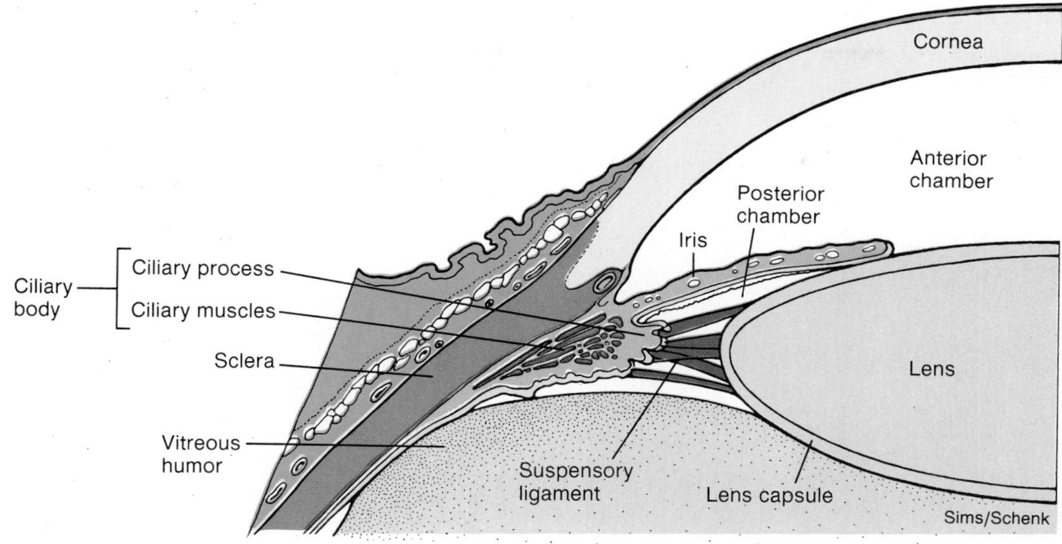


**Figure 9.34.** Aqueous humor maintains the intraocular pressure within the anterior and posterior chambers. It is secreted into the posterior chamber, flows through the pupil into the anterior chamber, and drains from the eyeball through the venous sinus (canal of Schlemm).

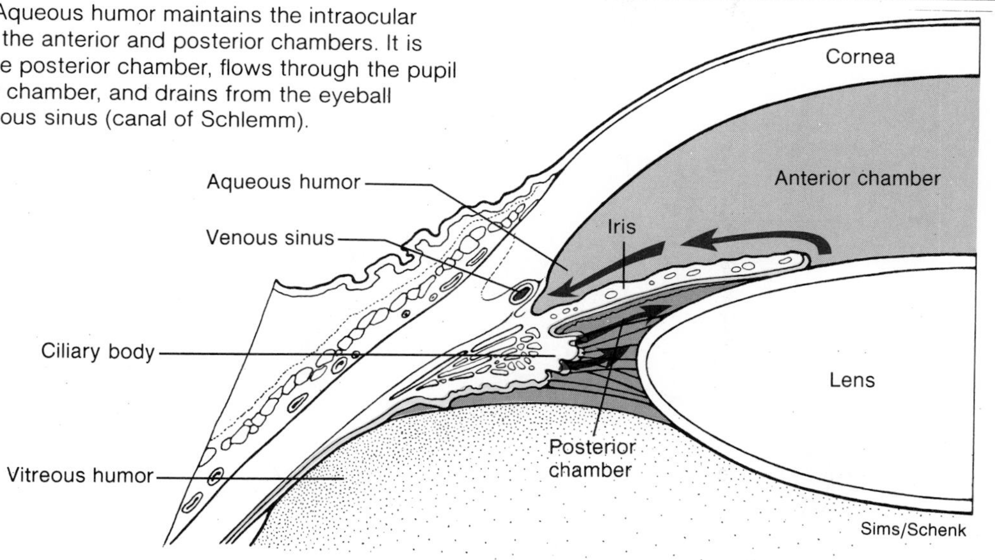

The portion of the eye located behind the lens is filled with a thick, viscous substance known as the **vitreous body.** Light from the lens that passes through the vitreous body enters the neural layer, which contains photoreceptors, at the back of the eye. This neural layer is called the **retina.** Light that passes through the retina is absorbed by a darkly pigmented *choroid* layer underneath. While passing through the retina, some of this light stimulates photoreceptors, which in turn activate other neurons. Neurons in the retina contribute fibers that are gathered together at a region called the *optic disc* (fig. 9.35) to exit the retina as the optic nerve. The optic disc is also the site of entry and exit of blood vessels.

## Refraction

Light that passes from a medium of one density into a medium of a different density is *refracted,* or bent. The degree of refraction depends on the comparative densities of the two media, as indicated by their *refractive index.* The refractive index of air is set at 1.00; the refractive index of the cornea, in comparison, is 1.38, and the refractive index of the lens is 1.45. Since the greatest difference in refractive index occurs at the air-cornea interface, the light is refracted most at the cornea.

**Figure 9.35.** A view of the retina as seen with an ophthalmoscope. Optic nerve fibers leave the eyeball at the optic disc to form the optic nerve. Note the blood vessels that can be seen entering the eyeball at the optic disc.

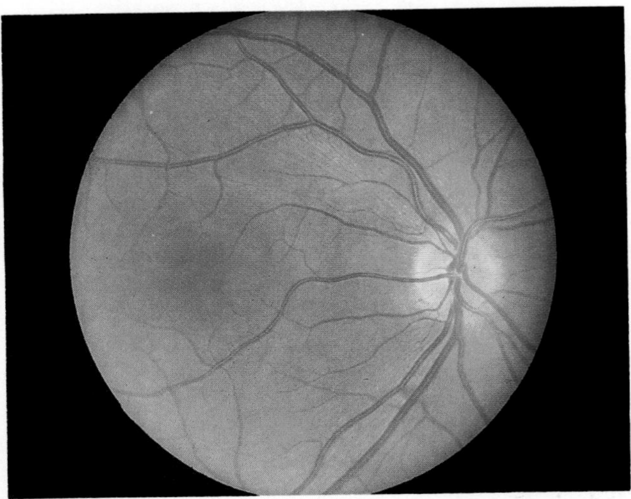

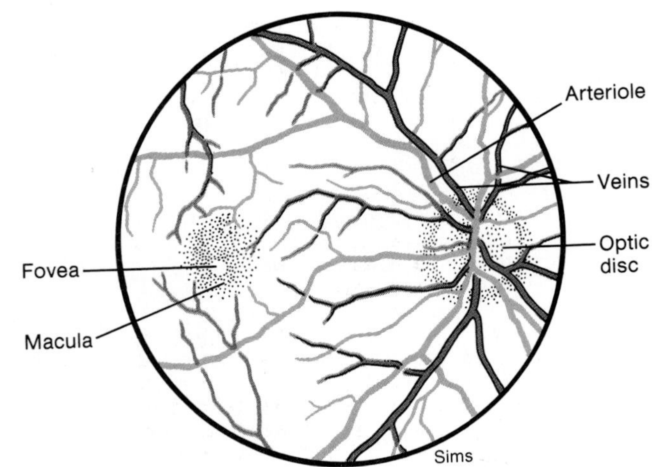


**Figure 9.36.** The refraction of light waves within the eyeball causes the image of an object to be inverted on the retina.

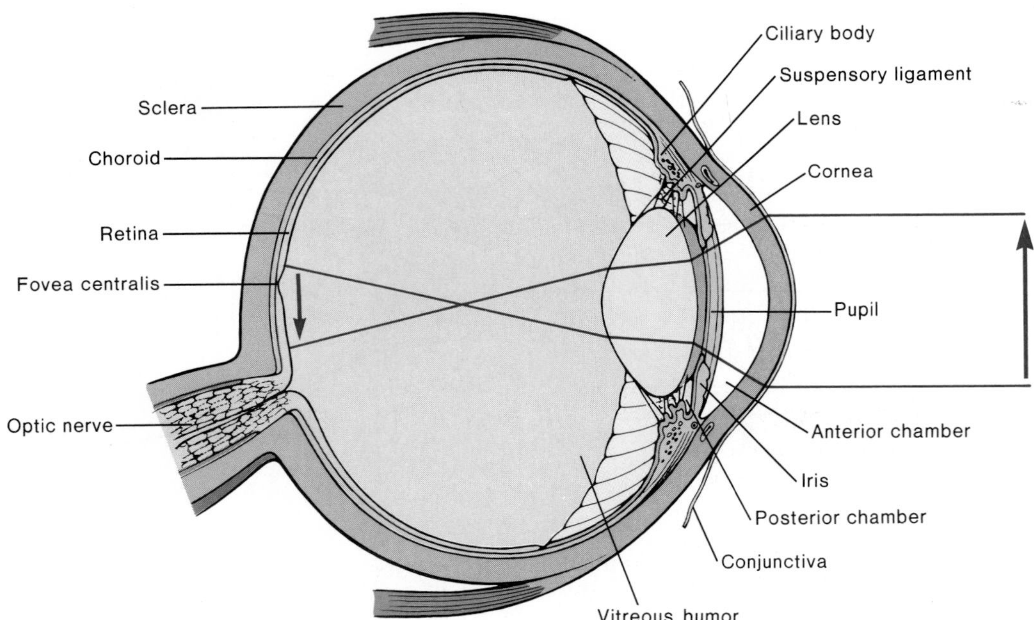


The degree of refraction also depends on the curvature of the interface between two media. The curvature of the cornea is constant, however, while the curvature of the lens can be varied. The refractive properties of the lens can thus provide fine control for focusing light on the retina. As a result of light refraction, the image formed on the retina is upside down and right to left (fig. 9.36).

The *visual field*—which is the part of the external world projected onto the retina—is thus reversed in each eye. The cornea and lens focus the right part of the visual field on the left half of the retina of each eye, while the left half of the visual field is focused on the right half of each retina (fig. 9.37). The medial (or nasal) half-retina of the left eye therefore receives the same image as the lateral (or temporal) half-retina of the right eye. The nasal half-retina of the right eye receives the same image as the temporal half-retina of the left eye.

**Figure 9.37.** Refraction of light in the cornea and lens produces a right-to-left image on the retina. The left side of the visual field is projected to the right half of each retina, while the right side of each visual field is projected to the left half of each retina.

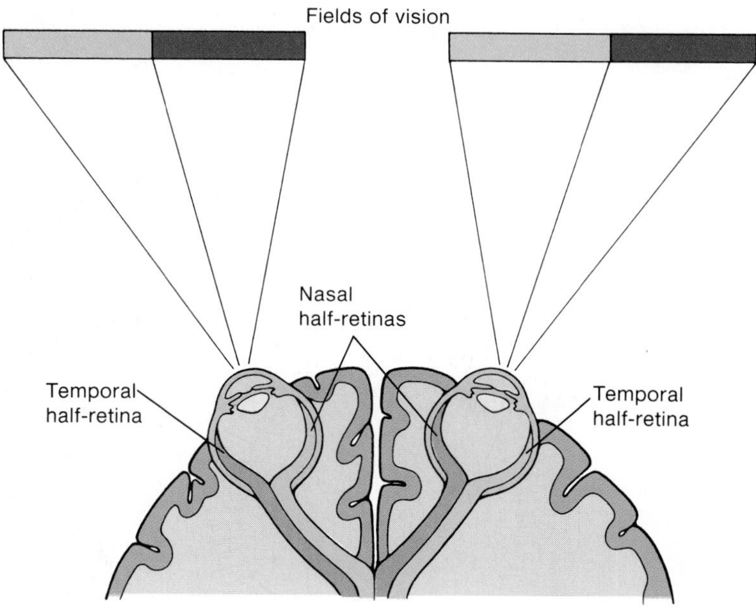


## Accommodation

When a normal eye views an object, parallel rays of light are refracted to a point, or *focus,* on the retina (see fig. 9.40). If the degree of refraction were to remain constant, movement of the object closer to or farther from the eye would cause corresponding movement of the focal point, so that the focus would either be behind or in front of the retina.

The ability of the eyes to keep the image focused on the retina as the distance between the eyes and object is changed is called **accommodation.** Accommodation results from contraction of the ciliary muscle, which is like a sphincter muscle that can vary its aperture (fig. 9.38). When the ciliary muscle is relaxed, its aperture is wide. Relaxation of the ciliary muscle thus places tension on the zonular fibers and pulls the lens taut. These are the conditions that prevail when viewing an object which is twenty feet or more from a normal eye; the image is focused on the retina and the lens is in its most flat, least convex form. As the object moves closer to the eyes the muscles of the ciliary body contract. This muscular contraction narrows the aperture of the ciliary body and thus reduces the tension on the zonular fibers which suspend the lens. When the tension is reduced, the lens becomes more round and convex as a result of its inherent elasticity (fig. 9.39).

The ability of a person's eyes to accommodate can be measured by the *near-point-of-vision* test, which is the minimum distance from the eyes that an object can be maintained in focus. This distance increases with age, and indeed accommodation in almost everyone over the age of forty-five is significantly impaired. Loss of accommodating ability with age is known as **presbyopia** (*presby* = old). This loss appears to have a number of causes, including thickening of the lens and a forward movement of the attachments of the zonular fibers to the lens. As a result of these changes, the zonular fibers and lens are pulled taut even when the ciliary muscle contracts. The lens is thus not able to thicken and increase its refraction when, for example, a printed page is brought close to the eyes.

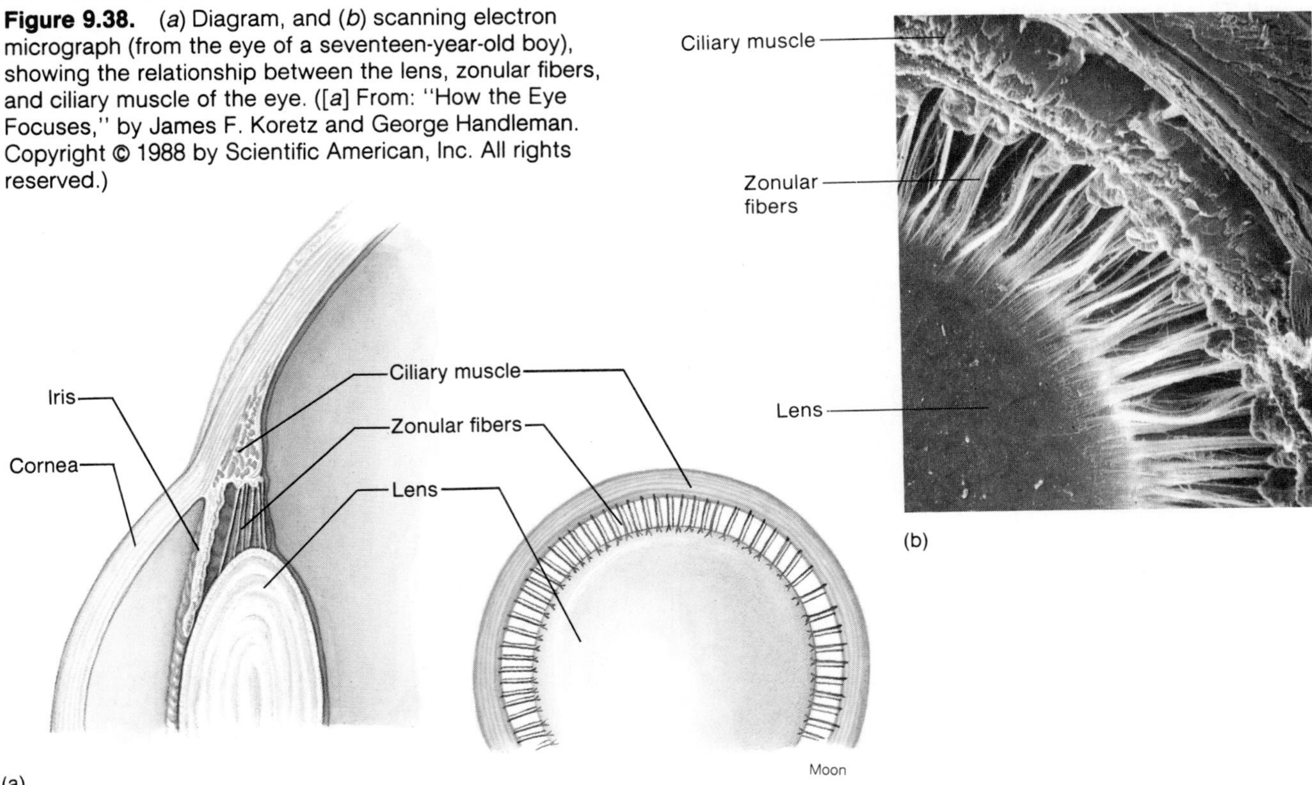


**Figure 9.38.** (*a*) Diagram, and (*b*) scanning electron micrograph (from the eye of a seventeen-year-old boy), showing the relationship between the lens, zonular fibers, and ciliary muscle of the eye. ([*a*] From: "How the Eye Focuses," by James F. Koretz and George Handleman. Copyright © 1988 by Scientific American, Inc. All rights reserved.)

## Visual Acuity

Visual acuity refers to the sharpness of vision. The sharpness of an image depends on the *resolving power* of the visual system—that is, on the ability of the visual system to distinguish (resolve) two closely spaced dots. The better the resolving power of the system is, the closer together these dots can be and still be seen as separate; when the resolving power of the system is exceeded, the dots are blurred together as a single image.

***Myopia and Hyperopia.*** When a person with normal visual acuity stands twenty feet from a *Snellen eye chart* (so that accommodation is not a factor influencing acuity), the line of letters marked "20/20" can be read. If a person has **myopia** (nearsightedness), this line will appear blurred because the focus of this image will be in front of the retina. This is usually caused by the fact that the eyeball is too long. Myopia is corrected by glasses with concave lenses that cause the light rays to diverge, so that the point of focus is farther from the lens and is thus pushed back to the retina (fig. 9.40).

If the eyeballs are too short, the line marked "20/20" will appear blurred because the focal length of the lens is longer than the distance to the retina. The image would thus be brought to a focus behind the retina, and the object must be placed farther from the eyes to be seen clearly. This condition is called **hyperopia** (farsightedness). Hyperopia is corrected by glasses with convex lenses that increase the convergence of light so that the focus is brought closer to the lens and falls on the retina.

***Astigmatism.*** The curvature of the cornea and lens is not perfectly symmetrical, so that light passing through some parts of these structures may be refracted to a different degree than light passing through other parts. When the asymmetry of the cornea and/or lens is significant, the person is said to have **astigmatism.** If a person with astigmatism views a circle of lines radiating from the center, like the spokes of a wheel, the image of these lines will not appear clear in all 360 degrees; the parts of the circle that appear blurred can thus be used to map the astigmatism. This condition is corrected by cylindrical lenses that compensate for the asymmetry in the cornea or lens of the eye.

---

1. *Using a line diagram, explain why an inverse image is produced on the retina. Also explain how the image in one eye corresponds to the image in the other eye.*
2. *Using a line diagram, show how parallel rays of light are brought to a focus on the retina. Explain how this focus is maintained as the distance from the object to the eye is increased or decreased (that is, explain accommodation).*
3. *Explain how a blurred image is produced when a person has presbyopia, myopia, or hyperopia.*

---

**Figure 9.39.** Changes in the shape of the lens during accommodation. (*a*) The lens is flattened for distant vision when the ciliary muscle fibers are relaxed and the suspensory ligaments are taut. (*b*) The lens is more spherical for closeup vision when the ciliary muscle fibers are contracted and the suspensory ligaments are relaxed.

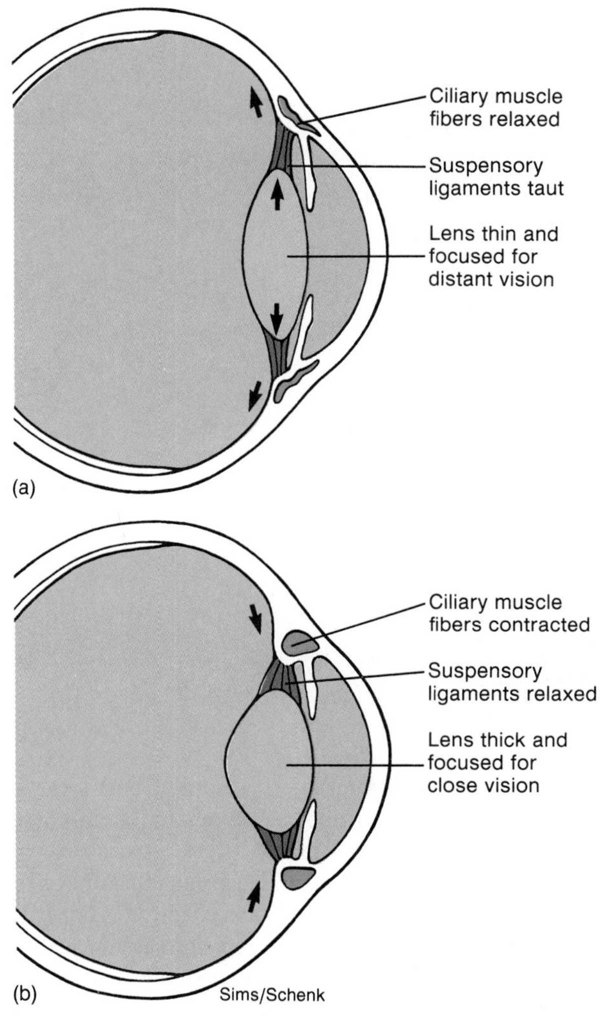


**Figure 9.40.** In a normal eye (*a*), parallel rays of light are brought to a focus on the retina by refraction in the cornea and lens. If the eye is too long, as in myopia (*b*), the focus is in front of the retina. This can be corrected by a concave lens. If the eye is too short (*c*), as in hyperopia, the focus is behind the retina. This is corrected by a convex lens. In astigmatism (*d*), light refraction is uneven due to an abnormal shape of the cornea or lens.

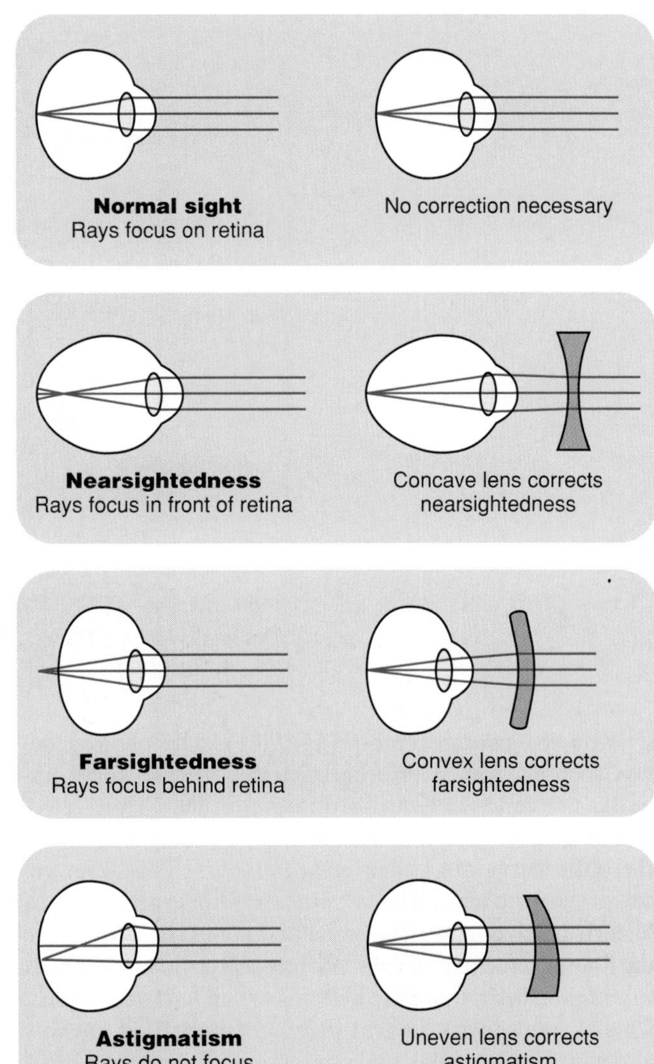


## The Retina

There are two types of photoreceptor neurons: rods and cones. Both receptor cell types contain pigment molecules which undergo dissociation in response to light, and it is this photochemical reaction that eventually results in the production of action potentials in sensory fibers of the optic nerve. Rods provide black-and-white vision under conditions of low light intensites whereas cones provide sharp color vision when light intensities are higher. The neural activity resulting from stimulation of optic nerve fibers involves a number of brain regions, and many levels of synaptic interactions are required to produce the image perceived by the brain.

The retina consists of a pigment epithelium, photoreceptor neurons called *rods* and *cones,* and layers of other neurons. The neural layers of the retina are actually a forward extension of the brain. In this sense the optic nerve can be considered a tract, and indeed the myelin sheaths of its fibers are derived from oligodendrocytes (like other CNS nerve fibers) rather than from Schwann cells.

Since the retina is an extension of the brain, the neural layers face outwards, toward the incoming light. Light, therefore, must pass through several neural layers before

**Figure 9.41.** The layers of the retina. The retina is inverted, so that light must pass through various layers of nerve cells before reaching the photoreceptors (rods and cones).

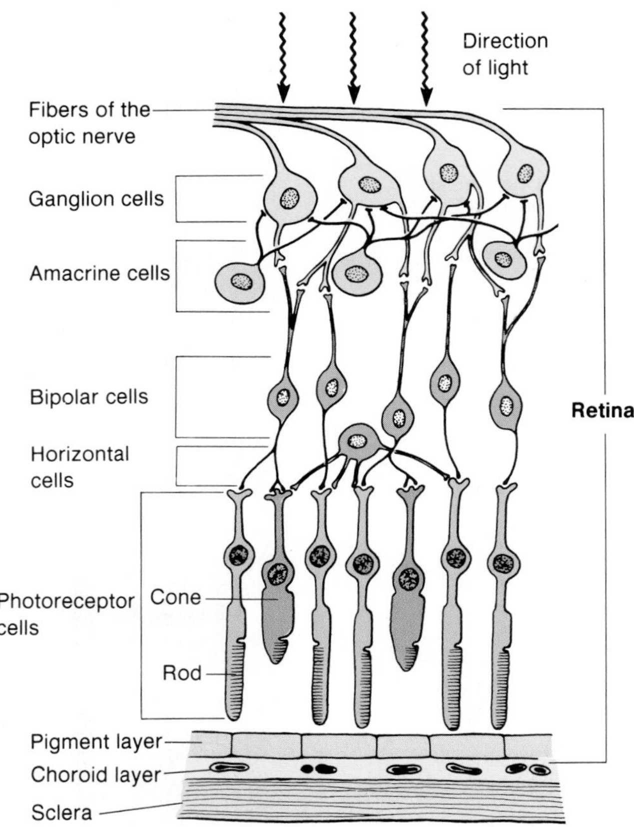


striking the photoreceptors (fig. 9.41). The photoreceptors then synapse with other neurons, so that nerve impulses are conducted outward in the retina.

The outer layers of neurons that contribute axons to the optic nerve are called *ganglion cells.* This layer receives synaptic input from *bipolar cells* underneath, which in turn receive input from rods and cones. In addition to the flow of information from photoreceptors to bipolar cells to ganglion cells, there are neurons called *horizontal cells,* which synapse with several photoreceptors (and possibly also with bipolar cells), and neurons called *amacrine cells,* which synapse with several ganglion cells.

## Effect of Light on the Rods

The photoreceptors—rods and cones (fig. 9.42)—are activated when light produces a chemical change in molecules of pigment contained within the membranous lamellae of the outer segments of the receptor cells. Rods contain a purple pigment known as **rhodopsin.** The pigment appears purple (a combination of red and blue), because it transmits light in the red and blue regions of the spectrum, while absorbing light energy in the green region. The wavelength of light that is absorbed best—the *absorption maximum*—is about 500 nm (a green-colored light).

Cars and other objects that are green in color are seen more easily at night (when rods are used for vision) than are red objects. This is because red light is not well absorbed by rhodopsin, and only absorbed light can produce the photochemical reaction that results in vision. In response to absorbed light, rhodopsin dissociates into its two components: a pigment called **retinene,** (or retinaldehyde) derived from vitamin A, and a protein called **opsin.** This reaction is known as the *bleaching reaction.*

Retinene can exist in two possible configurations (shapes)—one known as the all-*trans* form and one called the 11-*cis* form (fig. 9.43). The all-*trans* form is the most stable, but only the 11-*cis* form is found attached to opsin. In response to absorbed light energy, the 11-*cis* retinene is converted to the all-*trans* form, causing it to dissociate from the opsin. This dissociation reaction in response to light initiates changes in the ionic permeability of the rod cell membrane and ultimately results in the production of nerve impulses in the ganglion cells. As a result of these effects, rods provide black-and-white vision under conditions of low light intensity (as described in a later section).

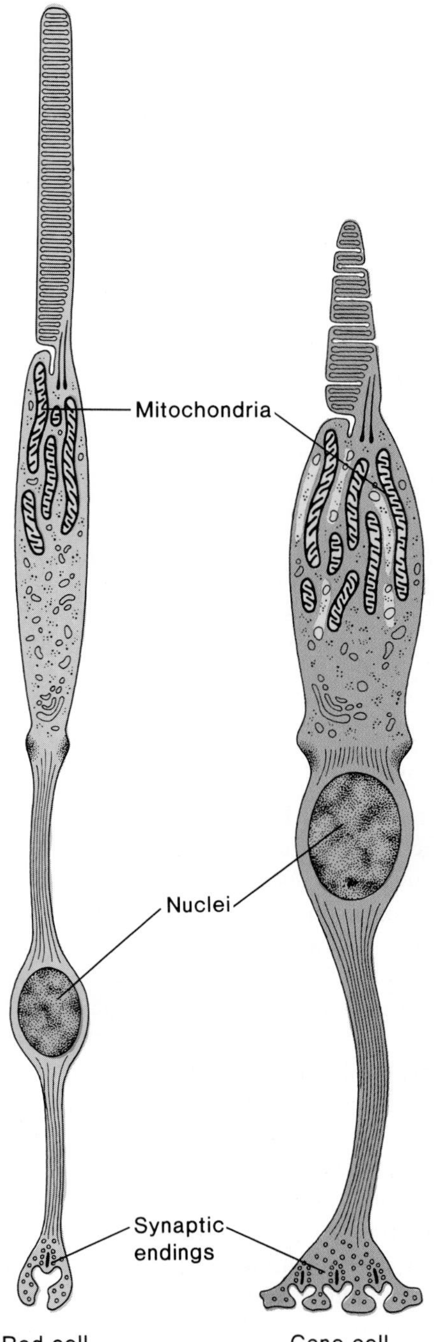

**Figure 9.42.** (*a*) Structure of a rod and cone. (*b*) A scanning electron micrograph of rods and cones.

***Dark Adaptation.*** The bleaching reaction that occurs in the light results in a lowered amount of rhodopsin in the rods and lowered amounts of visual pigments in the cones. When a light-adapted person first enters a darkened room, therefore, sensitivity to light is low and vision is poor. A gradual increase in photoreceptor sensitivity, known as *dark adaptation,* then occurs, reaching maximal sensitivity at about twenty minutes. The increased sensitivity to low light intensity is due partly to increased amounts of visual pigments produced in the dark. Increased pigments in the cones produce a slight dark adaptation in the first five minutes. Increased rhodopsin in the rods produces a much greater increase in sensitivity to low light levels and is partly responsible for the adaptation that occurs after about five minutes in the dark. In addition to the increased concentration of rhodopsin, other more subtle

**Figure 9.43.** The photopigment rhodopsin consists of the protein opsin combined with 11-*cis* retinene. In response to light the retinaldehyde is converted to a different form, called all-*trans,* and dissociates from the opsin. This photochemical reaction induces changes in ionic permeability that ultimately results in stimulation of ganglion cells in the retina.

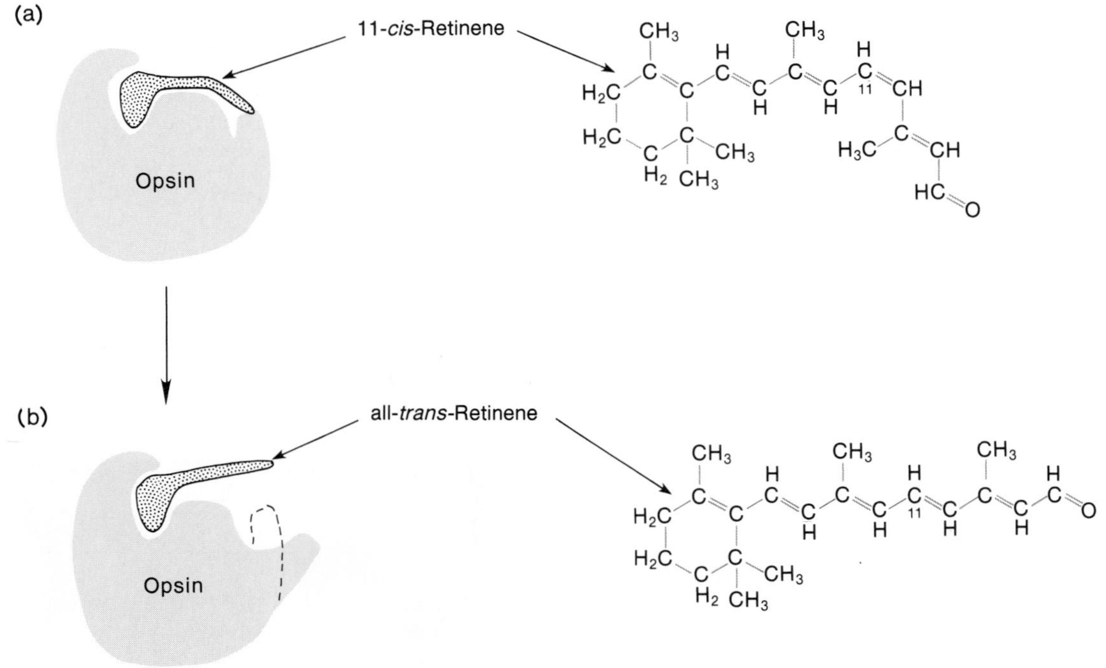

(a) Opsin — 11-*cis*-Retinene

(b) Opsin — all-*trans*-Retinene

(and less well-understood) changes occur in the rods that ultimately result in a 100,000-fold increase in light sensitivity in dark-adapted as compared to light-adapted eyes.

## Electrical Activity of Retinal Cells

The only neurons in the retina that produce all-or-none action potentials are ganglion cells and amacrine cells. The photoreceptors, bipolar cells, and horizontal cells instead produce only graded depolarizations or hyperpolarizations, analogous to EPSPs and IPSPs.

In the dark, the photoreceptors have a resting membrane potential that is less negative (closer to zero) than that of most other neurons. This is caused by a constant current of $Na^+$ into the cell, called a *dark current,* through special $Na^+$ channels. Light causes these $Na^+$ channels to become blocked; as a result, the photoreceptors become less depolarized than they are in the dark. Put another way, light causes the photoreceptors to become *hyperpolarized* in comparison to their membrane potential in the dark. Since light must ultimately have a stimulatory effect on the optic nerve, this hyperpolarization (which is associated with inhibition—chapter 7) is certainly surprising.

The translation of the effect of light on photoreceptors to the production of nerve impulses may be explained in the following manner. Photoreceptors in the dark release a neurotransmitter chemical at a constant rate at their synapses with bipolar cells. Hyperpolarization of the photoreceptors in response to light is similar to the hyperpolarization of other neurons that occurs during postsynaptic inhibition; it decreases the release of neurotransmitter by the photoreceptors. This neurotransmitter may be inhibitory to the bipolar cell (synaptic transmission between photoreceptors and bipolar cells can be either excitatory or inhibitory). If the neurotransmitter is inhibitory, the secretion of lower amounts of this chemical, as a result of the light-induced hyperpolarization of the photoreceptor neurons, will stimulate the bipolar cells. These neurons in turn activate ganglion cells, and action potentials will thus be produced on fibers of the optic nerve in response to light.

Light can also inhibit the production of action potentials in ganglion cells. The purpose of this inhibition is to sharpen the image produced by those ganglion cells that are stimulated in response to the light (this is similar to the process of lateral inhibition discussed earlier in this chapter). The inhibition is achieved when light causes a decrease in the amount of excitatory neurotransmitter released in the dark by photoreceptors. The light-induced decrease in excitatory neurotransmitter inhibits the activity of the bipolar cells and thus of the ganglion cells. These effects produce the characteristics of ganglion cell receptive fields, which is a topic discussed later in this chapter.

**Figure 9.44.** There are three types of cones. Each type of cone contains retinaldehyde combined with a different type of protein, producing a pigment that absorbs light maximally at a different wavelength. Color vision, according to the trichromatic theory, is produced by the activity of these blue cones, green cones, and red cones.

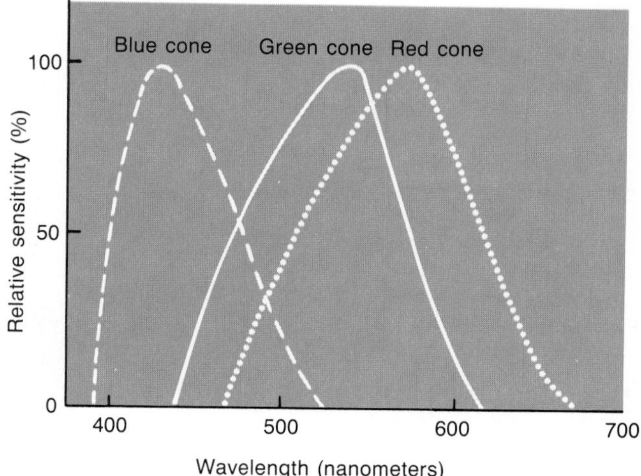


## Cones and Color Vision

Cones are less sensitive than rods to light, but provide color vision and greater visual acuity, as described in the next section. During the day, therefore, the high light intensity bleaches out the rods, and color vision with high acuity is provided by the cones. According to the **trichromatic theory** of color vision, our perception of a multitude of colors is due to stimulation of only three types of cones. Each type of cone contains retinene, as in rhodopsin, but this molecule is associated with a different protein than opsin. The protein is different for each of the three cone pigments, and as a result, each of the pigments absorbs light of a given wavelength (color) to a different degree. The three types of cones are designated blue, green, and red, according to the region of the visible spectrum in which each cone pigment absorbs light maximally (fig. 9.44). Our perception of any given color is produced by the relative degree to which each cone is stimulated by any given wavelength of visible light.

Suppose a person has become dark-adapted in a photographic darkroom over a period of twenty minutes or longer, but needs an increase in light to examine some prints. Since rods do not absorb red light but red cones do, a red light in a photographic darkroom allows vision (because of the red cones), but does not cause bleaching of the rods. When the light is turned off, therefore, the rods will still be dark-adapted and the person will still be able to see.

**Color blindness** is due to a congenital lack of one or more types of cones. People with normal color vision are *trichromats;* people with only two types of cones are *dichromats.* They may be missing red cones (have *protanopia*), or green cones (have *deuteranopia*), or blue cones (have *tritanopia*). Such a person, for example, may have difficulty distinguishing red from green. People who are *monochromats* have only one cone system and can only see black, white, and shades of gray. Color blindness is a trait carried on the X chromosome; since men have only one X chromosome per cell, whereas women have two X chromosomes (chapter 20), men are far more likely to be color blind than women (who can carry this trait in a recessive state).

## Visual Acuity and Sensitivity

While reading or similarly focusing visual attention on objects in daylight, each eye is oriented so that the image falls within a tiny area of the retina called the **fovea centralis.** The fovea is a pinhead-sized pit (*fovea* = pit) within a yellow area of the retina called the *macula lutea.* The pit is formed as a result of the displacement of neural layers around the fovea, so that light falls directly on photoreceptors in this region (fig. 9.45)—whereas light falling on other areas must pass through several layers of neurons, as previously described.

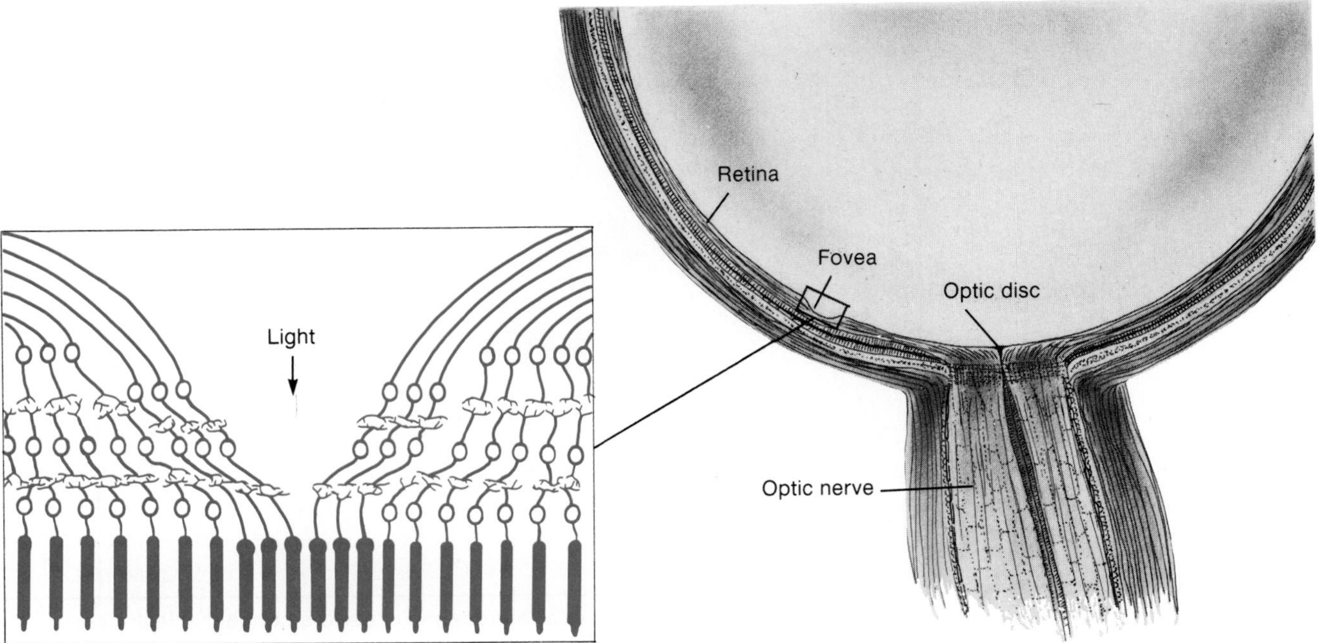


**Figure 9.45.** When the eyes "track" an object, the image is cast upon the fovea centralis of the retina. The fovea is literally a "pit" formed by parting of the neural layers, so that light falls directly on the photoreceptors (cones) in this region.

There are approximately 120 million rods and 6 million cones in each retina, but only about 1.2 million nerve fibers enter the optic nerve of each eye. This gives an overall convergence of photoreceptors on ganglion cells of about 105:1. This number is misleading, however, because the degree of convergence is much lower for cones than for rods, and it is 1:1 in the fovea.

The photoreceptors are distributed in such a way that the fovea contains only cones, whereas more peripheral regions of the retina contain a mixture of rods and cones. Approximately four thousand cones in the fovea provide input to approximately four thousand ganglion cells; each ganglion cell in this region, therefore, has a private line to the visual field. Each ganglion cell thus receives input from an area of retina corresponding to the diameter of one cone (about 2 $\mu$m). Peripheral to the fovea, however, many rods synapse with a single bipolar cell, and many bipolar cells synapse with a single ganglion cell. A single ganglion cell outside the fovea thus may receive input from large numbers of rods, corresponding to an area of about 1 mm$^2$ on the retina (fig. 9.46).

Since each cone in the fovea has a private line to a ganglion cell, and since each ganglion cell receives input from only a tiny region of the retina, visual acuity is greatest and sensitivity to low light is poorest when light falls on the fovea. In dim light only the rods are activated, and vision is best out of the corners of the eye when the image falls away from the fovea. Under these conditions, the convergence of many rods on a single bipolar cell and the convergence of many bipolar cells on a single ganglion cell increase sensitivity to dim light at the expense of visual acuity. Night vision is therefore less distinct than day vision.

### Neural Pathways from the Retina

As a result of light refraction by the cornea and lens, the right half of the visual field is projected to the left half of the retina of both eyes (the temporal half of the left retina and the nasal half of the right retina). The left half of the visual field is projected to the right half of the retina of both eyes. The temporal half of the left retina and the nasal half of the right retina therefore see the same image. Axons from ganglion cells in the left (temporal) half of the left retina pass to the left **lateral geniculate body** of the thalamus. Axons from ganglion cells in the nasal half of the

**Figure 9.46.** Since bipolar cells receive input from the convergence of many rods (*a*), and since a number of such bipolar cells converge on a single ganglion cell, rods provide high sensitivity to low levels of light at the expense of visual acuity. The 1:1:1 ratio of cones to bipolar cells to ganglion cells in the fovea, (*b*), in contrast, provides high visual acuity, but low sensitivity.

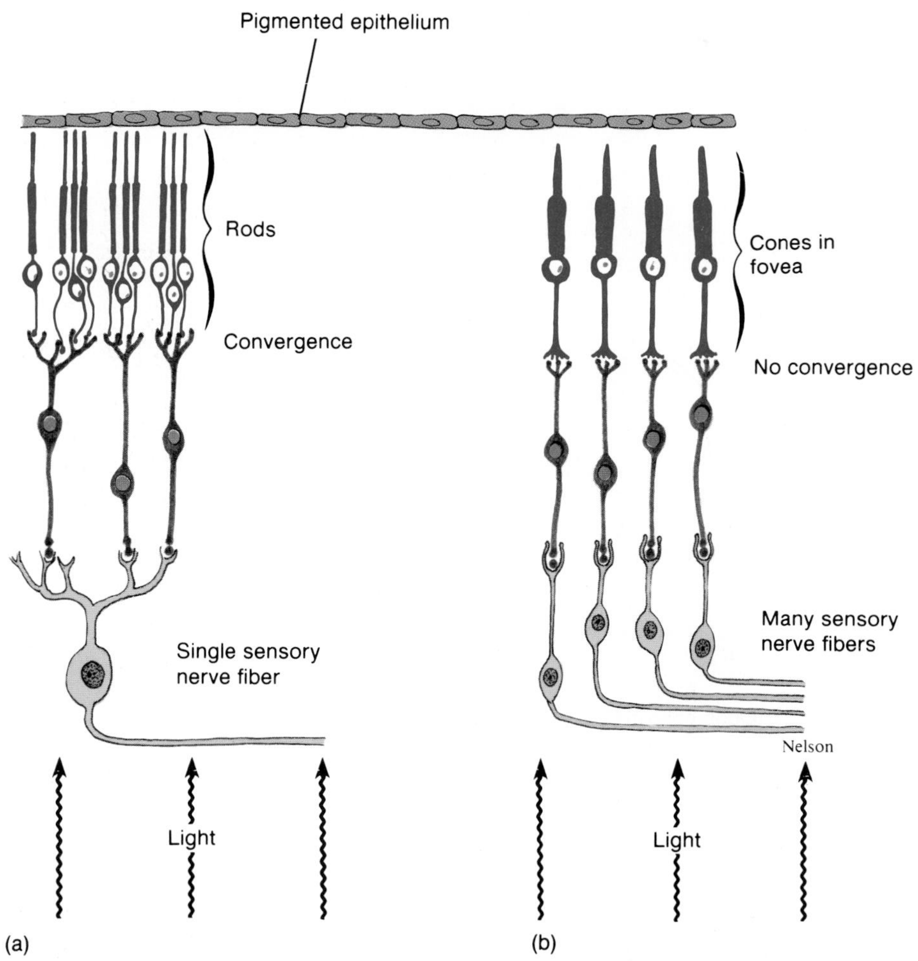


right retina cross (decussate) in the *optic chiasm* to synapse also in the left lateral geniculate body. The left lateral geniculate, therefore, receives input from both eyes that relates to the right half of the visual field (fig. 9.47).

The right lateral geniculate body, similarly, receives input from both eyes relating to the left half of the visual field. Neurons in both lateral geniculate bodies of the thalamus in turn project to the **striate cortex** of the occipital lobe in the cerebral cortex (fig. 9.47). This area is also called area 17, in reference to a numbering system developed by Brodmann. Neurons in area 17 synapse with neurons in areas 18 and 19 of the occipital lobe (fig. 9.48).

Approximately 70% to 80% of the axons from the retina pass to the lateral geniculate bodies and to the striate cortex. This **geniculostriate system** is involved in perception of the visual field. Put another way, the geniculostriate system is needed for answering the question, "What is it?" Approximately 20% to 30% of the fibers from the retina, however, follow a different path to the *superior colliculus* of the midbrain (also called the *optic tectum*). Axons from the superior colliculus activate motor pathways leading to eye and body movements. The **tectal system,** in other words, is needed for answering the question, "Where is it?"

**Figure 9.47.** The neural pathway leading from the retina to the lateral geniculate body to the visual cortex. As a result of the crossing of optic fibers, the visual cortex of each cerebral hemisphere receives input from the opposite (contralateral) visual field.

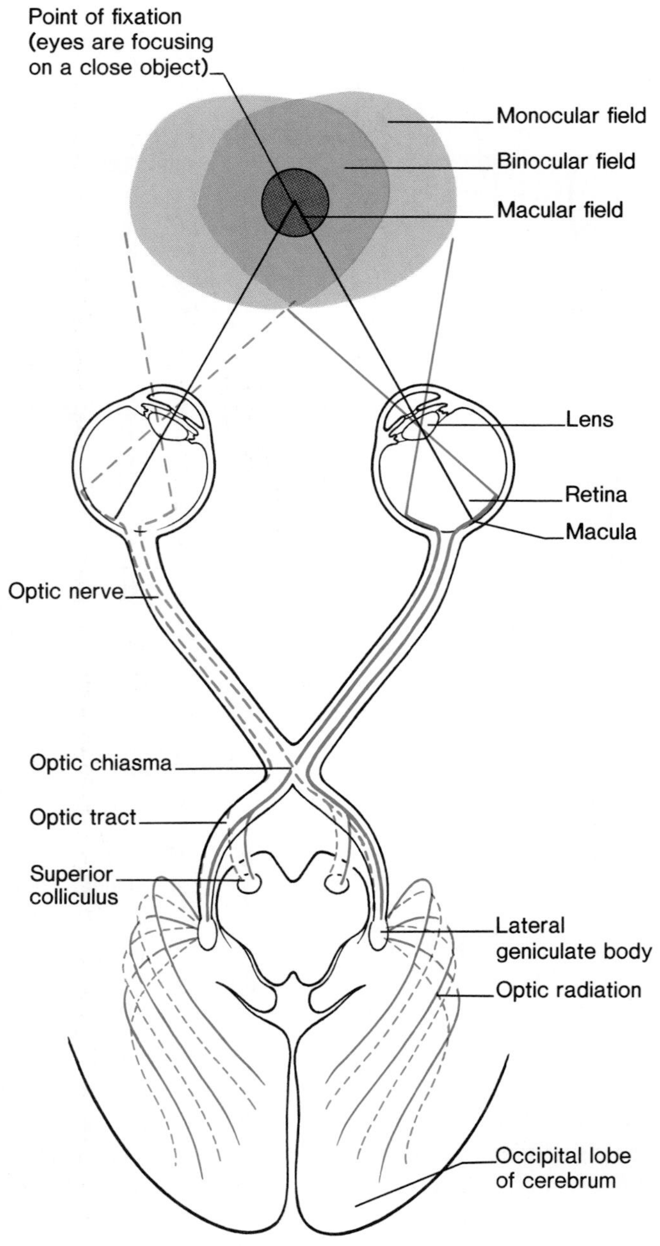

Figure 9.48. The striate cortex (area 17) and the visual association (areas 18 and 19).

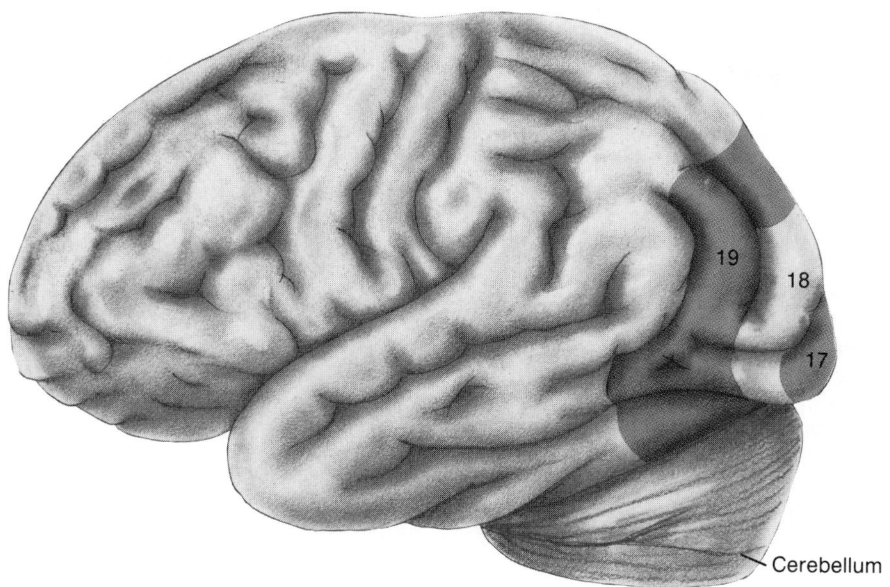


**Table 9.5** Muscles of the eye

| | | |
|---|---|---|
| **Extrinsic Muscles (striated)** | | |
| Superior rectus | Oculomotor nerve (III) | Rotates eye upward and toward midline |
| Inferior rectus | Oculomotor nerve (III) | Rotates eye downward and toward midline |
| Medial rectus | Oculomotor nerve (III) | Rotates eye toward midline |
| Lateral rectus | Abducens nerve (VI) | Rotates eye away from midline |
| Superior oblique | Trochlear nerve (IV) | Rotates eye downward and away from midline |
| Inferior oblique | Oculomotor nerve (III) | Rotates eye upward and away from midline |
| **Intrinsic Muscles (smooth)** | | |
| Ciliary muscles | Oculomotor nerve (III) parasympathetic fibers | Causes suspensory ligaments to relax |
| Iris, circular muscles | Oculomotor nerve (III) parasympathetic fibers | Causes size of pupil to decrease |
| Iris, radial muscles | Sympathetic fibers | Causes size of pupil to increase |

***Superior Colliculus and Eye Movements.*** Neural pathways from the superior colliculus to motor neurons in the spinal cord help mediate the startle response to the sight of an unexpected intruder. Other nerve fibers from the superior colliculus stimulate the **extrinsic eye muscles** (table 9.5), which are the striated muscles that move the eyes.

There are two types of eye movements coordinated by the superior colliculus. *Smooth pursuit movements* track moving objects and keep the image focused on the fovea centralis. *Saccadic eye movements* are short (lasting 20 to 50 msec), jerky movements that occur while the eyes appear to be still. These saccadic movements continuously move the image to different photoreceptors; if these movements were to stop, the image would disappear as the photoreceptors became bleached.

The tectal system is also involved in the control of the intrinsic eye muscles—the iris and the muscles of the ciliary body. Shining a light into one eye stimulates the *pupillary reflex* in which both pupils constrict. This is caused by activation of parasympathetic neurons by fibers from the superior colliculus. Postganglionic neurons in the ciliary ganglia behind the eyes, in turn, stimulate constrictor fibers in the iris. Contraction of the ciliary body during accommodation also involves stimulation of the superior colliculus.

1. *List the different layers of the retina, and describe the path of light and of nerve activity through these layers.*
2. *Describe the photochemical reaction in the rods, and explain the process of dark adaptation.*
3. *Describe the electrical state of photoreceptors in the dark, and explain how light affects the electrical activity of retinal cells.*
4. *Explain the trichromatic theory of color vision.*
5. *Compare the architecture of the fovea centralis with more peripheral regions of the retina, and explain how this architecture relates to visual acuity and sensitivity.*
6. *Describe how different parts of the visual field are projected on the retinas of both eyes, and describe the neural pathways of this information in the geniculostriate system.*
7. *Describe the neural pathways involved in the tectal system, and explain how these pathways are involved in the control of the extrinsic eye muscles, iris, and ciliary body.*

## *Neural Processing of Visual Information*

Electrical activity in ganglion cells of the retina, and neurons in the lateral geniculate nucleus and cerebral cortex, can be measured as a result of light on the retina. The way that each neuron type responds to light at a particular point of the retina, and the stimulus requirements for activation of each neuron, provides information about the way the brain interprets visual information.

Light that is cast on the retina directly affects the activity of photoreceptors and indirectly affects the neural activity in bipolar and ganglion cells. The part of the visual field that affects the activity of a particular ganglion cell can be considered to be its *receptive field.* Since each cone in the fovea has a private line to a ganglion cell, the receptive fields of these ganglion cells are equal to the width of one cone (about 2 $\mu$m). Ganglion cells in more peripheral parts of the retina receive input from hundreds of photoreceptors, and thus are influenced by a larger area of the retina (about 1 mm in diameter).

### Ganglion Cell Receptive Fields

Studies of the electrical activity of ganglion cells have yielded some surprising results. In the dark, each ganglion cell discharges spontaneously at a slow rate. When the room lights are turned on, the firing rate of many (but not all) ganglion cells increases slightly. A small spot of light that is directed at the center of some ganglion cells' receptive fields, however, stimulates a large increase in firing rate. A small spot of light can thus be a more effective stimulus than larger areas of light.

When the spot of light is moved only a short distance away from the center of the receptive field the ganglion cell responds in an opposite manner. The ganglion cell that was stimulated with light at the center of its receptive field is inhibited by light in the periphery of its field. The response produced by light in the center and by light in the "surround" of the visual field is *antagonistic.* Those ganglion cells that are stimulated by light at the center of their visual fields are said to have **on-center fields.** Ganglion cells that are inhibited by light in the center and stimulated by light in the surround have **off-center fields.**

The reason wide illumination of the retina has less effect than pinpoint illumination is now clear; diffuse illumination gives the ganglion cell conflicting orders—on and off. Because of the antagonism between the center and surround of ganglion cell receptive fields, the activity of each ganglion is a result of the *difference in light intensity* between the center and surround of its visual field. This is a form of *lateral inhibition* that helps to accentuate the contours of images and improve visual acuity.

### Lateral Geniculate Bodies

Each of the two lateral geniculate bodies receives input from ganglion cells in both eyes. The right lateral geniculate receives input from the right half of each retina (corresponding to the left half of the visual field); the left lateral geniculate receives input from the left half of each retina (corresponding to the right half of the visual field). Each neuron in the lateral geniculate, however, is activated only by input from one eye. Neurons that are activated by ganglion cells from the left eye are in separate layers within the lateral geniculate from those that are activated by the right eye (fig. 9.49).

The receptive field of each ganglion cell, as previously described, is the part of the retina it "sees" through its photoreceptor input. The receptive field of lateral geniculate neurons, similarly, is the part of the retina it "sees" through its ganglion cell input. Experiments in which the lateral geniculate receptive fields are mapped with a spot of light reveal that they are circular, with an antagonistic center and surround, much like the ganglion cell receptive fields.

**Figure 9.49.** The lateral geniculate nucleus. Each lateral geniculate consists of six layers (numbered 1 through 6 in this figure). Each of these layers receives input from only one eye, with right and left eyes alternating. An arrow through these six layers (see figure) of the left lateral geniculate, for example, encounters corresponding projections from a part of the right visual field in right and left eyes, alternatively, as it passes from the outer to the inner layers.

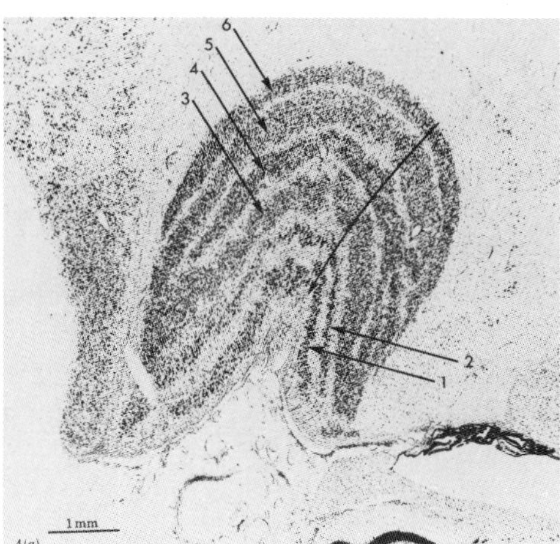


## The Cerebral Cortex

Projections of nerve fibers from the lateral geniculate bodies to area 17 of the occipital lobe form the *optic radiation* (see fig. 9.47). These fiber projections give area 17 a striped or striated appearance; this area is consequently also known as the striate cortex. Neurons in area 17, in turn, project to areas 18 and 19 of the occipital lobe. Cortical neurons in areas 17, 18, and 19 are thus stimulated indirectly by light on the retina. On the basis of their stimulus requirements, these cortical neurons are classified as simple, complex, and hypercomplex.

***Simple Neurons.*** The receptive fields of simple cortical neurons are rectangular rather than circular. This results from the fact that they receive input from lateral geniculate neurons whose receptive fields are aligned in a particular way (as illustrated in fig. 9.50). Simple cortical neurons are best stimulated by a slit or bar of light that is located in a precise part of the visual field (of either eye) at a precise orientation (fig. 9.51).

The striate cortex (area 17) contains simple, complex, and hypercomplex neurons. The other visual association areas, designated areas 18 and 19, contain only complex and hypercomplex cells. Complex neurons receive input from simple cells, and hypercomplex neurons receive input from complex cells.

**Figure 9.50.** Cortical neurons described as simple cells have rectangular receptive fields that are best stimulated by slits of light of particular orientations. This may be due to the fact that these simple cells receive input from ganglion cells that have circular receptive fields along a particular line.

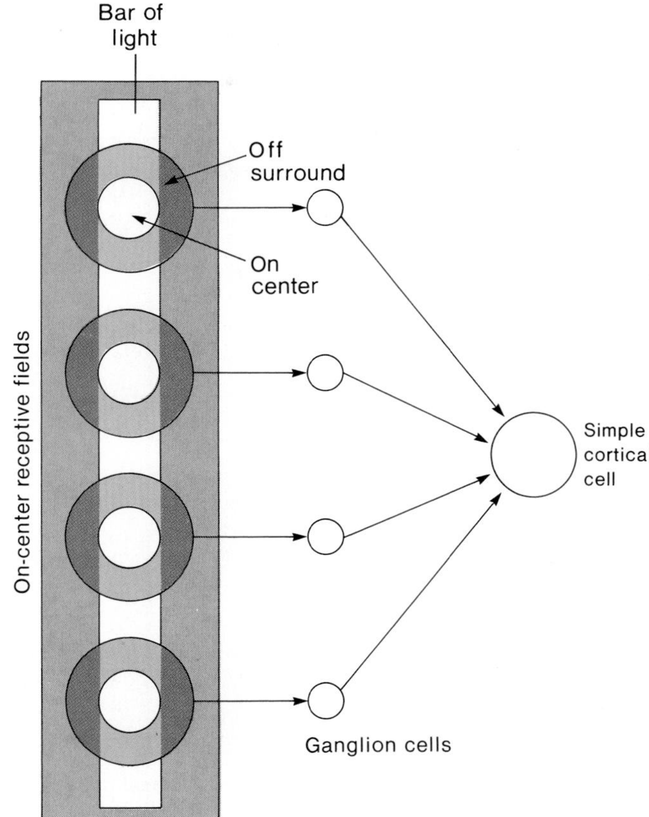


***Complex and Hypercomplex Neurons.*** Complex neurons respond best to straight lines with a specific orientation that move in a particular direction through the receptive field. Unlike simple neurons, complex neurons do not require that the stimulus have a particular position within the receptive field. Hypercomplex neurons require that the stimulus be of a particular length or have a particular bend or corner.

**Figure 9.51.** Simple cells are best stimulated by a slit or bar of light along a particular orientation within a particular region of the receptive field. The behavior of two different cortical cells (*a*) and (*b*) is illustrated. (Adapted from "Cellular Communication," by Stent, Gunther. Copyright © 1972 by Scientific American, Inc. All rights reserved.)

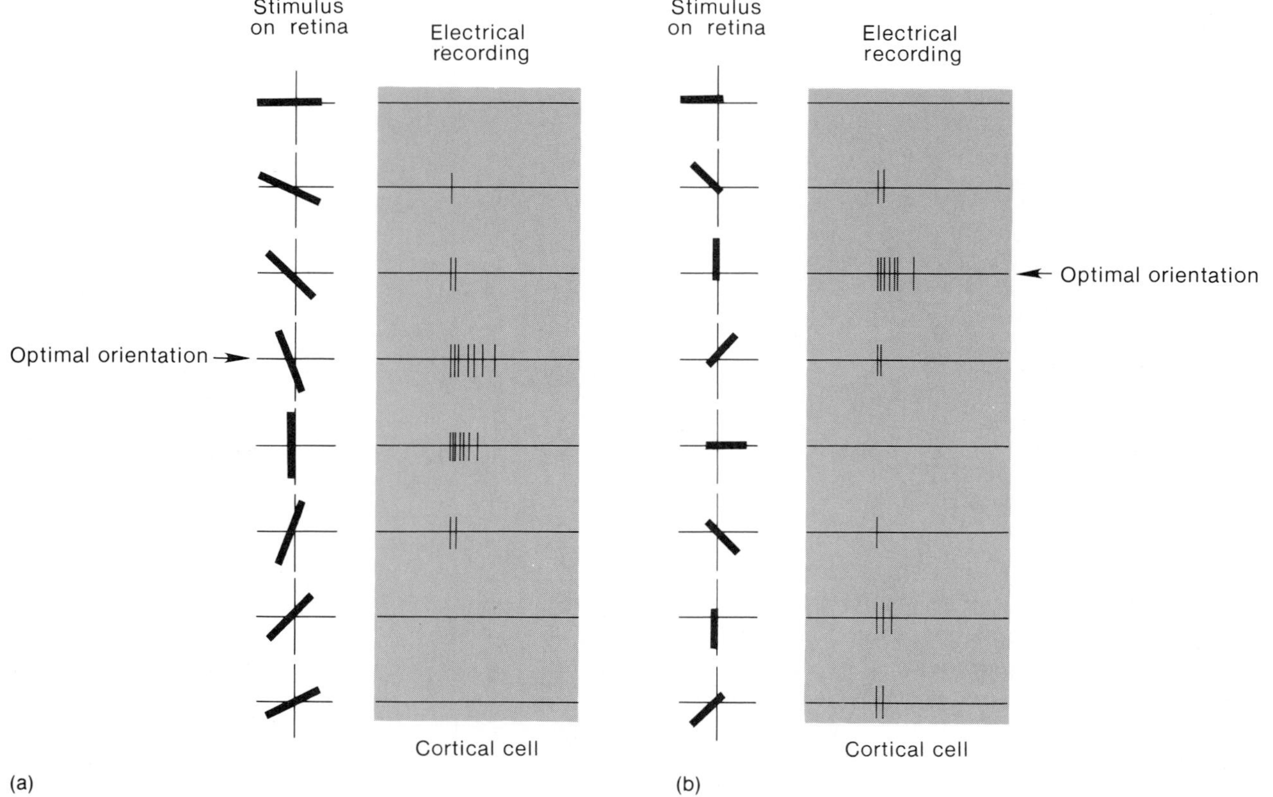


The dotlike information from ganglion and lateral geniculate cells is thus transformed in the occipital lobe into information about edges—their position, length, orientation, and movement. Although this represents a high degree of abstraction, the visual association areas of the occipital lobe probably serve as only an early stage in the integration of visual information. Other areas of the brain receive input from the visual association areas and provide meaning to visual perception.

1. *Describe the response characteristics of ganglion cells to light on the retina, and explain why a small spot of light is a more effective stimulus than general illumination of the retina.*
2. *Explain how the arrangement of ganglion cells' receptive fields can enhance visual acuity.*
3. *Describe the stimulus requirements of simple, complex, and hypercomplex cortical neurons.*

## Summary


### Characteristics of Sensory Receptors p. 212

I. Sensory receptors may be categorized on the basis of their structure, the stimulus energy they transduce, or the nature of their response.
  A. Receptors may be dendritic nerve endings, specialized neurons, or specialized epithelial cells associated with sensory nerve endings.
  B. Receptors may be chemoreceptors, photoreceptors, thermoreceptors, mechanoreceptors, or nocioreceptors.
    1. Proprioceptors include receptors in the muscles, tendons, and joints.
    2. The sense of sight, hearing, taste, olfaction, and equilibrium are grouped as special senses.
  C. Tonic receptors fire continuously as long as the stimulus is applied; these monitor the presence and intensity of a stimulus.
    1. Phasic receptors respond to stimulus changes; they do not respond to a sustained stimulus.
    2. Phasic receptors therefore help to provide sensory adaptation to sustained stimuli.

II. According to the law of specific nerve energies, each sensory receptor responds with lowest threshold to only one modality of sensation.
  A. That stimulus modality is called the adequate stimulus.
  B. Stimulation of the sensory nerve from a receptor by any means is interpreted in the brain as the adequate stimulus modality of that receptor.

III. Generator potentials are graded changes (usually depolarizations) in the membrane potential of the dendritic endings of sensory neurons.
  A. The magnitude of the potential change of the generator potential is directly proportional to the strength of the stimulus applied to the receptor.
  B. After the generator potential reaches a threshold value, increases in the magnitude of the depolarization results in increased frequency of action potential production in the sensory neuron.

### Cutaneous Sensations p. 215

I. Somatesthetic information—from cutaneous and proprioceptors—project by third-order neurons to the postcentral gyrus of the cerebrum.
  A. Proprioception and pressure sensations ascend on the ipsilateral side of the spinal cord, synapse in the medulla and cross to the contralateral side, and ascend in the medial lemniscus to the thalamus; neurons from the thalamus in turn project to the postcentral gyrus.
  B. Sensory neurons from other cutaneous receptors synapse and cross to the contralateral side in the spinal cord and ascend in the lateral and ventral spinothalamic tracts to the thalamus; neurons in the thalamus then project to the postcentral gyrus.

II. The receptive field of a cutaneous sensory neuron is the area of skin that, when stimulated, produces responses in the neuron.
  A. The receptive fields are smaller where the skin has a greater density of cutaneous receptors.
  B. The two-point touch threshold test reveals that the fingertips and mouth areas have a greater density of touch receptors and thus greater sensory acuity than other areas of the body.

III. Lateral inhibition acts to sharpen the sensation by inhibiting the activity of sensory neurons coming from areas of the skin around the area that is most greatly stimulated; the contrast between the most stimulated area and areas of lesser stimulation is thereby increased.

### Taste and Olfaction p. 219

I. The sense of taste is mediated by taste buds.
  A. The cells that compose the taste buds are innervated by the glossopharyngeal (IX) and facial (VII) nerves.
  B. A particular taste bud is most sensitive to one of four taste modalities: sweet, sour, bitter, and salty.
  C. Taste buds that are most sensitive to one modality of taste are located in characteristic regions of the tongue.

II. The olfactory receptors are dendritic endings of the olfactory (first cranial) nerve.

### Vestibular Apparatus and Equilibrium p. 221

I. The structures for equilibrium and hearing are located in the inner ear in the membranous labyrinth.
  A. The structure involved in equilibrium, known as the vestibular apparatus, consists of the otolith organs (utricle and saccule) and the semicircular canals.
  B. The utricle and saccule provide information about linear acceleration, whereas the semicircular canals provide information about angular acceleration.
  C. The sensory receptors for equilibrium are hair cells that have microvilli and one cilium (called a kinocilium).
    1. When the microvilli are bent in the direction of the kinocilium, the cell membrane becomes depolarized.
    2. When the microvilli are bent in the opposite direction, the membrane becomes hyperpolarized.

II. The hairs of the hair cells in the utricle and saccule protrude into the endolymph of the membranous labyrinth and are embedded within a gelatinous otolith membrane.
  A. When a person is upright, the hairs of the utricle are oriented vertically and those of the saccule are oriented horizontally.
  B. Linear acceleration causes a shearing force between the hairs and the otolith membrane, thus bending the hairs and electrically stimulating the sensory endings.

III. The three semicircular canals are oriented at right angles to each other, like the faces of a cube.
  A. The hair cells are embedded within a gelatinous membrane called the cupula, which projects into the endolymph.

B. Movement along one of the planes of a semicircular canal causes the endolymph to bend the cupula and stimulate the hair cells.
C. Stimulation of the hair cells in the vestibular apparatus activates sensory neurons of the vestibulocochlear nerve, which projects to the cerebellum and the vestibular nuclei of the medulla oblongata.
   1. The vestibular nuclei in turn send fibers to the oculomotor center, which controls eye movements.
   2. Spinning and then stopping abruptly can thus cause oscillatory movements of the eyes called nystagmus.

## The Ears and Hearing p. 227

I. The outer ear funnels sound waves of a given frequency (measured in hertz) and intensity (measured in decibels) to the tympanic membrane, causing it to vibrate.
II. Vibrations of the tympanic membrane cause movement of the middle ear ossicles—malleus, incus, and stapes—which in turn produces vibrations of the oval window of the cochlea.
III. Vibrations of the oval window set up a traveling wave of perilymph in the scala vestibuli.
   A. This wave can pass around the helicotrema to the scala tympani, or it can reach the scala tympani by passing through the scala media (cochlear duct).
   B. The scala media is filled with endolymph.
      1. The membrane of the cochlear duct that faces the scala vestibuli is called Reissner's membrane.
      2. The membrane that faces the scala tympani is called the basilar membrane.
IV. The sensory structure of the cochlea is called the organ of Corti.
   A. The organ of Corti consists of sensory hair cells on the basilar membrane.
      1. The hairs of the hair cells project upwards into an overhanging tectorial membrane.
      2. The hair cells are innervated by the vestibulocochlear (VIII) nerve.
   B. Sounds of high frequency cause maximum displacement of the basilar membrane closer to its base near the stapes; sounds of lower frequency produce maximum displacement of the basilar membrane closer to its apex near the helicotrema.
      1. Displacement of the basilar membrane causes the hairs to bend against the tectorial membrane and stimulate the production of nerve impulses.
      2. Pitch discrimination is thus dependent on the region of the basilar membrane that vibrates maximally to sounds of different frequencies.
      3. Pitch discrimination is enhanced by lateral inhibition.

## The Eyes and Vision p. 236

I. Light enters the cornea of the eye, passes through the pupil (the opening of the iris), and then through the lens, from which it is projected to the retina in the back of the eye.
   A. Light rays are bent, or refracted, by the cornea and lens.
   B. Because of refraction, the image on the retina is upside down and right to left.
   C. The right half of the visual field is projected to the left half of the retina in each eye, and vice versa.
II. Accommodation is the ability to maintain a focus on the retina when the distance between the object and the eyes is changed.
   A. Accommodation is produced by changes in the shape and refractive power of the lens.
   B. When the muscles of the ciliary body are relaxed, the suspensory ligament is tight and the lens is pulled to its least convex form.
      1. This gives the lens a low refractive power for distance vision.
      2. As an object is brought closer than 20 feet from the eyes, the ciliary body contracts, the suspensory ligament becomes less tight, and the lens becomes more convex and more powerful.
III. Visual acuity refers to the sharpness of the image and depends in part on the ability of the lens to bring the image to a focus on the retina.
   A. People with myopia have an eyeball that is too long, so that the image is brought to a focus in front of the retina; this is corrected by a concave lens.
   B. People with hyperopia have an eyeball that is too short, so that the image is brought to a focus behind the retina; this is corrected by a convex lens.
   C. Astigmatism is the uneven refraction of light around 360 degrees of a circle, caused by uneven curvature of the cornea and/or lens.

## The Retina p. 243

I. The retina contains photoreceptor neurons called rods and cones, which synapse with bipolar cells.
   A. When light strikes the rods, it causes the photodissociation of rhodopsin into retinene and opsin.
      1. This occurs maximally with a light wavelength of 500 nm.
      2. This reaction is called the bleaching reaction.
      3. Photodissociation is caused by the conversion of the 11-*cis* to the all-*trans* form of retinene, which cannot bond to opsin.
   B. In the dark, more rhodopsin can be produced and contained in the rods, so that the eyes become more sensitive to light; this is responsible for one component of dark adaptation.
   C. The rods provide black-and-white vision under conditions of low light intensity; at higher light intensity the rods are bleached out and the cones provide color vision.
II. In the dark, there is a constant movement of $Na^+$ into the rods that produces what is known as a "dark current."
   A. When light causes the dissociation of rhodopsin, the $Na^+$ channels become blocked and the rods become hyperpolarized in comparison to their membrane potential in the dark.
   B. Hyperpolarization of the rods results in their release of less neurotransmitter at their synapses with bipolar cells.
   C. Neurotransmitters from rods cause depolarization of bipolar cells in some cases, and hyperpolarization of bipolar cells in other cases; so when the rods are in light and release less neurotransmitter these effects are innverted.

III. According to the trichromatic theory of color vision, there are three systems of cones each of which responds to one of three colors: red, blue, and green.
  A. Each type of cone contains retinene attached to a different type of protein.
  B. The names for the cones signify the region of the spectrum in which the cones absorb light maximally.

IV. The fovea centralis contains only cones; more peripheral parts of the retina contain both cones and rods.
  A. Each cone in the fovea synapses with one bipolar cell, which in turn synapses with one ganglion cell.
    1. The ganglion cell that receives input from the fovea thus has a visual field equal to only that part of the retina which activated its cone.
    2. As a result of this 1:1 ratio of cones to bipolar cells, visual acuity is high in the fovea but sensitivity to low light levels is less than in other regions of the retina.
  B. In other regions of the retina, where rods predominate, there are a large number of rods that provide input to each ganglion cell (there is great convergence); visual acuity, as a result, is impaired, but sensitivity to low light levels is improved.

V. The right half of the visual field is projected to the left half of the retina of each eye.
  A. The left half of the left retina sends fibers to the left lateral geniculate body of the thalamus.
  B. The left half of the right retina also sends fibers to the left lateral geniculate body; this is because these fibers decussate in the optic chiasma.
  C. The left lateral geniculate body thus receives input from the left half of the retina of both eyes, corresponding to the right half of the visual field; the right lateral geniculate receives information about the left half of the visual field.
    1. Neurons in the lateral geniculate bodies send fibers to the striate cortex of the occipital lobes.
    2. The geniculostriate system is involved in providing meaning to the images that form on the retina.
  D. Instead of synapsing in the geniculate bodies, some fibers from the ganglion cells of the retina synapse in the superior colliculus of the midbrain, which controls eye movements.
    1. Since this brain region is also called the optic tectum, this pathway is called the tectal system.
    2. The tectal system enables the eyes to move and track an object; it is also responsible for the pupillary reflex and the changes in lens shape that are needed for accommodation.

## Neural Processing of Visual Information p. 252

I. The area of the retina that provides input to a ganglion cell is called the receptive field of the ganglion cell.
  A. The receptive field of a ganglion cell is roughly circular, with an "on" or "off" center and an antagonistic surround.
    1. A spot of light in the center of an "on" receptive field stimulates the ganglion cell; a spot of light in its surround inhibits the ganglion cell.
    2. The opposite is true for ganglion cells with "off" receptive fields.
    3. A larger light that stimulates both the center and the surround of a receptive field has less net effect on a ganglion cell than a smaller spot of light that only illuminates either the center or the surround.
  B. The antagonistic center and surround of the receptive field of ganglion cells provide lateral inhibition, which enhances contours and provides better visual acuity.

II. Each lateral geniculate body receives input from both eyes relating to the same part of the visual field.
  A. The neurons receiving input from each eye are arranged in layers within the lateral geniculate.
  B. The receptive fields of neurons in the lateral geniculate are circular with antagonistic center and surround, much like the receptive field of ganglion cells.

III. Cortical neurons involved in vision are either simple, complex, or hypercomplex.
  A. Simple neurons receive input from neurons in the lateral geniculate; complex neurons receive input from simple cells; and hypercomplex neurons receive input from complex cells.
  B. Simple neurons are best stimulated by a slit or bar of light that is located in a precise part of the visual field and has a precise orientation.
  C. Complex cells respond best to a straight line that has a particular orientation and moves in a particular direction; the position of the line in the visual field is not important.
  D. Hypercomplex cells respond best to lines that have a particular length or have a particular bend or corner.
  E. Information from these cells in the occipital lobe is integrated with information from other areas of the brain to provide meaningful visual perception.

## Review Activities

### Objective Questions

Match the following:

1. utricle and saccule
2. semicircular canals
3. cochlea

(a) cupula
(b) tectorial membrane
(c) basilar membrane
(d) otolith membrane

4. The dissociation of rhodopsin in the rods in response to light causes
   (a) the $Na^+$ channels to become blocked
   (b) the rods to secrete less neurotransmitter
   (c) the bipolar cells to become either stimulated or inhibited
   (d) all of the above
5. Tonic receptors
   (a) are fast-adapting
   (b) do not fire continuously to a sustained stimulus
   (c) produce action potentials at a greater frequency as the generator potential is increased
   (d) all of the above
6. Cutaneous receptive fields are smallest in
   (a) the fingertips
   (b) the back
   (c) the thighs
   (d) the arms
7. The process of lateral inhibition
   (a) increases the sensitivity of receptors
   (b) promotes sensory adaptation
   (c) increases sensory acuity
   (d) prevents adjacent receptors from being stimulated
8. The receptors for taste are
   (a) naked sensory nerve endings
   (b) encapsulated sensory nerve endings
   (c) modified epithelial cells
9. The utricle and saccule
   (a) are otolith organs
   (b) are located in the middle ear
   (c) provide a sense of linear acceleration
   (d) both *a* and *c*
   (e) both *b* and *c*
10. Since fibers of the optic nerve that originate in the nasal halves of each retina cross at the optic chiasma, each lateral geniculate receives input from
    (a) both the right and left sides of the visual field of both eyes
    (b) the ipsilateral visual field of both eyes
    (c) the contralateral visual field of both eyes
    (d) the ipsilateral field of one eye and the contralateral field of the other eye
11. When a person with normal vision views an object from a distance of at least 20 feet
    (a) the ciliary muscles are relaxed
    (b) the suspensory ligament is tight
    (c) the lens is in its most flat, least convex shape
    (d) all of the above
12. Glasses with concave lenses help correct
    (a) presbyopia
    (b) myopia
    (c) hyperopia
    (d) astigmatism
13. Parasympathetic nerves that stimulate constriction of the iris (in the pupillary reflex) are activated by neurons in
    (a) the lateral geniculate
    (b) the superior colliculus
    (c) the inferior colliculus
    (d) the striate cortex
14. A bar of light in a specific part of the retina, with a particular length and orientation, is the most effective stimulus for
    (a) ganglion cells
    (b) lateral geniculate cells
    (c) simple cortical cells
    (d) complex cortical cells
15. The ability of the lens to increase its curvature and maintain a focus at close distances is called
    (a) convergence
    (b) accommodation
    (c) astigmatism
    (d) ambylopia
16. Which of the following sensory modalities is transmitted directly to the cerebral cortex without being relayed through the thalamus?
    (a) taste
    (b) sight
    (c) smell
    (d) hearing
    (e) touch

### Essay Questions

1. Define the term *lateral inhibition,* and give examples of its effects in three sensory systems.
2. Describe the nature of the generator potential, and explain its relationship to stimulus intensity and to frequency of action potential production.
3. Explain how the vestibular apparatus provides information about changes in the position of our body in space.
4. Define accommodation, and explain how it is accomplished. Why is it more of a strain on the eyes to look at a small object close to the eyes than large objects far away?
5. Describe the effects of light on the photoreceptors, and explain how these effects influence the bipolar cells.
6. Explain why images that fall on the fovea centralis are seen more clearly than images that fall on the periphery of the retina. Why are the "corners of the eyes" more sensitive to light than the fovea?
7. Why do rods provide only black-and-white vision? Include a discussion of different types of color blindness in your answer.
8. Why are green-colored objects seen better at night than objects of other colors? What effect does red light in a darkroom have on a dark-adapted eye?
9. Describe the receptive fields of ganglion cells, and explain how the nature of these fields helps to improve visual acuity.

## Selected Readings


Boynton, R. M. 1979. *Human color vision.* New York: Holt, Rinehart & Winston.

Fireman, P. 1987. Newer concepts in otitis media. *Hospital Practice* 23:85.

Glickstein, M. September 1988. The discovery of the visual cortex. *Scientific American.*

Goldberg, J. M., and C. Fernandez. 1975. Vestibular mechanisms. *Annual Review of Physiology* 37:129.

Hubel, D. H. 1979. The visual cortex of normal and deprived monkeys. *American Scientist* 67:532.

Hubel, D. H., and T. Wiesel. September 1979. Brain mechanisms of vision. *Scientific American.*

Hudspeth, A. J. February 1983. The hair cells of the inner ear. *Scientific American.*

Koretz, J. F., and G. H. Handelman. July 1988. How the human eye focuses. *Scientific American.*

Loeb, G. E. February 1985. The functional replacement of the ear. *Scientific American.*

MacNichol, E. F., Jr. December 1964. Three-pigment color vision. *Scientific American.*

Nathans, J. February 1989. The genes for color vision. *Scientific American.*

O'Brian, D. F. 1982. The chemistry of vision. *Science* 218:961.

Parker, D. E. November 1980. The vestibular apparatus. *Scientific American.*

Pfaffmann, C., Frank, M., and R. Norgren. 1979. Neural mechanisms and behavioral aspects of taste. *Annual Review of Psychology* 30:283.

Rhode, W. S. 1984. Cochlear mechanics. *Annual Review of Physiology* 46:231.

Rushton, W. A. H. March 1975. Visual pigments and color blindness. *Scientific American.*

Van Essen, D. C. 1979. Visual areas of the mammalian cerebral cortex. *Annual Review of Neurosciences* 2:277.

# 10

# THE AUTONOMIC NERVOUS SYSTEM

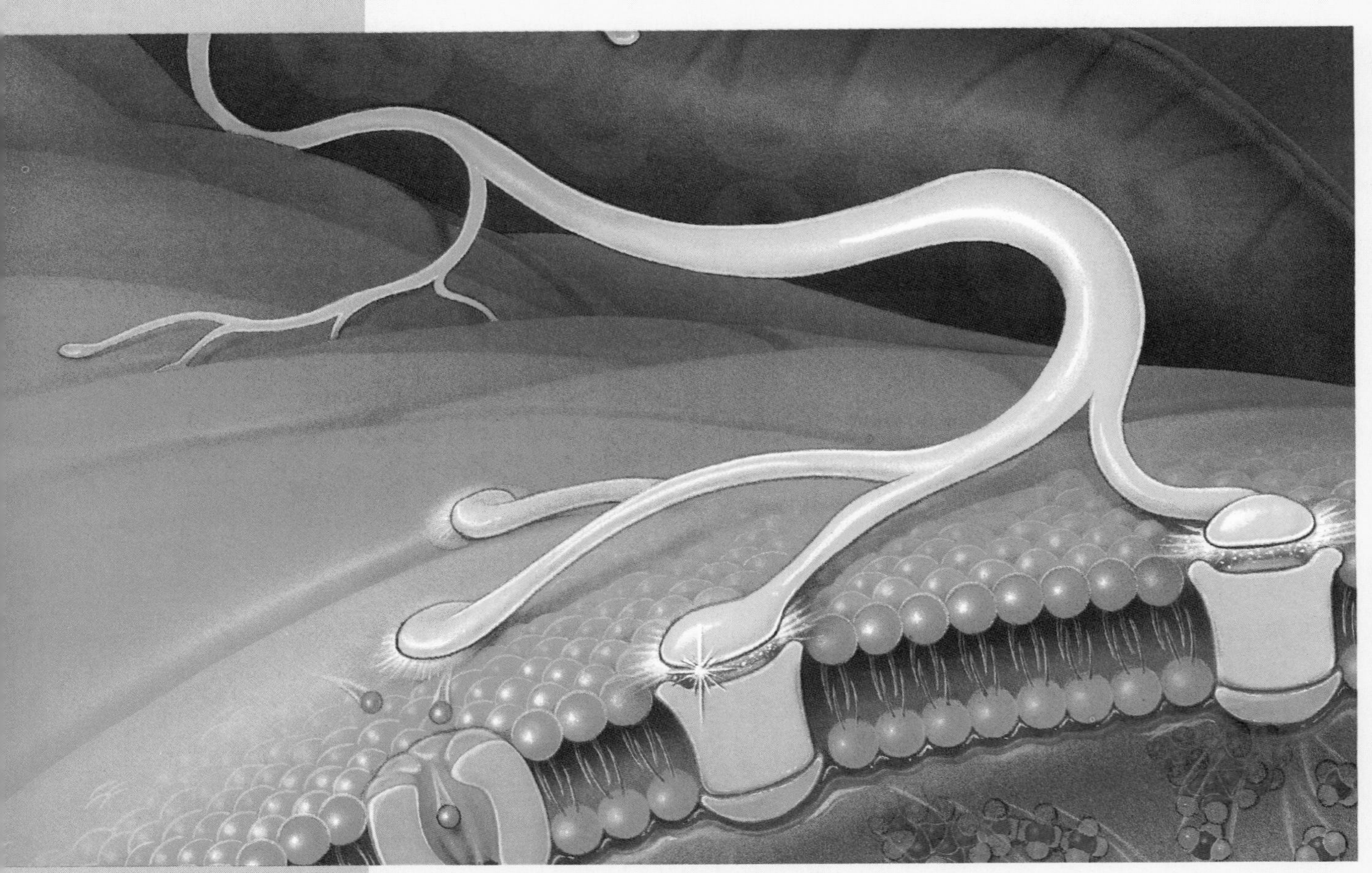

## Objectives

By studying this chapter, you should be able to

1. compare the structures and pathways of the autonomic system with those involved in the control of skeletal muscle
2. describe the structure and functional characteristics of cardiac muscle, smooth muscle, and exocrine glands
3. describe the structure of the sympathetic division of the autonomic system and the general functions of this division
4. describe the structure of the parasympathetic division of the autonomic system and the general functions of this division
5. indicate the neurotransmitters of the preganglionic and postganglionic neurons of the sympathetic and parasympathetic systems
6. describe the structural and functional relationship between the sympathetic system and the adrenal medulla
7. distinguish between the different types of adrenergic receptors, indicate their anatomic locations, and explain the physiological and clinical significance of these receptors
8. explain how the cholinergic receptors are divided into two categories and describe the effects produced by stimulation of these receptors
9. explain the antagonistic, complementary, and cooperative effects of sympathetic and parasympathetic innervation in different organs
10. describe the higher neural control of the autonomic system

## Outline

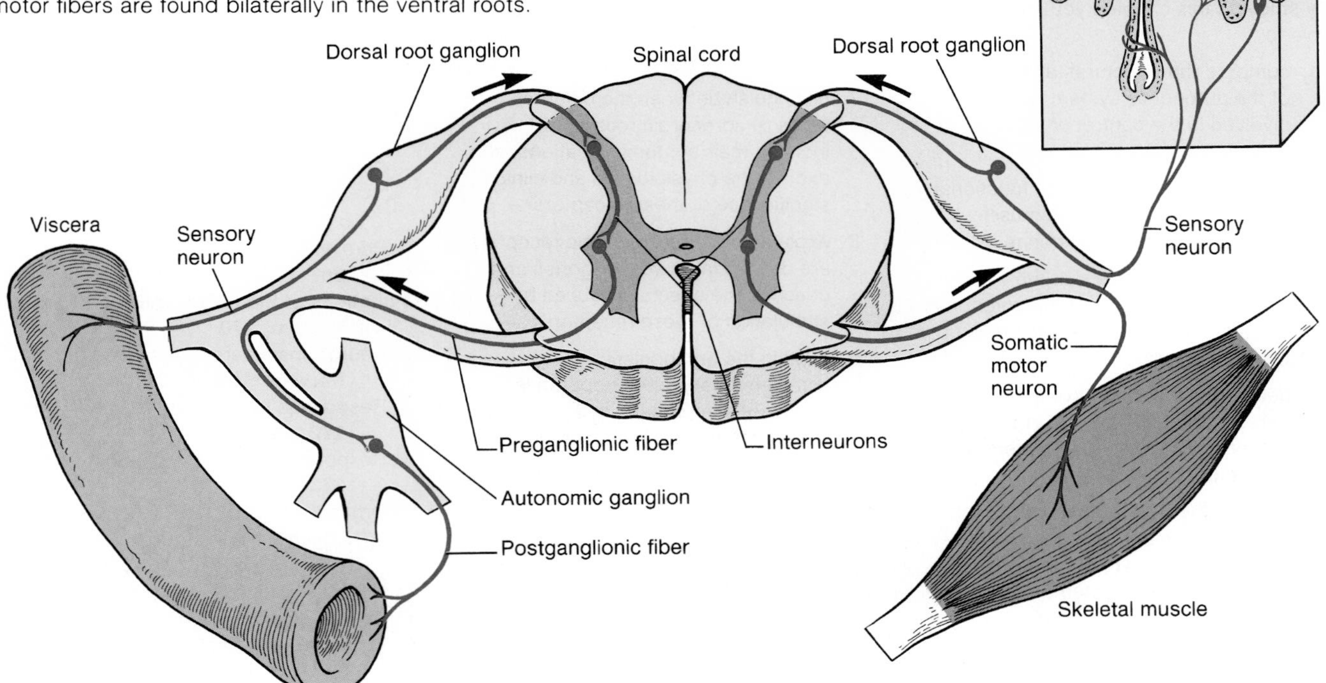


**Figure 10.1.** A comparison of a somatic motor reflex with an autonomic motor reflex. Although, for the sake of clarity, each is shown on different sides of the spinal cord, both visceral and somatic sensory neurons are found bilaterally in the dorsal roots of spinal nerves and somatic and autonomic motor fibers are found bilaterally in the ventral roots.

## ***Neural Control of Involuntary Effectors***

The autonomic nervous system helps to regulate the activities of cardiac muscle, smooth muscle, exocrine glands, and endocrine glands. In this regulation, impulses are conduced from the central nervous system by the axon of an autonomic motor neuron. This axon synapses with a second autonomic neuron, and it is the axon of this second neuron in the pathway that innervates the involuntary effector.

Autonomic motor nerves innervate organs whose functions are not usually under voluntary control. The effectors that respond to autonomic regulation include **cardiac muscle** (the heart), **smooth** (visceral) **muscles,** and **glands.** These are part of the organs of the *viscera* (organs within the body cavities) and of blood vessels. The involuntary effects of autonomic innervation contrast with the voluntary control of skeletal muscles by way of somatic motor neurons.

### Autonomic Neurons

As discussed in chapter 7, neurons of the peripheral nervous system (PNS) which conduct impulses away from the central nervous system (CNS) are known as *motor*, or *efferent*, neurons. There are two major categories of motor neurons: somatic and autonomic. Somatic motor neurons have their cell bodies within the CNS and send axons to skeletal muscles, which are usually under voluntary control. This was briefly described in chapter 8, under the topic of the reflex arc, and is reviewed in the right half of figure 10.1. The control of skeletal muscles by somatic motor neurons is discussed in depth in chapter 12.

Unlike somatic motor neurons, which conduct impulses along a single axon from the spinal cord to the neuromuscular junction, autonomic motor control involves two neurons in the efferent pathway (table 10.1). The first of these neurons has its cell body in the gray matter of the brain or spinal cord. The axon of this neuron does not directly innervate the effector organ but instead synapses with a second neuron within an *autonomic ganglion* (a ganglion is a collection of cell bodies outside the CNS). The first neuron is thus called a **preganglionic neuron.** The second neuron in this pathway, called a **postganglionic**

**Table 10.1** Comparisons of the somatic motor system with the autonomic motor system

| Feature | Somatic Motor | Autonomic Motor |
|---|---|---|
| Effector organs | Skeletal muscles | Cardiac muscle, smooth muscle, and glands |
| Presence of ganglia | No ganglia outside CNS | Cell bodies of postganglionic autonomic fibers located in paravertebral, prevertebral (collateral), and terminal ganglia |
| Number of neurons from CNS to effector | One | Two |
| Type of neuromuscular junction | Specialized motor end plate | No specialization of postsynaptic membrane; all areas of smooth muscle cells contain receptor proteins for neurotransmitters |
| Effect of nerve impulse on muscle | Excitatory only | Either excitatory or inhibitory |
| Type of nerve fibers | Fast-conducting, thick (9–13 $\mu$m), and myelinated | Slow-conducting; preganglionic fibers, lightly myelinated but thin (3 $\mu$m); postganglionic fibers, unmyelinated and very thin (about 1.0 $\mu$m) |
| Effect of denervation | Flaccid paralysis and atrophy | Much muscle tone and function remains; target cells show denervation hypersensitivity |

**neuron,** has an axon that extends from the autonomic ganglion and synapses with the cells of an effector organ (fig. 10.1, left half).

Preganglionic autonomic fibers originate in the midbrain and hindbrain and in the upper thoracic to the fourth sacral levels of the spinal cord. Autonomic ganglia are located in the head, neck, and abdomen; chains of autonomic ganglia also parallel the right and left sides of the spinal cord. The origin of the preganglionic fibers and the location of the autonomic ganglia help to differentiate the **sympathetic** and **parasympathetic** divisions of the autonomic system, which will be discussed in later sections of this chapter.

## Visceral Effector Organs

Since the autonomic nervous system helps to regulate the activities of glands, smooth muscles, and cardiac muscle, the story of autonomic control is integrated with the physiology of most of the body systems. Autonomic regulation is thus part of the explanation of endocrine regulation (chapter 11), smooth muscle function (chapter 12), functions of the heart and circulation (chapters 13 and 14), and of all the remaining systems to be discussed. Although the functions of the target organs of autonomic innervation are described in subsequent chapters, some of the common features of autonomic regulation are described in this section.

Unlike skeletal muscles, which enter a state of flaccid paralysis and atrophy when their motor nerves are severed, the involuntary effectors are somewhat independent of their innervation. Smooth muscles maintain a resting tone (tension) in the absence of nerve stimulation, for example. Damage to an autonomic nerve, in fact, makes its target tissue more sensitive than normal to stimulating agents. This phenomenon is called **denervation hypersensitivity.** Such compensatory changes can account, for example, for the observation that the ability of the mucosa of the stomach to secrete acid may be restored after its neural supply from the vagus nerve is severed (this procedure is called vagotomy, and is sometimes performed as a treatment for ulcers).

In addition to their intrinsic ("built-in") muscle tone, many smooth muscles and the cardiac muscle take their autonomy a step further. These muscles can contract rhythmically, even in the absence of nerve stimulation, in response to electrical waves of depolarization initiated by the muscles themselves. Autonomic innervation simply increases or decreases this intrinsic activity. Autonomic nerves also maintain a resting "tone," in the sense that they maintain a baseline firing rate that can be either increased or decreased. A decrease in the excitatory input to the heart, for example, will slow its rate of beat.

Unlike somatic motor neurons—where the release of the neurotransmitter (ACh) always stimulates the effector organ (skeletal muscles)—some autonomic nerves release transmitters that inhibit the activity of their effectors. An increase in the activity of the vagus, a nerve that supplies inhibitory fibers to the heart, for example, will slow the heart rate, whereas a decrease in this inhibitory input will increase the heart rate.

1. *Describe the preganglionic and postganglionic neurons in the autonomic system, and use a diagram to illustrate the difference in efferent outflow between somatic and autonomic nerves.*
2. *Describe how the regulation of the contraction of cardiac and smooth muscle cells differs from that of skeletal muscle cells. Explain how these muscles are affected by the experimental removal of their innervation.*

**Figure 10.2.** The sympathetic chain of paravertebral ganglia, showing its relationship to the vertebral column and the spinal cord.

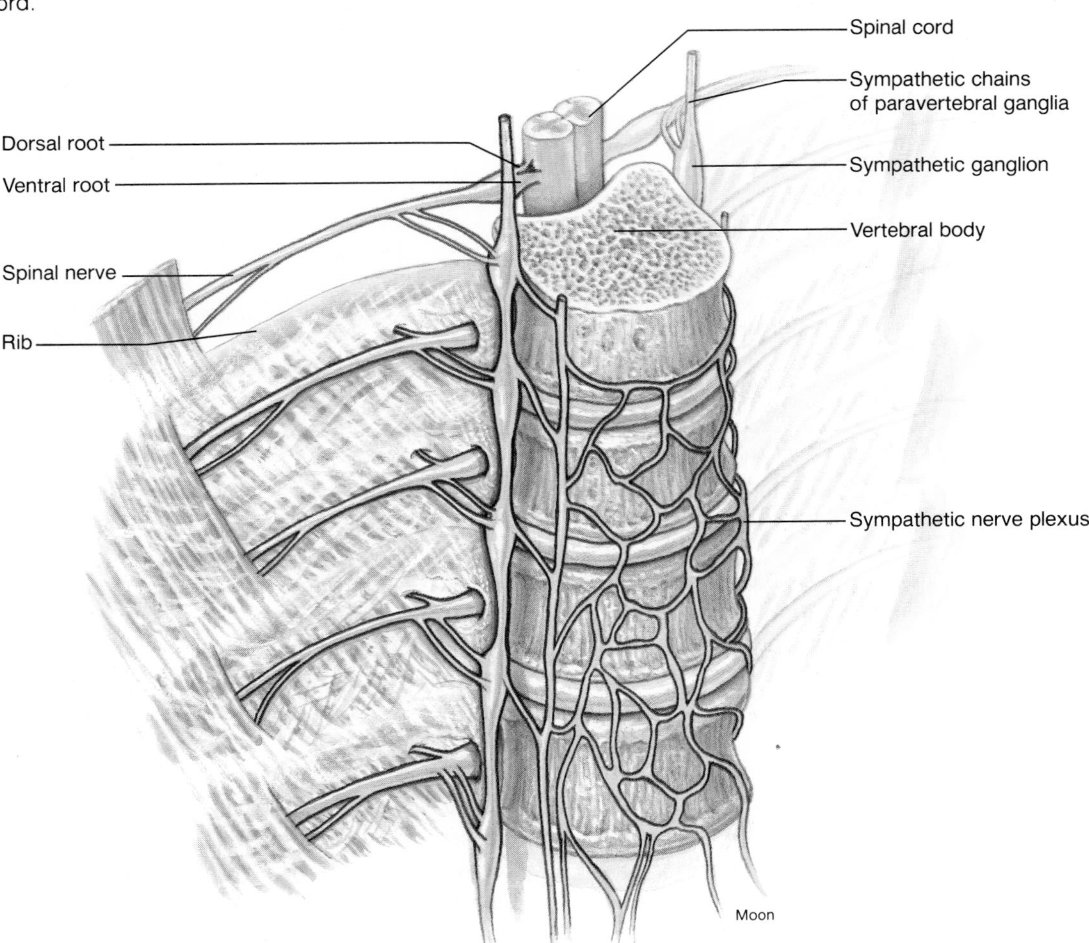


## Divisions of the Autonomic Nervous System

Preganglionic neurons of the sympathetic division of the autonomic system originate in the thoracic and lumbar levels of the spinal cord and send axons to sympathetic ganglia, which parallel the spinal cord. Preganglionic neurons of the parasympathetic division, in contrast, originate in the brain and in the sacral level of the spinal cord, and send axons to ganglia located in different regions of the body in or near the effector organs.

The sympathetic and parasympathetic divisions of the autonomic system share structural similarities: both consist of preganglionic neurons that originate in the CNS and postganglionic neurons that originate outside of the CNS in ganglia. The specific origin of the preganglionic fibers and the location of the ganglia, however, are different in the two divisions of the autonomic system.

### Sympathetic (Thoracolumbar) Division

The sympathetic system is also called the thoracolumbar division of the autonomic system because its preganglionic fibers exit the spinal cord from the first thoracic (T1) to the second lumbar (L2) levels. Most sympathetic nerve fibers, however, separate from the somatic motor fibers and synapse with postganglionic neurons within a double row of sympathetic ganglia, or **paravertebral ganglia,** located on either side of the spinal cord (fig. 10.2). Ganglia within each row are interconnected, forming a *sympathetic chain* of ganglia parallel to each side of the spinal cord.

Since the preganglionic sympathetic fibers are myelinated and thus appear white, they are called *white rami communicantes.* Some of these preganglionic sympathetic

**Figure 10.3.** Sympathetic chain ganglia, the sympathetic chain, and rami communicantes of the sympathetic division of the ANS. (Solid lines = preganglionic fibers; dashed lines = postganglionic fibers.)

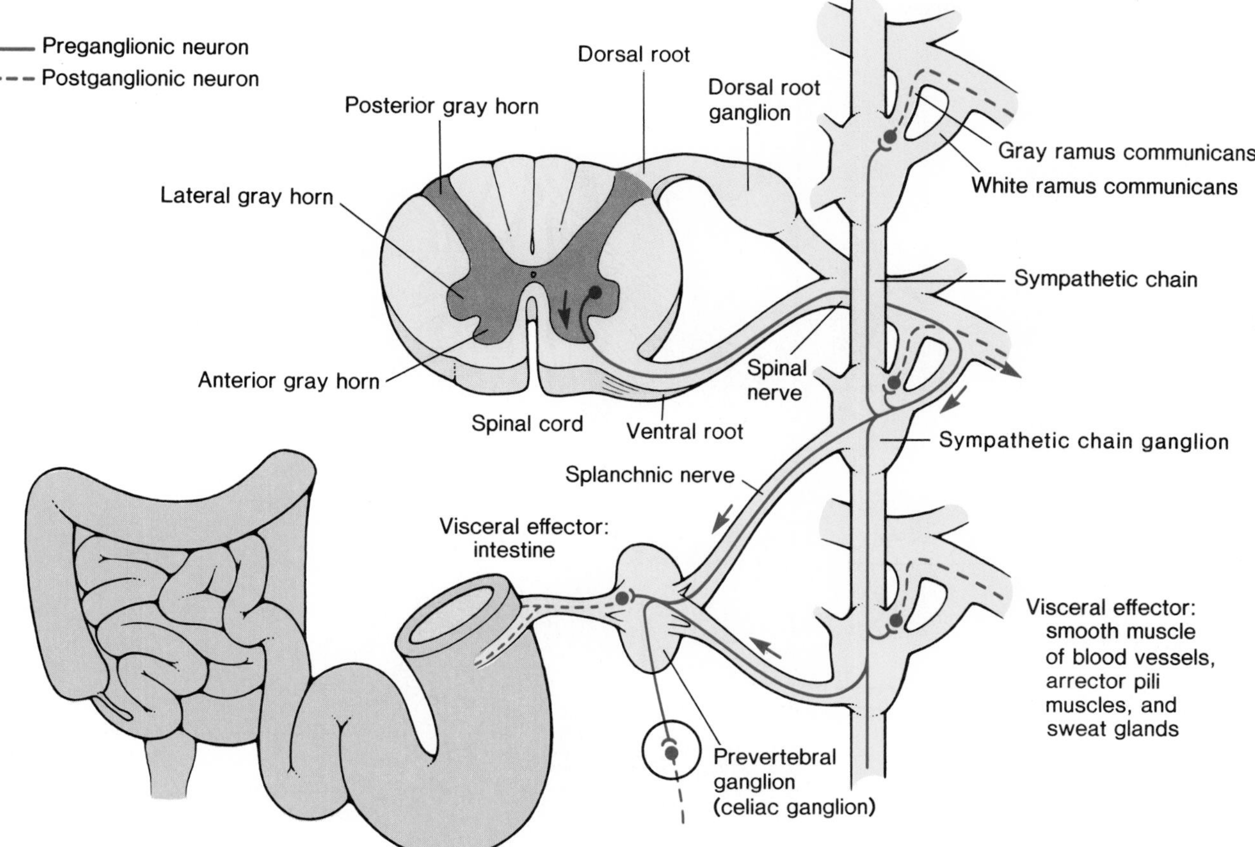


fibers synapse with postganglionic neurons located at their same level in the chain of sympathetic ganglia. Other preganglionic fibers travel up or down within the sympathetic chain before synapsing with postganglionic neurons. Since the fibers of the postganglionic sympathetic neurons are unmyelinated and thus appear gray, they form the *gray rami communicantes.* Postganglionic axons in the gray rami extend directly back to the ventral roots of the spinal nerves and travel distally within the spinal nerves to innervate their effector organs (fig. 10.3).

Within the chain of paravertebral ganglia, *divergence* is seen when preganglionic fibers branch to synapse with many postganglionic neurons located at different levels in the chain. *Convergence* is also seen when a given postganglionic neuron receives synaptic input from a large number of preganglionic fibers. The divergence of impulses from the spinal cord to the ganglia and the convergence of impulses within the ganglia usually result in the **mass activation** of almost all of the postganglionic fibers. This explains why the sympathetic system is usually activated as a unit and affects all of its effector organs at the same time.

Many preganglionic fibers that exit the spinal cord in the upper thoracic level travel into the neck, where they synapse in cervical sympathetic ganglia (fig. 10.4). Postganglionic fibers from here innervate the smooth muscles and glands of the head and neck.

***Collateral Ganglia.*** Many preganglionic fibers that exit the spinal cord below the level of the diaphragm pass through the sympathetic chain of ganglia without synapsing. Beyond the sympathetic chain, these preganglionic fibers form *splanchnic nerves.* Preganglionic fibers in the splanchnic nerves synapse in **collateral ganglia,** or *prevertebral ganglia.* These include the *celiac, superior*

**Figure 10.4.** The cervical sympathetic ganglia.

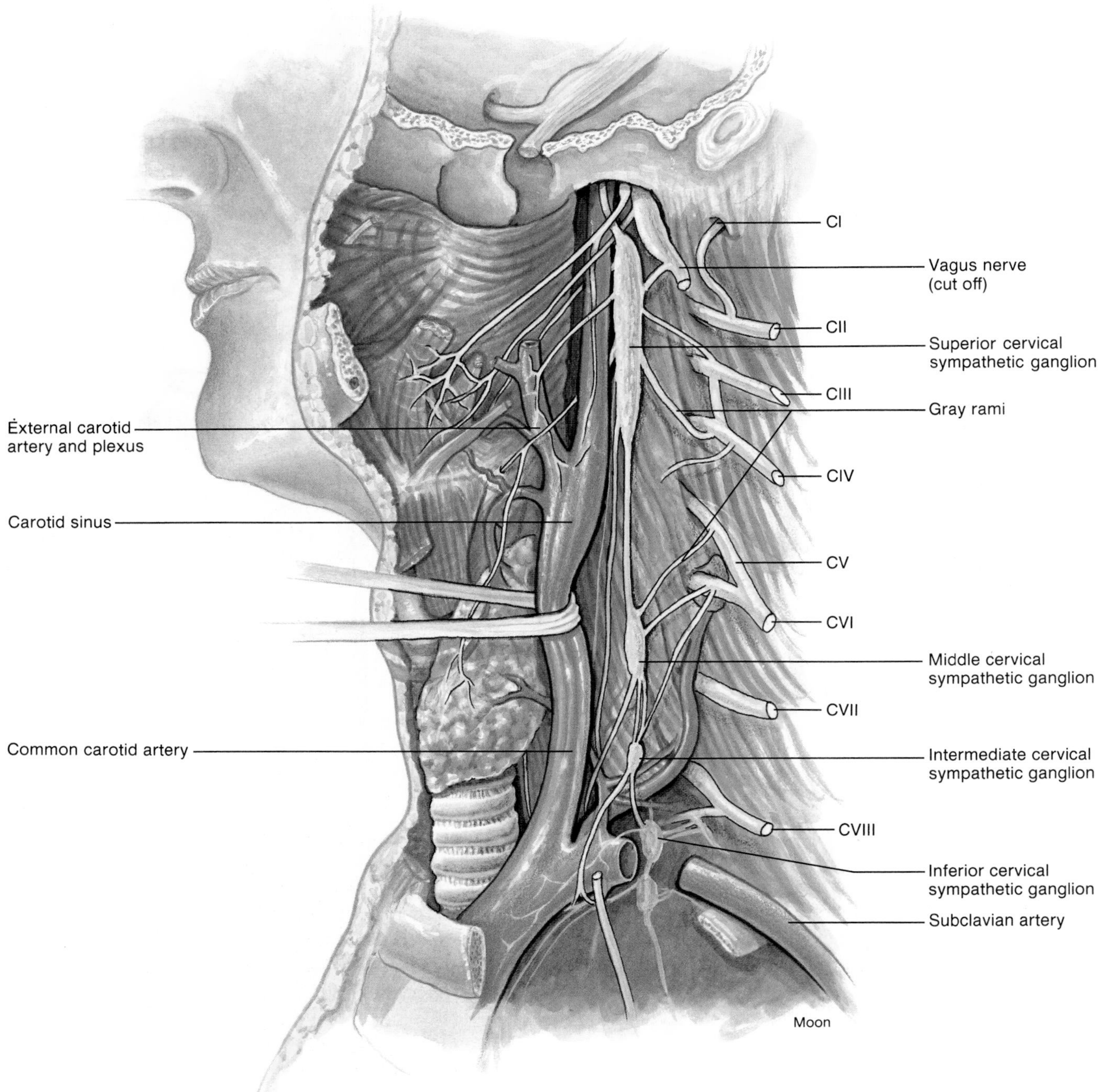


*mesenteric,* and *inferior mesenteric ganglia* (figs. 10.5 and 10.6). Postganglionic fibers that arise from the collateral ganglia innervate organs of the digestive, urinary, and reproductive systems.

***Adrenal Glands.*** The paired adrenal glands are located above each kidney. Each adrenal is composed of two parts: an outer **cortex** and an inner **medulla.** These two parts are really two functionally different glands with different embryonic origins, different hormones, and different regulatory mechanisms. The adrenal cortex secretes steroid hormones; the adrenal medulla secretes the hormone **epinephrine** (adrenalin) and, to a lesser degree, **norepinephrine,** when it is stimulated by the sympathetic system.

The adrenal medulla is a modified sympathetic ganglion whose cells are derived from postganglionic sympathetic neurons. The cells of the adrenal medulla are

Figure 10.5. The collateral sympathetic ganglia: the celiac and the superior and inferior mesenteric ganglia.

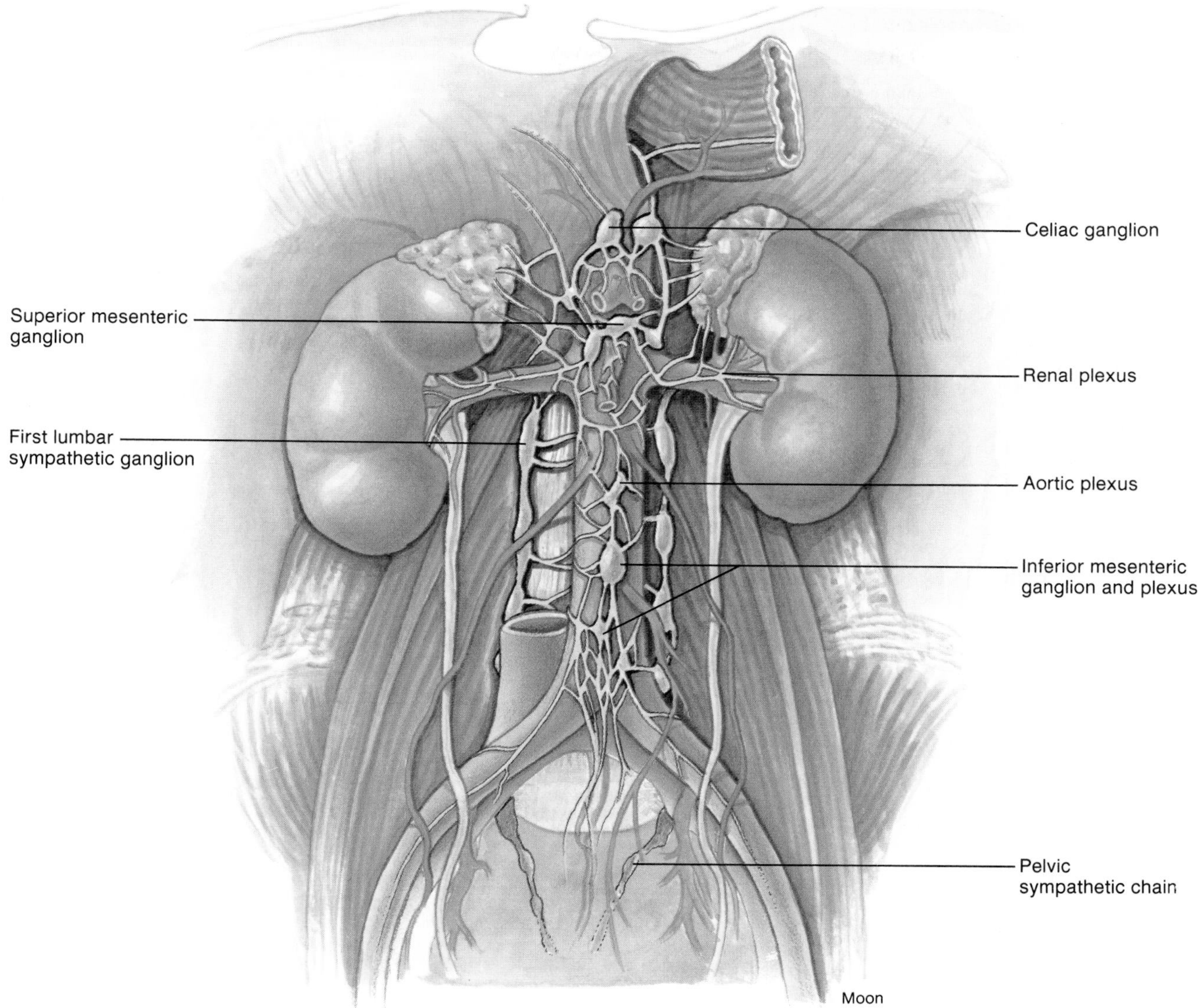


innervated by preganglionic sympathetic fibers originating in the thoracic level of the spinal cord; they secrete epinephrine into the blood in response to this neural stimulation. The effects of epinephrine are complementary to those of the neurotransmitter norepinephrine, which is released from postganglionic sympathetic nerve endings. For this reason and because the adrenal medulla is stimulated as part of the mass activation of the sympathetic system, the two are often grouped together as the **sympathoadrenal system.**

### Parasympathetic (Craniosacral) Division

The parasympathetic system is also known as the *craniosacral division* of the autonomic system. This is because its preganglionic fibers originate in the brain (specifically, the midbrain and the medulla oblongata of the brain stem) and in the second through fourth sacral levels of the spinal cord. These preganglionic parasympathetic fibers synapse in ganglia that are located next to—or actually within—the organs innervated. These parasympathetic ganglia, which are called **terminal ganglia,** supply the postganglionic fibers that synapse with the effector cells. Tables 10.2 and 10.3 show the comparative structures of the sympathetic and parasympathetic divisions. It should be noted that, unlike sympathetic fibers, most parasympathetic fibers do not travel within spinal nerves. As a result, cutaneous effectors (blood vessels, sweat glands, and arrector pili muscles) and blood vessels in skeletal muscles receive sympathetic but not parasympathetic innervation.

**Figure 10.6.** The autonomic nervous system. The sympathetic division is shown in color, the parasympathetic in black. The solid lines indicate preganglionic fibers, and the dotted lines indicate postganglionic fibers.

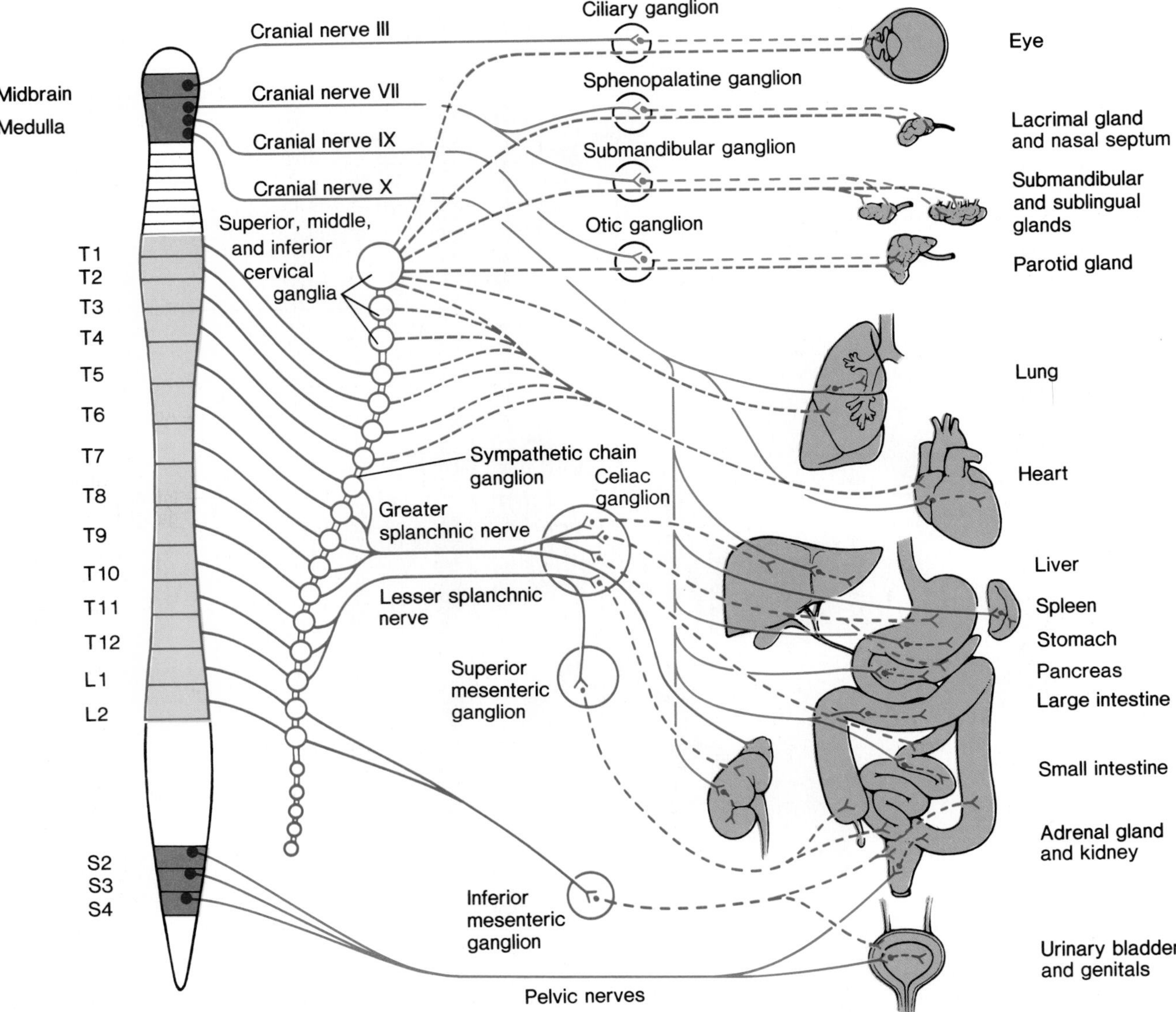


Four of the twelve pairs of cranial nerves (described in chapter 8) contain preganglionic parasympathetic fibers. These are the oculomotor (III), facial (VII), glossopharyngeal (IX), and vagus (X) nerves. Parasympathetic fibers within the first three of these cranial nerves synapse in ganglia located in the head; fibers in the vagus nerve synapse in terminal ganglia located in many regions of the body.

The oculomotor nerve contains somatic motor and parasympathetic fibers that originate in the oculomotor nuclei of the midbrain. These parasympathetic fibers synapse in the *ciliary ganglion,* whose postganglionic fibers innervate the ciliary muscle and constrictor fibers in the iris of the eye. Preganglionic fibers that originate in the medulla oblongata travel in the facial nerve to the *pterygopalatine ganglion,* also called the sphenopalatine ganglion, which sends postganglionic fibers to the nasal mucosa, pharynx, palate, and lacrimal glands. Another group of fibers in the facial nerve terminate in the *submandibular ganglion,* which sends postganglionic fibers to the submandibular and sublingual salivary glands. Preganglionic fibers of the glossopharyngeal nerve synapse in the *otic ganglion,* which sends postganglionic fibers to innervate the parotid salivary gland.

Other nuclei in the medulla oblongata contribute preganglionic fibers to the very long *tenth cranial,* or *vagus nerves* (the "vagrant" or "wandering" nerves). These preganglionic fibers travel through the neck to the thoracic

**Table 10.2** The sympathetic (thoracolumbar) system

| Parts of Body Innervated | Spinal Origin of Preganglionic Fibers | Origin of Postganglionic Fibers |
|---|---|---|
| Eye | C-8 and T-1 | Cervical ganglia |
| Head and neck | T-1 to T-4 | Cervical ganglia |
| Heart and lungs | T-1 to T-5 | Upper thoracic (paravertebral) ganglia |
| Upper extremity | T-2 to T-9 | Lower cervical and upper thoracic (paravertebral) ganglia |
| Upper abdominal viscera | T-4 to T-9 | Celiac and superior mesenteric (collateral) ganglia |
| Adrenal | T-10 and T-11 | — |
| Urinary and reproductive systems | T-12 to L-2 | Celiac and inferior mesenteric (collateral) ganglia |
| Lower extremities | T-9 to L-2 | Lumbar and upper sacral (paravertebral) ganglia |

**Table 10.3** The parasympathetic (craniosacral) system

| Effector Organs | Origin of Preganglionic Fibers | Nerve | Location of Terminal Ganglia |
|---|---|---|---|
| Eye (ciliary and iris muscles) | Midbrain (cranial) | Oculomotor (third cranial) nerve | Ciliary ganglion |
| Lacrimal, mucous, and salivary glands in head | Medulla oblongata (cranial) | Facial (seventh cranial) nerve | Sphenopalatine and submandibular ganglia |
| Parotid (salivary) gland | Medulla oblongata (cranial) | Glossopharyngeal (ninth cranial) nerve | Otic ganglion |
| Heart, lungs, gastrointestinal tract, liver, pancreas | Medulla oblongata (cranial) | Vagus (tenth cranial) nerve | Terminal ganglia in or near organ |
| Lower half of large intestine, rectum, urinary bladder, and reproductive organs | S-2 to S-4 (sacral) | Through pelvic spinal nerves | Terminal ganglia near organs |

**Table 10.4** Some comparisons between the structure of the sympathetic and parasympathetic systems

| | Sympathetic | Parasympathetic |
|---|---|---|
| Origin of Preganglionic Outflow | Thoracolumbar levels of spinal cord | Midbrain, hindbrain, and sacral levels of spinal cord |
| Location of Ganglia | Chain of paravertebral ganglia and prevertebral (collateral) ganglia | Terminal ganglia in or near effector organs |
| Distribution of Postganglionic Fibers | Throughout the body | Mainly limited to the head and the viscera of the chest, abdomen, and pelvis |
| Divergence of Impulses from Pre- to Postganglionic Fibers | Great divergence (one preganglionic may activate twenty postganglionic fibers) | Little divergence (one preganglionic only activates a few postganglionic fibers) |
| Mass Discharge of System as a Whole | Yes | Not normally |

cavity, and through the esophageal opening in the diaphragm to the abdominal cavity. In each region, some of these preganglionic fibers leave the main trunks of the vagus nerves and synapse with postganglionic neurons that are located *within* the innervated organs. The preganglionic vagus fibers are thus quite long, and provide parasympathetic innervation to the heart, lungs, esophagus, stomach, pancreas, liver, small intestine, and the upper half of the large intestine. Postganglionic parasympathetic fibers arise from terminal ganglia within these organs and synapse with effector cells (smooth muscles and glands).

Preganglionic fibers from the sacral levels of the spinal cord provide parasympathetic innervation to the lower half of the large intestine, rectum, and to the urinary and reproductive systems. These fibers, like those of the vagus, synapse with terminal ganglia located within the effector organs.

Parasympathetic nerves to the visceral organs thus consist of preganglionic fibers, whereas sympathetic nerves to these organs contain postganglionic fibers. A composite view of the sympathetic and parasympathetic systems is provided in figure 10.6, and these comparisons are summarized in table 10.4.

**Table 10.5** Effects of autonomic nerve stimulation on various visceral effector organs

| Effector Organ | Sympathetic Effect | Parasympathetic Effect |
|---|---|---|
| Eye | | |
| Iris (radial muscle) | Dilates pupil | ——— |
| Iris (sphincter muscle) | ——— | Constricts pupil |
| Ciliary muscle | Relaxes (for far vision) | Contracts (for near vision) |
| Glands | | |
| Lacrimal (tear) | ——— | Stimulates secretion |
| Sweat | Stimulates secretion | ——— |
| Salivary | Decreases secretion; saliva becomes thick | Increases secretion; saliva becomes thin |
| Stomach | ——— | Stimulates secretion |
| Intestine | ——— | Stimulates secretion |
| Adrenal medulla | Stimulates secretion of hormones | ——— |
| Heart | | |
| Rate | Increases | Decreases |
| Conduction | Increases rate | Decreases rate |
| Strength | Increases | ——— |
| Blood Vessels | Mostly constricts; affects all organs | Dilates in a few organs (e.g., penis) |
| Lungs | | |
| Bronchioles (tubes) | Dilates | Constricts |
| Mucous glands | Inhibits secretion | Stimulates secretion |
| Gastrointestinal Tract | | |
| Motility | Inhibits movement | Stimulates movement |
| Sphincters | Stimulates closing | Inhibits closing |
| Liver | Stimulates hydrolysis of glycogen | ——— |
| Adipose (fat) cells | Stimulates hydrolysis of fat | ——— |
| Pancreas | Inhibits exocrine secretions | Stimulates exocrine secretions |
| Spleen | Stimulates contraction | ——— |
| Urinary Bladder | Helps set muscle tone | Stimulates contraction |
| Piloerector Muscles | Stimulates erection of hair and goose bumps | ——— |
| Uterus | If pregnant: contraction<br>If not pregnant: relaxation | ——— |
| Penis | Erection; ejaculation | Erection (due to vasodilation) |

1. *Using a simple line diagram, illustrate the sympathetic pathway (a) from the spinal cord to the heart, and (b) from the spinal cord to the adrenal gland. Label the preganglionic and postganglionic fibers and the ganglion.*
2. *Describe the mass activation of the sympathetic system, and explain the significance of the term* sympathoadrenal system.
3. *Using a simple line diagram, illustrate the parasympathetic pathway from the brain to the heart. Compare the location of the pre- and postganglionic fibers and the location of their ganglia with those of the sympathetic division.*

## Functions of the Autonomic Nervous System

The sympathetic division of the autonomic system activates the body to "fight or flight," largely through the release of norepinephrine from postganglionic fibers and the secretion of epinephrine from the adrenal medulla. The parasympathetic division often has antagonistic effects through the release of acetylcholine from its postganglionic fibers. A balance between the actions of both divisions of the autonomic system is thus required for homeostasis to be preserved.

The sympathetic and parasympathetic divisions of the autonomic system affect the visceral organs in different ways. Mass activation of the sympathetic system prepares the body for intense physical activity in emergencies; the heart rate increases, blood glucose rises, and blood is diverted to the skeletal muscles (away from the visceral organs and skin). These and other effects are listed in table 10.5. The theme of the sympathetic system has been aptly summarized in a phrase: **"fight or flight."**

The effects of parasympathetic nerve stimulation are in many ways opposite to the effects of sympathetic stimulation. The parasympathetic system, however, is not normally activated as a whole. Stimulation of separate parasympathetic nerves can result in slowing of the heart, dilation of visceral blood vessels, and an increased activity of the digestive tract (table 10.5). The different responses of a visceral organ to sympathetic and parasympathetic nerve activity are due to the fact that the postganglionic fibers of these two divisions release different neurotransmitters.

**Figure 10.7.** Neurotransmitters of the autonomic motor system. ACh = acetylcholine; NE = norepinephrine; E = epinephrine. Those nerves that release ACh are called cholinergic; those nerves that release NE are called adrenergic. The adrenal medulla secretes both epinephrine (85%) and norepinephrine (15%) as hormones into the blood.

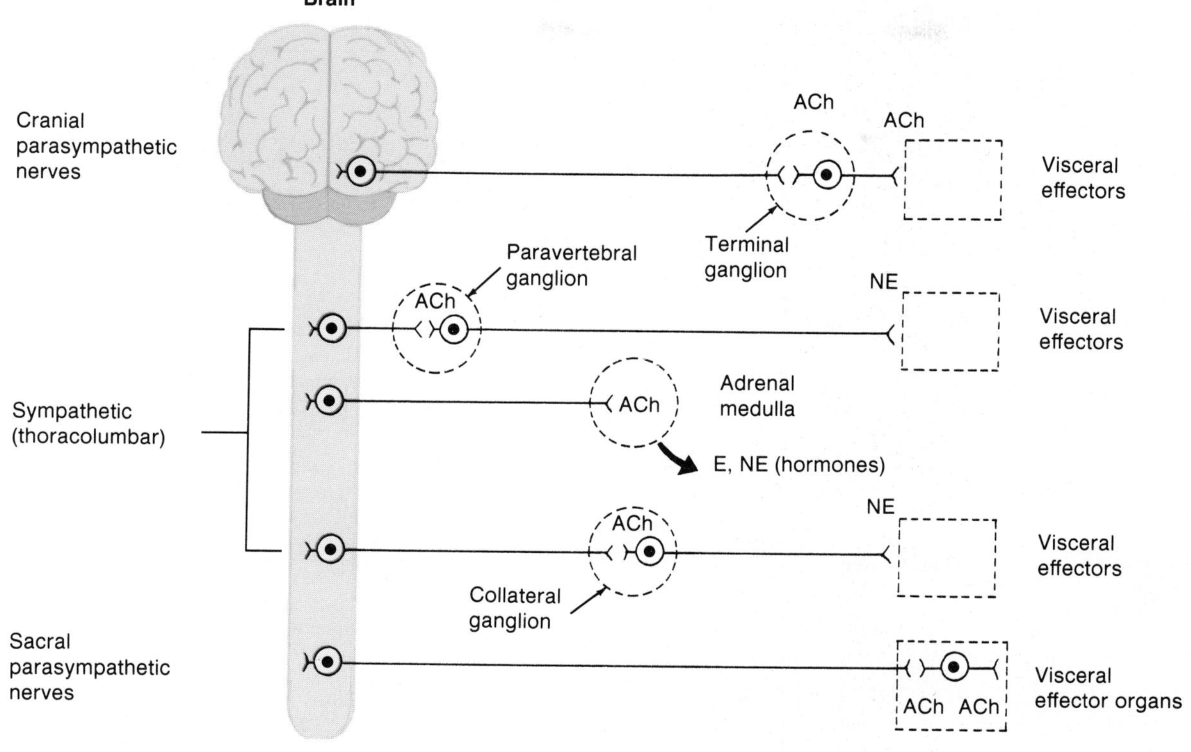


**Figure 10.8.** The structure of the catecholamines norepinephrine and epinephrine.

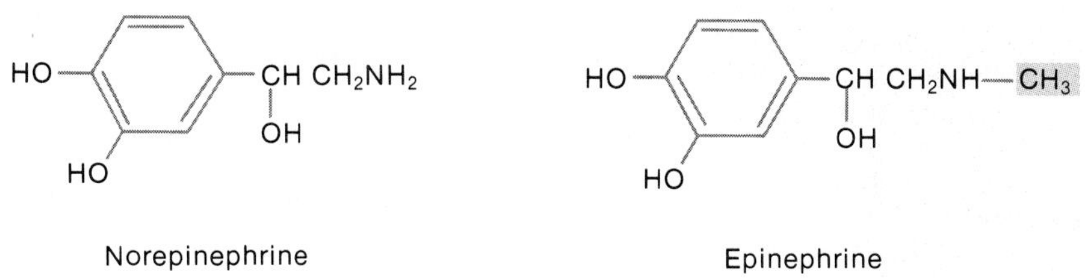

HO, HO — CH(OH) $CH_2NH_2$

Norepinephrine

HO, HO — CH(OH) $CH_2NH$—$CH_3$

Epinephrine

## Neurotransmitters of the Autonomic Nervous System

**Acetylcholine** (*ACh*) is the neurotransmitter of all preganglionic fibers (both sympathetic and parasympathetic). Acetylcholine is also the transmitter released by all parasympathetic postganglionic fibers at their synapses with effector cells (fig. 10.7). Transmission at these synapses is thus said to be **cholinergic.**

The neurotransmitter released by most postganglionic sympathetic nerve fibers is **norepinephrine** (*noradrenalin*). Transmission at these synapses is thus said to be **adrenergic.** There are a few exceptions to this rule: some sympathetic fibers that innervate blood vessels in skeletal muscles, as well as sympathetic fibers to sweat glands, release ACh (are cholinergic).

In view of the fact that the cells of the adrenal medulla are derived from postganglionic sympathetic neurons, it is not surprising that the hormones they secrete (normally about 85% epinephrine and 15% norepinephrine) are similar to the transmitter of postganglionic sympathetic neurons. Epinephrine differs from norepinephrine only in the presence of an additional methyl ($CH_3$) group, as shown in figure 10.8. Epinephrine, norepinephrine, and dopamine (a transmitter within the CNS) are collectively termed **catecholamines.** The catecholamines, together with serotonin (another transmitter within the CNS), are often referred to as *monoamines,* or *biogenic amines.*

**Table 10.6** Adrenergic and cholinergic effects of sympathetic and parasympathetc nerves

| | Effect of | | | |
|---|---|---|---|---|
| | Sympathetic | | Parasympathetic | |
| **Organ** | *Action* | *Receptor* | *Action* | *Receptor* |
| Eye | | | | |
| Iris | | | | |
| Radial muscle | Contracts | $\alpha$ | . . . | . . . |
| Circular muscle | . . . | . . . | Contracts | M |
| Heart | | | | |
| Sinoatrial node | Accelerates | $\beta_1$ | Decelerates | M |
| Contractility | Increases | $\beta_1$ | Decreases (atria) | M |
| Vascular smooth muscle | | | | |
| Skin, splanchnic vessels | Contracts | $\alpha$ | . . . | . . . |
| Skeletal muscle vessels | Relaxes | $\beta_2$ | . . . | . . . |
| | Relaxes | M* | . . . | . . . |
| Bronchiolar smooth muscle | Relaxes | $\beta_2$ | Contracts | M |
| Gastrointestinal tract | | | | |
| Smooth muscle | | | | |
| Walls | Relaxes | $\beta_2$ | Contracts | M |
| Sphincters | Contracts | $\alpha$ | Relaxes | M |
| Secretion | . . . | . . . | Increases | M |
| Myenteric plexus | Inhibits | $\alpha$ | . . . | . . . |
| Genitourinary smooth muscle | | | | |
| Bladder wall | Relaxes | $\beta_2$ | Contracts | M |
| Sphincter | Contracts | $\alpha$ | Relaxes | M |
| Uterus, pregnant | Relaxes | $\beta_2$ | . . . | . . . |
| | Contracts | $\alpha$ | . . . | . . . |
| Penis, seminal vesicles | Ejaculation | $\alpha$ | Erection | M |
| Skin | | | | |
| Pilomotor smooth muscle | Contracts | $\alpha$ | . . . | . . . |
| Sweat glands | | | | |
| Thermoregulatory | Increases | M | . . . | . . . |
| Apocrine (stress) | Increases | $\alpha$ | . . . | . . . |

*Vascular smooth muscle in skeletal muscle has sympathetic cholinergic dilator fibers.

Adrenergic receptors are indicated as alpha ($\alpha$) or beta ($\beta$); cholinergic receptors are indicated as muscarinic (M).

Reproduced, with permission, from Katzung, B. G.: *Basic and Clinical Pharmacology,* 4th edition, Copyright Appleton & Lange, 1989.

## Responses to Adrenergic Stimulation

Adrenergic stimulation—by epinephrine in the blood and by norepinephrine released from sympathetic nerve endings—has both excitatory and inhibitory effects. The heart, dilatory muscles of the iris, and the smooth muscles of many blood vessels are stimulated to contract. The smooth muscles of the bronchioles and of some blood vessels, however, are inhibited from contracting; adrenergic chemicals, therefore, cause these structures to dilate.

Since excitatory and inhibitory effects can be produced in different tissues by the same chemical, the responses clearly depend on the biochemistry of the tissue cells rather than on the intrinsic properties of the chemical. Included in the biochemical differences among the target tissues for catecholamines are differences in the *membrane receptor proteins* for these chemical agents. There are two major classes of these receptor proteins, designated **alpha** ($\alpha$) and **beta** ($\beta$) **adrenergic receptors.** (The interaction of neurotransmitters and receptor proteins in the postsynaptic membrane was described in chapter 7.)

Further experiments have revealed that there are two subtypes of each category of adrenergic receptor. These are designated by subscripts: $\alpha_1$ and $\alpha_2$; $\beta_1$ and $\beta_2$. Scientists have developed compounds that selectively bind more to one or the other type of adrenergic receptor and, by this means, either promote or inhibit the normal action produced when epinephrine or norepinephrine binds to the receptor. By using these selective compounds, it has been possible to determine the adrenergic receptor subtype in each organ and to correlate that receptor with the physiological effect of autonomic neurotransmitters in that organ (table 10.6). Both types of beta receptors appear to produce these effects by stimulating the production of cyclic AMP (discussed in chapter 7) within the target cells.

The activation of the $\alpha_2$ receptors has the opposite effect—cyclic AMP production is blocked. The response to $\alpha_1$ receptors may be mediated by a rise in the cytoplasmic concentration of $Ca^{++}$, which serves as a "second messenger" in the target cell instead of cyclic AMP. Each of these changes, it should be remembered, ultimately results in the characteristic response of the tissue to the autonomic neurotransmitter.

Review of table 10.6 reveals certain generalities about the actions of adrenergic receptors. The stimulation of alpha-adrenergic receptors consistently causes constriction of smooth muscles. One can thus state that the vasoconstrictor effect of sympathetic nerves always results from the activation of alpha-adrenergic receptors. The effects of beta-adrenergic activation are more complex; these receptors stimulate the relaxation of smooth muscles (in the digestive tract, brochioles, and uterus, for example), but stimulate contraction of cardiac muscle and promote an increase in cardiac rate.

The use of drugs that selectively stimulate or block (act as *agonists* or *antagonists*) adrenergic receptors is of great clinical benefit. People with hypertension, for example, may receive a beta-blocking drug, such as *propranolol,* which reduces the cardiac rate. People with asthma may be treated with epinephrine, which stimulates bronchodilation (a $\beta_2$ effect), but since epinephrine also stimulates the heart (a $\beta_1$ effect), it is advisable to use more selective $\beta_2$ agonist drugs for the treatment of asthma. Compounds that stimulate $\alpha_1$ adrenergic receptors, such as *phenylephrine,* are often part of medications to relieve a stuffy nose (by causing vasoconstriction in the nasal mucosa).

## Responses to Cholinergic Stimulation

Somatic motor neurons, all preganglionic autonomic neurons, and all postganglionic parasympathetic neurons are cholinergic—they release acetylcholine as a neurotransmitter. The cholinergic effects of somatic motor neurons and preganglionic autonomic neurons are always excitatory. The cholinergic effects of postganglionic parasympathetic fibers are usually excitatory, but there are notable exceptions—the parasympathetic fibers innervating the heart, for example, cause slowing of the heart rate. It is useful to remember that the effects of parasympathetic stimulation are, in general, opposite to the effects of sympathetic stimulation.

There are two known subtypes of cholinergic receptors. The drug *muscarine,* derived from some poisonous mushrooms, stimulates the cholinergic receptor proteins in the target organs of postganglionic parasympathetic nerve fibers (such as in the heart, eye, and digestive system). Muscarine does not stimulate ACh receptor proteins in autonomic ganglia or at the neuromuscular junction of skeletal muscle fibers. The ACh receptors stimulated by muscarine are therefore called **muscarinic receptors,** and the effects produced by parasympathetic nerves in their target organs are called *muscarinic effects* (indicated in table 10.6).

The drug *nicotine,* derived from the tobacco plant, specifically stimulates cholinergic transmission of preganglionic fibers at the autonomic ganglia, and also activation of the neuromuscular junction of skeletal muscles. These ACh receptors are thus called **nicotinic receptors** to distinguish them from the muscarinic receptors. The drug *curare,* used clinically to cause skeletal muscle relaxation, specifically blocks nicotinic receptors but has little effect on muscarinic receptors.

The muscarinic effects of ACh are specifically inhibited by the drug **atropine,** derived from the deadly nightshade plant (*Atropa belladonna*). Indeed, extracts of this plant were used by women during the middle ages to dilate their pupils (atropine inhibits parasympathetic stimulation of the iris). This was done to enhance their beauty (belladonna—beautiful woman). Atropine is used clinically today to dilate pupils during eye examinations, to dry mucous membranes of the respiratory tract prior to general anesthesia, and to inhibit spasmodic contractions of the lower digestive tract.

## Organs with Dual Innervation

Most visceral organs receive a dual innervation—they are innervated by both sympathetic and parasympathetic fibers. When this occurs, the effects of these two divisions may be antagonistic, complementary, or cooperative.

***Antagonistic Effects.*** The effects of sympathetic and parasympathetic innervation of the pacemaker region of the heart is the best example of the antagonism of these two systems. In this case sympathetic and parasympathetic fibers innervate the same cells. Adrenergic stimulation from sympathetic fibers increases the heart rate, and cholinergic stimulation from parasympathetic fibers inhibits the pacemaker cells and, thus, decreases the heart rate. A reverse of this antagonism is seen in the digestive tract, where sympathetic nerves inhibit and parasympathetic nerves stimulate intestinal movements and secretions.

The effects of sympathetic and parasympathetic stimulation on the diameter of the pupil of the eye are analogous to the reciprocal innervation of flexor and extensor skeletal muscles by somatic motor neurons (chapter 12). This is because the iris contains antagonistic muscle layers. Contraction of the radial muscles, which is stimulated by sympathetic nerves, causes dilation; contraction of the circular muscles, which are innervated by parasympathetic nerve endings, causes constriction of the pupils (fig. 10.9).

**Figure 10.9.** Reciprocal innervation of the iris muscles by the sympathetic and parasympathetic systems. Stimulation of sympathetic nerves produces contraction of the dilator (radial) muscles, which enlarges the pupil. Stimulation of the parasympathetic nerves produces contraction of the constrictor (circular) muscle layer, which makes the pupil smaller.

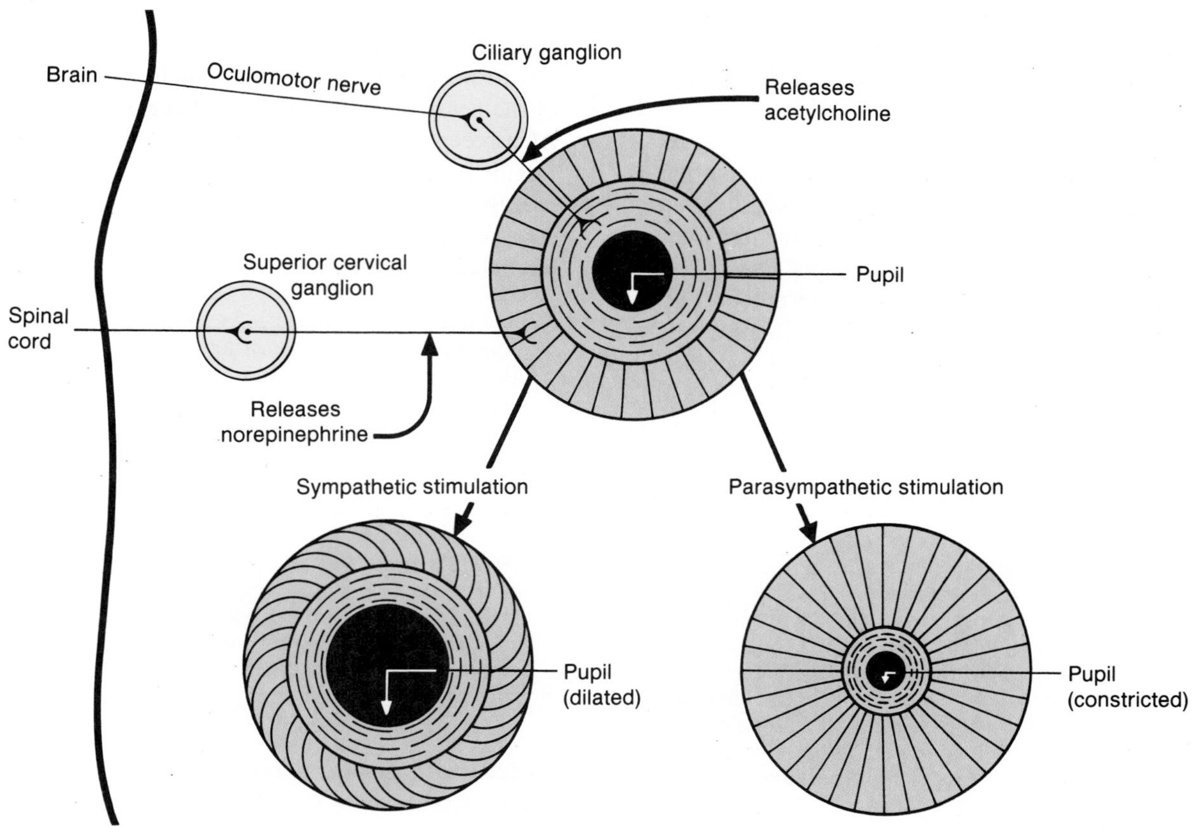


***Complementary Effects.*** The effects of sympathetic and parasympathetic stimulation on salivary gland secretion are complementary. The secretion of watery saliva is stimulated through parasympathetic nerves, which also stimulate the secretion of other exocrine glands in the digestive tract. Impulses through sympathetic nerves stimulate the constriction of blood vessels throughout the digestive tract. The resultant decrease in blood flow to the salivary glands causes the production of a thicker, more viscous saliva.

***Cooperative Effects.*** The effects of sympathetic and parasympathetic stimulation on the urinary and reproductive systems are cooperative. Erection of the penis, for example, is due to vasodilation resulting from parasympathetic nerve stimulation; ejaculation is due to stimulation through sympathetic nerves. Although the contraction of the urinary bladder is myogenic (independent of nerve stimulation), it is promoted in part by the action of parasympathetic nerves. This *micturition,* or urination, urge and reflex is also enhanced by sympathetic nerve activity, which increases the tone of the bladder muscles. Emotional states that are accompanied by high sympathetic nerve activity (such as extreme fear) may thus result in reflex urination at bladder volumes that are normally too low to trigger this reflex.

## Organs without Dual Innervation

Although most organs are innervated by both sympathetic and parasympathetic nerves, some—including the adrenal medulla, arrector pili muscles, sweat glands, and most blood vessels—receive only sympathetic innervation. In these cases regulation is achieved by increases or decreases in the tone (firing rate) of the sympathetic fibers. Constriction of the blood vessels, for example, is produced by increased sympathetic activity, which stimulates alpha-adrenergic receptors, and vasodilation results from decreased sympathetic nerve stimulation.

The sympathoadrenal system is required for *nonshivering thermogenesis:* animals deprived of their sympathetic system and adrenals cannot tolerate cold stress. The sympathetic system itself is required for proper thermoregulatory responses to heat. In a hot room, for example, decreased sympathetic stimulation produces dilation of the blood vessels in the surface of the skin, which increases cutaneous blood flow and provides better heat radiation.

**Table 10.7** Some effects which are stimulated by sensory input from afferent fibers in the vagus, which are transmitted to centers in the medulla oblongata

| Organs | Type of Receptors | Reflex Effects |
|---|---|---|
| Lungs | Stretch receptors | Inhibit further inhalation; stimulate an increase in cardiac rate and vasodilation |
| | Type J receptors | Stimulated by pulmonary congestion—produce feelings of breathlessness and cause a reflex fall in cardiac rate and blood pressure |
| Aorta | Chemoreceptors | Stimulated by rise in $CO_2$ and fall in $O_2$—produce increased rate of breathing, fall in heart rate, and vasoconstriction |
| | Baroreceptors | Stimulated by increased blood pressure—produce a reflex decrease in heart rate |
| Heart | Atrial stretch receptors | Inhibit antidiuretic hormone secretion, thus increasing the volume of urine excreted |
| | Stretch receptors in ventricles | Produce a reflex decrease in heart rate and vasodilation |
| Gastrointestinal tract | Stretch receptors | Feelings of satiety, discomfort, and pain |

During exercise, in contrast, there is increased sympathetic activity, which causes constriction of the blood vessels in the skin of the limbs and stimulation of sweat glands in the trunk.

The eccrine sweat glands in the trunk secrete a watery fluid in response to cholinergic sympathetic stimulation. Evaporation of this dilute sweat helps to cool the body. The eccrine sweat glands also secrete a chemical called **bradykinin** in response to sympathetic stimulation. Bradykinin stimulates dilation of the surface blood vessels near the sweat glands, helping to radiate some heat despite the fact that other cutaneous blood vessels are constricted. At the conclusion of exercise sympathetic stimulation is reduced, causing cutaneous blood vessels to dilate. This increases blood flow to the skin, which helps to eliminate metabolic heat. Notice that all of these thermoregulatory responses are achieved without the direct involvement of the parasympathetic system.

## Control of the Autonomic Nervous System by Higher Brain Centers

Visceral functions are regulated, to a large degree, by autonomic reflexes. In most autonomic reflexes sensory input is transmitted to brain centers, which integrate this information and appropriately respond by modifying the activity of efferent preganglionic autonomic neurons. The neural centers that directly control the activity of autonomic nerves are influenced by higher brain areas as well as by sensory input.

The **medulla oblongata** (chapter 8) of the brain stem is the area that most directly controls the activity of the autonomic system. Almost all autonomic responses can be elicited by experimental stimulation of the medulla, which contains centers for the control of the cardiovascular, pulmonary, urinary, reproductive, and digestive systems. Much of the sensory input to these centers travels in the afferent fibers of the vagus nerve, which is a mixed nerve containing both sensory and motor fibers. These reflexes are listed in table 10.7.

**Autonomic dysreflexia,** a serious condition producing rapid elevations in blood pressure that can lead to stroke (cerebrovascular accident), occurs in 85% of people with quadriplegia and others with spinal cord lesions above the sixth thoracic level. Lesions to the spinal cord first produce the symptoms of spinal shock, characterized by the loss of both skeletal muscle and autonomic reflexes. After a period of time both types of reflexes return in an exaggerated state; the skeletal muscles may become spastic, due to absence of higher inhibitory influences, and the visceral organs experience denervation hypersensitivity. Patients in this state have difficulty emptying their urinary bladders and must often be catheterized.

Noxious stimuli, such as overdistension of the urinary bladder, can result in reflex activation of the sympathetic nerves below the spinal cord lesion. This produces goose bumps, cold skin, and vasoconstriction in the regions served by the spinal cord below the level of the lesion. The rise in blood pressure resulting from this vasoconstriction activates pressure receptors that transmit impulses along sensory nerve fibers to the medulla. In response to this sensory input, the medulla directs a reflex slowing of the heart and vasodilation. Since descending impulses are blocked by the spinal lesion, however, the skin is warm and moist (due to vasodilation and sweat gland secretion) above the lesion, but cold below the level of spinal cord damage.

Although the medulla oblongata directly regulates the activity of autonomic motor fibers, the medulla is itself responsive to regulation by higher brain areas. One of these is the hypothalamus (chapter 8), which is the brain region that contains centers for the control of body temperature, hunger, thirst, regulation of the pituitary gland, and—together with the limbic system and cerebral cortex—various emotional states. Complex circuits between the hypothalamus and limbic system allow for autonomic responses to emotions. Blushing, pallor, fainting, breaking out in a cold sweat, and "butterflies in the stomach" are only some of the many visceral reactions that, as a result of autonomic activation, accompany emotions.

The autonomic correlates of motion sickness—nausea, sweating, and cardiovascular changes—are abolished by cutting the efferent tracts of the cerebellum. This demonstrates that impulses from the cerebellum to the medulla influence activity of the autonomic system. Experimental and clinical observations have also demonstrated that the frontal and temporal lobes of the cerebral cortex influence lower brain areas as part of their involvement in emotion and personality.

One of the most dramatic examples of the role of higher brain areas in personality and emotion is the famous crowbar accident, which occurred in 1848. A twenty-five-year-old railroad foreman, Phineas P. Gage, was tamping gunpowder into a hole in a rock when it exploded. The rod—three feet, seven inches long and one and one-fourth inches thick—was driven through his left eye, passed through his brain, and emerged in the back of his skull.

After a few minutes of convulsions, Gage got up, rode a horse three-quarters of a mile into town, and walked up a long flight of stairs to see a doctor. He recovered well, with no noticeable sensory or motor deficits. His associates, however, noted striking personality changes. Before the accident Gage was a responsible, capable, and financially prudent man. Afterwards, he appeared to lose social inhibitions, engaged in gross profanity (which he did not do before the accident), and seemed to be tossed about by chance whims. He was eventually fired from his job, and his previous friends remarked that he was "no longer Gage."

---

1. *Define the meaning of the terms* adrenergic *and* cholinergic, *and use these terms to describe the neurotransmitters of different autonomic nerve fibers.*
2. *List the effects of sympathoadrenal stimulation on different effector organs, and indicate which effects are due to alpha or beta receptor stimulation.*
3. *Describe the effects of the drug* atropine, *and explain these effects in terms of the actions of the parasympathetic system.*
4. *Explain how the sympathetic and parasympathetic systems can have antagonistic, cooperative, and complementary effects, and give examples.*
5. *Explain the structures and mechanisms involved when a person blushes.*

---

## Summary

### Neural Control of Involuntary Effectors p. 262

I. Preganglionic autonomic neurons originate from the brain or spinal cord; postganglionic neurons originate from ganglia located outside the CNS.

II. Smooth muscle, cardiac muscle, and glands receive autonomic innervation.
   - A. The involuntary effectors are somewhat independent of their innervation and become hypersensitive when their innervation is removed.
   - B. Autonomic nerves can have either excitatory or inhibitory effects on their target organs.

### Divisions of the Autonomic Nervous System p. 264

I. Preganglionic neurons of the sympathetic division originate in the spinal cord between the thoracic and lumbar levels.
   - A. Many of these fibers synapse with postganglionic neurons whose cell bodies are located in a double chain of sympathetic (paravertebral) ganglia outside the spinal cord.
   - B. Some preganglionic fibers synapse in collateral ganglia; these are the celiac, superior mesenteric, and the inferior mesenteric ganglia.
   - C. Some preganglionic fibers innervate the adrenal medulla, which secretes epinephrine (and some norepinephrine) into the blood in response to this stimulation.

II. Preganglionic parasympathetic fibers originate in the brain and in the sacral levels of the spinal cord.
   - A. Preganglionic parasympathetic fibers contribute to cranial nerves III, VII, IX, and X.
   - B. Preganglionic fibers of the vagus (X) nerve are very long and synapse in terminal ganglia located next to or within the innervated organ; short postganglionic fibers then innervate the effector cells.
   - C. The vagus provides parasympathetic innervation to the heart, lungs, esophagus, stomach, liver, small intestine, and upper half of the large intestine.
   - D. Parasympathetic outflow from the sacral levels of the spinal cord innervates terminal ganglia in the lower half of the large intestine, the rectum, and the urinary and reproductive systems.

### Functions of the Autonomic Nervous System p. 270

I. The sympathetic division of the autonomic system activates the body to "fight or flight" through adrenergic effects; the parasympathetic division often has antagonistic actions through cholinergic effects.

II. All preganglionic autonomic nerve fibers are cholinergic (use ACh as a neurotransmitter).
   - A. All postganglionic parasympathetic fibers are cholinergic.
   - B. Most postganglionic sympathetic fibers are adrenergic (use norepinephrine as a neurotransmitter).
   - C. Sympathetic fibers that innervate sweat glands and those that innervate blood vessels in skeletal muscles are cholinergic.

III. Adrenergic effects include stimulation of the heart, vasoconstriction in the viscera and skin, bronchodilation, and glycogenolysis in the liver.
   - A. There are two main groups of adrenergic receptor proteins: alpha and beta receptors.
   - B. Different organs have only alpha or beta receptors, and some organs (such as the heart) have both receptors.

C. There are two subtypes of alpha receptors ($\alpha_1$ and $\alpha_2$) and two subtypes of beta receptors ($\beta_1$ and $\beta_2$), which can be selectively stimulated or blocked by therapeutic drugs.

IV. Cholinergic effects of parasympathetic nerves are promoted by the drug muscarine and inhibited by atropine.

V. In organs with dual innervation, the actions of the sympathetic and parasympathetic divisions can be antagonistic, complementary, or cooperative.
   A. The effects are antagonistic in the heart and pupils.
   B. The actions are complementary in the regulation of salivary gland secretion and are cooperative in the regulation of the reproductive and urinary systems.

VI. In organs without dual innervation (such as most blood vessels), regulation is achieved by variations in sympathetic nerve activity.

VII. The medulla oblongata of the brain stem is the area that most directly controls the activity of the autonomic system.
   A. The medulla oblongata is in turn influenced by sensory input and by input from the hypothalamus.
   B. The hypothalamus orchestrates somatic, autonomic, and endocrine responses during various behavioral states.

## Review Activities

### Objective Questions

1. When a visceral organ is denervated
   (a) it ceases to function
   (b) it becomes less sensitive to subsequent stimulation by neurotransmitters
   (c) it becomes hypersensitive to subsequent stimulation
2. Parasympathetic ganglia are located
   (a) in a chain parallel to the spinal cord
   (b) in the dorsal roots of spinal nerves
   (c) next to or within the organs innervated
   (d) in the brain
3. The neurotransmitter of preganglionic sympathetic fibers is
   (a) norepinephrine
   (b) epinephrine
   (c) acetylcholine
   (d) dopamine
4. Which of the following results from stimulation of alpha-adrenergic receptors?
   (a) constriction of blood vessels
   (b) dilation of bronchioles
   (c) decreased heart rate
   (d) sweat gland secretion
5. Which of the following fibers release norepinephrine?
   (a) preganglionic parasympathetic fibers
   (b) postganglionic parasympathetic fibers
   (c) postganglionic sympathetic fibers in the heart
   (d) postganglionic sympathetic fibers in sweat glands
   (e) all of the above
6. The actions of sympathetic and parasympathetic fibers are cooperative in the
   (a) heart
   (b) reproductive system
   (c) digestive system
   (d) eyes
7. Propranolol is a "beta-blocker." It would therefore cause
   (a) vasodilation
   (b) slowing of the heart rate
   (c) increased blood pressure
   (d) secretion of saliva
8. Atropine blocks parasympathetic nerve effects. It would therefore cause
   (a) dilation of the pupils
   (b) decreased mucus secretion
   (c) decreased movements of the digestive tract
   (d) increased heart rate
   (e) all of the above
9. The area of the brain that is most directly involved in the reflex control of the autonomic system is the
   (a) hypothalamus
   (b) cerebral cortex
   (c) medulla oblongata
   (d) cerebellum
10. The two subtypes of cholinergic receptors are
   (a) adrenergic and nicotinic
   (b) dopaminergic and muscarinic
   (c) nicotinic and muscarinic
   (d) nicotinic and dopaminergic

### Essay Questions

1. Compare the sympathetic and parasympathetic systems in terms of the location of their ganglia and the distribution of their nerves.
2. Explain the anatomical and physiological relationship between the sympathetic nervous system and the adrenal glands.
3. Compare the effects of adrenergic and cholinergic stimulation on the cardiovascular and digestive systems.
4. Explain how effectors that receive only sympathetic innervation are regulated by the autonomic system.
5. Distinguish between the different types of adrenergic receptors, and explain how their differences are clinically exploited.

## Selected Readings


Decara, L. V. January 1970. Learning in the autonomic nervous system. *Scientific American.*

Goodman, L. S., and A. Gilman. 1980. *The pharmacological basis of therapeutics.* 6th ed. New York: The Macmillan Co.

Hoffman, B. B., and R. J. Lefkowitz. 1980. Alpha-adrenergic receptor subtypes. *New England Journal of Medicine* 302:1390.

Katzung, B. G., ed. 1984. *Basic and clinical pharmacology.* Los Altos: Lange Medical Publications.

Lefkowitz, B. B. 1976. Beta-adrenergic receptors: Recognition and regulation. *New England Journal of Medicine* 295:323.

Motulski, J. H., and P. A. Insel. 1982. Adrenergic receptors in man. *New England Journal of Medicine* 307:18.

Noback, C. E., and R. J. Demerest. 1975. *The human nervous system: basic principles of neurobiology.* 2d ed. New York: McGraw-Hill Book Company.

# 11

# ENDOCRINE GLANDS:

## *SECRETION AND ACTIONS OF HORMONES*

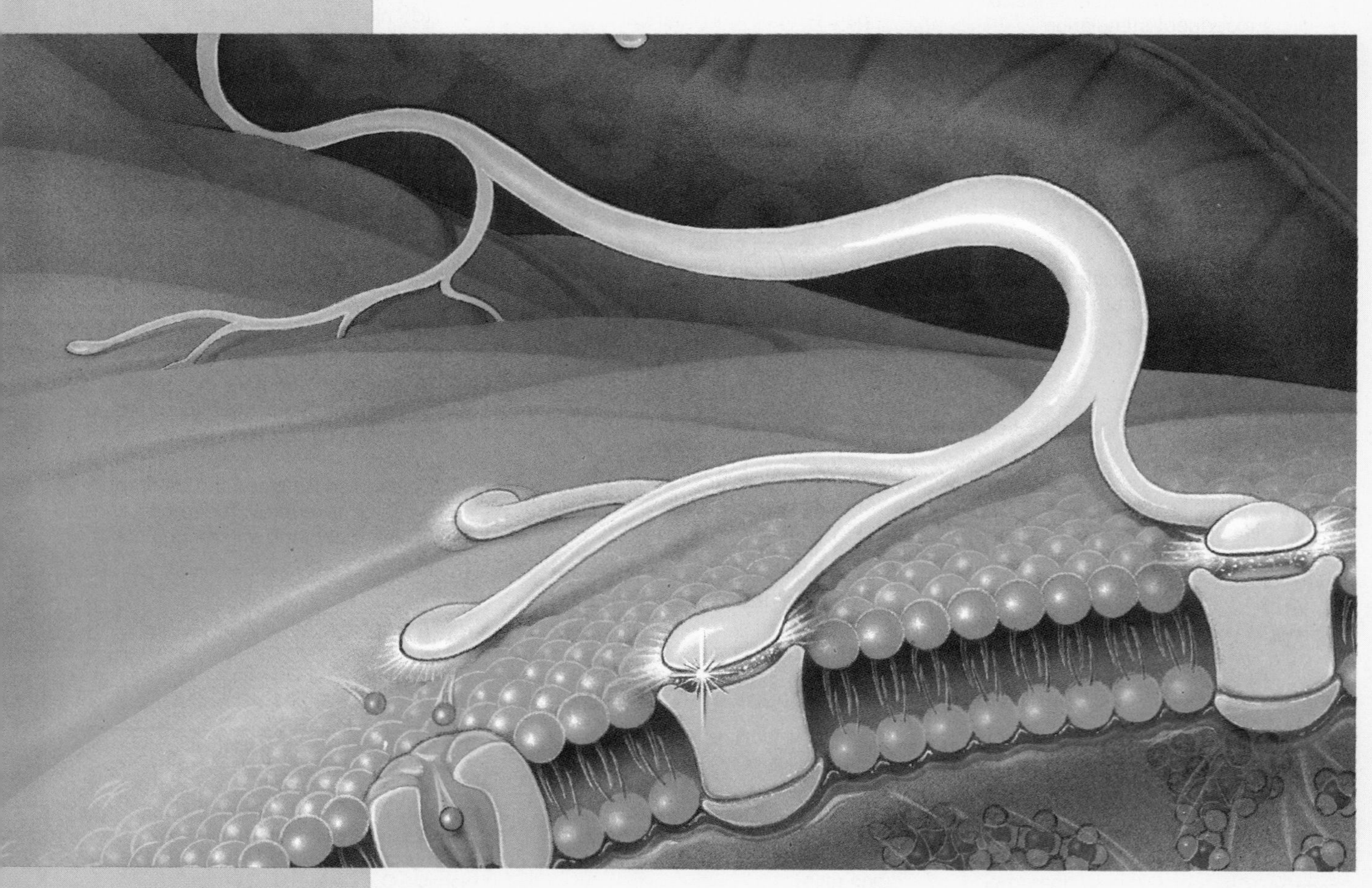

## Objectives

By studying this chapter, you should be able to

1. define the terms *hormone* and *endocrine gland,* and describe how chemical transformations in the endocrine gland or target cells can activate certain hormones
2. describe the different chemical classifications of hormones
3. explain how the concentrations of a hormone in the blood are regulated and how the effects of a hormone are influenced by its concentration
4. explain how different hormones can exert synergistic, permissive, or antagonistic effects
5. describe the mechanisms of hormone action in the case of steroid and thyroid hormones
6. describe the mechanism of hormone action in the case of amine, polypeptide, and glycoprotein hormones
7. describe the parts of the pituitary gland and the relationship between the pituitary gland and the hypothalamus
8. list the hormones secreted by the neurohypophysis (posterior pituitary) and explain the origin of these hormones and how the hypothalamus regulates their secretion
9. list the hormones of the adenohypophysis (anterior pituitary) and explain how their secretion is regulated by the hypothalamus
10. describe the production and actions of the thyroid hormones and explain how thyroid secretion is regulated
11. describe the location of the parathyroid glands and explain the actions of PTH and how secretion of this hormone is regulated
12. describe the types and actions of corticosteroids produced by the adrenal cortex and explain how the secretions of the adrenal cortex are regulated
13. describe the actions of epinephrine and norepinephrine, secreted by the adrenal medulla, and explain how the secretions of the adrenal medulla are regulated
14. explain why the pancreas is both an exocrine and an endocrine gland and describe the structure and functions of the islets of Langerhans
15. describe the actions of insulin and glucagon and explain how secretions of these hormones are regulated
16. identify briefly and describe the significance of hormones secreted by the pineal, thymus, gonads, and placenta
17. identify the chemical nature and physiological roles of the prostaglandins and describe some of the clinical benefits of drugs that inhibit prostaglandin synthesis

## Outline

**Figure 11.1.** (*a*) The anatomy of some of the endocrine glands. (*b*) The islets of Langerhans within the pancreas.

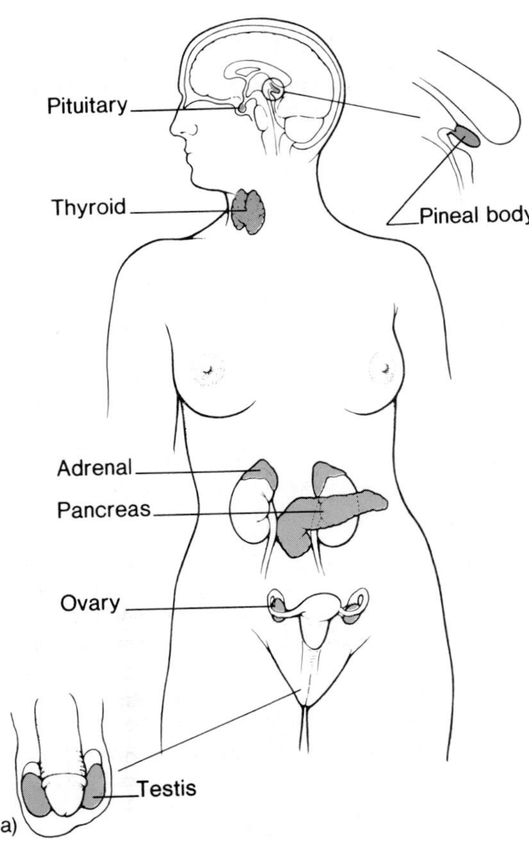


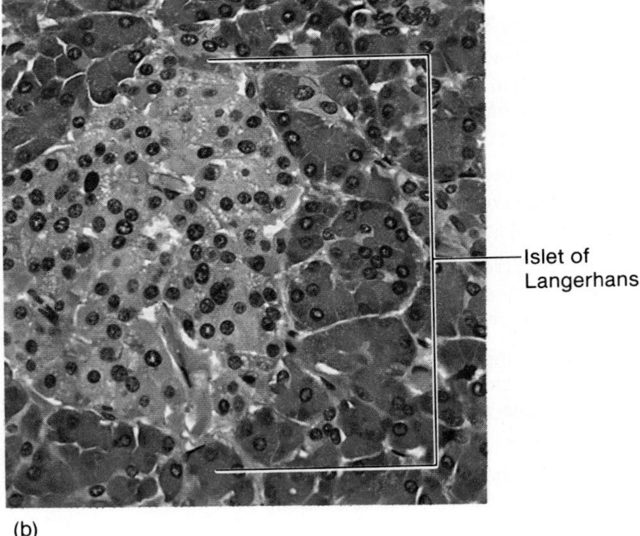


## *HORMONES: ACTIONS AND INTERACTIONS*

Hormones are regulatory molecules secreted into the blood by endocrine glands. The blood transports hormone molecules to the target organs, where each hormone exerts its characteristic regulatory function through events that occur at a cellular and molecular level. The particular effects of a given hormone are influenced by the concentration of that hormone in the blood, and by interactions with effects produced by other hormones in the target cells.

**Endocrine glands**, which lack the ducts present in exocrine glands (chapter 1), secrete biologically active chemicals called **hormones** into the blood. Many endocrine glands are discrete organs (fig. 11.1) whose primary functions are the production and secretion of hormones. The pancreas functions as both an exocrine and an endocrine gland; the endocrine portion of the pancreas is composed of microscopic structures called the islets of Langerhans (fig. 11.1*b*). The concept of the **endocrine system**, however, must be extended beyond these organs. In recent years, it has been discovered that many other organs in the body secrete hormones. When these hormones can be demonstrated to have significant physiological functions, the organs that produce these hormones may be categorized as endocrine glands in addition to their other functions. A partial list of the endocrine glands (table 11.1) should thus include the heart, liver, hypothalamus, and kidneys.

Hormones affect the metabolism of their target organs and, by this means, help regulate (1) total body metabolism, (2) growth, and (3) reproduction. The effects of hormones on body metabolism and growth are discussed in chapter 18; the regulation of reproductive functions by hormones is included in chapter 20.

### Chemical Classification of Hormones

Hormones secreted by different endocrine glands are diverse in chemical structure. All hormones, however, can be grouped into three general chemical categories: (1) **amines,** such as catecholamines (epinephrine and norepinephrine), and other derivatives of amino acids; (2) **polypeptides** and **glycoproteins,** including shorter chain polypeptides such as antidiuretic hormone and insulin and large glycoproteins such as thyroid-stimulating hormone (table 11.2); and (3) **steroids,** such as cortisol and testosterone.

**Table 11.1** A partial list of the endocrine glands

| Endocrine Gland | Major Hormones | Primary Target Organs | Primary Effects |
|---|---|---|---|
| Adrenal cortex | Cortisol<br>Aldosterone | Liver, muscles<br>Kidneys | Glucose metabolism; $Na^+$ retention, $K^+$ excretion |
| Adrenal medulla | Epinephrine | Heart, bronchioles, blood vessels | Adrenergic stimulation |
| Heart | Atrial natriuretic hormone | Kidneys | Promotes excretion of $Na^+$ in the urine |
| Hypothalamus | Releasing and inhibiting hormones | Anterior pituitary | Regulates secretion of anterior pituitary hormones |
| Intestine | Secretin and cholecystokinin | Stomach, liver, and pancreas | Inhibits gastric motility; stimulates bile and pancreatic juice secretion |
| Islets of Langerhans (pancreas) | Insulin<br>Glucagon | Many organs<br>Liver and adipose tissue | Insulin promotes cellular uptake of glucose and formation of glycogen and fat; glucagon stimulates hydrolysis of glycogen and fat |
| Kidneys | Erythropoietin | Bone marrow | Stimulates red blood cell production |
| Liver | Somatomedins | Cartilage | Stimulates cell division and growth |
| Ovaries | Estradiol-17$\beta$ and progesterone | Female genital tract and mammary glands | Maintains structure of genital tract; promotes secondary sexual characteristics |
| Parathyroids | Parathyroid hormone | Bone, intestine, and kidneys | Increases $Ca^{++}$ concentration in blood |
| Pineal | Melatonin | Hypothalamus and anterior pituitary | Affects secretion of gonadotrophic hormones |
| Pituitary, anterior | Trophic hormones | Endocrine glands and other organs | Stimulates growth and development of target organs; stimulates secretion of other hormones |
| Pituitary, posterior | Antidiuretic hormone<br>Oxytocin | Kidneys, blood vessels<br>Uterus, mammary glands | Antidiuretic hormone promotes water retention and vasoconstriction. Oxytocin stimulates contraction of uterus and mammary secretory units |
| Skin | 1,25-Dihydroxyvitamin $D_3$ | Intestine | Stimulates absorption of $Ca^{++}$ |
| Stomach | Gastrin | Stomach | Stimulates acid secretion |
| Testes | Testosterone | Prostate, seminal vesicles, other organs | Stimulates secondary sexual development |
| Thymus | Thymosin | Lymph nodes | Stimulates white blood cell production |
| Thyroid | Thyroxine ($T_4$) and triiodothyronine ($T_3$) | Most organs | Growth and development; stimulates basal rate of cell respiration (basal metabolic rate or BMR) |

**Table 11.2** Some examples of polypeptide and glycoprotein hormones

| Hormone | Structure | Gland | Primary Effects |
|---|---|---|---|
| Antidiuretic hormone | 8 amino acids | Posterior pituitary | Water retention and vasoconstriction |
| Oxytocin | 8 amino acids | Posterior pituitary | Uterine and mammary contraction |
| Insulin | 21 and 30 amino acids (double chain) | $\beta$ cells in islets of Langerhans | Cellular glucose uptake, lipogenesis, and glycogenesis |
| Glucagon | 29 amino acids | $\alpha$ cells in islets of Langerhans | Hydrolysis of stored glycogen and fat |
| ACTH | 39 amino acids | Anterior pituitary | Stimulation of adrenal cortex |
| Parathyroid hormone | 84 amino acids | Parathyroid | Increases blood $Ca^{++}$ concentration |
| FSH, LH, TSH | Glycoproteins | Anterior pituitary | Stimulates growth, development, and secretion of target glands |

**Figure 11.2.** Simplified biosynthetic pathways for steroid hormones. Notice that progesterone (a hormone secreted by the ovaries) is a common precursor in the formation of all other steroid hormones and that testosterone (the major androgen secreted by the testes) is a precursor in the formation of estradiol–17*β*, the major estrogen secreted by the ovaries.

Cholesterol

Pregnenolone

Progesterone

Secreted by ovaries

Cortisol (hydrocortisone)

Secreted by adrenal cortex

Androstenedione

Testosterone

Secreted by testes

Estradiol - 17β

Secreted by ovaries

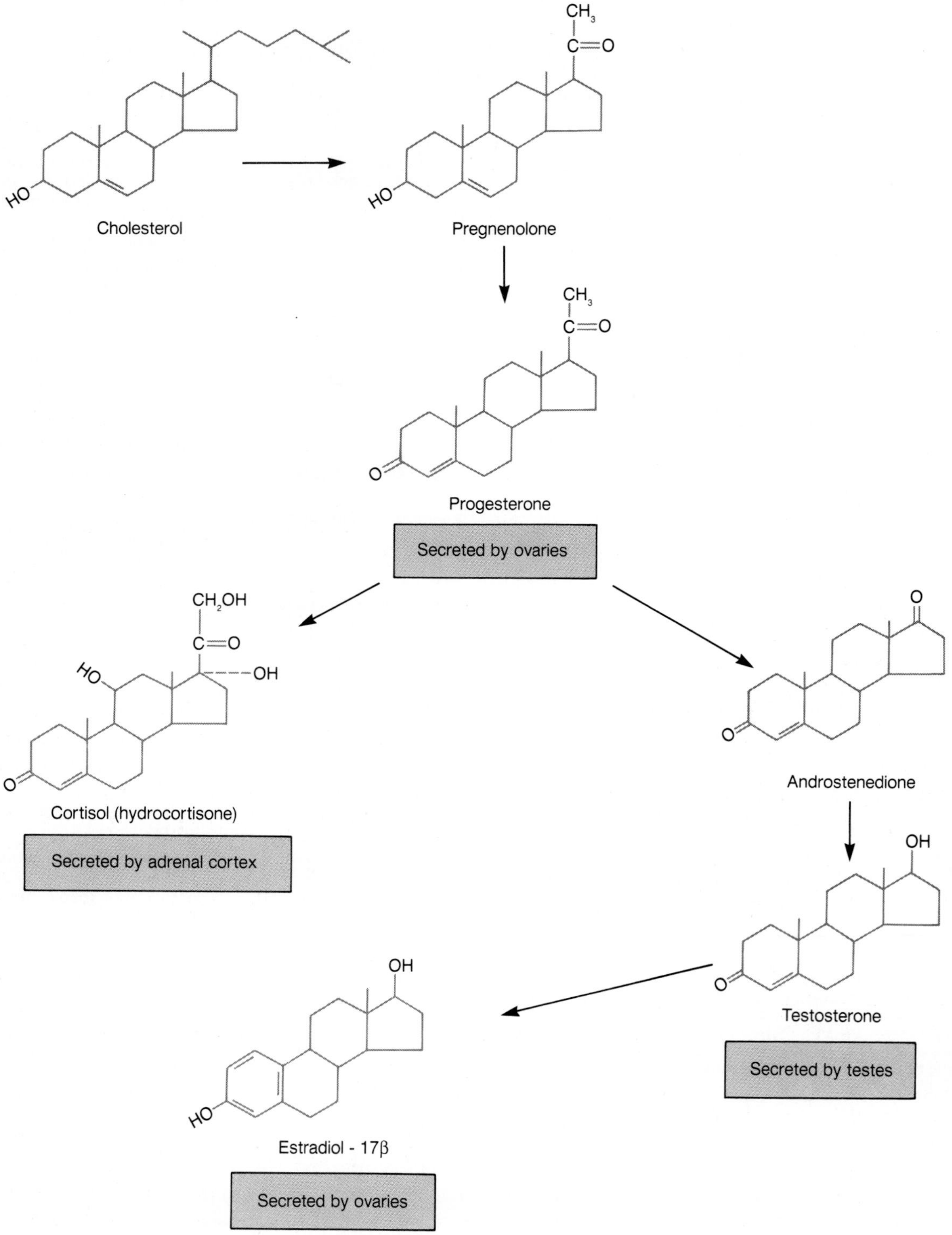

**Figure 11.3.** The thyroid hormones: thyroxine ($T_4$) and triiodothyronine ($T_3$) are secreted in a ratio of 9:1.

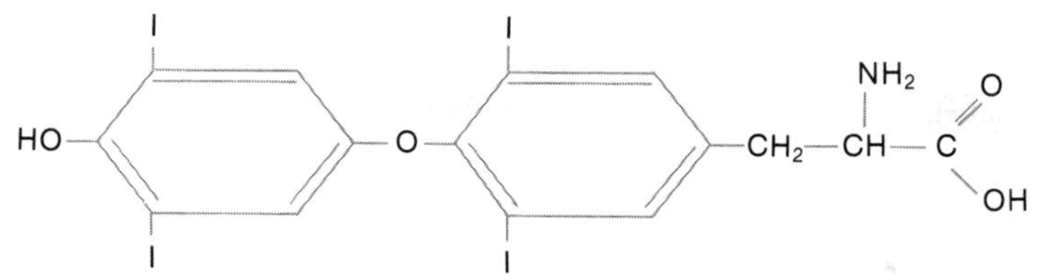

Thyroxine, or tetraiodothyronine ($T_4$)

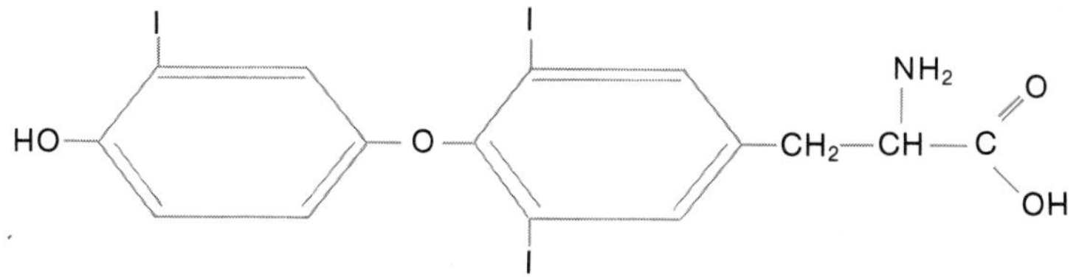

Triiodothyronine ($T_3$)

Steroid hormones, which are derived from cholesterol (fig. 11.2), are lipids and thus are not water-soluble. The gonads—testes and ovaries—secrete *sex steroids;* the adrenal cortex secretes *corticosteroids* such as cortisol, aldosterone, and others.

The major thyroid hormones are composed of two derivatives of the amino acid tyrosine bonded together. These hormones are unique because they contain iodine (fig. 11.3). When the hormone contains four iodine atoms, it is called *tetraiodothyronine* ($T_4$), or *thyroxine.* When it contains three atoms of iodine, it is called *triiodothyronine* ($T_3$). Although these hormones are not steroids, they are like steroids in that they are relatively small and nonpolar molecules. Steroid and thyroid hormones are active when taken orally (as a pill); sex steroids are contained in the contraceptive pill, and thyroid hormone pills are taken by people whose thyroid is deficient (who are hypothyroid). Other types of hormones cannot be taken orally because they would be digested into inactive fragments before they were absorbed into the blood.

## Common Aspects of Neural and Endocrine Regulation

It may be thought that endocrine regulation is chemical and therefore distinct from the electrical nature of neural control systems. This idea is incorrect. Electrical nerve impulses are in fact chemical events produced by the diffusion of ions through the neuron cell membrane (chapter 7). Interestingly, the action of some hormones (such as insulin) is accompanied by ion diffusion and electrical changes in the target cells, so changes in membrane potential are not unique to the nervous system. Also, most nerve fibers stimulate the cells they innervate through the release of a chemical neurotransmitter. Neurotransmitters differ from hormones in that they do not travel in the blood but instead diffuse only a very short distance across a synapse. In other respects, however, the actions of neurotransmitters and hormones are very similar.

Indeed, many polypeptide hormones, including those secreted by the pituitary gland and by the digestive tract, have been discovered in the brain. In certain locations in the brain, some of these compounds are produced and secreted as hormones. In other brain locations, some of these compounds apparently serve as neurotransmitters. The discovery of some of these polypeptides in unicellular organisms, which of course lack a nervous and endocrine system, suggests that these regulatory molecules appeared early in evolution and became incorporated into the function of nerve and endocrine tissue as these systems evolved. This fascinating theory would help to explain, for example, why insulin, a polypeptide hormone produced in the pancreas of vertebrates, is found in neurons of invertebrates (which lack a separate endocrine system).

Regardless of whether a particular chemical is acting as a neurotransmitter or as a hormone, in order for it to function in physiological regulation: (1) target cells must have specific **receptor proteins** that combine with the chemical; (2) the combination of the regulator molecule with its receptor proteins must cause a specific sequence of changes in the target cells; and (3) there must be a mechanism to rapidly turn off the action of the regulator; without an "off switch," physiological control is impossible. This last process involves rapid removal and/or chemical inactivation of the regulator molecules.

### Activation and Inactivation of Hormones

Hormone molecules that affect the metabolism of target cells are often derived from less active "parent," or *precursor,* molecules. In the case of polypeptide hormones, the precursor may be a longer chained **prohormone** that is cut and spliced together to make the hormone. Insulin, for example, is produced from *proinsulin* within the endocrine beta cells of the islets of Langerhans of the pancreas. In some cases the prohormone itself is derived from an even larger precursor molecule; in the case of insulin, this is called *pre-proinsulin.* Many authors use the term *prehormone* to indicate such precursors of prohormones.

In some cases the molecule secreted by the endocrine gland, and considered to be the hormone of that gland, is actually inactive in the target cells. In order to become active, the target cells must modify the chemical structure of the secreted hormone. Thyroxine ($T_4$), for example, must be changed into $T_3$ within the target cells to affect the metabolism of the target cells. Similarly, testosterone (secreted by the testes) and vitamin $D_3$ (secreted by the skin) are converted into more active molecules within their target cells (table 11.3). In this text, the term **prehormone** will be used to designate those molecules secreted by endocrine glands that are inactive until changed by their target cells.

The concentration of hormones in the blood primarily reflects the rate of secretion by the endocrine glands—hormones do not generally accumulate in the blood. This is because hormones are rapidly removed from the blood by target organs and by the liver. The **half-life** of a hormone, which is the time required for the plasma concentration of a given amount of a hormone to be reduced to half its reference level, ranges from minutes to hours for most hormones (thyroid hormone, however, has a half-life of several days). Hormones removed from the blood by the liver are converted by enzymatic reactions into less active products. Steroids, for example, are converted to more polar derivatives. These less active, more water-soluble polar derivatives are released into the blood and are excreted in the urine and bile.

### Effects of Hormone Concentrations on Tissue Response

The effects of hormones are very concentration dependent. Normal tissue responses are only produced when the hormones are present within their normal, or *physiological,* range of concentrations. When some hormones are taken in abnormally high, or *pharmacological,* concentrations (as when they are taken as drugs), they may produce noncharacteristic effects. This may be partly due to the fact that abnormally high concentrations of a hormone may cause the hormone to bond to tissue receptor proteins of different but related hormones. Also, since some steroid hormones can be converted by their target cells into products with other biological effects (such as the conversion of androgens into estrogens), the administration of large amounts of one steroid can result in the production of a significant amount of other steroids with different effects.

Pharmacological doses of hormones, particularly of steroids, can thus have widespread and often damaging "side effects." People with inflammatory diseases who are treated with high doses of cortisone over long periods of time, for example, may develop characteristic changes in bone and soft tissue structure. Contraceptive pills, which contain sex steroids, have a number of potential side effects that could not have been predicted at the time the pill was first introduced.

It is unfortunate that some bodybuilders and others wishing to quickly build large muscle mass take *anabolic steroids.* These compounds are synthetic androgens (male hormones), which promote protein synthesis in muscles; the presence of androgens secreted by the testes is responsible for the fact that males have, on the average, more muscle mass than females. Although administration of exogenous androgens may promote muscle growth, it also can produce a number of undesirable side effects. Since the liver and some other organs can change androgens into estrogens (females sex steroids), men who take exogenous androgens often develop **gynecomastia**—the development of femalelike mammary tissue! This abnormal tissue must be surgically removed. The androgenic effect of anabolic steroids inhibits the pituitary's secretion of gonadotropic hormones (FSH and LH), causing shrinkage of the testes and sterility. Other damaging effects, including liver cancer, kidney damage, heart disease, and antisocial behavior, may also occur as a result of the ill-advised taking of anabolic steroids.

***Priming Effects.*** Variations in hormone concentration within the normal, physiological range can affect the responsiveness of target cells. This is due in part to the effects of polypeptide and glycoprotein hormones on the number of their receptor proteins in target cells. Small amounts of gonadotropin-releasing hormone (GnRH), secreted by the hypothalamus, for example, increase the sensitivity of anterior pituitary cells to further GnRH stimulation. This is a priming effect, sometimes also called *upregulation.* Subsequent stimulation by GnRH thus causes a greater response from the anterior pituitary.

***Desensitization and Downregulation.*** Prolonged exposure to high concentrations of polypeptide hormones has been found to *desensitize* the target cells. Subsequent exposure to the same concentration of the same hormone thus produces less of a target tissue response. This desensitization may be partially due to the fact that high concentrations of these hormones cause a decrease in the number of receptor proteins in their target cells, a phenomenon

**Table 11.3** Conversion of some prehormones into biologically active derivatives

| Endocrine Gland | Prehormone | Active Products | Comments |
|---|---|---|---|
| Skin | Vitamin $D_3$ | 1,25-dihydroxyvitamin $D_3$ | Hydroxylation reactions occur in the liver and kidneys. |
| Testes | Testosterone | Dihydrotestosterone (DHT) | DHT and other 5$\alpha$-reduced androgens are formed in most androgen-dependent tissue. |
| | | Estradiol-17$\beta$ ($E_2$) | $E_2$ is formed in the brain from testosterone, where it is believed to affect both endocrine function and behavior; small amounts of $E_2$ are also produced in the testes. |
| Thyroid | Thyroxine ($T_4$) | Triiodothyronine ($T_3$) | Conversion of $T_4$ to $T_3$ occurs in almost all tissues. |

called *downregulation.* Such desensitization and downregulation of receptors has been shown to occur, for example, in adipose cells exposed to high concentrations of insulin and in testicular cells exposed to high concentrations of luteinizing hormone (LH).

In order to prevent desensitization from occurring under normal conditions, many polypeptide and glycoprotein hormones are secreted in pulses rather than continuously. This *pulsatile secretion* is an important aspect, for example, in the hormonal control of the reproductive system. The pulsatile secretion of GnRH and LH is needed to prevent desensitization; when these hormones are artificially presented in a continuous fashion, they produce a decrease (rather than the normal increase) in gonadal function. This has important clinical implications, as will be described in chapter 20.

### Hormone Interactions

A given target tissue is usually responsive to a number of different hormones, which may antagonize each other or work together to produce effects that are additive or complementary. The responsiveness of a target tissue to a particular hormone is thus affected not only by the concentration of that hormone, but also by the effects of other hormones on that tissue. Terms used to describe hormone interactions include *synergistic, permissive,* and *antagonistic.*

***Synergistic and Permissive Effects.*** When two or more hormones work together to produce a particular result their effects are said to be **synergistic.** These effects may be additive or complementary. The action of epinephrine and norepinephrine on the heart is a good example of an additive effect. Each of these hormones separately produces an increase in cardiac rate; if the same concentrations of these hormones are given together, the stimulation of cardiac rate is increased. The synergistic action of FSH and testosterone is an example of a complementary effect; each hormone separately stimulates a different stage of spermatogenesis during puberty, so that both hormones together are needed at that time to complete sperm development. Likewise, the ability of mammary glands to produce and secrete milk requires the synergistic action of many hormones—estrogen, cortisol, prolactin, oxytocin, and others.

A hormone is said to have a **permissive** effect on the action of a second hormone when it enhances the responsiveness of a target organ to the second hormone or when it increases the activity of the second hormone. Prior exposure of the uterus to estrogen, for example, induces the formation of receptor proteins for progesterone, which improves the response of the uterus when it is subsequently exposed to progesterone. Estrogen thus has a permissive effect on the responsiveness of the uterus to progesterone. Prior exposure of the gonads to FSH, as another example, stimulates the production of LH receptors, which enhances the responsiveness of the testes and ovaries to LH. Glucocorticoids (a class of corticosteroids including cortisol) exert permissive effects on the actions of catecholamines (epinephrine and norepinephrine). When there is an absence of these permissive effects due to abnormally low glucocorticoids, the catecholamines will not be as effective as they are normally. One symptom of this condition may be an abnormally low blood pressure.

Vitamin $D_3$ is a prehormone that must first be converted by enzymes in the kidneys and liver, which add two hydroxyl (OH) groups to form the active hormone 1,25-dihydroxyvitamin $D_3$. This hormone helps to raise blood calcium levels. Parathyroid hormone (PTH) has a permissive effect on the actions of vitamin $D_3$ because it stimulates the production of the hydroxylating enzymes in the kidneys and liver. By this means, an increased secretion of PTH results in an increase in the ability of vitamin $D_3$ to stimulate the intestinal absorption of calcium.

***Antagonistic Effects.*** In some situations the actions of one hormone antagonize the effects of another. Lactation during pregnancy, for example, is prevented because the high concentration of estrogen in the blood inhibits the secretion and action of prolactin. Another example of antagonism is the action of insulin and glucagon (two hormones from the islets of Langerhans) on adipose tissue; the formation of fat is promoted by insulin, whereas glucagon promotes fat breakdown.

**Table 11.4** Functional categories of hormones, based on the location of their receptor proteins and the mechanisms of their action

| Types of Hormones | Secreted by | Location of Receptors | Effects of Hormone–Receptor Interaction |
|---|---|---|---|
| Catecholamines, polypeptides, glycoproteins | All glands except adrenal cortex, gonads, and thyroid | Outer surface of cell membrane | Stimulates production of intracellular "second messenger," which activates previously inactive enzymes |
| Steroids | Adrenal cortex, testes, ovaries | Cytoplasm of target cells | Stimulates translocation of hormone-receptor complex to nucleus and activation of specific genes |
| Thyroxine ($T_4$) | Thyroid | Nucleus of target cells | After conversion to triiodothyronine ($T_3$), activates specific genes |

1. *Define the terms* prohormone *and* prehormone, *and give examples of each.*
2. *Distinguish between the effects of physiological and pharmacological concentrations of a hormone.*
3. *Explain how the response of the body to a given hormone can be affected by the concentration of that hormone in the blood.*
4. *Explain how the effects of one hormone can influence the response of a target organ to other hormones.*

## *Mechanisms of Hormone Action*

Each hormone exerts its characteristic effects on a target organ through the actions it has on the cells of these organs. These actions on the cellular level are the mechanisms by which the hormone works, and these mechanisms are similar for hormones that have similar chemical natures. Those hormones that are nonpolar pass through the target cell membrane and act directly within the target cell, whereas those that are polar act on the target cell membrane to cause the production of intracellular second messenger molecules.

Although each hormone has its own characteristic effects on specific target cells, hormones that are in the same chemical category have similar mechanisms of action. These similarities involve the location of cellular receptor proteins and the events that occur in the target cells after the hormone has combined with its receptor protein.

Hormones are delivered by the blood to every cell in the body, but only the **target cells** are able to respond to each hormone. In order to respond to a hormone, a target cell must have specific receptor proteins for that hormone. Receptor protein–hormone interaction is highly specific, much like the interaction of an enzyme with its substrate. Receptor proteins are not enzymes, however, and since they cannot be detected by the techniques used to assay enzymes, other methods must be used. These methods are based on the observation that hormones bond to receptors with a *high affinity* (high bond strength) and with a *low capacity.* The latter characteristic refers to the fact that there are only a limited number of receptors per target cell (usually a few thousand), so that it is possible to saturate receptors with hormones. Notice that the characteristics of specific bonding and saturation of receptor proteins are similar to the characteristics of enzyme and transport carrier proteins discussed in previous chapters.

The location of a hormone's receptor proteins in its target cells depends on the chemical nature of the hormone. Based on the location of the receptor proteins, hormones can be grouped into three categories: (1) receptor proteins within the nucleus of target cells—*thyroid hormones;* (2) receptor proteins within the cytoplasm of target cells—*steroid hormones;* and (3) receptor proteins in the outer surface of the target cell membrane—*amine, polypeptide,* and *glycoprotein hormones.* This information is summarized in table 11.4.

### Mechanisms of Steroid and Thyroid Hormone Action

Steroid and thyroid hormones are similar in size and in the fact that they are nonpolar and thus are not very water-soluble. Unlike other hormones, therefore, steroids and thyroid hormones (primarily thyroxine) do not travel dissolved in the aqueous portion of the plasma but instead are transported to their target cells attached to plasma carrier proteins. These hormones then dissociate from the carrier proteins in the blood and easily pass through the lipid component of the target cell's membrane.

**Figure 11.4.** The mechanism of the action of a steroid hormone (*H*) on the target cells.

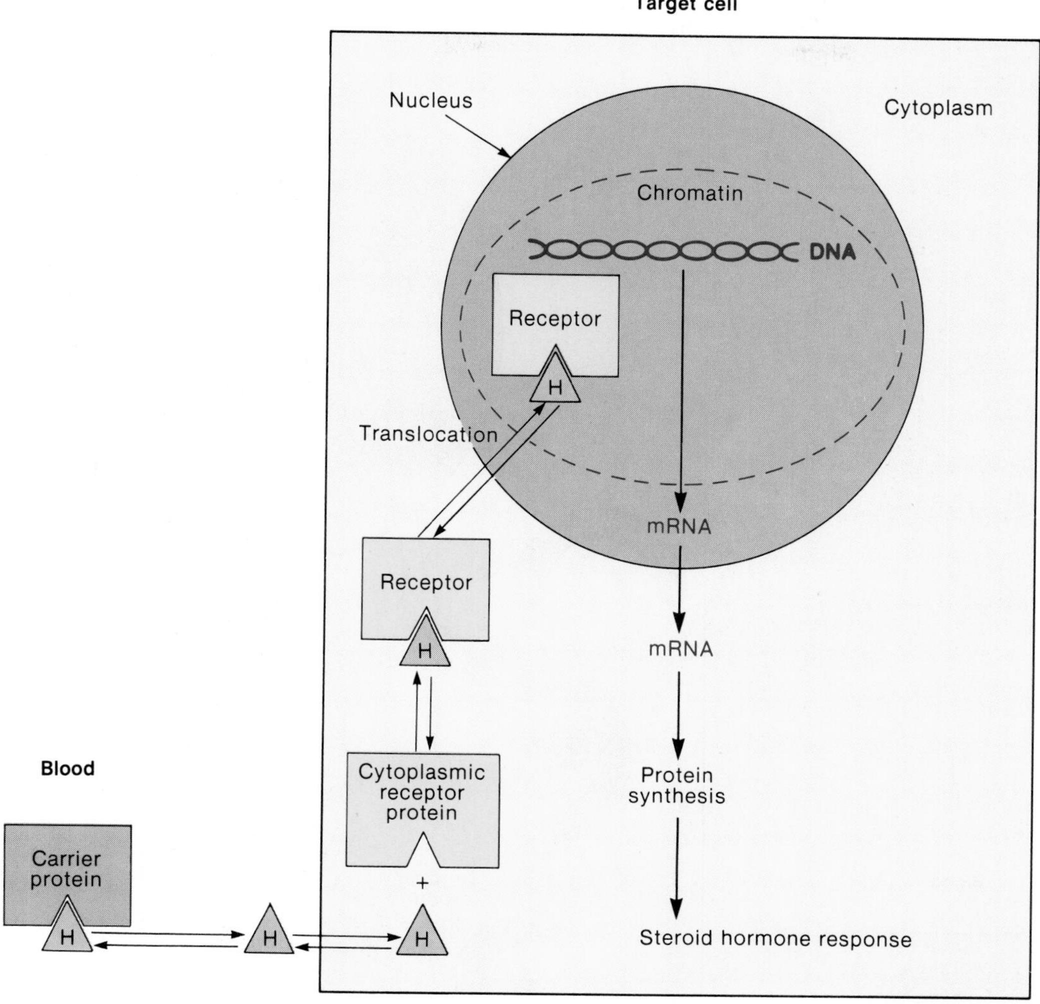


***Steroid Hormones.*** Once through the cell membrane, steroid hormones attach to *cytoplasmic receptor proteins* in the target cells. The steroid hormone–receptor protein complex then *translocates* to the nucleus and attaches by means of the receptor proteins to the chromatin. The sites of attachment in the chromatin, termed *acceptor sites,* are specific for the target tissue. This specificity is believed to be determined by acidic (nonhistone) proteins in the chromatin (chapter 3). According to one theory, part of the receptor bonds to an acidic protein while a different part of the receptor bonds to DNA.

The attachment of the receptor protein–steroid complex to the acceptor site "turns on" genes. Specific genes become activated by this process and produce nuclear RNA, which is then processed into messenger RNA (mRNA). This new mRNA enters ribosomes and codes for the production of new proteins. Since some of these newly synthesized proteins may be enzymes, the metabolism of the target cell is changed in a specific manner (fig. 11.4). Steroid hormones, in short, affect their target cells by stimulating genetic transcription (RNA synthesis) followed by genetic translation (protein synthesis).

***Thyroxine.*** The major hormone secreted by the thyroid gland is thyroxine, or **tetraiodothyronine ($T_4$).** Like steroid hormones, thyroxine travels in the blood attached to carrier proteins (primarily attached to *thyroxine-binding globulin,* or *TBG*). The thyroid also secretes a small amount of **triiodothyronine,** or **$T_3$.** The carrier proteins have a higher affinity for $T_4$ than for $T_3$, however, and as a result, the amount of unbound (or "free") $T_3$ is about ten times greater than the amount of free $T_4$ in the plasma.

**Figure 11.5.** The mechanism of the action of $T_3$ (triiodothyronine) on the target cells.

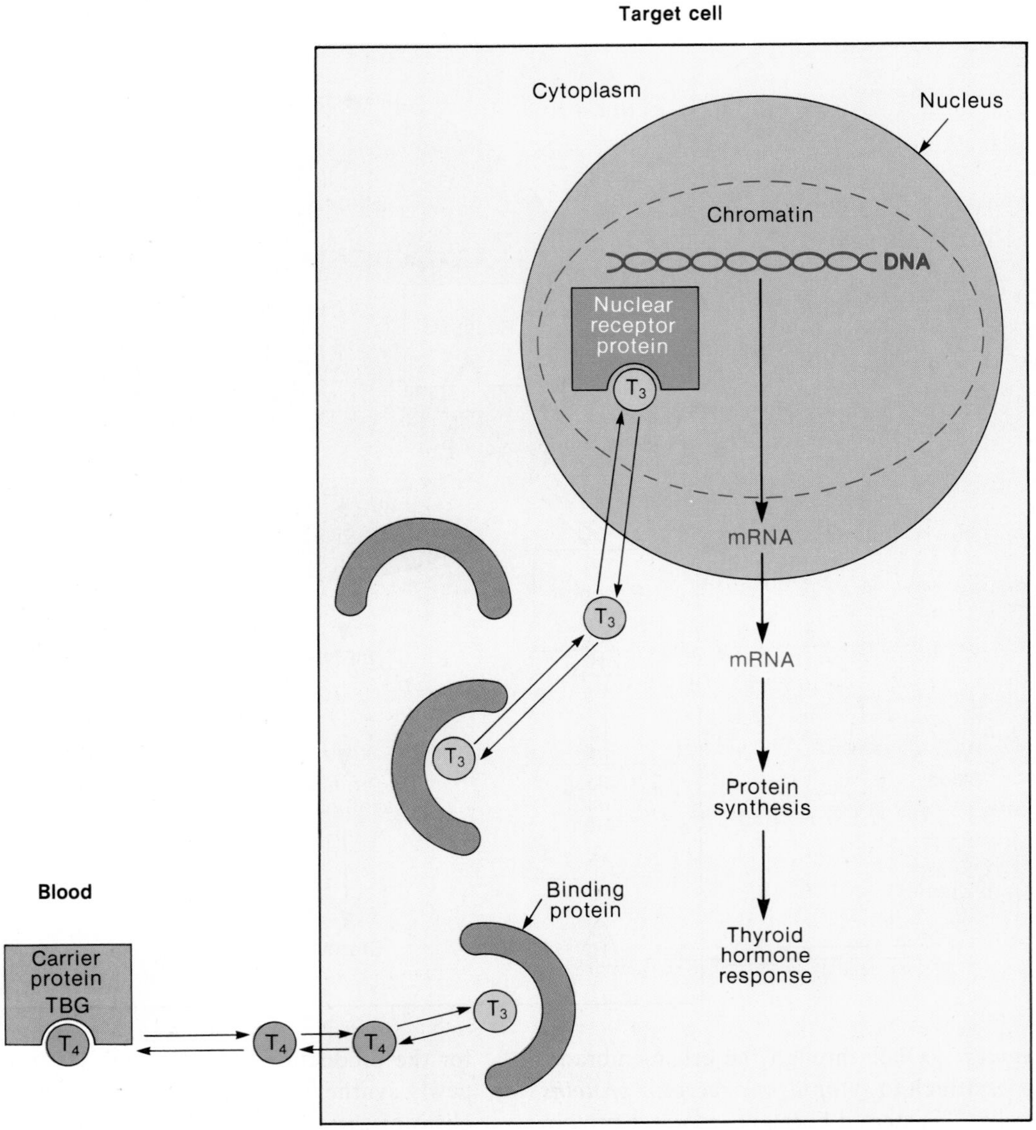

Approximately 99.96% of the thyroxine in the blood is attached to carrier proteins in the plasma; the rest is free. Only the free thyroxine and $T_3$ can enter target cells; the protein-bound thyroxine serves as a reservoir of this hormone in the blood (this is why it takes a couple of weeks after surgical removal of the thyroid for the symptoms of hypothyroidism to develop). Once the free thyroxine passes into the target cell cytoplasm, it is enzymatically converted into $T_3$. As previously discussed, it is the $T_3$ rather than $T_4$ which is active within the target cells.

Inactive $T_3$ receptor proteins are already in the nucleus attached to chromatin. These receptors are inactive until $T_3$ enters the nucleus from the cytoplasm. The attachment of $T_3$ to the chromatin-bound receptor proteins activates genes and results in the production of new mRNA and new proteins. This sequence of events is summarized in figure 11.5.

## Mechanisms of Amine, Polypeptide, and Glycoprotein Hormone Action: Second Messengers

Amine, polypeptide, and glycoprotein hormones cannot pass through the lipid barrier of the target cell membrane. Although some of these hormones may enter the cell by pinocytosis, most of the effects of these hormones are believed to result from interaction of these hormones with receptor proteins in the outer surface of the target cell membrane. Since these hormones do not have to enter the target cells to exert their effects, other molecules must mediate the actions of these hormones within the target cells. If you think of hormones as "messengers" from the endocrine glands, the intracellular mediators of the hormone's action can be called **second messengers.** (The concept of second messengers was discussed in relation to synaptic transmission in chapter 7.)

**Table 11.5** Hormones that activate adenylate cyclase and use cAMP as a second messenger and hormones that use other second messengers

| Hormones that Stimulate Adenylate Cyclase | Hormones that Use Other Second Messengers |
|---|---|
| Adrenocorticotropic hormone (ACTH) | Catecholamines ($\alpha$-adrenergic) |
| Calcitonin | Growth hormone (GH) |
| Epinephrine ($\beta$-adrenergic) | Insulin |
| Follicle-stimulating hormone (FSH) | Oxytocin |
| Glucagon | Prolactin |
| Luteinizing hormone (LH) | Somatomedin |
| Parathyroid hormone | Somatostatin |
| Thyrotropin-releasing hormone (TRH) | |
| Thyroid-stimulating hormone (TSH) | |
| Antidiuretic hormone | |

*Cyclic AMP.* Cyclic adenosine monophosphate (abbreviated cAMP) was the first "second messenger" to be discovered and is the best understood. The hormonal effects of epinephrine and norepinephrine, and the effects of norepinephrine released as a neurotransmitter, are due to cAMP production within the target cells. It was later discovered that the effects of many (but not all—see table 11.5) polypeptide and glycoprotein hormones are also mediated by cAMP.

The bonding of these hormones to their membrane receptor proteins activates an enzyme called **adenylate cyclase.** This enzyme is built into the cell membrane and, when activated, it catalyzes the following reaction:

$$ATP \rightarrow cAMP + PP_i$$

Adenosine triphosphate (ATP) is thus converted into cAMP plus two inorganic phosphates (*pyrophosphate,* abbreviated $PP_i$). As a result of the interaction of the hormone with its receptor and the activation of adenylate cyclase, therefore, the intracellular concentration of cAMP is increased. Cyclic AMP activates a previously inactive enzyme in the cytoplasm called **protein kinase.** The inactive form of this enzyme consists of two subunits: a catalytic subunit and an inhibitory subunit. The enzyme is produced in an inactive form and becomes active only when cAMP attaches to the inhibitory subunit. Bonding of cAMP to the inhibitory subunit causes it to dissociate from the catalytic subunit, which then becomes active (fig. 11.6). The hormone, in summary—acting through an increase in cAMP production—causes an increase in protein kinase enzyme activity within its target cells.

Active protein kinase catalyzes the attachment of phosphate groups to different proteins in the target cells. This causes some enzymes to become activated, and others to become inactivated. Cyclic AMP, acting through protein kinase, thus modulates the activity of enzymes that are already present in the target cell. This alters the metabolism of the target tissue in a manner characteristic of the actions of that specific hormone (table 11.6).

Like all biologically active molecules, cAMP must be rapidly inactivated for it to function effectively as a second messenger in hormone action. This function is served by an enzyme called **phosphodiesterase** within the target cells, which hydrolyzes cAMP into inactive fragments. Through the action of phosphodiesterase, the stimulatory effect of a hormone that uses cAMP as a second messenger depends upon the continuous generation of new cAMP molecules, and thus depends upon the level of secretion of the hormone.

Drugs that inhibit the activity of phosphodiesterase thus prevent the breakdown of cAMP and result in increased concentrations of cAMP within the target cells. The drug **theophylline and its derivatives,** for example, are used clinically to raise cAMP levels within bronchiolar smooth muscle. This duplicates and enhances the effect of epinephrine on the bronchioles (producing dilation) in people who suffer from asthma. **Caffeine,** a compound related to theophylline, is also a phosphodiesterase inhibitor, and thus produces its effects by raising the cAMP concentrations within tissue cells.

**Figure 11.6.** Cyclic AMP (cAMP) as a second messenger in the action of catecholamine, polypeptide, and glycoprotein hormones.

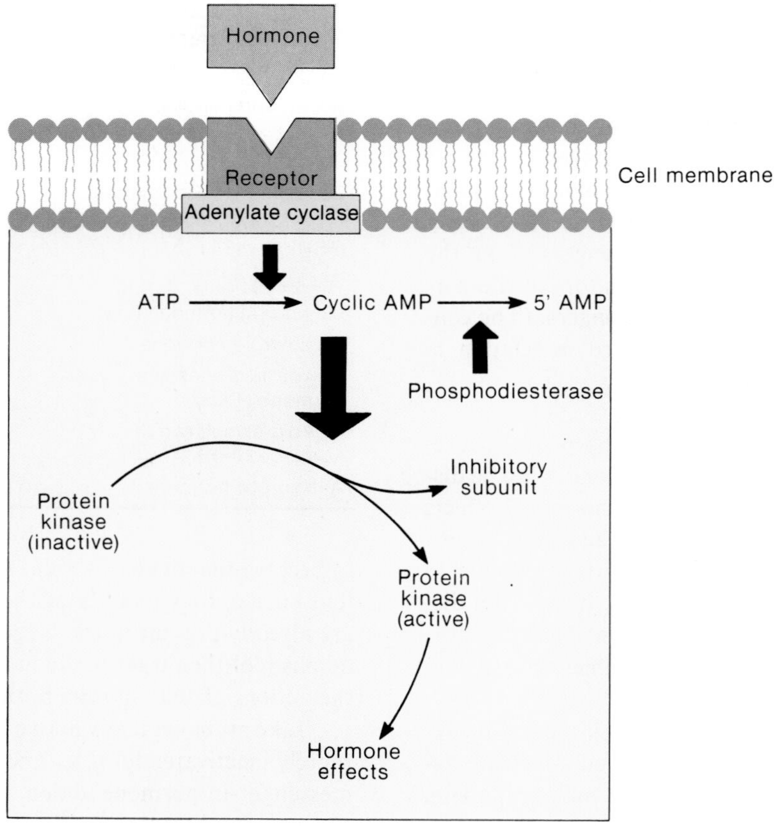


**Table 11.6** Sequence of events that occurs with cyclic AMP as a second messenger

1. The hormones combine with their receptors on the outer surface of target cell membranes.
2. Hormone-receptor interaction stimulates activation of adenylate cyclase on the cytoplasmic side of the membranes.
3. Activated adenylate cyclase catalyzes the conversion of ATP to cyclic AMP (cAMP) within the cytoplasm.
4. Cyclic AMP activates protein kinase enzymes that were already present in the cytoplasm in an inactive state.
5. Activated cAMP-dependent protein kinase transfers phosphate groups to (phosphorylates) other enzymes in the cytoplasm.
6. The activity of specific enzymes is either increased or inhibited by phosphorylation.
7. Altered enzyme activity mediates the target cell's response to the hormone.

In addition to cyclic AMP, other cyclic nucleotides may have second messenger functions. **Cyclic guanosine monophosphate (cGMP)**, in particular, has been shown to antagonize the action of cAMP in some cases, and to support its activity in other cases. For example, the control of cell division and the cell cycle (chapter 3) is related to the ratio of cAMP and cGMP in the cell.

***$Ca^{++}$ as a Second Messenger.*** The concentration of $Ca^{++}$ in the cytoplasm is kept very low by the action of active transport carriers—calcium pumps—in the cell membrane. Through the action of these pumps, the concentration of calcium in the cytoplasm is 5,000–10,000 times lower in the cytoplasm than in the extracellular fluid. In addition, the endoplasmic reticulum (chapter 3) of many cells contains calcium pumps that actively transport $Ca^{++}$ from the cytoplasm into the cisternae of the endoplasmic reticulum. The steep concentration gradient for $Ca^{++}$ that results allows various stimuli to evoke a rapid, though brief, diffusion of $Ca^{++}$ into the cell, which can serve as a signal in different control systems.

**Figure 11.7.** Hormones can stimulate the hydrolysis of liver glycogen by way of two second-messenger systems. In this instance, the action of epinephrine and glucagon via the cAMP system is physiologically more significant than the action of vasopressin and angiotensin II via the $Ca^{++}$ calmodulin system.

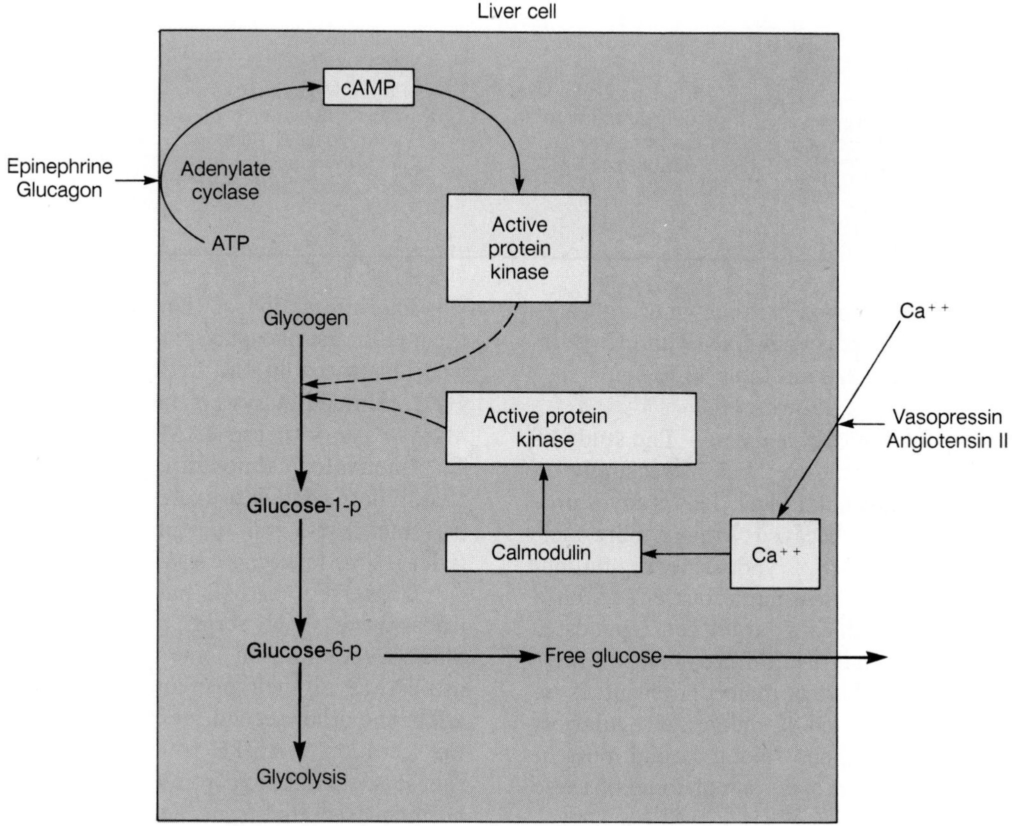


At the terminal boutons of axons, for example, the influx of $Ca^{++}$ serves as a signal for the release of neurotransmitters (chapter 7). Similarly, when skeletal muscles are stimulated to contract, the release of $Ca^{++}$ from the endoplasmic reticulum couples electrical excitation of the muscle cell to the mechanical processes that result in contraction (chapter 12). Additionally, there is mounting evidence that $Ca^{++}$ serves as a second messenger in the action of some hormones, including insulin.

Despite the fact that insulin is one of the most intensively studied hormones, its mechanism of action is still not completely understood. It is known, however, that insulin's actions do not depend on cAMP as a second messenger. This becomes obvious when one considers the fact that insulin's actions on adipose and liver cells are antagonistic to those of epinephrine (as previously discussed), which does use cAMP as a second messenger.

When insulin combines with its receptor protein on the surface of a target cell, there is a rapid and transient inflow of $Ca^{++}$. This $Ca^{++}$ binds to a cytoplasmic protein called **calmodulin**. Once $Ca^{++}$ binds to calmodulin, the now active calmodulin in turn activates specific protein kinase enzymes (those that add phosphate groups to proteins), which modify the actions of other enzymes in the cell. Activation of specific calmodulin-dependent enzymes by insulin then causes at least some of the effects of insulin in the target cell.

The activation of calmodulin-dependent protein kinase by hormones that use $Ca^{++}$ as a second messenger is analogous to the activation of cAMP-dependent protein kinase by hormones that use cAMP as a second messenger. This is illustrated in figure 11.7, in which the actions of two classes of hormones on liver cells are compared. In one case, epinephrine and glucagon stimulate glycogen breakdown using cAMP as a second messenger. In the other case (of lesser physiological significance), vasopressin and

**Table 11.7** Comparison of the effects of cyclic AMP (cAMP) and $Ca^{++}$ as second messengers in different tissues

| Second Messenger | Mechanism | Effects: Nerves | Effects: Muscles | Effects: Hormones |
|---|---|---|---|---|
| cAMP | Activation of cAMP-dependent protein kinase | Opens ion channels in postsynaptic membrane | Phosphorylation of myosin in skeletal muscle in response to depolarization (significance controversial) | Activation or inactivation of target cell enzymes in response to many polypeptide and glycoprotein hormones |
| $Ca^{++}$ | Activation of $Ca^{++}$-dependent regulatory protein (calmodulin) | Mediates release of neurotransmitter in response to depolarization | Phosphorylation of myosin activates actomyosin ATPase in smooth muscles and stimulates contraction | May mediate the effects of insulin on target cells; may increase cAMP by stimulation of adenylate cyclase or decrease cAMP by stimulation of phosphodiesterase |

angiogensin II stimulate glycogen breakdown using $Ca^{++}$ as a second messenger. The effects of cAMP and $Ca^{++}$ in different tissues are summarized in table 11.7.

***Interactions Among Second Messengers.*** The study of endocrine physiology has come to focus increasingly on events that occur at the molecular level. This study is producing a more complex—but because of this, a more complete and satisfying—picture of endocrine regulation. Since a given hormone can have numerous effects on a particular target cell, and since a target cell reponds to numerous hormones that may use different second messengers, a more complex picture of molecular events is required for adequate explanation of endocrine regulation. This brief section on interactions among second messengers is intended to provide an overview of some of these more advanced concepts.

Evidence has accumulated in recent years to support the presence of another class of second messengers that interacts with the calcium messenger system. Again, insulin can serve as an example. When insulin combines with its receptor protein, two events occur: (1) $Ca^{++}$ enters the cell and combines with calmodulin, as previously described; and (2) an enzyme in the cell membrane called *phospholipase C* becomes activated. The phospholipase C enzyme converts a class of membrane phospholipids (the phosphoinosides) into two additional second messengers. These phospholipid-derived second messengers are **inositol triphosphate** and **diacylglycerol** (the latter is composed of glycerol attached to two fatty acids—chapter 2).

Inositol triphosphate is released from the cell membrane and travels into the cytoplasm, where it stimulates the endoplasmic reticulum to release its store of $Ca^{++}$. This $Ca^{++}$ derived from the endoplasmic reticulum, like the $Ca^{++}$ that comes from the extracellular environment, combines with calmodulin and activates calmodulin-dependent protein kinase enzymes. The diacylglycerol, which was produced at the same time as the inositol triphosphate, stays in the cell membrane where its action is believed to be required for the sustained effects of the hormone.

The interaction of the $Ca^{++}$ messenger system and the membrane phospholipid-derived second messengers is well-illustrated in the action of insulin. Similarly, the $Ca^{++}$-calmodulin system has been shown to interact in many ways with the cAMP second messenger system. $Ca^{++}$-activated calmodulin, for example, has been shown to stimulate adenylate cyclase and thus raise cAMP levels in some cases, and to stimulate phosphodiesterase activity—thus lowering cAMP levels—in other cases.

Conversely, cAMP has been shown to modify the transport of $Ca^{++}$ across the cell membrane and endoplasmic reticulum in some tissues, and so hormones that use cAMP as their primary second messenger can also affect the other second messenger systems. For example, the secretion of ACTH from the anterior pituitary in response to stimulation from CRH (corticotropin-releasing hormone, secreted from the hypothalamus) is directly related to production of cAMP, but also involves the $Ca^{++}$-calmodulin system and the inositol-triphosphate/diacylglycerol system. These and other second messengers probably form the basis for the myriad effects of a given hormone on a target cell, and for the complex interactions that a number of hormones can have—synergistic, permissive, and antagonistic—on their target tissue.

---

1. *Using diagrams, describe how steroid hormones and thyroxine exert their effects on their target cells.*
2. *Using a diagram, describe how cyclic AMP is produced within a target cell in response to hormone stimulation. List three hormones that use cAMP as a second messenger.*
3. *Explain how cAMP functions as a second messenger in the action of some hormones.*
4. *Explain how $Ca^{++}$ can have regulatory functions in different tissues and why it can also be considered to be a second messenger in the action of some hormones.*
5. *Explain how the membrane-derived phospholipid second messengers interact with the $Ca^{++}$ messenger system.*

---

**Figure 11.8.** The structure of the pituitary gland as seen in sagittal view.

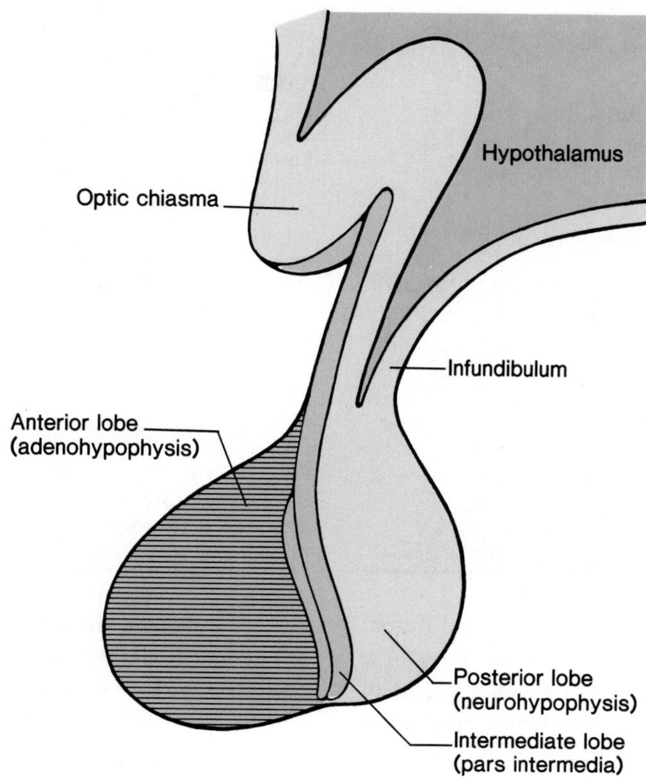


## Pituitary Gland

The pituitary gland is divided into an adenohypophysis and a neurohypophysis. The neurohypophysis secretes hormones that are actually produced by the hypothalamus, whereas the adenohypophysis produces and secretes its own hormones. The adenohypophysis is under the control of the hypothalamus, however, by way of releasing hormones secreted by the hypothalamus into a system of portal vessels. The adenohypophysis responds to the releasing hormones by secreting specific hormones which act on other endocrine glands, as well as on nonendocrine tissues. Feedback influences from many glands in the body usually act to inhibit the secretions of the adenohypophysis.

The pituitary gland, or hypophysis, is located on the inferior aspect of the brain in the region of the diencephalon (chapter 8). The pituitary is a rounded, pea-shaped gland measuring about 1.3 cm (0.5 in.) in diameter, and is attached to the hypothalamus by a stalklike structure called the infundibulum (fig. 11.8).

The pituitary gland is structurally and functionally divided into an anterior lobe, or **adenohypophysis,** and a posterior lobe called the **neurohypophysis.** These two parts have different embryonic origins. The adenohypophysis is derived from a pouch of epithelial tissue (*Rathke's pouch*) that migrates upwards from the embryonic mouth, whereas the neurohypophysis is formed as a downgrowth of the brain. The adenohypophysis consists of three parts: (1) the *pars distalis* is the rounded portion, and is the major endocrine part of the gland; (2) the *pars tuberalis* is the thin extension in contact with the infundibulum; and (3) the *pars intermedia* is between the anterior and posterior parts of the pituitary. These parts are illustrated in figure 11.8.

The neurohypophysis is the neural part of the pituitary gland. It consists of the *pars nervosa,* which is in contact with the pars intermedia and the pars distalis of the adenohypophysis, and the *infundibulum,* which is the connecting stalk to the hypothalamus. Nerve fibers extend through the infundibulum along with small neuroglialike cells, called *pituicytes.*

### Pituitary Hormones

The hormones secreted by the **anterior pituitary** (the pars distalis of the adenohypophysis) are called **trophic hormones.** The term *trophic* means "food." Although the anterior pituitary hormones are not food for their target organs, this term is used because high amounts of the anterior pituitary hormones cause their target organs to hypertrophy, while low amounts cause their target organs to atrophy. When names are applied to the hormones of the anterior pituitary, therefore, the "trophic" term—

**Table 11.8** Hormones secreted by the anterior pituitary

| Hormone | Target Tissue | Stimulated by Hormone | Regulation of Secretion |
|---|---|---|---|
| ACTH (adrenocorticotropic hormone) | Adrenal cortex | Secretion of glucocorticoids | Stimulated by CRH (corticotropin-releasing hormone); inhibited by glucocorticoids |
| TSH (thyroid-stimulating hormone) | Thyroid gland | Secretion of thyroid hormones | Stimulated by TRH (thyrotropin-releasing hormone); inhibited by thyroid hormones |
| GH (growth hormone) | Most tissue | Protein synthesis and growth; lipolysis and increased blood glucose | Inhibited by somatostatin; stimulated by growth hormone-releasing hormone |
| FSH (follicle-stimulating hormone) and LH (luteinizing hormone) | Gonads | Gamete production and sex steroid hormone secretion | Stimulated by GnRH (gonadotropin-releasing hormone); inhibited by sex steroids |
| Prolactin | Mammary glands and other sex accessory organs | Milk production Controversial actions in other organs | Inhibited by PIH (prolactin-inhibiting hormone) |
| LH (luteinizing hormone) | Gonads | Sex hormone secretion: ovulation and corpus luteum formation | Stimulated by GnRH |

conventionally shortened to *tropic* (which has a different meaning—"attracted to"), is incorporated into these names. This is also indicated by the shortened forms of the names for the anterior pituitary hormones, which end in the suffix—*tropin.* The hormones of the pars distalis (table 11.8) are:

**Growth hormone (GH, or somatotropin).** This hormone promotes the movement of amino acids into tissue cells and the incorporation of these amino acids into tissue proteins, thus stimulating growth of organs.

**Thyroid-stimulating hormone (TSH, or thyrotropin).** This hormone stimulates the thyroid gland to produce and secrete thyroxine (tetraiodothyronine, or $T_4$).

**Adrenocorticotropic hormone (ACTH, or corticotropin).** This hormone stimulates the adrenal cortex to secrete the glucocorticoids, such as hydrocortisone (cortisol).

**Follicle-stimulating hormone (FSH, or folliculotropin).** This hormone stimulates the growth and secretion of ovarian follicles in females and the production of sperm in the testes of males.

**Luteinizing hormone (LH, or luteotropin).** This hormone and FSH are collectively called **gonadotropic hormones.** In females, LH stimulates ovulation and the conversion of the ovulated ovarian follicle into an endocrine structure called a corpus luteum. In males, LH (which is sometimes also called interstitial cell-stimulating hormone, or ICSH) stimulates the secretion of male sex hormones (mainly testosterone) from the interstitial cells of Leydig in the testes.

**Prolactin.** This hormone is secreted in both males and females. Its best known function is the stimulation of milk production by the mammary glands of women after the birth of their babies. Prolactin plays a supporting role in the regulation of the male reproductive system by the gonadotropins (FSH and LH), and acts on the kidneys to help regulate water and electrolyte balance.

Inadequate growth hormone secretion during childhood causes *pituitary dwarfism.* Hyposecretion of growth hormone in an adult produces a rare condition called *pituitary cachexia (Simmonds' disease).* One of the symptoms of this disease is premature aging caused by tissue atrophy. Oversecretion of growth hormone during childhood, in contrast, causes *gigantism.* Excessive growth hormone secretion in an adult does not cause further growth in length because the epiphyseal discs have ossified. Hypersecretion of growth hormone in an adult causes *acromegaly,* in which the person's appearance gradually changes as a result of thickening of bones and the growth of soft tissues, particularly in the face, hands, and feet.

The pars intermedia of the adenohypophysis in an adult human is not well defined, and its function, if any, is poorly understood. It produces different forms of **melanocyte-stimulating hormone** (**MSH**), which in lower vertebrates (fish, amphibians, and reptiles) cause a darkening of the skin to provide camouflage against a dark background. This reaction involves a cell type that is absent in human skin. Although the exogenous administration of MSH to humans stimulates melanin (pigment) production in the melanocytes of skin, MSH does not appear to have this effect in normal physiology. It is interesting that

**Figure 11.9.** The posterior pituitary, or neurohypophysis, stores and secretes hormones (vasopressin and oxytocin) produced in neuron cell bodies within the supraoptic and paraventricular nuclei of the hypothalamus. These hormones are transported to the posterior pituitary by nerve fibers of the hypothalamo-hypophyseal tract.

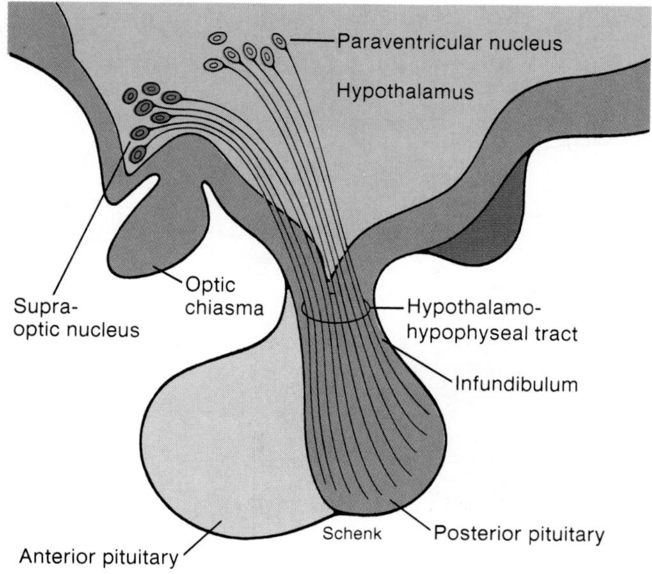


ACTH, secreted from the pars distalis, contains the amino acid sequence of MSH as part of its structure, and abnormally high amounts of ACTH secretion (as in Addison's disease) cause a darkening of the skin. The pars intermedia also produces large amounts of $\beta$-endorphin (chapter 7), as do other structures, but the physiological significance of this is not understood.

The posterior pituitary, or pars nervosa, secretes only two hormones, both of which are produced in the hypothalamus and merely stored in the posterior lobe of the pituitary:

**Antidiuretic hormone (ADH, or vasopressin).** Antidiuretic hormone stimulates the kidneys to retain water so that less water is excreted in the urine and more water is retained in the blood. This hormone also causes vasoconstriction in experimental animals, although the significance of this effect in humans is controversial.

**Oxytocin.** This hormone has no known function in males, but in females it is known to stimulate contractions of the uterus during labor and contractions of the mammary gland alveoli and ducts, which result in the milk-ejection reflex during lactation.

Injections of oxytocin may be given to a woman during labor if she is having difficulties in parturition. Increased amounts of oxytocin assist uterine contractions and generally speed up delivery. Oxytocin administration after parturition causes the uterus to regress in size and squeezes the blood vessels, thus minimizing the danger of hemorrhage.

## Hypothalamic Control of the Neurohypophysis

The **posterior pituitary** (pars nervosa of the neurohypophysis) secretes two hormones: antidiuretic hormone (ADH) and oxytocin. These two hormones, however, are actually produced in neuron cell bodies of the *supraoptic nuclei* and *paraventricular nuclei* of the hypothalamus. These nuclei within the hypothalamus are thus endocrine glands; the hormones they produce are transported along axons of the **hypothalamo-hypophyseal tract** (fig. 11.9) to the posterior pituitary, which stores and later secretes these hormones. The posterior pituitary is thus more a storage organ than a true gland.

The secretion of ADH and oxytocin from the posterior pituitary is controlled by **neuroendocrine reflexes.** In nursing mothers, for example, the stimulus of sucking acts via sensory nerve impulses to the hypothalamus to stimulate the reflex secretion of oxytocin. The secretion of ADH is stimulated by osmoreceptor neurons in the hypothalamus in response to a rise in blood osmotic pressure (chapter 6); its secretion is inhibited by sensory impulses from stretch receptors in the left atrium of the heart in response to a rise in blood volume. These reflexes are discussed in more detail in later chapters.

## Hypothalamic Control of the Adenohypophysis

At one time the anterior pituitary was called the "master gland" because it secretes hormones that regulate some other endocrine glands (fig. 11.10 and table 11.8). Adrenocorticotropic hormone (ACTH), thyroid-stimulating hormone (TSH), and the gonadotropic hormones (FSH

**Figure 11.10.** The hormones secreted by the anterior pituitary and the target organs for those hormones.

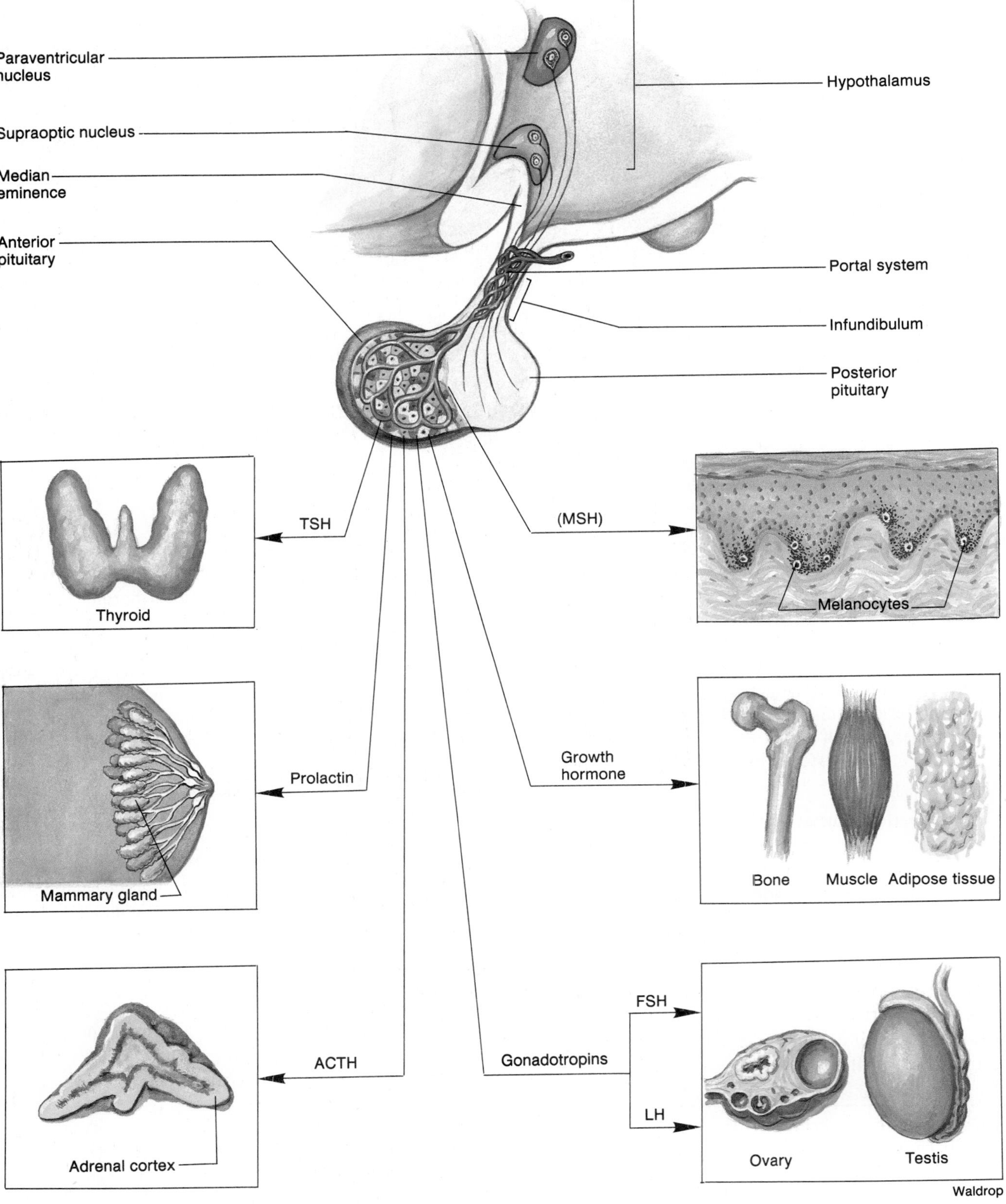

**Figure 11.11.** Neurons in the hypothalamus secrete releasing hormones (shown as dots) into the blood vessels of the hypothalamo-hypophyseal portal system. These releasing hormones stimulate the anterior pituitary to secrete its hormones into the general circulation.

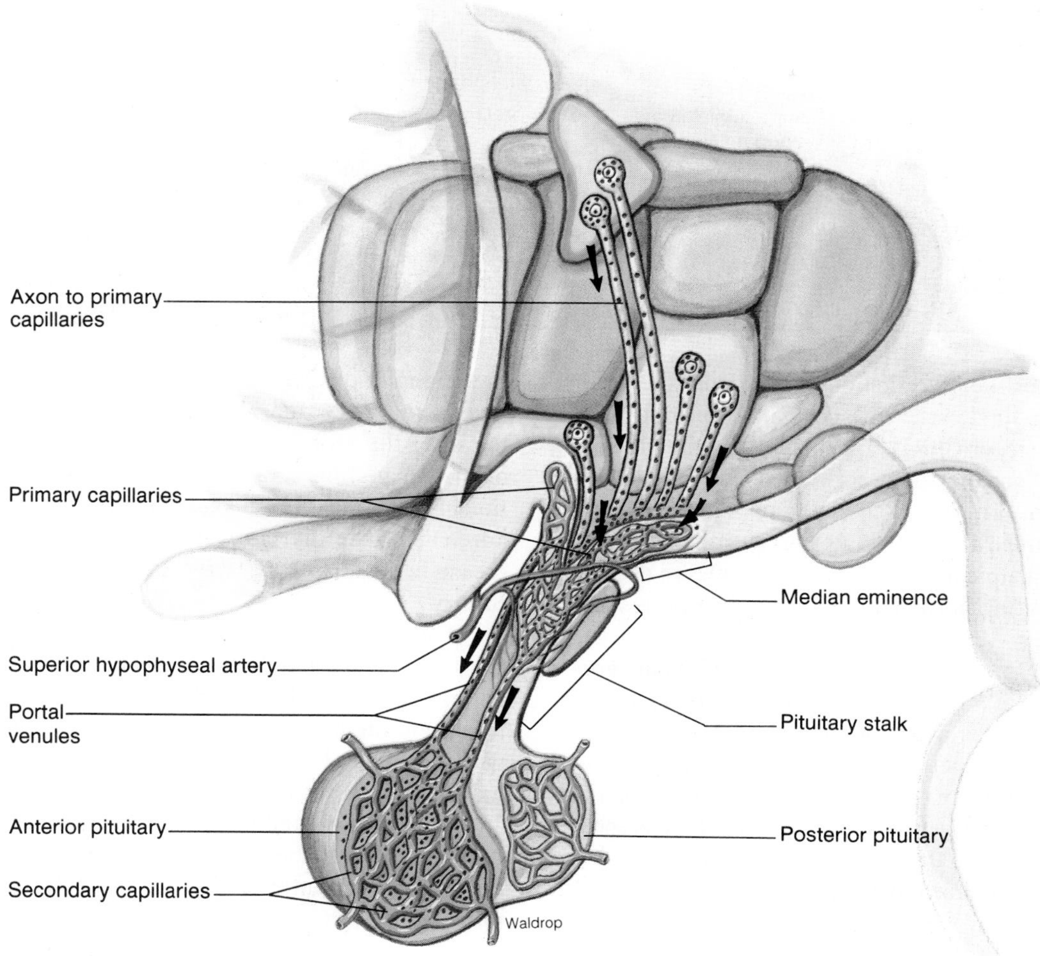


and LH) stimulate the adrenal cortex, thyroid, and gonads, respectively, to secrete their hormones. The anterior pituitary hormones also have a "trophic" effect on their target glands, in that the structure and health of the target glands depend on adequate stimulation by anterior pituitary hormones. The anterior pituitary, however, is not really the master gland, because secretion of its hormones is in turn controlled by hormones secreted by the hypothalamus.

***Releasing and Inhibiting Hormones.*** Since axons do not enter the anterior pituitary, hypothalamic control of the anterior pituitary is achieved through hormonal rather than neural regulation. Neurons in the hypothalamus produce releasing and inhibiting hormones, which are transported to axon endings in the basal portion of the hypothalamus. This region, known as the *median eminence,* contains blood capillaries that are drained by venules in the stalk of the pituitary.

The venules that drain the median eminence deliver blood to a second capillary bed in the anterior pituitary. Since this second capillary bed is downstream from the capillary bed in the median eminence and receives venous blood from it, the vascular link between the median eminence and the anterior pituitary comprises a *portal system.* (This is analogous to the hepatic portal system that delivers venous blood from the intestine to the liver—chapter 17.) The vascular link between the hypothalamus and the anterior pituitary is thus called the **hypothalamo-hypophyseal portal system.**

Neurons of the hypothalamus secrete polypeptide hormones into this portal system that regulate the secretions of the anterior pituitary (fig. 11.11 and table 11.9). Thyrotropin-releasing hormone (**TRH**) stimulates the secretion of TSH, and corticotropin-releasing hormone

**Table 11.9** Some hypothalamic hormones involved in the control of the anterior pituitary

| Hypothalamic Hormone | Structure | Effect on Anterior Pituitary | Action of Anterior Pituitary Hormone |
|---|---|---|---|
| Corticotropin-releasing hormone (CRH) | 41 amino acids | Stimulates secretion of adrenocorticotropic hormone (ACTH) | Stimulates secretions of adrenal cortex |
| Gonadotropin-releasing hormone (GnRH) | 10 amino acids | Stimulates secretion of follicle-stimulating hormone (FSH) and luteinizing hormone (LH) | Stimulates gonads to produce gametes (sperm and ova) and secrete sex steroids |
| Prolactin-inhibiting hormone (PIH) | Controversial (may be dopamine) | Inhibits prolactin secretion | Stimulates production of milk in mammary glands |
| Somatostatin | 14 amino acids | Inhibits secretion of growth hormone | Stimulates anabolism and growth in many organs |
| Thyrotropin-releasing hormone (TRH) | 3 amino acids | Stimulates secretion of thyroid-stimulating hormone (TSH) | Stimulates secretion of thyroid gland |
| Growth hormone–releasing hormone (GRH) | 44 amino acids | Stimulates growth hormone secretion | Stimulates anabolism and growth in many organs |

(**CRH**) stimulates the secretion of ACTH from the anterior pituitary. A single releasing hormone, gonadotropin-releasing hormone, or **GnRH,** appears to stimulate the secretion of both gonadotropic hormones (FSH and LH) from the anterior pituitary. The secretion of prolactin and of growth hormone from the anterior pituitary is regulated by hypothalamic inhibitory hormones, known as **PIH** (prolactin-inhibiting hormone) and **somatostatin,** respectively.

Recently, a specific **growth hormone-releasing hormone** (**GHRH**) that stimulates growth hormone secretion has been identified as a polypeptide consisting of forty-four amino acids. Experiments suggest that a releasing hormone for prolactin may also exist, but no such specific releasing hormone has yet been discovered. There is some indirect evidence that this proposed prolactin-releasing hormone may be produced by the posterior pituitary gland rather than by the hypothalamus.

### Feedback Control of the Adenohypophysis

In view of its secretion of releasing and inhibiting hormones, the hypothalamus might be considered the "master gland." The chain of command, however, is not linear; the hypothalamus and anterior pituitary are controlled by the effects of their own actions. In the endocrine system, to use an analogy, the general takes orders from the private. The hypothalamus and anterior pituitary are not master glands because their secretions are controlled by the target glands they regulate.

Anterior pituitary secretion of ACTH, TSH, and the gonadotropins (FSH and LH) is controlled by **negative feedback inhibition** from the target gland hormones. Secretion of ACTH is inhibited by a rise in corticosteroid secretion, for example, and TSH is inhibited by a rise in the secretion of thyroxine from the thyroid. These negative feedback relationships are easily demonstrated by removal of the target glands. Castration (surgical removal of the gonads), for example, produces a rise in the secretion of FSH and LH. In a similar manner, removal of the adrenals or the thyroid would result in an abnormal increase in ACTH or TSH secretion from the anterior pituitary.

These effects demonstrate that under normal conditions the target glands exert an inhibitory effect on the anterior pituitary. This inhibitory effect can occur at two levels: (1) the target gland hormones could act on the hypothalamus and inhibit the secretion of releasing hormones, and (2) the target gland hormones could act on the anterior pituitary and inhibit its response to the releasing hormones. Thyroxine, for example, appears to inhibit the response of the anterior pituitary to TRH and thus acts to reduce TSH secretion (fig. 11.12). Sex steroids, in contrast, reduce the secretion of gonadotropins by inhibiting both GnRH secretion and the ability of the anterior pituitary to respond to stimulation by GnRH (see fig. 11.13).

Recent evidence suggests that there may be retrograde transport of blood from the anterior pituitary to the hypothalamus. This may permit a *short feedback loop* where a particular trophic hormone inhibits the secretion of its releasing hormone from the hypothalamus. A high secretion of TSH, for example, may inhibit further secretion of TRH by this means.

In addition to negative feedback control of the anterior pituitary, there is one example where a hormone from a target organ actually stimulates the secretion of an anterior pituitary hormone. This occurs toward the middle of the menstrual cycle when the rising secretion of estradiol from the ovaries stimulates the anterior pituitary to secrete a "surge" of LH, which results in ovulation. This

**Figure 11.12.** The secretion of thyroxine from the thyroid is stimulated by the thyroid-stimulating hormone (TSH) from the anterior pituitary. The secretion of TSH is stimulated by the thyrotrophin-releasing hormone (TRH) secreted from the hypothalamus into the hypothalamo-hypophyseal portal system. This stimulation is balanced by the negative feedback inhibition of thyroxine, which decreases the responsiveness of the anterior pituitary to stimulation by TRH.

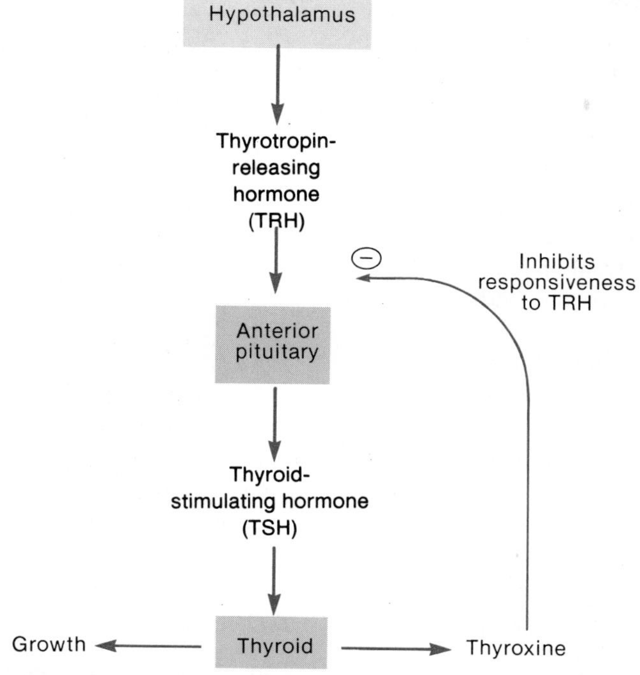


**Figure 11.13.** Negative feedback control of gonadotrophin secretion. The possible short negative feedback loop involving the inhibition of GnRH secretion of gonadotrophins is shown by a dotted arrow.

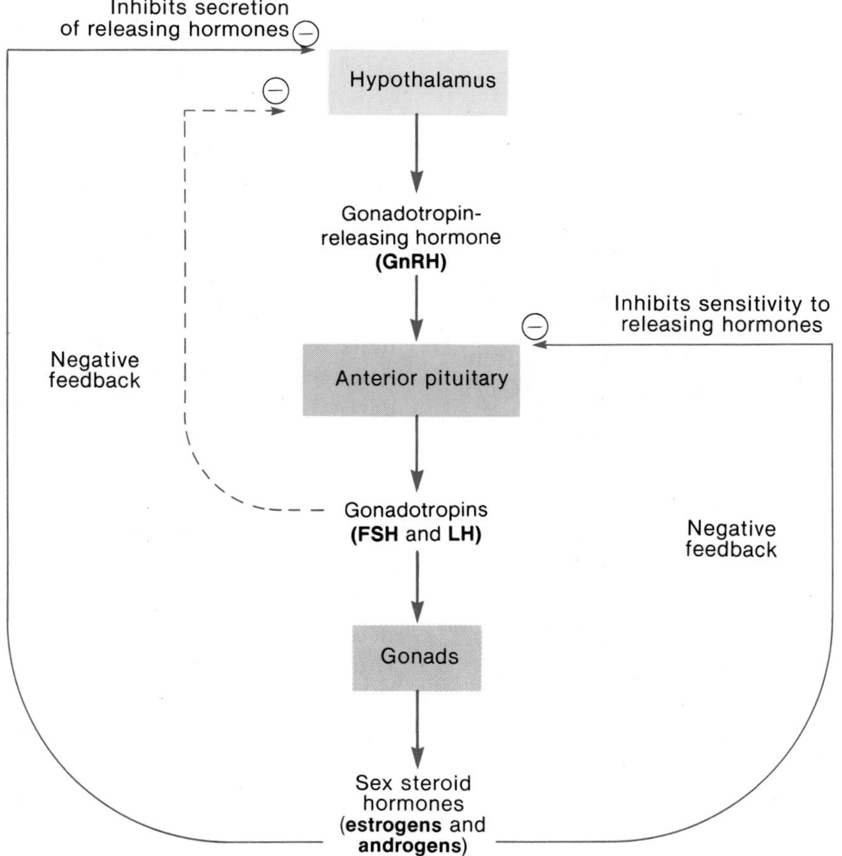

case is commonly referred to as a *positive feedback* effect to distinguish it from the more usual negative feedback inhibition of target gland hormones on anterior pituitary secretion. Interestingly, higher levels of estradiol at a later stage of the menstrual cycle exert the opposite effect—negative feedback inhibition—on LH secretion. The control of gonadotropin secretion is discussed in more detail in chapter 20.

### Higher Brain Function and Pituitary Secretion

The feedback effect of estradiol on the secretion of gonadotropic hormones is believed to be exerted at the level of the pituitary gland and hypothalamus. Since the hypothalamus receives neural input from "higher brain centers" (chapter 8), however, it is not surprising that the pituitary-gonad axis can be affected by emotions, so that intense emotions may alter the timing of ovulation or menstruation. The influences of higher brain centers on the pituitary-gonad axis also helps to explain the "dormitory effect," in which researchers have noted a tendency for the menstrual cycles to synchronize in women who room together.

This synchronization of menstrual cycles does not occur in a new roommate if her nasal cavity is plugged with cotton, suggesting that the dormitory effect is due to the action of **pheromones.** Pheromones are chemicals excreted by an individual which act through the olfactory sense and modify the physiology or behavior of another member of the same species. Pheromones are important regulatory molecules in the urine, and in vaginal and other secretions of lower mammals that help regulate their reproductive cycles and behavior. The importance of pheromones in human physiology and behavior is much more limited, but—because of our frequent cleansing, use of deodorants, and applications of artificial scents—is difficult to assess.

The effect of stress on the pituitary-adrenal axis is another good example of the influence of higher brain centers on pituitary function. Stressors, as described later in this chapter, produce an increase in CRH secretion from the hypothalamus, which in turn results in elevated ACTH and corticosteroid secretion. In addition, the influence of higher brain centers produces *circadian* ("about a day") *rhythms* in the secretion of many anterior pituitary hormones. The secretion of growth hormone, for example, is highest during sleep and decreases during wakefulness, although its secretion is also stimulated by the absorption of particular amino acids following a meal.

1. *Describe the embryonic origins of the adenohypophysis and neurohypophysis, and list the parts of each. Indicate which of these parts are also called the "anterior pituitary" and "posterior pituitary."*
2. *List the hormones secreted by the posterior pituitary. Describe the site of origin of these hormones and the mechanisms by which their secretions are regulated.*
3. *List the hormones secreted by the anterior pituitary, and describe how the hypothalamus controls the secretion of each hormone.*
4. *Draw a negative feedback loop showing the control of ACTH secretion. Explain how this system would be affected by (a) an injection of ACTH; (b) surgical removal of the pituitary; (c) an injection of corticosteroids; and (d) surgical removal of the adrenal glands.*

## ADRENAL GLANDS

The adrenal cortex and adrenal medulla are structurally and functionally different. The adrenal medulla secretes catecholamine hormones, which complement the sympathetic nervous system in the "fight or flight" reaction. The adrenal cortex secretes steroid hormones, which participate in the regulation of mineral balance, energy balance, and reproductive function. The adrenal medulla is stimulated by sympathetic neurons; the adrenal cortex is stimulated by ACTH as a nonspecific response to stress.

The adrenal glands are paired organs that cap the superior borders of the kidneys (fig. 11.14). Each adrenal consists of an outer cortex and inner medulla, which function as separate glands. The differences in structure and function of the adrenal cortex and medulla are related to the differences in their embryonic derivation; the adrenal medulla is derived from embryonic neural tissue (the same tissue that produces the sympathetic ganglia), whereas the adrenal cortex is derived from a different tissue (mesoderm).

As a result of its embryonic derivation, the adrenal medulla secretes catecholamine hormones (mainly epinephrine, with lesser amounts of norepinephrine) into the blood in response to stimulation by preganglionic sympathetic nerve fibers. The adrenal cortex does not receive neural innervation, and so must be stimulated hormonally (by ACTH secreted from the anterior pituitary). The cortex consists of three zones: an outer *zona glomerulosa*, a middle *zona fasciculata*, and an inner *zona reticularis* (fig. 11.14), which are believed to have different functions.

**Figure 11.14.** The structure of the adrenal gland showing the three zones of the adrenal cortex.

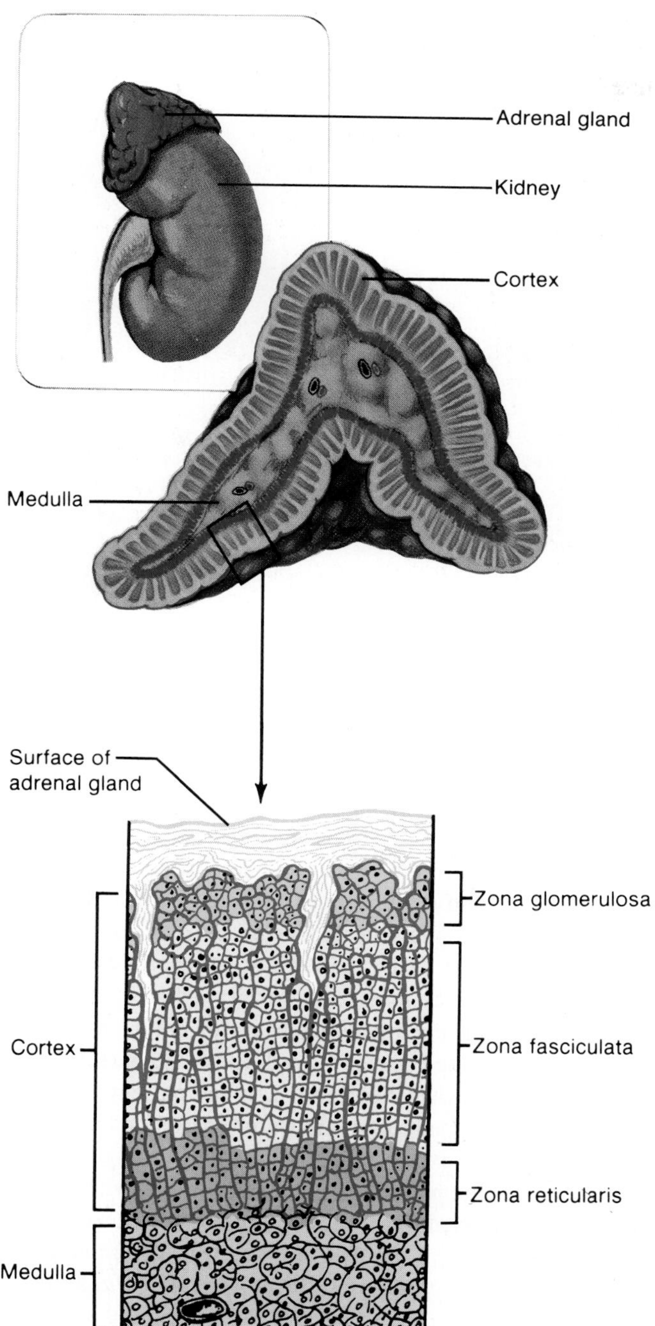


## Functions of the Adrenal Cortex

The adrenal cortex secretes steroid hormones called **corticosteroids,** or **corticoids,** for short. There are three functional categories of corticosteroids: (1) **mineralocorticoids,** which regulate $Na^+$ and $K^+$ balance; (2) **glucocorticoids,** which regulate the metabolism of glucose and other organic molecules; and (3) **sex steroids,** which are weak androgens (and lesser amounts of estrogens) that supplement the sex steroids secreted by the gonads. These hormones are secreted by the different zones of the adrenal cortex.

**Adrenogenital syndrome** is caused by the hypersecretion of adrenal sex hormones, particularly the androgens. Adrenogenital syndrome in young children causes premature puberty and enlarged genitals, especially the penis in a male and the clitoris in a female. An increase in body hair and a deeper voice are other characteristics. This condition in a mature woman can cause the growth of a beard. A related condition, called **congenital adrenal hyperplasia**, can cause masculinization of a female fetus prior to birth.

Aldosterone is the most potent mineralocorticoid. The mineralocorticoids are produced in the zona glomerulosa (fig. 11.15). The predominant glucocorticoid in humans is cortisol (hydrocortisone), which is secreted by the zona fasciculata and perhaps also by the zona reticularis. The secretion of aldosterone by the zona glomerulosa is stimulated by angiotensin II and by blood $K^+$. The secretion of cortisol by the zona fasciculata is stimulated by ACTH, which in turn is inhibited by negative feedback from cortisol (fig. 11.16).

Hypersecretion of corticosteroids results in **Cushing's syndrome.** This is generally caused by a tumor of the adrenal cortex or by oversecretion of ACTH from the adenohypophysis. Cushing's syndrome is characterized by changes in carbohydrate and protein metabolism, hyperglycemia, hypertension, and muscular weakness. Metabolic problems give the body a puffy appearance and can cause structural changes characterized as "buffalo hump" and "moon face." Similar effects are also seen when people with chronic inflammatory diseases receive prolonged treatment with corticosteroids, which are given to reduce inflammation and inhibit the immune response.

**Addison's disease** is caused by inadequate secretion of both glucocorticoids and mineralocorticoids, which results in hypoglycemia, sodium and potassium imbalance, dehydration, hypotension, rapid weight loss, and generalized weakness. A person with this condition who is not treated with corticosteroids will die within a few days because of the severe electrolyte imbalance and dehydration.

**Figure 11.15.** Simplified pathways for the synthesis of steroid hormones in the adrenal cortex. The adrenal cortex produces steroids that regulate $Na^+$ and $K^+$ balance (mineralocorticoids), steroids that regulate glucose balance (glucocorticoids), and small amounts of sex steroid hormones.

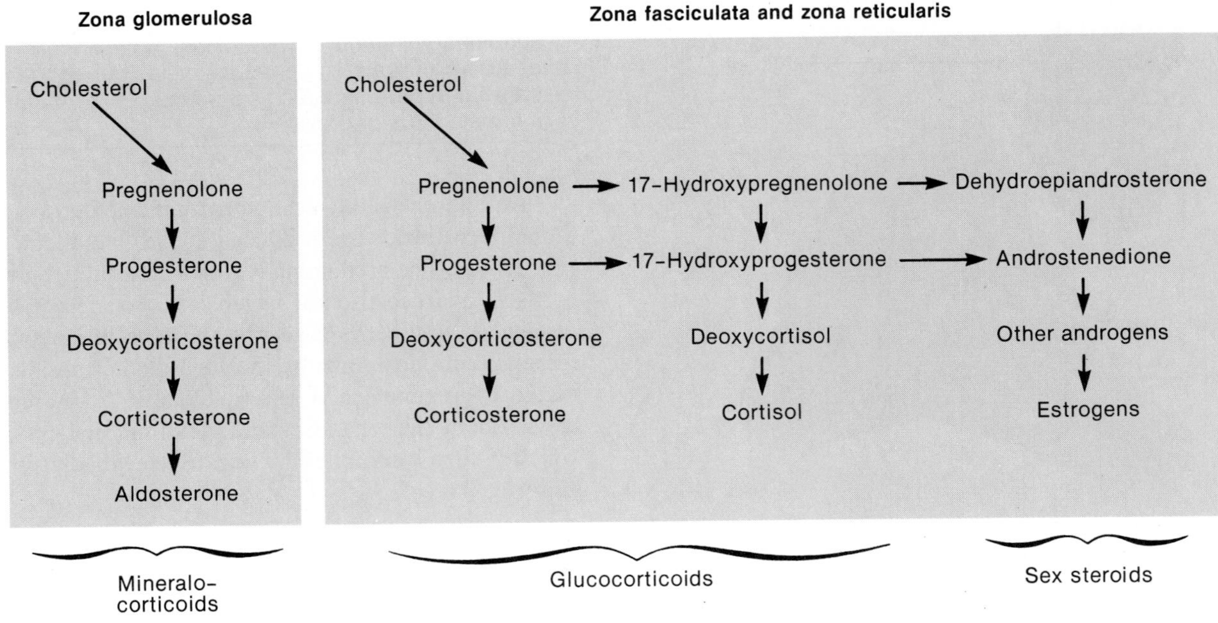


**Figure 11.16.** The activation of the pituitary-adrenal axis by nonspecific stress.

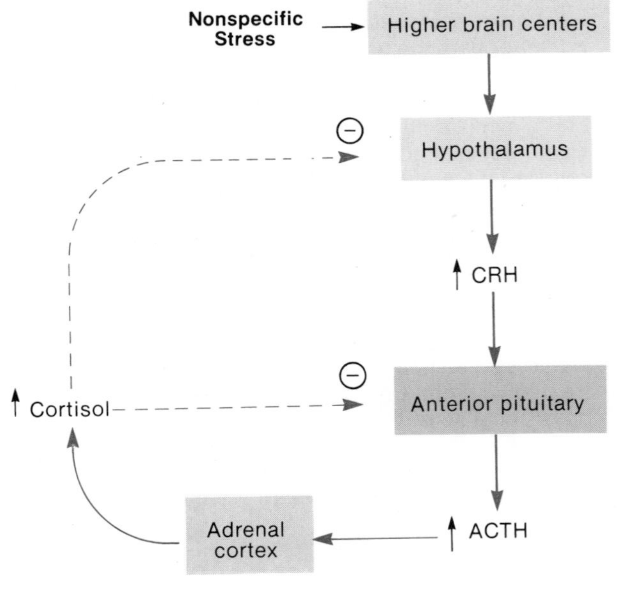


***Stress and the Adrenal Gland.*** In 1936, Hans Selye discovered that injections of cattle ovaries into rats (1) stimulated growth of the adrenal cortex; (2) caused atrophy of the lymphoid tissue of the spleen, lymph nodes, and thymus; and (3) produced bleeding peptic ulcers. At first he thought that these ovarian extracts contained a specific hormone that caused these effects. He later discovered that injections of a variety of substances, including foreign chemicals such as formaldehyde, could produce the same effects. Indeed, the same pattern of effects occurred when he placed rats in cold environments or when he dropped them into water and made them swim until they were exhausted.

The specific pattern of effects produced by these procedures suggested that these effects were the result of something that the procedures shared in common. Selye reasoned that all of the procedures were *stressful* and that the pattern of changes he observed represented a specific response to any stressful agent. He later discovered that all forms of stress produce these effects because all stressors stimulate the pituitary-adrenal axis. Under stressful conditions, there is increased secretion of ACTH and thus increased secretion of corticosteroids from the adrenal cortex.

Since stress is so very difficult to define, many people prefer to define stress operationally as any stimulus that activates the pituitary-adrenal axis. Using this criterion, it has been found that pleasant changes in one's life—such as marriage, a recent promotion, and so on—can be as stressful as unpleasant changes. On this basis, Selye has stated that there is "a nonspecific response of the body to readjust itself following any demand made upon it."

Selye termed this nonspecific response the **general adaptation syndrome (GAS).** Stress, in other words, produces GAS. There are three stages in the response to stress: (1) the *alarm reaction,* when the adrenal glands are activated; (2) the *stage of resistance,* in which readjustment occurs; and (3) if the readjustment is not complete, the *stage of exhaustion* may follow, leading to sickness and possibly death.

Excessive stimulation of the adrenal medulla can result in depletion of the body's energy reserves, and high levels of corticosteroid secretion from the adrenal cortex can significantly impair the immune system. It is reasonable to expect, therefore, that prolonged stress can result in increased susceptibility to disease. Indeed, many studies show that prolonged stress results in an increased incidence of cancer and other diseases.

## Functions of the Adrenal Medulla

The chromaffin cells of the adrenal medulla secrete **epinephrine** and **norepinephrine** in an approximate ratio of 4:1, respectively. These hormones are classified as amines (more specifically, as catecholamines) and are derived from the amino acid tyrosine.

The effects of these hormones are similar to those caused by stimulation of the sympathetic nervous system, except that the hormonal effect lasts about ten times longer. The hormones from the adrenal medulla increase cardiac output and heart rate, dilate coronary blood vessels, increase mental alertness, increase the respiratory rate, and elevate metabolic rate. A comparison of the effects of epinephrine and norepinephrine are presented in table 11.10.

The adrenal medulla is innervated by sympathetic nerve fibers. The impulses are initiated from the hypothalamus via the spinal cord when the sympathetic nervous system is stimulated. Many stressors, therefore, activate the adrenal medulla as well as the adrenal cortex. Activation of the adrenal medulla together with the sympathetic nervous system prepares the body for greater physical performance—the *fight-or-flight* response.

**Table 11.10** Comparison of the hormones from the adrenal medulla

| Epinephrine | Norepinephrine |
|---|---|
| Elevates blood pressure because of increased cardiac output and peripheral vasoconstriction | Elevates blood pressure because of generalized vasoconstriction |
| Accelerates respiratory rate and dilates respiratory passageways | Similar effect but to a lesser degree |
| Increases efficiency of muscular contraction | Similar effect but to a lesser degree |
| Increases rate of glycogen breakdown into glucose, so level of blood glucose rises | Similar effect but to a lesser degree |
| Increases rate of fatty acid released from fat, so level of blood fatty acids rises | Similar effect but to a lesser degree |
| Increases release of ACTH and TSH from the adenohypophysis of the pituitary gland | No effect |

From Kent M. Van De Graaff, *Human Anatomy,* 2d ed. 

A tumor of the adrenal medulla is referred to as a **pheochromocytoma.** This tumor causes hypersecretion of epinephrine and norepinephrine, which produces an effect similar to continuous sympathetic nerve stimulation. The symptoms of this condition are hypertension, elevated metabolism, hyperglycemia, sweating, and so on. It does not take long for the body to become totally fatigued under these conditions, making the patient susceptible to other diseases.

1. *List the categories of corticosteroids and the zones of the adrenal cortex that secrete these hormones.*
2. *List the hormones of the adrenal medulla, and describe their effects.*
3. *Explain how the secretions of the adrenal cortex and adrenal medulla are regulated.*
4. *Describe how stress affects the secretions of the adrenal cortex and medulla, and explain how excessive adrenal hormones may produce an increased susceptibility to disease.*

**Figure 11.17.** The thyroid gland. (*a*) Its relationship to the larynx and trachea. (*b*) A scan of the thyroid gland twenty-four hours after the intake of radioactive iodine.

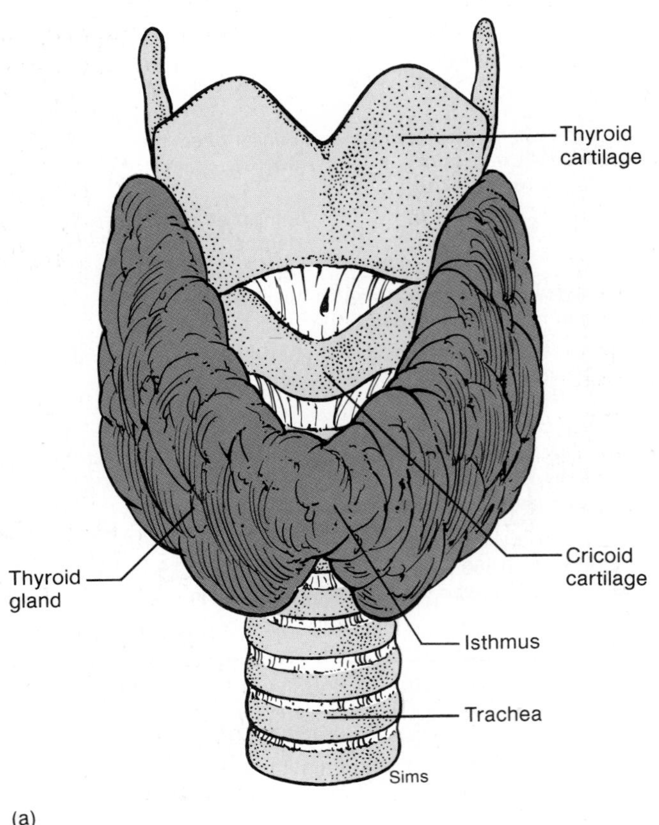


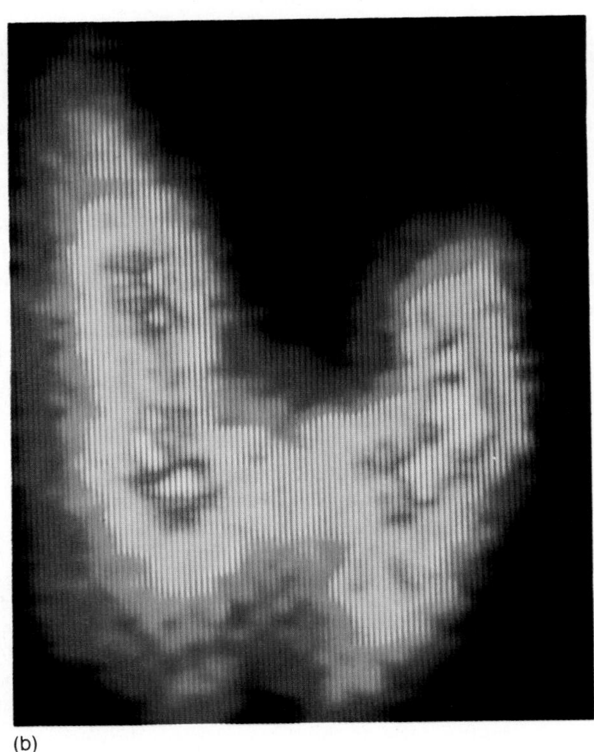


## THYROID AND PARATHYROIDS

The thyroid secretes thyroxine ($T_4$) and triiodothyronine ($T_3$), which are needed for proper growth and development and which are primarily responsible for determining the basal metabolic rate (BMR). Secretions of the thyroid are regulated by TSH from the anterior pituitary. The parathyroid glands secrete parathyroid hormone, which helps to raise the blood $Ca^{++}$ concentration. Low blood $Ca^{++}$ thus stimulates parathyroid hormone concentration in a negative feedback manner.

The thyroid gland is positioned just below the larynx (fig. 11.17). This gland consists of two lobes that lie on either side of the trachea and are connected anteriorly by a broad *isthmus.* The thyroid is the largest of the endocrine glands, weighing between 20 and 25 g.

On a microscopic level, the thyroid gland consists of many spherical hollow sacs called **thyroid follicles** (fig. 11.18). These follicles are lined with a simple cuboidal epithelium composed of *principal cells,* which synthesize the principal thyroid hormones. The interior of the follicles contains *colloid,* which is a protein-rich fluid. Between the follicles are epithelial cells called *parafollicular cells,* which produce a hormone called calcitonin (or thyrocalcitonin).

### Production and Action of Thyroid Hormones

The thyroid follicles actively accumulate iodide ($I^-$) from the blood and secrete it into the colloid. Once the iodide is in the colloid, it is oxidized to iodine and attached to specific amino acids (tyrosines) within the polypeptide chain of a protein called **thyroglobulin.** The attachment of one iodine to tyrosine produces *monoiodotyrosine (MIT);* the attachment of two iodines produces *diiodotyrosine (DIT).*

Within the colloid, enzymes modify the structure of MIT and DIT and couple them together (fig. 11.19). When two DIT molecules that are appropriately modified are coupled together, a molecule of **tetraiodothyronine, $T_4$,** or **thyroxine,** is produced. The combination of one MIT with one DIT forms **triiodothyronine, $T_3$.** Note that within the colloid $T_4$ and $T_3$ are still attached to thyroglobulin. Upon stimulation by TSH, the cells of the follicle take up a small volume of colloid by pinocytosis, hydrolyze the $T_3$ and $T_4$ from the thyroglobulin, and secrete the free hormones into the blood.

**Figure 11.18.** A photomicrograph (magnification ×250) of a thyroid gland, showing numerous thyroid follicles. Each follicle consists of follicular cells surrounding the fluid known as colloid, which contains thyroglobulin.

Follicles
Colloid
Follicular cells

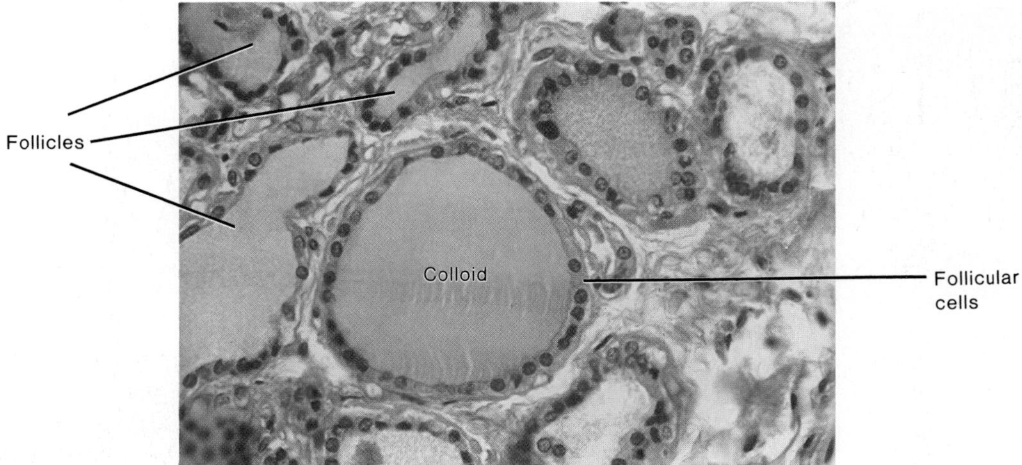

**Figure 11.19.** Stages in the formation and secretion of thyroid hormones. Iodide is actively accumulated by the follicular cells. In the colloid it is converted into iodine and attached to tyrosine amino acids within the thyroglobulin protein. Pinocytosis of iodinated thyroglobulin, coupling of MIT and DIT, and the release of thyroid hormones are stimulated by TSH from the anterior pituitary.

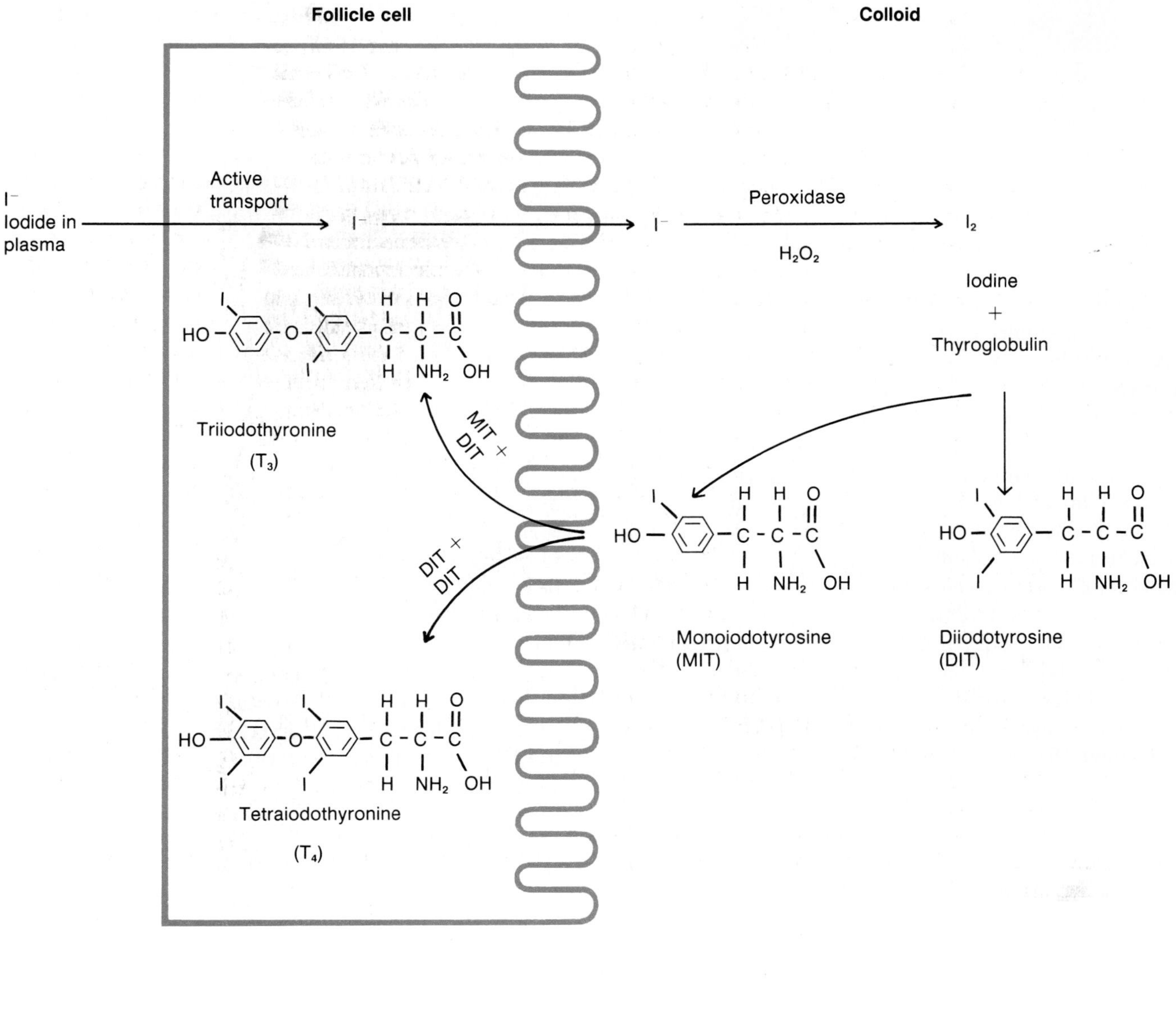

Follicle cell
Colloid
I⁻
Iodide in plasma
Active transport
I⁻
I⁻
Peroxidase
H₂O₂
I₂
Iodine
+
Thyroglobulin
Triiodothyronine
(T₃)
MIT + DIT
DIT + DIT
Monoiodotyrosine
(MIT)
Diiodotyrosine
(DIT)
Tetraiodothyronine
(T₄)

**Figure 11.20.** A simple or endemic goiter is caused by insufficient iodine in the diet.

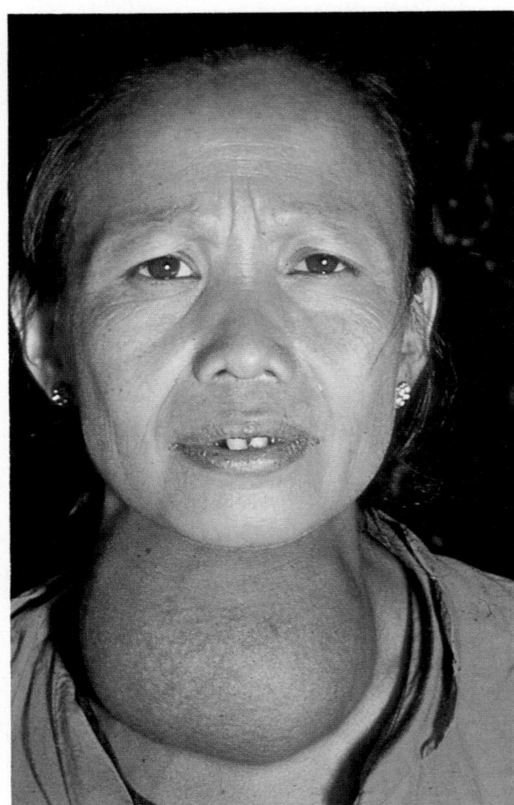

Most of the thyroid hormone molecules that enter the blood become attached to plasma carrier proteins. These hormone molecules can detach from the carrier proteins while others combine with the carriers, to produce an equilibrium state with only a very small percentage of hormone molecules in the free state. Only the thyroxine that is free in the plasma can enter the target cells, where it is converted to triiodothyronine and attached to nuclear receptor proteins, as described earlier in this chapter. Through the activation of genes, thyroid hormones stimulate protein synthesis, promote maturation of the nervous system, and increase the rate of energy utilization by the body. The development of the central nervous system is particularly dependent on thyroid hormones, and a deficiency of these hormones during development can cause serious mental retardation. The basal metabolic rate (BMR)—which is the minimum rate of caloric expenditure by the body—is determined to a large degree by the level of thyroid hormones in the blood. The physiological functions of thyroid hormones are described in more detail in chapter 18.

**Figure 11.21.** The mechanism of goiter formation in iodine deficiency. Low negative feedback inhibition results in excessive TSH secretion, which stimulates abnormal growth of the thyroid.

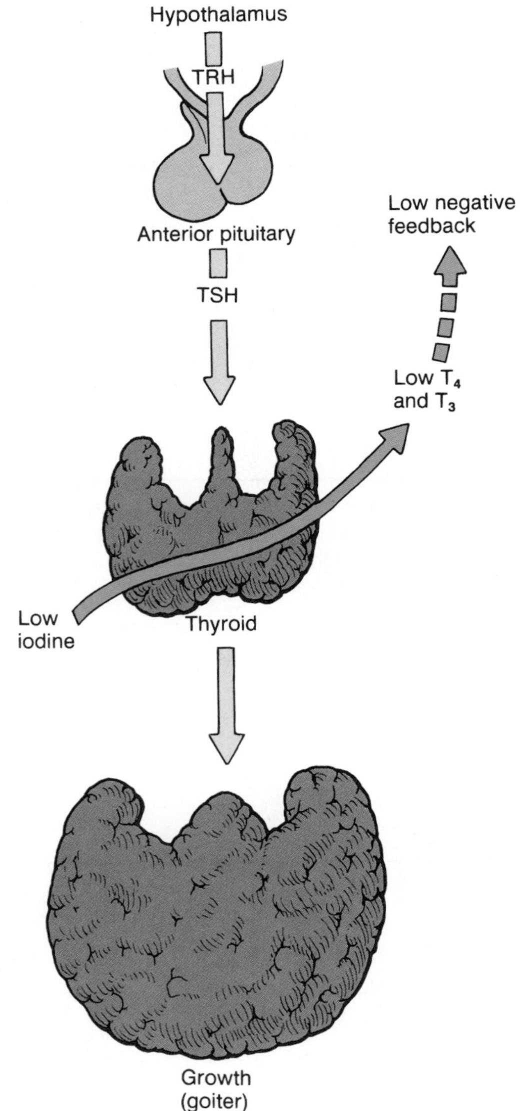


***Diseases of the Thyroid.*** Thyroid-stimulating hormone (TSH) from the anterior pituitary stimulates the thyroid to secrete thyroxine and exerts a trophic effect on the thyroid gland. This trophic effect is dramatically revealed in people who develop an **iodine-deficiency (endemic) goiter** (fig. 11.20). In the absence of sufficient dietary iodine the thyroid cannot produce adequate amounts of $T_4$ and $T_3$. The resulting lack of negative feedback inhibition causes abnormally high levels of TSH secretion, which in turn stimulate the abnormal growth of the thyroid (a goiter). These events are summarized in figure 11.21.

The infantile form of **hypothyroidism** is known as *cretinism*. An affected child usually appears normal at birth because thyroxine is received from the mother through the placenta. The clinical symptoms of cretinism are stunted growth, thickened facial features, abnormal bone development, mental retardation, low body temperature, and general lethargy. If cretinism is diagnosed early, it can be successfully treated by administering thyroxine.

Hypothyroidism in an adult causes *myxedema*. This disorder affects body fluids, causing an edema characterized by the accumulation of mucopolysaccharides. A person with myxedema has a low metabolic rate, lethargy, and a tendency to gain weight. This condition is treated with thyroxine or with triiodothyronine, which are taken orally (as pills).

**Graves' disease,** also called **toxic goiter,** involves growth of the thyroid associated with hypersecretion of thyroxine. This hyperthyroidism is produced by antibodies that act like TSH and stimulate the thyroid; it is an autoimmune disease. As a consequence of high levels of thyroxine secretion, the metabolic rate and heart rate increase, there is loss of weight, and the autonomic nervous system induces excessive sweating. In about half of the cases, *exophthalmos* (bulging of the eyes) also develops (fig. 11.22) because of edema in the tissues of the eye sockets and swelling of the extrinsic eye muscles.

**Figure 11.22.** Hyperthyroidism is characterized by an increased metabolic rate, weight loss, muscular weakness, and nervousness. The eyes may also protrude.

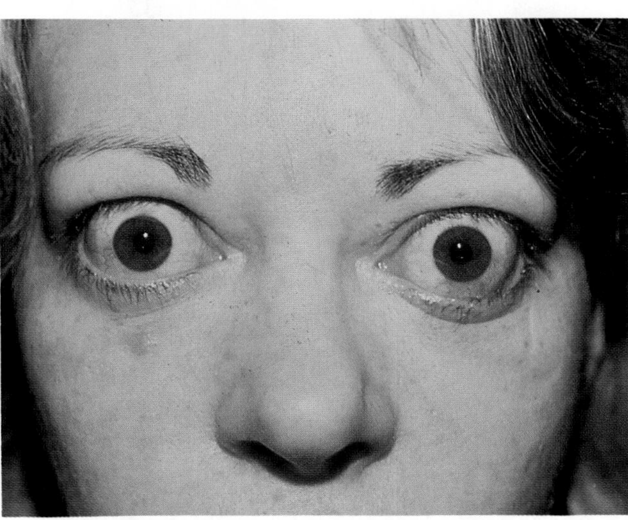

***Radiotherapy of Thyroid Cancer.*** The procedures involved in the treatment of thyroid cancer provide excellent examples of the practical applications of the principles previously presented. Thyroid cancer, particularly when it occurs in younger adults, has a more optimistic prognosis than most other forms of cancer. It is extremely slow growing and usually spreads only into the lymph nodes of the neck (although other sites can be invaded). Surgical treatment involves removing most of the thyroid and of the affected cervical lymph nodes. This surgical treatment is followed by swallowing solutions containing radioactive iodine ($^{131}I$), which is selectively transported only into cells of the thyroid gland and into cancerous thyroid cells that have metastasized (traveled) to other regions of the body. These cells are then killed by the radioactively labeled iodine.

Thyroid-stimulating hormone (TSH) stimulates the active accumulation of iodine into thyroid cells so that these cells can produce and secrete thyroxine. Therefore, in order for the ingested radioactive iodine to be maximally effective in the treatment of thyroid cancer, the blood levels of TSH must be raised to high levels.

When most of the patient's thyroid gland is surgically removed the blood levels of thyroxine gradually decline. Owing to the fact that most of the thyroxine is bound to protein in the plasma, it has an extremely long half-life. As the thyroxine levels in the blood decline, the negative feedback inhibition of TSH secretion lessens and TSH levels consequently rise. A period of about four to five weeks is required after removal of the thyroid for the TSH levels to rise sufficiently. At this point the high secretion of TSH can stimulate accumulation of radioactive iodine into thyroid cells and promote the destruction of both normal and metastasized cells.

## Parathyroid Glands

The small, flattened parathyroid glands are embedded in the posterior surfaces of the lateral lobes of the thyroid gland (see fig. 11.23). There are usually four parathyroid glands: a *superior* and an *inferior* pair. Each parathyroid gland is a small yellowish brown body measuring 3–8 mm (0.1–0.3 in.) in length, 2–5 mm (0.07–0.2 in.) in width, and about 1.5 mm (0.05 in.) in depth.

On a microscopic level, the parathyroids are composed of two types of epithelial cells. The cells that synthesize parathyroid hormone are called *chief cells* and are scattered among *oxyphil cells*. Oxyphil cells support the chief cells and are believed to produce reserve quantities of parathyroid hormone.

The parathyroid glands secrete one hormone called **parathyroid hormone (PTH).** This hormone promotes a rise in blood calcium levels by acting on the bones, kidneys, and intestine (fig. 11.24). Regulation of calcium balance is described in more detail in chapter 18.

1. *Describe the structure of the thyroid gland, and list the effect of thyroid hormones.*
2. *Describe how thyroid hormones are produced and how their secretion is regulated.*
3. *Explain the consequences of an inadequate dietary intake of iodine.*

**Figure 11.23.** A posterior view of the parathyroid glands.

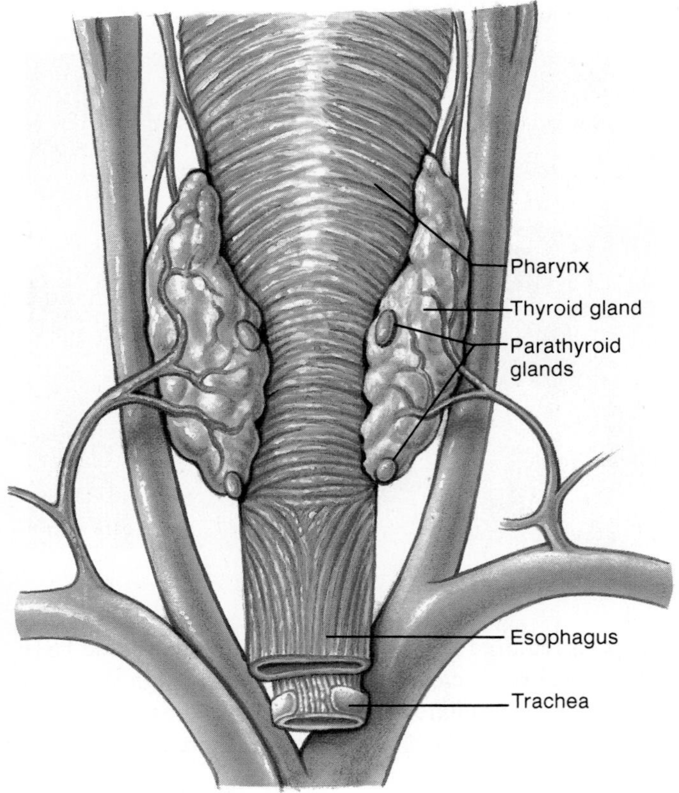


**Figure 11.24.** Actions of parathyroid hormone. An increased level of parathyroid hormone causes the bones to release calcium, the kidneys to conserve calcium loss through the urine, and the absorption of calcium through the intestinal wall. Negative feedback of increased calcium levels in the blood inhibits the secretion of this hormone.

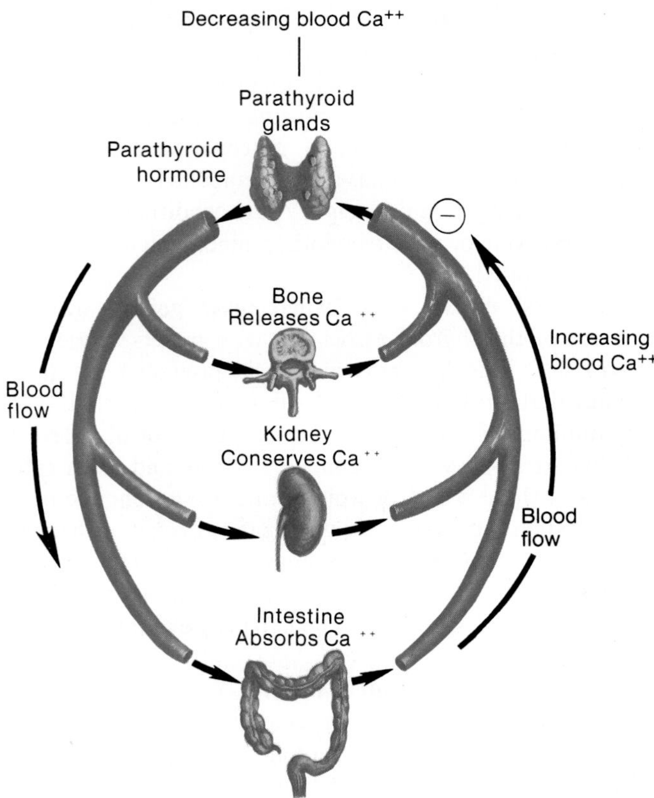


## *Pancreas and Other Endocrine Glands*

The islets of Langerhans in the pancreas secrete two hormones, insulin and glucagon, which are critically involved in the regulation of metabolic balance in the body. Insulin promotes the lowering of blood glucose and the storage of energy in the form of glycogen and fat. The hormone glucagon has antagonistic effects that act to raise the blood glucose. Additionally, many other organs secrete hormones that help regulate digestion, metabolism, growth, immune function, and reproduction.

The pancreas is both an endocrine and an exocrine gland. The gross structure of this gland and its exocrine functions in digestion are described in chapter 17. The endocrine portion of the pancreas consists of scattered clusters of cells called the **islets of Langerhans.** These endocrine structures are most common in the body and tail of the pancreas (fig. 11.25).

### Islets of Langerhans

On a microscopic level, the most conspicuous cells in the islets are the *alpha* and *beta* cells (fig. 11.26). The alpha cells secrete the hormone **glucagon,** and the beta cells secrete **insulin.**

Alpha cells secrete glucagon in response to a fall in the blood glucose concentrations. Glucagon stimulates the liver to convert glycogen to glucose **(glycogenolysis),** which causes the blood glucose level to rise. This effect represents the completion of a negative feedback loop. Glucagon also stimulates the hydrolysis of stored fat **(lipolysis)** and the consequent release of free fatty acids into the blood. This effect helps to provide energy substrates for the body during fasting, when blood glucose levels decrease. Glucagon, together with other hormones, also stimulates the conversion of fatty acids to ketone bodies, which can be secreted by the liver into the blood and used by many organs as an energy source. Glucagon is thus a hormone that helps to maintain homeostasis during times of fasting, when the body's energy reserves must be utilized (chapter 18).

**Figure 11.25.** The pancreas and the associated islets of Langerhans.

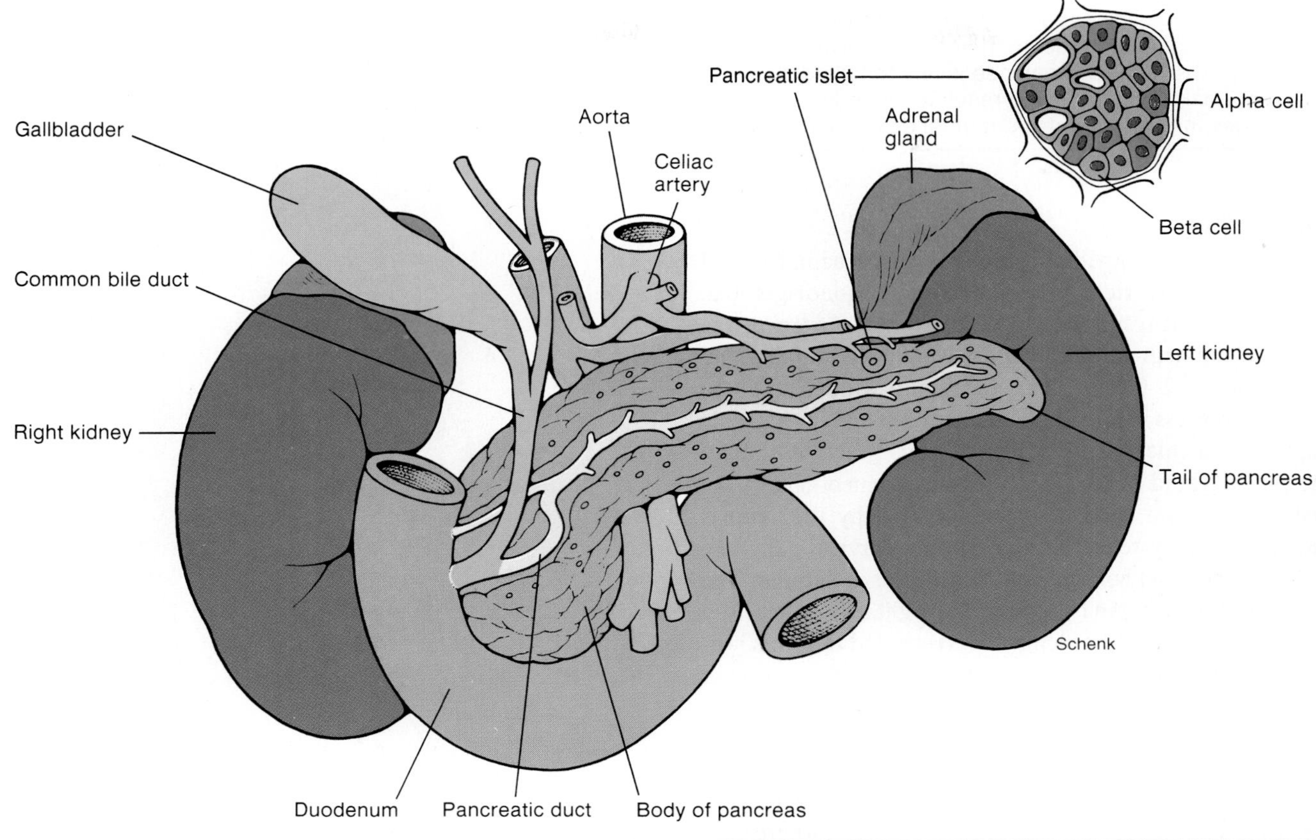


**Diabetes mellitus** is characterized by fasting hyperglycemia and the presence of glucose in the urine. There are two forms of this disease. *Type I*, or *insulin-dependent* diabetes mellitus, is caused by destruction of the beta cells and the resulting lack of insulin secretion. *Type II*, or *non-insulin-dependent* diabetes mellitus (which is the more common form), is caused by decreased tissue sensitivity to the effects of insulin, so that larger amounts of insulin are required to produce a normal effect. Both types of diabetes mellitus are also associated with abnormally high levels of glucagon secretion. The causes and symptoms of diabetes mellitus are described in more detail in chapter 18.

Beta cells secrete insulin in response to a rise in the blood glucose concentrations. Insulin promotes the entry of glucose into tissue cells, and the conversion of this glucose into energy storage molecules of glycogen and fat. Insulin also aids the entry of amino acids into cells and the production of cellular protein. The actions of insulin and glucagon are thus antagonistic. After a meal, insulin secretion is increased and glucagon secretion is decreased; fasting, in contrast, causes a rise in glucagon and a fall in insulin secretion.

**Figure 11.26.** The histology of the islets of Langerhans.

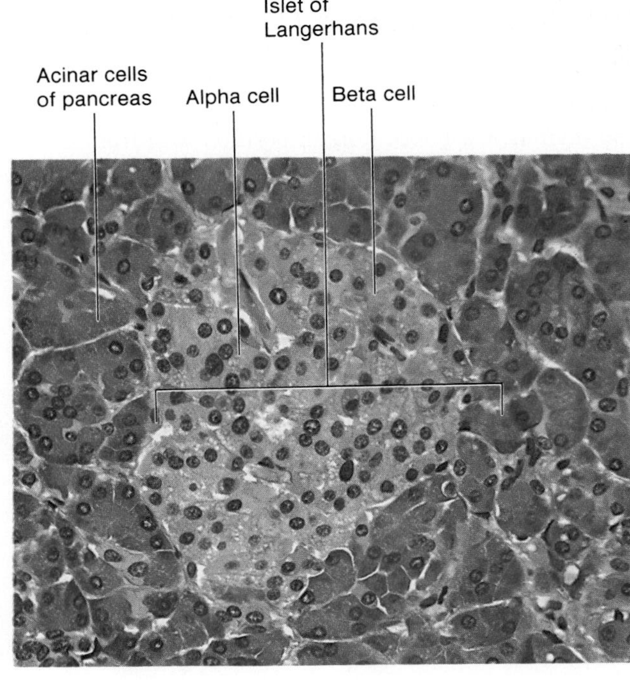

People who have a genetic predisposition for type II diabetes mellitus often first develop **reactive hypoglycemia.** In this condition, the rise in blood glucose that follows the ingestion of carbohydrates stimulates excessive secretion of insulin, which in turn causes the blood glucose levels to fall below the normal range. This can result in weakness, changes in personality, and mental disorientation.

## Pineal Gland

The small, cone-shaped pineal gland is located in the roof of the third ventricle near the corpora quadrigemina, where it is encapsulated by the meninges covering the brain. The pineal gland of a child weighs about 0.2 g and is 5–8 mm (0.2–0.3 in.) long and 9 mm wide. The gland begins to regress in size at about age seven and in the adult appears as a thickened strand of fibrous tissue. Although the pineal gland lacks direct nervous connections to the rest of the brain, it is highly innervated by the sympathetic nervous system from the superior cervical ganglion.

The principal hormone of the pineal is **melatonin.** The secretion of melatonin is inhibited by light and is therefore maximal at night. It has long been suspected that this hormone inhibits the pituitary-gonad axis. Indeed, a decrease in melatonin secretion in many lower vertebrates is responsible for maturation of the gonads during their reproductive season. Melatonin secretion is highest in children of ages one to five and decreases thereafter, reaching its lowest levels at the end of puberty, where concentrations are 75% lower than during early childhood. The secretion of melatonin is, therefore, believed by some researchers to play an important role in the onset of puberty, but this possibility is highly controversial.

## Thymus

The thymus is a bilobed organ positioned in front of the aorta and behind the manubrium of the sternum (fig. 11.27). Although the size of the thymus varies considerably from person to person, it is relatively large in newborns and children and sharply regresses in size after puberty. Besides decreasing in size, the thymus of adults becomes infiltrated with strands of fibrous and fatty connective tissue.

The thymus serves as the site of production of **T cells** *(thymus-dependent cells),* which are the lymphocytes involved in cell-mediated immunity (chapter 19). In addition to providing T cells, the thymus secretes a number of hormones that are believed to stimulate T cells after they leave the thymus.

**Figure 11.27.** The thymus is a bilobed organ within the mediastinum of the thorax.

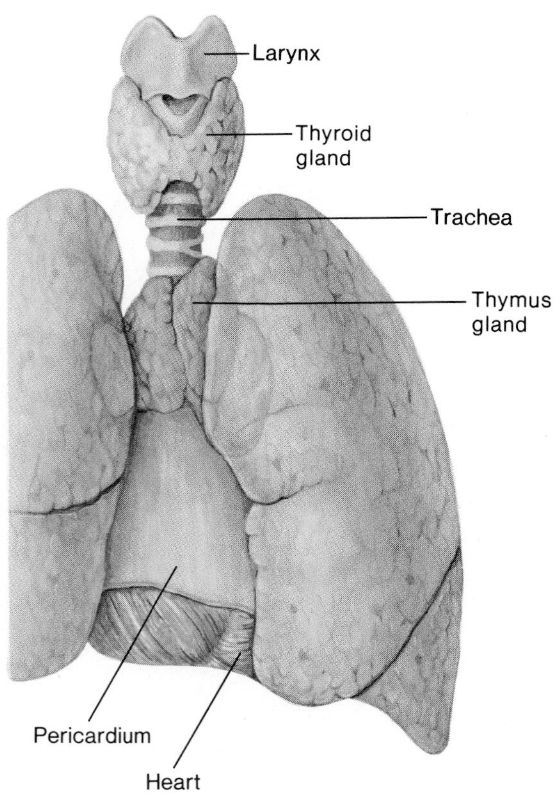


## Gastrointestinal Tract

The stomach and small intestine secrete a number of hormones that act on the gastrointestinal tract itself and on the pancreas and gallbladder (chapter 17). The effects of these hormones act together with regulation by the autonomic nervous system to coordinate the activities of different regions of the digestive tract and the secretions of pancreatic juice and bile.

## Gonads and Placenta

The gonads (**testes** and **ovaries**) secrete **sex steroids.** These include male sex hormones, or **androgens,** and female sex hormones—**estrogens** and **progestogens.** The principal hormones in each of these categories are *testosterone, estradiol-17β,* and *progesterone,* respectively.

The testes consist of two compartments: *seminiferous tubules,* which produce sperm, and *interstitial tissue* between the convolutions of the tubules. Within the interstitial tissue are *Leydig cells,* which secrete testosterone. Testosterone is needed for the development and maintenance of the male genitalia (penis and scrotum) and male sex accessory organs (prostate, seminal vesicles, epididymides, and ductus deferens), as well as for the development of male secondary sexual characteristics.

During the first half of the menstrual cycle, estrogen is secreted by many small structures within the ovary called *ovarian follicles*. These follicles contain the egg cell, or *ovum,* and *granulosa cells* that secrete estrogen. By about mid-cycle, one of these follicles grows very large and, in the process of ovulation, extrudes its ovum from the ovary. The empty follicle, under the influence of luteinizing hormone (LH) from the anterior pituitary, then becomes a new endocrine structure called a *corpus luteum*. The corpus luteum secretes progesterone as well as estradiol-17$\beta$.

The **placenta** is the organ responsible for nutrient and waste exchange between the fetus and mother. The placenta is also an endocrine gland; it secretes large amounts of estrogens and progesterone, as well as a number of polypeptide and protein hormones that are similar to some hormones secreted by the anterior pituitary gland. These latter hormones include **human chorionic gonadotropin (hCG)**, which is similar to LH, and **somatomammotropin**, which is similar in action to growth hormone and prolactin. The physiology of the placenta and other aspects of reproductive endocrinology are covered in chapter 20.

---

1. *Describe the structure of the endocrine pancreas, and indicate the sites of origin of insulin and glucagon.*
2. *Describe how insulin and glucagon secretion are affected by eating and by fasting, and explain the actions of these two hormones.*
3. *Describe the location of the pineal gland and the possible function of melatonin.*
4. *Describe the location and function of the thymus gland.*
5. *Explain how the hormones secreted by the gonads and placenta are categorized, and describe which hormones of these categories are secreted by each gland.*

---

## ***Prostaglandins***

Prostaglandins are a family of compounds that are autocrine regulators: they act within the same organ in which they are produced. There are a variety of prostaglandins, each derived from the precursor molecule arachidonic acid, which have different effects in many organs. The actions of prostaglandins supplement the regulatory effects of the endocrine and nervous systems.

In chapters 7 and 11, two types of regulatory molecules have been considered—neurotransmitters and hormones. These two classes of regulatory molecules cannot simply be defined by differences in chemical structure, since the same molecule (such as norepinephrine) may be in both categories, but they must rather be defined by function. Neurotransmitters are released by axons, travel across a narrow synaptic cleft, and affect a postsynaptic cell. Hormones are secreted into the blood by an endocrine gland and, through transport in the blood, come to influence the activities of one or more target organs.

There is yet another class of regulatory molecules, which are distinguished by the fact that they are produced in many different organs and are generally *active within the organ in which they are produced.* Molecules of this type are called *autocrine,* or *paracrine,* regulators. The major group of such molecules are the **prostaglandins.** Since prostaglandins do not generally function as hormones, they are not properly considered a part of the endocrine system. Nevertheless, as a result of their regulatory function and their close association with hormonal control mechanisms, prostaglandins will be considered in this chapter as a corollary of endocrine regulation.

A prostaglandin is a twenty-carbon-long fatty acid that contains a five-membered carbon ring. This molecule is derived from the precursor molecule *arachidonic acid,* which can be released from phospholipids in the cell membrane under hormonal or other stimulation. Arachidonic acid can then enter one of two possible metabolic pathways. In one case, the arachidonic acid may be converted by the enzyme *cyclo-oxygenase* into a prostaglandin, which can then be changed by other enzymes into other prostaglandins. In the other case, arachidonic acid may be converted by the enzyme *lipoxygenase* into **leukotrienes,** which are compounds that are closely related to the prostaglandins (fig. 11.28).

Prostaglandins are produced in almost every organ and have been implicated in a wide variety of regulatory functions. The study of prostaglandin function can be confusing, in part because a given prostaglandin may have opposite effects in different organs. Prostaglandins of the E series (PGE), for example, cause relaxation of smooth muscle in the bladder, bronchioles, intestine, and uterus, but the same molecules cause contraction of vascular smooth muscle. A different prostaglandin, designated $PGF_{2\alpha}$, has exactly the opposite effects.

The antagonistic effects of prostaglandins on blood clotting make good physiological sense. Blood platelets, which are required for blood clotting, produce *thromboxane $A_2$.* This prostaglandin promotes clotting by stimulating platelet aggregation and vasoconstriction. The endothelial cells of blood vessels, in contrast, produce a different prostaglandin, known as $PGI_2$, or *prostacyclin,* which has the opposite effects—it inhibits platelet aggregation and causes vasodilation. These antagonistic effects help to promote clotting while at the same time they ensure that clots do not normally form on the walls of intact blood vessels.

**Figure 11.28.** Formation and actions of leukotrienes and prostaglandins (PG = prostaglandin; TX = thromboxane).

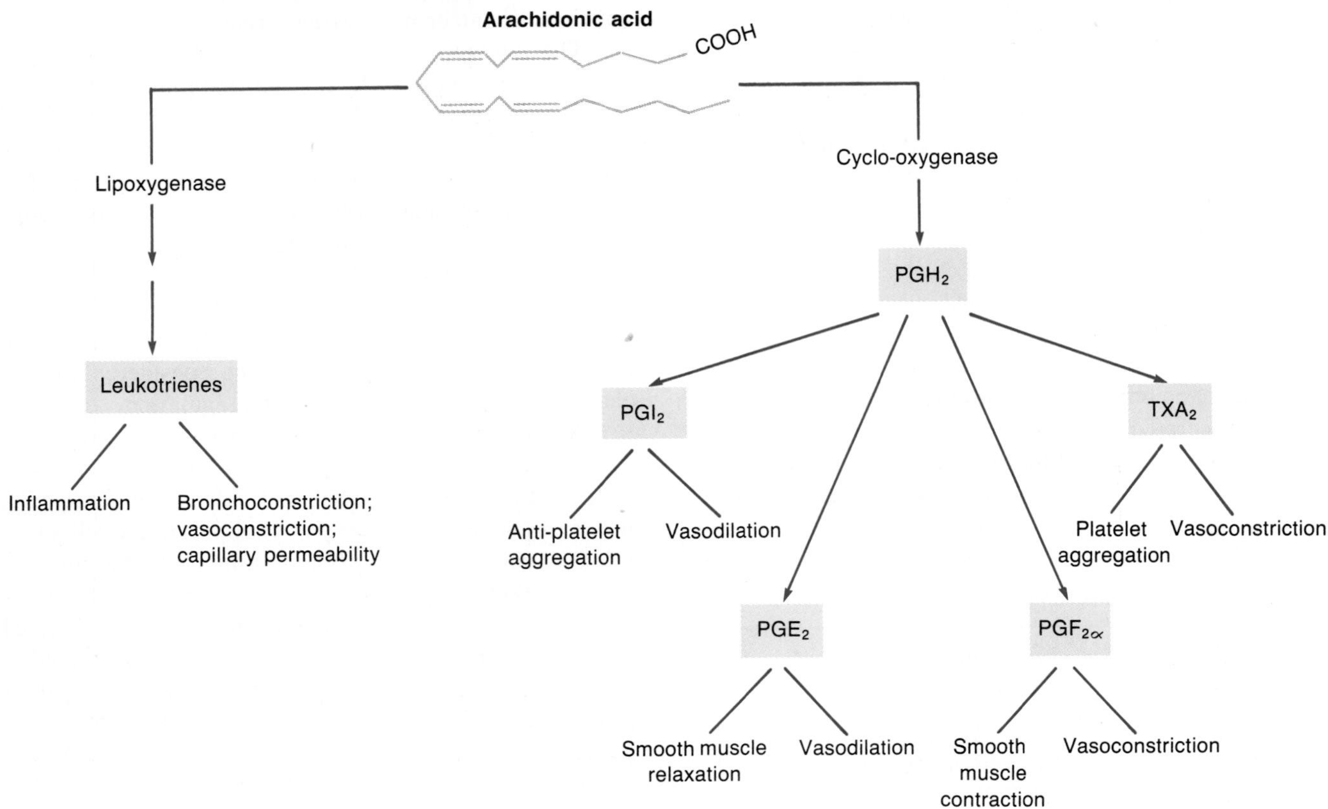


The following are some of the regulatory functions proposed for prostaglandins in different organs and systems:

**Inflammation.** Prostaglandins promote many aspects of the inflammatory process, including the development of pain and fever. Drugs that inhibit prostaglandin synthesis help to alleviate these symptoms.

**Reproductive system.** Prostaglandins may play a role in ovulation and corpus luteum function in the ovaries and in contraction of the uterus. Excessive prostaglandin production may be involved in premature labor, endometriosis, dysmenorrhea (painful menstrual cramps), and other gynecological disorders.

**Gastrointestinal tract.** The stomach and intestines produce prostaglandins, which are believed to inhibit gastric secretions and influence intestinal motility and fluid absorption. Since prostaglandins inhibit gastric secretion, drugs that suppress prostaglandin production may make a patient more susceptible to peptic ulcers.

**Respiratory system.** Some prostaglandins cause constriction whereas others cause dilation of blood vessels in the lungs and of bronchiolar smooth muscle. The leukotrienes are potent bronchoconstrictors, and these compounds, together with some prostaglandins, may cause respiratory distress and contribute to bronchoconstriction in asthma.

**Blood vessels.** Some prostaglandins are vasoconstrictors and others are vasodilators. In a fetus, $PGE_2$ is believed to promote dilation of the ductus arteriosus, a short vessel that connects the pulmonary artery with the aorta. After birth the ductus arteriosus normally closes as a result of a rise in blood oxygen when the baby breathes. If the ductus remains patent (open), however, it can be closed by the administration of drugs that inhibit prostaglandin synthesis.

**Blood clotting.** Thromboxane $A_2$, produced by blood platelets, promotes platelet aggregation and vasoconstriction. Prostacyclin, produced by vascular endothelial cells, inhibits platelet aggregation and promotes vasodilation.

**Kidneys.** Prostaglandins are produced in the medulla of the kidneys and cause vasodilation, resulting in increased renal blood flow and increased excretion of water and electrolytes in the urine.

**Aspirin** is the most widely used member of a class of drugs known as *nonsteroidal anti-inflammatory drugs.* Aspirin and other drugs of this class produce their effects because they specifically inhibit the cyclo-oxygenase enzyme that is needed for prostaglandin synthesis. Since prostaglandins promote inflammation, therefore, aspirin helps to reduce inflammation, pain, and fever. Acting through the inhibition of prostaglandin synthesis, which is needed for proper blood clotting, aspirin significantly increases the clotting time for eight days after administration (platelets are replaced every eight days). For this reason aspirin may be used to treat venous embolisms and arterial thromboses and is used in patients with myocardial ischemia in the hope that it will prevent the formation of blood clots in the coronary vessels.

1. *Describe the chemical nature of prostaglandins, and explain why these molecules are not considered to be hormones.*
2. *Describe the metabolic pathway by which prostaglandins are produced, and describe the different types of functions performed by different prostaglandins.*
3. *Explain how aspirin works, and describe the effects of aspirin on different body processes.*

## SUMMARY

### Hormones: Actions and Interactions p. 280

I. Precursors of active hormones may be called either prohormones or prehormones.
   A. Prohormones are relatively inactive precursor molecules made in the endocrine cells.
   B. Prehormones are the normal secretions of an endocrine gland that must be converted to other derivatives by target cells to be active.

II. The effects of a hormone in the body depend on its concentration.
   A. Abnormally high amounts of a hormone can result in effects that are not normally found.
   B. Target tissues can become desensitized by high hormone concentrations.

III. Hormones can interact in permissive, synergistic, or antagonistic ways.

### Mechanisms of Hormone Action p. 286

I. Steroid and thyroid hormones enter their target cells.
   A. Thyroid hormones attach to chromatin-bound receptors located in the nucleus.
   B. Steroid hormones bond to cytoplasmic receptor proteins and translocate to the nucleus.
   C. Attachment of the hormone-receptor protein complex to the chromatin activates genes and thereby stimulates RNA and protein synthesis.

II. Amine, polypeptide, and glycoprotein hormones bond to receptor proteins on the outer surface of the target cell membrane.
   A. In many cases, this leads to the intracellular production of cyclic AMP, which serves as a second messenger in the action of these hormones.
   B. In other cases, cyclic GMP, or $Ca^{++}$ may serve as a second messenger in the action of the hormone.

### Pituitary Gland p. 293

I. The pituitary secretes eight hormones.
   A. The anterior pituitary secretes growth hormone, thyroid-stimulating hormone, adrenocorticotropic hormone, follicle-stimulating hormone, luteinizing hormone, and prolactin.
   B. The posterior pituitary secretes antidiuretic hormone (also called vasopressin) and oxytocin.

II. The hormones of the posterior pituitary are produced in the hypothalamus and transported to the posterior pituitary by the hypothalamo-hypophyseal nerve tract.

III. Secretions of the anterior pituitary are controlled by hypothalamic hormones that stimulate or inhibit secretions of the anterior pituitary.
   A. Hypothalamic hormones include TRH, CRH, GnRH, PIH, somatostatin, and a growth hormone–releasing hormone.
   B. These hormones are carried to the anterior pituitary by the hypothalamo-hypophyseal portal system.

IV. Secretions of the anterior pituitary are also regulated by the feedback (usually negative feedback) of hormones from the target glands.

V. Higher brain centers, acting through the hypothalamus, can influence pituitary secretion.

### Adrenal Glands p. 300

I. The adrenal cortex secretes mineralocorticoids (mainly aldosterone), glucocorticoids (mainly cortisol), and sex steroids (primarily weak androgens).
   A. The glucocorticoids help regulate energy balance; they also can inhibit inflammation and suppress immune function.
   B. The pituitary-adrenal axis is stimulated by stress as part of the general adaptation syndrome.

II. The adrenal medulla secretes epinephrine and lesser amounts of norepinephrine, which complement the action of the sympathetic nervous system.

### Thyroid and Parathyroids p. 304

I. The thyroid follicles secrete tetraiodothyronine ($T_4$, or thyroxine) and lesser amounts of triiodothyronine ($T_3$).
   A. These hormones are formed within the colloid of the thyroid follicles.

B. The parafollicular cells of the thyroid secrete the hormone calcitonin, which may act to lower blood calcium levels.

II. The parathyroids are small structures embedded within the thyroid gland; the parathyroids secrete a hormone that promotes a rise in blood calcium levels.

## Pancreas and Other Endocrine Glands p. 308

I. Beta cells in the islets secrete insulin; alpha cells secrete glucagon.
   A. Insulin lowers blood glucose and stimulates the production of glycogen, fat, and protein.
   B. Glucagon raises blood glucose by stimulating the breakdown of liver glycogen; glucagon also promotes lipolysis and the formation of ketone bodies.
   C. The secretion of insulin is stimulated by a rise in blood glucose following meals; the secretion of glucagon is stimulated by a fall in blood glucose during periods of fasting.

II. The pineal gland, located on the roof of the third ventricle of the brain, secretes melatonin; this hormone may play a role in regulating reproductive function.

III. The thymus is the site of the production of T cell lymphocytes and secretes a number of hormones that may help regulate the immune system.

IV. The gastrointestinal tract secretes a number of hormones that help regulate functions of the digestive system.

V. The gonads secrete sex steroid hormones.
   A. Leydig cells in the interstitial tissue of the testes secrete testosterone and other androgens.
   B. Granulosa cells of the ovarian follicles secrete estrogen.
   C. The corpus luteum of the ovaries secretes progesterone as well as estrogen.

VI. The placenta secretes estrogen, progesterone, and a variety of polypeptide hormones that have actions similar to some anterior pituitary hormones.

## Prostaglandins p. 311

I. Prostaglandins are special, twenty-carbon-long fatty acids produced by many different organs that usually have regulatory functions within the organ in which they are produced.

---

# Review Activities

## Objective Questions

1. Hypothalamic-releasing hormones
   (a) are secreted into capillaries in the median eminence
   (b) are transported by portal veins to the anterior pituitary
   (c) stimulate the secretion of specific hormones from the anterior pituitary
   (d) all of the above
2. The hormone primarily responsible for setting the basal metabolic rate and for promoting the maturation of the brain is
   (a) cortisol
   (b) ACTH
   (c) TSH
   (d) thyroxine
3. Which of the following statements about the adrenal cortex is *true?*
   (a) It is not innervated by nerve fibers.
   (b) It secretes some androgens.
   (c) The zona granulosa secretes aldosterone.
   (d) The zona fasciculata is stimulated by ACTH.
   (e) All of the above are true.
4. The hormone insulin
   (a) is secreted by alpha cells in the islets of Langerhans
   (b) is secreted in response to a rise in blood glucose
   (c) stimulates the production of glycogen and fat
   (d) both *a* and *b*
   (e) both *b* and c

Match the hormone with the primary agent that stimulates its secretion.

5. Epinephrine
6. Thyroxine
7. Corticosteroids
8. ACTH

(a) TSH
(b) ACTH
(c) Growth hormone
(d) Sympathetic nerves
(e) CRH

9. Steroid hormones are secreted by
   (a) the adrenal cortex
   (b) the gonads
   (c) the thyroid
   (d) both *a* and *b*
   (e) both *b* and *c*
10. The secretion of which of the following hormones would be *increased* in a person with endemic goiter?
   (a) TSH
   (b) thyroxine
   (c) triiodothyronine
   (d) all of the above
11. Which of the following hormones use cAMP as a second messenger?
   (a) testosterone
   (b) cortisol
   (c) insulin
   (d) epinephrine
12. Which of the following terms best describes the type of interaction between the effects of insulin and glucagon?
   (a) synergistic
   (b) permissive
   (c) antagonistic
   (d) cooperative

## Essay Questions

1. Explain how the regulation of the neurohypophysis and adrenal medulla are related to their embryonic origins.
2. Compare steroid and polypeptide hormones in terms of their mechanism of action in target organs.
3. Explain the significance of the term *trophic* in regard to the actions of anterior pituitary hormones.
4. Suppose a drug blocks the conversion of $T_4$ to $T_3$. Explain what the effects of this drug would be on (*a*) TSH secretion, (*b*) thyroxine secretion, and (*c*) the size of the thyroid gland.
5. Explain why the phrase "master gland" is sometimes used to describe the anterior pituitary, and why this term is misleading.
6. Suppose a person's immune system made antibodies against insulin receptor proteins. Describe the possible effect of this condition on carbohydrate and fat metabolism.

## Selected Readings


Austin, L. A., and H. Heath, III. 1981. Calcitonin: Physiology and pathophysiology. *New England Journal of Medicine.* 304:269.

Axelrod, J., and T. D. Reisine. 1984. Stress hormones: their interaction and regulation. *Science* 224:452.

Baxter, J. D., and W. J. Funder. 1979. Hormone receptors. *New England Journal of Medicine.* 300:117.

Biglieri, E. G. 1989. ACTH effects on aldosterone, cortisol, and other steroids. *Hospital Practice* 24:145.

Brownstein, M. J. et al. 1980. Synthesis, transport and release of posterior pituitary hormones. *Science* 207:373.

Carmichael, S. W., and H. Winkler. August 1985. The adrenal chromaffin cell. *Scientific American.*

Demers, L. M. September 1984. The effects of prostaglandins. *Diagnostic Medicine,* p. 37.

Ganong, W. F., L. C. Alpert, and T. C. Lee. 1974. ACTH and the regulation of adrenocortical secretion. *New England Journal of Medicine* 290:1006.

Ganong, W. F. 1988. The stress concept—a dynamic overview. *Hospital Practice* 23:155.

Gershengorn, M. C. 1986. Mechanism of thyrotrophin releasing hormone stimulation of pituitary hormone secretion. *Annual Review of Physiology* 48:515.

Gillie, R. B. June 1971. Endemic goiter. *Scientific American.*

Guillemin, R., and R. Burgus. November 1972. The hormones of the hypothalamus. *Scientific American.*

Katzenellenbogen, B. S. 1980. Dynamics of steroid hormone receptor action. *Annual Review of Physiology* 42:17.

Krieger, D. T. 1984. Brain peptides: what, where, and why? *Science* 222:975.

Levitzki, A. 1988. From epinephrine to cyclic AMP. *Science* 241:800.

McEwen, B. S. July 1976. Interactions between hormones and nerve tissue. *Scientific American.*

Martin, C. R. *Endocrine physiology.* 1985. New York: Oxford University Press.

O'Malley, B., and W. T. Shrader. February 1976. The receptors of steroid hormones. *Scientific American.*

Rasmussen, H. 1986. The calcium messenger system. *New England Journal of Medicine* 314: First part, p. 1094; Second part, p. 1164.

Reichlin, S. et. al. 1976. Hypothalamic hormones. *Annual Review of Physiology* 39:389.

Reisine, T. 1988. Neurohumoral aspects of ACTH release. *Hospital Practice* 23:77.

Roth, J. 1980. Insulin receptors in diabetes. *Hospital Practice* 15:98.

Roth, J., and S. I. Taylor. 1982. Receptors for peptide hormones: alterations in diseases of humans. *Annual Review of Physiology* 44:639.

Schally, A. V. 1978. Aspects of the hypothalamic control of the pituitary gland. *Science* 202:18.

Selye, H. 1973. The evolution of the stress concept. *American Scientist* 61:692.

Spiegel, A. M. 1988. G proteins in clinical medicine. *Hospital Practice* 23:93.

Taylor, A. L., and L. M. Fishman. 1988. Corticotropin-releasing hormone. *New England Journal of Medicine* 319:213.

Turner, C. D. 1976. *General endocrinology.* 6th ed. Philadelphia: W. B. Saunders Co.

Ville, D. B. 1975. *Human endocrinology: A developmental approach.* Philadelphia: W. B. Saunders Co.

Williams, R. 1974. *Textbook of endocrinology.* 5th ed. Philadelphia: W. B. Saunders Co.

# 12 Muscle: Mechanisms of Contraction and Neural Control

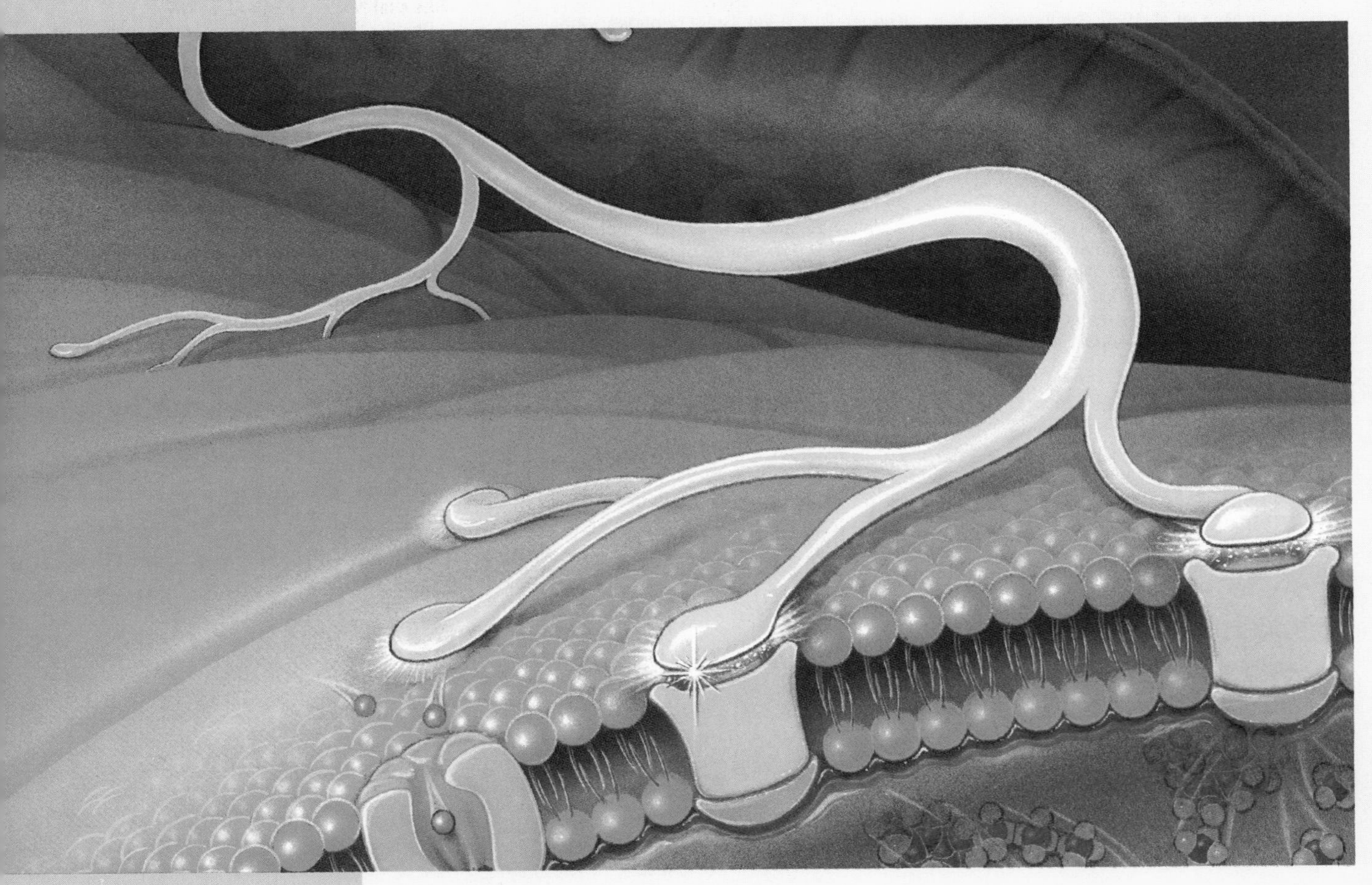

## Objectives

By studying this chapter, you should be able to

1. describe the gross and microscopic structure of skeletal muscles
2. describe the all-or-none contraction of skeletal muscle fibers and explain how muscles can produce graded and sustained contractions
3. define a motor unit, explain how motor units vary, and explain how the motor units are used to control muscle contraction
4. distinguish between isometric and isotonic contraction and describe how the series elastic component affects muscle contraction
5. describe the structure of sarcomeres and of myofibrils and explain how this structure accounts for the striated appearance of skeletal muscle fibers
6. explain the sliding filament theory of contraction
7. explain the events that occur during cross-bridge cycles and the role of ATP in muscle contraction
8. explain the physiological roles of tropomyosin and troponin and the role of $Ca^{++}$ in excitation-contraction coupling
9. describe how action potentials in a nerve and muscle fiber affect the contractile mechanism and explain the role of the sarcoplasmic reticulum in muscle contraction and relaxation
10. describe the structure and functions of muscle spindles and explain the mechanisms involved in a stretch reflex
11. explain the function of Golgi tendon organs and explain why a slow, gradual muscle stretch might avoid the spasm that could be caused by a rapid stretch
12. explain reciprocal innervation and describe the neural pathways involved in a crossed-extensor reflex
13. describe the significance of gamma motoneurons in the neural control of muscle contraction and in the maintenance of muscle tone
14. describe the neural pathways involved in the pyramidal and extrapyramidal systems
15. explain the differences in structure and function between slow-twitch, fast-twitch, and intermediate fibers
16. explain how muscles fatigue and what changes occur in muscle fibers in response to physical training
17. compare the structure and physiology of cardiac muscle with that of skeletal muscle
18. describe the structure of smooth muscle, and explain how its contraction is regulated

## Outline

**Figure 12.1.** Actions of antagonistic muscles in the leg.

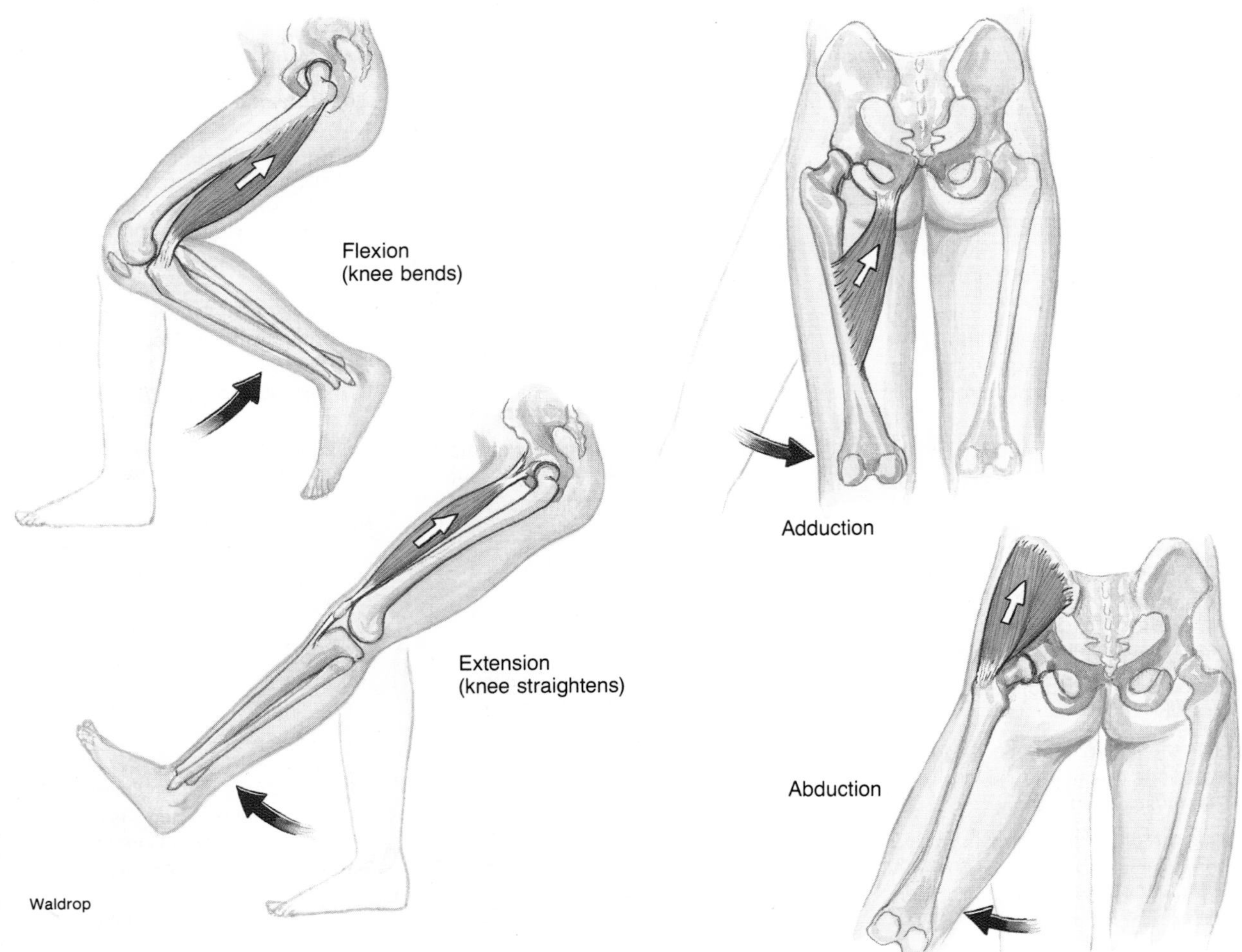


## Structure and Actions of Skeletal Muscles

Skeletal muscles are composed of individual muscle fibers which contract when they are stimulated by a motor neuron. Each motor neuron branches to innervate a number of muscle fibers, and all of these fibers contract when their motor neuron is activated. Activation of varying numbers of motor neurons, and thus varying numbers of muscle fibers, results in gradations in the strength of contraction of the whole muscle. In this way, the muscle can shorten and cause its characteristic movement of the skeleton.

Skeletal muscles are usually attached to bone on each end by tough connective-tissue *tendons.* When a muscle contracts it shortens, and this places tension on its tendons and attached bones. The muscle tension causes movement of the bones at a joint, where one of the attached bones generally moves more than the other. The more movable bony attachment of the muscle, known as its *insertion,* is pulled toward its less movable attachment (the *origin*). A variety of different skeletal movements are possible, depending on the type of joint involved and the attachments of the muscles (table 12.1 and fig. 12.1). When *flexor* muscles contract, for example, they decrease the angle of a joint. Contraction of *extensor* muscles increases the angle of their attached bones at the joint. Flexors and extensors that act on the same joint are thus *antagonistic muscles.*

**Table 12.1** Categories of skeletal muscle actions

| Category | Action |
|---|---|
| Extensor | Increases the angle at a joint |
| Flexor | Decreases the angle at a joint |
| Abductor | Moves limb away from the midline of the body |
| Adductor | Moves limb toward the midline of the body |
| Levator | Moves insertion upward |
| Depressor | Moves insertion downward |
| Rotator | Rotates a bone along its axis |
| Sphincter | Constricts an opening |

The position of the limbs, for example, is determined by the actions of a variety of antagonistic muscles. In addition to the movements of flexion and extension, a limb

**Figure 12.2.** The relationship between muscle fibers and the connective tissues of the tendon, epimysium, perimysium, and endomysium. Closeup shows an expanded view of a single muscle fiber.

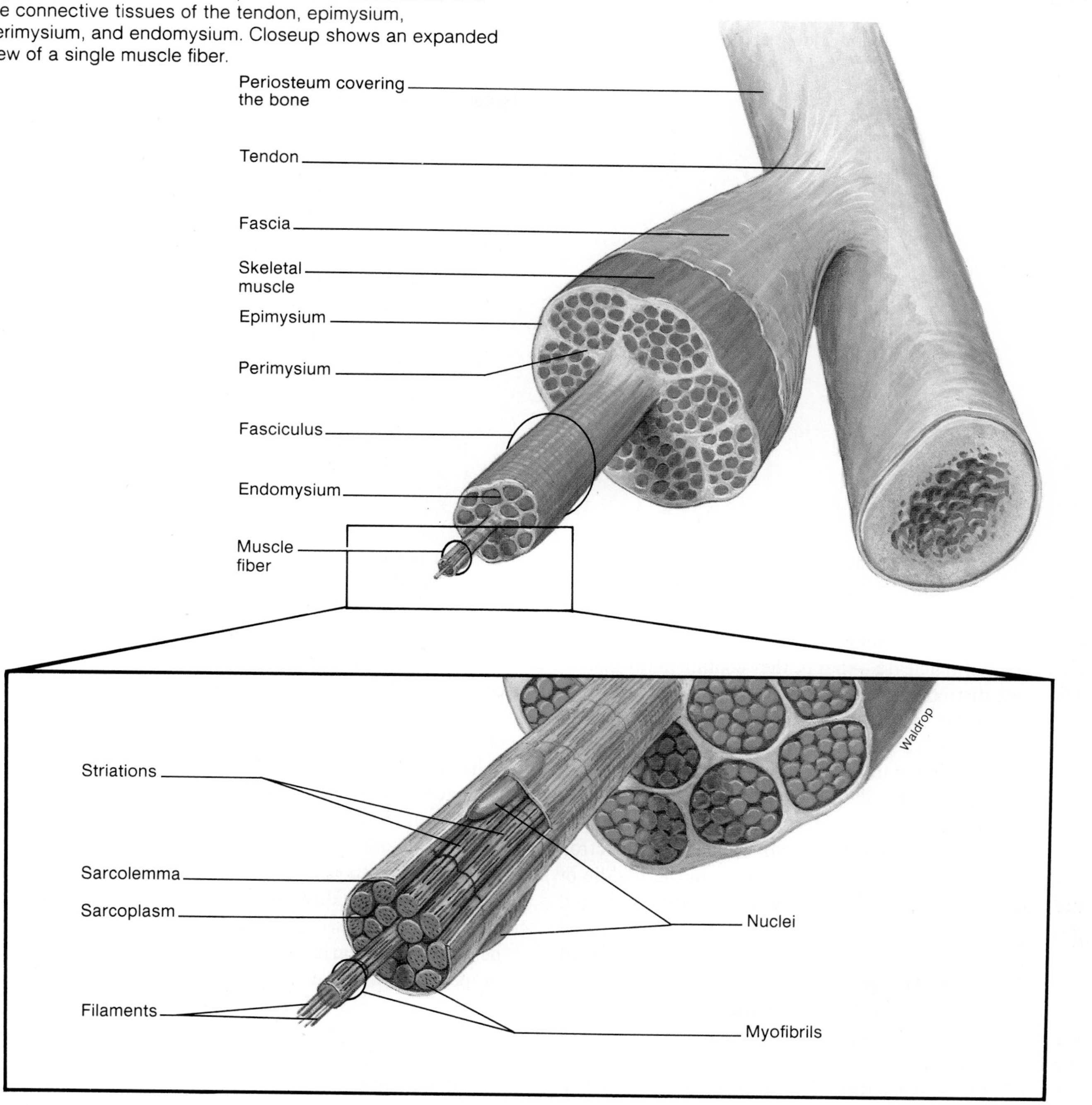

can be moved away from the midline of the body by *abductor* muscles, and brought inward toward the midline by contraction of *adductor* muscles. In all cases, these skeletal movements are produced by the shortening of the appropriate muscle groups—the *agonists*—while the antagonist muscles remain relaxed.

## Structure of Skeletal Muscles

The fibrous connective tissue proteins within the tendons continue in an irregular arrangement around the muscle to form a sheath known as the *epimysium* (*epi* = above; *my* = muscle). Connective tissue from this outer sheath extends into the body of the muscle, subdividing it into columns, or *fascicles* (e.g., the "strings" in stringy meat). Each of these fascicles is thus surrounded by its own connective tissue sheath, known as the *perimysium* (*peri* = around).

Dissection of a muscle fascicle under a microscope reveals that it, in turn, is composed of many **muscle fibers** (or *myofibers*) surrounded by wisps of connective tissue called *endomysium* (fig. 12.2). These fibers are actually the muscle cells. Since the connective tissue of the tendons, epimysium, perimysium, and endomysium is continuous, muscle fibers do not normally pull out of the tendons when they contract.

**Figure 12.3.** Appearance of skeletal muscle fibers in the light microscope (note the striated appearance produced by alternating dark A bands and light I bands).

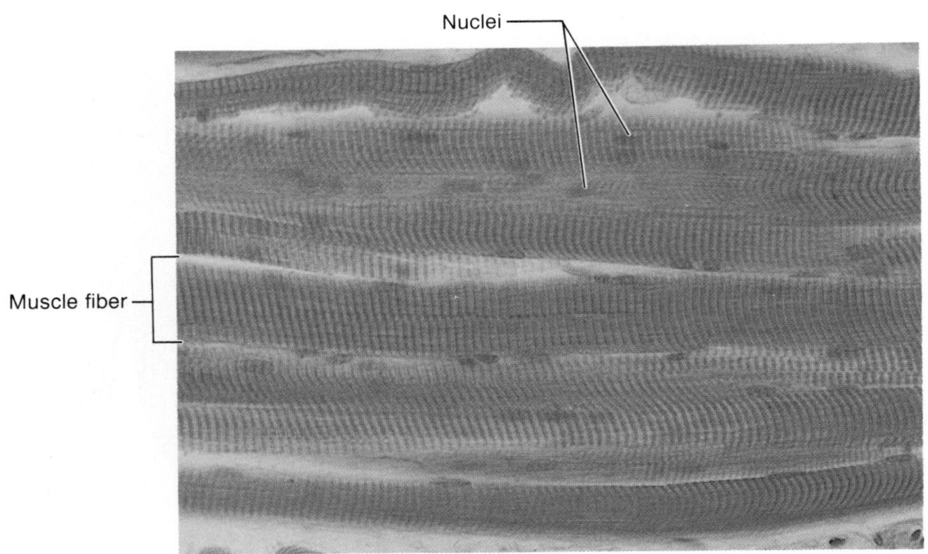


Despite their unusual fiber shape, muscle cells have the same organelles that are present in other cells: mitochondria, intracellular membranes, glycogen granules, and others. Unlike most other cells in the body, skeletal muscle fibers are multinucleate—that is, they contain many nuclei. The most distinctive feature of the microscopic appearance of skeletal muscle fibers, however, is their **striated** appearance (fig. 12.3). The striations (stripes) are produced by alternating dark and light bands that appear to cross the width of the fiber.

The dark bands are called **A bands** and the light bands are called **I bands**. At high magnification in an electron microscope, thin, dark lines can be seen in the middle of the I bands. These are called **Z lines**. The labels A, I, and Z are useful for describing the functional architecture of muscle fibers, and were derived during the history of muscle research. The letters A and I stand for *anisotropic* and *isotropic*, respectively, which indicate the behavior of polarized light as it passes through these regions; the letter Z comes from the German word *Zwischenscheibe*, which translates to "between disc." These derivations are of historical interest only.

## Twitch, Summation, and Tetanus

The contractile behavior of skeletal muscles is more easily studied *in vitro* (outside the body) than *in vivo* (within the body). In these studies a muscle, such as the gastrocnemius (calf muscle) of a frog, is usually mounted so that one end is fixed and the other is movable. In the classic studies of this kind, the movable end of the muscle directly produces deflections of a pen, which writes on a rotating drum recorder (fig. 12.4). When more modern equipment is used, the mechanical force of the muscle contraction is transduced into an electric current, which can be amplified and displayed as pen deflections in a multichannel recorder (fig. 12.5). In this way, the contractile behavior of the whole muscle in response to experimentally administered electric shocks can be studied.

When the muscle is stimulated with a single electric shock of a sufficient voltage, it quickly contracts and relaxes. This response is called a **twitch.** Increasing the stimulus voltage increases the strength of the twitch up to a maximum. The strength of a muscle contraction can thus be *graded,* or varied—an obvious requirement for the proper control of skeletal movements. If a second electric shock is delivered immediately after the first, it will produce a second twitch that may partially "ride piggyback" on the first. This response is called **summation.**

If the stimulator is set to deliver an increasing frequency of electric shocks automatically, the relaxation time between successive twitches will get shorter and shorter, as the strength of contraction increases in amplitude. This is **incomplete tetanus.** Finally, at a particular "fusion frequency" of stimulation, there is no visible relaxation between successive twitches (fig. 12.6). Contraction is smooth and sustained, as it is during normal muscle contraction

**Figure 12.4.** A setup for observing the contractile behavior of the isolated gastrocnemius muscle of a frog.

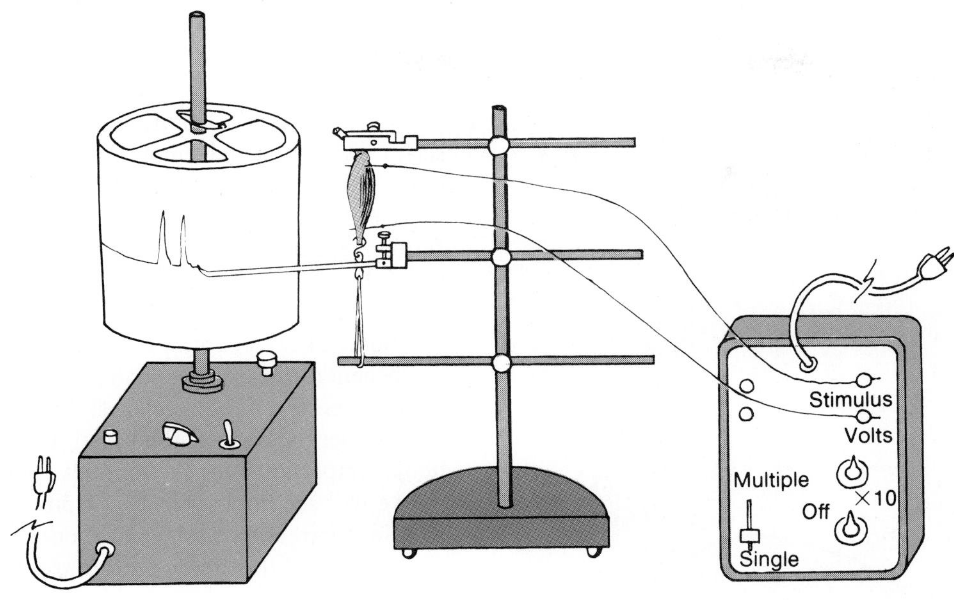


**Figure 12.5.** The physiograph Mark III recorder. This is one of many types of electronic recording devices that receive electrical signals produced by transducers and record these signals on moving paper.

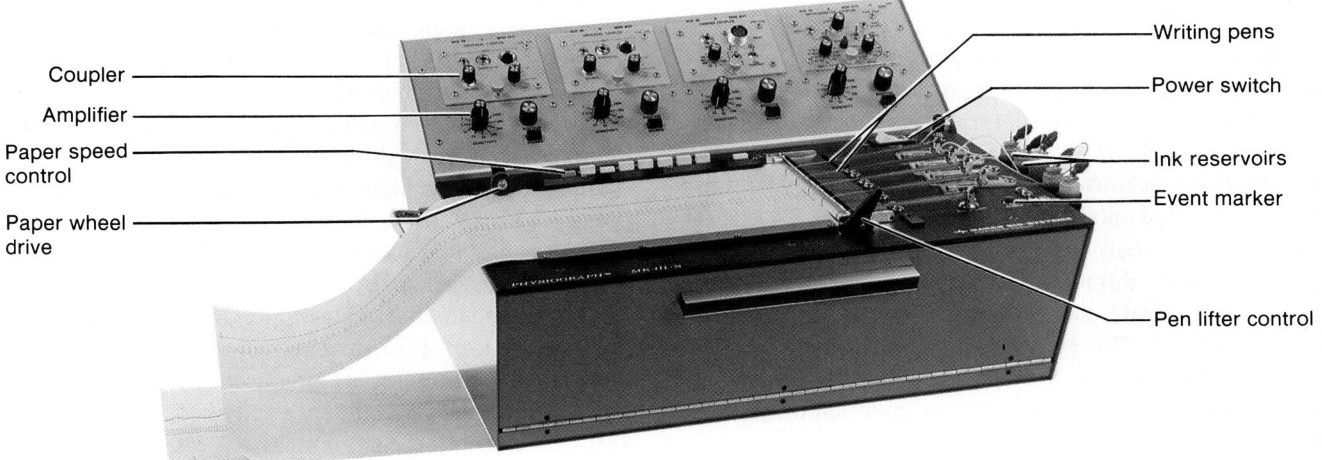


*in vivo.* This smooth, sustained contraction is called **complete tetanus.** The term *tetanus* should not be confused with the disease of the same name, which is accompanied by a painful state of muscle contracture, or *tetany.*

Research laboratories, using an experimental setup analogous to the one previously described, can study the contractile behavior of *an isolated muscle fiber.* The isolated fiber produces a twitch of submaximal strength in response to a single action potential. Since the fiber recovers from its refractory period before it has finished its twitch, it can be stimulated a number of times to produce a summation of twitches. Continued stimulation of the isolated fiber can then result in tetanus, analogous to the behavior of the whole muscle previously described.

**Figure 12.6.** (*a*) A recording of summation of muscle twitches using a physiograph. (*b*) An illustration of summation, tetanus, and fatigue in an isolated muscle.

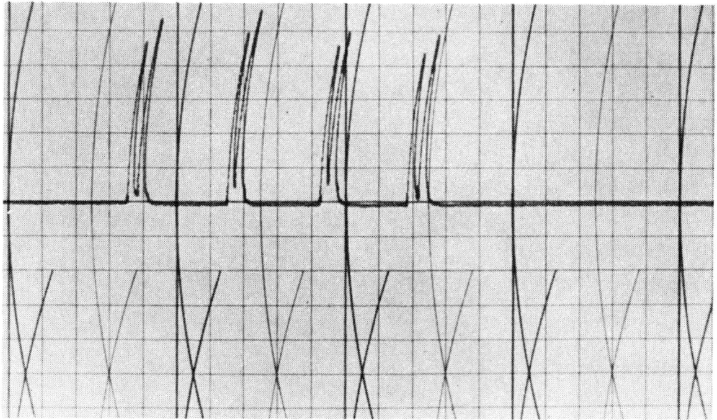

(a)

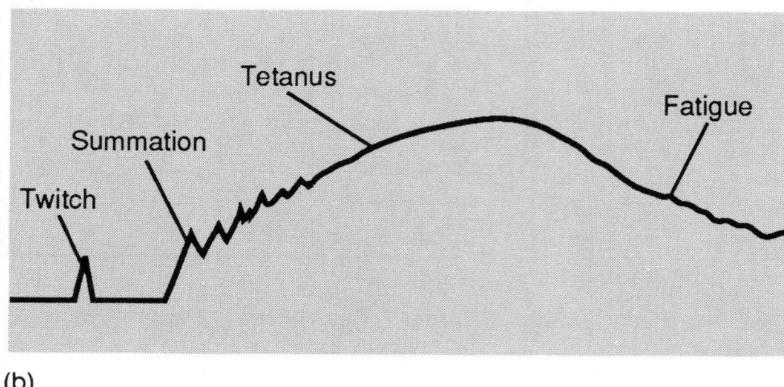


(b)

Stimulation of the whole muscle *in vitro* with an electric stimulator or stimulation of muscle fibers *in vivo* by motor axons, however, usually results in the production of many action potentials in the stimulated muscle fibers. Within the whole muscle, therefore, muscle fibers are normally stimulated maximally and thus produce *all-or-none* contractions. In this case, gradations in the strength of skeletal muscle contractions result from summation of the contractions of different *numbers* of muscle fibers and not from variations in the strength of the contractions of individual muscle fibers. Stronger muscle contractions, in other words, are produced by the stimulation of greater numbers of muscle fibers.

***Series Elastic Component.*** In order for a muscle to shorten when it contracts, and thus to move its insertion toward its origin, the noncontractile parts of the muscle and the connective tissue of its tendons must first be pulled tight. These structures, particularly the tendons, have elasticity—they resist distension, and when the distending force is released, they tend to spring back to their resting lengths. Since the tendons are in series with the force of muscle contraction, they provide what is called the **series elastic component** of muscle contraction. The series elastic component absorbs some of the tension as a muscle contracts, and must be pulled tight before muscle tension can result in muscle shortening.

When the gastrocnemius muscle was stimulated with a single electric shock in the previously described experiment, the amplitude of the twitch was reduced because some of the force of contraction was used to stretch the series elastic component. Delivery of a second shock quickly after the first thus produced a greater degree of muscle shortening than the first shock, culminating at the fusion frequency of stimulation with complete tetanus, in which the strength of contraction was much greater than that of individual twitches.

Some of the energy used to stretch the series elastic component during muscle contraction is released by elastic recoil when the muscle relaxes. This elastic recoil helps the muscles to return to their resting length, and is of particular importance for the muscles involved in breathing. As we will see in chapter 15, inspiration is produced by muscle contraction and expiration is produced by the elastic recoil of the thoracic structures that were stretched during inspiration.

## Motor Units

*In vivo,* each muscle fiber receives a single axon terminal from a somatic motor neuron (fig. 12.7), which stimulates it to contract by liberating acetylcholine at the neuromuscular junction (described in chapter 7). The cell body of a somatic motor neuron is located in the ventral horn of the gray matter of the spinal cord and gives rise to a single axon that emerges in the ventral root of a spinal nerve (chapter 8). Each axon, however, can produce a number of collateral branches to innervate an equal number of muscle fibers. Each somatic motor neuron together with all of the muscle fibers that it innervates is known as a **motor unit** (fig. 12.8).

Whenever a somatic motor neuron is activated, all of the muscle fibers that it innervates are stimulated to contract with all-or-none twitches. *In vivo*, graded contractions of whole muscles are produced by variations in the number of motor units that are activated, and the smooth contractions of complete tetanus are produced by rapid, asynchronous stimulation of different motor units.

Fine neural control over the strength of muscle contraction is optimal when there are many small motor units involved. In the extraocular muscles that position the eyes, for example, the *innervation ratio* of an average motor unit is one neuron per twenty-three muscle fibers, which affords a fine degree of control. The innervation ratio of the gastrocnemius, in contrast, averages one neuron per one thousand muscle fibers. Stimulation of these motor units results in more powerful contractions at the expense of finer gradations in contraction strength.

**Figure 12.7.** A motor end plate at the neuromuscular junction. (*a*) A neuromuscular junction is the site where the nerve fiber and muscle fiber meet. The motor end plate is the specialized portion of the sarcolemma of a muscle fiber surrounding the terminal end of the axon. Note the slight gap between the membrane of the axon and that of the muscle fiber. (*b*) A photomicrograph of muscle fibers and motor end plates. A motor neuron and the muscle fibers it innervates constitute a motor unit.

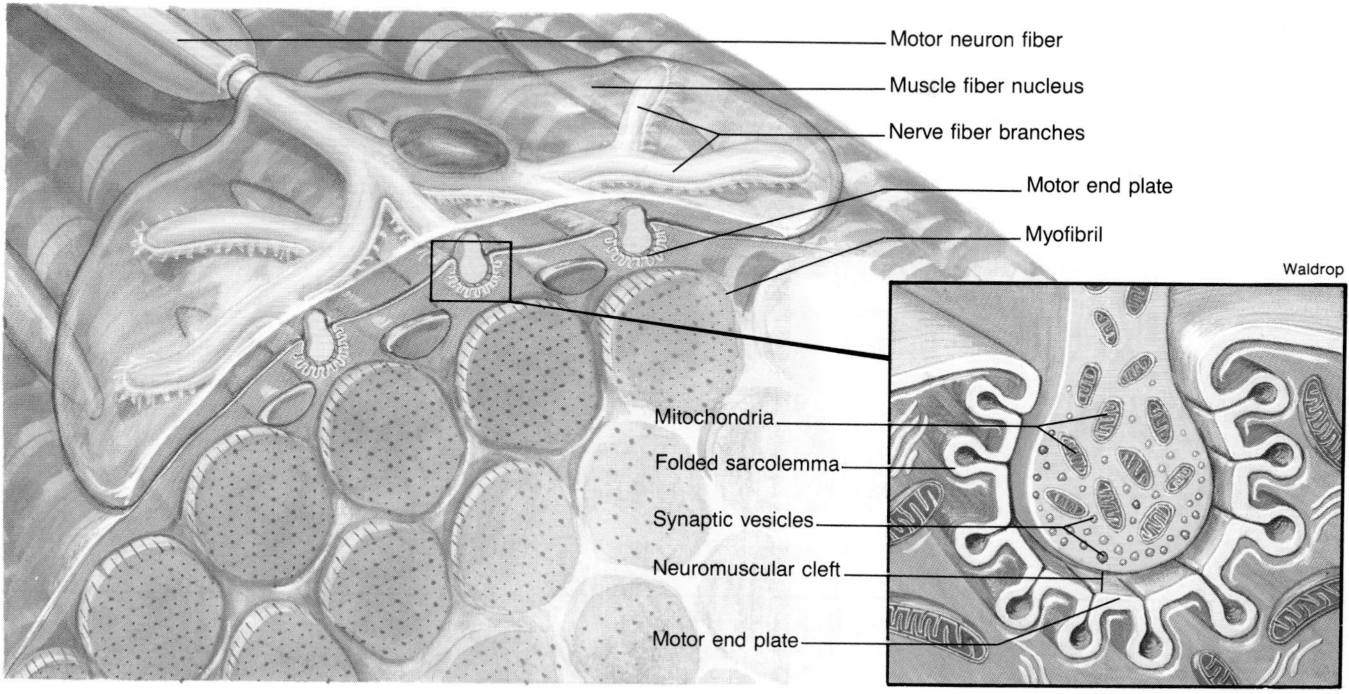


(a)

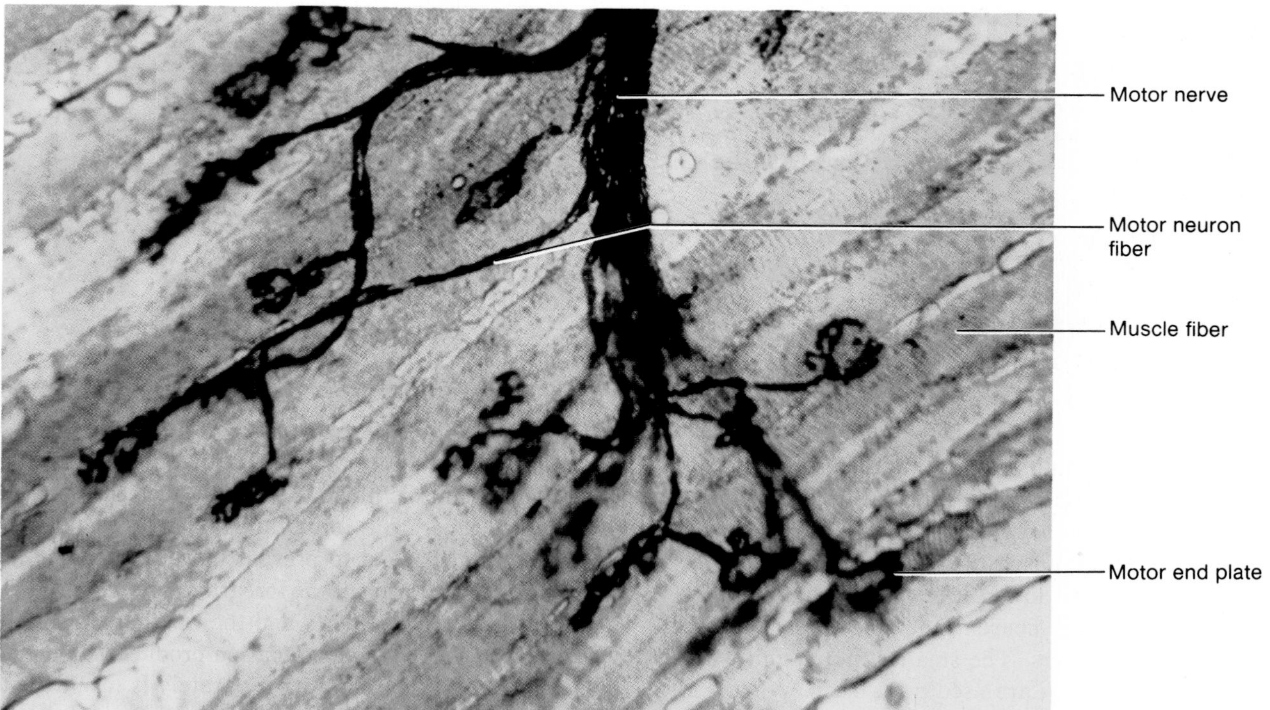


(b)

**Figure 12.8.** A diagram illustrating the innervation of muscle fibers by different motor units. (Actually, many more muscle fibers would be included in a single motor unit than is shown in this drawing.)

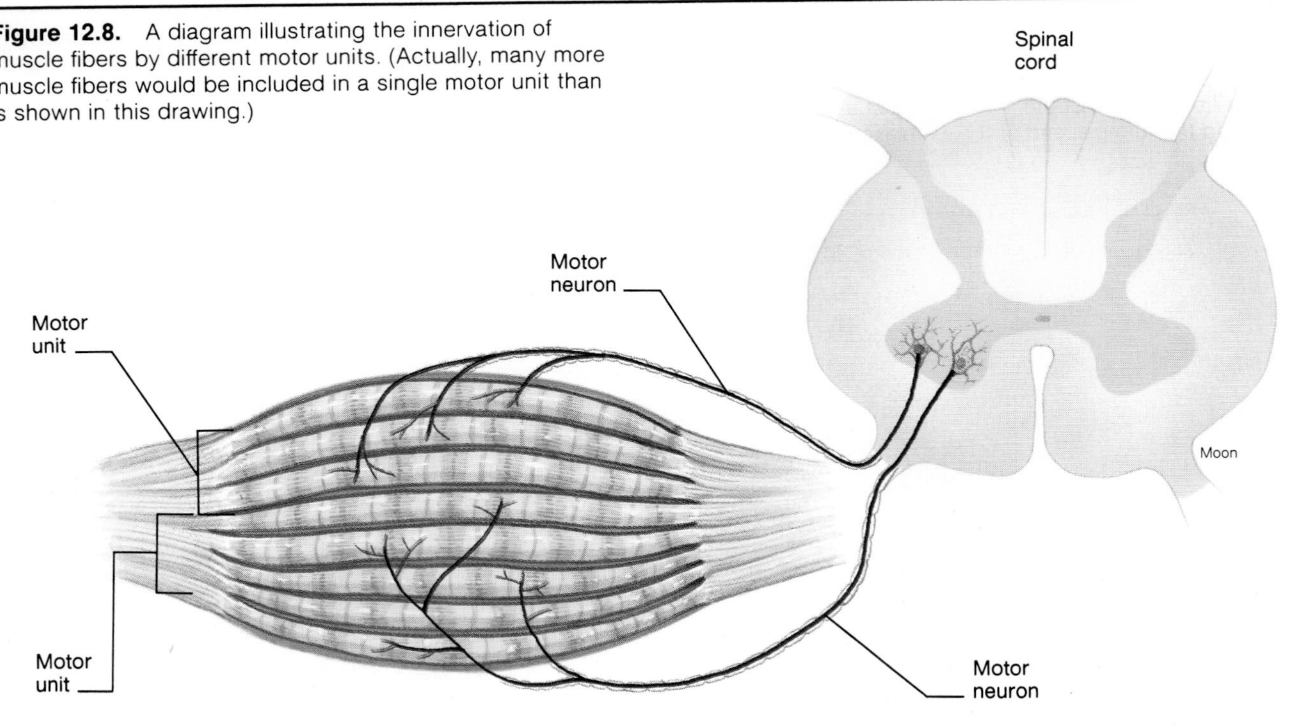


**Figure 12.9.** (*a*) Isometric and (*b*) isotonic contraction.

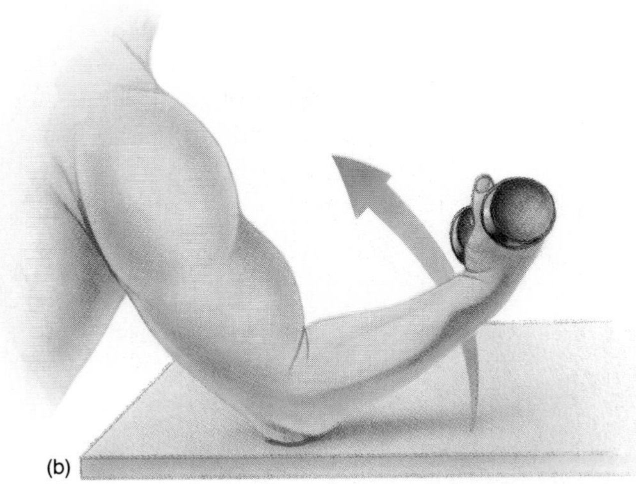


All of the motor units controlling the gastrocnemius, however, are not the same size. Innervation ratios (motor neuron: muscle fibers) vary from 1:100 to 1:2000. A neuron that innervates fewer muscle fibers has a smaller cell body and is stimulated by lower levels of excitatory input (EPSPs) than a larger neuron that innervates a greater number of muscle fibers. The smaller motor units, as a result, are the ones that are used most often. When contractions of greater strength are required, larger and larger motor units are activated; this is known as **recruitment** of motor units.

## Isotonic and Isometric Contractions

In order for muscle fibers to shorten when they contract, they must generate a force that is greater than the opposing forces that act to prevent movement of the muscle's insertion. Flexion of the forearm, for example, occurs against the force of gravity and the weight of the objects being lifted (fig. 12.9). The tension produced by the contraction of each muscle fiber separately is insufficient to overcome these opposing forces, but the combined contractions of many muscle fibers may be sufficient to overcome the opposing force and flex the forearm. In this case the muscle and all its fibers shorten in length.

**Figure 12.10.** A skeletal muscle fiber is composed of numerous myofibrils that contain the filaments of actin and myosin. A skeletal muscle fiber is striated and multinucleated.

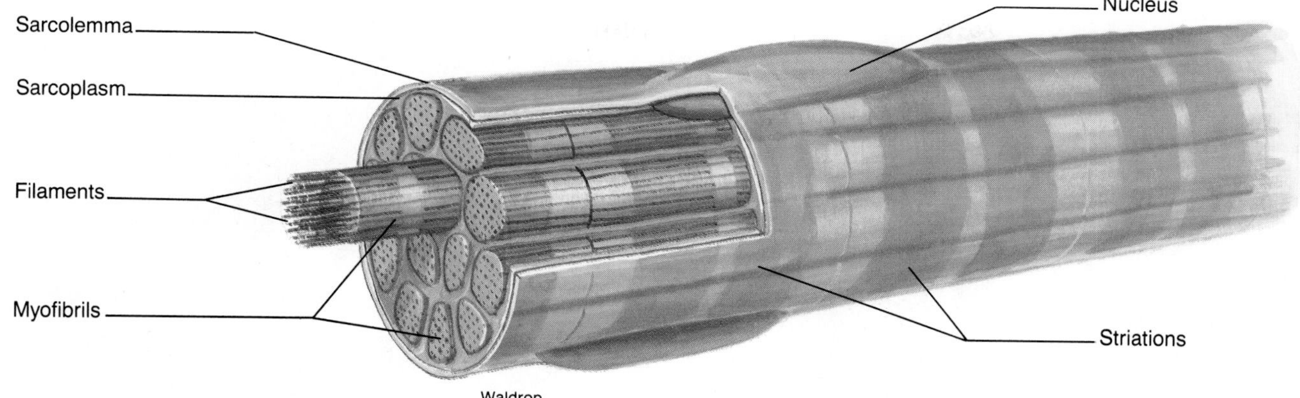


Contraction that results in muscle shortening is called *isotonic contraction,* so-called because the force of contraction remains relatively constant throughout the shortening process (iso = same; tonic = strength). If the opposing forces are too great or the number of motor units activated is too few to shorten the muscle, however, the contraction is called *isometric* (literally, "same length").

Isometric contraction can be voluntarily produced, for example, by lifting a weight and maintaining the forearm in a partially flexed position. One can then increase the amount of muscle tension produced by recruiting more motor units until the muscle begins to shorten; at this point, isometric contraction is converted to isotonic contraction.

1. *Illustrate and describe muscle twitch, summation, and tetanus.*
2. *Describe how graded muscle contractions result from fiber twitches that are all-or-none and how a smooth muscle contraction (tetanus) is produced* in vivo.
3. *Illustrate one motor unit with an innervation ratio of 1:5. Describe the functional significance of the innervation ratio.*
4. *Try to "make a muscle" with your arm and while doing so, feel your* biceps brachii *and your* triceps brachii. *Are these muscles contracting isotonically or isometrically? How is this type of contraction produced? (Include the antagonism of these two muscle groups in your answer.)*

## Mechanisms of Contraction

The A bands within each muscle fiber are composed of thick filaments and the I bands contain thin filaments. Each filament is composed of a number of protein molecules of a particular type. Movement of cross-bridges that extend from the thick to the thin filaments cause the sliding of the thin filaments over and between the thick filaments. This sliding of the filaments produces muscle tension and shortening. During muscle relaxation the cross-bridges are prevented from attaching to the thin filaments by additional proteins, which are moved out of the way when they bond to $Ca^{++}$. Since electrical excitation of a muscle fiber causes the release of $Ca^{++}$ into the muscle cell cytoplasm, and since $Ca^{++}$ allows the cross-bridges to attach to the thin filaments, electrical excitation results in muscle contraction.

When muscle cells are viewed in the electron microscope, which can produce images at several thousand times the magnification possible in an ordinary light microscope, each cell is seen to be composed of many subunits known as **myofibrils** (*fibrils* = little fibers—fig. 12.10). These myofibrils are approximately one micrometer (1 $\mu$m) in diameter and extend in parallel from one end of the muscle fiber to the other. The myofibrils are so densely packed that other organelles—such as mitochondria and intracellular membranes—are restricted to the narrow cytoplasmic spaces that remain between adjacent myofibrils.

With the electron microscope, it can be seen that the muscle fiber does not have striations that extend from one side of the fiber to the other. It is the myofibrils that are striated with dark (A) and light (I) bands (fig. 12.11). The striated appearance of the entire muscle fiber when seen with a light microscope is an illusion created by the fact that the dark and light bands of the myofibrils are in line with each other from one side of the fiber to the other. Since the separate myofibrils are not clearly seen at low magnification, the dark and light bands appear to be continuous across the width of the fiber.

When a myofibril is observed at high magnification in longitudinal section (side view), the A bands are seen to contain **thick filaments** (about 110 Å—angstroms—thick; 1 Å = $10^{-10}$m) that are stacked in register. It is

**Figure 12.11.** An electron micrograph of a longitudinal section of myofibrils, showing A, H, and I bands. Notice how the dark and light bands of each myofibril are stacked in register.

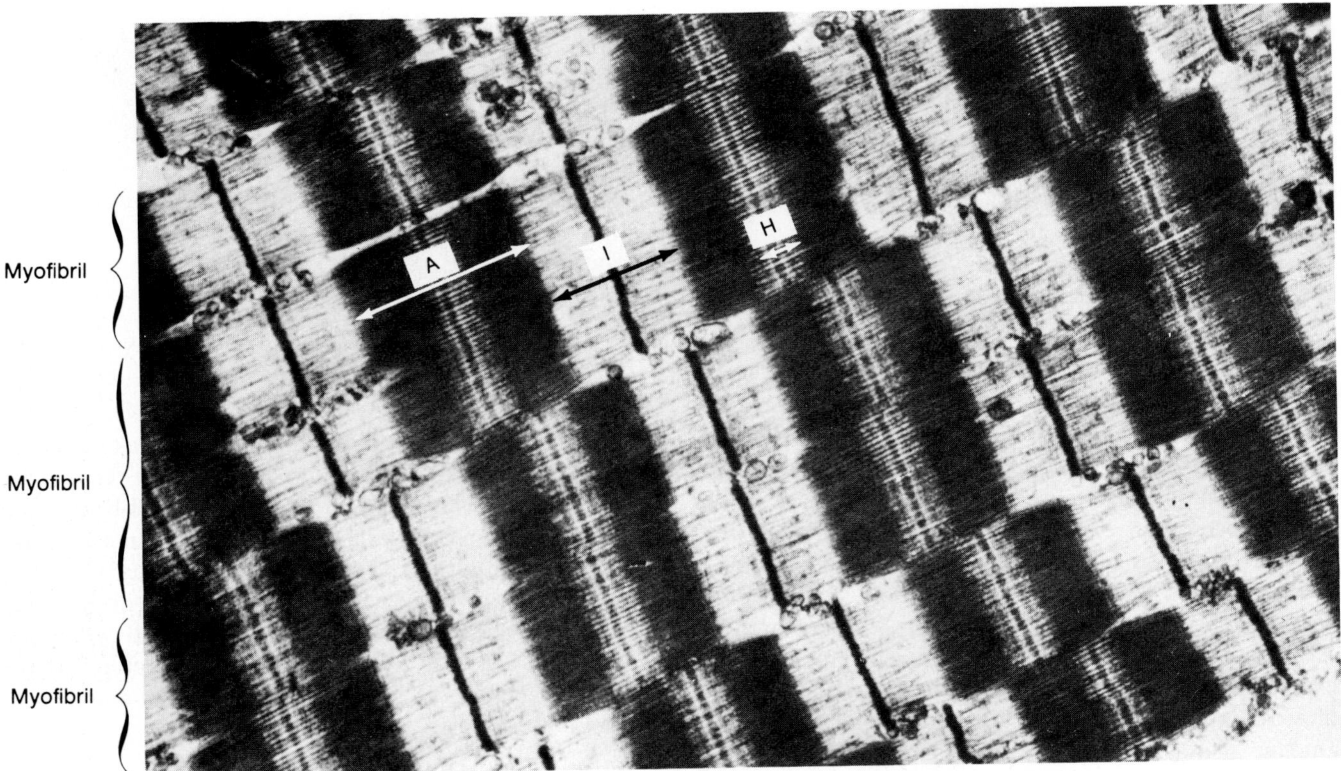


these thick filaments that give the A band its dark appearance. The lighter I band, in contrast, contains **thin filaments** (about 50–60 Å thick). The thick filaments are composed of the protein **myosin,** whereas the thin filaments are composed primarily of the protein **actin.**

The I bands within a myofibril are the lighter areas that extend from the edge of one stack of thick myosin filaments to the edge of the next stack of thick filaments. They are light in appearance because they contain only thin filaments. The thin filaments, however, do not end at the edges of the I bands. Instead, each thin filament continues partway into the A bands on each side (between the stack of thick filaments on each side of an I band). Since thick and thin filaments overlap at the edges of each A band, the edges of the A band are darker in appearance than the central region. These central lighter regions of the A bands are called the *H bands* (for *helle,* a German word for "bright"). The central H bands thus contain only thick filaments that are not overlapped with thin filaments.

In the center of each I band is a thin dark Z line. The arrangement of thick and thin filaments between a pair of Z lines forms a repeating pattern that serves as the basic subunit of striated muscle contraction. These subunits, from Z to Z, are known as **sarcomeres.** A longitudinal section of a myofibril thus presents a side view of successive sarcomeres.

This side view is, in a sense, misleading; there are numerous sarcomeres within each myofibril that are out of the plane of the section (and out of the picture). A better appreciation of the three-dimensional structure of a myofibril can be obtained by viewing the myofibril in cross section. In this view, it can be seen that the Z lines are in reality disc shaped and that the thin filaments that penetrate these Z discs surround the thick filaments in a hexagonal arrangement (fig. 12.12*c*). If one concentrates on a single row of dark thick filaments in this cross section, the alternating pattern of thick and thin filaments seen in longitudinal section becomes apparent.

## Sliding Filament Theory of Contraction

When a muscle contracts isotonically, it decreases in length as a result of the shortening of its individual fibers. Shortening of the muscle fibers, in turn, is produced by shortening of their myofibrils, which occurs as a result of the shortening of the distance from Z line to Z line. As the

**Table 12.2** Summary of the sliding filament theory of contraction

1. A myofiber, together with all its myofibrils, shortens by movement of the insertion toward the origin of the muscle.
2. Shortening of the myofibrils is caused by shortening of the sarcomeres—the distance between Z lines (or discs) is reduced.
3. Shortening of the sarcomeres is accomplished by sliding of the myofilaments—each filament remains the same length during contraction.
4. Sliding of the filaments is produced by asynchronous power strokes of myosin cross-bridges, which pull the thin filaments (actin) over the thick filaments (myosin).
5. The A bands remain the same length during contraction, but are pulled toward the origin of the muscle.
6. Adjacent A bands are pulled closer together as the I bands between them shorten.
7. The H bands shorten during contraction as the thin filaments from each end of the sarcomeres are pulled toward the middle.

**Figure 12.12.** Electron micrographs of myofibrils of a muscle fiber. (*a*) At low power (1,600×), a single muscle fiber containing numerous myofibrils. (*b*) At high power (53,000×), myofibrils in longitudinal section. Notice the sarcomeres and overlapping thick and thin filaments. (*c*) The hexagonal arrangement of thick and thin filaments as seen in cross section (arrows point to cross-bridges; SR = sarcoplasmic reticulum; M = mitochondria). From Kessel, R. G., and Kardon, R. H.: *Tissues and Organs: A Text-Atlas of Scanning Electron Microscopy.* © 1979 W. H. Freeman and Company.

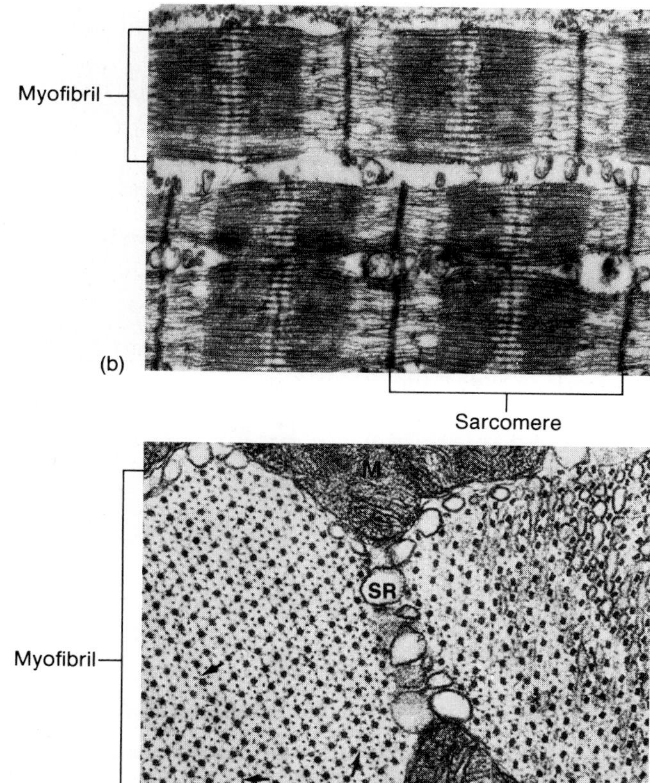


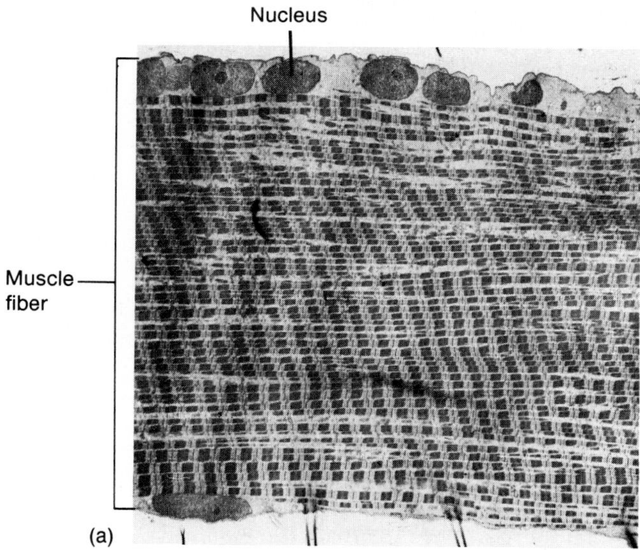


sarcomeres shorten in length, however, the A bands do *not* shorten but instead move closer together. The I bands—which represent the distance between A bands of successive sarcomeres—decrease in length (table 12.2).

The thin actin filaments composing the I band, however, do not shorten. Close examination reveals that the thick and thin filaments remain the same length during muscle contraction. Shortening of the sarcomeres is produced, not by shortening of the filaments, but rather by the *sliding* of thin filaments over and between the thick filaments. In the process of contraction, the thin filaments on either side of each A band slide deeper and deeper toward the center, producing increasing amounts of overlap with the thick filaments. The I bands (containing only thin filaments) and H bands (containing only thick filaments) thus get shorter during contraction (fig. 12.13).

***Cross-Bridges.*** Sliding of the filaments is produced by the action of numerous **cross-bridges** that extend out from the myosin toward the actin. These cross-bridges are part

**Figure 12.13.** The sliding filament model of contraction. As the filaments slide, the Z lines are brought closer together. The A bands remain the same length during contraction, but the I and H bands get progressively shorter and may eventually become obliterated.

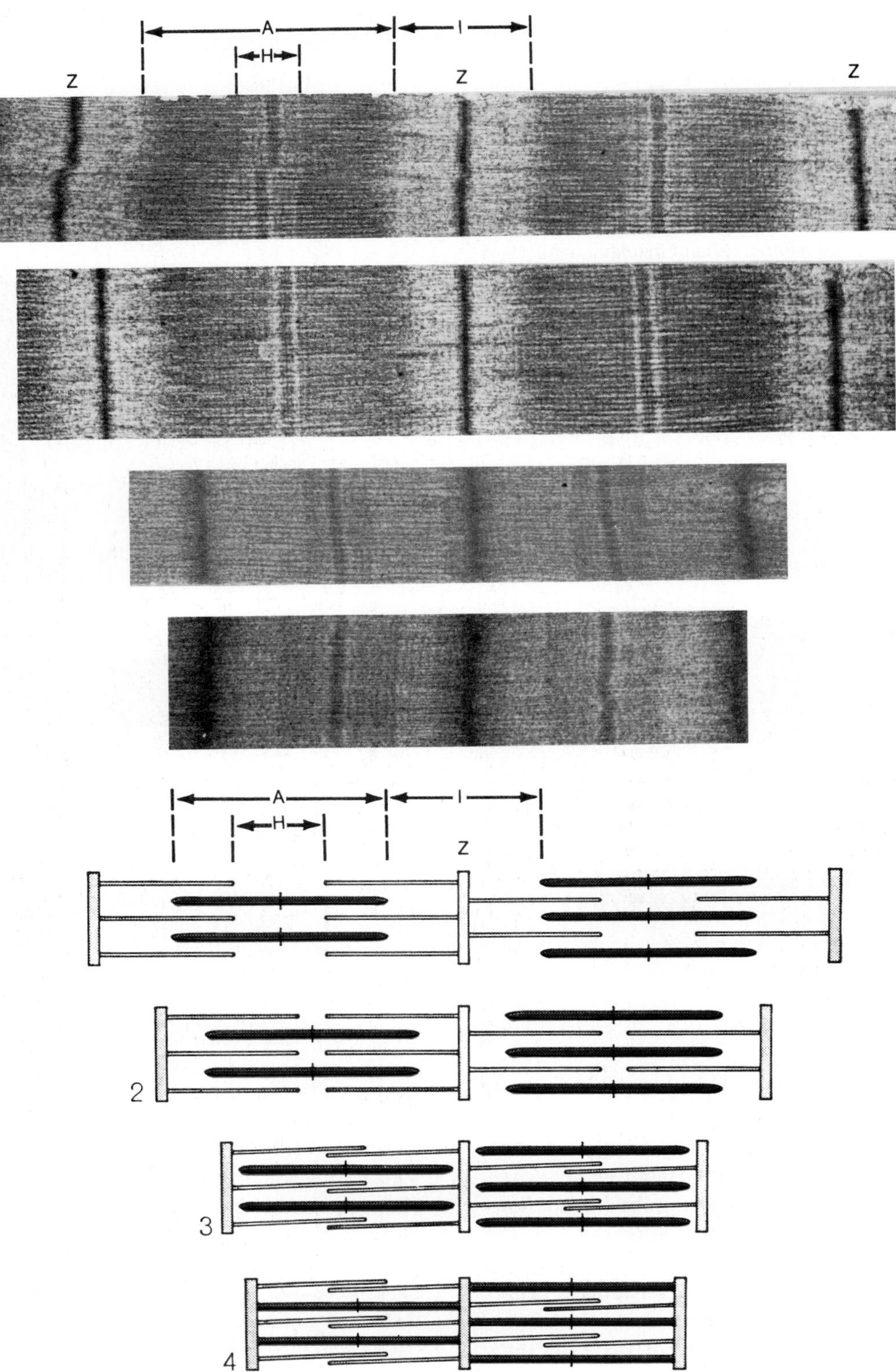

**Figure 12.14.** Myosin cross-bridges are oriented in opposite directions on either side of a sarcomere.

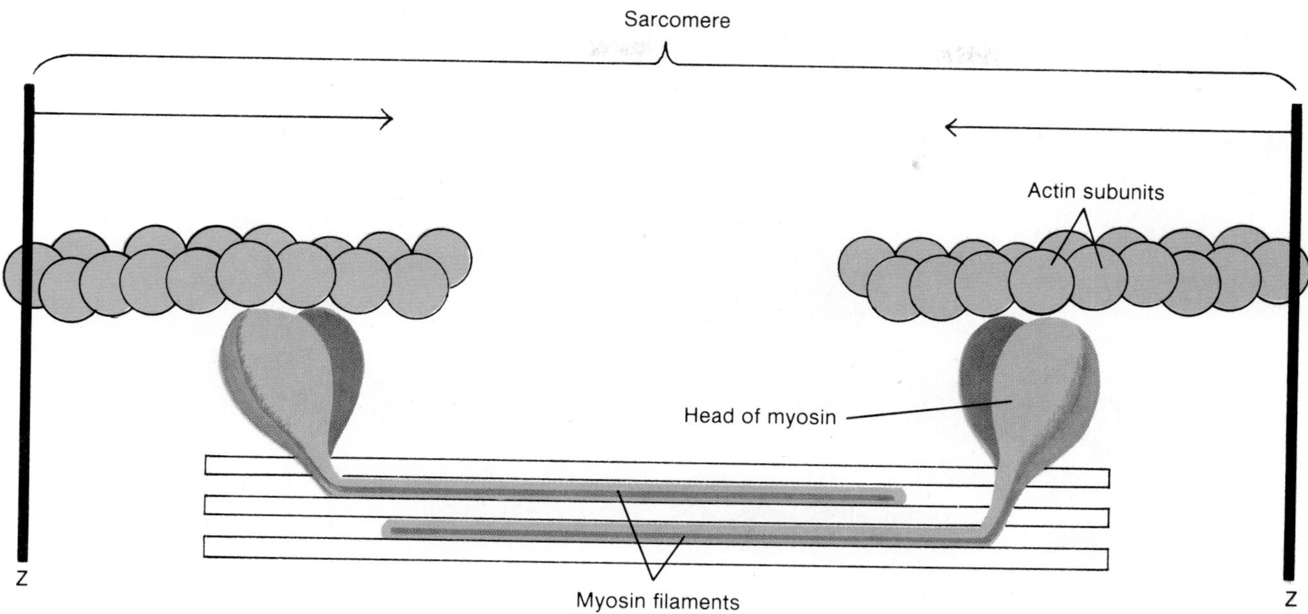


of the myosin proteins that extend from the axis of the thick filaments to form "arms" that terminate in globular "heads" (fig. 12.14). The orientation of cross-bridges on one side of a sarcomere is opposite to that on the other side, so that when they attach to actin on each side of the sarcomere they can pull the actin from each side toward the center.

Isolated muscles *in vitro* are easily stretched (although this is opposed *in vivo* by the stretch reflex, described in a later section), demonstrating that the myosin cross-bridges are not attached to actin when the muscle is at rest. Each globular head of a cross-bridge contains an ATP-bonding site closely associated with an actin-binding site (fig. 12.15). The globular heads function as **myosin ATPase** enzymes, splitting ATP into ADP and $P_i$. This reaction occurs before the cross-bridges combine with actin, and indeed is required for activating the cross-bridges so that they can attach to actin. The ADP and $P_i$ remain bonded to the myosin heads until the cross-bridges attach to the actin.

The myosin heads are able to bond to specific attachment sites in the actin subunits. When the cross-bridges bond to actin they undergo a conformation change. This has two effects: (1) ADP and $P_i$ are released; and (2) the cross-bridges change their orientation, resulting in a *power stroke,* which pulls the thin filaments toward the center of the A bands. At the end of the power stroke each cross-bridge bonds to a fresh ATP molecule. This bonding of the cross-bridge to a new ATP causes the cross-bridge to break its bond with actin and resume its resting orientation. The myosin ATPase will then split ATP and become activated as in the previous cycle. Note that the splitting of ATP is required *before* a cross-bridge can attach to actin and undergo a power stroke and that the attachment of a *new ATP* is needed for the cross-bridge to release from actin at the end of a power stroke.

Because the cross-bridges are quite short, a single contraction cycle and power stroke of all the cross-bridges in a muscle would shorten the muscle by only about 1% of its resting length. Since muscles can shorten up to 60% of their resting lengths, it is obvious that the contraction cycles must be repeated many times. In order for this to occur, the cross-bridges must detach from the actin at the end of a power stroke, reassume their resting orientation, and then reattach to the actin and repeat the cycle.

During normal contraction, however, only about 50% of the cross-bridges are attached at any given time. The power strokes are thus not in synchrony, as the strokes of a competitive rowing team are. Rather, they are like the actions of a team engaged in tug-of-war in which the pulling action of the members is asynchronous. Some cross-bridges are engaged in power strokes at all times during the contraction.

The detachment of a cross-bridge from actin at the end of a power stroke requires that a new ATP molecule bind to the myosin ATPase. The importance of this process is illustrated by the muscular contracture called **rigor mortis,** which occurs due to lack of ATP when the muscle dies. This results in the formation of "rigor complexes" between myosin and actin that cannot detach. In rigor mortis, all of the cross-bridges are attached to actin at the same time.

**Figure 12.15.** The structure of myosin, showing its binding sites for ATP and for actin.

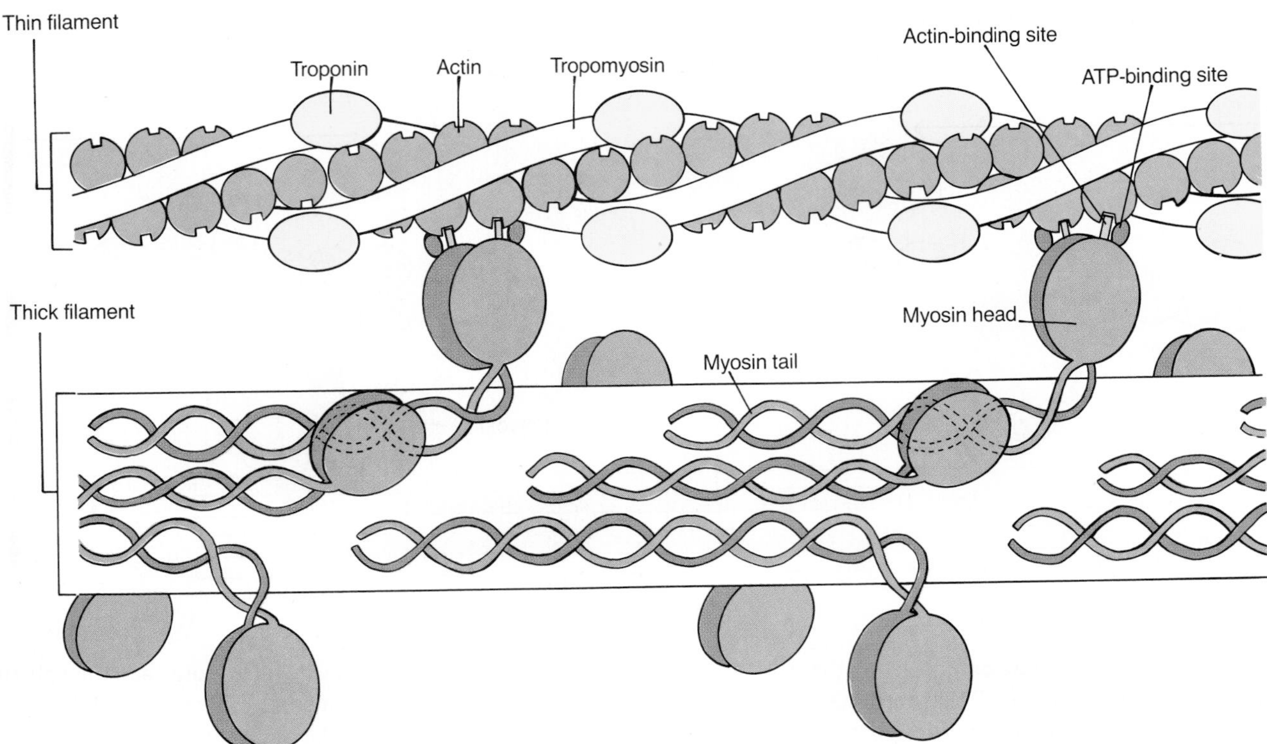


***Length-Tension Relationship.*** The strength of a muscle's contraction is affected by a number of factors. These include the number of muscle fibers within the muscle that are stimulated to contract, the thickness of each muscle fiber (thicker fibers have more myofibrils and thus can exert more power), and the initial length of the muscle fibers when they are at rest.

There is an "ideal" resting length of muscle fibers. When the resting length is more than this ideal, the overlap between actin and myosin is so little that few cross-bridges can attach. When the muscle is stretched to the point that there is no overlap of actin with myosin, no cross-bridges can attach to the thin filaments and the muscle cannot contract. When the muscle is shortened to about 60% of its resting length, the Z lines abut against the thick filaments so that further contraction cannot occur.

The strength of a muscle's contraction can be measured by the force required to prevent it from shortening. Under these isometric conditions, the strength of contraction, or *tension,* can be measured when the muscle length at rest is varied. Maximum tension is produced when the muscle is at its normal resting length *in vivo* (fig. 12.16). If the muscle were any shorter or longer than its normal length, in other words, its strength of contraction would be reduced. This resting length is maintained by reflex contraction in response to passive stretching, as described in a later section of this chapter.

## Regulation of Contraction

When the cross-bridges attach to actin they undergo power strokes and cause muscle contraction. In order for a muscle to relax, therefore, the attachment of myosin cross-bridges to actin must be prevented. The regulation of cross-bridge attachment to actin is a function of two proteins found associated with actin in the thin filaments.

The actin filament—or *F-actin*—is a polymer formed of three hundred to four hundred globular subunits (*G-actin*), arranged in a double row and twisted to form a helix (fig. 12.17). A different type of protein, known as **tropomyosin,** lies within the groove between the double row of G-actin. There are forty to sixty tropomyosin molecules per thin filament, with each tropomyosin spanning a distance of approximately seven actin subunits.

Attached to the tropomyosin, rather than directly to the actin, is a third type of protein, called **troponin,** within the thin filaments. Troponin and tropomyosin work together to regulate the attachment of cross-bridges to actin and thus serve as a switch for muscle contraction and relaxation. In a relaxed muscle, the position of the tropomyosin in the thin filaments is such that it physically blocks the cross-bridges from bonding to specific attachment sites in the actin. Thus, in order for the myosin cross-bridges to attach to actin, the tropomyosin must be moved. This requires the interaction of troponin with $Ca^{++}$, as described in the next section.

**Figure 12.16.** The length-tension relationship in skeletal muscles. Maximum relative tension (1.0) is achieved when the muscle is 100%–200% of its resting length (sarcomere lengths from 2.0 to 2.25 $\mu$m). Increases or decreases in muscle (and sarcomere) lengths result in rapid decreases in tension.

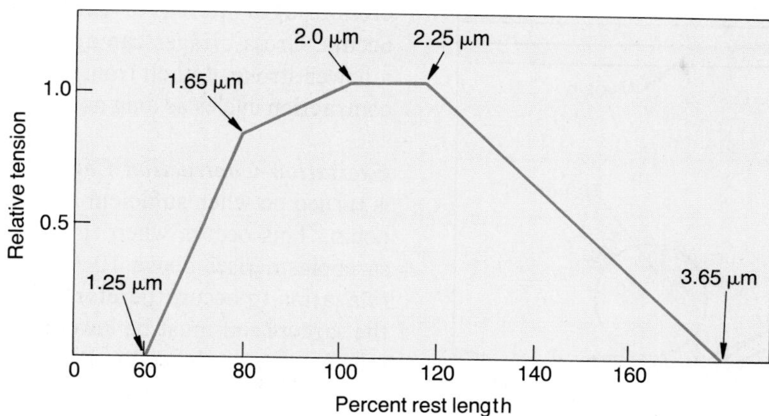


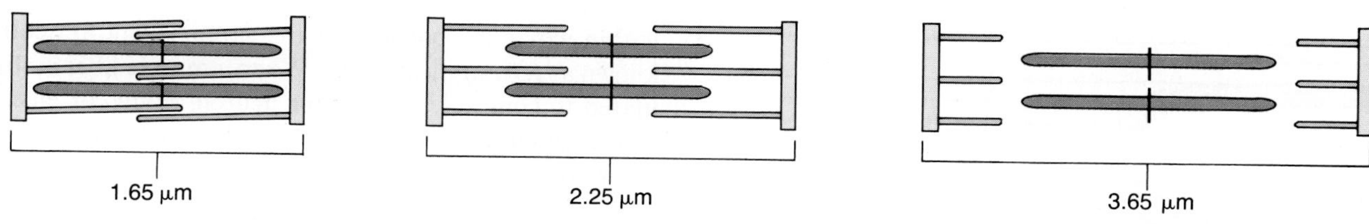


**Figure 12.17.** The relationship of troponin and tropomyosin to actin in the thin filaments. The tropomyosin is attached to actin, whereas the troponin complex of three subunits is attached to tropomyosin (not directly to actin).

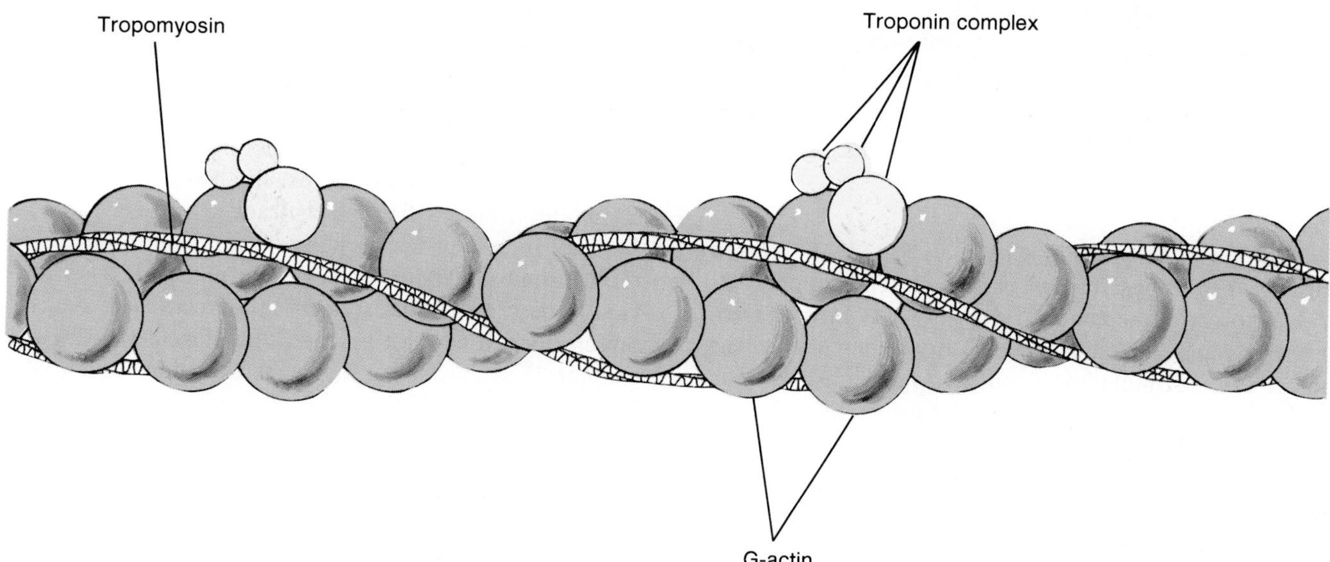

**Figure 12.18.** The attachment of $Ca^{++}$ to troponin causes movement of the troponin-tropomyosin complex, which exposes binding sites on the actin. The myosin cross-bridges can then attach to actin and undergo a power stroke.

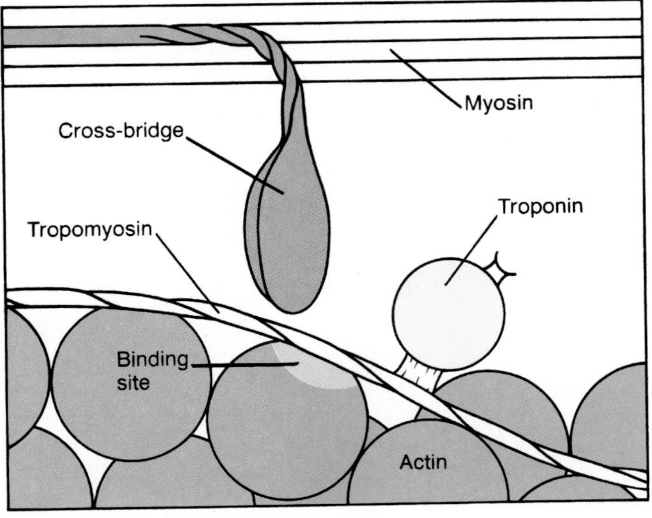


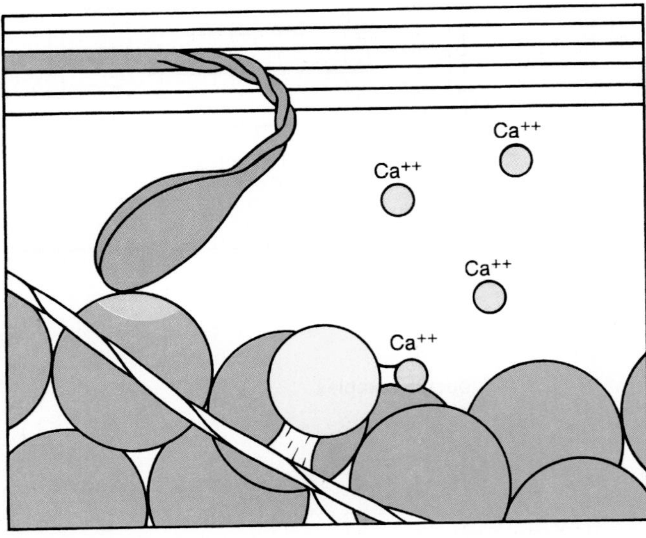


***Role of $Ca^{++}$ in Muscle Contraction.*** In a relaxed muscle, when tropomyosin blocks the attachment of cross-bridges to actin, the concentration of $Ca^{++}$ in the sarcoplasm (cytoplasm of muscle cells) is very low. When the muscle cell is stimulated to contract, mechanisms which will be discussed shortly cause the concentration of $Ca^{++}$ in the sarcoplasm to quickly rise. Some of this $Ca^{++}$ attaches to a subunit of troponin, causing a conformational change that moves the troponin *and* its attached tropomyosin out of the way so that the cross-bridges can attach to actin (fig. 12.18). Once the attachment sites on the actin are exposed, the cross-bridges can bind to actin, undergo power strokes, and produce muscle contraction.

The position of the troponin-tropomyosin complexes in the thin filaments is thus adjustable. When $Ca^{++}$ is not attached to troponin, the tropomyosin is in a position that inhibits attachment of cross-bridges to actin; muscle contraction is prevented. When $Ca^{++}$ attaches to troponin, the troponin-tropomyosin complexes shift position—the cross-bridges attach to actin, and muscle contraction occurs. Cross-bridges can again attach to actin, produce a power stroke, detach from actin, and can continue these contraction cycles as long as $Ca^{++}$ is attached to troponin.

***Excitation-Contraction Coupling.*** Muscle contraction is turned on when sufficient amounts of $Ca^{++}$ bind to troponin. This occurs when the $Ca^{++}$ concentration of the sarcoplasm rises above $10^{-6}$ molar. In order for muscle relaxation to occur, therefore, the $Ca^{++}$ concentration of the sarcoplasm must be lowered below this level. Muscle relaxation is produced by the active transport of $Ca^{++}$ out of the sarcoplasm into the **sarcoplasmic reticulum** (fig. 12.19). The sarcoplasmic reticulum is a modified endoplasmic reticulum, consisting of interconnected sacs and tubes that surround each myofibril within the muscle cell.

Most of the $Ca^{++}$ in a relaxed muscle fiber is stored within expanded portions of the sarcoplasmic reticulum known as *terminal cisternae.* When a muscle fiber is stimulated to contract, by a motor neuron *in vivo* or electric shocks *in vitro*, the stored $Ca^{++}$ is released from the sarcoplasmic reticulum so that it can attach to troponin. When a muscle fiber is no longer stimulated, the $Ca^{++}$ from the sarcoplasm is actively accumulated into the sarcoplasmic reticulum. In order to understand how the release and uptake of $Ca^{++}$ is regulated, one more organelle within the muscle fiber must be described.

The terminal cisternae of the sarcoplasmic reticulum are separated by only a very narrow gap from **transverse tubules** (or **T tubules**), which are narrow membranous "tunnels" formed from and continuous with the cell membrane (the *sarcolemma*). The transverse tubules thus open to the extracellular environment through pores in the cell surface, and are capable of conducting action potentials. The stage is now set to explain exactly how a motor neuron stimulates a muscle fiber to contract.

The release of acetylcholine from axon terminals at the neuromuscular junctions (motor end plates), as previously described, causes electrical activation of skeletal muscle fibers. End-plate potentials (analogous to EPSPs—chapter 7) are produced that generate action potentials. Muscle action potentials, like the action potentials of axons, are all-or-none events that are regenerated along the cell membrane. It must be remembered that action potentials involve the flow of ions between the extracellular and intracellular environments across a cell membrane that separates these two compartments. In muscle cells, therefore, action potentials can be conducted into the interior of the fiber across the membrane of the transverse tubules.

**Figure 12.19.** A drawing of myofibrils, showing their relationship to the sarcoplasmic reticulum, transverse tubules, and sarcolemma.

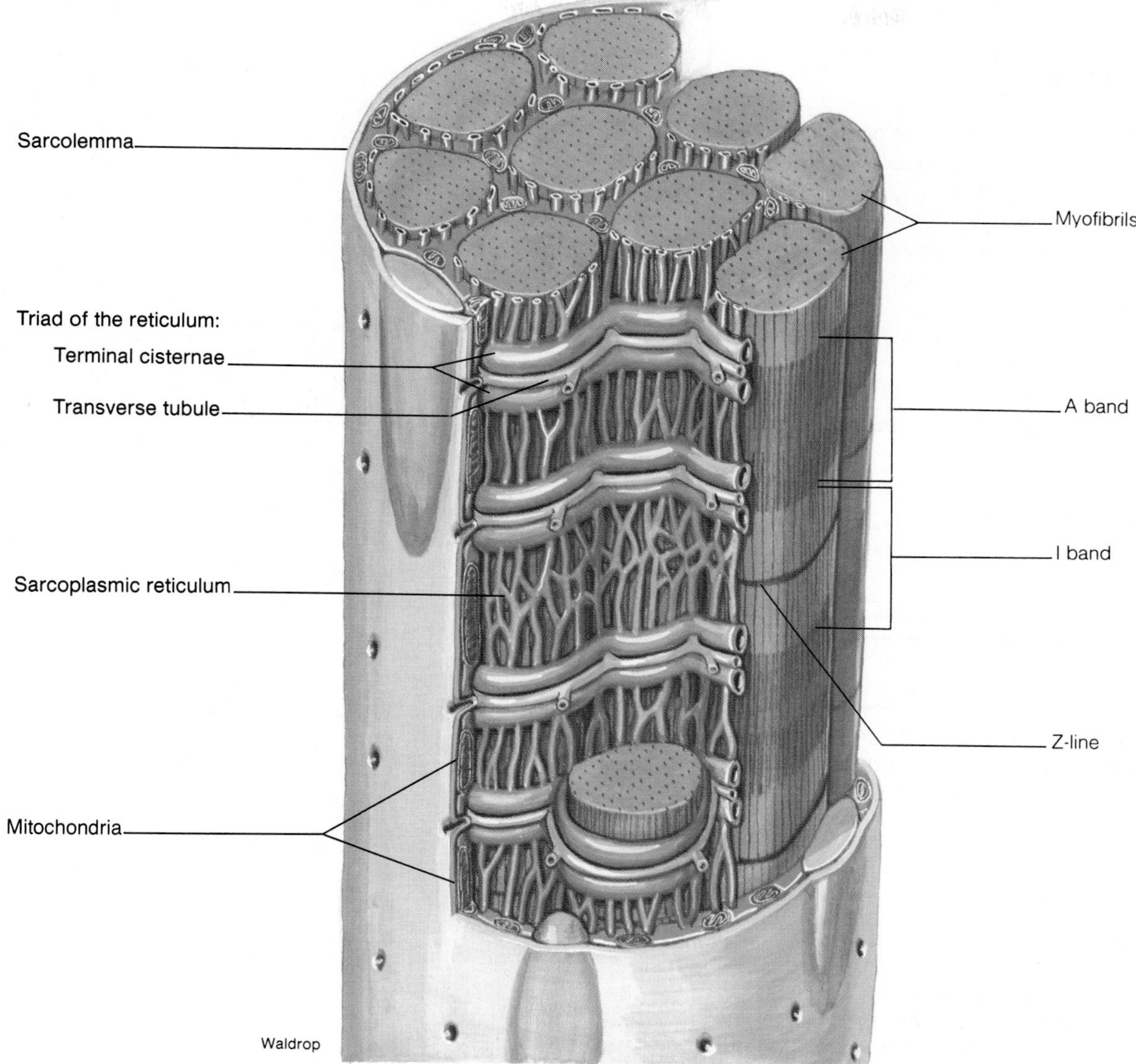


*Action potentials in the transverse tubules cause the release of $Ca^{++}$ from the sarcoplasmic reticulum.* Since the transverse tubules do not physically touch the sarcoplasmic reticulum, a second messenger is required for communication between these two organelles. Recent evidence suggests that inositol triphosphate, a molecule that also serves as a second messenger in the action of some hormones (chapter 11), may be secreted by the transverse tubules and stimulate the sarcoplasmic reticulum to release its stored $Ca^{++}$. The released $Ca^{++}$ diffuses into the sarcomeres and binds to troponin, causing the displacement of tropomyosin and allowing the actin to bind to the myosin cross-bridges. Muscle contraction is thus stimulated.

As long as action potentials continue to be produced—which is as long as the neural stimulation of the muscle is continued—$Ca^{++}$ will remain attached to troponin and cross-bridges will be able to undergo contraction cycles. When neural activity and action potentials in the muscle fiber cease, the sarcoplasmic reticulum actively accumulates $Ca^{++}$ and muscle relaxation occurs. Note that the return of $Ca^{++}$ to the sarcoplasmic reticulum involves active transport and thus requires the hydrolysis of ATP. ATP is therefore needed for muscle relaxation as well as for muscle contraction.

**Table 12.3** Summary of events that occur during excitation-contraction coupling

1. Action potentials in a somatic motor nerve cause the release of acetylcholine neurotransmitter at the myoneural junction (one myoneural junction per myofiber).
2. Acetylcholine, through its interaction with receptors in the muscle cell membrane (sarcolemma), produces action potentials that are regenerated across the sarcolemma.
3. The membranes of the transverse tubules (T tubules) are continuous with the sarcolemma and conduct action potentials deep into the muscle fiber.
4. Action potentials in the T tubules, by a mechanism that is poorly understood, stimulate the release of $Ca^{++}$ from the terminal cisternae of the sarcoplasmic reticulum.
5. $Ca^{++}$ released into the sarcoplasm attaches to troponin, causing a change in its structure.
6. The shape change in troponin causes its attached tropomyosin to shift position in the actin filament, thus exposing bonding sites for the myosin cross-bridges.
7. Myosin cross-bridges, previously activated by the hydrolysis of ATP, attach to actin.
8. Once the previously activated cross-bridges attach to actin, they undergo a power stroke and pull the thin filaments over the thick filaments.
9. Attachment of fresh ATP allows the cross-bridges to detach from actin and repeat the contraction cycle as long as $Ca^{++}$ remains attached to troponin.
10. When action potentials stop being produced, the sarcoplasmic reticulum actively accumulates $Ca^{++}$ and tropomyosin moves again to its inhibitory position.

Neural regulation of skeletal muscle contraction is thus mediated by calcium ions. The mechanisms of electrical excitation and the mechanisms of muscle contraction (sliding of the filaments) are "coupled" through adjustments of the sarcoplasmic $Ca^{++}$ concentration. This is known as *excitation-contraction coupling* and is summarized in table 12.3.

1. *Describe how the lengths of the A, I, and H bands change during contraction, and explain these changes according to the sliding filament theory.*
2. *Describe a cycle of cross-bridge activity during contraction, and explain the role of ATP in this cycle.*
3. *Draw a sarcomere in a relaxed muscle and a sarcomere in a contracted muscle, label the bands in each, and explain the significance of the differences observed.*
4. *Describe the molecular structure of myosin and actin, and explain the positions and functions of tropomyosin and troponin.*
5. *Draw a flowchart (using arrows) of the events that occur between the time that ACh is released from a nerve ending and the time that $Ca^{++}$ is released from the sarcoplasmic reticulum.*
6. *Explain the requirements for $Ca^{++}$ and ATP in muscle contraction.*
7. *Explain how muscle relaxation is produced.*

## *Neural Control of Skeletal Muscles*

Skeletal muscles contain stretch receptors known as muscle spindles, which stimulate the production of nerve impulses in sensory neurons when the muscle is stretched. This can evoke activity in alpha motoneurons, which stimulate the muscle to contract and shorten. Other motor neurons, called gamma motoneurons, stimulate the tightening of the spindles and thus increase their sensitivity. These motor neurons can be activated at the spinal cord level as a result of a stretch reflex, or they may be activated consciously by synapses with descending motor tracts from the brain.

Motor neurons in the spinal cord, or **lower motor neurons** (often shortened to *motoneurons*), are those previously described that have cell bodies in the spinal cord and axons within nerves that stimulate muscle contraction (table 12.4). The activity of these neurons is influenced by (1) sensory feedback from the muscles and tendons and (2) facilitory and inhibitory effects from **higher motor neurons** in the brain that contribute axons to descending motor tracts. Lower motor neurons are thus said to be the *final common pathway* by which sensory stimuli and higher brain centers exert control over skeletal movements.

The cell bodies of lower motor neurons are located in the ventral horn of the gray matter of the spinal cord (chapter 8). Axons from these cell bodies leave the ventral side of the spinal cord to form the *ventral roots* of spinal nerves (fig. 12.20). The *dorsal roots* of spinal nerves contain sensory fibers whose cell bodies are located in the *dorsal root ganglia.* Both sensory (*afferent*) and motor (*efferent*) fibers join in a common connective tissue sheath to form the spinal nerves at each segment of the spinal cord. In the lumbar region (the small of the back), there are about twelve thousand sensory and six thousand motor fibers per spinal nerve.

About 375,000 cell bodies have been counted in a lumbar segment—a number far larger than can be accounted for by the number of motor neurons. Most of these neurons do not contribute fibers to the spinal nerve, but rather serve as *interneurons* whose fibers conduct impulses up, down, and across the central nervous system.

**Table 12.4** Some terms used to describe the neural control of skeletal muscles

| Term | Description |
|---|---|
| 1. Lower motoneurons | Neurons whose axons innervate skeletal muscles—also called the "final common pathway" in the control of skeletal muscles |
| 2. Higher motoneurons | Neurons in the brain that are involved in the control of skeletal movements and that act by facilitating or inhibiting (usually by way of interneurons) the activity of the lower motoneurons |
| 3. Alpha motoneurons | Lower motoneurons whose fibers innervate ordinary (extrafusal) muscle fibers |
| 4. Gamma motoneurons | Lower motoneurons whose fibers innervate the muscle spindle fibers (intrafusal fibers) |
| 5. Agonist/antagonist | Pair of muscles or muscle groups that insert on the same bone, the agonist being the muscle of reference |
| 6. Synergist | A muscle whose action facilitates the action of the agonist |
| 7. Ipsilateral/contralateral | Ipsilateral—to the same side, or the side of reference; contralateral—the opposite side |
| 8. Afferent/efferent | Afferent neurons—sensory; efferent neurons—motor |

**Figure 12.20.** A sensory neuron, an association neuron (interneuron), and a somatic motor neuron at the spinal cord level.

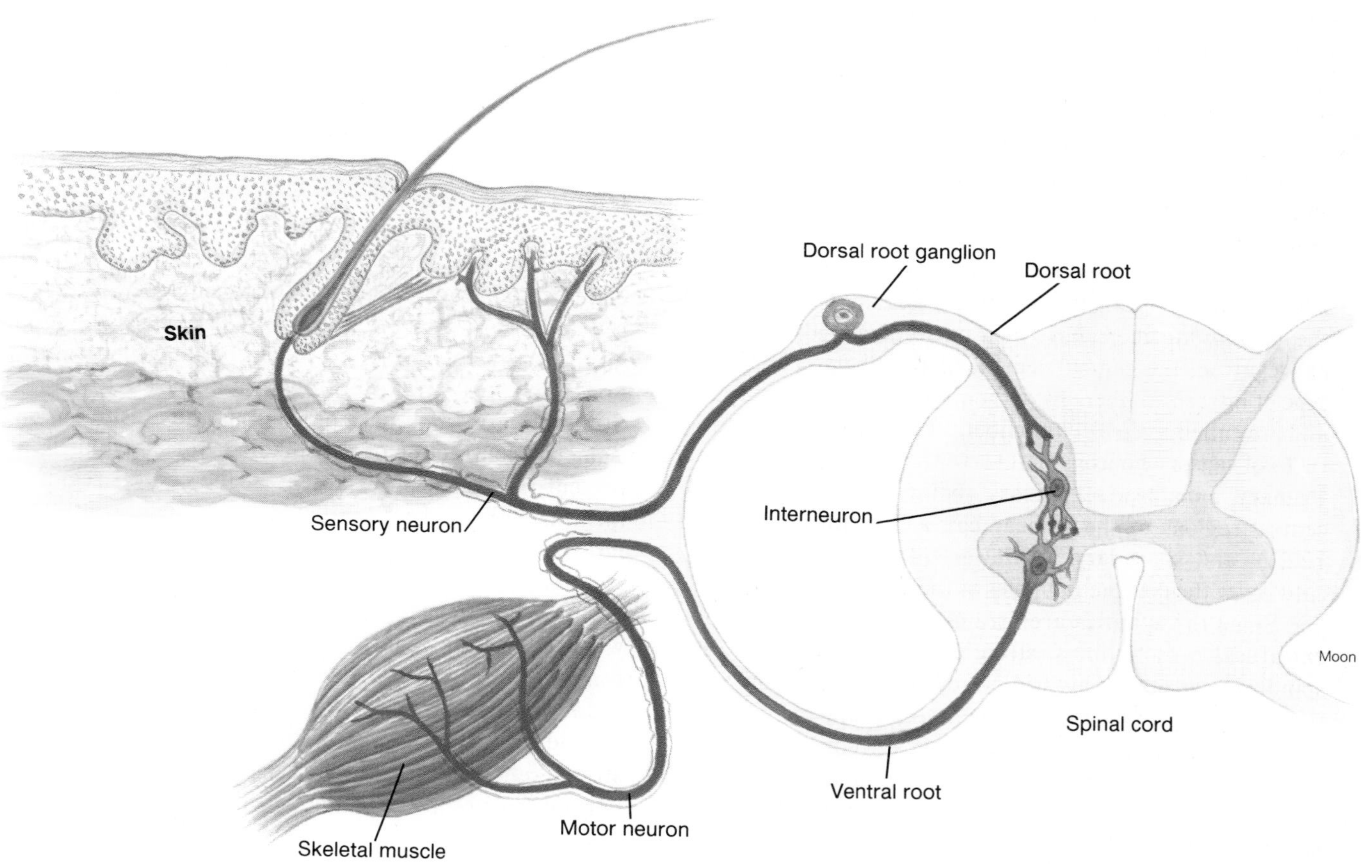


Those fibers that conduct impulses to higher spinal cord segments and the brain form *ascending tracts.* Those fibers that conduct to lower spinal segments contribute to *descending tracts.* And those that cross the midline of the CNS to synapse on the opposite side are part of *commissural tracts.* Interneurons can thus conduct impulses up and down on the same, or *ipsilateral,* side, and can affect neurons on the opposite, or *contralateral,* side of the central nervous system.

## Muscle Spindle Apparatus

In order for the nervous system to control skeletal movements properly, it must receive continuous sensory feedback information concerning the effects of its actions. This sensory information includes (1) the tension that the muscle exerts on its tendons, provided by the **Golgi tendon organs** and (2) muscle length, provided by the **muscle spindle apparatus.** The spindle apparatus, so-called because it is wider in the center and tapers toward the ends,

**Table 12.5** Spindle apparatus content of selected skeletal muscles

| Muscle | | Muscle Weight (g) | Average Number of Spindles | Number of Spindles per Gram Muscle |
|---|---|---|---|---|
| Gastrocnemius | | 7.6 | 35 | 5 |
| Rectus femoris | | 8.36 | 104 | 12 |
| Tibialis anterior | leg | 4.57 | 71 | 15 |
| Semitendinosis | | 6.41 | 114 | 18 |
| Soleus | | 2.49 | 56 | 23 |
| Fifth interossei—foot | | 0.33 | 29 | 88 |
| Fifth interossei—hand | | 0.21 | 25 | 119 |

functions as a length detector. Muscles that require the finest degree of control, such as the muscles of the hand, have the highest density of spindles (table 12.5).

Each spindle apparatus contains several thin muscle cells, called *intrafusal fibers* (*fusus* = spindle), packaged within a connective tissue sheath. Like the stronger and more numerous "ordinary" muscle fibers outside the spindles—the *extrafusal fibers*—the spindles insert into tendons on each end of the muscle. Spindles are therefore said to be in parallel with the extrafusal fibers.

Unlike the extrafusal fibers, which contain myofibrils along their entire length, the contractile apparatus is absent from the central regions of the intrafusal fibers. The central, noncontracting part of an intrafusal fiber contains nuclei. There are two types of intrafusal fibers. One type, the *nuclear bag* fibers, have their nuclei arranged in a loose aggregate in the central regions of the fibers. The other type of intrafusal fibers have their nuclei arranged in rows and are called *nuclear chain* fibers. There are likewise two types of sensory neurons that serve these intrafusal fibers. **Primary, annulospiral, sensory endings** wrap around the central regions of the nuclear bag and chain fibers (fig. 12.21), and **secondary,** or **flower-spray, endings** are located over the contracting poles of the nuclear chain fibers.

Since the spindles are arranged in parallel with the extrafusal muscle fibers, stretching a muscle causes its spindles to stretch. This stimulates both the primary and secondary sensory endings. The spindle apparatus thus serves as a length detector because the frequency of impulses produced in the primary and secondary endings is proportional to the length of the muscle. The primary endings, however, are most stimulated at the onset of stretch, whereas the secondary endings respond in a more tonic (sustained) fashion as stretch is maintained. Sudden, rapid stretching of a muscle activates both types of sensory endings and is thus a more powerful stimulus for the muscle spindles than slower, more gradual stretching, which has less effect on the primary sensory endings. Since the activation of the sensory endings in muscle spindles produces a reflex contraction, the force of this reflex contraction is greater in response to rapid stretch than to gradual stretch.

### Alpha and Gamma Motoneurons

There are two types of lower motor neurons in the spinal cord that innervate skeletal muscles. The motor neurons that innervate the extrafusal muscle fibers are called **alpha motoneurons,** and those that innervate the intrafusal fibers are called **gamma motoneurons** (fig. 12.21). The alpha motoneurons are larger and more rapidly conducting (60–90 meters per second) than the thinner, more slowly conducting (10–40 meters per second) gamma motoneurons. Since only the extrafusal muscle fibers are sufficiently strong and numerous to cause a muscle to shorten, only stimulation by the alpha motoneurons can cause muscle contraction that results in skeletal movements.

The intrafusal fibers of the muscle spindle are stimulated to contract by gamma motoneurons, which comprise one-third of all efferent fibers in spinal nerves. The intrafusal fibers are too few in number and their contraction is too weak, however, to cause a muscle to shorten. Stimulation by gamma motoneurons thus results in only isometric contraction of the spindles. Since myofibrils are present in the poles but absent in the central regions of intrafusal fibers, the more distensible central region of the intrafusal fiber is pulled toward the ends in response to stimulation by gamma motoneurons. As a result, the spindle is tightened. This effect of gamma motoneurons, which is sometimes termed *active stretch* of the spindles, functions to increase the sensitivity of the spindles when the entire muscle is passively stretched by external forces. The activation of gamma motoneurons thus enhances the stretch reflex (described in the next section), and is an important feature in the voluntary control of skeletal movements.

### Co-activation of Alpha and Gamma Motoneurons

Most of the fibers in the descending motor tracts synapse with interneurons in the spinal cord; only about 10% of the descending fibers synapse directly with the lower (alpha and gamma) motor neurons. It is likely that very rapid movements are produced by direct synapses with the lower motor neurons, whereas most other movements are produced indirectly via synapses with spinal interneurons, which in turn stimulate the motor neurons.

**Figure 12.21.** The structure of muscle spindles and their relationship to skeletal muscles.

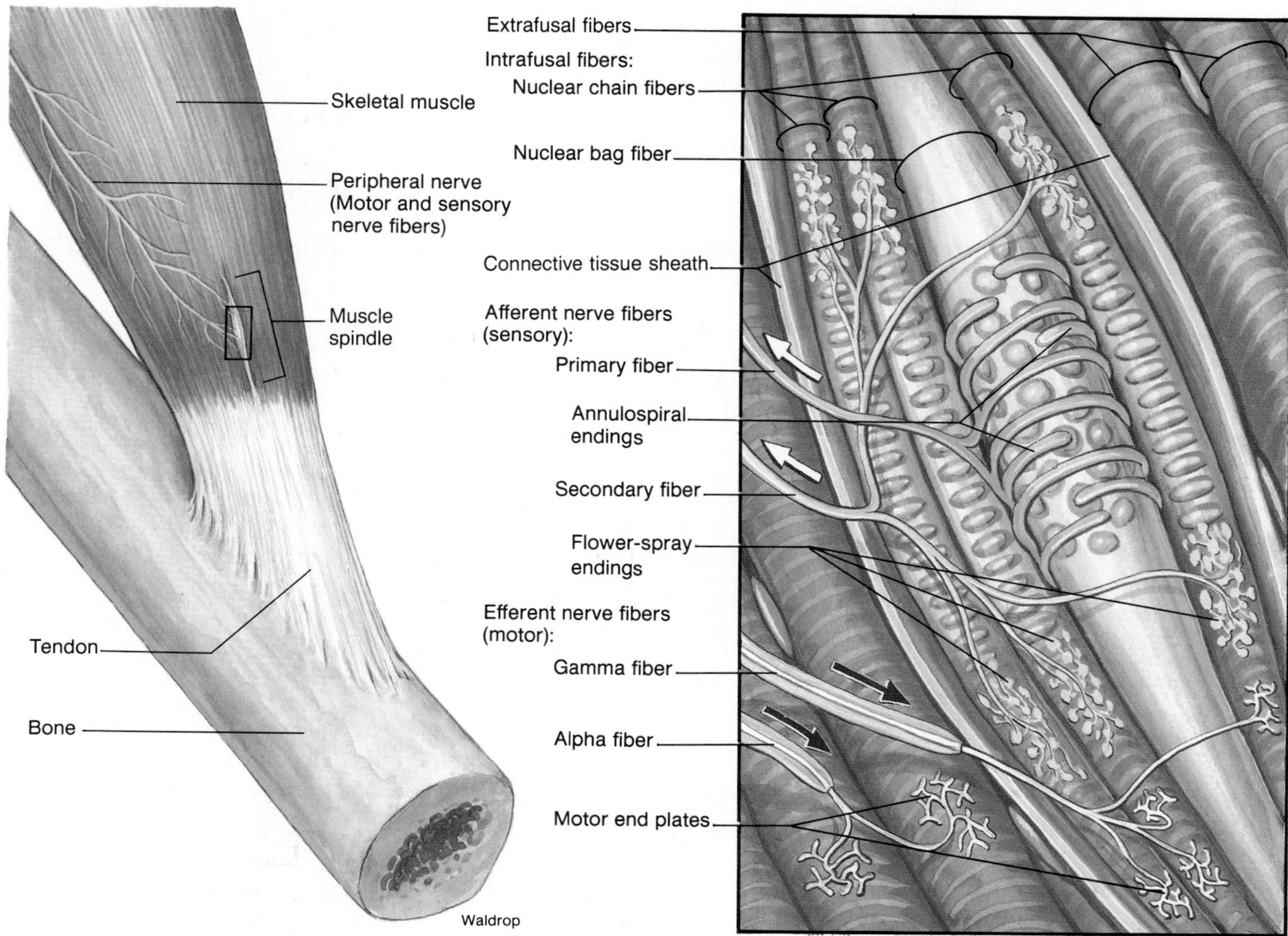


*Higher motor neurons*—neurons in the brain that contribute fibers to descending motor tracts—usually stimulate both alpha and gamma motoneurons simultaneously. This is known as **co-activation** of the alpha and gamma motoneurons. Stimulation of alpha motoneurons results in muscle contraction and shortening; stimulation of gamma motoneurons stimulates contraction of the intrafusal fibers and thus "takes out the slack" that would otherwise be present in the spindles as the muscles shorten. In this way the spindles remain under tension and provide information about the length of the muscle even while the muscle is shortening.

When higher motor neurons stimulate the contraction of one muscle, they inhibit, via synapses with spinal interneurons, the alpha and gamma motoneurons of the antagonist muscles. Inhibition of the gamma motoneurons makes the antagonist less sensitive to stretch, so that its stretch reflex will not oppose the intended action. Flexion of the arm by contraction of the biceps, for example, stretches the triceps muscle. The triceps is not normally stimulated to contract by this action, however, because its stretch reflex is dampened by inhibition of its gamma motoneurons.

Under normal conditions the activity of gamma motoneurons is maintained at the level needed to keep the muscle spindles under proper tension while the muscles are relaxed. Undue relaxation of the muscles is prevented by stretch and activation of the spindles, which in turn elicits a reflex contraction (described in the next section). This mechanism produces a normal resting muscle length and state of tension, or **muscle tone**.

## Skeletal Muscle Reflexes

Although skeletal muscles are often called voluntary muscles, because they are controlled by descending motor pathways that are under conscious control, they often contract in an unconscious, reflex fashion in response to particular stimuli. In the simplist type of reflex, a skeletal muscle contracts in response to the stimulus of muscle stretch. More complex reflexes involve inhibition of antagonistic muscles and regulation of a number of muscles on both sides of the body.

**Table 12.6** Summary of events that occur during a monosynaptic stretch reflex

1. Passive stretch of a muscle (produced by tapping its tendon) stretches the spindle (intrafusal) fibers.
2. Stretching of a spindle distorts its central (bag or chain) region, which stimulates dendritic endings of sensory nerves.
3. Action potentials are conducted by afferent (sensory) nerve fibers into the spinal cord on the dorsal roots of spinal nerves.
4. Axons of sensory neurons synapse with dendrites and cell bodies of somatic motor neurons located in the ventral horn gray matter of the spinal cord.
5. Efferent nerve impulses in the axons of somatic motor neurons (which form the ventral roots of spinal nerves) are conducted to the ordinary (extrafusal) muscle fibers. These neurons are alpha motoneurons.
6. Release of acetylcholine from the endings of alpha motoneurons stimulates contraction of the extrafusal fibers and thus of the whole muscle.
7. Contraction of the muscle relieves the stretch of its spindles, thus decreasing electrical activity in the spindle afferent nerve fibers.

**Table 12.7** Effects of lesions (damage) at different levels on the neural control of skeletal muscles

| | |
|---|---|
| Dorsal roots of spinal nerves (sensory) | Coordinated movements difficult; difficulty walking (ataxia) |
| Transsection of spinal cord | First—*spinal shock,* characterized by lack of stretch reflexes and low muscle tone. After a few weeks—*decerebrate rigidity,* characterized by hyperactive stretch reflexes, flexion of the arms, and extension of the legs (spasticity); also, forced flexion of hand or foot, producing oscillating extension and flexion (clonus) |
| Cutting of pyramids | Voluntary movements requiring great effort; movements lacking precision; low muscle tone on side opposite lesion |
| Transsection between midbrain and cerebral cortex | Animal placed on side can right itself, stand, and walk; clumsy movements and some rigidity of extensor muscles |
| Removal of motor cortex | No paralysis, but uncoordinated movements |

***The Monosynaptic Stretch Reflex.*** Reflex contraction of skeletal muscles occurs in response to sensory input and is not dependent upon the activation of higher motor neurons. The **reflex arc,** which describes the nerve impulse pathway from sensory to motor endings in such reflexes, involves only a few synapses within the CNS. The simplest of all reflexes—the muscle *stretch reflex*—consists of only one synapse within the CNS. The sensory neuron directly synapses with the motor neuron, without involving spinal cord interneurons. The stretch reflex is thus *monosynaptic* in terms of the individual reflex arcs (many sensory neurons, of course, are activated at the same time, leading to the activation of many motor neurons). Resting skeletal muscles are maintained at an optimal length, as previously described under the heading "length-tension relationship," by stretch reflexes.

The stretch reflex is present in all muscles, but it is most dramatic in the extensor muscles of the limbs. The **knee jerk reflex**—the most commonly evoked stretch reflex—is initiated by striking the patellar tendon with a rubber mallet. This stretches the entire body of the muscle and thus passively stretches the spindles within the muscle. Sensory nerves with primary (annulospiral) endings in the spindles are thus activated, which synapse within the ventral gray matter of the spinal cord with *alpha motoneurons.* These large, rapidly conducting motor nerve fibers stimulate the extrafusal fibers of the extensor muscle, resulting in isotonic contraction and the knee jerk. This is an example of negative feedback—stretching of the muscles (and spindles) stimulates shortening of the muscles (and spindles). These events are summarized in table 12.6 and in figure 12.22.

Damage to spinal nerves, or to the cell bodies of lower motor neurons (by poliovirus, for example), produces a *flaccid paralysis* characterized by reduced muscle tone, depressed stretch reflexes, and atrophy. Damage to higher motor neurons or descending motor tracts (table 12.7) at first produces *spinal shock* in which there is a flaccid paralysis. This is followed in a few weeks by *spastic paralysis,* which is characterized by increased muscle tone, exaggerated stretch reflexes, and other signs of hyperactive lower motor neurons.

The appearance of spastic paralysis suggests that higher motor neurons normally exert an inhibitory effect on lower alpha and gamma motor neurons. When this inhibition is removed the gamma motoneurons become hyperactive, and the spindles thus become overly sensitive to stretch. This can be shown dramatically in such a patient by forcefully dorsiflecting a foot (pushing it up) and then releasing it. Forced extension stretches the antagonistic flexor muscles, which contract and produce the opposite movement (plantar flexion). Alternating activation of antagonistic stretch reflexes produces a flapping motion known as **clonus.**

**Figure 12.22.** The knee jerk reflex, an example of a monosynaptic stretch reflex.

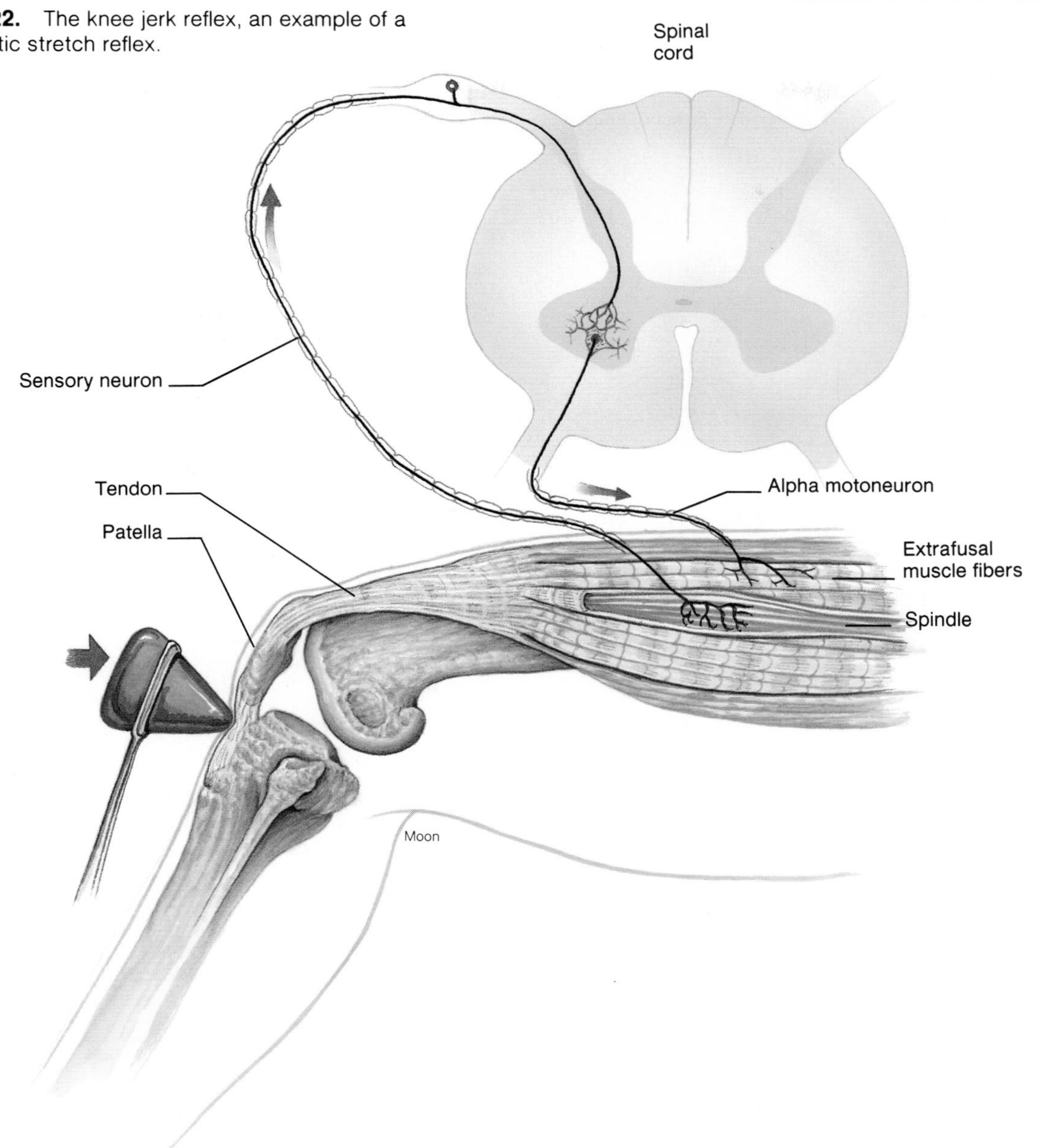


***Golgi Tendon Organs.*** The Golgi tendon organs continuously monitor tension in the tendons, produced by muscle contraction or passive stretching of a muscle. Sensory neurons from these receptors synapse with interneurons in the spinal cord; these interneurons, in turn, have *inhibitory synapses* (via IPSPs and postsynaptic inhibition—chapter 7) with motor neurons that innervate the muscle (fig. 12.23). This is called an inhibitory **disynaptic reflex** (because two synapses are crossed in the CNS), and it helps prevent excessive muscle contractions or excessive passive muscle stretching. Indeed, if a muscle is stretched extensively, it will actually relax as a result of the inhibitory effects of the Golgi tendon organs.

Rapid stretching of skeletal muscles produces very forceful muscle contractions as a result of the activation of primary and secondary endings in the muscle spindles and the monosynaptic stretch reflex. This can result in painful muscle spasms, as may occur, for example, when muscles are forcefully pulled in the process of setting broken bones. Painful muscle spasms may be avoided in physical exercise by stretching slowly and thereby stimulating mainly the secondary endings in the muscle spindles. A slower rate of stretch also provides time for the inhibitory Golgi tendon organ reflex to occur and promote muscle relaxation.

**Figure 12.23.** An increase in muscle tension stimulates the activity of sensory nerve endings in the Golgi tendon organ. This sensory input stimulates ⊕ an interneuron, which in turn, inhibits ⊖ the activity of a motor neuron innervating that muscle. This is therefore a disynaptic reflex.

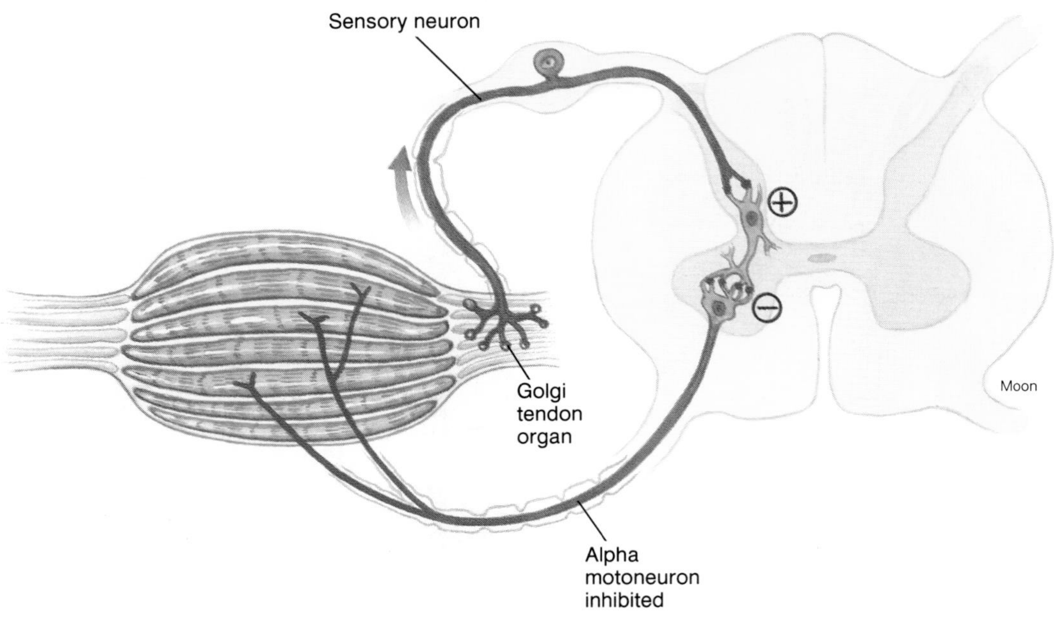


**Figure 12.24.** A diagram of reciprocal innervation. Afferent impulses from muscle spindles stimulate alpha motoneurons to the agonist muscle (the extensor) directly, but (via an inhibitory interneuron) inhibit activity in the alpha motoneuron to the antagonist muscle.

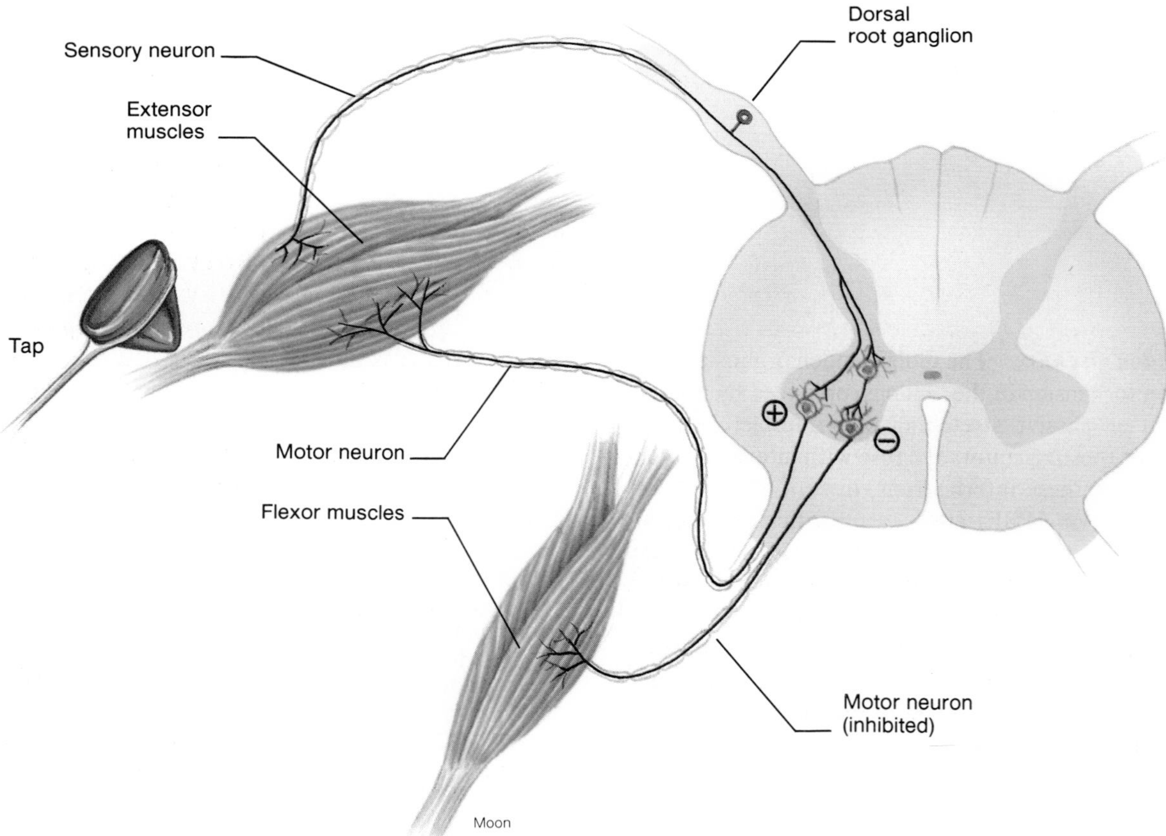

**Figure 12.25.** The crossed-extensor reflex, demonstrating double reciprocal innervation.

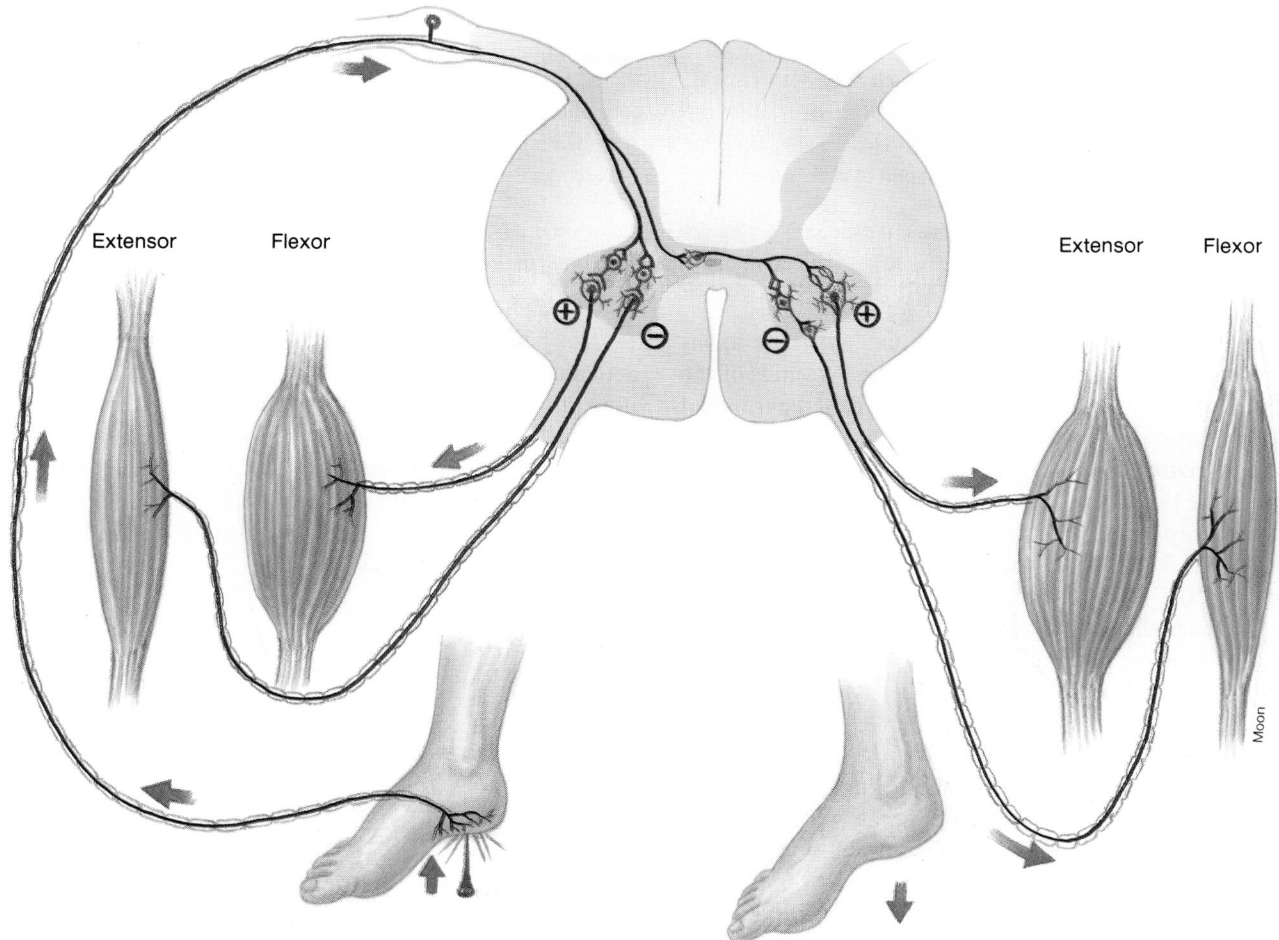


***Reciprocal Innervation and the Crossed-Extensor Reflex.*** When a limb is flexed, the antagonistic extensor muscles are passively stretched. Extension of a limb similarly stretches the antagonistic flexor muscles. If the monosynaptic stretch reflexes were not inhibited, reflex contraction of the antagonistic muscles would always interfere with the intended movement (or produce clonus, as previously described). Fortunately, whenever the "intended," or agonist muscles, are stimulated to contract, the alpha and gamma motoneurons that stimulate the antagonist muscles are inhibited.

This inhibition of the motoneurons that stimulate antagonistic muscles is accomplished by interneurons, which inhibit motor neurons via inhibitory postsynaptic potentials (IPSPs). In the knee jerk and other stretch reflexes, the sensory neuron that stimulates the motor neuron of the agonist muscle also stimulates, via collateral branches, the interneurons that inhibit the motor neurons of antagonist muscles. This dual stimulatory and inhibitory activity is called **reciprocal innervation** (fig. 12.24).

The stretch reflex, with its reciprocal innervations, involves the muscles of one limb only and is controlled by only one segment of the spinal cord. More complex reflexes involve muscles controlled by a number of spinal cord segments and affect muscles on the contralateral side of the cord. Such reflexes involve **double reciprocal innervation** of muscles.

Double reciprocal innervation is illustrated by the **crossed-extensor reflex.** If one steps on a tack with the right foot, for example, this foot is withdrawn by contraction of its flexors and relaxation of its extensors. The contralateral left leg, in contrast, extends to help support the body during this withdrawal reflex. The extensors of the left leg contract while its flexors relax. These events are shown in figure 12.25.

## Higher Motor Neuron Control of Skeletal Muscles

Higher motor neurons are neurons in the brain which influence the control of skeletal muscle by lower motor neurons (alpha and gamma motoneurons). Neurons in the

**Table 12.8** Some symptoms of higher motor neuron damage

| |
|---|
| **Babinski reflex**—extension of the big toe when the sole of the foot is rubbed along the lateral border |
| **Spastic paralysis**—high muscle tone and hyperactive stretch reflexes; flexion of arms and extension of legs |
| **Hemiplegia**—paralysis of upper and lower limbs on one side—commonly produced by damage to motor tracts as they pass through internal capsule (such as by cerebrovascular accident—stroke) |
| **Paraplegia**—paralysis of lower limbs on both sides, by lower spinal cord damage |
| **Quadriplegia**—paralysis of both upper and lower limbs on both sides, by damage to the upper region of the spinal cord or brain |
| **Chorea**—random uncontrolled contractions of different muscle groups (such as Saint Vitus' dance), produced by damage to basal ganglia |
| **Resting tremor**—shaking of limbs at rest; disappears during voluntary movements; produced by damage to basal ganglia |
| **Intention tremor**—oscillations of arm following voluntary reaching movements; produced by damage to cerebellum |

precentral gyrus of the cerebral cortex contribute axons that cross to the contralateral sides in the pyramids of the medulla oblongata; these tracts are thus called **pyramidal tracts** (chapter 8). The pyramidal tracts include the *lateral* and *ventral corticospinal tracts.* Neurons in other areas of the brain produce the **extrapyramidal tracts**. The major extrapyramidal tract is the *reticulospinal tract* (review fig. 8.21), which originates in the reticular formation of the medulla oblongata and pons. Brain areas that influence the activity of extrapyramidal tracts are believed to produce the inhibition of lower motor neurons previously described.

***Cerebellum.*** The cerebellum, like the cerebrum, receives sensory input from muscle spindles and Golgi tendon organs. It also receives fibers from areas of the cerebral cortex devoted to vision, hearing, and equilibrium (the latter areas are called the vestibular nuclei because they process sensory input from the vestibular apparatus in the inner ear).

There are no descending tracts from the cerebellum. The cerebellum can only influence motor activity indirectly, through its output to the vestibular nuclei, red nucleus, and basal ganglia. These structures, in turn, affect lower motor neurons via the vestibulospinal tract, rubrospinal tract, and reticulospinal tract. It is interesting that all output from the cerebellum is inhibitory; these inhibitory effects aid motor coordination by eliminating inappropriate neural activity. Damage to the cerebellum interferes with the ability to coordinate movements with spatial judgment. Under or overreaching for an object may occur, followed by *intention tremor,* in which the limb moves back and forth in a pendulumlike motion.

***Basal Ganglia.*** The basal ganglia, or cerebral nuclei, include the *caudate nucleus, putamen,* and *globus pallidus* (chapter 8). Often included in this group are other nuclei of the *thalamus, subthalamus, substantia nigra,* and *red nucleus.* Acting directly via the rubrospinal tract and indirectly via synapses in the reticular formation and thalamus, the basal ganglia have profound effects on the activity of lower motor neurons.

The basal ganglia, acting through synapses in the reticular formation particularly, appear normally to exert an inhibitory influence on the activity of lower motor neurons. Damage to the basal nuclei thus results in increased muscle tone, as previously described. People with such damage display *akinesia* (lack of desire to use the affected limb) and *chorea*—sudden and uncontrolled random movements (table 12.8).

**Parkinson's disease** (or *paralysis agitans*) is a disorder of the basal ganglia involving degeneration of fibers from the substantia nigra. These fibers, which use dopamine as a neurotransmitter, are required to antagonize the effects of other fibers that use acetylcholine (ACh) as a transmitter. The relative deficiency of dopamine compared to ACh is believed to produce the symptoms of Parkinson's disease, including *resting tremor.* This "shaking" of the limbs tends to disappear during voluntary movements and then reappear when the limb is again at rest.

1. *Draw a muscle spindle surrounded by a few extrafusal fibers, and show the location of primary and secondary sensory endings. Describe the way these endings respond to muscle stretch.*
2. *Describe all of the events that occur between the time that the patellar tendon is struck with a mallet and the time that the leg kicks.*
3. *Explain how a Golgi tendon organ is stimulated, and describe the disynaptic reflex that occurs.*
4. *Explain the significance of reciprocal innervation and double reciprocal innervation in muscle reflexes.*
5. *Explain how spindles serve as length detectors, and explain the functions of gamma motoneurons. Explain why gamma motoneurons are stimulated at the same time as alpha motoneurons during voluntary muscle contractions.*
6. *Explain how a person with spinal cord damage might develop clonus.*
7. *Explain the functions of the cerebral cortex, basal ganglia, and cerebellum in the control of skeletal muscle.*

**Figure 12.26.** The production and utilization of phosphocreatine in muscles.

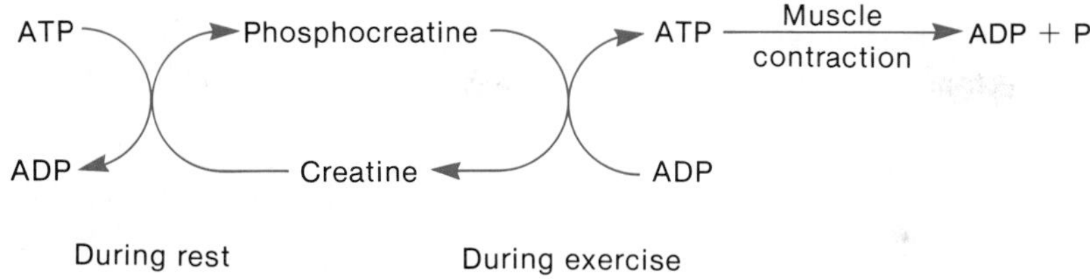


## *Skeletal Muscle Types and Energy Usage*

Skeletal muscles generate ATP through aerobic and anaerobic respiration, and through the use of phosphate groups donated by creatine phosphate. The aerobic and anaerobic abilities of skeletal muscle fibers differ, depending on the type of muscle fiber. Fast-twitch (type I) fibers are adapted for anaerobic respiration, and slow-twitch (type II) fibers are adapted for aerobic respiration. The former fibers can produce more power, but the latter have a greater resistance to fatigue.

The metabolism of skeletal muscles was described in chapter 5 ("Metabolism of Skeletal Muscles") as a case study in the basic processes of metabolism, and should be reviewed at this time. As described in this previous chapter, skeletal muscles at rest obtain most of their energy from the aerobic respiration of fatty acids. During exercise, muscle glycogen and blood glucose are also used as energy sources. Energy obtained by cell respiration is used to make ATP, which serves as the immediate source of energy for (1) the movement of the cross-bridges for muscle contraction and (2) the pumping of $Ca^{++}$ into the sarcoplasmic reticulum for muscle relaxation.

The ability of muscles to convert the potential energy present in fuel molecules to the mechanical energy of muscle contraction—called the *mechanical efficiency*—averages about 24%, although some top endurance athletes have a mechanical efficiency of up to 30%. The reason for this variation is not entirely known, but is believed to be largely due to genetic differences in the amount of myosin ATPase within the muscles. The remaining energy liberated by cell respiration—76% of the total energy in the average example—is lost as body heat.

During sustained muscle contraction, ATP may be utilized faster than the rate of ATP production through cell respiration. At these times the rapid renewal of ATP is extremely important. This is accomplished by combining ADP with phosphate derived from another high-energy phosphate compound called **phosphocreatine** (or **creatine phosphate**).

The phosphocreatine concentration within muscle cells is greater than three times the concentration of ATP and represents a ready reserve of high-energy phosphate that can be donated directly to ADP (fig. 12.26). During times of rest, the depleted reserve of phosphocreatine can be restored by the reverse reaction—phosphorylation of creatine with phosphate derived from ATP.

**Figure 12.27.** Creatine kinase in plasma is separated from other proteins and identified by the rate of its movement in an electric field. The height of each peak indicates concentration; and the position of each peak indicates the isoenzyme form. As can be seen in the *lower figure,* the plasma concentration of creatine kinase is elevated in both heart disease (myocardial infarction) and skeletal muscle disease, but the two types of disease may be distinguished in this test by the appearance of a different isoenzyme in myocardial infarction than in muscle disease.

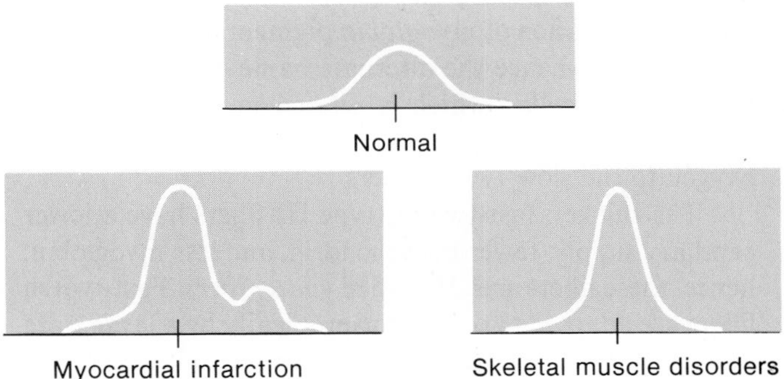


The enzyme that transfers phosphate between creatine and ATP is called **creatine kinase,** or **creatine phosphokinase**. Skeletal muscle and heart muscle have two different forms of this enzyme (they have different isoenzymes, as described in chapter 4). The skeletal muscle isoenzyme is found to be elevated in the blood of people with *muscular dystrophy* (degenerative disease of skeletal muscles). The plasma concentration of the isoenzyme characteristic of heart muscle is elevated as a result of *myocardial infarction* (damage to heart muscle), and measurements of this enzyme are thus used as a means of diagnosing this condition (fig. 12.27).

## Slow- and Fast-Twitch Fibers

Skeletal muscle fibers can be divided on the basis of their contraction speed (time required to reach maximum tension) into **slow-twitch,** or **type I fibers,** and **fast-twitch,** or **type II fibers.** These differences are associated with different myosin ATPase isoenzymes, designated "slow" and "fast," by which the two fiber types can be distinguished when they are appropriately stained (fig. 12.28). The extraocular muscles that position the eyes, for example, have a high proportion of fast-twitch fibers and reach maximum tension in about 7.3 msec (milliseconds—thousandths of a second); the soleus muscle in the leg, in contrast, has a high proportion of slow-twitch fibers and requires about 100 msec to reach maximum tension (fig. 12.29).

Muscles like the soleus are *postural muscles* that must be able to sustain a contraction for a long period of time without fatigue. This is aided by other characteristics of slow-twitch (type I) fibers that endow them with a high capacity for aerobic respiration. Slow-twitch fibers are served by large numbers of capillaries, have numerous mitochondria and aerobic respiratory enzymes, and have a high concentration of *myoglobin* pigment. Myoglobin is a red pigment—hence the alternate name of *red fibers* for these muscle cells—which is related to the hemoglobin pigment of blood and serves to improve the delivery of oxygen to the slow-twitch fibers.

The thicker, fast-twitch (type II) fibers have a lower capillary supply, fewer mitochondria, and less myoglobin; hence, these fibers are also called *white fibers.* Fast-twitch fibers are adapted to respire anaerobically by a large store of glycogen and a high concentration of glycolytic enzymes, which enable these fibers to be distinguished when appropriately stained (fig. 12.30). In addition to the type I (slow-twitch) and type II (fast-twitch) fibers, human muscles may also have an intermediate form of fibers. These intermediate fibers are fast-twitch but also have a high aerobic capability. These are sometimes called type IIA, to distinguish them from the anaerobically adapted fast-twitch fibers (which are then labeled IIB). A comparison of these fiber types is summarized in table 12.9.

Interestingly, the conduction rate of motor neurons that innervate fast-twitch fibers is faster (80–90 meters per second) than the conduction rate to slow-twitch fibers (60–70 meters per second). The fiber type indeed seems to be determined by the motor neuron. When the motor neurons to different fiber types are switched in experimental animals, the previously fast-twitch fibers become slow and the slow-twitch fibers become fast. As expected from these observations, all of the muscle fibers innervated by the same motor neuron (that are part of the same motor unit) are of the same type.

A muscle such as the gastrocnemius (calf muscle) contains both fast- and slow-twitch fibers, although fast-twitch fibers predominate. A given somatic motor axon, however, only innervates muscle fibers of one type. The size of these motor units differ; the motor units composed of slow-twitch fibers tend to be smaller (have fewer fibers) than the motor units of fast-twitch fibers. Since motor units are recruited from smaller to larger when increasing effort is required, as previously described, the smaller motor units with slow-twitch fibers would be used most often in routine activities. Larger motor units with fast-twitch fibers, which can exert a great deal of force but which respire anaerobically and thus fatigue quickly, would be used relatively infrequently and for only short periods of time.

**Figure 12.28.** Skeletal muscle (of a cat) stained to indicate the activity of myosin ATPase. Type II fibers contain higher ATPase activity than type I fibers.

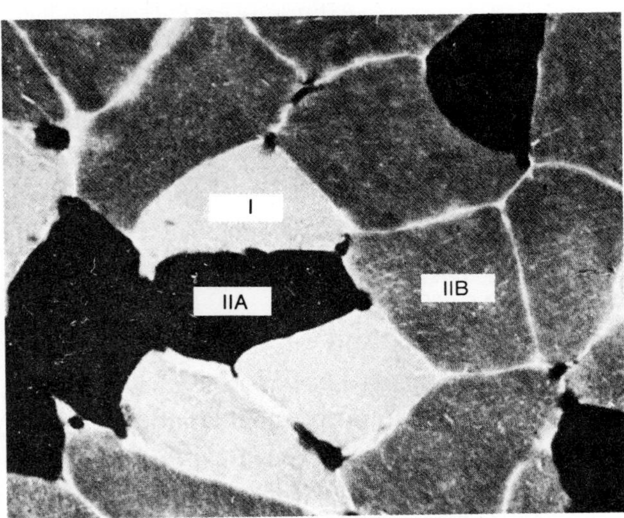


## Muscle Fatigue

Muscle fatigue may be defined as the inability to maintain a particular muscle tension when the contraction is sustained or to reproduce a particular tension during rhythmic contraction over time. Fatigue during a sustained maximal contraction, when all the motor units are used and the rate of neural firing is maximal—as when lifting an extremely heavy weight—appears to be due to an accumulation of extracellular $K^+$. (Remember that $K^+$ leaves axons and muscle fibers during the repolarization phase of action potentials.) This reduces the membrane potential of muscle fibers and interferes with their ability to produce action potentials. Fatigue under these circumstances lasts only a short time, and maximal tension can again be produced after less than a minute rest.

Fatigue during moderate exercise involving rhythmical contractions occurs as the slow-twitch fibers deplete their reserve glycogen and fast-twitch fibers are increasingly recruited. Fast-twitch fibers obtain their energy through anaerobic respiration, converting glucose to lactic acid, and this results in a rise in intracellular $H^+$ and a fall in pH. The decrease in muscle pH, in turn, inhibits

**Figure 12.29.** A comparison of the rates with which maximum tension is developed in three muscles. These include the relatively fast-twitch extraocular (*a*) and gastrocnemius (*b*) muscles and the slow-twitch soleus muscle (*c*).

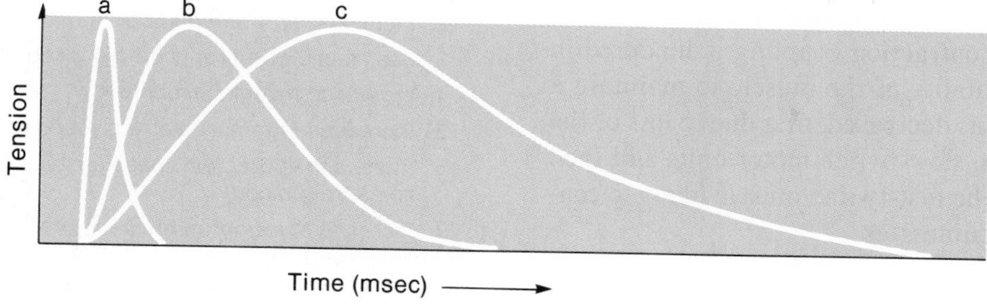


**Figure 12.30.** Skeletal muscle (of a cat) stained to show the activity of glycolytic enzymes in different fiber types. Type IIB (white) fibers have the fastest rate of glycolysis, while type I (red) fibers have the slowest.

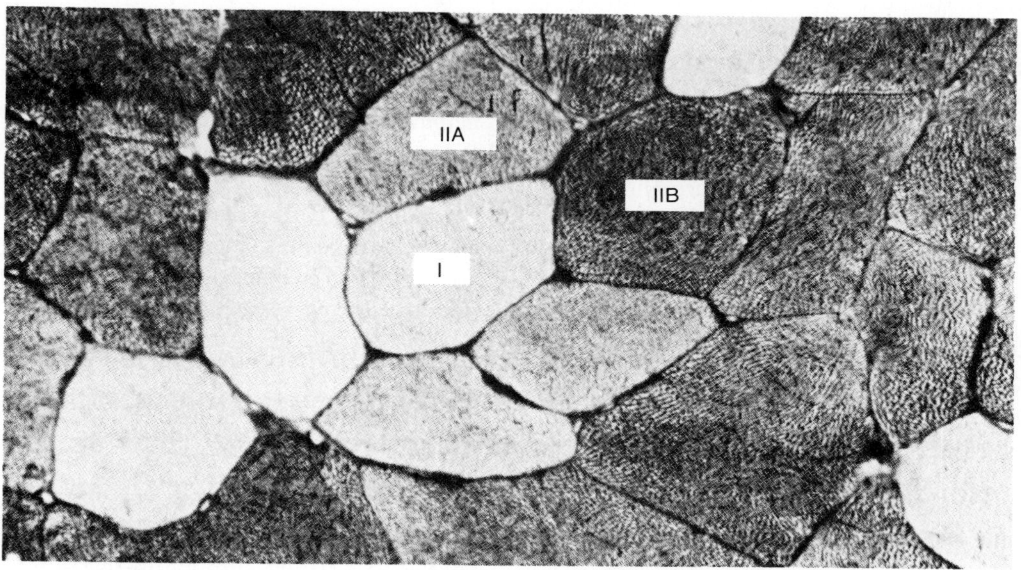


**Table 12.9** Characteristics of red, intermediate, and white muscle fibers

| | Red (Type I) | Intermediate (Type IIA) | White (Type IIB) |
|---|---|---|---|
| Diameter | Small | Intermediate | Large |
| Z-line thickness | Wide | Intermediate | Narrow |
| Glycogen content | Low | Intermediate | High |
| Resistance to fatigue | High | Intermediate | Low |
| Capillaries | Many | Many | Few |
| Myoglobin content | High | High | Low |
| Respiration type | Aerobic | Aerobic | Anaerobic |
| Twitch rate | Slow | Fast | Fast |
| Myosin ATPase content | Low | High | High |

the activity of key glycolytic enzymes so that the rate of ATP production is reduced. Since ATP is needed for the active accumulation of $Ca^{++}$ into the sarcoplasmic reticulum, the decrease in ATP is believed to result in a loss of this intracellular $Ca^{++}$ to the extracellular environment. With inadequate stores of $Ca^{++}$ in the sarcoplasmic reticulum, excitation-contraction coupling is hindered in the muscle cells. The ability of the muscle to maintain a particular tension is thus decreased, first due to loss of the contribution of smaller, slow-twitch motor units and then due to the inability of the fast-twitch muscle fibers to contract as lactic acid accumulates.

Note that the decrease in ATP within the muscle cell causes it to fatigue due to interference with excitation-contraction coupling, not due to direct interference with the cross-bridge cycle. This is supported by the observation that, even at maximal exhaustion, the amount of ATP in muscle fibers is only reduced by about 25% (although the amount of phosphocreatine may be completely depleted). Actually, the interference with excitation-contraction coupling can be thought of as a protective mechanism, because if the ATP were significantly depleted the muscle would enter a state analogous to rigor mortis. This does not occur in a living muscle, even during maximal exercise, because (1) ATP can be quickly reformed, using the phosphate group from creatine phosphate; and (2) if ATP does become depleted by a certain amount, the muscle fatigues due to interference with excitation-contraction coupling.

### Adaptations to Exercise

The maximal oxygen uptake, obtained during very strenuous exercise, averages 50 ml $O_2$ per minute per kilogram body weight in males between twenty and twenty-five years old (females average 25% lower). Top endurance athletes, however, can have maximal oxygen uptakes as high as 86 ml $O_2$ per minute per kilogram. When exercise is performed at low levels of effort, such that the oxygen consumption rate is less than 50% of its maximum, the energy for muscle contraction is obtained almost entirely from aerobic cell respiration. Anaerobic cell respiration, with its consequent production of lactic acid, contributes to the energy requirements as the exercise level rises and more than 60% of the maximal oxygen uptake is required. Top endurance-trained athletes, however, can continue to respire aerobically, with little lactic acid production, at up to 80% of their maximal oxygen uptake. These athletes thus produce less lactic acid at a given level of exercise than the average person and, therefore, are less subject to fatigue than the average person.

Since the fiber types are determined by their innervations, endurance training cannot change fast-twitch (type II) fibers to slow-twitch (type I) fibers. All fiber types, however, adapt to endurance training by an increase in myoglobin and aerobic respiratory enzymes, so that the maximal oxygen uptake can be increased by up to 20% through this training. In addition to changes in aerobic capacity, fibers show an increase in their content of triglycerides, which serves as an alternate energy source helping to spare their stores of glycogen. A summary of the changes that occur as a result of endurance training is presented in table 12.10.

**Table 12.10** Summary of the effects of endurance training (long-distance running, swimming, bicycling, etc.) on skeletal muscles

1. Improved ability to obtain ATP from oxidative phosphorylation
2. Increased size and number of mitochondria
3. Less lactic acid produced per given amount of exercise
4. Increased myoglobin content
5. Increased intramuscular triglyceride content
6. Increased lipoprotein lipase (enzyme needed to utilize lipids from blood)
7. Increased proportion of energy derived from fat, less from carbohydrates
8. Lower rate of glycogen depletion during exercise
9. Improved efficiency in extracting oxygen from blood

Endurance training does not increase the size of muscles. Muscle enlargement is produced only by frequent bouts of muscle contraction against a high resistance—as in weight lifting. As a result of this latter type of training, type II muscle fibers become thicker, and the muscle therefore grows by hypertrophy (increase in cell size, rather than number). This happens as the myofibrils within a muscle fiber get thicker, due to the addition of new sarcomeres and myofibrils. After a myofibril attains a certain thickness it may split into two myofibrils, which may then each become thicker due to the addition of sarcomeres. Muscle hypertrophy, in summary, is associated with an increase in the number and size of myofibrils within the muscle fibers.

1. *Draw a figure illustrating the relationship between ATP and creatine phosphate, and explain the physiological significance of this relationship.*
2. *Describe the characteristics of slow- and fast-twitch fibers (including intermediate fibers). Explain how the fiber types are determined and the functions of different fiber types.*
3. *Explain the different causes of muscle fatigue, and relate the causes of fatigue to the fiber types.*
4. *Describe the effects of endurance training and strength training on the fiber characteristics of the muscles.*

## Cardiac and Smooth Muscle

Cardiac muscle, like skeletal muscle, is striated and contains sarcomeres that shorten by sliding of thin and thick filaments. Unlike skeletal muscle, however, cardiac muscle can produce action potentials spontaneously and conduct these impulses from one cell to another. Smooth muscles lack striations but appear to also contract by means of cross-bridge interaction between myosin and actin. The characteristics of this interaction, and the mechanisms of its regulation, differ in some respects from those of striated muscle.

Unlike skeletal muscles, which are voluntary effectors regulated by somatic motor neurons, cardiac and smooth muscles are involuntary effectors regulated by autonomic motor neurons. Although there are important differences between skeletal muscle and cardiac and smooth muscle, there are also significant similarities. All types of muscle are believed to contract by means of sliding of thin actin filaments over thick myosin filaments; the sliding of the filaments is produced by the action of myosin cross-bridges in all types of muscles; and excitation-contraction coupling in all types of muscles involves $Ca^{++}$. This section serves only as an introduction to the properties of cardiac and smooth muscles. The functions of these muscles in the internal organs are explored in subsequent chapters.

### Cardiac Muscle

Like skeletal muscle cells, cardiac (heart) muscle cells, or *myocardial cells,* are striated; they contain actin and myosin filaments arranged in the form of sarcomeres, and they contract by means of the sliding filament mechanism. The long, fibrous skeletal muscle cells, however, are structurally and functionally separated from each other, whereas the myocardial cells are short, branched, and interconnected. Adjacent myocardial cells are joined by electrical synapses, or **gap junctions** (described in chapter 7). Gap junctions in cardiac muscle have an affinity for stain that gives them the appearance of dark lines between adjacent cells when viewed in the light microscope. These dark staining lines are known as *intercalated discs* (fig. 12.31).

**Figure 12.31.** Cardiac muscle. Note that the cells are short, branched, and striated and that they are interconnected by intercalated discs.

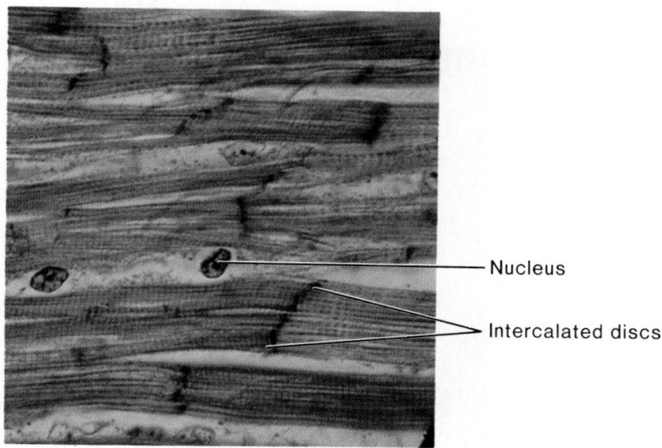


Electrical impulses that originate at any point in the mass of myocardial cells, called the *myocardium,* can spread to all cells in the mass that are joined by gap junctions. Because all cells in a myocardium are electrically joined, the myocardium behaves as a single functional unit, or a *functional syncytium* (*sin-sish'e-um*). Thus, unlike skeletal muscles which can produce contractions that are graded depending on the number of cells stimulated, the heart contracts with an **all-or-none** contraction.

Unlike skeletal muscles, which require external stimulation by somatic motor nerves before they can produce action potentials and contract, cardiac muscle is able to produce action potentials automatically. Cardiac action potentials normally originate in a specialized group of cells called the *pacemaker.* However, the rate of this spontaneous depolarization and, thus, the rate of the heartbeat, are regulated by autonomic innervation. The properties of cardiac muscle in relation to the heart are explained in more detail in chapter 13.

### Smooth Muscle

Smooth (visceral) muscles are arranged in circular layers around the walls of blood vessels, bronchioles (small air passages in the lungs), and in the sphincter muscles of the digestive tract. Both circular and longitudinal smooth muscle layers are found in the tubular digestive tract, the ureters (which transport urine), the ductus deferens (which transport sperm), and the uterine tubes (which transport ova). The alternate contraction of circular and longitudinal smooth muscle layers in the intestine produces **peristaltic waves,** which propel the contents of these tubes in one direction.

Smooth muscle cells do not contain sarcomeres (which produce striations in skeletal and cardiac muscle). Smooth muscle cells do, however, contain a great amount of actin and some myosin, which produces a ratio of thin-to-thick filaments of about 16:1 (in striated muscles the ratio is 2:1). Unlike striated muscles, in which the thick filaments are short and stacked between Z discs in sarcomeres, the myosin filaments in smooth muscle cells are quite long (fig. 12.32).

The long length of myosin filaments and the fact that they are not organized into sarcomeres may be advantageous for the function of smooth muscles. This is because smooth muscles must be able to contract even when greatly stretched—in the urinary bladder, for example, the smooth muscle cells may be stretched up to two and a half times

**Figure 12.32.** An electron micrograph of the thick and thin filaments of smooth muscle. A longitudinal section of a complete long myosin filament is shown between the arrows (32,000×).

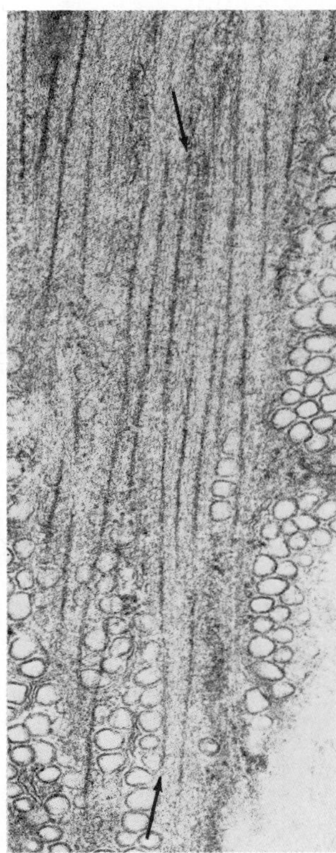

their resting length. The smooth muscle cells of the uterus may be stretched up to eight times their original length by the end of pregnancy. Striated muscles, because of their structure, lose their ability to contract when the sarcomeres are stretched to the point where actin and myosin no longer overlap.

As in striated muscles, the contraction of smooth muscles is triggered by a sharp rise in the $Ca^{++}$ concentration within the cytoplasm of the muscle cells. Unlike skeletal muscles, the sarcoplasmic reticulum of smooth muscles is poorly developed, and $Ca^{++}$ released from this organelle may account for only the initial phase of smooth muscle contraction. Extracellular $Ca^{++}$, diffusing into the smooth muscle cell through its plasma membrane, is responsible for sustained contractions. This $Ca^{++}$ enters primarily through voltage-sensitive $Ca^{++}$ gates. The opening of these gates is graded by the amount of depolarization; the greater the depolarization, the more $Ca^{++}$ will enter the cell and the stronger will be the smooth muscle contraction.

Drugs such as *verapamil* and related compounds are **calcium channel blockers,** or calcium antagonists, which act directly on smooth muscle cells of blood vessels to cause vasodilation. This effect is useful in the treatment of such conditions as angina pectoris (pain resulting from heart disease) and hypertension (high blood pressure). Vasodilation results from relaxation of the smooth muscle cells within the walls of blood vessels. This relaxation is apparently produced by the fact that these drugs inhibit the opening of $Ca^{++}$ channels in the smooth muscle cell membrane.

The events that follow the entry of $Ca^{++}$ into the cytoplasm are somewhat different in smooth muscles than in striated muscles. In striated muscles $Ca^{++}$ combines with troponin. Troponin, however, is not present in smooth muscle cells. In smooth muscles, $Ca^{++}$ combines with a protein in the cytoplasm called **calmodulin**, which is structurally similar to troponin and which was previously discussed in relation to the function of $Ca^{++}$ as a second messenger in hormone action (chapter 11). The calmodulin-$Ca^{++}$ complex thus formed combines with and activates an enzyme called *myosin light chain kinase,* which catalyzes the phosphorylation of (addition of phosphate groups to) the myosin cross-bridges. Unlike the situation in striated muscles, the cross-bridges in smooth muscle cells must be phosphorylated before they can bond to actin.

In this way, the concentration of $Ca^{++}$ in the smooth muscle cell cytoplasm determines how many cross-bridges will combine with actin, and thus determines the strength of contraction. The concentration of $Ca^{++}$ is in turn regulated by the degree of depolarization. Unlike the situation in striated muscle cells, which produce all-or-none action potentials and thus usually produce all-or-none contractions, smooth muscle cells may produce graded depolarizations and contractions without producing action potentials. (This is discussed in more detail in conjunction with intestinal contractions in chapter 17.) Indeed, only these graded depolarizations are conducted from cell-to-cell in many smooth muscles.

In addition to being graded, the contractions of smooth muscle cells are slow and sustained. The slowness of contraction is related to the fact that myosin ATPase in smooth muscle is slower in its action (splitting ATP for the cross-bridge cycle) than it is in striated muscle. The sustained nature of smooth muscle contraction is explained by the theory that cross-bridges in smooth muscles can enter a *latch* state.

The latch state allows smooth muscle to maintain its contraction in a very energy-efficient manner. This ability is obviously important for smooth muscles, which encircle

**Table 12.11** Some comparisons of skeletal, cardiac, and smooth muscles

| Skeletal Muscle | Cardiac Muscle | Smooth Muscle |
|---|---|---|
| Striated; actin and myosin arranged in sarcomeres | Striated; actin and myosin arranged in sarcomeres | Not striated; more actin than myosin; actin inserts into dense bodies and cell membrane |
| Well-developed sarcoplasmic reticulum and transverse tubules | Moderately developed sarcoplasmic reticulum and transverse tubules | Poorly developed sarcoplasmic reticulum; no transverse tubules |
| Contains troponin in the thin filaments | Contains troponin in the thin filaments | Contains a $Ca^{++}$ binding protein; may be located in thick filaments |
| $Ca^{++}$ released into cytoplasm from sarcoplasmic reticulum | $Ca^{++}$ enters cytoplasm from sarcoplasmic reticulum and extracellular fluid | $Ca^{++}$ enters cytoplasm from extracellular fluid, sarcoplasmic reticulum, and perhaps mitochondria |
| Cannot contract without nerve stimulation; denervation results in muscle atrophy | Can contract without nerve stimulation; action potentials originate in pacemaker cells of heart | Maintains tone in absence of nerve stimulation; visceral smooth muscle produces pacemaker potentials; denervation results in hypersensitivity to stimulation |
| Muscle fibers stimulated independently; no gap junctions | Gap junctions present as intercalated discs | Gap junctions present in most smooth muscles |

the walls of hollow organs and must sustain contractions for very long time periods. The mechanisms by which the latch state are produced, however, are not completely understood. One current hypothesis is that latched cross-bridges are produced when the phosphate group is removed from some previously phosphorylated cross-bridges which are already attached to actin. The dephosphorylated cross-bridges detach from actin at a much slower rate than do the phosphorylated cross-bridges, resulting in the association of the cross-bridges with the thin filaments for a longer period of time. This latch mechanism allows the smooth muscles to sustain a contraction while hydrolyzing much less ATP than would otherwise be required.

Despite their differences, it is currently believed that both smooth muscles and striated muscles contract by means of a *sliding filament* mechanism. The sliding filament mechanism in smooth muscle, however, is not as well understood as it is in striated muscle. Some comparisons of skeletal muscle, cardiac muscle, and smooth muscle are given in table 12.11.

***Single-Unit and Multi-Unit Smooth Muscles.*** Smooth muscles are often grouped into two functional categories: **single-unit** and **multi-unit.** Single-unit smooth muscles have numerous gap junctions (electrical synapses) between adjacent cells that weld them together electrically; they thus behave as a single unit, much like cardiac muscle (previously discussed). Multi-unit smooth muscles have few, if any, gap junctions; the individual cells must thus be stimulated separately by impulses through nerve endings. This is similar to the numerous motor units required for the control of skeletal muscles.

Single-unit smooth muscles display *pacemaker* activity in which certain cells stimulate others in the mass. This is similar to the situation in cardiac muscle. Single-unit smooth muscles also display intrinsic, or *myogenic,* electrical activity and contraction in response to stretch. For example, the stretch induced by an increase in the volume of a small artery or a section of the gastrointestinal tract can stimulate myogenic contraction. Such contraction does not require stimulation by autonomic nerves. Contraction of multi-unit smooth muscles, in contrast, requires nerve stimulation. Comparisons of single-unit and multi-unit smooth muscles are given in table 12.12.

***Autonomic Innervation of Smooth Muscles.*** There are significant differences between the neural control of skeletal muscles and that of smooth muscles. A skeletal muscle fiber has only one junction with a somatic nerve fiber, and the receptors for the neurotransmitter are located only at the neuromuscular junction. In contrast, the entire surface of smooth muscle cells contains neurotransmitter receptor proteins. Neurotransmitter molecules are released along a stretch of an autonomic nerve fiber that is located some distance from the smooth muscle cells. The regions of the autonomic fiber that release transmitters appear as bulges, or *varicosities,* and the neurotransmitters released from these varicosities stimulate a number of smooth muscle cells. Since there are a number of varicosities along a stretch of an autonomic nerve ending, they form synapses "in passing"—or *synapses en passant*—with the smooth muscle cells.

---

1. *Explain how cardiac muscle differs from skeletal muscle in its structure and regulation of contraction.*
2. *Contrast the structure of a smooth muscle cell with that of a skeletal muscle fiber, and explain the advantages of each type of structure.*
3. *Explain how depolarization of a smooth muscle cell results in contraction, and explain why smooth muscle contractions are slow and sustained.*
4. *Distinguish between single-unit and multi-unit smooth muscles.*

---

**Table 12.12** Some comparisons between single-unit and multi-unit smooth muscles

| | Single-Unit Muscle | Multi-Unit Muscle |
|---|---|---|
| Location | Gastrointestinal tract; uterus; ureter; small arteries (arterioles) | Arrector pili muscles of hair follicles; ciliary muscle (attached to lens); iris; vas deferens; large arteries |
| Origin of electrical activity | Spontaneous activity by pacemakers (myogenic) | Not spontaneously active; potentials are neurogenic |
| Type of potentials | Action potentials | Graded depolarizations |
| Response to stretch | By contraction; not dependent on nerve stimulation | No inherent response |
| Presence of gap junctions | Numerous; join all cells together electrically | Few (if any) |
| Type of contraction | Slow and sustained | Slow and sustained |

## SUMMARY

### Structure and Actions of Skeletal Muscles p. 318

I. Skeletal muscles are attached by tendons to bones.
   A. Skeletal muscles are composed of separate cells or fibers that are attached in parallel to the tendons.
   B. Skeletal muscle fibers are striated.
      1. The dark striations are called A bands and the light regions are called I bands.
      2. Z lines are located in the middle of each I band.

II. Muscles *in vitro* can exhibit twitch, summation, and tetanus.
   A. The rapid contraction and relaxation of muscle fibers is called a twitch.
   B. A whole muscle also produces a twitch in response to a single electrical pulse *in vitro*.
      1. The stronger the electric shock, the stronger the muscle twitch—whole muscle can give a graded contraction.
      2. The graded contraction of whole muscles is due to different numbers of fibers participating in the contraction.
   C. The summation of fiber twitches can occur so rapidly that the muscle produces a smooth, sustained contraction known as tetanus.

III. The contraction of muscle fibers *in vivo* is stimulated by somatic motor neurons.
   A. Each somatic motor axon branches to innervate a number of muscle fibers.
   B. The motor neuron and the muscle fibers it innervates is called a motor unit.
      1. When a muscle is composed of many motor units (such as in the hand), there is fine control of muscle contraction.
      2. The large muscles of the leg have relatively few motor units, which are correspondingly large.
      3. Sustained contractions are produced by the asynchronous stimulation of different motor units.

IV. When a muscle exerts tension without shortening, the contraction is termed isometric; when shortening does occur, the contraction is isotonic.

### Mechanisms of Contraction p. 325

I. Skeletal muscle cells, or fibers, contain structures called myofibrils.
   A. Each myofibril is striated with dark (A) and light (I) bands; there are Z lines in the middle of each I band.
   B. The A bands contain thick filaments composed of myosin.
      1. The edges of each A band also contain thin filaments overlapped with the thick filaments.
      2. The central regions of the A bands contain only thick filaments—these regions are the H bands.
   C. The I bands contain only thin filaments, which are composed primarily of the protein called actin.
   D. Thin filaments are formed of globular actin subunits known as G-actin. A protein known as tropomyosin is also located periodically in the thin filaments; another protein—troponin—is attached to the tropomyosin.

II. Myosin cross-bridges extend out from the thick filaments to the thin filaments.
   A. At rest, the cross-bridges are not attached to actin.
      1. The cross-bridge heads function as ATPase enzymes.
      2. ATP is split into ADP and $P_i$, activating the cross-bridge.
   B. When the activated cross-bridges attach to actin, they undergo a power stroke and in the process release ADP and $P_i$.
   C. At the end of a power stroke, the cross-bridge bonds to a new ATP.
      1. This allows the cross-bridge to detach from actin and repeat the cycle.
      2. Rigor mortis is caused by the inability of cross-bridges to detach from actin due to a lack of ATP.

III. The activity of the cross-bridges causes the thin filaments to slide toward the centers of the sarcomeres.
   A. The filaments slide—they do not shorten—during muscle contraction.
   B. The lengths of the H and I bands decrease, whereas the A bands stay the same length during contraction.

IV. When a muscle is at rest, the $Ca^{++}$ concentration of the sarcoplasm is very low and cross-bridges are prevented from attaching to actin.
   A. The $Ca^{++}$ is actively transported into the sarcoplasmic reticulum.
   B. The sarcoplasmic reticulum is a modified endoplasmic reticulum that surrounds the myofibrils.

V. Action potentials are conducted by transverse tubules into the muscle fiber.
   A. Transverse tubules are invaginations of the cell membrane that almost touch the sarcoplasmic reticulum.
   B. Action potentials in the transverse tubules stimulate the release of $Ca^{++}$ from the sarcoplasmic reticulum.

VI. When action potentials cease, $Ca^{++}$ is removed from the sarcoplasm and stored in the sarcoplasmic reticulum.

## Neural Control of Skeletal Muscles p. 334

I. The somatic motor neurons that innervate the muscles are called the lower motor neurons.
   A. Alpha motoneurons innervate the ordinary, or extrafusal, muscle fibers—these are the fibers that produce muscle shortening during contraction.
   B. Gamma motoneurons innervate the intrafusal fibers of the muscle spindles.

II. Muscle spindles are length detectors in the muscle.
   A. Spindles consist of several intrafusal fibers wrapped together and in parallel with the extrafusal fibers.
   B. Stretching of the muscle stretches the spindles; this excites sensory endings in the spindle apparatus.
      1. Impulses in the sensory neurons travel into the spinal cord in the dorsal roots of spinal nerves.
      2. The sensory neuron makes a synapse directly with an alpha motoneuron within the spinal cord—this produces a monosynaptic reflex.
      3. The alpha motoneuron stimulates the extrafusal muscle fibers to contract, thus relieving the stretch—this is called the stretch reflex.
   C. The activity of gamma motoneurons tightens the spindles, thus making them more sensitive to stretch and better able to monitor the length of the muscle, even during muscle shortening.

III. The Golgi tendon organs monitor the tension that the muscle exerts on its tendons.
   A. As the tension increases, sensory neurons from Golgi tendon organs inhibit the activity of alpha motoneurons.
   B. This is a disynaptic reflex, because the sensory neurons synapse with interneurons, which in turn make inhibitory synapses with motoneurons.

IV. A crossed-extensor reflex occurs when a foot steps on a tack.
   A. Sensory input from the injured foot causes stimulation of flexor muscles and inhibition of the antagonistic extensor muscles.
   B. The sensory input also crosses the spinal cord to cause stimulation of extensor and inhibition of flexor muscles in the contralateral leg.

V. Most of the fibers of descending tracts synapse with spinal interneurons, which in turn synapse with the lower motor neurons.
   A. Alpha and gamma motoneurons are usually stimulated at the same time, or co-activated.
   B. The stimulation of gamma motoneurons keeps the muscle spindles under tension and sensitive to stretch.
   C. Higher motor neurons, primarily in the basal ganglia, also exert inhibitory effects on gamma motoneurons.
      1. Damage to the spinal cord at first causes muscles to become flaccid as stimulation by higher motor neurons is blocked.
      2. After a period of time, the gamma motoneurons become hyperexcitable, since they are freed from higher motor neuron inhibition and cause muscle spasticity and clonus.

VI. Neurons in the brain that affect the lower motor neurons are called higher motor neurons.
   A. The fibers of neurons in the precentral gyrus, or motor cortex, descend to the lower motor neurons as the lateral and ventral corticospinal tracts.
      1. Most of these fibers cross to the contralateral side in the brain stem, forming structures called the pyramids; this system is therefore called the pyramidal system.
      2. The left side of the brain thus controls the musculature on the right side, and vice versa.
      3. The pyramidal system is responsible for fine control of voluntary movements.
   B. Other descending motor tracts are part of the extrapyramidal system.
      1. The neurons of the extrapyramidal system make numerous synapses in different areas of the brain, including the midbrain, brain stem, basal ganglia, and cerebellum.
      2. Damage to the cerebellum produces intention tremor.
      3. The degeneration of fibers in the basal ganglia that use dopamine as a transmitter produces Parkinson's disease.

### Skeletal Muscle Types and Energy Usage p. 343

I. Aerobic cell respiration is ultimately required for the production of ATP needed for cross-bridge activity.
  A. New ATP can be quickly produced, however, from the combination of ADP with phosphate derived from phosphocreatine.
    1. The phosphocreatine represents a ready reserve of high-energy phosphate during sustained muscle contractions.
    2. Phosphocreatine is produced at rest from creatine, and phosphate that is derived from ATP.
  B. There are three types of muscle fibers.
    1. Slow-twitch red fibers are adapted for aerobic respiration and are resistant to fatigue.
    2. Fast-twitch white fibers are adapted for anaerobic respiration.
    3. Intermediate fibers are fast-twitch but adapted for aerobic respiration.

II. Muscle fatigue may be caused by a number of mechanisms.
  A. Fatigue during sustained maximal contraction may be produced by the accumulation of extracellular $K^+$ as a result of high levels of nerve activity.
  B. Fatigue during rhythmic, moderate exercise is primarily a result of anaerobic respiration by fast-twitch fibers.
    1. The production of lactic acid lowers the intracellular pH, which inhibits glycolysis and decreases ATP concentrations.
    2. Decreased ATP inhibits excitation-contraction coupling, possibly due to a cellular loss of $Ca^{++}$.

III. Physical training affects the characteristics of the muscle fibers.
  A. Endurance training increases the aerobic capacity of all muscle fiber types, so that their reliance on anaerobic respiration—and thus their susceptibility to fatigue—is decreased.
  B. Strength training causes hypertrophy of the muscle fibers due to an increase in the size and number of myofibrils.

### Cardiac and Smooth Muscle p. 347

I. Cardiac muscle is striated and contains sarcomeres.
  A. Unlike skeletal muscle, action potentials in the heart originate in myocardial cells; stimulation by neurons is not required.
  B. Also unlike the situation in skeletal muscles, action potentials can cross from one myocardial cell to another.

II. Smooth muscle cells lack sarcomeres and are not striated.
  A. Smooth muscle cells contain myosin and actin, but these are not arranged in sarcomeres.
  B. Myosin filaments are very long, and because of this, smooth muscle cells can contract even when they are greatly stretched.
  C. When stimulated by graded depolarizations, $Ca^{++}$ enters smooth muscle cells and combines with calmodulin; this activates an enzyme that phosphorylates myosin cross-bridges.
  D. Unlike the situation in striated muscles, phosphorylation of cross-bridges is required for their bonding to actin.
  E. Depending upon their neural regulation, smooth muscles can be classified as single-unit or multi-unit.

## Review Activities

### Objective Questions

1. A graded whole muscle contraction is produced *in vivo* primarily by variations in
   (a) the strength of the fiber's contraction
   (b) the number of fibers that are contracting
   (c) both of the above
   (d) neither of the above
2. The series elastic component of muscle contraction is responsible for
   (a) increased muscle shortening to successive twitches
   (b) time delay between contraction and shortening
   (c) the lengthening of muscle after contraction has ceased
   (d) all of the above
3. Which of the following muscles have motor units with the highest innervation ratio?
   (a) leg muscles
   (b) arm muscles
   (c) muscles that move the fingers
   (d) muscles of the torso
4. The stimulation of gamma motoneurons produces
   (a) isotonic contraction of intrafusal fibers
   (b) isometric contraction of intrafusal fibers
   (c) either isotonic or isometric contraction of intrafusal fibers
   (d) contraction of extrafusal fibers
5. In a single reflex arc involved in the knee jerk reflex, how many synapses are activated within the spinal cord?
   (a) thousands
   (b) hundreds
   (c) dozens
   (d) two
   (e) one
6. Spastic paralysis may occur when there is damage to
   (a) the lower motor neurons
   (b) the higher motor neurons
   (c) either the lower or the higher motor neurons
7. When a skeletal muscle shortens during contraction, which of the following statements is *false?*
   (a) The A bands shorten.
   (b) The H bands shorten.
   (c) The I bands shorten.
   (d) The sarcomeres shorten.
8. Electrical excitation of a muscle fiber *most directly* causes the
   (a) movement of tropomyosin
   (b) attachment of the cross-bridges to actin

(c) release of $Ca^{++}$ from the sarcoplasmic reticulum
(d) splitting of ATP

9. The energy for muscle contraction is *most directly* obtained from
   (a) phosphocreatine
   (b) ATP
   (c) anaerobic respiration
   (d) aerobic respiration
10. Which of the following statements about cross-bridges is *false*?
   (a) They are composed of myosin.
   (b) They bond to ATP after they detach from actin.
   (c) They contain an ATPase.
   (d) They split ATP before they attach to actin.
11. When a muscle is stimulated to contract, $Ca^{++}$ bonds to
   (a) myosin
   (b) tropomyosin
   (c) actin
   (d) troponin
12. Which of the following statements about muscle fatigue is *false*?
   (a) It may result when ATP is no longer available for the cross-bridge cycle.
   (b) It may be caused by a loss of muscle cell $Ca^{++}$.
   (c) It may be caused by the accumulation of extracellular $K^{+}$.
   (d) It may be a result of lactic acid production.
13. Which of the following types of muscle cells are *not* capable of spontaneous depolarization?
   (a) single-unit smooth muscle
   (b) multi-unit smooth muscle
   (c) cardiac muscle
   (d) skeletal muscle
   (e) both *b* and *d*
   (f) both *a* and *c*
14. Which of the following types of muscle is striated and contains gap junctions?
   (a) single-unit smooth muscle
   (b) multi-unit smooth muscle
   (c) cardiac muscle
   (d) skeletal muscle
15. In an isotonic muscle contraction
   (a) the length of the muscle remains constant
   (b) the muscle tension remains constant
   (c) both muscle length and tension are changed
   (d) movement of bones does not occur

## Essay Questions

1. Using the concept of motor units, explain how skeletal muscles *in vivo* produce graded and sustained contractions.
2. Describe how an isometric contraction can be converted into an isotonic contraction, using the concepts of motor unit recruitment and the series elastic component of muscles.
3. Why don't the cross-bridges attach to the thin filaments when a muscle is relaxed? Trace the sequence of events that allows the cross-bridges to attach to the thin filaments when a muscle is stimulated by a nerve.
4. Using the sliding filament theory of contraction, explain why the contraction strength of a muscle is maximal at a particular muscle length.
5. How is muscle tone maintained? Explain why muscle tone is first decreased and then increased when descending motor tracts are damaged.
6. Explain the role of ATP in muscle contraction and in muscle relaxation.
7. Why are all the muscle fibers of a given motor unit all of the same type? Why are smaller motor units and slow-twitch muscle fibers used more frequently than larger motor units and fast-twitch fibers?
8. Explain how endurance training raises the level of exercise that can be performed before muscle fatigue occurs.
9. Compare the mechanism of excitation-coupling in striated muscle with that of smooth muscle.

## SELECTED READINGS


Astrand, P. O., and K. Rodahl. 1977. *Textbook of work physiology: physiological basis of exercise.* New York: McGraw-Hill Book Company.

Bourne, G. H., ed. 1973. *The structure and function of muscle.* 2d ed. 4 vols. New York: Academic Press.

Cohen, C. November 1975. The protein switch of muscle contraction. *Scientific American.*

Drachman, D. B. 1978. Myasthenia gravis. *New England Journal of Medicine* 298:136.

Eisenberg, E., and L. E. Greene. 1980. The relation of muscle biochemistry to muscle physiology. *Annual Review of Physiology* 42:293.

Emont-Denand, F., C. C. Hunt, and Y. Laporte. 1988. How muscle spindles signal changes in muscle length. *News in Physiological Sciences* 3:105.

Felig, P., and J. Wahren. 1975. Fuel homeostasis in exercise. *New England Journal of Medicine* 293:1078.

Grinnel, A. D., and M. A. B. Brazier, eds. 1981. *Regulation of muscle contraction: excitation-contraction coupling.* New York: Academic Press.

Hoyle, G. April 1970. How is muscle turned on and off? *Scientific American.*

Huxley, H. E. 1969. The mechanism of muscular contraction. *Science* 164:1356.

Margaria, R. March 1972. The sources of muscular energy. *Scientific American.*

Merton, P. A. May 1972. How we control the contraction of our muscles. *Scientific American.*

Murphy, R. A. 1988. Muscle cells of hollow organs. *News in Physiological Sciences* 3:124.

Murray, J. H., and A. Weber. February 1974. The cooperative action of muscle proteins. *Scientific American.*

Nadel, E. R. 1985. Physiological adaptations to aerobic training. *American Scientist* 73:334.

Porter, K. R., and C. Franzini-Armstrong. March 1965. The sarcoplasmic reticulum. *Scientific American.*

Shepard, R. J. 1982. *Physiology and biochemistry of exercise.* New York: Praeger Publishers.

Stein, R. B., ed. 1980. *Nerve and muscle: membranes, cells, and systems.* New York: Plenum Publishing Corp.

# 13

# Heart and Circulation

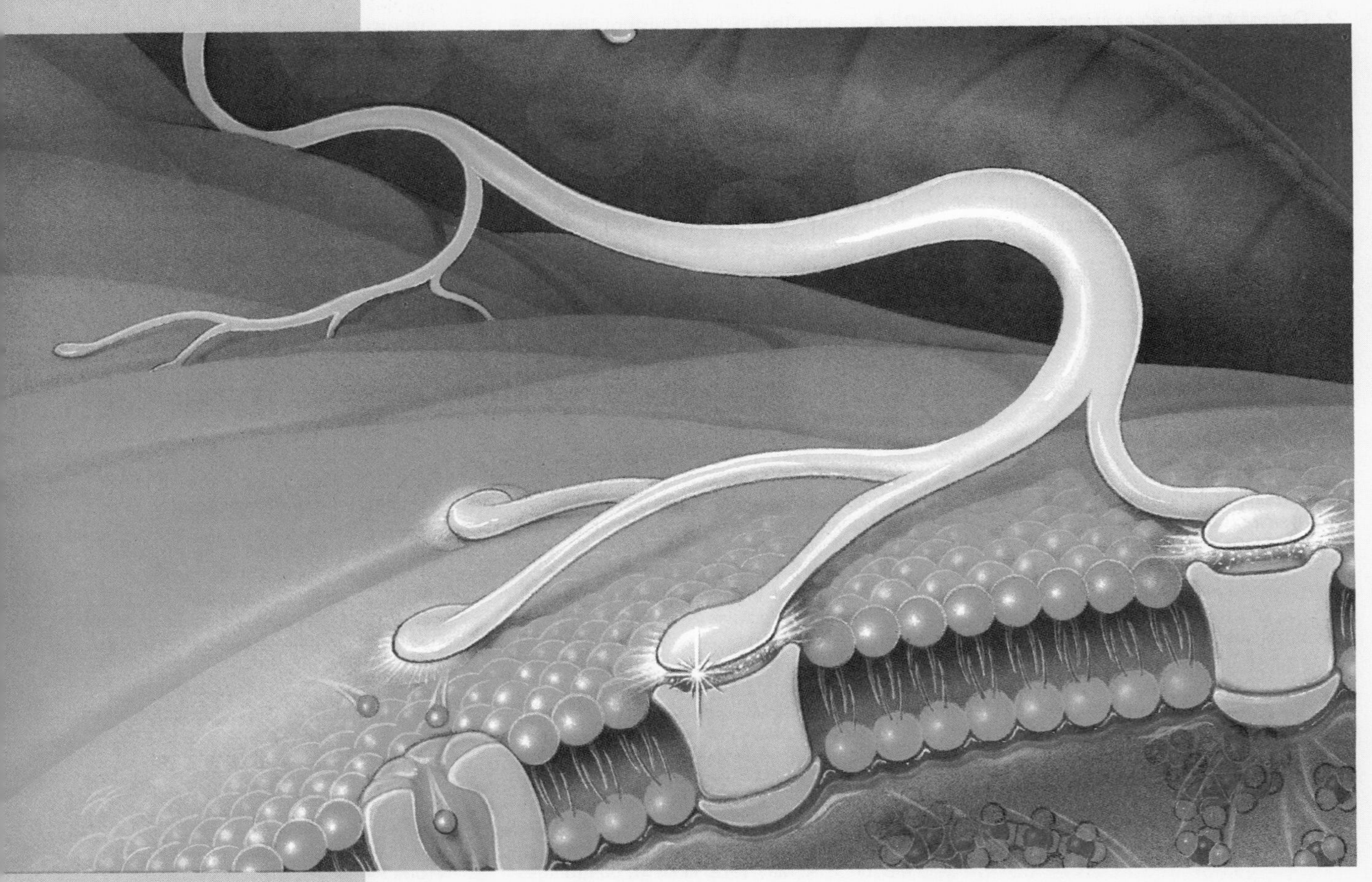

## Objectives

By studying this chapter, you should be able to

1. describe the composition of blood plasma and the classification of the formed elements of the blood
2. explain the meaning and significance of the blood types
3. explain how a blood clot is formed and how it is ultimately destroyed
4. describe the path of the blood through the heart and the function of the atrioventricular and semilunar valves
5. describe the structures and pathways of the pulmonary and systemic circulation
6. describe the structures and pathways of electrical impulse conduction in the heart
7. explain the electrical activity in the sinoatrial node and why this tissue functions as the normal pacemaker of the heart
8. relate the time involved in the production of an action potential to the time involved in the contraction of myocardial cells and explain the significance of this relationship
9. describe the pressure changes that occur in the ventricles during the cardiac cycle and relate these changes to the action of the valves and the flow of blood
10. explain the origin of the heart sounds and describe when these sounds are produced in the cardiac cycle
11. explain the cause of each wave in an electrocardiogram and relate these waves to other events in the cardiac cycle
12. describe some common arrhythmias that can be detected with an ECG
13. compare the structure of an artery and vein and explain how the structure of each type of vessel relates to its function
14. describe the structure of capillaries and explain how this structure aids the function of these vessels
15. explain the nature and significance of atherosclerosis and describe how this condition may be produced
16. describe the components and functions of the lymphatic system

## Outline

## Functions and Components of the Circulatory System

Blood serves numerous functions, including the transport of respiratory gases, nutritive molecules, metabolic wastes, and hormones. Blood is transported through the body in a closed system of vessels from and to the heart. The heart consists of four chambers; two of these receive blood from veins, and two pump blood into arteries. The arterial and venous systems are continuous by way of tiny vessels known as blood capillaries.

A unicellular organism can provide for its own maintenance and continuity by performing the wide variety of functions needed for life. By contrast, the complex human body is composed of trillions of specialized cells that demonstrate a division of labor. Cells of a multicellular organism depend on one another for the very basics of their existence. The majority of the cells of the body are firmly implanted in tissues and are incapable of procuring food and oxygen or even moving away from their own wastes. Therefore, a highly specialized and effective means of transporting materials within the body is needed.

The blood serves this transportation function. An estimated 60,000 miles of vessels throughout the body of an adult ensure that continued sustenance reaches each of the trillions of living cells. The blood can also, however, serve to transport disease-causing viruses, bacteria, and their toxins. To ensure against this, the circulatory system has protective mechanisms: the white blood cells and lymphatic system. In order to perform its various functions, the circulatory system works together with the respiratory, urinary, digestive, endocrine, and integumentary systems in maintaining homeostasis.

### Functions of the Circulatory System

The functions of the circulatory system can be divided into three areas: transportation, regulation, and protection.

**Transportation.** All of the substances essential for cellular metabolism are transported by the circulatory system. These substances can be categorized as follows:

a. Respiratory. Red blood cells, called **erythrocytes,** transport oxygen to the tissue cells. In the lungs, oxygen from the inhaled air attaches to hemoglobin molecules within the erythrocytes and is transported to the cells for aerobic respiration. Carbon dioxide produced by cell respiration is carried by the blood to the lungs for elimination in the exhaled air.
b. Nutritive. The digestive system is responsible for the mechanical and chemical breakdown of food so that it can be absorbed through the intestinal wall and into the blood vessels of the circulatory system. The blood then carries these absorbed products of digestion through the liver and to the cells of the body.
c. Excretory. Metabolic wastes, excessive water and ions, as well as other molecules in plasma (the fluid portion of blood) are filtered through the capillaries of the kidneys and excreted in urine.

**Regulation.** The blood carries hormones and other regulatory molecules from their site of origin to distant target tissues.

**Protection.** The circulatory system protects against injury and foreign microbes or toxins introduced into the body. The clotting mechanism protects against blood loss when vessels are damaged, and **leukocytes** (white blood cells) provide immunity from many disease-causing agents.

### Major Components of the Circulatory System

The circulatory system is frequently divided into the **cardiovascular system,** which consists of the heart and blood vessels, and the **lymphatic system,** which consists of lymph vessels and lymph nodes.

The **heart** is a four-chambered, double pump. Its pumping action creates the pressure head needed to push blood in the vessels to the lungs and body cells. At rest, the heart of an adult pumps about 5 liters of blood per minute. It takes about one minute for blood to be circulated to the most distal extremity and back to the heart.

**Blood vessels** form a tubular network that permits blood to flow from the heart to all the living cells of the body and then back to the heart. *Arteries* carry blood away from the heart, whereas *veins* return blood to the heart. Arteries and veins are continuous with each other through smaller blood vessels.

Arteries branch extensively to form a "tree" of ever smaller vessels. Those that are microscopic in diameter are called *arterioles.* Conversely, microscopic-sized veins, called *venules,* deliver blood into ever larger vessels that empty into the large veins. Blood passes from the arterial to the venous system in *capillaries,* which are the thinnest and most numerous blood vessels. All exchanges of fluid, nutrients, and wastes between the blood and tissues occur across the walls of capillaries.

As blood plasma passes through capillaries, the hydrostatic pressure of the blood forces some of this fluid out of the capillary walls. Fluid derived from plasma that passes out of capillary walls into the surrounding tissues is called *tissue fluid,* or *interstitial fluid.* Some of this fluid returns directly to capillaries, and some enters into **lymph vessels** located in the connective tissues around the blood vessels. Fluid in lymph vessels is called *lymph.* This fluid is returned to the venous blood at particular sites. **Lymph nodes,** positioned along the way, cleanse the lymph

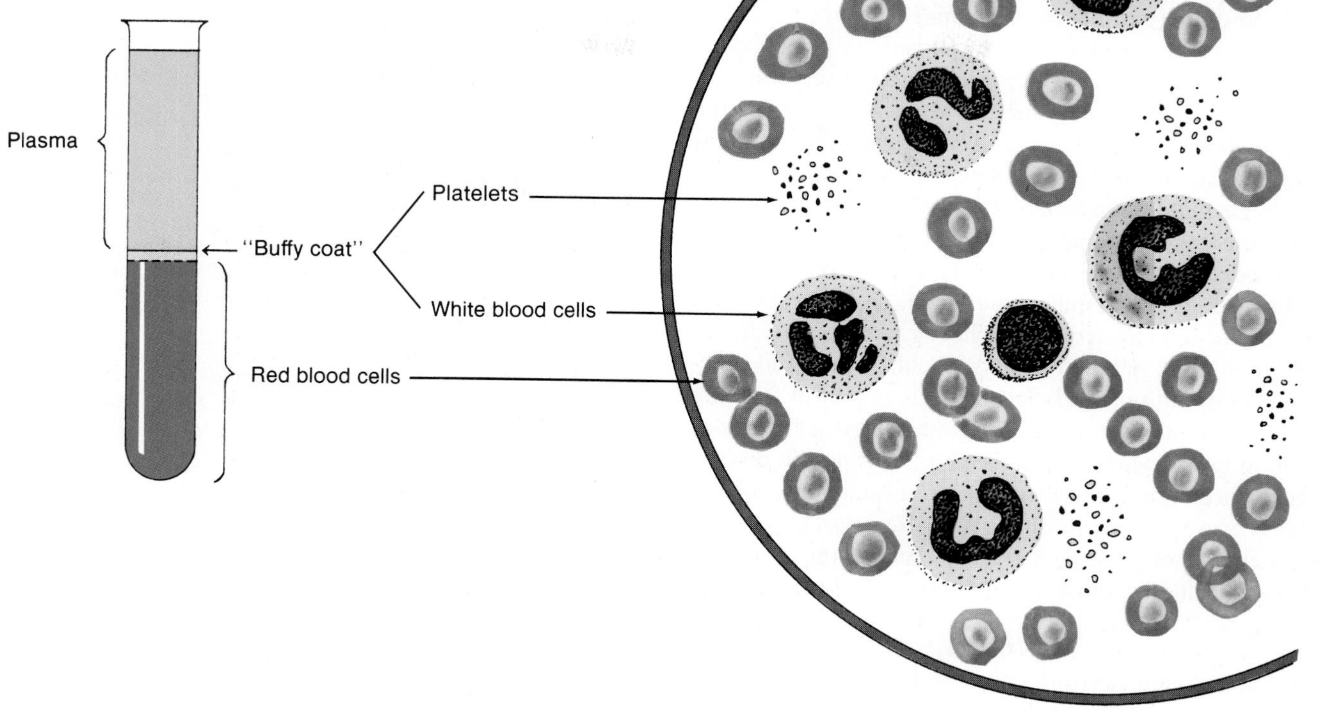

**Figure 13.1.** Blood cells become packed at the bottom of the test tube when whole blood is centrifuged, leaving the fluid plasma at the top of the tube. Red blood cells are the most abundant of the blood cells—white blood cells and platelets form only a thin, light-colored "buffy coat" at the interface between the packed red blood cells and the plasma.



prior to its return to the venous blood. The lymphatic system is thus considered a part of the circulatory system and is discussed at the end of this chapter.

1. *Name the components of the circulatory system that function in oxygen transport, in the transport of nutrients from the digestive system, and in protection.*
2. *Define and describe the function of arteries, veins, and capillaries.*
3. *Define the terms* interstitial fluid *and* lymph. *Describe their relationships to blood plasma.*

## Composition of the Blood

Blood consists of formed elements that are suspended and carried in the plasma. These formed elements and their major functions include erythrocytes (oxygen transport), leukocytes (immune defense), and platelets (blood clotting). Plasma contains different types of proteins and many water-soluble molecules.

The average-sized adult has about 5 liters of blood, which constitute about 8% of the total body weight. Blood leaving the heart is referred to as *arterial blood.* Arterial blood, with the exception of that going to the lungs, is bright red in color due to a high concentration of oxyhemoglobin (the combination of oxygen with hemoglobin) in the erythrocytes. *Venous blood* is blood returning to the heart and, except for the venous blood from the lungs, has a darker color (due to hemoglobin that is no longer combined with oxygen). Blood has a viscosity that ranges between 4.5 and 5.5. This means that it flows thicker than water, which has a viscosity of 1. Blood has a pH range of 7.35 to 7.45 and a temperature within the thorax of the body of about 38° C (100.4° F).

Blood is composed of a cellular portion, called **formed elements,** and a fluid portion, called **plasma.** When a blood sample is centrifuged, the heavier formed elements are packed into the bottom of the tube, leaving plasma at the top (fig. 13.1). The formed elements constitute approximately 45% of the total blood volume (the *hematocrit*), and the plasma accounts for the remaining 55%.

### Plasma

Plasma is a straw-colored liquid consisting of water and dissolved solutes. Sodium ion is the major solute of the plasma in terms of its concentration (and therefore contributes more than other solutes to the total osmolality of plasma). In addition to $Na^+$, plasma contains many other salts and ions, as well as organic molecules such as metabolites, hormones, enzymes, antibodies, and other proteins. The values of some of these constituents of plasma are shown in table 13.1.

***Plasma Proteins.*** Plasma proteins constitute 7%–9% of the plasma. The three types of proteins are albumins, globulins, and fibrinogen. **Albumins** account for most (60%–80%) of the plasma proteins and are the smallest in size. They are produced by the liver and serve to provide the osmotic pressure needed to draw water from the surrounding tissue fluid into the capillaries. This action is needed to maintain blood volume and pressure. **Globulins** are divided into three subtypes: **alpha globulins, beta globulins,** and **gamma globulins.** The alpha and beta globulins are produced by the liver and function to transport lipids and fat-soluble vitamins in the blood. Gamma globulins are antibodies produced by lymphocytes (one of the formed elements found in blood and lymphoid tissues) and function in immunity. **Fibrinogen,** which accounts for only about 4% of the total plasma proteins, is an important clotting factor produced by the liver. During the process of clot formation (described later in this chapter), fibrinogen is converted into insoluble threads of *fibrin.* The fluid from clotted blood, which is called **serum,** thus does not contain fibrinogen but is otherwise identical to plasma.

## The Formed Elements of Blood

The formed elements of blood include two types of blood cells: **erythrocytes,** or red blood cells, and **leukocytes,** or white blood cells. Erythrocytes are by far the most numerous of these two types: a cubic millimeter of blood contains 5.1 million to 5.8 million erythrocytes in males and 4.3 million to 5.2 million erythrocytes in females. The same volume of blood, in contrast, contains only 5,000 to 9,000 leukocytes.

***Erythrocytes.*** Erythrocytes are flattened, biconcave discs, about 7 μm in diameter and 2.2 μm thick. Their unique shape relates to their function of transporting oxygen and provides an increased surface area through which gas can diffuse (fig. 13.2). Erythrocytes lack a nucleus and mitochondria (they get energy from anaerobic respiration). Because of these deficiencies, erythrocytes have a circulating life span of only about 120 days before they are destroyed by phagocytic cells in the liver, spleen, and bone marrow.

Each erythrocyte contains approximately 280 million *hemoglobin* molecules, which give blood its red color. Each hemoglobin molecule consists of a protein, called globin, and an iron-containing pigment, called heme. The iron group of heme is able to combine with oxygen in the lungs and release oxygen in the tissues.

**Table 13.1** Representative normal plasma values

| Measurement | Normal Range |
|---|---|
| Blood volume | 80–85 ml/kg body weight |
| Blood osmolality | 280–296 mOsm |
| Blood pH | 7.35–7.45 |
| Enzymes | |
| Creatine phosphokinase (CPK) | Female: 10–79 U/L<br>Male: 17–148 U/L |
| Lactic dehydrogenase (LDH) | 45–90 U/L |
| Phosphatase (acid) | Female: 0.01–0.56 Sigma U/ml<br>Male: 0.13–0.63 Sigma U/ml |
| Hematology values | |
| Hematocrit | Female: 37%–48%<br>Male: 45%–52% |
| Hemoglobin | Female: 12–16 g/100 ml<br>Male: 13–18 g/100 ml |
| Red blood cell count | 4.2–5.9 million/mm$^3$ |
| White blood cell count | 4,300–10,880/mm$^3$ |
| Hormones | |
| Testosterone | Male: 300–1,100 ng/100 ml<br>Female: 25–90 ng/100 ml |
| Adrenocorticotropic hormone (ACTH) | 15–70 pg/ml |
| Growth hormone | Children: over 10 ng/ml<br>Adult male: below 5 ng/ml |
| Insulin | 6–26 μU/ml (fasting) |
| Ions | |
| Bicarbonate | 24–30 mmol/l |
| Calcium | 2.1–2.6 mmol/l |
| Chloride | 100–106 mmol/l |
| Potassium | 3.5–5.0 mmol/l |
| Sodium | 135–145 mmol/l |
| Organic molecules (other) | |
| Cholesterol | 120–220 mg/100 ml |
| Glucose | 70–110 mg/100 ml (fasting) |
| Lactic acid | 0.6–1.8 mmol/l |
| Protein (total) | 6.0–8.4 g/100 ml |
| Triglyceride | 40–150 mg/100 ml |
| Urea nitrogen | 8–25 mg/100 ml |
| Uric acid | 3–7 mg/100 ml |

Excerpted from material originally appearing in *The New England Journal of Medicine,* "Case Records of the Massachusetts General Hospital," Vol. 302, No. 1, pp. 37-48, January 3, 1980 and "Case Records of the Massachusetts General Hospital," Vol. 314, No. 1, pp. 39-49. 

Anemia is present when there is an abnormally low hemoglobin concentration and/or red blood cell count. The most common cause of this condition is a deficiency in iron **(iron-deficiency anemia),** which is an essential component of the hemoglobin molecule. In **pernicious anemia,** there is inadequate availability of vitamin $B_{12}$, which is needed for red blood cell production. This results from atrophy of the glandular mucosa of the stomach, which normally secretes a substance, called *intrinsic factor,* that is needed for absorption of vitamin $B_{12}$ obtained in the diet. **Aplastic anemia** is anemia due to destruction of the bone marrow, which may be caused by chemicals (including benzene and arsenic) and by X rays.

**Figure 13.2.** The structure of an erythrocyte.

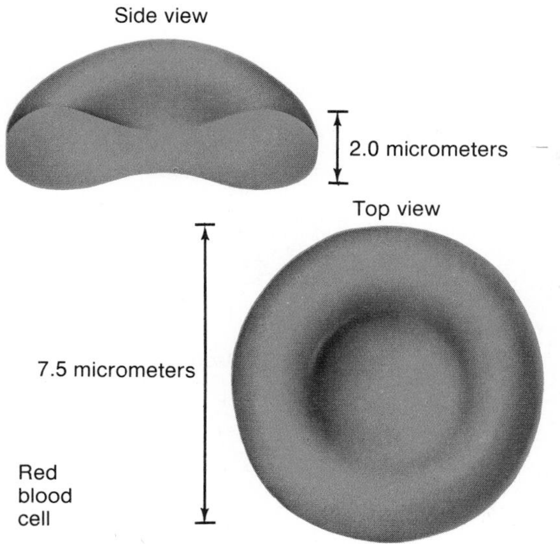


***Leukocytes.*** Leukocytes, or white blood cells, differ from erythrocytes in several ways. Leukocytes contain nuclei and mitochondria and can move in an amoeboid fashion (erythrocytes are not able to move independently). Because of their ameboid ability, leukocytes can squeeze through pores in capillary walls and get to a site of infection, whereas erythrocytes usually remain confined within blood vessels. The movement of leukocytes through capillary walls is called *diapedesis*.

Leukocytes, which are almost invisible under the microscope unless they are stained, are classified according to their stained appearance. Those leukocytes that have granules in their cytoplasm are called **granular leukocytes,** and those that do not are called **agranular** (or nongranular) **leukocytes.** The granular leukocytes are also identified by their oddly shaped nuclei, which in some cases are contorted into lobes separated by thin strands. The granular leukocytes are therefore known as *polymorphonuclear leukocytes* (abbreviated PMN).

The stain used to identify leukocytes is usually a mixture of a pink-to-red stain called eosin and a blue-to-purple stain called a "basic stain." Granular leukocytes with pink-staining granules are therefore called *eosinophils,* and those with blue-staining granules are called *basophils.* Those with granules that have little affinity for either stain are *neutrophils.* Neutrophils are the most abundant type of leukocyte, comprising 50%–70% of the leukocytes in the blood.

There are two types of agranular leukocytes: lymphocytes and monocytes. *Lymphocytes* are usually the second most numerous type of leukocyte; they are small cells with round nuclei and little cytoplasm. *Monocytes,* in contrast, are the largest of the leukocytes and generally have kidney- or horseshoe-shaped nuclei. In addition to these two cell types, there are smaller numbers of large lymphocytes, or *plasma cells,* that may be difficult to distinguish from monocytes. Plasma cells produce and secrete large amounts of antibodies.

***Platelets.*** Platelets, or **thrombocytes,** are the smallest of the formed elements and are actually fragments of large cells called *megakaryocytes,* found in bone marrow. (This is why the term *formed elements* is used rather than *blood cells* to describe erythrocytes, leukocytes, and platelets.) The fragments that enter the circulation as platelets lack nuclei but, like leukocytes, are capable of ameboid movement. The platelet count per cubic millimeter of blood is 130,000 to 360,000. Platelets survive about five to nine days and then are destroyed by the spleen and liver.

Platelets play an important role in blood clotting. They constitute the major portion of the mass of the clot, and phospholipids in their cell membranes serve to activate the clotting factors in plasma that result in threads of fibrin, which reinforce the platelet plug. Platelets that attach together in a blood clot also release a chemical called *serotonin,* which stimulates constriction of the blood vessels and thus reduces the flow of blood to the injured area.

The appearance of the formed elements of the blood is shown in figure 13.3, and a summary of the characteristics of these formed elements is presented in table 13.2.

Blood cell counts are an important source of information in determining the health of a person. An abnormal increase in erythrocytes, for example, is termed **polycythemia** and is indicative of several dysfunctions. An abnormally low red blood cell count is **anemia.** An elevated leukocyte count, called **leukocytosis,** is often associated with localized infection. A large number of immature leukocytes within a blood sample is diagnostic of the disease **leukemia.**

## Hemopoiesis

Blood cells are constantly formed through a process called **hemopoiesis.** The term *erythropoiesis* refers to the formation of erythrocytes, and *leukopoiesis* to the formation of leukocytes. These processes occur in two classes of tissues. **Myeloid tissue** is the red bone marrow of the long bones, ribs, sternum, bodies of vertebrae, and portions of the skull. **Lymphoid tissue** includes the lymph nodes, tonsils, spleen, and thymus. The bone marrow produces all of the different types of blood cells, while one type—lymphocytes—is also produced in the lymphoid tissue.

**Figure 13.3.** Types of formed elements.

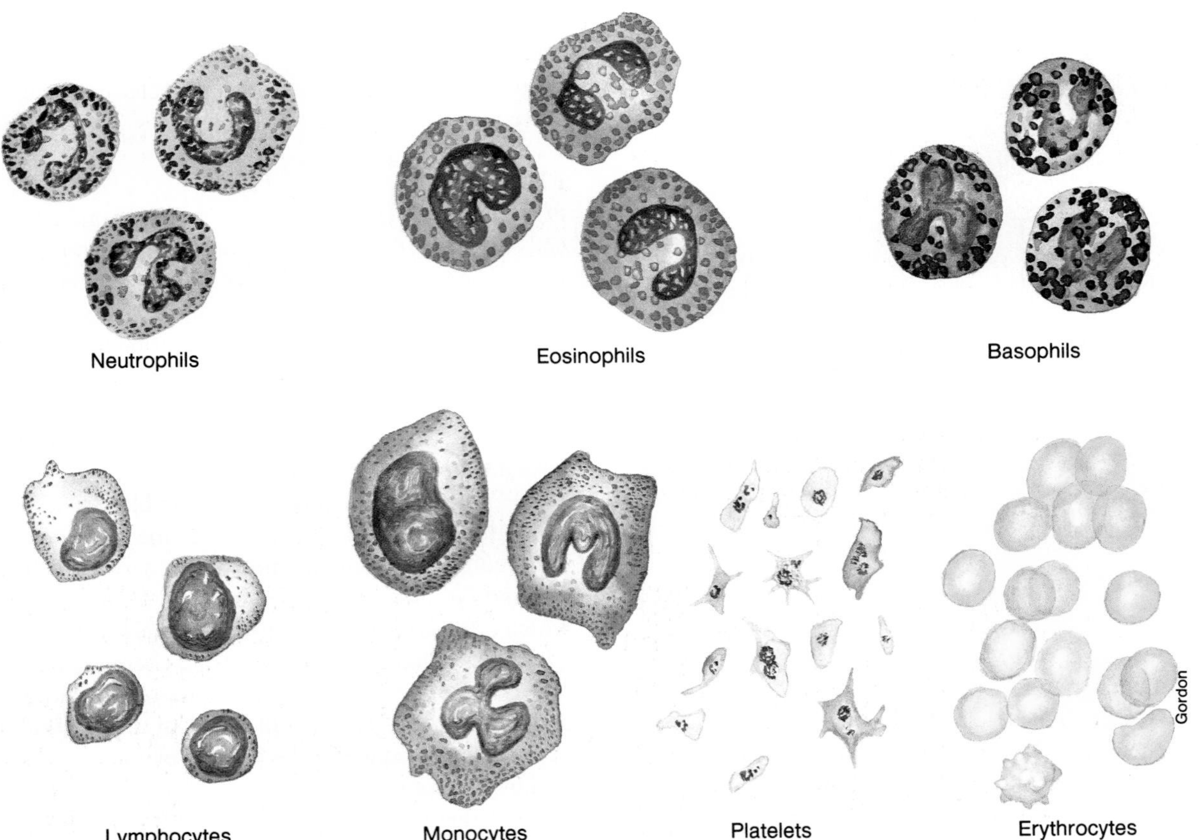


**Table 13.2** Formed elements of the blood

| Component | Description | Number Present | Function |
|---|---|---|---|
| Erythrocyte (red blood cell) | Biconcave disc without nucleus; contains hemoglobin; survives 100–120 days | 4,000,000 to 6,000,000/mm³ | Transports oxygen and carbon dioxide |
| Leukocytes (white blood cells) | | 5,000 to 10,000/mm³ | Aid in defense against infections by microorganisms |
| Granulocytes | About twice the size of red blood cells; cytoplasmic granules present; survive 12 hours to 3 days | | |
| 1. Neutrophil | Nucleus with 2–5 lobes; cytoplasmic granules stain slightly pink | 54%–62% of white cells present | Phagocytic |
| 2. Eosinophil | Nucleus bilobed; cytoplasmic granules stain red in eosin stain | 1%–3% of white cells present | Helps to detoxify foreign substances; secretes enzymes that break down clots |
| 3. Basophil | Nucleus lobed; cytoplasmic granules stain blue in hematoxylin stain | Less than 1% of white cells present | Releases anticoagulant heparin |
| Agranulocytes | Cytoplasmic granules absent; survive 100–300 days | | |
| 1. Monocyte | 2–3 times larger than red blood cell; nuclear shape varies from round to lobed | 3%–9% of white cells present | Phagocytic |
| 2. Lymphocyte | Only slightly larger than red blood cell; nucleus nearly fills cell | 25%–33% of white cells present | Provides specific immune response (including antibodies) |
| Platelet (thrombocyte) | Cytoplasmic fragment; survives 5–9 days | 130,000 to 360,000/mm³ | Clotting |

Erythropoiesis is an extremely active process. It is estimated that about 2.5 million erythrocytes are produced every second in order to replace the number that are continuously destroyed by the spleen and liver. During the destruction of erythrocytes, iron is salvaged and returned to the red bone marrow, where it is used again in the formation of erythrocytes. The life span of an erythrocyte is approximately 120 days. Agranular leukocytes remain functional for 100–300 days under normal conditions. Granular leukocytes, in contrast, have an extremely short life span of twelve hours to three days.

Hemopoiesis begins the same way in both myeloid and lymphoid tissue. A population of undifferentiated (unspecialized) cells gradually differentiate (specialize) to become "stem cells," which give rise to the blood cells. At each step along the way, stem cells can duplicate themselves by mitosis, thus ensuring that the parent population never becomes depleted. As the cells become differentiated they develop membrane receptors for chemical signals that cause further development along particular lines. The earliest stem cells that can be distinguished under a microscope are the *erythroblasts* (which become erythrocytes), *myeloblasts* (which become granular leukocytes), *lymphoblasts* (which form lymphocytes), and *monoblasts* (which form monocytes).

The production of different subtypes of lymphocytes is stimulated by chemicals called **lymphokines**, which are discussed as part of the immune system in chapter 19. The production of red blood cells is stimulated by a hormone called **erythropoietin**, which is secreted by the kidneys. Erythropoietin, a glycoprotein hormone containing 166 amino acids, stimulates cell division and differentiation of erythrocyte stem cells in bone marrow. The secretion of erythropoietin by the kidneys is stimulated whenever the delivery of oxygen to the kidneys and other organs is lower than normal. Under these conditions—which can occur, for example, when a person lives at high altitude—the increased production of red blood cells allows the blood to carry a higher concentration of oxygen to the tissues.

Considering the fact that the kidneys produce erythropoietin, and that erythropoietin is needed to stimulate red blood cell production, it is not surprising that patients with kidney disease may also suffer from anemia. It has recently been shown that this anemia can be effectively treated by giving these patients human erythropoietin. This exciting therapeutic breakthrough was made possible by the commercial production of erythropoietin, using cultured mammalian cells that had incorporated the human gene for erythropoietin through genetic engineering techniques.

## Red Blood Cell Antigens and Blood Typing

All cells in the body contain, on their surfaces, certain molecules that can be recognized as foreign by the immune system of another individual. These molecules are known as *antigens*. As part of the immune response, particular lymphocytes secrete a class of proteins, called *antibodies*, which bond in a specific fashion to antigens. The specificity of antibodies for antigens is analogous to the specificity of enzymes for their substrates, and of receptor proteins for neurotransmitters and hormones. A complete description of antibodies and antigens is provided in chapter 19.

***ABO System.*** The distinguishing antigens on other cells are far more varied than the antigens on red blood cells. Red blood cell antigens, however, are of extreme clinical importance because their types must be matched between donors and recipients for blood transfusions. There are several groups of red blood cell antigens, but the major group is known as the ABO system. In terms of the antigens present on the red blood cell surface, a person may be *type A* (with only A antigens), *type B* (with only B antigens), *type AB* (with both A and B antigens), or *type O* (with neither A nor B antigens). It should again be noted that the blood type denotes the class of antigens (chemically, a type of glycolipid) found on the red blood cell surface.

Each person inherits two genes (one from each parent) that control the production of the ABO antigens. The genes for A or B antigens are dominant to the gene for O, since the latter simply means the absence of A or B. A person who is type A, therefore, may have inherited the A gene from each parent (may have the genotype AA), or the A gene from one parent and the O gene from the other parent (and thus have the genotype AO). Likewise, a person who is type B may have the genotype BB or BO. It follows that a type O person inherited the O gene from each parent (has the genotype OO), whereas a type AB person inherited the A gene from one parent and the B gene from the other (there is no dominance-recessive relationship between A and B).

The immune system is tolerant of its own red blood cell antigens. A person who is type A, for example, does not produce anti-A antibodies. Surprisingly, however, people with type A blood do make antibodies against the B antigen, and conversely, people with blood type B make antibodies against the A antigen. This is believed to result from the fact that antibodies made in response to some common bacteria cross-react with the A or B antigens. A

**Figure 13.4.** The agglutination (clumping) of red blood cells occurs when cells with A-type antigens are mixed with anti-A antibodies and when cells with B-type antigens are mixed with anti-B antibodies. No agglutination would occur with type O blood (not shown).

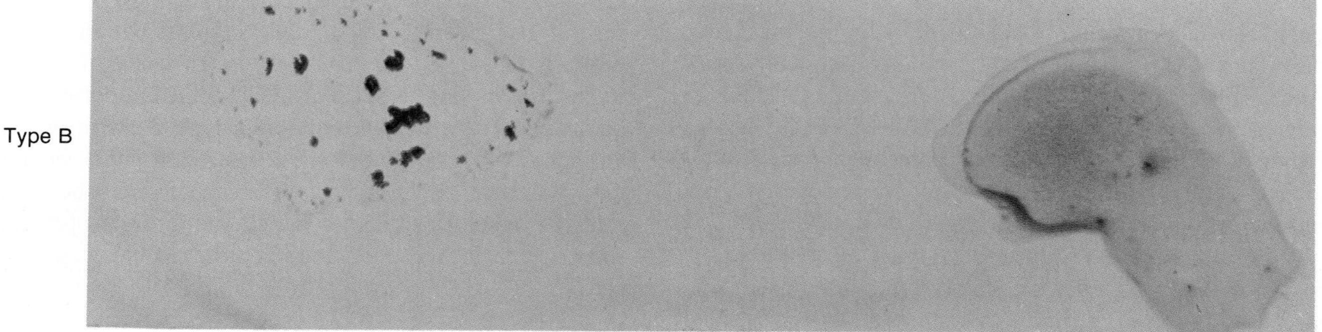


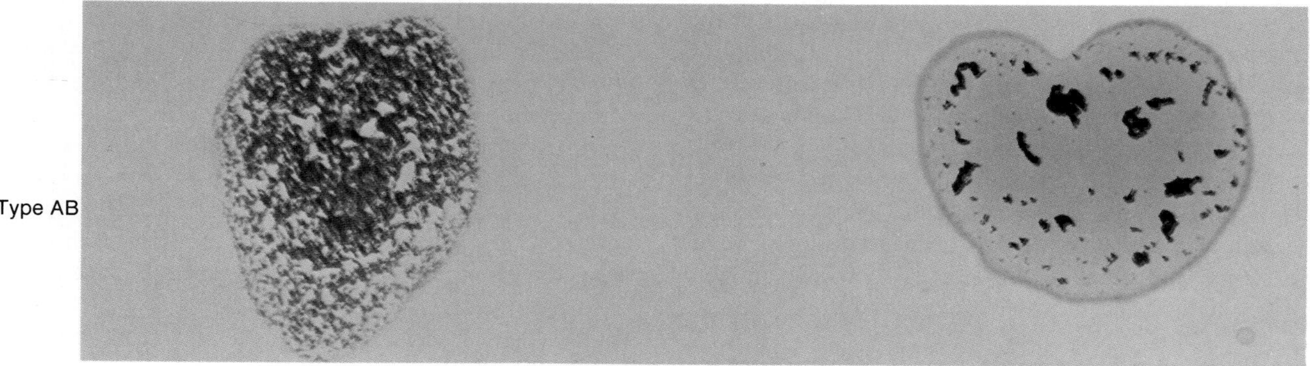


**Table 13.3** The ABO system of antigens on red blood cells

| Antigen on RBCs | Antibody in Plasma |
|---|---|
| A | Anti-B |
| B | Anti-A |
| O | Anti-A and anti-B |
| AB | Neither anti-A nor anti-B |

person who is type A, therefore, acquires antibodies that can react with B antigens by exposure to these bacteria but does not develop antibodies that can react with A antigens, because this is prevented by tolerance mechanisms.

People who are type AB develop tolerance to both these antigens and thus do not produce either anti-A or anti-B antibodies. Those who are type O, in contrast, do not develop tolerance to either antigen and, therefore, have both anti-A and anti-B antibodies in their plasma (table 13.3).

***Transfusion Reactions.*** Before transfusions are performed, a *major cross-match* is made by mixing serum from the recipient with blood cells from the donor. If the types do not match—if the donor is type A, for example, and the recipient is type B—the recipient's antibodies attach to the donor's red blood cells and form bridges that cause the cells to clump together, or **agglutinate** (fig. 13.4).

Because of this agglutination reaction, the A and B antigens are sometimes called *agglutinogens*, and the antibodies against them are called *agglutinins*. Transfusion errors that result in such agglutination in the blood can produce a blockage of small blood vessels and cause organ damage.

In emergencies, type O blood has been given to people who are type A, B, AB, or O. Since type O red blood cells lack A or B antigens, the recipient's antibodies cannot cause agglutination of the donor red blood cells. Type O is, therefore, a *universal donor*, but only as long as the volume of plasma donated is small, because plasma from a type O person would agglutinate type A, type B, and type AB red blood cells. Likewise, type AB people are *universal recipients* because they lack anti-A and anti-B antibodies and thus cannot agglutinate donor red blood cells. (Donor plasma could agglutinate recipient red blood cells if the transfusion volume were too large.) Because of the dangers involved, the universal donor and recipient concept in blood transfusions is strongly discouraged.

***Rh Factor.*** Another group of antigens found in most red blood cells is the *Rh factor* (Rh stands for Rhesus monkey, in which these antigens were first discovered). People who have these antigens are said to be **Rh positive**, whereas those who do not are **Rh negative**. There are fewer Rh negative people because this condition is recessive to Rh positive. The Rh factor is of particular significance when Rh negative mothers give birth to Rh positive babies.

Since the fetal and maternal blood are normally kept separate across the placenta (chapter 20), the Rh negative mother is not usually exposed to the Rh antigen of the fetus during the pregnancy. At the time of birth, however, a variable degree of exposure may occur, and the mother's immune system may become sensitized and produce antibodies against the Rh antigen. This does not always occur, however, because the exposure may be minimal and because Rh negative women vary in their sensitivity to the Rh factor. If the woman does produce antibodies against the Rh factor, these antibodies can cross the placenta in subsequent pregnancies and cause hemolysis of the Rh positive red blood cells of the fetus. The baby is therefore born anemic, with a condition called *erythroblastosis fetalis*.

Erythroblastosis fetalis can be prevented by injecting the Rh negative mother with antibodies against the Rh factor (one trade name for this is RhoGAM—the GAM is short for gamma globulin, the class of plasma proteins in which antibodies are found) within seventy-two hours after the birth of each Rh positive baby. This is a type of passive immunization in which the injected antibodies inactivate the Rh antigens and thus prevent the mother from becoming actively immunized to them.

---

1. *List the different classes of plasma proteins, and describe their functions.*
2. *Distinguish between the different types of formed elements of the blood in terms of their origin, appearance, and function.*
3. *Describe how the rate of erythropoiesis is regulated.*
4. *Explain what "type A positive" means, and explain what can happen in a blood transfusion if donor and recipient are not properly matched.*

---

## Blood Clotting

Trauma to a blood vessel initiates a sequence of events that leads to the formation of a blood clot. These events require the presence of blood platelets and a number of proteins called clotting factors in the plasma. Ultimately, an active enzyme called thrombin is formed which catalyzes the conversion of fibrinogen into insoluble threads of fibrin. A delayed effect of the events that cause clot formation is the activation of plasmin, an enzyme that digests fibrin and promotes dissolution of the blood clot.

When a blood vessel is injured, a number of physiological mechanisms are activated that promote **hemostasis,** or the cessation of bleeding (*hemo* = blood; *stasis* = standing). Breakage of the endothelial lining of a vessel exposes collagen proteins from the subendothelial connective tissue to the blood. This initiates three separate, but overlapping, hemostatic mechanisms: (1) vasoconstriction; (2) the formation of a platelet plug; and (3) the production of a web of fibrin proteins around the platelet plug.

### Functions of Platelets

In the absence of vessel damage, platelets are repelled from each other and from the endothelial lining of vessels. The repulsion of platelets from an intact endothelium is believed to be due to *prostacyclin,* a derivative of prostaglandins, produced within the endothelium. Mechanisms that prevent platelets from sticking to the blood vessels and to each other are obviously needed to prevent inappropriate blood clotting.

Damage to the endothelium of vessels exposes subendothelial tissue to the blood. Platelets are able to stick to exposed collagen proteins that have become coated with a protein (*von Willebrand factor*) secreted by endothelial cells. Platelets contain secretory granules; when platelets stick to collagen, they *degranulate* as the secretory granules release their products. These products include *ADP* (adenosine diphosphate), *serotonin,* and a prostaglandin called *thromboxane* $A_2$. This event is known as the **platelet release reaction.**

**Figure 13.5.** A scanning electron micrograph showing threads of fibrin.

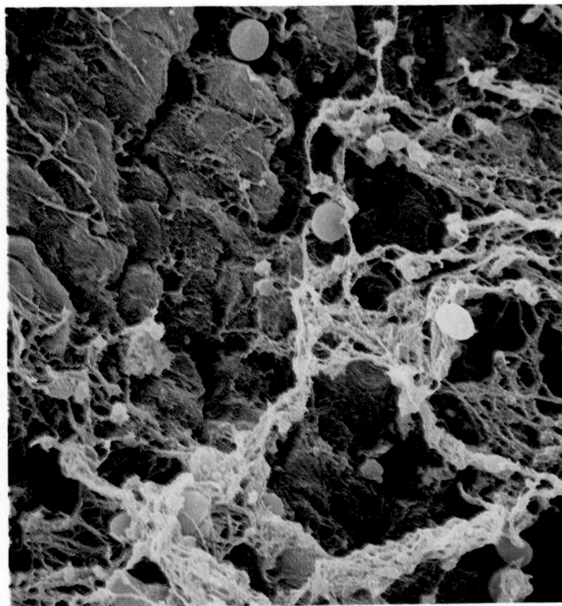

Serotonin and thromboxane $A_2$ stimulate vasoconstriction, which helps to decrease blood flow to the injured vessel. Phospholipids that are exposed on the platelet membrane participate in the activation of clotting factors.

The release of ADP and thromboxane $A_2$ from platelets that are stuck to exposed collagen makes other platelets in the vicinity "sticky," so that they adhere to those stuck to the collagen. The second layer of platelets, in turn, undergoes a platelet release reaction, and the ADP and thromboxane $A_2$ that are secreted cause additional platelets to aggregate at the site of injury. This produces a **platelet plug** in the damaged vessel, which is strengthened by the activation of plasma-clotting factors.

In order to undergo a release reaction, the production of prostaglandins (chapter 7) by the platelets is required. **Aspirin** inhibits the conversion of arachidonic acid (a cyclic fatty acid) into prostaglandins and thus inhibits the release reaction and consequent formation of a platelet plug. The ingestion of excessive amounts of aspirin can thus significantly prolong bleeding time, which is why blood donors and women in the last trimester of pregnancy are advised to avoid aspirin. The inhibition of clot formation is beneficial, however, in people who have atherosclerosis, and thus aspirin is recommended for patients who have survived myocardial infarction and may be in danger of a recurrence. (Atherosclerosis and myocardial infarction are described later in this chapter.)

**Figure 13.6.** The sequence of events leading to platelet aggregation and the formation of a blood clot.

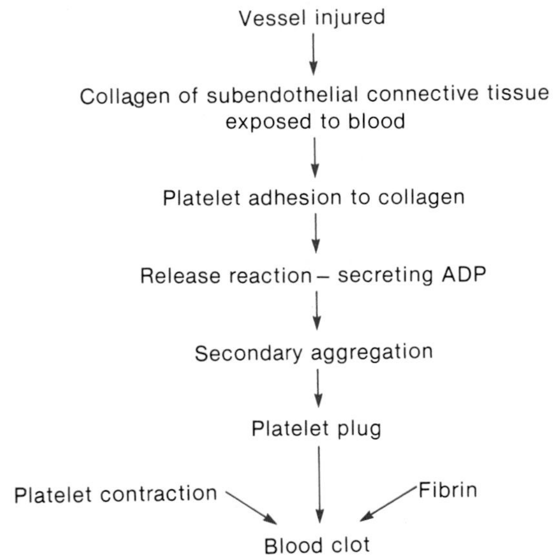


## Clotting Factors: Formation of Fibrin

The platelet plug is strengthened by a meshwork of insoluble protein fibers known as **fibrin** (fig. 13.5). Blood clots therefore contain platelets, fibrin, and usually trapped red blood cells that give the clot a red color (clots formed in arteries generally lack red blood cells and are gray in color). Finally, contraction of the platelets in the process of *clot retraction* forms a more compact and effective plug (fig. 13.6). Fluid squeezed from the clot as it retracts is called *serum,* which is plasma without fibrinogen (the soluble precursor of fibrin).

There are two pathways that result in the conversion of fibrinogen into fibrin. Blood left in a test tube will clot without the addition of any external chemicals; the pathway that produces this clot is thus called the **intrinsic pathway.** The intrinsic pathway also produces clots in damaged blood vessels when collagen is exposed to plasma. Damaged tissues, however, release a chemical that initiates a "shortcut" to the formation of fibrin. Since this chemical is not part of blood, the shorter pathway is called the **extrinsic pathway.**

***Intrinsic Pathway.*** The intrinsic pathway is initiated by the exposure of plasma to negatively charged surfaces, such as that provided by collagen or glass. This activates a plasma protein called **factor XII** (table 13.4), which is a protein-digesting enzyme (protease). Active factor XII in turn activates another plasma protein—**factor XI**—by cleaving part of its inactive precursor. Activated factor XI, in turn, activates **factor IX.**

**Figure 13.7.** The extrinsic and intrinsic clotting pathways that lead to the formation of insoluble fibrin polymers.

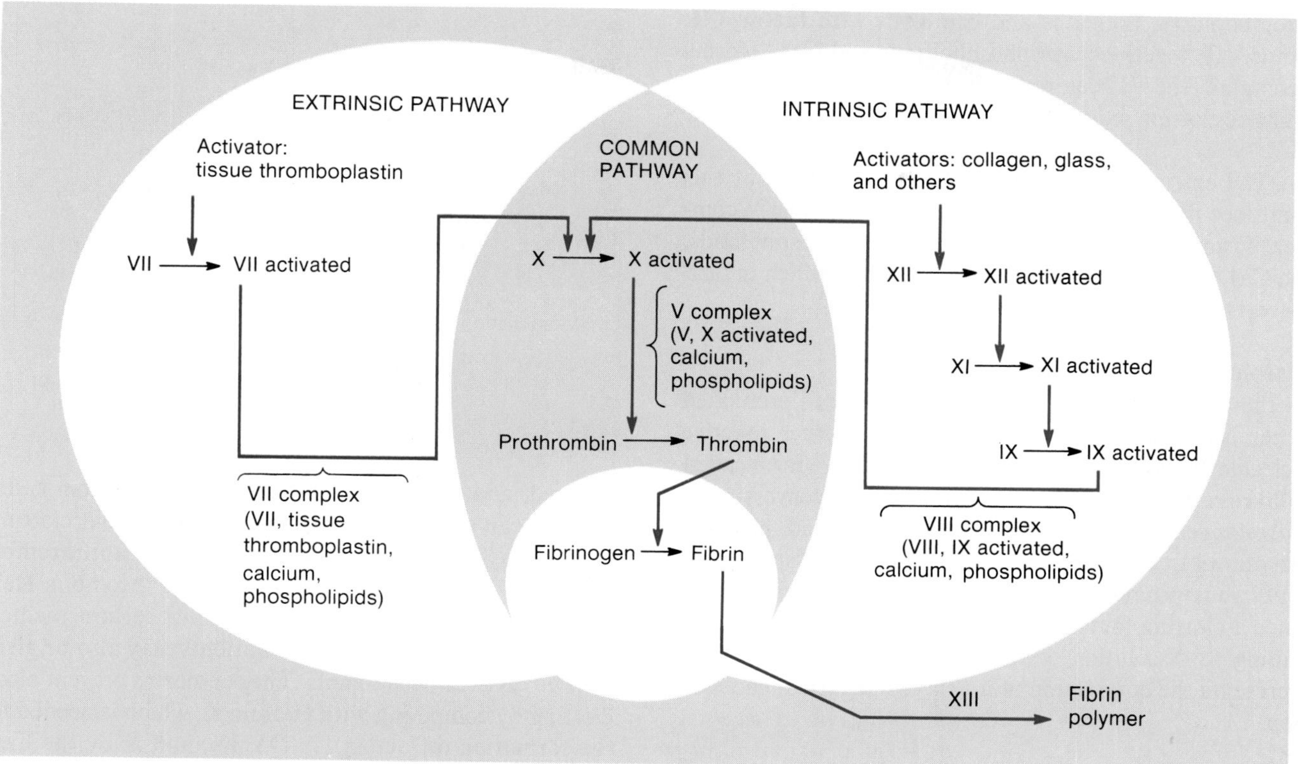


**Table 13.4** The plasma clotting factors

| Factor | Name | Function | Pathway |
|---|---|---|---|
| I | Fibrinogen | converted to fibrin | common |
| II | Prothrombin | enzyme | common |
| III | Tissue thromboplastin | cofactor | extrinsic |
| IV | Calcium ions ($Ca^{++}$) | cofactor | intrinsic, extrinsic, and common |
| V | Proaccelerin | cofactor | common |
| VII | Proconvertin | enzyme | extrinsic |
| VIII | Antihemophilic factor | cofactor | intrinsic |
| IX | Plasma thromboplastin component; Christmas factor | enzyme | intrinsic |
| X | Stuart-Prower factor | enzyme | common |
| XI | Plasma thromboplastin antecedent | enzyme | intrinsic |
| XII | Hageman factor | enzyme | intrinsic |
| XIII | Fibrin stabilizing factor | enzyme | common |

The next steps in the sequence require the presence of phospholipids, which are provided by platelets, and $Ca^{++}$. The combination of these with active factor IX and **factor VIII** forms a complex that activates **factor X.** A complex is then formed between active factor X, **factor V,** platelet phospholipids, and $Ca^{++}$ that converts **prothrombin** (inactive **factor II**) to **thrombin.**

Thrombin is a protease that converts the soluble protein **fibrinogen** (**factor I**) into **fibrin** monomers. These monomers are joined together to form insoluble fibrin polymer by the action of **factor XIII.** The intrinsic clotting sequence is shown on the right side of figure 13.7.

***Extrinsic Pathway.*** The formation of fibrin can occur more rapidly as a result of the release of **tissue thromboplastin,** or **factor III,** from damaged tissue cells. Tissue thromboplastin activates and combines with **factor VII;** factor VII, together with phospholipids and $Ca^{++}$, forms a complex (the VII complex) that activates factor X. The extrinsic clotting sequence is shown on the left side of figure 13.7.

The extrinsic and intrinsic pathways overlap at this point (see fig. 13.7) into a common pathway. The V complex, formed by active factor X, factor V, phospholipids, and $Ca^{++}$ converts prothrombin to thrombin, which in turn converts fibrinogen to fibrin.

## Dissolution of Clots

As the damaged blood vessel wall is repaired, activated factor XII promotes the conversion of another inactive molecule in plasma, *prekallikrein,* to the active form called **kallikrein.** Kallikrein, in turn, catalyzes the conversion of *plasminogen* into the active molecule called **plasmin.** Plasmin is an enzyme that digests fibrin into "split products," thus promoting dissolution of the clot (fig. 13.8). Since a clotting factor (factor XII) initiates the pathway leading to dissolution of the clot, the action of plasmin represents the completion of a delayed negative feedback loop.

There are a number of plasminogen activators besides kallikrein that are used clinically to promote dissolution of clots. Some of these are endogenous activators produced by tissues and leukocytes, and some are exogenous compounds that are normal bacterial products. An exciting recent development in genetic engineering technology is the commercial availability of one of the endogenous compounds, called **tissue plasminogen activator (TPA)**, which is the product of human genes introduced into bacteria. **Streptokinase** is a potent and more widely used activator of plasminogen but is classified as exogenous because it is a normal bacterial product. Streptokinase and TPA may be injected into the general circulation or injected specifically into a coronary vessel that has become occluded by a thrombus (blood clot).

Kallikrein, in addition to its function in the formation of plasmin, also stimulates the formation of a molecule called **bradykinin** from its precursor *(kininogen).* Bradykinin stimulates vasodilation, which helps to counter the vasoconstrictor effects of prostaglandins and serotonin released by platelets during the formation of a clot.

***Anticoagulants.*** Clotting of the blood in test tubes can be prevented by the addition of *citrate* or *EDTA,* which chelates (binds to) calcium. By this means $Ca^{++}$ levels in the blood that can participate in the clotting sequence are lowered, and clotting is inhibited. A mucoprotein called *heparin* can also be added to the tube to prevent clotting. Heparin activates a plasma protein called antithrombin III, which combines with and inactivates thrombin. Heparin is also given intravenously during certain medical procedures to prevent clotting. Patients may also be given *coumarins* as anticoagulants. The coumarins prevent blood clotting by competing with vitamin K, which is needed for the formation of factors II, VII, IX, and X by the liver. In contrast to the immediate effects of heparin, therefore, coumarin must be given to a patient for several days to be effective.

**Figure 13.8.** Events that produce dissolution of the blood clot and vasodilation.

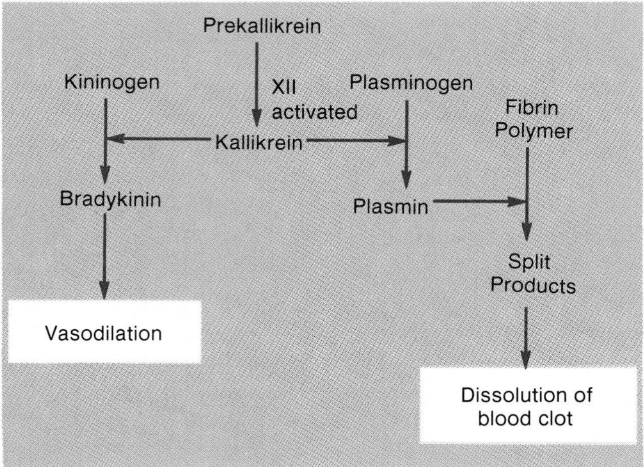


***Hereditary Clotting Disorders.*** Examples of hereditary clotting disorders include two different genetic defects in factor VIII. A defect in one subunit of factor VIII prevents this factor from participating in the intrinsic clotting pathway. This genetic disease, called **hemophilia A,** is an X-linked recessive trait that is prevalent in the royal families of Europe. A defect in another subunit of factor VIII results in **von Willebrand's disease.** In this disease, rapidly circulating platelets are unable to stick to collagen and a platelet plug cannot be formed. Some acquired and inherited defects in the clotting system are summarized in table 13.5.

1. *Describe how a platelet plug is formed when a vessel is cut.*
2. *List the steps shared in common by the intrinsic and extrinsic clotting pathways.*
3. *Explain the meaning of the terms* intrinsic *and* extrinsic *in terms of the clotting pathways, and describe how these pathways differ from each other.*
4. *Describe the steps that lead to the formation of plasmin, and explain how this might be regarded as a negative feedback mechanism.*

**Table 13.5** Some acquired and inherited defects in the clotting mechanism

| Category | Cause of Disorder | Comments |
|---|---|---|
| Acquired clotting disorders | Vitamin K deficiency | Inadequate formation of prothrombin and other clotting factors in the liver |
| | Aspirin | Inhibits prostaglandin production, resulting in defective platelet release reaction |
| Anticoagulants | Coumarin | Competes with the action of vitamin K |
| | Heparin | Inhibits activity of thrombin |
| | Citrate | Combines with $Ca^{++}$ and thus inhibits the activity of many clotting factors |
| Inherited clotting disorders | Hemophilia A (defective factor $VIII_{AHF}$) | Recessive trait carried on X chromosome; results in delayed formation of fibrin |
| | Von Willebrand's disease (defective factor $VIII_{VWF}$) | Dominant trait carried on autosomal chromosome; impaired ability of platelets to adhere to collagen in subendothelial connective tissue |
| | Hemophilia B (defective factor IX), also called Christmas disease | Recessive trait carried on X chromosome; results in delayed formation of fibrin |

## Structure of the Heart

The heart contains four chambers: two upper atria, which receive venous blood, and two lower ventricles, which eject blood into arteries. The right ventricle pumps blood to the lungs, where the blood becomes oxygenated, whereas the left ventricle pumps oxygenated blood to the entire body. The flow of blood from atria to ventricles, and then from ventricles to arteries, is assured by the action of two pairs of one-way valves within the heart.

The heart is divided into four chambers. The right and left **atria** (singular, *atrium*) receive blood from the venous system; the right and left **ventricles** pump blood into the arterial system. The right atrium and ventricle (sometimes called the *right pump*) are separated from the left atrium and ventricle (the *left pump*) by a muscular wall, or *septum*. This septum normally prevents mixture of the blood from the two sides of the heart.

There is a layer of dense connective tissue between the atria and ventricles known as the **fibrous skeleton** of the heart. Bundles of myocardial cells (described in chapter 12) in the atria attach to the upper margin of this fibrous skeleton, and the myocardial cell bundles of the ventricles attach to the lower margin. As a result, the myocardia of the atria and ventricles are structurally and functionally separated from each other, and special conducting tissue is needed to carry action potentials from the atria to the ventricles (this will be described in a later section). The connective tissue of the fibrous skeleton also forms rings, called *annuli fibrosi*, around the four heart valves, providing a foundation for the support of the valve flaps.

### Pulmonary and Systemic Circulation

Blood that has become partially depleted of its oxygen content and increased in carbon dioxide content, as a result of gas exchange across tissue capillaries, returns to the right atrium. This blood then enters the right ventricle, which pumps it into the *pulmonary trunk* and *pulmonary arteries*. The pulmonary arteries branch to transport blood to the lungs, where gas exchange occurs between the lung capillaries and the air sacs (alveoli) of the lungs. Oxygen diffuses from the air to the capillary blood, while carbon dioxide diffuses in the opposite direction.

The blood that returns to the left atrium by way of the *pulmonary veins* is therefore enriched in oxygen and partially depleted in carbon dioxide. The path of blood from the heart (right ventricle), through the lungs, and back to the heart (left atrium) completes one circuit: the **pulmonary circulation.**

Oxygen-rich blood in the left atrium enters the left ventricle and is pumped into a very large, elastic artery—the *aorta*. The aorta ascends for a short distance, makes a U-turn, and then descends through the thoracic (chest) and abdominal cavities. Arterial branches from the aorta supply oxygen-rich blood to all of the organ systems and are thus part of the **systemic circulation.**

As a result of cellular respiration, the oxygen concentration is lower and the carbon dioxide concentration is higher in the tissues than in the capillary blood. Blood that drains into the systemic veins is thus partially depleted in oxygen and increased in carbon dioxide content. These veins ultimately empty into two large veins—the *superior* and *inferior venae cavae*—that return the oxygen-poor blood to the right atrium. This completes the systemic circulation: from the heart (left ventricle), through the organ systems, and back to the heart (right atrium). The characteristics of the systemic and pulmonary circulations are summarized in table 13.6 and illustrated in figure 13.9.

The numerous small muscular arteries and arterioles of the systemic circulation present greater resistance to blood flow than that in the pulmonary circulation. The amount of work performed by the left ventricle is greater (by a factor of 5 to 7) than that performed by the right

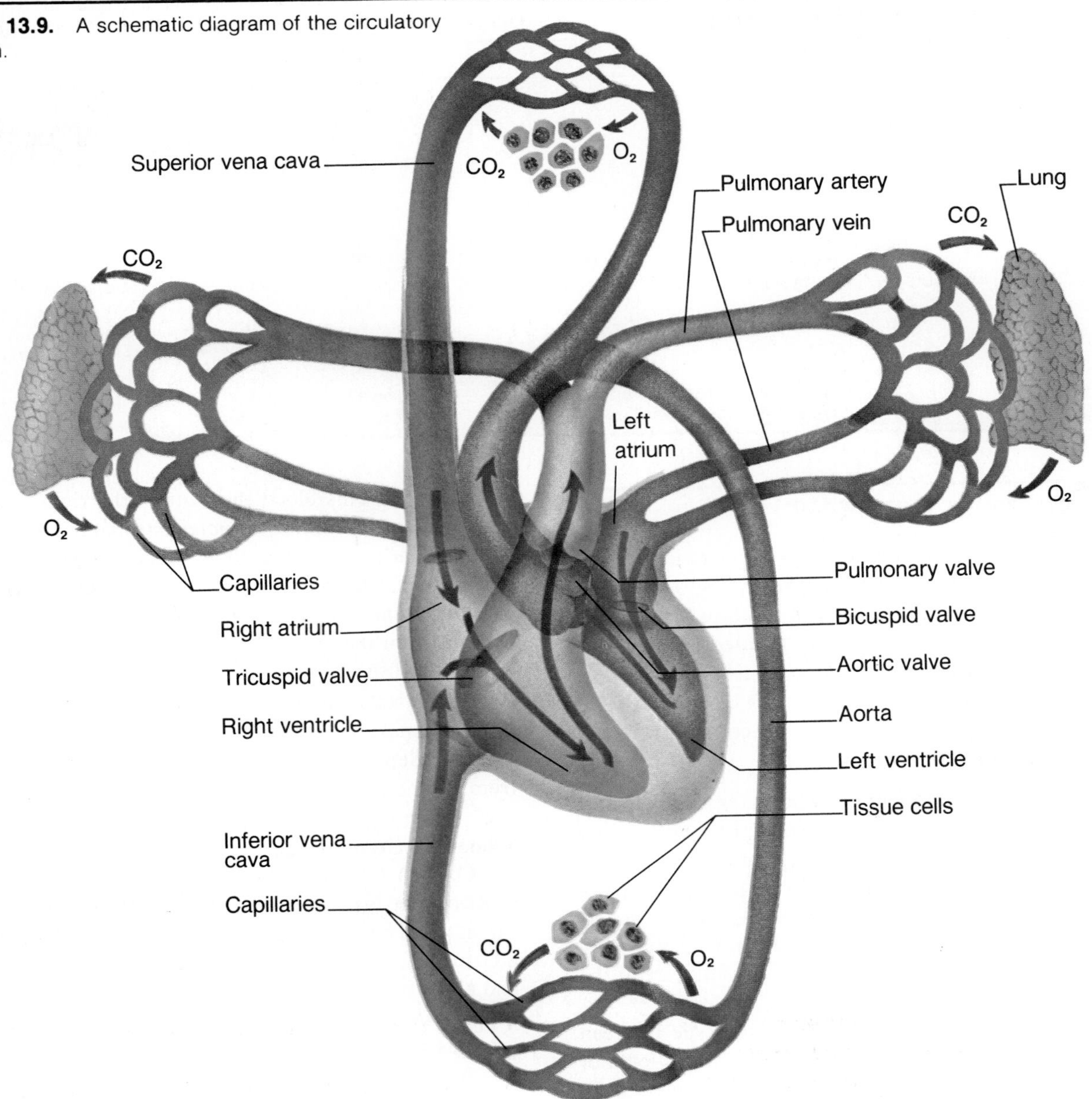


**Figure 13.9.** A schematic diagram of the circulatory system.

**Table 13.6** Summary of the pulmonary and systemic circulations

| | Source | Arteries | $O_2$ Content of Arteries | Veins | Termination |
|---|---|---|---|---|---|
| Pulmonary Circulation | Right ventricle | Pulmonary arteries | Low | Pulmonary veins | Left atrium |
| Systemic Circulation | Left ventricle | Aorta and its branches | High | Superior and inferior venae cavae and their branches* | Right atrium |

*Blood from the coronary circulation does not enter the venae cavae, but instead returns directly to the right atrium via the coronary sinus.

ventricle, because the rate of blood flow through the systemic circulation must be matched to the flow rate of the pulmonary circulation despite the differences in resistance. It is not surprising, therefore, that the muscular wall of the left ventricle is thicker (8–10 mm) than that of the right ventricle (2–3 mm).

### Atrioventricular and Semilunar Valves

Although adjacent myocardial cells are joined together mechanically and electrically by intercalated discs, the atria and ventricles are separated into two functional units by a sheet of connective tissue located between them. Embedded within this sheet of tissue are one-way **atrioventricular (AV) valves** (fig. 13.10). The AV valve located

**Figure 13.10.** A diagram of the structure of the heart (*a*) and photograph of the aortic and pulmonary semilunar valves (*b*).

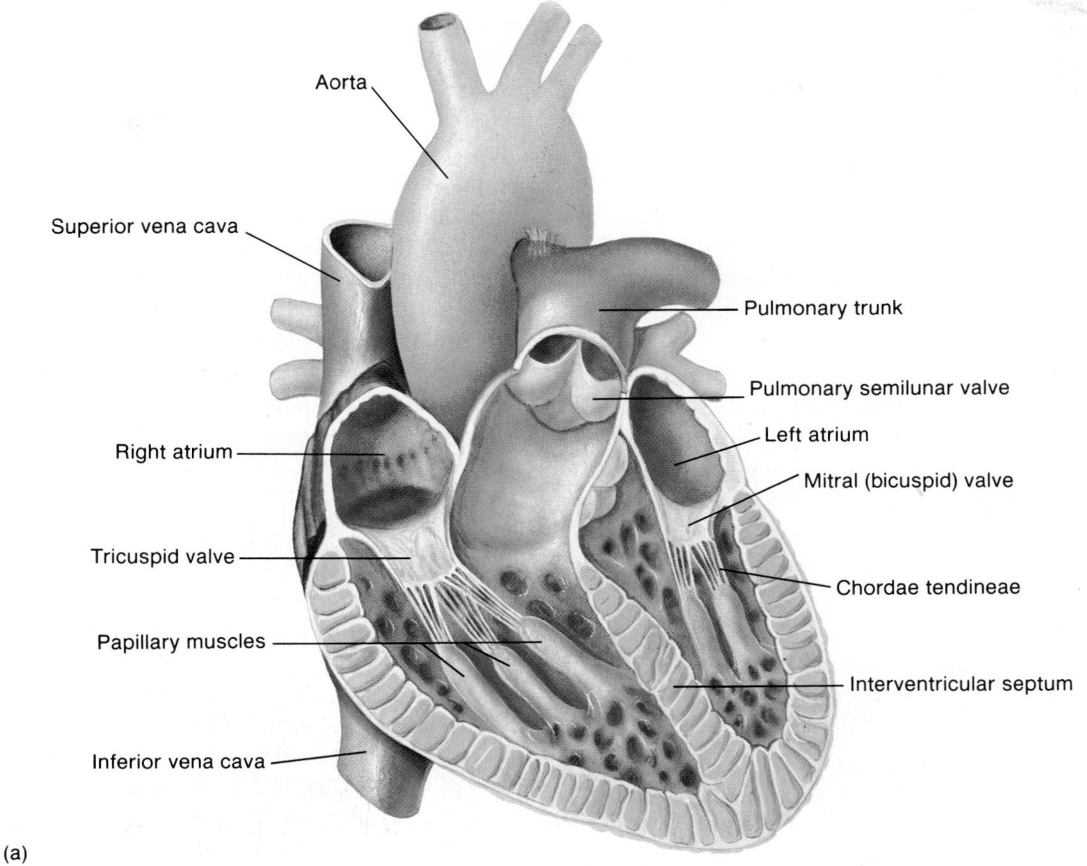


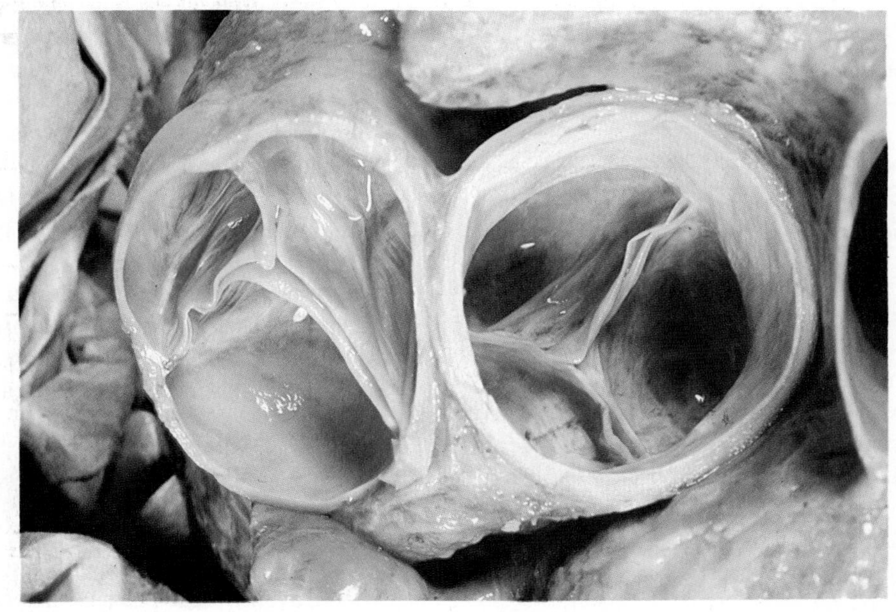


between the right atrium and right ventricle has three flaps, and is therefore called the *tricuspid valve.* The AV valve between the left atrium and left ventricle has two flaps and is thus called the *bicuspid valve;* this is also known as the *mitral valve.*

The AV valves allow blood to flow from the atria to the ventricles, but normally prevent the backflow of blood into the atria. Opening and closing of these valves occur as a result of pressure differences between the atria and ventricles. When the ventricles are relaxed, the venous return of blood to the atria causes the pressure in the atria to exceed that in the ventricles. The AV valves therefore open, allowing blood to enter the ventricles. As the ventricles contract, the intraventricular pressure rises above the pressure in the atria and pushes the AV valves closed.

There is a danger, however, that the high pressure produced by contraction of the ventricles could push the valve flaps too much and evert them. This is normally prevented by contraction of the *papillary muscles* within the ventricles, which are connected to the AV valve flaps by the *chordae tendineae.* Contraction of the papillary muscles occurs at the same time as contraction of the muscular walls of the ventricles and serves to keep the valve flaps tightly closed.

One-way **semilunar valves** are located at the origin of the pulmonary artery and aorta. These valves open during ventricular contraction, allowing blood to enter the pulmonary and systemic circulations. During ventricular relaxation, when the pressure in the arteries is greater than the pressure in the ventricles, the semilunar valves snap shut and thus prevent the backflow of blood into the ventricles.

1. *Using a flow diagram (arrows), describe the pathway of the pulmonary circulation. Indicate the relative amounts of oxygen and carbon dioxide in the vessels involved.*
2. *Use a flow diagram to describe the systemic circulation, and indicate the relative amounts of oxygen and carbon dioxide in the blood vessels.*
3. *Name the AV valves and the valves of the pulmonary artery and aorta. Describe how these valves ensure a one-way flow of blood.*
4. *Explain the structure and functional significance of the fibrous skeleton of the heart.*

## The Cardiac Cycle and the Heart Sounds

The two atria fill with blood and then contract simultaneously. This is then followed by simultaneous contraction of both ventricles, which sends blood through the pulmonary and systemic circulations. Contraction of the ventricles causes a rise in pressure, which closes the AV valves and opens the semilunar valves; relaxation of the ventricles produces a fall in pressure, which closes the semilunar valves and opens the AV valves. The closing of first the AV valves and then the semilunar valves produces the "lub-dub" sounds heard with a stethoscope.

The cardiac cycle refers to the repeating pattern of contraction and relaxation of the heart. The phase of contraction is called **systole,** and the phase of relaxation is called **diastole.** When these terms are used alone they refer to contraction and relaxation of the ventricles. It should be noted, however, that the atria also contract and relax. There is an atrial systole and diastole. Atrial contraction occurs toward the end of diastole, when the ventricles are relaxed; when the ventricles contract during systole, the atria are relaxed.

The heart thus has a two-step pumping action. The right and left atria contract almost simultaneously, followed about 0.1–0.2 seconds later by contraction of the right and left ventricles. During the time when both the atria and ventricles are relaxed the venous return of blood fills the atria. The buildup of pressure that results causes the AV valves to open and blood to flow from atria to ventricles. It has been estimated that the ventricles are about 80% filled with blood even before the atria contract. Contraction of the atria adds the final 20% to the *end-diastolic volume*, which is the total volume of blood in the ventricles at the end of diastole.

It is interesting that the blood contributed by contraction of the atria does not appear to be essential for life. Elderly people who have atrial fibrillation (a condition in which the atria fail to contract) do not appear to have a higher mortality than those who have normally functioning atria. People with atrial fibrillation, however, become fatigued more easily during exercise because the reduced filling of the ventricles compromises the ability of the heart to sufficiently increase its output and blood flow during exercise. Cardiac output and blood flow during rest and exercise are discussed in chapter 14.

Contraction of each ventricle in systole ejects about two-thirds of the blood that they contain—an amount called the *stroke volume*—leaving one-third of the initial amount left in the ventricles as the *end-systolic volume.* The ventricles then fill with blood during the next cycle. At an average **cardiac rate** of 75 beats per minute, each cycle lasts 0.8 second; 0.5 second is spent in diastole, and systole takes 0.3 second (fig. 13.11).

### Pressure Changes during the Cardiac Cycle

When the heart is in diastole, pressure in the systemic arteries averages about 80 mm Hg (millimeters of mercury). The following events in the cardiac cycle then occur:

1. As the ventricles begin their contraction the intraventricular pressure rises, causing the AV valves to snap shut. At this time, the ventricles are neither being filled with blood (because the AV valves are closed) nor ejecting blood (because the intraventricular pressure has not risen sufficiently to open the semilunar valves). This is the phase of *isovolumetric contraction.*
2. When the pressure in the left ventricle becomes greater than the pressure in the aorta the phase of *ejection* begins as the semilunar valves open. The

**Figure 13.11.** The cardiac cycle of ventricular systole and diastole. Contraction of the atria occurs in the last 0.1 second of ventricular diastole. Relaxation of the atria occurs during ventricular systole. The durations of systole and diastole given are accurate for a cardiac rate of 75 beats per minute.

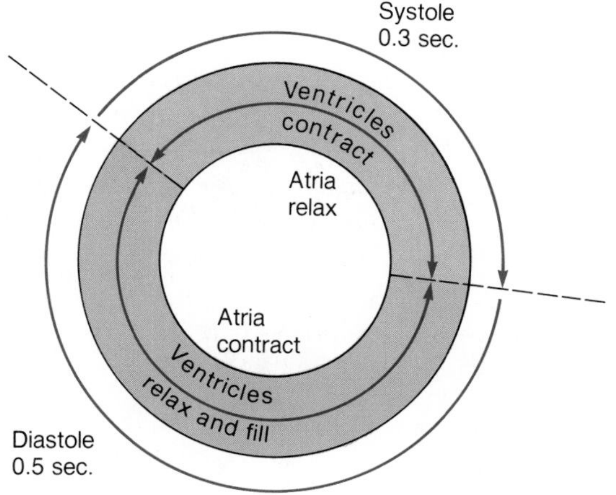


pressure in the left ventricle and aorta rises to about 120 mm Hg (fig. 13.12) as the ventricular volume decreases.

3. As the pressure in the left ventricle falls below the pressure in the aorta, the back pressure causes the semilunar valves to snap shut. The pressure in the aorta falls to 80 mm Hg while pressure in the left ventricle falls to 0 mm Hg.
4. During *isovolumetric relaxation* the AV and semilunar valves are closed. This phase lasts until the pressure in the ventricles falls below the pressure in the atria.
5. When the pressure in the ventricles falls below the pressure in the atria, a phase of *rapid filling* of the ventricles occurs.
6. *Atrial contraction* (*atrial systole*) empties the final amount of blood into the ventricles immediately prior to the next phase of isovolumetric contraction of the ventricles.

Similar events occur in the right ventricle and pulmonary circulation, but the pressures are lower. The maximum pressure produced at systole in the right ventricle is 25 mm Hg, which falls to a low of 8 mm Hg at diastole.

## Heart Sounds

Closing of the AV and semilunar valves produces sounds that can be heard at the surface of the chest with a stethoscope. These sounds are often described phonetically as *lub-dub*. The "lub," or **first sound,** is produced by closing of the AV valves during isovolumetric contraction of the ventricles. The "dub," or **second sound,** is produced by closing of the semilunar valves when the pressure in the ventricles falls below the pressure in the arteries. The first sound is thus heard when the ventricles contract at systole, and the second sound is heard when the ventricles relax at the beginning of diastole.

**Figure 13.12.** The relationship between the heart sounds and the intraventricular pressure and volume. Closing of the AV valves occurs during the early part of contraction, when the intraventricular pressure rises prior to ejection of blood. Closing of the semilunar valves occurs at the beginning of ventricular relaxation, just prior to filling. The first and second sounds thus appear during the stages of isovolumetric contraction and isovolumetric relaxation (*iso* = same). Numbers indicate steps described in text.

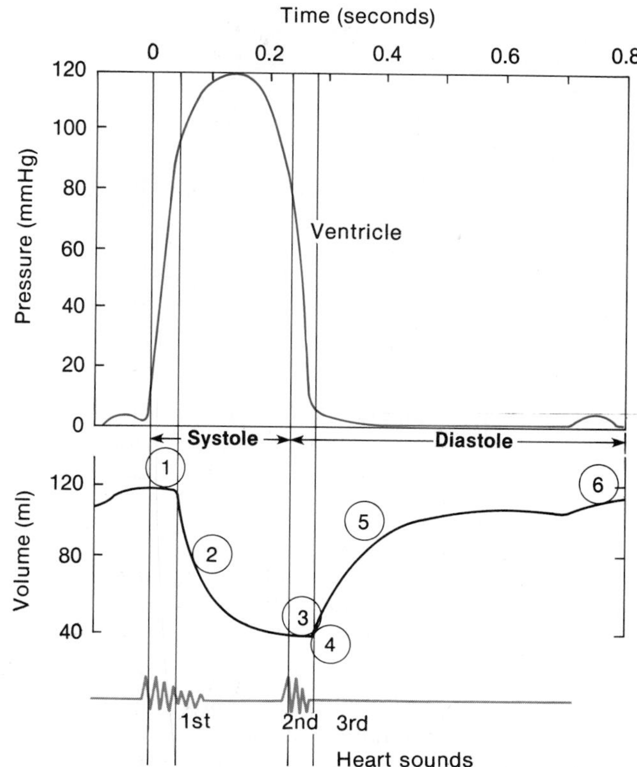


The first sound may, particularly during inhalation, be heard to split into tricuspid and mitral components. Closing of the tricuspid is best heard at the fifth intercostal space (between the ribs) just to the left of the sternum, and closing of the mitral valve is best heard at the fifth left intercostal space at the apex of the heart (fig. 13.13). The second sound may also be split under certain conditions. Closing of the pulmonary and aortic semilunar valves is best heard at the second left and right intercostal spaces, respectively.

***Heart Murmurs.*** Murmurs are abnormal heart sounds produced by abnormal patterns of blood flow in the heart. Many murmurs are caused by defective heart valves.

**Figure 13.13.** Routine stethoscope positions for listening to the heart sounds.

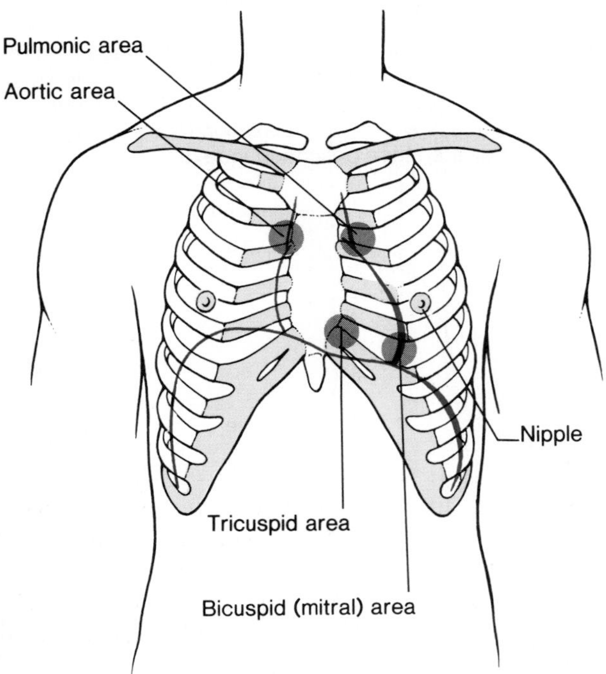


Defective heart valves may be congenital, or more commonly, they can occur as a result of *rheumatic endocarditis,* associated with rheumatic fever. In this disease the valves become damaged by antibodies made in response to an infection caused by streptococcus bacteria (the same bacteria that produce strep throat). Many people have small defects that produce detectable murmurs but do not seriously compromise the pumping ability of the heart. Larger defects, however, may have dangerous consequences and thus may require surgical correction.

In *mitral stenosis,* for example, the mitral valve becomes thickened and calcified. This can impair the blood flow from the left atrium to the left ventricle. An accumulation of blood in the left atrium may cause a rise in left atrial and pulmonary vein pressure, resulting in pulmonary hypertension. As a compensation for the increased pulmonary pressure, the right ventricle grows thicker and stronger.

Valves are said to be incompetent when they do not close properly, and murmurs may be produced as blood regurgitates through the valve flaps. One important cause of incompetent AV valves is damage to the papillary muscles (fig. 13.10). When this occurs, the tension in the chordae tendineae may not be sufficient to prevent the valve from everting as pressure in the ventricle rises during systole.

Murmurs can also be produced by the flow of blood through *septal defects*—holes in the septum between the right and left sides of the heart. These are usually congenital and may occur either in the interatrial or interventricular septum (fig. 13.14). When a septal defect is not accompanied by other abnormalities, blood will usually pass through the defect from the left to the right side, due to the higher pressure on the left side. The buildup of blood and pressure on the right side of the heart that results may lead to pulmonary hypertension and edema (fluid in the lungs).

**Figure 13.14.** Abnormal patterns of blood flow due to septal defects. Left-to-right shunting of blood is shown (*circled areas*) because the left pump is at a higher pressure than the right pump. Under certain conditions, however, the pressure in the right atrium may exceed that of the left, causing right-to-left shunting of blood through a septal defect in the atria.

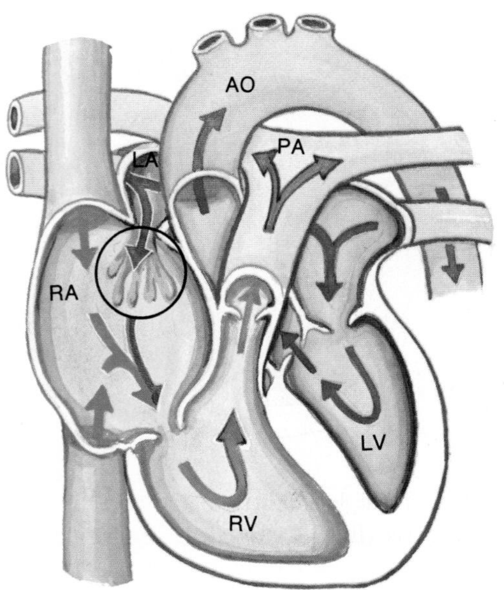


Septal defect in atria

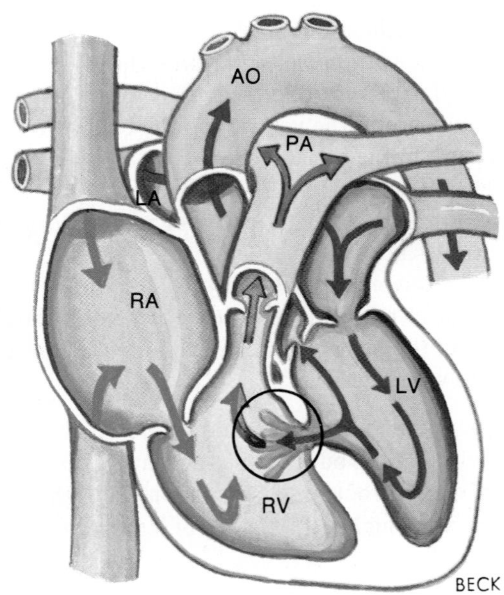


Septal defect in ventricles

**Figure 13.15.** The flow of blood through a patent (open) ductus arteriosus.

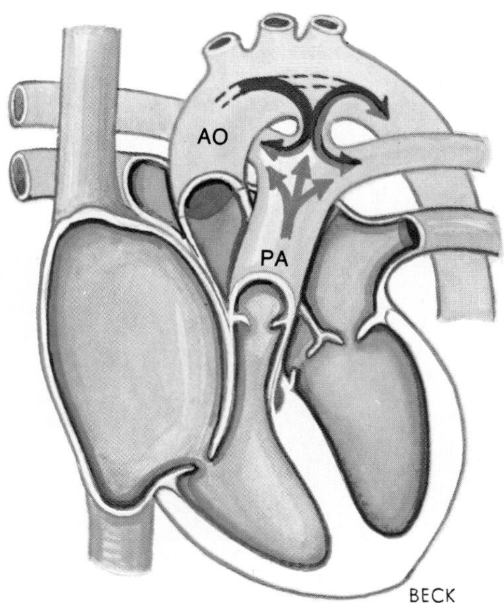


The lungs of a fetus are collapsed, and blood is routed away from the pulmonary circulation by an opening in the interatrial septum called the **foramen ovale** and by a connection between the pulmonary trunk and aorta called the **ductus arteriosus** (fig. 13.15). These shunts normally close after birth, but when they remain open (are *patent*), murmurs can result. Since blood usually goes from left to right through these shunts, the left ventricle still pumps blood that is high in oxygen. When other defects are present, however (as in the *tetralogy of Fallot*), a significant amount of oxygen-depleted blood from the right side of the heart may enter the left side and the systemic circulation through these defects. The mixture of oxygen-poor blood from the right side with oxygen-rich blood in the left side of the heart lowers the oxygen concentration of the blood ejected into the systemic circulation. Since blood low in oxygen imparts a bluish tinge to the skin, the baby may be born *cyanotic* (blue).

1. *Using a figure or an outline, describe the sequence of events that occurs during the cardiac cycle. Indicate when atrial and ventricular filling occur and when atrial and ventricular contraction occur.*
2. *Describe, in words, how the pressure in the left ventricle and in the systemic arteries varies during the cardiac cycle.*
3. *Draw a figure to illustrate the pressure variations described in question 2, and indicate in your figure when the AV and semilunar valves close. Discuss the origin of the heart sounds.*
4. *Explain why blood usually flows from left-to-right through a septal defect. What must occur through a septal defect to produce cyanosis?*

## Electrical Activity of the Heart and the Electrocardiogram

The pacemaker region of the heart (SA node) exhibits a spontaneous depolarization that causes action potentials and results in the automatic beating of the heart. Electrical impulses are conducted by myocardial cells in the atria and are transmitted to the ventricles by special conducting tissue. The characteristics of the action potentials of myocardial cells ensure that the atria and ventricles each contract and relax in a manner that allows them to function as a pump. Electrocardiogram waves correspond to the electical events in the heart as follows: P wave (depolarization of the atria); QRS wave (depolarization of the ventricles); and T wave (repolarization of the ventricles).

As described in chapter 12, myocardial cells are short, branched, and interconnected by *gap junctions* which function as electrical synapses. The entire mass of cells interconnected by gap junctions is known as a *myocardium*. A myocardium is potentially a single functioning unit, or *functional syncitium*, since action potentials that originate in any cell in the mass can be transmitted to all the other cells. The atria and ventricles are different myocardia that are separated by the fibrous skeleton of the heart, as previously described. Since the impulse normally originates in the atria, the atrial myocardium is excited before that of the ventricles.

### Electrical Activity of the Heart

If the heart of a frog is removed from the body, and all neural innervations are severed, it will still continue to beat as long as the myocardial cells remain alive. The automatic nature of the heartbeat is referred to as *automaticity*. Experiments with isolated myocardial cells, and clinical experience with patients who have specific heart disorders, has revealed that there are many regions within the heart that are capable of originating action potentials and functioning as a pacemaker. In a normal heart, however, only one region demonstrates spontaneous electrical activity and by this means functions as a pacemaker. This pacemaker region is called the **sinoatrial node**, or **SA node**. The SA node is located in the right atrium near the opening of the superior vena cava, and is about 5 mm wide, 15 mm long, and 1.5 mm deep.

The cells of the SA node do not keep a resting membrane potential in the manner of resting neurons or skeletal muscle cells. Instead, during the period of diastole, the SA node exhibits a spontaneous depolarization called the **pacemaker potential**. The membrane potential begins at about $-60$ mV and gradually depolarizes to $-40$ mV, which is the threshold for producing an action potential in these cells. This spontaneous depolarization is produced by the diffusion of $Ca^{++}$ through openings in the membrane called *slow calcium channels*. At the threshold level

**Figure 13.16.** Pacemaker potentials and action potentials in the SA node.

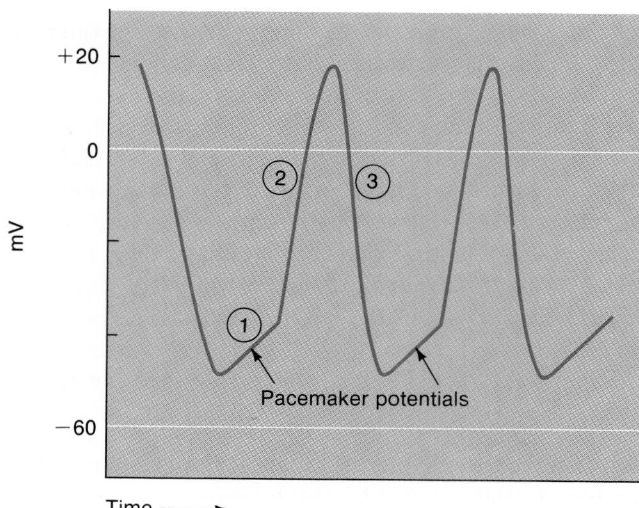


**Figure 13.17.** An action potential in a myocardial cell from the ventricles. The plateau phase of the action potential is maintained by a slow inward diffusion of $Ca^{++}$. The cardiac action potential, as a result, has a duration that is about one hundred times longer than the "spike potential" of an axon.

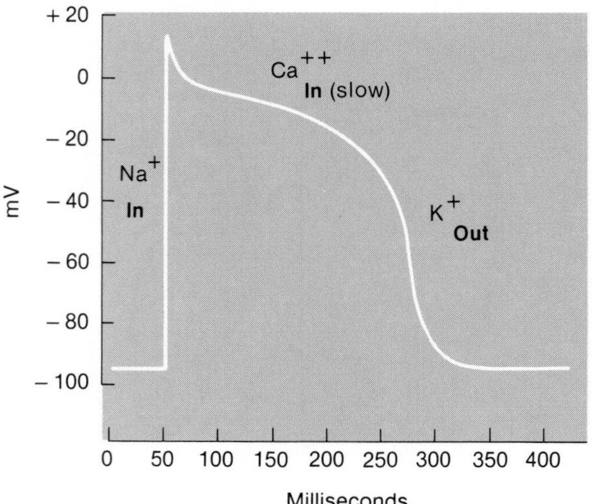


of depolarization other channels, called *fast calcium channels*, open, and $Ca^{++}$ rapidly diffuses into the cells. The opening of voltage-regulated $Na^+$ gates, and inward diffusion of $Na^+$ that results, may also contribute to the upshoot phase of the action potential in pacemaker cells (fig. 13.16). Repolarization is produced by the opening of $K^+$ gates and outward diffusion of $K^+$, as in the other excitable tissues previously discussed. Once repolarization to −60 mV has been achieved, a new pacemaker potential begins that again culminates, at the end of diastole, with a new action potential.

Some other regions of the heart, including the area around the SA node and the atrioventricular bundle, can potentially produce pacemaker potentials. The rate of spontaneous depolarization of these cells, however, is slower than that of the SA node. The potential pacemaker cells are, therefore, stimulated by action potentials from the SA node before they can stimulate themselves through their own pacemaker potentials. If action potentials from the SA node are prevented from reaching these areas (through blockage of conduction), they will generate pacemaker potentials at their own rate and serve as sites for the origin of action potentials; they will function as pacemakers. A pacemaker other than the SA node is called an *ectopic pacemaker* (also called an *ectopic focus*). From this discussion, it is clear that the rhythm set by such an ectopic pacemaker is usually slower than that normally set by the SA node.

Once another myocardial cell has been stimulated by action potentials originating in the SA node it produces its own action potential. The majority of myocardial cells have resting membrane potentials of about −90 mV. When stimulated by action potentials from a pacemaker region, these cells become depolarized to threshold, at which point their voltage-regulated $Na^+$ gates open. The upshoot phase of the action potential of nonpacemaker cells is due to the inward diffusion of $Na^+$. Following the rapid reversal of the membrane polarity, the membrane potential quickly declines to about −10 to −20 mV. Unlike the action potential of other cells, however, this level of depolarization is maintained for 200–300 msec (milliseconds) before repolarization (fig. 13.17). This *plateau phase* results from a slow inward diffusion of $Ca^{++}$, which balances a slow outward diffusion of cations. Rapid repolarization at the end of the plateau phase is achieved, as in other cells, by the opening of $K^+$ gates and the rapid outward diffusion of $K^+$ that results.

***Conducting Tissues of the Heart.*** Action potentials that originate in the SA node spread to adjacent myocardial cells of the right and left atria through the gap junctions between these cells. Since the myocardium of the atria are separated from the myocardium of the ventricles by the fibrous skeleton of the heart, however, the impulse cannot be conducted directly from the atria to the ventricles. Specialized conducting tissue, composed of modified myocardial cells, is thus required.

Once the impulse spreads through the atria, it passes to the **atrioventricular node (AV node)**, which is located on the inferior portion of the interatrial septum (fig. 13.18). The AV node is smaller than the SA node, averaging 5 mm in length, 2 mm in width, and 0.5 mm in depth. From

**Figure 13.18.** The conduction system of the heart.

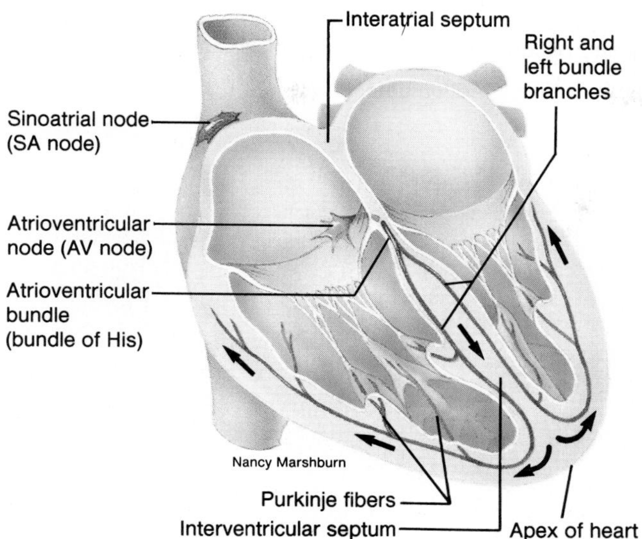


here, the impulse continues through the **atrioventricular bundle,** or **bundle of His,** beginning at the top of the interventricular septum. This conducting tissue pierces the fibrous skeleton of the heart and continues to descend along the interventricular septum. The atrioventricular bundle divides into right and left bundle branches, which are continuous with the **Purkinje fibers** within the ventricular walls. Stimulation of these fibers causes both ventricles to contract simultaneously and eject blood into the pulmonary and systemic circulation.

***Conduction of the Impulse.*** Action potentials from the SA node spread very quickly—at a rate of 0.8 to 1.0 meter (m) per second (sec)—across the myocardial cells of both atria. The conduction rate then slows considerably as the impulse passes into the AV node. Slow conduction of impulses (0.03 to 0.05 m/sec) through the AV node accounts for over half of the time delay between excitation of the atria and ventricles. After the impulses spread through the AV node, the conduction rate increases greatly in the atrioventricular bundle and reaches very high velocities (5 m/sec) in the Purkinje fibers. As a result of this rapid conduction of impulses, ventricular contraction begins about 0.1–0.2 second after the contraction of the atria.

Unlike skeletal muscles, the heart cannot sustain a contraction. This is because the atria and ventricles behave as if each were composed of only one muscle cell; the entire myocardium of each is electrically stimulated as a single unit and contracts as a unit. This contraction, which corresponds in time to the long action potential of myocardial cells and lasts almost 300 m/sec, is analogous to the twitch produced by a single skeletal muscle fiber (which lasts only 20–100 m/sec in comparison). The heart normally cannot be stimulated again until after it has relaxed from its previous contraction, because myocardial cells have *long refractory periods* (fig. 13.19), which correspond to the long duration of their action potentials. Summation of contractions is thus prevented, and the myocardium must relax after each contraction. The rhythmic pumping action of the heart is thus ensured.

Abnormal patterns of electrical conduction in the heart can produce abnormalities of the cardiac cycle and seriously compromise the function of the heart. These *arrhythmias* may be treated with a variety of drugs that inhibit specific aspects of the cardiac action potentials and, in this way, inhibit the production or conduction of impulses along abnormal pathways. Drugs used to treat arrhythmias may (1) block the fast $Na^+$ channel (quinidine, procainamide, lidocaine); (2) block the slow $Ca^{++}$ channel (verapamil); or (3) block $\beta$-adrenergic receptors (propranolol), since the rate of impulse production and conduction are stimulated by catecholamines.

**Figure 13.19.** The time course for the myocardial action potential (*A*) is compared with the duration of contraction (*B*). Notice that the long action potential results in a correspondingly long absolute refractory period (*ARP*) and relative refractory period (*RRP*). These refractory periods last almost as long as the contraction, so that the myocardial cells cannot be stimulated a second time until they have finished their contraction from the first stimulus.

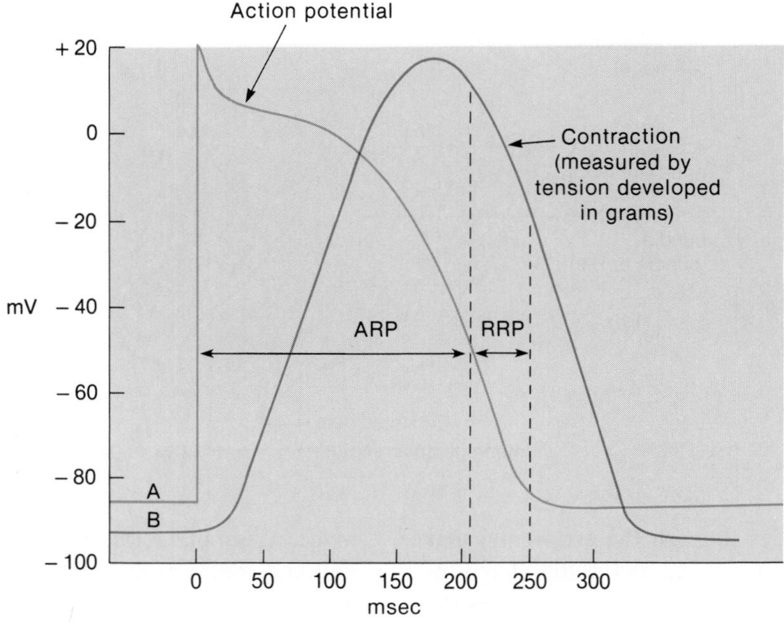


**Figure 13.20.** The electrocardiogram indicates the conduction of electrical impulses through the heart (*a*) and measures and records both the intensity of this electrical activity (in millivolts) and the time intervals involved (*b*).

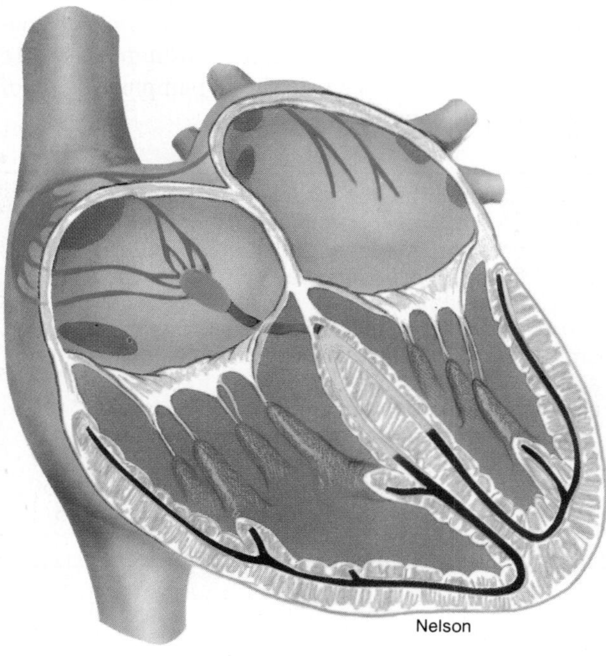


(a)

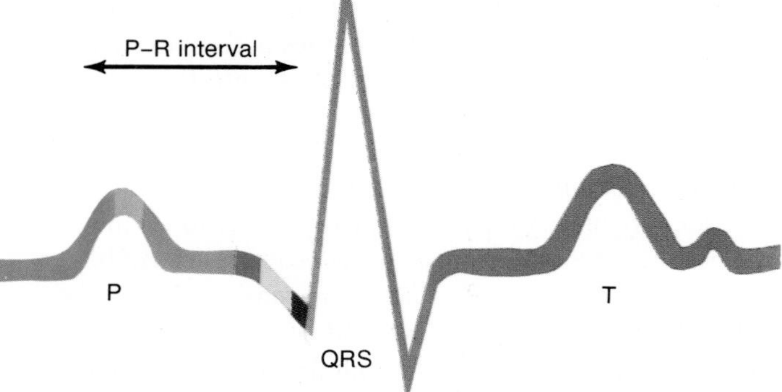


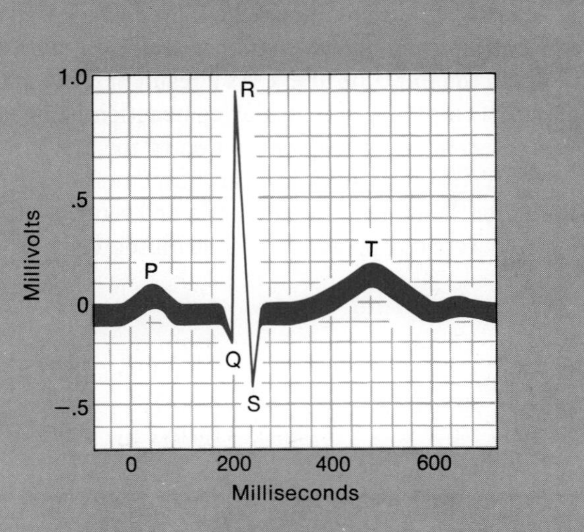


(b)

**Figure 13.21.** The placement of the bipolar limb leads and the exploratory electrode for the unipolar chest leads in an electrocardiogram (ECG). (RA = right arm, LA = left arm, LL = left leg.)

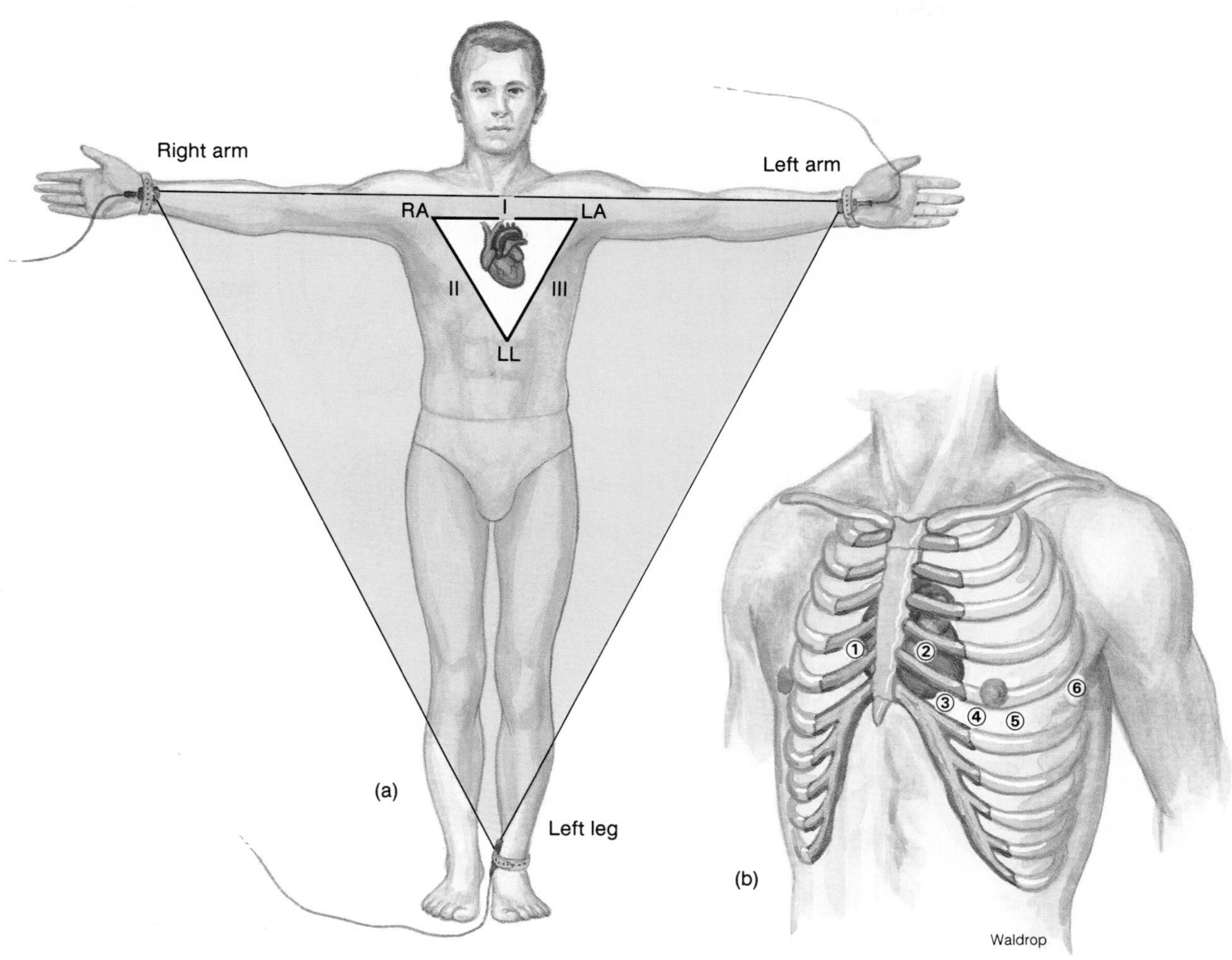


## The Electrocardiogram

A pair of surface electrodes placed directly on the heart will record a repeating pattern of potential changes. As action potentials spread from the atria to the ventricles, the voltage measured between these two electrodes will vary in a way that provides a "picture" of the electrical activity of the heart. The nature of this picture can be varied by changing the position of the recording electrodes; different positions provide different perspectives, enabling an observer to gain a more complete picture of the electrical events.

The body is a good conductor of electricity because tissue fluids contain a high concentration of ions that move (creating a current) in response to potential differences. Potential differences generated by the heart are thus conducted to the body surface, where they can be recorded by surface electrodes placed on the skin. The recording thus obtained is called an **electrocardiogram (ECG or EKG)** (fig. 13.20).

There are two types of ECG recording electrodes, or "leads." The *bipolar limb leads* record the voltage between electrodes placed on the wrists and legs. These bipolar leads include lead I (right arm to left arm), lead II (right arm to left leg), and lead III (left arm to left leg). In the *unipolar leads,* voltage is recorded between a single "exploratory electrode" placed on the body and an electrode that is built into the electrocardiograph and maintained at zero potential (ground).

The unipolar limb leads are placed on the right arm, left arm, and left leg, and are abbreviated AVR, AVL, and AVF, respectively. The unipolar chest leads are labeled one through six, starting from the midline position (fig. 13.21). There are thus a total of twelve standard ECG leads that "view" the changing pattern of the heart's electrical activity from different perspectives (table 13.7). This is important because certain abnormalities are best seen with particular leads and may not be visible at all with other leads.

**Figure 13.22.** The conduction of electrical impulses in the heart, as indicated by the electrocardiogram (ECG). The direction of the arrows in (*e*) indicates that depolarization of the ventricles occurs from the inside (endocardium) out (to the epicardium), whereas the arrows in (*g*) indicate that repolarization of the ventricles occurs in the opposite direction.

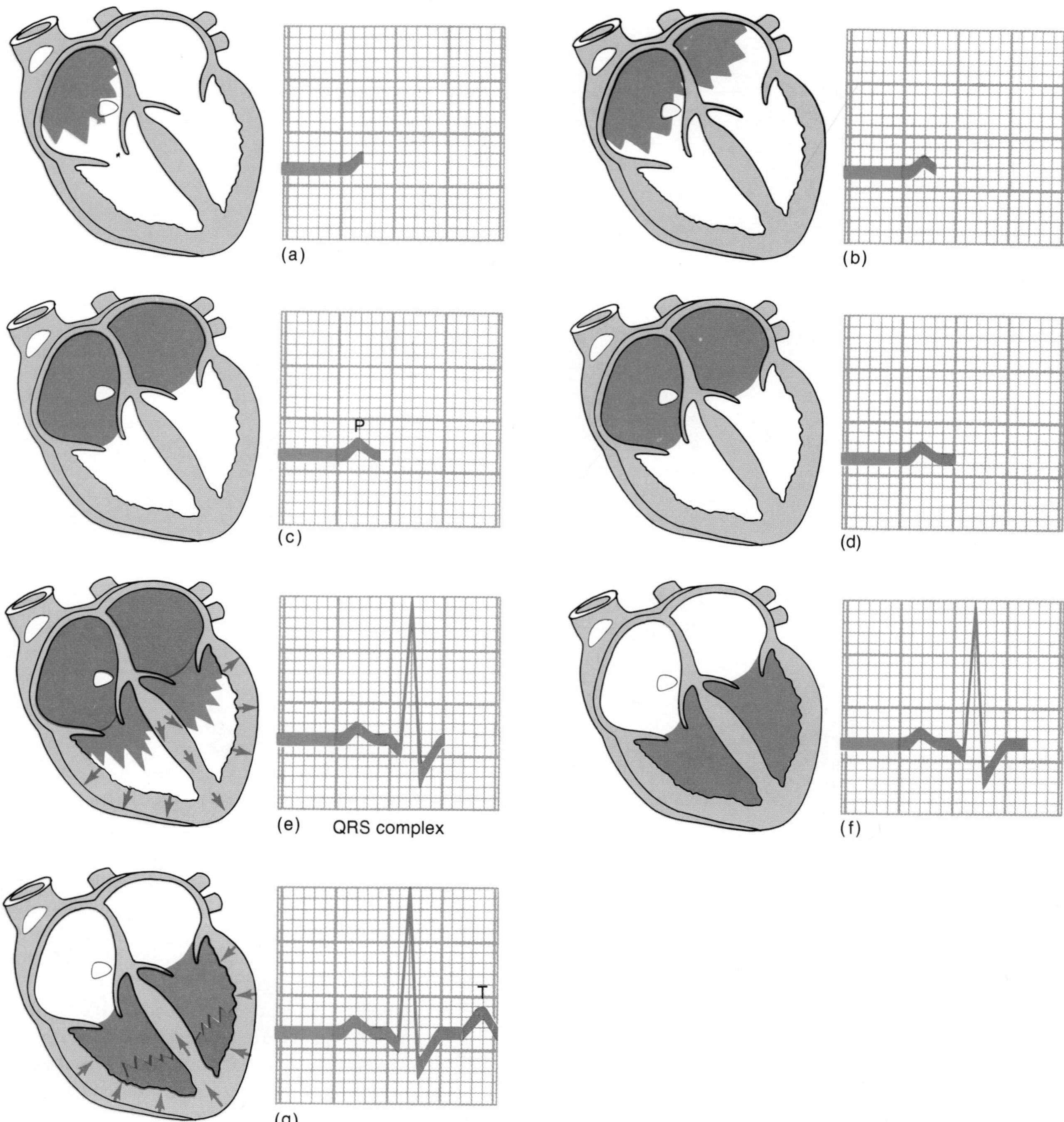

**Table 13.7** The electrocardiograph (ECG) leads

| Name of Lead | Placement of Electrodes |
|---|---|
| Bipolar limb leads | |
| I | Right arm and left arm |
| II | Right arm and left leg |
| III | Left arm and left leg |
| Unipolar limb leads | |
| AVR | Right arm |
| AVL | Left arm |
| AVF | Left leg |
| Unipolar chest leads | |
| $V_1$ | 4th intercostal space right of sternum |
| $V_2$ | 4th intercostal space left of sternum |
| $V_3$ | 5th intercostal space to the left of the sternum |
| $V_4$ | 5th intercostal space in line with the middle of the clavicle (collarbone) |
| $V_5$ | 5th intercostal space to the left of $V_4$ |
| $V_6$ | 5th intercostal space in line with the middle of the axilla (underarm) |

Each cardiac cycle produces three distinct ECG waves, designated P, QRS, and T. It should be noted that these waves are not action potentials; they represent changes in potential between two regions on the surface of the heart which are produced by the composite effects of action potentials in many myocardial cells. For example, the spread of depolarization through the atria causes a potential difference that is indicated by an upward deflection of the ECG line. When about half the mass of the atria is depolarized this upwards deflection reaches a maximum value, because the potential difference between the depolarized and unstimulated portions of the atria is at a maximum. When the entire mass of the atria is depolarized, the ECG returns to baseline because all regions of the atria have the same polarity. The spread of atrial depolarization thus creates the **P wave**.

Conduction of the impulse into the ventricles similarly creates a potential difference that results in a sharp upward deflection of the ECG line, which then returns to the baseline as the enire mass of the ventricles becomes depolarized. The spread of the depolarization into the ventricles is thus represented by the **QRS wave**. During this time the atria repolarize, but this event is hidden by the greater depolarization occurring in the ventricles. Finally, repolarization of the ventricles produces the **T wave** (fig. 13.22).

***Correlation of the ECG with Heart Sounds.*** Depolarization of the ventricles, as indicated by the QRS wave, stimulates contraction by promoting the uptake of $Ca^{++}$ into the regions of the sarcomeres. The QRS wave is thus seen to occur at the beginning systole. The rise in intraventricular pressure that results causes the AV valves to close, so that the first heart sound ($S_1$, or lub) is produced immediately after the QRS wave (fig. 13.23).

**Figure 13.23.** The relationship between changes in intraventricular pressure and the electrocardiogram during the cardiac cycle. The QRS wave (representing depolarization of the ventricles) occurs at the beginning of systole, whereas the T wave (representing repolarization of the ventricles) occurs at the beginning of diastole.

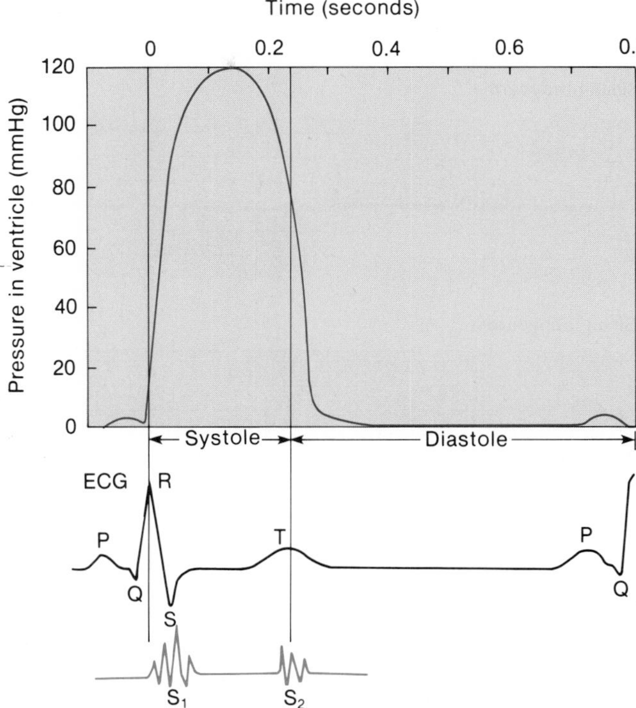


Repolarization of the ventricles, as indicated by the T wave, occurs at the same time that the ventricles relax at the beginning of diastole. The resulting fall in intraventricular pressure causes the aortic and pulmonary semilunar valves to close, so that the second heart sound ($S_2$, or dub) is produced shortly after the T wave in an electrocardiogram begins.

## Arrhythmias Detected by the Electrocardiograph

Arrhythmias, or abnormal heart rhythms, can be detected and described by the abnormal ECG tracings they produce. Although proper clinical interpretation of electrocardiograms requires more information than is presented in this chapter, some knowledge of abnormal rhythms is interesting in itself and is useful in gaining an understanding of normal physiology.

**Figure 13.24.** In (*a*) the heartbeat is paced by the normal pacemaker—the SA node (hence the name sinus rhythm). This can be abnormally slow (bradycardia—46 beats per minute in this example) or fast (tachycardia—136 beats per minute in this example). Compare the pattern of tachycardia in (*a*) with the tachycardia in (*b*), which is produced by an ectopic pacemaker in the ventricles. This dangerous condition can quickly lead to ventricular fibrillation, also shown in (*b*).

Sinus bradycardia

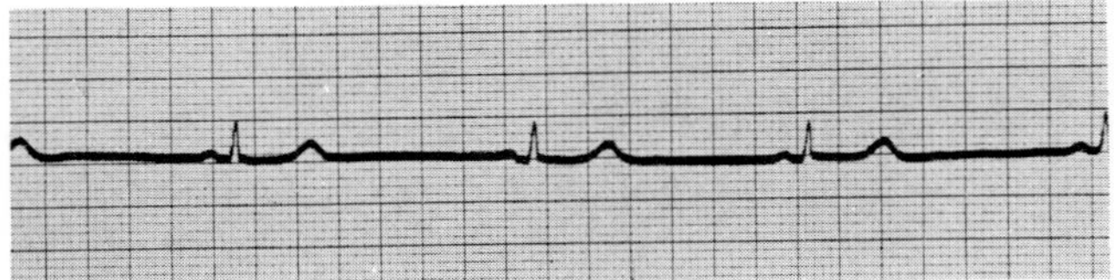

Sinus tachycardia

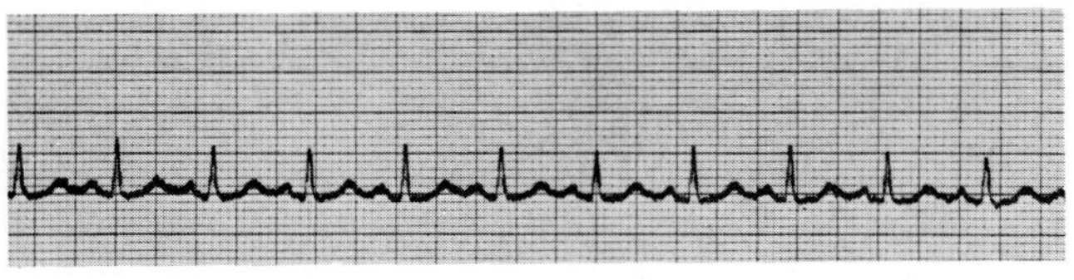

(a)

Ventricular tachycardia

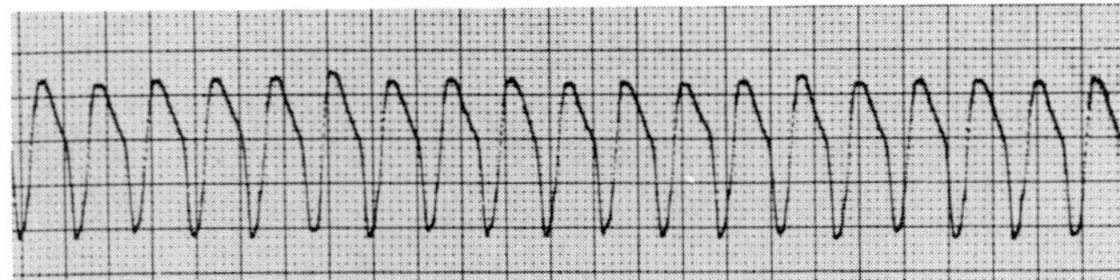

Ventricular fibrillation

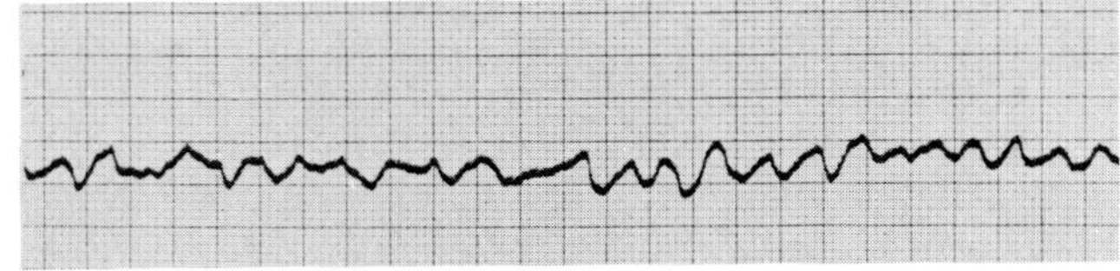

(b)

Since a heartbeat occurs whenever a normal QRS complex is seen and since the ECG chartpaper moves at a known speed, so that its x-axis indicates time, the cardiac rate (beats per minute) can be easily obtained from examination of the ECG recording. A cardiac rate slower than 60 beats per minute indicates **bradycardia;** a rate faster than 100 beats per minute is described as **tachycardia** (fig. 13.24).

Both bradycardia and tachycardia can occur normally. Endurance-trained athletes, for example, commonly have a slower heart rate than the general population. This *athlete's bradycardia* occurs as a result of higher levels of parasympathetic inhibition of the SA node and is a beneficial adaptation. Activation of the sympathetic system, during exercise or emergencies ("fight or flight"), causes a normal tachycardia to occur.

Abnormal tachycardia occurs when a person is at rest. This may result from abnormally fast pacing by the atria, due to drugs or to the development of abnormally fast *ectopic pacemakers*—cells located outside the SA node that assume a pacemaker function. This abnormal atrial tachycardia thus differs from normal "sinus" (SA node) tachycardia. *Ventricular tachycardia* results when abnormally fast ectopic pacemakers in the ventricles cause them to beat rapidly and independently of the atria. This is very dangerous because it can quickly degenerate into a lethal condition known as ventricular fibrillation.

***Flutter and Fibrillation.*** Extremely rapid rates of electrical excitation and contraction of either the atria or the ventricles may occur. In *flutter,* the contractions are very rapid (200–300 per minute) but are coordinated. In *fibrillation,* contractions of different groups of myocardial fibers occur at different times so that a coordinated pumping action of the chambers is impossible.

*Atrial flutter* usually degenerates quickly into *atrial fibrillation.* This causes the pumping action of the atria to stop. Since the ventricles fill to about 80% of their end-diastolic volume before atrial contraction normally occurs, however, the heart is still able to eject a sufficient quantity of blood into the circulation. People who have atrial fibrillation can thus live for many years. People who have *ventricular fibrillation,* in contrast, can live for only a few minutes before the brain and heart—which are very dependent upon oxygen for their metabolism—cease to function.

Fibrillation is caused by a continuous recycling of electrical waves, known as **circus rhythms,** through the myocardium. This recycling is normally prevented by the fact that the entire myocardium enters a simultaneous refractory period (due to the long duration of action potentials, as previously discussed). If some cells emerge from their refractory period before others, however, electrical waves can be continuously regenerated and conducted. Recycling of electrical waves along continuously changing pathways produces uncoordinated contraction and an impotent pumping action.

Circus rhythms are thus produced whenever impulses can be conducted without interruption by nonrefractory tissue. This may occur when the conduction pathway is longer than normal, as in a dilated heart. It can also be produced by an electric shock delivered during the middle of the T wave, when different myocardial cells are in different stages of recovery from their refractory period. Finally, circus rhythms and fibrillation may be produced by damage to the myocardium, which slows the normal rate of impulse conduction.

Fibrillation can sometimes be stopped by a strong electric shock delivered to the chest. This procedure is called **electrical defibrillation.** The electric shock depolarizes all the myocardial cells at the same time, causing them all to enter a refractory state. Conduction of circus rhythms thus stops, and—within a couple of minutes—the SA node can begin to stimulate contraction in a normal fashion. This does not correct the initial problem that caused circus rhythms and fibrillation, of course, but it does keep the person alive long enough to take other corrective measures.

***AV Node Block.*** The time interval between the beginning of atrial depolarization—indicated by the P wave—and the beginning of ventricular depolarization (as shown by the Q part of the QRS complex) is called the *P-R interval.* In the normal heart this time interval is 0.12 to 0.20 second. Damage to the AV node causes slowing of impulse conduction and is reflected by changes in the P-R interval. This condition is known as *AV node block* (fig. 13.25).

**First-degree AV node block** occurs when the rate of impulse conduction through the AV node (as reflected by the P-R interval) is greater than 0.20 second. **Second-degree AV node block** occurs when the AV node is damaged so severely that only one out of every two, three, or four atrial electrical waves can pass through to the ventricles. This is indicated in an ECG by the presence of P waves without associated QRS waves.

In **third-degree,** or **complete, AV node block,** none of the atrial waves can pass through the AV node to the ventricles. The atria are paced by the SA node (follow a normal "sinus rhythm"), but the ventricles are paced by a different, ectopic pacemaker (usually located in the bundle of His or Purkinje fibers). Since the SA node is the normal pacemaker by virtue of the fact that it has the fastest cycle of electrical activity, the ectopic pacemaker in the ventricles causes the ventricles to beat at an abnormally slow rate. The bradycardia that results is usually corrected by insertion of an artificial pacemaker.

A number of abnormal conditions, including a blockage in conduction of the impulse along the bundle of His, require the insertion of an **artificial pacemaker.** These are battery powered devices, about the size of a locket, which may be placed in permanent position under the skin. The electrodes from the pacemaker are guided by means of a fluoroscope through a vein to the right atrium, through the tricuspid valve, and into the right ventricle. The electrodes are fixed to the trabeculae carnae and are in contact with the wall of the ventricle. When these electrodes deliver shocks—either at a continuous pace or on demand (when the heart's own impulse doesn't arrive on time)—both ventricles are depolarized and contract and then repolarize and relax just as they do in response to endogenous stimulation.

1. *Describe the electrical activity of the cells of the SA node, and explain how the SA node functions as the normal pacemaker.*
2. *Using a line diagram, illustrate a myocardial action potential and the time course for myocardial contraction. Describe how the relationship between these two events prevents the heart from sustaining a contraction and how it normally prevents circus rhythms of electrical activity.*
3. *Draw an ECG and label the waves. Indicate the electrical events in the heart that produce these waves.*
4. *Draw a figure illustrating the relationship between ECG waves and the heart sounds. Explain this relationship.*
5. *Using a flowchart (arrows), describe the pathway of electrical conduction of the heart, starting with the SA node. Explain how damage to the AV node affects this conduction pathway and the ECG.*
6. *Describe normal and pathological examples of bradycardia and tachycardia. Also, describe how flutter and fibrillation are produced.*

**Figure 13.25.** Atrioventricular (AV) node block. In first-degree block the P-R interval is greater than 0.20 second (in the example here, the P-R interval is 0.26–0.28 second). In second-degree block P waves are seen that are not accompanied by QRS waves; in this example, the atria are beating 90 times per minute (as represented by the P waves), while the ventricles are beating at 50 times per minute (as represented by the QRS waves). In third-degree block the ventricles are paced independently of the atria by an ectopic pacemaker. Ventricular depolarization (QRS) and repolarization (T) therefore have a variable position in the electrocardiogram in relationship to the P waves (atrial depolarization).

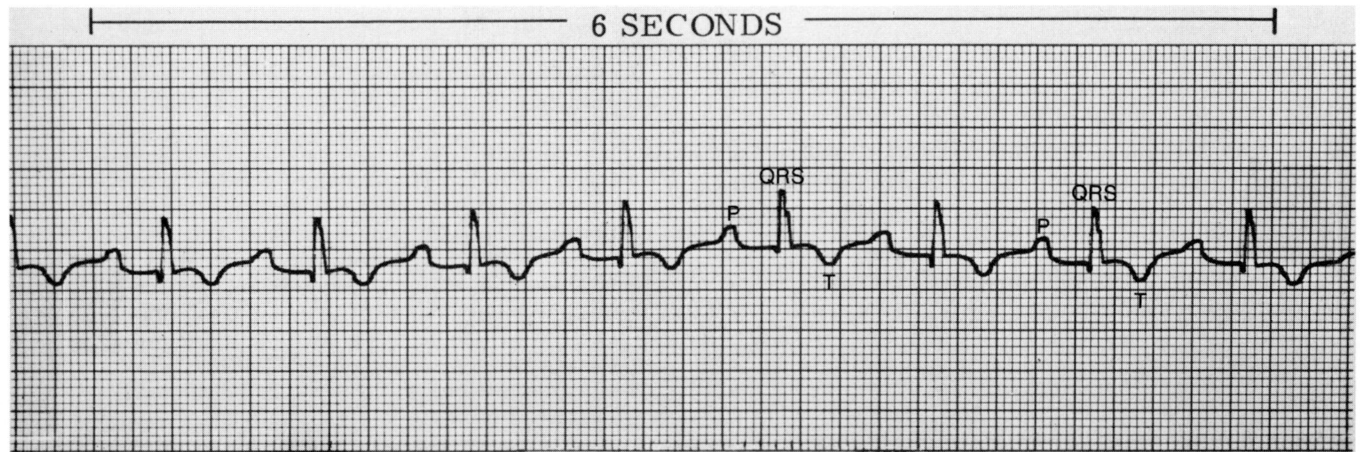


First-degree AV block

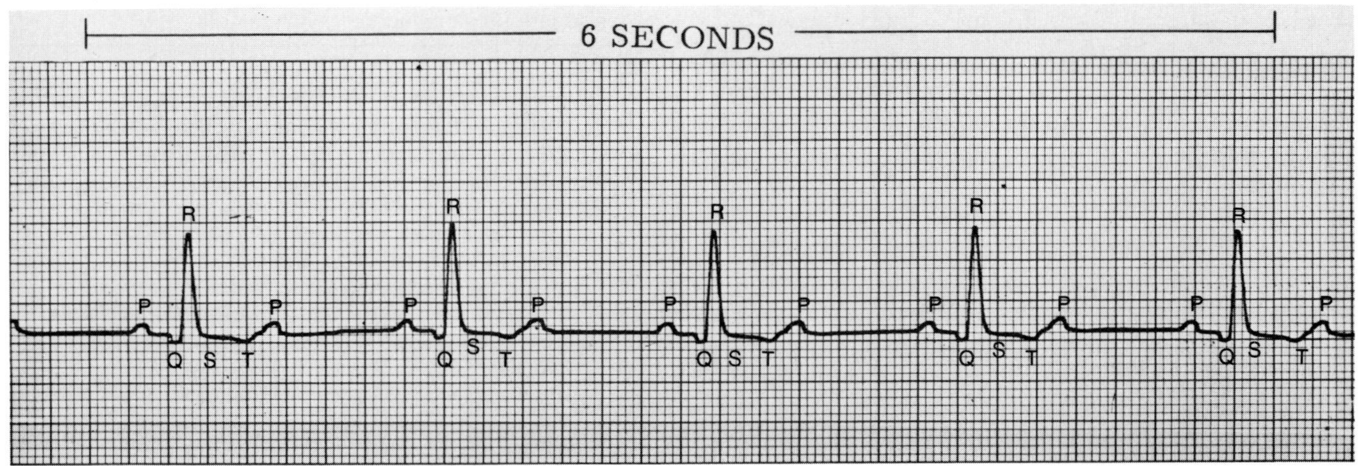


Second-degree AV block

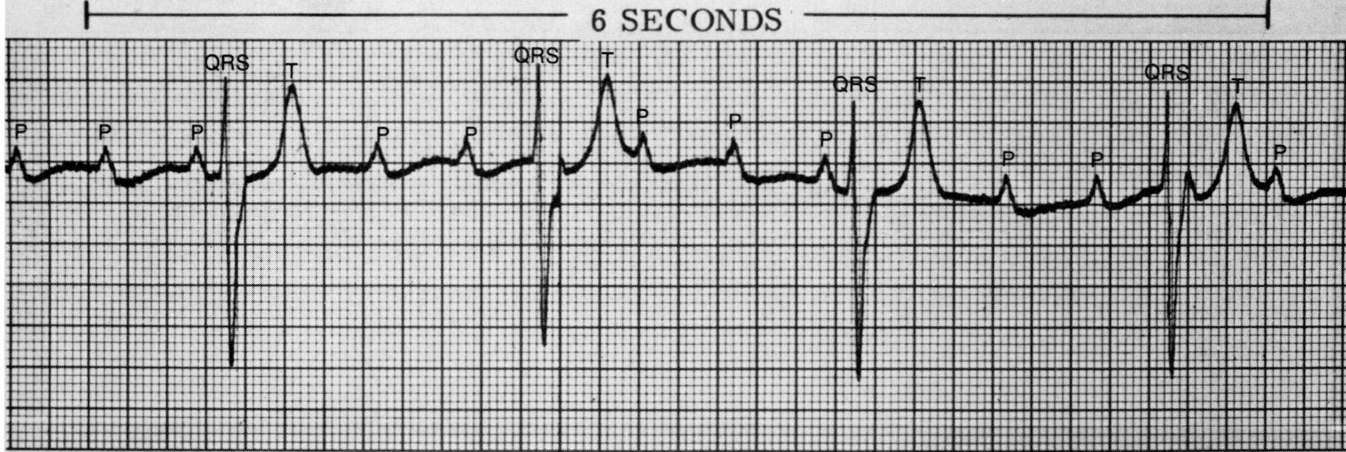


Third-degree AV block

**Figure 13.26.** The structure of a medium-sized artery and vein showing the relative thickness and composition of the tunicas.

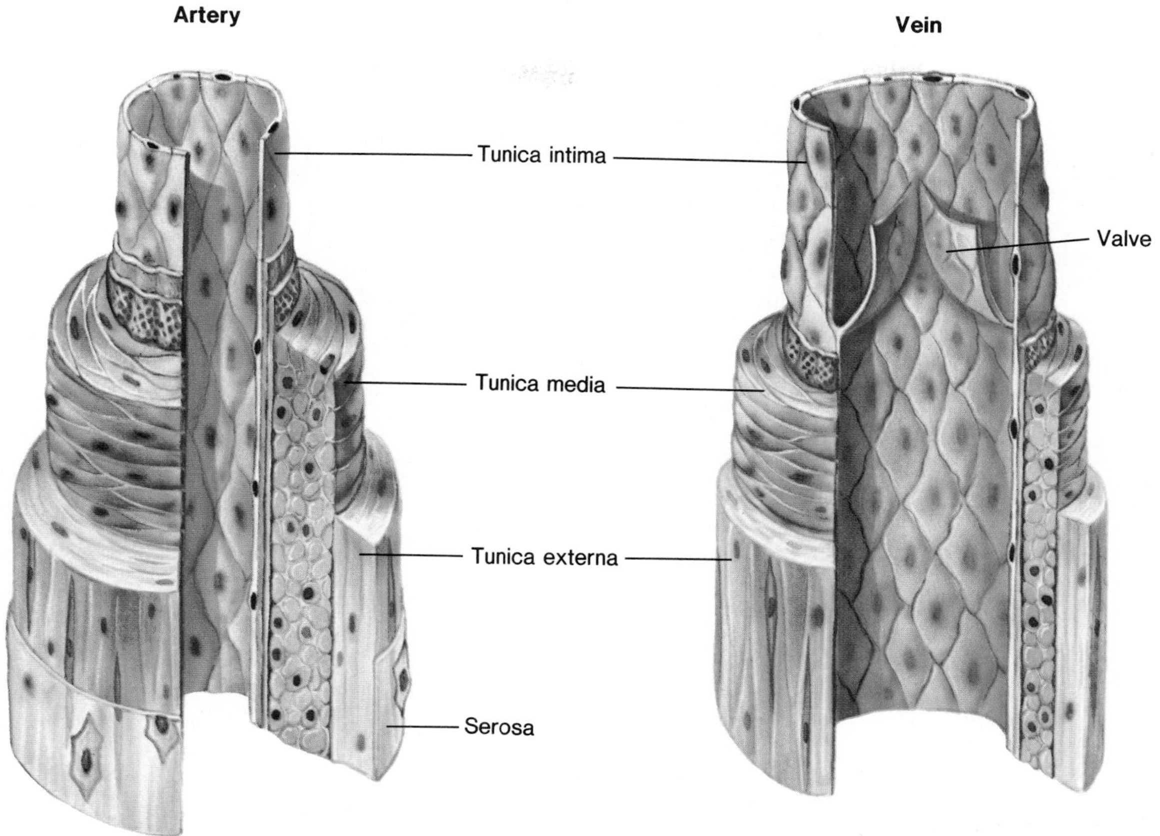


## Blood Vessels

The thick muscle layer of arteries allows them to transmit blood under high pressure ejected from the heart, and the elastic recoil of the large arteries further contributes to blood flow. The thinner muscle layer of veins allows them to distend when an increased amount of blood enters them, and their one-way valves ensure that blood flows back to the heart. Capillaries are composed of only a single layer of endothelium, which allows water and other molecules to move across the capillary walls and thus permits exchanges between the blood and tissue fluid.

Blood vessels form a tubular network throughout the body that permits blood to flow from the heart to all the living cells of the body and then back to the heart. Blood leaving the heart passes through vessels of progressively smaller diameters, referred to as **arteries, arterioles,** and **capillaries.** Capillaries are microscopic vessels that join the arterial flow to the venous flow. Blood returning to the heart from the capillaries passes through vessels of progressively larger diameters, called **venules** and **veins.**

The walls of arteries and veins are composed of three coats, or "tunics." The outermost layer is the **tunica externa,** the middle layer is the **tunica media,** and the inner layer is the **tunica intima.** The tunica externa is composed of connective tissue. The tunica media is composed primarily of smooth muscle. The tunica intima consists of three parts: (1) an innermost simple squamous epithelium, the *endothelium,* which lines the lumina of all blood vessels; (2) the basement membrane (a layer of glycoproteins) overlying some connective tissue fibers; and (3) a layer of elastic fibers, or *elastin,* forming an *internal elastic lamina.*

Although both arteries and veins have the same basic structure (fig. 13.26), there are some important differences between these vessels. Arteries have more muscle for their diameters than do comparably sized veins. As a result, arteries appear more round in cross section, whereas veins are usually partially collapsed. In addition, many veins have valves, which are absent in arteries.

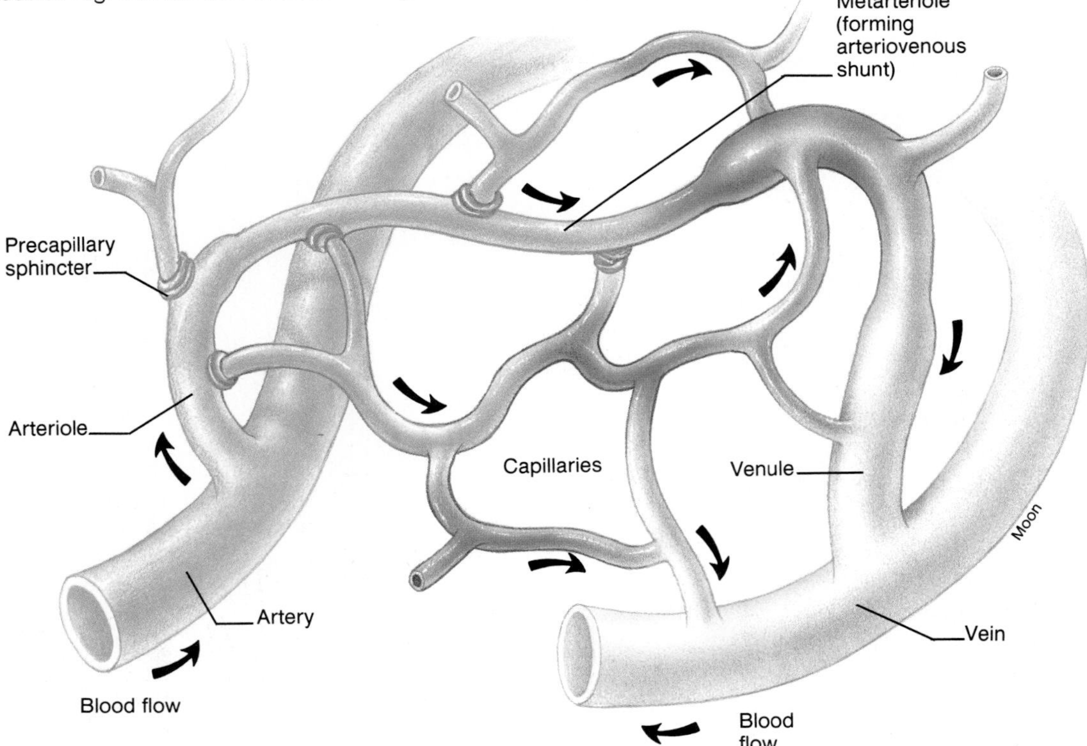


**Figure 13.27.** The microcirculation. Metarterioles (arteriovenous anastomoses) provide a path of least resistance between arterioles and venules. Precapillary sphincter muscles regulate the flow of blood through the capillaries.

## Arteries

The aorta and other large arteries contain numerous layers of elastin fibers between smooth muscle cells in the tunica media. These large arteries expand when the pressure of the blood rises as a result of the heart's contraction; they recoil, like a stretched rubber band, when the blood pressure falls during relaxation of the heart. This elastic recoil helps to produce a smoother, less pulsatile flow of blood through the smaller arteries and arterioles.

The small arteries and arterioles are less elastic than the larger arteries and have a thicker layer of smooth muscle for their diameters. Unlike the larger **elastic arteries,** therefore, the smaller **muscular arteries** retain almost the same diameter as the pressure of the blood rises and falls during the heart's pumping activity. Since arterioles and small muscular arteries have narrow lumina, they provide the greatest resistance to blood flow through the arterial system.

Small muscular arteries that are 100 $\mu$m or less in diameter branch to form smaller arterioles (20–30 $\mu$m in diameter). In some tissues, blood from the arterioles can enter the venules directly through *arteriovenous anastomoses.* In most cases, however, blood from arterioles passes into capillaries (fig. 13.27). Capillaries are the narrowest of blood vessels (7–10 $\mu$m in diameter), and serve as the "business end" of the circulatory system in which exchanges of gases and nutrients between the blood and the tissues occur.

## Capillaries

The arterial system branches extensively (table 13.8) to deliver blood to over forty billion capillaries in the body. The extensiveness of these branchings is indicated by the fact that all tissue cells are located within a distance of only 60–80 $\mu$m of a capillary and by the fact that capillaries provide a total surface area of 1,000 square miles for exchanges between blood and tissue fluid.

Despite their large number, capillaries contain only about 250 ml of blood at any time, out of a total blood volume of about 5,000 ml (most is contained in the venous system). The amount of blood flowing through a particular capillary bed is determined in part by the action of the **precapillary sphincter muscles.** These muscles allow only 5%–10% of the capillary beds in skeletal muscles, for example, to be open at rest. Blood flow to an organ is regulated by the action of these precapillary sphincters and by the degree of resistance to blood flow (due to constriction or dilation) provided by the small arteries and arterioles in the organ.

**Table 13.8** Characteristics of the vascular supply to the mesenteries in a dog

| Kind of Vessel | Diameter (mm) | Number | Total Cross-Sectional Area (cm²) | Length (cm) | Total Volume (cm³) |
|---|---|---|---|---|---|
| Aorta | 10 | 1 | 0.8 | 40 | 30 |
| Large arteries | 3 | 40 | 3.0 | 20 | 60 |
| Main artery branches | 1 | 600 | 5.0 | 10 | 50 |
| Terminal branches | 0.6 | 1,800 | 5.0 | 1 | 25 |
| Arterioles | 0.02 | 40,000,000 | 125 | 0.2 | 25 |
| Capillaries | 0.008 | 1,200,000,000 | 600 | 0.1 | 60 |
| Venules | 0.03 | 80,000,000 | 570 | 0.2 | 110 |
| Terminal veins | 1.5 | 1,800 | 30 | 1 | 30 |
| Main venous branches | 2.4 | 600 | 27 | 10 | 270 |
| Large veins | 6.0 | 40 | 11 | 20 | 220 |
| Vena cava | 12.5 | 1 | 1.2 | 40 | 50 |
| | | | | | 930 |



Unlike the vessels of the arterial and venous systems, the walls of capillaries are composed of only one cell layer—a simple squamous epithelium, or endothelium (fig. 13.28). The absence of smooth muscle and connective-tissue layers permits a more rapid rate of transport of materials between the blood and the tissues.

***Types of Capillaries.*** Different organs have different types of capillaries, which are distinguished by significant differences in structure. In terms of their endothelial lining, these capillary types include those that are *continuous,* those that are *discontinuous,* and those that are *fenestrated.*

Continuous capillaries are those in which adjacent endothelial cells are closely joined together. These are found in muscles, lungs, adipose tissue, and in the central nervous system. Continuous capillaries in the CNS lack intercellular channels; this fact contributes to the blood-brain barrier. Continuous capillaries in other organs have narrow intercellular channels (about 40–45 Å in width), which allow the passage of molecules other than protein between the capillary blood and tissue fluid (fig. 13.29).

Examination of endothelial cells with an electron microscope has revealed the presence of pinocytotic vesicles (fig. 13.28), which suggests that the intracellular transport of material may occur across the capillary walls. This type of transport appears to be the only mechanism of capillary exchange available within the central nervous system and may account, in part, for the selective nature of the blood-brain barrier.

The kidneys, endocrine glands, and intestines have *fenestrated capillaries,* characterized by wide intercellular pores (800–1,000 Å) that are covered by a layer of mucoprotein which may serve as a diaphragm. In the bone marrow, liver, and spleen, the distance between endothelial cells is so great that these *discontinuous* capillaries appear as little cavities (*sinusoids*) in the organ.

**Figure 13.28.** An electron micrograph of a capillary in a coronary muscle. Notice the thin intercellular channel and the fact that the capillary wall is only one cell thick. Arrows show some of the many pinocytotic vesicles.

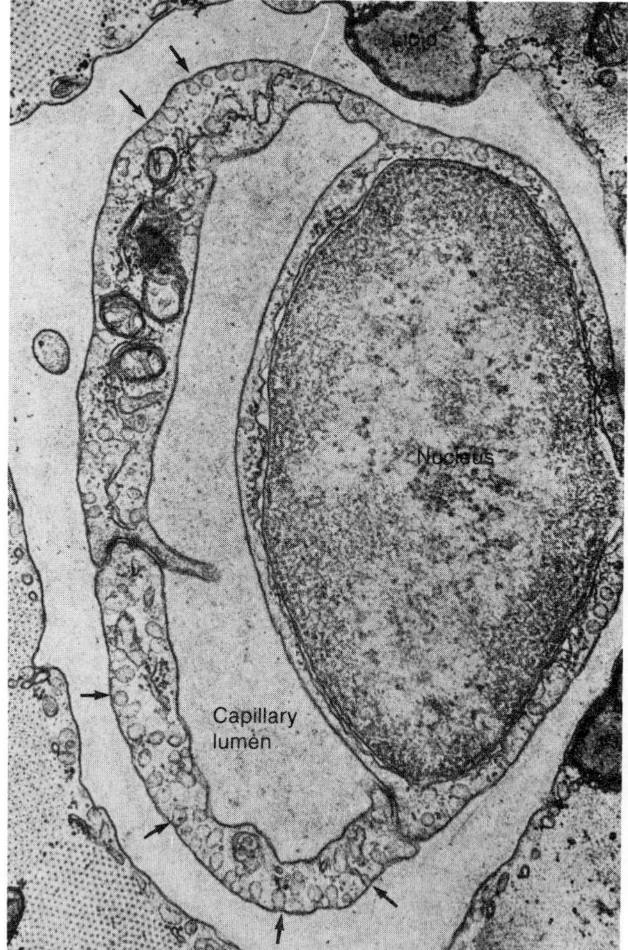

**Figure 13.29.** Diagrams of continuous, fenestrated, and discontinuous capillaries as they appear in the electron microscope. This classification is derived from the continuity of the endothelial layer.

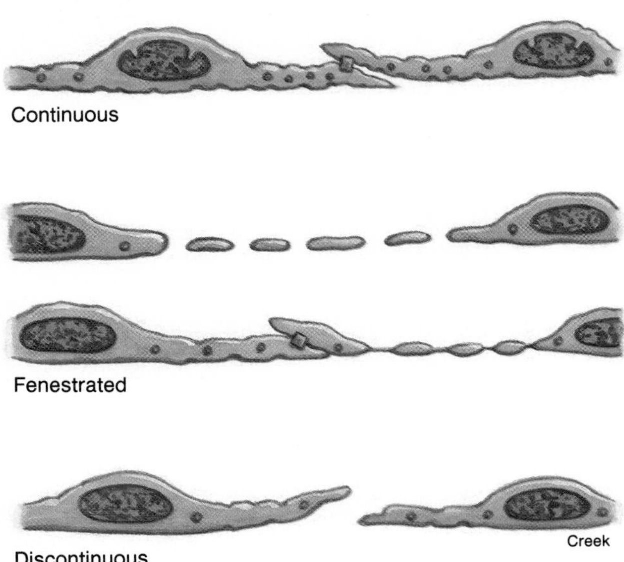


## Veins

The average pressure in the veins is only 2 mm Hg, compared to a much higher average arterial pressure of about 100 mm Hg. These pressures represent the hydrostatic pressure that the blood exerts on the walls of the vessels, and the numbers indicate the differences from atmospheric pressure.

The low venous pressure is insufficient to return blood to the heart, particularly from the lower limbs. Veins, however, pass between skeletal muscle groups that produce a massaging action as they contract (fig. 13.30). As the veins are squeezed by contracting skeletal muscles, a one-way flow of blood to the heart is ensured by the presence of **venous valves.** The ability of these valves to prevent the flow of blood away from the heart was demonstrated in the seventeenth century by William Harvey (fig. 13.31). After applying a tourniquet to a subject's arm, Harvey found that he could push the blood in a bulging vein toward the heart but not in the reverse direction.

The effect of the massaging action of skeletal muscles on venous blood flow is often described as the **skeletal muscle pump.** The rate of venous return to the heart is dependent, in large part, on the action of skeletal muscle pumps. When these pumps are less active, as when a person stands still or is bedridden, blood accumulates in the veins and causes them to bulge. When a person is more active, blood returns to the heart at a faster rate and less is left in the venous system.

The accumulation of blood in the veins of the legs over a long period of time, as may occur in people with occupations that require standing still all day, can cause the veins to stretch to the point where the venous valves are no longer efficient. This can produce **varicose veins.** During walking the movements of the foot activate the soleus muscle pump. This effect can be produced in bedridden people by upward and downward manipulations of the feet.

Action of the skeletal muscle pumps aid the return of venous blood from the lower limbs to the large abdominal veins. Movement of venous blood from abdominal to thoracic veins, however, is aided by an additional mechanism: breathing. When a person inhales, the diaphragm—a muscular sheet separating the thoracic and abdominal cavities—contracts. As the diaphragm contracts it changes from dome-shaped to a more flattened form and thus protrudes more into the abdomen. This has two effects; it increases the pressure in the abdomen, thus squeezing the abdominal veins, and it decreases the pressure in the thoracic cavity. The pressure difference in the veins created by this inspiratory movement of the diaphragm forces blood into the thoracic veins that return the venous blood to the heart.

## Atherosclerosis

Atherosclerosis is the most common form of arteriosclerosis (hardening of the arteries) and, through its contribution to heart disease and stroke, is responsible for about 50% of the deaths in the United States. In atherosclerosis, localized *plaques,* or **atheromas,** protrude into the lumen of the artery and thus reduce blood flow. The atheromas additionally serve as sites for *thrombus* (blood clot) formation, which can further occlude the blood supply to an organ (fig. 13.32).

It is currently believed that atheromas begin as "fatty streaks," which are gray-white areas that protrude into the lumen of arteries, particularly at arterial branch points. These are universally present in the aorta and coronary arteries by the age of ten, but progress to more advanced stages at different rates in different people. As this progression occurs, smooth muscle cells and others fill with lipids to give a "foamy cell" appearance. Later, damage to the endothelial lining of the artery causes blood platelets to adhere to subendothelial tissue. Platelets are believed to release chemicals that cause smooth muscle cells to proliferate and migrate from the tunica media to the intima, resulting in a tumorlike growth. The intercellular space of the atheroma later becomes filled with lipids and cholesterol crystals, and then is hardened by deposits of calcium. The damage to the endothelial covering, and

Figure 13.30. The action of the one-way venous valves. Contraction of skeletal muscles helps to pump blood toward the heart, but is prevented from pushing blood away from the heart by closure of the venous valves.

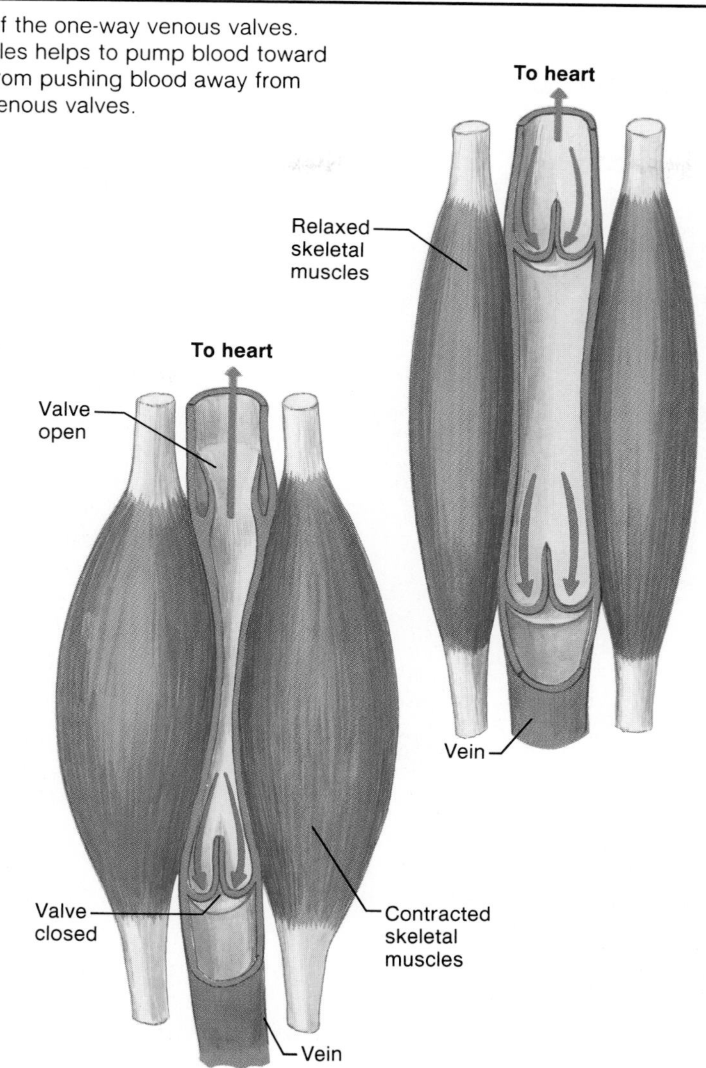


Figure 13.31. A classical demonstration by William Harvey of the existence of venous valves that prevent the flow of blood away from the heart.

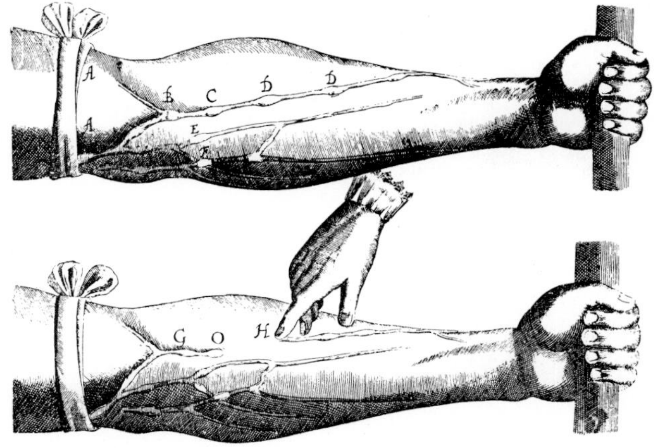

**Figure 13.32.** (*a*) The lumen (cavity) of a human coronary artery is partially occluded by an atheroma and (*b*) almost completely occluded by a thrombus. (*c*) The structure of an atheroma is diagrammed.

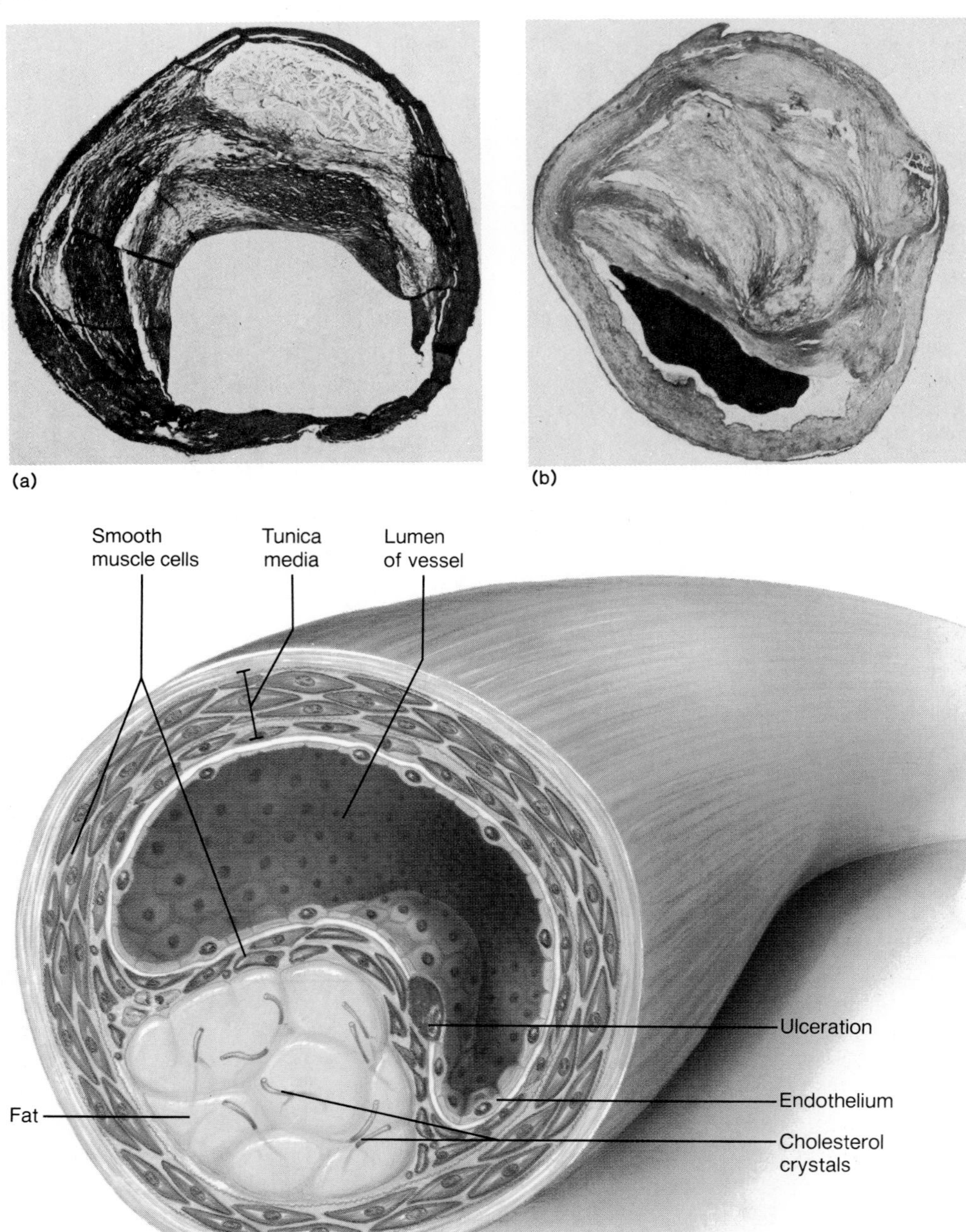

consequent adherence of platelets to the subendothelial tissue, also promotes the formation of blood clots at the atheromas which further occlude the artery.

Risk factors in the development of atherosclerosis include advanced age, smoking, hypertension, and high blood cholesterol concentrations. It is currently believed that an intact and properly functioning endothelium protects against atherosclerosis. If the endothelium in a particular region of an artery is removed, or if it is damaged in some way, growth factors, or *mitogens* (chemicals that stimulate mitosis), may stimulate proliferation of the smooth muscle cells and the growth of an atheroma. These growth factors may be derived from blood platelets, endothelial cells, and/or monocytes.

***Cholesterol and Plasma Lipoproteins.*** There is good evidence that high blood cholesterol is associated with an increased risk of atherosclerosis. This high blood cholesterol can be produced by a diet rich in cholesterol and saturated fat, or it may be the result of an inherited condition known as *familial hypercholesteremia.* This condition is inherited as a single dominant gene; individuals who inherit two of these genes have extremely high cholesterol concentrations (regardless of diet) and usually suffer heart attacks during childhood.

Lipids, including cholesterol, are carried in the blood attached to protein carriers (this topic is covered in detail in chapter 17). Cholesterol is carried to the arteries by plasma proteins called **low-density lipoproteins (LDL).** These particles, produced by the liver, consist of a core of cholesterol surrounded by a layer of phospholipids (to make the particle water-soluble) and a protein. Cells in various organs contain receptors for the protein in LDL; when LDL attaches to its receptors, the cell engulfs the LDL by receptor-mediated endocytosis (described in chapter 6) and utilizes the cholesterol for different purposes. Most of the LDL in the blood is removed in this way by the liver.

Once LDL has passed through the endothelium of an artery, it may stimulate monocytes to enter the area and engulf the cholesterol (thereby becoming "foam cells"). The monocytes may then be stimulated by LDL to secrete a growth factor that either begins or contributes to the development of an atheroma. A high blood concentration of LDL favors these events. Recent evidence shows that people who eat a diet high in cholesterol and saturated fat, and people with familial hypercholesteremia, have a high blood LDL concentration because their tissues (principally the liver) have a low number of LDL receptors. With fewer LDL receptors the liver is less able to remove the LDL from the blood, the blood LDL concentration is raised, and the risk of atherosclerosis is greatly increased.

Excessive cholesterol may be released from cells and travel in the blood as **high-density lipoproteins (HDL),** which are removed by the liver. The cholesterol in HDL is not taken into the artery wall because these cells lack the membrane receptor required for endocytosis of HDL, and therefore this cholesterol does not contribute to atherosclerosis. Indeed, a high proportion of cholesterol in HDL as compared to LDL is beneficial, since it indicates that cholesterol may be traveling away from the blood vessels to the liver. The concentration of HDL-cholesterol appears to be higher and the risk of atherosclerosis lower in people who exercise regularly. The HDL-cholesterol concentration, for example, is higher in marathon runners than in joggers and is higher in joggers than in inactive men. Women in general have higher HDL-cholesterol concentrations and a lower risk of atherosclerosis than men.

---

Most people can significantly lower their blood cholesterol concentration through a regimen of exercise and diet. Since saturated fat in the diet raises blood cholesterol, such foods as fatty meat, egg yolks, and internal animal organs (liver, brain, etc.) should be eaten only sparingly. Cholesterol is also lowered by food containing omega-3-fatty acids (fatty acids with a double bond at the third carbon from the methyl end of the molecule). These molecules are derived from the essential fatty acid alpha-linoleic acid, produced in the chloroplasts of plants. Fish and other seafood provide a rich source of omega-3-fatty acids, and authorities have thus urged people to eat two or three fish meals per week (in place of beef) as a means of protecting against atherosclerosis.

---

***Ischemic Heart Disease.*** A tissue is said to be **ischemic** when it receives an inadequate supply of oxygen because of an inadequate blood flow. The most common cause of myocardial ischemia is atherosclerosis of the coronary arteries. The adequacy of blood flow is relative—it depends on the metabolic requirements of the tissue for oxygen. An obstruction in a coronary artery, for example, may allow sufficient blood flow at rest but may produce ischemia when the heart is stressed by exercise or emotional conditions.

**Figure 13.33.** Depression of the S-T segment of the electrocardiogram as a result of myocardial ischemia.

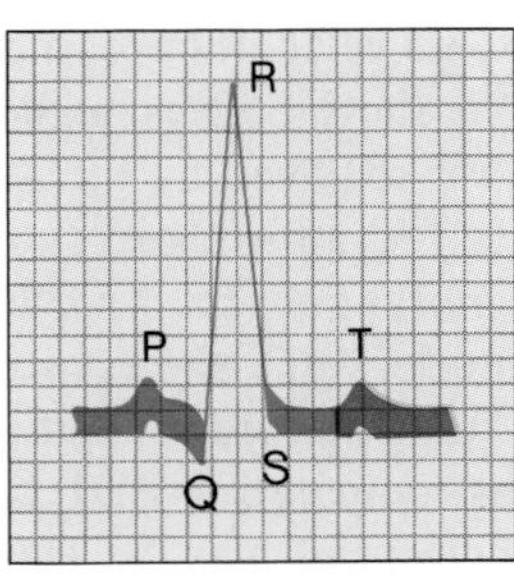

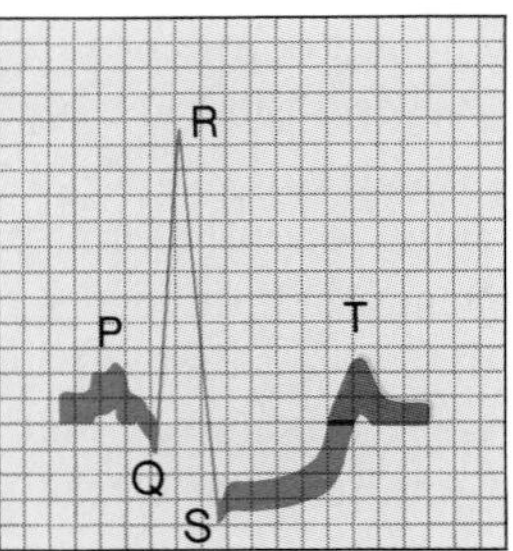


Normal    Ischemia

**Table 13.9** Changes in the enzyme activity in plasma following a myocardial infarction

| Serum Enzyme | Earliest Increase (hr) | Maximum Concentration (hr) | Return to Normal (days) | Amplitude of Increase (× normal) |
|---|---|---|---|---|
| Creatine kinase | 3 to 6 | 24 to 36 | 3 | 7 |
| Malate dehydrogenase | 4 to 6 | 24 to 48 | 5 | 4 |
| AST | 6 to 8 | 24 to 48 | 4 to 6 | 5 |
| Lactate dehydrogenase | 10 to 12 | 48 to 72 | 11 | 3 |
| $\alpha$-Hydroxybutyrate dehydrogenase | 10 to 12 | 48 to 72 | 13 | 3 to 4 |
| Aldolase | 6 to 8 | 24 to 48 | 4 | 4 |
| ALT | Usually normal, unless there are other complications | | | |
| Isocitrate dehydrogenase | Usually normal | | | |

From Montgomery, Rex, Dryer, Robert L., Conway, Thomas W., and Spector, Arthur A.: *Biochemistry: a case-oriented approach,* ed. 4, St. Louis, 1983, The C. V. Mosby Co.

Myocardial ischemia is associated with increased concentrations of blood lactic acid produced by anaerobic respiration of the ischemic tissue. This condition often causes substernal pain, which may also be referred to the left shoulder and arm, as well as other areas. This referred pain is called **angina pectoris.** People with angina frequently take nitroglycerin or related drugs that help relieve the ischemia and pain. These drugs are effective because they stimulate vasodilation, which improves circulation to the heart and decreases the work that the heart must perform to eject blood into the arteries.

Myocardial cells are adapted to respire aerobically and cannot respire anaerobically for more than a few minutes. If ischemia and anaerobic respiration continue for more than a few minutes, *necrosis* (cellular death) may occur in the areas most deprived of oxygen. A sudden, irreversible injury of this kind is called a **myocardial infarction,** or **MI.** The lay term "heart attack," though imprecise, usually refers to a myocardial infarction.

Myocardial ischemia may be detected by changes in the S-T segment of the electrocardiogram (fig. 13.33). The diagnosis of myocardial infarction is aided by measurement of the concentration of enzymes in the blood that are released by the infarcted tissue. Plasma concentrations of *creatine phosphokinase* (*CPK*), for example, increase within three to six hours after the onset of symptoms and return to normal after three days. Plasma levels of *lactate dehydrogenase* (*LDH*) reach a peak within forty-eight to seventy-two hours after the onset of symptoms and remain elevated for about eleven days (table 13.9).

1. *Describe the basic structural pattern of arteries and veins. Describe how arteries and veins differ in structure and how these differences contribute to the resistance function of arteries and the capacitance function of veins.*
2. *Describe the functional significance of the "skeletal muscle pump," and illustrate the action of venous valves.*
3. *Explain the functions of capillaries, and describe the structural differences between capillaries in different organs.*
4. *Explain how cholesterol is carried in the plasma and how the concentrations of cholesterol carriers are related to the risk of developing atherosclerosis.*
5. *Explain how angina pectoris is produced and the relationship of this symptom to conditions in the heart.*

**Figure 13.34.** A schematic diagram showing the structural relationship of a capillary bed and a lymph capillary.

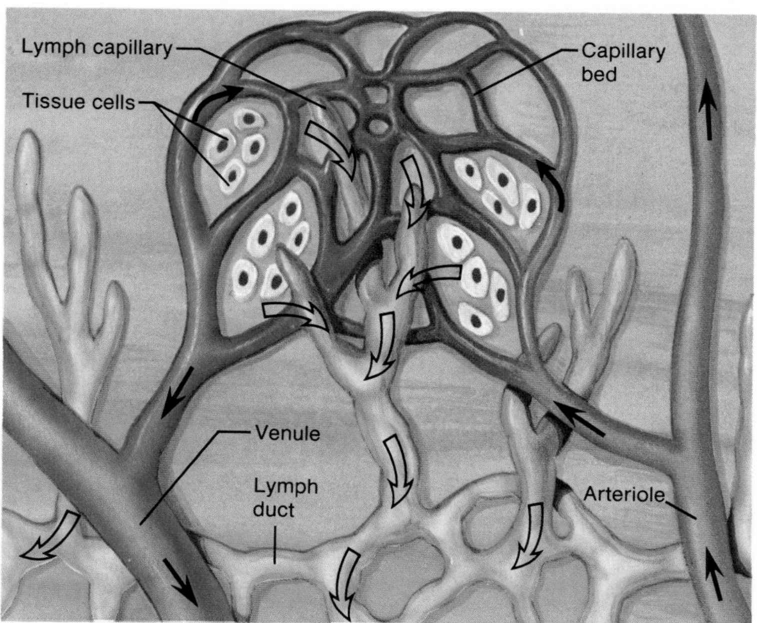


## LYMPHATIC SYSTEM

Lymphatic vessels absorb excess tissue fluid and transport this fluid—now called lymph—to ducts that drain into veins. Lymph nodes, and lymphoid tissue in the thymus, spleen, and tonsils produce lymphocytes, which are white blood cells involved in immunity.

The lymphatic system has three basic functions: (1) it transports interstitial (tissue) fluid, which was initially formed as a blood filtrate, back to the blood; (2) it transports absorbed fat from the small intestine to the blood; and (3) its cells—called *lymphocytes*—help provide immunological defenses against disease-causing agents.

The smallest vessels of the lymphatic system are the *lymph capillaries* (fig. 13.34). Lymph capillaries are microscopic, closed-ended tubes that form vast networks in the intercellular spaces within most organs. Because the walls of lymph capillaries are composed of endothelial cells with porous junctions, interstitial fluid, proteins, microorganisms, and absorbed fat (in the intestine—chapter 17) can easily enter. Once fluid enters the lymphatic capillaries, it is referred to as **lymph**.

From merging lymph capillaries, the lymph is carried into larger *lymphatic vessels*. The walls of lymphatic vessels are similar to those of veins in that they have the same three layers and contain valves to prevent the backflow of lymph. The lymphatic vessels eventually empty into one of two principal vessels: the *thoracic duct* and the *right lymphatic duct*. These ducts drain the lymph into the left and right subclavian veins, respectively. Thus interstitial fluid, which is formed by filtration of plasma out of blood capillaries (a process described in chapter 14), is ultimately returned back to the cardiovascular system (fig. 13.35).

Before the lymph is returned to the cardiovascular system, it is filtered through **lymph nodes** (fig. 13.36). Lymph nodes contain phagocytic cells, which help to remove pathogens, and *germinal centers*, which are the sites of lymphocyte production. The tonsils, thymus, and spleen—together called *lymphoid organs*—likewise contain germinal centers and are sites of lymphocyte production. Lymphocytes are the cells of the immune system which respond in a specific fashion to antigens, and their functions are described as part of the immune system in chapter 19.

1. *Compare the composition of lymph and blood, and describe the relationship between blood capillaries and lymphatic capillaries.*
2. *Explain how the lymphatic system differs from the cardiovascular system, and how these two systems are related.*
3. *Describe the functions of lymph nodes and lymphoid organs.*

**Figure 13.35.** The schematic relationship of the circulatory and lymphatic systems. Lymphatic vessels transport lymph fluid from interstitial spaces to the venous bloodstream.

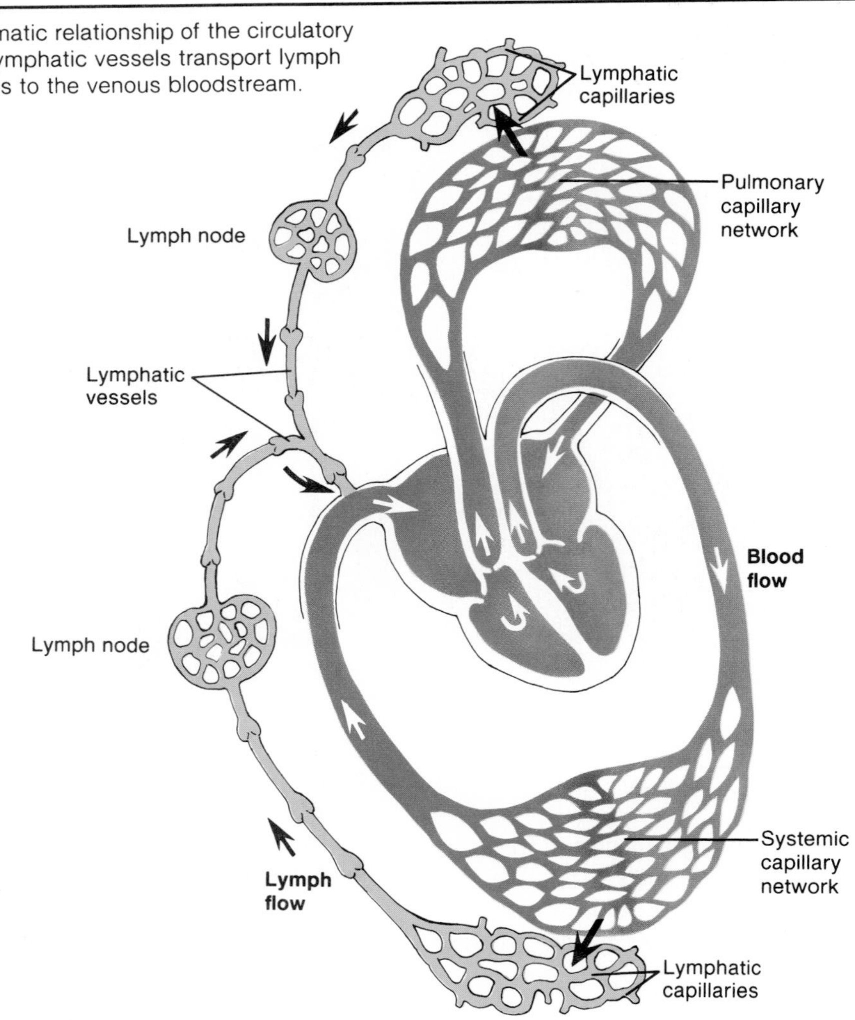


**Figure 13.36.** Photograph of a lymph node and lymphatic vessels.

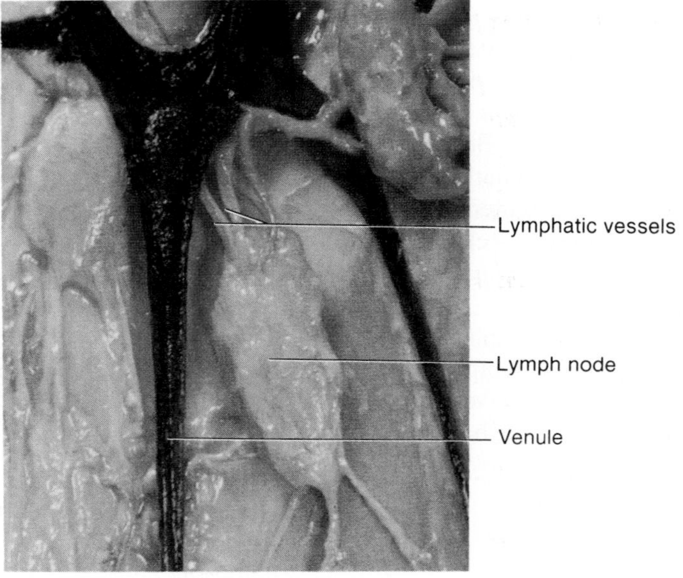

## Summary


### Functions and Components of the Circulatory System p. 356

I. The blood transports oxygen and nutrients to the tissue cells and removes waste products from the tissues; the blood also serves a regulatory function through its transport of hormones.
   A. Oxygen is carried by red blood cells, or erythrocytes.
   B. The white blood cells, or leukocytes, serve to protect the body from disease.

II. The circulatory system consists of the cardiovascular system (heart and blood vessels) and the lymphatic system.

### Composition of the Blood p. 357

I. Plasma is the fluid part of the blood, containing dissolved ions and various organic molecules.
   A. Hormones are found in the plasma portion of the blood.
   B. Plasma proteins are divided into albumins and alpha, beta, and gamma globulins.

II. The formed elements of the blood include erythrocytes, leukocytes, and platelets.
   A. Erythrocytes, or red blood cells, contain hemoglobin and transport oxygen.
   B. Leukocytes may be granular (also called polymorphonuclear) or agranular; they function in immunity.
   C. Thrombocytes, or platelets, are required for blood clotting.

III. Production of red blood cells is stimulated by the hormone erythropoietin, and development of different kinds of white blood cells is controlled by chemicals called lymphokines.

IV. The major blood typing groups are the ABO system and the Rh system.
   A. The type refers to the kind of antigens found on the red blood cells' surface.
   B. When different types of blood are mixed, antibodies against the red blood cell antigens cause the red blood cells to agglutinate

### Blood Clotting p. 363

I. When a blood vessel is damaged, platelets adhere to the exposed subendothelial collagen proteins.
   A. Platelets that stick to collagen undergo a release reaction, in which they secrete ADP, serotonin, and thromboxane $A_2$.
   B. Serotonin and thromboxane $A_2$ cause vasoconstriction; ADP and thromboxane $A_2$ attract other platelets and make them stick to the growing mass of platelets that are stuck to the collagen in the broken vessel.

II. In the formation of a blood clot, a soluble protein called fibrinogen is converted into insoluble threads of fibrin.
   A. This reaction is catalyzed by the enzyme thrombin.
   B. Thrombin is derived from its inactive precursor, called prothrombin, by either an intrinsic or an extrinsic pathway.
      1. The intrinsic pathway is the longer of the two and requires the activation of more clotting factors.
      2. The shorter extrinsic pathway is initiated by the secretion of tissue thromboplastin.
   C. The clotting sequence requires $Ca^{++}$ and phospholipids present in the platelet cell membranes.

III. Dissolution of the clot eventually occurs by the digestive action of plasmin, which cleaves fibrin into split products.

### Structure of the Heart p. 367

I. The right and left sides of the heart pump blood through the pulmonary and systemic circulations.
   A. The right ventricle pumps blood to the lungs; this blood then returns to the left atrium.
   B. The left ventricle pumps blood into the aorta and systemic arteries; this blood then returns to the right atrium.

II. The heart contains two pairs of one-way valves.
   A. The atrioventricular valves allow blood to go from the atria to the ventricles, but not in the reverse direction.
   B. The semilunar valves allow blood to leave the ventricles and enter the pulmonary and systemic circulations, but these valves prevent blood from returning from the arteries to the ventricles.

III. The electrical impulse begins in the sinoatrial node and spreads through both atria by electrical conduction from one myocardial cell to another.
   A. The impulse then excites the atrioventricular node, from which it is conducted by the bundle of His into the ventricles.
   B. The Purkinje fibers transmit the impulse into the ventricular muscle and cause it to contract.

### The Cardiac Cycle and the Heart Sounds p. 370

I. The heart is a two-step pump; first the atria contract and then the ventricles contract.
   A. During diastole, first the atria and then the ventricles fill with blood.
   B. The ventricles are about 80% filled before the atria contract and add the final 20% to the end-diastolic volume.
   C. Contraction of the ventricles ejects about two-thirds of their blood, leaving about one-third as the end-systolic volume.

II. When the ventricles contract at systole, the pressure within them first rises sufficiently to close the AV valves and then rises sufficiently to open the semilunar valves.
   A. Blood is ejected from the ventricles until the pressure within them falls below the pressure in the arteries; at this point the semilunar valves close and the ventricles begin relaxation.
   B. When the pressure in the ventricles falls below the pressure in the atria, a phase of rapid filling of the ventricles occurs, followed by the final filling caused by contraction of the atria.

III. Closing of the AV valves produces the first heart sound, or "lub," at systole; closing of the semilunar valves produces the second heart sound, or "dub," at diastole. Abnormal valves can cause abnormal sounds called murmurs.

### Electrical Activity of the Heart and the Electrocardiogram p. 373

I. In the normal heart the impulse originates in the SA node, due to a spontaneous depolarization called the pacemaker potential.
   A. When this spontaneous depolarization reaches a threshold value, an action potential is produced due to opening of the voltage-regulated $Na^+$ gates.
   B. Repolarization is produced by the outward diffusion of $K^+$, but a stable resting membrane potential is not attained because spontaneous depolarization once more occurs.

C. Other myocardial cells are capable of spontaneous activity, but the SA node is the normal pacemaker because its rate of spontaneous depolarization is the fastest.
D. When the action potential produced by the SA node reaches other myocardial cells, they produce action potentials that are quite long due to a slow, inward diffusion of $Ca^{++}$.
E. The long action potential and long refractory period of myocardial cells allows the entire mass of cells to be in a refractory period while it contracts; this prevents the myocardium from being stimulated again until after it relaxes.

II. The regular pattern of conduction in the heart produces a changing pattern of potential differences between two points on the body surface.
A. Measurements of the voltage between two points on the surface of the body caused by the electrical activity of the heart is called an electrocardiogram (ECG).
B. The P wave is caused by depolarization of the atria; the QRS wave is caused by depolarization of the ventricles; the T wave is produced by repolarization of the ventricles.
C. The ECG can be used to detect abnormal cardiac rates, abnormal conduction between the atria and ventricles, and other abnormal patterns of electrical conduction in the heart.

## Blood Vessels p. 383

I. Arteries contain three layers, or tunics: the intima, media, and externa.
A. The tunica intima consists of a layer of endothelium, which is separated from the tunica media by a band of elastin fibers.
B. The tunica media consists of smooth muscle.
C. Large arteries have many layers of elastin and are elastic; medium and small arteries and arterioles are less distensible and thus provide greater resistance to blood flow.

II. Capillaries are the narrowest but the most numerous of the blood vessels.
A. Capillary walls consist of only a single layer of endothelial cells, which provides for the exchange of molecules between the blood and the surrounding tissues.
B. The flow of blood from arterioles to capillaries is regulated by precapillary sphincter muscles.
C. The capillary wall may be continuous, fenestrated, or discontinuous.

III. Veins have the same three tunics as arteries, but generally have a thinner muscular layer than comparably sized arteries.
A. Veins are more distensible than arteries and can expand to hold a larger quantity of blood.
B. Many veins have venous valves that permit a one-way flow of blood to the heart.
C. The flow of blood back to the heart is aided by contraction of the skeletal muscles that surround veins; this action is called the skeletal muscle pump.

IV. Atherosclerosis of arteries can occlude blood flow to the heart and brain.
A. Atheromas begin as smooth muscle cells migrate into the tunica intima and proliferate; eventually the atheroma accumulates cholesterol and the endothelium frays, which promotes blood clot formation.
B. Atherosclerosis appears to be promoted by a high blood level of low-density lipoproteins (LDL), which carry cholesterol into the artery wall.
C. Occulsion of blood flow to the heart can lead to myocardial ischemia and myocardial infarction.

## Lymphatic System p. 391

I. Lymphatic capillaries are blind-ended but highly permeable, and can drain excess tissue fluid into lymphatic vessels.
II. Lymph passes through lymph nodes and is returned by way of lymph ducts to the venous blood.

# REVIEW ACTIVITIES

## Objective Questions

1. Which of the following statements is *false*?
   (a) Most of the total blood volume is contained in veins.
   (b) Capillaries have a greater total surface area than any other type of vessel.
   (c) Exchanges between blood and tissue fluid occur across the walls of venules.
   (d) Small arteries and arterioles present great resistance to blood flow.
2. All arteries in the body contain oxygen-rich blood with the exception of
   (a) the aorta
   (b) the pulmonary artery
   (c) the renal artery
   (d) the coronary arteries
3. The "lub," or first heart sound, is produced by closing of
   (a) the aortic semilunar valve
   (b) the pulmonary semilunar valve
   (c) the tricuspid valve
   (d) the bicuspid valve
   (e) both AV valves
4. The first heart sound is produced at
   (a) the beginning of systole
   (b) the end of systole
   (c) the beginning of diastole
   (d) the end of diastole
5. Changes in the cardiac rate primarily reflect changes in the duration of
   (a) systole
   (b) diastole
6. The QRS wave of an ECG is produced by
   (a) depolarization of the atria
   (b) repolarization of the atria
   (c) depolarization of the ventricles
   (d) repolarization of the ventricles
7. The second heart sound immediately follows the occurrence of
   (a) the P wave
   (b) the QRS wave
   (c) the T wave
8. The cells that normally have the fastest rate of spontaneous diastolic depolarization are located in
   (a) the SA node
   (b) the AV node
   (c) the bundle of His
   (d) the Purkinje fibers

9. Which of the following statements is *true?*
   (a) The heart can produce a graded contraction.
   (b) The heart can produce a sustained contraction.
   (c) The action potentials produced at each cardiac cycle normally travel around the heart in circus rhythms.
   (d) All of the myocardial cells in the ventricles are normally in a refractory period at the same time.
10. An ischemic injury to the heart that results in death of some myocardial cells is
    (a) angina pectoris
    (b) a myocardial infarction
    (c) fibrillation
    (d) heart block
11. The activation of factor X is
    (a) part of the intrinsic pathway only
    (b) part of the extrinsic pathway only
    (c) part of both the intrinsic and extrinsic pathways
    (d) not part of either the intrinsic or extrinsic pathways
12. Platelets
    (a) form a plug by sticking to each other
    (b) release chemicals that stimulate vasoconstriction
    (c) provide phospholipids needed for the intrinsic pathway
    (d) all of the above
13. Antibodies against both type A and type B antigens are found in the plasma of a person who is
    (a) type A
    (b) type B
    (c) type AB
    (d) type O
    (e) all of these
14. Production of which of the following blood cells is stimulated by a hormone secreted by the kidneys?
    (a) lymphocytes
    (b) monocytes
    (c) erythrocytes
    (d) neutrophils
    (e) thrombocytes
15. Which of the following statements about plasmin is *true*?
    (a) It is involved in the intrinsic clotting system.
    (b) It is involved in the extrinsic clotting system.
    (c) It functions in fibrinolysis.
    (d) It promotes the formation of emboli.
16. During the phase of isovolumetric relaxation of the ventricles, the pressure in the ventricles is
    (a) rising
    (b) falling
    (c) first rising, then falling
    (d) constant

### Essay Questions

1. Explain why the beat of the heart is automatic and why the SA node functions as the normal pacemaker.
2. Compare the duration of the heart's contraction with the myocardial action potential and refractory period. Explain the significance of these relationships.
3. Describe the pressure changes that occur during the cardiac cycle, and relate these changes to the occurrence of the heart sounds.
4. Describe the causes of the P, QRS, and T waves of an ECG, and indicate when each of these waves occurs in the cardiac cycle. Explain why the first heart sound occurs immediately after the QRS wave and why the second sound occurs at the time of the T wave.
5. Can a defective valve be detected by an ECG? Can a partially damaged AV node be detected by auscultation (listening) with a stethoscope? Explain.
6. Explain how a cut in the skin initiates both the intrinsic and extrinsic clotting pathways. Which pathway finishes first? Why?

## SELECTED READINGS


Benditt, D. G. *et al.* 1988. Supraventricular tachycardias: mechanisms and therapies. *Hospital Practice* 23:161.

Benditt, E. P. February 1977. The origin of atherosclerosis. *Scientific American.*

Berne, R. M., and M. N. Levy. 1981. *Cardiovascular physiology.* 4th ed. St. Louis: The C. V. Mosby Co.

Brown, M. S., and J. L. Goldstein. 1984. How LDL receptors influence cholesterol and atherosclerosis. *Scientific American.*

Conover, M. B. 1980. *Understanding electrocardiography.* 3d. ed. St. Louis: The C. V. Mosby Co.

Del Zoppo, G. J., and L. A. Harker. 1984. Blood/vessel interaction in coronary disease. *Hospital Practice* 19:163.

Dubin, D. 1981. *Rapid interpretation of EKG's.* 3d. ed. Tampa, Fla.: Cover Publishing Co.

Fitzgerald, D., and R. Lazzara. 1988. Functional anatomy of the conducting system. *Hospital Practice* 23:81.

Fulkow, B., and E. Neill. 1971. *Circulation.* London: Oxford University Press.

Garrard, J. M. 1988. Platelet aggregation: Cellular regionation and physiologic role. *Hospital Practice* 23:89.

Glasser, S. P., and R. G. Zoble. 1985. Management of cardiac arrhythmias. *Hospital Practice* 20:127.

Golde, D. W., and J. C. Gasson. July 1988. Hormones that stimulate the growth of blood cells. *Scientific American.*

Hills, D., and E. Braunwald. 1977. Myocardial ischemia. *New England Journal of Medicine* 296: first part, p. 971; second part, p. 1033; third part, p. 1093.

Knopp, R. H. 1988. New approaches to cholesterol lowering: efficacy and safety. *Hospital Practice, Supplement 1* 23:22.

Lawn, R. M., and G. A. Vehar. March 1986. The molecular genetics of hemophilia. *Scientific American.*

Leon, A. S. 1983. Exercise and coronary heart disease. *Hospital Practice* 18:38.

Little, R. C. 1981. *Physiology of the heart and circulation.* 2d. ed. Chicago: Year Book Medical Publishers.

Robinson, T. F., S. M. Factor, and E. H. Sonnenblick. June 1986. The heart as a suction pump. *Scientific American.*

Rosenberg, R. D., and K. A. Bauer. 1986. New insights into hypercoaguable states. *Hospital Practice* 21:131.

Ross, R. 1986. The pathogenesis of atherosclerosis: An update. *New England Journal of Medicine* 314:488.

Shen, W. K., and H. C. Strauss. 1988. Mechanisms of bradyarrythmias and blocks. *Hospital Practice* 23:93.

Smith, J. J., and J. P. Kampine. 1980. *Circulatory physiology: The essentials.* Baltimore: Williams & Wilkins Co.

Spear, J. F., and E. N. Moore. 1982. Mechanisms of cardiac arrhythmias. *Annual Review of Physiology* 44:485.

Stein, B. *et al.* 1988. Pathogenesis of coronary occlusion. *Hospital Practice* 23:87.

Wood, J. E. January 1968. The venous system. *Scientific American.*

Zucker, M. B. June 1980. The function of blood platelets. *Scientific American.*

# 14 Cardiac Output, Blood Flow, and Blood Pressure

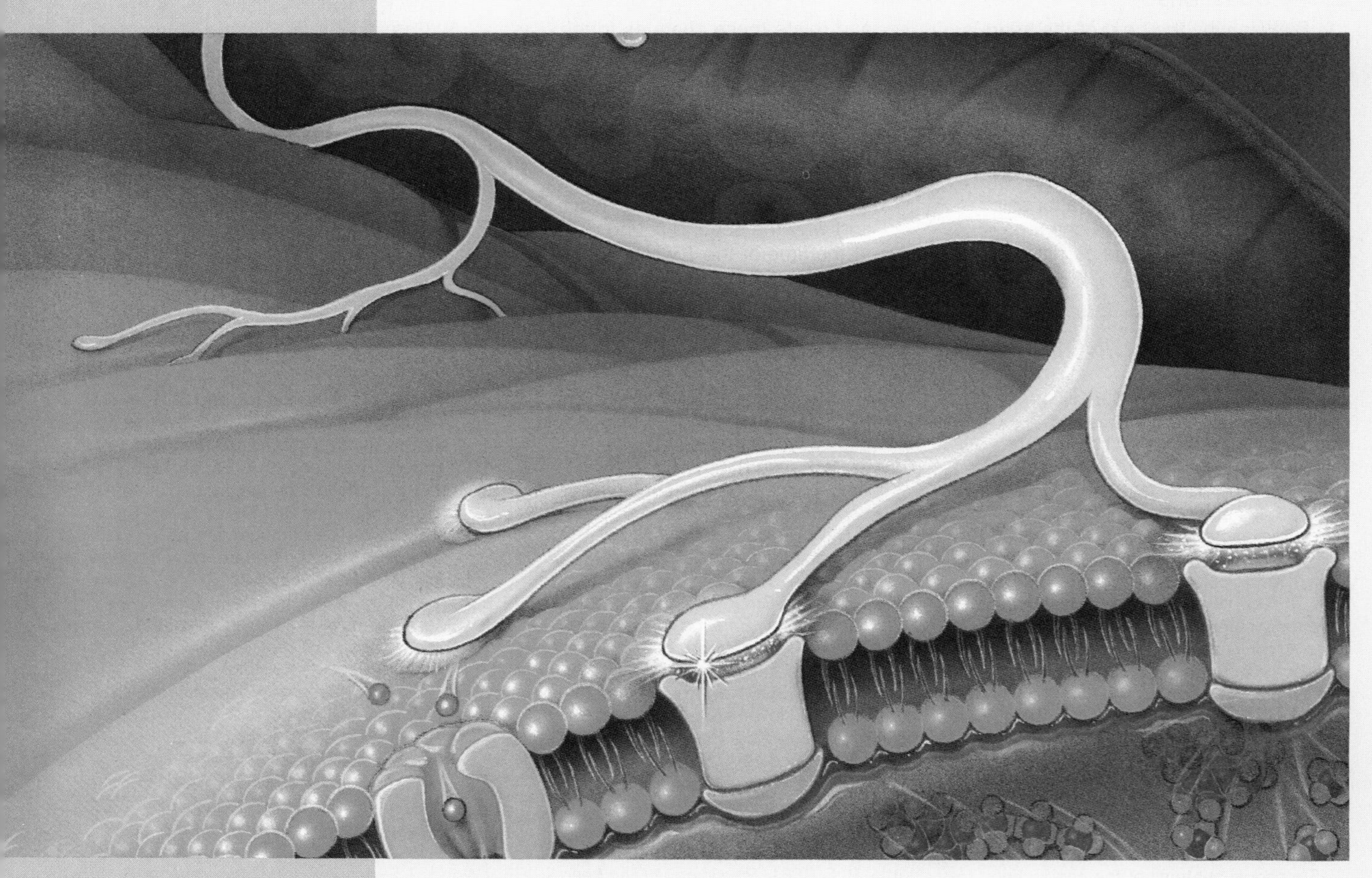

## Objectives

By studying this chapter, you should be able to

1. define cardiac output and describe how it is affected by cardiac rate and stroke volume
2. explain how autonomic nerves regulate the cardiac rate and the strength of ventricular contraction
3. explain the intrinsic regulation of stroke volume (the Frank-Starling law of the heart)
4. describe the factors that affect the venous return of blood to the heart
5. explain how tissue fluid is formed and how it is returned to the capillary blood
6. explain how edema may be produced
7. explain how antidiuretic hormone helps to regulate the blood volume, plasma osmolality, and the blood pressure
8. explain the role of aldosterone in the regulation of blood volume and pressure
9. describe the renin-angiotensin system and its significance in cardiovascular regulation
10. explain, using Poiseuille's law, how blood flow is regulated
11. define total peripheral resistance, and explain how vascular resistance is regulated by extrinsic control mechanisms
12. describe the mechanisms involved in the autoregulation of blood flow
13. explain the mechanisms by which blood flow to the heart and skeletal muscles is regulated
14. describe the changes that occur in the cardiac output and in the distribution of blood flow in the body during exercise
15. describe the cutaneous circulation and explain how circulation in the skin is regulated
16. describe the factors that regulate the arterial blood pressure
17. describe the baroreceptor reflex and explain its significance in blood pressure regulation
18. explain how the sounds of Korotkoff are produced and how these sounds are used to measure blood pressure
19. describe how the pulse pressure and mean arterial pressure are calculated and explain the significance of these measurements
20. explain the mechanisms that contribute to and that help compensate for the conditions of hypertension, circulatory shock,and congestive heart failure

## Outline

## Cardiac Output

The pumping ability of the heart is a function of the number of beats per minute (cardiac rate) and the volume of blood ejected per beat (stroke volume). The cardiac rate is regulated primarily by autonomic nerves. The stroke volume is dependent on the volume of blood in the ventricle, the strength of ventricular contraction, and the resistance to blood flow through the arteries and arterioles.

The cardiac output is equal to the volume of blood pumped per minute by each ventricle. The average resting **cardiac rate** in an adult is 70 beats per minute; the average **stroke volume** (volume of blood pumped per beat by each ventricle) is 70–80 ml per beat. The product of these two variables gives an average cardiac output of 5,000 ml per minute:

$$\text{cardiac output (ml/min)} = \text{stroke volume (ml/beat)} \times \text{cardiac rate (beats/min)}$$

The **total blood volume** is also equal to about 5–6 liters. This means that each ventricle pumps the equivalent of the total blood volume each minute under resting conditions. Put another way, it takes about a minute for a drop of blood to complete the systemic and pulmonary circuits. An increase in cardiac output, as occurs during exercise, must thus be accompanied by an increased rate of blood flow through the circulation. This is accomplished by factors that regulate the cardiac rate and stroke volume.

### Regulation of Cardiac Rate

In the complete absence of neural influences, the heart will continue to beat according to the rhythm set by the SA node (discussed in chapter 13). This automatic rhythm is produced by the spontaneous decay of the resting membrane potential to a threshold depolarization, at which point voltage-regulated membrane gates are opened and action potentials are produced. As described in chapter 13, $Ca^{++}$ enters the myocardial cytoplasm during the action potential, attaches to troponin, and causes contraction.

Normally, however, sympathetic and vagus (parasympathetic) nerve fibers to the heart are continuously active and modify the rate of spontaneous depolarization of the SA node. Norepinephrine, released primarily by sympathetic nerve endings, and epinephrine, secreted by the adrenal medulla, stimulate an increase in the spontaneous rate of firing of the SA node. Acetylcholine, released from parasympathetic endings, hyperpolarizes the SA node and thus decreases the rate of its spontaneous firing (fig. 14.1). The actual pace set by the SA node at any time depends on the net effect of these antagonistic influences. Mechanisms that affect the cardiac rate are said to have a **chronotropic effect** (*chrono* = time); those that increase cardiac rate have a positive chronotropic effect, and those that decrease the rate have a negative chronotropic effect.

**Table 14.1** Effects of autonomic nerve activity on the heart

| Region Affected | Sympathetic Nerve Effects | Parasympathetic Nerve Effects |
|---|---|---|
| SA node | Increased rate of diastolic depolarization; increased cardiac rate | Decreased rate of diastolic depolarization; decreased cardiac rate |
| AV node | Increased conduction rate | Decreased conduction rate |
| Atrial muscle | Increased strength of contraction | Decreased strength of contraction |
| Ventricular muscle | Increased strength of contraction | No significant effect |

Autonomic innervation of the SA node represents the major means by which cardiac rate is regulated. Autonomic nerves do, however, affect cardiac rate by other mechanisms to a lesser degree. Sympathetic endings in the musculature of the atria and ventricles increase the strength of contraction and decrease slightly the time spent in systole when the cardiac rate is high (table 14.1).

In exercise, the cardiac rate increases as a result of decreased vagus nerve inhibition of the SA node. Further increases in cardiac rate are achieved by increased sympathetic nerve stimulation. The resting bradycardia (slow heart rate) of endurance-trained athletes is due largely to high vagus nerve activity. The activity of the autonomic innervation of the heart is coordinated by **cardiac control centers** in the *medulla oblongata* of the brain stem. There may be separate cardioaccelerator and cardioinhibitory centers in the medulla, but this is currently controversial. These centers, in turn, are affected by higher brain areas and by sensory feedback from pressure receptors, or *baroreceptors,* in particular places in the aorta and carotid arteries. In this way, a rise in blood pressure can produce a reflex slowing of the heart. The *baroreceptor reflex* is discussed in more detail in relation to blood pressure regulation later in this chapter.

### Regulation of Stroke Volume

The stroke volume is regulated by three variables: (1) the **end-diastolic volume (EDV),** which is the volume of blood in the ventricles at the end of diastole; (2)the **total peripheral resistance**, which is the frictional resistance, or impedance, to blood flow in the arteries; and (3) the **contractility,** or strength, of ventricular contraction.

The end-diastolic volume is the amount of blood in the ventricles just prior to contraction. This is a work load imposed on the ventricles prior to contraction and is thus

**Figure 14.1.** The rhythm set by the pacemaker potentials (*colored lines*) in the SA node. Sympathetic nerve effects decrease the rate of spontaneous depolarization, thus influencing the rate at which action potentials are produced.

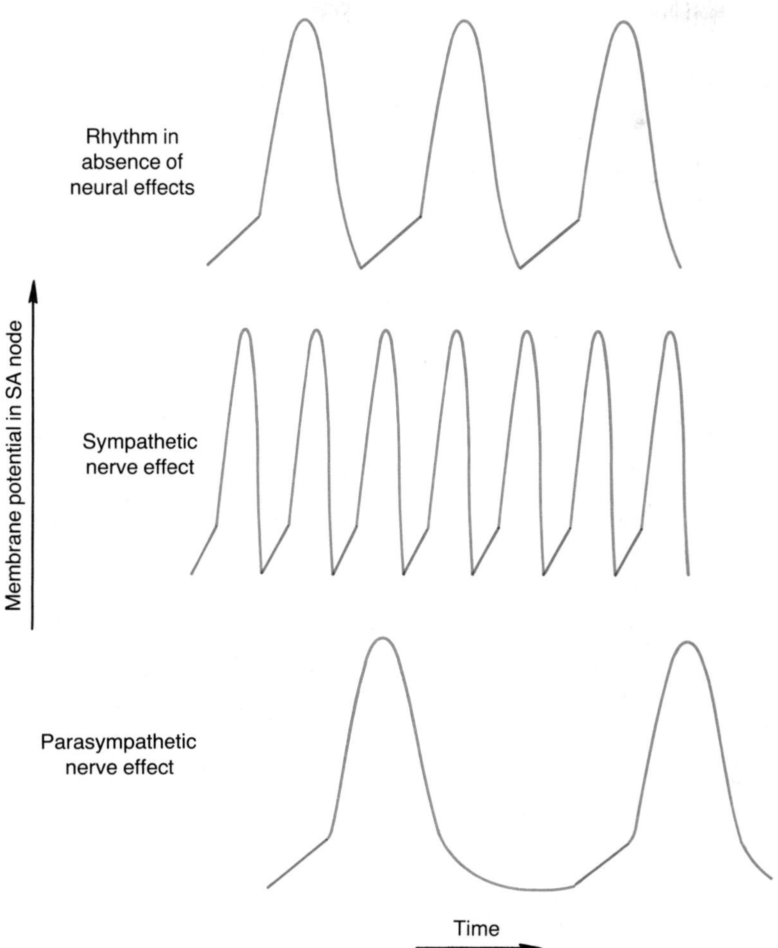


sometimes called a *preload.* The stroke volume is directly proportional to the preload; an increase in EDV results in an increase in stroke volume. The stroke volume is also directly proportional to contractility; when the ventricles contract more strongly, they pump more blood.

In order to eject blood, the pressure generated in a ventricle when it contracts must be greater than the pressure in the arteries (since blood only flows from a location of higher to one of lower pressure). The pressure in the arterial system before the ventricle contracts is in turn a function of the total peripheral resistance—the higher the peripheral resistance, the higher the pressure. As blood begins to be ejected from the ventricle, the added volume of blood in the arteries causes a rise in pressure against the "bottleneck" presented by the peripheral resistance; ejection of blood stops shortly after the aortic pressure becomes equal to the intraventricular pressure. The total peripheral resistance thus presents an impedance to the ejection of blood from the ventricle, or an *afterload* imposed on the ventricle after contraction has begun.

In summary, the stroke volume is inversely proportional to the total peripheral resistance; the greater the peripheral resistance, the lower the stroke volume. It should be noted that this lowering of stroke volume in response to a raised peripheral resistance only occurs for a few beats. Thereafter, a healthy heart is able to compensate for the increased peripheral resistance by beating more strongly. This compensation occurs by means of a mechanism called the Frank-Starling law, to be described shortly.

The proportion of the end-diastolic volume that is ejected against a given afterload depends upon the strength of ventricular contraction. Normally, contraction strength is sufficient to eject 70–80 ml of blood out of a total end-diastolic volume of 110–130 ml. The *ejection fraction* is

**Figure 14.2.** The Frank-Starling mechanism (law of the heart). When the heart muscle is subject to increased amounts of stretch, it contracts more strongly. As a result of the increased contraction strength (shown as tension), the time required to reach maximum contraction is the same regardless of the degree of stretch.

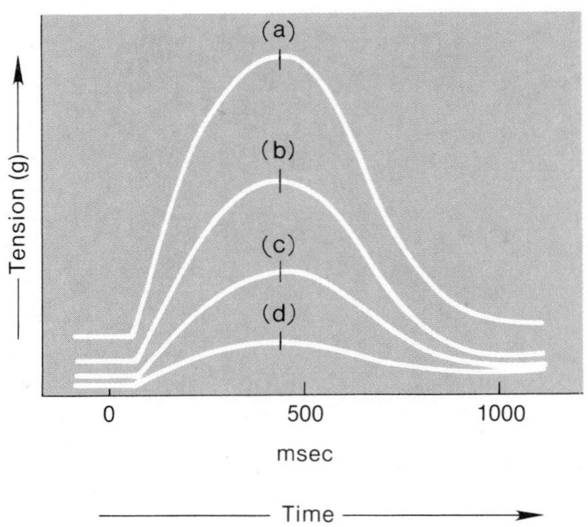


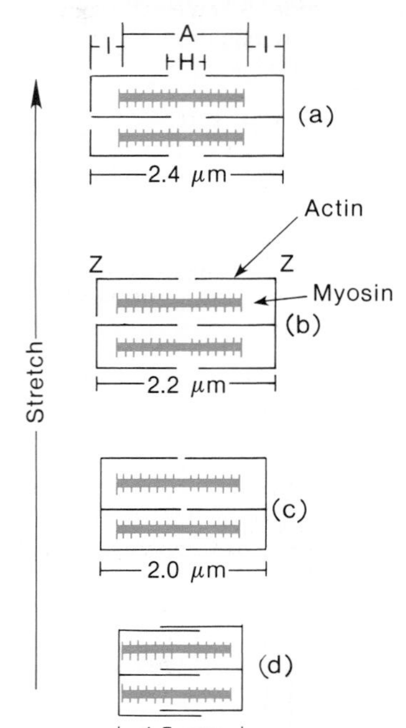


thus about two-thirds. More blood is pumped per beat as the EDV increases, and thus the ejection fraction remains relatively constant over a range of end-diastolic volumes. In order for this to be true, the strength of ventricular contraction must increase as the end-diastolic volume increases.

***Frank-Starling Law of the Heart.*** Two physiologists, Frank and Starling, demonstrated that normally the strength of ventricular contraction varies directly with the end-diastolic volume. Even in experiments where the heart is removed from the body (and is thus not subject to neural or hormonal regulation) and where the heart is filled with blood flowing from a reservoir, an increase in EDV within the physiological range results in increased contraction strength and, therefore, in increased stroke volume. This relationship between EDV, contraction strength, and stroke volume is thus a built-in, or *intrinsic,* property of heart muscle, and is known as the **Frank-Starling law of the heart**.

***Intrinsic Control of Contraction Strength.*** The intrinsic control of contraction strength and stroke volume is due to variations in the degree to which the myocardium is stretched by the end-diastolic volume. As the EDV rises within the physiological range, the myocardium is increasingly stretched and, as a result, contracts more forcibly.

Stretch can also increase the contraction strength of skeletal muscles. The resting length of skeletal muscles, however, is close to ideal, so that significant stretching decreases contraction strength. This is not true of the heart. Prior to filling with blood during diastole, the sarcomere lengths of myocardial cells are only about 1.5 μm. At this length the actin filaments from each side overlap in the middle of the sarcomeres, and the cells can only contract weakly (fig. 14.2).

As the ventricles fill with blood the myocardium stretches, so that the actin filaments overlap with myosin only at the edges of the A bands (fig. 14.2). This allows more force to be developed during contraction. Since this more advantageous overlapping of actin and myosin is produced by stretching of the ventricles and since the degree of stretching is controlled by the degree of filling (the end-diastolic volume), the strength of contraction is intrinsically adjusted by the end-diastolic volume.

The Frank-Starling law explains how the heart can adjust to a rise in total peripheral resistance: (1) a rise in peripheral resistance causes the stroke volume of the ventricle to be decreased, so that (2) more blood remains in the ventricle and the end-diastolic volume is greater for the next cycle; as a result, (3) the ventricle is more greatly stretched in the next cycle, and contracts more strongly to eject more blood. This allows a healthy ventricle to sustain a normal cardiac output. A very important consequence of these events is that the cardiac output (ml/min)

**Figure 14.3.** The regulation of cardiac output. Factors that stimulate cardiac output are shown as solid arrows; factors that inhibit cardiac output are shown as dotted arrows.

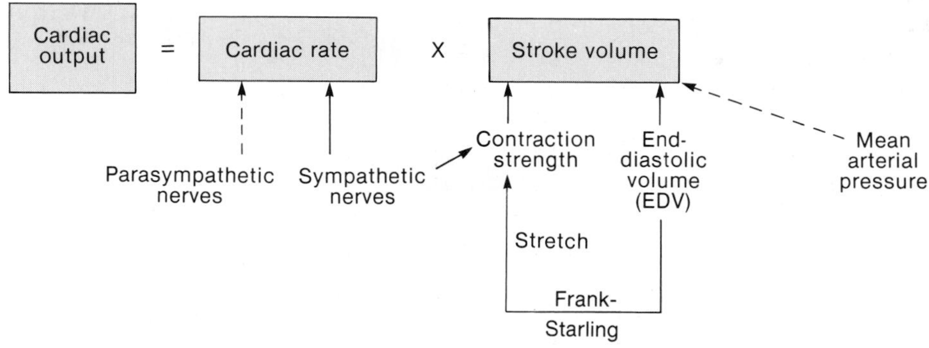


of the left ventricle, which pumps blood into the ever-changing resistances of the systemic circulation, can be maintained to match the output of the right ventricle (which pumps blood into the pulmonary circulation). This may seem to be a subtle point, but a moment's reflection will reveal that the rate of blood flow through the pulmonary and systemic circulations must be equal in order to prevent fluid accumulation in the lungs and to deliver fully oxygenated blood to the body.

***Extrinsic Control of Contractility.*** The *contractility* is the contraction strength at any given fiber length. At any given degree of stretch, the strength of ventricular contraction depends on the activity of the sympathoadrenal system. Norepinephrine from sympathetic endings and epinephrine from the adrenal medulla produce an increase in contraction strength. This is often called a **positive inotropic effect,** and is believed to result from an increase in the amount of $Ca^{++}$ available to the sarcomeres.

The cardiac output is thus affected in two ways by the activity of the sympathoadrenal system: the positive inotropic effect on contractility and the positive chronotropic effect that increases cardiac rate (fig. 14.3). Stimulation through the parasympathetic nerve to the heart has a negative chronotropic effect, but does not directly affect the contraction strength of the ventricles.

## Venous Return

The end-diastolic volume—and thus the stroke volume and cardiac output—is controlled by factors that affect the *venous return*, which is the return of blood via veins to the heart. The rate at which the atria and ventricles are filled with venous blood depends on the total blood volume and the venous pressure (pressure in the veins), which serves as the driving force for the return of blood to the heart.

**Figure 14.4.** The distribution of blood within the circulatory system at rest. (From *Circulation* by Bjorn Folkow and Eric Neil. Copyright © 1971 by Oxford University Press, Inc. Reprinted by permission.)

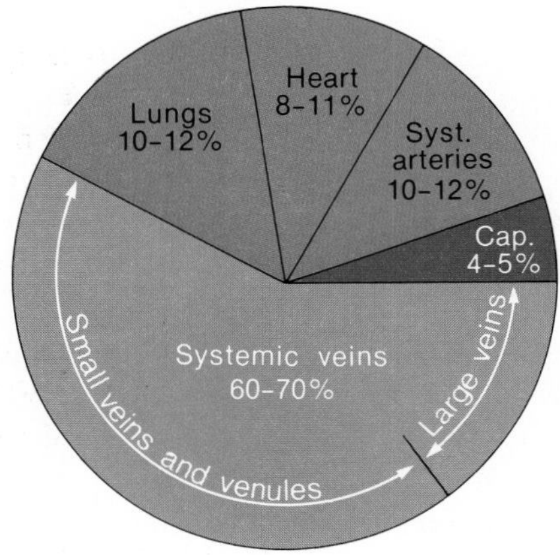


Veins have thinner, less muscular walls than do arteries and thus have a higher *compliance*. This means that a given amount of pressure will cause more distension (expansion) in veins than in arteries, so that the veins can hold more blood. Approximately two-thirds of the total blood volume is located in the veins (fig. 14.4). Veins are therefore called *capacitance vessels,* by analogy with capacitors in electronics, which can accumulate electrical charges. Muscular arteries and arterioles expand less under pressure (are less compliant) and thus are called *resistance vessels.*

Although veins contain almost 70% of the total blood volume, the mean venous pressure is only 2 mm Hg, compared to a mean arterial pressure of 90–100 mm Hg. The lower venous pressure is due in part to a pressure drop between arteries and capillaries and in part to the high venous compliance.

**Figure 14.5.** Variables that affect venous return and thus end-diastolic volume. Direct relationships are indicated by solid arrows; inverse relationships are shown with dashed arrows.

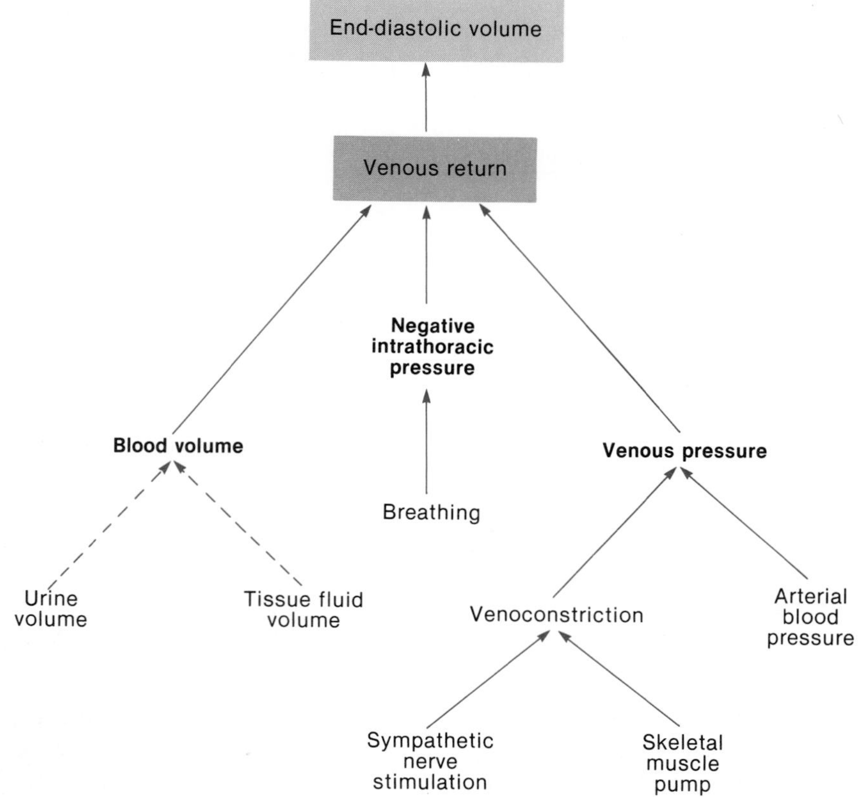


The venous pressure is highest in the venules (10 mm Hg) and lowest at the junction of the venae cavae with the right atrium (0 mm Hg). In addition to this pressure difference, the venous return to the heart is aided by (1) sympathetic nerve activity, which stimulates smooth muscle contraction in the venous walls and thus reduces compliance; (2) the skeletal muscle pump, which squeezes veins during muscle contraction; and (3) the pressure difference between the thoracic and abdominal cavities, which promotes the flow of venous blood back to the heart.

Contraction of the skeletal muscles functions as a "pump" by virtue of its squeezing action on veins (described in chapter 13). Contraction of the diaphragm during inhalation also improves venous return. This results from the fact that as the diaphragm contracts, it lowers to increase the thoracic volume and decrease the abdominal volume. This creates a partial vacuum in the thoracic cavity and a higher pressure in the abdominal cavity. The pressure difference thus produced favors blood flow from abdominal to thoracic veins (fig. 14.5).

1. *Describe how the stroke volume is intrinsically regulated by the end-diastolic volume, and explain the significance of this regulation.*
2. *Describe the effects of autonomic nerve stimulation on the cardiac rate and stroke volume.*
3. *Define the terms* preload *and* afterload, *and explain how these affect the cardiac output.*
4. *List the factors that affect venous return, and draw a flowchart to show how an increased venous return can result in an increased cardiac output.*

## Blood Volume

Fluid in the extracellular environment of the body is distributed between the blood and the tissue fluid compartments by filtration and osmotic forces acting across the walls of capillaries. The kidneys have a major role in regulating the blood volume, because urine is derived from blood plasma. Acting through their effects on the kidneys, a number of hormones regulate the amount of urine excreted and by this means influence the volume of blood remaining in the vascular system.

**Figure 14.6.** The daily intake and excretion of body water and its distribution between different intracellular and extracellular compartments. (From *Circulation* by Bjorn Folkow and Eric Neil. Copyright © 1971 by Oxford University Press, Inc. Reprinted by permission.)

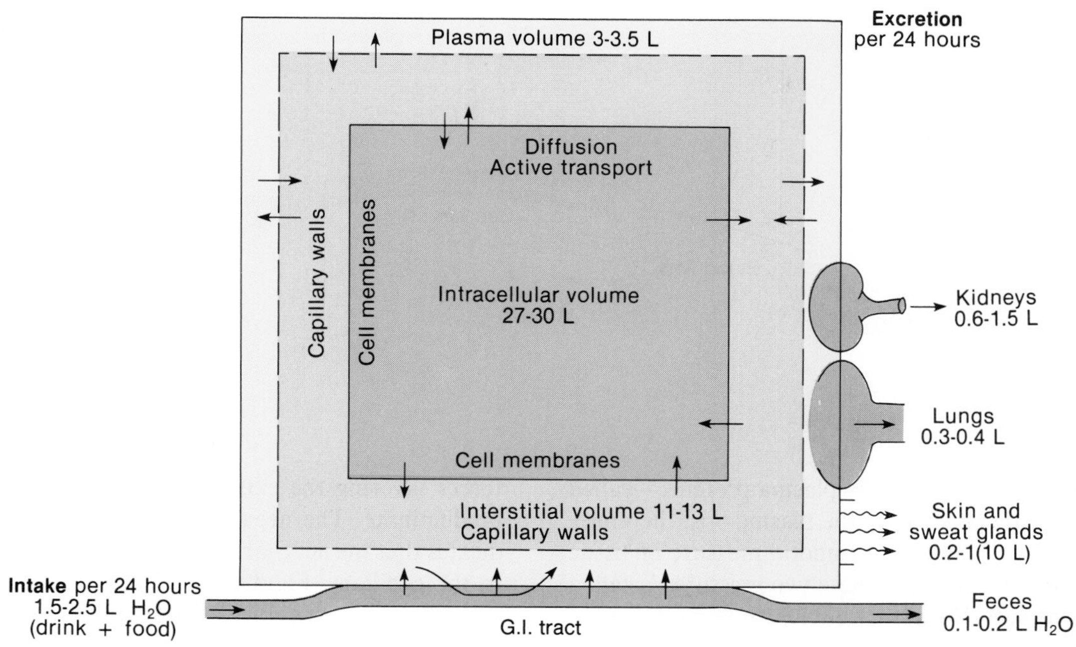


Blood volume represents one part, or compartment, of the total body water. Approximately two-thirds of the total body water is located within cells—in the **intracellular compartment.** The remaining one-third is in the **extracellular compartment.** This extracellular fluid is normally distributed so that about 80% is in the tissues—as tissue or *interstitial fluid*—and 20% is in the blood plasma (fig. 14.6).

The distribution of water between these compartments is determined by a balance between opposing forces acting at the capillaries. Blood pressure, for example, promotes the formation of tissue fluid from plasma, whereas osmotic forces draw water from the tissues into the vascular system. The total volume of intracellular and extracellular fluid is normally maintained constant by a balance between water loss and water gain. Mechanisms that affect drinking, urine volume, and the distribution of water between plasma and tissue fluid thus help regulate blood volume and, by this means, help regulate cardiac output and blood flow.

## Exchange of Fluid between Capillaries and Tissues

The distribution of extracellular fluid between the plasma and interstitial compartments is in a state of dynamic equilibrium. Tissue fluid is not normally a "stagnant pond" but is rather a continuously circulating medium, formed from and returning to the vascular system. In this way, the tissue cells receive a continuously fresh supply of glucose and other plasma solutes that are filtered through tiny endothelial channels in the capillary walls.

Filtration results from the blood pressure within the capillaries. This hydrostatic pressure, which is exerted against the inner capillary wall, is equal to about 37 mm Hg at the arteriolar end of systemic capillaries and drops to about 17 mm Hg at the venular end of the capillaries. The **net filtration pressure** is equal to the hydrostatic pressure of the blood in the capillaries minus the hydrostatic pressure of tissue fluid outside the capillaries, which opposes filtration. If, as an extreme example, these two values were equal, there would be no filtration. The magnitude of the tissue hydrostatic pressure varies from organ to organ. If the hydrostatic pressure in the tissue fluid is 1 mm Hg, as it is outside the capillaries of skeletal muscles, the net filtration pressure would be $37 - 1 = 36$ mm Hg at the arteriolar end of the capillary and $17 - 1 = 16$ mm Hg at the venular end.

Glucose, comparably sized organic molecules, inorganic salts, and ions are filtered along with water through the capillary channels. The concentrations of these substances in tissue fluid are thus the same as in plasma. The protein concentration of tissue fluid (2 g/100 ml), however, is less than the protein concentration of plasma (6–8 g/100 ml). This difference is due to the fact that the filtration of proteins is restricted by the capillary pores.

**Figure 14.7.** Tissue or interstitial fluid is formed by filtration as a result of blood pressures at the arteriolar ends of capillaries and is returned to venular ends of capillaries by the colloid osmotic pressure of plasma proteins. Solid arrows indicate the net direction of flow.

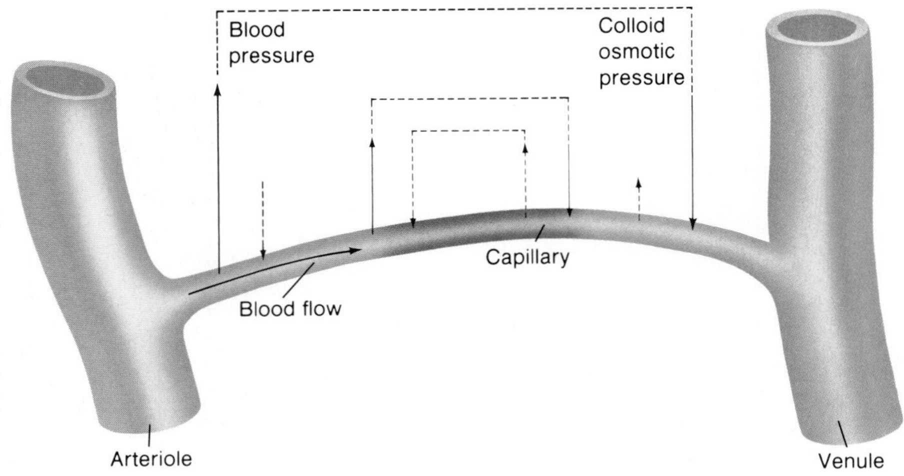


The osmotic pressure exerted by plasma proteins—called the **colloid osmotic pressure** of the plasma—is, therefore, much greater than the colloid osmotic pressure of tissue fluid. The difference between these two pressures is called the **oncotic pressure.** Since the colloid osmotic pressure of the tissue fluid is so low it can be neglected, the oncotic pressure is essentially equal to the colloid osmotic pressure of the plasma. This value has been estimated to be 25 mm Hg. Since water will move by osmosis from the solution of lower to the solution of higher osmotic pressure (chapter 6), this oncotic pressure favors the movement of water into the capillaries.

Whether fluid will move out of or into the capillary depends on the magnitudes of the net filtration pressure, which varies from the arteriolar to the venular end of the capillary, and on the oncotic pressure. These opposing forces that affect the distribution of fluid across the capillary are known as **Starling forces**, and their effects can be calculated according to the following equation:

Fluid movement is proportional to: $(P_c + \pi_i) - (P_i + \pi_p)$

Where:

$P_c$ = hydrostatic pressure in the capillary
$\pi_i$ = colloid osmotic pressure of the interstitial (tissue) fluid
$P_i$ = hydrostatic pressure of interstitial fluid
$\pi_p$ = colloid osmotic pressure of the plasma

The variables of this equation in the first group, enclosed by parentheses, provide the sum of forces acting to move fluid out of the capillary. The variables in the second group provide the sum of forces acting to move fluid into the capillary. In the capillaries of skeletal muscles, the values are as follows: at the arteriolar end of the capillary, $(37 + 0) - (1 + 25) = 11$ mm Hg; at the venular end of the capillary, $(17 + 0) - (1 + 25) = -9$ mm Hg. The positive value at the arteriolar end indicates that the forces favoring the extrusion of fluid from the capillary predominate. The negative value at the venular end indicates that the net Starling forces favor the return of fluid to the capillary. Fluid thus leaves the capillaries at the arteriolar end and returns to the capillaries at the venular end (fig. 14.7).

This "classic" view of capillary dynamics has recently been challenged by some investigators who believe that when capillaries are open, the net filtration force exceeds the force for the osmotic return of water throughout the length of the capillary. They believe that the opposite is true in closed capillaries (capillaries can be opened or closed by the action of precapillary sphincter muscles). Net filtration, in summary, would occur in open capillaries, whereas net absorption of water would occur in closed capillaries.

By either proposed mechanism, plasma and tissue fluid are thus continuously interchanged. The return of fluid to the vascular system at the venular ends of the capillaries, however, does not exactly equal the amount filtered at the arteriolar ends. According to some estimates, approximately 85% of the capillary filtrate is returned directly to the capillaries; the remaining 15% (amounting to at least 2 L per day) is returned to the vascular system by way of the lymphatic system. Lymphatic capillaries, it may be recalled from chapter 13, drain excess tissue fluid and proteins and, by way of lymphatic vessels, ultimately return this fluid to the venous system.

***Causes of Edema.*** Excessive accumulation of tissue fluid is known as **edema.** This condition is normally prevented by a proper balance between capillary filtration and osmotic uptake of water and by proper lymphatic drainage. Edema may thus result from (1) *high arterial blood pressure,* which increases capillary pressure and causes excessive filtration; (2) *venous obstruction*—as in phlebitis

**Table 14.2** The causes of edema

| Variables that Cause Edema | Effects |
|---|---|
| Increased blood pressure or venous obstruction | Increases capillary filtration pressure so that more tissue fluid is formed at the arteriolar ends of capillaries. |
| Increased tissue protein concentration | Decreases osmosis of water into the venular ends of capillaries. Usually a localized tissue edema due to leakage of plasma proteins through capillaries during inflammation and allergic reactions. Myxedema due to hypothyroidism is also in this category. |
| Decreased plasma protein concentration | Decreases osmosis of water into the venular ends of capillaries. May be caused by liver disease (which can be associated with insufficient plasma protein production), kidney disease (due to leakage of plasma protein into urine), or protein malnutrition. |
| Obstruction of lymphatic vessels | Infections by filaria roundworms (nematodes) transmitted by a certain species of mosquito block lymphatic drainage, causing edema and tremendous swelling of the affected areas. |

(where a thrombus forms in a vein) or mechanical compression of veins (during pregnancy, for example)—which produces a congestive increase in capillary pressure; (3) *leakage of plasma proteins into tissue fluid,* which causes reduced osmotic flow of water into the capillaries (this occurs during inflammation and allergic reactions as a result of increased capillary permeability); (4) *myxedema*—the excessive production of particular glycoproteins (mucin) in the interstitial spaces caused by hypothyroidism; (5) *decreased plasma protein concentration* as a result of liver disease (the liver makes most of the plasma proteins), or as a result of kidney disease in which proteins are excreted in the urine; and (6) *obstruction of the lymphatic drainage* (table 14.2).

In the tropical disease *filariasis,* mosquitoes transmit a nematode worm parasite to humans. The larvae of these worms invade lymphatic vessels and block lymphatic drainage. The edema that results can be so severe that the tissues swell to produce an elephantlike appearance—a condition aptly termed **elephantiasis** (fig. 14.8).

## Regulation of Blood Volume by the Kidneys

The formation of urine begins in the same manner as the formation of tissue fluid—by filtration of plasma through capillary pores. These capillaries are known as *glomeruli,* and the filtrate they produce enters a system of tubules which transports and modifies the filtrate (by mechanisms discussed in chapter 16). The kidneys produce about 180 L per day of blood filtrate, but since there is only 5.5 L of blood in the body, it is clear that most of this filtrate must be returned to the vascular system and recycled. Only about 1–2 L per day of urine is excreted; 98%–99% of the amount filtered is **reabsorbed** back into the vascular system.

**Figure 14.8.** Parasitic larvae that block lymphatic drainage produce tissue edema, resulting in elephantiasis.

The volume of urine excreted can be varied by changes in the reabsorption of filtrate. If 99% of the filtrate is reabsorbed, for example, 1% must be excreted. Decreasing the reabsorption by only 1%, to 98%, would double the volume of urine excreted (increasing to 2% of the amount filtered). Carrying the logic further, a doubling of urine volume from, for example, one to two liters would result in the loss of one additional liter of blood volume. The percent of the glomerular filtrate reabsorbed—and thus the urine volume and blood volume—is adjusted according to the needs of the body by the action of specific hormones on the kidneys. Through their effects on the kidneys, and the resulting changes in blood volume, these hormones serve important functions in the regulation of the cardiovascular system.

***Regulation by Antidiuretic Hormone (ADH).*** One of the major hormones involved in the regulation of blood volume is **antidiuretic hormone (ADH),** also known as *vasopressin.* As described in chapter 11, this hormone is produced by neurons in the hypothalamus, transported by axons into the neurohypophysis (posterior pituitary), and released from this storage gland in response to hypothalamic stimulation. The secretion of ADH from the posterior pituitary occurs when neurons called **osmoreceptors** in the hypothalamus detect an increase in plasma osmolality (osmotic pressure).

An increase in plasma osmolality occurs when the plasma becomes more concentrated (chapter 6). This can be produced either by *dehydration* or by *excessive salt*

**Figure 14.9.** The role of antidiuretic hormone (ADH) in the regulation of blood osmolality and volume. Both dehydration and salt ingestion stimulate drinking and increased water retention. In dehydration, this provides a negative feedback mechanism (shown by dashed arrows) that helps to counter the loss of blood volume. As a result of salt ingestion, this same mechanism produces a rise in blood volume. Numbers indicate the sequence of cause-and-effect steps.

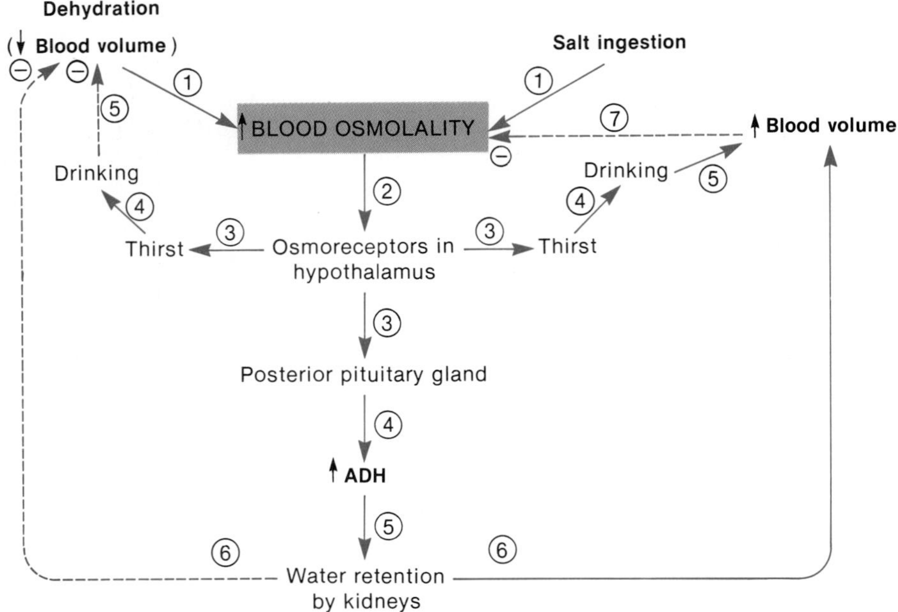


*intake*. Stimulation of osmoreceptors produces sensations of thirst, leading to increased water intake, and increased ADH secretion from the posterior pituitary. Through mechanisms that will be discussed in conjunction with kidney physiology in chapter 16, ADH stimulates water reabsorption from the filtrate. A smaller volume of urine is thus excreted as a result of the action of ADH (fig. 14.9).

A person who is dehydrated or who eats excessive amounts of salt thus drinks more and urinates less. This raises the blood volume and, in the process, dilutes the plasma to lower its previously elevated osmolality. The rise in blood volume that results from these mechanisms is extremely important in stabilizing the condition of a dehydrated person with low blood volume and pressure. In a person who eats excessive amounts of salt, however, the increased blood volume could produce an abnormally high blood pressure (hypertension).

Drinking excessive amounts of water without excessive amounts of salt does not result in hypertension. The water does enter the blood from the intestine and momentarily raises the blood volume; at the same time, however, it dilutes the blood. Dilution of the blood decreases the plasma osmolality and thus inhibits ADH secretion. With less ADH there is less reabsorption of filtrate in the kidneys—a larger volume of urine is excreted. Water is therefore a *diuretic* —a substance that promotes urine formation—because it inhibits the secretion of antidiuretic hormone.

There is another mechanism—in addition to the osmoreceptor effects—by which drinking excessive amounts of water can inhibit ADH secretion. An excessively high blood volume stimulates stretch receptors located in the left atrium of the heart. Stimulation of these stretch receptors, in turn, activates a reflex inhibition of ADH secretion, which in turn promotes a lowering of blood volume through increased urine production. This inhibition of ADH secretion is independent of the inhibitory effects that a decrease in plasma osmolality would produce; experimental stimulation of the stretch receptors by inflation of a balloon in the left atrium duplicates the inhibition of ADH secretion produced by high blood volume.

During prolonged exercise, particularly on a warm day, a substantial amount of water may be lost from the body through sweating (up to 900 ml per hour or more). The lowering of blood volume that results decreases the ability of the body to dissipate heat, and the consequent overheating of the body can cause ill effects that require an end to the exercise. It is thus important for an athlete under these conditions to remain well hydrated, but this cannot be achieved by drinking pure water. This is because blood sodium is lost in sweat, so that a lesser amount of water is required to dilute the blood osmolality back to normal. Under these conditions the blood volume is abnormally low although the blood osmolality is in the normal range. When the blood osmolality is normal the urge to drink is extinguished and any additional water ingested is eliminated in the urine, due to inhibition of ADH secretion. For these reasons, athletes performing prolonged endurance exercise should drink solutions containing sodium (as well as carbohydrates for energy) and should drink at a predetermined rate rather than at a rate determined only by thirst.

**Figure 14.10.** The renin-angiotensin-aldosterone system. Negative feedback effects are shown with dashed arrows. Numbers indicate the sequence of cause-and-effect steps.

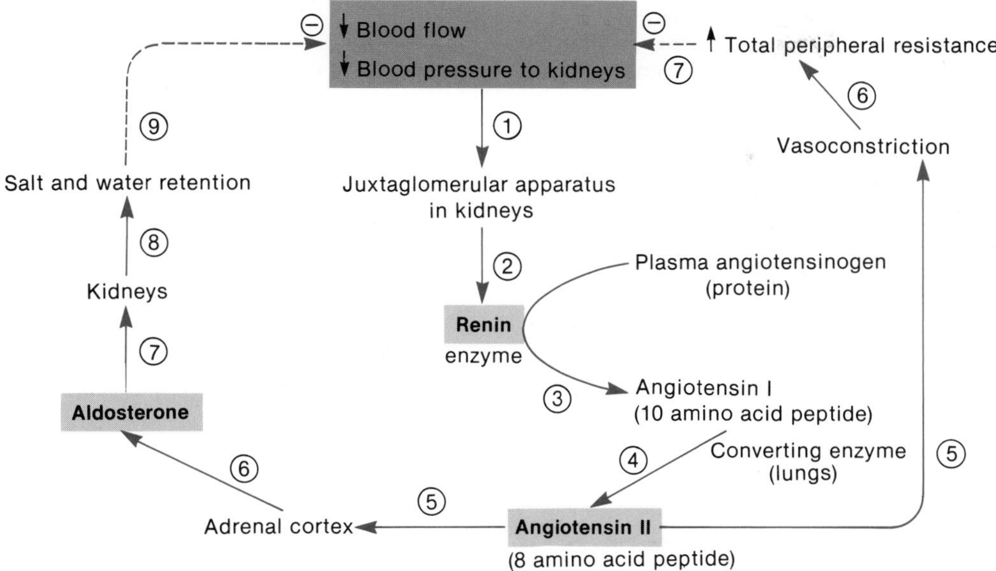


***Regulation by Aldosterone.*** From the preceding discussion, it is clear that a certain amount of dietary salt is required to maintain blood volume and pressure. Throughout most of human history, salt was in short supply and was thus highly valued. Moorish merchants in the sixth century traded an ounce of salt for an ounce of gold; salt cakes were used as money in Abyssinia; part of a Roman soldier's pay was given in salt—it is from this practice that the word salary (*sal* = salt) is derived. The phrase "worth his salt" is derived from the fact that Greeks and Romans sometimes purchased slaves with salt.

Since $Na^+$ and $Cl^-$ are easily filtered in the kidneys, a mechanism must exist to promote the reabsorption and retention of salt when the dietary salt intake is too low. **Aldosterone,** a steroid hormone secreted by the adrenal cortex, stimulates the reabsorption of salt by the kidneys. Aldosterone is thus a "salt-retaining hormone." Retention of salt indirectly promotes retention of water (in part, by the action of ADH, as previously discussed). The action of aldosterone thus causes an increase in blood volume, but—unlike the effects of ADH—does not cause a change in plasma osmolality. This is because aldosterone promotes the reabsorption of salt and water in proportionate amounts, whereas ADH only promotes the reabsorption of water. Thus aldosterone, unlike ADH, does not act to dilute the blood. Since regulatory mechanisms generally follow the principles of negative feedback, it makes sense that ADH secretion, but not aldosterone secretion, should be stimulated by a rise in plasma osmolality. The secretion of aldosterone is stimulated during salt deprivation, when the blood volume and pressure are reduced. The adrenal cortex, however, is not directly stimulated to secrete aldosterone by these conditions. Instead, a decrease in blood volume and pressure activates an intermediate mechanism, described in the next section.

***Renin-Angiotensin System.*** When the blood flow and pressure are reduced in the renal artery (as they would be in the low blood volume state of salt deprivation), a group of cells in the kidneys called the *juxtaglomerular apparatus* secretes the enzyme **renin** into the blood. This enzyme cleaves a ten-amino-acid polypeptide called *angiotensin I* from a plasma protein called *angiotensinogen.* As angiotensin I passes through the capillaries of the lungs and other organs, an *angiotensin-converting enzyme* removes two amino acids. This leaves an eight-amino-acid polypeptide called **angiotensin II** (fig. 14.10). Conditions of salt deprivation, low blood volume, and low blood pressure, in summary, cause increased production of angiotensin II in the blood.

Angiotensin II has numerous effects that cause a rise in blood pressure. This rise in pressure is due both to constriction of muscular arteries and arterioles (vasoconstriction) and to increases in blood volume. Vasoconstriction is produced directly by the effects of angiotensin II on blood vessels and indirectly by the augmentation of sympathetic nerve activity by angiotensin II (causing increased release of norepinephrine in the blood vessels).

One of the newer classes of drugs that can be used to treat hypertension (high blood pressure) are the **angiotensin-converting enzyme** (abbreviated **ACE) inhibitors.** These drugs block the formation of angiotension II and thus promote vasodilation, which decreases the total peripheral resistance. This reduces the afterload of the heart, and is thus also beneficial in the treatment of left ventricular hypertrophy and congestive heart failure.

Angiotensin II promotes a rise in blood volume by means of two mechanisms: (1) thirst centers in the hypothalamus are stimulated by angiotensin II, so that more water is ingested; and (2) secretion of aldosterone from the adrenal cortex is stimulated by angiotensin II, and higher aldosterone secretion causes more salt and water to be retained by the kidneys. The relationship between angiotensin II and aldosterone is sometimes described as the *renin-angiotensin-aldosterone system.*

This mechanism can also work in the opposite direction: high salt intake, leading to high blood volume and pressure, normally inhibits renin secretion. With less angiotensin II formation and less aldosterone secretion, less salt is retained by the kidneys and more is excreted in the urine. Unfortunately, many people with chronically high blood pressure may have normal or even elevated levels of renin secretion. In these cases, the intake of salt must be lowered to match the impaired ability to excrete salt in the urine.

***Atrial Natriuretic Hormone.*** As described in the previous section, a fall in blood volume is compensated, through activation of the renin-angiotensin-aldosterone system, by renal retention of fluid. An increase in blood volume, conversely, is compensated by renal excretion of a larger volume of urine. Experiments suggest that the increase in water excretion under conditions of high blood volume is at least partly due to an increase in the excretion of $Na^+$ in the urine, or *natriuresis* (*natrium* = sodium; *uresis* = making water).

Increased $Na^+$ excretion (natriuresis) may be produced by a decline in aldosterone secretion, but there is evidence that there is a separate hormone which stimulates natriuresis. This **natriuretic hormone** would thus be antagonistic to aldosterone and would promote $Na^+$ and water excretion in the urine in response to a rise in blood volume. The atria of the heart have recently been shown to produce a hormone with these properties, which is now identified as *atrial natriuretic hormone.* By promoting salt and water excretion in the urine, atrial natriuretic hormone can act to lower the blood volume and pressure. This is analogous to the action of diuretic drugs taken by people with hypertension, as described later in this chapter.

1. *Describe the composition of tissue fluid. Using a flow diagram, explain how tissue fluid is formed and how it is returned to the vascular system.*
2. *Define the term* edema, *and describe four different mechanisms that can produce this condition.*
3. *Describe the effects of dehydration on blood and urine volumes, and explain the cause-and-effect mechanism involved.*
4. *Describe how salt deprivation causes increased salt and water retention by the kidneys. Explain how a high-salt diet can be compensated by increased urinary excretion of salt.*

## *Vascular Resistance and Blood Flow*

The rate of blood flow to an organ is dependent on the resistance to flow in the small arteries and arterioles that serve the organ. Vasodilation decreases resistance and increases flow, whereas vasoconstriction increases resistance and decreases flow. Vasodilation and vasoconstriction occur in response to extrinsic neural and hormonal signals, as well as to intrinsic control mechanisms within the organ. These intrinsic mechanisms include direct responses of the vascular smooth muscle to pressure or to local chemical conditions.

The amount of blood that the heart pumps per minute is equal to the rate of venous return and thus is equal to the rate of blood flow through the entire circulation. The cardiac output of 5–6 L per minute is distributed unequally to the different organs. At rest, blood flow is about 2,500 ml per minute through the liver, kidneys, and gastrointestinal tract, 1,200 ml per minute through the skeletal muscles, 750 ml per minute through the brain, and 250 ml per minute through the coronary arteries of the heart. The balance of the cardiac output (500–1,100 ml per minute) is distributed to the other organs (table 14.3).

### Physical Laws Describing Blood Flow

The flow of blood through the vascular system, like the flow of any fluid through a tube, depends in part on the difference in pressure at the two ends of the tube. If the pressure at both ends of the tube is the same there will be no flow. If the pressure at one end is greater than at the other, blood will flow from the region of higher to the region of lower pressure. The rate of blood flow is proportional

**Table 14.3** Estimated distribution of the cardiac output at rest

| Organs | Blood Flow | |
|---|---|---|
| | Milliliters per minute | Percent total |
| Gastrointestinal tract and liver | 1,400 | 24 |
| Kidneys | 1,100 | 19 |
| Brain | 750 | 13 |
| Heart | 250 | 4 |
| Skeletal muscles | 1,200 | 21 |
| Skin | 500 | 9 |
| Other organs | 600 | 10 |
| Total organs | 5,800 | 100 |

From O. L. Wade and J. M. Bishop, *Cardiac Output and Regional Blood Flow.* Copyright © 1962 Blackwell Scientific Publications, Ltd., England. Used with permission.

**Figure 14.11.** The flow of blood in the systemic circulation is ultimately dependent on the pressure difference ($\Delta P$) between the origin of the flow (mean pressure of about 100 mm Hg in the aorta) and the end of the circuit—zero mm Hg in the vena cava where it joins the right atrium (*RA*). (LA = left atrium, RV = right ventricle, LV = left ventricle.)

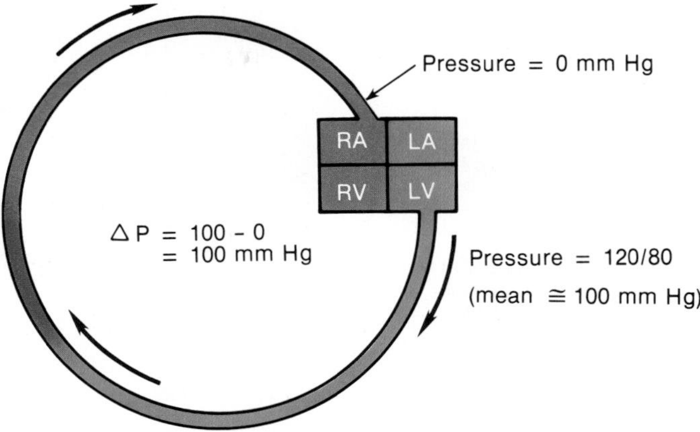


to the pressure difference ($P_1 - P_2$) between the two ends of the tube. The term **pressure difference** is abbreviated $\Delta P$, in which the Greek letter $\Delta$ (*delta*) means "change in" (fig. 14.11).

If the systemic circulation is pictured as a single tube leading from and back to the heart (fig. 14.11), blood flow through this system would occur as a result of the pressure difference between the beginning of the tube (the aorta) and the end of the tube (the junction of the venae cavae with the right atrium). The average, or mean, arterial pressure is about 100 mm Hg; the pressure at the right atrium is 0 mm Hg. The "pressure head," or driving force ($\Delta P$), is therefore about $100 - 0 = 100$ mm Hg.

Blood flow is directly proportional to the pressure difference between the two ends of the tube ($\Delta P$) but is *inversely proportional* to the frictional resistance to blood flow through the vessels. Inverse proportionality is expressed by showing one of the factors in the denominator of a fraction, since a fraction decreases when the denominator increases:

$$\text{Blood flow} \propto \frac{\Delta P}{\text{Resistance}}$$

The **resistance** to blood flow through a vessel is directly proportional to the length of the vessel and to the viscosity of the blood (the "thickness," or ability of molecules to "slip over" each other). Vascular resistance is inversely proportional to the fourth power of the radius of the vessel:

$$\text{Resistance} \propto \frac{L\eta}{r^4}$$

where: L = length of vessel
$\eta$ = viscosity of blood
r = radius of vessel

For example, if one vessel has half the radius of another and if all other factors are the same, the smaller vessel would have sixteen times ($2^4$) the resistance of the larger vessel. Blood flow through the larger vessel, as a result, would be sixteen times greater than in the smaller vessel (fig. 14.12).

When physical constants are added to this relationship, the rate of blood flow can be calculated according to **Poiseuille's** *(pwah-zuh'yez)* **law:**

$$\text{Blood flow} = \frac{\Delta P\, r^4\, (\pi)}{\eta\, L\, (8)}$$

Vessel length and blood viscosity do not vary significantly in normal physiology, although blood viscosity is increased in severe dehydration and in the polycythemia (high red blood cell count) that occurs as an adaptation to life at high altitude. The major physiological regulators of blood flow through an organ are the mean arterial pressure (driving the flow) and the vascular resistance to flow. At a given mean arterial pressure, blood can be diverted from one organ to another by variations in the degree of vasoconstriction and vasodilation. Vasoconstriction in one organ and vasodilation in another results in a diversion of blood to the second organ. Since arterioles are the smallest arteries and can become narrower by vasoconstriction, they provide the greatest resistance to blood flow (fig. 14.13). Blood flow to an organ is thus largely determined by the degree of vasoconstriction or vasodilation of its arterioles. The rate of blood flow to an organ can be increased by dilation of its arterioles and can be decreased by constriction of its arterioles.

**Figure 14.12.** Relationships between blood flow, vessel radius, and resistance. (*a*) The resistance and blood flow is equally divided between two branches of a vessel. (*b*) A doubling of the radius of one branch and halving of the radius of the other produces a sixteenfold increase in blood flow in the former and a sixteenfold decrease of blood flow in the latter.

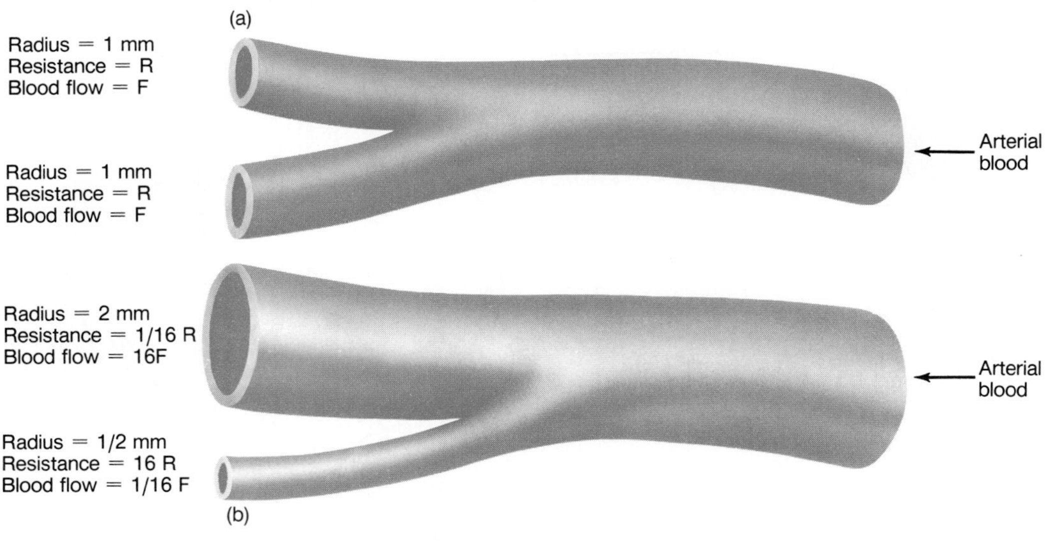


**Figure 14.13.** Blood pressure in different vessels of the systemic circulation.

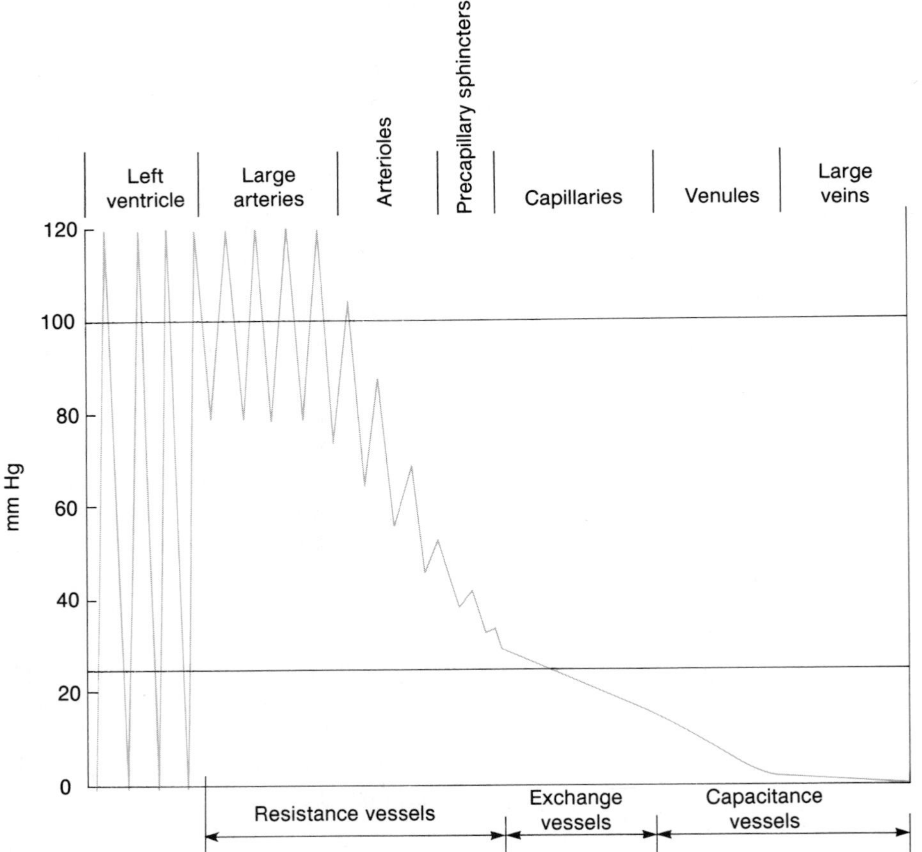

**Figure 14.14.** A diagram of the systemic and pulmonary circulations. Notice that with few exceptions (such as blood flow in the renal circulation), the flow of arterial blood is in parallel rather than in series (arterial blood does not flow from one organ to another).

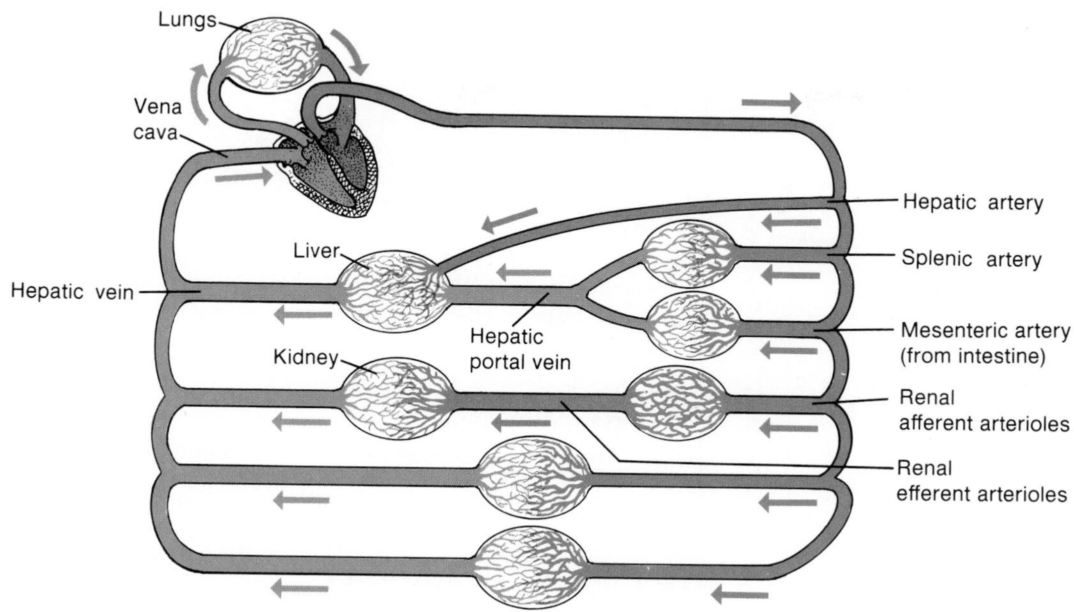


***Total Peripheral Resistance.*** The sum of all the vascular resistances within the systemic circulation is called the *total peripheral resistance.* The arterial supplies to the organs, which contribute to this total, are generally in parallel rather than in series with each other. That is, arterial blood passes through only one set of resistance vessels (arterioles) before returning to the heart (fig. 14.14). Since one organ is not "downstream" from another in terms of its arterial supply, changes in resistance within one organ directly affect blood flow in only that organ.

Vasodilation in a large organ might, however, significantly decrease the total peripheral resistance and, by this means, might decrease the mean arterial pressure. If this is not prevented by compensations, the driving force for blood flow through all organs might be reduced. This is normally prevented by an increase in the cardiac output and by vasoconstriction in other areas. During exercise of the large muscles, for example, the arterioles in the exercising muscles are dilated. This would cause a great fall in mean arterial pressure if there were no compensations. The blood pressure actually rises during exercise, however, because the cardiac output is increased, and because there is constriction of the arterioles in the viscera and skin.

## Extrinsic Regulation of Blood Flow

The term *extrinsic regulation* refers to control by the autonomic nervous system and endocrine system. **Angiotensin II,** for example, directly stimulates vascular smooth muscle to produce generalized vasoconstriction. Antidiuretic hormone (ADH) also has a vasoconstrictor effect at high concentrations; this is why it is also called **vasopressin.** This vasopressor effect of ADH is not believed to be significant under physiological conditions in humans.

***Regulation by Sympathetic Nerves.*** Stimulation of the sympathoadrenal system produces an increase in the cardiac output (as previously discussed) and an increase in total peripheral resistance. The latter effect is due to alpha-adrenergic stimulation (chapter 10) of vascular smooth muscle by norepinephrine and, to a lesser degree, by epinephrine. This produces vasoconstriction of the arterioles in the viscera and skin.

Even when a person is calm, the sympathoadrenal system is active to a certain degree and helps set the "tone" of vascular smooth muscles. In this case, **adrenergic sympathetic fibers** (those that release norepinephrine) cause a basal level of vasoconstriction throughout the body. During the fight-or-flight reaction the activity of adrenergic fibers increases, so that vasoconstriction is produced in the digestive tract, kidneys, and skin.

Arterioles in skeletal muscles receive **cholinergic sympathetic fibers,** which release acetylcholine as a neurotransmitter. During the fight-or-flight reaction, the activity of these cholinergic fibers increases. This causes

**Table 14.4** Extrinsic control of vascular resistance and blood flow

| Extrinsic Agent | Effect | Comments |
|---|---|---|
| Sympathetic nerves | | |
| Alpha-adrenergic | Vasoconstriction | It occurs throughout the body. This is the dominant effect of sympathetic nerve stimulation on the vascular system. |
| Beta-adrenergic | Vasodilation | There is some activity in arterioles in skeletal muscles and in coronary vessels, but effects are masked by dominant alpha-receptor-mediated constriction. |
| Cholinergic | Vasodilation | Effects are localized to arterioles in skeletal muscles and are only activated during defense (fight-or-flight) reaction. |
| Parasympathetic nerves | Vasodilation | Effects are primarily restricted to the gastrointestinal tract, external genitals, and salivary glands and have little effect on total peripheral resistance. |
| Angiotensin II | Vasoconstriction | A powerful vasoconstrictor produced as a result of secretion of renin from the kidneys, it may function to help maintain adequate filtration pressure in kidneys when systemic blood flow and pressure are reduced. |
| ADH (vasopressin) | Vasoconstriction | Although the effects of this hormone on vascular resistance and blood pressure in anesthetized animals are well documented, the importance of these effects in conscious humans is controversial. |
| Histamine | Vasodilation | It promotes localized vasodilation during inflammation and allergic reactions. |
| Bradykinins | Vasodilation | Bradykinins are polypeptides secreted by sweat glands that promote local vasodilation. |
| Prostaglandins | Vasodilation or vasoconstriction | Prostaglandins are cyclic fatty acids that can be produced by most tissues, including blood vessel walls. Prostaglandin $I_2$ is a vasodilator, whereas thromboxane $A_2$ is a vasoconstrictor. The physiological significance of these effects is presently controversial. |

vasodilation. Vasodilation in skeletal muscles is also produced by epinephrine secreted by the adrenal medulla, which stimulates beta-adrenergic receptors. During the fight-or-flight reaction, therefore, blood flow is decreased to the viscera and skin, due to the alpha-adrenergic effects of vasoconstriction in these organs, and increased to the skeletal muscles. This diversion of blood flow to the skeletal muscles during emergency conditions may produce an "extra edge" for the skeletal muscles' response to the emergency.

***Parasympathetic Control of Blood Flow.*** Parasympathetic endings in arterioles are always cholinergic and always promote vasodilation. Parasympathetic innervation of blood vessels, however, is limited to the digestive tract, external genitals, and salivary glands. Because of this limited distribution, the parasympathetic system is less important than the sympathetic system in the control of total peripheral resistance. The extrinsic control of blood flow is summarized in table 14.4.

## Intrinsic Regulation of Blood Flow

Extrinsic control mechanisms affect resistance and flow in many regions of the body. In contrast to these more generalized effects, intrinsic ("built-in") mechanisms within individual organs provide a more localized regulation of vascular resistance and blood flow. Intrinsic mechanisms are classified as (1) *myogenic* and (2) *metabolic.* Some organs, the brain and kidneys in particular, utilize these intrinsic mechanisms to maintain relatively constant flow rates despite wide fluctuations in blood pressure. This ability is termed **autoregulation.**

***Myogenic Control Mechanisms.*** If the arterial blood pressure and flow through an organ are inadequate—if the organ is inadequately *perfused* with blood—the metabolism of the organ cannot be maintained beyond a limited period of time. Excessively high blood pressure can also be dangerous, particularly in the brain, because this may cause fine blood vessels to burst (causing cerebrovascular accident—CVA, or stroke).

Changes in systemic arterial pressure are compensated in the brain and some other organs by the appropriate responses of vascular smooth muscle. A decrease in arterial pressure causes cerebral vessels to dilate, so that adequate rates of blood flow can be maintained despite the decreased pressure. High blood pressure, in contrast, causes cerebral vessels to constrict so that finer vessels downstream are protected from the elevated pressure. These responses are myogenic; they are direct responses by the vascular smooth muscle to changes in pressure.

**Table 14.5** The intrinsic control of vascular resistance and blood flow

| Category | Agent (↑ = increase; ↓ = decrease) | Mechanisms | Comments |
|---|---|---|---|
| Myogenic | ↑ Blood pressure | Stretching of the arterial wall as the blood pressure rises directly stimulates increased smooth muscle tone (vasoconstriction). | It helps to maintain relatively constant rates of blood flow and pressure within an organ despite changes in systematic arterial pressure (autoregulation). |
| Metabolic | ↓ Oxygen<br>↑ Carbon dioxide<br>↓ pH<br>↑ Adenosine<br>↑ K | Local changes in gas and metabolic concentrations act directly on vascular smooth muscle walls to produce vasodilation in the systemic circulation. The importance of different agents varies in different organs. | It aids in autoregulation of blood flow and also helps to shunt increased amounts of blood to organs with higher metabolic rates (active hyperemia). |

***Metabolic Control Mechanisms.*** Local vasodilation within an organ can occur as a result of the chemical environment created by the organ's metabolism. The localized chemical conditions that promote vasodilation include (1) *decreased oxygen concentrations* that result from increased metabolic rate; (2) *increased carbon dioxide concentrations;* (3) *decreased tissue pH* (due to $CO_2$, lactic acid, and other metabolic products); and (4) the *release of adenosine or* $K^+$ from the tissue cells.

The vasodilation that occurs in response to tissue metabolism can be demonstrated by constricting the blood supply to an area for a short time and then removing the constriction. The constriction allows metabolic products to accumulate by preventing venous drainage of the area. When the constriction is removed and blood flow resumes, the metabolic products that have accumulated cause vasodilation. The tissue thus appears red. This response is called **reactive hyperemia.** A similar increase in blood flow occurs in skeletal muscles and other organs as a result of increased metabolism. This is called **active hyperemia.** Intrinsic control mechanisms are summarized in table 14.5.

1. *Describe the relationship between blood flow rate, arterial blood pressure, and vascular resistance.*
2. *Describe the relationship between vascular resistance and the radius of a vessel. Explain how blood flow can be diverted from one organ to another.*
3. *Explain how vascular resistance and blood flow are regulated by (1) sympathetic adrenergic fibers, (2) sympathetic cholinergic fibers, and (3) parasympathetic fibers.*
4. *Define* autoregulation, *and explain how this is accomplished through myogenic and metabolic mechanisms.*

## Blood Flow to the Heart and Skeletal Muscles

Blood flow to the heart and skeletal muscles is regulated by both extrinsic and intrinsic mechanisms. These mechanisms provide increased rates of blood flow when the metabolic requirements of these tissues are increased during exercise.

Survival requires that the heart and brain receive adequate blood flow at all times. The ability of skeletal muscles to respond quickly in emergencies and to maintain continued high levels of activity may also be critically important for survival. During such times, high rates of blood flow to the skeletal muscles must be maintained without compromising blood flow to the heart and brain. This is accomplished by mechanisms that increase the cardiac output and that divert a higher proportion of the cardiac output to the heart, skeletal muscles, and brain and away from the viscera and skin.

### Aerobic Requirements of the Heart

The coronary arteries supply an enormous number of capillaries, which are packed within the myocardium at a density of about 2,500–4,000 per cubic millimeter of tissue. Fast-twitch skeletal muscles, in contrast, have a capillary density of 300–400 per cubic millimeter of tissue. Each myocardial cell, as a consequence, is within 10 $\mu m$ of a capillary (compared to an average distance in other organs of 60–80 $\mu m$). The exchange of gases by diffusion between myocardial cells and capillary blood thus occurs very quickly.

**Table 14.6** Some vascular and metabolic comparisons between heart and skeletal muscle

| | Cardiac Muscle | Skeletal Muscle |
|---|---|---|
| Number of capillaries | 4× | 1× |
| Mean blood flow | 10–20× | 1× |
| Myolemma (sarcolemma) | Thin, low resistance | Thicker, higher resistance |
| Sarcomere length | 1× | 1.7× |
| Glycogen concentration | Maintained | Depleted by fasting, diabetes |
| Glycolytic enzyme systems | 1× | 2×, strongly developed |
| Creatine phosphate concentration | 1× | 6× |
| Anaerobic energy production | Beating: 2 min Arrested: 30–90 min | Up to 40% total energy |
| Lactic acid production | Terminal mechanism | Frequent when incurring $O_2$ debt |
| Ability to incur $O_2$ debt | 1× | 4× |
| Increased $O_2$ requirement met primarily by | Increased flow | Increased extraction |
| Oxygen consumption | 3× | 1× |
| Oxygen extraction at rest | Near maximal | Significant reserve |
| Increased $O_2$ consumption with increased work | 2× | 30× |
| Myoglobin | Present | Present in red skeletal muscle |
| Mitochondria | Abundant, giant | Fewer, smaller |
| Krebs cycle enzymes | 2–3× | 1× |
| Cytochrome-C | 6× | 1× |
| Myosin ATPase activity | 1× | 3× |

From Brachfeld, N., "The Physiology of Muscular Exercise," *Primary Cardiology,* June, 1979, p. 112. Reproduced with permission.

Contraction of the myocardium squeezes the coronary arteries. Unlike blood flow in all other organs, flow in the coronary vessels thus decreases in systole and increases during diastole. The myocardium, however, contains large amounts of *myoglobin,* a pigment related to hemoglobin (the molecules in red blood cells which carry oxygen). Myoglobin in the myocardium stores oxygen during diastole and releases its oxygen during systole. In this way, the myocardial cells can receive a continuous supply of oxygen even though coronary blood flow is temporarily reduced during systole.

In addition to containing large amounts of myoglobin, heart muscle contains numerous mitochondria and aerobic respiratory enzymes. This indicates that—even more than slow-twitch skeletal muscles—the heart is extremely specialized for aerobic respiration (table 14.6). The normal heart always respires aerobically, even during heavy exercise when the metabolic demand for oxygen can rise to five times resting levels. This increased oxygen requirement is met by a corresponding increase in coronary blood flow, from about 80 ml at rest to about 400 ml per minute per 100 g tissue during heavy exercise.

### Regulation of Coronary Blood Flow

Sympathetic nerve fibers, through stimulation of alpha-adrenergic receptors in the coronary arterioles, produce a relatively high vascular resistance in the coronary circulation at rest. Vasodilation of coronary vessels may be produced in part by sympathetic nerve activation of beta-adrenergic receptors. Most of the vasodilation that occurs during exercise, however, is due to intrinsic metabolic control mechanisms. As the metabolism of the myocardium increases, local accumulation of carbon dioxide, $K^+$, and adenosine acts directly on the vascular smooth muscle to cause relaxation and vasodilation.

Under abnormal conditions, the blood flow to the myocardium may be inadequate, resulting in myocardial ischemia (described in chapter 13). This can result from blockage by atheromas and/or blood clots or from muscular spasm of a coronary artery (fig. 14.15). Occlusion of a coronary artery can be visualized by a technique called *selective coronary arteriography.* In this procedure, a catheter (plastic tube) is inserted into a brachial or femoral artery all the way to the opening of the coronary arteries in the aorta, and radiographic contrast material is injected. The picture thus obtained is called an **angiogram.**

If the occlusion is sufficiently great a coronary bypass operation may be performed. In this procedure a length of blood vessel, usually taken from the saphenous vein in the leg, is sutured to the aorta and to the coronary artery at a location beyond the site of the occlusion (fig. 14.16).

**Figure 14.15.** An angiogram of the left coronary artery in a patient (*a*) when the ECG was normal and (*b*) when the ECG showed evidence of myocardial ischemia. Notice that a coronary artery spasm—see arrow in (*b*)—appears to accompany the ischemia.

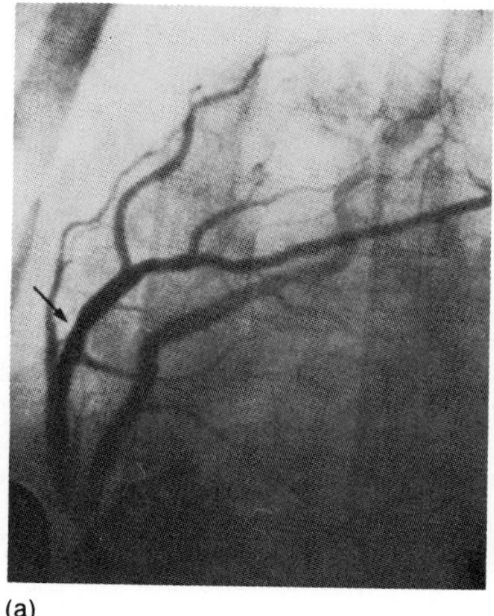
(a)

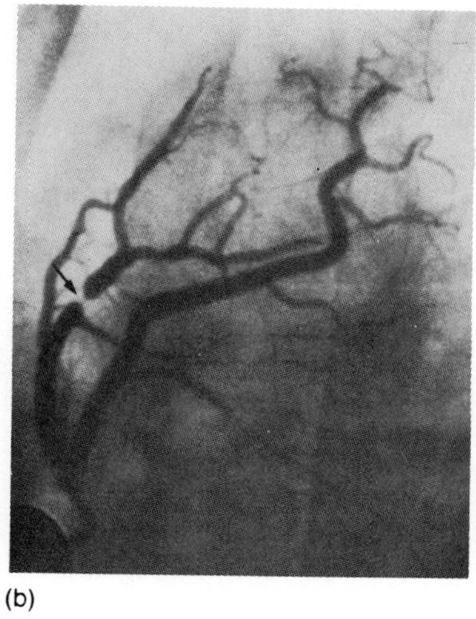
(b)

## Regulation of Blood Flow through Skeletal Muscles

The arterioles in skeletal muscles, like those of the coronary circulation, have a high vascular resistance at rest as a result of alpha-adrenergic sympathetic stimulation. This produces a relatively low rate of blood flow per tissue weight (4–6 ml per minute per 100 g tissue), but because muscles have such a large mass, this still accounts for 20%–25% of the total blood flow in the body at rest. Also like in the heart, blood flow in a skeletal muscle decreases when the muscle contracts and squeezes its arterioles, and in fact blood flow stops entirely when the muscle contracts beyond about 70% of its maximum. Pain and fatigue thus occur much more quickly when an isometric contraction is sustained than when rhythmic isotonic contractions are performed.

In addition to adrenergic fibers, which promote vasoconstriction by stimulation of alpha-adrenergic receptors, there are also sympathetic cholinergic fibers in skeletal muscles. These cholinergic fibers, together with the stimulation of beta-adrenergic receptors by the hormone epinephrine, stimulate vasodilation as part of the "fight-or-flight" response to any stressful state, including that existing just prior to exercise (table 14.7). These extrinsic controls have been previously discussed and function to regulate blood flow through muscles at rest and at the anticipation of exercise.

**Figure 14.16.** A diagram of coronary artery bypass surgery.

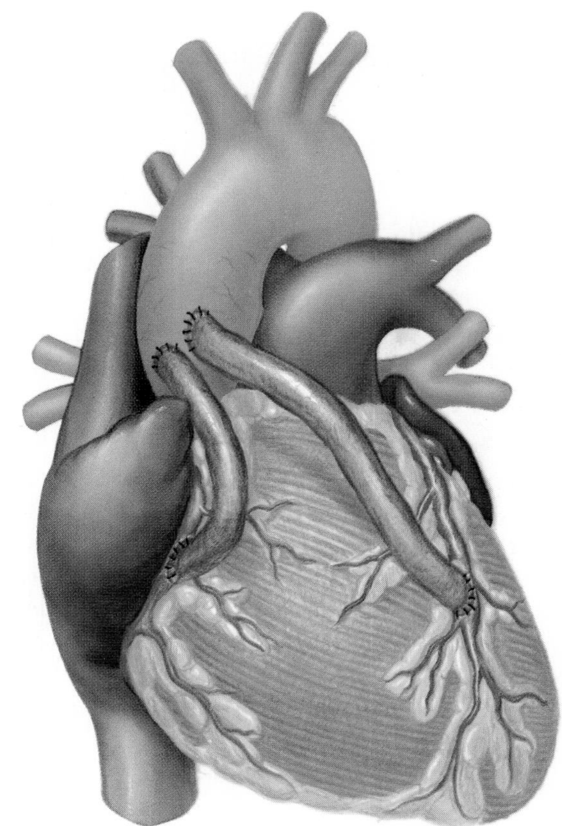

**Table 14.7** Changes in skeletal muscle blood flow under conditions of rest and exercise

| Condition | Blood Flow (ml/min) | Mechanism |
|---|---|---|
| Rest | 1,000 | High adrenergic sympathetic stimulation of vascular alpha-receptors causing vasoconstriction |
| Beginning exercise | Increased | Dilation of arterioles in skeletal muscles due to cholinergic sympathetic nerve activity and stimulation of β-adrenergic receptors by the hormone epinephrine |
| Heavy exercise | 20,000 | Fall in alpha-adrenergic activity<br>Increased sympathetic cholinergic activity<br>Increased metabolic rate of exercising muscles producing intrinsic vasodilation |

**Figure 14.17.** The distribution of blood flow (cardiac output) during rest and heavy work. At rest, the cardiac output is 5 L per minute (*bottom of figure*); during heavy work the cardiac output increases to 25 L per minute (*top of figure*). During rest, for example, the brain receives 15% of 5 L per minute (= 750 ml/min.), whereas during exercise it receives 3% to 4% of 25 L per minute (0.03 × 25 = 750 ml/min.). Flow to the skeletal muscles increases more than twentyfold, because the total cardiac output increases (from 5 L/min. to 25 L/min.) and because the percent of the total received by the muscles increases from 15% to 80%.

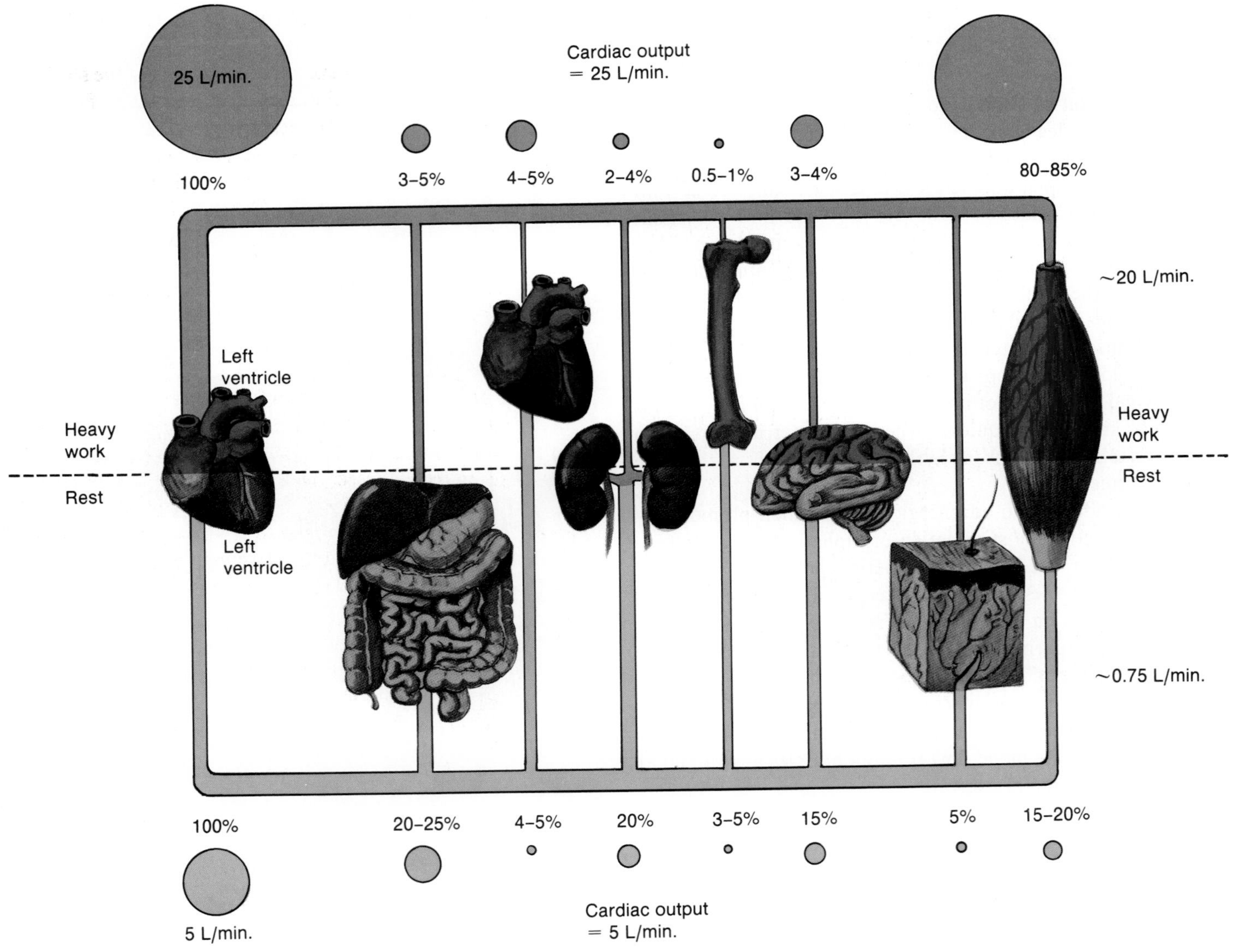

**Table 14.8** Relationship between age and average maximum cardiac rate

| Age | Maximum Cardiac Rate |
|---|---|
| 20–29 | 190 beats/min |
| 30–39 | 160 beats/min |
| 40–49 | 150 beats/min |
| 50–59 | 140 beats/min |
| 60 and above | 130 beats/min |

As exercise progresses, the vasodilation and increased skeletal muscle blood flow that occur are almost entirely due to intrinsic metabolic control. The high metabolic rate of skeletal muscles during exercise causes local changes such as increased carbon dioxide concentrations, decreased pH (due to carbonic acid and lactic acid), decreased oxygen, increased extracellular $K^+$, and the secretion of adenosine. Like the intrinsic control of the coronary circulation, these changes cause vasodilation of arterioles in skeletal muscles. This decreases the vascular resistance and increases the rate of blood flow. This effect is combined with the recruitment of capillaries by the opening of precapillary sphincter muscles (only 5%–10% of the skeletal muscle capillaries are open at rest). As a result of these changes, skeletal muscles can receive as much as 85% of the total blood flow in the body during maximal exercise, amounting to as much as a 100-fold increase in the rate of blood flow over resting conditions.

While the vascular resistance in skeletal muscles decreases during exercise, the resistance to flow through visceral organs and skin increases. This increased resistance occurs because of vasoconstriction stimulated by adrenergic sympathetic fibers and results in decreased rates of blood flow through these organs. During exercise, therefore, the blood flow to skeletal muscles increases because of three simultaneous changes: (1) increased total blood flow (cardiac output); (2) metabolic vasodilation in the exercising muscles; and (3) the diversion of blood away from the viscera and skin. Blood flow to the heart also increases during exercise, whereas blood flow to the brain does not appear to change significantly (fig. 14.17).

Although the skeletal muscles consume large amounts of oxygen during exercise, the oxygen concentration of arterial blood (going to the tissues) is usually not reduced. This is due to the fact that the rate and depth of breathing is also increased during exercise and to the fact that the pulmonary blood flow increases to match the increased systemic blood flow. The oxygen concentration of venous blood (draining from the tissues), however, can be decreased to one-half or one-third the resting levels. This is because the exercising muscles are extracting a much higher proportion of the oxygen delivered to them in the arterial blood. The ability of exercising muscles to extract oxygen from arterial blood is improved by endurance training, which results in increased capillary density and increased amounts of aerobic enzymes in the trained muscles.

### Circulatory Changes during Exercise

During exercise the cardiac output can increase five-fold, from about 5 L per minute to about 25 L per minute. This is primarily due to an increase in cardiac rate. The cardiac rate, however, can only increase up to a maximum value (table 14.8), which is determined mainly by the person's age. In very well-trained athletes the stroke volume can also increase significantly, allowing these individuals to achieve a cardiac output during strenuous exercise as much as six or seven times greater than their resting values.

In most people the stroke volume can only increase from 10% to 35% during exercise. The fact that the stroke volume can increase at all during exercise may at first be surprising, in view of the fact that the heart has less time to fill with blood between beats when it is pumping faster. Despite the faster beat, however, the end-diastolic volume during exercise is not decreased. This is because the venous return is aided by the improved action of the skeletal muscle pumps and by increased respiratory movements during exercise (fig. 14.18). Since the end-diastolic volume is not significantly changed during exercise, any increase in stroke volume that occurs must be due to an increase in the proportion of blood ejected per stroke.

The proportion of the end-diastolic volume ejected per stroke can increase from 67% at rest to as much as 90% during heavy exercise. This increased *ejection fraction* is produced by the increased contractility that results from sympathoadrenal stimulation. There may also be a decrease in total peripheral resistance as a result of vasodilation in the exercising skeletal muscles, which decreases the afterload and thus further augments the increase in stroke volume. The cardiovascular changes that occur during exercise are summarized in table 14.9.

Endurance training often results in a lowering of the resting cardiac rate and an increase in the resting stroke volume. The lowering of the resting cardiac rate results from a greater degree of inhibition of the SA node by the

**Figure 14.18.** Cardiovascular adaptations to exercise.

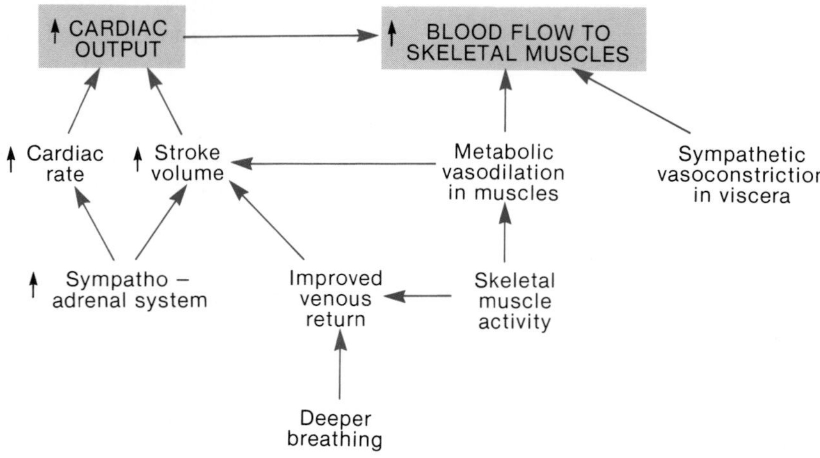


**Table 14.9** Cardiovascular changes that occur during moderate exercise

| Variable | Change | Mechanisms |
|---|---|---|
| Cardiac output | Increased | Cardiac rate and stroke volume increased |
| Cardiac rate | Increased | Increased sympathetic nerve activity; decreased activity of the vagus nerve |
| Stroke volume | Increased | Increased myocardial contractility due to stimulation by sympathoadrenal system; decreased total peripheral resistance |
| Total peripheral resistance | Decreased | Vasodilation of arterioles in skeletal muscles (and in skin when thermoregulatory adjustments are needed) |
| Arterial blood pressure | Increased | Increased systolic and pulse pressure due primarily to increased cardiac output; diastolic pressure rises less due to decreased total peripheral resistance |
| End-diastolic volume | Unchanged | Decreased filling time at high cardiac rates is compensated by increased venous pressure, increased activity of the skeletal muscle pump, and decreased intrathoracic pressure aiding the venous return |
| Blood flow to heart and muscles | Increased | Increased muscle metabolism produces intrinsic vasodilation; aided by increased cardiac output and increased vascular resistance in visceral organs |
| Blood flow to visceral organs | Decreased | Vasoconstriction in digestive tract, liver, and kidneys due to sympathetic nerve stimulation |
| Blood flow to skin | Increased | Metabolic heat produced by exercising muscles produces reflex (involving hypothalamus) that reduces sympathetic constriction of arteriovenous shunts and arterioles |
| Blood flow to brain | Unchanged | Autoregulation of cerebral vessels, which maintains constant cerebral blood flow despite increased arterial blood pressure |

vagus nerve. The increased resting stroke volume is believed to be due to an increase in blood volume; indeed, studies have shown that the blood volume can increase by about 500 ml after only eight days of training. These adaptations enable the trained athlete to produce a larger proportionate increase in cardiac output and achieve a higher absolute cardiac output during exercise. This large cardiac output is the major factor in the improved oxygen delivery to skeletal muscles that occurs as a result of endurance training.

1. *Describe blood flow and oxygen delivery to the myocardium during systole and diastole.*
2. *Describe how blood flow to the heart is affected by exercise, and explain how blood flow to the heart is regulated at rest and during exercise.*
3. *Describe the mechanisms that produce vasodilation of the arterioles in skeletal muscles during exercise, and describe two other reasons why the blood flow to muscles increases during exercise.*
4. *Explain how the stroke volume can increase during exercise despite the fact that the filling times are reduced at high cardiac rates.*

## Blood Flow to the Brain and Skin

Myogenic vasodilation and constriction help to maintain a relatively constant rate of cerebral blood flow. Within the brain, blood is diverted toward areas that are most active by metabolite-induced vasodilation. Blood flow to the skin is controlled by sympathetic nerve fibers, and can be greatly reduced in order to conserve body heat when the ambient temperature is cold.

The examination of cerebral and cutaneous blood flow is a study in contrasts. Cerebral blood flow is regulated primarily by intrinsic mechanisms; cutaneous blood flow is regulated by extrinsic mechanisms. Cerebral blood flow is relatively constant; cutaneous blood flow exhibits more variation than blood flow in any other organ. The brain is the organ that can least tolerate and the skin is the organ that can most tolerate low rates of blood flow.

### Cerebral Circulation

When the brain is deprived of oxygen for a few seconds, the person loses consciousness; irreversible brain injury may occur after a few minutes. For these reasons, the cerebral blood flow is remarkably constant at about 750 ml per minute. This amounts to about 15% of the total cardiac output at rest.

Unlike the coronary and skeletal muscle blood flow, cerebral blood flow is not normally influenced by sympathetic nerve activity. Only when the mean arterial pressure rises to about 200 mm Hg do sympathetic nerves cause a significant degree of vasoconstriction in the cerebral circulation. This vasoconstriction helps to protect small, thin-walled arterioles from bursting under the pressure and thus helps to prevent cerebrovascular accident (stroke).

In the normal range of arterial pressures, cerebral blood flow is regulated almost exclusively by intrinsic mechanisms. These mechanisms help ensure a constant rate of blood flow despite changes in systemic arterial pressure—a process called *autoregulation.* The autoregulation of cerebral blood flow is achieved by both myogenic and metabolic mechanisms.

Myogenic regulation occurs when there is variation in systemic arterial pressure. The cerebral arteries automatically dilate when the blood pressure falls and constrict when the pressure rises. This helps to maintain a constant flow rate during the normal pressure variations that occur during rest, exercise, and emotional states.

The cerebral vessels are also sensitive to the carbon dioxide concentration of arterial blood. When the carbon dioxide concentration rises, as a result of inadequate ventilation (hypoventilation), the cerebral arterioles dilate. This is believed to be due to decreases in the pH of cerebrospinal fluid rather than to a direct effect of $CO_2$ on the cerebral vessels. Conversely, when the arterial $CO_2$ falls below normal during hyperventilation, the cerebral vessels constrict. The resulting decrease in cerebral blood flow is responsible for the dizziness that occurs during hyperventilation.

The cerebral arterioles are exquisitely sensitive to local changes in metabolic activity, so that those brain regions with the highest metabolic activity get the most blood. Indeed, areas of the brain that control specific processes have been mapped by the changing patterns of blood flow that result when these areas are activated. Visual and auditory stimuli, for example, increase blood flow to the appropriate sensory areas of the cerebral cortex, whereas motor activities such as movements of the eyes, arms, and organs of speech result in different patterns of blood flow (fig. 14.19).

The exact mechanisms whereby increases in neural activity in a particular area of the brain elicit local vasodilation are not completely understood. There is evidence, however, that local cerebral vasodilation may be caused by $K^+$, which is released from active neurons during repolarization. It has been proposed that astrocytes may take up this extruded $K^+$ near the active neurons and then release the $K^+$ through their perivascular feet (chapter 7) surrounding arterioles, thereby causing the arterioles to dilate.

### Cutaneous Blood Flow

The skin is the outer covering of the body and as such serves as the first line of defense against invasion by disease-causing organisms. The skin, as the interface between the internal and external environments, also serves to help maintain a constant deep-body temperature despite changes in the ambient (external) temperature, a process called *thermoregulation.* The small thickness and large size of the skin (1.0–1.5 mm thick; 1.7–1.8 square meters in surface area) make it an effective radiator of heat when the body temperature is greater than the ambient temperature. The transfer of heat from the body to the external environment is aided by the flow of warm blood through capillary loops near the surface of the skin.

Blood flow through the skin is adjusted to maintain deep-body temperature at about 37° C (98.6° F). These adjustments are made by variations in the degree of constriction or dilation of ordinary arterioles and of unique **arteriovenous anastomoses** (fig. 14.20). These latter vessels, found predominantly in the fingertips, palms of the hands, toes, soles of the feet, ears, nose, and lips, shunt (divert) blood directly from arterioles to deep venules, thus bypassing superficial capillary loops. Both the ordinary arterioles and the arteriovenous anastomoses are innervated by sympathetic nerve fibers. When the ambient temperature is cold, sympathetic nerves stimulate cutaneous vasoconstriction; cutaneous blood flow is thus decreased, so that less heat will be lost from the body. Since

**Figure 14.19.** Computerized picture of blood flow distribution in the brain after injecting the carotid artery with a radioactive isotope. In (*a*), on the left, the subject followed a moving object with his eyes. High activity is seen over the occipital lobe of the brain. In (*a*), on the right, the subject listened to spoken words. Notice that the high activity is seen over the temporal lobe (the auditory cortex). In (*b*), on the left, the subject moved his fingers on the side of the body opposite to the cerebral hemisphere being studied. In (*b*), on the right, the subject counted to twenty. High activity is shown over the mouth area of the motor cortex, the supplementary motor area, and the auditory cortex.

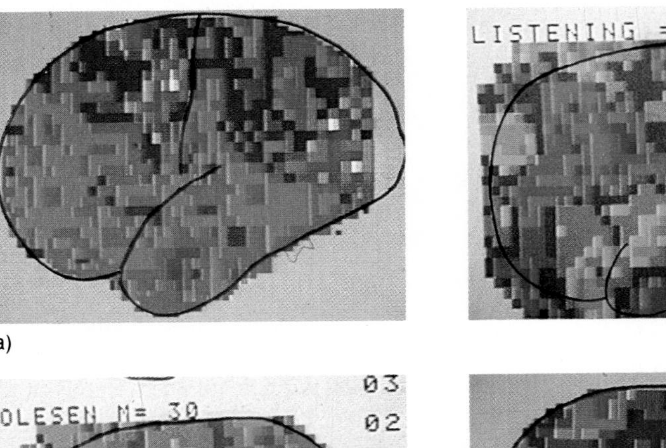


(a)

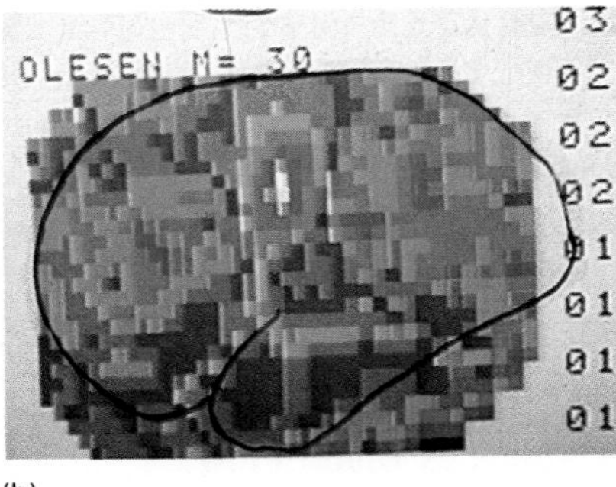


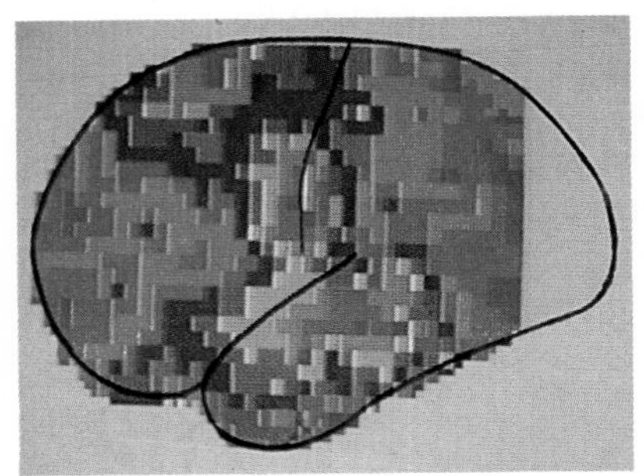

(b)

**Figure 14.20.** Circulation in the skin showing arteriovenous anastomoses, which function as shunts allowing blood to be diverted directly from the arteriole to the venule, thus bypassing superficial capillary loops.

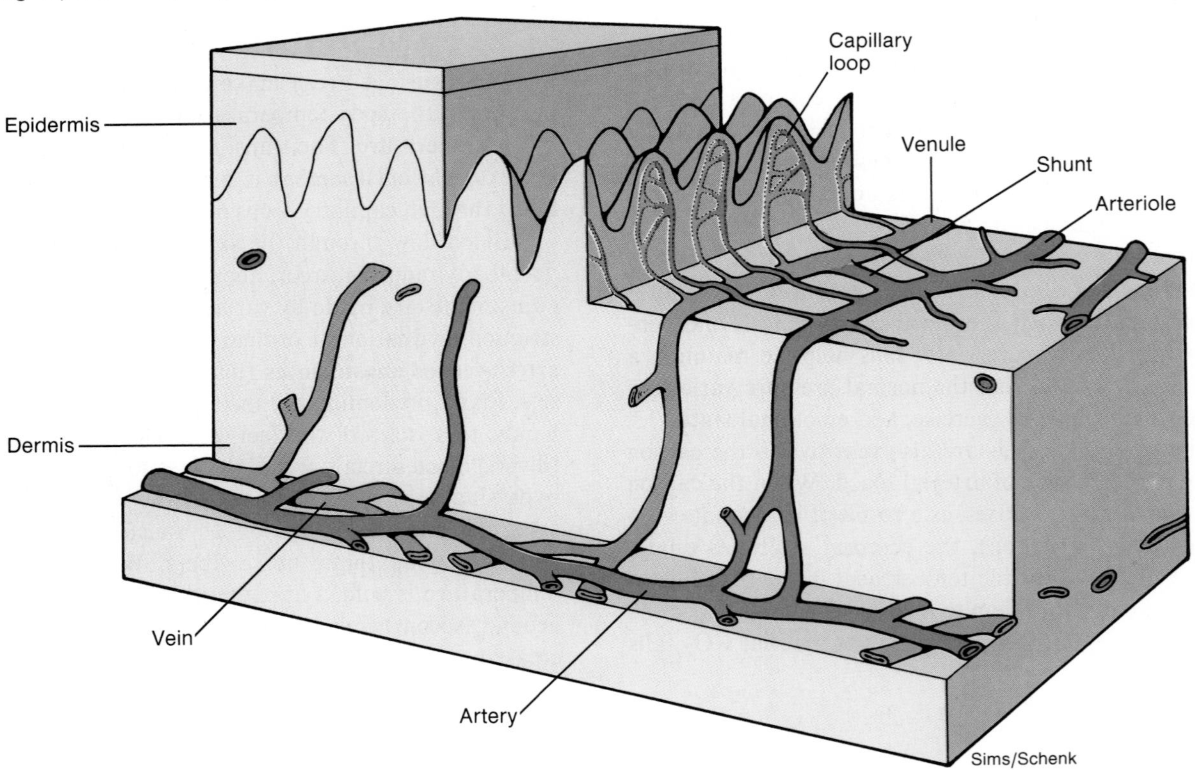

the arteriovenous anastomoses also constrict, the skin may appear rosy as a result of the fact that blood is diverted to the superficial capillary loops. In spite of this rosy appearance, however, the total cutaneous blood flow and rate of heat loss is lower than under usual conditions.

Skin can tolerate an extremely low blood flow in cold weather because its metabolic rate decreases when the ambient temperature decreases. In cold weather, therefore, the skin requires less blood. During severe cold exposure, however, blood flow to the skin can be so severely restricted that the tissue dies: this is *frostbite.* Blood flow to the skin can vary from less than 20 ml per minute at maximal vasoconstriction to as much as 3–4 L per minute at maximal vasodilation.

As the temperature warms, cutaneous arterioles in the hands and feet dilate as a result of decreased sympathetic nerve activity. Continued warming causes dilation of arterioles in other areas of the skin. If the resulting increase in cutaneous blood flow is not sufficient to cool the body, secretion of the sweat glands may be stimulated. Sweat helps to cool the body as it evaporates from the surface of the skin. Also, the sweat glands secrete **bradykinin,** a polypeptide that stimulates vasodilation. This increases blood flow to the skin and to the sweat glands, so that larger volumes of more dilute sweat are produced.

Under the usual conditions of ambient temperature, the cutaneous vascular resistance is high and the blood flow is low when a person is not exercising. In the pre-exercise state of "fight or flight," sympathetic nerve activity reduces cutaneous blood flow still further. During exercise, however, the need to maintain a deep-body temperature takes precedence over the need to maintain an adequate systemic blood pressure. As the body temperature rises during exercise, vasodilation in cutaneous vessels occurs together with vasodilation in the exercising muscles. This can produce an even greater lowering of total peripheral resistance. If exercise is performed in hot and humid weather and if restrictive clothing is worn that increases skin temperature and cutaneous vasodilation, a dangerously low blood pressure may be produced after exercise has ceased and the cardiac output has declined. People have lost consciousness and even died as a result.

Changes in cutaneous blood flow occur as a result of changes in sympathetic nerve activity. Since the activity of the sympathetic nervous system is controlled by the brain, emotional states, acting through control centers in the medulla oblongata, can affect sympathetic activity and cutaneous blood flow. During fear reactions, for example, vasoconstriction in the skin together with activation of the sweat glands can produce a pallor and a "cold sweat." Other emotions may cause vasodilation and blushing.

1. *Define the term* autoregulation, *and describe how this is accomplished in the cerebral circulation.*
2. *Describe how hyperventilation can cause dizziness.*
3. *Explain how cutaneous blood flow is adjusted to maintain a constant deep-body temperature.*

## Blood Pressure

The pressure of the arterial blood is regulated by the blood volume, total peripheral resistance, and the cardiac rate. Changes in these factors function to maintain a correct arterial pressure by feedback responses to the activity of pressure receptors, or baroreceptors. Arterial pressure rises and falls as the heart goes through systole and diastole, and values of systolic and diastolic pressure can be determined by a technique which exploits the sounds made by the turbulent flow of blood through a constriction of an artery.

Resistance to flow in the arterial system is greatest in the arterioles because these vessels have the smallest diameters. Although the total blood flow through a system of arterioles must be equal to the flow in the larger vessel that gave rise to those arterioles, the narrow diameter of each arteriole reduces the flow rate in each according to Poiseuille's law. Blood flow rate and pressure is thus reduced in the capillaries, which are located downstream of the high resistance imposed by the arterioles. The blood pressure upstream of the arterioles—in the medium and large arteries—is correspondingly increased (fig. 14.21).

The blood pressure and flow rate within the capillaries are further reduced by the fact that their total cross-sectional area is much greater, due to their large number, than the cross-sectional areas of arteries and arterioles (fig. 14.22). Thus, although each capillary is much narrower than each arteriole, the capillary beds served by arterioles do not provide as great a resistance to blood flow as do the arterioles.

Variations in the diameter of arterioles due to vasoconstriction and vasodilation thus simultaneously affect both blood flow through capillaries and the *arterial blood pressure* "upstream" from the capillaries. An increase in total peripheral resistance due to vasoconstriction of arterioles can raise arterial blood pressure. Blood pressure can also be raised by an increase in the cardiac output. This may be due to elevations in cardiac rate or stroke volume, which in turn is affected by other factors. The three most important variables affecting blood pressure are the **cardiac rate, stroke volume** (determined primarily

**Figure 14.21.** A constriction increases blood pressure upstream (analogous to the arterial pressure) and decreases pressure downstream (analogous to capillary and venous pressure).

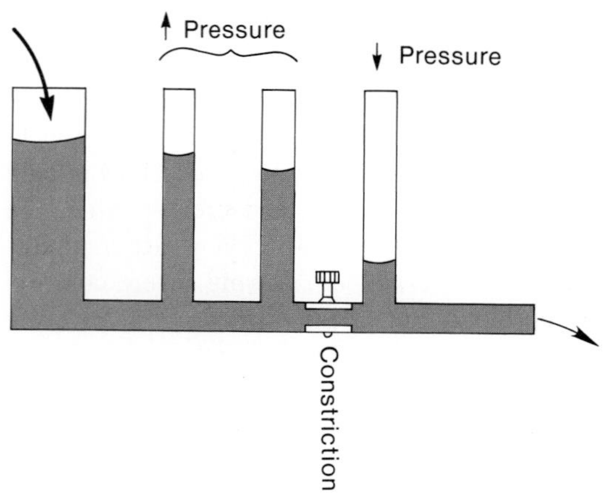


**Figure 14.22.** As blood passes from the aorta to the smaller arteries, arterioles, and capillaries, the cross-sectional area increases as the pressure decreases.

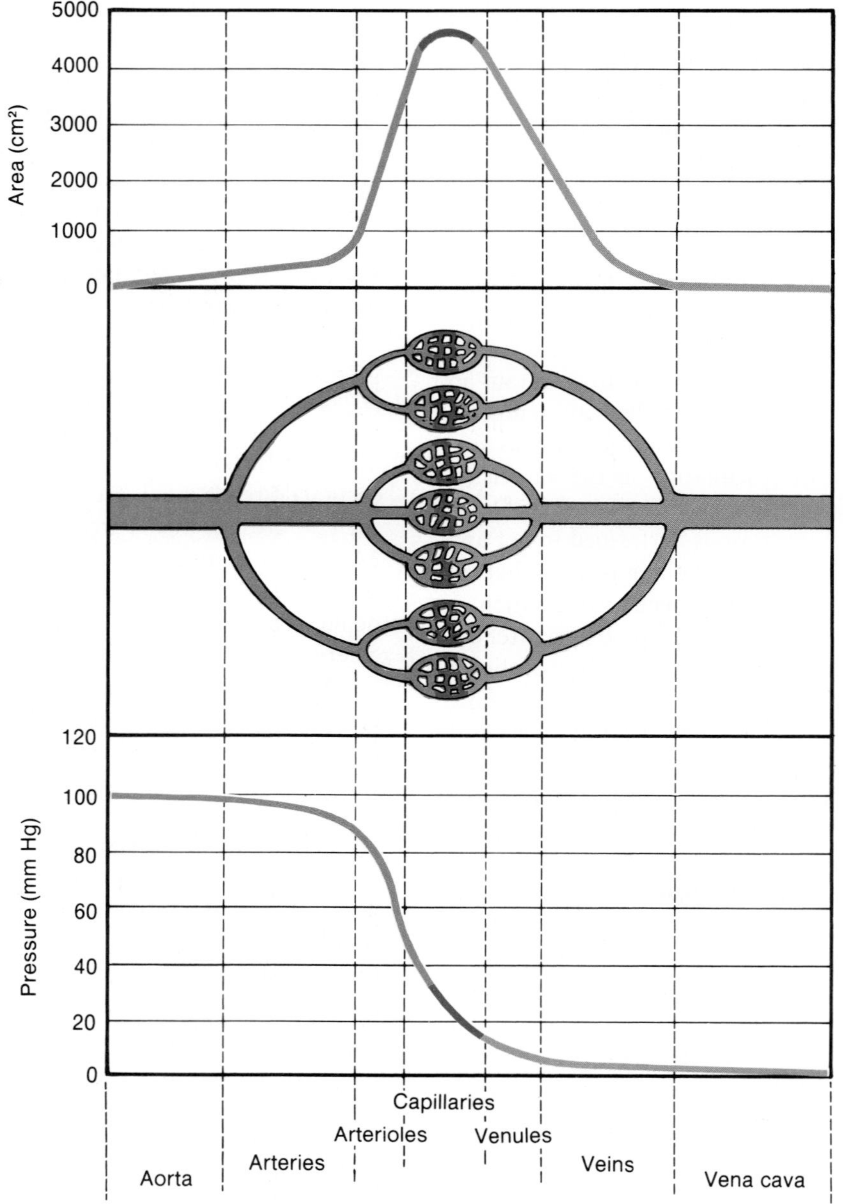

**Figure 14.23.** Action potential frequency in sensory nerve fibers from baroreceptors in the carotid sinus and aortic arch. As the blood pressure increases, the baroreceptors become increasingly stretched. This results in an increase in the frequency of action potentials that are transmitted to the cardiac and vasomotor control centers in the medulla oblongata.

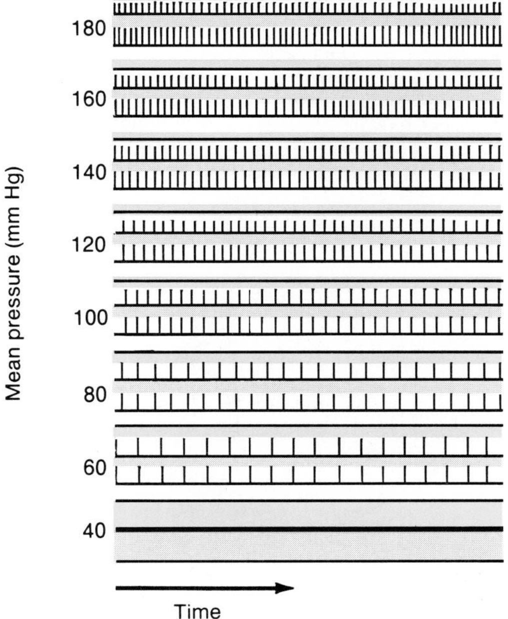


by the **blood volume**), and **total peripheral resistance.** An increase in any of these, if not compensated by a decrease in another variable, will result in an increased blood pressure.

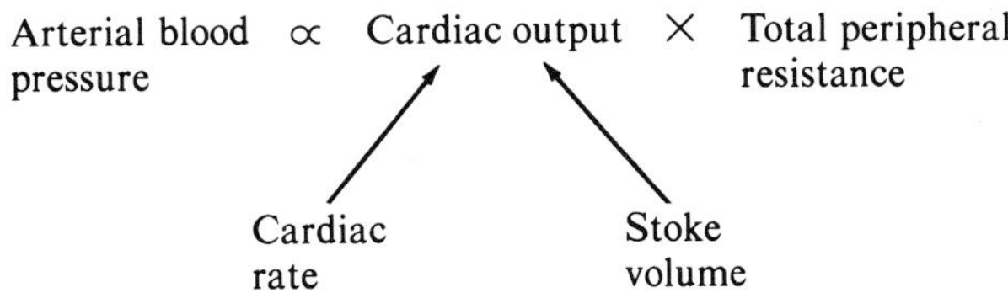


Blood pressure can thus be regulated by the kidneys, which control blood volume and thus stroke volume, and by the sympathoadrenal system. Increased activity of the sympathoadrenal system can raise blood pressure by stimulating vasoconstriction of arterioles (thus raising total peripheral resistance) and by promoting an increased cardiac output. Sympathetic stimulation can also affect blood volume indirectly, by stimulating constriction of renal blood vessels and thus reducing urine output.

## Regulation of Blood Pressure

In order for blood pressure to be maintained within limits, specialized receptors for pressure are needed. These **baroreceptors** are stretch receptors located in the *aortic arch* and in the *carotid sinuses.* An increase in pressure causes the walls of these arterial regions to stretch and stimulate the activity of sensory nerve endings (fig. 14.23). A fall in pressure below the normal range, in contrast, causes a decrease in the frequency of action potentials produced by these sensory nerve fibers.

Sensory nerve activity from the baroreceptors ascends via the vagus and glossopharyngeal nerves to the medulla oblongata, which directs the autonomic system to respond appropriately. **Vasomotor control centers** in the medulla control vasoconstriction/vasodilation and hence help regulate total peripheral resistance. **Cardiac control centers** in the medulla regulate the cardiac rate (fig. 14.24).

***Baroreceptor Reflex.*** When a person goes from a lying to a standing position, there is a shift of 500–700 ml of blood from the veins of the thoracic cavity to veins in the lower extremities, which expand to contain the extra volume of blood. This pooling of blood reduces the venous return and cardiac output. The resulting fall in blood pressure is almost immediately compensated by the baroreceptor reflex. A decrease in baroreceptor sensory information, traveling in the ninth and tenth cranial nerves to the medulla oblongata, inhibits parasympathetic activity and promotes sympathetic nerve activity, resulting in increased cardiac rate and vasoconstriction. These responses help maintain an adequate blood pressure upon standing (fig. 14.25).

An effective baroreceptor reflex helps to maintain adequate blood flow to the brain upon standing. Since the baroreceptor reflex may require a few seconds to be fully effective, however, many people feel dizzy and disoriented if they stand up too rapidly. If the baroreceptor sensitivity is abnormally reduced, perhaps by atherosclerosis, an uncompensated fall in pressure may occur upon standing. This condition—called **postural,** or **orthostatic, hypotension** (hypotension = low blood pressure)—can make a person feel extremely dizzy or even faint because of inadequate perfusion of the brain.

The baroreceptor reflex can also mediate the opposite response. When the blood pressure rises above an individual's normal range, the baroreceptor reflex causes a slowing of the cardiac rate and vasodilation. Manual massage of the carotid sinus, a procedure sometimes employed by physicians to reduce tachycardia and lower blood pressure, also evokes this reflex. Such carotid massage should be used cautiously, however, because the intense vagus-nerve-induced slowing of the cardiac rate could cause loss of consciousness (as occurs in emotional fainting). Manual massage of both carotid sinuses simultaneously can even cause cardiac arrest in susceptible people.

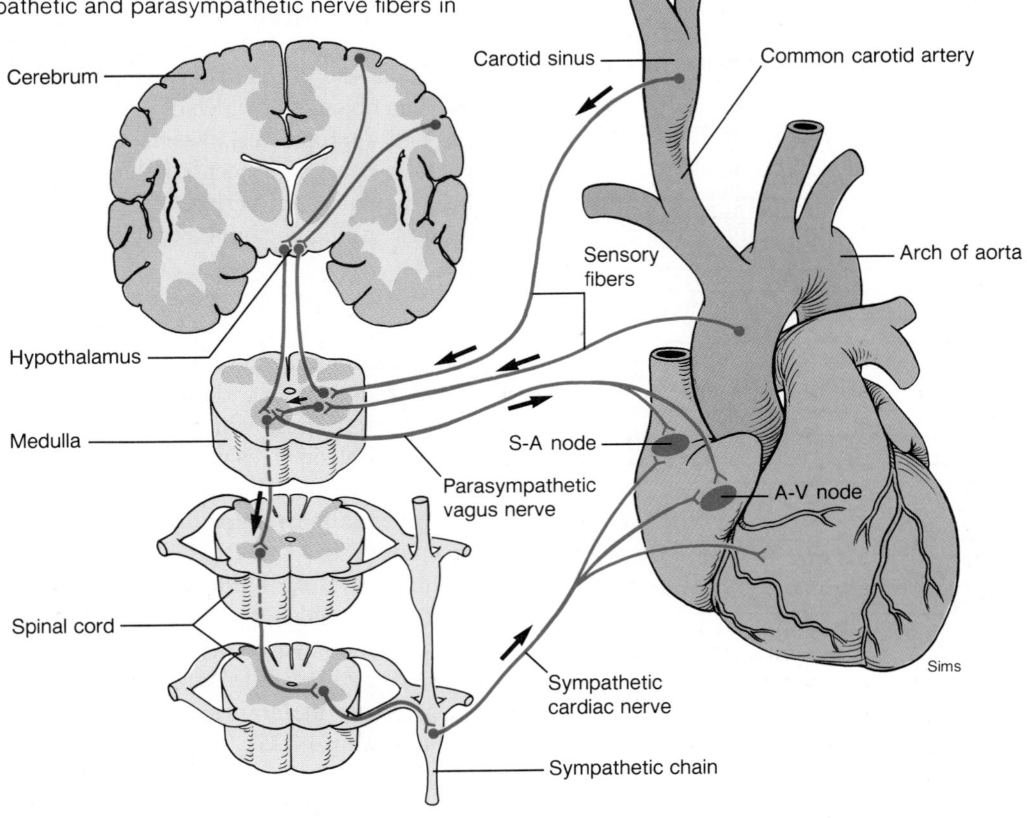


**Figure 14.24.** The baroreceptor reflex. Sensory stimuli from baroreceptors in the carotid sinus and the aortic arch, acting via control centers in the medulla oblongata, affect the activity of sympathetic and parasympathetic nerve fibers in the heart.

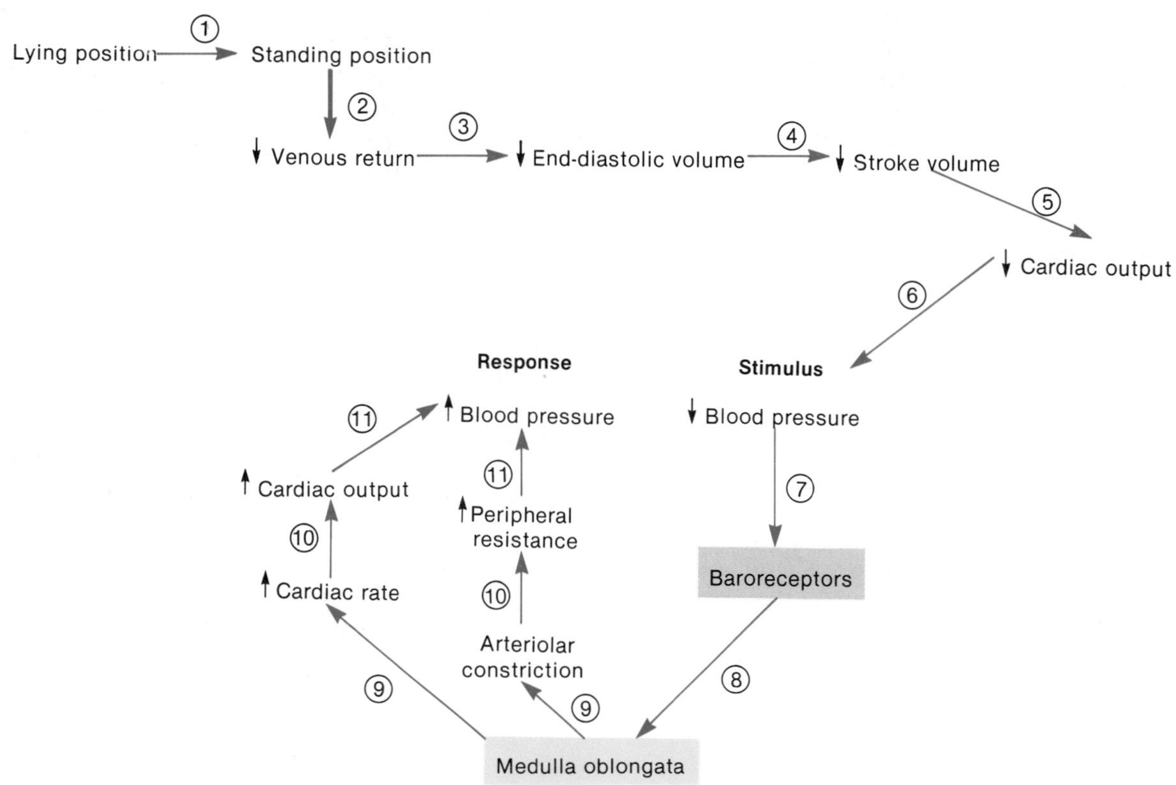


**Figure 14.25.** Compensations for the upright posture induced by the baroreceptor reflex. Numbers indicate the sequence of cause-and-effect steps.

**Figure 14.26.** The use of a pressure cuff and sphygmomanometer to measure blood pressure.

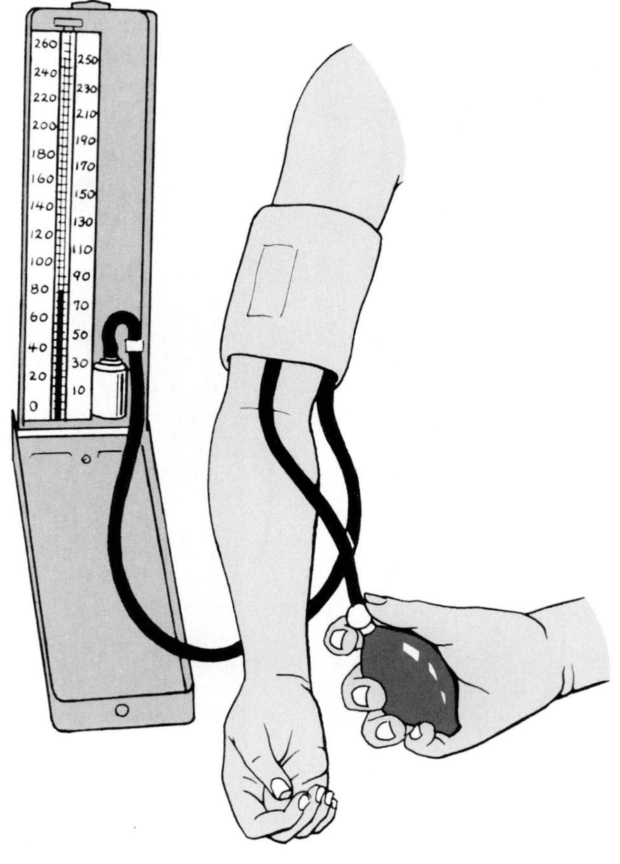


A person is said to perform the **Valsalva maneuver** when exhalation is attempted against a closed glottis (which prevents the air from escaping—see chapter 15). This maneuver, commonly performed during forceful defecation or when lifting heavy weights, increases the intrathoracic pressure. This at first causes the arterial pressure to rise due to compression of the aorta. The thoracic veins are also compressed, however, and since this results in a fall in venous return and cardiac output, the arterial blood pressure subsequently falls. The lowering of arterial pressure stimulates the baroreceptor reflex, resulting in tachycardia and increased total peripheral resistance. When the glottis is finally opened and the air is exhaled, the cardiac output returns to normal but the total peripheral resistance is still elevated, causing a rise in blood pressure. Acting through the baroreceptors, the rise in blood pressure next results in bradycardia and a lowering of the blood pressure to normal values.

The decreased cardiac output that occurs due to constriction of thoracic veins, and the fluctuations in blood pressure that occur during the Valsalva maneuver, are undesirable effects that can be particularly dangerous for a person with heart disease. Even for healthy people, it is advisable to exhale normally when lifting weights.

***Other Reflexes Controlling Blood Pressure.*** The reflex control of ADH secretion by osmoreceptors in the hypothalamus and the control of angiotensin II production and aldosterone secretion by the juxtaglomerular apparatus of the kidneys have been previously discussed. Antidiuretic hormone and aldosterone increase blood volume, and angiotensin II stimulates vasoconstriction to cause an increase in blood pressure.

Another reflex that may be important to blood pressure regulation is initiated by stretch receptors located in the left atrium of the heart. These receptors are activated by increased venous return to the heart and produce (1) reflex tachycardia, as a result of increased sympathetic nerve activity; and (2) inhibition of ADH secretion, resulting in larger volumes of urine excretion and a lowering of blood volume.

## Measurement of Blood Pressure

Stephen Hales (1677–1761) accomplished the first documented measurement of blood pressure by inserting a cannula into the artery of a horse and measuring the height to which blood would rise in the vertical tube. Modern clinical blood pressure measurements, fortunately, are less direct. The indirect, or **auscultatory,** method of blood pressure measurement is based on the correlation of blood pressure and arterial sounds.

In the auscultatory method, an inflatable rubber bladder within a cloth cuff is wrapped around the upper arm and a stethoscope is applied over the brachial artery (fig. 14.26). The artery is normally silent before inflation of the cuff because blood normally travels in a smooth *laminar flow* through the arteries. The term laminar means layered—blood in the central axial stream moves the fastest, and blood flowing closer to the artery wall moves more slowly. There is little transverse movement between these layers that would produce mixing.

The laminar flow that normally occurs in arteries produces little vibration and is thus silent. When the artery is pinched, however, blood flow through the constriction becomes turbulent, which causes the artery to vibrate and produce sounds (much like the sounds produced by water through a kink in a garden hose). The tendency of the cuff pressure to constrict the artery is opposed by the blood pressure. Thus, in order to constrict the artery, the cuff pressure must be greater than the diastolic blood pressure. If the cuff pressure is also greater than the systolic blood pressure the artery would be pinched off and silent. *Turbulent flow* and sounds produced by vibrations of the artery as a result of this flow, therefore, only occur when the cuff pressure is greater than the diastolic and less than the systolic blood pressure.

Suppose that a person has a systolic pressure of 120 mm Hg and a diastolic pressure of 80 mm Hg (the average normal values). When the cuff pressure is between

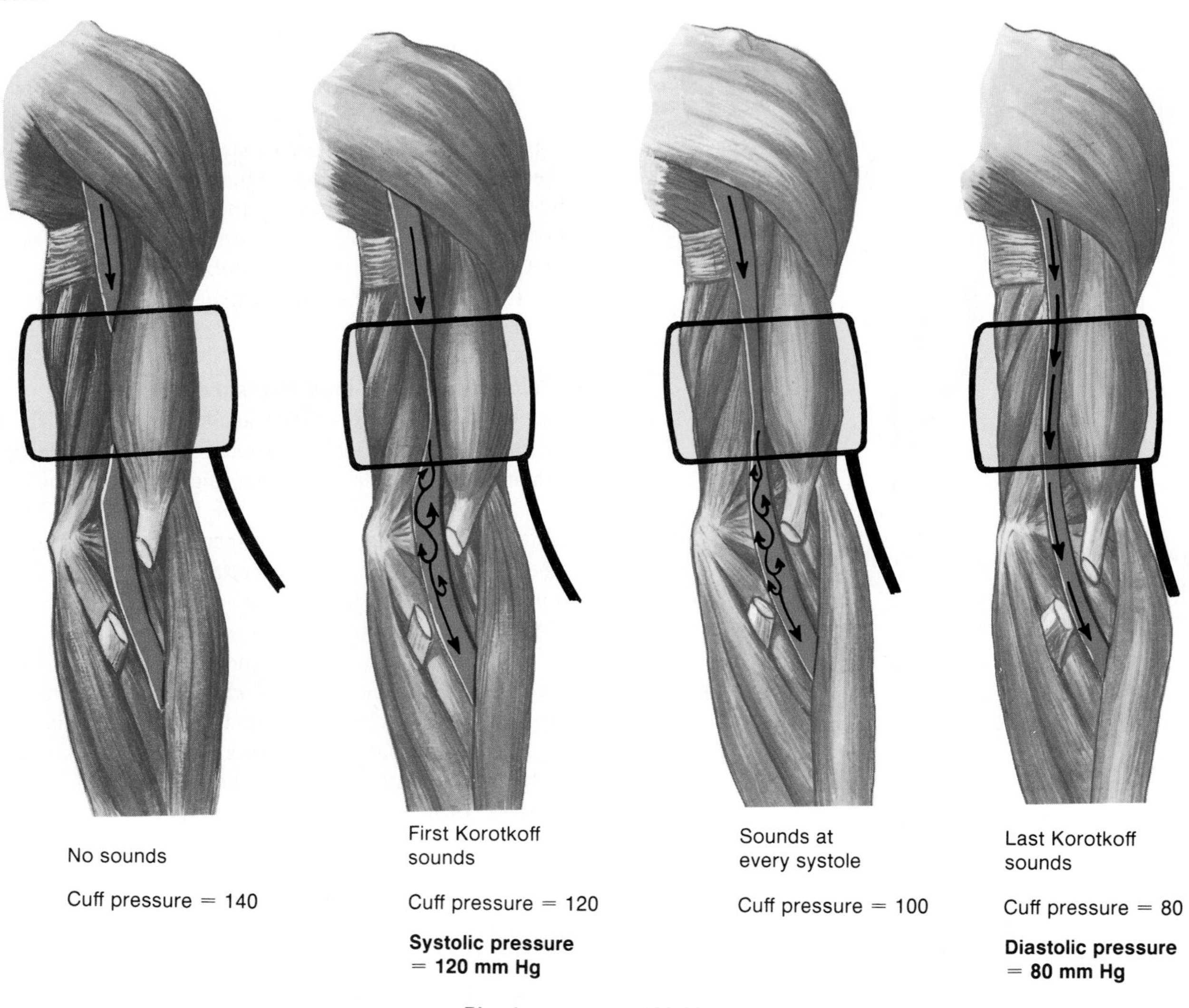


**Figure 14.27.** Korotkoff sounds are produced by the turbulent flow of blood through the partially constricted brachial artery. This occurs when the cuff pressure is greater than the diastolic pressure but less than the systolic pressure.

80 and 120 mm Hg, the artery will be closed during diastole and open during systole. As the artery begins to open with every systole, turbulent flow of blood through the constriction will create vibrations that are heard as the **sounds of Korotkoff,** as shown in figure 14.27. These are usually "tapping" sounds because the artery becomes constricted, blood flow stops, and silence resumes with every diastole. It should be understood that the sounds of Korotkoff are *not* "lub-dub" sounds produced by closing of the heart valves (those sounds can only be heard on the chest, not on the brachial artery).

Initially, the cuff is usually inflated to produce a pressure greater than the systolic pressure so that the artery is pinched off and silent. The pressure in the cuff is read from an attached meter called a *sphygmomanometer.* A valve is then turned to allow the release of air from the cuff, causing a gradual decrease in cuff pressure. When the cuff pressure is equal to the systolic pressure, the **first sound** of Korotkoff is heard as blood passes in a turbulent flow through the constricted opening of the artery.

Korotkoff sounds will continue to be heard at every systole as long as the cuff pressure remains greater than the diastolic pressure. When the cuff pressure becomes equal to or less than the diastolic pressure, the sounds disappear because the artery remains open, laminar flow occurs, and the vibrations of the artery stop (fig. 14.28). The **last sound** of Korotkoff thus occurs when the cuff pressure is equal to the diastolic pressure.

Different phases in the measurement of blood pressure are identified on the basis of the quality of the Korotkoff sounds (fig. 14.29). In some people, the Korotkoff

**Figure 14.28.** The indirect, or auscultatory, method of blood-pressure measurement. Korotkoff sounds, produced by turbulent blood flow through a constricted artery, occur whenever the cuff pressure is less than the systolic blood pressure and greater than the diastolic blood pressure. As a result, the first Korotkoff sound is heard when the cuff pressure is equal to the systolic blood pressure, and the last sound is heard when the cuff pressure and diastolic blood pressures are equal.

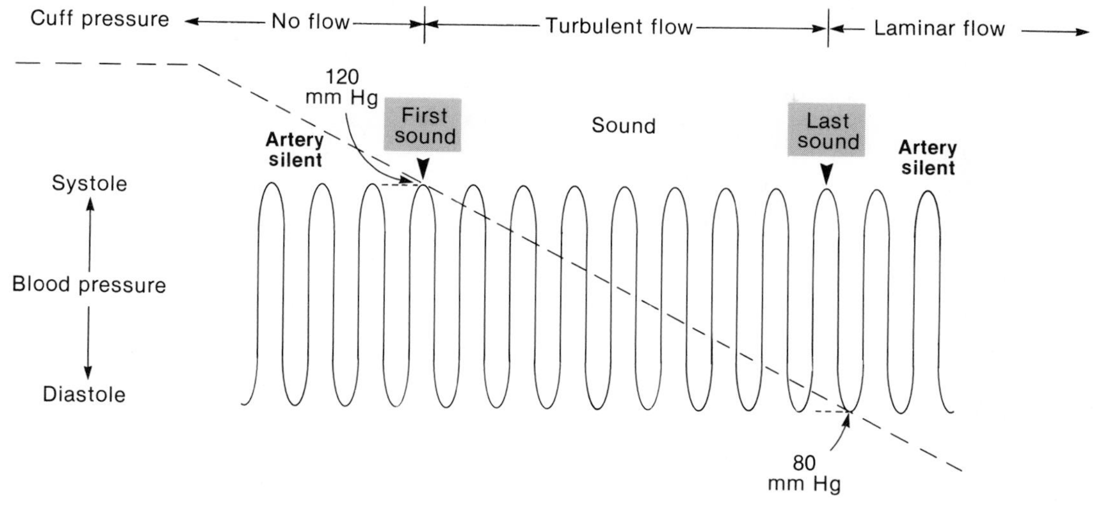


**Figure 14.29.** The five phases of blood-pressure measurement.

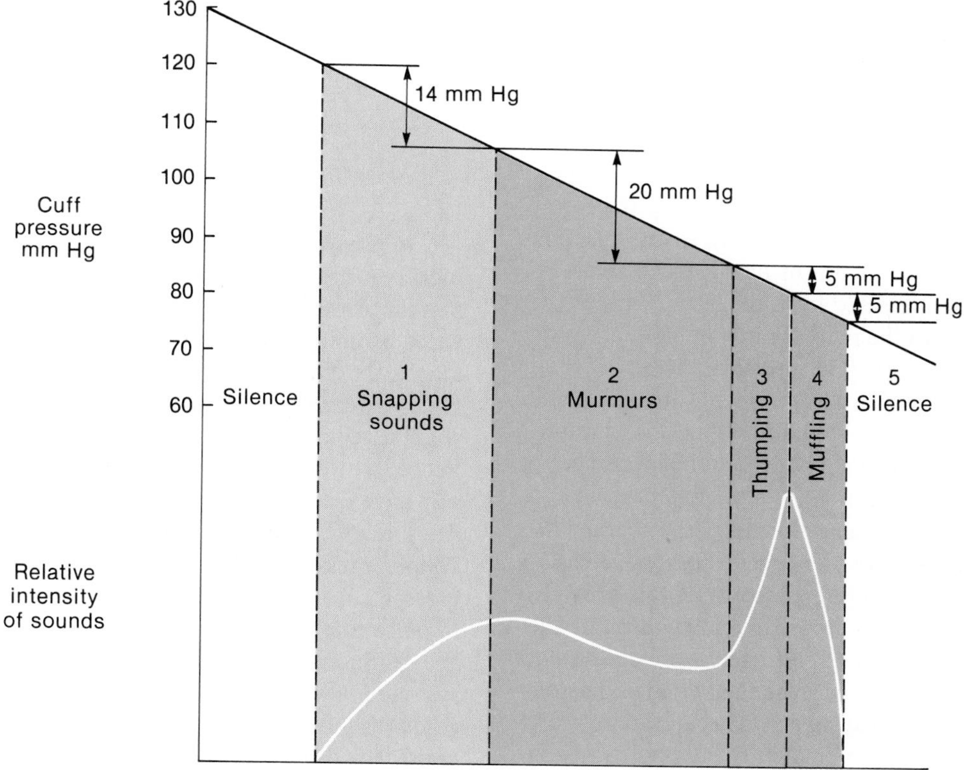


sounds do not disappear even when the cuff pressure is reduced to zero (zero pressure means that it is equal to atmospheric pressure). In these cases—and often routinely—the onset of muffling of the sounds (phase 4 in fig. 14.29) is used rather than the onset of silence (phase 5) as an indication of diastolic pressure. Normal blood pressure values are shown in table 14.10.

## Pulse Pressure and Mean Arterial Pressure

The **pulse pressure** is equal to the difference between the systolic and diastolic pressures. If a person has a blood pressure of 120/80 (systolic/diastolic), therefore, the pulse pressure would be 40 mm Hg.

$$\text{Pulse pressure} = \text{Systolic pressure} - \text{Diastolic pressure}$$

**Table 14.10** Normal arterial blood pressure at different ages

| Age | Systolic | | Diastolic | | Age | Systolic | | Diastolic | |
|---|---|---|---|---|---|---|---|---|---|
| | Men | Women | Men | Women | | Men | Women | Men | Women |
| 1 day | 70 | | | | 16 years | 118 | 116 | 73 | 72 |
| 3 days | 72 | | | | 17 years | 121 | 116 | 74 | 72 |
| 9 days | 73 | | | | 18 years | 120 | 116 | 74 | 72 |
| 3 weeks | 77 | | | | 19 years | 122 | 115 | 75 | 71 |
| 3 months | 86 | | | | 20–24 years | 123 | 116 | 76 | 72 |
| 6–12 months | 89 | 93 | 60 | 62 | 25–29 years | 125 | 117 | 78 | 74 |
| 1 year | 96 | 95 | 66 | 65 | 30–34 years | 126 | 120 | 79 | 75 |
| 2 years | 99 | 92 | 64 | 60 | 35–39 years | 127 | 124 | 80 | 78 |
| 3 years | 100 | 100 | 67 | 64 | 40–44 years | 129 | 127 | 81 | 80 |
| 4 years | 99 | 99 | 65 | 66 | 45–49 years | 130 | 131 | 82 | 82 |
| 5 years | 92 | 92 | 62 | 62 | 50–54 years | 135 | 137 | 83 | 84 |
| 6 years | 94 | 94 | 64 | 64 | 55–59 years | 138 | 139 | 84 | 84 |
| 7 years | 97 | 97 | 65 | 66 | 60–64 years | 142 | 144 | 85 | 85 |
| 8 years | 100 | 100 | 67 | 68 | 65–69 years | 143 | 154 | 83 | 85 |
| 9 years | 101 | 101 | 68 | 69 | 70–74 years | 145 | 159 | 82 | 85 |
| 10 years | 103 | 103 | 69 | 70 | 75–79 years | 146 | 158 | 81 | 84 |
| 11 years | 104 | 104 | 70 | 71 | 80–84 years | 145 | 157 | 82 | 83 |
| 12 years | 106 | 106 | 71 | 72 | 85–89 years | 145 | 154 | 79 | 82 |
| 13 years | 108 | 108 | 72 | 73 | 90–94 years | 145 | 150 | 78 | 79 |
| 14 years | 110 | 110 | 73 | 74 | 95–106 years | 145 | 149 | 78 | 81 |
| 15 years | 112 | 112 | 75 | 76 | | | | | |

From Diem, K., and Lentner, C., Eds., *Documenta Geigy Scientific Tables,* 7th ed., J. R. Geigy S. A., Basle, Switzerland, 1970. With permission.

At diastole in this example the aortic pressure equals 80 mm Hg. When the left ventricle contracts, the intraventricular pressure rises above 80 mm Hg and ejection begins. As a result, the amount of blood in the aorta increases by the amount ejected from the left ventricle (the stroke volume). Due to the increase in volume there is an increase in blood pressure. The pressure in the brachial artery, where blood pressure measurements are commonly taken, therefore increases to 120 mm Hg in this example. The rise in pressure from diastolic to systolic levels (pulse pressure) is thus a reflection of the stroke volume.

The **mean arterial pressure** represents the average arterial pressure during the cardiac cycle. This value is significant because it is the difference between this pressure and the venous pressure that drives blood through the capillary beds of organs. The mean arterial pressure is not a simple arithmetic average because the period of diastole is longer than the period of systole. Mean arterial pressure can most correctly be approximated by adding one-third of the pulse pressure to the diastolic pressure. If a person has a blood pressure of 120/80, for example, the mean arterial pressure would be $80 + 1/3\ (40) = 93$ mm Hg.

$$\text{Mean arterial pressure} = \text{Diastolic pressure} + \tfrac{1}{3} \times \text{Pulse pressure}$$

A rise in total peripheral resistance and cardiac rate increases the diastolic pressure more than it increases the systolic pressure. When the baroreceptor reflex is activated by going from a lying to a standing position, for example, the diastolic pressure usually increases by 5–10 mm Hg, whereas the systolic pressure is either unchanged or slightly reduced (as a result of decreased venous return). People with hypertension (high blood pressure) who usually have elevated total peripheral resistance and cardiac rates, likewise have a greater increase in diastolic than in systolic pressure. Dehydration or blood loss results in decreased cardiac output and thus also produces a decrease in pulse pressure.

An increase in cardiac output, in contrast, raises the systolic pressure more than it raises the diastolic pressure (though both pressures do rise). This occurs during exercise, for example, when the blood pressure may rise to values as high as 200/100 (yielding a pulse pressure of 100 mm Hg). The effects of different variables on blood pressure are summarized in table 14.11.

**Table 14.11** Effects of different variables on blood pressure measurements

| Variable | Diastolic Pressure | Systolic Pressure | Pulse Pressure |
|---|---|---|---|
| ↑Cardiac rate | Increases (more) | Increases (less) | Decreases |
| ↑Peripheral resistance | Increases (more) | Increases (less) | Decreases |
| ↑Blood volume* | Increases (less) | Increases (more) | Increases |
| ↑Stroke volume | Increases (less) | Increases (more) | Increases |

*Conversely, a fall in blood volume (as in dehydration) produces a decrease in pulse pressure.

1. *Describe the relationship between blood pressure and the total cross-sectional area of arteries, arterioles, and capillaries. Describe how arterioles influence blood flow through capillaries and arterial blood pressure.*
2. *Describe how the baroreceptor reflex helps to compensate for a fall in blood pressure. Explain how a person who is severely dehydrated would have a rapid pulse.*
3. *Describe how the sounds of Korotkoff are produced and how these sounds are used to measure blood pressure.*
4. *Define* pulse pressure *and explain its physiological significance.*

## HYPERTENSION, SHOCK, AND CONGESTIVE HEART FAILURE

An understanding of the normal function, or physiology, of the cardiovascular system is prerequisite to the study of its *pathophysiology,* or mechanisms of abnormal function. Since the mechanisms that regulate cardiac output, blood flow, and blood pressure are highlighted in particular disease states, a study of pathophysiology at this time can strengthen the students' understanding of the mechanisms involved in normal function.

### Hypertension

Approximately 20% of all adults in the United States have *hypertension*—blood pressure in excess of the normal range for the person's age and sex. Hypertension that is a result of (secondary to) known disease processes is logically called **secondary hypertension.** Secondary hypertension comprises only about 10% of the hypertensive population. Hypertension that is the result of complex and poorly understood processes is not-so-logically called **primary,** or **essential, hypertension.**

Diseases of the kidneys and arteriosclerosis of the renal arteries can cause secondary hypertension because of high blood volume. More commonly, the reduction of renal blood flow can raise blood pressure by stimulating the secretion of vasoactive chemicals from the kidneys. Experiments in which the renal artery is pinched, for example, produce hypertension that is associated (at least initially) with elevated renin secretion. These and other causes of secondary hypertension are summarized in table 14.12.

***Essential Hypertension.*** The vast majority of people with hypertension have essential hypertension. An increased total peripheral resistance is a universal characteristic of this condition. Cardiac rate and the cardiac output are elevated in many, but not all, of these cases.

The secretion of renin, which is correlated with angiotensin II production and aldosterone secretion, is likewise variable. Although some people with essential hypertension have low renin secretion, most have either normal or elevated levels of renin secretion. Renin secretion in the normal range is inappropriate for people with hypertension, since high blood pressure should inhibit renin secretion and, through a lowering of aldosterone, result in greater excretion of salt and water. Inappropriately high levels of renin secretion could thus contribute to hypertension by promoting (via stimulation of aldosterone secretion) salt and water retention and high blood volume.

Sustained high stress (acting via the sympathetic nervous system) and high salt intake appear to act synergistically in the development of hypertension. There is some evidence that $Na^+$ enhances the vascular response to sympathetic stimulation. Further, sympathetic nerves can cause constriction of the renal blood vessels and thus decrease the excretion of salt and water. In one study, hypertensive students given an IQ test (which was apparently stressful) had decreased renal blood flow, increased renin secretion, decreased excretion of salt and water, and a greater increase in blood pressure than did students with normal blood pressure. There also appears to be a genetic component to these responses; students who were normotensive but who had a family history of hypertension had responses that were intermediate between hypertensive students and students who were normotensive with no family history of hypertension.

As an adaptive response to prolonged high blood pressure, the arterial wall becomes thickened. This response can lead to arteriosclerosis and results in an even greater increase in total peripheral resistance, thus raising blood pressure still more in a positive feedback fashion.

The interactions between salt intake, sympathetic nerve activity, cardiovascular responses to sympathetic nerve activity, kidney function, and genetics make it difficult to sort out the cause-and-effect sequence that leads to essential hypertension. Many people have suggested that

**Table 14.12** Some possible causes of secondary hypertension

| System Involved | Examples | Mechanisms |
|---|---|---|
| Kidneys | Kidney disease<br>Renal artery disease | Decreased urine formation<br>Secretion of vasoactive chemicals |
| Endocrine | Excess catecholamines (tumor of adrenal medulla)<br>Excess aldosterone (Conn's syndrome) | Increased cardiac output and total peripheral resistance<br>Excess salt and water retention by the kidneys |
| Nervous | Increased intracranial pressure<br>Damage to vasomotor center | Activation of sympathoadrenal system<br>Activation of sympathoadrenal system |
| Cardiovascular | Complete heart block; patent ductus arteriosus<br>Arteriosclerosis of aorta; coarctation of aorta | Increased stroke volume<br>Decreased distensibility of aorta |

**Figure 14.30.** The sequence of events that has been proposed as a possible cause of essential hypertension.

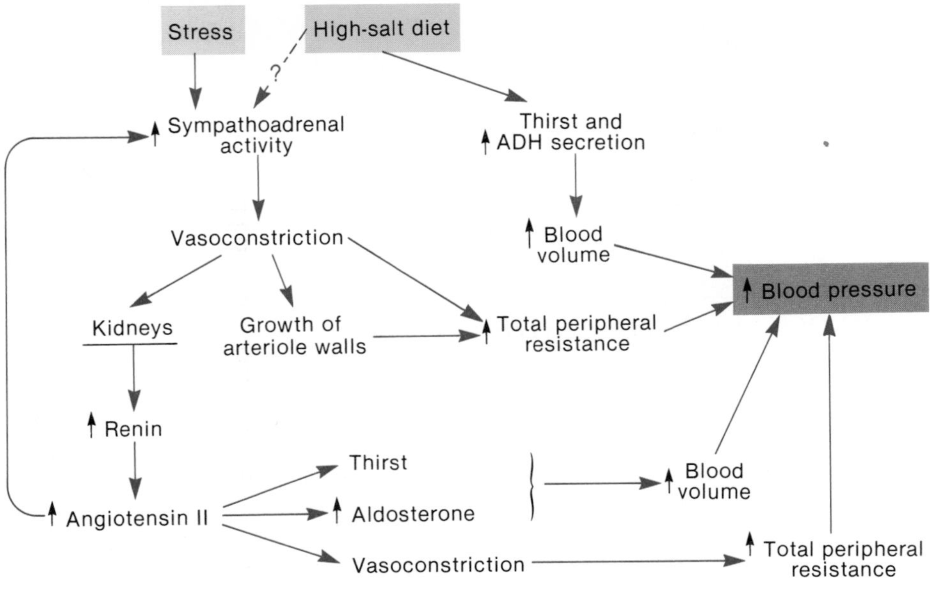


there is no single cause-and-effect sequence but rather a web of causes and effects (fig. 14.30). This is currently controversial.

***Dangers of Hypertension.*** If other factors remain constant, blood flow increases as arterial blood pressure increases. People with hypertension thus have adequate perfusion of their organs with blood until the hypertension causes vascular damage. Hypertension, as a result, is usually without symptoms until a dangerous amount of vascular damage is produced.

Hypertension is dangerous because (1) high arterial pressure increases the afterload, making it more difficult for the ventricles to eject blood; this increases the amount of work that the heart must perform and may result in pathological changes in heart structure and function, leading to congestive heart failure; (2) high pressure may damage cerebral blood vessels, leading to cerebrovascular accident (stroke); and (3) it contributes to the development of atherosclerosis, which can itself lead to heart disease and stroke as previously described.

***Treatment of Hypertension.*** Hypertension is usually treated by restricting salt intake and by taking drugs that act in a variety of ways. Most commonly, these drugs are *diuretics* that increase urine volume, thus decreasing blood volume and pressure. Sympathetic-blocking drugs are also often used; drugs that block beta-adrenergic receptors (such as propranolol) decrease cardiac rate. Various vasodilators (table 14.13) may also be used to decrease total peripheral resistance.

## Circulatory Shock

Circulatory shock occurs when there is inadequate blood flow and/or oxygen utilization by the tissues. Some of the signs of shock (table 14.14) are a result of inadequate tissue perfusion; other signs of shock are produced by cardiovascular responses that help to compensate for the poor

**Table 14.13** Examples and mechanisms of action of some antihypertensive drugs

| Category of Drugs | Examples | Mechanisms |
|---|---|---|
| Extracellular fluid volume depletors | Thiazide diuretics | Increase volume of urine excreted, thus lowering blood volume |
| Sympathoadrenal system inhibitors | Clonidine; alpha-methyldopa | Acts on brain to decrease sympathoadrenal stimulation |
| | Guanethidine; reserpine | Depletes norepinephrine from sympathetic nerve endings |
| | Propranolol | Blocks beta-adrenergic receptors, decreasing cardiac output and/or renin secretion |
| | Phentolamine | Blocks alpha-adrenergic receptors, decreasing sympathetic vasoconstriction |
| Vasodilators | Hydralazine; sodium nitroprusside | Causes vasodilation by acting directly on vascular smooth muscle |
| Calcium channel blockers | Verapamil | Inhibits diffusion of $Ca^{++}$ into vascular smooth muscle cells, causing vasodilation and reduced peripheral resistance. |
| Angiotension II inhibitors | Captopril | Inhibits the conversion of angiotension I into angiotension II |

**Table 14.14** Signs of shock

| | Early Sign | Late Sign |
|---|---|---|
| Blood pressure | Decreased pulse pressure<br>Increased diastolic pressure | Decreased systolic pressure |
| Urine | Decreased $Na^+$ concentration<br>Increased osmolality | Decreased volume |
| Blood pH | Increased pH (alkalosis) due to hyperventilation | Decreased pH (acidosis) due to "metabolic" acids |
| Effects of poor tissue perfusion | Slight restlessness; occasionally warm, dry skin | Cold, clammy skin<br>"Cloudy" senses |

From R. F. Wilson, editor, *Principles and Techniques of Critical Care,* vol. 1. Copyright © 1977 Upjohn Company. Reprinted by permission of F. A. Davis Company, Philadelphia, PA.

**Table 14.15** Cardiovascular reflexes that help to compensate for circulatory shock

| Organ(s) | Compensatory Mechanisms |
|---|---|
| Heart | Sympathoadrenal stimulation produces increased cardiac rate and increased stroke volume, due to "positive inotropic effect" on myocardial contractility |
| Digestive tract and skin | Decreased blood flow due to vasoconstriction as a result of sympathetic nerve stimulation (alpha-adrenergic effect) |
| Kidneys | Decreased urine production as a result of sympathetic-nerve-induced constriction of renal arterioles; increased salt and water retention due to increased aldosterone and antidiuretic hormone (ADH) secretion |

tissue perfusion (table 14.15). When these compensations are effective, they (together with emergency medical care) are able to reestablish adequate tissue perfusion. In some cases, however, and for reasons that are not clearly understood, the shock may progress to an irreversible stage and death may result.

***Hypovolemic Shock.*** The term *hypovolemic shock* refers to circulatory shock due to low blood volume, as might be caused by hemorrhage (bleeding), dehydration, or burns. This is accompanied by decreased blood pressure and decreased cardiac output. In response to these changes, the sympathoadrenal system is activated by means of the baroreceptor reflex. As a result, tachycardia is produced and vasoconstriction occurs in the skin, digestive tract, kidneys, and muscles. Decreased blood flow through the kidneys stimulates renin secretion and activation of the renin-angiotensin-aldosterone system. A person in hypovolemic shock thus has low blood pressure, rapid pulse, cold, clammy skin, and little urine excretion.

Since the resistance in the coronary and cerebral circulations is not increased, blood is diverted to the heart and brain at the expense of other organs. Interestingly, a similar response occurs in diving mammals and, to a lesser degree, in Japanese pearl divers during prolonged submersion. These responses help to deliver blood to the two organs that have the highest requirements for aerobic metabolism.

Vasoconstriction in organs other than the brain and heart raises total peripheral resistance, which helps (along with the reflex increase in cardiac rate) to compensate for the drop in blood pressure due to low blood volume. Constriction of arterioles also decreases capillary blood flow and capillary filtration pressure. Less filtrate is formed, as a result, while the osmotic return of fluid to the capillary due to the plasma colloid osmotic pressure is either unchanged or increased (during dehydration). The blood volume is thus raised at the expense of tissue fluid volume. Blood volume is also conserved by decreased urine production, which occurs as a result of vasoconstriction in the kidneys and the water-conserving effects of ADH and aldosterone, which are secreted in increased amounts during shock.

***Other Causes of Circulatory Shock.*** A rapid fall in blood pressure occurs in **anaphylactic shock** as a result of severe allergic reaction (usually to bee stings or penicillin). This results from the widespread release of histamine, which causes vasodilation and thus decreases total peripheral resistance. A rapid fall in blood pressure also occurs in **neurogenic shock,** in which sympathetic tone is decreased, usually because of upper spinal cord damage or spinal anesthesia. **Cardiogenic shock** results from cardiac failure, as defined by a cardiac output that is inadequate to maintain tissue perfusion.

## Congestive Heart Failure

Cardiac failure occurs when the cardiac output is insufficient to maintain the blood flow required by the body. This may be due to heart disease—resulting from myocardial infarction or congenital defects—or to hypertension, which increases the afterload of the heart. The most common causes of left pump failure are myocardial infarction, aortic valve stenosis, and incompetence of the aortic and bicuspid (mitral) valves. Failure of the right pump is usually caused by prior failure of the left pump.

Heart failure can also result from disturbance in the electrolyte concentrations of the blood. Excessive plasma $K^+$ concentration decreases the resting membrane potential of myocardial cells; low blood $Ca^{++}$ reduces excitation-contraction coupling. High blood $K^+$ and low blood $Ca^{++}$ can thus cause the heart to stop in diastole. Conversely, low blood $K^+$ and high blood $Ca^{++}$ can arrest the heart in systole.

**Figure 14.31.** The relationship between end-diastolic volume and ventricular pressure (a measure of contraction strength). At a given end-diastolic volume the ventricular pressure is decreased during heart failure and increased by sympathetic stimulation. Heart failure may be compensated by sympathetic stimulation.

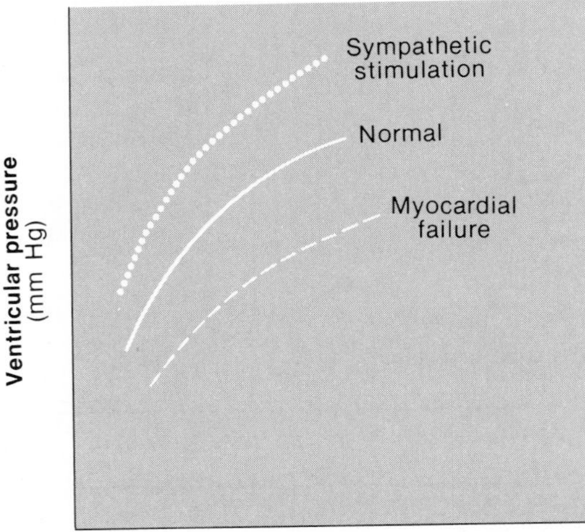


The term *congestive* is often used in describing heart failure because of the increased venous volume and pressure that results. Failure of the left pump, for example, raises the left atrial pressure and produces pulmonary congestion and edema. This causes shortness of breath and fatigue; if severe, pulmonary edema can be fatal. Failure of the right pump results in increased right atrial pressure, which produces congestion and edema in the systemic circulation.

People with congestive heart failure are often treated with the drug *digitalis*. Digitalis appears to bind to and inhibit the action of $Na^+/K^+$ pumps in the cell membranes, causing a rise in the intracellular concentrations of $Na^+$. The increased availability of $Na^+$, in turn, stimulates the activity of another membrane transport carrier, which exchanges $Na^+$ for extracellular $Ca^{++}$. As a result, the intracellular concentrations of $Ca^{++}$ are increased, which strengthens the contractions of the heart.

The compensatory responses that occur during congestive heart failure are similar to those that occur during hypovolemic shock. Activation of the sympathoadrenal system stimulates cardiac rate, contractility of the ventricles (fig. 14.31), and constriction of arterioles. As in hypovolemic shock, renin secretion is increased and urine output is reduced.

As a result of these compensations, chronically low cardiac output is associated with elevated blood volume and dilation and hypertrophy of the ventricles. These changes can themselves be dangerous. Elevated blood volume places a work overload on the heart, and the enlarged ventricles have a higher metabolic requirement for oxygen. These problems are often treated with drugs that increase myocardial contractility (such as digitalis), drugs that are vasodilators (such as nitroglycerin), and diuretic drugs that lower blood volume by increasing the volume of urine excreted.

1. *Explain how stress and a high-salt diet can contribute to hypertension.*
2. *Using a flowchart to show cause and effect, explain why a person in hypovolemic shock may have a fast pulse and cold, clammy skin.*
3. *Describe the compensatory mechanisms that act to raise blood volume during cardiovascular shock.*
4. *Describe congestive heart failure, and explain the compensatory responses that occur during this condition.*

## SUMMARY


### Cardiac Output p. 398

I. Cardiac rate is increased by sympathoadrenal stimulation and decreased by the effects of parasympathetic fibers that innervate the SA node.

II. Stroke volume is regulated both extrinsically and intrinsically.
   A. The Frank-Starling law of the heart describes the way the end-diastolic volume, through various degrees of myocardial stretching, influences the contraction strength of the myocardium and thus the stroke volume.
   B. The end-diastolic volume is called the preload; the total peripheral resistance, through its effect on arterial blood pressure, provides an afterload that acts to reduce the stroke volume.
   C. At a given end-diastolic volume, the amount of blood ejected depends on contractility; strength of contraction is increased by sympathoadrenal stimulation.

III. The venous return of blood to the heart is dependent largely on the total blood volume and mechanisms that improve the flow of blood in veins.
   A. The total blood volume is regulated by the kidneys.
   B. The venous flow of blood to the heart is aided by the action of skeletal muscle pumps and the effects of breathing.

### Blood Volume p. 402

I. Tissue fluid is formed from and returns to the blood.
   A. The hydrostatic pressure of the blood forces fluid from the arteriolar ends of capillaries into the interstitial spaces of the tissues.
   B. Since the colloid osmotic pressure of plasma is greater than tissue fluid, water returns by osmosis to the venular ends of capillaries.
   C. Excess tissue fluid is returned to the venous system by lymphatic vessels.
   D. Edema occurs when there is an accumulation of tissue fluid.

II. The kidneys control the blood volume by regulating the amount of filtered fluid that will be reabsorbed.
   A. Antidiuretic hormone stimulates reabsorption of water from the kidney filtrate and thus acts to maintain the blood volume.
   B. A decrease in blood flow through the kidneys activates the renin-angiotensin system.
   C. Angiotensin II stimulates vasoconstriction and the secretion of aldosterone by the adrenal cortex.
   D. Aldosterone acts on the kidneys to promote the retention of salt and water.

### Vascular Resistance and Blood Flow p. 408

I. Poiseuille's law describes the fact that blood flow is directly related to the pressure difference between the two ends of a vessel and is inversely related to the resistance to blood flow through the vessel.

II. Extrinsic regulation of vascular resistance is provided mainly by the sympathetic nervous system, which stimulates vasoconstriction of arterioles in the viscera and skin.

III. Intrinsic control of vascular resistance allows organs to autoregulate their own blood flow rates.
   A. Myogenic regulation occurs when vessels constrict or dilate as a direct response to a rise or fall in blood pressure.
   B. Metabolic regulation occurs when vessels dilate in response to the local chemical environment within the organ.

### Blood Flow to the Heart and Skeletal Muscles p. 413

I. The heart normally respires aerobically because of its high capillary supply, myoglobin content, and enzyme content.

II. During exercise, when the heart's metabolism increases, intrinsic metabolic mechanisms stimulate vasodilation of the coronary vessels and thus increase coronary blood flow.

III. Just prior to exercise and at the start of exercise, blood flow through skeletal muscles increases due to vasodilation caused by cholinergic sympathetic nerve fibers; during exercise, intrinsic metabolic vasodilation occurs.

IV. Since cardiac output can increase by a factor of five or more during exercise, the heart and skeletal muscles receive an increased proportion of a higher total blood flow.
   A. The cardiac rate increases due to lower activity of the vagus nerve and higher activity of the sympathetic nerve.
   B. The venous return is greater because of higher activity of the skeletal muscle pumps and increased breathing.
   C. Increased contractility of the heart, combined with a decrease in total peripheral resistance, can result in a higher stroke volume.

### Blood Flow to the Brain and Skin p. 419

I. Cerebral blood flow is regulated both myogenically and metabolically.

A. Cerebral vessels automatically constrict if the systemic blood pressure rises too high.
B. Metabolic products cause local vessels to dilate and supply more active areas with more blood.

II. The skin has unique arteriovenous anastomoses, which can shunt the blood away from surface capillary loops.
A. Sympathetic nerve fibers cause constriction of cutaneous arterioles.
B. As a thermoregulatory response, there is increased cutaneous blood flow and increased flow through surface capillary loops when body temperature rises.

### Blood Pressure p. 421

I. Baroreceptors in the aortic arch and carotid sinuses affect, via the sympathetic nervous system, the cardiac rate and the total peripheral resistance.
A. The baroreceptor reflex causes pressure to be maintained when an upright posture is assumed and can cause a lowered pressure when the carotid sinuses are massaged.
B. Other mechanisms that affect blood volume help to regulate blood pressure.

II. Blood pressure is commonly measured indirectly by auscultation of the brachial artery when a pressure cuff is inflated and deflated.
A. The first sound of Korotkoff, caused by turbulent flow of blood through a constriction in the artery, occurs when the cuff pressure equals the systolic pressure.
B. The last sound of Korotkoff is heard when the cuff pressure equals the diastolic blood pressure.

III. The mean arterial pressure represents the driving force for blood flow through the arterial system.

### Hypertension, Shock, and Congestive Heart Failure p. 429

I. Hypertension, or high blood pressure, is classified as either primary or secondary.
A. Primary hypertension, also called essential hypertension, may be the result of the interaction of many mechanisms that raise the blood volume, cardiac output, and/or peripheral resistance.
B. Secondary hypertension is the direct result of known, specific diseases.

II. Circulatory shock occurs when there is inadequate delivery of oxygen to the organs of the body.
A. In hypovolemic shock, low blood volume causes low blood pressure; this may progress to an irreversible state that cannot be stopped by compensations.
B. The fall in blood volume and pressure stimulates various reflexes that produce a rise in cardiac rate, shift of fluid from the tissues into the vascular system, decrease in urine volume, and vasoconstriction.

III. Congestive heart failure occurs when the cardiac output is insufficient to supply the blood flow required by the body; the term "congestive" is used to describe the increased venous volume and pressure that results.

---

## REVIEW ACTIVITIES

### Objective Questions

1. According to the Frank-Starling law, the strength of ventricular contraction is
   (a) directly proportional to the end-diastolic volume
   (b) inversely proportional to the end-diastolic volume
   (c) independent of the end-diastolic volume
2. In the absence of compensations, the stroke volume will decrease when
   (a) blood volume increases
   (b) venous return increases
   (c) contractility increases
   (d) arterial blood pressure increases
3. Which of the following statements about tissue fluid is *false?*
   (a) It contains the same glucose and salt concentration as plasma.
   (b) It contains a lower protein concentration than plasma.
   (c) Its colloid osmotic pressure is greater than that of plasma.
   (d) Its hydrostatic pressure is less than that of plasma.
4. Edema may be caused by
   (a) high blood pressure
   (b) decreased plasma protein concentration
   (c) leakage of plasma protein into tissue fluid
   (d) blockage of lymphatic vessels
   (e) all of the above
5. Both ADH and aldosterone act to
   (a) increase urine volume
   (b) increase blood volume
   (c) increase total peripheral resistance
   (d) all of the above
6. The greatest resistance to blood flow occurs in the
   (a) large arteries
   (b) medium-sized arteries
   (c) arterioles
   (d) capillaries
7. If a vessel were to dilate to twice its previous radius and if pressure remained constant, blood flow through this vessel would
   (a) increase by a factor of 16
   (b) increase by a factor of 4
   (c) increase by a factor of 2
   (d) decrease by a factor of 2
8. The sounds of Korotkoff are produced by
   (a) closing of the semilunar valves
   (b) closing of the AV valves
   (c) the turbulent flow of blood through an artery
   (d) elastic recoil of the aorta
9. Vasodilation in the heart and skeletal muscles during exercise is primarily due to the effects of
   (a) alpha-adrenergic stimulation
   (b) beta-adrenergic stimulation
   (c) cholinergic stimulation
   (d) products released by the exercising muscle cells
10. Blood flow in the coronary circulation is
    (a) increased during systole
    (b) increased during diastole
    (c) constant throughout the cardiac cycle
11. Blood flow in the cerebral circulation
    (a) varies with systemic arterial pressure
    (b) is regulated primarily by the sympathetic system
    (c) is maintained constant within physiological limits
    (d) is increased during exercise

12. Which of the following organs is able to tolerate the greatest restriction in blood flow? The
    (a) brain
    (b) heart
    (c) skeletal muscles
    (d) skin
13. Arteriovenous shunts in the skin
    (a) divert blood to superficial capillary loops
    (b) are closed when the ambient temperature is very cold
    (c) are closed when the deep-body temperature rises much above 37° C
    (d) all of the above
14. An increase in blood volume will cause
    (a) a decrease in ADH secretion
    (b) an increase in $Na^+$ excretion in the urine
    (c) a decrease in renin secretion
    (d) all of these
15. The volume of blood pumped by the left ventricle per minute is
    (a) greater than the volume pumped by the right ventricle
    (b) less than the volume pumped by the right ventricle
    (c) the same as the volume pumped by the right ventricle
    (d) either less or greater than the volume pumped by the right ventricle, depending on the strength of contraction
16. Blood pressure is lowest in the
    (a) arteries
    (b) arterioles
    (c) capillaries
    (d) venules
    (e) veins

## Essay Questions

1. Define the terms *contractility, preload,* and *afterload,* and explain how these factors affect the cardiac output.
2. Explain, using the Frank-Starling law, how the stroke volume is affected by (a) bradycardia and (b) a "missed beat."
3. Which part of the cardiovascular system contains the most blood? Which part provides the greatest resistance to blood flow? Which part provides the greatest cross-sectional area? Explain.
4. Explain how the kidneys regulate blood volume.
5. A person who is dehydrated drinks more and urinates less. Explain the mechanisms involved.
6. Explain, using Poiseuille's law, how arterial blood flow can be diverted from one organ system to another.
7. Describe the mechanisms that increase the cardiac output during exercise and that increase the rate of blood flow to the heart and skeletal muscles.
8. Explain how an anxious person may have a cold, clammy skin and how the skin becomes hot and flushed on a hot, humid day.

## SELECTED READINGS


Atlas, S. A. 1986. Atrial natriuretic factor: Renal and systemic effects. *Hospital Practice* 21:67.

Berne, R. M., and M. N. Levy. 1981. *Cardiovascular physiology.* 4th ed. St. Louis: The C. V. Mosby Company.

Braunwald, E. 1974. Regulation of the circulation. *New England Journal of Medicine* 290: first part, p. 1124; second part, p. 1420.

Brody, M. J., J. R. Haywood, and K. B. Toun. 1980. Neural mechanisms in hypertension. *Annual Review of Physiology* 42:441.

Cantin, M., and J. Genest. February 1986. The heart as an endocrine gland. *Scientific American.*

Carafol, E., and J. T. Penniston. November 1985. The calcium signal. *Scientific American.*

Del Zoppo, G. J., and L. A. Harker. 1984. Blood vessel interaction in coronary disease. *Hospital Practice* 19:163.

Donald, D. E., and J. T. Shepard. 1980. Autonomic regulation of the peripheral circulation. *Annual Review of Physiology* 42:429.

Folkow, B., and E. Neil. 1971. *Circulation.* London: Oxford University Press.

Franciosa, J. A. 1981. Hypertensive left heart failure: Pathogenesis and therapy. *Hospital Practice* 16:165.

Granger, D. E., and P. R. Kvietys. 1981. The splanchnic circulation: Intrinsic regulation. *Annual Review of Physiology* 43:409.

Herd, J. A. 1984. Cardiovascular response to stress in man. *Annual Review of Physiology* 46:177.

Hilton, P. J. 1986. Cellular sodium transport in essential hypertension. *New England Journal of Medicine* 314:222.

Hilton, S. M., and K. M. Spyer. 1980. Central nervous regulation of vascular resistance. *Annual Review of Physiology* 42:399.

Kaplan, N. M. 1980. The control of hypertension: A therapeutic breakthrough. *American Scientist* 68:537.

Katz, A. M. 1975. Congestive heart failure. *New England Journal of Medicine* 293:1184.

Kayz, M. H. 1987. A physiologic approach to the treatment of heart failure. *Hospital Practice* 22:117.

Kontos, H. A. 1981. Regulation of the cerebral circulation. *Annual Review of Physiology* 43:397.

Laragh, J. H. 1985. Atrial natriuretic hormone, the renin-aldosterone axis, and blood pressure-electrolyte homeostasis. *New England Journal of Medicine* 313:1330.

Light, K. C., J. P. Loepke, P. A. Obrist, and P. W. Willis, IV. 1983. Psychological stress induces sodium and fluid retention in men at high risk for hypertension. *Science* 220:249.

McCarron, D. A. et al. 1984. Blood pressure and nutrient intake in the United States. *Science* 224:1392.

Nadel, E. R. 1985. Physiological adaptations to aerobic training. *American Scientist* 73:334.

Needleman, P., and J. E. Greenwald. 1986. Atriopeptin: A cardiac hormone intimately involved in fluid, electrolyte, and blood pressure homeostasis. *New England Journal of Medicine* 314:828.

Olsson, R. A. 1981. Local factors regulating cardiac and skeletal muscle blood flow. *Annual Review of Physiology* 43:385.

Robinson, T. F., S. M. Factor, and E. H. Sonnenblick. June 1986. The heart as a suction pump. *Scientific American.*

Ross, J., Jr. 1983. The failing heart and circulation. *Hospital Practice* 18:151.

Schatz, I. J. 1983. Orthostatic hypotension: Diagnosis and treatment. *Hospital Practice* 18:59.

Stephensen, R. B. 1984. Modification of reflex regulation of blood pressure by behavior. *Annual Review of Physiology* 46:133.

Vatner, S. F., and E. Braunwald. 1975. Cardiovascular control mechanisms in the conscious state. *New England Journal of Medicine* 293:970.

Weber, K. T., J. S. Janicki, and W. Laskey. 1983. The mechanics of ventricular function. *Hospital Practice* 18:113.

Weinberger, M. H. 1986. Dietary sodium and blood pressure. *Hospital Practice* 21:55.

Zelis, R., S. F. Flaim, A. J. Liedke, and S. H. Nellis. 1981. Cardiovascular dynamics in the normal and failing heart. *Annual Review of Physiology* 43:455.

# Respiratory Physiology

# 15

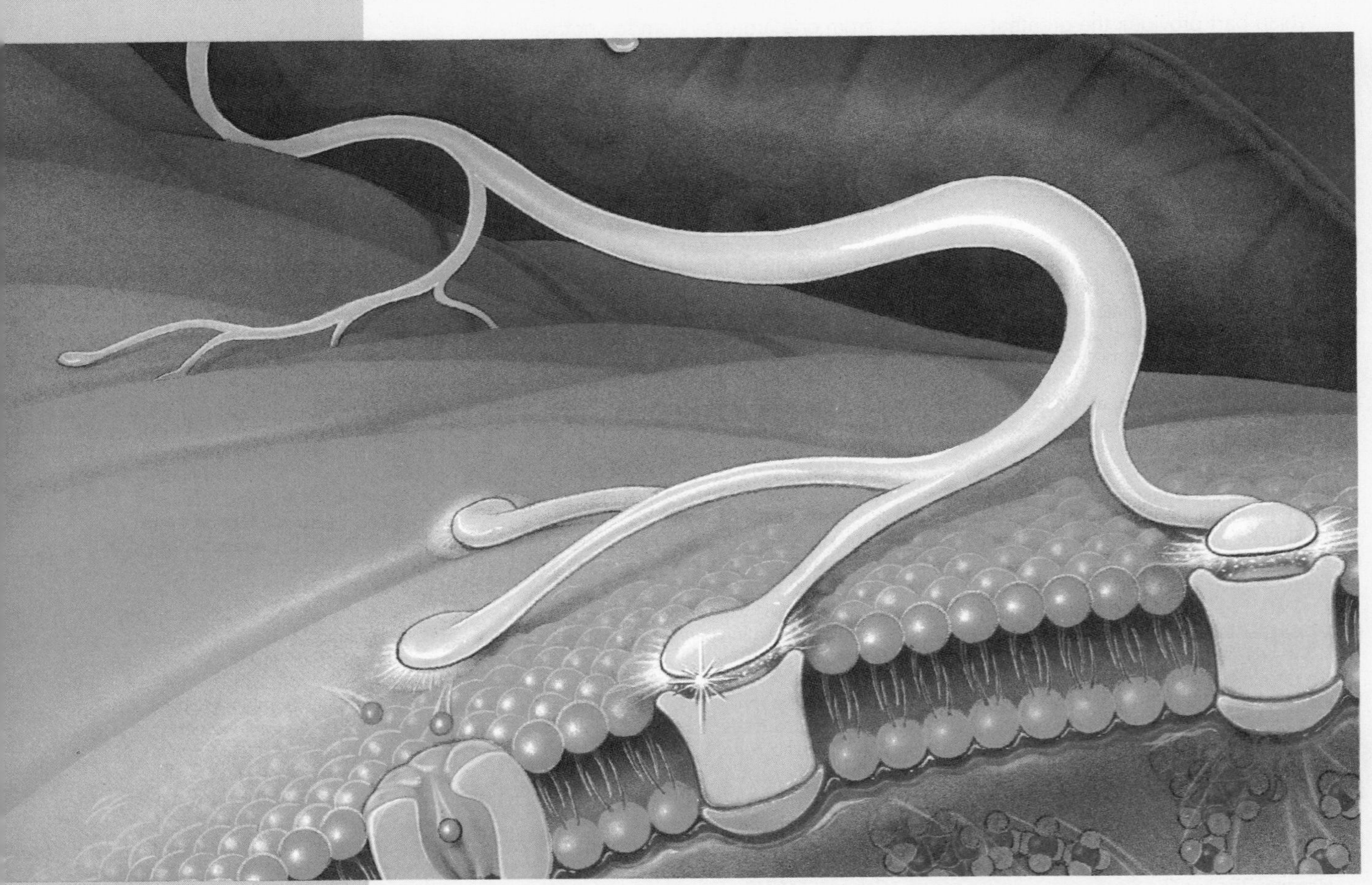

## OBJECTIVES

By studying this chapter, you should be able to

1. describe the structures that comprise the lungs and explain their functions; describe the compartmentation of the thoracic cavity
2. explain how the intrapulmonary and intrapleural pressures vary during ventilation and how Boyle's law applies to these pressure changes
3. define the terms *compliance* and *elasticity* and explain how these lung properties affect ventilation
4. explain the significance of surface tension in lung mechanics, how the law of LaPlace applies to lung function, and the role of pulmonary surfactant
5. explain how inspiration and expiration are accomplished in unforced breathing and describe the accessory respiratory muscles that are used in forced breathing
6. define the various lung volumes and capacities that can be measured by spirometry and explain how obstructive diseases may be detected by the FEV test
7. describe the nature of some pulmonary disorders, including asthma, bronchitis, emphysema, and fibrosis
8. explain Dalton's law and describe how the partial pressure of a gas in a mixture of gases is calculated
9. explain Henry's law, describe how the partial pressure of oxygen and carbon dioxide in a fluid (such as blood) is measured, and explain the clinical significance of these measurements
10. describe the roles of the medulla oblongata, pons, and cerebral cortex in the regulation of breathing
11. explain why the $P_{CO_2}$ and pH of blood, rather than its $P_{O_2}$, serve as the primary chemical stimuli for breathing
12. explain how the chemoreceptors in the medulla oblongata and the peripheral chemoreceptors in the aortic and carotid bodies respond to changes in $P_{CO_2}$, pH, and $P_{O_2}$
13. describe the Hering-Breuer reflex and its significance
14. describe the different forms of hemoglobin and their significances
15. describe the loading and unloading reactions and explain how the extent of these reactions is influenced by the $P_{O_2}$ and affinity of hemoglobin for oxygen
16. describe the oxyhemoglobin dissociation curve, explain the reason for and the significance of its shape, and demonstrate how the curve is used to derive the percent unloading of oxygen
17. explain how oxygen transport is influenced by changes in blood pH and temperature and explain the effect and physiological significance of 2,3-DPG on oxygen transport
18. list the different forms of carbon dioxide transport in the blood and explain the chloride shift in the tissues and the reverse chloride shift in the lungs
19. explain how carbon dioxide affects blood pH and describe how hypoventilation and hyperventilation affect acid-base balance
20. describe the hyperpnea of exercise and explain how the anaerobic threshold is affected by endurance training
21. explain the compensations of the respiratory system to life at a high altitude

## OUTLINE

**Figure 15.1.** A diagram showing the relationship between lung alveoli and pulmonary capillaries.

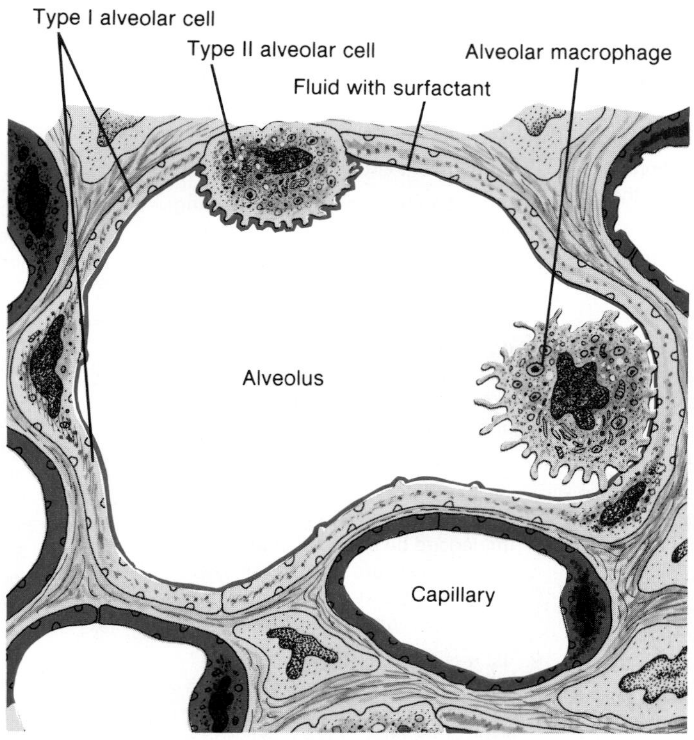


## The Respiratory System

The respiratory system is divided into a respiratory zone, where gas exchange between air and blood occur, and a conducting zone, which conducts the air to the respiratory zone. The exchange of gases between air and blood occur across the walls of tiny air sacs called alveoli, which are only one cell across in thickness to permit very rapid rates of gas diffusion. The lungs are covered with a thin, wet membrane which is against a similar membrane that lines the wall of the thorax.

The term *respiration* includes three separate but related functions: (1) **ventilation** (breathing); (2) **gas exchange,** which occurs between the air and blood in the lungs and between the blood and other tissues of the body; and (3) **oxygen utilization** by the tissues in the energy-liberating reactions of cell respiration. Ventilation and the exchange of gases (oxygen and carbon dioxide) between the air and blood are together called *external respiration.* Gas exchange between the blood and other tissues and oxygen utilization by the tissues are together known as *internal respiration.*

Ventilation is the mechanical process that moves air into and out of the lungs. Since air in the lungs has a higher oxygen concentration than in the blood, oxygen diffuses from air to blood. Carbon dioxide, conversely, moves from the blood to the air within the lungs by diffusing down its concentration gradient. As a result of this gas exchange, the inspired air contains more oxygen and less carbon dioxide than the expired air. More importantly, blood leaving the lungs (in the pulmonary veins) contains a higher oxygen and a lower carbon dioxide concentration than the blood delivered to the lungs in the pulmonary arteries. This results from the fact that the lungs function to bring the blood into gaseous equilibrium with the air.

Gas exchange between the air and blood occurs entirely by diffusion through lung tissue. This diffusion occurs very rapidly because there is a high surface area within the lungs and a very short diffusion distance between blood and air. The fact that blood in the pulmonary veins and systemic arteries is almost in complete equilibrium with the inspired air is testimony to the high efficiency of normal lung function.

### Structure of the Respiratory System

Gas exchange in the lungs occurs across about 300 million tiny (0.25–0.50 mm in diameter) air sacs, known as **alveoli.** The enormous number of these structures provides a high surface area—60–80 square meters, or about 760 square feet—for diffusion of gases. The diffusion rate is further increased by the fact that each alveolus is only one cell-layer thick, so that the total "air-blood barrier" is only two cells across (an alveolar cell and a capillary endothelial cell), or about 2 $\mu$m. This is an average distance because the type II alveolar cells are thicker than the type I cells (fig. 15.1). Where the basement membranes of cap-

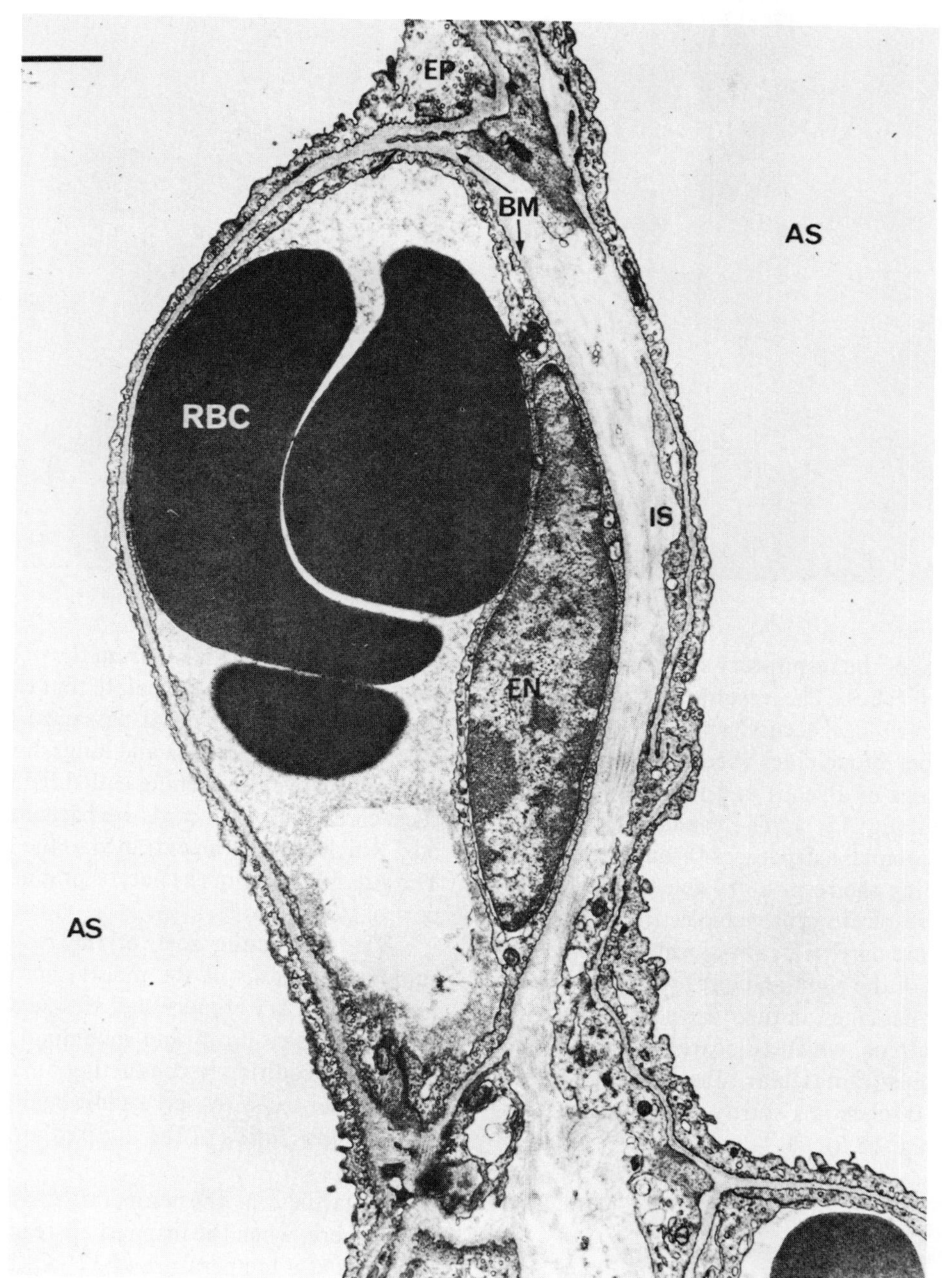


**Figure 15.2.** An electron micrograph of a capillary within the thin interalveolar septum that separates two adjacent alveoli. Note the short distance separating the alveolar space on one side (left, in this figure) from the capillary. RBC = red blood cell; BM = basement membrane; IS = interstitial connective tissue.

illary endothelial cells fuse with those of type I alveolar cells, the diffusion distance is less than 1 $\mu$m (fig. 15.2).

Alveoli are polyhedral in shape and are usually clustered together, like the units of a honeycomb. Air within one member of a cluster can enter other members through tiny pores. These clusters of alveoli usually occur at the ends of *respiratory bronchioles,* which are the very thin air tubes that end blindly in alveolar sacs. Individual alveoli also occur as separate outpouchings along the length of respiratory bronchioles. Although the distance between each respiratory bronchiole and its terminal alveoli is only about 0.5 mm, these units together comprise most of the mass of the lungs.

**Figure 15.3.** (*a*) A scanning electron micrograph showing lung alveoli and a small bronchiole. (*b*) The alveoli under higher power (the arrow points to an alveolar pore through which air can pass from one alveolus to another).

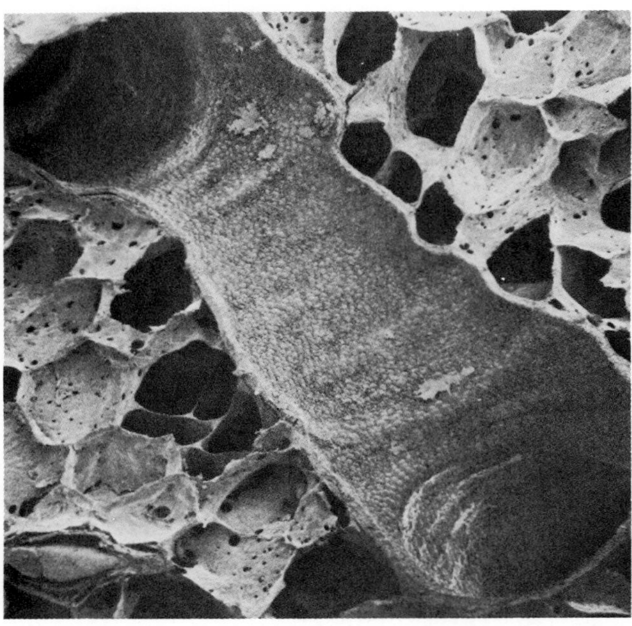

(a)

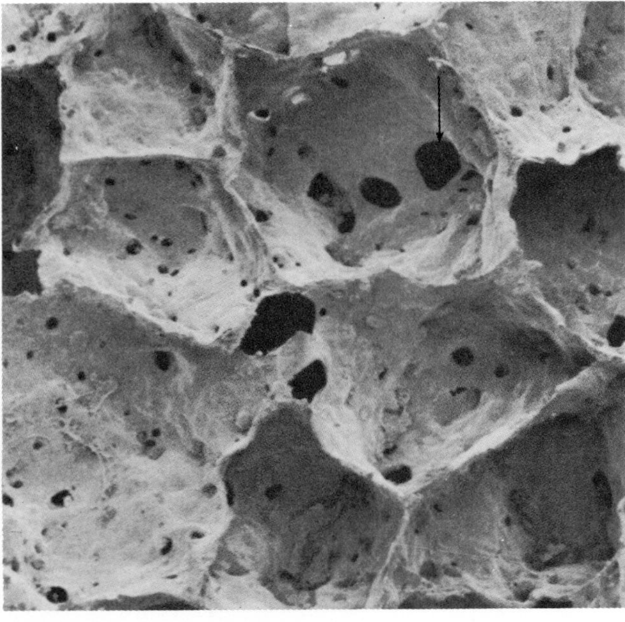

(b)

The air passages of the respiratory system are divided into two functional zones. The **respiratory zone** is the region where gas exchange occurs, and it therefore includes the respiratory bronchioles (because they contain separate outpouchings of alveoli) and the terminal clusters of alveolar sacs (fig. 15.3). The **conducting zone** includes all of the anatomical structures through which air passes before reaching the respiratory zone (fig. 15.4).

Air enters the respiratory bronchioles from *terminal bronchioles,* which are narrow airways formed from many successive divisions of the right and left *primary bronchi.* These two large air passages, in turn, are continuous with the *trachea,* or windpipe, which is located in the neck in front of the esophagus (a muscular tube carrying food to the stomach). The trachea is a sturdy tube supported by rings of cartilage (fig. 15.5).

If the trachea becomes occluded through inflammation, excessive secretion, trauma, or aspiration of a foreign object, it may be necessary to create an emergency opening into this tube so that ventilation can still occur. A **tracheotomy** is the process of surgically opening the trachea, and a **tracheostomy** is the procedure of inserting a tube into the trachea to permit breathing and to keep the passageway open. A tracheotomy should be performed only by a competent physician, however, as there is great risk of cutting a recurrent laryngeal nerve or jugular vein, which is also in this area.

Air enters the trachea from the *pharynx,* which is the cavity located behind the palate that receives the contents of both the oral and nasal passages. In order for air to enter or leave the trachea and lungs, however, it must pass through a valvelike opening called the *glottis* between the vocal cords. The vocal cords are part of the *larynx,* or voice box, which guards the entrance to the trachea (fig. 15.6). The Adam's apple in the neck is produced by a protruding part of the larynx.

The conducting zone of the respiratory system, in summary, consists of the mouth, nose, pharynx, larynx, trachea, primary bronchi, and all successive branchings of the bronchioles up to and including the terminal bronchioles. In addition to conducting air into the respiratory zone, these structures serve additional functions: *warming* and *humidification* of the inspired air and *filtration* and *cleaning.*

Regardless of the temperature and humidity of the atmosphere, when the inspired air reaches the respiratory zone it is at a temperature of 37° C (body temperature), and it is saturated with water vapor. The first function is needed to maintain a constant internal body temperature, and the latter function is needed to protect delicate lung tissue from desiccation.

Mucus secreted by cells of the conducting zone serves to trap small particles in the inspired air and thereby performs a filtration function. This mucus is moved along at

**Figure 15.4.** The conducting and respiratory zones of the respiratory system.

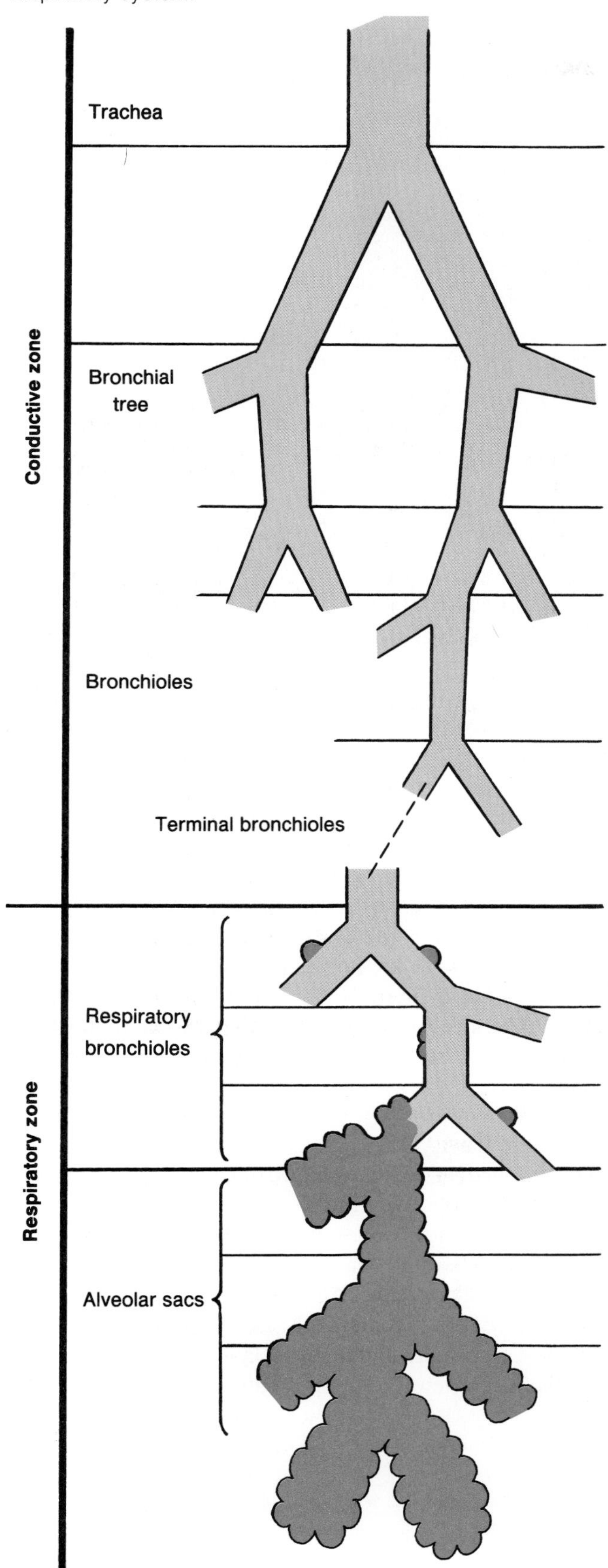


**Figure 15.5.** The conducting zone of the respiratory system. (*a*) is illustrated from the larynx to the primary bronchi, and (*b*) is shown by a plastic cast of the airway from the trachea to the terminal bronchioles.

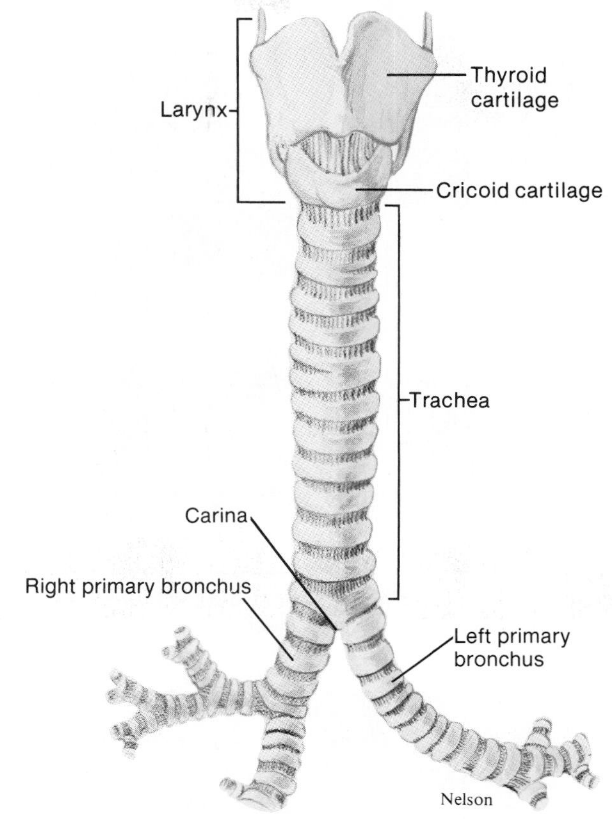


(a)

(b)

Figure 15.6. A photograph of the larynx, showing the true and false vocal cords and the glottis.

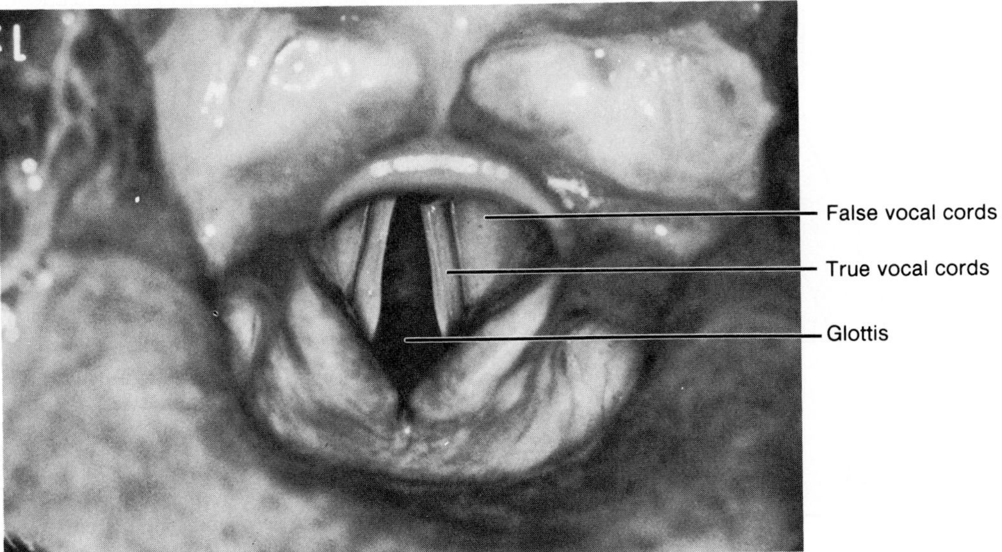


Figure 15.7. A scanning electron micrograph of a bronchial wall showing cilia, which help to cleanse the lung by moving trapped particles.

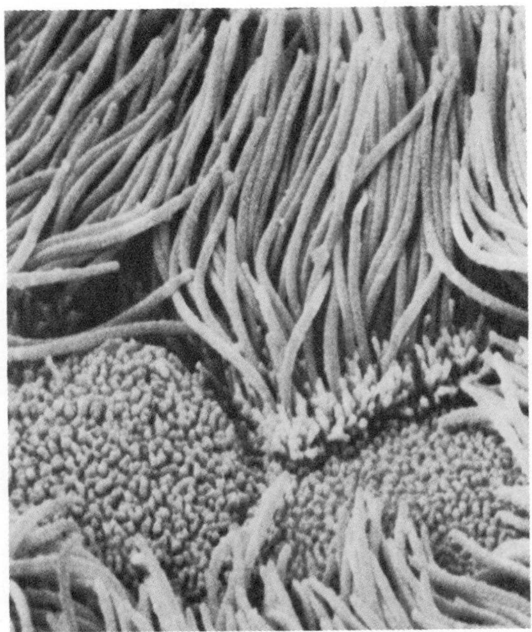

a rate of 1–2 centimeters per minute by cilia projecting from the tops of epithelial cells that line the conducting zone (fig. 15.7). There are about three hundred cilia per cell that beat in a coordinated fashion to move mucus toward the pharynx, where it can either be swallowed or expectorated.

As a result of this filtration function, particles larger than about 6 μm do not normally enter the respiratory zone of the lungs. The importance of this function is evidenced by the disease called *black lung,* which occurs in miners who inhale too much carbon dust and therefore develop pulmonary fibrosis (as described in a later section). The alveoli themselves are normally kept clean by the action of *macrophages* (literally, "big eaters") that reside within them. The cleansing action of cilia and macrophages in the lungs has been shown to be diminished by cigarette smoke.

## Thoracic Cavity

The *diaphragm,* a dome-shaped sheet of striated muscle, divides the body cavity into two parts. The area below the diaphragm, or the *abdominal cavity,* contains the liver, pancreas, gastrointestinal tract, spleen, genitourinary tract, and other organs. Above the diaphragm, the chest, or *thoracic cavity,* contains the heart, large blood vessels, trachea, esophagus, and thymus gland in the central region and is filled elsewhere by the right and left lungs.

**Figure 15.8.** A cross section of the thoracic cavity, showing the mediastinum and pleural membranes.

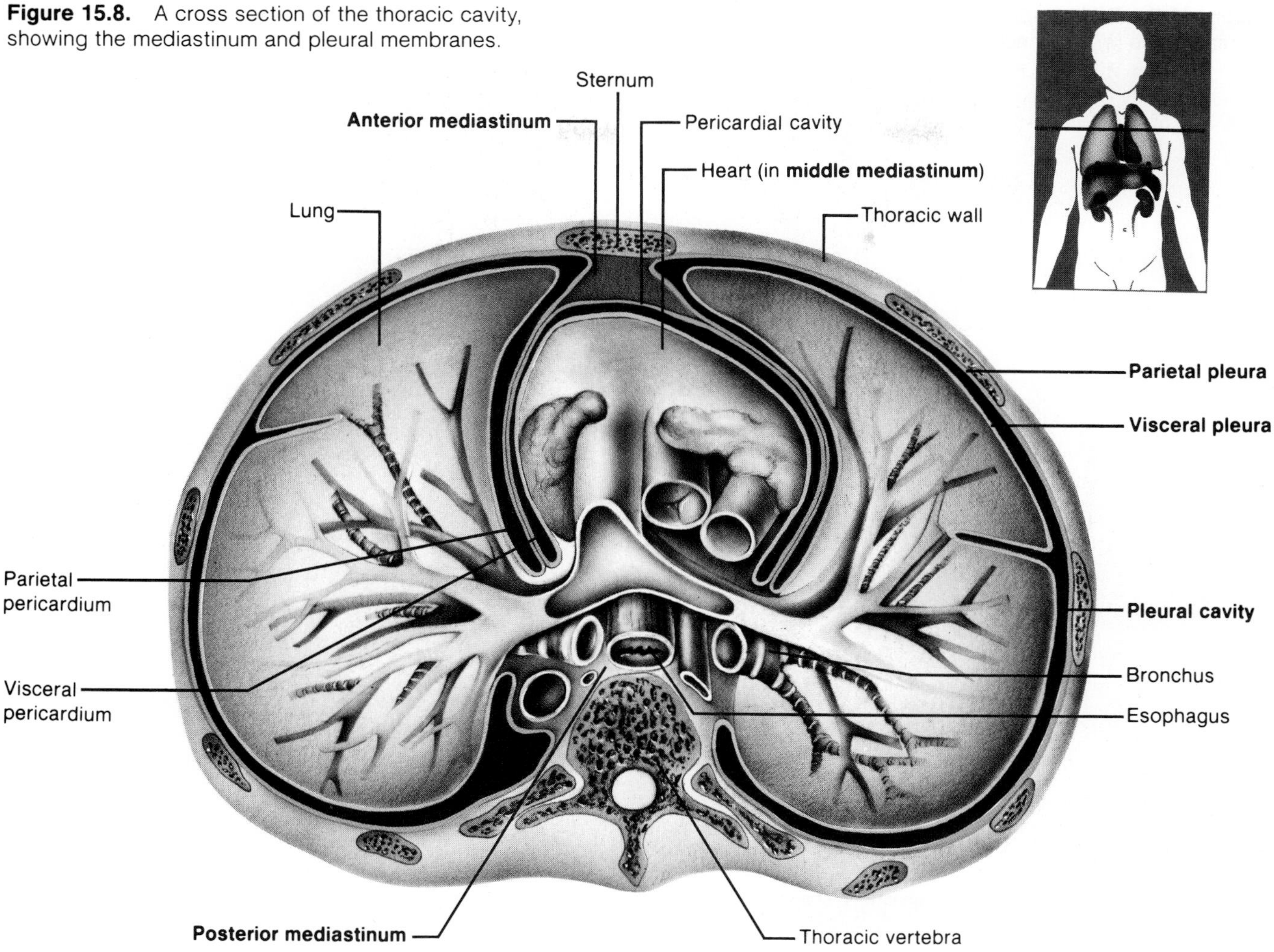

The structures in the central region—or *mediastinum*—are enveloped by a double layer of wet epithelial membranes, called the *pleural membranes.* One membrane of this double layer is continuous with the *parietal pleural membrane,* a wet epithelial membrane that lines the inside of the thoracic wall. The other layer is continuous with the *visceral pleural membranes* that cover the surface of the lungs (fig. 15.8).

The lungs normally fill the thoracic cavity so that the visceral pleural membranes covering the lungs are pushed against the parietal pleural membrane lining the thoracic wall. There is thus, under normal conditions, little or no air between the visceral and parietal pleural membranes. There is, however, a "potential space"—called the *intrapleural space*—that can become a real space if the visceral and parietal pleural membranes separate when a lung collapses. The normal position of the lungs in the thoracic cavity is shown in the radiograph in figure 15.9.

1. *Describe the structures involved in gas exchange in the lungs and how this process occurs.*
2. *Describe the structures and functions of the conducting zone of the respiratory system.*
3. *Describe how each lung is packaged separately in pleural membranes, and describe the relationship between the visceral and parietal pleural membranes.*

**Figure 15.9.** Radiographic (X ray) views of the chest of a normal female (*a*) and a normal male (*b*).

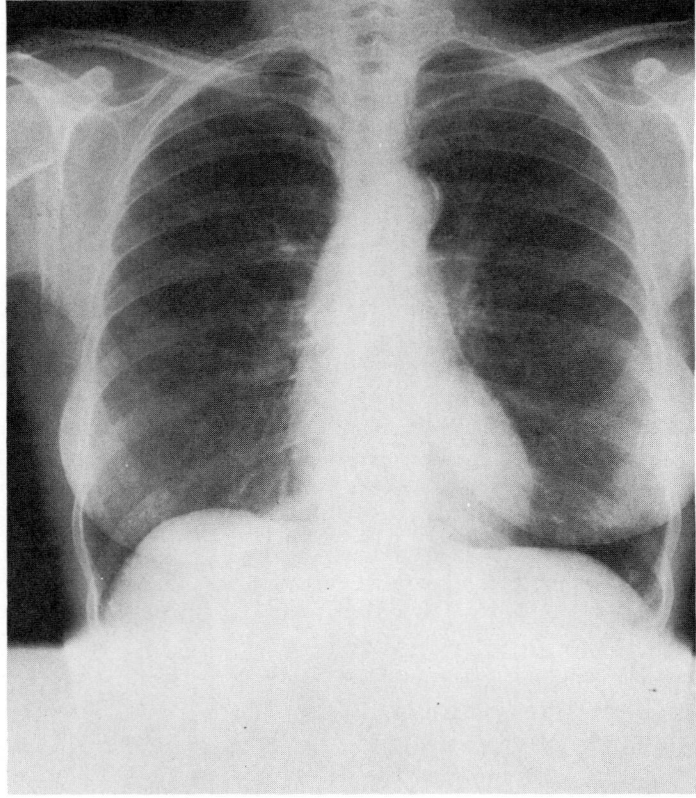

(a)

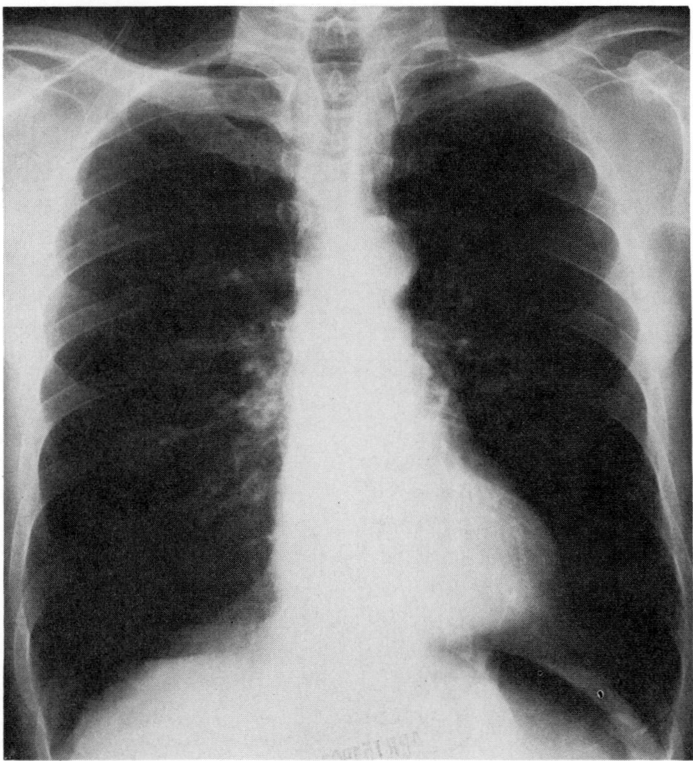

(b)

## *Physical Aspects of Ventilation*

During inspiration, the volume of the lungs is increased and the pressure of the air within the lungs is decreased below the pressure of the atmosphere. The greater pressure of the atmosphere then causes air to enter the lungs. The distension of the lungs is resisted by surface tension within the alveoli, which is kept within manageable limits by the presence of pulmonary surfactant. Elastic recoil of the lungs when the distending force is removed causes the lung volume to decrease and results in expiration.

Movement of air from the conducting zone to the terminal bronchioles occurs as a result of the pressure difference between the two ends of the airways. Air flow through bronchioles, like blood flow through blood vessels, is directly proportional to the pressure difference and inversely proportional to the frictional resistance to flow. The pressure differences in the pulmonary system are induced by changes in lung volumes. Ventilation is thus influenced by the physical properties of the lungs, including their compliance, elasticity, and surface tension.

### Intrapulmonary and Intrapleural Pressures

The wet, serous membranes of the visceral and parietal pleurae are normally against each other, so that the lungs are stuck to the chest wall in the same manner that two wet pieces of glass stick to each other. The *intrapleural space* between the two wet membranes contains only a thin layer of fluid secreted by the pleural membranes. The pleural cavity in a healthy, living organism is thus potential rather than real; it can become real only in abnormal situations when air enters the intrapleural space. Since the lungs normally remain against the chest wall, they get larger and smaller together with the thoracic cavity during respiratory movements.

Air enters the lungs during inspiration because the atmospheric pressure is greater than the **intrapulmonary,** or **alveolar, pressure.** Since the atmospheric pressure does not usually change, the intrapulmonary pressure must fall below atmospheric pressure to cause inspiration. A pressure below that of the atmosphere is called a *subatmospheric pressure,* or *negative pressure.* During quiet inspiration, for example, the intrapulmonary pressure may become 3 mm Hg less than the pressure of the atmosphere. This subatmospheric pressure is commonly shown as −3 mm Hg. Expiration, conversely, occurs when the intrapulmonary pressure is greater than the atmospheric pressure. During quiet expiration, for example, the intrapulmonary pressure may rise to at least +3 mm Hg over the atmospheric pressure.

The lack of air in the intrapleural space produces a subatmospheric **intrapleural pressure,** which is lower than the intrapulmonary pressure (table 15.1). There is thus a pressure difference across the wall of the lung—called the **transpulmonary pressure**—which is the difference between the intrapulmonary pressure and the intrapleural pressure. Since the pressure within the lungs (intrapulmonary pressure) is greater than outside the lungs (intrapleural pressure), the difference in pressure (transpulmonary pressure) acts to expand the lungs as the thoracic volume expands during inspiration.

**Table 15.1** Intrapulmonary and intrapleural pressures in normal, quiet breathing, and the transpulmonary pressure acting to expand the lungs

| | Inspiration | Expiration |
|---|---|---|
| Intrapulmonary pressure (mm Hg) | −3 (to zero) | +3 (to zero) |
| Intrapleural pressure (mm Hg) | −6 | −3 |
| Transpulmonary pressure (mm Hg) | +3 | +6 |

Note: Pressures indicate mm Hg below or above atmospheric pressure. Intrapleural pressure is normally always negative (subatmospheric).

***Boyle's Law.*** Changes in intrapulmonary pressure occur as a result of changes in lung volume. This follows from **Boyle's law,** which states that *the pressure of a given quantity of gas is inversely proportional to its volume.* An increase in lung volume during inspiration decreases intrapulmonary pressure to subatmospheric levels; air therefore goes in. A decrease in lung volume raises the intrapulmonary pressure above that of the atmosphere, thus pushing air out. These changes in lung volume occur as a consequence of changes in thoracic volume, as will be described in a later section on the mechanics of breathing.

### Physical Properties of the Lungs

In order for inspiration to occur, the lungs must be able to expand when stretched; they must have high *compliance.* In order for expiration to occur, the lungs must get smaller when this stretching force is released; they must have *elasticity.* The tendency to get smaller is also aided by *surface tension* forces within the alveoli.

***Compliance.*** The lungs are very distensible—they are, in fact, about one hundred times more distensible than a toy balloon. Another term for distensibility is **compliance,** which is defined as *the change in lung volume per change in transpulmonary pressure.* A given transpulmonary pressure, in other words, will cause greater or lesser expansion, depending on the compliance of the lungs.

The compliance of the lungs is reduced by factors that produce a resistance to distension. If the lungs were filled with concrete (as an extreme example), a given transpulmonary pressure would produce no increase in lung volume and no air would enter; the compliance would be zero. The infiltration of lung tissue with connective tissue proteins, a condition called *pulmonary fibrosis,* similarly decreases lung compliance (fig. 15.10). In *emphysema,* in which alveolar tissue is destroyed, the lungs are less resistant to distension and have a greater compliance (if no fibrosis is present). Because of the high lung compliance, a person with emphysema may be able to inhale easily. Exhalation, however, is more difficult, because the loss of alveoli decreases the elasticity of the lungs.

***Elasticity.*** The term **elasticity** refers to the tendency of a structure to return to its initial size after being distended. The lungs are very elastic, due to a high content of elastin proteins, and resist distension. Since the lungs are normally stuck to the chest wall, they are always in a state of elastic tension. This tension increases during inspiration when the lungs are stretched and is reduced by elastic recoil during expiration. The elasticity of the lungs and of other thoracic structures thus aids in pushing the air out during expiration.

The elastic nature of lung tissue is revealed when air enters the intrapleural space (as a result of an open chest wound, for example). This condition is called a **pneumothorax,** which is shown in figure 15.11. As air enters the intrapleural space, the intrapleural pressure rises until it is equal to the atmospheric pressure. When the intrapleural pressure is the same as the intrapulmonary pressure the lung can no longer expand. Not only does the lung not expand during inspiration, it actually collapses away from the chest wall as a result of elastic recoil. Fortunately, a pneumothorax usually occurs only in one lung since each lung is in a separate pleural compartment.

***Surface Tension.*** The forces that act to resist distension include elastic resistance and **surface tension** that is exerted by fluid in the alveoli. Although the alveoli are relatively dry, they do contain a very thin film of fluid, much like soap bubbles. Surface tension is created by the fact that water molecules at the surface are attracted more to other water molecules than to air. As a result, the surface water molecules are pulled tightly together by attractive forces from underneath (fig. 15.12).

The surface tension of an alveolus produces a force that is directed inward and, as a result, creates pressure within the alveolus. As described by the **Law of LaPlace,** the pressure thus created is directly proportional to the surface tension and *inversely proportional to the radius* of the alveolus (fig. 15.13). According to this law, the pressure in a smaller alveolus would be greater than in a larger alveolus if the surface tension is the same in both.

**Figure 15.10.** The volume of air inspired as a function of the transpulmonary pressure. The slope of each line (change in volume per unit change in pressure) in the linear regions is called the *compliance* and is a measure of the distensibility of the lungs.

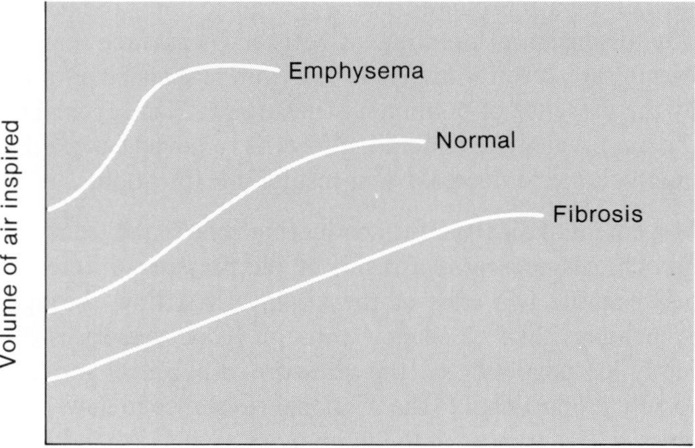


The greater pressure of the smaller alveolus would then cause it to empty its air into the larger one (fig. 15.13). This does not normally occur because, as an alveolus decreases in size, its surface tension (the numerator in the equation) is decreased as its radius (the denominator) is reduced. The cause of the reduced surface tension, which prevents the alveoli from collapsing, is described in the next section.

## Surfactant and the Respiratory Distress Syndrome

Alveolar fluid contains a phospholipid known as dipalmitoyl lecithin, probably attached to a protein, which functions to lower surface tension. This compound is called **lung surfactant,** which is a contraction of the term *surface active agent.* Because of the presence of surfactant, the surface tension in the alveoli is lower than would be predicted if surfactant were absent. Further, the ability of surfactant to lower surface tension improves as the alveoli get smaller during expiration. This may be because the surfactant molecules become more concentrated as the alveoli get smaller. Surfactant thus prevents the alveoli from collapsing during expiration, as would be predicted from the Law of LaPlace. Even after a forceful expiration, the alveoli remain open and a *residual volume* of air remains in the lungs. Since the alveoli do not collapse, less surface tension has to be overcome to inflate them at the next inspiration.

Surfactant is produced by type II alveolar cells (fig. 15.14) in late fetal life. Since surfactant does not start to be produced until about the eighth month, premature babies are sometimes born with lungs that lack sufficient surfactant, and their alveoli are collapsed as a result. This condition is called **respiratory distress syndrome.** It is also called **hyaline membrane disease,** because the high surface

**Figure 15.11.** A pneumothorax of the right lung. The right side of the thorax appears uniformly dark because it is filled with air; the space between the ribs is also greater than on the left due to release from the elastic tension of the lungs. The left lung appears denser (less dark) because of shunting of blood from the right to the left lung.

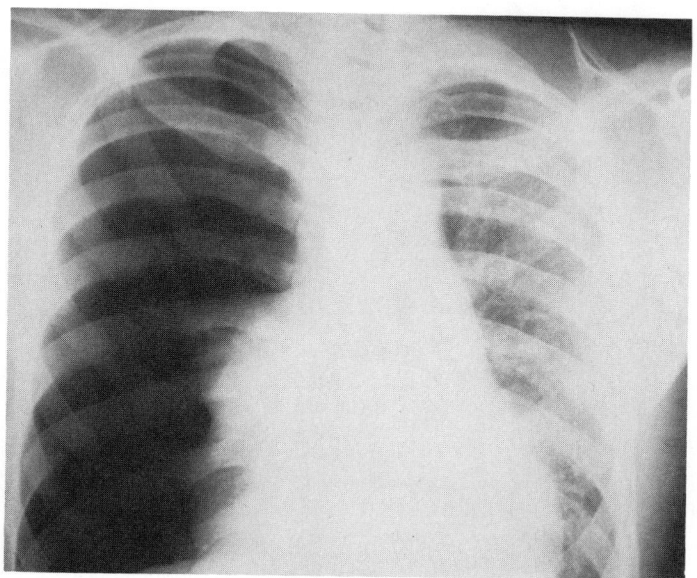

**Figure 15.12.** Water molecules at the surface have a greater attraction for other water molecules than for air. The surface molecules are thus attracted to each other and pulled tightly together by the attractive forces of water underneath. This produces surface tension.

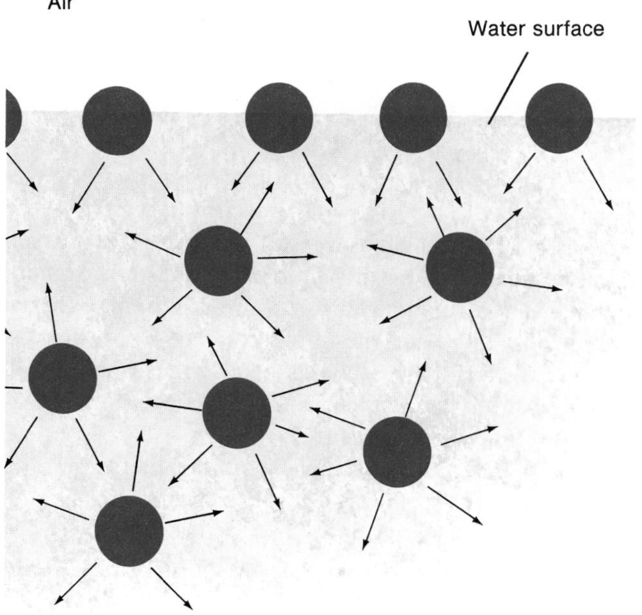


**Figure 15.13.** According to the Law of LaPlace, the pressure created by surface tension should be greater in the smaller alveolus (*right*) than in the larger alveolus (*left*). This implies that (without surfactant) smaller alveoli would collapse and empty their air into larger alveoli.

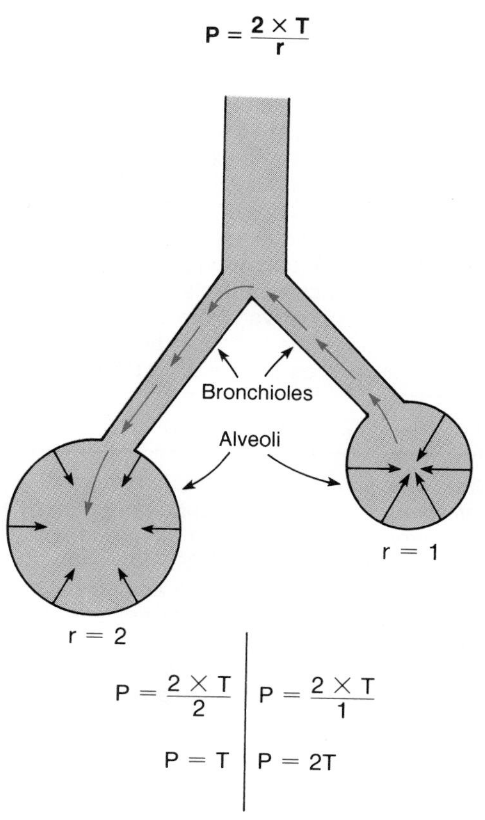

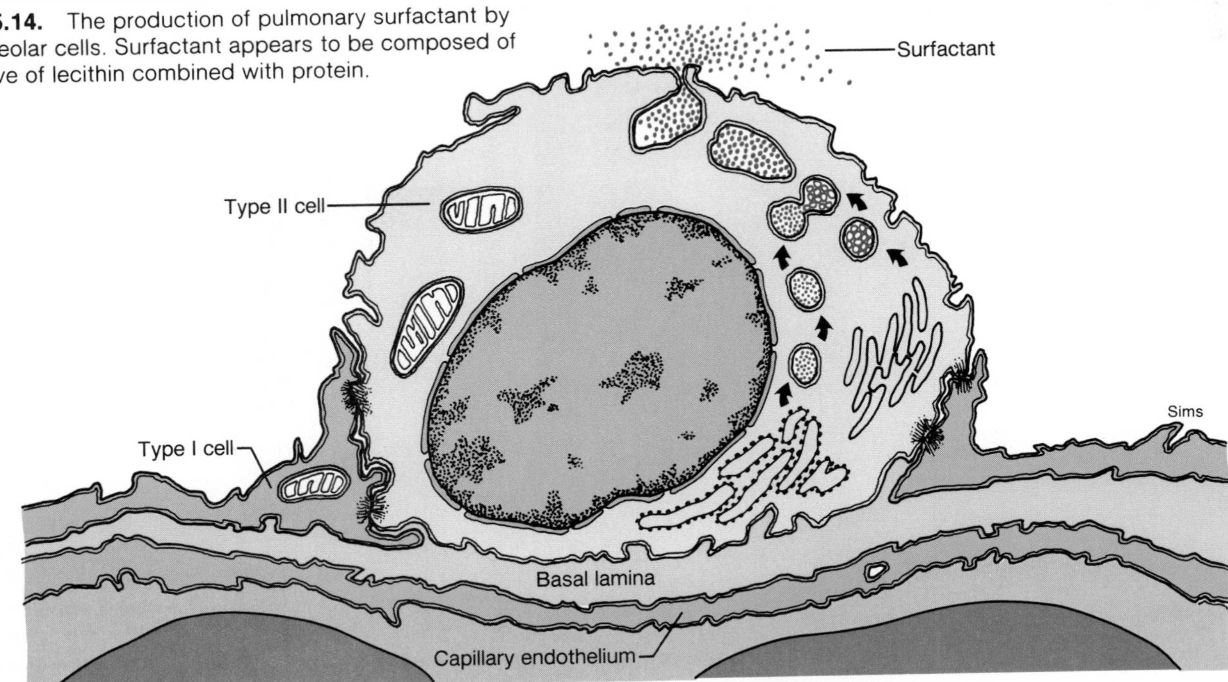


**Figure 15.14.** The production of pulmonary surfactant by type II alveolar cells. Surfactant appears to be composed of a derivative of lecithin combined with protein.

tension causes plasma fluid to leak into the alveoli, producing a glistening "membrane" appearance (and pulmonary edema). This condition does not occur in all premature babies; the rate of lung development depends on hormonal conditions (thyroxine and hydrocortisone primarily) and on genetic factors.

Even under normal conditions, the first breath of life is a difficult one because the newborn must overcome great surface tension forces in order to inflate its partially collapsed alveoli. The transpulmonary pressure required for the first breath is fifteen to twenty times that required for subsequent breaths, and an infant with respiratory distress syndrome must duplicate this effort with every breath. Fortunately, many babies with this condition can be saved by mechanical ventilators that keep them alive long enough for their lungs to mature and manufacture sufficient surfactant. In addition, surfactant derived from cow lungs or from the amniotic fluid of the baby can be delivered into the baby's lungs through an endotracheal tube.

1. *Describe how the intrapulmonary and intrapleural pressures change during inspiration and expiration, and explain the reasons for these changes in terms of Boyle's law.*
2. *Define the terms* compliance *and* elasticity, *and explain how these lung properties affect inspiration and expiration.*
3. *Describe lung surfactant, and explain why the alveoli would collapse in the absence of surfactant.*

## Mechanics of Breathing

Inspiration results from contraction of muscles that expand the volume of the thorax, while unforced expiration occurs as a result of elastic recoil. Contractions of accessory respiratory muscles produce forced inspiration and expiration. Pulmonary function testing allows lung disorders to be classified into two major groups: restrictive and obstructive. Restrictive disorders are characterized by an abnormally low vital capacity, whereas the rate of forced expiration ($FEV_{1.0}$) is abnormally slow in obstructive disorders.

Breathing, or **pulmonary ventilation,** refers to the movement of air into and out of the respiratory system. This movement of air occurs as a result of differences between the atmospheric and the intrapulmonary pressures; air goes in when the intrapulmonary pressure is subatmospheric, and air goes out when the intrapulmonary pressure rises above that of the atmosphere. As previously discussed, these pressure changes within the lungs occur as a consequence of changes in thoracic volume.

The thorax must be semirigid yet flexible. It must be sufficiently rigid so that it can protect vital organs and provide attachments for many short, powerful muscles. It must be flexible to function as a bellows during the ventilation cycle. The rigidity and the surfaces for muscle attachment are provided by the bony composition of the rib cage. The rib cage is pliable, however, because the ribs are separate from one another and because most ribs (the upper ten of the twelve pairs) are attached to the sternum by resilient costal cartilages. The vertebral attachments

Figure 15.15. A change in lung volume, as shown by radiographs, during expiration (*a*) and inspiration (*b*). The increase in lung volume during full inspiration is shown by comparison with the lung volume in full expiration (*dashed lines*).

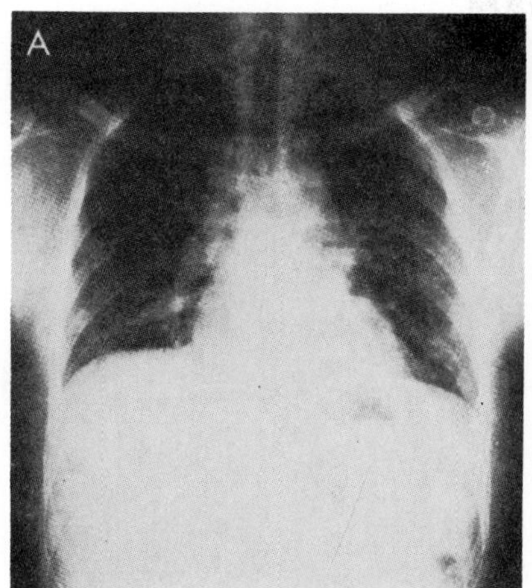


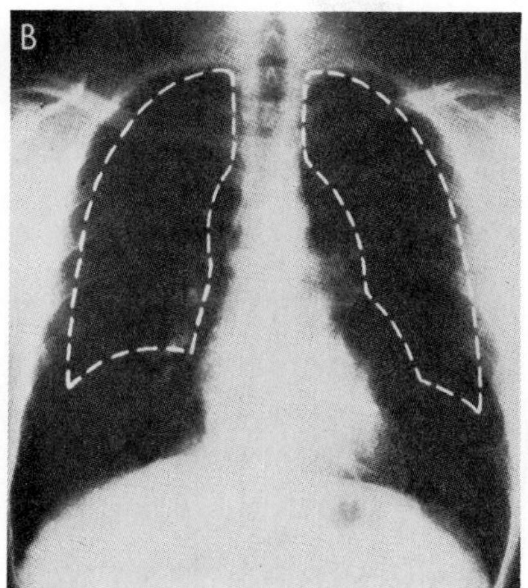


Figure 15.16. The position of the principal muscles of inspiration. Contraction of these muscles enlarges the size of the thoracic cavity.

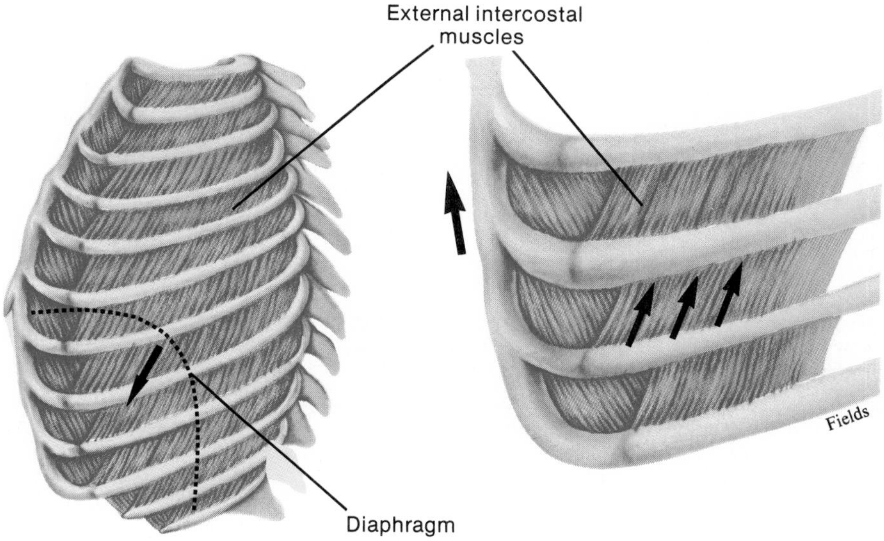


likewise provide considerable mobility. The structure of the rib cage and associated cartilages provide continuous elastic tension, so that when stretched by muscle contraction during inspiration, the rib cage can return passively to its resting dimensions when the muscles relax. This elastic recoil is greatly aided by the elasticity of the lungs.

Pulmonary ventilation consists of two phases, inspiration and expiration. Inspiration (inhalation) and expiration (exhalation) are accomplished by alternately increasing and decreasing the volumes of the thorax and lungs (fig. 15.15).

## Inspiration and Expiration

The thoracic cavity increases in size during inspiration in three directions: anteroposteriorly, laterally, and vertically. This is accomplished by contractions of the **diaphragm** and the **external intercostal** muscles (fig. 15.16). A contraction of the dome-shaped diaphragm downward increases the thoracic volume vertically. A simultaneous contraction of the external intercostals increases the lateral and anteroposterior dimensions of the thorax.

**Figure 15.17.** The mechanics of pulmonary ventilation. (*a*) The lungs and pleural cavities prior to inspiration. (*b*) During inspiration, the diaphragm is contracted and the rib cage elevated; intrapleural pressure is reduced and air inflates the lungs. (*c*) Inspiration is completed as intrapulmonary pressure is equal to atmospheric pressure. (*d*) As the muscles of inspiration are relaxed, the rib cage recoils and the intrapleural and intrapulmonary pressures are raised until the intrapulmonary pressure equals the atmospheric pressure (*a*).

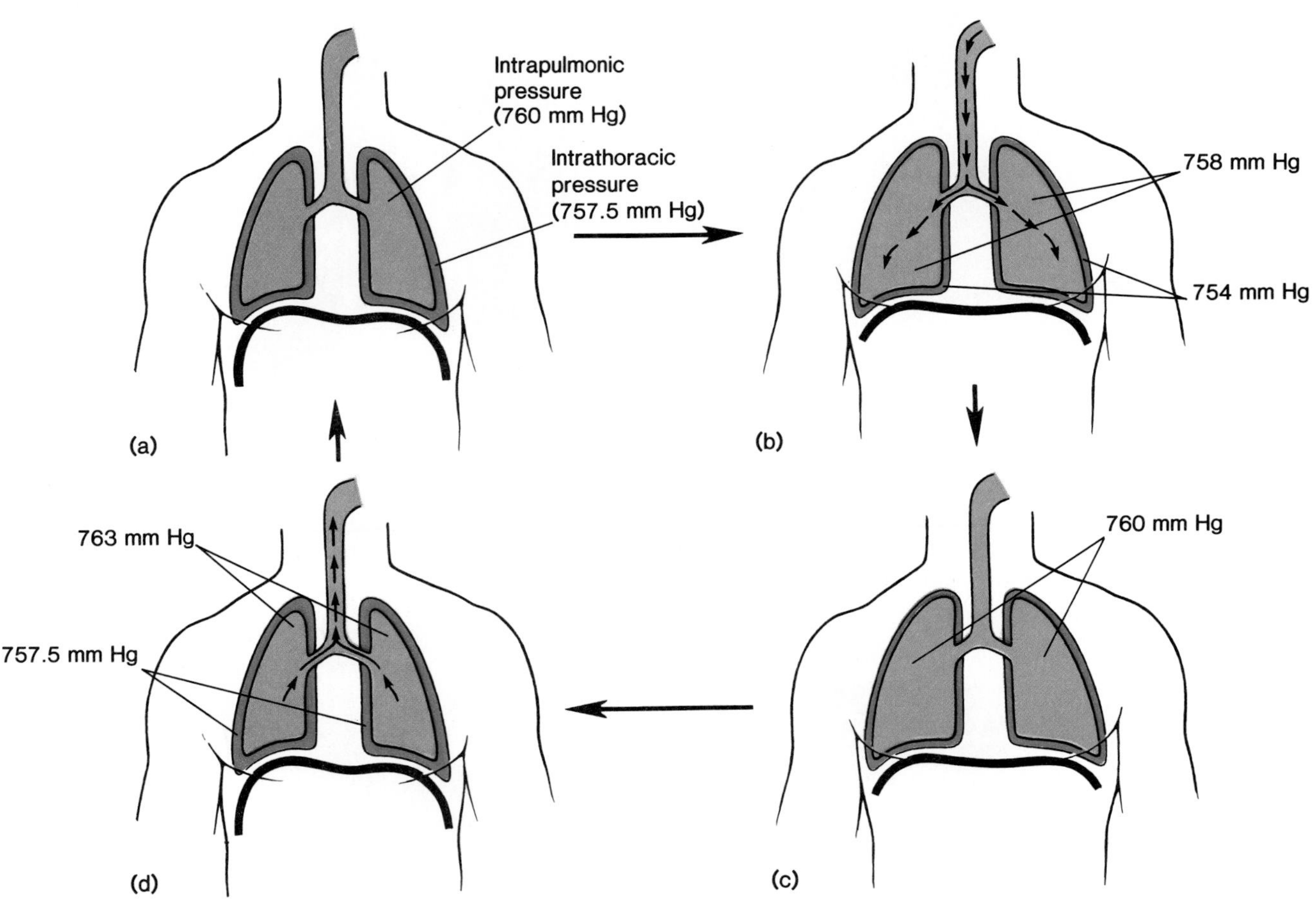


**Table 15.2** Summary of the mechanisms involved in normal, quiet ventilation and forced ventilation

| | Inspiration | Expiration |
|---|---|---|
| Normal, quiet breathing | Contraction of the diaphragm and external intercostal muscles increases the thoracic and lung volume, decreasing intrapulmonary pressure to about −3 mm Hg. | Relaxation of the diaphragm and external intercostals, plus elastic recoil of lungs, decreases lung volume and increases intrapulmonary pressure to about +3 mm Hg. |
| Forced ventilation | Inspiration, aided by contraction of accessory muscles such as the scalenes and sternocleidomastoid, decreases intrapulmonary pressure to −20 mm Hg or less. | Expiration, aided by contraction of abdominal muscles and internal intercostal muscles, increases intrapulmonary pressure to +30 mm Hg or more. |

Other thoracic muscles become involved in forced (deep) inspiration. The most important of these is the scalenus, followed by the pectoralis minor, and in extreme cases the sternocleidomastoid. Contraction of these muscles elevates the ribs in an anteroposterior direction, while at the same time the upper rib cage is stabilized so that the intercostals become more effective.

Quiet expiration is a passive process. After becoming stretched by contractions of the diaphragm and thoracic muscles, the thorax and lungs recoil as a result of their elastic tension when the respiratory muscles relax. The decrease in lung volume raises the pressure within the alveoli above the atmospheric pressure and pushes the air out.

During forced expiration the *internal intercostal* muscles contract and depress the rib cage. The abdominal muscles may also aid expiration, because when they contract they force abdominal organs up against the diaphragm and further decrease the volume of the thorax. By this means the intrapulmonary pressure can rise 20 or 30 mm Hg above the atmospheric pressure. The events that occur during inspiration and expiration are summarized in table 15.2 and shown in figure 15.17.

Figure 15.18. A spirometer (Collins 9L respirometer) used to measure lung volumes and capacities.

Figure 15.19. A spirogram showing respiratory air volumes.

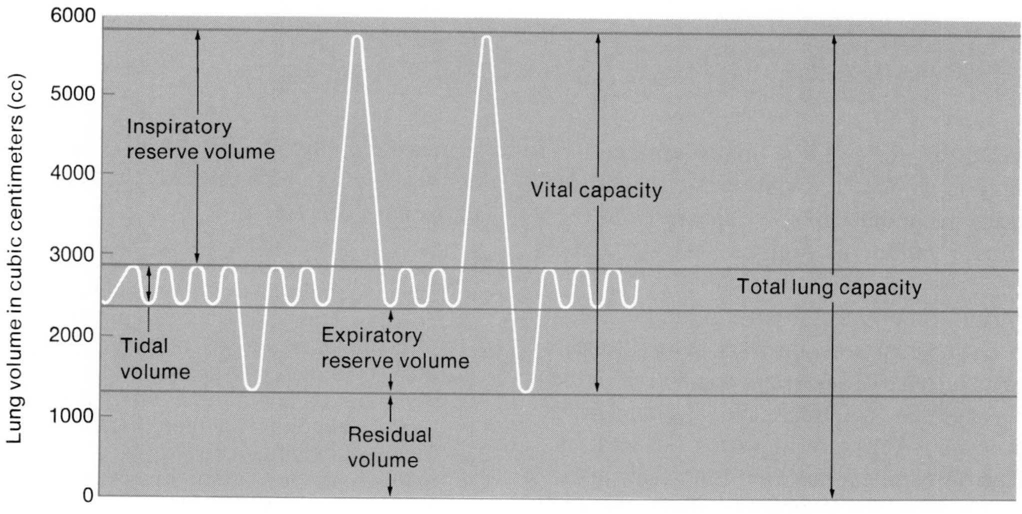


## Pulmonary Function Tests

Pulmonary function may be assessed clinically by means of a technique known as *spirometry.* In this procedure, a subject breathes in a closed system in which air is trapped within a light plastic bell floating in water. The bell moves up when the subject exhales and down when the subject inhales. The movements of the bell cause corresponding movements of a pen, which traces a record of the breathing on a rotating drum recorder (fig. 15.18).

The appearance of a spirogram taken during quiet, resting breathing is shown in figure 15.19, and the various lung volumes and capacities are defined in table 15.3. During quiet breathing, the amount of air expired in each breath (the **tidal volume**) is equal to about 500 ml. This is about 12% of the **vital capacity** (the maximum amount of air that can be expired after a maximum inspiration). Multiplying the tidal volume at rest times the number of breaths per minute yields a **total minute volume** of about

**Table 15.3** Definitions of terms used to describe lung volumes and capacities

| Term | Definition |
|---|---|
| Lung volumes | The four nonoverlapping components of the total lung capacity |
| Tidal volume | The volume of gas inspired or expired in an unforced respiratory cycle |
| Inspiratory reserve volume | The maximum volume of gas that can be inspired from the end of a tidal inspiration |
| Expiratory reserve volume | The maximum volume of gas that can be expired from the end of a tidal expiration |
| Residual volume | The volume of gas remaining in the lungs after a maximum expiration |
| Lung capacities | Measurements that are the sum of two or more lung volumes |
| Total lung capacity | The total amount of gas in the lungs at the end of a maximum inspiration |
| Vital capacity | The maximum amount of gas that can be expired after a maximum inspiration |
| Inspiratory capacity | The maximum amount of gas that can be inspired at the end of a tidal expiration |
| Functional residual capacity | The amount of gas remaining in the lungs at the end of a tidal expiration |

**Figure 15.20.** An illustration of the one-second forced expiratory volume ($FEV_{1.0}$) spirometry test for detecting obstructive pulmonary disorders.

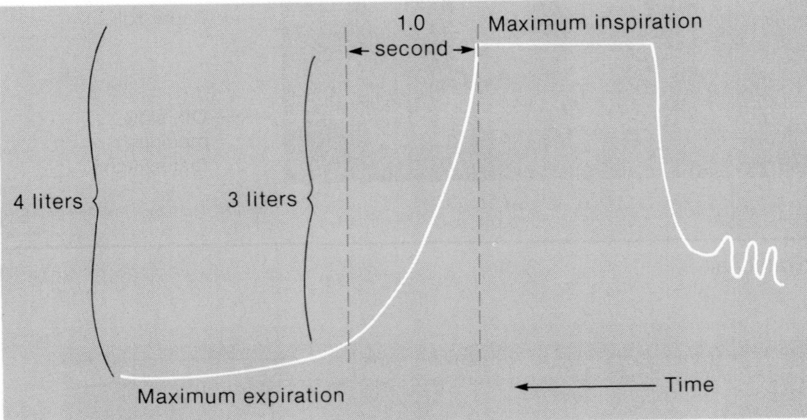


$$FEV_{1.0} = \frac{3L}{4L} \times 100\% = 75\%$$

6 L per minute. During exercise, the tidal volume can increase to as much as 50% of the vital capacity, and the number of breaths per minute likewise increases to produce a total minute volume as high as 100–200 L per minute.

Spirometry is useful in the diagnosis of lung diseases. On the basis of pulmonary function tests, lung disorders can be classified as *restrictive* or *obstructive*. In restrictive disorders, such as pulmonary fibrosis, the vital capacity is reduced below normal. The rate at which the vital capacity can be forcibly exhaled, however, is normal. In disorders that are only obstructive, such as asthma, the vital capacity is normal because lung tissue is not damaged. In asthma the bronchioles constrict, and this bronchoconstriction increases the resistance to air flow. Though the vital capacity is normal, the increased airway resistance makes expiration more difficult and takes a longer time. Obstructive disorders are thus diagnosed by tests that measure the rate of expiration. One such test is the **forced expiratory volume (FEV),** in which the percent of the vital capacity that can be exhaled in the first second ($FEV_{1.0}$) is measured (fig. 15.20). An $FEV_{1.0}$ that is significantly less than 75% to 85% suggests the presence of obstructive pulmonary disease.

Bronchoconstriction often occurs in response to the inhalation of noxious agents in the air, such as from smoke or smog. The $FEV_{1.0}$ has, therefore, been used by researchers to determine the effects of various components of smog and of passive cigarette smoke inhalation on pulmonary function. These studies have shown that it is unhealthy to exercise on very smoggy days and that inhalation of smoke from other people's cigarettes in a closed environment can measurably affect pulmonary function.

## Pulmonary Disorders

People with pulmonary disorders frequently complain of **dyspnea,** which is a subjective feeling of "shortness of breath." Dyspnea may occur even when ventilation is normal, however, and may not occur even when total minute volume is very high as in exercise. Some of the terms used to define ventilation are defined in table 15.4.

**Table 15.4** Definitions of some terms used to describe ventilation

| Term | Definition |
|---|---|
| Air spaces | Alveolar ducts, alveolar sacs, and alveoli |
| Airways | Structures that conduct air from the mouth and nose to the respiratory bronchioles |
| Alveolar ventilation | Removal and replacement of gas in alveoli; equal to the tidal volume minus the volume of dead space times the ventilation product of rate |
| Anatomical dead space | Volume of the conducting airways to the zone where gas exchange occurs |
| Apnea | Cessation of breathing |
| Dyspnea | Unpleasant, subjective feeling of difficult or labored breathing |
| Eupnea | Normal, comfortable breathing at rest |
| Hyperventilation | Alveolar ventilation that is excessive in relation to metabolic rate; results in abnormally low alveolar $CO_2$ |
| Hypoventilation | An alveolar ventilation that is low in relation to metabolic rate; results in abnormally high alveolar $CO_2$ |
| Physiological dead space | Combination of anatomical dead space and underventilated or underperfused alveoli that do not contribute normally to blood-gas exchange |
| Pneumothorax | Presence of gas in the pleural space (the space between the visceral and parietal pleural membranes) causing lung collapse |
| Torr | Synonymous with millimeters of mercury (760 mm Hg = 760 torr) |

***Asthma.*** The obstruction of air flow through the bronchioles may occur as a result of excessive mucous secretion, inflammation, and/or contraction of the smooth muscles in the bronchioles. **Asthma** results from bronchiolar constriction, which increases airway resistance and makes breathing difficult. Constriction of the bronchiolar smooth muscles is stimulated by leukotrienes (molecules related to prostaglandins) and to a lesser degree by histamine, released by leukocytes and mast cells. This can be provoked by an allergic reaction or by the release of acetylcholine from parasympathetic nerve endings.

Epinephrine and related compounds stimulate beta-adrenergic receptors in the bronchioles and by this means promote bronchodilation. Asthmatics can, therefore, take epinephrine as an inhaled spray to relieve the symptoms of an asthma attack. It has been learned that there are two subtypes of beta receptors for epinephrine and that the subtype in the heart (called $\beta_1$) is different from the one in the bronchioles ($\beta_2$). By utilizing these differences, compounds have been developed that can more selectively stimulate the $\beta_2$-adrenergic receptors and cause bronchodilation, without having as great an effect on the heart as epinephrine does.

***Emphysema.*** Alveolar tissue is destroyed in emphysema, resulting in fewer but larger alveoli (fig. 15.21). This produces a decreased surface area for gas exchange and a decreased ability of the bronchioles to remain open during expiration. Collapse of the bronchioles as a result of the compression of the lungs during expiration produces *air trapping,* which further decreases the efficiency of gas exchange in the alveoli.

**Figure 15.21.** Photomicrographs of tissue from a normal lung and from the lung of a person with emphysema. In emphysema, lung tissue is destroyed, resulting in the presence of fewer and larger alveoli.

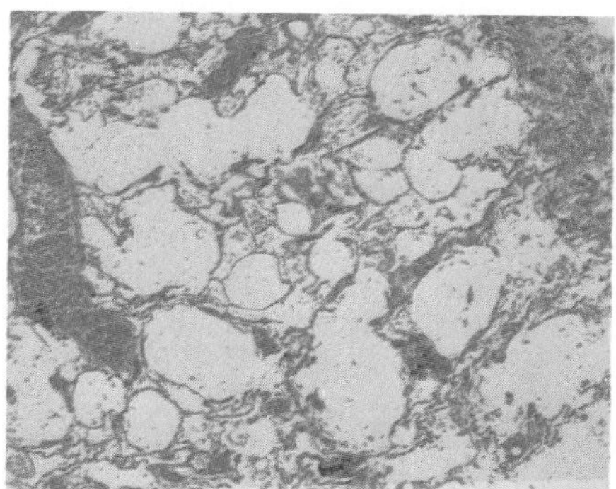

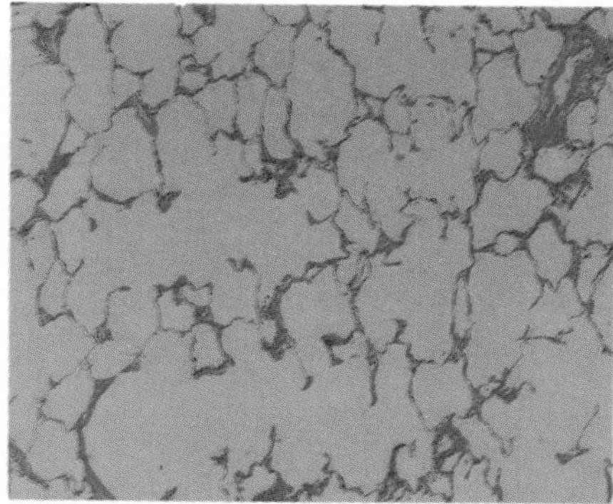

**Table 15.5** Pulmonary disorders caused by inhalation of noxious agents containing particles less than 6 μm in size

| Condition | Agent | Pulmonary Disorders |
|---|---|---|
| Silicosis | Silica | Fibrosis, obstructive emphysema, tuberculosis, cor pulmonale |
| Anthracosis (Black lung) | Coal dust | Similar to silicosis (most coal dust contains silica) |
| Asbestosis | Asbestos fibers | Diffuse fibrosis, cor pulmonale, pulmonary malignancy |

There are different types of emphysema. The most common type occurs almost exclusively in people who have smoked cigarettes heavily over a period of years. A component of cigarette smoke apparently stimulates the macrophages and leukocytes to secrete proteolytic (protein-digesting) enzymes that destroy lung tissues. A less common type of emphysema results from the genetic inability to produce a plasma protein called $\alpha_1$-antitrypsin. This protein normally inhibits proteolytic enzymes such as trypsin and thus normally protects the lungs against the effects of enzymes that are released from alveolar macrophages.

Chronic bronchitis and emphysema, the most common causes of respiratory failure, together are called **chronic obstructive pulmonary disease (COPD).** In addition to the more direct obstructive and restrictive aspects of these conditions, other pathological changes may occur. These include edema, inflammation, hyperplasia (increased cell number), zones of pulmonary fibrosis, pneumonia, pulmonary emboli (traveling blood clots), and heart failure. Patients with severe emphysema may eventually develop *cor pulmonale*—pulmonary hypertension with hypertrophy and the eventual failure of the right ventricle.

***Pulmonary Fibrosis.*** Under certain conditions, for reasons that are poorly understood, lung damage leads to pulmonary fibrosis instead of emphysema. In this condition the normal structure of the lungs is disrupted by the accumulation of fibrous connective tissue proteins. Fibrosis can result, for example, from the inhalation of particles less than 6 μm in size, which are able to accumulate in the respiratory zone of the lungs. Included in this category is *anthracosis,* or black lung, which is produced by the inhalation of carbon particles from coal dust (table 15.5).

1. *Describe the actions of the diaphragm and external intercostal muscles during inspiration, and explain how quiet expiration is produced.*
2. *Describe how forced inspiration and forced expiration are produced.*
3. *Define the terms* tidal volume *and* vital capacity. *Describe how the total minute volume is calculated, and explain how this value changes during exercise.*
4. *Describe how the vital capacity and the forced expiratory volume measurements are affected by asthma and pulmonary fibrosis, and explain why these measurements are affected.*

## Gas Exchange in the Lungs

Gas exchange between the alveolar air and the blood in pulmonary capillaries results in an increased concentration of oxygen, and a decreased concentration of carbon dioxide, in the blood leaving the lungs. This blood enters the systemic arteries, where blood gas measurements are taken to assess the effectiveness of lung function. Measurements of the concentrations of oxygen and carbon dioxide dissolved in plasma are taken with electrodes and indicated using units of partial pressure.

The atmosphere is an ocean of gas that exerts pressure on all objects within it. The amount of this pressure can be measured with a glass U-tube filled with fluid. One end of the U-tube is exposed to the atmosphere, while the other side is continuous with a sealed vacuum tube. Since the atmosphere presses on the exposed side, but not on the side connected to the vacuum tube, atmospheric pressure pushes fluid in the U-tube up on the vacuum side to a height determined by the atmospheric pressure and the density of the fluid. Water, for example, would be pushed up to a height of 33.9 feet (10,332 mm) at sea level, whereas mercury (Hg)—which is more dense—is raised to a height of 760 mm. As a matter of convenience, therefore, devices used to measure atmospheric pressure (barometers) use mercury rather than water. The atmospheric pressure at sea level is thus said to be equal to *760 mm Hg* (or *760 Torr*), which is also described as a pressure of *one atmosphere* (fig. 15.22).

According to **Dalton's law,** the total pressure of a gas mixture (such as air) is equal to the sum of the pressures that each gas in the mixture would exert independently. The pressure that a particular gas in the mixture would exert independently is equal to the product of the total pressure times the percent of that gas in the mixture. This is known as the **partial pressure** of that gas in the mixture. The total pressure of the gas mixture is thus equal to the sum of the partial pressures of the constituent gases. Since oxygen comprises about 21% of the atmosphere, for example, its partial pressure (abbreviated $P_{O_2}$) is 21% of 760, or about 159 mm Hg. Since nitrogen comprises about 78% of the atmosphere, its partial pressure is equal to 0.78 × 760 = 593 mm Hg. These two gases thus contribute about 99% of the total pressure of 760 mm Hg:

$$P_{\text{dry atmosphere}} = P_{N_2} + P_{O_2} + P_{CO_2} = 760 \text{ mm Hg}$$

**Figure 15.22.** Atmospheric pressure at sea level can push a column of mercury to a height of 760 millimeters. This is also described as 760 torr, or one atmospheric pressure.

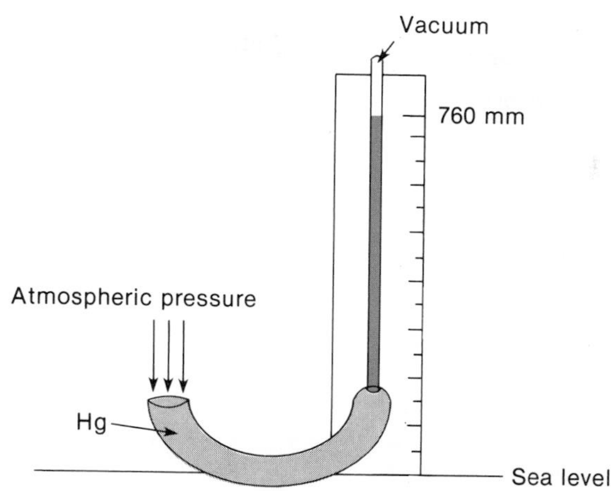


**Table 15.6** The effect of altitude on $P_{O_2}$

| **Changes in $P_{O_2}$ at Various Altitudes** | | | | |
|---|---|---|---|---|
| Altitude (Feet above Sea Level) | Atmospheric Pressure (mm Hg) | $P_{O_2}$ in Air (mm Hg) | $P_{O_2}$ in Alveoli (mm Hg) | $P_{O_2}$ in Arterial Blood (mm Hg) |
| 0 | 760 | 159 | 105 | 100 |
| 2,000 | 707 | 148 | 97 | 92 |
| 4,000 | 656 | 137 | 90 | 85 |
| 6,000 | 609 | 127 | 84 | 79 |
| 8,000 | 564 | 118 | 79 | 74 |
| 10,000 | 523 | 109 | 74 | 69 |
| 20,000 | 349 | 73 | 40 | 35 |
| 30,000 | 226 | 47 | 21 | 19 |

### Calculation of $P_{O_2}$

As one goes to a higher altitude, the total atmospheric pressure and the partial pressure of the constituent gases decrease (table 15.6). At Denver, for example (5,000 feet above sea level), the atmospheric pressure is decreased to 619 mm Hg, and the $P_{O_2}$ is therefore reduced to 619 × 0.21 = 130 mm Hg. At the peak of Mount Everest (at 29,000 feet), the $P_{O_2}$ is only 42 mm Hg. As one descends below sea level, as in ocean diving, the total pressure increases by one atmosphere for every 33 feet. At 33 feet therefore, the pressure equals 2 × 760 = 1,520 mm Hg. At 66 feet, the pressure equals three atmospheres.

Inspired air contains variable amounts of moisture. By the time the air has passed into the respiratory zone of the lungs, however, it is normally saturated with water vapor (has a relative humidity of 100%). The capacity of air to contain water vapor depends on its temperature; since the temperature of the respiratory zone is constant at 37° C, its water vapor pressure is also constant (47 mm Hg).

Water vapor, like the other constituent gases, contributes a partial pressure to the total atmospheric pressure. Since the total atmospheric pressure is constant (depending only on the height of the air mass), the water vapor "dilutes" the contribution of other gases to the total pressure:

$$P_{\text{wet atmosphere}} = P_{N_2} + P_{O_2} + P_{CO_2} + P_{H_2O}$$

When the effect of water vapor pressure is considered, the partial pressure of oxygen in the inspired air is decreased at sea level to

$$P_{O_2} \text{ (sea level)} = 0.21 \quad (760 - 47) = 150 \text{ mm Hg}$$

As a result of gas exchange in the alveoli, the $P_{O_2}$ of *alveolar air* is further diminished to about 105 mm Hg. A comparison of the partial pressures of the inspired air with the partial pressures of alveolar air is shown in figure 15.23.

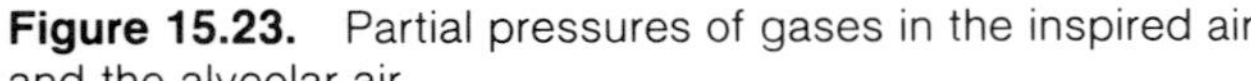

**Figure 15.23.** Partial pressures of gases in the inspired air and the alveolar air.

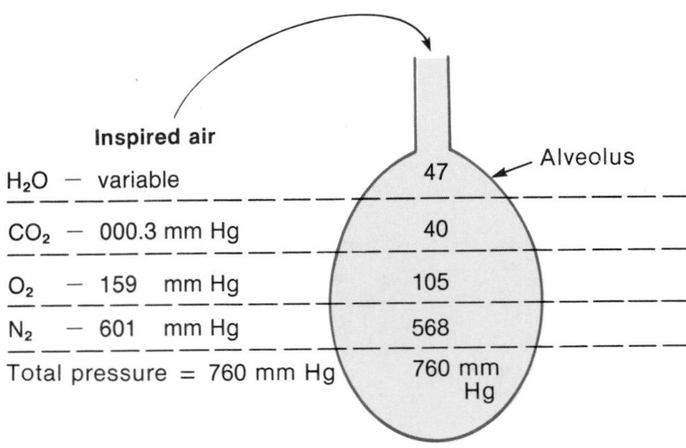


| **Table 15.7** Some of the physical laws that are important in ventilation and gas exchange | |
|---|---|
| **Physical Law** | **Description** |
| Boyle's law | The volume of a gas is inversely proportional to its pressure when the temperature and mass are constant (PV = constant). |
| Dalton's law | The total pressure of a gas mixture is equal to the sum of the partial pressures of its constituent gases. The partial pressure of each gas is the pressure it would exert if it alone occupied the total volume of the mixture. |
| Graham's law | The rate of diffusion of a gas through a liquid is directly proportional to its solubility and inversely proportional to its density (or gram molecular weight). |
| Henry's law | The weight of a gas dissolved in a liquid at a given temperature is proportional to the partial pressure of the gas. |
| LaPlace's law | The inward pressure tending to collapse a bubble is equal to two times its surface tension divided by the radius of the bubble. |

## Partial Pressures of Gases in Blood

The enormous surface area of alveoli and the short diffusion distance between alveolar air and the capillary blood quickly help to bring the blood into gaseous equilibrium with the alveolar air. This function is further aided by the fact that each alveolus is surrounded by so many capillaries that they form an almost continuous sheet of blood around the alveoli (fig. 15.24).

When a fluid and a gas, such as blood and alveolar air, are at equilibrium, the amount of gas dissolved in the fluid reaches a maximum value. This value depends, according to **Henry's law** (table 15.7), on (1) the solubility of the gas in the fluid, which is a physical constant; (2) the temperature of the fluid—more gas can be dissolved in cold water than warm water; and (3) the partial pressure of the gas. Since the temperature of the blood does not vary significantly, *the concentration of a gas dissolved in a fluid (such as plasma) depends directly on its partial pressure in the gas mixture.* When water—or plasma—is brought into equilibrium with air at a $P_{O_2}$ of 100 mm Hg, for example, the fluid will contain 0.3 ml $O_2$ per 100 ml of fluid at 37° C. If the $P_{O_2}$ of the gas were reduced by half, the amount of dissolved oxygen would also be reduced by half.

***Blood-Gas Measurements.*** Measurement of the oxygen content of blood (in ml of $O_2$ per 100 ml blood) is a laborious procedure. Fortunately, an **oxygen electrode** that produces an electric current in proportion to the amount of *dissolved oxygen* has been developed. If this electrode is placed in a fluid while oxygen is artificially bubbled into it, the current produced by the oxygen electrode will increase up to a maximum value. At this maximum value the fluid is *saturated* with oxygen—that is, all of the oxygen that can be dissolved at that temperature and $P_{O_2}$ is dissolved. At a constant temperature the amount dissolved, and thus the electric current, depend only on the $P_{O_2}$ of the gas.

As a matter of convenience, it can now be said that the *fluid has the same $P_{O_2}$ as the gas.* If it is known that the gas has a $P_{O_2}$ of 152 mm Hg, for example, the deflection of the needle produced on a scale by the oxygen electrode can be calibrated at 152 mm Hg (fig. 15.25). The actual amount of dissolved oxygen under these circumstances is not known or very important (it can be looked up in solubility tables, if desired), since this is simply a linear function of the $P_{O_2}$. A lower $P_{O_2}$ indicates that less oxygen is dissolved, and a higher $P_{O_2}$ indicates that more oxygen is dissolved.

**Figure 15.24.** The high surface area of contact between the pulmonary capillaries and the alveoli allow rapid exchange of gases between the air and blood.

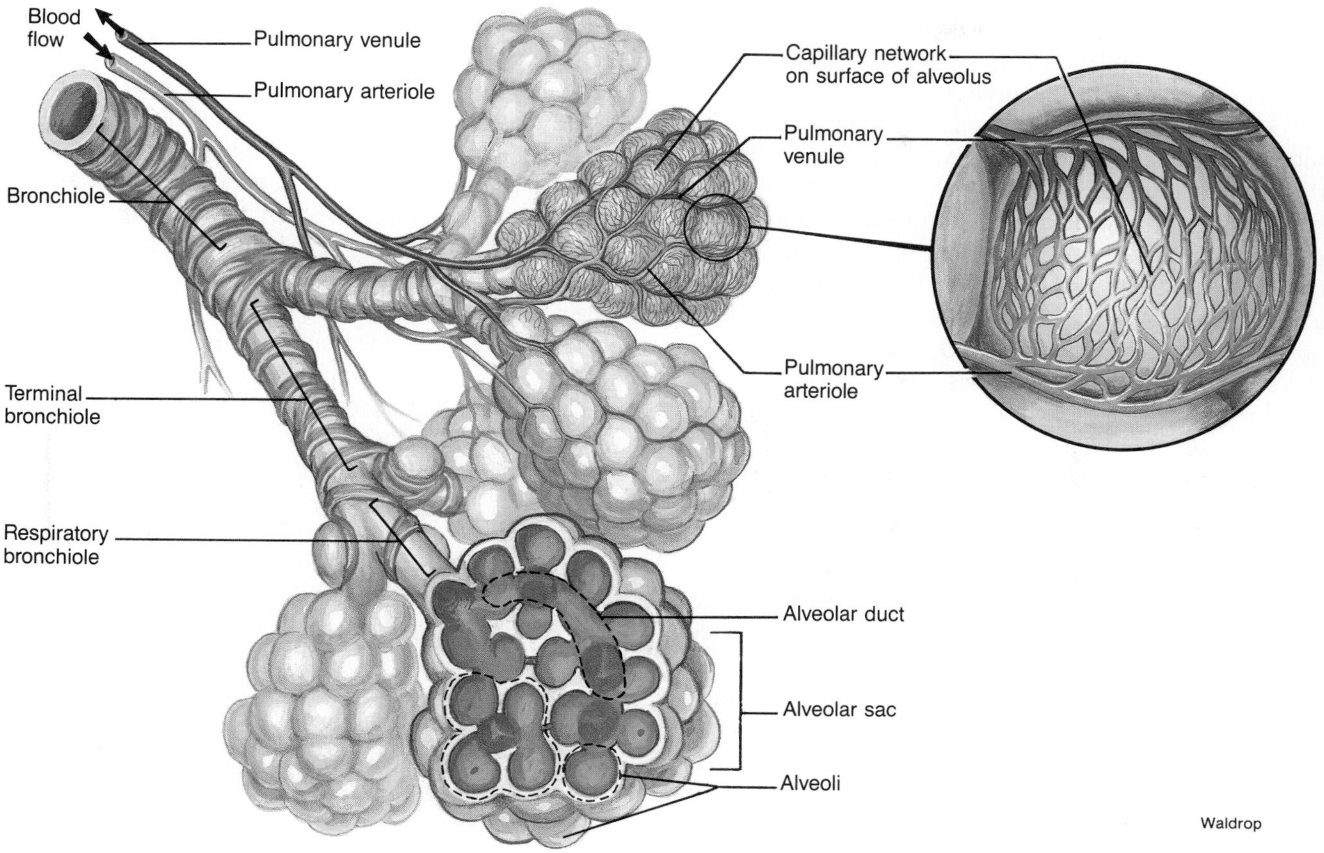

Blood flow
Pulmonary venule
Pulmonary arteriole
Bronchiole
Terminal bronchiole
Respiratory bronchiole
Capillary network on surface of alveolus
Pulmonary venule
Pulmonary arteriole
Alveolar duct
Alveolar sac
Alveoli
Waldrop

**Figure 15.25.** Blood-gas measurements using a $P_{O_2}$ electrode. (*a*) The electrical current generated by the oxygen electrode is calibrated so that the needle of the blood-gas machine points to the $P_{O_2}$ of the gas with which the fluid is in equilibrium. (*b*) Once standardized in this way, the electrode can be inserted in a fluid, such as blood, and the $P_{O_2}$ of this solution can be measured.

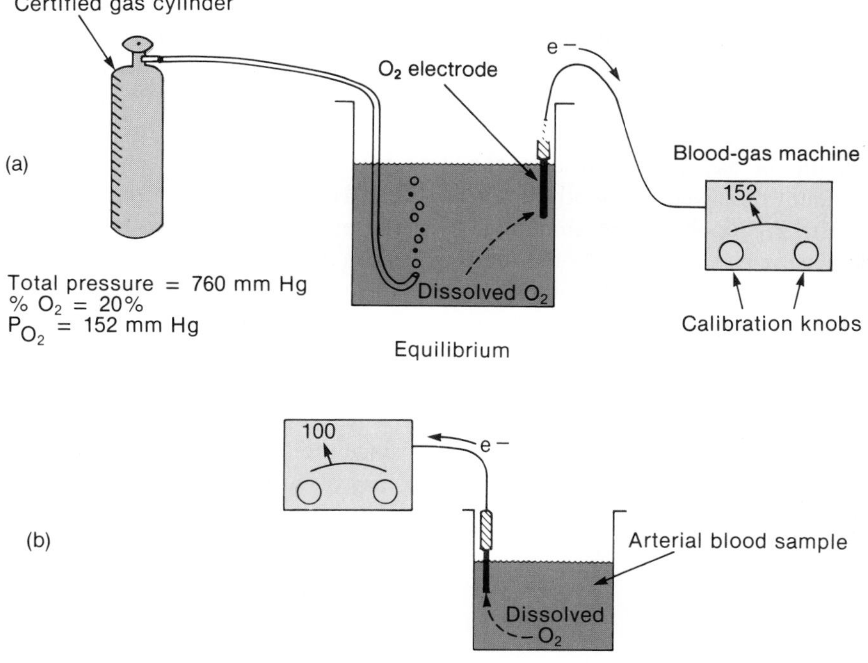

If the oxygen electrode is next inserted into an unknown sample of blood, the $P_{O_2}$ of that sample can be read directly from the previously calibrated scale. Suppose, as illustrated in figure 15.25, the blood sample has a $P_{O_2}$ of 100 mm Hg. Since alveolar air has a $P_{O_2}$ of about 105 mm Hg, this reading indicates that the blood is almost in complete equilibrium with the alveolar air.

The oxygen electrode responds only to the oxygen dissolved in water or plasma; it cannot respond to the oxygen that is hidden in red blood cells. Most of the oxygen in blood, however, is located in the red blood cells attached to hemoglobin. The oxygen content of whole blood thus depends on both its $P_{O_2}$ and its red blood cell and hemoglobin content. At a $P_{O_2}$ of about 100 mm Hg, whole blood normally contains almost 20 ml $O_2$ per 100 ml blood; of this amount, only 0.3 ml $O_2$ is dissolved in the plasma and 19.7 ml $O_2$ is within the red blood cells. Since only the 0.3 ml $O_2$ per 100 ml blood affects the $P_{O_2}$ measurement, this measurement would be unchanged if the red blood cells were removed from the sample.

### Significance of Blood $P_{O_2}$ and $P_{CO_2}$ Measurements

Since blood $P_{O_2}$ measurements are not directly affected by the oxygen in red blood cells, the $P_{O_2}$ does not provide a measurement of the total oxygen content of whole blood. It does, however, provide a good index of *lung function.* If the inspired air has a normal $P_{O_2}$ but the arterial $P_{O_2}$ is below normal, for example, gas exchange in the lungs must be impaired. Measurements of arterial $P_{O_2}$ thus provide valuable information in the treatment of people with pulmonary diseases, during surgery (when breathing may be depressed by anesthesia), and in the care of premature babies with respiratory distress syndrome.

When the lungs are functioning properly, the $P_{O_2}$ of systemic arterial blood is only 5 mm Hg less than the $P_{O_2}$ of alveolar air. Hyperventilation, therefore, cannot significantly increase the blood $P_{O_2}$. At a normal $P_{O_2}$ of about 100 mm Hg, hemoglobin is almost completely loaded with oxygen. An increase in blood $P_{O_2}$—produced, for example, by breathing 100% oxygen from a gas tank—thus cannot significantly increase the amount of oxygen contained in the red blood cells. This can, however, significantly increase the amount of oxygen dissolved in the plasma (because the amount dissolved is directly determined by the $P_{O_2}$). If the $P_{O_2}$ doubles, the amount of oxygen dissolved in the plasma also doubles, but the total oxygen content of whole blood increases only slightly since most of the oxygen by far is not in plasma but in the red blood cells.

Since the oxygen carried by red blood cells must first dissolve in plasma before it can diffuse to the tissue cells, however, a doubling of the blood $P_{O_2}$ means that the *rate of oxygen delivery* to the tissues would double under these conditions. For this reason, breathing from a tank of 100% oxygen (with a $P_{O_2}$ of 760 mm Hg) would significantly increase oxygen delivery to the tissues, although it would have little effect on the total oxygen content of blood.

An electrode that produces a current in response to dissolved carbon dioxide has also been developed, so that the $P_{CO_2}$ of blood can be measured together with its $P_{O_2}$. Blood in the systemic veins, which is delivered to the lungs by the pulmonary arteries, usually has a $P_{O_2}$ of 40 mm Hg and a $P_{CO_2}$ of 46 mm Hg. After gas exchange in the alveoli of the lungs, blood in the pulmonary veins and systemic arteries has a $P_{O_2}$ of about 100 mm Hg and a $P_{CO_2}$ of 40 mm Hg (fig. 15.26). The values in arterial blood are relatively constant and clinically significant, because they reflect lung function. Blood-gas measurements of venous blood are not performed clinically because these values are far more variable; venous $P_{O_2}$ is much lower and $P_{CO_2}$ much higher after exercise, for example, than at rest, whereas arterial values are not significantly affected by usual changes in physical activity.

### Pulmonary Circulation and Ventilation/Perfusion Ratios

In a fetus, the pulmonary circulation has a high vascular resistance as a result of the fact that the lungs are partially collapsed. This high vascular resistance helps to shunt blood from the right to the left atrium through the foramen ovale, and from the pulmonary artery to the aorta through the ductus arteriosus (described in chapter 13). After birth, the foramen ovale and ductus arteriosus close, and the vascular resistance of the pulmonary circulation falls sharply. This fall in vascular resistance at birth is due to (1) opening of the vessels as a result of the subatmospheric intrapulmonary pressure and physical stretching of the lungs during inspiration and (2) to dilation of the pulmonary arterioles in response to increased alveolar $P_{O_2}$.

In the adult, the right ventricle (like the left) has an output of about 5.5 L per minute. The rate of blood flow through the pulmonary circulation is thus equal to the flow rate through the systemic circulation. Blood flow, as described in chapter 14, is directly proportional to the pressure difference between the two ends of a vessel and inversely proportional to the vascular resistance. In the systemic circulation, the mean arterial pressure is 90–100 mm Hg, and the pressure of the right atrium is 0 mm Hg; the pressure difference is thus about 100 mm Hg. The mean pressure of the pulmonary artery, in contrast, is only 15 mm Hg and the pressure of the left atrium is 5 mm Hg. The driving pressure in the pulmonary circulation is thus 15 − 5, or 10 mm Hg.

Since the driving pressure in the pulmonary circulation is only one-tenth that of the systemic circulation, and since the flow rates are equal, it follows that the pulmonary vascular resistance must be one-tenth that of the systemic vascular resistance. The pulmonary circulation, in other words, is a low resistance, low pressure pathway. The

**Figure 15.26.** The $P_{O_2}$ and $P_{CO_2}$ of blood as a result of gas exchange in the lung alveoli and gas exchange between systemic capillaries and body cells.

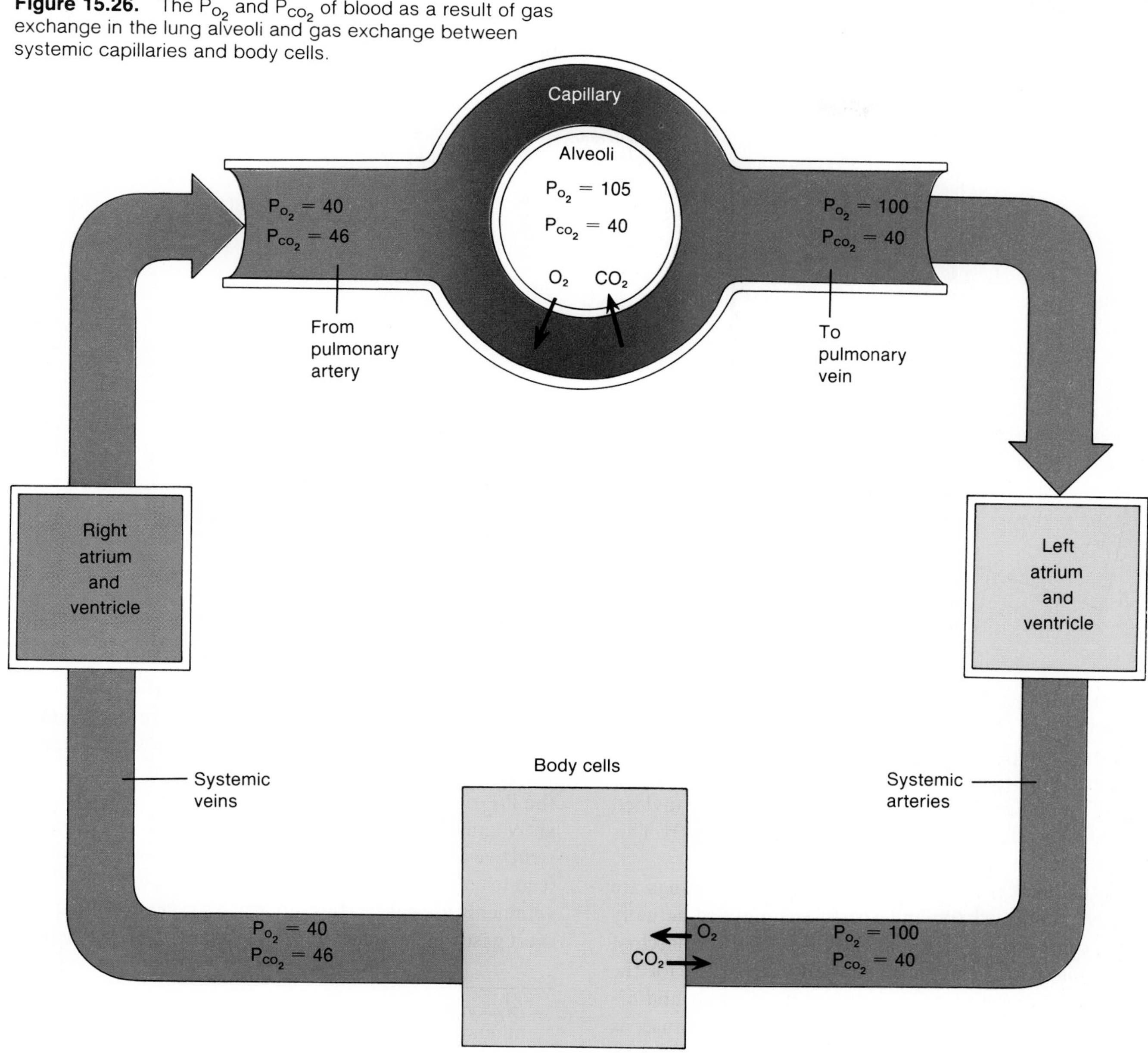


low pulmonary blood pressure produces less filtration pressure than in the systemic capillaries, and thus helps to protect against *pulmonary edema,* which is a dangerous condition that can impede ventilation and gas exchange. Pulmonary edema can occur when there is pulmonary hypertension produced, for example, by left heart failure.

Pulmonary arterioles constrict when the alveolar $P_{O_2}$ is low and dilate as the alveolar $P_{O_2}$ is raised. This response is opposite to that of systemic arterioles, which dilate in response to low tissue $P_{O_2}$ (as described in chapter 14). Dilation of the systemic arterioles when the $P_{O_2}$ is low helps to supply more blood and oxygen to the tissues; constriction of the pulmonary arterioles when the alveolar $P_{O_2}$ is low helps to decrease blood flow to alveoli that are inadequately ventilated.

Constriction of the pulmonary arterioles where the alveolar $P_{O_2}$ is low and dilation of arterioles where the alveolar $P_{O_2}$ is high helps to *match ventilation to perfusion* (the term *perfusion* refers to blood flow). If this autoregulation of blood flow did not occur, blood from poorly ventilated alveoli would mix with blood from well-ventilated alveoli and the blood leaving the lungs would have a lowered $P_{O_2}$ as a result of this dilution effect.

Dilution of the $P_{O_2}$ of pulmonary vein blood actually does occur to some degree despite autoregulation of pulmonary blood flow. When a person stands upright, gravity causes a greater blood flow to the base of the lungs than to the apex (top). Ventilation likewise decreases from base

**Figure 15.27.** Ventilation, blood flow, and ventilation/perfusion ratios at the apex and base of the lungs. The ratios indicate that the apex is relatively overventilated and the base underventilated in relation to their blood flows. Such uneven matching of ventilation to perfusion results in the fact that the blood leaving the lungs has a $P_{O_2}$ that is slightly less (by 4–5 mm Hg) than the $P_{O_2}$ of alveolar air.

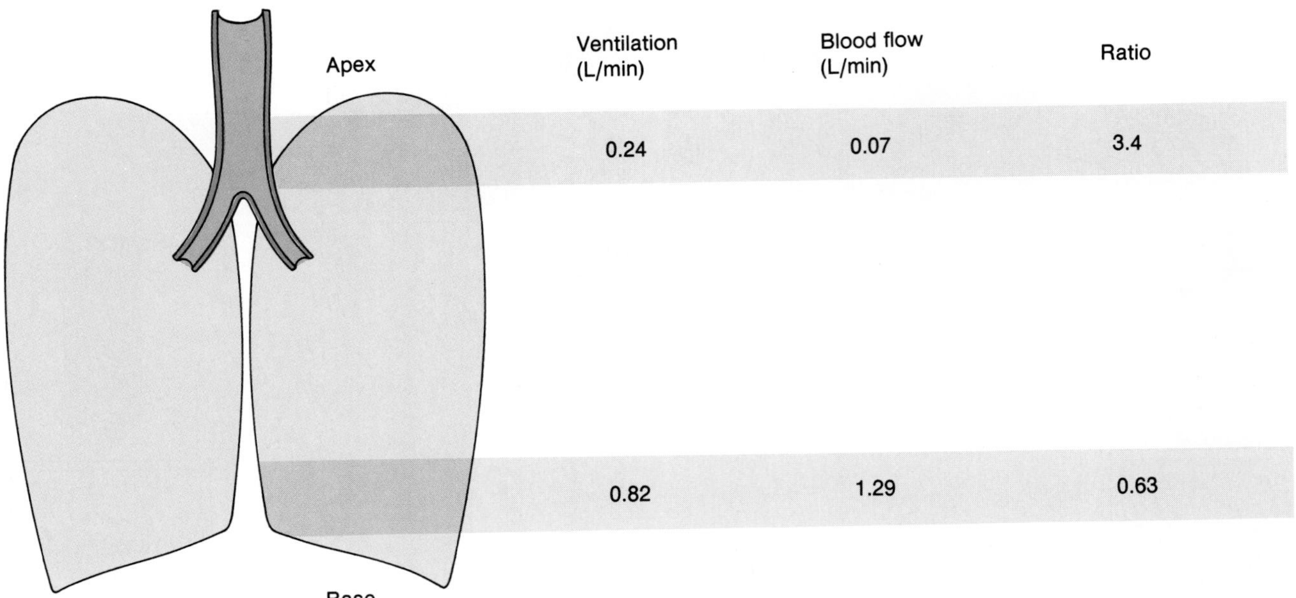


to apex, but this decrease is not proportionate to the decrease in blood flow. The *ventilation/perfusion ratio* at the apex is thus high (0.24 L air per minute divided by 0.07 L blood per minute equals a ratio of 3.4/1.0), while at the base of the lungs it is low (0.82 L air per minute divided by 1.29 L blood per minute equals a ratio of 0.6/1.0). This is illustrated in figure 15.27.

Functionally, the alveoli at the apex of the lungs are thus overventilated (or underperfused) and are actually larger, compared to alveoli at the base. This mismatch of ventilation/perfusion ratios is largely responsible for the 5 mm Hg difference in $P_{O_2}$ between alveolar air and arterial blood. This difference is not significant because, as previously described, the hemoglobin of normal arterial blood is almost completely saturated with oxygen.

## Disorders Caused by High Partial Pressures of Gases

The total atmospheric pressure increases by one atmosphere (760 mm Hg) for every 10 m (33 feet) below sea level. If one dives 10 m below sea level, therefore, the partial pressures and amounts of dissolved gases in the plasma are twice that at sea level. At 66 feet they are three times, and at 100 feet they are four times the values at sea level. The increased amounts of nitrogen and oxygen dissolved in the blood plasma under these conditions can have serious effects on the body.

***Oxygen Toxicity.*** Although breathing 100% oxygen at one or two atmospheres pressure can be safely tolerated for a few hours, higher partial oxygen pressures can be very dangerous. *Oxygen toxicity* develops rapidly when the $P_{O_2}$ rises above about 2.5 atmospheres. This is apparently caused by the oxidation of enzymes and other destructive changes that can damage the nervous system and lead to coma and death. For these reasons, deep sea divers commonly use gas mixtures in which oxygen is diluted with inert gases such as nitrogen (as in ordinary air) or helium.

*Hyperbaric oxygen*—oxygen at greater than one atmosphere pressure—is often used to treat conditions such as carbon monoxide poisoning, circulatory shock, and gas gangrene. Before the dangers of oxygen toxicity were realized these hyperbaric oxygen treatments sometimes resulted in tragedy. Particularly tragic were the cases of **retrolental fibroplasia,** in which damage to the retina and blindness resulted from hyperbaric oxygen treatment of premature babies with hyaline membrane disease.

***Nitrogen Narcosis.*** Although at sea level nitrogen is physiologically inert, larger amounts of dissolved nitrogen under hyperbaric conditions have deleterious effects. Since it takes time for the nitrogen to dissolve, these effects usually don't appear until the person has remained submerged over an hour. *Nitrogen narcosis* resembles alcohol intoxication; depending on the depth of the dive, the diver may experience "rapture of the deep" or may become so drowsy that he or she is totally incapacitated.

***Decompression Sickness.*** The amount of nitrogen dissolved in the plasma decreases as the diver ascends to sea level, due to a progressive decrease in the $P_{N_2}$. If the diver surfaces slowly, a large amount of nitrogen can diffuse through the alveoli and be eliminated in the expired breath. If decompression occurs too rapidly, however, bubbles of nitrogen gas ($N_2$) can form in the blood and block small blood channels, producing muscle and joint pain as well as more serious damage. These effects are known as *decompression sickness,* or the bends.

The cabins of airplanes that fly long distances at high altitudes (30,000 to 40,000 feet) are pressurized so that the passengers and crew are not exposed to the very low atmospheric pressures of these altitudes. If a cabin were to become rapidly depressurized at high altitude, much less nitrogen could remain dissolved at the greatly lowered pressure. People in this situation, like the divers that ascend too rapidly, would thus experience decompression sickness.

1. *Describe how the $P_{O_2}$ of air is calculated and how this value is affected by altitude, diving, and water vapor pressure.*
2. *Describe how blood $P_{O_2}$ measurements are taken, and explain the physiological and clinical significance of these measurements.*
3. *Explain how the arterial $P_{O_2}$ and the oxygen content of whole blood is affected by (1) hyperventilation; (2) breathing from a tank containing 100% oxygen; (3) anemia (low red blood cell count and hemoglobin concentration); and (4) high altitude.*
4. *Describe the ventilation/perfusion ratios of the lungs, and explain why systemic arterial blood has a slightly lower $P_{O_2}$ than alveolar air.*
5. *Explain how decompression sickness is produced in divers who ascend too rapidly.*

## REGULATION OF BREATHING

The motor neurons that stimulate the muscles involved in breathing are controlled by two major descending pathways: one from the cerebral cortex, which controls voluntary breathing, and one from the medulla oblongata, which controls involuntary breathing. The unconscious, rhythmic control of breathing results from spontaneous activities of neurons within the medulla, but these neurons are influenced by neurons in the pons and by sensory feedback from receptors sensitive to the $P_{CO_2}$, pH, and $P_{O_2}$ of arterial blood. Of these stimuli, a rise in blood $P_{CO_2}$ and fall in pH are the most effective in provoking increased ventilation.

Inspiration and expiration are produced by the contraction and relaxation of skeletal muscles in response to activity in somatic motor neurons in the spinal cord. The activity of these motor neurons, in turn, is controlled by descending tracts from neurons in the respiratory control centers in the medulla oblongata and from neurons in the cerebral cortex.

The automatic control of breathing is regulated by nerve fibers that descend in the lateral and ventral white matter of the spinal cord from the medulla oblongata. The voluntary control of breathing is a function of the cerebral cortex and involves nerve fibers that descend in the corticospinal tracts (chapter 8). The separation of the voluntary and involuntary pathways is dramatically illustrated in the condition called **Ondine's curse,** in which neurological damage abolishes the automatic but not the voluntary control of breathing. People with this condition must consciously force themselves to breathe and be put on artificial respirators when they sleep.

### Brain Stem Respiratory Centers

A loose aggregation of neurons in the reticular formation of the *medulla oblongata* forms the **rhythmicity center** that controls automatic breathing. The rhythmicity center consists of interacting pools of neurons that fire either during inspiration (*I neurons*) or expiration (*E neurons*). The I neurons project to and stimulate spinal motoneurons that innervate the respiratory muscles. Expiration is a passive process that occurs when the I neurons are inhibited by the activity of the E neurons. The activity of I and E neurons varies in a reciprocal way, so that a rhythmic pattern of breathing is produced. The cycle of inspiration and expiration is thus intrinsic to the neural activity of the medulla. The rhythmicity center in the medulla is divided into a dorsal group of neurons, which regulates the activity of the phrenic nerves to the diaphragm, and a ventral group, which controls the motor neurons to the intercostal muscles.

The activity of the medullary rhythmicity center is influenced by centers in the *pons*. As a result of research in which the brain stem is destroyed at different levels, two respiratory control centers have been identified in the pons. One area—the **apneustic center**—appears to promote inspiration by stimulating the I neurons in the medulla. The other pons area—called the **pneumotaxic center**—seems to antagonize the apneustic center and inhibit inspiration (fig. 15.28). The apneustic center is believed to provide a tonic, or constant, stimulus for inspiration which is cyclically inhibited by the activity of the pneumotaxic center.

**Figure 15.28.** Approximate locations of the brain stem respiratory centers.

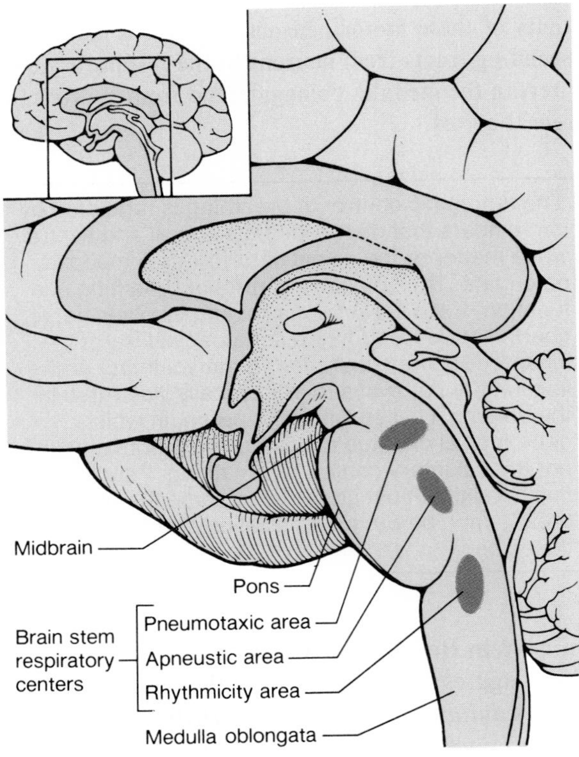


The automatic control of breathing is also influenced by input from receptors sensitive to the chemical composition of the blood. There are two groups of *chemoreceptors* that respond to changes in blood $P_{CO_2}$, pH, and $P_{O_2}$. These are the **central chemoreceptors** in the medulla oblongata and the **peripheral chemoreceptors.** The peripheral chemoreceptors are contained within small nodules associated with the aorta and the carotid arteries, and receive blood from these critical arteries via small arterial branches. The peripheral chemoreceptors include the **aortic bodies**, located around the aortic arch, and the **carotid bodies**, located just beyond the branching of each common carotid into the internal and external carotid arteries (fig. 15.29). The aortic and carotid bodies should not be confused with the aortic and carotid sinuses (chapter 14), which are located within these arteries and contain receptors that monitor the blood pressure.

The peripheral chemoreceptors control breathing indirectly via sensory nerve fibers to the medulla. The aortic bodies send sensory information to the medulla in the vagus (tenth) cranial nerve; the carotid bodies stimulate sensory fibers in the glossopharyngeal (ninth) cranial nerve. The neural and sensory control of ventilation is summarized in figure 15.30.

**Table 15.8** The effect of ventilation, as measured by total minute volume (breathing rate × tidal volume), on the $P_{CO_2}$ of arterial blood

| Total Minute Volume | Arterial $P_{CO_2}$ | Type of Ventilation |
|---|---|---|
| 2 L/min | 80 mm Hg | Hypoventilation |
| 4–5 L/min | 40 mm Hg | Normal ventilation |
| 8 L/min | 20–25 mm Hg | Hyperventilation |

## Effects of Blood $P_{CO_2}$ and pH on Ventilation

Chemoreceptor input to the brain stem modifies the rate and depth of breathing so that, under normal conditions, arterial $P_{CO_2}$, pH, and $P_{O_2}$ remain relatively constant. If hypoventilation (inadequate ventilation) occurs, $P_{CO_2}$ quickly rises and pH falls. The fall in pH is due to the fact that carbon dioxide can combine with water to form carbonic acid, which in turn can release $H^+$ to the solution. This is shown in the following equations:

$$CO_2 + H_2O \rightarrow H_2CO_3$$

$$H_2CO_3 \rightarrow HCO_3^- + H^+$$

The oxygen content of the blood decreases much more slowly, because there is a large "reservoir" of oxygen attached to hemoglobin. During hyperventilation, conversely, blood $P_{CO_2}$ quickly falls and pH rises due to the excessive elimination of carbonic acid. The oxygen content of blood, on the other hand, is not significantly increased by hyperventilation (hemoglobin in arterial blood is 97% saturated with oxygen during normal ventilation).

The blood $P_{CO_2}$ and pH are, therefore, more immediately affected by changes in ventilation than is the oxygen content. Indeed, changes in $P_{CO_2}$ provide a sensitive index of ventilation, as shown in table 15.8. In view of these facts, it is not surprising that changes in $P_{CO_2}$ provide the most potent stimulus for the reflex control of ventilation. Ventilation, in other words, is adjusted to maintain a constant $P_{CO_2}$; proper oxygenation of the blood occurs naturally as a side product of this reflex control.

The rate and depth of ventilation are normally adjusted to maintain an arterial $P_{CO_2}$ of 40 mm Hg. Hypoventilation causes a rise in $P_{CO_2}$—a condition called *hypercapnia*. Hyperventilation, conversely, results in *hypocapnia*. Chemoreceptor regulation of breathing in response to changes in $P_{CO_2}$ is illustrated in figure 15.31.

***Chemoreceptors in the Medulla.*** The chemoreceptors most sensitive to changes in the arterial $P_{CO_2}$ are located in the ventral area of the medulla, near the exit of the ninth and tenth cranial nerves. These chemoreceptor neurons are anatomically separate from, but synaptically communicate with, the neurons of the respiratory control center in the medulla.

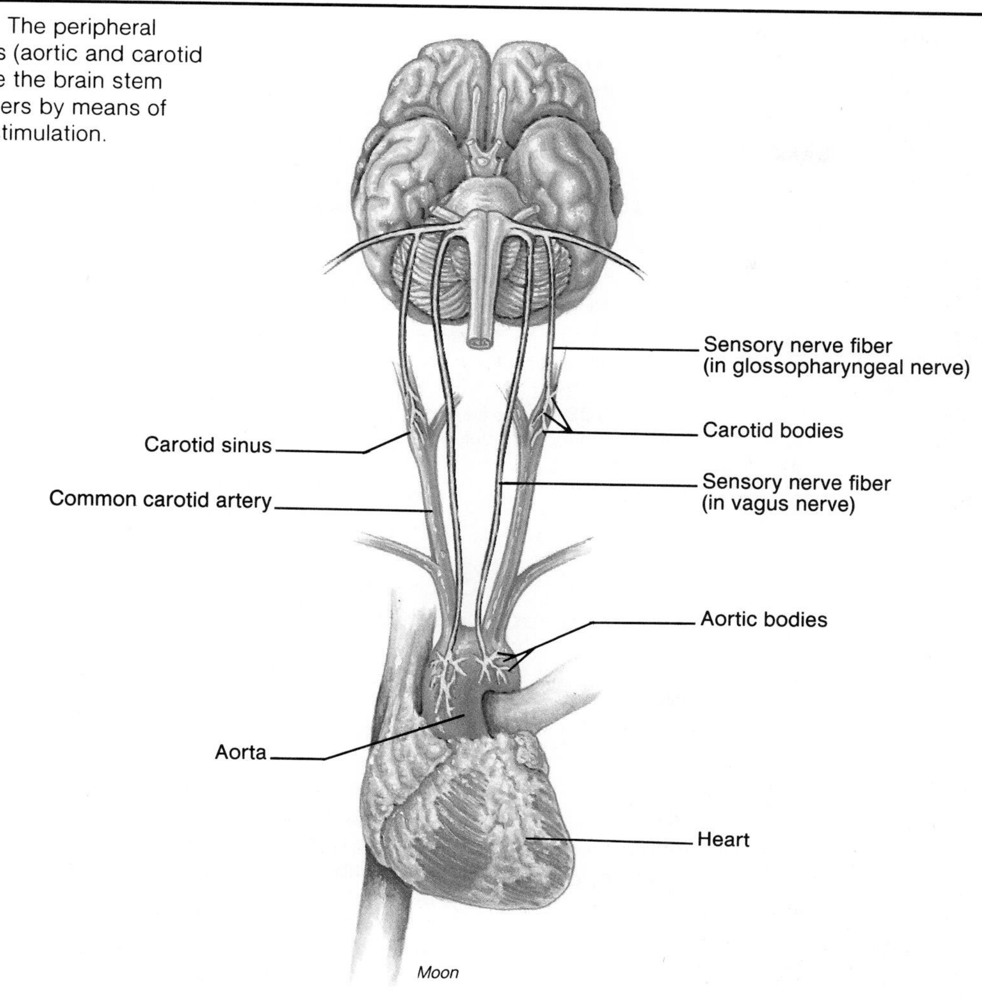


**Figure 15.29.** The peripheral chemoreceptors (aortic and carotid bodies) regulate the brain stem respiratory centers by means of sensory nerve stimulation.

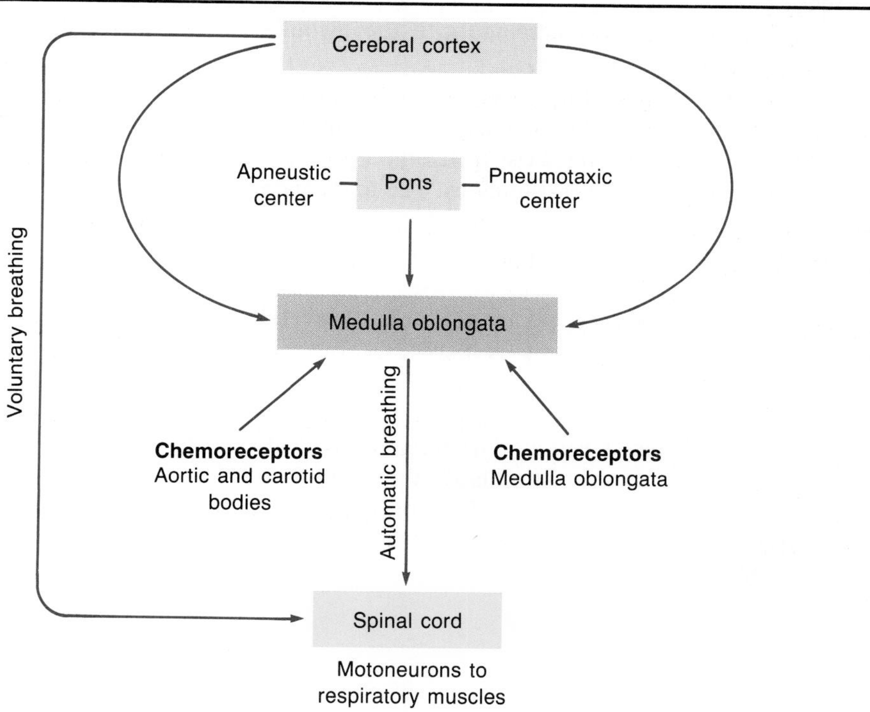


**Figure 15.30.** A schematic illustration of the control of ventilation by the central nervous system. The feedback effects of pulmonary stretch receptors and "irritant" receptors are not shown.

**Figure 15.31.** Negative feedback control of ventilation through changes in blood $P_{CO_2}$ and pH. The box represents the blood-brain barrier, which allows $CO_2$ to pass into the cerebrospinal fluid but prevents the passage of $H^+$.

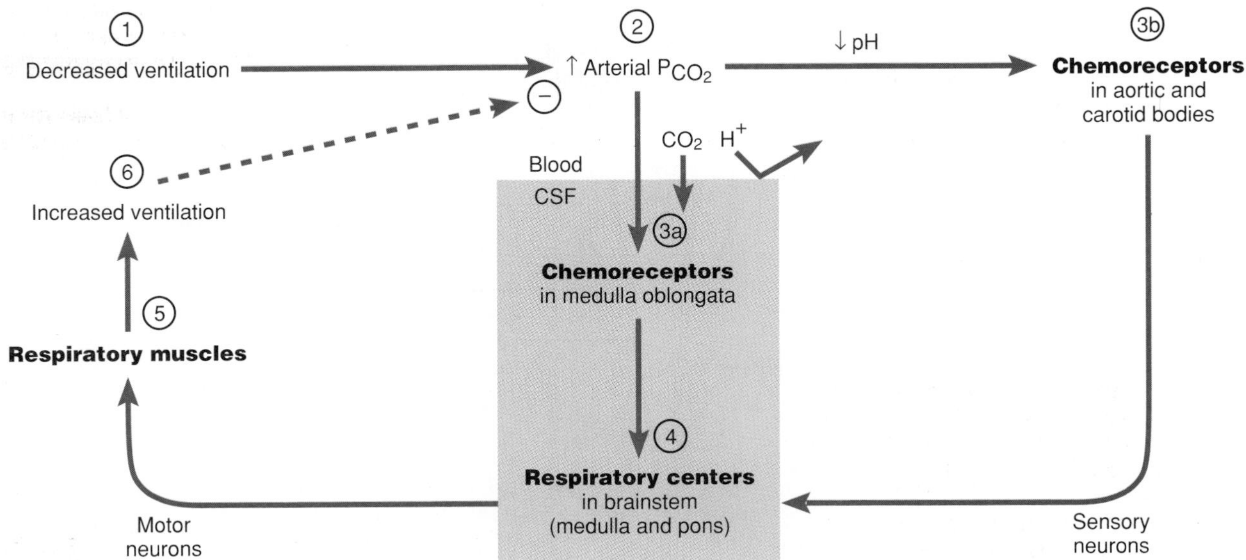


An increase in arterial $P_{CO_2}$ causes a rise in the $H^+$ concentration of the blood as a result of increased carbonic acid concentrations. The $H^+$ in the blood, however, cannot cross the blood-brain barrier and, therefore, cannot influence the medullary chemoreceptors. Carbon dioxide in the arterial blood *can* cross the blood-brain barrier and, through the formation of carbonic acid, can lower the pH of cerebrospinal fluid. This fall in cerebrospinal fluid pH directly stimulates the chemoreceptors in the medulla when there is a rise in arterial $P_{CO_2}$.

The chemoreceptors in the medulla are ultimately responsible for 70%–80% of the increased ventilation that occurs in response to a sustained rise in arterial $P_{CO_2}$. This response, however, takes several minutes. The immediate increase in ventilation that occurs when $P_{CO_2}$ rises is produced by stimulation of the peripheral chemoreceptors.

***Peripheral Chemoreceptors.*** The aortic and carotid bodies are not stimulated directly by blood $CO_2$. Instead, they are stimulated by a rise in the $H^+$ concentration (fall in pH) of arterial blood, which occurs when the blood $CO_2$, and thus carbonic acid is raised. The retention of $CO_2$ during hypoventilation thus stimulates the medullary chemoreceptors through a lowering of cerebrospinal fluid pH and stimulates peripheral chemoreceptors through a lowering of blood pH.

People who hyperventilate during psychological stress are sometimes told to breathe into a paper bag so that they rebreathe their expired air that is enriched in $CO_2$. This procedure helps to raise their blood $P_{CO_2}$ back up to the normal range. This is needed because hypocapnia causes cerebral vasoconstriction. In addition to producing dizziness, the cerebral ischemia that results can lead to acidotic conditions in the brain that, through stimulation of the medullary chemoreceptors, can cause further hyperventilation. Breathing into a paper bag can thus relieve the hypocapnia and stop the hyperventilation.

## Effects of Blood $P_{O_2}$ on Ventilation

Under normal conditions, blood $P_{O_2}$ affects breathing only indirectly, by influencing the chemoreceptor sensitivity to changes in $P_{CO_2}$. Chemoreceptor sensitivity to $P_{CO_2}$ is augmented by a low $P_{O_2}$ (so ventilation is increased at a high altitude, for example) and is decreased by a high $P_{O_2}$. If the blood $P_{O_2}$ is raised by breathing 100% oxygen, therefore, the breath can be held longer because the response to increased $P_{CO_2}$ is blunted.

When the blood $P_{CO_2}$ is held constant by experimental techniques, the $P_{O_2}$ of arterial blood must fall from 100 mm Hg to below 50 mm Hg before ventilation is significantly stimulated (fig. 15.32). This stimulation is apparently due to a direct effect of $P_{O_2}$ on the carotid bodies.

**Figure 15.32.** Comparison of the effects of increasing concentrations of $CO_2$ in air with decreasing concentrations of $O_2$ in air on respiration. Note that respiration increases linearly with increasing $CO_2$ concentration, whereas $O_2$ concentrations must decrease to half the normal value before respiration is stimulated.

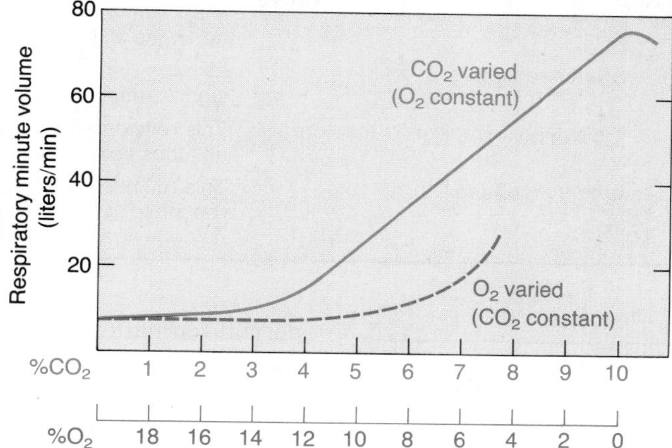


**Table 15.9** Definitions of some terms used to describe blood oxygen and carbon dioxide levels

| Term | Definition |
|---|---|
| Hypoxemia | The oxygen content or $P_{O_2}$ is lower than normal in arterial blood. |
| Hypoxia | The oxygen content or $P_{O_2}$ is lower than normal in the lungs, blood, or tissues. This is a more general term than hypoxemia. Tissues can be hypoxic, for example, even though there is no hypoxemia (as when the blood flow is occluded). |
| Hypercapnia, or hypercarbia | The $P_{CO_2}$ of systemic arteries is higher than 40 mm Hg. Usually this occurs when the ventilation is inadequate for a given metabolic rate (hypoventilation). Antonyms are *hypocapnia* and *hypocarbia* (usually produced by hyperventilation). |

**Table 15.10** The sensitivity of chemoreceptors to changes in blood gases and pH

| Stimulus | Chemoreceptor(s) | Comments |
|---|---|---|
| ↑ $P_{CO_2}$ | Medulla oblongata; aortic and carotid bodies | Medullary chemoreceptors are sensitive to the pH of cerebrospinal fluid (CSF). Diffusion of $CO_2$ from the blood into the CSF lowers the pH of CSF by forming carbonic acid. Similarly, the aortic and carotid bodies are stimulated by a fall in blood pH induced by increases in blood $CO_2$. |
| ↓ pH | Aortic and carotid bodies | Peripheral chemoreceptors are stimulated by decreased blood pH independent of the effect of blood $CO_2$. Chemoreceptors in the medulla are not affected by changes in blood pH because $H^+$ cannot cross the blood-brain barrier. |
| ↓ $P_{O_2}$ | Carotid bodies | Low blood $P_{O_2}$ (hypoxemia) augments the chemoreceptor response to blood $P_{CO_2}$ and can stimulate ventilation directly when the $P_{O_2}$ falls below 50 mm Hg. |

Since this degree of *hypoxemia,* or low blood oxygen (table 15.9), does not normally occur even in breath-holding, $P_{O_2}$ does not normally exert this direct effect on breathing.

In emphysema, when there is a chronic retention of carbon dioxide, the chemoreceptors eventually lose their ability to respond to increases in the arterial $P_{CO_2}$. The abnormally high $P_{CO_2}$, however, enhances the sensitivity of the carotid bodies to a fall in $P_{O_2}$. For people with emphysema, breathing may thus be stimulated by a *hypoxic drive* rather than by increases in blood $P_{CO_2}$.

The effect of changes in the blood $P_{CO_2}$, pH, and $P_{O_2}$ on chemoreceptors and the regulation of ventilation are summarized in table 15.10.

**Table 15.11** Pulmonary receptors and reflexes served by sensory nerve fibers in the vagus (tenth cranial nerve)

| Stimulus | Receptors | Comments |
|---|---|---|
| Stretch of lungs at inspiration | Stretch receptors | The Hering-Breuer reflex. This reflex is believed to be important in the control of breathing in the newborn but not in the adult at normal tidal volumes. |
| Stretch of lungs at expiration | Stretch receptors | Like the Hering-Breuer reflex, this is not believed to be significant in adults at normal tidal volumes. |
| Pulmonary congestion | Type J (juxta-capillary) receptors | This reflex may produce feelings of dyspnea at high altitudes and during severe exercise. |
| Irritation | Irritant receptors | This reflex causes reflex constriction of bronchioles in response to irritation from smoke, smog, and other noxious agents. |

There are a variety of disease processes that can produce cessation of breathing during sleep, or *sleep apnea.* **Sudden infant death syndrome (SIDS)** is an especially tragic form of sleep apnea that claims the lives of about ten thousand babies annually in the United States. Victims of this condition are apparently healthy two-to-five-month-old babies who die in their sleep without apparent reason—hence, the layman's term of *crib death.* These deaths seem to be caused by failure of the respiratory control mechanisms in the brain stem and/or by failure of the carotid bodies to be stimulated by reduced arterial oxygen.

Abnormal breathing patterns often appear prior to death by brain damage or heart disease. The most common of these abnormal patterns is **Cheyne-Stokes breathing,** in which the depth of breathing progressively increases and then progressively decreases. These cycles of increasing and decreasing tidal volumes may be followed by periods of apnea of varying durations. Cheyne-Stokes breathing may be caused by neurological damage or by insufficient oxygen delivery to the brain. The latter may result from heart disease or from a brain tumor that diverts a large part of the vascular supply from the respiratory centers.

### Pulmonary Stretch and Irritant Reflexes

The lungs contain various types of receptors that influence the brain stem respiratory control centers via sensory fibers in the vagus (table 15.11). Irritant receptors in the lungs, for example, stimulate reflex constriction of the bronchioles in response to smoke and smog. Similarly, sneezing, sniffing, and coughing may be stimulated by irritant receptors in the nose, larynx, and trachea.

The **Hering-Breuer reflex** is stimulated by pulmonary stretch receptors. The activation of these receptors during inspiration inhibits the respiratory control centers, making further inspiration increasingly difficult. This helps to prevent undue distension of the lungs and may contribute to the smoothness of the ventilation cycles. A similar inhibitory reflex may occur during expiration. The Hering-Breuer reflex appears to be important in the control of normal ventilation in the newborn. Pulmonary stretch receptors in adults, however, are probably not active at normal resting tidal volumes (500 ml per breath) but may contribute to respiratory control at high tidal volumes, as during exercise.

1. *Describe the effects of voluntary hyperventilation and breath-holding on arterial $P_{CO_2}$, pH, and oxygen content. Indicate the relative degree of changes in these values.*
2. *Using a flowchart to show a negative feedback loop, explain the relationship between ventilation and arterial $P_{CO_2}$.*
3. *Explain the effect of increased arterial $P_{CO_2}$ on (1) chemoreceptors in the medulla oblongata and (2) chemoreceptors in the aortic and carotid bodies.*
4. *Explain the role of arterial $P_{O_2}$ in the regulation of breathing. Explain why ventilation increases when a person goes to a high altitude.*

## Hemoglobin and Oxygen Transport

Hemoglobin without oxygen, or deoxyhemoglobin, can bond to oxygen to form oxyhemoglobin. This "loading" reaction occurs in the capillaries of the lungs. The dissociation of oxyhemoglobin, or "unloading" reaction, occurs in the tissue capillaries. Dissociation of oxyhemoglobin occurs to a greater extent, providing more oxygen to the tissue, when the $P_{O_2}$ of tissue capillaries is lower and when the bond strength (affinity) between hemoglobin and oxygen is weaker. A weaker affinity, and thus improved unloading of oxygen, occurs when there is a fall in pH or an increase in temperature. The flexibility in oxygen transport and delivery by hemoglobin is a result of the fact that hemoglobin is composed of four specific polypeptide chains.

If the lungs are functioning properly, blood leaving in the pulmonary veins and traveling in the systemic arteries has a $P_{O_2}$ of about 100 mm Hg, indicating a plasma oxygen concentration of about 0.3 ml $O_2$ per 100 ml blood. The

**Figure 15.33.** Plasma and whole blood that are brought into equilibrium with the same gas mixture have the same $P_{O_2}$ and thus the same amount of dissolved oxygen molecules (shown as black dots). The oxygen content of whole blood, however, is much higher than that of plasma because of the binding of oxygen to hemoglobin.

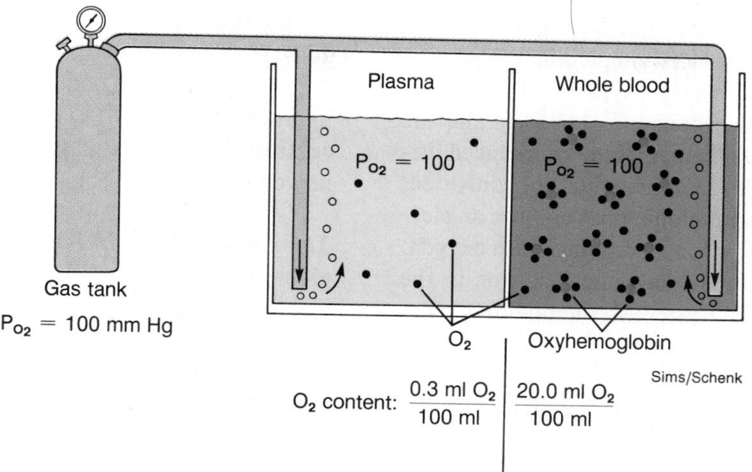


**Figure 15.34.** (*a*) An illustration of the three-dimensional structure of hemoglobin, in which the two alpha and two beta polypeptide chains are shown; the four heme groups are represented as flat structures with iron (*dark spheres*) in the centers. (*b*) The chemical structure of heme.

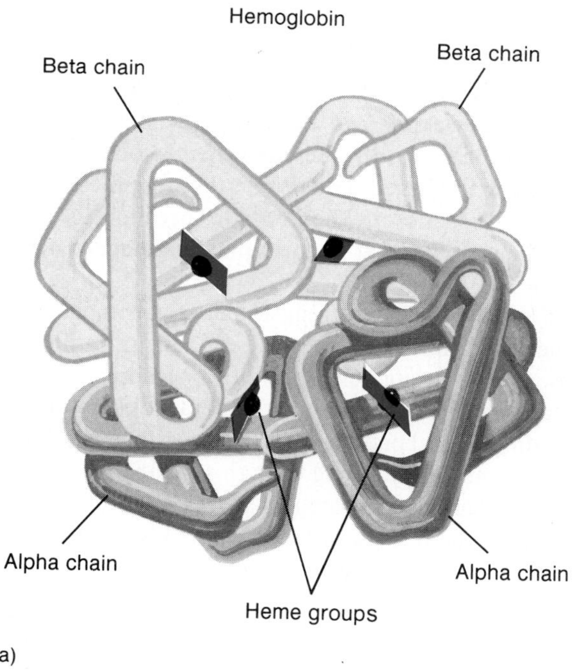


(a)

$CH_3$ $CH=CH_2$
$CH_3-C$ $C=C-CH_3$
N ---- Fe — N
$CH_2-C$ $C=C-CH=CH_2$
$CH_2$ $HC=C$ $C=CH$
COOH
$CH_2$ $CH_3$
$CH_2$
COOH

(b)

total oxygen content of the blood, however, cannot be derived from knowing only the $P_{O_2}$ of plasma. The total oxygen content depends not only on the $P_{O_2}$ but also on the hemoglobin concentration. If the $P_{O_2}$ and hemoglobin concentration are normal, arterial blood contains about 20 ml of $O_2$ per 100 ml of blood (fig. 15.33).

## Hemoglobin

Most of the oxygen in the blood is contained within the red blood cells, where it is chemically bonded to **hemoglobin.** Each hemoglobin molecule consists of (1) a protein globin part, composed of four polypeptide chains; and (2) four nitrogen-containing, disc-shaped organic pigment molecules, called *hemes* (fig. 15.34).

The protein part of hemoglobin is composed of two identical *alpha chains,* which are each 141 amino acids long, and two identical *beta chains,* which are each 146 amino acids long. Each of the four polypeptide chains is combined with one heme group. In the center of each heme group is one atom of iron, which can combine with one molecule of oxygen ($O_2$). One hemoglobin molecule can thus combine with four molecules of oxygen; since there are about 280 million hemoglobin molecules per red blood cell, each red blood cell can carry over a billion molecules of oxygen.

Normal heme contains iron in the reduced form ($Fe^{++}$, or ferrous iron). In this form, the iron can share electrons and bond to oxygen to form **oxyhemoglobin.** When oxyhemoglobin dissociates to release oxygen to the tissues, the heme iron is still in the reduced ($Fe^{++}$) form and the hemoglobin is called **deoxyhemoglobin,** or **reduced hemoglobin.** The term *oxyhemoglobin* is thus not equivalent to *oxidized* hemoglobin; hemoglobin does not lose an electron (and become oxidized) when it combines with oxygen. Oxidized hemoglobin, or **methemoglobin,** has iron in the oxidized ($Fe^{+++}$, or ferric) state. Methemoglobin thus lacks the electron it needs to form a bond with oxygen and cannot participate in oxygen transport. Blood normally contains only a small amount of methemoglobin, but certain drugs can increase this amount.

**Carboxyhemoglobin** is another abnormal form of hemoglobin, in which the reduced heme is combined with *carbon monoxide* instead of oxygen. Since the bond with carbon monoxide is about 210 times stronger than the bond with oxygen, carbon monoxide tends to displace oxygen in hemoglobin and remains attached to hemoglobin as the blood passes through systemic capillaries. The transport of oxygen to the tissues is thus reduced in carbon monoxide poisoning.

According to federal standards, the percent carboxyhemoglobin in the blood of active nonsmokers should not be higher than 1.5%. Studies have shown, however, that these values can range as high as 3% or more in nonsmokers and as high as 10% or more in smokers in some cities. Although these high levels may not cause immediate problems in healthy people, long-term effects on health might occur. People with respiratory or cardiovascular diseases would be particularly vulnerable to the negative effects of carboxyhemoglobin on oxygen transport.

***Hemoglobin Concentration.*** The *oxygen-carrying capacity* of whole blood is determined by its concentration of normal hemoglobin. If the hemoglobin concentration is below normal—a condition called **anemia**—the oxygen concentration of the blood is reduced below normal. Conversely, when the hemoglobin concentration is increased above the normal range—as occurs in **polycythemia** (high red blood cell count)—the oxygen-carrying capacity of blood is increased accordingly. This can occur as an adaptation to life at a high altitude.

The production of hemoglobin and red blood cells in bone marrow is controlled by a hormone called **erythropoietin,** produced primarily by the kidneys (as described in chapter 13). The production of erythropoietin—and thus the production of red blood cells—is stimulated when the delivery of oxygen to the kidneys and other organs is lower than normal. Red blood cell production is also promoted by androgens, which explains why the hemoglobin concentration in men averages 1–2 g per 100 ml higher than in women.

***The Loading and Unloading Reactions.*** Deoxyhemoglobin and oxygen combine to form oxyhemoglobin; this is called the **loading reaction.** Oxyhemoglobin, in turn, dissociates to yield deoxyhemoglobin and free oxygen molecules; this is the **unloading reaction.** The loading reaction occurs in the lungs and the unloading reaction occurs in the systemic capillaries.

Loading and unloading can thus be shown as a reversible reaction:

$$\text{Deoxyhemoglobin} + O_2 \underset{\text{(tissues)}}{\overset{\text{(lungs)}}{\rightleftarrows}} \text{Oxyhemoglobin}$$

The extent that the reaction will go in each direction depends on two factors: (1) the $P_{O_2}$ of the environment and (2) the *affinity,* or bond strength, between hemoglobin and oxygen. High $P_{O_2}$ drives the equation to the right (favors the loading reaction); at the high $P_{O_2}$ of the pulmonary capillaries almost all the deoxyhemoglobin molecules combine with oxygen. Low $P_{O_2}$ in the systemic capillaries drives the reaction in the opposite direction to promote unloading. The extent of this unloading depends on how low the $P_{O_2}$ values are.

The affinity (bond strength) between hemoglobin and oxygen also influences the loading and unloading reactions. A very strong bond would favor loading but inhibit unloading; a weak bond would hinder loading but improve unloading. The bond strength between hemoglobin and oxygen is normally strong enough so that 97% of the hemoglobin leaving the lungs is in the form of oxyhemoglobin, yet the bond is sufficiently weak so that adequate amounts of oxygen are unloaded to sustain aerobic respiration in the tissues. Under normal resting conditions only about 22% of the oxygen is unloaded; this satisfies the tissue needs for oxygen yet maintains an oxygen reserve in the blood for emergency conditions.

## The Oxyhemoglobin Dissociation Curve

Blood in the systemic arteries, at a $P_{O_2}$ of 100 mm Hg, has a *percent oxyhemoglobin saturation* of 97% (which means that 97% of the hemoglobin is in the form of oxyhemoglobin). This blood is delivered to the systemic capillaries, where oxygen diffuses into the tissue cells and is consumed in aerobic respiration. Blood leaving in the systemic veins is thus reduced in oxygen; it has a $P_{O_2}$ of about

**Figure 15.35.** The percent of oxyhemoglobin saturation and the blood oxygen content are shown at different values of $P_{O_2}$. Notice that there is about a 25% decrease in percent oxyhemoglobin as the blood passes through the tissue from arteries to veins, resulting in the unloading of approximately 5 ml $O_2$ per 100 ml to the tissues.

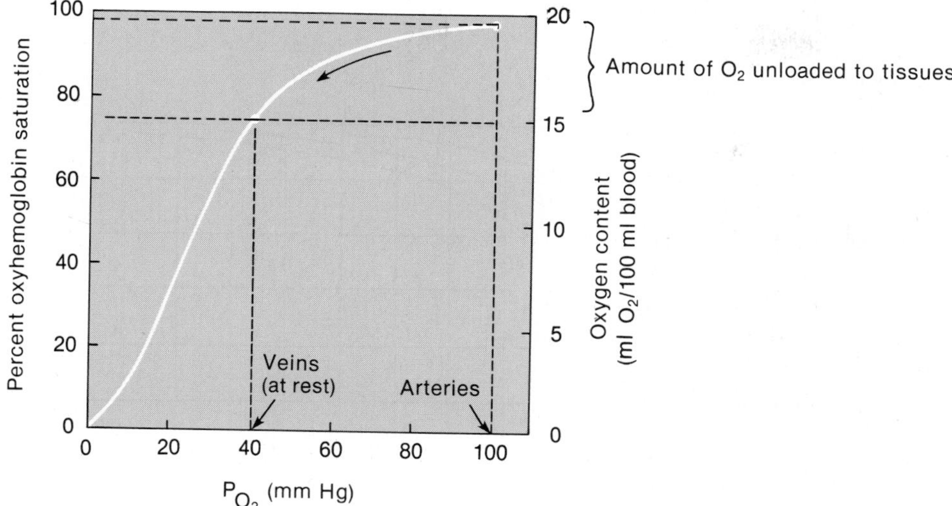


**Table 15.12** The relationship between percent oxyhemoglobin saturation and $P_{O_2}$ (at pH = 7.40 and temperature = 37°C)

| | | | | | | | | | | | |
|---|---|---|---|---|---|---|---|---|---|---|---|
| **$P_{O_2}$(mm Hg)** | 100 | 80 | 61 | 45 | 40 | 36 | 30 | 26 | 23 | 21 | 19 |
| **Percent oxyhemoglobin** | 97 | 95 | 90 | 80 | 75 | 70 | 60 | 50 | 40 | 35 | 30 |
| | Arterial blood | | | | Venous blood | | | | | | |

40 mm Hg and a percent saturation of about 75% (table 15.12). In other words, blood entering the tissues contains 20 ml $O_2$ per 100 ml blood, and blood leaving the tissues contains 15.5 ml $O_2$ per 100 ml blood (fig. 15.35). Thus, 22%, or 4.5 ml of $O_2$ out of 20 ml $O_2$ per 100 ml blood, is unloaded to the tissues.

A graphic illustration of the percent oxyhemoglobin saturation at different values of $P_{O_2}$ is called an **oxyhemoglobin dissociation curve** (fig. 15.35). The values in this graph are obtained by subjecting samples of blood *in vitro* (outside the body) to different partial oxygen pressures. The percent oxyhemoglobin saturations obtained by this procedure, however, can be used to predict what the percent unloading would be *in vivo* (within the body) with a given difference in arterial and venous $P_{O_2}$ values.

Figure 15.35 shows the difference between the arterial and venous $P_{O_2}$ and the percent oxyhemoglobin saturation at rest. The relatively large amount of oxyhemoglobin remaining in the venous blood at rest functions as an oxygen reserve. If a person stops breathing, there is a sufficient reserve of oxygen in the blood to keep the brain and heart alive for approximately four to five minutes in the absence of cardiopulmonary resuscitation (CPR) techniques. This reserve supply of oxygen can also be tapped when the tissue's requirements for oxygen are raised.

The oxyhemoglobin dissociation curve is S-shaped, or *sigmoidal.* The fact that it is relatively flat at high $P_{O_2}$ values indicates that changes in $P_{O_2}$ within this range have little effect on the loading reaction. One would have to ascend as high as 10,000 feet, for example, before the oxyhemoglobin saturation of arterial blood would decrease from 97% to 93%. At more common elevations the percent oxyhemoglobin saturation would not be significantly different from the 97% value at sea level.

At the steep part of the sigmoidal curve, however, small changes in $P_{O_2}$ values produce large differences in percent saturation. A decrease in *venous* $P_{O_2}$ from 40 mm Hg to 30 mm Hg, as might occur during mild exercise, corresponds to a change in percent saturation from 75% to 58%. Since the *arterial* percent saturation is usually still 97% during exercise, this change in venous percent saturation indicates that more oxygen has been unloaded to the tissues. The difference between the arterial and venous percent saturations indicates the percent unloading: in these examples, 97% minus 75% = 22% unloading at rest, and 97% minus 58% = 39% unloading during mild exercise. During heavier exercise, the venous $P_{O_2}$ can drop to 20 mm Hg or less, indicating a percent unloading in excess of 70%.

**Figure 15.36.** A decrease in blood pH (an increase in $H^+$ concentration) decreases the affinity of hemoglobin for oxygen at each $P_{O_2}$ value, resulting in a "shift to the right" of the oxyhemoglobin dissociation curve. A curve that is shifted to the right has a lower percent oxyhemoglobin saturation at each $P_{O_2}$, but the effect is more marked at lower $P_{O_2}$ values. This is called the *Bohr effect.*

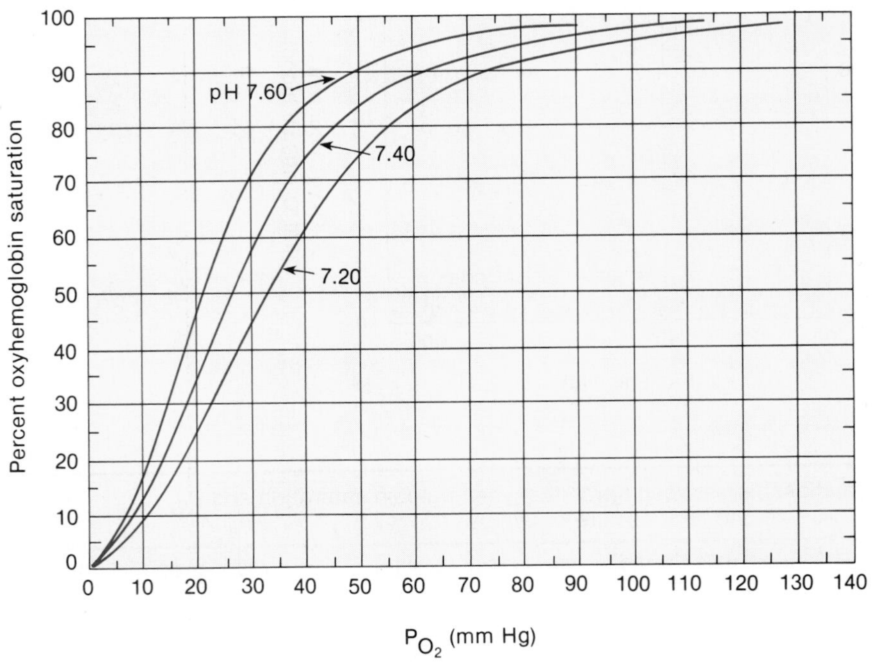


**Table 15.13** Effect of pH on hemoglobin affinity for oxygen and unloading of oxygen to the tissues

| pH | Affinity | Arterial $O_2$ Content per 100 ml | Venous $O_2$ Content per 100 ml | $O_2$ Unloaded to Tissues per 100 ml |
|---|---|---|---|---|
| 7.40 | Normal | 19.8 ml $O_2$ | 14.8 ml $O_2$ | 5.0 ml $O_2$ |
| 7.60 | Increased | 20.0 ml $O_2$ | 17.0 ml $O_2$ | 3.0 ml $O_2$ |
| 7.20 | Decreased | 19.2 ml $O_2$ | 12.6 ml $O_2$ | 6.6 ml $O_2$ |

## Effect of pH and Temperature on Oxygen Transport

In addition to changes in $P_{O_2}$, the loading and unloading reactions are influenced by changes in the bond strength, or affinity, of hemoglobin for oxygen. The affinity is decreased when the pH is lowered and increased when the pH is raised; this is called the **Bohr effect.** When the affinity of hemoglobin for oxygen is reduced, there is less loading of the blood with oxygen in the lungs but greater unloading of oxygen in the tissues. A weakening of the bond between hemoglobin and oxygen, however, has less effect at the higher $P_{O_2}$ values in the lungs than at the lower $P_{O_2}$ values of the tissues. A lowering of pH thus results in slightly less oxygen loading in the lungs but significantly more oxygen unloading in the tissues. The net effect is that the tissues receive more oxygen when the blood pH is lowered (table 15.13). Since the pH can be decreased by carbon dioxide (through the formation of carbonic acid), the Bohr effect helps to provide a little more oxygen to the tissues when their carbon dioxide output (and metabolism) is increased.

When the percent oxyhemoglobin saturation at different pH values is graphed as a function of $P_{O_2}$, the dissociation curve is shown to be shifted to the right by a lowering of pH and shifted to the left by a rise in pH (fig. 15.36). If the percent unloading is calculated by subtracting the percent oxyhemoglobin saturation at given $P_{O_2}$ values for arterial and venous blood, it will be clear that a *shift to the right* of the curve indicates a greater oxygen unloading, whereas a *shift to the left* indicates less unloading but slightly more oxygen loading in the lungs.

When oxyhemoglobin dissociation curves are constructed at constant pH values but at different temperatures, it can be seen that the affinity of hemoglobin for oxygen is decreased by a rise in temperature. An increase in temperature weakens the bond between hemoglobin and oxygen and thus has the same effect as a fall in pH; the

**Figure 15.37.** The oxyhemoglobin dissociation curve is shifted to the right as the temperature increases, indicating a lowered affinity of hemoglobin for oxygen at each $P_{O_2}$. This effect, like the Bohr effect (see figure 15.35), is more marked at lower $P_{O_2}$ values.

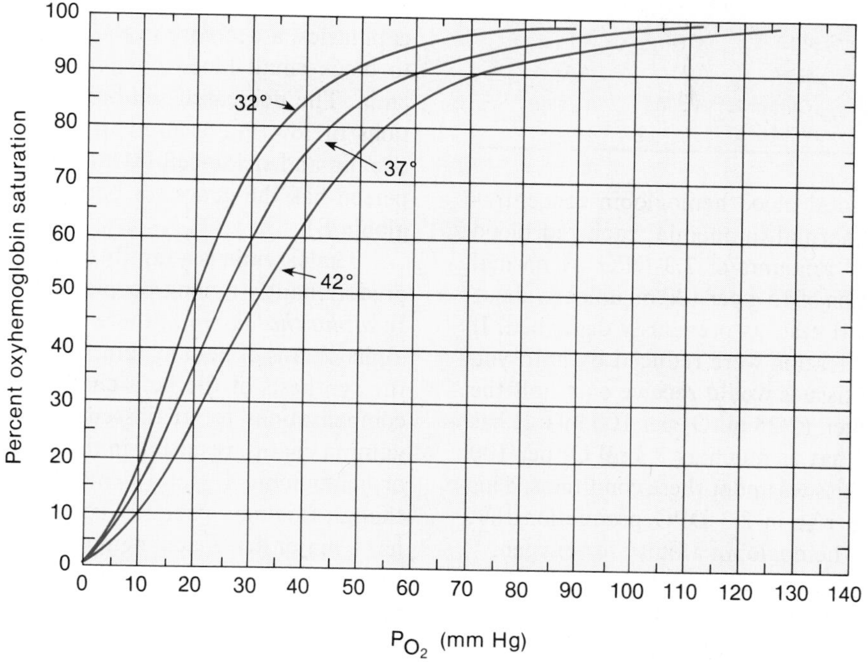


**Table 15.14** Factors that affect the affinity of hemoglobin for oxygen and the position of the oxyhemoglobin dissociation curve

| Factor | Affinity | Position of Curve | Comments |
|---|---|---|---|
| ↓ pH | Decreased | Shift to the right | Called the Bohr effect; increases oxygen delivery during hypercapnia |
| ↑ Temperature | Decreased | Shift to the right | Increases oxygen unloading during exercise and fever |
| ↑ 2,3-DPG | Decreased | Shift to the right | Increases oxygen unloading when there is a decrease in total hemoglobin or total oxygen content; an adaptation to anemia and high-altitude living |

oxyhemoglobin dissociation curve is shifted to the right (fig. 15.37). At higher temperatures, therefore, more oxygen is unloaded to the tissues than would be the case if the bond strength were constant. This effect can significantly increase the delivery of oxygen to muscles that are warmed during exercise.

## Effect of 2,3-DPG on Oxygen Transport

Mature red blood cells lack both nuclei and mitochondria. Without mitochondria they cannot respire aerobically; the very cells that carry oxygen are the only cells in the body that cannot use it! Red blood cells, therefore, must obtain energy through the anaerobic respiration of glucose. At a certain point in the glycolytic pathway there is a "side reaction" in red blood cells that results in a unique product—**2,3-diphosphoglyceric acid (2,3-DPG).**

The enzyme that produces 2,3-DPG is inhibited by oxyhemoglobin. When the oxyhemoglobin concentration is decreased, therefore, the production of 2,3-DPG is increased. This increase in 2,3-DPG production can occur when the total hemoglobin concentration is low (in anemia) or when the $P_{O_2}$ is low (at a high altitude, for example). 2,3-DPG combines with deoxyhemoglobin and makes it more stable. At the $P_{O_2}$ values in the tissue capillaries, therefore, a higher proportion of the oxyhemoglobin will be converted to deoxyhemoglobin by unloading its oxygen. An increased concentration of 2,3-DPG in red blood cells thus increases oxygen unloading (table 15.14) and shifts the oxyhemoglobin dissociation curve to the right.

The importance of 2,3-DPG within red blood cells is now recognized in blood banking. Old, stored red blood cells can lose their ability to produce 2,3-DPG as they lose their ability to metabolize glucose. Modern techniques for blood storage, therefore, include the addition of energy substrates for respiration and phosphate sources needed for the production of 2,3-DPG.

**Table 15.15** Some differences between normal hemoglobin A and two mutant hemoglobin forms

| | Hemoglobin | | |
|---|---|---|---|
| | A | S | C |
| DNA Base Triplet | CTT | CAT | TTT |
| mRNA Codon | GAA | GUA | AAA |
| Amino Acid in Position 6 | Glutamic acid | Valine | Lysine |

***Anemia.*** When the total blood hemoglobin concentration is reduced below normal in anemia, each red blood cell produces increased amounts of 2,3-DPG. A normal hemoglobin concentration of 15 g per 100 ml unloads about 4.5 ml $O_2$ per 100 ml at rest, as previously described. If the hemoglobin concentration were reduced by half, you might expect that the tissues would receive only half the normal amount of oxygen (2.25 ml $O_2$ per 100 ml). It has been shown, however, that as much as 3.3 ml $O_2$ per 100 ml are unloaded to the tissues under these conditions. This occurs as a result of a rise in 2,3-DPG production that produces a decrease in hemoglobin affinity for oxygen.

***Hemoglobin F.*** The effects of 2,3-DPG are also important in the transfer of oxygen from maternal to fetal blood. The mother has hemoglobin molecules composed of two alpha and two beta chains, as previously described, whereas the fetal hemoglobin contains two alpha and two *gamma chains* in place of beta chains (gamma chains differ from beta chains in 37 of their amino acids). Normal adult hemoglobin in the mother (*hemoglobin A*) is able to bond to 2,3-DPG. Fetal hemoglobin (*hemoglobin F*), in contrast, cannot bond to 2,3-DPG, and thus has a higher affinity for oxygen at a given $P_{O_2}$ than does hemoglobin A. Since hemoglobin F can have a higher percent oxyhemoglobin saturation than hemoglobin A at a given $P_{O_2}$, oxygen is transferred from the maternal to the fetal blood as these two come into close proximity in the placenta.

## Inherited Defects in Hemoglobin Structure and Function

There are a number of hemoglobin diseases that are produced by inherited (congenital) defects in the protein part of hemoglobin. **Sickle-cell anemia**—a disease found in one out of 625, and carried in a recessive state by 8%–11% of the black population of the United States—for example, is caused by an abnormal form of hemoglobin called *hemoglobin S.* Hemoglobin S differs from normal hemoglobin A in only one amino acid: valine is substituted for glutamic acid in position 6 on the beta chains. This amino acid substitution is caused by a single base change in the region of DNA that codes for the beta chains (table 15.15).

Under conditions of low blood $P_{O_2}$, hemoglobin S comes out of solution and cross links to form a "paracrystalline gel" within the red blood cells. This causes the characteristic sickle shape of red blood cells (fig. 15.38) and makes them less flexible. Since red blood cells must be able to bend in the middle to pass through many narrow capillaries, a decrease in their flexibility may cause them to block small blood channels and produce organ ischemia. The decreased solubility of hemoglobin S in solutions of low $P_{O_2}$ is used in the diagnosis of sickle-cell anemia and sickle-cell trait (the carrier state, in which a person has the genes for both hemoglobin A and hemoglobin S).

**Thalassemia** is a family of hemoglobin diseases found predominantly among people of Mediterranean ancestry. In *alpha thalassemia,* there is decreased synthesis of the alpha chains of hemoglobin, whereas in *beta thalassemia* the synthesis of the beta chains is impaired. One of the compensations for thalassemia is increased synthesis of gamma chains, resulting in the retention of large amounts of hemoglobin F (fetal hemoglobin) into adulthood. Although this may partially compensate for the anemia, it has a major drawback; hemoglobin F has a higher affinity for oxygen than hemoglobin A and thus cannot unload as much oxygen to the tissues.

Some types of abnormal hemoglobins have been shown to be advantageous in the environments in which they evolved. A person who is a carrier for sickle-cell anemia, for example (and who therefore has both hemoglobin A and hemoglobin S), has a high resistance to malaria. This is because the parasite that causes this disease cannot live in red blood cells that contain hemoglobin S.

## Muscle Myoglobin

**Myoglobin** is a red pigment found exclusively in striated muscle fibers. In particular, slow-twitch, aerobically respiring skeletal fibers and cardiac muscle fibers are rich in myoglobin. Myoglobin is similar to hemoglobin, but it has one rather than four hemes and, therefore, can combine with only one molecule of oxygen.

Myoglobin has a higher affinity for oxygen than does hemoglobin, and its dissociation curve is therefore to the left of the oxyhemoglobin dissociation curve (fig. 15.39). The shape of the myoglobin curve is also different from the oxyhemoglobin dissociation curve; it is rectangular, indicating that oxygen will be released only when the $P_{O_2}$ gets very low. The differences in behavior between myoglobin and hemoglobin highlight the importance of the tetrameric (four-subunit) structure of hemoglobin. When one heme group of hemoglobin combines with oxygen a shape change occurs that allows the other heme groups to combine with oxygen more easily. The same interaction beween the four subunits of hemoglobin allows the hemoglobin to unload its oxygen at higher values of $P_{O_2}$ than can myoglobin.

**Figure 15.38.** (*a*) A sickled red blood cell as seen in the light microscope. (*b*) Normal cells. (*c*). Sickled red blood cells as seen in the scanning electron microscope.

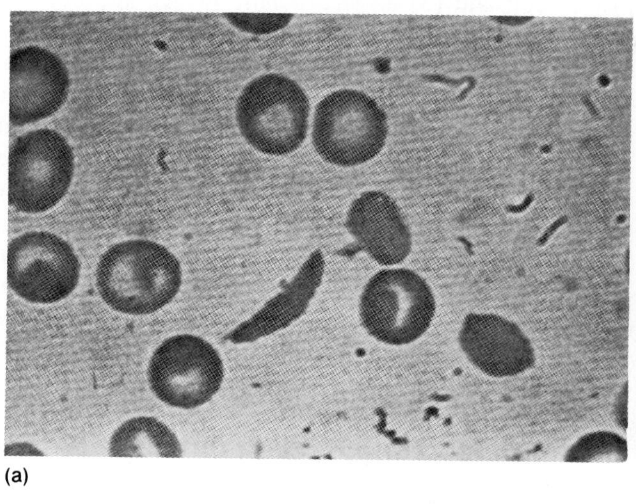

(a)

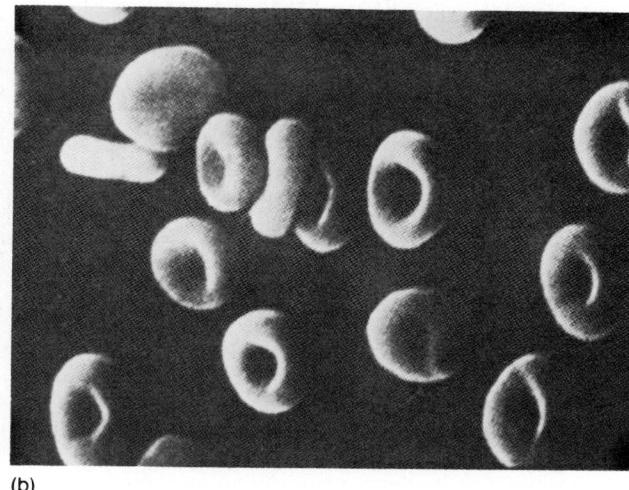

(b)

(c)

**Figure 15.39.** A comparison of the dissociation curves for hemoglobin and for myoglobin. At the $P_{O_2}$ of venous blood, the myoglobin retains almost all of its oxygen, indicating a higher affinity than hemoglobin for oxygen. The myoglobin does, however, release its oxygen at the very low $P_{O_2}$ values found inside the mitochondria.

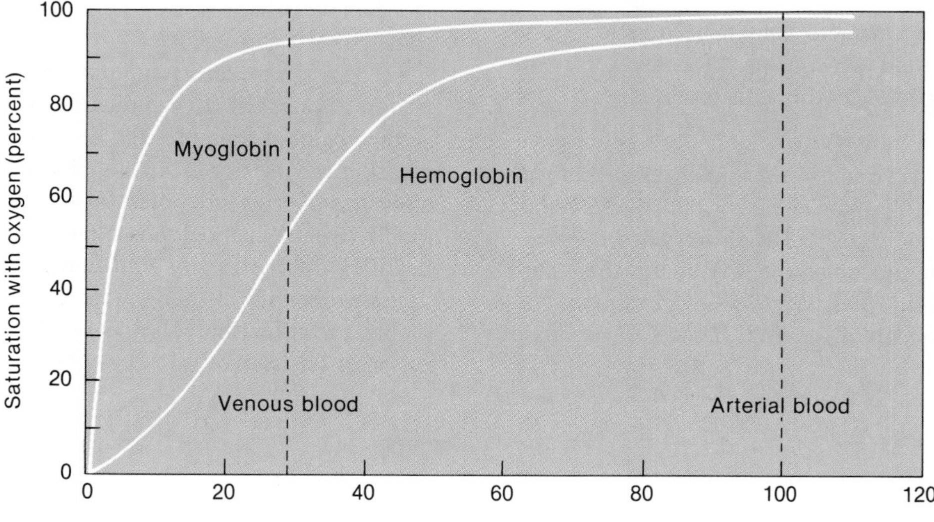

Since the $P_{O_2}$ in mitochondria is very low (because oxygen is incorporated into water here), myoglobin may act as a "middleman" in the transfer of oxygen from blood to the mitochondria within muscle cells. Myoglobin may also have an oxygen storage function, which is of particular importance in the heart. During diastole, when the coronary blood flow is greatest, the myoglobin can load up with oxygen. This stored oxygen can then be released during systole, when the coronary arteries are squeezed closed by the contracting myocardium.

---

1. *Describe the effect of $P_{O_2}$ on the loading and unloading reactions. Illustrate this with a graph, and label the regions of the graph that show when loading and unloading occur.*
2. *Draw an oxyhemoglobin dissociation curve, and label the values found in the arterial blood and venous blood under resting conditions. Use this graph to show an increase in the percent unloading that can occur during exercise.*
3. *Describe how changes in pH and temperature affect the hemoglobin affinity for oxygen and the position of the oxyhemoglobin dissociation curve. Explain the effect of these changes on oxygen transport.*
4. *Explain how a person who is anemic or a person at high altitude could have an increase in the percent unloading of oxygen by hemoglobin.*

---

## *Carbon Dioxide Transport and Acid-Base Balance*

Carbon dioxide is transported in the blood primarily in the form of bicarbonate ($HCO_3^-$), which is released when carbonic acid dissociates. Bicarbonate is produced in the red blood cells from carbonic acid, which is formed as a result of the catalytic action of an enzyme found in the red blood cells but not in plasma. Bicarbonate is an anion that can buffer $H^+$, and thus helps to maintain a normal arterial blood pH. Hypoventilation raises, and hyperventilation lowers, the carbonic acid concentration of the blood. Hypoventilation can thus produce respiratory acidosis, whereas hyperventilation can produce respiratory alkalosis.

Carbon dioxide is carried by the blood in three forms: (1) as *dissolved $CO_2$*—since carbon dioxide is about twenty-one times more soluble than oxygen in water, a substantial portion (about one-tenth) of the total blood $CO_2$ is dissolved in plasma; (2) as *carbaminohemoglobin*—about one-fifth of the total blood $CO_2$ is carried attached to an amino acid in hemoglobin (carbaminohemoglobin should not be confused with carboxyhemoglobin, which is a combination of hemoglobin with carbon monoxide); and (3) as *bicarbonate,* which accounts for most of the $CO_2$ carried by the blood.

Carbon dioxide is able to combine with water to form carbonic acid. This reaction occurs spontaneously in the plasma at a slow rate but occurs much more rapidly within the red blood cells due to the catalytic action of the enzyme **carbonic anhydrase.** Since this enzyme is confined to the red blood cells, most of the carbonic acid is produced there rather than in the plasma. The formation of carbonic acid from $CO_2$ and water is favored by the high $P_{CO_2}$ found in tissue capillaries (this is an example of the *law of mass action,* described in chapter 4).

$$CO_2 + H_2O \xrightarrow[\text{high } P_{CO_2}]{\text{carbonic anhydrase}} H_2CO_3$$

### The Chloride Shift

As a result of catalysis by carbonic anhydrase within the red blood cells, large amounts of carbonic acid are produced as blood passes through the systemic capillaries. The buildup of carbonic acid concentrations within the red blood cells favors the dissociation of these molecules into *$H^+$* (protons, which contribute to the acidity of a solution) and *$HCO_3^-$* (bicarbonate). The equation describing this reaction has been previously introduced.

The $H^+$ ions released by the dissociation of carbonic acid are largely buffered by their combination with deoxyhemoglobin within the red blood cells. Although the unbuffered $H^+$ ions are free to diffuse out of the red blood cells, more bicarbonate diffuses outward into the plasma than does $H^+$. As a result of the "trapping" of $H^+$ ions within the red blood cells by their attachment to hemoglobin and the outward diffusion of bicarbonate, the inside of the red blood cell gains a net positive charge. This attracts chloride ions ($Cl^-$), which move into the red blood cells as $HCO_3^-$ moves out. This exchange of anions as blood moves through the tissue capillaries is called the **chloride shift** (fig. 15.40).

There is an interesting interaction between the transport of oxygen and carbon dioxide. The bonding of $H^+$ by oxyhemoglobin promotes the unloading of oxygen by decreasing the affinity of hemoglobin for oxygen (the Bohr effect), and thus promotes the conversion of oxyhemoglobin to deoxyhemoglobin. Now, since deoxyhemoglobin bonds $H^+$ more strongly than does oxyhemoglobin, the act of unloading its oxygen improves the ability of hemoglobin to buffer the $H^+$ released by carbonic acid. Removal of $H^+$ from solution by combining with hemoglobin

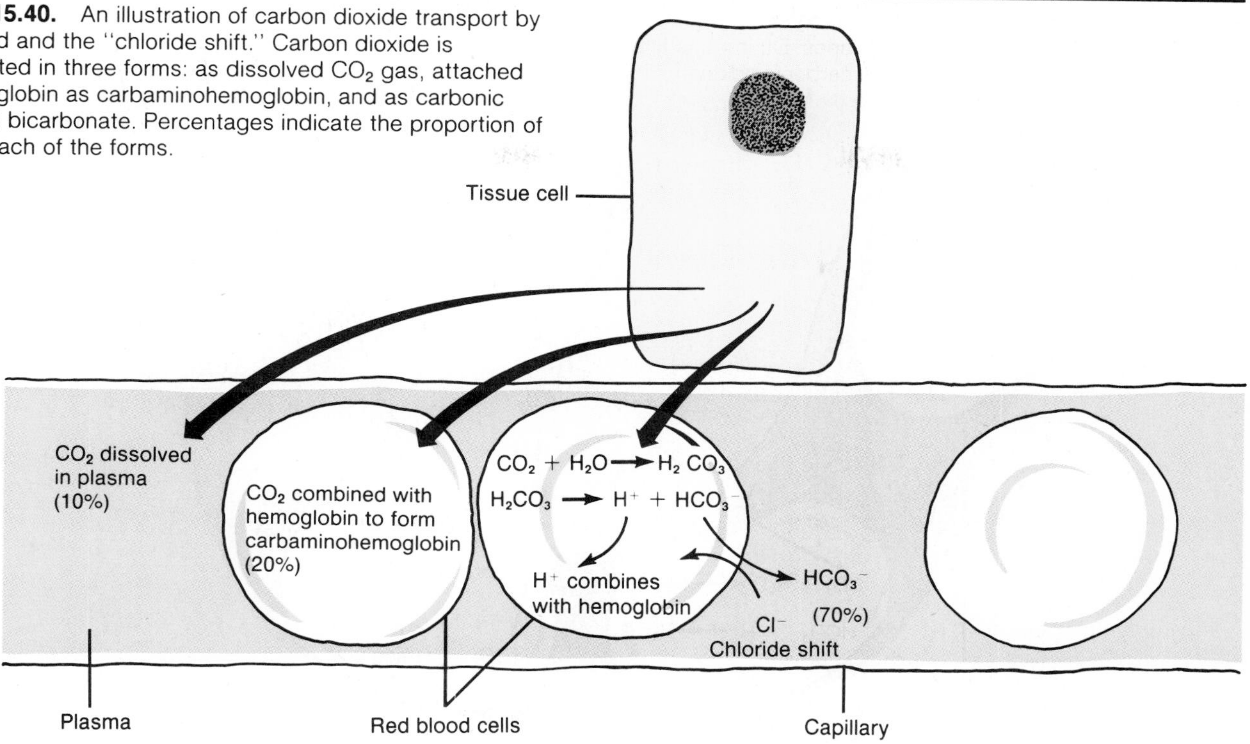

**Figure 15.40.** An illustration of carbon dioxide transport by the blood and the "chloride shift." Carbon dioxide is transported in three forms: as dissolved $CO_2$ gas, attached to hemoglobin as carbaminohemoglobin, and as carbonic acid and bicarbonate. Percentages indicate the proportion of $CO_2$ in each of the forms.



(by the law of mass action), in turn, favors the continued production of carbonic acid. The formation of carbonic acid, in summary, enhances oxygen unloading to the tissues (the Bohr effect), while oxygen unloading (through the increased formation of deoxyhemoglobin) improves the ability of the blood to form carbonic acid and thus transport carbon dioxide.

When blood reaches the pulmonary capillaries, deoxyhemoglobin is converted to oxyhemoglobin. Since oxyhemoglobin has a lower affinity for $H^+$ than does deoxyhemoglobin, $H^+$ are released within the red blood cells. This attracts $HCO_3^-$ from the plasma, which combines with $H^+$ to form carbonic acid:

$$H^+ + HCO_3^- \longrightarrow H_2CO_3$$

Under conditions of low $P_{CO_2}$, as occurs in the pulmonary capillaries, carbonic anhydrase catalyzes the conversion of carbonic acid to carbon dioxide and water:

$$H_2CO_3 \xrightarrow[\text{low } P_{CO_2}]{\text{carbonic anhydrase}} CO_2 + H_2O$$

In summary, the carbon dioxide produced by the tissue cells is converted within the systemic capillaries, mostly through the action of carbonic anhydrase in the red blood cells, to carbonic acid. With the buildup of carbonic acid concentrations in the RBCs, the carbonic acid dissociates into bicarbonate and $H^+$, which results in the chloride shift.

A *reverse chloride shift* operates in the pulmonary capillaries to convert carbonic acid to $CO_2$ gas, which is eliminated in the expired breath (fig. 15.41). The $P_{CO_2}$, carbonic acid, $H^+$, and bicarbonate concentrations in the systemic arteries are thus maintained relatively constant by normal ventilation.

## Ventilation and Acid-Base Balance

Normal systemic arterial blood has a pH of 7.35 to 7.45. Using the definition of pH described in chapter 2, this means that arterial blood has a $H^+$ concentration of about $10^{-7.4}$ molar. Some of these $H^+$ are derived from carbonic acid, and some are derived from nonvolatile *metabolic acids* (fatty acids, ketone bodies, lactic acid, and others) that cannot be eliminated in the expired breath.

Under normal conditions the $H^+$ released by metabolic acids do not affect blood pH because these $H^+$ combine with buffers and are thereby removed from solution. The buffers of the blood include hemoglobin within the red blood cells (as previously described), and proteins and bicarbonate within the plasma. These buffers, however, would eventually become saturated if the body were a closed system that could not eliminate acids. The elimination of acids from the body is accomplished by the lungs, through exhalation of carbon dioxide, and by the kidneys, through excretion of $H^+$ in the urine (chapter 16).

**Figure 15.41.** Carbon dioxide is released from the blood as it travels through the pulmonary capillaries. During this time a "reverse chloride shift" occurs and carbonic acid is transformed into $CO_2$ and $H_2O$.

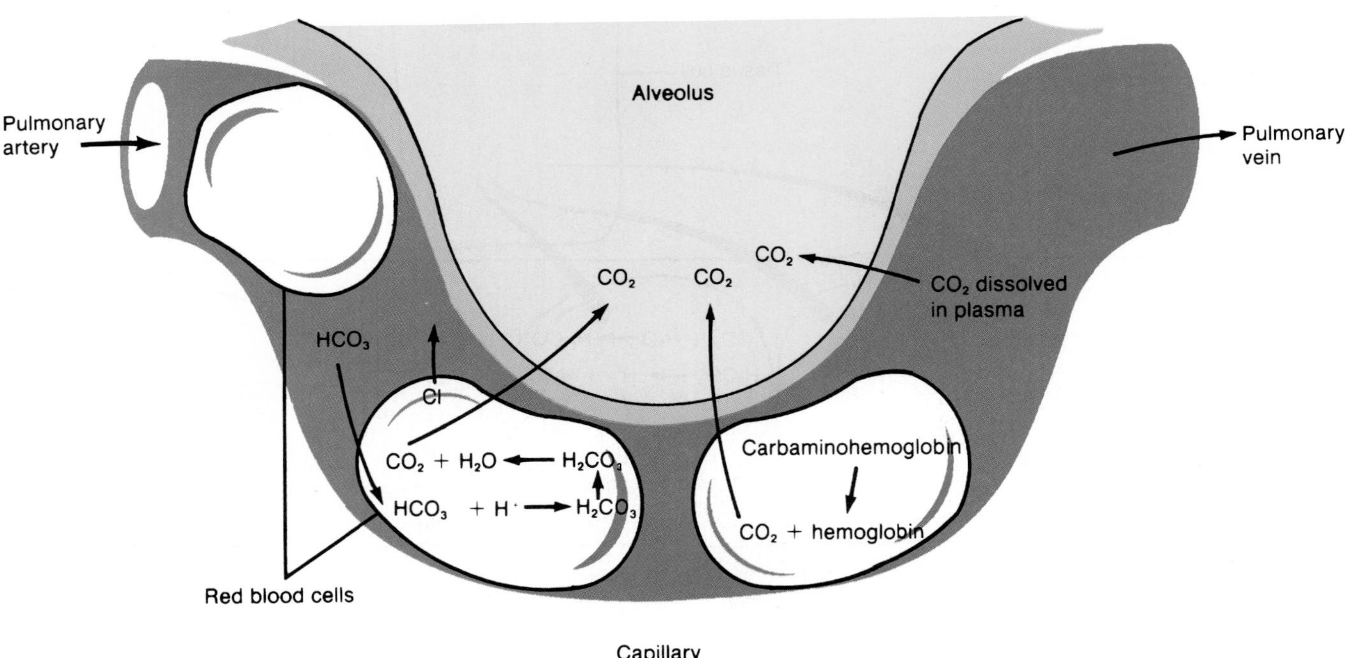


**Table 15.16** Definition of terms used to describe acid-base balance

| Terms | Definitions |
|---|---|
| Acidosis, respiratory | Increased carbon dioxide retention (due to hypoventilation), which can result in the accumulation of carbonic acid and thus a fall in blood pH below normal |
| Acidosis, metabolic | Increased production of "nonvolatile" acids such as lactic acid, fatty acids, and ketone bodies, or loss of blood bicarbonate (such as by diarrhea) resulting in a fall in blood pH below normal |
| Alkalosis, respiratory | A rise in blood pH due to loss of $CO_2$ and carbonic acid (through hyperventilation) |
| Alkalosis, metabolic | A rise in blood pH produced by loss of nonvolatile acids (as in excessive vomiting) or by excessive accumulation of bicarbonate base |
| Compensated acidosis or alkalosis | Metabolic acidosis or alkalosis are partially compensated by opposite changes in blood carbonic acid levels (through changes in ventilation). Respiratory acidosis or alkalosis are partially compensated by increased retention or excretion of bicarbonate in the urine. |

*Bicarbonate is the major buffer in the plasma* and acts to maintain a blood pH of 7.4 despite the constant production of nonvolatile metabolic acids by the tissues. In this buffering process, some of the $HCO_3^-$ released from the red blood cells during the chloride shift is converted into $H_2CO_3$ in the plasma. Normally, however, there is still a buffer reserve of free bicarbonate that can help protect against unusually large additions of metabolic acids to the blood. These processes are illustrated in figure 15.42.

Normal plasma, therefore, contains free bicarbonate, carbonic acid, and a $H^+$ concentration indicated by a pH of 7.4. If the $H^+$ concentration of the blood should fall, the carbonic acid produced by the buffering reaction can dissociate and serve as a source of additional $H^+$. If the $H^+$ concentration should rise, bicarbonate can remove this excess $H^+$ from solution. Carbonic acid and bicarbonate are thus said to function as a *buffer pair.*

***Acidosis and Alkalosis.*** A fall in blood pH below 7.35 is called **acidosis,** because the pH is to the acid-side of normal. Acidosis does not mean acidic (pH less than 7); a blood pH of 7.2, for example, represents serious acidosis. Similarly, a rise in blood pH above 7.45 is known as **alkalosis.** There are two components to acid-base balance: *respiratory* and *metabolic.* Respiratory acidosis and alkalosis are due to abnormal concentrations of carbonic acid, as a result of abnormal ventilation. Metabolic acidosis and alkalosis result from abnormal amounts of $H^+$ derived from nonvolatile metabolic acids (table 15.16).

**Figure 15.42.** Bicarbonate released into the plasma from red blood cells functions to buffer $H^+$ produced by the ionization of metabolic acids (lactic acid, fatty acids, ketone bodies, and others). Binding of $H^+$ to hemoglobin also promotes the unloading of $O_2$.

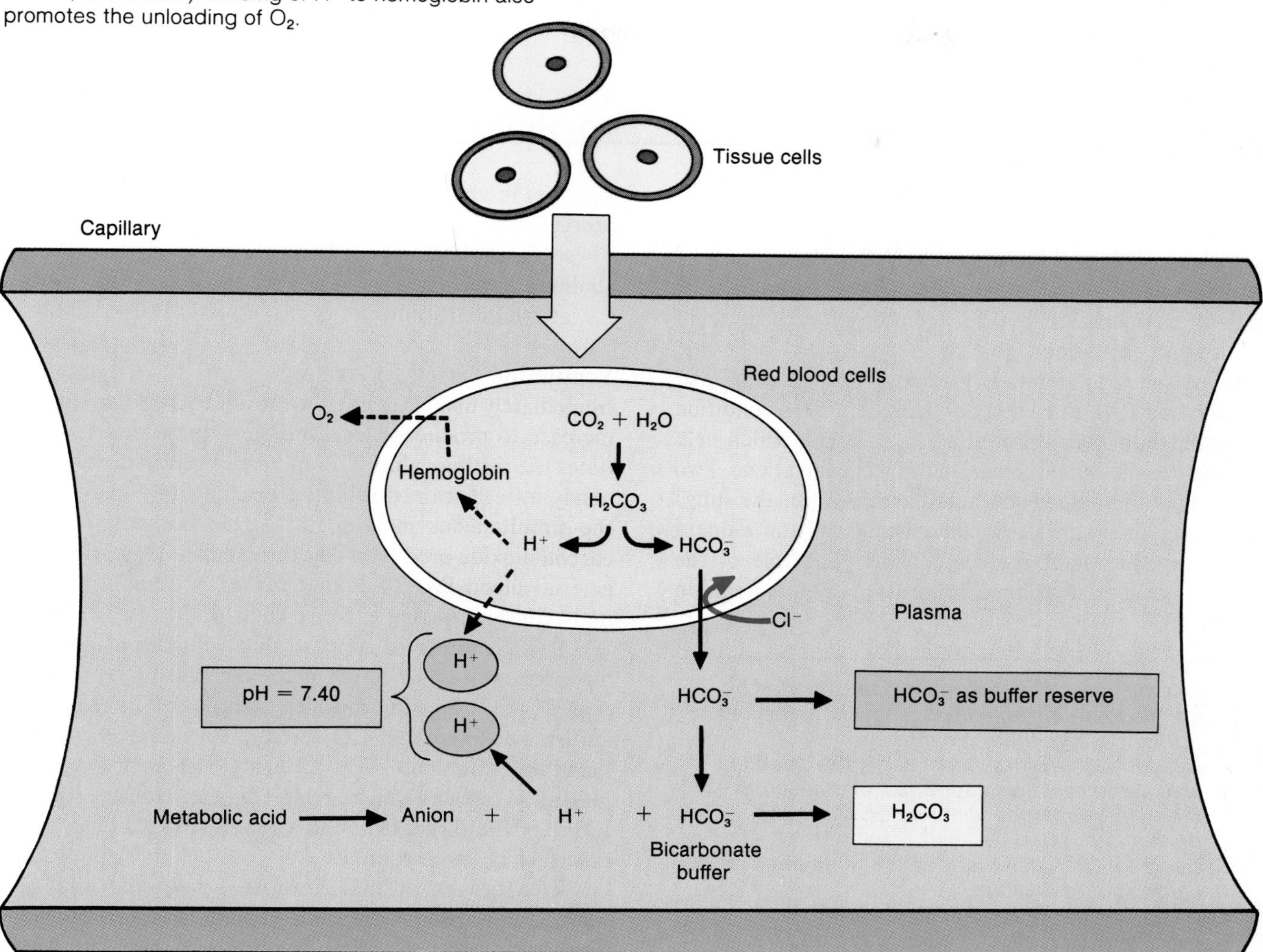


Ventilation is normally adjusted to keep pace with the metabolic rate. *Hypoventilation* occurs when the rate of $CO_2$ production exceeds the rate at which it is "blown off" in ventilation. Indeed, hypoventilation can be operationally defined as an abnormally high arterial $P_{CO_2}$. Under these conditions, carbonic acid production is excessively high and **respiratory acidosis** occurs. In *hyperventilation,* conversely, the rate of ventilation is greater than the rate of $CO_2$ production. Arterial $P_{CO_2}$ therefore decreases, causing less formation of carbonic acid than under normal conditions. The depletion of carbonic acid raises the pH, and **respiratory alkalosis** occurs.

**Metabolic acidosis** can occur when the production of nonvolatile acids is abnormally increased. In uncontrolled diabetes mellitus, for example, ketone bodies (derived from fatty acids) may accumulate and produce *ketoacidosis.* In order for metabolic acidosis to occur, however, the buffer reserve of bicarbonate must first be depleted (this is why metabolic acidosis can also be produced by the excessive loss of bicarbonate, as in diarrhea). Until the buffer reserve is depleted the pH remains normal—ketosis can occur, for example, without ketoacidosis. **Metabolic alkalosis,** a less common condition than metabolic acidosis, can result when there is a loss of acidic gastric juice by vomiting or when there is an excessive intake of bicarbonate (from stomach antacids or from an intravenous solution). Since the kidneys are responsible for regulating the metabolic component of acid-base balance, this topic will be covered in more detail in chapter 16.

***Compensations for Acidosis and Alkalosis.*** A change in blood pH, produced by alterations in either the respiratory or metabolic component of acid-base balance, can be compensated by a change in the other component. Metabolic acidosis, for example, stimulates hyperventilation (because the aortic and carotid bodies are sensitive to blood $H^+$) and thus causes a secondary respiratory alkalosis to be produced. The person is still acidotic, but not as much

**Table 15.17** The effect of lung function on blood acid-base balance

| Condition | pH | $P_{CO_2}$ | Ventilation | Cause or Compensation |
|---|---|---|---|---|
| Normal | 7.35–7.45 | 39–41 mm Hg | Normal | Not applicable |
| Respiratory acidosis | Low | High | Hypoventilation | Cause of the acidosis |
| Respiratory alkalosis | High | Low | Hyperventilation | Cause of the alkalosis |
| Metabolic acidosis | Low | Low | Hyperventilation | Compensation for acidosis |
| Metabolic alkalosis | High | High | Hypoventilation | Compensation for alkalosis |

so as would be the case without the compensation. People with compensated metabolic acidosis would thus have a low pH accompanied by a low blood $P_{CO_2}$ as a result of the hyperventilation. Metabolic alkalosis, similarly, is partially compensated by the retention of carbonic acid due to hypoventilation (table 15.17).

A person with respiratory acidosis has a low pH and a high blood $P_{CO_2}$ due to hypoventilation. This condition can be partially compensated by the kidneys, which help to regulate the blood bicarbonate concentration. Two organs thus regulate blood acid-base balance: the lungs (regulating the respiratory component) and the kidneys (regulating the metabolic component). The role of the kidneys in acid-base balance is discussed in more detail in chapter 16.

1. *List the ways that carbon dioxide is carried by the blood. Using equations, show how carbonic acid and bicarbonate are formed.*
2. *Describe the events that occur in the chloride shift in the systemic capillaries, and describe the reverse chloride shift that occurs in the pulmonary capillaries.*
3. *Describe the functions of bicarbonate and carbonic acid in blood.*
4. *Describe how hyperventilation and hypoventilation affect the blood pH, and explain the mechanisms involved.*
5. *Explain the mechanisms whereby a person with ketoacidosis hyperventilates. Explain the potential benefits of hyperventilation under these conditions.*

## *Effect of Exercise and High Altitude on Respiratory Function*

The arterial blood gases and pH do not significantly change during moderate exercise because ventilation increases to keep pace with increased metabolism. This increased ventilation requires neural feedback from the exercising muscles and chemoreceptor stimulation. Trained muscles can extract a higher proportion of the oxygen from the arterial blood, and thus lower the venous $P_{O_2}$ more than can untrained muscles. Adaptations occur at high altitude in both the control of ventilation and the oxygen transport abilities of blood to permit an adequate delivery of oxygen to the tissues.

Changes in ventilation and oxygen delivery occur during exercise and during acclimatization to a high altitude. These changes help to compensate for the increased metabolic rate during exercise and for the decreased arterial $P_{O_2}$ at a high altitude.

### Ventilation during Exercise

Immediately upon exercise, the rate and depth of breathing increase to produce a total minute volume that is many times the resting value. This increased ventilation, particularly in well-trained athletes, is exquisitely matched to the simultaneous increase in oxygen consumption and carbon dioxide production by the exercising muscles. The arterial blood $P_{O_2}$, $P_{CO_2}$, and pH, thus, remain surprisingly constant during exercise (fig. 15.43).

It is tempting to suppose that ventilation increases during exercise as a result of the increased $CO_2$ production by the exercising muscles. Ventilation increases together with increased $CO_2$ production, however, so that blood measurements of $P_{CO_2}$ during exercise are not significantly higher than at rest. The mechanisms responsible for the increased ventilation during exercise must therefore be more complex.

Two groups of mechanisms—*neurogenic* and *humoral*—have been proposed to explain the increased ventilation that occurs during exercise. Possible neurogenic mechanisms include the following: (1) sensory nerve activity from the exercising limbs may stimulate the respiratory muscles, either through spinal reflexes or via the brain stem respiratory centers; and/or (2) input from the cerebral cortex may stimulate the brain stem centers to modify ventilation. These neurogenic theories help to explain the immediate increase in ventilation that occurs at the beginning of exercise.

Rapid and deep ventilation continues after exercise has stopped, suggesting that humoral (chemical) factors in the blood may also stimulate ventilation during exercise. Since the $P_{O_2}$, $P_{CO_2}$, and pH of the blood samples from exercising subjects are within the resting range, these humoral theories propose that (1) the $P_{CO_2}$ and pH in the region of the chemoreceptors may be different from these values "downstream" where blood samples are taken; and/or (2) there may be cyclic variations in these values that stimulate the chemoreceptors but cannot be detected by blood samples. The evidence suggests that both neurogenic and humoral mechanisms are involved in the *hyperpnea,* or increased ventilation, of exercise. (Note that

**Figure 15.43.** The effect of moderate and heavy exercise on arterial blood gases and pH. Notice that there are no consistent and significant changes in these measurements during the first several minutes of moderate and heavy exercise and that only the $P_{CO_2}$ changes (actually decreases) during more prolonged exercise.

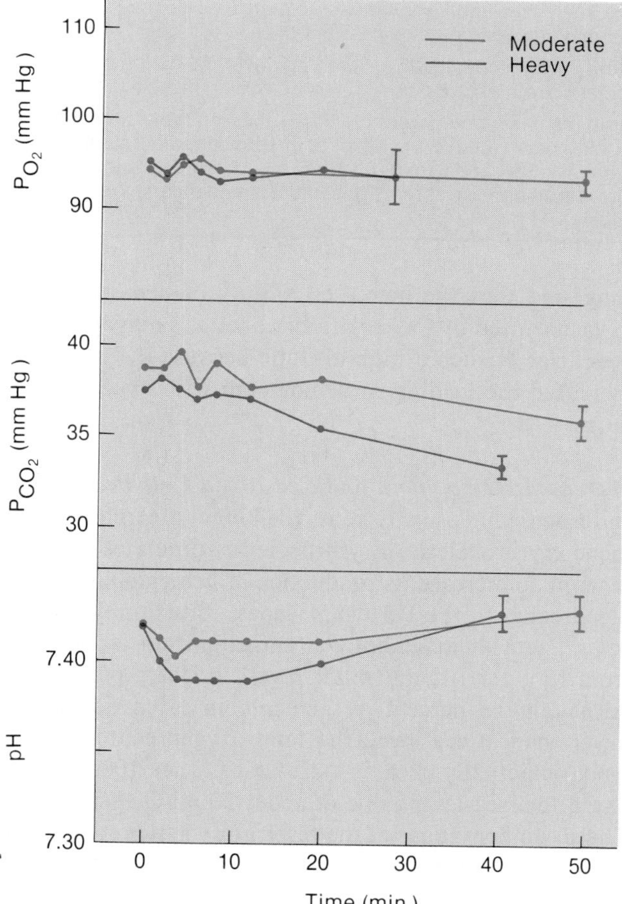


hyperpnea differs from hyperventilation in that the blood $P_{CO_2}$ remains in the normal range during hyperpnea but is decreased in hyperventilation.)

***Anaerobic Threshold and Endurance Training.*** The ability of the cardiopulmonary system to deliver adequate amounts of oxygen to the exercising muscles at the beginning of exercise may be insufficient, as a result of the time lag required to make proper cardiovascular adjustments. During this time, therefore, the muscles respire anaerobically and a "stitch in the side"—probably due to hypoxia of the diaphragm—may develop. After the cardiovascular adjustments have been made, a person may experience a "second wind" when the muscles receive sufficient oxygen for their needs.

Continued heavy exercise can cause a person to reach the **anaerobic threshold,** which is the maximum rate of oxygen consumption that can be attained before blood lactic acid levels rise as a result of anaerobic respiration. This occurs when 50%–60% of the maximal oxygen uptake of the person has been reached. The rise in lactic acid levels is due to the aerobic limitations of the muscles; it is not due to a malfunction of the cardiopulmonary system. Indeed, the arterial oxygen hemoglobin saturation remains at 97% and venous blood draining the muscles contains unused oxygen. The anaerobic threshold, however, is higher in endurance-trained athletes than it is in other people.

The rise in blood lactic acid that occurs when the anaerobic threshold is exceeded is due to the inability of the exercising muscles to increase their oxygen consumption rate sufficiently to prevent anaerobic respiration. Endurance training increases the skeletal muscle content of myoglobin, mitochondria, and Krebs cycle enzymes. These muscles, therefore, are able to utilize more of the oxygen delivered to them by the arterial blood. At a given level of exercise, consequently, the venous blood that drains from muscles in endurance-trained people contains a lower percent oxyhemoglobin than in other people. The effects of exercise and endurance training on respiratory function are summarized in table 15.18.

## Acclimatization to High Altitude

When a person who is from a region near sea level moves to a significantly higher elevation, several adjustments in the respiratory system must be made to compensate for the decreased atmospheric pressure and $P_{O_2}$ at the higher altitude. These adjustments include changes in ventilation, in the hemoglobin affinity for oxygen, and in the total hemoglobin concentration.

Reference to table 15.6 indicates that at an altitude of 8,000 feet, for example, the $P_{O_2}$ of arterial blood is 74 mm Hg (compared to 100 mm Hg at sea level). Reference to table 15.12 indicates that the percent oxyhemoglobin saturation decreases from 97% at sea level to about 94% at this lower $P_{O_2}$. The amount of oxygen attached to hemoglobin, and thus the total oxygen content of blood, is thus decreased. In addition, the rate at which oxygen can be delivered to the tissue cells (by the plasma-derived tissue fluid) after it dissociates from oxyhemoglobin is reduced at the higher altitude. This is because the maximum concentration of oxygen that can be dissolved in the plasma decreases in a linear fashion with the fall in $P_{O_2}$. People may thus experience rapid fatigue even at more moderate elevations, such as 5,000–6,000 feet, in which the oxyhemoglobin saturation is only slightly decreased. Compensations made by the respiratory system gradually reduce the amount of fatigue caused by a given amount of exertion at high altitudes.

***Changes in Ventilation.*** A decrease in arterial $P_{O_2}$ makes the chemoreceptors more sensitive to increases in $P_{CO_2}$; a decrease in $P_{O_2}$ below 50 mm Hg directly stimulates the carotid bodies to increase ventilation. These

**Table 15.18** Changes in respiratory function during exercise

| Variable | Change | Comments |
| --- | --- | --- |
| Ventilation | Increased | This is not hyperventilation because ventilation is matched to increased metabolic rate. Mechanisms responsible for increased ventilation are not well understood. |
| Blood gases | No change | Blood-gas measurements during light, moderate, and heavy exercise show little change because ventilation is increased to match increased muscle oxygen consumption and carbon dioxide production. |
| Oxygen delivery to muscles | Increased | Although the total oxygen content and $P_{O_2}$ do not increase during exercise, there is an increased rate of blood flow to the exercising muscles. |
| Oxygen extraction by muscles | Increased | Increased oxygen consumption lowers the tissue $P_{O_2}$ and lowers the affinity of hemoglobin for oxygen (due to the effect of increased temperature). More oxygen, as a result, is unloaded so that venous blood contains a lower oxyhemoglobin saturation than at rest. This effect is enhanced by endurance training. |

changes cause *hyperventilation,* which becomes stabilized after a few days at about 2–3 L per minute more than the total minute volume at sea level. As a result of this hyperventilation, the arterial $P_{CO_2}$ decreases from 40 mm Hg (its value at sea level) to about 29 mm Hg.

Hyperventilation cannot, of course, increase blood $P_{O_2}$ above that of the inspired air. The $P_{O_2}$ of arterial blood is therefore low at high altitudes. In the Peruvian Andes, for example, the normal arterial $P_{O_2}$ is reduced from 100 mm Hg (at sea level) to 45 mm Hg. The loading of hemoglobin with oxygen is therefore incomplete, producing an oxyhemoglobin saturation that is decreased from 97% (at sea level) to 81%.

***Hemoglobin Affinity for Oxygen.*** Normal arterial blood at sea level only unloads about 22% of its oxygen to the tissues at rest; the percent saturation is reduced from 97% in arterial blood to 75% in venous blood. As a partial compensation for the decrease in oxygen content at high altitude, the affinity of hemoglobin for oxygen is reduced, so that a higher proportion of oxygen is unloaded. This occurs because the low oxyhemoglobin content of red blood cells stimulates the production of 2,3-DPG, which in turn decreases the hemoglobin affinity for oxygen.

At very high altitudes, however, the story becomes more complex. At the summit of Mount Everest, in one study, the very low arterial $P_{O_2}$ (28 mm Hg) stimulated intense hyperventilation, so that the arterial $P_{CO_2}$ was decreased to 7.5 mm Hg. The resultant respiratory alkalosis (arterial pH greater than 7.7) caused a shift to the left of the oxyhemoglobin dissociation curve (indicating greater affinity of hemoglobin for oxygen) despite the antagonistic effects of increased 2,3-DPG concentrations. It was suggested that the increased affinity of hemoglobin for oxygen caused by the respiratory alkalosis may have been beneficial at such a high altitude because it would have increased the loading of hemoglobin with oxygen in the lungs.

***Increased Hemoglobin and Red Blood Cell Production.*** In response to tissue hypoxia, the kidneys secrete the hormone erythropoietin. Erythropoietin stimulates the bone marrow to increase its production of hemoglobin and red blood cells. In the Peruvian Andes, for example, people have a total hemoglobin concentration that is increased from 15 g per 100 ml (at sea level) to 19.8 g per 100 ml. Although the percent oxyhemoglobin saturation is still lower than at sea level, the total oxygen content of the blood is actually greater—22.4 ml $O_2$ per 100 ml compared to a sea level value of about 20 ml $O_2$ per 100 ml. These compensations of the respiratory system to high altitude are summarized in table 15.19. It should be noted that these changes are not unalloyed benefits. Polycythemia (high red blood cell count) increases the viscosity of blood, and pulmonary hypertension, which is more common at high altitude, can cause accompanying edema and ventricular hypertrophy which may lead to heart failure.

1. *Describe the effect of exercise on the blood values of $P_{O_2}$, $P_{CO_2}$, and pH, and explain how ventilation might be increased during exercise.*
2. *Explain why endurance-trained athletes have a higher anaerobic threshold than other people.*
3. *Describe the changes that occur in the respiratory system during acclimatization to life at a high altitude.*

**Table 15.19** Changes in respiratory function during acclimatization to high altitude

| Variable | Change | Comments |
|---|---|---|
| Partial pressure of oxygen | Decreased | Due to decreased total atmospheric pressure |
| Percent oxyhemoglobin saturation | Decreased | Due to lower $P_{O_2}$ in pulmonary capillaries |
| Ventilation | Increased | Due to low arterial $P_{O_2}$; ventilation usually returns to normal after a few days |
| Total hemoglobin | Increased | Due to stimulation by erythropoietin; raises oxygen capacity of blood to partially or completely compensate for the reduced partial pressure |
| Oxyhemoglobin affinity | Decreased | Due to increased DPG within the red blood cells; results in a higher percent unloading of oxygen to the tissues, which may partially or completely compensate for the reduced arterial oxyhemoglobin saturation |

## Summary

### The Respiratory System p. 438

I. Alveoli are small, numerous, thin-walled air sacs that provide an enormous surface area for gas diffusion.
   A. The region of the lungs where gas exchange with the blood occurs is known as the respiratory zone.
   B. The trachea, bronchi, and bronchioles that deliver air to the respiratory zone comprise the conducting zone.

II. The thoracic cavity is limited by the chest wall and diaphragm.
   A. The structures of the thoracic cavity are covered by thin, wet pleural membranes.
   B. The lungs are covered by a visceral pleura, which is normally against the parietal pleura, which lines the chest wall.
   C. The potential space between the two pleural membranes is called the intrapleural space.

### Physical Aspects of Ventilation p. 445

I. The intrapleural and intrapulmonary pressures vary during ventilation.
   A. The intrapleural pressure is always less than the intrapulmonary pressure.
   B. The intrapulmonary pressure is subatmospheric during inspiration and greater than the atmospheric pressure during expiration.
   C. Pressure changes in the lungs are produced by variations in lung volume, in accordance with the inverse relationship between the volume and pressure of a gas described by Boyle's law.

II. The mechanics of ventilation are influenced by the physical properties of the lungs.
   A. The compliance of the lungs is the change in lung volume per change in transpulmonary pressure (the difference between intrapulmonary and intrapleural pressure).
   B. The elasticity of the lungs refers to their tendency to recoil after distension.
   C. The surface tension of the fluid in the alveoli exerts a force directed inwards, which acts to resist distension.

III. On first consideration, it would seem that the surface tension in the alveoli would create a pressure that would cause smaller alveoli to collapse and empty their air into large alveoli.
   A. This would occur because the pressure caused by a given amount of surface tension would be greater in smaller alveoli than in larger alveoli, as described by the law of LaPlace.
   B. Collapse of alveoli due to surface tension does not normally occur, however, because the presence of pulmonary surfactant (a combination of phospholipid and protein) acts to lower surface tension.
   C. In hyaline membrane disease, the lungs of premature infants collapse due to the absence of surfactant.

### Mechanics of Breathing p. 448

I. Inspiration and expiration are accomplished by contraction and relaxation of striated muscles.
   A. During quiet inspiration, the diaphragm and external intercostal muscles contract and thus increase the volume of the thorax.
   B. During quiet expiration, these muscles relax, and the elastic recoil of the lungs and thorax causes a decrease in thoracic volume.
   C. Forced inspiration and expiration are aided by contraction of the accessory respiratory muscles.

II. Spirometry aids the diagnosis of a number of pulmonary disorders.
   A. In restrictive disease, such as pulmonary fibrosis, the vital capacity measurement is decreased below normal.
   B. In obstructive disease, such as asthma and bronchitis, the forced expiratory volume is reduced below normal because of increased airway resistance to air flow.

III. Asthma results from bronchoconstriction; emphysema and chronic bronchitis are frequently referred to as chronic obstructive pulmonary disease.

## Gas Exchange in the Lungs p. 454

I. According to Dalton's law, the total pressure of a gas mixture is equal to the sum of the pressures that each gas in the mixture would exert independently.
  A. The partial pressure of a gas in a dry gas mixture is thus equal to the total pressure times the percent composition of that gas in the mixture.
  B. Since the total pressure of a gas mixture decreases with altitude above sea level, the partial pressures of the constituent gases likewise decrease with altitude.
  C. When the partial pressure of a gas in a wet gas mixture is calculated, the water vapor pressure must be taken into account.

II. According to Henry's law, the amount of gas that can be dissolved in a fluid is directly proportional to the partial pressure of that gas in contact with the fluid.
  A. The concentrations of oxygen and carbon dioxide that are dissolved in plasma are proportional to an electric current generated by special electrodes that react with these gases.
  B. Normal arterial blood has a $P_{O_2}$ of 100 mm Hg, indicating a concentration of dissolved oxygen of 0.3 ml per 100 ml of blood; the oxygen contained in red blood cells (about 19.7 ml per 100 ml of blood) does not affect the $P_{O_2}$ measurement.

III. The $P_{O_2}$ and $P_{CO_2}$ measurements of arterial blood provide information about lung function.

IV. In addition to proper ventilation of the lungs, blood flow (perfusion) in the lungs must be adequate and matched to air flow (ventilation) in order for adequate gas exchange to occur.

V. Abnormally high partial pressures of gases in blood can cause a variety of disorders, including oxygen toxicity, nitrogen narcosis, and decompression sickness.

## Regulation of Breathing p. 461

I. The rhythmicity center in the medulla oblongata directly controls the muscles of respiration.
  A. Activity of the inspiratory and expiratory neurons varies in a reciprocal way to produce an automatic breathing cycle.
  B. Activity in the medulla is influenced by the apneustic and pneumotaxic centers in the pons, as well as by sensory feedback information.
  C. Conscious breathing involves direct control by the cerebral cortex via corticospinal tracts.

II. Breathing is affected by chemoreceptors sensitive to the $P_{CO_2}$, pH, and $P_{O_2}$ of the blood.
  A. The $P_{CO_2}$ of the blood and consequent changes in pH are usually of greater importance than the blood $P_{O_2}$ in the regulation of breathing.
  B. Central chemoreceptors in the medulla oblongata are sensitive to changes in blood $P_{CO_2}$, because these changes cause the pH of cerebrospinal fluid to change.
  C. The peripheral chemoreceptors in the aortic and carotid bodies are sensitive to changes in blood $P_{CO_2}$ indirectly, because of consequent changes in blood pH.

III. Decreases in blood $P_{O_2}$ directly stimulate breathing only when the blood $P_{O_2}$ is less than 50 mm Hg; a drop in $P_{O_2}$ also stimulates breathing indirectly, by making the chemoreceptors more sensitive to changes in $P_{CO_2}$ and pH.

IV. At tidal volumes of one liter or more, inspiration is inhibited by stretch receptors in the lungs (the Hering-Breuer reflex); there is a similar deflation reflex.

## Hemoglobin and Oxygen Transport p. 466

I. Hemoglobin is composed of two alpha and two beta polypeptide chains and four heme groups that contain a central atom of iron.
  A. When the iron is in the reduced form and not attached to oxygen, the hemoglobin is called deoxyhemoglobin, or reduced hemoglobin; when it is attached to oxygen, it is called oxyhemoglobin.
  B. If the iron is attached to carbon monoxide, the hemoglobin is called carboxyhemoglobin; when the iron is in an oxidized state and unable to transport any gas, the hemoglobin is called methemoglobin.
  C. Deoxyhemoglobin combines with oxygen in the lungs (the loading reaction) and breaks its bonds with oxygen in the tissue capillaries (the unloading reaction); the extent of each reaction is determined by the $P_{O_2}$ and the affinity of hemoglobin for oxygen.

II. A graph of percent oxyhemoglobin saturation at different values of $P_{O_2}$ is called an oxyhemoglobin dissociation curve.
  A. At rest, the difference between arterial and venous oxyhemoglobin saturations indicates that about 22% of the oxyhemoglobin unloads its oxygen to the tissues.
  B. During exercise, the venous $P_{O_2}$ and percent oxyhemoglobin saturation are decreased, indicating that a higher percent of the oxyhemoglobin unloaded its oxygen to the tissues.

III. The pH and temperature of the blood influence the affinity of hemoglobin for oxygen and thus the extent of loading and unloading.
  A. A fall in pH decreases the affinity, and a rise in pH increases the affinity of hemoglobin for oxygen; this is called the Bohr effect.
  B. A rise in temperature decreases the affinity of hemoglobin for oxygen.
  C. When the affinity is decreased, the oxyhemoglobin dissociation curve is shifted to the right; this indicates a greater percentage unloading of oxygen to the tissues.

IV. The affinity of hemoglobin for oxygen is also decreased by an organic molecule in the red blood cells called 2,3-diphosphoglyceric acid (2,3-DPG).

A. Since oxyhemoglobin inhibits 2,3-DPG production, there will be higher 2,3-DPG concentrations when the oxyhemoglobin is decreased due to anemia or low $P_{O_2}$ (as in high altitude).
B. If a person is anemic, the lowered hemoglobin concentration is partially compensated by the fact that a higher percent of the oxyhemoglobin will unload its oxygen due to the effect of 2,3-DPG.
C. Since fetal hemoglobin cannot bond 2,3-DPG, it has a higher affinity for oxygen than the mother's hemoglobin; this facilitates the transfer of oxygen to the fetus.

V. Inherited defects in the amino acid composition of hemoglobin are responsible for such diseases as sickle-cell anemia and thalassemia.

VI. Striated muscles have myoglobin, a pigment related to hemoglobin, which can combine with oxygen and deliver it to the muscle cell mitochondria at low values of $P_{O_2}$.

### Carbon Dioxide Transport and Acid-Base Balance p. 474

I. Red blood cells contain an enzyme called carbonic anhydrase, which catalyzes the reversible reaction whereby carbon dioxide and water are used to form carbonic acid.
A. This reaction is favored by the high $P_{CO_2}$ in the tissue capillaries, and as a result, carbon dioxide produced by the tissues is converted into carbonic acid in the red blood cells.
B. Carbonic acid then ionizes to form $H^+$ and $HCO_3^-$ (bicarbonate).
C. Since much of the $H^+$ is buffered by hemoglobin, but more bicarbonate is free to diffuse outward, an electrical gradient is established which draws $Cl^-$ into the red blood cells; this is called the chloride shift.
D. A reverse chloride shift occurs in the lungs; in this process, the low $P_{CO_2}$ favors the conversion of carbonic acid to carbon dioxide, which can be exhaled.

II. By adjusting the blood concentration of carbon dioxide and thus of carbonic acid, the process of ventilation helps to maintain proper acid-base balance of the blood.
A. Normal arterial blood pH is 7.40; a pH less than 7.35 is termed acidosis, and a pH greater than 7.45 is termed alkalosis.
B. Hyperventilation causes respiratory alkalosis, and hypoventilation causes respiratory acidosis.
C. Metabolic acidosis stimulates hyperventilation, which can cause a respiratory alkalosis as a partial compensation.

### Effect of Exercise and High Altitude on Respiratory Function p. 478

I. During exercise there is increased ventilation, or hyperpnea, which can be matched to the increased metabolic rate so that the arterial blood $P_{CO_2}$ remains normal.
A. This hyperpnea may be caused by proprioceptor information, cerebral input, and/or changes in arterial $P_{CO_2}$ and pH.
B. During heavy exercise the anaerobic threshold may be reached at about 50%–60% of the maximal oxygen uptake; at this point, lactic acid is released into the blood by the muscles.
C. Endurance training enables the muscles to utilize oxygen more effectively, so that greater levels of exercise can be performed before the anaerobic threshold is reached.

II. Acclimatization to a high altitude involves changes that help to deliver oxygen more effectively to the tissues despite reduced arterial $P_{O_2}$.
A. Hyperventilation occurs in response to the low $P_{O_2}$.
B. The red blood cells produce more 2,3-DPG, which lowers the affinity of hemoglobin for oxygen and improves the unloading reaction.
C. The kidneys produce the hormone erythropoietin, which stimulates the bone marrow to increase its production of red blood cells, so that more oxygen can be carried by the blood at given values of $P_{O_2}$.

## Review Activities

### Objective Questions

1. Which of the following statements about intrapulmonary and intrapleural pressure is true?
   (a) The intrapulmonary pressure is always subatmospheric.
   (b) The intrapleural pressure is always greater than the intrapulmonary pressure.
   (c) The intrapulmonary pressure is greater than the intrapleural pressure.
   (d) The intrapleural pressure equals the atmospheric pressure.
2. If the transpulmonary pressure equals zero,
   (a) a pneumothorax has probably occurred
   (b) the lungs cannot inflate
   (c) elastic recoil causes the lungs to collapse
   (d) all of the above
3. The maximum amount of air that can be expired after a maximum inspiration is the
   (a) tidal volume
   (b) forced expiratory volume
   (c) vital capacity
   (d) maximum expiratory flow rate
4. If the blood lacked red blood cells but the lungs were functioning normally,
   (a) the arterial $P_{O_2}$ would be normal
   (b) the oxygen content of arterial blood would be normal
   (c) both a and b
   (d) neither a nor b
5. If a person were to dive with scuba equipment to a depth of 66 feet, which of the following statements would be *false?*
   (a) The arterial $P_{O_2}$ would be three times normal.

(b) The oxygen content of plasma would be three times normal.
(c) The oxygen content of whole blood would be three times normal.

6. Which of the following would be most affected by a decrease in the affinity of hemoglobin for oxygen? The
   (a) arterial $P_{O_2}$
   (b) arterial percent oxyhemoglobin saturation
   (c) venous oxyhemoglobin saturation
   (d) arterial $P_{CO_2}$
7. If a person with normal lung function were to hyperventilate for several seconds, there would be a significant
   (a) increase in the arterial $P_{O_2}$
   (b) decrease in the arterial $P_{CO_2}$
   (c) increase in the arterial percent oxyhemoglobin saturation
   (d) decrease in the arterial pH
8. Erythropoietin is produced by the
   (a) kidneys
   (b) liver
   (c) lungs
   (d) bone marrow
9. The affinity of hemoglobin for oxygen is decreased under conditions of
   (a) acidosis
   (b) fever
   (c) anemia
   (d) acclimatization to a high altitude
   (e) all of the above
10. Most of the carbon dioxide in the blood is carried in the form of
   (a) dissolved $CO_2$
   (b) carbaminohemoglobin
   (c) bicarbonate
   (d) carboxyhemoglobin
11. The bicarbonate concentration of the blood would be decreased during
   (a) metabolic acidosis
   (b) respiratory acidosis
   (c) metabolic alkalosis
   (d) respiratory alkalosis
12. The chemoreceptors in the medulla are directly stimulated by
   (a) $CO_2$ from the blood
   (b) $H^+$ from the blood
   (c) $H^+$ in cerebrospinal fluid that is derived from blood $CO_2$
   (d) decreased arterial $P_{O_2}$
13. The rhythmic control of breathing is produced by the activity of inspiratory and expiratory neurons in the
   (a) medulla oblongata
   (b) apneustic center of the pons
   (c) pneumotaxic center of the pons
   (d) cerebral cortex
14. Which of the following occurs during hypoxemia?
   (a) increased ventilation
   (b) increased production of 2,3-DPG
   (c) increased production of erythropoietin
   (d) all of the above
15. During exercise, which of the following statements is true?
   (a) The arterial percent oxyhemoglobin saturation is decreased.
   (b) The venous percent oxyhemoglobin saturation is decreased.
   (c) The arterial $P_{CO_2}$ is measurably increased.
   (d) The arterial pH is measurably decreased.
16. All of the following can bond to hemoglobin *except*
   (a) $HCO_3^-$
   (b) $O_2$
   (c) $H^+$
   (d) $CO_2$
17. The partial pressure of carbon dioxide is higher in
   (a) the alveoli than in the pulmonary arteries
   (b) the systemic arteries than in the tissues
   (c) the systemic veins than in the systemic arteries
   (d) the pulmonary veins than in the pulmonary arteries

## Essay Questions

1. Using a flow diagram to show cause and effect, explain how contraction of the diaphragm produces inspiration.
2. Radiographic (X-ray) pictures show that the ribs of a person with a pneumothorax are expanded and farther apart. Explain why this occurs.
3. Explain, using a flowchart, how a rise in blood $P_{CO_2}$ stimulates breathing. Include both the central and peripheral chemoreceptors in your answer.
4. A person with ketoacidosis may hyperventilate. Explain why this occurs, and explain why this hyperventilation can be stopped by an intravenous fluid containing bicarbonate.
5. What blood measurements can be performed to detect (*a*) anemia, (*b*) carbon monoxide poisoning, and (*c*) poor lung function?
6. Explain how measurements of blood $P_{CO_2}$, bicarbonate, and pH are affected by hypoventilation and hyperventilation.
7. Explain how blood pH and bicarbonate concentrations are affected by respiratory and metabolic acidosis.
8. How would an increase in the red blood cell content of 2,3-DPG affect the $P_{O_2}$ of venous blood? Explain your answer.

## SELECTED READINGS


Avery, M. E., N. S. Wang, and H. W. Taeusch, Jr. March 1975. The lung of the newborn infant. *Scientific American.*

Berger, A. J., R. A. Mitchel, and J. W. Severinghaus. 1977. Regulation of respiration. *New England Journal of Medicine* 297: first part, p. 92; second part, p. 138; third part, p. 194.

Brample, D. M., and D. R. Carrier. 1983. Running and breathing in mammals. *Science* 219:251.

Browning, R. J. 1982. Pulmonary disease: Back to basics (part 1); Putting blood gases to work (part 2). *Diagnostic Medicine.* First part: Jan/Feb, p. 39; second part: March/April, p. 59.

Cherniak, N. S. 1986. Breathing disorders during sleep. *Hospital Practice* 21:81.

Dantzker, D. R. 1986. Physiology and pathophysiology of pulmonary gas exchange. *Hospital Practice* 21:135.

Davis, J. M., *et al.* 1988. Changes in pulmonary mechanics after the administration of surfactant to infants with respiratory distress syndrome. *New England Journal of Medicine* 319:476.

Finch, C. A., and C. Lenfant. 1972. Oxygen transport in man. *New England Journal of Medicine* 286:407.

Flenley, D. C., and P. M. Warren. 1983. Ventilatory response to $O_2$ and $CO_2$ during exercise. *Annual Review of Physiology* 45:415.

Fraser, R. G., and J. A. P. Pare. 1977. *Structure and function of the lung.* 2d ed. Philadelphia: W. B. Saunders Co.

Guz, A. 1975. Regulation of respiration in man. *Annual Review of Physiology* 37:303.

Haddad, G. G., and R. B. Mellins. 1984. Hypoxia and respiratory control in early life. *Annual Review of Physiology* 46:629.

Irsigler, G. B., and J. W. Severinghaus. 1980. Clinical problems of ventilatory control. *Annual Review of Medicine* 31:109.

Kassirer, J. P., and N. E. Madias. 1980. Respiratory acid-base disorders. *Hospital Practice* 15:57.

Macklem, P. T. 1986. Respiratory muscle dysfunction. *Hospital Practice* 21:83.

Massaro, D. 1986. Oxygen: Toxicity and tolerance. *Hospital Practice* 21:95.

Murray, J. F. 1985. The lungs and heart failure. *Hospital Practice* 20:55.

Nadel, E. R. 1985. Physiological adaptations to aerobic training. *American Scientist* 73:334.

Naeye, R. L. April 1980. Sudden infant death. *Scientific American.*

Naquet, R., and J. C. Rostain. 1988. Deep-diving humans. *News in Physiological Sciences* 3:72.

Perutz, M. F. December 1978. Hemoglobin structure and respiratory transport. *Scientific American.*

Rigatto, H. 1984. Control of ventilation in the newborn. *Annual Review of Physiology* 46:661.

Roussos, C., and P. T. Macklem. 1982. The respiratory muscles. *New England Journal of Medicine* 307:786.

Shannon, D. C., and D. H. Kelly. 1982. SIDS and near-SIDS. *New England Journal of Medicine* 306: first part, p. 959; second part, p. 1022.

Snapper, J. R., and K. L. Bingham. 1986. Pulmonary edema. *Hospital Practice* 21:87.

Tobin, M. J. 1986. Update on strategies in mechanical ventilation. *Hospital Practice* 21:69.

Walker, D. W. 1984. Peripheral and central chemoreceptors in the fetus and newborn. *Annual Review of Physiology* 46:687.

West, J. B. 1984. Human physiology at extreme altitudes on Mount Everest. *Science* 223:784.

Whipp, B. J. 1983. Ventilatory control during exercise in humans. *Annual Review of Physiology* 45:393.

# 16

# Physiology of the Kidneys

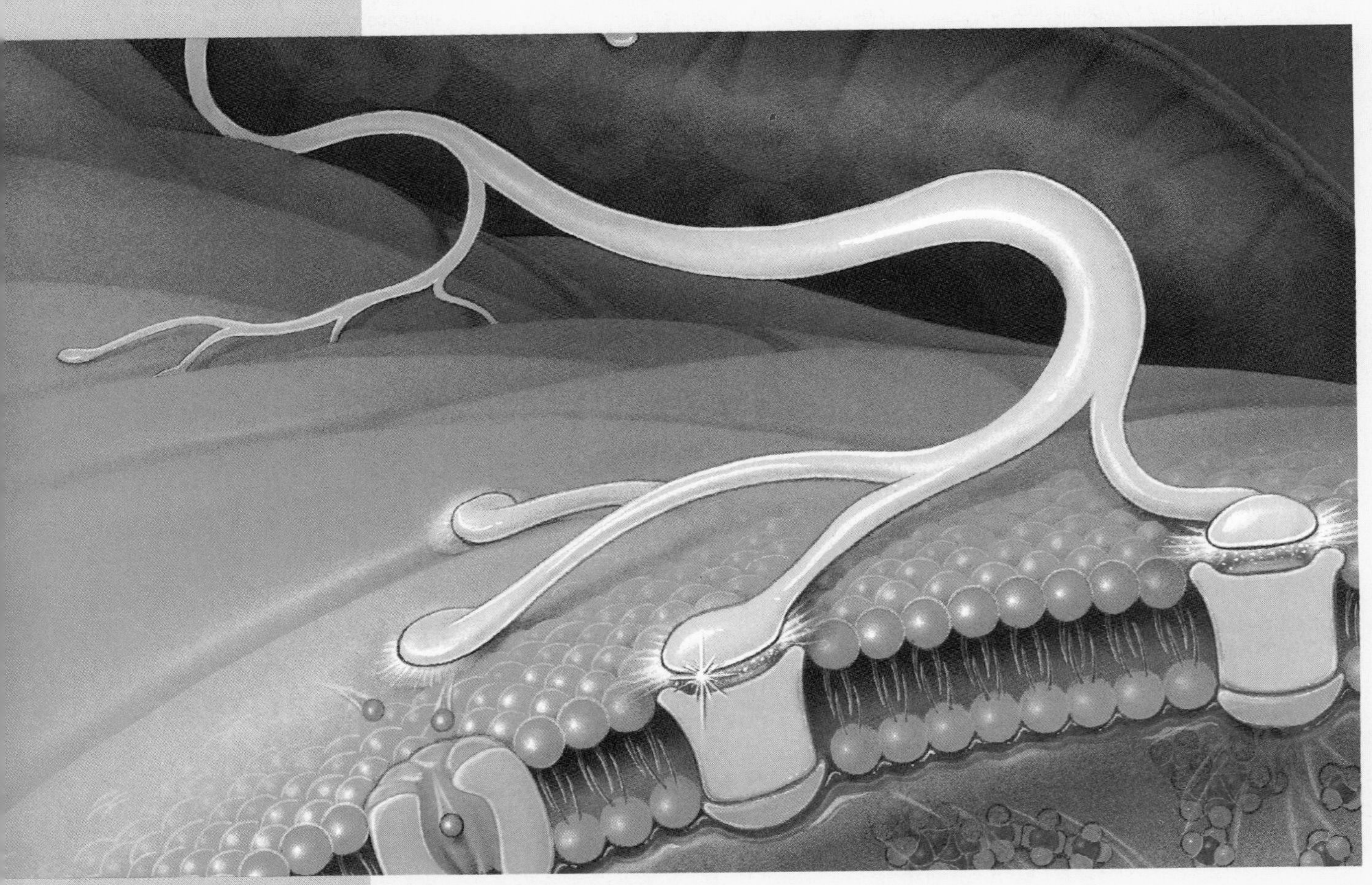

## Objectives

By studying this chapter, you should be able to

1. describe the different regions of the nephron tubules and explain the anatomic relationship between the tubules and the gross structure of the kidney
2. describe the structural and functional relationships between the nephron tubules and their associated blood vessels
3. describe the composition of glomerular ultrafiltrate and explain how it is produced, how it enters Bowman's capsule, and how the glomerular filtration rate is regulated
4. explain how the proximal convoluted tubule reabsorbs salt and water and relate the mechanisms of reabsorption to the structure of the tubular epithelial cells
5. describe the processes of active transport and osmosis in the ascending and descending limbs of the loop of Henle and explain how these processes function to produce a countercurrent multiplier system
6. explain how the vasa recta function in countercurrent exchange to maintain a high osmolality in the tissue fluid of the renal medulla, and state the significance of this high osmolality
7. explain how antidiuretic hormone (ADH) functions to regulate the final urine volume
8. describe the mechanisms of glucose and amino acid reabsorption and explain the meanings of the terms *transport maximum* and *renal plasma threshold*
9. describe the meaning of the term *renal plasma clearance,* explain why the clearance of inulin is equal to the glomerular filtration rate (GFR), and explain how the GFR is calculated
10. explain how the clearance of different molecules is determined and how the processes of reabsorption and secretion affect the clearance measurement
11. explain how the renal plasma clearance of para-aminohippuric acid (PAH) can be used to measure the total renal blood flow
12. explain the mechanism of $Na^+$ reabsorption in the distal tubule and why this reabsorption occurs together with the secretion of $K^+$
13. describe the effects of aldosterone on the distal convoluted tubule and how aldosterone secretion is affected by the blood $Na^+$ and $K^+$ concentration
14. describe the structure of the juxtaglomerular apparatus and explain how activation of the renin-angiotensin system results in the stimulation of aldosterone secretion
15. describe the interaction between plasma $K^+$ and $H^+$ concentrations and explain how this affects the tubular secretion of these ions
16. distinguish between the respiratory and metabolic components of acid-base balance and describe the role of the kidneys in the regulation of acid-base balance
17. describe the different mechanisms by which substances can act as diuretics and explain why some cause excessive loss of $K^+$ whereas others are classified as potassium-sparing diuretics

## Outline

**Figure 16.1.** The anatomical locations of the kidneys, ureters, and urinary bladder.

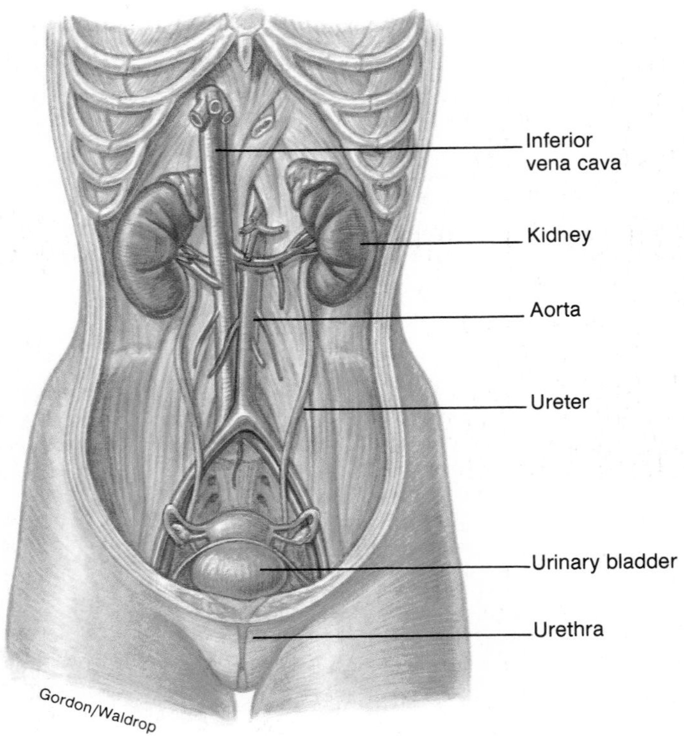


## Structure and Function of the Kidneys

Each kidney contains many tiny tubules which open to a cavity that is drained by the ureter. Each of these tubules receives a blood filtrate, similar to tissue fluid, which is modified as it passes through different regions of the tubule and is thereby changed into urine. The tubules and associated blood vessels thus form the functioning units of the kidneys known as nephrons. Urine is transported by the ureter of each kidney to the urinary bladder, and is excreted from the body through the urethra.

The primary function of the kidneys is regulation of the extracellular fluid (plasma and tissue fluid) environment in the body. This function is accomplished through the formation of urine, which is a modified filtrate of plasma. In the process of urine formation, the kidneys regulate (1) the volume of blood plasma and thus contribute significantly to the regulation of blood pressure; (2) the concentration of waste products in the blood; (3) the concentration of electrolytes ($Na^+$, $K^+$, $HCO_3^-$, and other ions) in the plasma; and (4) the pH of plasma. In order to understand how these functions are performed by the kidneys, a knowledge of kidney structure is required.

### Gross Structure of the Urinary System

The paired right and left kidneys are located in the abdominal cavity below the diaphragm and the liver. Each kidney in an adult weighs approximately 160–175 g, and is about 10–12 cm long and 5–6 cm wide—about the size of a fist. Urine produced in the kidneys is drained into a cavity known as the *renal pelvis* (= basin), and from there it is channeled via two long ducts—the *ureters*—to the single *urinary bladder* (fig. 16.1).

A coronal section of the kidney shows two distinct regions (fig. 16.2). The outer **cortex,** in contact with the capsule, is reddish brown and granular in appearance because of its many capillaries. The deeper region, or **medulla,** is lighter in color and striped in appearance because of the presence of microscopic tubules and blood vessels. The medulla is composed of eight to fifteen conical *renal pyramids,* separated by *renal columns.*

The cavity of the kidney collects and transports urine from the kidney to the ureter. It is divided into several portions. Each pyramid projects into a small depression called a *minor calyx* (the plural form is *calyces*). Several minor calyces unite to form a *major calyx.* In turn, the major calyces join to form the funnel-shaped *renal pelvis.* The renal pelvis is actually an expanded portion of the ureter in the kidney and serves to collect urine from the calyces and transport it to the ureter and urinary bladder (fig. 16.3)

**Figure 16.2.** The internal structures of a kidney. (*a*) A coronal section showing the structure of the cortex, medulla, and renal calyces. (*b*) A diagrammatic magnification of a renal pyramid and cortex to depict the tubules. (*c*) A diagrammatic view of a single nephron.

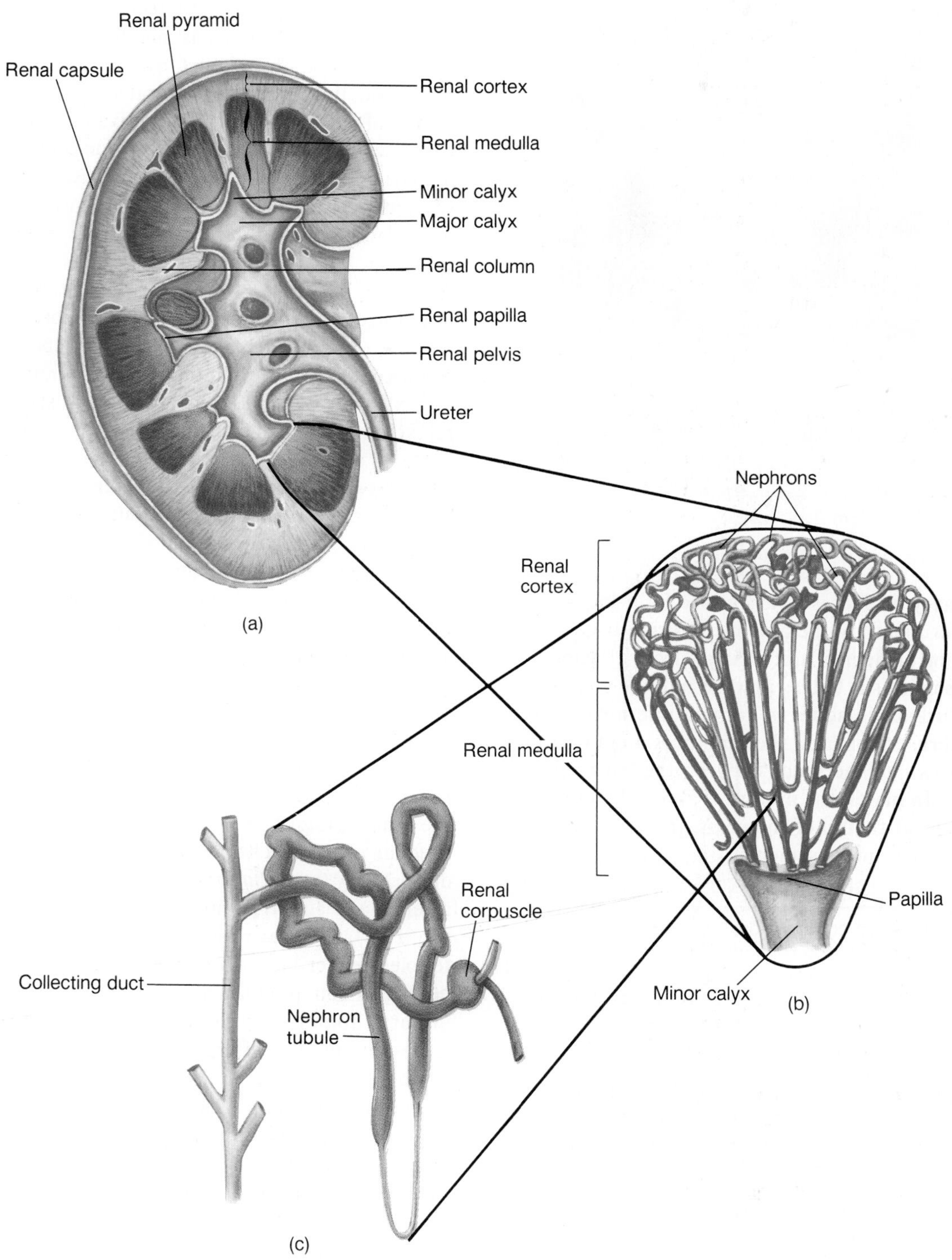

**Figure 16.3.** A false color X ray of the calyces and renal pelvises of the kidneys, the ureters, and the urinary bladder.

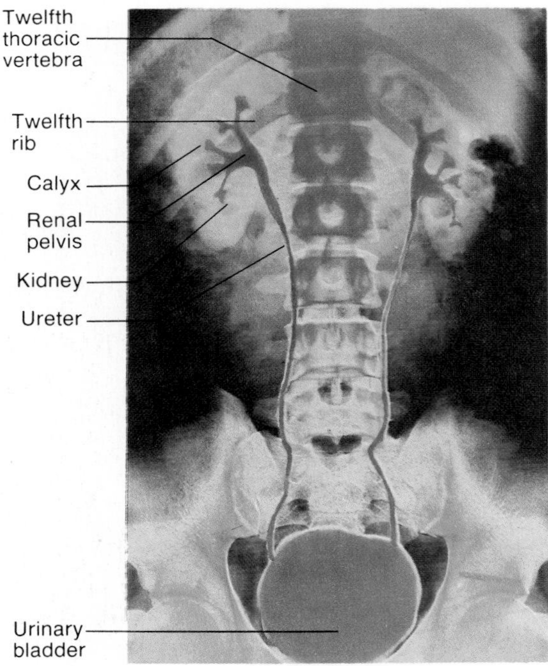


The **urinary bladder** is a storage sac for urine, and its shape is determined by the amount of urine it contains. As the urinary bladder fills, it changes from a triangular to an ovoid shape as it bulges upward into the abdominal cavity. The urinary bladder is drained inferiorly by the tubular **urethra**. In females, the urethra is 4 cm (1.5 in.) long and opens into the space between the labia minora (chapter 20). In males, the urethra is about 20 cm (8 in.) long and opens at the tip of the penis, from which it can discharge either urine or semen.

**Kidney stones** are composed of crystals and proteins that grow in the renal medulla until they break loose and pass into the urine collection system. Small stones that are anchored in place are not usually noticed, but large stones in the calyces or pelvis may obstruct the flow of urine. When a stone breaks loose and passes into a ureter it produces a steadily increasing sensation of pain, which often becomes so intense that the patient requires narcotic drugs. Most kidney stones contain crystals of calcium oxalate, but stones may also be composed of crystals of calcium phosphate, uric acid, or cystine. These substances are normally present in urine in a supersaturated state from which they can crystalize for a variety of reasons.

***Micturition Reflex.*** The urethra has two muscular sphincters. The upper sphincter is composed of smooth muscle and is called the *internal urethral sphincter*; the lower sphincter is composed of voluntary skeletal muscle and is called the *external urethral sphincter*. The actions of these sphincters are regulated in the process of urination, which is also known as **micturition**.

Micturition is controlled by a reflex center located in the second, third, and fourth sacral levels of the spinal cord. Filling of the urinary bladder activates stretch receptors which send impulses to this micturition center. As a result, parasympathetic neurons are activated which stimulate rhythmic contractions of the detrusor muscle of the urinary bladder and cause relaxation of the internal urethral sphincter. At this point, a sense of urgency is perceived by the brain, but there is still voluntary control over the external urethral sphincter. When urination is consciously allowed to occur, descending motor tracts to the micturition center cause inhibition of somatic motor fibers to the external urethral sphincter. This muscle, as a result, relaxes, and urine is expelled.

## Microscopic Structure of the Kidney

The **nephron** is the functional unit of the kidney that is responsible for the formation of urine. Each kidney contains more than a million nephrons. A nephron consists of **tubules** and associated small blood vessels. Fluid formed by capillary filtration enters the tubules and is subsequently modified by transport processes; the resulting fluid that leaves the tubules is urine.

***Renal Blood Vessels.*** Arterial blood enters the kidney through the *renal artery,* which divides into *interlobar arteries* (fig. 16.4), that pass between the pyramids through the renal columns. *Arcuate arteries* branch from the interlobar arteries at the boundary of the cortex and medulla. Many *interlobular arteries* radiate from the arcuate arteries and subdivide into numerous **afferent arterioles** (fig. 16.5), which are microscopic in size. The afferent arterioles deliver blood into capillary networks, called **glomeruli,** which produce a blood filtrate that enters the urinary tubules. The blood remaining in the glomerulus leaves through an **efferent arteriole**, which delivers the blood into another capillary network, the **peritubular capillaries**, that surrounds the tubules. This arrangement of blood vessels, in which a capillary bed (the glomerulus) is drained by an arteriole rather than by a venule, and delivered to a second capillary bed located downstream (the peritubular capillaries) is unique. Blood from the peritubular capillaries is drained into veins that parallel the course of the arteries in the kidney and are named the *interlobular veins, arcuate veins,* and *interlobar veins.* The interlobar veins descend between the pyramids, converge, and leave the kidney as a single *renal vein* that empties into the inferior vena cava.

Figure 16.4. The vascular structure of the kidneys. (*a*) An illustration of the major arterial supply and (*b*) a scanning electron micrograph of the glomeruli. (From *Tissues and Organs: A Text-Atlas of Scanning Electron Microscopy* by R. G. Kessel and R. Kardon. W. H. Freeman and Company © 1979.)

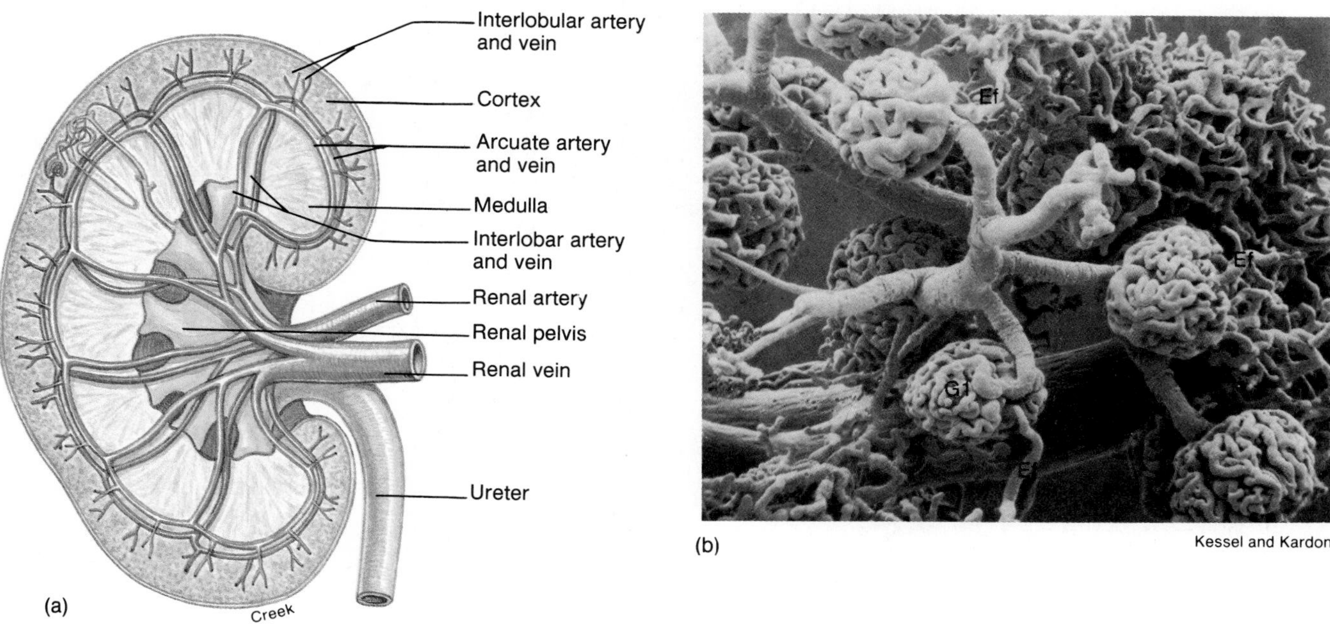


Figure 16.5. A simplified illustration of blood flow from a glomerulus to an efferent arteriole, to the peritubular capillaries, and to the venous drainage of the kidneys.

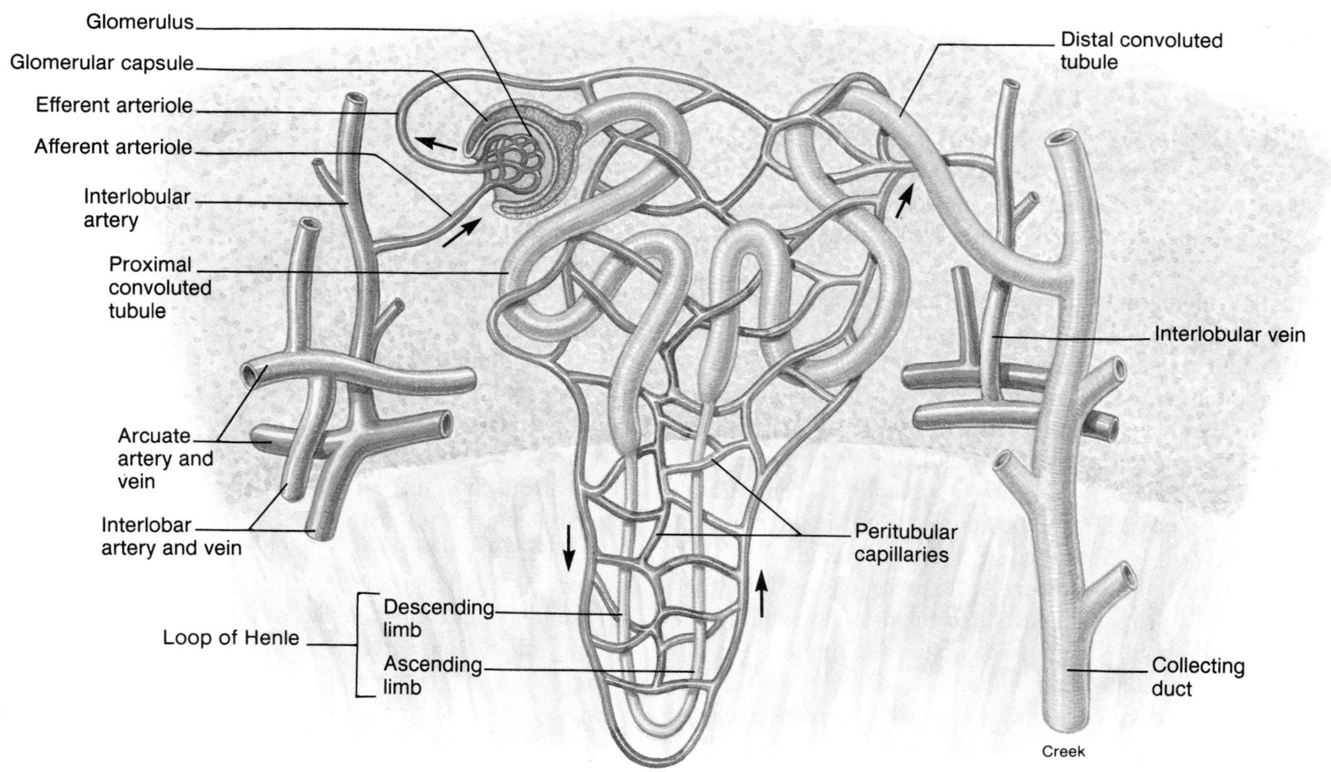

**Figure 16.6.** Cortical nephrons are located almost exclusively within the renal cortex, and juxtamedullary nephrons are located, for the most part, within the outer portion of the renal medulla.

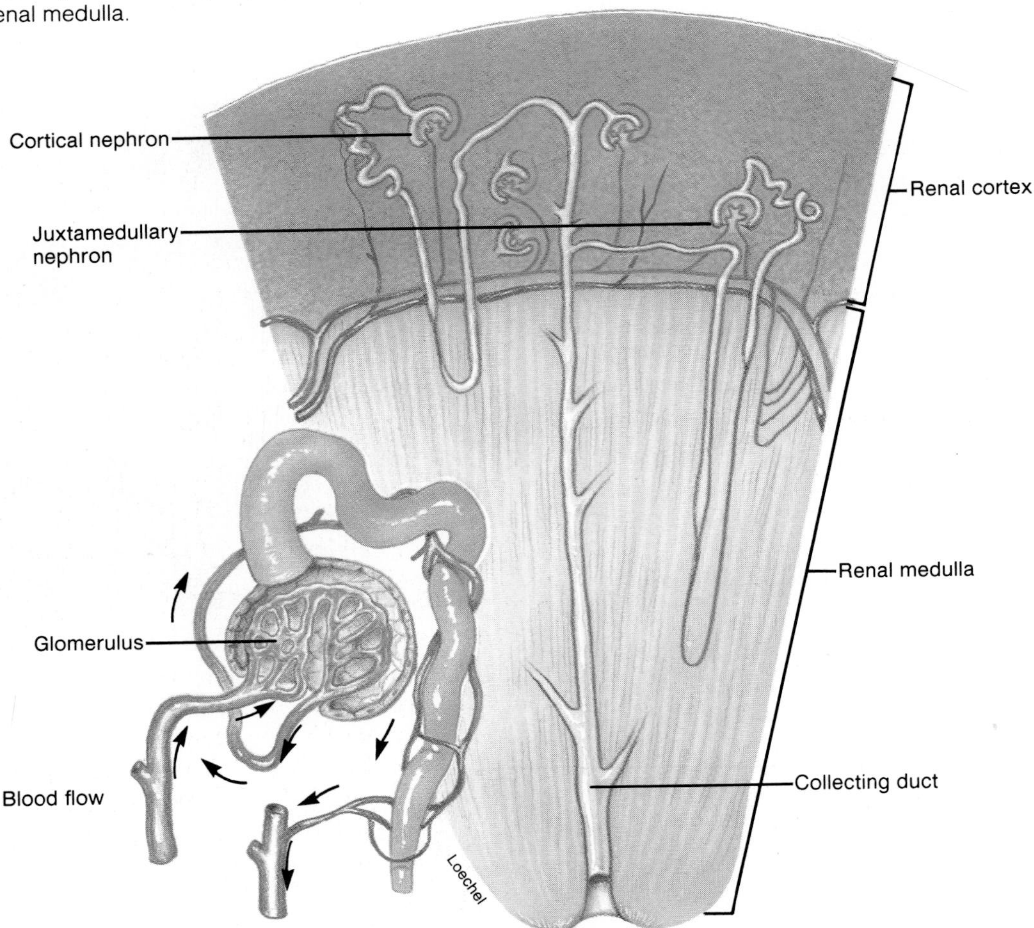


***Nephron Tubules.*** The tubular portion of a nephron consists of a glomerular capsule, a proximal convoluted tubule, a descending limb of the loop of Henle, an ascending limb of the loop of Henle, and a distal convoluted tubule (fig. 16.5).

The **glomerular (Bowman's) capsule** surrounds the glomerulus. The glomerular capsule and its associated glomerulus are located in the cortex of the kidney and together constitute the *renal corpuscle.* The glomerular capsule contains an inner visceral layer of epithelium around the glomerular capillaries and an outer parietal layer. The space between these two layers is continuous with the lumen of the tubule and receives the glomerular filtrate, as will be described in the next section.

Filtrate in the glomerular capsule passes into the lumen of the **proximal convoluted tubule.** The wall of the proximal convoluted tubule consists of a single layer of cuboidal cells, containing millions of microvilli; these microvilli serve to increase the surface area for reabsorption. In the process of reabsorption, salt, water, and other molecules needed by the body are transported from the lumen, through the tubular cells, and into the surrounding peritubular capillaries.

The glomerulus, glomerular capsule, and proximal convoluted tubule are located in the renal cortex. Fluid passes from the proximal convoluted tubule to the **loop of Henle.** This fluid is carried into the medulla in the **descending limb** of the loop and returns to the cortex in the **ascending limb** of the loop. Back in the cortex, the tubule becomes coiled again and is called the **distal convoluted tubule.** In contrast to the proximal tubule, the distal convoluted tubule is shorter and has relatively few microvilli. The distal convoluted tubule is the last segment of the nephron and terminates as it empties into a collecting duct.

There are two types of nephrons, which are classified according to their position in the kidney and the lengths of their loops of Henle. Nephrons that originate in the inner one-third of the cortex—called *juxtamedullary nephrons*—have longer loops of Henle than the more numerous *cortical nephrons,* which originate in the outer two-thirds of the cortex (fig. 16.6). The juxtamedullary nephrons are involved in the process of concentrating the urine by means of a mechanism described in a later section.

The distal convoluted tubules of several nephrons drain into a **collecting duct.** Fluid is then drained by the collecting duct from the cortex to the medulla as the collecting duct passes through a renal pyramid. This fluid, now called urine, passes into a minor calyx. Urine is then funneled through the renal pelvis and out of the kidney in the ureter.

**Polycystic kidney disease** is a condition inherited as an autosomal dominant trait (chapter 20) which affects one in 600–1,000 people. This disease is thus more common than such genetic diseases as sickle-cell anemia, cystic fibrosis, or muscular dystrophy. In 50% of the people who inherit the defective gene (located on the short arm of chromosome 16), progressive renal failure develops during middle age to the point that dialysis or kidney transplants are required. The cysts that develop are expanded portions of the renal tubule. Cysts which originate in the proximal tubule contain fluid that resembles glomerular filtrate and plasma. Cysts which originate in the distal tubule contain a lower NaCl concentration and a higher potassium and urea concentration than plasma, as a result of the transport processes that occur during the passage of fluid in the nephron.

1. *Describe the "theme" of kidney function in a single sentence, and list the components of this functional theme.*
2. *Trace the course of blood flow through the kidney from the renal artery to the renal vein.*
3. *Trace the course of tubular fluid from the glomerular capsules to the ureter.*
4. *Draw a diagram of the tubular component of a nephron. Label the segments, and indicate which parts are in the cortex and which are in the medulla.*

## Glomerular Filtration

The glomerular capillaries have large pores, and the layer of Bowman's capsule in contact with the glomerulus has filtration slits. Water together with dissolved solutes—but not proteins—can thus pass from the blood plasma to the inside of the capsule and the lumen of the nephron tubules. The volume of this filtrate produced per minute by both kidneys is called the glomerular filtration rate (GFR). The GFR is regulated by sympathetic nerves and by intrinsic mechanisms.

The glomerular capillaries have extremely large pores (200–500 Å in diameter) called *fenestrae*, and are thus said to be *fenestrated*. As a result of these large pores, glomerular capillaries are one hundred to four hundred times more permeable to plasma water and dissolved solutes than are the capillaries of skeletal muscles. Although the pores of glomerular capillaries are large, they are still small enough to prevent the passage of red blood cells, white blood cells, and platelets into the filtrate.

Before the filtrate can enter the interior of the glomerular capsule it must pass through the capillary pores, the basement membrane (a thin layer of glycoproteins immediately outside the endothelial cells), and the inner, visceral layer of the glomerular capsule. The inner layer of the glomerular capsule is composed of unique cells, called *podocytes,* with numerous cytoplasmic extensions known as *pedicels* (fig. 16.7). These pedicels interdigitate, like fingers wrapped around the glomerular capillaries. The narrow slits between adjacent pedicels provide the passageways through which filtered molecules must pass to enter the interior of the glomerular capsule (fig. 16.8).

Although the glomerular capillary pores are apparently large enough to permit the passage of proteins, the fluid that enters the capsular space is almost completely free of plasma proteins. This exclusion of plasma proteins from the filtrate is partially a result of their negative charges, which hinder their passage through the negatively charged glycoproteins in the basement membrane of the capillaries. The large size and negative charges of plasma proteins may also restrict their movement through the filtration slits between pedicels.

### Glomerular Ultrafiltrate

The fluid that enters the glomerular capsule is called *ultrafiltrate* (fig. 16.9), because it is formed under pressure (the hydrostatic pressure of the blood). This is similar to the formation of tissue fluid by other capillary beds in the body in response to Starling forces (chapter 14). The force favoring filtration is opposed by a counterforce developed by the hydrostatic pressure of fluid in the glomerular capsule. Also, since the protein concentration of the tubular fluid is low (less than 2–5 mg per 100 ml) compared to that of plasma (6–8 g per 100 ml), the greater colloid osmotic pressure of plasma promotes the osmotic return of filtered water. When these opposing forces are subtracted from the hydrostatic pressure of the glomerular capillaries, a *net filtration pressure* of approximately 10 mm Hg is obtained.

Because glomerular capillaries are extremely permeable and have a high surface area, this modest net filtration pressure produces an extraordinarily large volume of filtrate. The **glomerular filtration rate (GFR)** is the volume of filtrate produced per minute by both kidneys. The GFR averages 115 ml per minute in women and 125 ml per minute in men. This is equivalent to 7.5 L per hour or 180 L per day (about 45 gallons)! Since the total blood volume averages about 5.5 L, this means that the total blood volume is filtered into the urinary tubules every forty minutes. Most of the filtered water must obviously be returned immediately to the vascular system, or a person would literally urinate to death within several minutes.

**Figure 16.7.** (*a*) The inner (visceral) layer of Bowman's capsule is composed of podocytes, as shown in this scanning electron micrograph. Very fine extensions of these podocytes form foot processes, or pedicels, that interdigitate around the glomerular capillaries. Spaces between adjacent pedicels form the "filtration slits." (*b*) An illustration of the relationship between glomerular capillaries and the inner layer of Bowman's capsule.

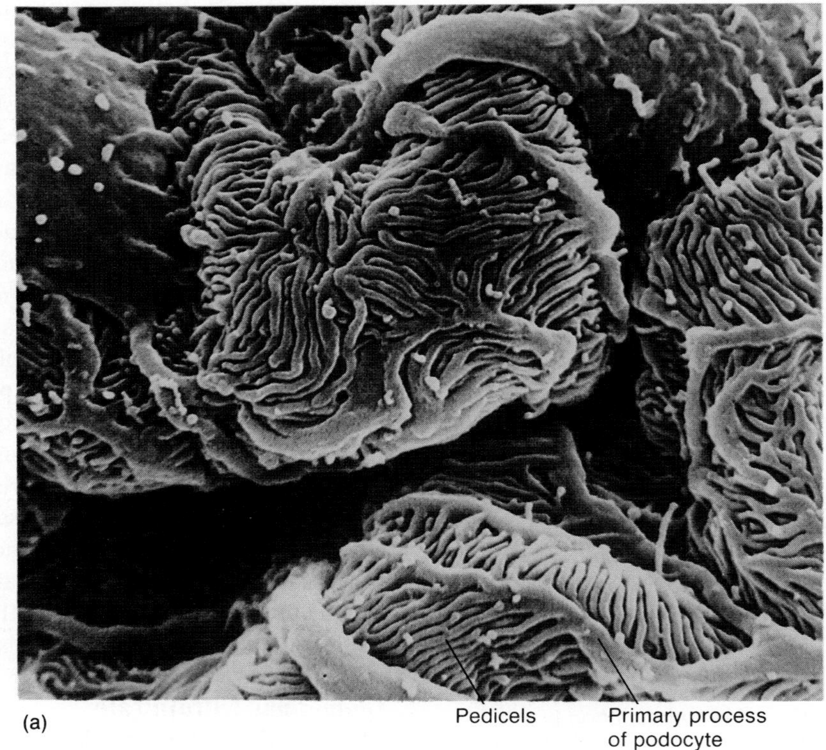


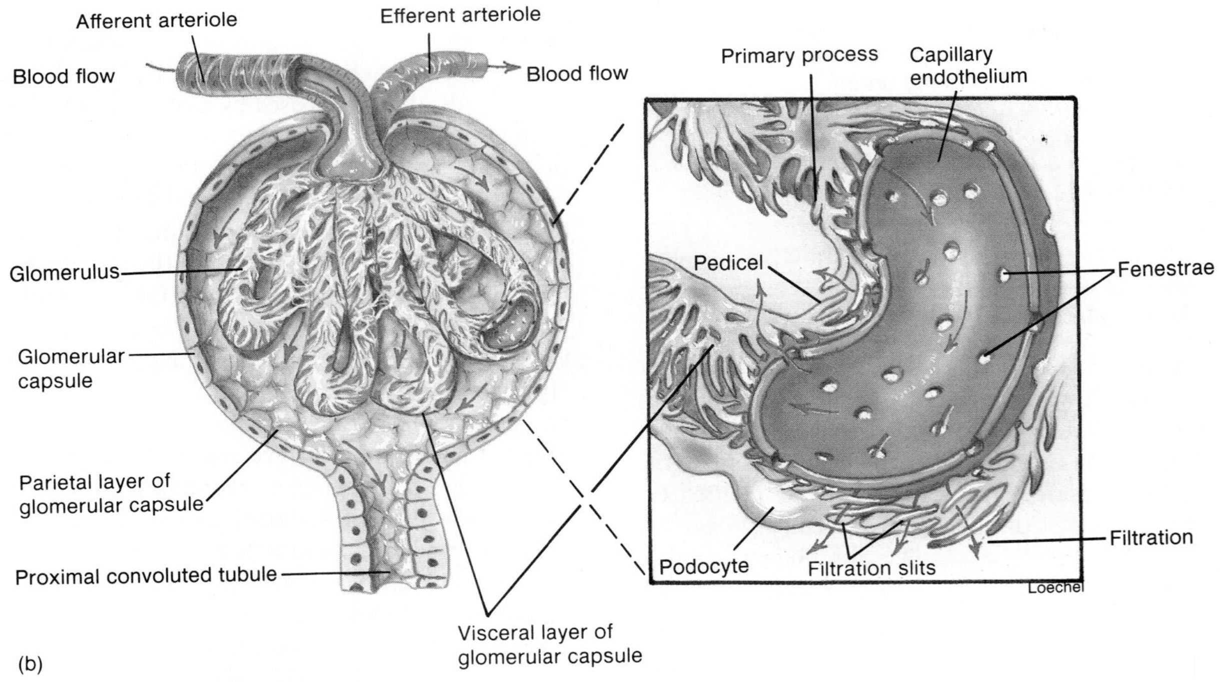

**Figure 16.8.** (*a*) An electron micrograph and (*b*) an illustration of the "filtration barrier" between the capillary lumen and the cavity of the glomerular (Bowman's) capsule.

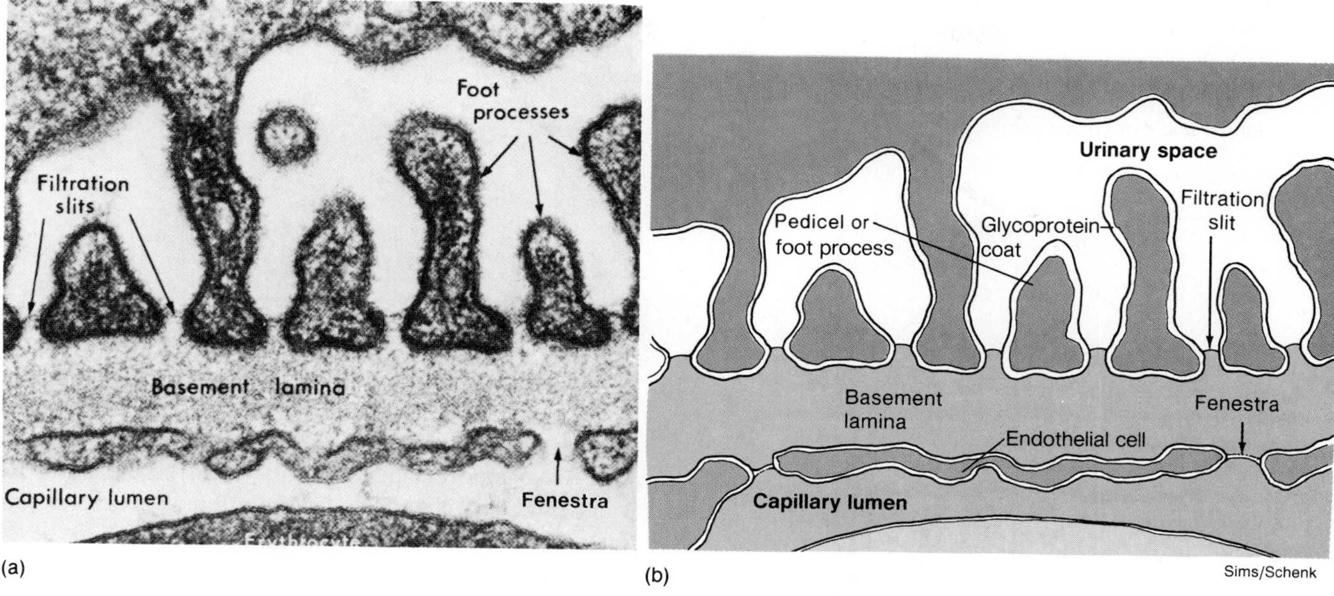


**Figure 16.9.** The formation of glomerular ultrafiltrate. Proteins (*large circles*) are not filtered, but smaller plasma solutes (*dots*) easily enter the glomerular ultrafiltrate. Arrows indicate the direction of filtration.

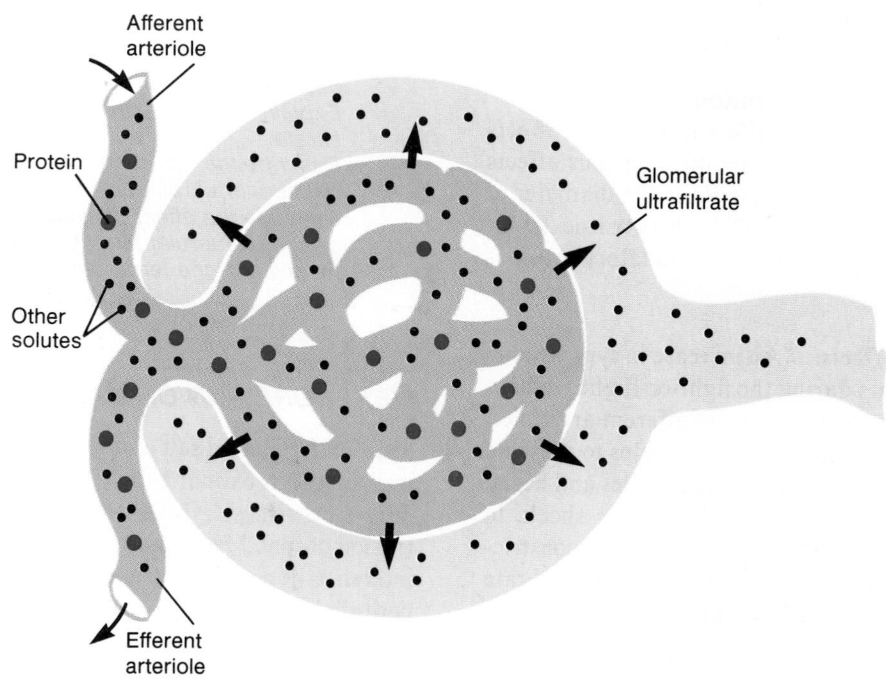

**Figure 16.10.** The effect of increased sympathetic nerve activity on kidney function and other physiological processes.

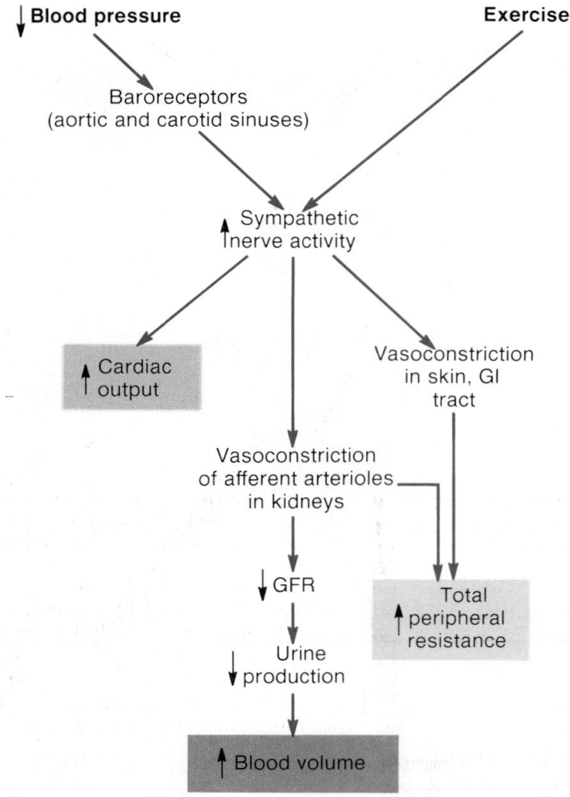


## Regulation of Glomerular Filtration Rate

Vasoconstriction or dilation of afferent arterioles affects the rate of blood flow to the glomerulus and thus affects the glomerular filtration rate. Changes in the diameter of the afferent arterioles result from both extrinsic (sympathetic innervation) and intrinsic regulatory mechanisms.

***Sympathetic Nerve Effects.*** An increase in sympathetic nerve activity, as occurs during the fight-or-flight reaction and exercise, stimulates constriction of afferent arterioles. This is an alpha-adrenergic effect, which helps to preserve blood volume and to divert blood to the muscles and heart. A similar effect occurs during cardiovascular shock, in which sympathetic nerve activity stimulates vasoconstriction. The decreased GFR and the resulting decreased rate of urine formation help to compensate for the rapid drop of blood pressure under these circumstances (fig. 16.10).

***Renal Autoregulation.*** When the direct effect of sympathetic stimulation is experimentally removed, the effect of systemic blood pressure on GFR can be observed. Under these conditions, surprisingly, the GFR remains relatively constant despite changes in mean arterial pressure within a range of 70–180 mm Hg (normal mean arterial pressure is 100 mm Hg). The ability of the kidneys to maintain a relatively constant GFR in the face of fluctuating blood pressures is called **renal autoregulation.**

Renal autoregulation results from the effects of locally produced chemicals on the afferent arterioles (effects on the efferent arterioles are believed to be of secondary importance). When systemic arterial pressure falls toward a mean of 70 mm Hg, the afferent arterioles dilate, and when the pressure rises, the afferent arterioles constrict. Blood flow to the glomeruli and GFR can thus remain relatively constant within the autoregulatory range of blood pressure values. The effects of different regulatory mechanisms on GFR are summarized in table 16.1.

**Table 16.1** Regulation of glomerular filtration rate (GFR)

| Regulation | Stimulus | Afferent Arteriole | GFR |
|---|---|---|---|
| Sympathetic nerves | Activation by aortic and carotid baroreceptors or by higher brain centers | Constricts | Decreases |
| Autoregulation | Decreased blood pressure | Dilates | No change |
| Autoregulation | Increased blood pressure | Constricts | No change |

1. *Describe the structures that plasma fluid must pass through to enter the glomerular capsule. Explain how proteins are excluded from the filtrate.*
2. *Describe the forces that affect the formation of glomerular ultrafiltrate.*
3. *Describe the effect of sympathetic innervation on the glomerular filtration rate, and explain the meaning of the term "renal autoregulation."*

# Reabsorption of Salt and Water

Most of the filtered salt and water is reabsorbed across the wall of the proximal tubule. The reabsorption of water occurs by osmosis, in which water follows the active extrusion of NaCl from the tubule and into the surrounding interstitial spaces. Salt and water then diffuse into the peritubular capillaries. Most of the remaining water in the filtrate is reabsorbed across the wall of the collecting duct. This also occurs by osmosis as a result of the fact that the interstitial fluid in the renal medulla has a higher osmotic pressure than the filtrate. The high osmotic pressure of the renal medulla is produced by transport processes that operate in the loop of Henle. Since antidiuretic hormone (ADH) increases the permeability of the collecting duct to water, more ADH secretion results in more water reabsorption.

**Figure 16.11.** Plasma water and its dissolved solutes (except proteins) enter the glomerular ultrafiltrate, but most of these filtered molecules are reabsorbed. The term *reabsorption* refers to the transport of molecules out of the tubular filtrate back into the blood.

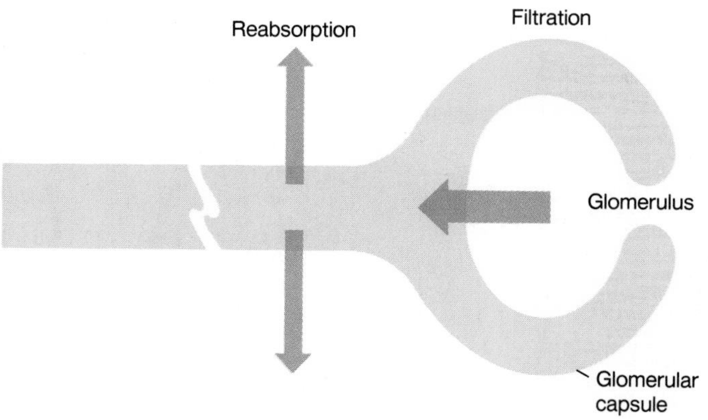


Although about 180 L per day of glomerular ultrafiltrate are produced, the kidneys normally excrete only 1–2 L per day of urine. Approximately 99% of the filtrate must thus be returned to the vascular system, while 1% is excreted in the urine. The urine volume, however, varies according to the needs of the body. When a well-hydrated person drinks a liter or more of water, urine volume increases to 16 ml per minute (the equivalent of 23 L per day if this were to continue for twenty-four hours). In severe dehydration, when the body needs to conserve water, only 0.3 ml per minute, or 400 ml per day, of urine are produced. A volume of 400 ml per day of urine is needed to excrete the amount of metabolic wastes produced by the body; this is called the *obligatory water loss*. When water in excess of this amount is excreted, the urine volume is increased and its concentration is decreased.

Regardless of the body's state of hydration, it is clear that most of the filtered water must be returned to the vascular system to maintain blood volume and pressure. The return of filtered molecules from the tubules to the blood is called **reabsorption** (fig. 16.11). It is important to realize that the transport of water always occurs passively by *osmosis;* there is no such thing as active transport of water. A concentration gradient must thus be created between tubular fluid and blood that favors the osmotic return of water to the vascular system.

## Reabsorption in the Proximal Tubule

Since all plasma solutes, with the exception of proteins, are able to enter the glomerular ultrafiltrate freely, the osmolality of the filtrate is essentially the same as that of plasma (300 milliosmoles per liter, or 300 mOsm). The filtrate is thus said to be isosmotic with the plasma. Osmosis thus cannot occur unless the concentration of plasma in the peritubular capillaries and the concentration of filtrate are altered by active transport processes. This is achieved by the active transport of $Na^+$ from the filtrate to the peritubular blood.

***Active and Passive Transport.*** The epithelial cells that compose the wall of the proximal tubule are joined together by gap junctions only on their apical sides—that is, the sides of each cell that are closest to the lumen of the tubule (fig. 16.12). Each cell therefore has four exposed surfaces: the apical side facing the lumen, which contains microvilli, the opposite, basal side facing the peritubular capillaries, and the lateral sides, facing the narrow clefts between adjacent epithelial cells.

The concentration of $Na^+$ in the glomerular ultrafiltrate—and thus in the fluid entering the proximal tubule—is the same as in plasma. The epithelial cells of the tubule, however, have a much lower $Na^+$ concentration. This lower $Na^+$ concentration is partially due to the low permeability of the cell membrane to $Na^+$ and partially due to the active transport of $Na^+$ out of the cell by $Na^+/K^+$ pumps, as described in chapter 6. In the cells of the proximal tubule, the $Na^+/K^+$ pumps are located in the basal and lateral sides of the cell membrane but not in the apical membrane. As a result of the action of these active transport pumps, a concentration gradient is created that favors the diffusion of $Na^+$ from the tubular fluid, across the apical cell membranes, and into the epithelial cells of the proximal tubule. The $Na^+$ is then extruded into the surrounding tissue fluid by the $Na^+/K^+$ pumps.

The transport of $Na^+$ from the tubular fluid to the interstitial (tissue) fluid surrounding the epithelial cells of the proximal tubule creates a potential difference across the wall of the tubule. This electrical gradient favors the passive transport of $Cl^-$ toward the higher $Na^+$ concentration in the tissue fluid. Chloride ions, therefore, passively follow sodium ions out of the filtrate to the interstitial

**Figure 16.12.** An illustration of the appearance of tubule cells in the electron microscope. Molecules that are reabsorbed pass through the tubule cells from the apical membrane (facing the filtrate) to the basolateral membrane (facing the blood).

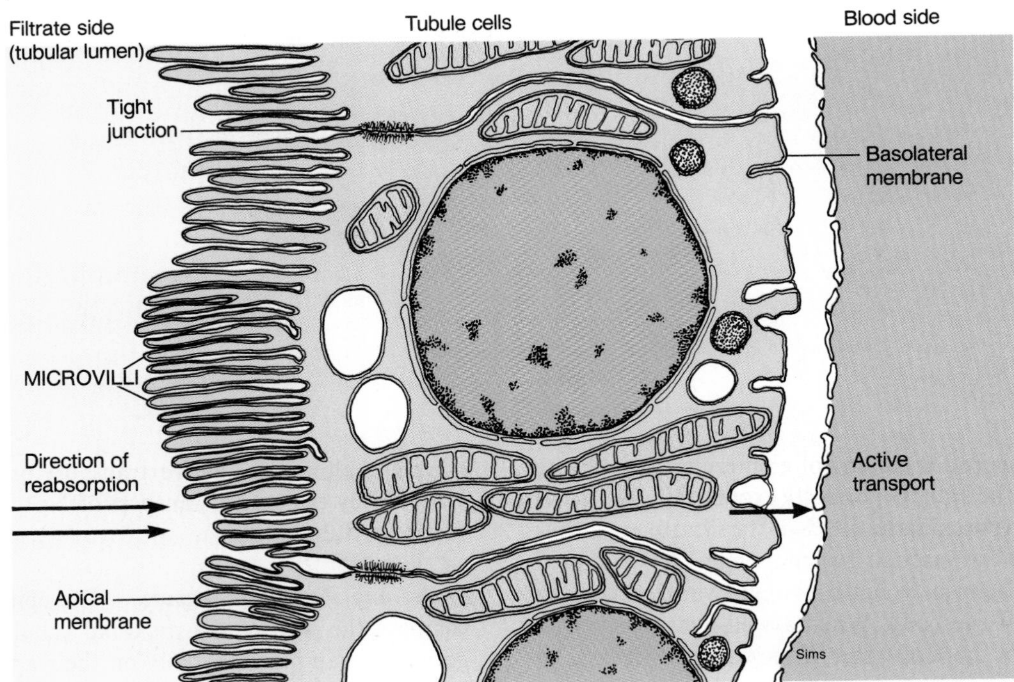


fluid. As a result of the accumulation of NaCl, the osmolality and osmotic pressure of the tissue fluid surrounding the epithelial cells is increased above that of the tubular fluid. This is particularly true of the tissue fluid between the lateral membranes of adjacent epithelial cells, where the narrow spaces permit the accumulated NaCl to achieve a higher concentration.

An osmotic gradient is thus created beween the tubular fluid and the tissue fluid surrounding the proximal tubule. Since the cells of the proximal tubule are permeable to water, water moves by osmosis from the tubular fluid into the epithelial cells and then across the basal and lateral sides of the epithelial cells into the tissue fluid. The salt and water which were reabsorbed from the tubular fluid can then move passively into the surrounding peritubular capillaries and in this way be returned to the blood (fig. 16.13).

***Significance of Proximal Tubule Reabsorption.*** Approximately 65% of the salt and water in the original glomerular ultrafiltrate is reabsorbed across the proximal tubule and returned to the vascular system. The volume of tubular fluid remaining is reduced accordingly, but this fluid is still isosmotic with the blood (has a concentration of 300 mOsm). This results from the fact that the cell membranes in the proximal tubule are freely permeable to water so that water and salt are removed in proportionate amounts.

An additional smaller amount of salt and water is returned to the vascular system by reabsorption in the loop of Henle. This reabsorption, like that in the proximal tubule, occurs constantly regardless of the person's state of hydration. Unlike reabsorption in later regions of the nephron, it is not subject to hormonal regulation. Approximately 85% of the filtered salt and water is, therefore, reabsorbed in a constant, unregulated fashion in the early regions of the nephron (proximal tubule and loop of Henle). This reabsorption is very costly in terms of energy expenditures, accounting for as much as 6% of the calories consumed by the body at rest.

Since 85% of the original glomerular ultrafiltrate is immediately reabsorbed in the early region of the nephron, only 15% of the initial filtrate remains to enter the distal convoluted tubule and collecting duct. This is still a large volume of fluid—15% × GFR (180 L per day) = 27 L

**Figure 16.13.** Mechanisms of salt and water reabsorption in the proximal tubule. Sodium is actively transported out of the filtrate, and chloride follows passively by electrical attraction. Water follows the salt out of the tubular filtrate into the peritubular capillaries by osmosis.

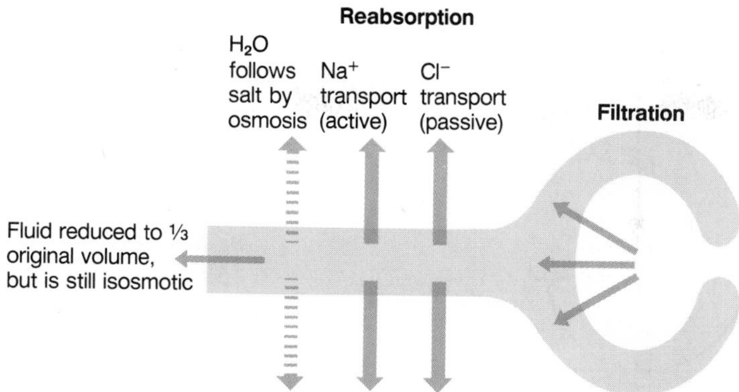


per day—that must be reabsorbed to varying degrees in accordance with the body's state of hydration. This "fine tuning" of the percent reabsorption and urine volume is accomplished by the action of hormones on the later regions of the nephron.

## The Countercurrent Multiplier System

Water cannot be actively transported across the tubule wall, and osmosis of water cannot occur if the tubular fluid and surrounding tissue fluid are isotonic to each other. In order for water to be reabsorbed by osmosis, the surrounding tissue fluid must be hypertonic. The osmotic pressure of the tissue fluid in the renal medulla is, in fact, raised to over four times that of plasma. This results partly from the fact that the tubule bends; the geometry of the loop of Henle allows interaction to occur between the descending and ascending limbs. Since the ascending limb is the active partner in this interaction, its properties will be described before those of the descending limb.

***Ascending Limb of the Loop of Henle.*** Salt (NaCl) is actively extruded from the ascending limb into the surrounding tissue fluid. This is not accomplished, however, by the same process that occurs in the proximal tubule. Instead, $Na^+$, $K^+$, and $Cl^-$ passively diffuse from the filtrate into the ascending limb cells, in a ratio of 1 $Na^+$ to 1 $K^+$ to 2 $Cl^-$. The $Na^+$ is then actively transported across the basolateral membrane to the tissue fluid by the $Na^+$/$K^+$ pump. $Cl^-$ follows the $Na^+$ passively because of electrical attraction, and $K^+$ diffuses back into the filtrate (fig. 16.14).

The ascending limb is structurally divisible into two regions: a *thin segment,* nearest to the tip of the loop, and a *thick segment* of varying lengths, which carries the filtrate outwards into the cortex and into the distal convoluted tubule. It is currently believed that only the cells of the thick segments of the ascending limb are capable of actively transporting NaCl from the filtrate into the surrounding tissue fluid.

Though the mechanism of NaCl transport is different in the ascending limb than in the proximal tubule, the net effect is the same: salt (NaCl) is extruded into the surrounding tissue fluid. Unlike the epithelial walls of the proximal tubule, however, the walls of the ascending limb of the loop of Henle are *not permeable to water.* The tubular fluid thus becomes increasingly dilute as it ascends toward the cortex, whereas the tissue fluid around the loops of Henle in the medulla becomes increasingly more concentrated. By means of these processes, the tubular fluid that enters the distal tubule in the cortex is made hypotonic (with a concentration of about 100 mOsm), whereas the tissue fluid in the medulla is made hypertonic.

***Descending Limb of the Loop of Henle.*** The deeper regions of the medulla, around the tips of the loops of juxtamedullary nephrons, reach a concentration of 1200–1400 mOsm. In order to reach this high a concentration, the salt pumped out of the ascending limb must accumulate in the tissue fluid of the medulla. This occurs as a result of the properties of the descending limb, to be discussed

**Figure 16.14.** In the thick segment of the ascending limb of the loop, $Na^+$ and $K^+$, together with two $Cl^-$, enter the tubule cells. $Na^+$ is then actively transported out into the interstitial space, and $Cl^-$ follows passively. The $K^+$ diffuses back into the filtrate, and some also enters the interstitial space.

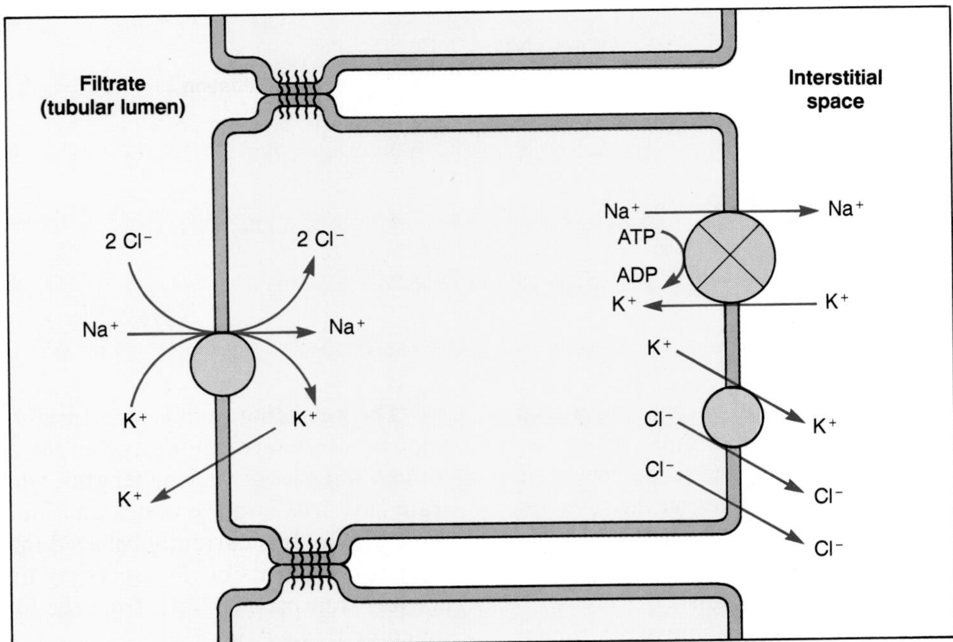


next, and as a result of the fact that blood vessels around the loop do not carry back all of the extruded salt to the general circulation. The capillaries in the medulla are uniquely arranged to trap NaCl in the tissue fluid, as will be discussed in a later section.

The descending limb does not actively transport salt, and indeed is believed to be impermeable to the passive diffusion of salt. It is, however, permeable to water. Since the surrounding interstitial fluid is hypertonic to the filtrate in the descending limb, water is drawn out of the descending limb by osmosis and enters blood capillaries. The concentration of tubular fluid is thus increased, and its volume is decreased, as it descends toward the tips of the loops.

As a result of these passive transport processes in the descending limb, the fluid that "rounds the bend" at the tip of the loop has the same osmolality as the surrounding tissue fluid (1200–1400 mOsm). There is, therefore, a higher salt concentration arriving in the ascending limb than there would be if the descending limb simply delivered isotonic fluid. Salt transport by the ascending limb is increased accordingly, so that the "saltiness" of the tissue fluid is multiplied (fig. 16.15).

***Countercurrent Multiplication.*** Countercurrent flow (flow in opposite directions) in the ascending and descending limbs and the close proximity of the two limbs allow interaction to occur. Since the concentration of the tubular fluid in the descending limb reflects the concentration of surrounding tissue fluid, and since the concentration of this tissue fluid is raised by the active extrusion of salt from the ascending limb, a *positive feedback* mechanism is created. The more salt the ascending limb extrudes, the more concentrated will be the fluid that returns to it from the descending limb. This positive feedback mechanism multiplies the concentration of tissue fluid and descending limb fluid and is thus called the **countercurrent multiplier system.**

The countercurrent multiplier system recirculates salt and thus traps some of the salt that enters the loop of Henle in the tissue fluid of the renal medulla. This system results in a gradually increasing concentration of renal tissue fluid from the cortex to the inner medulla; the osmolality of tissue fluid increases from 300 mOsm (isotonic) in the cortex to 1200–1400 mOsm in the deepest part of the medulla.

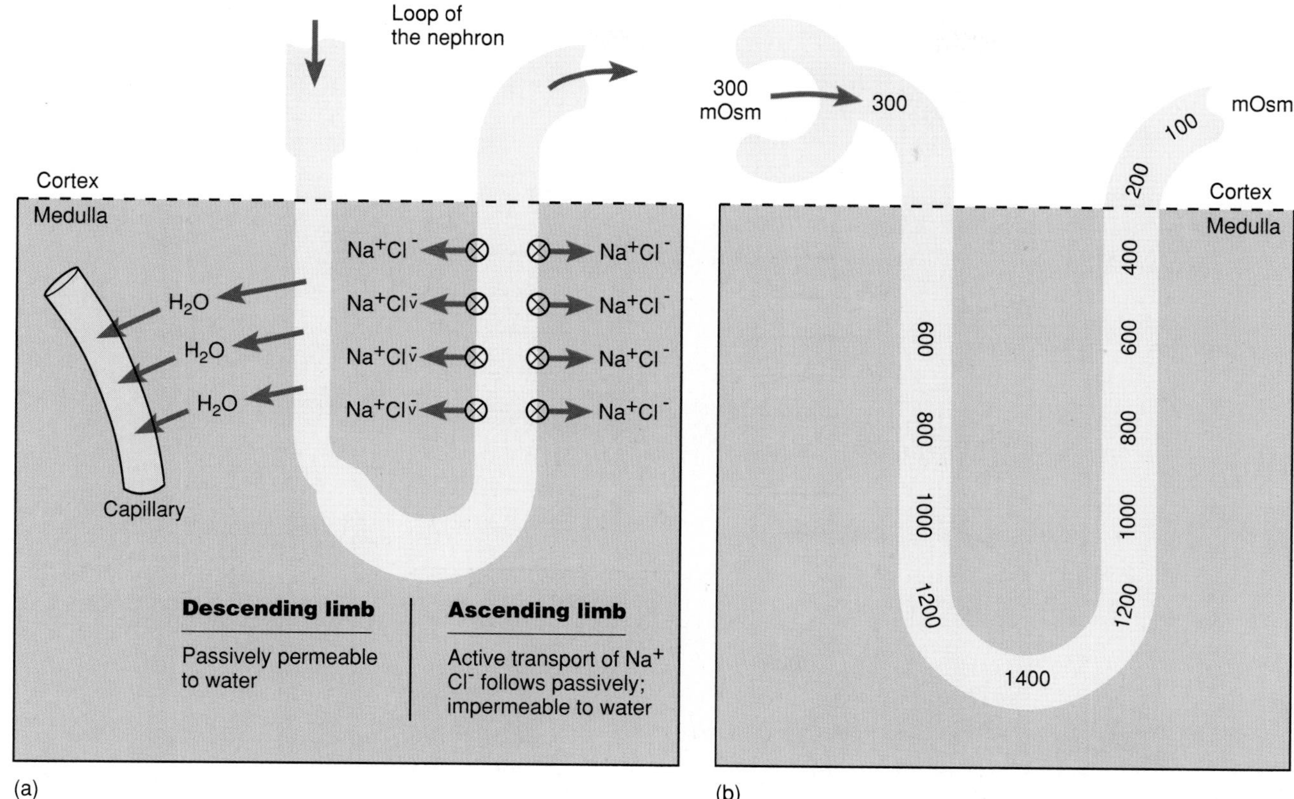


**Figure 16.15.** The countercurrent multiplier system. The extrusion of sodium chloride from the ascending limb makes the surrounding tissue fluid more concentrated. This concentration is multiplied by the fact that the descending limb is passively permeable so that its fluid increases in concentration as the surrounding tissue fluid becomes more concentrated. The transport properties of the loop and their effect on tubular fluid concentration is shown in (*a*). The values of these changes in osmolality, together with the effect on surrounding tissue fluid concentration, are shown in (*b*).

***Vasa Recta.*** In order for the countercurrent multiplier system to be effective, most of the salt that is extruded from the ascending limbs must remain in the tissue fluid of the medulla, while most of the water that leaves the descending limbs must be removed by the blood. This is accomplished by vessels known as the **vasa recta,** which form long capillary loops that parallel the long loops of Henle of the juxtamedullary nephrons (see fig. 16.18).

The vasa recta maintain the hypertonicity of the renal medulla by means of a mechanism known as **countercurrent exchange.** Salt and other solutes (such as urea, described in the next section) that are present at high concentrations in the medullary tissue fluid diffuse into the blood as the blood descends into the capillary loops of the vasa recta, but then passively diffuse out of the ascending vessels and back into the descending vessels (where the concentration is lower). Solutes are thus recirculated and trapped within the medulla. Since the walls of the vasa recta are freely permeable to dissolved solutes, the concentration of these solutes becomes the same inside the vasa recta as in the surrounding interstitial fluid at each level within the medulla. The colloid osmotic pressure within the vasa recta, however, is higher than in the interstitial fluid because plasma proteins do not easily pass through the capillary walls. This is similar to the situation in other capillary beds (chapter 14), and results in the osmotic movement of water into both the descending and ascending limbs of the vasa recta. The vasa recta thus trap salt and urea within the interstitial fluid but transport water out of the renal medulla (fig. 16.16).

***Effects of Urea.*** Countercurrent multiplication of the NaCl concentration is the most well-established mechanism to explain the hypertonicity of the interstitial fluid in the medulla. It is currently believed, however, that urea also contributes significantly to the total osmolality of the interstitial fluid.

**Figure 16.16.** Countercurrent exchange in the vasa recta. The diffusion of salt and water first into and then out of these blood vessels helps to maintain the "saltiness" (hypertonicity) of the interstitial fluid in the renal medulla (numbers indicate osmolality).

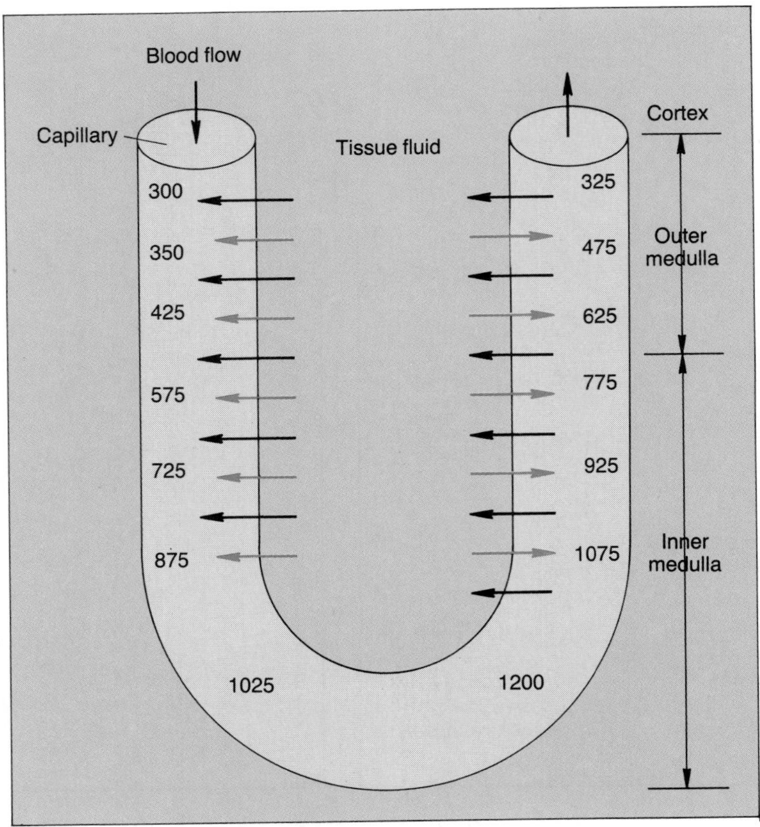


The role of urea was inferred from experimental evidence showing that active transport of $Na^+$ only occurs in the thick segments of the ascending limbs. The thin segments of the ascending limbs, which are located in the deeper regions of the medulla, are not able to extrude salt actively. Since salt does indeed leave the thin segments, a diffusion gradient for salt must exist even though the surrounding tissue fluid has the same osmolality as the tubular fluid. Investigators therefore concluded that molecules other than salt—specifically urea—contribute to the hypertonicity of the tissue fluid.

It was later shown that the ascending limb of the loop of Henle and the collecting duct are permeable to urea. Urea can thus diffuse out of the collecting duct and into the ascending limb (fig. 16.17). In this way, a certain amount of urea is recycled through these two segments of the nephron and is thus trapped in the interstitial fluid. The transport properties of different tubule segments are summarized in table 16.2.

## Collecting Duct: Effect of Antidiuretic Hormone (ADH)

As a result of the recycling of salt between the ascending and descending limbs and possibly of the recycling of urea between the collecting duct and the loop of Henle, the medullary tissue fluid is made very hypertonic. The collecting ducts must transport their fluid through this hypertonic environment in order to empty their contents of urine into the calyces. Whereas the fluid surrounding the collecting ducts in the medulla is hypertonic, the fluid that passes into the collecting ducts in the cortex is hypotonic as a result of the active extrusion of salt by the ascending limbs of the loops.

The walls of the collecting ducts are *permeable to water but not to salt*. Since the surrounding tissue fluid in the renal medulla is very hypertonic, as a result of the countercurrent multiplier system, water is drawn out of the collecting ducts by osmosis. This water does not dilute the surrounding tissue fluid because it is transported by

**Figure 16.17.** According to some authorities, urea diffuses out of the collecting duct and contributes significantly to the concentration of the interstitial fluid in the renal medulla. The active transport of $Na^+$ out of the thick segments of the ascending limbs also contributes to the hypertonicity of the medulla so that water is reabsorbed by osmosis from the collecting ducts.

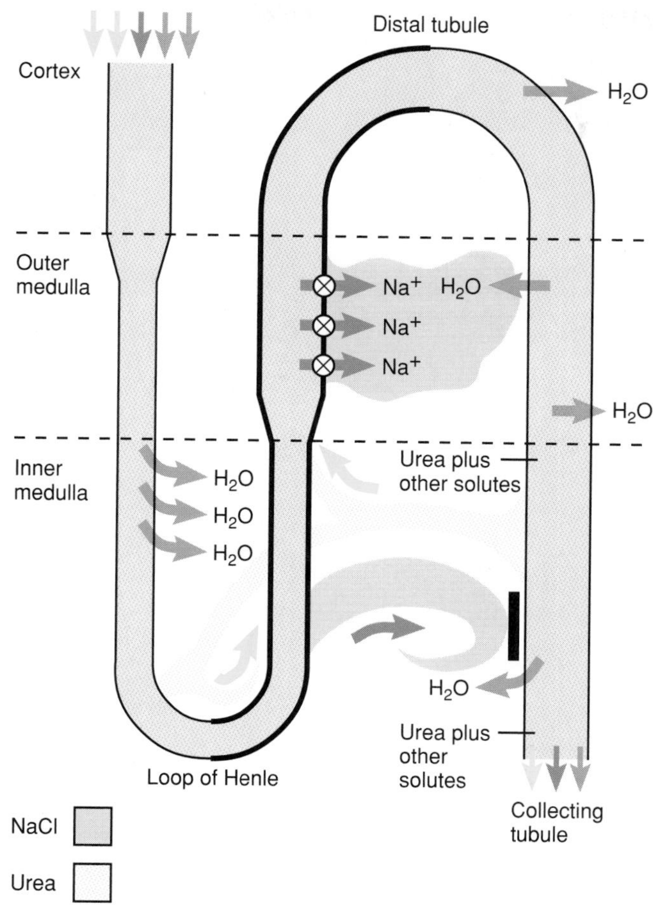


**Table 16.2** Properties of different nephron segments in regard to the concentrating-diluting mechanisms of the kidney

| Nephron Segment | Active Transport | Passive Transport | | |
|---|---|---|---|---|
| | | Salt | Water | Urea |
| Proximal tubule | $Na^+$ | $Cl^-$ | Yes | Yes |
| Descending limb of Henle's loop | None | Maybe | Yes | No |
| Thin segment of ascending limb | None | NaCl | No | Yes |
| Thick segment of ascending limb | $Na^+$ | $Cl^-$ | No | No |
| Distal tubule | $Na^+$ | No | No | No |
| Collecting duct* | Slight $Na^+$ | No | Yes (ADH) or Slight (no ADH) | Yes |

*The permeability of the collecting duct to water depends on the presence of ADH.

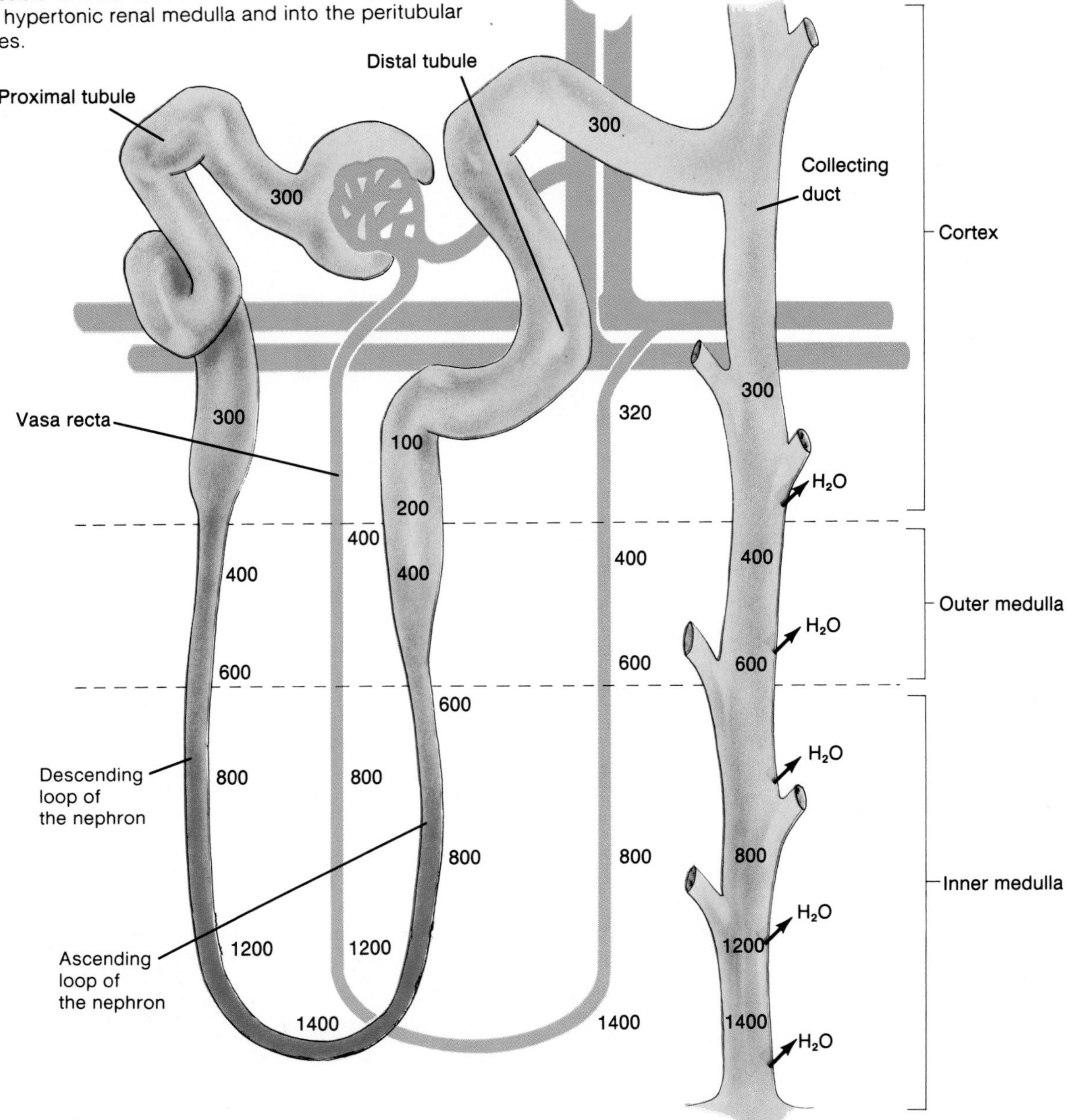


**Figure 16.18.** The countercurrent multiplier system in the loop of the nephron and countercurrent exchange in the vasa recta help create a hypertonic renal medulla. Under the influence of antidiuretic hormone (ADH), the collecting duct is permeable to water so that water is drawn by osmosis out into the hypertonic renal medulla and into the peritubular capillaries.

capillaries to the general circulation. In this way, most of the water remaining in the filtrate is returned to the vascular system (fig. 16.18).

The osmotic gradient created by the countercurrent multiplier system provides the force for water reabsorption through the collecting ducts. The rate of this reabsorption, however, is determined by the permeability of the collecting duct cell membranes to water. The permeability of the collecting duct to water, in turn, is determined by the concentration of **antidiuretic hormone (ADH)** in the blood. When the concentration of ADH is increased, the collecting ducts become more permeable to water, and more water is reabsorbed. A decrease in ADH, conversely, results in less reabsorption of water and thus in the excretion of a larger volume of more dilute urine.

ADH is produced by neurons in the hypothalamus and is secreted from the posterior pituitary gland (chapter 11).

**Table 16.3** Antidiuretic hormone secretion and action

| Stimulus | Receptors | Secretion of ADH | Effects on | |
|---|---|---|---|---|
| | | | Urine Volume | Blood |
| ↑Osmolality (dehydration) | Osmoreceptors in hypothalamus | Increased | Decreased | Increased water retention; decreased blood osmolality |
| ↓Osmolality | Osmoreceptors in hypothalamus | Decreased | Increased | Water loss increases blood osmolality |
| ↑Blood volume | Stretch receptors in left atrium | Decreased | Increased | Decreased blood volume |
| ↓Blood volume | Stretch receptors in left atrium | Increased | Decreased | Increased blood volume |

The secretion of ADH is stimulated when osmoreceptors in the hypothalamus respond to an increase in blood osmotic pressure. During dehydration, therefore, when the plasma becomes more concentrated, increased secretion of ADH promotes increased permeability of the collecting ducts to water. In severe dehydration only the minimal amount of water needed to eliminate the body's wastes is excreted. This minimum, about 400 ml per day, is limited by the fact that urine cannot become more concentrated than the medullary tissue fluid surrounding the collecting ducts. Under these conditions about 99.8% of the initial glomerular ultrafiltrate is reabsorbed.

A person in a state of normal hydration excretes about 1.5 L per day of urine, indicating that 99.2% of the glomerular ultrafiltrate volume is reabsorbed. Notice that small changes in percent reabsorption translate into large changes in urine volume. Increasing water ingestion—and thus decreasing ADH secretion (table 16.3)—results in correspondingly larger volumes of urine excretion. It should be noted that even in the complete absence of ADH some water is still reabsorbed through the collecting ducts.

**Diabetes insipidus** is a disease associated with the inadequate secretion or action of ADH. The collecting ducts are thus not very permeable to water and, therefore, a large volume (5–10 L per day) of dilute urine is produced. The dehydration that results causes intense thirst, but a person with this condition has difficulty drinking enough to compensate for the large volumes of water lost in the urine.

1. *Describe the mechanisms for salt and water reabsorption in the proximal tubule.*
2. *Compare the transport of $Na^+$, $Cl^-$, and water across the walls of the proximal tubule, ascending and descending limbs of the loop of Henle, and collecting duct.*
3. *Explain the interaction of the ascending and descending limbs of the loop and how this interaction results in a hypertonic renal medulla.*
4. *Explain how ADH helps the body to conserve water, and describe how variations in ADH secretion affect the volume and concentration of urine.*

## Renal Plasma Clearance

As blood passes through the kidneys, some of the constituents of the plasma are removed and excreted in the urine. The blood is thus "cleared," to one degree or another, of particular solutes in the process of urine formation. These solutes may be removed from the blood by filtration through the glomerular capillaries and by secretion by the tubular cells into the filtrate. At the same time, certain molecules in the tubular fluid can be reabsorbed back into the blood. The process of reabsorption thus reduces the renal clearance of these substances. Calculations of the renal plasma clearance of a substance indicate whether or not the kidneys reabsorb or secrete that substance.

One of the major functions of the kidneys is the excretion of waste products such as urea, creatinine, and other molecules. These molecules are filtered through the glomerulus into the glomerular capsule along with water, salt, and other plasma solutes. In addition, some waste products can gain access to the urine by a process called **secretion** (fig. 16.19). Secretion is the opposite of reabsorption. Molecules that are secreted move out of the peritubular capillaries and into the tubular cells, from which they are actively transported into the tubular lumen. In this way, molecules that were not filtered out of the blood in the glomerulus, but instead passed through the efferent arterioles to the peritubular capillaries, can still be excreted in the urine.

Although most (about 99%) of the filtered water is returned to the vascular system by reabsorption, most of the unneeded molecules that are filtered or secreted are eliminated in the urine. The concentration of these substances in the renal vein leaving the kidney is therefore lower than their concentrations in the blood entering the kidney in the renal artery. Some of the blood that passes through the kidneys, in other words, is "cleared" of these waste products.

### Renal Excretion of Inulin: Measurement of GFR

If a substance is neither reabsorbed nor secreted by the tubules, the amount excreted per minute in the urine will be equal to the amount that is filtered out of the glomeruli.

**Figure 16.19.** Secretion refers to the active transport of substances from the peritubular capillaries into the tubular fluid. This transport is in a direction opposite to that of reabsorption.

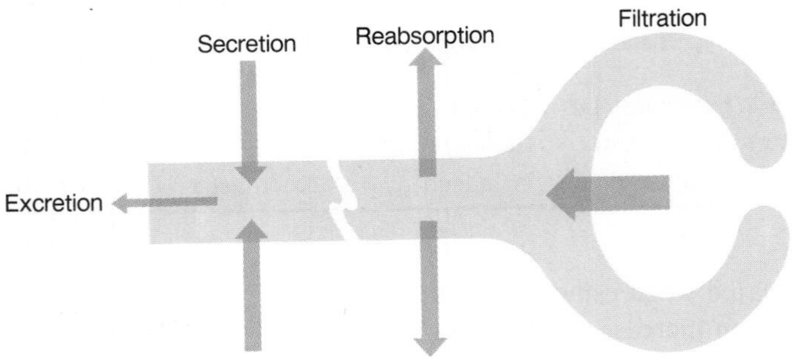


**Figure 16.20.** The renal clearance of inulin. (*a*) Inulin is present in the blood entering the glomeruli, and (*b*) some of this blood, together with its dissolved inulin, is filtered. All of this filtered inulin enters the urine, whereas most of the filtered water is returned to the vascular system (is reabsorbed). (*c*) The blood leaving the kidneys in the renal vein, therefore, contains less inulin than the blood that entered the kidneys in the renal artery. Since inulin is filtered but neither reabsorbed nor secreted, the inulin clearance rate equals the glomerular filtration rate (GFR).

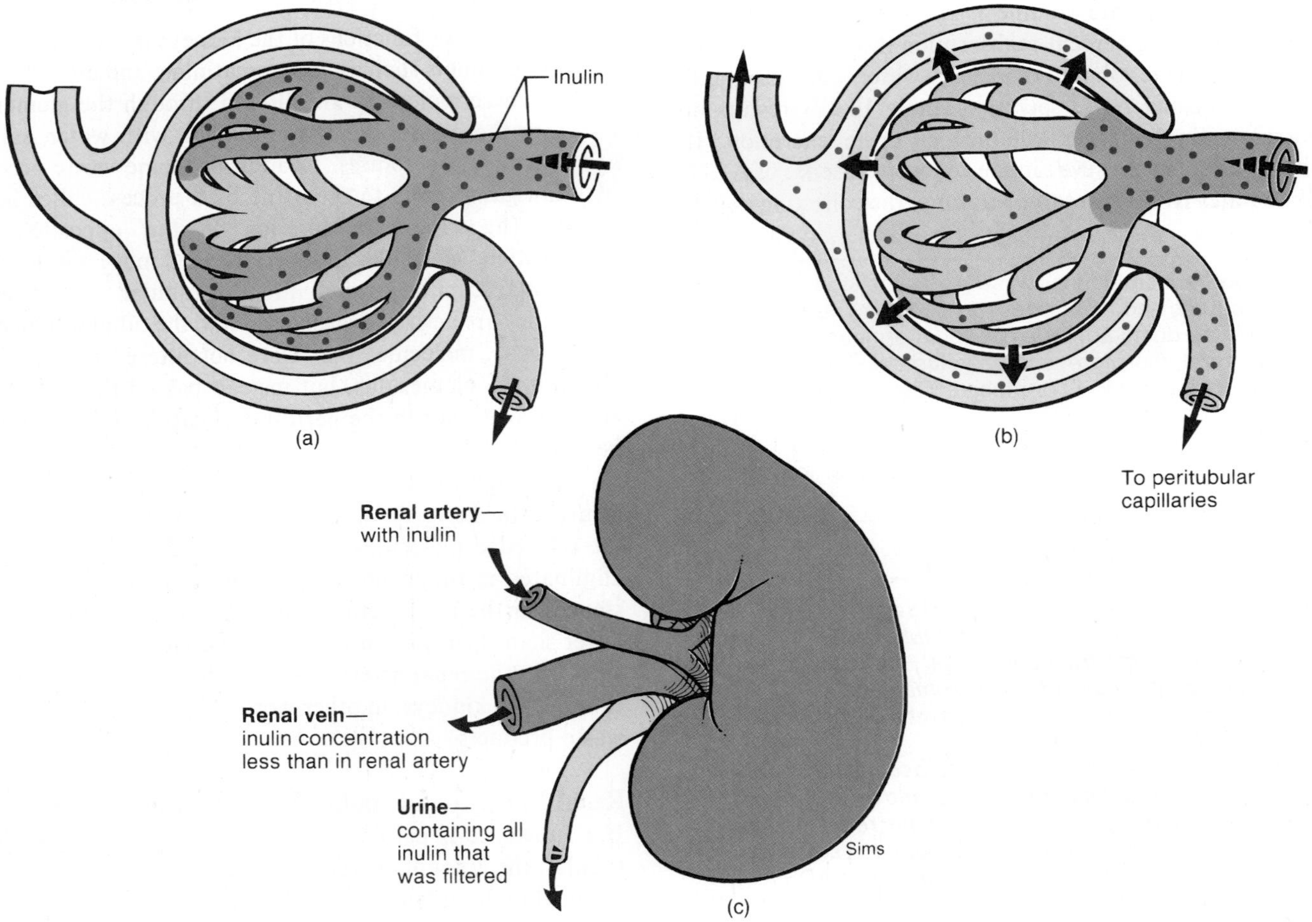

Inulin

(a)

(b)

To peritubular capillaries

**Renal artery**— with inulin

**Renal vein**— inulin concentration less than in renal artery

**Urine**— containing all inulin that was filtered

Sims

(c)

**Table 16.4** The effects of filtration, reabsorption, and secretion on renal clearance rates

| Term | Means | Effect on Renal Clearance |
|---|---|---|
| Filtered | A substance enters the glomerular ultrafiltrate | Some or all of a filtered substance may enter the urine and be "cleared" from the blood. |
| Reabsorbed | The transport of a substance from the filtrate, through tubular cells, and into the blood | Reabsorption decreases the rate at which a substance is cleared; clearance rate is less than the glomerular filtration rate (GFR). |
| Secreted | The transport of a substance from peritubular blood, through tubular cells, and into the filtrate | When a substance is secreted by the nephrons, its clearance rate is greater than the GFR. |

There does not seem to be a single substance produced by the body, however, that is not reabsorbed or secreted to some degree. Plants such as artichokes, dahlias, onions, and garlic, fortunately, do produce such a compound. This compound, a polymer of the monosaccharide fructose, is *inulin.* Once injected into the blood, inulin is filtered by the glomeruli, and the amount of inulin excreted per minute is exactly equal to the amount that was filtered per minute (fig. 16.20).

If the concentration of inulin in urine is measured and the rate of urine formation is determined, the rate of inulin excretion can easily be calculated:

$$\frac{\textit{Quantity excreted per minute}}{(\text{mg/min})} = \underset{\left(\frac{\text{ml}}{\text{min}}\right)}{V} \times \underset{\left(\frac{\text{mg}}{\text{ml}}\right)}{U}$$

Where $V$ = rate of urine formation
$U$ = inulin concentration in urine

The rate at which a substance is filtered by the glomeruli (in mg per minute) can be calculated by multiplying the ml per minute of plasma that is filtered (the **glomerular filtration rate,** or **GFR**) by the concentration of that substance in the plasma. This is shown in the following equation:

$$\frac{\textit{Quantity filtered per minute}}{(\text{mg/min})} = \underset{\left(\frac{\text{ml}}{\text{min}}\right)}{GFR} \times \underset{\left(\frac{\text{mg}}{\text{ml}}\right)}{P}$$

Where $P$ = inulin concentration in plasma

Since inulin is neither reabsorbed nor secreted, the amount filtered equals the amount excreted:

$$\underset{\text{(amount filtered)}}{GFR \times P} = \underset{\text{(amount excreted)}}{V \times U}$$

If the last equation is now solved for the glomerular filtration rate,

$$GFR_{(\text{ml/min})} = \frac{V_{(\text{ml/min})} \; U_{(\text{mg/ml})}}{P_{(\text{mg/ml})}}$$

Suppose, for example, that inulin is infused into a vein and its concentration in the urine and plasma are found to be 30 mg per ml and 0.5 mg per ml, respectively. If the rate of urine formation is 2 ml per minute, the GFR can be calculated as follows:

$$GFR = \frac{2\ \text{ml/min} \;\; 30\ \text{mg/ml}}{0.5\ \text{mg/ml}} = 120\ \text{ml/min}$$

This equation states that 120 ml of plasma per minute must have been filtered in order to excrete the measured amount of inulin that appeared in the urine. The glomerular filtration rate is thus 120 ml per minute in this example.

Measurements of the plasma concentration of **creatinine** are often used clinically as an index of kidney function. Creatinine, produced as a waste product of muscle creatine, is secreted to a slight degree by the renal tubules so that its excretion rate is a little above that of inulin. Since it is released into the blood at a constant rate and since its excretion is closely matched to the GFR, an abnormal decrease in GFR causes the plasma creatinine concentration to rise. A simple measurement of blood creatinine concentration can thus indicate if the GFR is normal and thus provide information about the health of the kidneys.

***Clearance Calculations.*** The **renal plasma clearance** is *the volume of plasma from which a substance is completely removed in one minute* by excretion in the urine. Notice that the units for renal plasma clearance are ml/min. In the case of inulin, which is filtered but neither reabsorbed nor secreted, the amount of inulin that enters the urine is that which is contained in the volume of plasma filtered. The clearance of inulin is thus equal to the GFR (120 ml/min in the previous example). This volume of filtered plasma, however, also contains other solutes which may be reabsorbed to varying degrees. If a portion of a filtered solute is reabsorbed, the amount excreted in the urine is less than that which was contained in the 120 ml of plasma filtered. The renal plasma clearance of a substance that is reabsorbed must thus be less than the GFR (table 16.4).

If a substance is not reabsorbed, all of the filtered amount will be cleared. If this substance is, in addition, secreted by active transport into the renal tubules from the peritubular blood, an additional amount of plasma can be cleared of that substance. The renal plasma clearance of a substance that is filtered and secreted is therefore greater than the GFR (table 16.5). Thus, in order to compare the renal "handling" of various substances in terms

**Table 16.5** Renal "handling" of different plasma molecules

| If Substance is | Example | Concentration in Renal Vein | Renal Clearance Rate |
|---|---|---|---|
| Not filtered | Proteins | Same as in renal artery | Zero |
| Filtered, not reabsorbed nor secreted | Inulin | Less than in renal artery | Equal to GFR (115–125 ml/min) |
| Filtered, partially reabsorbed | Urea | Less than in renal artery | Less than GFR |
| Filtered, completely reabsorbed | Glucose | Same as in renal artery | Zero |
| Filtered and secreted | PAH | Less than in renal artery; approaches zero | Greater than GFR; up to total plasma flow rate (~625 ml/min) |
| Filtered, reabsorbed, and secreted | $K^+$ | Variable | Variable |

of their reabsorption or secretion, the renal plasma clearance is calculated using the same formula used for the determination of the GFR:

$$\text{Clearance Rate} = \frac{V \times U}{P}$$

Where $V$ = urine volume per minute
$U$ = concentration of substance in urine
$P$ = concentration of substance in plasma

***Clearance of Urea.*** Urea may be used as an example of how the clearance calculations can reveal the way the kidneys handle a molecule. Urea is a waste product of amino acid metabolism that is secreted by the liver into the blood and is filtered into the glomerular capsules. Using the formula for renal clearance previously described, and sample values, the following clearance may be calculated:

If $V = 2$ ml/min
$U = 7.5$ mg/ml of urea
$P = 0.2$ mg/ml of urea

$$\text{urea clearance} = \frac{2 \text{ ml/min} \times 7.5 \text{ mg/ml}}{0.2 \text{ mg/ml}} = 75 \text{ ml/min}$$

The clearance of urea in this example is less (75 ml/min) than the clearance of inulin (120 ml/min). Thus, even though 120 ml of plasma filtrate entered the nephrons, only the amount of urea contained in 75 ml of filtrate is excreted per minute. The kidneys therefore must reabsorb some of the urea that is filtered. Despite the fact that it is a waste product, a significant portion of the filtered urea (ranging from 40%–60%) is always reabsorbed. This is a passive reabsorption that cannot be avoided because of the high permeability of cell membranes to urea.

## Clearance of PAH: Measurement of Renal Blood Flow

Not all of the blood delivered to the glomeruli is filtered into the glomerular capsules; most of the glomerular blood passes through to the efferent arterioles and peritubular capillaries. The inulin and urea in this unfiltered blood are not excreted but instead return to the general circulation. Blood must therefore make many passes through the kidneys before it can be completely cleared of a given amount of inulin or urea.

In order for compounds in the unfiltered renal blood to be cleared, they must be secreted into the tubules by active transport from the peritubular capillaries. In this way all of the blood going to the kidneys can potentially be cleared of a secreted compound in a single pass. This is the case for a molecule called **para-aminohippuric acid,** or **PAH,** (fig. 16.21). The clearance (in ml/min) of PAH can be used to measure the *total renal blood flow* (in ml/min). The normal PAH clearance has been found to average 625 ml/min. Since the glomerular filtration rate averages about 120 ml/min, this indicates that only about 120/625, or roughly 20%, of the renal plasma flow is filtered. The remaining 80% passes on to the efferent arterioles.

Since filtration and secretion clear only the molecules dissolved in plasma, the PAH clearance measures the renal plasma flow. In order to convert this to the total renal blood flow, the volume of blood occupied by erythrocytes must be taken into account. If the hematocrit (chapter 13) is 45, for example, erythrocytes occupy 45% of the blood volume and plasma accounts for the remaining 55% of the blood volume. The **total renal blood flow** is calculated by dividing the PAH clearance rate by the fractional blood volume occupied by plasma (0.55, in this example). The total renal blood flow in this example is thus 625 ml/min divided by 0.55, or 1.1 L/min.

Many antibiotics, like penicillin, are secreted by the renal tubules and thus have clearance rates greater than the glomerular filtration rate. Because penicillin is rapidly removed from the blood by renal clearance, large amounts must be administered to be effective. The ability of the kidneys to be visualized in radiographs is improved by the injection of Diodrast, a material that is secreted into the tubules and improves contrast by absorbing X rays. Many drugs and some hormones are inactivated in the liver by chemical transformations and are rapidly cleared from the blood by active secretion in the nephrons.

**Figure 16.21.** Some of the para-aminohippuric acid (PAH) in glomerular blood (*a*) is filtered into Bowman's capsules (*b*). The PAH present in the unfiltered blood is secreted from peritubular capillaries into the nephron (*c*), so that all of the blood leaving the kidneys is free of PAH (*d*). The clearance rate of PAH therefore equals the total plasma flow to the glomeruli.

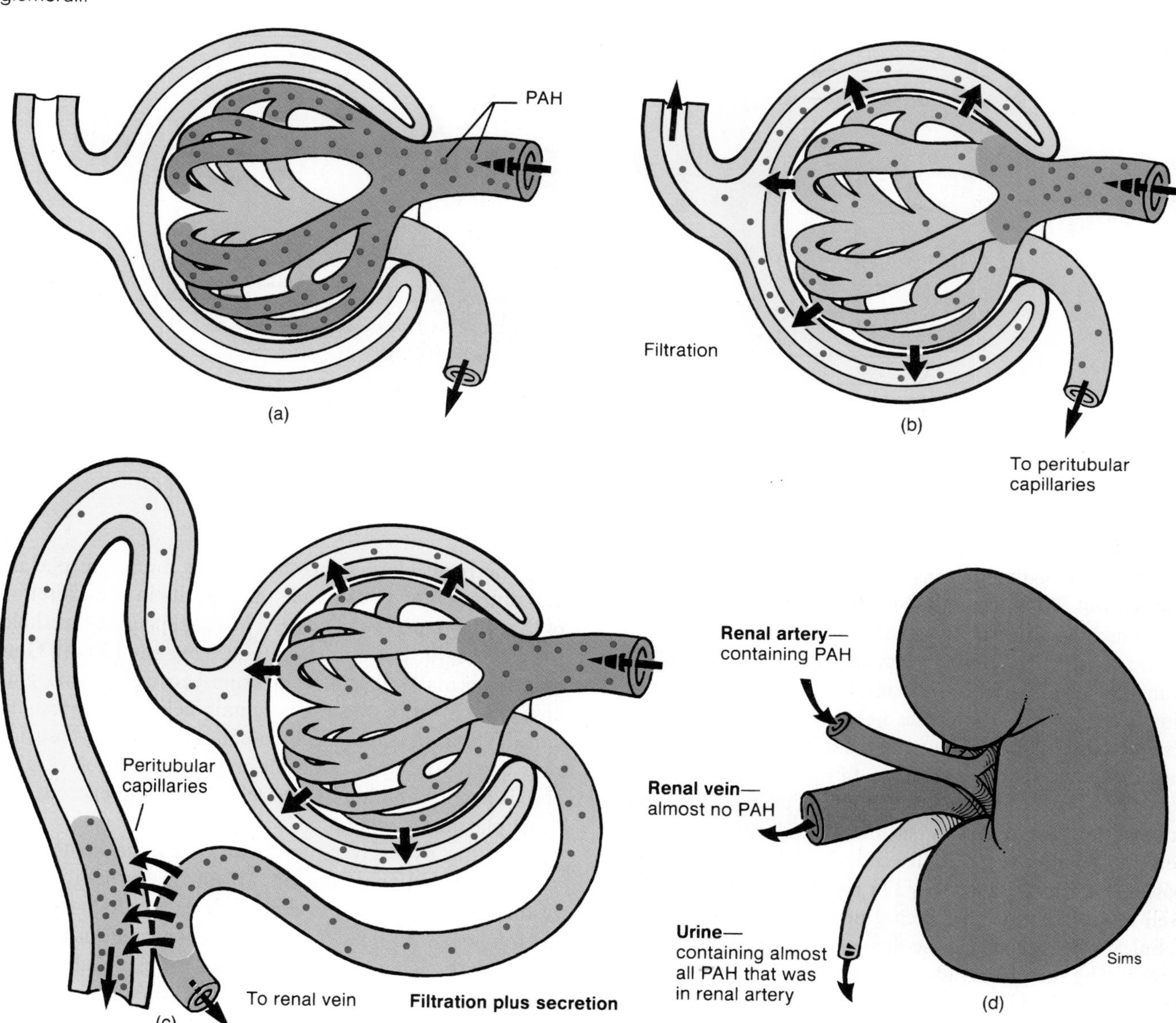


## Reabsorption of Glucose and Amino Acids

Glucose and amino acids in the blood are easily filtered by the glomeruli into the renal tubules. These molecules, however, are usually not present in the urine. It can be concluded, therefore, that filtered glucose and amino acids are normally completely reabsorbed by the nephrons.

The reabsorption of glucose and amino acids is an energy-requiring process, which occurs primarily in the proximal convoluted tubules. The energy required for movement of these compounds from the filtrate into the tubule cells is provided by cotransport (chapter 6) with $Na^+$, which diffuses down its electrochemical gradient when it enters the cells. This cotransport occurs because the glucose and amino acids appear to share carriers with $Na^+$ in the apical membrane. The extrusion of glucose and amino acids from the other side of the cell (across the basolateral membranes) occurs by means of a $Na^+$-independent active transport carrier (fig. 16.22).

**Figure 16.22.** The reabsorption of glucose in the proximal tubule. Glucose is reabsorbed by cotransport with $Na^+$ from the tubular fluid and then transported by $Na^+$-independent carriers through the basolateral membrane into the peritubular blood. By this means normally all of the filtered glucose is reabsorbed.

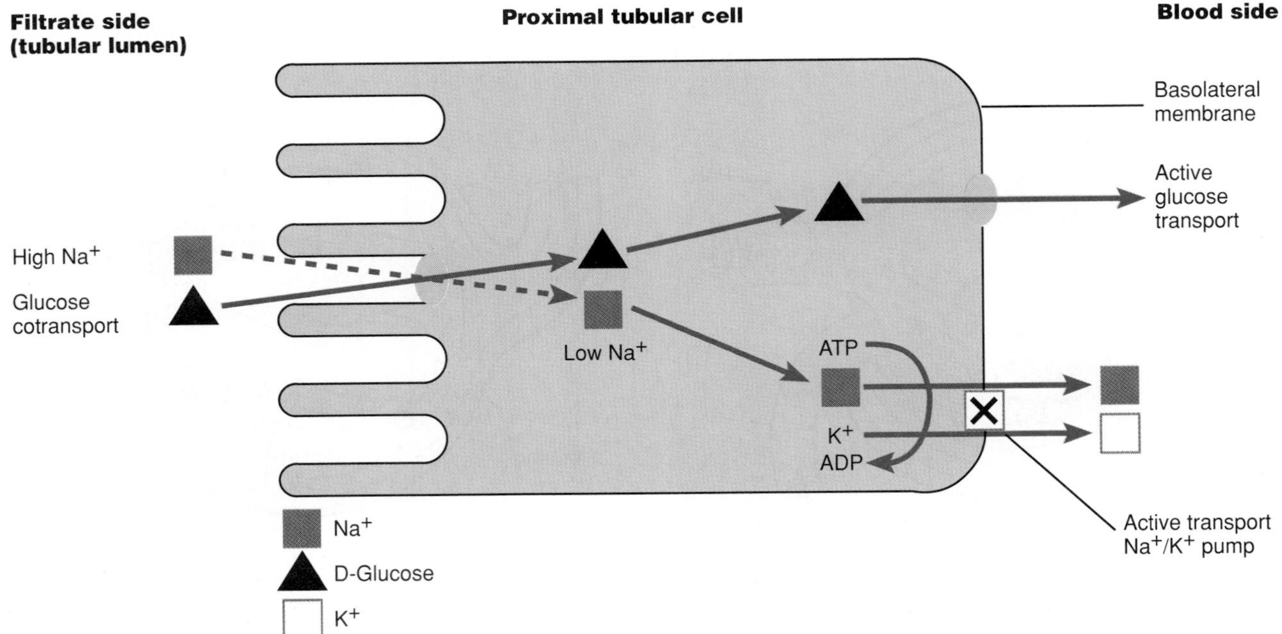


Carrier-mediated transport displays the property of *saturation.* This means that when the transported molecule (such as glucose) is present in sufficiently high concentrations, all of the carriers become "busy," and the transport rate reaches a maximal value. The concentration of transported molecules needed to just saturate the carriers and to just achieve the maximal transport rate is called the **transport maximum** (abbreviated $T_m$).

The carriers for glucose and amino acids in the renal tubules are not normally saturated and so are able to remove the filtered molecules completely. The $T_m$ for glucose, for example, averages 375 mg per minute, which is well above the rate at which glucose is delivered to the tubules. The rate of glucose delivery can be calculated by multiplying the plasma glucose concentration (about 1 mg per ml) by the GFR (about 125 ml per minute). Approximately 125 mg per minute are thus delivered to the tubules, whereas a rate of 375 mg per minute are required to reach saturation.

***Glycosuria.*** Glucose appears in the urine—a condition called glycosuria—when more glucose passes through the tubules than can be reabsorbed. This occurs when the plasma glucose concentration reaches 180–200 mg per 100 ml. Since the rate of glucose delivery under these conditions is still below the average $T_m$ for glucose, one must conclude that some nephrons have considerably lower $T_m$ values than the average.

The **renal plasma threshold** is the minimum plasma concentration of a substance that results in the excretion of that substance in the urine. The renal plasma threshold for glucose, for example, is 180–200 mg per 100 ml. Glucose is normally absent from urine because plasma glucose concentrations normally remain below this threshold value. The appearance of glucose in the urine (glycosuria) thus occurs only when the plasma concentration of glucose is abnormally high (hyperglycemia) and exceeds the renal plasma threshold.

Fasting hyperglycemia is caused by the inadequate secretion or action of insulin. When this hyperglycemia results in glycosuria, the disease is called **diabetes mellitus.** A person with uncontrolled diabetes mellitus also excretes a large volume of urine, because the excreted glucose carries water with it as a result of the osmotic pressure it generates in the tubules. This condition should not be confused with diabetes insipidus, in which a large volume of dilute urine is excreted as a result of inadequate ADH secretion.

The excretion of amino acids in the urine is not usually due to the presence of a high amino acid concentration in the blood. Rather, specific amino acids are excreted when their carriers are missing or defective due to a genetic disease. Since different classes of amino acids are reabsorbed by different carriers, the types of amino acids that "spill" into the urine are characteristic of the genetic defect (table 16.6).

**Table 16.6** Inherited diseases associated with the presence of specific amino acids in the urine

| Disease | Cause of Disease | Effect of Defect | Treatment |
|---|---|---|---|
| Cystinuria | Renal carriers for cystine and related amino acids are defective. | Kidney stones | Bicarbonate and diuretic administration |
| Hartnup disease | Renal carriers for tryptophane are defective. | Decreased NAD and NADP within body cells | Nicotinamide administration |
| Homocystinuria | Enzyme defect results in excessive blood levels of homocystine. | Speech defects, mental retardation | Diet low in methionine, high in cystine |
| Phenylketonuria | Enzyme defect results in excessive accumulation of phenylalanine and in urinary excretion of phenylpyruvic acid. | Severe mental retardation | Diet low in phenylalanine |

1. *Define renal plasma clearance, and describe how it is measured. Explain why the glomerular filtration rate is equal to the clearance rate of inulin.*
2. *Define the terms* reabsorption *and* secretion. *Describe how the renal plasma clearance is affected by the processes of reabsorption and secretion, and give examples.*
3. *Explain why the total renal blood flow can be measured by the clearance rate of PAH.*
4. *Define transport maximum and renal plasma threshold. Explain why people with diabetes mellitus have glycosuria.*

## Renal Control of Electrolyte Balance

Aldosterone stimulates the reabsorption of $Na^+$ and the secretion of $K^+$. Indeed, $K^+$ cannot be eliminated from the blood without the aldosterone-induced stimulation of $K^+$ secretion into the nephron tubules. Secretion of aldosterone from the adrenal cortex is thus stimulated by a high blood $K^+$ concentration. A low blood $Na^+$ also stimulates aldosterone secretion, but this is an indirect effect that is mediated by the renin-angiotensin system. Since the blood concentration of $H^+$ is influenced by the concentration of $K^+$, renal control of electrolyte balance also influences the acid-base balance of the blood.

The kidneys help regulate the concentrations of plasma electrolytes—sodium, potassium, chloride, bicarbonate, and phosphate—by matching the urinary excretion of these compounds to the amounts ingested. The control of plasma $Na^+$ is important in the regulation of blood volume and pressure; the control of plasma $K^+$ is required to maintain proper function of cardiac and skeletal muscles. The regulation of $Na^+/K^+$ balance is also intimately related to renal control of acid-base balance.

### Role of Aldosterone in $Na^+/K^+$ Balance

Approximately 90% of the filtered $Na^+$ and $K^+$ is reabsorbed in the early part of the nephron before the filtrate reaches the distal tubule. This reabsorption occurs at a constant rate and is not subject to hormonal regulation. The final concentration of $Na^+$ and $K^+$ in the urine is varied according to the needs of the body by processes that occur in the late distal tubule and in the cortical region of the collecting duct (the portion of the collecting duct within the medulla does not participate in this regulation). Renal excretion and retention of $Na^+$ and $K^+$ are regulated by **aldosterone,** a steroid hormone secreted by the adrenal cortex.

***Sodium Reabsorption.*** Although 90% of the filtered sodium is reabsorbed in the early region of the nephron, the amount left in the filtrate delivered to the distal convoluted tubule is still quite large. In the absence of aldosterone, 80% of this amount is automatically reabsorbed through the wall of the distal tubule into the peritubular blood; this is 8% of the amount filtered. The amount of sodium excreted without aldosterone is thus 2% of the amount filtered. Although this percentage seems small, the actual amount of sodium this represents is an impressive 30 g per day excreted in the urine. When aldosterone is secreted in maximal amounts, in contrast, all of the sodium delivered to the distal tubule is reabsorbed. Under these conditions urine contains no $Na^+$ at all.

***Potassium Secretion.*** About 90% of the filtered $K^+$ is reabsorbed in the early regions of the nephron (mainly in the proximal tubule). When aldosterone is absent, all of the filtered $K^+$ that remains is reabsorbed in the distal tubule. In the absence of aldosterone, therefore, no $K^+$ is excreted in the urine. The presence of aldosterone stimulates the *secretion of $K^+$* from the peritubular blood into the late distal tubule and cortical collecting duct (fig. 16.23). This aldosterone-induced secretion is thus the only

**Figure 16.23.** Potassium is almost completely reabsorbed in the proximal tubule, but under aldosterone stimulation, it is secreted into the distal tubule. All of the $K^+$ in urine is derived from secretion rather than from filtration.

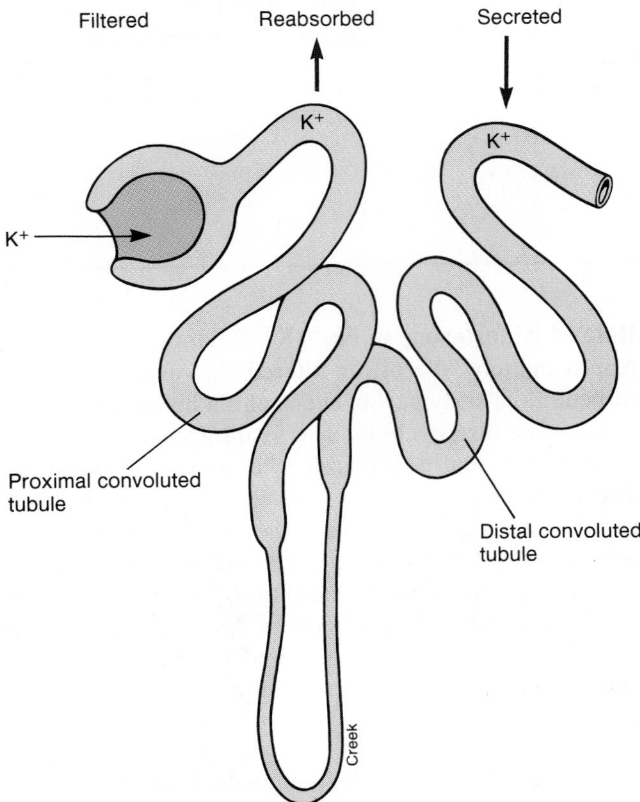


means by which $K^+$ can be eliminated in the urine. When aldosterone secretion is maximal, as much as fifty times more $K^+$ is excreted in the urine, because of secretion into the distal tubule, than was originally filtered through the glomeruli.

In summary, aldosterone promotes sodium retention and potassium loss from the blood by stimulating the reabsorption of $Na^+$ and the secretion of $K^+$ across the wall of the late distal convoluted tubules and cortical portions of the collecting ducts. Since aldosterone promotes the retention of $Na^+$, it contributes to an increased blood volume and pressure.

The body cannot get rid of excess $K^+$ in the absence of aldosterone-stimulated secretion of $K^+$ into the distal tubules. Indeed, when both adrenal glands are removed from an experimental animal, the **hyperkalemia** (high blood $K^+$) that results can produce fatal cardiac arrhythmias. Abnormally low plasma $K^+$ concentrations, as might result from excessive aldosterone secretion, can also produce arrhythmias as well as muscle weakness.

## Control of Aldosterone Secretion

Since aldosterone promotes $Na^+$ retention and $K^+$ loss, one might predict (on the basis of negative feedback) that aldosterone secretion will be increased when there is a low $Na^+$ or a high $K^+$ concentration in the blood. This indeed is the case. A rise in blood $K^+$ *directly* stimulates the secretion of aldosterone from the adrenal cortex. Decreases in plasma $Na^+$ concentrations also promote aldosterone secretion, but they do so indirectly.

***Juxtaglomerular Apparatus.*** The juxtaglomerular apparatus is the region in each nephron where the afferent arteriole and distal tubule come into contact (fig. 16.24). The microscopic appearance of the afferent arteriole and distal tubule in this small region differs from the appearance in other regions. *Granular cells* within the afferent arteriole secrete the enzyme **renin** into the blood; this enzyme catalyzes the conversion of angiotensinogen (a protein) into angiotensin I (a ten-amino-acid polypeptide).

Secretion of renin into the blood thus results in the formation of angiotensin I, which is then converted to **angiotensin II** by *angiotensin converting enzyme* as blood passes through the lungs and other organs. Angiotensin II, in addition to other effects (described in chapter 14), stimulates the adrenal cortex to secrete aldosterone. Secretion of renin from the granular cells of the juxtaglomerular apparatus is thus said to initiate the **renin-angiotensin-aldosterone system.** Conditions that result in renin secretion thus cause increased aldosterone secretion and, by this means, promote the reabsorption of $Na^+$ in the distal convoluted tubules.

***Regulation of Renin Secretion.*** A fall in plasma $Na^+$ concentration is always accompanied by a fall in blood volume (chapter 14). This is because ADH secretion is inhibited by the decreased plasma concentration (osmolality); with less ADH, less water is reabsorbed through the collecting ducts and more is excreted in the urine. The fall in blood volume and the fall in renal blood flow that results causes increased renin secretion. Increased renin secretion is believed to be due in part to the direct effect of blood flow on the granular cells, which may function as baroreceptors in the afferent arterioles. Renin secretion is also stimulated by sympathetic nerve activity, which is increased when the blood volume and pressure fall.

An increased secretion of renin acts, via the increased production of angiotensin II, to stimulate aldosterone secretion. In consequence, there is less sodium excreted in the urine and more retained in the blood. This negative feedback system is illustrated in figure 16.25.

**Figure 16.24.** The juxtaglomerular apparatus (*a*) includes the region of contact of the afferent arteriole with the distal tubule. The afferent arterioles in this region contain granular cells with renin, and the distal tubule cells in contact with the granular cells form an area called the macula densa (*b*). The granular cells of the afferent arteriole are innervated by renal sympathetic nerve fibers.

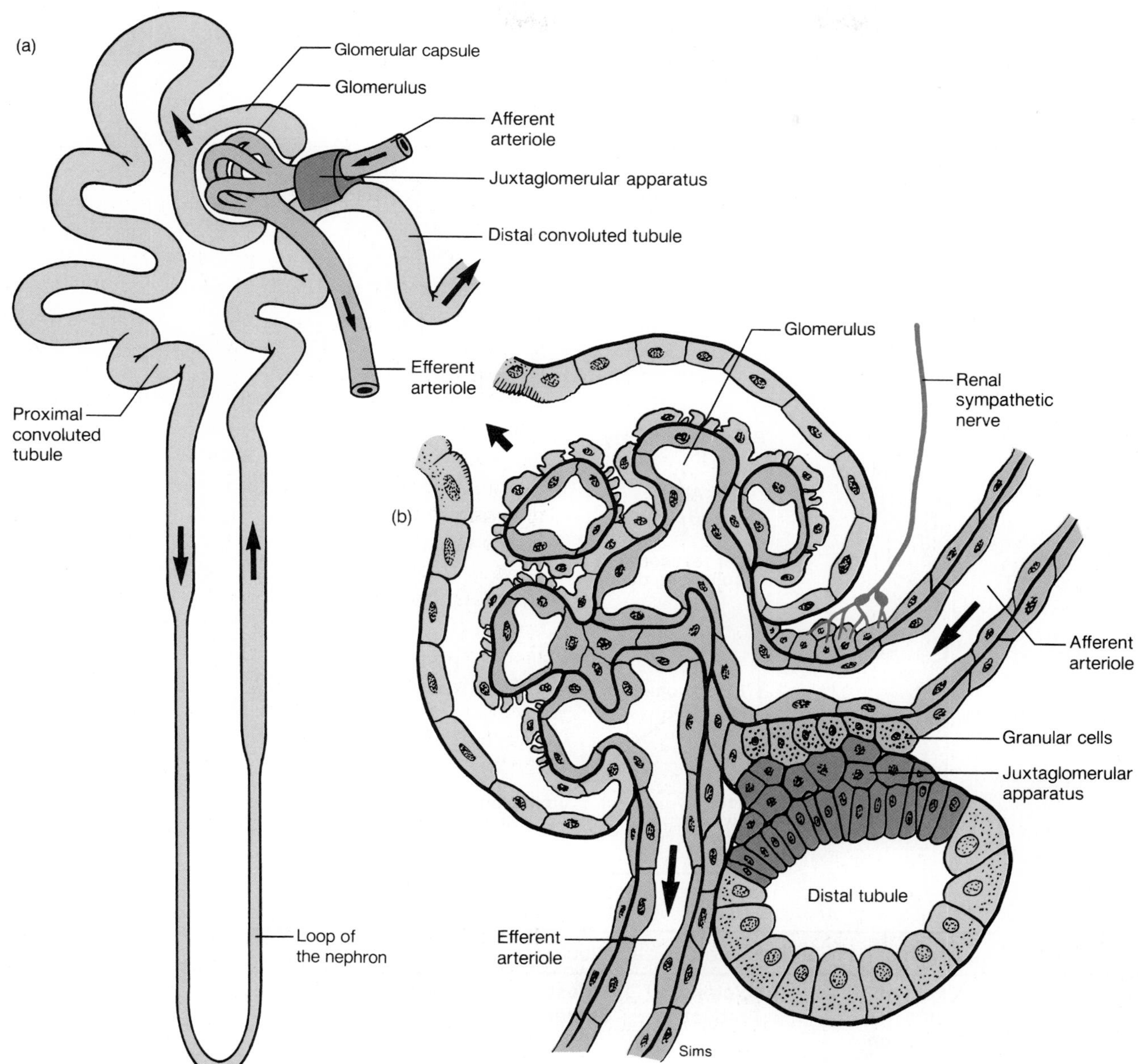


***Role of the Macula Densa.*** The region of the distal tubule in contact with the granular cells of the afferent arteriole is called the **macula densa** (fig. 16.24). There is evidence that this region helps to inhibit renin secretion when the blood $Na^+$ concentration is raised.

According to the proposed mechanism, the cells of the macula densa respond to $Na^+$ within the filtrate delivered to the distal tubule. When the plasma $Na^+$ concentration is raised, the rate of $Na^+$ delivered to the distal tubule is also increased. Through an effect on the macula densa, this increase in filtered $Na^+$ may inhibit the granular cells from secreting renin. Aldosterone secretion thus decreases, and since less $Na^+$ is reabsorbed in the distal tubule, more $Na^+$ is excreted in the urine. The regulation of renin and aldosterone secretion is summarized in table 16.7.

**Figure 16.25.** The sequence of events by which a low sodium (salt) intake leads to increased sodium reabsorption by the kidneys. The dotted arrow and negative sign show the completion of the negative feedback loop.

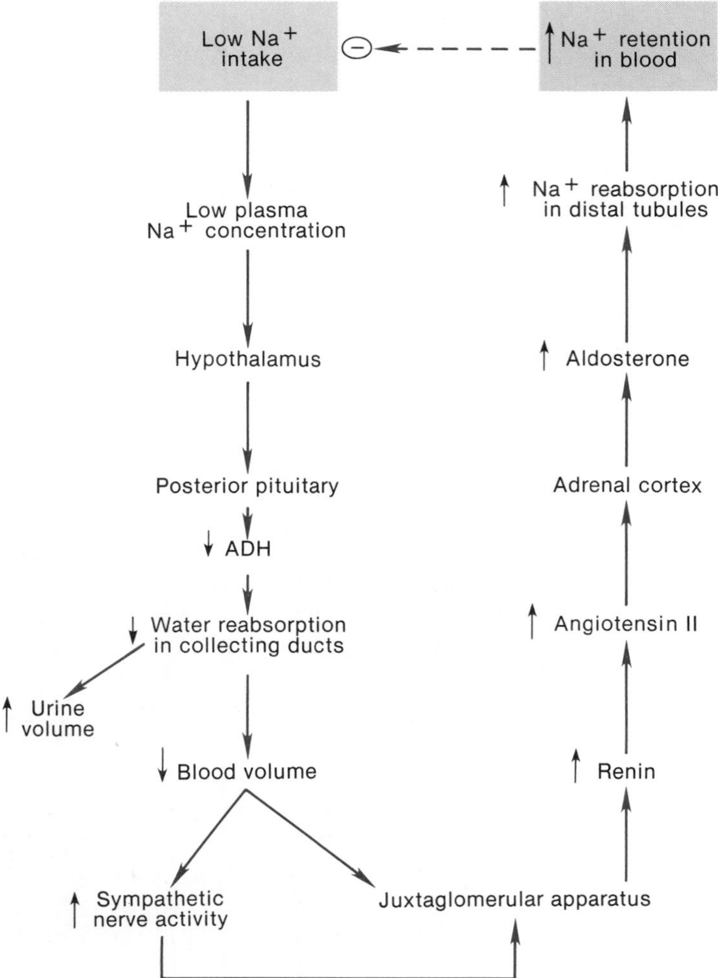


**Table 16.7** Regulation of renin and aldosterone secretion

| Stimulus | Effect on Renin Secretion | Angiotensin II Production | Aldosterone Secretion | Mechanisms |
|---|---|---|---|---|
| ↓$Na^+$ | Increased | Increased | Increased | Low blood volume stimulates renal baroreceptors; granular cells release renin. |
| ↑$Na^+$ | Decreased | Decreased | Decreased | Increased blood volume inhibits baroreceptors; increased $Na^+$ in distal tubule acts via macula densa to inhibit release of renin from granular cells. |
| ↑$K^+$ | None | Not changed | Increased | Direct stimulation of adrenal cortex |
| ↑Sympathetic nerve activity | Increased | Increased | Increased | $\alpha$-adrenergic effect stimulates constriction of afferent arterioles; $\beta$-adrenergic effect stimulates renin secretion directly. |

***Natriuretic Hormone.*** Expansion of the blood volume causes increased salt and water excretion in the urine. This is due in part to an inhibition of aldosterone secretion, as previously described. There is much experimental evidence, however, that the increased salt excretion that occurs under these conditions is due not only to the inhibition of aldosterone secretion, but also to the increased secretion of another substance with hormone properties. This other substance is called **natriuretic hormone** and is so named because it stimulates salt excretion (the opposite of aldosterone's action). The source and chemical nature of natriuretic hormone remained elusive for many years, but recent evidence has shown that the atria of the heart produce a polypeptide that appears to fit the description of the natriuretic hormone proposed by renal physiologists. This polypeptide is currently known as *atrial natriuretic hormone.*

## Relationship between $Na^+$, $K^+$, and $H^+$

Hormones, as a general rule, alter the rate of already existing processes. In the absence of aldosterone, the distal tubule reabsorbs 80% of the $Na^+$ delivered to it; aldosterone can increase this reabsorption to 100%. At a given level of aldosterone, the amount of $Na^+$ reabsorbed in the distal tubule is a given proportion of the $Na^+$ delivered to it. Some diuretic drugs inhibit $Na^+$ reabsorption in the loop of Henle and, therefore, increase the delivery of $Na^+$ to the distal tubule. As a result, there is an increased reabsorption of $Na^+$ in the distal convoluted tubule when a person takes these types of diuretics. It should be noted that the increased reabsorption of $Na^+$ in the distal region of the nephron cannot completely compensate for the effect of the diuretic drug (fortunately), so that a person who takes this drug does have a higher salt and water excretion in the urine and a lower blood volume as a result.

***Relationship between $Na^+$ and $K^+$.*** The reabsorption of $Na^+$ in the distal convoluted tubules occurs together with $K^+$ secretion. This occurs because the aldosterone-stimulated reabsorption of $Na^+$ creates a large potential difference between the two sides of the tubular wall, with the lumen side very negative ($-50$ mV) in comparison to the basolateral side. The secretion of $K^+$ into the tubular fluid is driven by this electrical gradient. Because of the $Na^+/K^+$ exchange in the distal tubule, an increase in $Na^+$ reabsorption in the distal tubule results in an increase in $K^+$ secretion. People who take diuretics that inhibit $Na^+$ reabsorption in the loop of Henle, for these reasons, tend to have excessive $K^+$ secretion into the distal tubules and, therefore, excessive $K^+$ loss in the urine. The actions of different diuretics and their side effects on blood $K^+$ are discussed in the last section of this chapter.

The $K^+$ loss that occurs with many diuretics may present serious side effects to these medications. If $K^+$ secretion into the distal convoluted tubules is significantly increased, a condition of **hypokalemia** (low blood $K^+$) may be produced, which must be compensated for by the increased ingestion of potassium. People who take diuretics for the treatment of high blood pressure are usually on a low-sodium diet and often must supplement their meals with potassium chloride (KCl).

***Relationship between $K^+$ and $H^+$.*** The plasma $K^+$ concentration indirectly affects the plasma $H^+$ concentration (pH). Changes in plasma pH likewise affect the $K^+$ concentration of the blood. These effects serve to stabilize the *ratio* of $K^+$ to $H^+$. When the extracellular $H^+$ concentration increases, for example, some of the $H^+$ moves into the tissue cells and causes cellular $K^+$ to diffuse outward into the extracellular fluid. The plasma concentration of $H^+$ is thus decreased while the $K^+$ increases, helping to reestablish the proper ratio of these ions in the extracellular fluid. A similar effect occurs in the cells of the distal region of the nephron.

In the cells of the late distal tubule and cortical collecting duct, positively charged ions ($K^+$ and $H^+$) are secreted in response to the negative polarity produced by reabsorption of $Na^+$ (fig. 16.26). When a person has severe acidosis, there is an increased amount of $H^+$ secretion at the expense of a decrease in the amount of $K^+$ secreted. Acidosis may thus be accompanied by a rise in blood $K^+$. When a person has hyperkalemia as the primary problem, conversely, there is an increased secretion of $K^+$ and thus a decreased secretion of $H^+$. Hyperkalemia can thus cause an increase in the blood concentration of $H^+$ and acidosis.

Aldosterone stimulates the secretion of $H^+$ as well as $K^+$ into the distal tubules. Abnormally high aldosterone secretion, in **primary aldosteronism,** or **Conn's syndrome,** therefore, results in both hypokalemia and metabolic alkalosis. Conversely, abnormally low aldosterone secretion, as occurs in **Addison's disease,** can produce hyperkalemia which is accompanied by metabolic acidosis.

**Figure 16.26.** In the distal tubule, $K^+$ and $H^+$ are secreted in exchange for $Na^+$. High concentrations of $H^+$ may therefore decrease $K^+$ secretion, and vice versa.

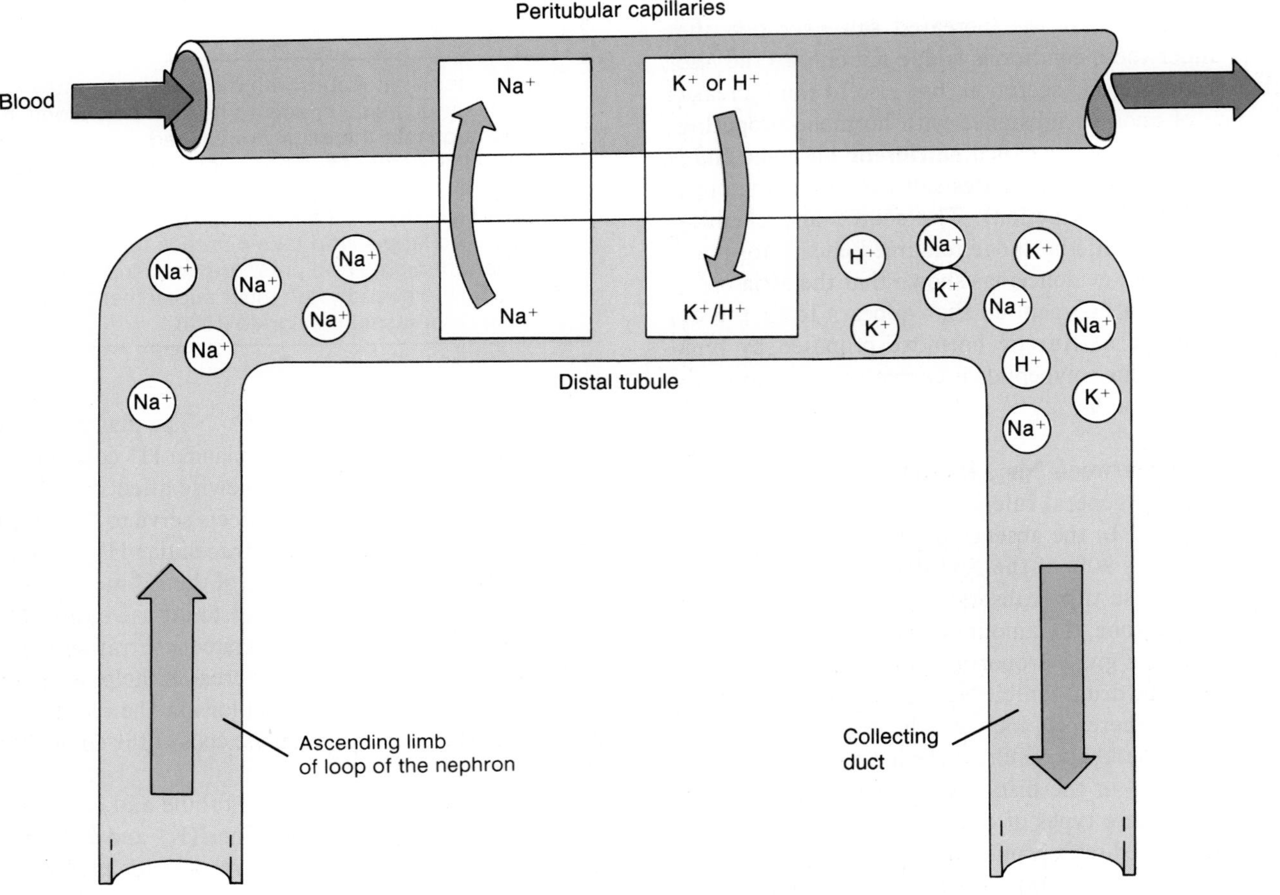


1. *Describe the effects of aldosterone on the renal nephrons, and explain how aldosterone secretion is regulated.*
2. *Describe how changes in plasma $Na^+$ concentrations regulate renin secretion, and explain how the secretion of renin acts to help regulate the plasma $Na^+$ concentration.*
3. *Explain why people who take some diuretic drugs may suffer from hypokalemia and why they must supplement their diets with potassium.*
4. *Explain the mechanisms by which the distal tubule secretes $K^+$ and $H^+$, and describe how hyperkalemia may affect the blood pH.*

## Renal Control of Acid-Base Balance

The kidneys help regulate the acid-base balance of the blood through their excretion of $H^+$ and their reabsorption of $HCO_3^-$. By means of these processes, the kidneys are responsible for the metabolic component of acid-base balance.

Normal arterial blood has a pH of 7.35–7.45. It should be recalled that the pH number is *inversely* related to the $H^+$ concentration. An increase in $H^+$, derived from carbonic acid or the nonvolatile metabolic acids, can lower blood pH. An increase in $H^+$ derived from metabolic acids, however, is normally prevented from changing the blood pH by **bicarbonate buffer,** which combines with and thus removes from solution the excess $H^+$.

### Respiratory and Metabolic Components of Acid-Base Balance

Acid-base balance has both respiratory and metabolic components. The respiratory component describes the effect of ventilation on arterial $P_{CO_2}$ and thus on the production of carbonic acid ($H_2CO_3$). The metabolic component describes the effect of nonvolatile metabolic acids—lactic acid, fatty acids, and ketone bodies—on blood pH. Since these acids are normally buffered by bicarbonate ($HCO_3^-$), the metabolic component can be described in terms of the free $HCO_3^-$ concentration. An increase in metabolic acids "uses up" free bicarbonate, as $HCO_3^-$ is converted to $H_2CO_3$, and is thus associated with a fall in plasma $HCO_3^-$ concentrations. A decrease in metabolic acids, conversely, is associated with a rise in free $HCO_3^-$.

**Table 16.8** Classification of metabolic and respiratory components of acidosis and alkalosis

| $P_{CO_2}$ | $HCO_3^-$ | Condition | Causes |
|---|---|---|---|
| Normal | Low | Metabolic acidosis | Increased production of "nonvolatile" acids (lactic acid, ketone bodies, and others), or loss of $HCO_3^-$ in diarrhea |
| Normal | High | Metabolic alkalosis | Vomiting of gastric acid; hypokalemia; excessive steroid administration |
| Low | Low | Respiratory alkalosis | Hyperventilation |
| High | High | Respiratory acidosis | Hypoventilation |

**Table 16.9** The effect of changes in $P_{CO_2}$ on the blood pH

| $P_{CO_2}$ (mm Hg) | $H_2CO_3$ (mEq/L) | $HCO_3^-$ (mEq/L) | $HCO_3^-/H_2CO_3$ ratio | pH | Condition |
|---|---|---|---|---|---|
| 20 | 0.6 | 24 | 40/1 | 7.70 | Respiratory alkalosis |
| 30 | 0.9 | 24 | 26.7/1 | 7.53 | Respiratory alkalosis |
| 40 | 1.2 | 24 | 20/1 | 7.40 | Normal |
| 50 | 1.5 | 24 | 16/1 | 7.30 | Respiratory acidosis |
| 60 | 1.8 | 24 | 13.3/1 | 7.22 | Respiratory acidosis |

Note: A blood pH of less than 7.35 due to high $P_{CO_2}$ is called respiratory acidosis. A blood pH greater than 7.45 due to low $P_{CO_2}$ is called respiratory alkalosis.

Since the respiratory component of acid-base balance is represented by the plasma $CO_2$ concentration and the metabolic component is represented by the free bicarbonate concentration, the study of acid-base balance can be considerably simplified. A normal arterial pH is obtained when both the lungs and kidneys are functioning correctly, and thus when there is a proper ratio of bicarbonate to carbon dioxide concentrations. Indeed, the pH can actually be calculated given these values (the $CO_2$ concentration is obtained indirectly by measuring the plasma $P_{CO_2}$ with an electrode). Using the **Henderson-Hasselbalch** equation, a normal pH is obtained when the ratio of these concentrations is 20 to 1:

$$pH = 6.1 + \log \frac{[HCO_3^-]}{[CO_2]}$$

A change in this ratio results in an abnormal blood pH. Pure respiratory acidosis or alkalosis occurs when the $HCO_3^-$ concentration is normal but the $P_{CO_2}$ and $H_2CO_3$ concentrations are altered. Pure metabolic acidosis or alkalosis occurs when the $P_{CO_2}$ and $H_2CO_3$ are normal but the $HCO_3^-$ concentration is abnormal. This classification is summarized in table 16.8.

A normal blood pH is produced when the $P_{CO_2}$ is 40 mm Hg and the bicarbonate concentration is 24 milliequivalents (mEq) per liter (one mEq is equal to one millimole times the valence of the ion; in this case, since the valence is 1, mEq = mmole). Table 16.9 shows how changes in $P_{CO_2}$ alter the ratio of $HCO_3^-$ to $H_2CO_3$ and thus affect blood pH. For the sake of simplicity, this table shows the effect of changing $P_{CO_2}$ when the bicarbonate concentration remains at a constant, normal value. Such changes in $P_{CO_2}$ are produced by hyperventilation and hypoventilation and result in respiratory alkalosis and acidosis, respectively.

## Mechanisms of Renal Acid-Base Regulation

The kidneys help regulate the blood pH by excreting $H^+$ in the urine and by the reabsorption of bicarbonate. These two mechanisms are interdependent—the reabsorption of bicarbonate occurs as a result of the filtration and secretion of $H^+$. The kidneys normally reabsorb all of the filtered bicarbonate and excrete about 40–80 mEq of $H^+$ per day. The amount depends upon how much the body produces; under normal conditions, exactly the same amount of $H^+$ is excreted as is produced. Normal urine, therefore, is free of bicarbonate and is slightly acidic (with a pH range between 5 and 7). A summary of the mechanisms involved in acidification of the urine and reabsorption of bicarbonate is provided in figure 16.27.

***Reabsorption of Bicarbonate in the Proximal Tubule.*** The apical membranes of the tubule cells (facing the lumen) are impermeable to bicarbonate. The reabsorption of bicarbonate must therefore occur indirectly. When the urine is acidic, $HCO_3^-$ combines with $H^+$ to form carbonic acid. Carbonic acid in the filtrate is then converted to $CO_2$ and $H_2O$ by the action of **carbonic anhydrase.** This enzyme is located in the apical cell membrane of the proximal tubule in contact with the filtrate. Notice that the reaction that occurs in the filtrate is the same one that occurs within the red blood cells in pulmonary capillaries (discussed in chapter 14).

The tubule cell cytoplasm also contains carbonic anhydrase. Under the conditions of high $CO_2$ that prevail within the cytoplasm, carbonic anhydrase catalyzes the reverse reaction (similar to that which occurs within red blood cells in tissue capillaries). The $CO_2$ that enters the tubule cells is converted to carbonic acid, which in turn dissociates to $HCO_3^-$ and $H^+$ within the tubule cells. The

**Figure 16.27.** Schematic diagram summarizing how the urine becomes acidified and how bicarbonate is reabsorbed from the filtrate.

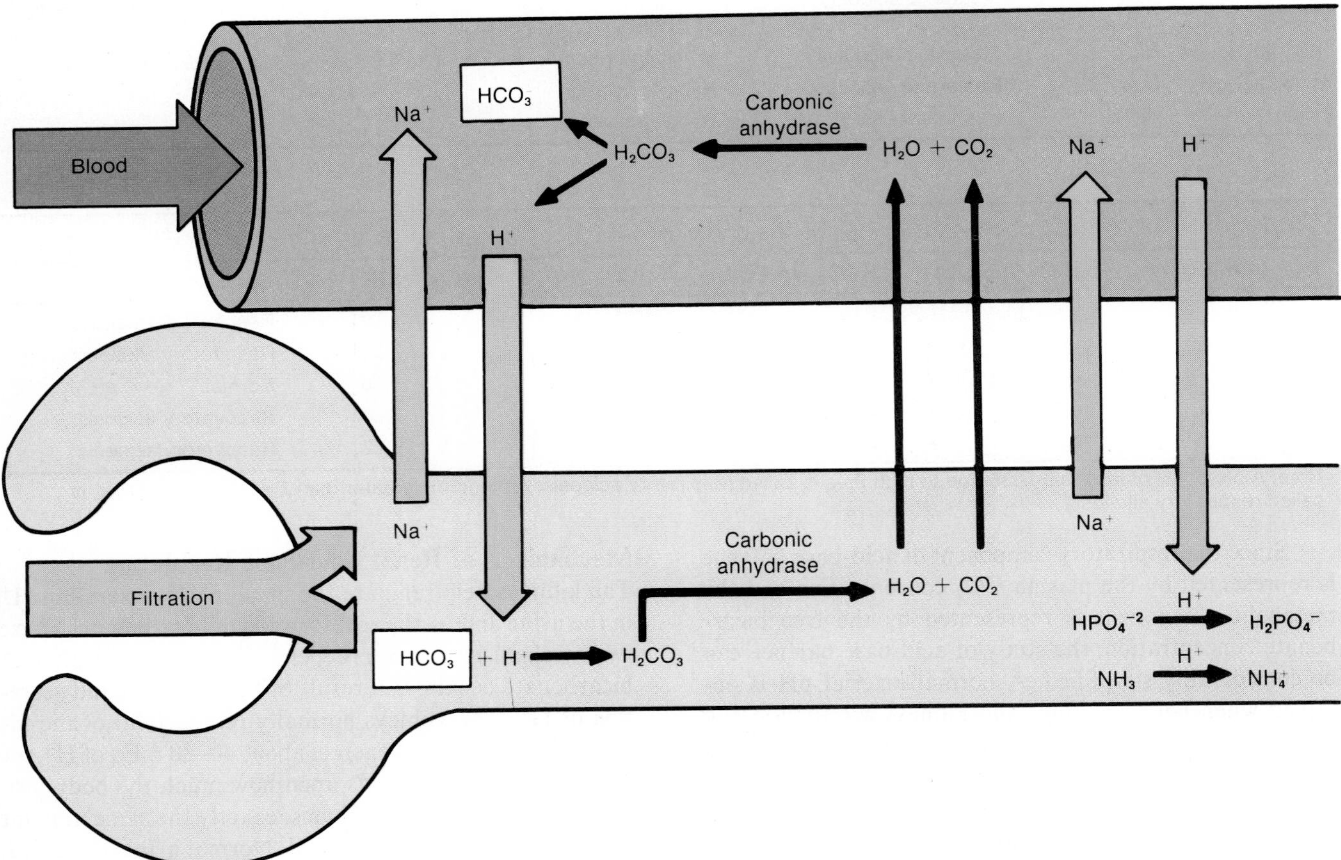


bicarbonate within the tubule cell can then diffuse through the basolateral membrane and enter the blood (fig. 16.28). When conditions are normal, the same amount of $HCO_3$ passes into the blood as was removed from the filtrate. The $H^+$, which was produced at the same time as $HCO_3$ in the cytoplasm of the tubule cell, can either pass back into the filtrate or into the blood. Under acidotic conditions, almost all of the $H^+$ goes back into the filtrate and is used to help reabsorb all of the filtered bicarbonate.

During alkalosis there is less $H^+$ secreted into the filtrate. Since the reabsorption of filtered bicarbonate requires the combination of $HCO_3^-$ with $H^+$ to form carbonic acid, less bicarbonate is reabsorbed. This results in urinary excretion of bicarbonate, which helps to partially compensate for the alkalosis.

In this way, disturbances in acid-base balance caused by respiratory problems can be partially compensated by changes in plasma bicarbonate concentrations. Metabolic acidosis or alkalosis—in which changes in bicarbonate concentrations occur as the primary disturbance—can be similarly compensated in part by changes in ventilation. These interactions of the respiratory and metabolic components of acid-base balance are shown in table 16.10.

***Urinary Buffers.*** When a person has a blood pH less than 7.35 (acidosis), the urine pH almost always falls below 5.5. The nephron, however, cannot produce a urine pH significantly less than 4.5. In order for more $H^+$ to be excreted, the acid must be buffered. Actually, most of the $H^+$ excreted even in normal urine is in a buffered form. Bicarbonate cannot serve this function because it is normally completely reabsorbed. Instead, the buffering action of phosphates (mainly $HPO_4^{-2}$) and ammonia ($NH_3$) provide the means for excreting most of the $H^+$ in the urine. Phosphate enters the urine by filtration. Ammonia (whose presence is strongly evident in a diaper pail or kitty litter box) is produced in the tubule cells by deamination of amino acids. These molecules buffer $H^+$ as described in the following equations:

$$NH_3 + H^+ \rightarrow NH_4^+ \text{ (ammonium ion)}$$

$$HPO_4^{-2} + H^+ \rightarrow H_2PO_4^-$$

**Figure 16.28.** The mechanism of bicarbonate reabsorption. Through this mechanism, the cells of the proximal tubule can reabsorb bicarbonate while secreting $H^+$. (CA = carbonic anhydrase.)

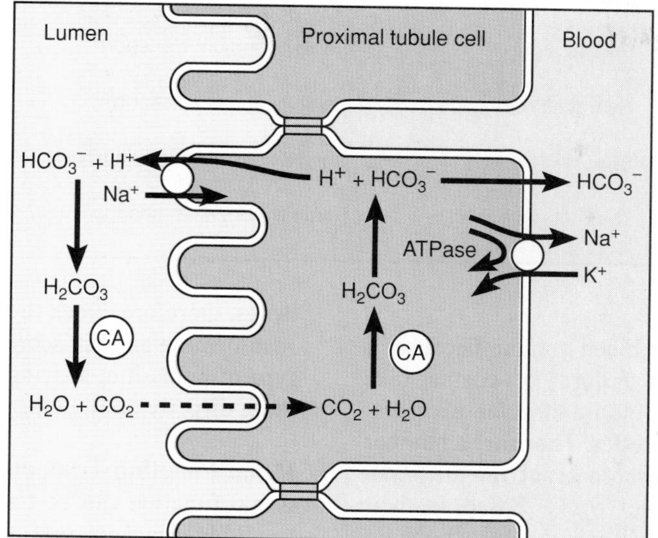


**Table 16.10** Categories of disturbances in acid-base balance, including those that involve mixtures of respiratory and metabolic components

| $P_{CO_2}$ (mmHg) | Bicarbonate (mEq/L) | | |
|---|---|---|---|
| | Less than 21 | 21–26 | More than 26 |
| More than 45 | Combined metabolic and respiratory acidosis | Respiratory acidosis | Metabolic alkalosis and respiratory acidosis |
| 35–45 | Metabolic acidosis | Normal | Metabolic alkalosis |
| Less than 35 | Metabolic acidosis and respiratory alkalosis | Respiratory alkalosis | Combined metabolic and respiratory alkalosis |

1. *Describe the respiratory and metabolic components of acid-base balance, and explain how these components are represented by the carbonic acid and bicarbonate concentration of blood.*
2. *Explain how the kidneys reabsorb filtered bicarbonate and how this process is affected by acidosis and alkalosis.*
3. *Suppose a person with diabetes mellitus has an arterial pH of 7.30, an abnormally low arterial $P_{CO_2}$, and an abnormally low bicarbonate concentration. Identify the type of acid-base disturbance, and explain how these values might have been produced.*

## Clinical Applications

Different types of diuretic drugs act on specific segments of the nephron tubule to indirectly inhibit the reabsorption of water and thus to promote the lowering of blood volume. A knowledge of the mechanism of action of diuretics thus promotes better understanding of the physiology of the nephron. Clinical analysis of the urine, similarly, is meaningful only when the mechanisms that produce normal urine composition are understood.

The importance of kidney function in maintaining homeostasis and the ease with which urine can be collected and used as a mirror of the plasma's chemical composition make the clinical study of renal function and urine composition particularly significant. Further, the ability of the kidneys to regulate blood volume is exploited clinically in the management of high blood pressure.

**Table 16.11** Actions of different classes of diuretics

| Category of Diuretic | Example | Mechanism of Action | Major Site of Action |
|---|---|---|---|
| Carbonic anhydrase inhibitors | Acetazolamide | Inhibits reabsorption of bicarbonate | Proximal tubule |
| Loop diuretics | Furosemide | Inhibits sodium transport | Thick segments of ascending limbs |
| Thiazides | Hydrochlorothiazide | Inhibits sodium transport | Last part of ascending limb and first part of distal tubule |
| Potassium-sparing diuretics | Spironolactone | Inhibits action of aldosterone | Last part of distal tubule and cortical collecting duct |
| | Triamterene | Inhibits $Na^+/K^+$ exchange | Last part of distal tubule and cortical collecting duct |

## Use of Diuretics

People who need to lower their blood volume because of hypertension, congestive heart failure, or edema take medications that increase the volume of urine excreted. Such medications are called **diuretics.** There are a number of diuretic drugs in clinical use that act on the renal nephron in different ways (table 16.11). Based on their chemical structure or aspects of their actions, the commonly used diuretics are categorized as carbonic acid inhibitors, loop diuretics, thiazides, osmotic diuretics, and potassium-sparing diuretics.

The most powerful diuretics, inhibiting salt and water reabsorption by as much as 25%, are the drugs that act to inhibit active salt transport out of the ascending limb of the loop of Henle. Examples of these loop diuretics include *furosemide* and *ethacrynic acid.* The thiazide diuretics, like *hydrochlorothiazide,* inhibit salt and water reabsorption by as much as 8% through inhibition of salt transport by the first segment of the distal convoluted tubule. The carbonic anhydrase inhibitors (*acetazolamide*) are much weaker diuretics and act, primarily in the proximal tubule, to prevent the water reabsorption that occurs when bicarbonate is reabsorbed.

When extra solutes are present in the filtrate, they increase the osmotic pressure of the filtrate and in this way decrease the osmotic reabsorption of water throughout the nephron. *Mannitol* is sometimes used clinically for this purpose. Osmotic diuresis can occur in diabetes mellitus due to the presence of glucose in the filtrate and urine; this extra solute causes the excretion of excessive amounts of water in the urine and can result in severe dehydration of the person with uncontrolled diabetes.

The previously mentioned diuretics can, as discussed in an earlier section, result in the excessive secretion of $K^+$ into the filtrate and its excessive elimination in the urine. For this reason, potassium-sparing diuretics are sometimes used. *Spironolactones* are aldosterone antagonists, which compete with aldosterone for cytoplasmic receptor proteins in the cells of the late distal tubule. These drugs, therefore, block the aldosterone stimulation of $Na^+$ reabsorption and $K^+$ secretion. *Triamterene* is a different type of potassium-sparing diuretic, which appears to act more directly on the $Na^+/K^+$ pumps in the distal tubule.

## Renal Function Tests and Kidney Disease

Renal function can be tested by techniques that include the PAH clearance rate, which measures total blood flow to the kidneys, and the measurement of GFR by the inulin clearance rate. The plasma creatinine concentration, as previously described, also provides an index of renal function. These tests aid the diagnosis of kidney diseases such as glomerulonephritis and renal insufficiency.

***Glomerulonephritis.*** Inflammation of the glomeruli, or glomerulonephritis, is currently believed to be an *autoimmune disease* which involves the person's own antibodies (as described in chapter 19). These antibodies may have been raised against the basement membrane of the glomerular capillaries, but more commonly, they appear to have been produced in response to streptococcus infections (such as strep throat). A variable number of glomeruli are destroyed in this condition, and the remaining glomeruli become more permeable to plasma proteins. Leakage of proteins into the urine results in decreased plasma colloid osmotic pressure and can therefore lead to edema.

***Renal Insufficiency.*** When nephrons are destroyed—as in chronic glomerulonephritis, infection of the renal pelvis (*pyelonephritis*), and loss of a kidney—or when kidney function is reduced by either arteriosclerosis or blockage by kidney stones, a condition of *renal insufficiency* may develop. This can cause hypertension, due primarily to the retention of salt and water, and *uremia* (high plasma urea concentrations). The inability to excrete urea is accompanied by elevated plasma $H^+$ (acidosis) and elevated $K^+$, which are more immediately dangerous than the high urea. Uremic coma appears to result from these associated changes.

Patients with uremia or the potential for developing uremia are often placed on *dialysis* machines. The term *dialysis* refers to the separation of molecules on the basis of size by their ability to diffuse through an artificial semipermeable membrane. Urea and other wastes can easily pass through the membrane pores, whereas plasma proteins are left behind (just as occurs across glomerular capillaries). The plasma is thus cleansed of these wastes as they pass from the blood into the dialyzing fluid. Unlike the tubules, however, the dialysis membrane cannot reabsorb $Na^+$, $K^+$, glucose, and other needed molecules. These substances are prevented from diffusing through the membrane by including them in the dialysis fluid. More recent techniques include the use of the patient's own peritoneal membranes (which line the abdominal cavity) for dialysis.

The difficulties encountered in attempting to compensate for renal insufficiency and the many dangers presented by this condition are stark reminders of the importance of kidney function in maintaining homeostasis. The ability of the kidneys to regulate blood volume and chemical composition in a way that can be adjusted according to the body's needs requires great complexity of function. Homeostasis is maintained in large part by coordination of these functions with those of the cardiovascular and pulmonary systems, as described in the preceding chapters.

1. *List the different categories of clinical diuretics, and explain how each exerts its diuretic effect.*
2. *Explain why most diuretics can cause excessive loss of $K^+$, and explain how this is prevented by the potassium-sparing diuretics.*
3. *Define uremia, describe its dangers, and explain how this condition may be corrected by the use of renal dialysis.*

## SUMMARY

### Structure and Function of the Kidneys p. 488

I. The kidney is divided into an outer cortex and inner medulla.
   A. The medulla is composed of renal pyramids, separated by renal columns.
   B. The renal pyramids empty urine into the calyces that drain into the renal pelvis, and from there to the ureter.

II. Each kidney contains more than a million microscopic functional units called nephrons, which consist of vascular and tubular components.
   A. Filtration occurs in the glomerulus, which receives blood from an afferent arteriole.
   B. Glomerular blood is drained by an efferent arteriole, which delivers blood to peritubular capillaries that surround the nephron tubules.
   C. The glomerular capsule and the proximal and distal convoluted tubules are located in the cortex.
   D. The loop of Henle is located in the medulla.
   E. Filtrate from the distal convoluted tubule is drained into collecting ducts, which plunge through the medulla to empty urine into the calyces.

### Glomerular Filtration p. 493

I. A filtrate derived from plasma in the glomerulus must pass through a basement membrane of the glomerular capillaries and through slits in the processes of the podocytes, which comprise the inner layer of the glomerular capsule.
   A. The glomerular ultrafiltrate is formed under the force of blood pressure and has a low protein concentration.
   B. The glomerular filtration rate is 115–125 ml/min.

II. The glomerular filtration rate can be regulated by constriction or dilation of the afferent arterioles.
   A. Sympathetic innervation causes constriction of the afferent arterioles.
   B. Intrinsic mechanisms help to autoregulate the rate of renal blood flow and the glomerular filtration rate.

### Reabsorption of Salt and Water p. 496

I. Approximately 65% of the filtered salt and water is reabsorbed across the proximal convoluted tubules.
   A. Sodium is actively transported, chloride follows passively by electrical attraction, and water follows the salt out of the proximal tubule.
   B. Salt transport in the proximal tubules is not under hormonal regulation.

II. The reabsorption of most of the remaining water occurs as a result of the action of the countercurrent multiplier system.
   A. Sodium is actively extruded from the ascending limb, followed passively by chloride.
   B. Since the ascending limb is impermeable to water, the remaining filtrate becomes hypotonic.
   C. Because of this salt transport and because of countercurrent exchange in the vasa recta, the tissue fluid of the medulla becomes hypertonic.
   D. The hypertonicity of the medulla is multiplied by a positive feedback mechanism involving the descending limb, which is passively permeable to water and perhaps to salt.

III. The collecting duct is permeable to water but not to salt.
   A. As the collecting ducts pass through the hypertonic renal medulla, water leaves by osmosis and is carried away in surrounding capillaries.
   B. The permeability of the collecting ducts to water is stimulated by antidiuretic hormone (ADH).

### Renal Plasma Clearance p. 505

I. Inulin is filtered but neither reabsorbed nor secreted; its clearance is thus equal to the glomerular filtration rate.
II. Some of the filtered urea is reabsorbed; its clearance is therefore less than the glomerular filtration rate.
III. Since almost all the PAH in blood going through the kidneys is cleared by filtration and secretion, the PAH clearance is a measure of the total renal blood flow.
IV. Normally all of the filtered glucose and amino acids are reabsorbed; glycosuria occurs when the transport carriers for glucose become saturated due to hyperglycemia.

### Renal Control of Electrolyte Balance p. 511

I. Aldosterone stimulates sodium reabsorption and potassium secretion in the distal convoluted tubule.
II. Aldosterone secretion is stimulated directly by a rise in blood potassium and indirectly by a fall in blood sodium.
   A. Decreased blood flow through the kidneys stimulates the secretion of the enzyme renin from the juxtaglomerular apparatus.
   B. Renin catalyzes the formation of angiotensin I, which is then converted to angiotensin II.
   C. Angiotensin II stimulates the adrenal cortex to secrete aldosterone.
III. Aldosterone stimulates the secretion of $H^+$ as well as potassium into the filtrate in exchange for sodium.

### Renal Control of Acid-Base Balance p. 516

I. The lungs regulate the $P_{CO_2}$ and carbonic acid concentration of the blood, whereas the kidneys regulate the bicarbonate concentration.
II. Filtered bicarbonate combines with $H^+$ to form carbonic acid in the filtrate.
   A. Carbonic anhydrase in the membranes of microvilli in the tubules catalyzes the conversion of carbonic acid to carbon dioxide and water.
   B. Carbon dioxide is reabsorbed and converted in either the tubule cells or the red blood cells to carbonic acid, which dissociates to bicarbonate and $H^+$.
   C. In addition to reabsorbing bicarbonate, the kidneys excrete $H^+$, which is buffered by ammonium and phosphate buffers.

### Clinical Applications p. 519

I. Diuretic drugs are used clinically to increase the urine volume and thus to lower the blood volume and pressure.
   A. Loop diuretics and the thiazides inhibit active $Na^+$ transport in the ascending limb and early portion of the distal tubule, respectively.
   B. Osmotic diuretics are extra solutes in the filtrate that increase the osmotic pressure of the filtrate and inhibit the osmotic reabsorption of water.
   C. The potassium-sparing diuretics act on the distal tubule to inhibit the reabsorption of $Na^+$ and secretion of $K^+$.
II. In glomerulonephritis the glomeruli can permit the leakage of plasma proteins into the urine.
III. People with renal insufficiency may be treated by the technique of renal dialysis.

---

## *Review Activities*

### Objective Questions

1. Which of the following statements about the renal pyramids is *false?*
   (a) They are located in the medulla.
   (b) They contain glomeruli.
   (c) They contain collecting ducts.
   (d) They empty urine into the calyces.

Match the following:

2. Active transport of sodium; water follows passively
3. Active transport of sodium; impermeable to water
4. Passively permeable to water and maybe salt
5. Passively permeable to water only

(a) proximal tubule
(b) descending limb
(c) ascending limb
(d) distal tubule
(e) collecting duct

6. Antidiuretic hormone promotes the retention of water by stimulating the
   (a) active transport of water
   (b) active transport of chloride
   (c) active transport of sodium
   (d) permeability of the collecting duct to water
7. Aldosterone stimulates sodium reabsorption and potassium secretion in the
   (a) proximal convoluted tubule
   (b) descending limb of the loop
   (c) ascending limb of the loop
   (d) distal convoluted tubule
   (e) collecting duct
8. Substance *X* has a clearance greater than zero but less than that of inulin. What can you conclude about substance *X?*
   (a) It is not filtered.
   (b) It is filtered, but neither reabsorbed nor secreted.
   (c) It is filtered and partially reabsorbed.
   (d) It is filtered and secreted.
9. Substance *Y* has a clearance greater than that of inulin. What can you conclude about *Y?*
   (a) It is not filtered.
   (b) It is filtered, but neither reabsorbed nor secreted.
   (c) It is filtered and partially reabsorbed.
   (d) It is filtered and secreted.
10. About 65% of the glomerular ultrafiltrate is reabsorbed in the
   (a) proximal tubule
   (b) distal tubule
   (c) loop of Henle
   (d) collecting duct
11. Diuretic drugs that act in the loop of Henle
   (a) inhibit active sodium transport
   (b) result in the increased flow of filtrate to the distal convoluted tubule
   (c) cause the increased secretion of potassium into the tubule
   (d) promote the excretion of salt and water
   (e) all of the above

12. The appearance of glucose in the urine
    (a) occurs normally
    (b) indicates the presence of kidney disease
    (c) occurs only when the transport carriers for glucose become saturated
    (d) is a result of hypoglycemia
13. Reabsorption of water through the tubules occurs by
    (a) osmosis
    (b) active transport
    (c) facilitated diffusion
    (d) all of the above
14. Which of the following factors oppose(s) filtration from the glomerulus?
    (a) plasma oncotic pressure
    (b) hydrostatic pressure in Bowman's capsule
    (c) plasma hydrostatic pressure
    (d) both a and b
    (e) both b and c
15. The countercurrent exchanger in the vasa recta
    (a) removes $Na^+$ from the extracellular fluid
    (b) maintains high concentrations of NaCl in the extracellular fluid
    (c) raises the concentration of $Na^+$ in the blood leaving the kidneys
    (d) causes large quantities of $Na^+$ to enter the filtrate
    (e) all of these
16. The kidneys help maintain acid-base balance by
    (a) $K^+/H^+$ exchange in the distal regions of the nephron
    (b) the action of carbonic anhydrase within the apical cell membranes
    (c) the action of carbonic anhydrase within the cytoplasm of the tubule cells
    (d) the buffering action of phosphates and ammonia in the urine
    (e) all of these

## Essay Questions

1. Explain how glomerular ultrafiltrate is produced and why it has a low protein concentration.
2. Explain how the countercurrent multiplier system works, and describe its functional significance.
3. Explain how countercurrent exchange occurs in the vasa recta, and describe its functional significance.
4. Explain the mechanisms whereby diuretic drugs may cause an excessive loss of potassium; also explain how the potassium-sparing diuretics work.
5. Explain how the structure of the epithelial wall of the proximal tubule and the distribution of $Na^+/K^+$ pumps in the epithelial cell membranes contribute to the ability of the proximal tubule to reabsorb salt and water.

## Selected Readings


Alexander, E. 1986. Metabolic acidosis: Recognition and etiologic diagnosis. *Hospital Practice* 21:100E.

Anderson, B. 1977. Regulation of body fluids. *Annual Review of Physiology* 39:185.

Bauman, J. W., and F. P. Chinard. 1975. *Renal function: physiological and medical aspects.* St. Louis: C. V. Mosby Co.

Beeuwkes, R., III. 1980. The vascular organization of the kidney. *Annual Review of Physiology* 42:531.

Brenner, B. M., and R. Beeuwkes, III. 1978. The renal circulation. *Hospital Practice* 13:35.

Brenner, B. M., T. H. Hostetter, and H. D. Humes. 1978. Molecular basis of proteinuria of glomerular origin. *New England Journal of Medicine* 298:826.

Buckalew, V. M., Jr., and K. A. Gruber. 1984. Natriuretic hormone. *Annual Review of Physiology* 46:343.

Coe, F. L., and J. H. Parks. 1988. Pathophysiology of kidney stones and strategies for treatment. *Hospital Practice* 23:185.

deBold, A. 1985. Atrial natriuretic factor: A hormone produced by the heart. *Science* 230:767.

Epstein, F. H., and R. S. Brown. 1988. Acute renal failure: A collection of paradoxes. *Hospital Practice* 23:171.

Galla, J. H., and R. G. Luke. 1987. Pathophysiology of metabolic alkalosis. *Hospital Practice* 22:123.

Giebisch, G. H., and B. Stanton. 1979. Potassium transport in the nephron. *Annual Review of Physiology* 41:241.

Glassock, R. J. 1987. Pathophysiology of acute glomerulonephritis. *Hospital Practice* 22:163.

Gluck, S. L. 1989. Cellular and molecular aspects of renal $H^+$ transport. *Hospital Practice* 24:149.

Hays, R. M. 1978. Principles of ion and water transport in the kidneys. *Hospital Practice* 13:79.

Hollenberg, N. K. 1986. The kidney in heart failure. *Hospital Practice* 21:81.

Jacobson, H. R. 1987. Diuretics: Mechanisms of action and uses. *Hospital Practice* 22:129.

Kokko, J. S. 1979. Renal concentrating and diluting mechanisms. *Hospital Practice* 14:110.

Kurtzman, N. A. 1987. Renal tubular acidosis: A constellation of syndromes. *Hospital Practice* 22:173.

Peart, W. S. 1975. Renin-angiotensin system. *New England Journal of Medicine* 292:302.

Rector, F. C., Jr., and M. G. Cogan. 1980. The renal acidoses. *Hospital Practice* 15:99.

Reid, I. A., B. J. Morris, and W. F. Ganong. 1978. The renin-angiotensin system. *Annual Review of Physiology* 40:377.

Renkin, E. M., and R. R. Robinson. 1974. Glomerular filtration. *New England Journal of Medicine* 290:79.

Steinmetz, P. R., and B. M. Koeppen. 1984. Cellular mechanisms of diuretic action along the nephron. *Hospital Practice* 19:125.

Tanner, R. L. 1980. Control of acid excretion by the kidney. *Annual Review of Medicine* 31:35.

Their, S. O. 1987. Diuretic mechanisms as a guide to therapy. *Hospital Practice* 22:81.

Thompson, C. 1988. The spectrum of renal cystic diseases. *Hospital Practice* 23:165.

Vander, A. J. 1980. *Renal physiology.* 2d ed. New York: McGraw-Hill Book Company.

Walker, L. A., and H. Valtin. 1982. Biological importance of nephron heterogeneity. *Annual Review of Physiology* 44:203.

Warnock, D. G., and F. C. Rector, Jr. 1979. Proton secretion by the kidney. *Annual Review of Physiology* 41:197.

Weinman, E. J., W. P. Dubinsky, Jr., and S. Shenolikar. 1989. Regulation of the renal $Na^+$ - $H^+$ exchanger. *Hospital Practice* 24:157.

# THE DIGESTIVE SYSTEM

# 17

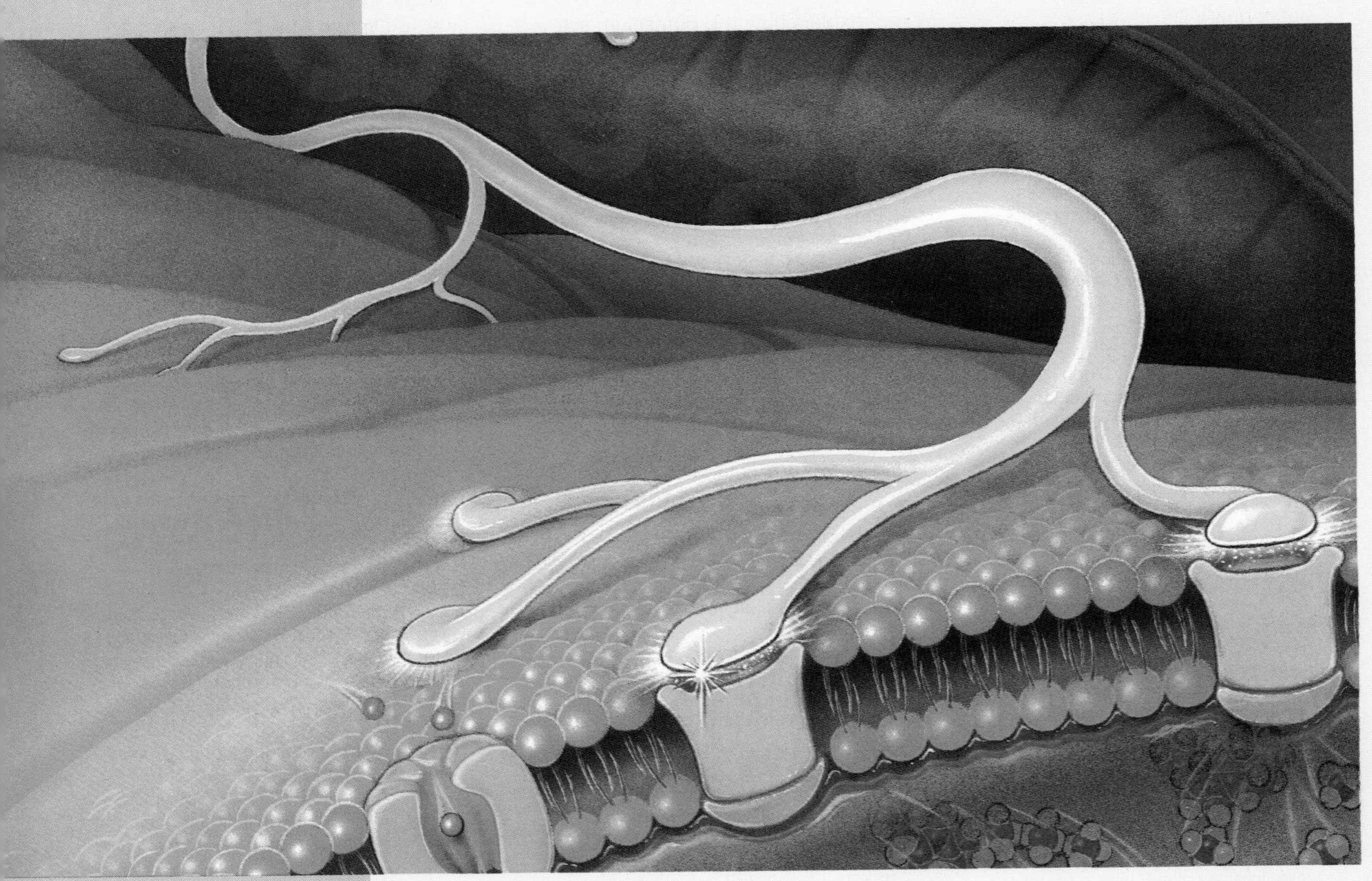

## Objectives

By studying this chapter, you should be able to

1. list the structures of the digestive system and enumerate the functions of different organs of the digestive system
2. describe the layers of the gastrointestinal tract and their function
3. describe the mechanism and function of peristalsis and explain the significance of the lower esophageal sphincter
4. describe the structure of the gastric mucosa, list the secretions of the mucosa and their functions, and identify the cells that produce each of these secretions
5. explain the roles of HCl and pepsin and explain why the stomach does not normally digest itself
6. describe the structure and function of the villi, microvilli, and crypts of Lieberkühn in the small intestine
7. describe the location and functions of the brush border enzymes of the intestine
8. explain the electrical activity that occurs in the intestine and describe the nature of peristalsis and segmentation
9. explain how the large intestine absorbs fluid and electrolytes
10. describe the flow of blood in the liver and explain how the liver functions to modify the chemical composition of the blood
11. describe the composition and functions of bile and explain how bile is kept separate from blood in the liver
12. trace the pathway of the formation, conjugation, and excretion of bilirubin and explain how jaundice may be produced
13. explain the significance of the enterohepatic circulation of various compounds and describe the enterohepatic circulation of bile pigment
14. describe the endocrine and exocrine structures of the pancreas and describe the composition and functions of pancreatic juice
15. describe the enzymes and locations involved in the digestion of carbohydrates, lipids, and proteins and explain how monosaccharides and amino acids are absorbed
16. describe the roles of bile and pancreatic lipase in fat digestion and trace the pathways and structures involved in the absorption of lipids
17. explain how gastric secretion is regulated during the cephalic, gastric, and intestinal phases
18. explain how pancreatic juice and bile secretion is regulated by nerves and hormones
19. discuss the trophic effects of gastrointestinal hormones on the digestive tract

## Outline

**Figure 17.1.** The digestion of food molecules occurs by means of hydrolysis reactions.

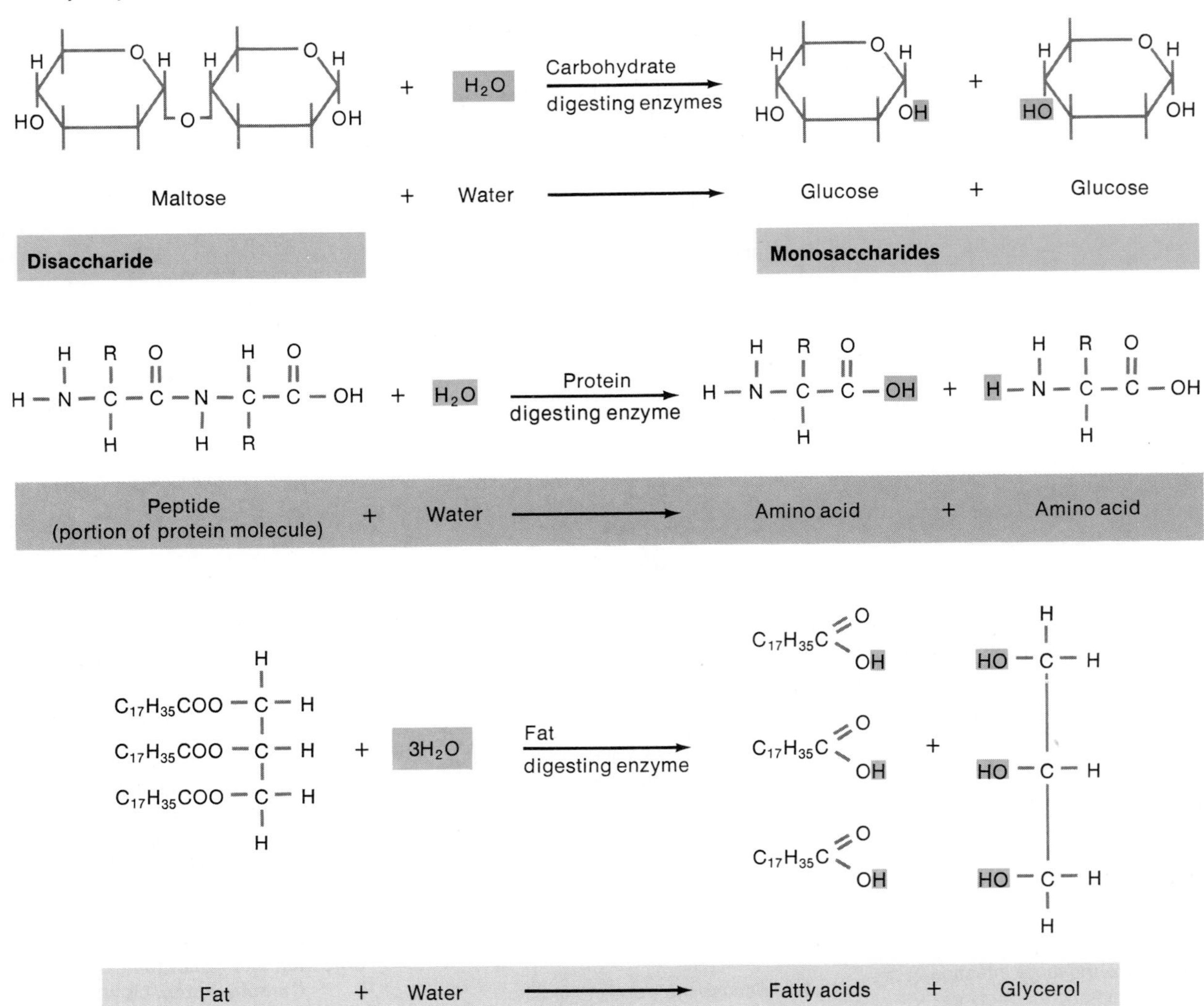


## INTRODUCTION TO THE DIGESTIVE SYSTEM

Within the lumen of the digestive tract, large food molecules are hydrolyzed into their monomers. These monomers pass through the inner layer, or mucosa, of the intestine to enter the blood or lymph in a process called absorption. Digestion and absorption are aided by specializations of the mucosa and by characteristic movements caused by contractions of the muscle layers of the digestive tract. The movements and secretions of the digestive system are stimulated by parasympathetic and inhibited by sympathetic innervations.

Unlike plants, which can form organic molecules using inorganic compounds such as carbon dioxide, water, and ammonia, humans and other animals must obtain their basic organic molecules from food. Some of the ingested food molecules are needed for their energy (caloric) value—obtained by the reactions of cell respiration and used in the production of ATP—and the balance is used to make additional tissue.

Most of the organic molecules that are ingested are similar to the molecules that form the structure of human tissues. These are generally large molecules *(polymers)*, which are composed of subunits *(monomers)*. Within the gastrointestinal tract, the process of **digestion** occurs in which these large molecules are broken down into their monomers by *hydrolysis* reactions (reviewed in fig. 17.1). The monomers thus formed are transported across the wall of the intestine into the blood and lymph; this process is called **absorption.** Digestion and absorption are the primary functions of the digestive system.

Since the composition of food is similar to the composition of body tissues, enzymes that digest food could also digest the person's own tissues. This does not normally occur because there is a variety of protective devices that inactivate digestive enzymes in the body and keep these enzymes separated from the cytoplasm of tissue cells. The fully active digestive enzymes are normally limited in their location to the lumen (cavity) of the gastrointestinal tract.

The lumen of the gastrointestinal tract is continuous with the environment because it is open at both ends (mouth and anus). Indigestible materials, such as cellulose from plant walls, pass from one end to the other without crossing the epithelial lining of the digestive tract (that is, without being absorbed). In this sense, these indigestible materials never enter the body, and the harsh conditions required for digestion thus occur *outside* the body.

In *planaria* (a type of flatworm) the gastrointestinal tract has only one opening—the mouth is also the anus. Each cell that lines the gastrointestinal tract is thus exposed to food, absorbable digestion products, and waste products. The two-ended digestive tract of higher organisms, in contrast, allows one-way transport, which is ensured by wavelike muscle contractions and by the action of sphincter muscles. This one-way transport allows different regions of the gastrointestinal tract to be specialized for different functions, as a "dis-assembly line." There are seven functions:

**Ingestion.** The taking of food into the digestive system by way of the mouth.

**Mastication.** Chewing movements to pulverize food and mix it with saliva.

**Deglutition.** The swallowing of food to move it from the mouth to the stomach.

**Digestion.** Mechanical and chemical breakdown of food into absorbable molecules.

**Absorption.** The passage of monomers of food molecules through the mucous membrane of the intestine into blood or lymph.

**Peristalsis.** Rhythmic, wavelike contractions that move food through the digestive tract.

**Defecation.** The discharge of indigestible wastes, called feces, from the body.

Anatomically and functionally the digestive system can be divided into a tubular *gastrointestinal (GI) tract,* or an *alimentary canal,* and *accessory organs.* The GI tract is approximately 9 m (30 ft) long and extends from the mouth to the anus. It traverses the thoracic cavity and enters the abdominal cavity at the level of the diaphragm. The anus is located at the inferior portion of the pelvic cavity. The organs of the GI tract include the *oral (buccal) cavity, pharynx, esophagus, stomach, small intestine,* and *large intestine* (fig. 17.2). The accessory digestive organs include the *teeth, tongue, salivary glands, liver, gallbladder,* and *pancreas.* The term *viscera* is frequently used to refer to the abdominal organs of digestion, but this term can actually be used to indicate any organ in the thoracic and abdominal cavities. *Gut* is an anatomical term that generally refers to the developing stomach and intestine in the embryo.

**Figure 17.2.** The digestive system, including the digestive tract and the accessory digestive organs.

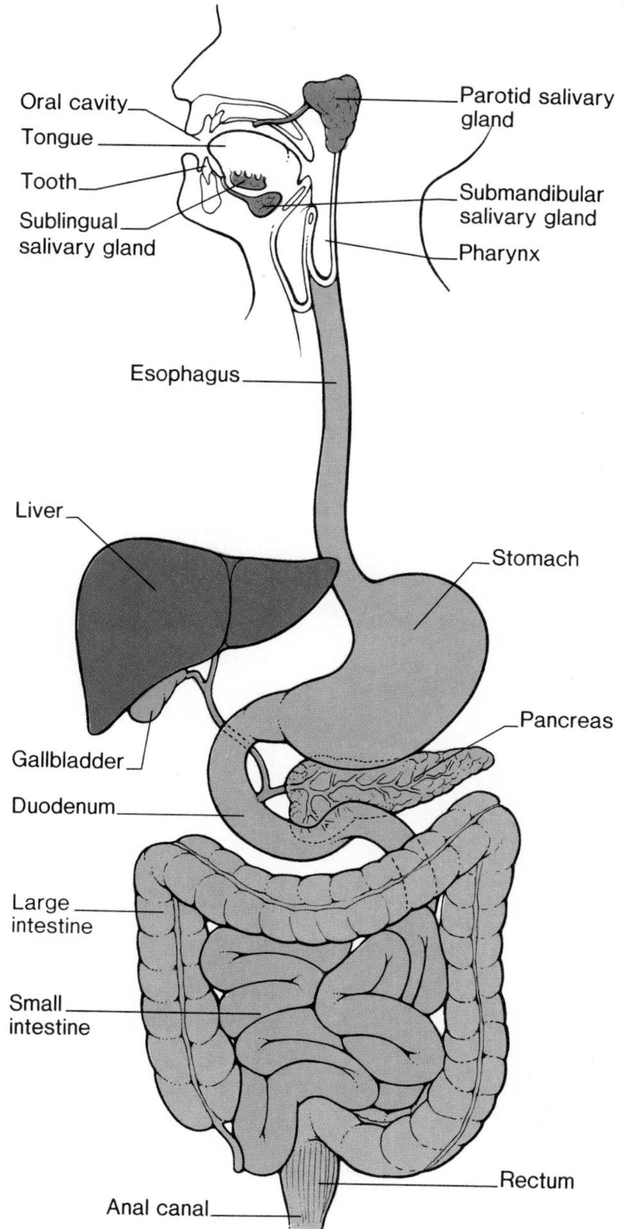

**Figure 17.3.** (*a*) An illustration of the major layers of the intestine. The insert shows how folds of mucosa form projections called *villi*. (*b*) An illustration of a cross section of the intestine showing layers and glands.

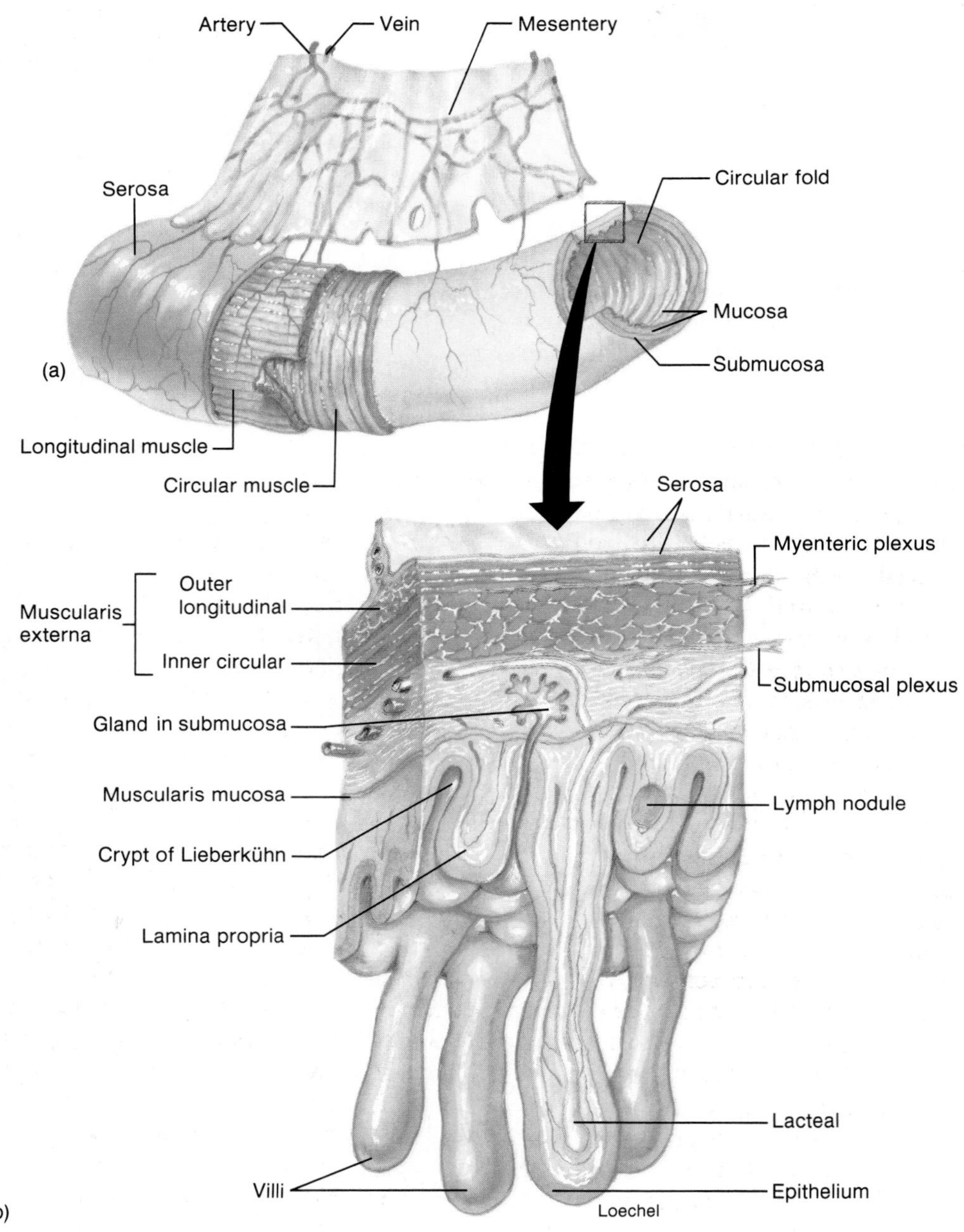


## Layers of the Gastrointestinal Tract

The GI tract from the esophagus to the anal canal is composed of four layers, or *tunics*. Each tunic contains a dominant tissue type that performs specific functions in the digestive process. The four tunics of the GI tract, from the inside out, are the **mucosa, submucosa, muscularis,** and **serosa** (fig. 17.3a).

***Mucosa.*** The mucosa surrounds the lumen of the GI tract and is the absorptive and major secretory layer. It consists of a simple columnar *epithelium* supported by the *lamina propria,* which is a thin layer of connective tissue. The lamina propria contains numerous lymph nodules that are important in protecting against disease (fig. 17.3b). External to the lamina propria is a thin layer of smooth muscle called the *muscularis mucosa.* This is the muscle layer that causes portions of the GI tract to have numerous small folds, thus greatly increasing the absorptive surface area. Specialized goblet cells in the mucosa secrete mucus throughout most of the GI tract.

***Submucosa.*** The relatively thick submucosa is a highly vascular layer of connective tissue serving the mucosa. Absorbed molecules that pass through the columnar epithelial cells of the mucosa enter into blood vessels of the submucosa. In addition to blood vessels, the submucosa contains glands and nerve plexuses. The *submucosal plexus (Meissner's plexus)* (fig. 17.3b) provides an autonomic nerve supply to the muscularis mucosa.

***Muscularis.*** The muscularis (also called the muscularis externa) is responsible for segmental contractions and peristaltic movement through the GI tract. The muscularis has an inner circular and an outer longitudinal layer of smooth muscle. Contractions of these layers move the food through the tract and physically pulverize and churn the food with digestive enzymes. The *myenteric plexus (Auerbach's plexus)* located between the two muscle layers provides the major nerve supply to the GI tract and includes fibers and ganglia from both the sympathetic and parasympathetic divisions of the autonomic nervous system.

***Serosa.*** The outer serosa layer completes the wall of the GI tract. It is a binding and protective layer consisting of areolar connective tissue covered with a layer of simple squamous epithelium.

### Innervation of the Gastrointestinal Tract

The GI tract is innervated by the sympathetic and parasympathetic divisions of the autonomic nervous system. The vagus nerve is the source of parasympathetic activity in the esophagus, stomach, pancreas, gallbladder, small intestine, and upper portion of the large intestine. The lower portion of the large intestine receives parasympathetic innervation from spinal nerves in the sacral region. The submucosal plexus and myenteric plexus are the sites where parasympathetic preganglionic fibers synapse with postganglionic neurons that innervate the smooth muscle of the GI tract. Stimulation of these parasympathetic fibers increases peristalsis and the secretions of the GI tract.

Postganglionic sympathetic fibers pass through the submucosal and myenteric plexuses and innervate the GI tract. Sympathetic nerve stimulation acts antagonistically to the effects of parasympathetic nerves by reducing peristalsis and secretions and stimulating the contraction of sphincter muscles along the GI tract.

---

1. *Define the terms* digestion *and* absorption, *describe how molecules are digested, and indicate which molecules are absorbed.*
2. *Describe the structure and function of the mucosa, submucosa, and muscularis layers of the gastrointestinal tract.*
3. *Describe the location and composition of the submucosal and myenteric plexuses, and explain the actions of autonomic nerves on the gastrointestinal tract.*

---

## *ESOPHAGUS AND STOMACH*

Swallowed food is passed through the esophagus to the stomach by wavelike contractions known as peristalsis. The mucosa of the stomach contains gastric glands with parietal cells, which secrete hydrochloric acid, and chief cells, which secrete pepsinogen. The latter molecules are an inactive form of the protein-digesting enzyme pepsin, which becomes activated in the acidic environment within the lumen of the stomach. As a result, the stomach partially digests proteins and functions to store its contents, called chyme, for later processing by the intestine.

Swallowed food is passed from the esophagus to the stomach, where it is churned and mixed with hydrochloric acid and pepsin, a protein-digesting enzyme. The mixture thus produced is pushed by muscular contractions of the stomach past the pyloric sphincter, which guards the junction of the stomach to the duodenum of the small intestine.

### Esophagus

The **esophagus** is the portion of the GI tract that connects the pharynx to the stomach. It is a collapsible muscular tube approximately 25 cm (10 in.) long, originating at the level of the larynx and located posterior to the trachea.

The esophagus is located within the mediastinum of the thorax (chapter 15) and passes through an opening in the diaphragm called the *esophageal hiatus* just above the stomach. The esophagus is lined with a nonkeratinized stratified squamous epithelium; its walls contain either striated or smooth muscle, depending on the location. The upper third of the esophagus contains striated muscle; the middle third contains a mixture of striated and smooth muscle, and the terminal portion contains only smooth muscle.

Swallowed food is pushed from one end of the esophagus to the other by a wavelike muscular contraction called **peristalsis** (fig. 17.4). Peristaltic waves are produced by constriction of the lumen as a result of circular muscle

**Figure 17.4.** (*a*) A diagram and (*b*) an X ray of peristalsis in the esophagus.

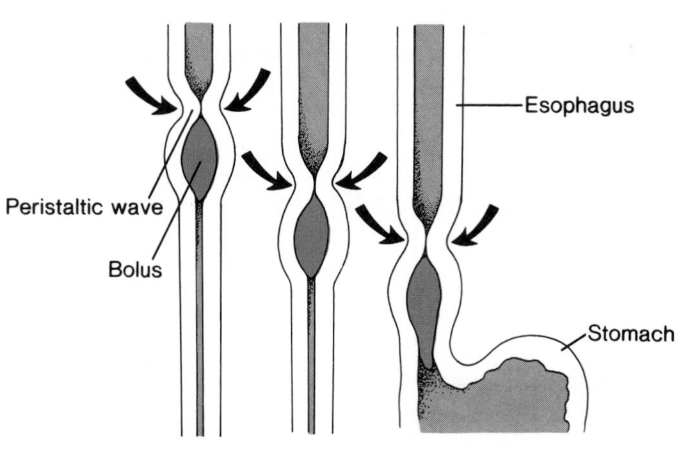


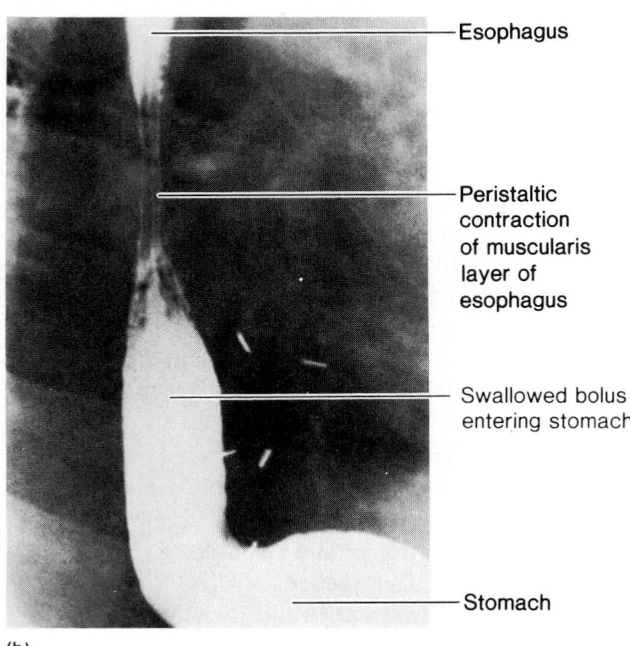


contraction, followed by shortening of the tube by longitudinal muscle contraction. These contractions progress from the superior end of the esophagus to the *gastroesophageal junction* at a rate of about 2–4 cm per second, as they empty the contents of the esophagus into the cardiac region of the stomach.

The lumen of the terminal portion of the esophagus is slightly narrowed because of a thickening of the circular muscle fibers in its wall. This portion is referred to as the **lower esophageal (gastroesophageal) sphincter.** The muscle fibers of this region constrict after food passes into the stomach to help prevent the stomach contents from regurgitating into the esophagus. Regurgitation would occur because the pressure in the abdominal cavity is greater than the pressure in the thoracic cavity as a result of respiratory movements. The lower esophageal sphincter must remain closed, therefore, until food is pushed through it by peristalsis into the stomach.

The lower esophageal sphincter is not a true sphincter muscle that can be identified histologically, and it does at times permit the acidic contents of the stomach to enter the esophagus. This can create a burning sensation commonly called **heartburn,** although the heart is not involved. Infants under one year old have difficulty controlling their lower esophageal sphincter and thus often "spit up" following meals. Certain mammals, such as rodents, have a true gastroesophageal sphincter and cannot regurgitate, which is why poison grains can kill mice and rats effectively.

## Stomach

The stomach is the most distensible part of the GI tract. It is a J-shaped pouch that is continuous with the esophagus superiorly and empties into the duodenum of the small intestine inferiorly. The functions of the stomach are to store food; to initiate the digestion of proteins; and to move the food into the small intestine as a pasty material called **chyme**.

Swallowed food is delivered from the esophagus to the *cardiac region* of the stomach (figs. 17.5 and 17.6). An imaginary horizontal line drawn through the cardiac region divides the stomach into an upper *fundus* and a lower *body,* which comprises about two-thirds of the stomach. The distal portion of the stomach is called the *pyloric region,* or *antrum.* Contractions of the stomach push partially digested food from the antrum, through the pyloric sphincter, and thus into the first part of the small intestine.

The inner surface of the stomach is thrown into long folds called *rugae,* which can be seen with the unaided eye. Microscopic examination of the gastric mucosa shows that it is likewise folded. The openings of these folds into the stomach lumen are called **gastric pits.** The cells that line the folds deeper in the mucosa secrete various products into the stomach; these form the exocrine **gastric glands** (figs. 17.7 and 17.8).

There are several different types of cells in the gastric glands that secrete different products: (1) **goblet cells,** which secrete mucus; (2) **parietal cells,** which secrete hydrochloric acid; (3) **chief cells,** which secrete pepsinogen, an inactive form of the protein-digesting enzyme pepsin; (4) **argentaffin cells,** which secrete serotonin and histamine; and (5) **G cells,** which secrete the hormone gastrin.

**Figure 17.5.** The major regions and structures of the stomach.

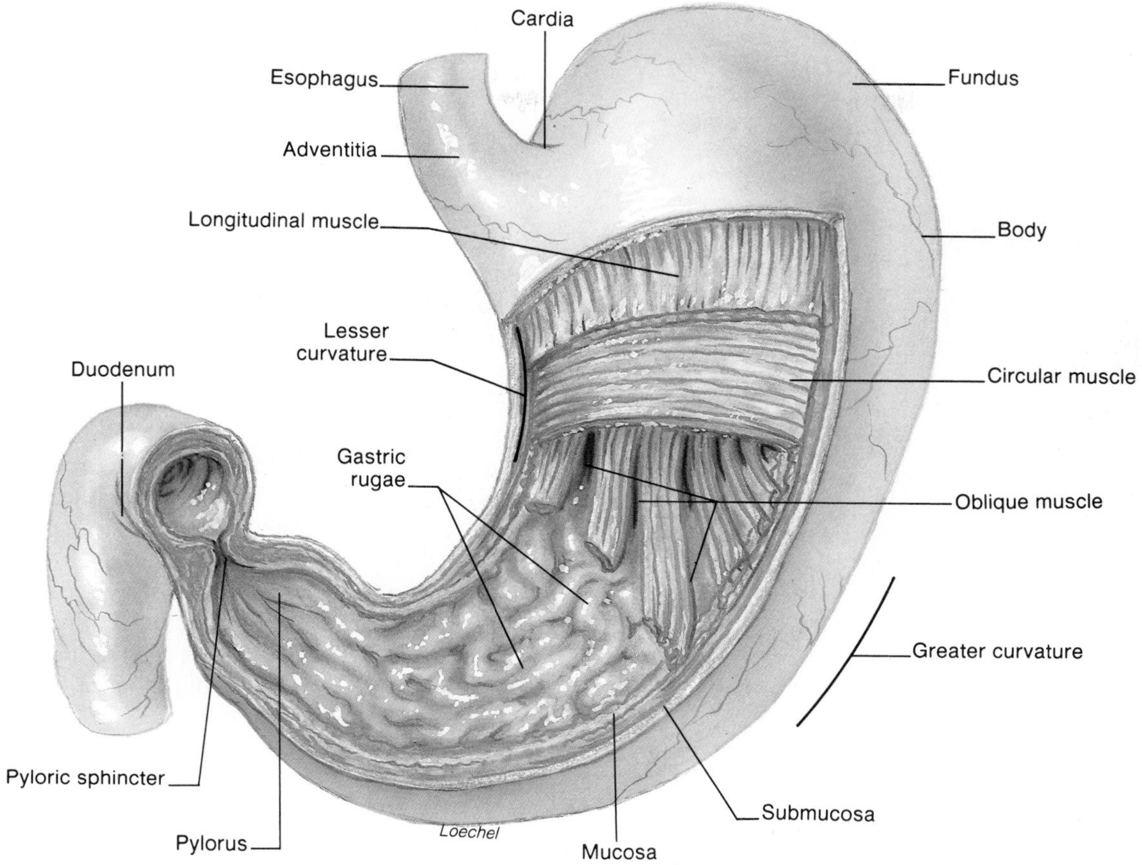

Cardia
Esophagus
Fundus
Adventitia
Longitudinal muscle
Body
Lesser curvature
Duodenum
Circular muscle
Gastric rugae
Oblique muscle
Greater curvature
Pyloric sphincter
Submucosa
Pylorus
Loechel
Mucosa

**Figure 17.6.** An X ray of a stomach.

Fundus
Rugae
Pyloric sphincter
Duodenum
Body
Pylorus

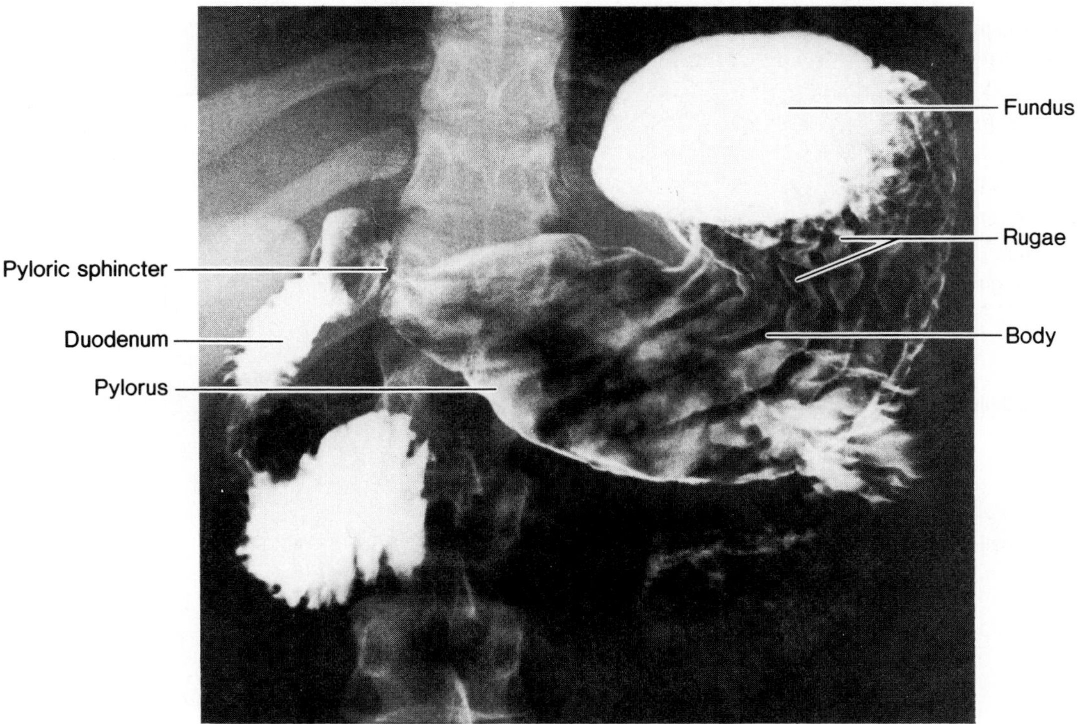

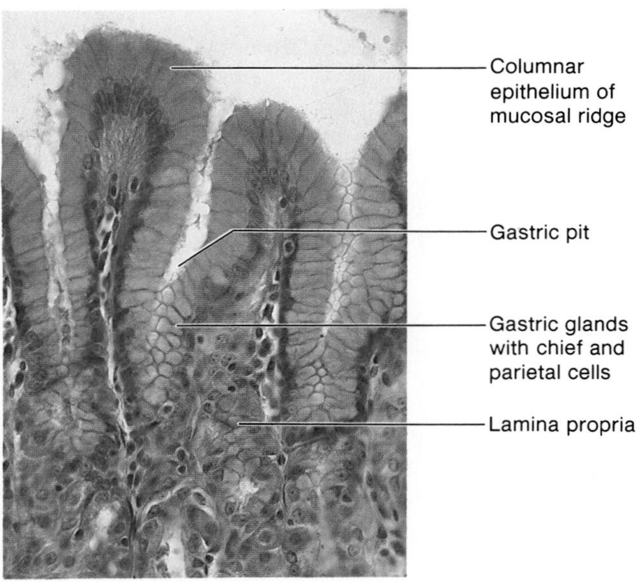


**Figure 17.7.** Microscopic structures of the mucosa of the stomach.

In addition to these products (table 17.1), the gastric mucosa (probably the parietal cells) secretes a polypeptide called *intrinsic factor,* which is required for absorption of vitamin $B_{12}$ in the intestine.

***Pepsin and Hydrochloric Acid Secretion.*** The secretion of hydrochloric acid by parietal cells makes gastric juice very acidic, with a pH less than 2. This strong acidity serves three functions: (1) ingested proteins are denatured at low pH—that is, their tertiary structure (chapter 2) is altered so that they become more digestible; (2) under acidic conditions, weak pepsinogen enzymes partially digest each other—this frees the active pepsin enzyme as small peptide fragments are removed (fig. 17.9); and (3) pepsin is more active under acidic conditions—it has a pH optimum (chapter 4) of about 2.0. The peptide bonds of ingested protein are broken (through hydrolysis reactions) by pepsin under acidic conditions; the HCl itself does not directly digest proteins.

***Digestion and Absorption in the Stomach.*** Proteins are only partially digested in the stomach by the action of pepsin. Carbohydrates and fats are not digested at all in the stomach. The complete digestion of food molecules

**Figure 17.8.** Gastric pits and gastric glands of the mucosa. (*a*) Gastric pits are the openings of the gastric glands. (*b*) Gastric glands consist of mucous cells, chief cells, and parietal cells, each of which produces a specific secretion.

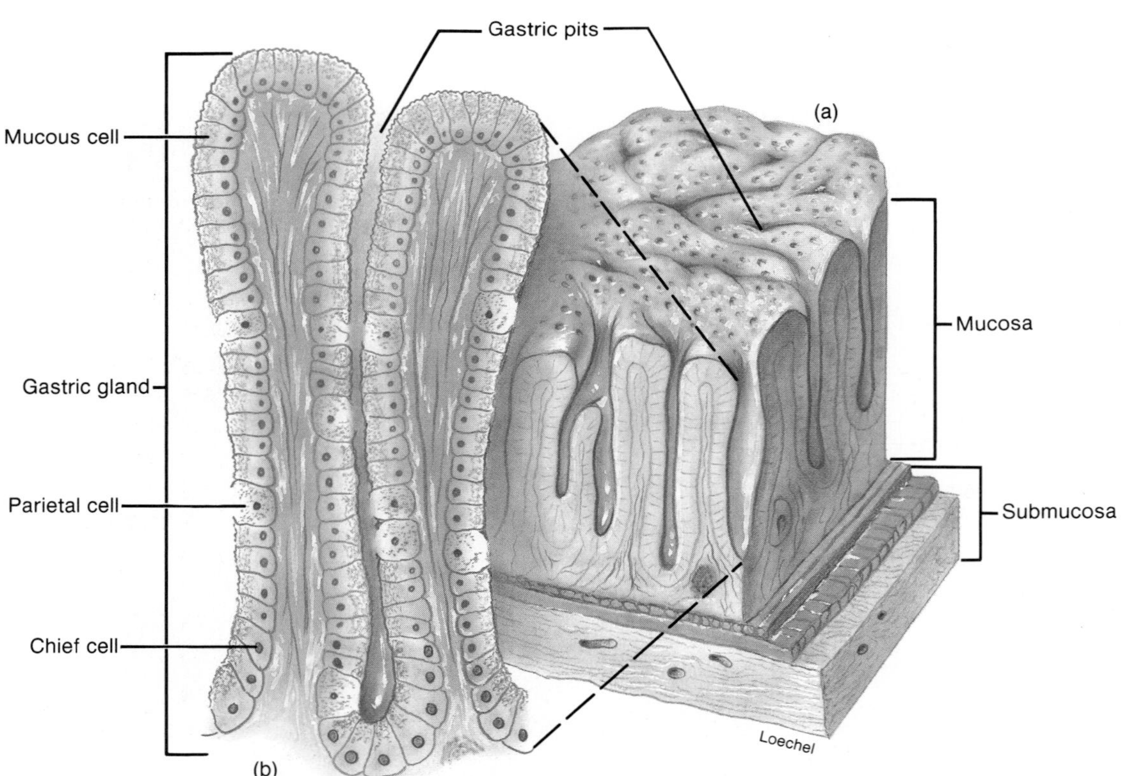

**Table 17.1** Secretions of the fundus and pyloric regions of the stomach

| Stomach Region | Cell Type | Secretions |
|---|---|---|
| Fundus | Parietal cells | Hydrochloric acid; intrinsic factor |
| | Chief cells | Pepsinogen |
| | Goblet cells | Mucus |
| | Argentaffin cells | Histamine, serotonin |
| Pyloric | G cells | Gastrin |
| | Chief cells | Pepsinogen |
| | Goblet cells | Mucus |

occurs later, when chyme enters the small intestine. Patients with partial gastric resections, therefore, and even those with complete gastrectomies (removal of the stomach), can still adequately digest and absorb their food.

Almost all of the products of digestion are absorbed through the wall of the intestine; the only commonly ingested substances that can be absorbed across the stomach wall are alcohol and aspirin. Absorption occurs as a result of the lipid solubility of these molecules. The passage of aspirin through the gastric mucosa has been shown to cause bleeding, which may be significant if large amounts of aspirin are taken.

The only function of the stomach that appears to be essential for life is the secretion of intrinsic factor. This polypeptide is needed for the absorption of vitamin $B_{12}$ in the terminal portion of the ileum in the small intestine, and vitamin $B_{12}$ is required for maturation of red blood cells in the bone marrow. A patient with a gastrectomy has to receive $B_{12}$ orally (together with intrinsic factor) or through injections to prevent the development of **pernicious anemia.**

***Gastritis and Peptic Ulcers.*** The gastric mucosa is normally resistant to the harsh conditions of low pH and pepsin activity in gastric juice. The reasons for its resistance to pepsin digestion are not well understood. Its resistance to hydrochloric acid appears to be due to three interrelated mechanisms: (1) the stomach lining is covered with a thin layer of alkaline mucus, which may offer limited protection; (2) the epithelial cells of the mucosa are joined together by tight junctions, which prevent acid from leaking into the submucosa; and (3) the epithelial cells that are damaged are exfoliated (shed) and replaced by new cells. This latter process results in the loss and replacement of about one-half million cells a minute, so that the entire epithelial lining is replaced every three days.

**Figure 17.9.** The gastric mucosa secretes the inactive enzyme pepsinogen and hydrochloric acid (HCl). In the presence of HCl, the active enzyme pepsin is produced. Pepsin digests proteins into shorter polypeptides.

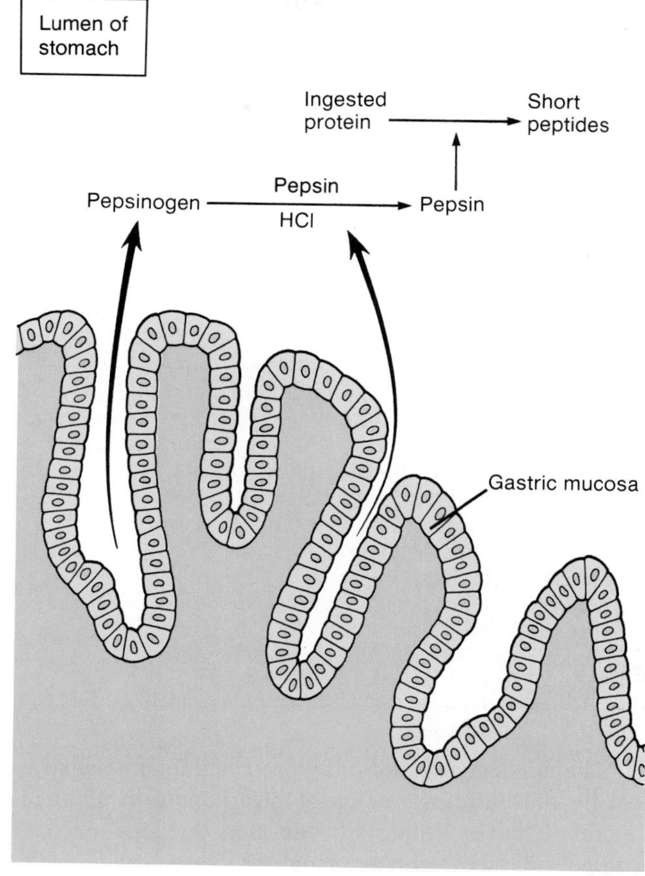


These three mechanisms provide a barrier that prevents self-digestion of the stomach. Breakdown of the gastric mucosa barrier may occur at times, however, perhaps as a result of the detergent action of bile salts that are regurgitated from the small intestine through the pyloric sphincter. When this occurs, acid can leak through the mucosal barrier to the submucosa, which causes direct damage and stimulates inflammation. The histamine released from mast cells (chapter 19) during inflammation may stimulate further acid secretion and result in further damage to the mucosa. The inflammation that occurs during these events is called **acute gastritis.**

Once the mucosal barrier is broken, the acid can produce craterlike holes and even complete perforations of the stomach wall. These are called gastric ulcers. More

**Table 17.2** Gastric secretion of hydrochloric acid, measured in milliequivalents (mEq) (one milliequivalent is equal to one millimole of $H^+$)

| Condition | Day (mEq/hr) | Night (mEq/12 hrs) | Maximal (mEq/hr) |
|---|---|---|---|
| Normal | 2–3 | 18 | 16–20 |
| Gastric ulcer | 2–4 | 8 | 16–20 |
| Duodenal ulcer | 4–10 | 60 | 25–40 |
| Zollinger-Ellison syndrome (G cell tumor) | 30 | 120 | 45 |

commonly encountered, however, are ulcers of the first 3–4 cm of the duodenum. Ulcers of the stomach and duodenum are known as **peptic ulcers.** It has been estimated that about 10% of American men and 4% of American women will develop peptic ulcers.

There is evidence that acute gastritis, and perhaps peptic ulcers, may be aggravated or even caused by a bacterial infection. The bacterium, *Campylobacter pylori,* appears to penetrate between cells in the gastric and duodenal mucosa and has been found in 90% of patients with duodenal ulcers and 70% of patients with gastric ulcers. Although the association of these bacteria with the presence of an ulcer does not prove a causal relationship, it is interesting that treatment of ulcer patients with antibiotics appears to reduce the chances of a relapse occurring.

The duodenum is normally protected against gastric acid by the buffering action of bicarbonate in alkaline pancreatic juice. People who develop duodenal ulcers, however, produce excessive amounts of gastric acid (table 17.2) that cannot be neutralized by pancreatic juice. This excessive gastric acid secretion is apparently due to excessive parasympathetic stimulation of the stomach. People with gastritis and peptic ulcers must avoid substances that stimulate acid secretion, including caffeine and alcohol, and often must take antacids.

The secretion of gastric acid is stimulated by *acetylcholine* (from parasympathetic nerve endings), *gastrin* (a hormone secreted by the stomach), and *histamine* (secreted by mast cells in the lamina propria). In response to these stimulants, the parietal cells secrete $H^+$ into the gastric lumen by active transport against a million-to-one concentration gradient, using a carrier that functions as an **$H^+/K^+$ ATPase pump.** People with ulcers, consequently, can be treated with drugs that block ACh (atropine; pirenzepine) or gastrin (proglumide) or histamine (cimetidine) action. Experimental drugs that may be important in the future treatment of ulcers include one that specifically blocks the $H^+/K^+$ pump (omeprazole), and various prostaglandin analogues that appear to block secretion of $H^+$.

1. *Describe the structure and function of the lower esophageal sphincter.*
2. *List the secretory cells of the gastric mucosa and the products they secrete.*
3. *Explain the roles of hydrochloric acid in the stomach and how pepsinogen is activated.*
4. *Explain the cause of peptic ulcers and why they are more likely to be duodenal than gastric in location.*
5. *Describe the factors that stimulate the parietal cells to secrete $H^+$ and explain how various drugs function to inhibit acid secretion.*

## Small Intestine

The mucosa of the small intestine is folded into villi which project into the lumen. The cells that line these villi, further, have foldings of their plasma membrane called microvilli. This arrangement greatly increases the surface area for absorption. It also improves digestion, because the digestive enzymes of the small intestine are embedded within the cell membrane of the microvilli. Contractions of muscles within the small intestine produce movements that mix the chyme and propel it to the large intestine.

The *small intestine* is that portion of the GI tract between the pyloric sphincter of the stomach and the ileocecal valve opening into the large intestine. It is called "small" because of its smaller diameter compared to that of the large intestine. The small intestine is approximately 3.7 m (12 ft) long in a living person, but it will measure nearly twice this length in a cadaver when the muscle wall is relaxed. The first 20–30 cm (10 in.) from the pyloric sphincter is the **duodenum** (fig. 17.10). The next two-fifths of the small intestine is the **jejunum,** and the last three-fifths is the **ileum.** The ileum empties into the large intestine through the ileocecal valve.

The products of digestion are absorbed across the epithelial lining of intestinal mucosa. Absorption of carbohydrates, lipids, protein, calcium, and iron occurs primarily in the duodenum and jejunum. Bile salts, vitamin $B_{12}$, water, and electrolytes are absorbed primarily in the ileum. Absorption occurs at a rapid rate as a result of the large mucosal surface area in the small intestine, provided by folds. The mucosa and submucosa form large folds called

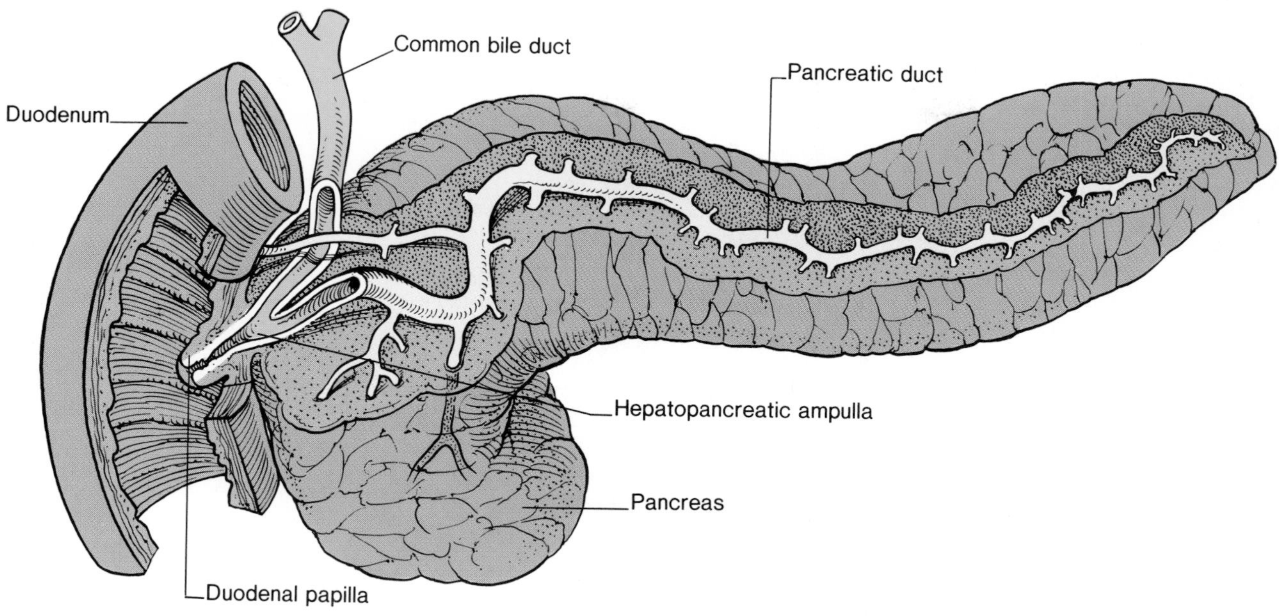


**Figure 17.10.** The duodenum and associated structures.

**Figure 17.11.** The histology of the duodenum.

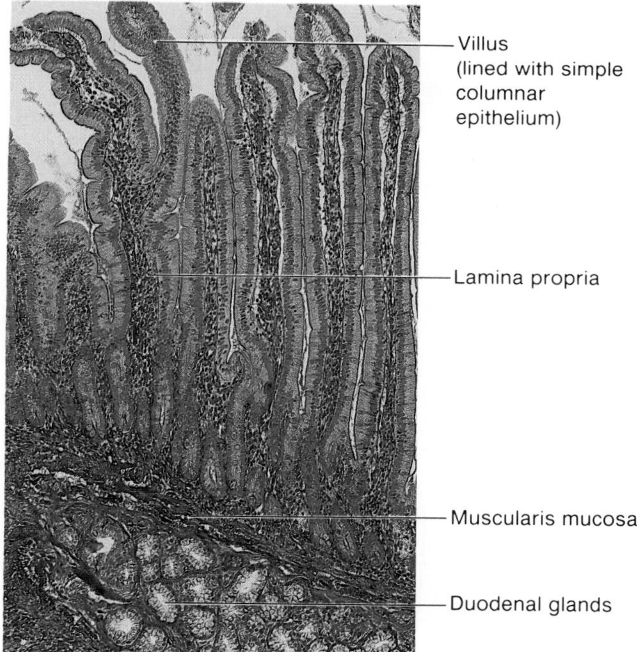


the *plicae circularis,* which can be observed with the unaided eye. The surface area is further increased by the microscopic folds of mucosa, called villi, and by the foldings of the apical cell membrane of epithelial cells (which can only be seen with an electron microscope), called microvilli.

## Villi and Microvilli

Each **villus** is a fingerlike fold of mucosa that projects into the intestinal lumen (fig. 17.11). The villi are covered with columnar epithelial cells and interspersed among these are the mucus-secreting goblet cells. The lamina propria forms a connective core of each villus and contains numerous lymphocytes, blood capillaries, and a lymphatic vessel called the *central lacteal* (fig. 17.12). Absorbed monosaccharides and amino acids are secreted into the blood capillaries; absorbed fat enters the central lacteals.

Epithelial cells at the tips of the villi are continuously exfoliated (shed) and are replaced by cells that are pushed up from the bases of the villi. The epithelium at the base of the villi invaginates downwards at various points to form narrow pouches that open through pores to the intestinal lumen. These structures are called the **crypts of Lieberkühn** (fig. 17.13).

Microvilli are fingerlike projections formed by foldings of the cell membrane, which can only be clearly seen in an electron microscope. In a light microscope, the microvilli produce a somewhat vague **brush border** on the edges of the columnar epithelial cells. The term *brush border* is thus often used synonymously with microvilli in descriptions of the intestine (fig. 17.14).

## Intestinal Enzymes

In addition to providing a large surface area for absorption, the cell membranes of the microvilli contain digestive enzymes. These enzymes are not secreted into the lumen, but instead remain attached to the cell membrane with their active sites exposed to the chyme. These **brush border enzymes** hydrolyze disaccharides, polypeptides, and other substrates (table 17.3). One brush border enzyme, *enterokinase,* is required for activation of the protein-digesting enzyme *trypsin,* which enters the intestine in pancreatic juice.

**Figure 17.12.** A diagram of the structure of an intestinal villus.

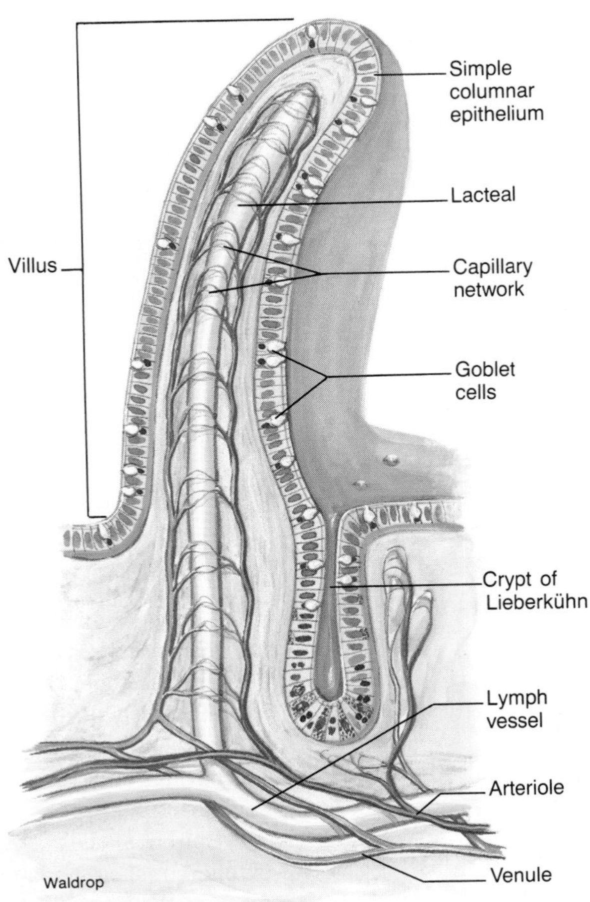


**Figure 17.13.** (*a*) Intestinal villi and crypts of Lieberkühn. The crypts serve as sites for production of new epithelial cells. The time required for migration of these new cells to the tip of the villi is shown in (*b*). Epithelial cells are exfoliated from the tips of the villi.

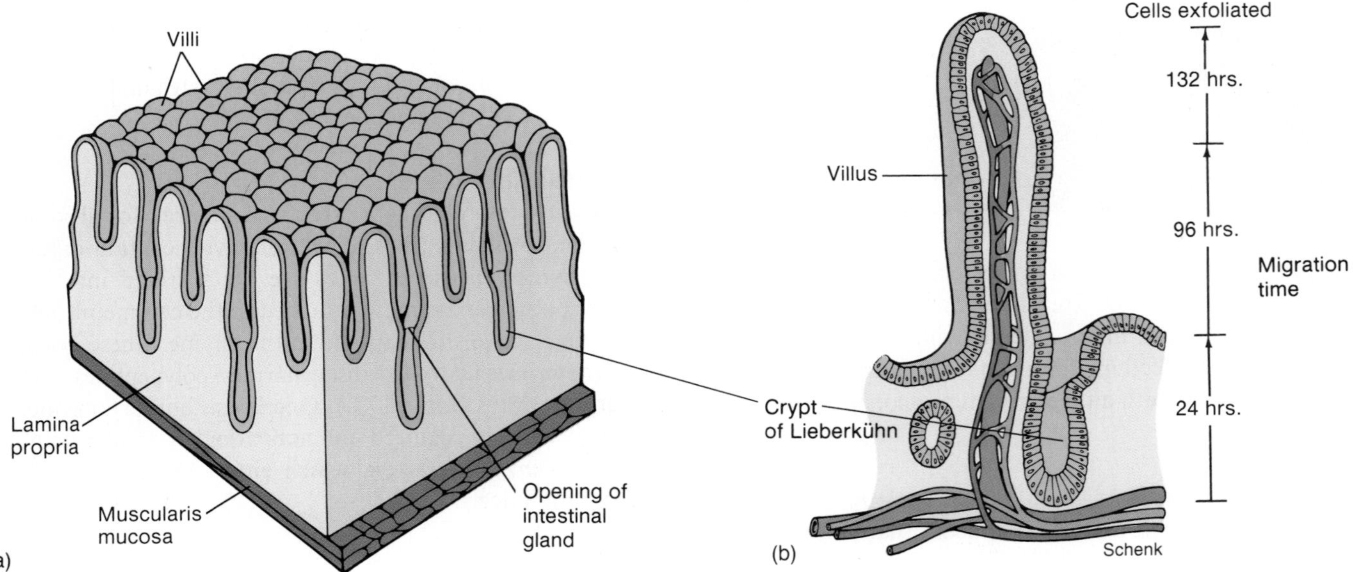

**Figure 17.14.** (*a*) Glycoproteins within the membranes of the microvilli contain polysaccharides that extend out into the lumen and form the glycocalyx, which covers the brush border epithelium. (*b*) The cell membrane of epithelial cells that line the small intestine are folded into microvilli. (*c*) The microvilli contain actin filaments attached to a terminal web of myosin filaments. This allows the microvilli to shorten. (*d*) A diagram of an intestinal epithelial cell.

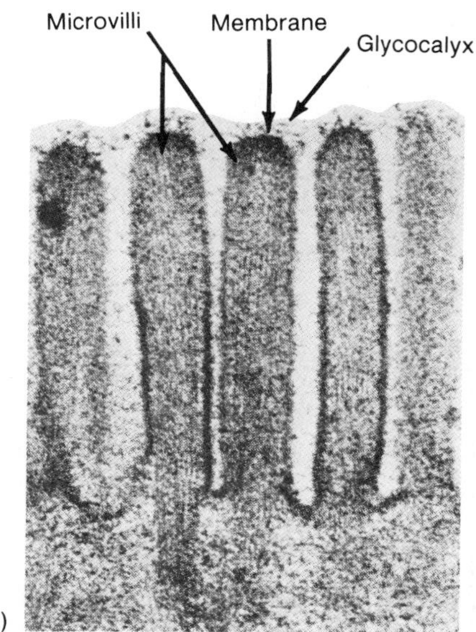


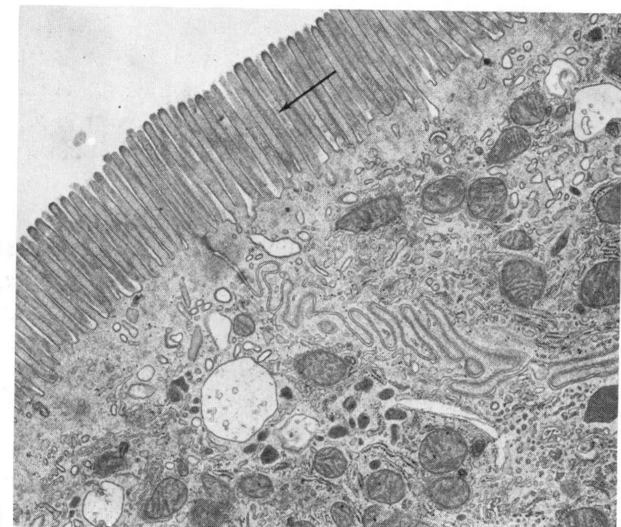


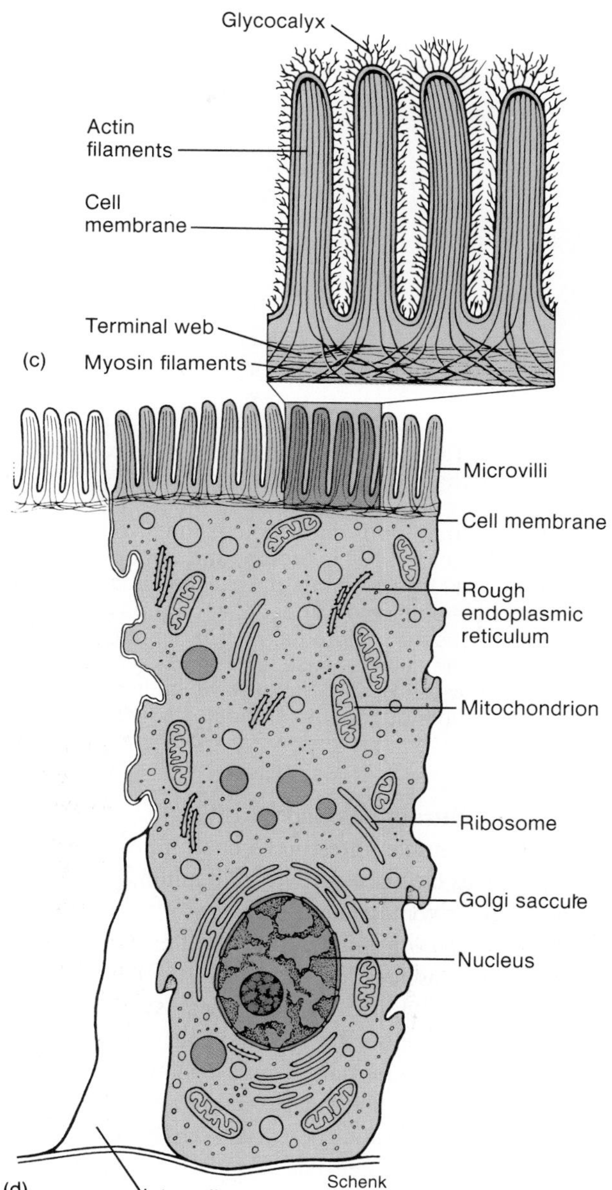


**Table 17.3** Brush border enzymes attached to the cell membrane of microvilli in the small intestine

| Category | Enzyme | Comments |
|---|---|---|
| Disaccharidase | Sucrase | Digests sucrose to glucose and fructose; deficiency produces gastrointestinal disturbances |
| | Maltase | Digests maltose to glucose |
| | Lactase | Digests lactose to glucose and galactose; deficiency produces gastrointestinal disturbances (lactose intolerance) |
| Peptidase | Aminopeptidase | Produces free amino acids, dipeptides, and tripeptides |
| | Enterokinase | Activates trypsin (and indirectly other pancreatic juice enzymes); deficiency results in protein malnutrition |
| Phosphatase | $Ca^{++}$, $Mg^{++}$—ATPase | Needed for absorption of dietary calcium; enzyme activity regulated by vitamin D |
| | Alkaline phosphatase | Removes phosphate groups from organic molecules; enzyme activity may be regulated by vitamin D |

**Figure 17.15.** The smooth muscle of the gastrointestinal tract produces and conducts spontaneous pacesetter potentials. As these potential changes reach a threshold level of depolarization, they stimulate the production of action potentials, which in turn stimulate smooth muscle contraction.

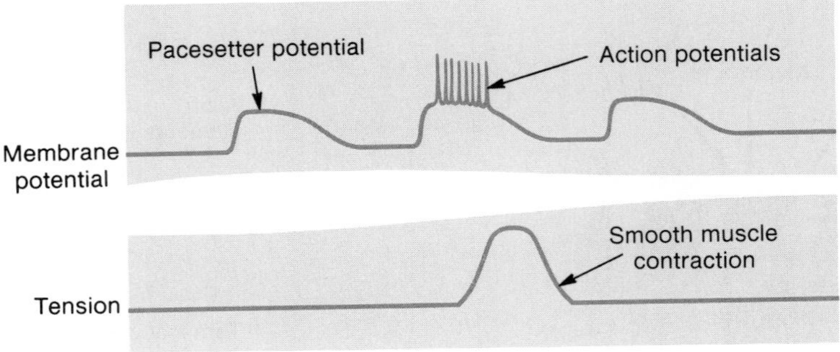


The ability to digest milk sugar, or lactose, depends on the presence of a brush border enzyme called lactase. This enzyme is present in most children under the age of four but becomes inactive in most adults. This can result in **lactose intolerance.** The presence of large amounts of undigested lactose in the intestine causes diarrhea, gas, cramps, and other unpleasant symptoms. Yogurt is better tolerated than milk because it contains lactase produced by the yogurt bacteria, which becomes activated in the duodenum and digests lactose.

**Figure 17.16.** Segmentation of the small intestine. Simultaneous contractions of many segments of the intestine help to mix the chyme with digestive enzymes and mucus.

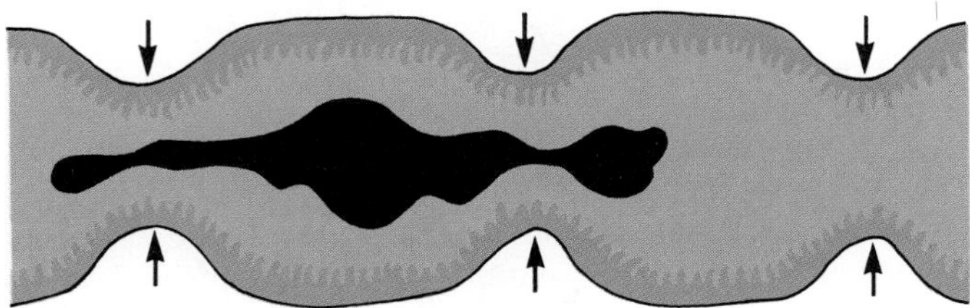

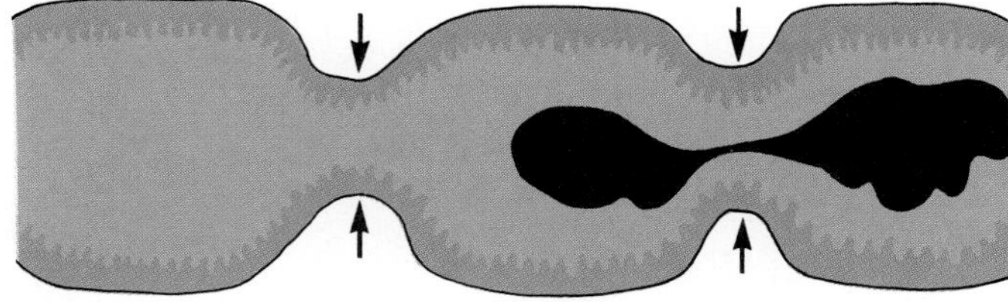

## Intestinal Contractions and Motility

Like cardiac muscle, intestinal smooth muscle is capable of spontaneous electrical activity and automatic, rhythmic contractions. Spontaneous depolarization begins in the longitudinal smooth muscle and is conducted to the circular smooth muscle layer across *nexuses*. The term *nexus* is used here to indicate an electrical synapse between smooth muscle cells. The spontaneous depolarizations, called **pacesetter potentials,** decrease in amplitude as they are conducted from one muscle cell to another, much like excitatory postsynaptic potentials (EPSPs). Also like EPSPs, pacesetter potentials stimulate the production of action potentials in the smooth muscle cells through which they are conducted (fig. 17.15).

The nexuses conduct the pacesetter potentials, not the action potentials. Action potentials are, therefore, limited to those smooth muscle cells that are depolarized to threshold by the spreading pacesetter potentials. When action potentials are produced, they stimulate smooth muscle contraction. The rate at which this automatic activity occurs is influenced by autonomic nerves. Contraction is stimulated by parasympathetic (vagus nerve) innervation and is reduced by sympathetic nerve activity.

The small intestine has two major types of contractions: peristalsis and segmentation. Peristalsis is much weaker in the small intestine than in the esophagus and stomach. **Intestinal motility**—the movement of chyme through the intestine—is relatively slow and is due primarily to the fact that the pressure at the pyloric end of the small intestine is greater than at the distal end.

The major contractile activity of the small intestine is **segmentation.** This term refers to muscular constrictions of the lumen, which occur simultaneously at different intestinal segments (fig. 17.16). This action serves to mix the chyme more thoroughly.

1. *Describe the intestinal structures that increase surface area, and explain the function of the crypts of Lieberkühn.*
2. *Define the term* brush border enzymes, *and list examples. Explain why many adults cannot tolerate milk.*
3. *Describe how smooth muscle contraction in the small intestine is regulated, and explain the function of segmentation.*

**Figure 17.17.** The large intestine.

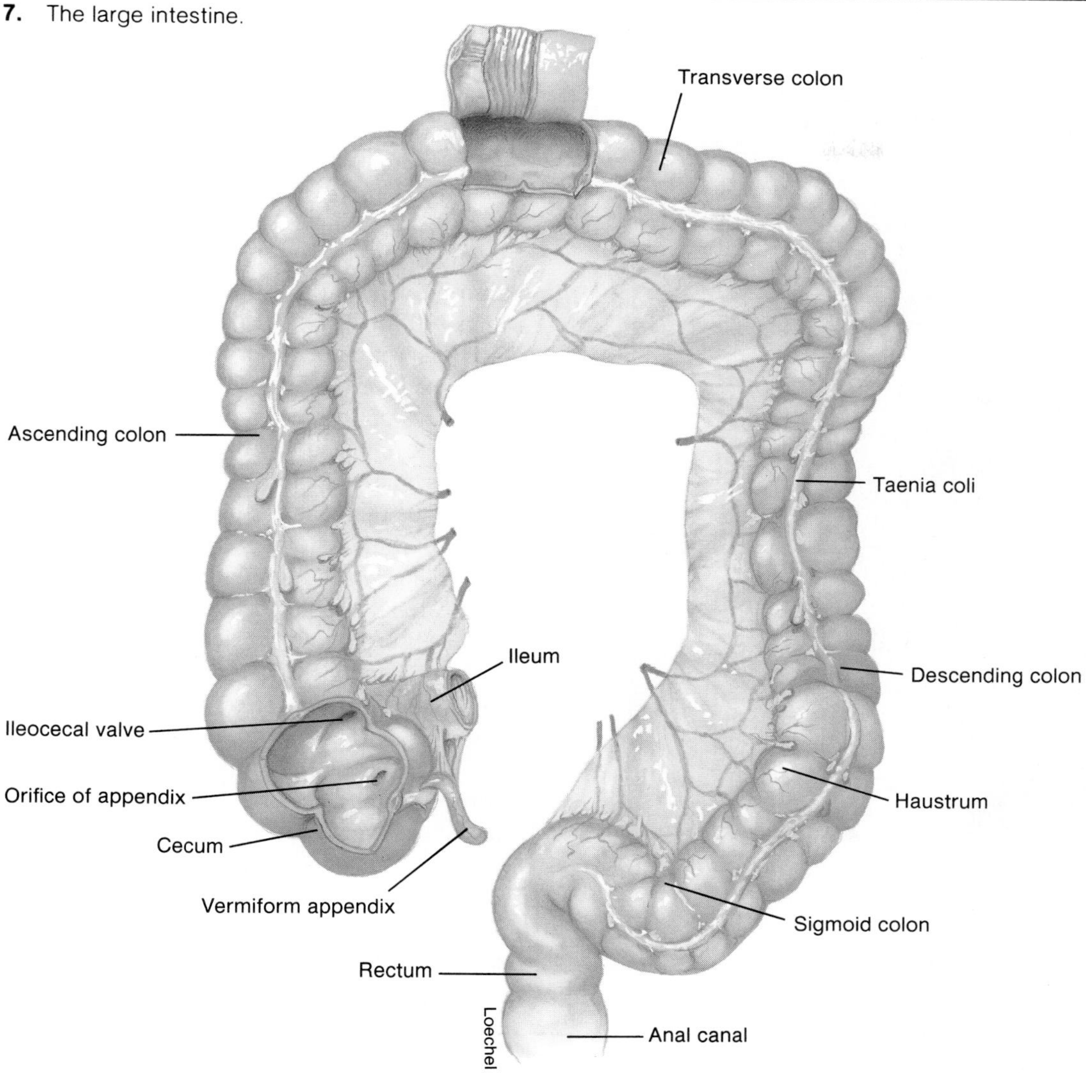

## LARGE INTESTINE

The large intestine absorbs water and electrolytes from the chyme it receives from the small intestine and, in a process regulated by the action of sphincter muscles, passes waste products out of the body through the rectum and anal canal.

Chyme from the ileum passes into the *cecum,* which is a blind-ending pouch at the beginning of the large intestine, or *colon.* Waste material then passes in sequence through the ascending colon, transverse colon, descending colon, sigmoid colon, rectum, and anal canal (fig. 17.17). Waste material (feces) is excreted through the anus.

As in the small intestine, the mucosa of the large intestine contains many scattered lymphocytes and lymphatic nodules and is covered by columnar epithelial cells and mucous-secreting goblet cells. Although this epithelium does form crypts of Lieberkühn, there are no villi in the large intestine—the intestinal mucosa therefore presents a flat appearance. The outer surface of the colon bulges outwards to form pouches, or **haustra** (fig. 17.18). Occasionally, the muscularis externa of the haustra may become so weakened that the wall forms a more elongated outpouching, or diverticulum (*divert* = turned aside). Inflammation of these structures is called *diverticulitis.*

The *vermiform appendix* is a short, thin outpouching of the cecum. It does not function in digestion, but like the tonsils, it contains numerous lymphatic nodules (fig. 17.19) and is subject to inflammation—a condition called **appendicitis.** This is commonly detected in its later stages by pain in the lower right quadrant of the abdomen. Rupture of the appendix can cause inflammation of the surrounding body cavity—*peritonitis.* This dangerous event may be prevented by surgical removal of the inflamed appendix (appendectomy).

**Figure 17.18.** A radiograph after a barium enema showing the haustra of the large intestine.

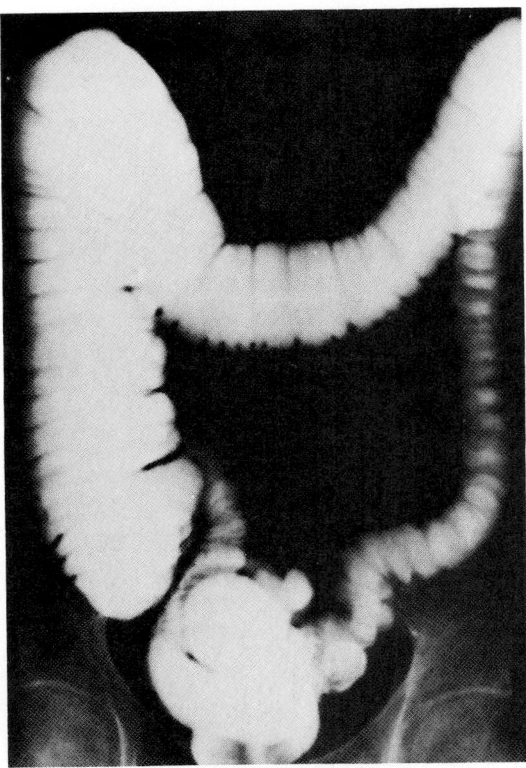

## Fluid and Electrolyte Absorption in the Intestine

Most of the fluid and electrolytes in the lumen of the digestive tract are absorbed by the small intestine. Although a person may only drink about 1.5 L/day of water, the small intestine receives 7–9 L/day as a result of the fluid secreted into the digestive tract by the salivary glands, stomach, pancreas, liver, and gallbladder. The small intestine absorbs most of this fluid and passes about 1.5–2.0 L/day of fluid to the large intestine. The large intestine absorbs about 90% of this remaining volume, leaving less than 200 ml/day of fluid to be excreted in the feces.

Absorption of water in the intestine occurs passively as a result of the osmotic gradient created by the active transport of ions. The epithelial cells of the intestinal mucosa are joined together much like those of the kidney tubules and, like the kidney tubules, contain $Na^+/K^+$ pumps in the basolateral membrane. The analogy with kidney tubules is emphasized by the observation that aldosterone, which stimulates salt and water reabsorption in the renal tubules, also appears to stimulate salt and water absorption in the ileum.

The handling of salt and water transport in the large intestine is made more complex by the fact that the large intestine can secrete, as well as absorb, water. The secretion of water by the mucosa of the large intestine occurs by osmosis as a result of the active transport of $Na^+$, followed passively by $Cl^-$, out of the epithelial cells into the intestinal lumen. Secretion in this way is normally minor compared to the far greater amount of salt and water absorption, but this balance may be altered in some disease states.

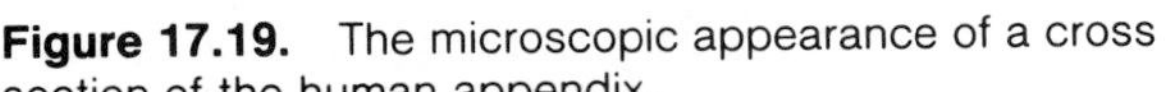

**Figure 17.19.** The microscopic appearance of a cross section of the human appendix.

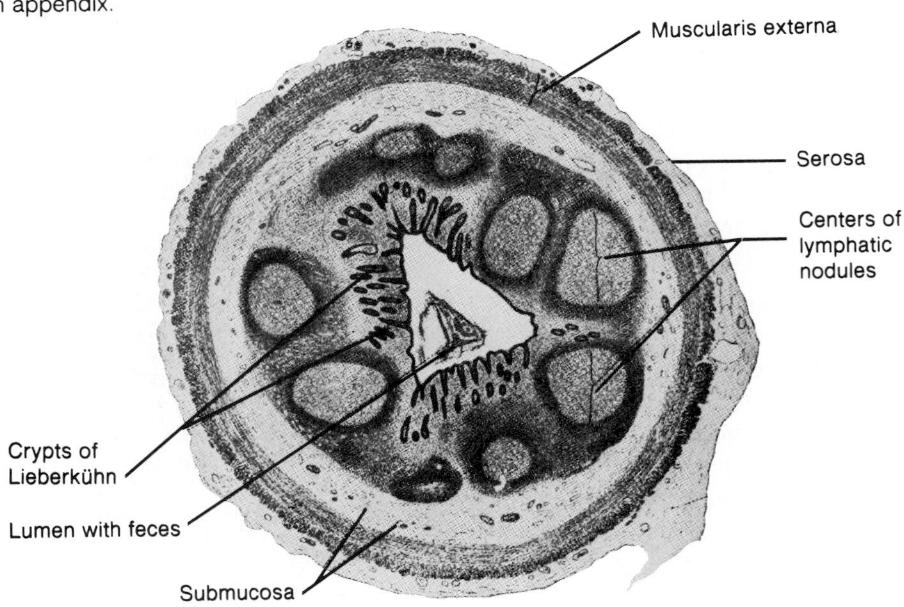

**Diarrhea** is characterized by excessive fluid excretion in the feces. There are three different mechanisms, illustrated by three different diseases, which can cause diarrhea. In *cholera,* severe diarrhea results from a chemical called *enterotoxin,* released from the infecting bacteria. Enterotoxin, acting via stimulation of adenylate cyclase and cAMP production, stimulates active $Na^+$ and water secretion. In *celiac sprue,* a disease produced in susceptible people by eating foods that contain gluten (proteins from grains such as wheat), diarrhea results from inadequate absorption of fluid due to damage to the intestinal mucosa. In *lactose intolerance,* diarrhea is produced by the increased osmolarity of the contents of the intestinal lumen as a result of the presence of undigested lactose.

### Defecation

After electrolytes and water have been absorbed, the waste material that is left passes to the rectum, leading to an increase in rectal pressure and the urge to defecate. If the urge to defecate is denied, feces are prevented from entering the anal canal by the external anal sphincter. In this case the feces remain in the rectum and may even back up into the sigmoid colon. The **defecation reflex** normally occurs when the rectal pressure rises to a particular level that is determined, to a large degree, by habit. At this point the internal anal sphincter relaxes to admit feces into the anal canal.

During the act of defecation the longitudinal rectal muscles contract to increase rectal pressure and the internal and external anal sphincter muscles relax. Excretion is aided by contractions of abdominal and pelvic skeletal muscles, which raise the intra-abdominal pressure and help push the feces from the rectum through the anal canal and out the anus.

1. *Describe how electrolytes and water are absorbed in the large intestine, and explain how diarrhea may be produced.*
2. *Describe the structures and mechanisms involved in defecation.*

## *Liver, Gallbladder, and Pancreas*

The primary digestive function of the liver is the production of bile, which is secreted into ducts that drain to the gallbladder and duodenum. The gallbladder simply stores and concentrates bile, which contains bile salts that aid fat digestion and absorption. Blood from the intestine, which contains the absorbed products of digestion, flows to the liver where it is chemically modified. Waste products and toxins are removed by the liver, and proteins, glucose, and lipids can be secreted by the liver into the blood which returns to the general circulation. The exocrine portion of the pancreas secretes bicarbonate, which neutralizes the acidity of the chyme, and critically important digestive enzymes into the small intestine.

The liver, which is the largest internal organ, lies immediately beneath the diaphragm in the abdominal cavity. Attached to the inferior surface of the liver, between its right and quadrate lobes, is the pear-shaped gallbladder. This organ is approximately 10 cm long by 3.5 cm wide. The pancreas is located behind the stomach along the posterior abdominal wall.

### Structure of the Liver

Although the liver is the largest internal organ, it is, in a sense, only one to two cells thick. This is because the liver cells, or **hepatocytes,** form **plates** that are one to two cells thick and separated from each other by large capillary spaces, called **sinusoids** (fig. 17.20). The sinusoids are lined with phagocytic **Kupffer cells,** but the large intercellular gaps between adjacent Kupffer cells make these sinusoids more highly permeable than other capillaries. The plate structure of the liver and the high permeability of the sinusoids allow each hepatocyte to have direct contact with the blood.

In **cirrhosis,** large numbers of liver lobules are destroyed and replaced with permanent connective tissue and "regenerative nodules" of hepatocytes. These regenerative nodules don't have the platelike structure of normal liver tissue and are therefore less functional. One indication of this decreased function is the entry of ammonia from the hepatic portal blood into the general circulation. Cirrhosis may be caused by chronic alcohol abuse, viral hepatitis, and other agents that attack liver cells.

***Portal System.*** The products of digestion that are absorbed into blood capillaries in the intestine do not directly enter the general circulation. Instead, this blood is delivered first to the liver. Capillaries in the digestive tract drain into the *hepatic portal vein,* which carries this blood to capillaries in the liver. It is not until the blood has passed through this second capillary bed that it enters the general circulation through the *hepatic vein* that drains the liver. The term **portal system** is used to describe this unique pattern of circulation: capillaries → vein → capillaries → vein. In addition to receiving venous blood from the intestine, the liver also receives arterial blood via the *hepatic artery.*

***Liver Lobules.*** The hepatic plates are arranged into functional units called **liver lobules** (fig. 17.21). In the middle of each lobule is a *central vein,* and at the periphery of each lobule are branches of the hepatic portal vein and of the hepatic artery, which open into the spaces *between* hepatic plates. Arterial blood and portal venous

**Figure 17.20.** The structure of the liver. (*a*) A scanning electron micrograph of the liver. Hepatocytes are arranged in plates so that blood that passes through sinusoids (*b*) will be in contact with each liver cell. (Photo from: *Tissues and Organs: A Text Atlas of Scanning Electron Microscopy* by R. G. Kessel and R. Kardon. W. H. Freeman and Company. © 1979.)

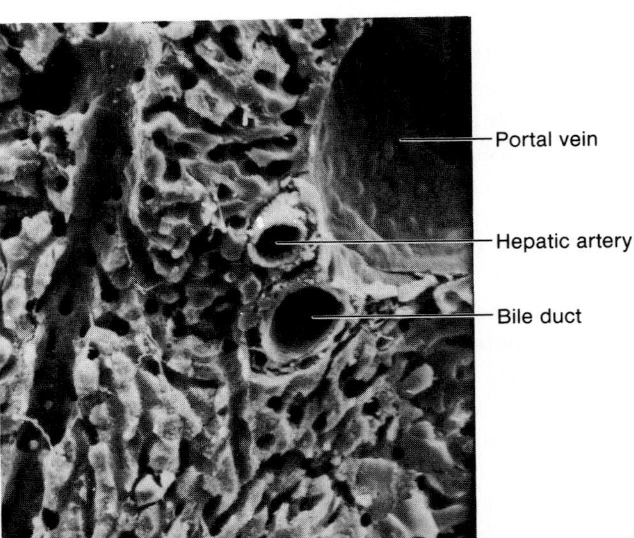


(a)

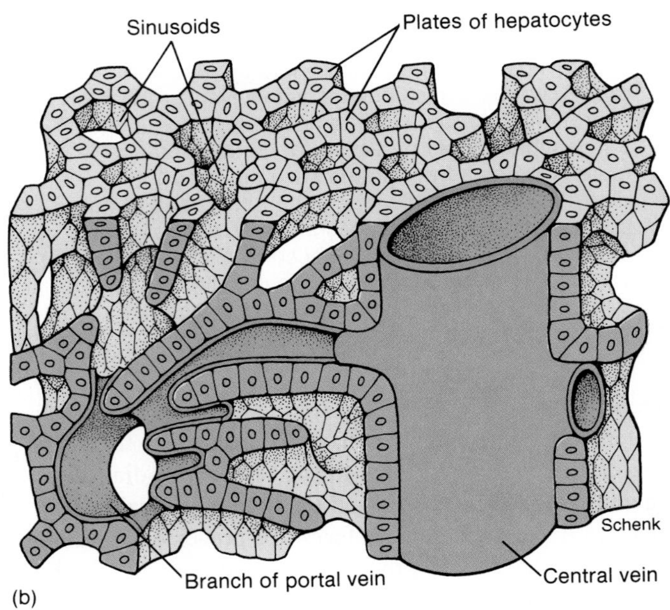


(b)

**Figure 17.21.** The flow of blood and bile in a liver lobule. Blood flows within sinusoids from a portal vein to the central vein (from the periphery to the center of a lobule). Bile flows within hepatic plates from the center to bile ducts at the periphery of a lobule.

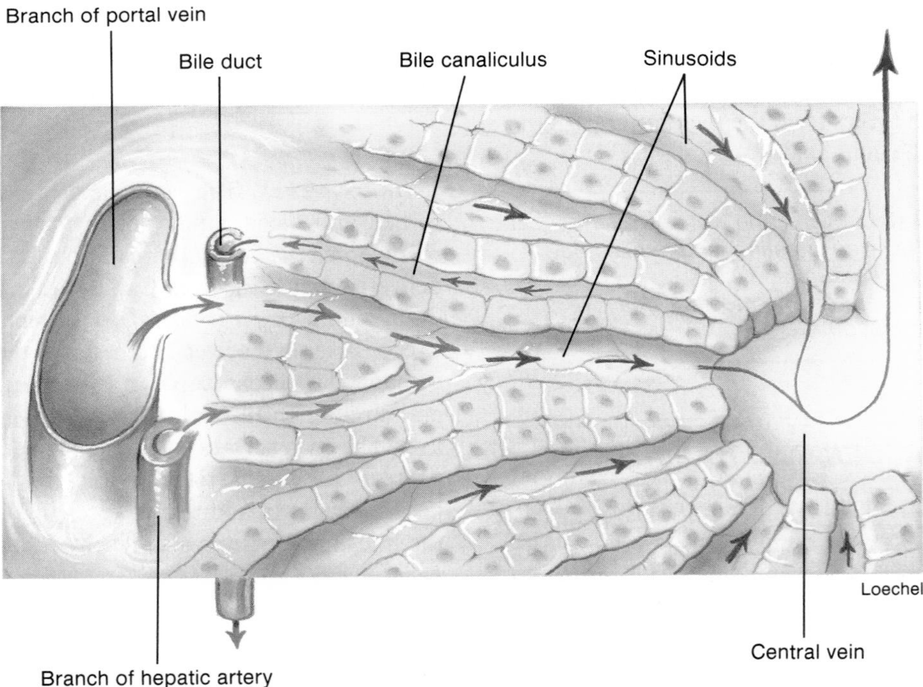

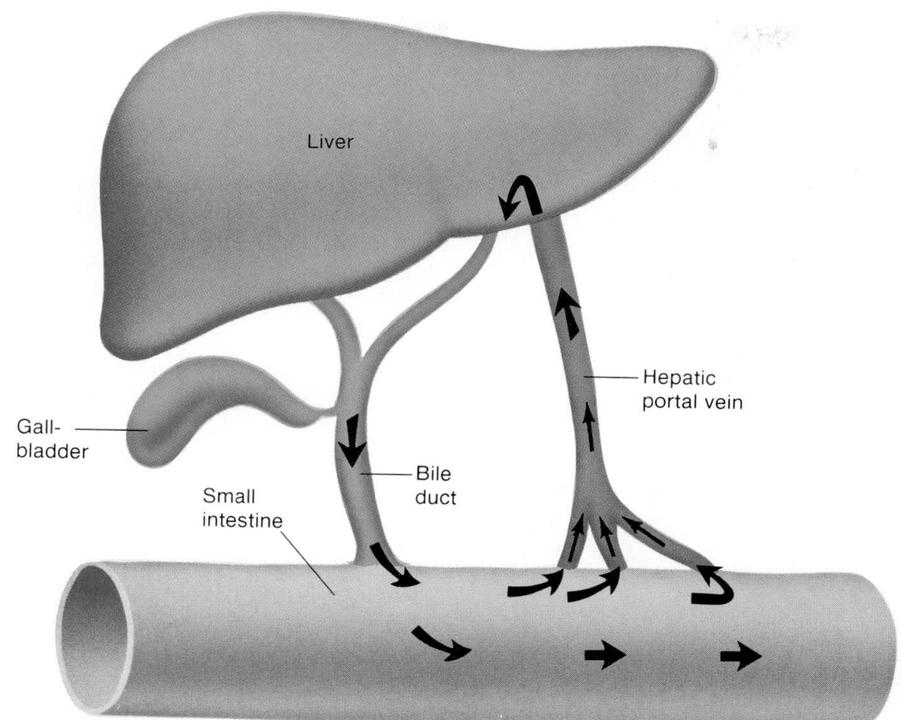


**Figure 17.22.** Enterohepatic circulation. Substances secreted in the bile may be absorbed by the intestinal epithelium and recycled to the liver via the hepatic portal vein.

blood, containing molecules absorbed in the GI tract, thus mix as the blood flows within the sinusoids from the periphery of the lobule to the central vein. The central veins of different liver lobules converge to form the hepatic vein, which carries blood from the liver to the inferior vena cava.

Bile is produced by the hepatocytes and secreted into thin channels called **bile canaliculi,** located *within* each hepatic plate (fig. 17.21). These bile canaliculi are drained at the periphery of each lobule by *bile ducts,* which in turn drain into *hepatic ducts* that carry bile away from the liver. Since blood travels in the sinusoids and bile travels in the opposite direction within the hepatic plates, blood and bile do not mix in the liver lobules.

***Enterohepatic Circulation.*** In addition to the normal constituents of bile, a wide variety of exogenous compounds (drugs) are secreted by the liver into the bile ducts (table 17.4). The liver can thus "clear" the blood of particular compounds by removing them from the blood and excreting them into the intestine with the bile. (The liver can also clear the blood by other mechanisms that will be described in a later section.) Molecules that are cleared from the blood by secretion into the bile are excreted in the feces; this is analogous to renal clearance of blood through excretion in the urine (chapter 16).

**Table 17.4** Compounds that the liver excretes into the bile

| | Compound | Comments |
|---|---|---|
| Endogenous (naturally occurring) | Bile salts, urobilinogen, cholesterol | High percentage is absorbed and has an enterohepatic circulation[1] |
| | Lecithin | Small percentage is absorbed and has an enterohepatic circulation |
| | Bilirubin | No enterohepatic circulation |
| Exogenous (drugs) | Ampicillin, streptomycin, tetracycline | High percentage is absorbed and has an enterohepatic circulation |
| | Sulfonamides, penicillin | Small percentage is absorbed and has an enterohepatic circulation |

[1]Compounds with an enterohepatic circulation are absorbed to some degree by the intestine and are returned to the liver in the hepatic portal vein.

**Table 17.5** Summary of the major categories of liver functions

| Functional Category | Actions |
|---|---|
| Detoxification of blood | Phagocytosis by Kupffer cells |
| | Chemical alteration of biologically active molecules (hormones and drugs) |
| | Production of urea, uric acid, and other molecules that are less toxic than parent compounds |
| | Excretion of molecules in bile |
| Carbohydrate metabolism | Conversion of blood glucose to glycogen and fat |
| | Production of glucose from liver glycogen and from other molecules (amino acids, lactic acid) by gluconeogenesis |
| | Secretion of glucose into the blood |
| Lipid metabolism | Synthesis of triglyceride and cholesterol |
| | Excretion of cholesterol in bile |
| | Production of ketone bodies from fatty acids |
| Protein synthesis | Production of albumin |
| | Production of plasma transport proteins |
| | Production of clotting factors (fibrinogen, prothrombin, and others) |
| Secretion of bile | Synthesis of bile salts |
| | Conjugation and excretion of bile pigment (bilirubin) |

Many compounds that are released with the bile into the intestine are not excreted with the feces, however. Some of these can be absorbed through the small intestine and enter the hepatic portal blood. These absorbed molecules are thus carried back to the liver, where they can be again secreted by hepatocytes into the bile ducts. Compounds that recirculate between the liver and intestine in this way are said to have an **enterohepatic circulation** (fig. 17.22).

## Functions of the Liver

As a result of its very large and diverse enzymatic content, its unique structure, and the fact that it receives venous blood from the intestine, the liver has a wider variety of functions than any other organ in the body. A summary of the major categories of liver function is presented in table 17.5.

**Table 17.6** Composition of the bile

| Component | Concentration |
|---|---|
| pH | 5.7–8.6 |
| Bile salts | 140–2,230 mg/100 ml |
| Lecithin | 140–810 mg/100 ml |
| Cholesterol | 97–320 mg/100 ml |
| Bilirubin | 12–70 mg/100 ml |
| Urobilinogen | 5–45 mg/100 ml |
| Sodium | 145–165 mEq/L |
| Potassium | 2.7–4.9 mEq/L |
| Chloride | 88–115 mEq/L |
| Bicarbonate | 27–55 mEq/L |

***Bile Production and Secretion.*** The liver produces and secretes 250–1,500 ml of bile per day. The major constituents of bile include bile salts, bile pigment (bilirubin), phospholipids (mainly lecithin), cholesterol, and inorganic ions (table 17.6).

Bile pigment, or bilirubin, is produced in the spleen, liver, and bone marrow as a derivative of the heme groups (minus the iron) from hemoglobin. The **free bilirubin** is not very water-soluble and thus most is carried in the blood attached to albumin proteins. This protein-bound bilirubin can neither be filtered by the kidneys into the urine, nor can it be directly excreted by the liver into the bile.

The liver can take some of the free bilirubin out of the blood and conjugate (combine) it with glucuronic acid. This **conjugated bilirubin** is water-soluble and can be secreted into the bile. Once in the bile, the conjugated bilirubin can enter the intestine where it is converted by bacteria into another pigment—**urobilinogen**—which is partially responsible for the color of the feces. About 30% to 50% of the urobilinogen, however, is absorbed by the intestine and enters the hepatic portal vein. Of the urobilinogen that enters the liver sinusoids, some is secreted into the bile and is thus returned to the intestine in an enterohepatic circulation, while the rest enters the general circulation (fig. 17.23). The urobilinogen in plasma, unlike free bilirubin, is not attached to albumin and, therefore, is easily filtered by the kidneys into the urine, giving urine its characteristic yellow color.

The **bile salts** are derivatives of cholesterol that have two to four polar groups on each molecule. The principal bile salts in humans are *cholic acid* and *chenodeoxycholic acid* (fig. 17.24). In aqueous solutions these molecules "huddle" together to form aggregates known as **micelles.** The nonpolar parts are located in the central region of the micelle (away from water), whereas the polar groups face water around the periphery of the micelle. Lecithin, cholesterol, and other lipids in the small intestine enter these micelles in a process that aids the digestion and absorption of fats (described in a later section).

**Figure 17.23.** The enterohepatic circulation of urobilinogen. Bacteria in the intestine convert bile pigment (bilirubin) into urobilinogen. Some of this pigment goes out in the feces; some is absorbed by the intestine and is recycled through the bile. A portion of the uroglobin that is absorbed goes into the general circulation and is therefore filtered by the kidneys into the urine.

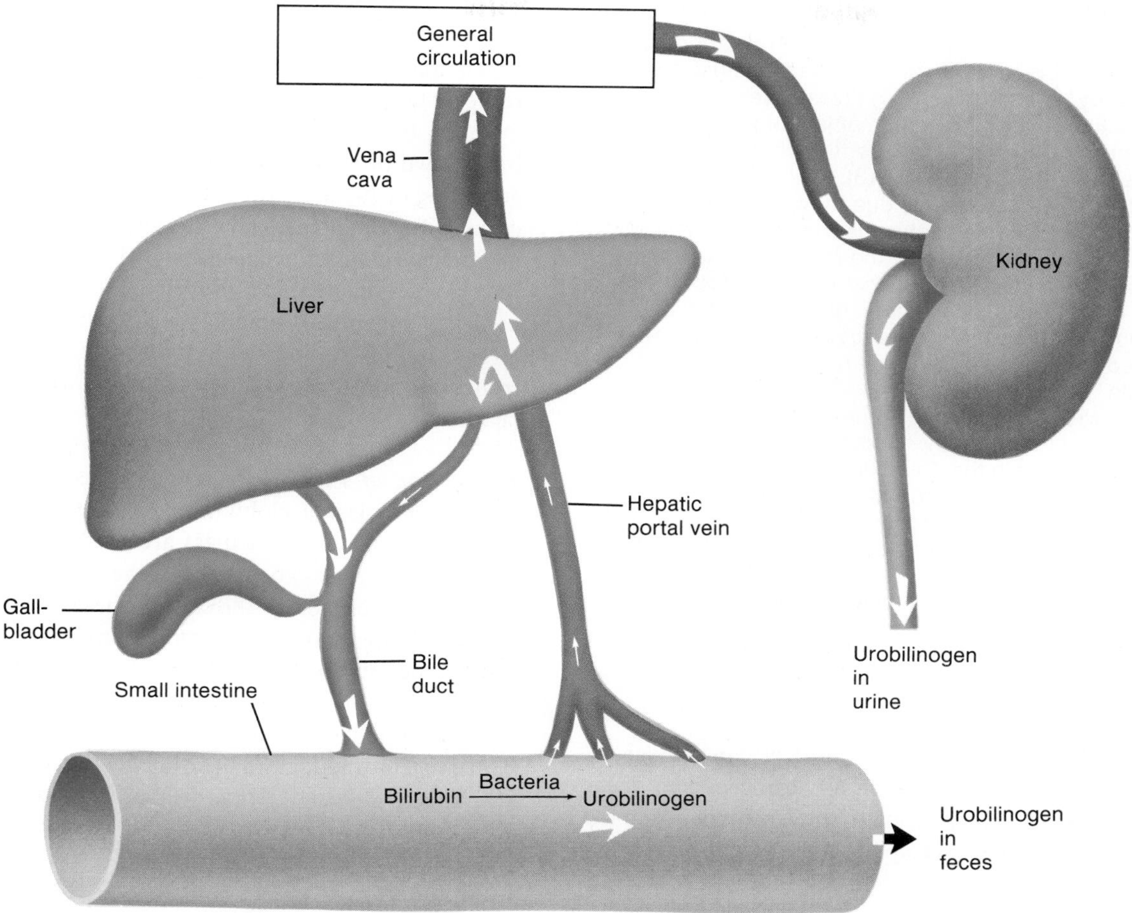


**Figure 17.24.** The two major bile acids (which form bile salts) in humans.

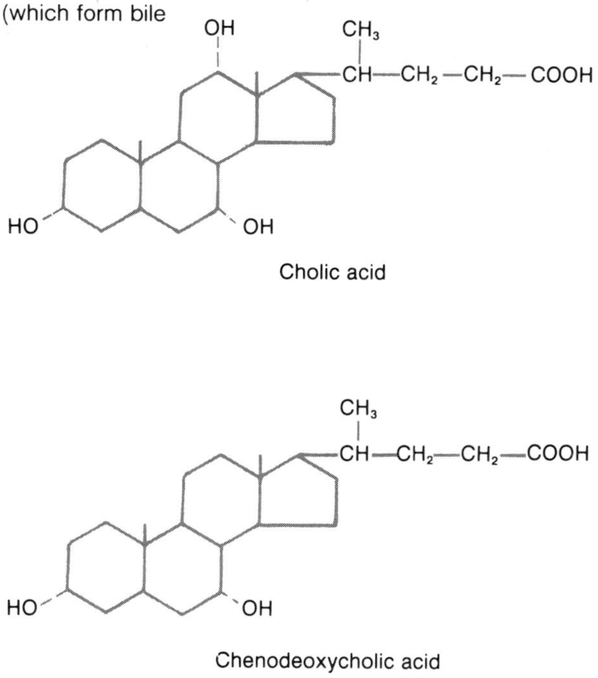

Cholic acid

Chenodeoxycholic acid

**Jaundice** is a yellow staining of the tissues produced by high blood concentrations of either free or conjugated bilirubin. Jaundice due to high blood levels of conjugated bilirubin in adults may result when bile excretion is blocked by gallstones. Since free bilirubin is derived from heme, jaundice due to high blood levels of free bilirubin is usually caused by an excessively high rate of red blood cell destruction. This is the cause of jaundice in newborn babies who suffer from erythroblastosis fetalis (described in chapter 13). *Physiological jaundice of the newborn* is due to high levels of free bilirubin in otherwise healthy neonates. This type of jaundice may be caused by the rapid fall in blood hemoglobin concentrations that normally occurs at birth, or in premature infants it may be caused by inadequate amounts of hepatic enzymes that are needed to conjugate bilirubin and thus excrete it in the bile.

Newborn infants with jaundice are usually treated by *phototherapy,* in which they are placed under blue light in the 400–500 nm wavelength range. This light is absorbed by bilirubin in cutaneous vessels and results in the conversion of bilirubin to a more polar isomer, which is soluble in plasma without having to be conjugated with glucuronic acid. The more water-soluble photoisomer of bilirubin can then be excreted in the bile.

***Detoxification of the Blood.*** The liver can remove biologically active molecules such as hormones and drugs from the blood by (1) excretion of these compounds in the bile; (2) phagocytosis by Kupffer cells, which line the sinusoids; and (3) chemical alteration of these molecules within the hepatocytes.

Ammonia, for example, is a very toxic molecule produced by the action of bacteria in the intestine. The observation that portal vein blood has an ammonia concentration that is four to fifty times greater than in the hepatic vein means that this compound is removed by the liver. The liver has the enzymes needed to convert ammonia into less toxic *urea* molecules, which are secreted by the liver into the blood and excreted by the kidneys in the urine. Similarly, the liver converts toxic porphyrins into *bilirubin* and toxic purines into *uric acid.*

Steroid hormones and other nonpolar compounds, such as many drugs, are inactivated in their passage through the liver by modifications of their chemical structures. The liver has enzymes that convert these molecules into more polar (more water-soluble) forms by **hydroxylation** (the addition of $OH^-$ groups) and by **conjugation** with highly polar groups such as sulfate and glucuronic acid. Polar derivatives of steroid hormones and drugs are less biologically active and, because of their increased water solubility, are more easily excreted by the kidneys into the urine.

***Secretion of Glucose, Triglycerides, and Ketone Bodies.*** The liver helps to regulate the blood glucose concentration by either removing glucose from or adding glucose to the blood, according to the needs of the body. After a carbohydrate-rich meal, the liver can remove some glucose from the hepatic portal blood and convert it into glycogen and triglycerides (**glycogenesis** and **lipogenesis,** respectively). During fasting, the liver secretes glucose into the blood. This glucose can be derived from the breakdown of stored glycogen in a process called **glycogenolysis,** or it can be produced by the conversion of noncarbohydrate molecules (such as amino acids) into glucose in a process called **gluconeogenesis.** The liver also contains the enzymes required to convert free fatty acids into ketone bodies (**ketogenesis**), which are secreted into the blood in large amounts during fasting. These processes are controlled by hormones and are explained in more detail in chapter 18.

***Production of Plasma Proteins.*** Plasma albumin and most of the plasma globulins (with the exception of immunoglobulins, or antibodies) are produced by the liver. Albumin comprises about 70% of the total plasma protein and contributes most to the colloid osmotic pressure of the blood (chapter 14). The globulins produced by the liver have a wide variety of functions, including the transport of cholesterol and triglycerides, transport of steroid and thyroid hormones, inhibition of trypsin activity, and blood clotting. Clotting factors I (fibrinogen), II (prothrombin), III, V, VII, IX, and XI, as well as angiotensinogen, are all produced by the liver.

## Gallbladder

The gallbladder is a saclike organ attached to the inferior surface of the liver. This organ stores and concentrates bile, which drains to it from the liver by way of the hepatic ducts, bile duct, and *cystic duct,* respectively. A sphincter valve at the neck of the gallbladder allows a storage capacity of about 35 to 50 ml. The inner mucosal layer of the gallbladder is arranged in rugae similar to those of the stomach. When the gallbladder fills with bile, it expands to the size and shape of a small pear. Bile is a yellowish-green fluid containing bile salts, bilirubin, cholesterol, and other compounds as previously discussed. Contraction of the muscularis layer of the gallbladder ejects bile through the cystic duct into the *common bile duct,* which conveys bile into the duodenum (fig. 17.25).

Bile is continuously produced by the liver and drains through the hepatic and common bile ducts to the duodenum. When the small intestine is empty of food, the *sphincter of Oddi* at the end of the common bile duct closes, and bile is forced up to the cystic duct and then to the gallbladder for storage.

**Figure 17.25.** The pancreatic duct joins the common bile duct to empty its secretions through the duodenal papilla into the duodenum. The release of bile and pancreatic juice into the duodenum is controlled by the sphincter of Oddi.

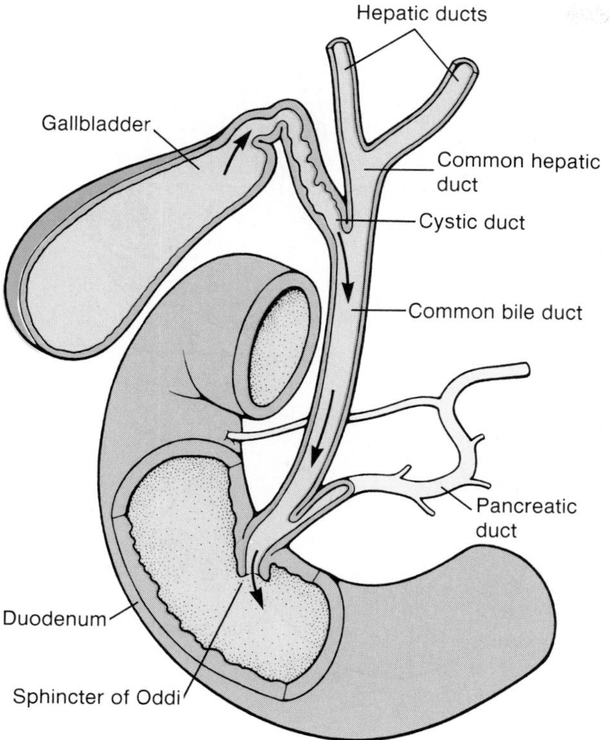


Approximately twenty million Americans have **gallstones,** which can produce painful symptoms by obstructing the cystic or common bile ducts. Gallstones commonly contain cholesterol as their major component, and become hardened by the precipitation of inorganic salts (fig. 17.26). Cholesterol normally has an extremely low water solubility (20 $\mu$g/L), but it can be present in bile at two million times its water solubility (40 g/L) because it enters the hydrophobic centers of mixed micelles of bile salts and lecithin. In order for gallstones to be produced, the liver must secrete enough cholesterol to create a supersaturated solution, and some substance must serve as a nucleus for the formation of cholesterol crystals. There is evidence that the gallbladder of people with gallstones secretes a glycoprotein that may serve as the nucleating factor in the formation of cholesterol crystals. Gallstones may sometimes be dissolved by treatment with the bile salt chenodeoxycholic acid, or they may have to be removed surgically.

**Figure 17.26.** (*a*) An X ray of a gallbladder that contains gallstones. (*b*) A posterior view of a gallbladder that has been removed (cholecystectomy) and cut open to reveal its gallstones (bilary calculi). A dime is placed in the photo to show relative size.

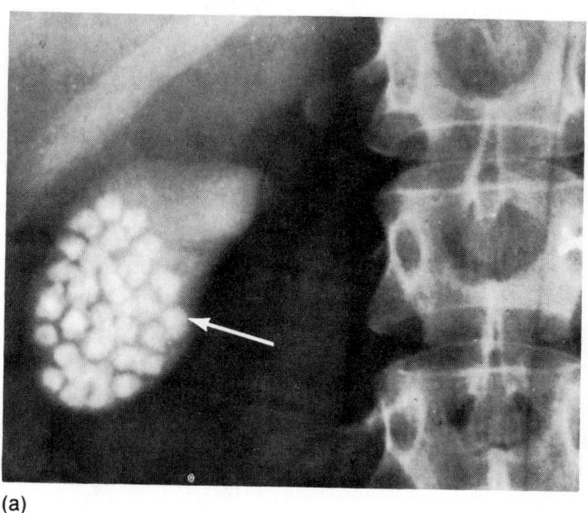

(a)

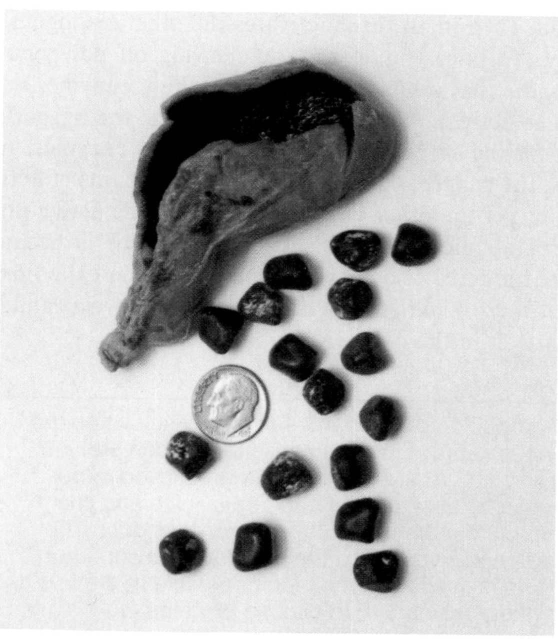

(b)

## Pancreas

The pancreas is a soft, lobulated, glandular organ that has both exocrine and endocrine functions. The endocrine function is performed by clusters of cells, called the **islets of Langerhans,** that secrete the hormones insulin and glucagon into the blood (chapter 18). As an exocrine gland, the pancreas secretes pancreatic juice through the pancreatic duct (fig. 17.27) into the duodenum. Within the lobules of the pancreas are the exocrine secretory units, called **acini.** Each acinus consists of a single layer of epithelial cells surrounding a lumen, into which the constituents of pancreatic juice are secreted.

***Pancreatic Juice.*** Pancreatic juice contains water, bicarbonate, and a wide variety of digestive enzymes that are secreted into the duodenum. These enzymes include (1) **amylase,** which digests starch; (2) **trypsin,** which digests protein; and (3) **lipase,** which digests triglycerides; other pancreatic enzymes are indicated in table 17.7. It should be noted that the complete digestion of food molecules in the small intestine requires the action of both pancreatic enzymes and brush border enzymes.

Most pancreatic enzymes are produced as inactive molecules, or *zymogens,* which help to minimize the risk of self-digestion within the pancreas. The inactive form of trypsin, called trypsinogen, is activated within the small intestine by the catalytic action of the brush border enzyme *enterokinase.* Enterokinase converts trypsinogen to active trypsin. Trypsin, in turn, activates the other zymogens of pancreatic juice (fig. 17.28) by cleaving off polypeptide sequences that inhibit the activity of these enzymes.

The activation of trypsin is, therefore, the triggering event for the activation of other pancreatic enzymes. Actually, the pancreas does produce small amounts of active trypsin, yet the other enzymes don't become active until pancreatic juice enters the duodenum. This is because pancreatic juice also contains a small protein called *pancreatic trypsin inhibitor,* which attaches to trypsin and inactivates it in the pancreas.

Inflammation of the pancreas may result when the various safeguards against self-digestion are insufficient. **Acute pancreatitis** is believed to be caused by the reflux of pancreatic juice and bile from the duodenum into the pancreatic duct. The leakage of trypsin into the blood also occurs, but trypsin is inactive in the blood because of the inhibitory action of two plasma proteins, $\alpha_1$-antitrypsin and $\alpha_2$-macroglobulin. Pancreatic amylase may also leak into the blood, but it is not active because its substrate (glycogen) is not present in blood. Pancreatic amylase activity can be measured *in vitro,* however, and these measurements are commonly performed to assess the health of the pancreas.

**Figure 17.27.** The pancreas is both an exocrine and an endocrine gland. Pancreatic juice—the exocrine product—is secreted by acinar cells into the pancreatic duct. Scattered "islands" of cells, called the islets of Langerhans, secrete the hormones insulin and glucagon into the blood.

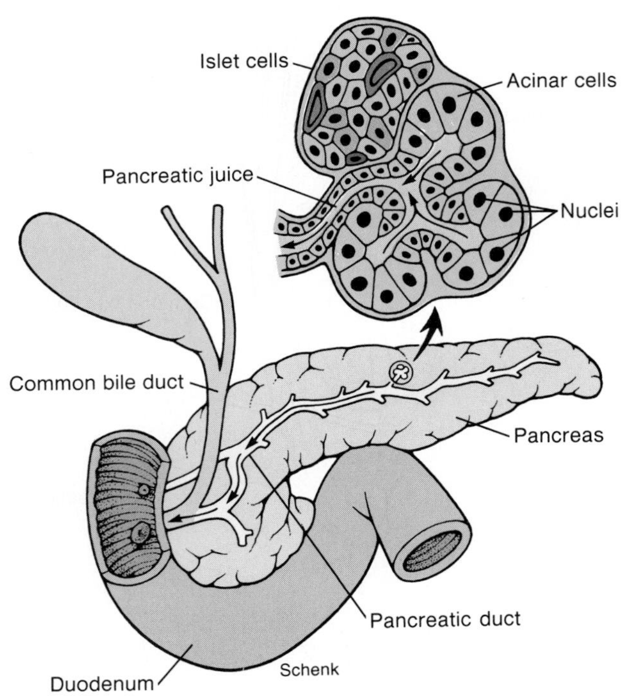


1. *Describe the structure of liver lobules, and trace the pathways for the flow of blood and bile in the lobules.*
2. *Describe the composition and function of bile, and trace the flow of bile from the liver and gallbladder to the duodenum.*
3. *Explain how the liver inactivates and excretes compounds such as hormones and drugs.*
4. *Describe the enterohepatic circulation of bilirubin and urobilinogen.*
5. *Explain how the liver helps maintain a constant blood glucose concentration and how the pattern of venous blood flow permits this task to be performed.*
6. *Describe the endocrine and exocrine structures and functions of the pancreas, and explain why the pancreas does not normally digest itself.*

## Digestion and Absorption of Carbohydrates, Lipids, and Proteins

Polysaccharides and polypeptides are hydrolyzed into their monomers of monosaccharides and amino acids, respectively, which enter the epithelial cells of the intestinal villi

Figure 17.28. The pancreatic protein-digesting enzyme *trypsin* is secreted in an inactive form known as trypsinogen. This inactive enzyme (zymogen) is activated by a brush border enzyme, enterokinase (*EN*), located in the cell membrane of microvilli. Active trypsin in turn activates other zymogens in pancreatic juice.

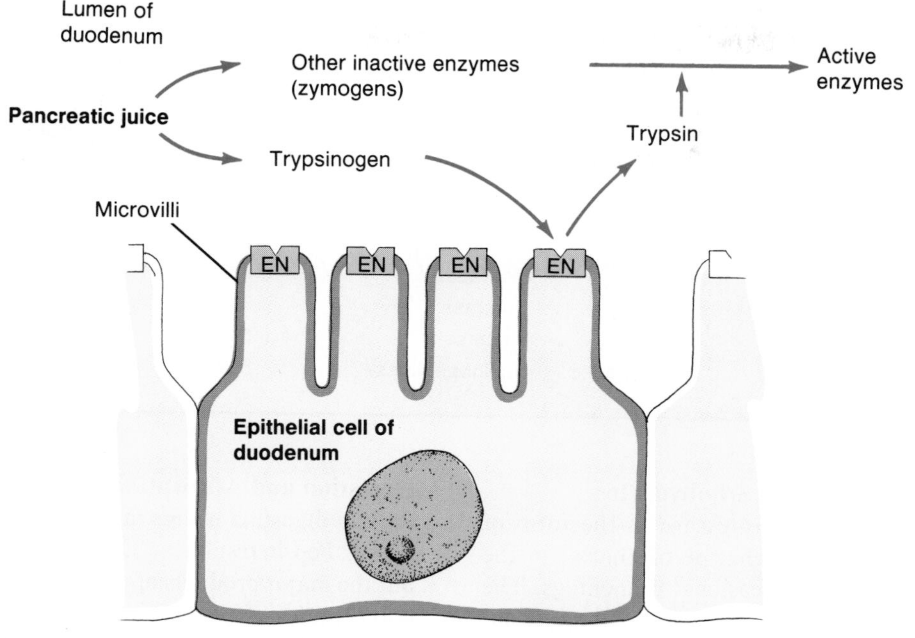


**Table 17.7** Enzymes in pancreatic juice

| Enzyme | Zymogen | Activator | Action |
|---|---|---|---|
| Trypsin | Trypsinogen | Enterokinase | Cleaves internal peptide bonds |
| Chymotrypsin | Chymotrypsinogen | Trypsin | Cleaves internal peptide bonds |
| Elastase | Proelastase | Trypsin | Cleaves internal peptide bonds |
| Carboxypeptidase | Procarboxypeptidase | Trypsin | Cleaves last amino acid from carboxyl-terminal end of polypeptide |
| Phospholipase | Prophospholipase | Trypsin | Cleaves fatty acids from phospholipids such as lecithin |
| Lipase | None | None | Cleaves fatty acids from glycerol |
| Amylase | None | None | Digests starch to maltose and short chains of glucose molecules |
| Cholesterolesterase | None | None | Releases cholesterol from its bonds with other molecules |
| Ribonuclease | None | None | Cleaves RNA to form short chains |
| Deoxyribonuclease | None | None | Cleaves DNA to form short chains |

and are secreted into blood capillaries. Fat is emulsified by the action of bile salts, hydrolyzed into fatty acids and monoglycerides, and absorbed into the intestinal epithelial cells. Once in the cells, triglycerides are resynthesized and combined with proteins to form particles called chylomicrons. These particles are secreted into the lymphatic fluid of the central lacteals, and thus enter the blood by way of the lymph.

The caloric (energy) value of food is found predominantly in its content of carbohydrates, lipids, and proteins. In the average American diet, carbohydrates account for approximately 50% of the total calories, protein accounts for 11% to 14%, and lipids make up the balance. These food molecules consist primarily of long combinations of subunits (monomers), which must be digested by hydrolysis reactions into the free monomers before absorption can occur. The characteristics of the major digestive enzymes are summarized in table 17.8.

**Table 17.8** Summary of the sources and activities of the major digestive enzymes

| Region or Source | | | | | |
|---|---|---|---|---|---|
| **Organ** | **Source** | **Substrate** | **Enzymes** | **Optimum pH** | **Products** |
| Mouth | Saliva | Starch | Salivary amylase | 6.7 | Maltose |
| Stomach | Gastric glands | Protein | Pepsin | 1.6–2.4 | Shorter polypeptides |
| Duodenum | Pancreatic juice | Starch | Pancreatic amylase | 6.7–7.0 | Maltose, maltriose, and oligosaccharides |
| | | Polypeptides | Trypsin, chymotrypsin, carboxypeptidase | 8.0 | Amino acids, dipeptides, and tripeptides |
| | | Triglycerides | Pancreatic lipase | 8.0 | Fatty acids and monoglycerides |
| | Epithelial membranes | Maltose | Maltase | 5.0–7.0 | Glucose |
| | | Sucrose | Sucrase | 5.0–7.0 | Glucose + fructose |
| | | Lactose | Lactase | 5.8–6.2 | Glucose + galactose |
| | | Polypeptides | Aminopeptidase | 8.0 | Amino acids, dipeptides, tripeptides |

## Digestion and Absorption of Carbohydrates

Most of the ingested carbohydrates are in the form of starch, which is a long polysaccharide of glucose in the form of straight chains with occasional branchings. The most commonly ingested sugars are the disaccharides sucrose (table sugar, consisting of glucose and fructose) and lactose (milk sugar, consisting of glucose and galactose). The digestion of starch begins in the mouth with the action of **salivary amylase,** or **ptyalin.** This enzyme cleaves some of the bonds between adjacent glucose molecules, but most people don't chew their food long enough for sufficient digestion to occur in the mouth. The digestive action of salivary amylase stops when the swallowed bolus enters the stomach because this enzyme is inactivated at the low pH of gastric juice.

The digestion of starch, therefore, occurs mainly in the duodenum as a result of the action of **pancreatic amylase.** This enzyme cleaves the straight chains of starch to produce the disaccharide *maltose* and the trisaccharide *maltriose.* Pancreatic amylase, however, cannot hydrolyze the bond between glucose molecules at the branch points in the starch. As a result, short, branched chains of glucose molecules, called *oligosaccharides,* are released together with maltose and maltriose by the activity of this enzyme (fig. 17.29).

Maltose, maltriose, and oligosaccharides of glucose released from partially digested starch, together with the disaccharides sucrose and lactose, are hydrolyzed to their monosaccharides by brush border enzymes, located on the microvilli of the epithelial cells in the small intestine. The absorption of these monosaccharides across the membrane of the microvilli occurs by means of **coupled transport.** In this process, glucose binds to the same carrier as $Na^+$ and enters the epithelial cell as $Na^+$ diffuses down its electrochemical gradient. This is a type of active transport, because energy from ATP is needed to maintain the $Na^+$ gradient. Glucose is then secreted from the epithelial cells into capillaries within the villi.

## Digestion and Absorption of Proteins

Protein digestion begins in the stomach with the action of pepsin. Pepsin results in the liberation of some amino acids, but the major products of pepsin digestion are short-chain polypeptides. This activity helps to produce a more homogenous chyme, but it is not essential for the complete digestion of protein that occurs—even in people with total gastrectomies—in the small intestine.

Most protein digestion occurs in the duodenum and jejunum. The pancreatic juice enzymes **trypsin, chymotrypsin,** and **elastase** cleave peptide bonds within the interior of the polypeptide chains. These enzymes are thus grouped together as *endopeptidases.* Enzymes that remove amino acids from the ends of polypeptide chains, in contrast, are *exopeptidases.* These include the pancreatic juice enzyme **carboxypeptidase,** which removes amino acids from the carboxyl-terminal end of polypeptide chains, and the brush border enzyme **aminopeptidase.** Aminopeptidase cleaves amino acids from the amino-terminal end of polypeptide chains.

As a result of the action of these enzymes, polypeptide chains are digested into free amino acids, dipeptides, and tripeptides. The free amino acids are absorbed through the epithelial cells of the intestinal mucosa and secreted into blood capillaries. This absorption is carrier-mediated, involving the coupled transport of free amino acids with $Na^+$, and uses four different carrier systems for different classes of amino acids. The dipeptides and tripeptides may enter epithelial cells by a different carrier system, but they are then digested within these cells into amino acids, which are secreted into the blood (fig. 17.30). Newborn babies appear to be capable of absorbing a substantial amount of undigested proteins (hence they can absorb antibodies from their mother's first milk); in adults, however, only the free amino acids enter the portal vein. Foreign food protein, which would be very antigenic, does not normally

**Figure 17.29.** Pancreatic amylase digests starch into maltose, maltriose, and short oligosaccharides containing branch points in the chain of glucose molecules.

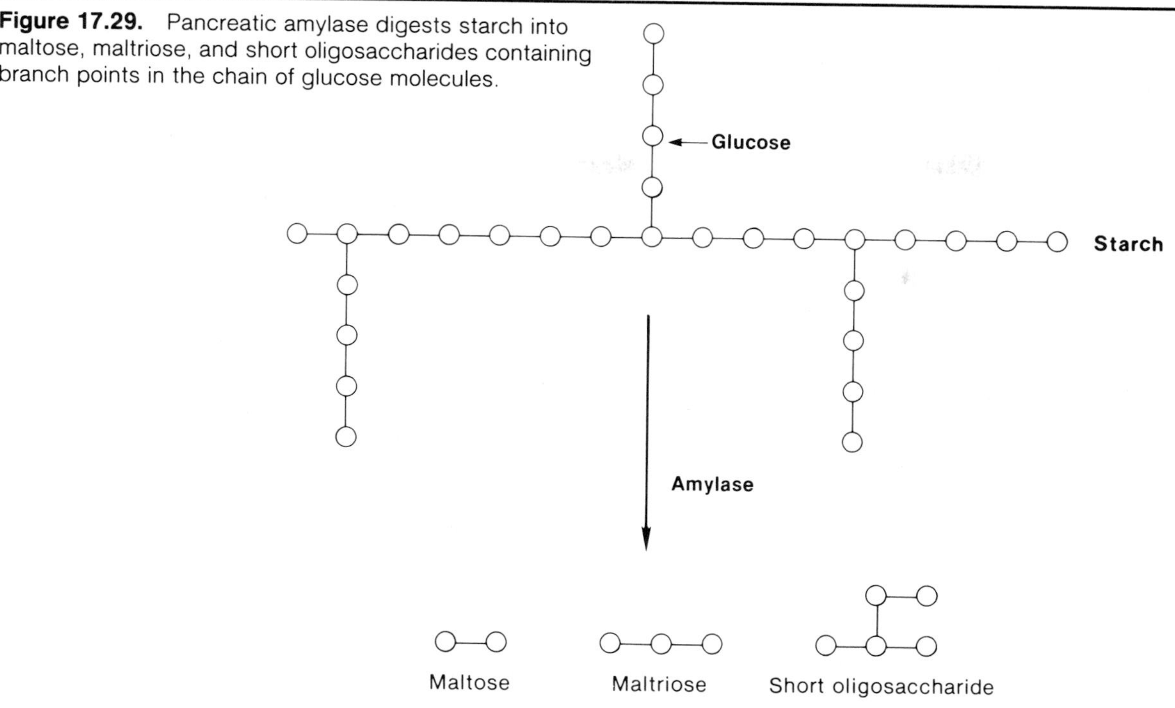


**Figure 17.30.** Polypeptide chains are digested into free amino acids, dipeptides, and tripeptides by the action of pancreatic juice enzymes and brush border enzymes. The amino acids, dipeptides, and tripeptides enter duodenal epithelial cells. Dipeptides and tripeptides are hydrolyzed into free amino acids within the epithelial cells, and these products are secreted into capillaries that carry them to the hepatic portal vein.

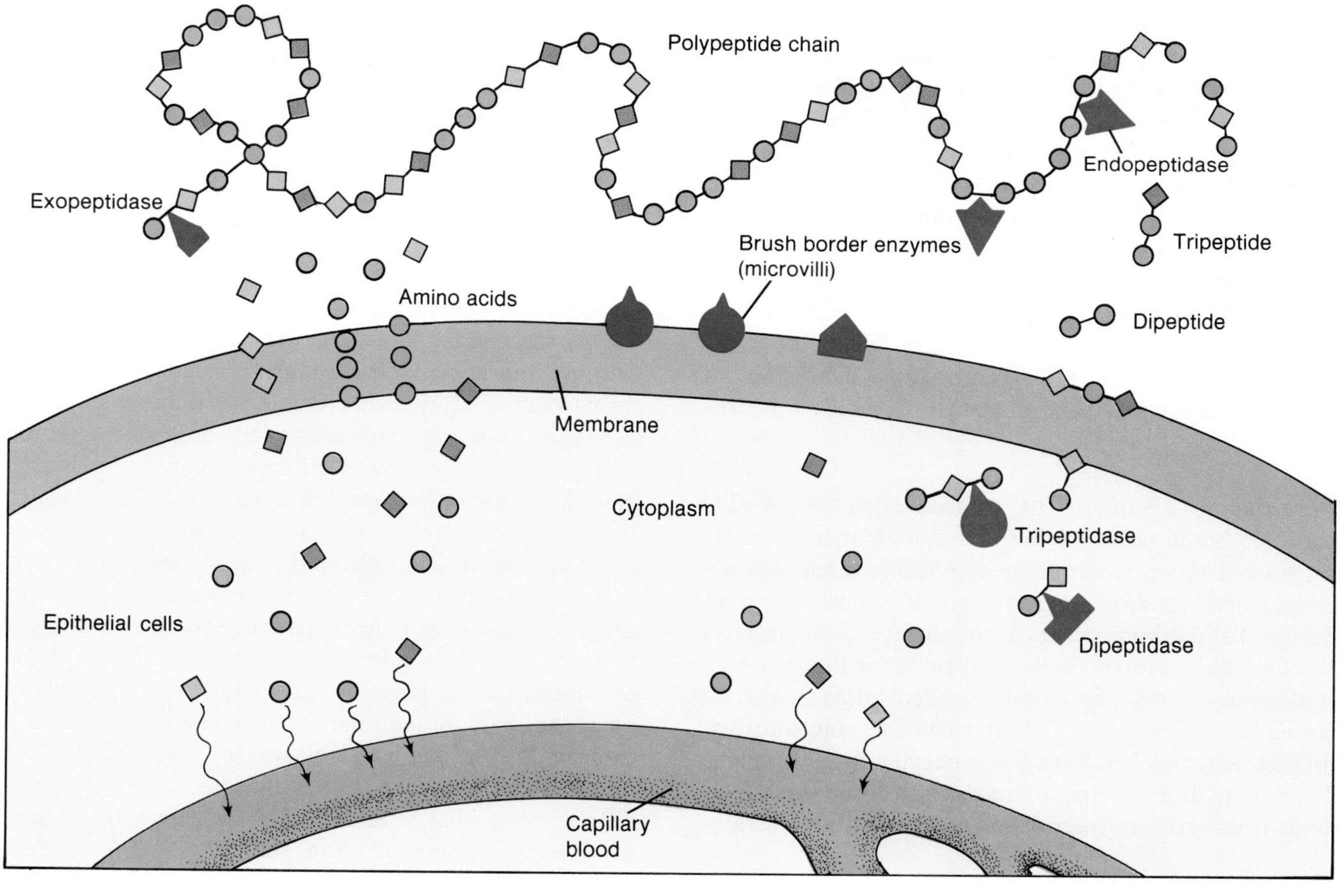

**Figure 17.31.** Pancreatic lipase digests fat (triglycerides) by cleaving off the first and third fatty acids. This produces free fatty acids and monoglycerides. Sawtooth structures indicate hydrocarbon chains in the fatty acids.

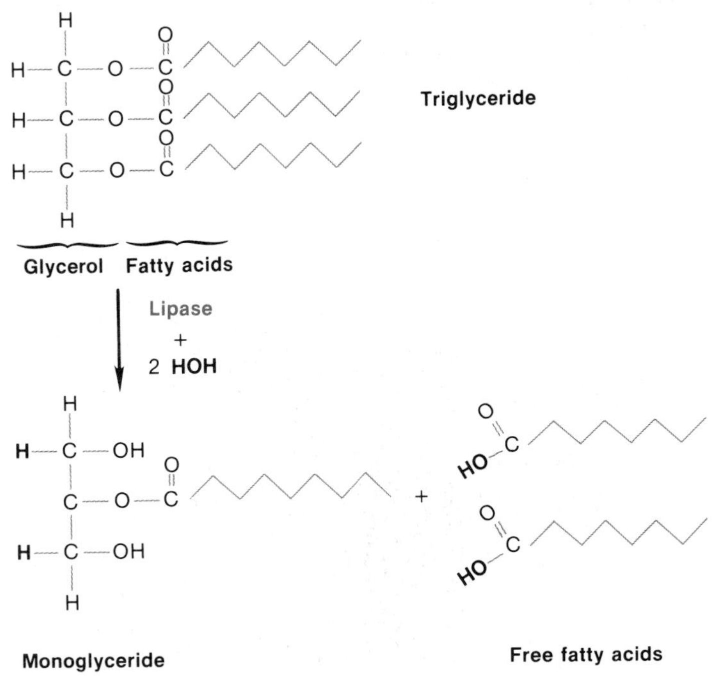

enter the blood. An interesting exception to this is the protein toxin that causes botulism, produced by *Clostridium botulinum* bacteria, which is resistant to digestion and is absorbed into the blood.

## Digestion and Absorption of Lipids

Although the salivary glands and stomach produce lipases, there is very little fat digestion until the fat in chyme arrives in the duodenum in the form of fat globules. Through mechanisms described in the next section, the arrival of fat in the duodenum serves as a stimulus for the secretion of bile. Mixed micelles of bile salts, lecithin, and cholesterol are secreted into the duodenum and act to break up the fat droplets into much finer droplets. This process, called **emulsification,** results in the formation of tiny *emulsification droplets* of triglycerides. Note that emulsification is not chemical digestion—the bonds joining glycerol and fatty acids are not hydrolyzed by this process.

***Digestion of Lipids.*** The emulsification of fat aids digestion because the smaller and more numerous emulsification droplets present a greater surface area than the unemulsified fat droplets that originally entered the duodenum. Fat digestion occurs at the surface of the droplets through the enzymatic action of **pancreatic lipase,** which is aided in its action by a protein called *colipase,* also secreted by the pancreas, which coats the emulsification droplets and "anchors" the lipase enzyme to the droplets. Through hydrolysis, lipase removes two of the three fatty acids from each triglyceride molecule and thus liberates *free fatty acids* and *monoglycerides* (fig. 17.31). **Phospholipase A** likewise digests phospholipids such as lecithin into fatty acids and lysolecithin (the remainder of the lecithin molecule after two fatty acids are removed).

Free fatty acids, monoglycerides, and lysolecithin are more polar than the undigested lipids and are able to move more easily into the mixed micelles of bile salts, lecithin, and cholesterol (fig. 17.32). These micelles then move to the brush border of the intestinal epithelium where absorption occurs.

***Absorption of Lipids.*** Free fatty acids, monoglycerides, and lysolecithin can leave the micelles and pass through the membrane of the microvilli to enter the intestinal epithelial cells. There is also some evidence that the micelles may be transported intact into the epithelial cells and that the lipid digestion products may be removed intracellularly from the micelles. In either event, these products are used to *resynthesize* triglycerides and phospholipids within the epithelial cells. This is different from the absorption of amino acids and monosaccharides, which pass through the epithelial cells without being altered.

Triglycerides, phospholipids, and cholesterol are then combined with protein inside the epithelial cells to form small particles called **chylomicrons.** These tiny combinations of lipid and protein are secreted into the lymphatic capillaries of the intestinal villi (fig. 17.33). Absorbed lipids thus pass through the lymphatic system, eventually entering the venous blood by way of the thoracic duct (chapter 13). The absorption of lipids is thus significantly different from that of amino acids and monosaccharides, which enter the hepatic portal vein.

**Figure 17.32.** Steps in the digestion of fat (triglycerides) and the entry of fat digestion products (fatty acids and monoglycerides) into micelles of bile salts secreted by the liver into the duodenum.

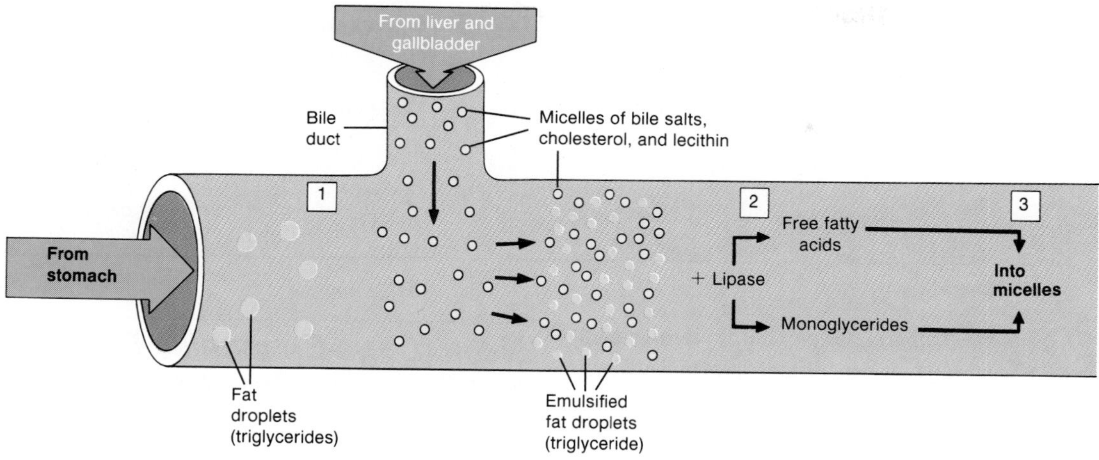


Step 1 Emulsification of fat droplets by bile salts

Step 2 Hydrolysis of triglycerides in emulsified fat droplets into fatty acid and monoglycerides

Step 3 Dissolving of fatty acids and monoglycerides into micelles to produce "mixed micelles"

**Figure 17.33.** Fatty acids and monoglycerides from the micelles within the small intestine are absorbed by epithelial cells and converted intracellularly into triglycerides. These are then combined with protein to form chylomicrons, which enter the lymphatic vessels (lacteals) of the villi. These lymphatic vessels transport the chylomicrons to the thoracic duct, which empties them into the venous blood.

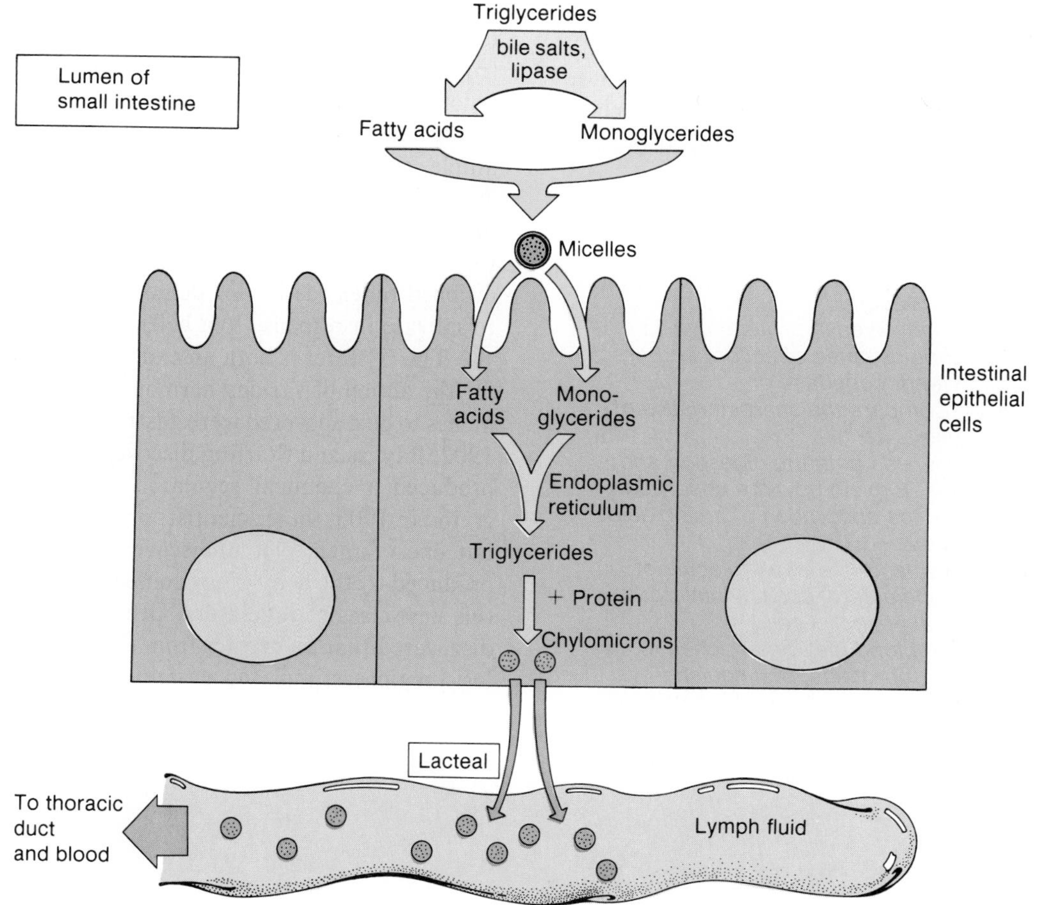

**Table 17.9** Summary of the characteristics of the lipid carrier proteins (lipoproteins) found in plasma

| Lipoprotein class | Origin | Destination | Major lipids | Functions |
|---|---|---|---|---|
| Chylomicrons | Intestine | Many organs | Triglycerides, other lipids | Delivers lipids of dietary origin to body cells |
| Very-low density lipoproteins (VLDL) | Liver | Many organs | Triglycerides, cholesterol | Delivers endogenously produced triglycerides to body cells |
| Low-density lipoproteins (LDL) | Intravascular removal of triglycerides from VLDL | Blood vessels, liver | Cholesterol | Delivers endogenously produced cholesterol to various organs |
| High-density lipoproteins (HDL) | Liver and intestine | Liver and steroid hormone producing glands | Cholesterol | Removes and degrades cholesterol |

***Transport of Lipids in Blood.*** Once the chylomicrons are in the blood, their triglyceride content is removed by the enzyme **lipoprotein lipase,** which is attached to the endothelium of blood vessels. This enzyme hydrolyzes triglycerides and thus provides free fatty acids and glycerol for use by the tissue cells. The remaining *remnant particles,* containing cholesterol, are taken up by the liver. This is a process of endocytosis (chapter 6), which requires membrane receptors for the protein part (or *apoprotein*) of the remnant particle.

Cholesterol and triglycerides produced by the liver are combined with other apoproteins and secreted into the blood as *very-low-density lipoproteins (VLDL),* which serve to deliver triglycerides to different organs. Once the triglycerides are removed, the VLDL are converted to *low-density lipoproteins (LDL),* which transport cholesterol to various organs, including blood vessels (chapter 13). Excess cholesterol is returned from these organs to the liver attached to *high-density lipoproteins (HDL).* The characteristics of these lipoproteins are summarized in table 17.9.

1. *List the enzymes involved in carbohydrate digestion, indicating their origin, sites of action, substrates, and products.*
2. *List the enzymes involved in protein digestion, indicating their origins, sites of action, and mode of action (endopeptidase or exopeptidase). Compare the characteristics of pepsin and trypsin.*
3. *Describe how bile aids both the digestion and absorption of fats. Explain how the absorption of fat differs from the absorption of amino acids and monosaccharides.*
4. *Trace the pathway and fate of a molecule of triglyceride and cholesterol in a chylomicron within an intestinal epithelial cell.*
5. *Cholesterol in the blood may be attached to any of four possible lipoproteins. Distinguish between these proteins in terms of the origin and destination of the cholesterol they carry.*

## Neural and Endocrine Regulation of the Digestive System

The activities of different regions of the GI tract are coordinated by the actions of the vagus nerve and various hormones secreted by the stomach and intestine. The stomach begins to increase its secretion in anticipation of a meal, and further increases its activities in response to the arrival of chyme. The entry of chyme into the duodenum stimulates secretion of hormones that promote contractions of the gallbladder, secretion of pancreatic juice, and inhibition of gastric activity. Each step in the regulatory process occurs in proportion to the amount and composition of the chyme.

The motility and glandular secretions of the GI tract are, to a large degree, automatic. Neural and endocrine control mechanisms, however, can stimulate or inhibit these automatic functions to help coordinate the different stages of digestion. The sight, smell, or taste of food, for example, can stimulate salivary and gastric secretions via activation of the vagus nerve, which helps to "prime" the digestive system in preparation for a meal. Stimulation of the vagus, in this case, originates in the brain and is a conditioned reflex (as Pavlov demonstrated by training dogs to salivate in response to a bell).

The GI tract is both an endocrine gland and a target for the action of various hormones. Indeed, the first hormones to be discovered were gastrointestinal hormones. In 1902, Bayliss and Starling discovered that the duodenum produced a chemical regulator, which they named **secretin;** in 1905, these scientists proposed that secretin was but one of many yet undiscovered chemical regulators produced by the body. They coined the term *hormones* for this new class of regulators. Other investigators in 1905 discovered that an extract from the stomach antrum (pyloric region) stimulated gastric secretion. The hormone **gastrin** was thus the second hormone to be discovered.

**Table 17.10** Summary of the physiological effects of gastrointestinal hormones

| Secreted by | Hormone | Effects |
|---|---|---|
| Stomach | Gastrin | Stimulates parietal cells to secrete HCl<br>Stimulates chief cells to secrete pepsinogen<br>Maintains structure of gastric mucosa |
| Small intestine | Secretin | Stimulates water and bicarbonate secretion in pancreatic juice<br>Potentiates actions of cholecystokinin on pancreas |
| Small intestine | Cholecystokinin (CCK) | Stimulates contraction of the gallbladder<br>Stimulates secretion of pancreatic juice enzymes<br>Potentiates action of secretin on pancreas<br>Maintains structure of exocrine pancreas (acini) |
| Small intestine | Gastric inhibitory peptide (GIP) | Inhibits gastric emptying<br>Inhibits gastric acid secretion<br>Stimulates secretion of insulin from endocrine pancreas (islets of Langerhans) |

The chemical structures of gastrin, secretin, and the duodenal hormone **cholecystokinin** (*CCK*) were determined in the 1960s. More recently, a fourth hormone produced by the small intestine, **gastric inhibitory peptide** *(GIP),* has been added to the list of proven GI tract hormones. The effects of these hormones are summarized in table 17.10.

## Regulation of Gastric Function

Gastric motility and secretion are, to some extent, automatic. Waves of contraction that serve to push chyme through the pyloric sphincter, for example, are initiated spontaneously by pacesetter cells in the greater curvature of the stomach. The secretion of hydrochloric acid (HCl) and pepsinogen, likewise, can be stimulated in the absence of neural and hormonal influences by the presence of cooked or partially digested protein in the stomach. The effects of autonomic nerves and hormones are superimposed on this automatic activity. This extrinsic control of gastric function is conveniently divided into three phases: (1) the cephalic phase; (2) the gastric phase; and (3) the intestinal phase. These are summarized in table 17.11.

***Cephalic Phase.*** The cephalic phase of gastric regulation refers to control by the brain via the vagus nerve. As previously discussed, various conditioned stimuli can evoke gastric secretion. This conditioning in humans is, of course, more subtle than that shown in response to a bell by Pavlov's dogs. A conversation about appetizing food was actually shown, in one study, to be a more potent stimulus for gastric acid secretion than was the sight and smell of food!

Activation of the vagus nerve can stimulate HCl and pepsinogen secretion by two mechanisms: (1) direct vagus stimulation of the gastric parietal and chief cells (the primary mechanism); and (2) vagus stimulation of gastrin secretion by the G cells, which in turn stimulates the parietal and chief cells to secrete HCl and pepsinogen, respectively. This cephalic phase stimulation of gastric secretion continues into the first 30 minutes of a meal, but gradually declines in importance as the next phase becomes predominant.

**Table 17.11** The cephalic, gastric, and intestinal phases in the regulation of gastric acid secretion

| Phase of Regulation | Description |
|---|---|
| Cephalic phase | 1. Sight, smell, and taste of food cause stimulation of vagus nuclei in brain<br>2. Vagus stimulates acid secretion<br>a) Direct stimulation of parietal cells (major effect)<br>b) Stimulation of gastrin secretion; gastrin stimulates acid secretion (lesser effect) |
| Gastric phase | 1. Distension of stomach stimulates vagus nerve; vagus stimulates acid secretion<br>2. Amino acids and peptides in stomach lumen stimulate acid secretion<br>a) Direct stimulation of parietal cells (lesser effect)<br>b) Stimulation of gastrin secretion; gastrin stimulates acid secretion (major effect)<br>3. Gastrin secretion inhibited when pH of gastric juice falls below 2.5 |
| Intestinal phase | 1. Neural inhibition of gastric emptying and acid secretion<br>a) Arrival of chyme in duodenum causes distension, increase in osmotic pressure<br>b) These stimuli activate a neural reflex that inhibits gastric activity<br>2. Gastric inhibitory peptide (GIP) secreted by duodenum in response to fat in chyme; GIP inhibits gastric acid secretion |

***Gastric Phase.*** The arrival of chyme into the stomach stimulates the gastric phase of regulation. Gastric secretion is stimulated in response to two factors: (1) distension of the stomach, which is determined by the amount of

**Figure 17.34.** The stimulation of gastric acid (HCl) secretion by the presence of proteins in the stomach lumen and by the hormone gastrin. The secretion of gastrin is inhibited by gastric acidity. This forms a negative feedback loop.

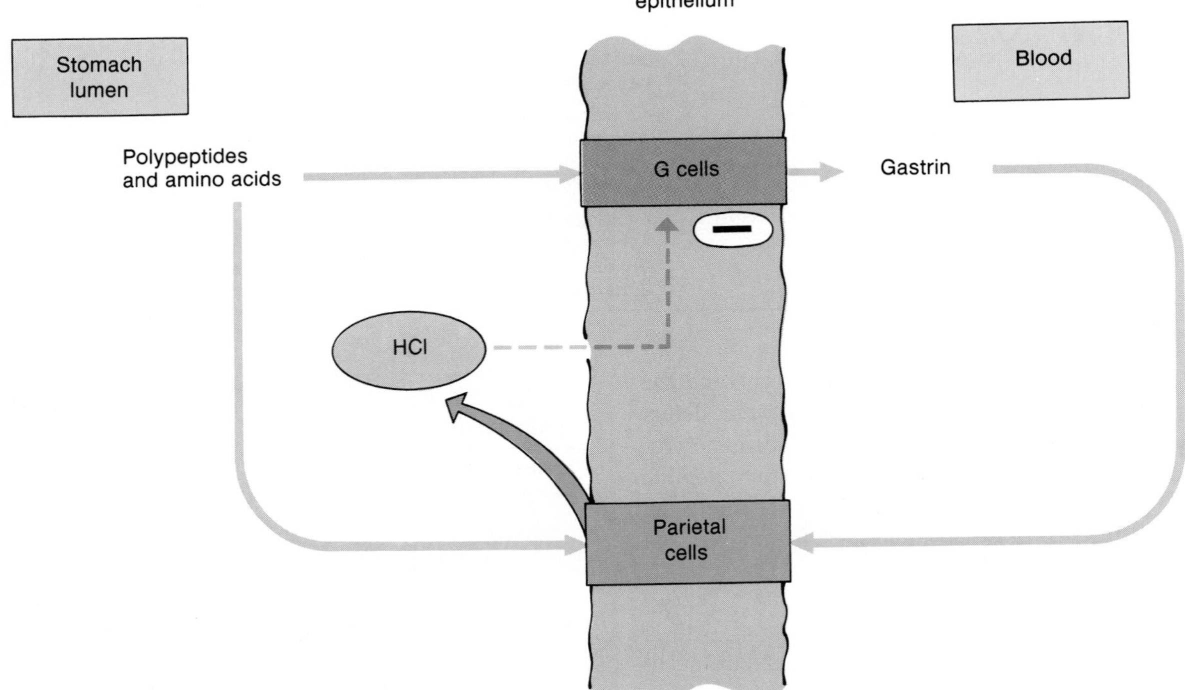


chyme; and (2) the chemical nature of the chyme. While intact proteins have little stimulatory effect, the presence of short polypeptides and amino acids in the stomach stimulates the G cells to secrete gastrin and the parietal and chief cells to secrete HCl and pepsinogen, respectively. Since gastrin also stimulates HCl and pepsinogen secretion, a *positive feedback mechanism* develops: as more HCl and pepsinogen are secreted, more short polypeptides and amino acids are released from the ingested protein, thus stimulating more secretion of gastrin and, therefore, more secretion of HCl and pepsinogen (fig. 17.34). Glucose in the chyme, in contrast, has no effect on gastric secretion, whereas the presence of fat actually inhibits acid secretion.

Secretion of HCl during the gastric phase is also regulated by a *negative feedback mechanism.* As the pH of gastric juice drops, so does the secretion of gastrin—at a pH of 2.5 gastrin secretion is reduced, and at a pH of 1.0 gastrin secretion is totally abolished. The secretion of HCl, which is largely under the control of gastrin, thus declines accordingly. The presence of proteins and polypeptides in the stomach help to buffer the acid and thus to prevent a rapid fall in gastric pH; more acid can thus be secreted when proteins are present than when they are absent. The arrival of protein into the stomach thus stimulates acid secretion two ways—by the positive feedback mechanism previously discussed and by lessening of the negative feedback control of acid secretion. The amount of acid secreted is, by means of these effects, closely matched to the amount of protein ingested. As the stomach is emptied, the protein buffers leave, the pH thus falls, and the secretion of gastrin and HCl is accordingly inhibited.

The amino acids tryptophane and phenylalanine have been shown to be the most potent stimulators of gastrin and acid secretion. People who have peptic ulcers must avoid foods that strongly stimulate acid secretion, such as milk, cola, beer, coffee (including decaffeinated coffee), tea, and wine. Such people should also avoid tryptophane tablets commonly sold in health food stores to promote sleep.

***Intestinal Phase.*** The intestinal phase of gastric regulation refers to the inhibition of gastric activity when chyme enters the small intestine. Investigators in 1886 demonstrated that the addition of olive oil to a meal inhibits gastric emptying, and in 1929 it was shown that the

**Table 17.12** Regulation of pancreatic juice and bile secretion by the hormones *secretin* and *cholecystokinin* (CCK) and by the neurotransmitter *acetylcholine*, released from parasympathetic nerve endings

| Description | Secretin | CCK | Acetylcholine (Vagus Nerve) |
|---|---|---|---|
| Stimulus for release | Acidity of chyme decreases duodenal pH below 4.5 | Fat and protein in chyme | Sight, smell of food; distension of stomach |
| Second messenger | Cyclic AMP | $Ca^{++}$ | $Ca^{++}$ |
| Effect on pancreatic juice | Stimulates water and bicarbonate secretion; potentiates action of CCK | Stimulates enzyme secretion; potentiates action of secretin | Stimulates enzyme secretion |
| Effect on bile | Stimulates secretion | Potentiates action of secretin; stimulates contraction of gallbladder | Stimulates contraction of gallbladder |

presence of fat inhibits gastric juice secretion. This inhibitory intestinal phase of gastric regulation is due to both a neural reflex originating from the duodenum and to a chemical hormone secreted by the duodenum.

The arrival of chyme into the duodenum increases its osmolality. This stimulus, together with stretch of the duodenum and possibly other stimuli, produces a neural reflex that results in the inhibition of gastric motility and secretion. The presence of fat in the chyme also stimulates the duodenum to secrete a hormone that inhibits gastric function. This inhibitory hormone is believed to be gastric inhibitory peptide (GIP), although the ability of GIP to inhibit gastric acid secretion is currently controversial.

The inhibitory neural and endocrine mechanisms during the intestinal phase prevent the further passage of chyme from the stomach to the duodenum. This gives the duodenum time to process the load of chyme that it has previously received. Since GIP is stimulated by fat in the chyme, a breakfast of bacon and eggs takes longer to pass through the stomach—and makes one feel "fuller" for a longer time—than does a breakfast of pancakes and syrup.

## Regulation of Pancreatic Juice and Bile Secretion

The arrival of chyme into the duodenum stimulates the intestinal phase of gastric regulation and, at the same time, stimulates reflex secretion of pancreatic juice and bile. The entry of new chyme is thus retarded as the previous load is digested. The secretion of pancreatic juice and bile is stimulated by both neural reflexes initiated in the duodenum and by secretion of the duodenal hormones cholecystokinin (CCK) and secretin.

***Pancreatic Juice.*** The secretion of pancreatic juice is stimulated by both secretin and CCK. Secretion of secretin and CCK, however, occurs in response to different stimuli and these two hormones have different effects on the composition of pancreatic juice. The release of secretin occurs in response to a fall in duodenal pH below 4.5; this pH fall occurs for only a short time, however, because the acidic chyme is rapidly neutralized by alkaline pancreatic juice. The secretion of CCK occurs in response to the fat content of chyme in the duodenum.

Secretin stimulates the production of bicarbonate by the pancreas. Since bicarbonate neutralizes the acidic chyme and since secretin is released in response to the low pH of chyme, this completes a negative feedback loop in which secretin indirectly inhibits its secretion. Cholecystokinin, in contrast, stimulates the production of pancreatic enzymes such as trypsin, lipase, and amylase. Secretin and CCK can have different effects on the same cells (the pancreatic acinar cells) because their actions are mediated by different intracellular compounds that act as second messengers. The second messenger of secretin action is cyclic AMP, whereas the second messenger for CCK is $Ca^{++}$ (table 17.12).

***Secretion of Bile.*** The liver secretes bile continuously, but this secretion is greatly augmented following a meal. This increased secretion is due to the release of secretin and CCK from the duodenum. Secretin is the major stimulator of bile secretion by the liver, and CCK enhances this effect. The arrival of chyme in the duodenum also causes the gallbladder to contract and eject bile. Contraction of the gallbladder occurs in response to neural reflexes from the duodenum and in response to stimulation by CCK (but not secretin).

## Trophic Effects of Gastrointestinal Hormones

Patients with tumors of the stomach pylorus have high acid secretion and hyperplasia (growth) of the gastric mucosa. Surgical removal of the pylorus reduces gastric secretion and prevents growth of the gastric mucosa. Patients with peptic ulcers are sometimes treated by vagotomy—cutting of the vagus nerve. Vagotomy also reduces acid secretion but has no effect on the gastric mucosa. These observations suggest that the hormone gastrin, secreted by the pyloric mucosa, may exert stimulatory, or *trophic,* effects on the gastric mucosa. The structure of the gastric mucosa, in other words, is dependent on the effects of gastrin.

In the same way, the structure of the acinar (exocrine) cells of the pancreas is dependent upon the trophic effects of CCK. Perhaps this explains why the pancreas, as well as the GI tract, atrophies during starvation. Since neural reflexes appear to be capable of regulating digestion, perhaps the primary function of the GI hormones is trophic—that is, maintenance of the structure of their target organs.

1. *Describe the positive and negative feedback mechanisms that operate during the gastric phase of HCl and pepsinogen secretion.*
2. *Describe the mechanisms involved in the intestinal phase of gastric regulation, and explain why a fatty meal takes longer to leave the stomach than a meal low in fat.*
3. *Explain the hormonal mechanisms involved in the production and release of pancreatic juice and bile.*

## SUMMARY

### Introduction to the Digestive System p. 526

I. The digestion of food molecules involves the hydrolysis of these molecules into their subunits.
   A. The digestion of food occurs in the lumen of the GI tract and is catalyzed by specific enzymes.
   B. The digestion products are absorbed through the intestinal mucosa and enter the blood or lymph.

II. The layers (tunics) of the GI tract are, from the inside outward, mucosa, submucosa, muscularis, and serosa.
   A. The mucosa consists of a simple columnar epithelium, a thin layer of connective tissue, called the lamina propria, and a thin layer of smooth muscle, called muscularis mucosa.
   B. The submucosa is composed of connective tissue, the muscularis consists of layers of smooth muscles, and the serosa is connective tissue covered with the visceral peritoneum.
   C. The submucosa contains the submucosal plexus, and the muscularis contains the myenteric plexus of autonomic nerves.

### Esophagus and Stomach p. 529

I. Peristaltic waves of contraction push food through the lower esophageal sphincter into the stomach.

II. The stomach consists of a cardia, fundus, body, and pyloris (antrum), which ends with the pyloric sphincter.
   A. The lining of the stomach is thrown into folds, or rugae, and the mucosa is formed into gastric pits and gastric glands.
   B. The parietal cells of the gastric glands secrete HCl, and the chief cells secrete pepsinogen.
   C. In the acidic environment of gastric juice, pepsinogen is converted into the active protein-digesting enzyme called pepsin.
   D. Some digestion of protein occurs in the stomach; the most important function of the stomach is the secretion of intrinsic factor, which is needed for the absorption of vitamin $B_{12}$ in the intestine.

### Small Intestine p. 534

I. Regions of the small intestine include the duodenum, jejunum, and ileum; the common bile duct and pancreatic duct empty into the duodenum.

II. Fingerlike extensions of mucosa called villi project into the lumen, and at the bases of the villi the mucosa forms narrow pouches called the crypts of Lieberkühn.
   A. New epithelial cells are formed in the crypts.
   B. The membrane of intestinal epithelial cells is folded to form microvilli; this is called the brush border of the mucosa and serves to increase surface area.

III. Digestive enzymes, called brush border enzymes, are located in the membranes of the microvilli.

IV. The small intestine exhibits two major types of movements—peristalsis and segmentation.

### Large Intestine p. 539

I. The large intestine is divided into the cecum, colon, rectum, and anal canal.
   A. The vermiform appendix is attached to the inferior medial margin of the cecum.
   B. The colon consists of ascending, transverse, descending, and sigmoid portions.
   C. Bulges in the walls of the large intestine are called haustra.

II. Three types of movements occur in the large intestine: peristalsis, haustral churning, and mass movement.

III. The large intestine absorbs water and electrolytes.
   A. Although most of the water that enters the GI tract is absorbed in the small intestine, about 1–1.5 L/day pass to the large intestine, which absorbs about 90% of this amount.
   B. $Na^+$ is actively absorbed and water follows passively, in a manner analogous to the reabsorption of NaCl and water in the renal tubules.
IV. Defecation occurs when the anal sphincters relax and contraction of other muscles raises the rectal pressure.

## Liver, Gallbladder, and Pancreas p. 541

I. The liver, the largest internal organ, is composed of functional units called lobules.
   A. Liver lobules consist of plates of hepatic cells separated by capillary sinusoids.
   B. Blood flows from the periphery of each lobule, where the hepatic artery and portal vein empty through the sinusoids and out the central vein.
   C. Bile flows within the hepatocyte plates, in canaliculi, to the bile ducts.
   D. Substances excreted in the bile can be returned to the liver in the hepatic portal blood; this is called an enterohepatic circulation.
   E. Bile consists of a pigment called bilirubin, bile salts, cholesterol, and other molecules.
   F. The liver detoxifies the blood by excreting substances in the bile, by phagocytosis, and by chemical inactivation.
   G. The liver modifies the plasma concentrations of proteins, glucose, triglycerides, and ketone bodies.
II. The gallbladder serves to store and concentrate the bile, and it releases bile through the cystic duct and common bile duct to the duodenum.
III. The pancreas is both an exocrine and an endocrine gland.
   A. The endocrine portion is known as the islets of Langerhans and secretes the hormones insulin and glucagon.
   B. The exocrine acini of the pancreas produce pancreatic juice, which contains various digestive enzymes and bicarbonate.

## Digestion and Absorption of Carbohydrates, Lipids, and Proteins p. 548

I. The digestion of starch begins in the mouth through the action of salivary amylase.
   A. Pancreatic amylase digests starch into disaccharides and short-chain oligosaccharides.
   B. Complete digestion into monosaccharides is accomplished by brush border enzymes.
II. Protein digestion begins in the stomach by the action of pepsin.
   A. Pancreatic juice contains protein-digesting enzymes, including trypsin, chymotrypsin, and others.
   B. The brush border contains digestive enzymes that help to complete the digestion of proteins into amino acids.
   C. Amino acids, like monosaccharides, are absorbed and secreted into capillary blood entering the portal vein.
III. Lipids are digested in the small intestine after being emulsified by bile salts.
   A. Free fatty acids and monoglycerides enter particles called micelles, formed in large part by bile salts, and they are absorbed in this form or as free molecules.
   B. Once inside the mucosal epithelial cells, these subunits are used to resynthesize triglycerides.
   C. Triglycerides in the epithelial cells, together with proteins, form chylomicrons, which are secreted into the central lacteals of the villi.
   D. Chylomicrons are transported by lymph to the thoracic duct and there enter the blood.

## Neural and Endocrine Regulation of the Digestive System p. 554

I. The regulation of gastric function occurs in three phases.
   A. In the cephalic phase, the activity of higher brain centers, acting via the vagus nerve, stimulates gastric juice secretion.
   B. In the gastric phase, the secretion of HCl and pepsin is controlled by the gastric contents and by the hormone gastrin, secreted by the gastric mucosa.
   C. In the intestinal phase, the activity of the stomach is inhibited by neural reflexes from the duodenum and by gastric inhibitory peptide (GIP), secreted by the duodenum.
II. The secretion of the hormones secretin and cholecystokinin (CCK) regulate pancreatic juice and bile secretion.
   A. Secretin secretion is stimulated by the arrival of acidic chyme into the duodenum.
   B. CCK secretion is stimulated by the presence of fat in the chyme arriving in the duodenum.
   C. Contraction of the gallbladder occurs in response to a neural reflex and to the secretion of CCK by the duodenum.
III. Gastrointestinal hormones may be needed for the maintenance of the GI tract and accessory digestive organs.

## Review Activities

### Objective Questions

1. Intrinsic factor
   (a) is secreted by the stomach
   (b) is a polypeptide
   (c) promotes absorption of vitamin $B_{12}$ in the intestine
   (d) helps prevent pernicious anemia
   (e) all of the above
2. Intestinal enzymes such as lactase are
   (a) secreted by the intestine into the chyme
   (b) produced by the crypts of Lieberkühn
   (c) produced by the pancreas
   (d) attached to the cell membrane of microvilli in the epithelial cells of the mucosa
3. Which of the following statements about gastric secretion of HCl is *false?*
   (a) HCl is secreted by parietal cells.
   (b) HCl hydrolyzes peptide bonds.
   (c) HCl is needed for the conversion of pepsinogen to pepsin.
   (d) HCl is needed for maximum activity of pepsin.
4. Most digestion occurs in the
   (a) mouth
   (b) stomach
   (c) small intestine
   (d) large intestine
5. Which of the following statements about trypsin is true?
   (a) Trypsin is derived from trypsinogen by the digestive action of pepsin.
   (b) Active trypsin is secreted into the pancreatic acini.
   (c) Trypsin is produced in the crypts of Lieberkühn.
   (d) Trypsinogen is converted to trypsin by the brush border enzyme enterokinase.
6. During the gastric phase, the secretion of HCl and pepsinogen is stimulated by
   (a) vagus nerve stimulation that originates in the brain
   (b) polypeptides in the gastric lumen and by gastrin secretion
   (c) secretin and cholecystokinin from the duodenum
   (d) all of the above
7. The secretion of HCl by the stomach mucosa is inhibited by
   (a) neural reflexes from the duodenum
   (b) the secretion of gastric inhibitory peptide from the duodenum
   (c) the lowering of gastric pH
   (d) all of the above
8. The first organ to receive the blood-borne products of digestion is the
   (a) liver
   (b) pancreas
   (c) heart
   (d) brain
9. Which of the following statements about hepatic portal blood is true?
   (a) It contains absorbed fat.
   (b) It contains ingested proteins.
   (c) It is mixed with bile in the liver.
   (d) It is mixed with blood from the hepatic artery in the liver.
10. Absorption of salt and water is the principal function of which region of the GI tract?
    (a) esophagus
    (b) stomach
    (c) duodenum
    (d) jejunum
    (e) large intestine
11. Cholecystokinen (CCK) is a hormone that stimulates the
    (a) production of bile
    (b) release of pancreatic enzymes
    (c) contraction of the gallbladder
    (d) both a and b
    (e) both b and c
12. Which of the following statements about vitamin $B_{12}$ is *false*?
    (a) Lack of this vitamin can produce pernicious anemia.
    (b) Intrinsic factor is needed for absorption of vitamin $B_{12}$.
    (c) Damage to the gastric mucosa may lead to a deficiency in vitamin $B_{12}$.
    (d) Vitamin $B_{12}$ is absorbed primarily in the jejunum.

### Essay Questions

1. Explain how the gastric secretion of HCl and pepsin is regulated during the cephalic, gastric, and intestinal phases.
2. Describe how pancreatic enzymes become activated in the lumen of the intestine, and explain the need for these mechanisms.
3. What is the function of bicarbonate in pancreatic juice? Explain why ulcers are more likely to be located in the duodenum than in the stomach.
4. Explain why the pancreas is considered to be both an exocrine and an endocrine gland. Given this information, predict what effects tying of the pancreatic duct would have on pancreatic structure and function.
5. Explain how jaundice is produced when (*a*) the person has gallstones, (*b*) the person has a high rate of red blood cell destruction, and (*c*) the person has liver disease. In which of these cases would phototherapy for the jaundice be effective? Explain.

## SELECTED READINGS


Binder, H. J. 1984. The pathophysiology of diarrhea. *Hospital Practice* 19:107.

Bleich, H. L., and E. S. Boro. 1979. Protein digestion and absorption. *New England Journal of Medicine* 300:659.

Bortoff, A. 1972. Digestion. *Annual Review of Physiology* 28:201.

Carey, M. C., D. M. Small, and C. M. Bliss. 1983. Lipid digestion and absorption. *Annual Review of Physiology* 45:651.

Chou, C. C. 1982. Relationship between intestinal blood flow and motility. *Annual Review of Physiology* 44:29.

Christensen, R. R. 1971. The controls of gastrointestinal movements: Some old and new views. *New England Journal of Medicine* 285:483.

Cohen, S. 1983. Neuromuscular disorders of the gastrointestinal tract. *Hospital Practice* 18:121.

Davenport, H. W. 1982. *Physiology of the digestive tract.* 5th ed. Chicago: Year Book Medical Publishers.

Dockray, G. J. 1979. Comparative biochemistry and physiology of gut hormones. *Annual Review of Physiology* 41:83.

Freeman, H. J., and Y. S. Kim. 1978. Digestion and absorption of proteins. *Annual Review of Physiology* 29:99.

Gardner, J. D., and R. T. Jensen. 1986. Receptors and cell activation associated with pancreatic enzyme secretion. *Annual Review of Physiology* 48:103.

Gollan, J. L., and A. B. Knapp. 1985. Bilirubin metabolism and congenital jaundice. *Hospital Practice* 20:83.

Gray, G. M. 1975. Carbohydrate digestion and absorption: Role of the small intestine. *New England Journal of Medicine* 292:1225.

Grossman, M. I. 1979. Neural and hormonal regulation of gastrointestinal function: An overview. *Annual Review of Physiology* 41:27.

Guth, P. H. 1982. Stomach blood flow and acid secretion. *Annual Review of Physiology* 44:3.

Hersey, S. J., S. H. Norris, and A. J. Gilbert. 1984. Cellular control of pepsinogen secretion. *Annual Review of Physiology* 46:393.

Holt, K. M., and J. I. Isenberg. 1985. Peptic ulcer disease: Physiology and pathophysiology. *Hospital Practice* 20:89.

Kappas, A., and A. P. Alvarez. June 1975. How the liver metabolizes foreign substances. *Scientific American.*

McGuigan, J. E. 1978. Gastrointestinal hormones. *Annual Review of Physiology* 29:99.

Moog, F. November 1981. The lining of the small intestine. *Scientific American.*

Salen, G., and S. Shefer. 1983. Bile acid synthesis. *Annual Review of Physiology* 45:679.

Sanders, M. J., and A. H. Soll. 1986. Characterization of receptors regulating secretory function in the fundic mucosa. *Annual Review of Physiology* 48:89.

Smith, B. F., and T. Lamont. 1984. The pathogenesis of gallstones. *Hospital Practice* 19:93.

Soll, A. and J. H. Walsh. 1979. Regulation of gastric acid secretion. *Annual Review of Physiology* 41:35.

Walsh, J. H. 1988. Peptides as regulators of gastric acid secretion. *Annual Review of Physiology* 50:41.

Walsh, J. H., and M. I. Grossman. 1975. Gastrin. *New England Journal of Medicine* 292: first part, p. 1324; second part, p. 1377.

Weisbrodt, N. W. 1981. Patterns of intestinal motility. *Annual Review of Physiology* 43:21.

Williams, J. A. 1984. Regulatory mechanisms in pancreas and salivary acini. *Annual Review of Physiology* 46:361.

Wolfe, M. M., and A. H. Soll. 1988. The physiology of gastric acid secretion. *New England Journal of Medicine* 319:1707.

# Regulation of Metabolism

# 18

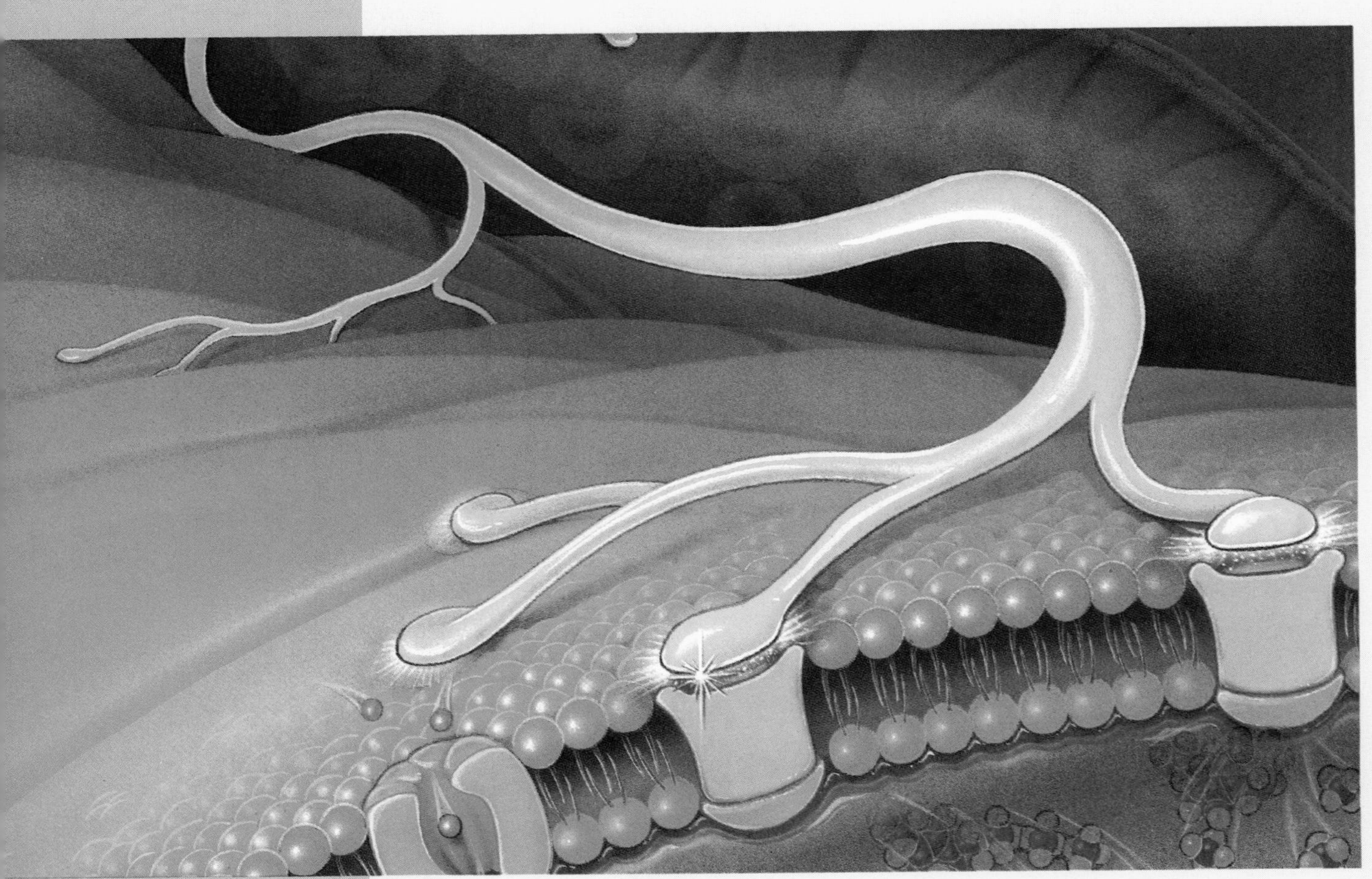

## Objectives

By studying this chapter, you should be able to

1. describe the significance of the basal metabolic rate, and explain how the metabolic rate is affected by physical activity, temperature, and eating
2. distinguish between the caloric and anabolic requirements for food, and explain the meaning of the terms *essential amino acids* and *essential fatty acids*
3. distinguish between fat-soluble and water-soluble vitamins, and describe some of the functions of different vitamins
4. define the terms *energy reserves* and *circulating energy substrates* and describe how these interact during anabolism and catabolism
5. describe the regulation of eating and describe in general terms the endocrine control of metabolism
6. describe the actions of insulin and glucagon and explain how the secretion of these hormones is regulated
7. explain how insulin and glucagon regulate metabolism during feeding and fasting
8. explain the causes and symptoms of type I and type II diabetes mellitus and of reactive hypoglycemia
9. describe the metabolic effects of epinephrine and the glucocorticoids
10. describe the effects of thyroxine on cell respiration and the relationship between thyroxine levels and the basal metabolic rate
11. explain the causes and symptoms of hypothyroidism and hyperthyroidism in terms of the actions of thyroid hormones
12. describe the metabolic effects of growth hormone and the body's requirements for growth hormone and thyroxine for proper growth
13. describe the actions of parathyroid hormone, 1,25-dihydroxyvitamin $D_3$, and calcitonin and explain how the secretion of these hormones is regulated
14. describe how 1,25-dihydroxyvitamin $D_3$ is produced and explain why this compound is needed to prevent osteomalacia and rickets

## Outline

## *Nutritional Requirements*

The body consumes energy continuously, but the extent of energy consumption can be increased by a variety of factors. These include muscular activity, digestion, and the actions of epinephrine, norepinephrine, and thyroxine. The energy needs of the body must be met by the caloric value of food to prevent catabolism of the body's own fat, carbohydrates, and protein. Additionally, food molecules—particularly the essential amino acids and fatty acids—are needed for replacement of molecules in the body that are continuously degraded. Vitamins and elements do not directly provide energy, but instead are required for diverse enzymatic reactions.

Living tissue is maintained by the constant expenditure of energy. This energy is obtained directly from ATP and indirectly from the cell respiration of glucose, fatty acids, ketone bodies, amino acids, and other organic molecules. These molecules are ultimately obtained from food, but they can also be obtained from the glycogen, fat, and protein stored in the body. The energy value of food is commonly measured in *kilocalories,* which are also called "big calories" and spelled with a capital letter (*C*alories). One kilocalorie is equal to 1,000 calories; one calorie is defined as the amount of heat required to raise the temperature of one cubic centimeter of water from 15° to 16° C. As described in chapter 5, the amount of energy released as heat when a quantity of food is combusted *in vitro* is equal to the amount of energy released within cells through the process of aerobic respiration. This is four calories per gram for carbohydrate or proteins, and nine calories per gram for fat. When this energy is released by cell respiration, some is transferred to the high-energy bonds of ATP, and some is lost as heat.

### Metabolic Rate and Caloric Requirements

The total rate of body metabolism, or the **metabolic rate,** can be measured by either the amount of heat generated by the body or by the amount of oxygen consumed by the body per minute. This rate is influenced by a variety of factors. For example, metabolic rate is increased by physical activity and by eating. The energy cost of eating (which must be subtracted from the caloric value of the food to produce a correct value for the energy the body actually obtains from the food) is called the *thermal effect of food (TEF).* The increased rate of metabolism that accompanies the assimilation of food can last more than six hours after a meal.

Temperature is also an important factor determining metabolic rate. This occurs for two reasons: (1) temperature itself is a determinant of the rate of chemical reactions; and (2) the hypothalamus contains *temperature control centers*, as well as temperature-sensitive cells that act as sensors for changes in body temperature. In response to deviations from a "set point" for body temperature (chapter 1), the control areas of the hypothalamus can direct physiological responses that help to correct the deviations and maintain a constant body temperature. Changes in body temperature are thus accompanied by physiological responses that influence the total metabolic rate.

If the body temperature is too hot the rate of all chemical reactions is raised, and metabolic rate is accordingly increased. The temperature control centers in the hypothalamus, however, are stimulated by a rise in body temperature. As a result, the person experiences a loss of appetite, decreased physical activity, increased sweating and respiration, and cutaneous vasodilation. A cold body temperature, in contrast, would itself decrease metabolic rate. In response to a fall in body temperature, however, the hypothalamus directs a number of compensatory responses: cutaneous vasoconstriction, shivering, increased hunger and food consumption, and increased activity of the sympathoadrenal system. The latter reaction results in more epinephrine and norepinephrine release, which act to increase the rate of cell respiration. Through these responses, a cold ambient temperature is accompanied by an increase in metabolic rate.

Hypothermia (low body temperature)—where the core body temperature is dropped to 21°–24° C—is often induced during open heart or brain surgery. The compensatory responses are dampened by the general anesthetic, and the lower body temperature drastically reduces the needs of the tissues for $O_2$. Under these conditions, the circulation can be stopped and bleeding is significantly reduced.

The metabolic rate of an awake, relaxed person 12–14 hours after eating and at comfortable temperature is known as the **basal metabolic rate (BMR).** The BMR is determined primarily by the age, sex, and body surface area of the person, but is also strongly influenced by the level of thyroid secretion. A person with hyperthyroidism has an abnormally high BMR, and a person with hypothyroidism has a low BMR. An interesting recent finding is that the BMR may be influenced by genetic inheritance, and that at least some families that are prone to obesity may have a genetically determined low BMR. They are thus more "fuel efficient." This ability may have evolved over many generations in response to chronic food shortage, and, in the words of one scientist, may represent a "thrifty gene rendered detrimental by progress."

The differences in energy requirements among most people, however, are due primarily to differences in physical activity. Average daily energy expenditures may range

**Table 18.1** Desirable body weights, according to sex, age, height, and body frame

| | Height Feet | Height Inches | Small Frame | Medium Frame | Large Frame |
|---|---|---|---|---|---|
| Men | 5 | 2 | 128–134 | 131–141 | 138–150 |
| | 5 | 3 | 130–136 | 133–143 | 140–153 |
| | 5 | 4 | 132–138 | 135–145 | 142–156 |
| | 5 | 5 | 134–140 | 137–148 | 144–160 |
| | 5 | 6 | 136–142 | 139–151 | 146–164 |
| | 5 | 7 | 138–145 | 142–154 | 149–168 |
| | 5 | 8 | 140–148 | 145–157 | 152–172 |
| | 5 | 9 | 142–151 | 148–160 | 155–176 |
| | 5 | 10 | 144–154 | 151–163 | 158–180 |
| | 5 | 11 | 146–157 | 154–166 | 161–184 |
| | 6 | 0 | 149–160 | 157–170 | 164–188 |
| | 6 | 1 | 152–164 | 160–174 | 168–192 |
| | 6 | 2 | 155–168 | 164–178 | 172–197 |
| | 6 | 3 | 158–172 | 167–182 | 176–202 |
| | 6 | 4 | 162–176 | 171–187 | 181–207 |

Weights at ages 25–59 based on lowest mortality. Weight in pounds according to frame (in indoor clothing weighing 5 lbs., shoes with 1″ heels).

| | Height Feet | Height Inches | Small Frame | Medium Frame | Large Frame |
|---|---|---|---|---|---|
| Women | 4 | 10 | 102–111 | 109–121 | 118–131 |
| | 4 | 11 | 103–113 | 111–123 | 120–134 |
| | 5 | 0 | 104–115 | 113–126 | 122–137 |
| | 5 | 1 | 106–118 | 115–129 | 125–140 |
| | 5 | 2 | 108–121 | 118–132 | 128–143 |
| | 5 | 3 | 111–124 | 121–135 | 131–147 |
| | 5 | 4 | 114–127 | 124–138 | 134–151 |
| | 5 | 5 | 117–130 | 127–141 | 137–155 |
| | 5 | 6 | 120–133 | 130–144 | 140–159 |
| | 5 | 7 | 123–136 | 133–147 | 143–163 |
| | 5 | 8 | 126–139 | 136–150 | 146–167 |
| | 5 | 9 | 129–142 | 139–153 | 149–170 |
| | 5 | 10 | 132–145 | 142–156 | 152–173 |
| | 5 | 11 | 135–148 | 145–159 | 155–176 |
| | 6 | 0 | 138–151 | 148–162 | 158–179 |

Weights at ages 25–59 based on lowest mortality. Weight in pounds according to frame (in indoor clothing weighing 3 lbs., shoes with 1″ heels).

Reprinted with permission of the Metropolitan Insurance Companies.

from about 1,300 to 5,000 kilocalories per day. The average values for people not engaged in heavy manual labor but who are active during their leisure time is about 2,900 kilocalories per day for a man and 2,100 kilocalories per day for a woman. People engaged in office work, the professions, sales, and comparable occupations consume up to 5 kilocalories per minute during work. More physically demanding occupations may require energy expenditures of 7.5 to 10 kilocalories per minute.

When the caloric intake is greater than the energy expenditures, excess calories are stored primarily as fat. This is true regardless of the source of the calories—carbohydrates, protein, or fat—because these molecules can be converted to fat by the metabolic pathways described in chapter 5. Appropriate body weights are indicated in table 18.1.

Obesity is a risk factor for cardiovascular diseases, renal disease, diabetes mellitus, gallbladder disease, the development of kidney stones, and some malignancies (particularly endometrial and breast cancer). Obesity in childhood is due to an increase in both the size and number of adipose cells; weight gain in adulthood is due mainly to an increase in adipose cell size, although the number of these cells may also increase in extreme weight gains. When weight is lost, the size of the adipose cells decreases, but the number of adipose cells does not decrease. It is thus important to prevent further increases in weight in all overweight people but particularly so in children.

The degree of obesity can be most conveniently determined by the **body mass index.** This is equal to the nude body weight (in kilograms) divided by the square of the barefoot height (in meters). For example, a person who weighs 170 pounds (77.1 kg) and is five-feet-nine-inches tall (1.75 m), has a body mass index of $77.1 \div 1.75^2 = 25.1$. The normal body mass index is 20–25 for males and 19–24 for females. Statistically, the morbidity and mortality rates are significantly increased when the body mass index is greater than 30. When the body mass index is over 40, the risk factor for cardiovascular disease due to obesity is comparable to risk factors due to smoking, hypertension, or hyperlipidemia.

Weight is lost when the caloric value of the food ingested is less than the amount required in cell respiration over a period of time. Weight loss, therefore, can be achieved by dieting alone or in combination with an exercise program to raise the metabolic rate. A summary of the caloric expenditure produced by different types of exercises is provided in table 18.2. In addition to the calories expended directly during the exercise, additional calories may be consumed through the course of the day due to an increased BMR as a result of the exercise program. The ability of exercise to increase the BMR, however, is controversial, and appears to occur only in people who exercise extensively rather than moderately and on a regular basis. Since vigorous exercise in obese people may place dangerous demands on the cardiovascular system, an exercise program under these conditions should be attempted only under medical supervision.

**Table 18.2** Energy consumed (in kilocalories per minute) by different types of activities

| Activity | Weight in pounds 105–115 | 127–137 | 160–170 | 182–192 |
|---|---|---|---|---|
| **Bicycling** | | | | |
| 10 mph | 5.41 | 6.16 | 7.33 | 7.91 |
| Stationary, 10 mph | 5.50 | 6.25 | 7.41 | 8.16 |
| **Calisthenics** | 3.91 | 4.50 | 7.33 | 7.91 |
| **Dancing** | | | | |
| Aerobic | 5.83 | 6.58 | 7.83 | 8.58 |
| Square | 5.50 | 6.25 | 7.41 | 8.00 |
| **Gardening, weeding, and digging** | 5.08 | 5.75 | 6.83 | 7.50 |
| **Jogging** | | | | |
| 5.5 mph | 8.58 | 9.75 | 11.50 | 12.66 |
| 6.5 mph | 8.90 | 10.20 | 12.00 | 13.20 |
| 8.0 mph | 10.40 | 11.90 | 14.10 | 15.50 |
| 9.0 mph | 12.00 | 13.80 | 16.20 | 17.80 |
| **Rowing, machine** | | | | |
| Easily | 3.91 | 4.50 | 5.25 | 5.83 |
| Vigorously | 8.58 | 9.75 | 11.50 | 12.66 |
| **Skiing** | | | | |
| Downhill | 7.75 | 8.83 | 10.41 | 11.50 |
| Cross-country, 5 mph | 9.16 | 10.41 | 12.25 | 13.33 |
| Cross-country, 9 mph | 13.08 | 14.83 | 17.58 | 19.33 |
| **Swimming, crawl** | | | | |
| 20 yards per minute | 3.91 | 4.50 | 5.25 | 5.83 |
| 40 yards per minute | 7.83 | 8.91 | 10.50 | 11.58 |
| 55 yards per minute | 11.00 | 12.50 | 14.75 | 16.25 |
| **Walking** | | | | |
| 2 mph | 2.40 | 2.80 | 3.30 | 3.60 |
| 3 mph | 3.90 | 4.50 | 5.30 | 5.80 |
| 4 mph | 4.50 | 5.20 | 6.10 | 6.80 |

## Anabolic Requirements

In addition to providing the body with energy, food also supplies the raw materials for synthesis reactions—collectively termed **anabolism**—that occur constantly within the cells of the body. Anabolic reactions include those that synthesize DNA and RNA, protein, glycogen, triglycerides, and other polymers. These anabolic reactions must occur constantly to replace those molecules that are hydrolyzed into their component monomers. These hydrolysis reactions, together with the reactions of cell respiration which break the monomers down ultimately to carbon dioxide and water, are collectively termed **catabolism.**

Exercise and fasting, acting through changes in hormonal secretion, cause an increase in the catabolism of stored glycogen, fat, and body protein. These molecules are also broken down at a certain rate in a person who is neither exercising nor fasting. Some of the monomers thus formed (amino acids, glucose, and fatty acids) are used to immediately resynthesize body protein, glycogen, and fat. However, some of the glucose derived from stored glycogen, for example, or fatty acids derived from stored triglycerides, undergo cell respiration for energy. For this reason, new monomers must be obtained from food to prevent a continual decline in the amount of protein, glycogen, and fat in the body.

The *turnover rate* of a particular molecule is the rate at which it is broken down and resynthesized. For example, the average daily turnover of carbohydrates is 250 g/day. Since some of the glucose in the body is reused to form glycogen, the average daily dietary requirement for carbohydrate is somewhat less than this amount, about 150 g/day. The average daily turnover for protein is 150 g/day, but since many of the amino acids derived from the catabolism of body proteins can be reused in protein synthesis, a person only needs about 35 g/day of protein in the diet. It should be noted that these are average figures and will differ in people as a result of differences in size, sex, age, genetics, and physical activity. The average daily turnover of fat is about 100 g/day, but very little is

**Table 18.3** Summary descriptions of the characteristics of vitamins

| Vitamin | Action | Deficiency symptoms | Sources |
|---|---|---|---|
| A | Constituent of visual pigment; strengthens epithelial membranes | Night blindness; dry skin | Yellow vegetables and fruit |
| $B_1$ (Thiamine) | Cofactor for enzymes that catalyze decarboxylation | Beriberi; neuritis | Liver, unrefined cereal grains |
| $B_2$ (Riboflavin) | Part of flavoproteins (such as FAD) | Glossitis; cheilosis | Liver, milk |
| $B_6$ (Pyridoxine) | Coenzyme for decarboxylase and transaminase enzymes | Convulsions | Liver, corn, wheat, and yeast |
| $B_{12}$ (Cyanocobalamin) | Coenzyme for amino acid metabolism; needed for erythropoiesis | Pernicious anemia | Liver, meat, eggs, milk |
| Biotin | Needed for fatty acid synthesis | Dermatitis; enteritis | Egg yolk, liver, tomatoes |
| C | Needed for collagen synthesis in connective tissues | Scurvy | Citrus fruits, green leafy vegetables |
| D | Needed for intestinal absorption of calcium and phosphate | Rickets; osteomalacia | Fish liver |
| E | Antioxidant | Muscular dystrophy | Milk, eggs, meat, leafy vegetables |
| Folates | Needed for reactions that transfer one carbon | Sprue: anemia | Green leafy vegetables |
| K | Promotes reactions needed for function of clotting factors | Hemorrhage; inability to form clot | Green leafy vegetables |
| Niacin | Part of NAD and NADP | Pellagra | Liver, meat, yeast |
| Pantothenic acid | Part of coenzyme A | Dermatitis; enteritis; adrenal insufficiency | Liver, eggs, yeast |

required in the diet (other than that which supplies sufficient fat-soluble vitamins and essential fatty acids), because fat can easily be produced from excess carbohydrates.

The minimal amounts of dietary protein and fat required to meet the turnover rate are only adequate if they supply sufficient amounts of the essential amino acids and fatty acids. These molecules are termed *essential* because the body cannot make them and thus is forced to obtain them in the diet for proper protein and fat synthesis. The eight **essential amino acids** are lysine, methionine, valine, leucine, isoleucine, tryptophane, phenylalanine, and threonine. The **essential fatty acids** are linoleic acid and linolenic acid.

## Vitamins and Elements

Vitamins are small organic molecules that cannot be made by the body and that serve as coenzymes in metabolic reactions (table 18.3). There are two groups of vitamins—the fat-soluble vitamins (A, D, E, and K) and the water-soluble vitamins. Water-soluble vitamins include thiamine ($B_1$), riboflavin ($B_2$), niacin ($B_3$), pyridoxine ($B_6$), pantothenic acid, biotin, folic acid, vitamin $B_{12}$, and vitamin C (ascorbic acid). Recommended daily allowances for these vitamins are shown in table 18.4.

Many of the water-soluble vitamins serve as coenzymes in the metabolism of carbohydrates, lipids, and proteins. Thiamine, for example, is needed for the activity of the enzyme that converts pyruvic acid to acetyl coenzyme A. Riboflavin and niacin are needed for the production of FAD and NAD, respectively; these latter compounds serve as coenzymes that transfer hydrogens during cell respiration. Pyridoxine is a cofactor for the enzymes involved in amino acid metabolism. Deficiencies of the water-soluble vitamins can, for obvious reasons, have widespread effects in the body.

Many fat-soluble vitamins have highly specialized functions. Vitamin K, for example, is required for the production of prothrombin and for clotting factors VII, IX, and X. Vitamin D is converted into a hormone that participates in the regulation of calcium balance. The visual pigments in the rods and cones of the retina are derived from vitamin A. Vitamin A and related compounds, called

**Table 18.4** Recommended daily allowances for vitamins and elements

| | Infants | Children | Adolescents (15–18 yrs) | | Adults (23–50 yrs) | |
|---|---|---|---|---|---|---|
| | 0–6 mos | 4–6 yrs | Males | Females | Males | Females |
| Weight, kg (lb.) | 6 (13) | 20 (44) | 66 (145) | 55 (120) | 70 (154) | 55 (120) |
| Height, cm (in.) | 60 (24) | 112 (44) | 176 (69) | 163 (64) | 178 (70) | 163 (64) |
| Protein, g | kg × 2.2 | 30 | 56 | 46 | 56 | 44 |
| Fat-soluble vitamins | | | | | | |
| Vitamin A, μg | 420 | 500 | 1000 | 800 | 1000 | 800 |
| Vitamin D, μg | 10 | 10 | 10 | 10 | 5 | 5 |
| Vitamin E activity, mg | 3 | 6 | 10 | 8 | 10 | 8 |
| Water-soluble vitamins | | | | | | |
| Ascorbic acid, mg | 35 | 45 | 60 | 60 | 60 | 60 |
| Folic acid, μg | 30 | 200 | 400 | 400 | 400 | 400 |
| Niacin, mg | 6 | 11 | 18 | 14 | 18 | 13 |
| Riboflavin, mg | 0.4 | 1.0 | 1.7 | 1.3 | 1.6 | 1.2 |
| Thiamine, mg | 0.3 | 0.9 | 1.4 | 1.1 | 1.4 | 1.0 |
| Vitamin $B_6$, mg | 0.3 | 1.3 | 2.0 | 2.0 | 2.2 | 2.0 |
| Vitamin $B_{12}$, μg | 0.5 | 2.5 | 3.0 | 3.0 | 3.0 | 3.0 |
| Elements | | | | | | |
| Calcium, mg | 360 | 800 | 1200 | 1200 | 800 | 800 |
| Phosphorus, mg | 240 | 800 | 1200 | 1200 | 800 | 800 |
| Iodine, μg | 40 | 90 | 150 | 150 | 150 | 150 |
| Iron, mg | 10 | 10 | 18 | 18 | 10 | 18 |
| Magnesium, mg | 50 | 200 | 400 | 300 | 350 | 300 |
| Zinc, mg | 3 | 10 | 15 | 15 | 15 | 15 |

*Source:* Food and Nutrition Board, *Recommended Dietary Allowances,* 9th ed., National Academy of Sciences—National Research Council, Washington, D.C., 1980.

**Table 18.5** Recommended daily intake of trace elements

| Element | Safe and Adequate Intake (mg/day) |
|---|---|
| Iron (males) | 10 |
| Iron (females) | 18 |
| Zinc | 15 |
| Manganese | 2.5 to 5.0 |
| Fluorine | 1.5 to 4.0 |
| Copper | 2.0 to 3.0 |
| Molybdenum | 0.15 to 0.5 |
| Chromium | 0.05 to 0.2 |
| Selenium | 0.05 to 0.2 |
| Iodine | 0.15 |

Source: Food and Nutrition Board, National Academy of Sciences, Washington, D.C. 1980.

*retinoids,* also have effects on genetic expression in epithelial cells; these compounds are now used clinically in the treatment of some skin conditions, and researchers are attempting to derive related compounds that may aid the treatment of some cancers.

Elements are needed as cofactors for specific enzymes and for a wide variety of other critical functions. Elements that are required in relatively large amounts per day include sodium, potassium, magnesium, calcium, phosphorus, and chlorine (table 18.4). In addition, the following **trace elements** are recognized as essential: iron, zinc, manganese, fluorine, copper, molybdenum, chromium, and selenium. These must be ingested in amounts ranging from 50 micrograms to 18 milligrams per day (table 18.5).

1. *Explain how the metabolic rate is influenced by exercise, ambient temperature, and the assimilation of food.*
2. *Distinguish between the caloric and anabolic requirements of the diet.*
3. *List the fat-soluble vitamins, and describe some of their functions.*
4. *Explain the roles of vitamins $B_1$, $B_2$, and $B_3$ in energy metabolism.*

**Figure 18.1.** A schematic flowchart of energy pathways in the body.

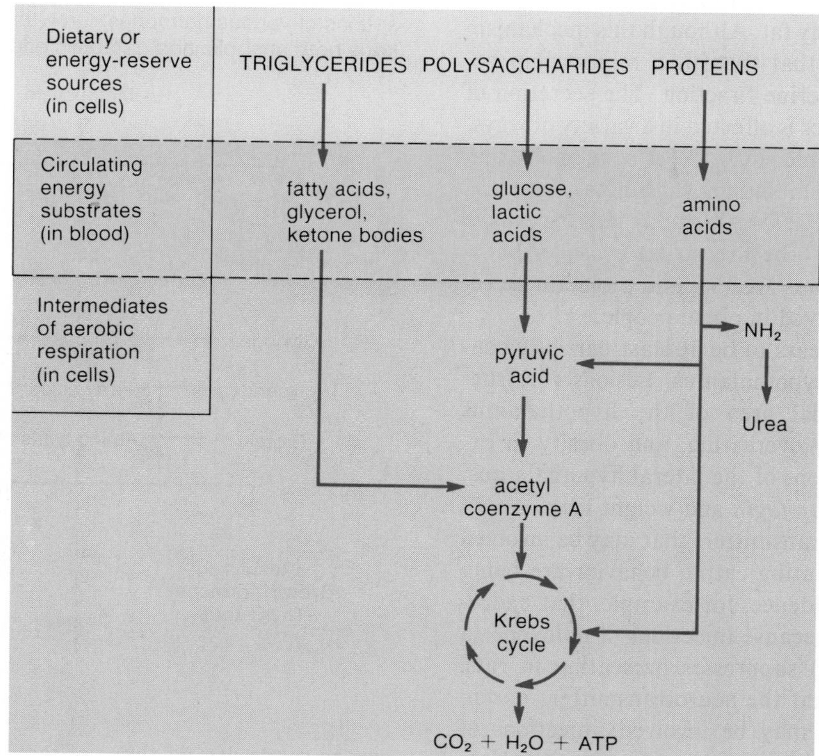


## Regulation of Energy Metabolism

The blood plasma contains circulating glucose, fatty acids, amino acids, and other molecules that can be used by the body tissues for cell respiration. These circulating molecules may be derived from the food or from breakdown of the body's glycogen and fat reserves, or from body protein. The building of the body's energy reserves following a meal, and the utilization of these reserves between meals, is regulated by the action of a number of hormones that act to promote either anabolism or catabolism of these reserves.

The molecules that can be oxidized for energy by the processes of cell respiration may be derived from the **energy reserves** of glycogen, fat, or protein. Glycogen and fat function primarily as energy reserves, whereas this represents a secondary, emergency function for proteins. Although body protein can provide amino acids for energy, this requires the breakdown of proteins needed for muscle contraction, structural strength, enzymatic activity, and other functions. Alternatively, the molecules used for cell respiration can be derived from the products of digestion that are absorbed through the small intestine. Since these molecules—glucose, fatty acids, amino acids, and others—are carried by the blood to the tissue cells for use in cell respiration, they can be called **circulating energy substrates** (fig. 18.1).

Because of differences in cellular enzyme content, different organs have different *preferred energy sources.* The brain has an almost absolute requirement for blood glucose as its energy source, for example. A fall in the plasma concentration of glucose below about 50 mg per 100 ml can thus "starve" the brain and have disastrous consequences. Resting skeletal muscles, in contrast, use fatty acids as their preferred energy source. Similarly, ketone bodies (derived from fatty acids), lactic acid, and amino acids can be used to different degrees as energy sources by various organs. The plasma normally contains adequate concentrations of all of these circulating energy substrates to meet the energy needs of the body.

### Eating

Ideally, one should eat the kinds and amounts of foods that provide adequate vitamins, minerals, essential amino acids and fatty acids, and calories. Proper caloric intake is that which maintains energy reserves (primarily fat and glycogen) and results in a body weight within an optimum range for health.

There is a tendency for body weight to be stable despite short-term changes in caloric intake. It has thus been proposed that there may be some mechanism that is sensitive to the amount of body fat. Although this mechanism is not known, it is clear that there is a relationship between body fat and endocrine function. The secretion of anterior pituitary hormones is affected in a variety of ways. Obese women, for example, may experience menstrual cycle abnormalities and hirsutism (hairiness), whereas women who drop below 15%–20% body fat may have amenorrhea (cessation of the menstrual cycle). Abnormalities in growth hormone, ACTH, and prolactin secretion have also been observed in obese people.

Eating behavior appears to be at least partially controlled by areas of the hypothalamus. Lesions (destruction) in the ventromedial area of the hypothalamus produce *hyperphagia,* or overeating, and obesity in experimental animals. Lesions of the lateral hypothalamus, in contrast, produce *hypophagia* and weight loss.

The chemical neurotransmitters that may be involved in neural pathways mediating eating behavior are being investigated. There is evidence, for example, that endorphins may be involved, because injections of naloxone (a morphine-blocking drug) suppresses overeating in rats. There is also evidence that the neurotransmitters norepinephrine and serotonin may be involved; injections of norepinephrine into the brain of rats cause overeating, whereas injections of serotonin have the opposite effect. Interestingly, the intestinal hormone cholecystokinin (CCK) also appears to function as a neurotransmitter in the brain, and it has been shown that injections of CCK cause experimental animals to stop eating. More research is needed to elucidate the structure and processes involved in the regulation of eating.

## Hormonal Regulation of Metabolism

The absorption of energy carriers from the intestine is not continuous; it rises to high levels following meals (the *absorptive state*) and tapers toward zero between meals (the *postabsorptive* or *fasting* state). Despite this, the plasma concentration of glucose and other energy substrates does not remain high during periods of absorption and does not normally fall below a certain level during periods of fasting. During the absorption of digestion products from the intestine, energy substrates are removed from the blood and deposited as energy reserves from which withdrawals can be made during times of fasting (fig. 18.2). This assures that there will be an adequate plasma concentration of energy substrates to sustain tissue metabolism at all times.

**Figure 18.2.** The balance of metabolism can be tilted toward anabolism (synthesis of energy reserves) or catabolism (utilization of energy reserves) by the combined actions of various hormones. Growth hormone and thyroxine have both anabolic and catabolic effects.

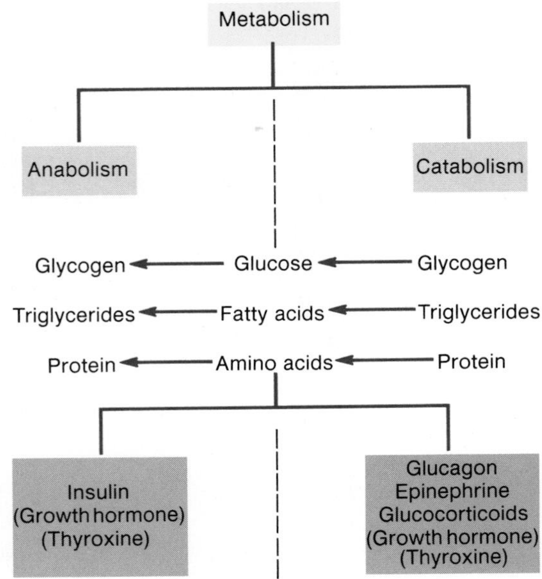


The rate of deposit and withdrawal of energy substrates into and from the energy reserves and the conversion of one type of energy substrate into another are regulated by the actions of hormones. The balance between anabolism and catabolism is determined by the antagonistic effects of hormones such as insulin, glucagon, growth hormone, thyroxine, and others (fig. 18.2). The specific metabolic effects of these hormones are summarized in table 18.6, and some of these actions are illustrated in figure 18.3.

1. *Define the terms* energy reserves *and* circulating energy carriers, *and give examples.*
2. *Describe the structures and neurotransmitters that may be involved in the regulation of eating.*
3. *Which hormones promote an increase in blood glucose? Which promote a decrease? List the hormones that stimulate fat synthesis (lipogenesis) and fat breakdown (lipolysis).*

**Table 18.6** Summary of the endocrine regulation of metabolism

| Hormone | Blood Glucose | Carbohydrate Metabolism | Protein Metabolism | Lipid Metabolism |
|---|---|---|---|---|
| Insulin | Decreased | ↑Glycogen formation<br>↓Glycogenolysis<br>↓Gluconeogenesis | ↑Amino acid transport | ↑Lipogenesis<br>↓Lipolysis<br>↓Ketogenesis |
| Glucagon | Increased | ↓Glycogen formation<br>↑Glycogenolysis<br>↑Gluconeogenesis | No direct effect | ↑Lipolysis<br>↑Ketogenesis |
| Growth hormone | Increased | ↑Glycogenolysis<br>↑Gluconeogenesis<br>↓Glucose utilization | ↑Protein synthesis | ↓Lipogenesis<br>↑Lipolysis<br>↑Ketogenesis |
| Glucocorticoids | Increased | ↑Glycogen formation<br>↑Gluconeogenesis | ↓Protein synthesis | ↓Lipogenesis<br>↑Lipolysis<br>↑Ketogenesis |
| Epinephrine | Increased | ↓Glycogen formation<br>↑Glycogenolysis<br>↑Gluconeogenesis | No direct effect | ↑Lipolysis<br>↑Ketogenesis |
| Thyroxine | No effect | ↑Glucose utilization | ↑Protein synthesis | No direct effect |

**Figure 18.3.** Different hormones participate both synergistically and antagonistically in the regulation of metabolism.

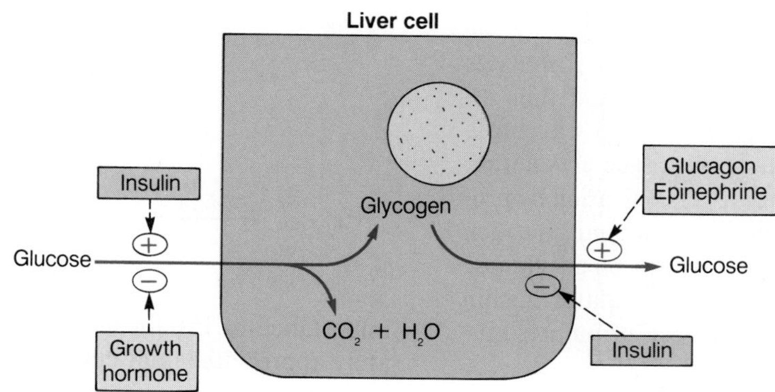


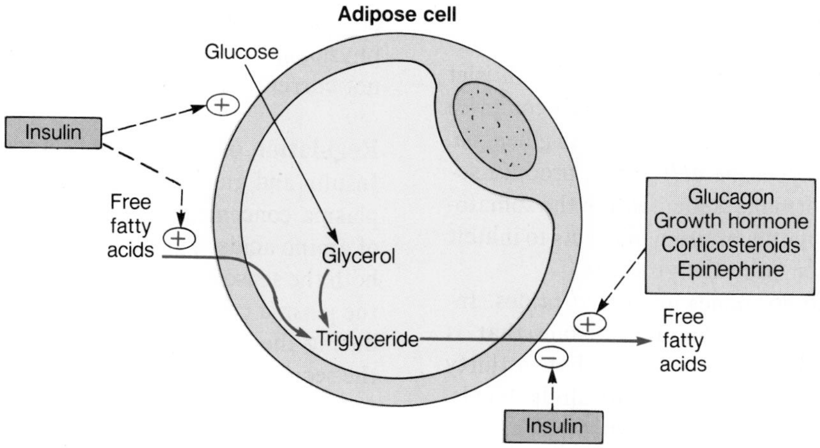

**Figure 18.4.** The cellular composition of a normal pancreatic islet.

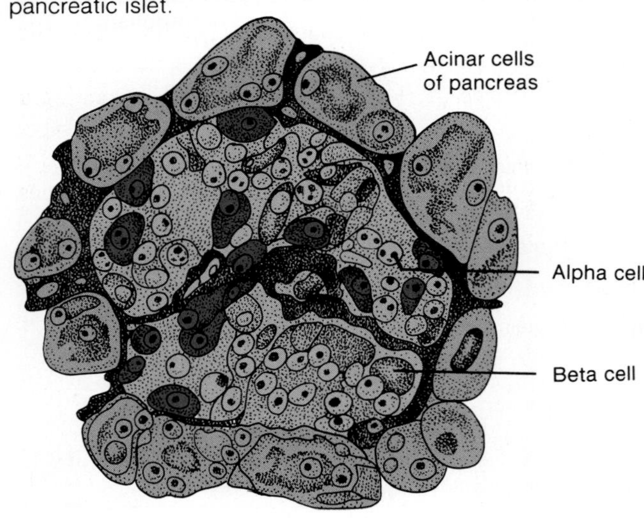


## *Energy Regulation by the Islets of Langerhans*

Insulin and glucagon are two hormones secreted by the islets of Langerhans in the pancreas. Insulin stimulates the entry of blood glucose into tissue cells and thus increases the storage of glycogen and fat while causing a fall in blood glucose concentration. Glucagon acts antagonistically to raise the blood glucose concentration by promoting glycogenolysis. Secretion of insulin and glucagon is under negative feedback control by the blood glucose concentration; a rise in blood glucose stimulates insulin secretion, whereas a fall in blood glucose stimulates glucagon secretion.

Scattered within a "sea" of pancreatic exocrine tissue (the acini) are islands of hormone-secreting cells. These **islets of Langerhans** contain three distinct cell types that secrete different hormones (fig. 18.4). The most numerous are the *beta cells;* these cells comprise 60% of each islet and secrete the hormone **insulin.** The *alpha cells* comprise about 25% of each islet and secrete the hormone **glucagon.** The least numerous cell type, the *delta cells,* produce **somatostatin.** The latter hormone is identical to the somatostatin produced by the hypothalamus, which acts to inhibit growth hormone secretion from the pituitary.

All three pancreatic hormones are polypeptides. Insulin consists of two polypeptide chains—one that is twenty-one amino acids long and another that is thirty amino acids long—joined together by disulfide bonds. Glucagon is a twenty-one-amino-acid polypeptide, and somatostatin contains fourteen amino acids. Insulin was the first of these hormones to be discovered (in 1921). The importance of insulin in diabetes mellitus was immediately appreciated, and clinical use of insulin in the treatment of this disease began almost immediately after its discovery. The physiological role of glucagon was discovered later, and the importance of glucagon in the development of diabetes has only recently been suspected. The physiological significance of islet-secreted somatostatin is not currently known.

**Figure 18.5.** The secretion from the B (beta) cells and A (alpha) cells of the islets of Langerhans is regulated to a large degree by the blood glucose concentration. (*a*) A high blood glucose concentration stimulates insulin and inhibits glucagon secretion. (*b*) A low blood glucose concentration, conversely, stimulates glucagon and inhibits insulin secretion. The actions of insulin and glucagon provide negative feedback control of the blood glucose concentration.

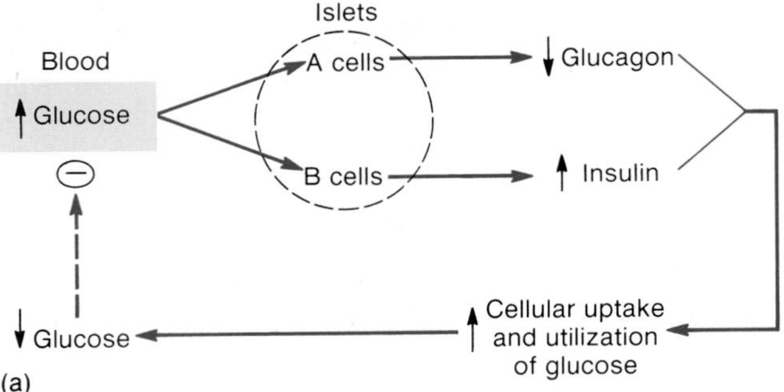


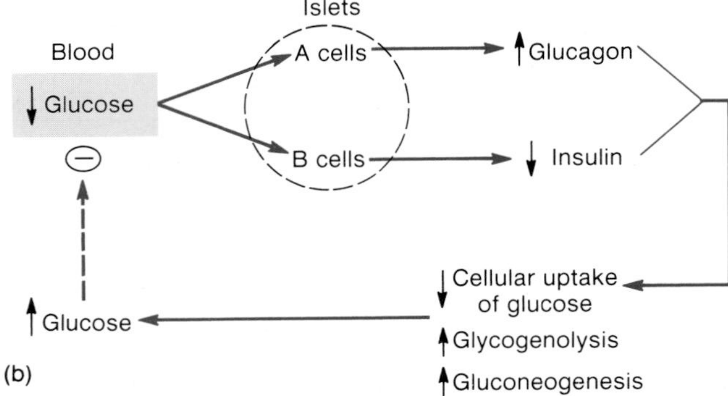


### Regulation of Insulin and Glucagon Secretion

Insulin and glucagon secretion is largely regulated by the plasma concentrations of glucose and, to a lesser degree, of amino acids. The alpha and beta cells, therefore, act as both the sensors and effectors in this control system. Since the plasma concentration of glucose and amino acids rises during the absorption of a meal and falls during fasting, the secretion of insulin and glucagon likewise changes between the absorptive and postabsorptive states. These changes in insulin and glucagon secretion, in turn, cause changes in plasma glucose and amino acid concentrations and thus help to maintain homeostasis via negative feedback loops (fig. 18.5).

**Figure 18.6.** Changes in blood glucose and plasma insulin concentrations after the ingestion of 100 grams of glucose in an oral glucose tolerance test.

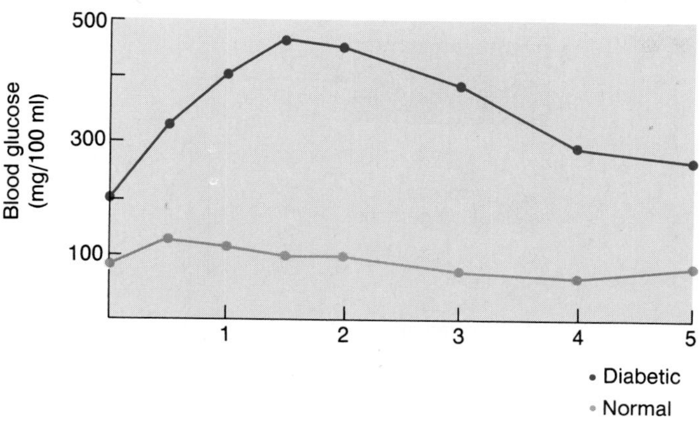


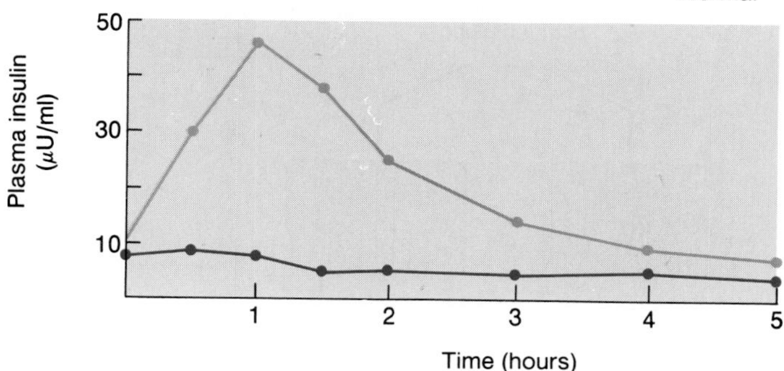


***Effects of Glucose.*** During the absorption of a carbohydrate meal, the plasma glucose concentration rises. This rise in plasma glucose (1) stimulates the beta cells to secrete insulin and (2) inhibits the secretion of glucagon from the alpha cells. Insulin acts to *stimulate the cellular uptake of plasma glucose.* A rise in insulin secretion therefore lowers the plasma glucose concentration. Since glucagon has the antagonistic effect of raising the plasma glucose concentration, by stimulating glycogenolysis in the liver, the inhibition of glucagon secretion complements the effect of increased insulin during the absorption of a carbohydrate meal. A rise in insulin and a fall in glucagon secretion thus help to lower the high plasma glucose concentration that occurs during periods of absorption.

During fasting, the plasma glucose concentration falls. At this time, therefore, (1) insulin secretion decreases and (2) glucagon secretion increases. These changes in hormone secretion prevent the cellular uptake of blood glucose into organs such as the muscles, liver, and adipose tissue and promote the release of glucose from the liver (through the actions of glucagon). A negative feedback loop is therefore completed (fig. 18.5), which helps to retard the fall in plasma glucose concentration that occurs during fasting.

The ability of the beta cells to secrete insulin, as well as the ability of insulin to lower blood glucose, is measured clinically by the **oral glucose tolerance test** (fig. 18.6). In this procedure, a person drinks a glucose solution and blood samples are taken periodically for plasma glucose measurements. In a normal person the rise in blood glucose produced by drinking this solution is reversed to normal levels within two hours following glucose ingestion.

People with **diabetes mellitus**—due to the inadequate secretion or action of insulin—maintain a state of high plasma glucose concentration (hyperglycemia) during the oral glucose tolerance test (fig. 18.6). People who have **reactive hypoglycemia** (low plasma glucose concentration due to excessive insulin secretion) have lower-than-normal blood glucose concentrations five hours following glucose ingestion. These conditions are described in more detail in a later section.

***Effects of Ingested Protein.*** Although the ingestion of carbohydrates causes a rise in insulin and a fall in glucagon secretion, the ingestion of proteins stimulates the secretion of both hormones. This rise in insulin secretion helps to complete a negative feedback loop because insulin promotes the uptake of amino acids into the tissues and the incorporation of these amino acids into proteins. The importance of a simultaneous rise in glucagon and insulin secretions can be understood in terms of the need to maintain a constant blood glucose concentration. A rise in insulin secretion after a protein-rich meal, if not accompanied by a simultaneous rise in glucagon secretion, would cause hypoglycemia in the absence of ingested carbohydrates. A rise in both insulin and glucagon secretions thus promotes lowering of the amino acid concentration of the blood, while the blood glucose concentration remains relatively constant due to the antagonistic effects of insulin and glucagon on glucose metabolism.

Since most meals contain both carbohydrates and proteins, insulin secretion is normally raised following meals, but the effects on glucagon secretion are variable. In an average meal that is rich in carbohydrates, the suppressive effects of high plasma glucose are more potent than the stimulatory effects of amino acids on glucagon secretion. In general, therefore, the insulin secretion is increased and the glucagon secretion is decreased during the period of absorption.

***Effects of Autonomic Nerves.*** The islets of Langerhans receive both parasympathetic and sympathetic innervation. The activation of the parasympathetic system during meals stimulates insulin secretion at the same time that

**Table 18.7** Regulation of insulin and glucagon secretion.

| Regulator | Effect on Insulin Secretion | Effect on Glucagon Secretion |
|---|---|---|
| Hyperglycemia | Stimulates | Inhibits |
| Hypoglycemia | Inhibits | Stimulates |
| Gastric inhibitory peptide | Stimulates | Stimulates (?) |
| Sympathetic nerve impulses | Inhibits | Stimulates |
| Parasympathetic nerve impulses | Stimulates | Inhibits |
| Amino acids | Stimulates | Stimulates |
| Somatostatin | Inhibits* | Inhibits* |

*Inhibitory effects of somatostatin on insulin and glucagon secretion may not occur under normal conditions.

gastrointestinal function is stimulated. The activation of the sympathetic system, in contrast, stimulates glucagon secretion and inhibits insulin secretion. The effects of glucagon, together with those of epinephrine, produce a "stress hyperglycemia" when the sympathoadrenal system is activated.

***Effects of Gastric Inhibitory Peptide (GIP).*** Surprisingly, insulin secretion increases more rapidly following glucose ingestion than it does following an intravenous injection of glucose. This is due to the fact that the intestine, in response to glucose ingestion, secretes a hormone that stimulates insulin secretion before the glucose is absorbed. Insulin secretion thus begins to rise in anticipation of a rise in blood glucose. The intestinal hormone that mediates this effect is believed to be gastric inhibitory peptide (GIP— chapter 17).

Table 18.7 summarizes the effects of various factors that regulate insulin and glucagon secretion.

The mechanisms that regulate insulin and glucagon secretion and the actions of these hormones normally prevent the plasma glucose concentration from rising above 170 mg per 100 ml after a meal or from falling below about 50 mg per 100 ml between meals. This regulation is important because abnormally high blood glucose can damage tissue cells (as may occur in diabetes mellitus), and abnormally low blood glucose can damage the brain. The later effect results from the fact that glucose enters the brain by facilitated diffusion; when the rate of this diffusion is too low, due to low plasma glucose concentrations, the supply of metabolic energy for the brain may become inadequate. This can result in weakness, dizziness, personality changes, and ultimately in coma and death.

**Figure 18.7.** A rise in blood glucose concentration stimulates insulin secretion. Insulin promotes a fall in blood glucose by stimulating the cellular uptake of glucose and the conversion of glucose to glycogen and fat.

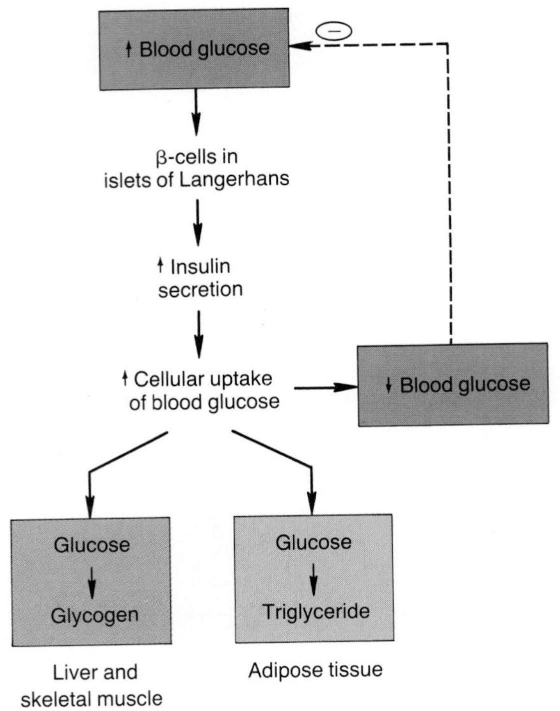


## Insulin and Glucagon: Absorptive State

The lowering of plasma glucose by insulin is, in a sense, a side effect of the primary action of this hormone. Insulin is the major hormone that promotes anabolism in the body. During absorption of the products of digestion and the subsequent rise in the plasma concentrations of circulating energy substrates, insulin promotes the cellular uptake of plasma glucose and its incorporation into glycogen and fat (fig. 18.7). As previously described, insulin also promotes the cellular uptake of amino acids and their incorporation into proteins. The stores of large, energy reserve molecules are thus increased while the plasma concentrations of glucose and amino acids are decreased.

The synthesis of triglycerides (fat) within adipose cells depends upon insulin-stimulated glucose uptake from plasma. Once inside the adipose cells, glucose can be converted into $\alpha$-glycerol phosphate and acetyl coenzyme A (acetyl CoA). The formation of fatty acids from acetyl CoA and condensation of three fatty acids with glycerol yields triglycerides. Although the adipose cells can also utilize free fatty acids from the blood, they cannot incorporate glycerol from the blood into triglycerides. This is because adipose cells lack the enzymes needed to convert

**Figure 18.8.** The synthesis of triglycerides (fat) within adipose cells. Notice that fat can be produced from glucose and from the fatty acids released by the hydrolysis of plasma lipoproteins.

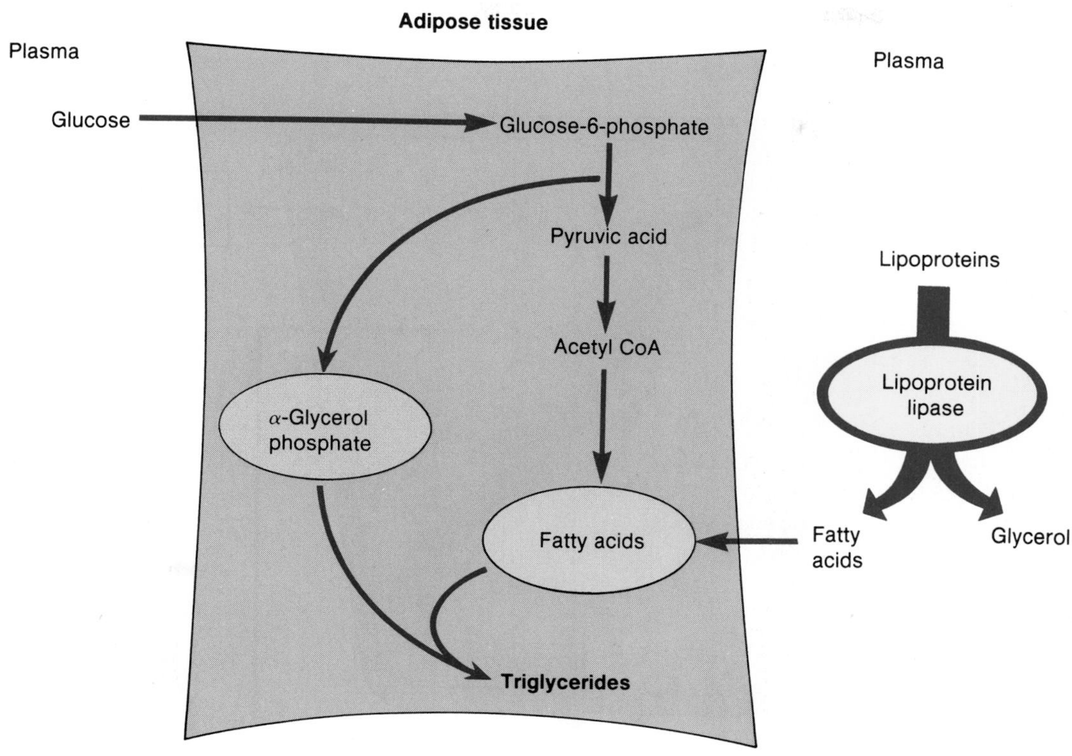


glycerol to α-glycerol phosphate, which is the precursor needed in triglyceride synthesis (fig. 18.8). Entry of blood glucose into adipose cells—which is directly dependent on insulin—thus determines the rate at which fat is produced.

A nonobese 70 kg man has approximately 10 kg (about 82,500 kcal) of stored fat. Since 250 g of fat can supply the energy requirements for one day, this reserve fuel is sufficient for about forty days. Glycogen is less efficient as an energy reserve, and less is stored in the body; there is about 100 g (400 kcal) of glycogen stored in the liver and 375–400 g (1,500 kcal) in skeletal muscles. Insulin promotes the cellular uptake of glucose into the liver and muscles and the conversion of glucose into glucose-6-phosphate. In the liver and muscles, this can be changed into glucose-1-phosphate, which is used as the precursor of glycogen. Once the stores of glycogen are filled, the continued ingestion of excess calories results in the continued production of fat rather than of glycogen.

## Insulin and Glucagon: Postabsorptive State

Glucagon stimulates and insulin suppresses the hydrolysis of liver glycogen, or **glycogenolysis.** Thus during times of fasting, when glucagon secretion is high and insulin secretion is low, liver glycogen is used as a source of additional blood glucose. This process is essentially the reverse of that which formed glycogen and results in the liberation of free glucose from glucose-6-phosphate by the action of an enzyme called *glucose-6-phosphatase* (chapter 5). Only the liver has this enzyme and therefore only the liver can use its stored glycogen as a source of additional blood glucose. Since muscles lack glucose-6-phosphatase, the glucose-6-phosphate produced from muscle glycogen can only be used for glycolysis by the muscle cells themselves.

Since there are only about 100 g of stored glycogen in the liver, adequate blood glucose levels could not be maintained for very long during fasting using this source alone. The low levels of insulin secretion during fasting, together with elevated glucagon secretion, however, promote **gluconeogenesis:** the formation of glucose from noncarbohydrate molecules. Low insulin allows the release of amino acids from skeletal muscles, while glucagon stimulates the production of enzymes in the liver that convert amino acids to pyruvic acid and pyruvic acid to glucose. During prolonged fasting and exercise, gluconeogenesis in the liver using amino acids from muscles may be the only source of blood glucose.

The secretion of glucose from the liver during fasting thus compensates for the low plasma glucose concentrations and helps to provide the glucose needed by the brain.

**Figure 18.9.** Increased glucagon secretion and decreased insulin secretion during fasting favors catabolism. These hormonal changes result in elevated release of glucose, fatty acids, ketone bodies, and amino acids into the blood. Notice that the liver secretes glucose that is derived both from the breakdown of liver glycogen and from the conversion of amino acids in gluconeogenesis.

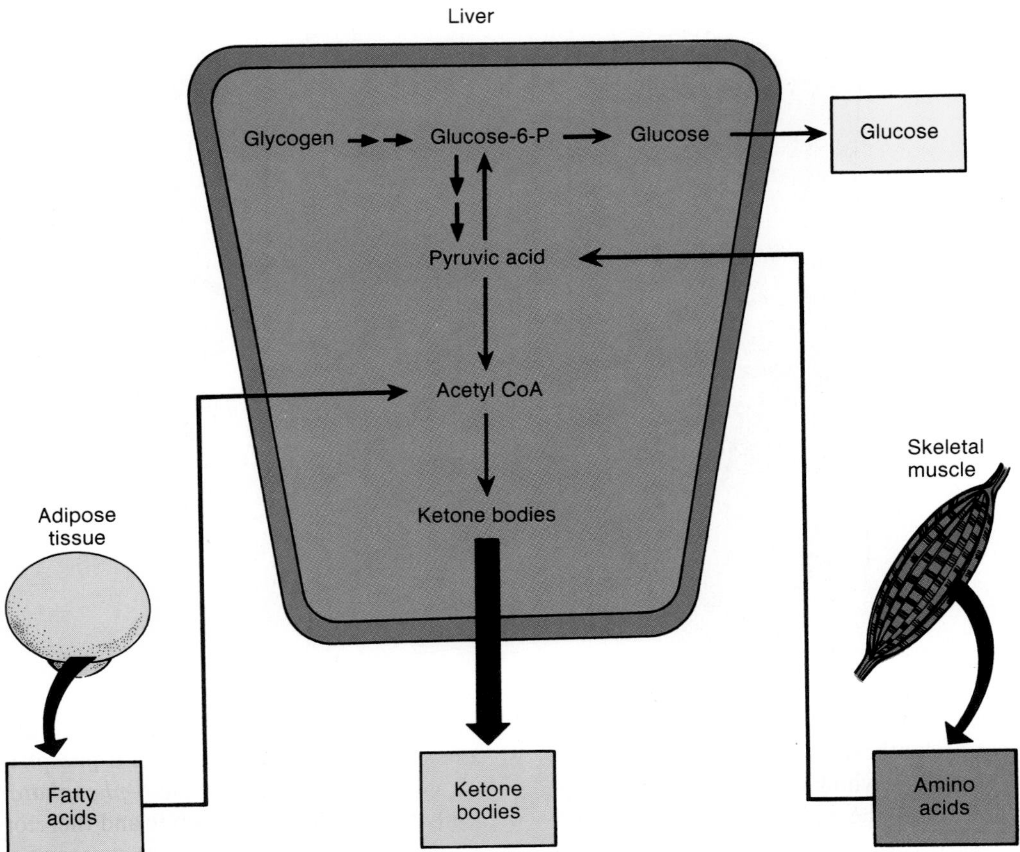


In the presence of low insulin secretion, however, other organs cannot utilize blood glucose as an energy source. This helps to spare glucose for the brain, but alternative energy sources are needed for these other organs. The major alternative energy sources that are used by the liver and skeletal muscles are free fatty acids and ketone bodies.

Glucagon, in the presence of low insulin levels, stimulates in adipose cells an enzyme called hormone-sensitive lipase. This enzyme catalyzes the hydrolysis of stored triglycerides and the release of free fatty acids and glycerol into the blood. Glucagon also stimulates enzymes in the liver that convert some of these fatty acids into ketone bodies. These derivatives of fatty acids—including acetoacetic acid, $\beta$-hydroxybutyric acid, and acetone—can also be released into the blood (fig. 18.9). Several organs in the body can use ketone bodies, like fatty acids, as a source of acetyl CoA in aerobic respiration.

Through the stimulation of **lipolysis** (the breakdown of fat) and **ketogenesis** (the formation of ketone bodies) the high glucagon and low insulin levels that are found during fasting provide circulating energy substrates for use by the muscles, liver, and other organs. Through liver glycogenolysis and gluconeogenesis, these hormonal changes help to provide adequate levels of blood glucose to sustain the metabolism of the brain. The antagonistic action of insulin and glucagon (fig. 18.10) thus promotes appropriate metabolic responses during periods of fasting and periods of absorption. The actions of insulin and glucagon are summarized in table 18.8.

**Figure 18.10.** The inverse relationship between insulin and glucagon secretion during the absorption of a meal and during fasting. Changes in the insulin:glucagon ratio tilts metabolism toward anabolism during the absorption of food and toward catabolism during fasting.

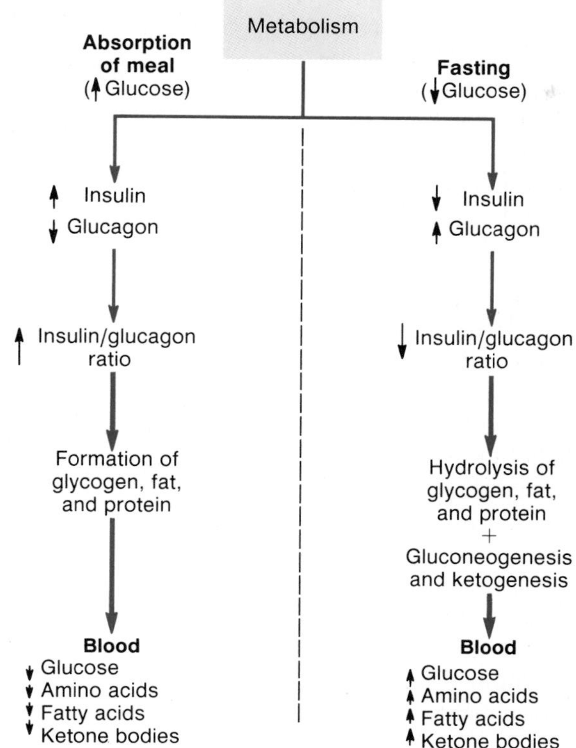


**Table 18.8** Comparison of the metabolic effects of insulin and glucagon

| Actions | Insulin | Glucagon |
|---|---|---|
| Cellular glucose transport | ↑ | 0 |
| Glycogen synthesis | ↑ | ↓ |
| Glycogenolysis in liver | ↓ | ↑ |
| Gluconeogenesis | ↓ | ↑ |
| Amino acid uptake; protein synthesis | ↑ | 0 |
| Inhibition of amino acid release; protein degradation | ↓ | 0 |
| Lipogenesis | ↑ | 0 |
| Lipolysis | ↓ | ↑ |
| Ketogenesis | ↓ | ↑ |

Key:
↑ = Increased
↓ = Decreased
0 = No direct effect

1. *Describe how the secretions of insulin and glucagon change during periods of absorption and periods of fasting. Explain how these changes in hormone secretion are produced.*
2. *Describe how the synthesis of fat in adipose cells is regulated by insulin. Explain how fat metabolism is regulated by insulin and glucagon during periods of absorption and fasting.*
3. *Define the following terms:* glycogenolysis, gluconeogenesis, *and* ketogenesis. *Explain how insulin and glucagon affect each of these processes during periods of absorption and fasting.*
4. *Describe two pathways used by the liver to produce glucose for secretion into the blood. Explain why skeletal muscles cannot secrete glucose into the blood.*

## Diabetes Mellitus and Hypoglycemia

Since insulin has a central role in the regulation of metabolism, defects in the actions of insulin produce predictable metabolic disturbances that are characteristic of the disease diabetes mellitus. A person with type I diabetes requires injections of insulin; a person with type II diabetes can control the condition by other methods. In both types, hyperglycemia and glycosuria result from the inability of glucose to move out of the blood and into the tissue cells. A person with reactive hypoglycemia, in contrast, secretes excessive amounts of insulin and thus experiences hypoglycemia in response to the stimulus of a carbohydrate meal.

Chronic high blood glucose, or hyperglycemia, is the hallmark of the disease **diabetes mellitus.** The name of this disease is derived from the fact that glucose "spills over" into the urine when the blood glucose concentration is too high. The hyperglycemia of diabetes mellitus results from either the insufficient secretion of insulin by the beta cells of the islets of Langerhans or the inability of secreted insulin to stimulate the cellular uptake of glucose from the blood. Diabetes mellitus, in short, results from the inadequate secretion or action of insulin.

There are two forms of diabetes mellitus. In **type I,** or **insulin-dependent,** diabetes the beta cells are destroyed and secrete little or no insulin. This form of the disease accounts for only about 10% of the cases of diabetes in the country. About 90% of the people who have diabetes have **type II,** or **non-insulin-dependent,** diabetes mellitus. Type I diabetes has also been called *juvenile-onset diabetes,* because this condition is usually diagnosed in people under the age of thirty. Type II, or *maturity-onset, diabetes* is usually diagnosed in people over the age of thirty. Some comparisons of these two forms of diabetes mellitus are shown in table 18.9.

**Table 18.9** Comparison of juvenile-onset and maturity-onset diabetes mellitus

| Characteristics | Juvenile-Onset (Type I) | Maturity-Onset (Type II) |
|---|---|---|
| Usual age at onset | Under 20 years | Over 40 years |
| Development of symptoms | Rapid | Slow |
| Percent of diabetic population | About 10% | About 90% |
| Development of ketoacidosis | Common | Rare |
| Association with obesity | Rare | Common |
| Beta cells of islets | Destroyed | Usually not destroyed |
| Insulin secretion | Decreased | Normal or increased |
| Autoantibodies to islet cells | Present | Absent |
| Associated with particular HLA antigens | Yes | No |
| Usual treatment | Insulin injections | Diet; oral stimulators of insulin secretion |

### Type I Diabetes Mellitus

Type I diabetes mellitus results when the beta cells of the islets of Langerhans are destroyed by the effects of a virus or another environmental agent, which may act directly or by provoking the action of the body's own antibodies (chapter 19) that attack their beta cells. Removal of the insulin-secreting cells in this way causes hyperglycemia and the appearance of glucose in the urine. Without insulin, glucose cannot enter the adipose cells; the rate of fat synthesis thus lags behind the rate of fat breakdown and large amounts of free fatty acids are released from the adipose cells.

In a person with uncontrolled type I diabetes, many of the fatty acids released from adipose cells are converted into ketone bodies in the liver. This may result in an elevated ketone body concentration in the blood (ketosis), and if the buffer reserve of bicarbonate is neutralized, it may also result in *ketoacidosis.* During this time, the glucose and excess ketone bodies that are excreted in the urine act as osmotic diuretics and cause the excessive excretion of water in the urine. This can produce severe dehydration, which, together with ketoacidosis and associated disturbances in electrolyte balance, may lead to coma and death (fig. 18.11).

In addition to the lack of insulin, people with type I diabetes have an abnormally high secretion of glucagon from the alpha cells of the islets. Glucagon stimulates glycogenolysis in the liver and thus helps to raise the blood glucose concentration. Glucagon also stimulates the production of enzymes in the liver that convert fatty acids into ketone bodies. Some researchers believe that the full symptoms of diabetes result from high glucagon secretion as well as from the absence of insulin. The lack of insulin may be largely responsible for hyperglycemia and for the release of large amounts of fatty acids into the blood. The high glucagon secretion may contribute to the hyperglycemia and be largely responsible for the development of ketoacidosis.

### Insulin Resistance: Obesity and Type II Diabetes

The effects produced by insulin, or any hormone, depend on the concentration of that hormone in the blood and on the sensitivity of the target tissue to given amounts of the hormone. Tissue responsiveness to insulin, for example, varies under normal conditions. For reasons that are incompletely understood, exercise increases insulin sensitivity and obesity decreases insulin sensitivity of the target

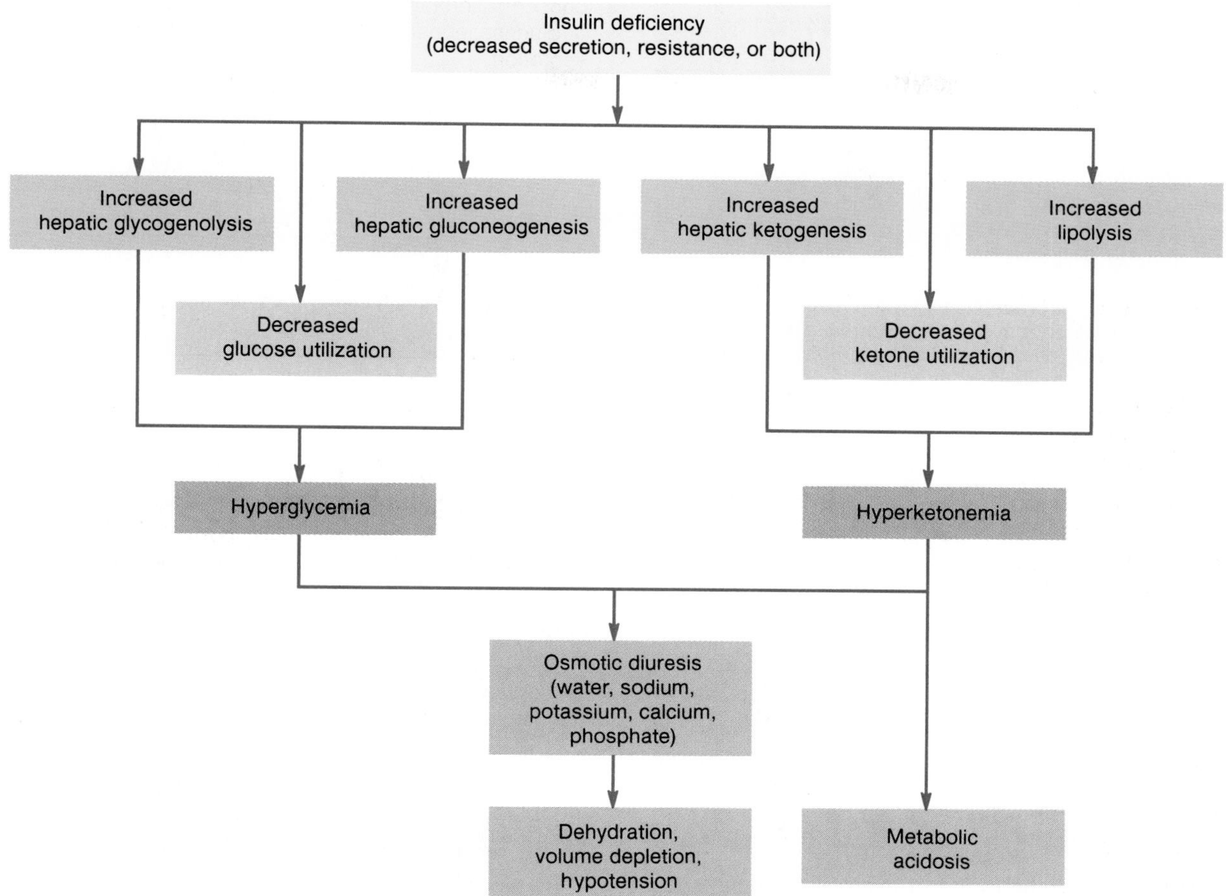


**Figure 18.11.** The sequence of events by which an insulin deficiency may lead to coma and death.

tissues. The islets of a nondiabetic obese person, therefore, must secrete high amounts of insulin to maintain the blood glucose concentration in the normal range. Conversely, nondiabetic people who are thin and exercise regularly require lower insulin secretion to maintain the proper blood glucose concentration.

Type II diabetes is usually slow to develop, is hereditary, and occurs most often in people who are overweight. Unlike type I diabetes mellitus, most people who have type II diabetes have normal or even elevated levels of insulin in their blood. Despite this, people with type II diabetes have hyperglycemia if untreated. This must mean that, even though the insulin levels may be in the normal range, the amount of insulin secreted is inadequate.

Much evidence has been obtained to show that people with type II diabetes have an abnormally low tissue sensitivity to insulin. This is true even if the person is not obese, but the problem is compounded by the decreased tissue sensitivity that accompanies obesity. There is also evidence, however, that the beta cells are not functioning correctly: whatever amount of insulin they secrete is inadequate to the task. Recent evidence also suggests that the beta cells of a person with type II diabetes secretes insulin in an abnormal pattern. Further research is needed to determine if the primary disturbance in type II diabetes is in the beta cells or in the target tissues or both of these.

Since obesity decreases insulin sensitivity, people who are genetically predisposed to insulin resistance may develop symptoms of diabetes when they gain weight. Conversely, this type of diabetes mellitus can usually be controlled by increasing tissue sensitivity to insulin through diet and exercise. If this is not sufficient, oral drugs (generically known as the sulfonylureas) are available that increase insulin secretion and also stimulate tissue responsiveness to insulin.

People with type II diabetes don't usually develop ketoacidosis. The hyperglycemia itself, however, can be dangerous on a long-term basis. Diabetes is the second leading cause of blindness in the United States, and people with diabetes frequently have circulatory problems that increase the tendency to develop gangrene and increase the risk of atherosclerosis. The causes of damage to the retina and lens of the eyes and to blood vessels are not well

**Table 18.10** Comparison of coma due to diabetic ketoacidosis and to hypoglycemia

| | Diabetic Ketoacidosis | Hypoglycemia |
|---|---|---|
| Onset | Hours to days | Minutes |
| Causes | Insufficient insulin; other diseases | Excess insulin; insufficient food; excessive exercise |
| Symptoms | Excessive urination and thirst; headache, nausea, and vomiting | Hunger, headache, confusion, stupor |
| Physical Findings | Deep, labored breathing; breath has acetone odor; blood pressure decreased, pulse weak; skin is dry | Pulse, blood pressure, and respiration are normal; skin is pale and moist |
| Laboratory Findings | Urine: glucose present, ketone bodies increased<br>Plasma: glucose and ketone bodies increased, bicarbonate decreased | Urine: normal<br>Plasma: glucose concentration low, bicarbonate normal |

understood. It is believed, however, that these problems may result from a long-term exposure to high blood glucose, which (1) causes water to leave tissue cells by osmosis, and thus produces dehydration of capillary endothelial cells; and (2) results in glycosylation of tissue proteins.

### Hypoglycemia

People with type I diabetes mellitus are dependent upon insulin injections to prevent hyperglycemia and ketoacidosis. If inadequate insulin is injected, the person may enter a coma as a result of the ketoacidosis, electrolyte imbalance, and dehydration that develop. An overdose of insulin, however, can also produce a coma as a result of the hypoglycemia (abnormally low blood glucose levels) produced. The physical signs and symptoms of diabetic and hypoglycemic coma are sufficiently different (table 18.10) to allow hospital personnel to distinguish between these two types.

Less severe symptoms of hypoglycemia are usually produced by an oversecretion of insulin from the islets of Langerhans after a carbohydrate meal. This **reactive hypoglycemia** is caused by an exaggerated response of the beta cells to a rise in blood glucose and is most commonly seen in adults who are genetically predisposed to type II diabetes. People with reactive hypoglycemia must, therefore, limit their intake of carbohydrates and eat small meals at frequent intervals, rather than two or three meals per day.

The symptoms of reactive hypoglycemia include tremor, hunger, weakness, hypoglycemia, blurred vision, and impaired mental ability. The appearance of some of these symptoms, however, does not necessarily indicate reactive hypoglycemia and a given level of blood glucose does not always produce these symptoms. For these reasons a number of tests must be performed, including the oral glucose tolerance test, to confirm the diagnosis of reactive hypoglycemia. In the glucose tolerance test, reactive hypoglycemia is shown when the initial rise in blood glucose produced by the ingestion of a glucose solution triggers excessive insulin secretion, so that the blood glucose levels fall below normal within five hours (fig. 18.12).

**Figure 18.12.** An idealized oral glucose tolerance test in a person with reactive hypoglycemia. The blood glucose concentration falls below the normal range within five hours of glucose ingestion as a result of excessive insulin secretion.

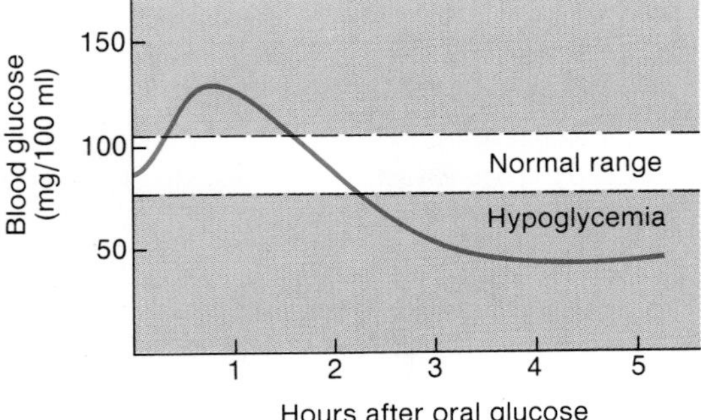


1. *Explain how ketoacidosis and dehydration are produced in a person with type I diabetes mellitus.*
2. *Describe the causes of hyperglycemia in a person with type II diabetes, and explain how weight loss may help to control this condition.*
3. *Explain how reactive hypoglycemia is produced, and describe the dangers of this condition.*

## Metabolic Regulation by Adrenal Hormones, Thyroxine, and Growth Hormone

Epinephrine, the glucocorticoids, thyroxine, and growth hormone stimulate the catabolism of carbohydrates and lipids. These hormones are thus antagonistic to insulin in their regulation of carbohydrate and lipid metabolism. Thyroxine and growth hormone, however, stimulate protein synthesis and are needed for body growth and proper development of the central nervous system. These hormones thus have an anabolic effect, which is complementary to that of insulin, on protein synthesis.

The anabolic effects of insulin are antagonized by glucagon, as previously described, and by the actions of a variety of other hormones. The hormones of the adrenals,

**Figure 18.13.** Cyclic AMP (cAMP) serves as a second messenger in the actions of epinephrine and glucagon on liver and adipose tissue metabolism.

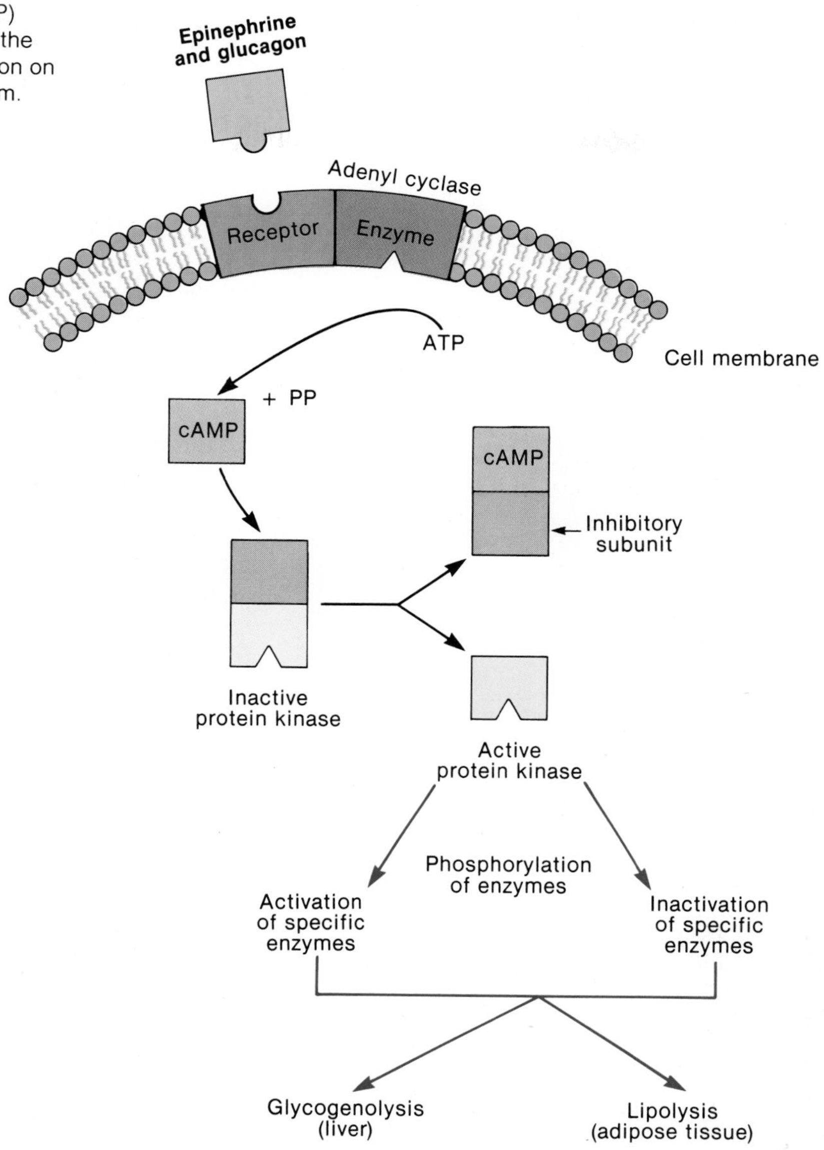

thyroid, and anterior pituitary (specifically growth hormone) antagonize the action of insulin on carbohydrate and lipid metabolism. The actions of insulin, thyroxine, and growth hormone, however, can act synergistically in the stimulation of protein synthesis.

## Adrenal Hormones

The adrenal gland consists of two different parts that have different embryonic origins, that secrete different hormones, and that are regulated by different control systems. The **adrenal medulla** secretes catecholamine hormones—epinephrine and lesser amounts of norepinephrine—in response to sympathetic nerve stimulation. The **adrenal cortex** secretes corticosteroid hormones. These are grouped into two functional categories: **mineralocorticoids,** such as aldosterone, which regulate $Na^+$ and $K^+$ balance, and **glucocorticoids,** such as hydrocortisone (cortisol), which participate in metabolic regulation. The secretion of glucocorticoids is stimulated by adrenocorticotropic hormone (ACTH) from the anterior pituitary.

***Metabolic Effects of Epinephrine.*** The metabolic effects of epinephrine are similar to those of glucagon. Both stimulate glycogenolysis and the release of glucose from the liver and lipolysis and the release of fatty acids from adipose tissue. These actions occur in response to glucagon during fasting, when low blood glucose stimulates glucagon secretion, and in response to epinephrine during the "fight-or-flight" reaction to stress. The latter effect provides circulating energy substrates in anticipation of the need for intense physical activity. Glucagon and epinephrine have similar mechanisms of action; the actions of both are mediated by cyclic AMP (fig. 18.13).

**Figure 18.14.** The catabolic actions of glucocorticoids help raise the blood concentration of glucose and other energy-carrier molecules.

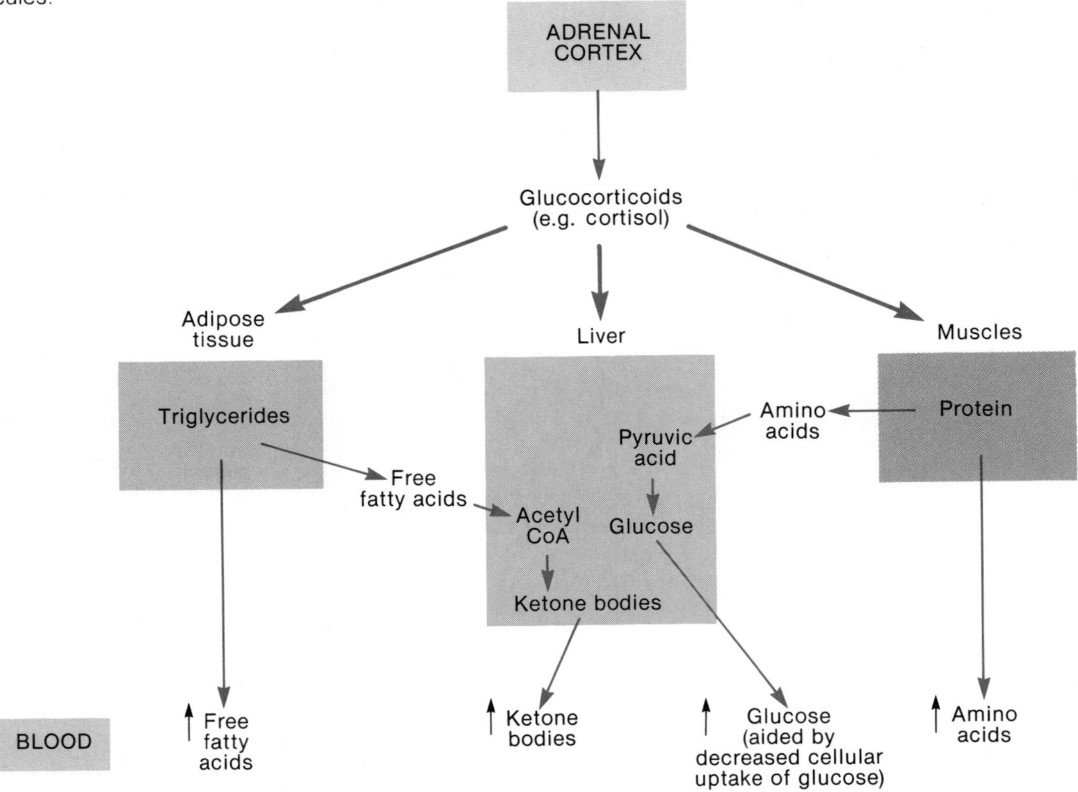


***Metabolic Effects of Glucocorticoids.*** Hydrocortisone and other glucocorticoids are secreted by the adrenal cortex in response to ACTH stimulation. The secretion of ACTH from the anterior pituitary occurs as part of the general adaptation syndrome in response to stress (chapter 11). Since prolonged fasting or prolonged exercise certainly qualify as stressors, ACTH—and thus glucocorticoid secretion—is stimulated under these conditions. The increased secretion of glucocorticoids during prolonged fasting or exercise supports the effects of increased glucagon and decreased insulin secretion from the islets.

Like glucagon, hydrocortisone promotes lipolysis and ketogenesis; it also stimulates the synthesis of hepatic enzymes that promote gluconeogenesis. Although it stimulates enzyme (protein) synthesis in the liver, hydrocortisone promotes protein breakdown in the muscles. This latter effect increases the blood levels of amino acids and thus provides the substrates needed by the liver for gluconeogenesis. The release of circulating energy substrates—amino acids, glucose, fatty acids, and ketone bodies—into the blood in response to hydrocortisone (fig. 18.14) helps to compensate for a state of prolonged fasting or exercise. Whether these metabolic responses are beneficial in other stressful states is open to question.

## Thyroxine

The thyroid follicles secrete thyroxine, also called tetraiodothyronine ($T_4$), in response to stimulation by thyroid-stimulating hormone (TSH) from the anterior pituitary. Almost all organs in the body are targets of thyroxine action. Thyroxine itself, however, is not the active form of the hormone within the target cells; thyroxine is a prehormone that must first be converted to triiodothyronine ($T_3$) within the target cells to be active (chapter 11). Acting via its conversion to $T_3$, thyroxine (1) regulates the rate of cell respiration and (2) contributes to proper growth and development, particularly during early childhood.

***Thyroxine and Cell Respiration.*** Thyroxine stimulates the rate of cell respiration in almost all cells in the body. This is believed to be due to thyroxine-induced lowering of cellular ATP concentrations. ATP exerts an end-product inhibition (chapter 4) of cell respiration so that when ATP concentrations increase, the rate of cell respiration decreases. Conversely, a lowering of ATP concentrations, as may occur in response to thyroxine, stimulates cell respiration.

The mechanisms of thyroxine action are not entirely understood. One effect of thyroxine, however—the stimulation of $Na^+/K^+$ pump activity in cell membranes—could account for the thyroxine-induced lowering of cell

**Figure 18.15.** A mechanism proposed to explain the effects of thyroid hormones on basal metabolic rate. Through the activation of genes and the stimulation of protein synthesis, thyroid hormones increase the activity of the $Na^+/K^+$ pump. This active transport carrier accounts for a large percentage of the energy expenditures in the cell. The concentration of ATP, therefore, declines as a result of this increased energy usage, and the decreased ATP concentrations stimulate increased cellular respiration.

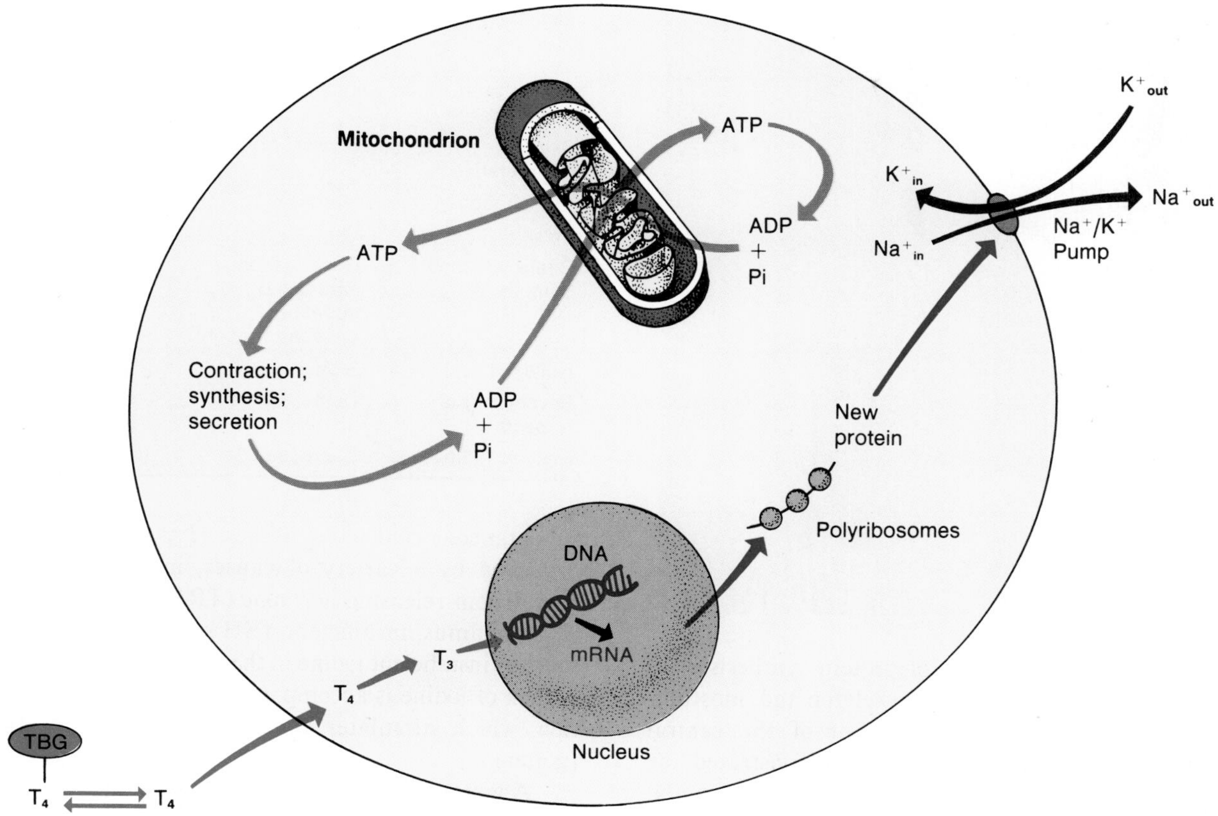


ATP concentrations. The active transport of $Na^+$ and $K^+$ represents a significant energy "sink" in the cell, accounting for about 12% of the calories consumed at rest. Through stimulation of $Na^+/K^+$ pumps, therefore, thyroxine could significantly decrease ATP concentrations and thus stimulate the rate of cell respiration (fig. 18.15).

The metabolic rate under resting and carefully defined conditions is known as the basal metabolic rate (BMR), as previously described. Most commonly measured by the rate of oxygen consumption, the BMR indicates the "idling speed" of the body. Since the activity of the $Na^+/K^+$ pumps contributes significantly to the energy consumed in the basal state and since the activity of these pumps is set by thyroxine secretion, the BMR can be used as an index of thyroid function. Indeed, such measurements were used clinically to evaluate thyroid function prior to the development of direct chemical determinations of $T_4$ and $T_3$ in the blood.

The coupling of energy-releasing reactions to energy-requiring reactions is never 100% efficient; a proportion of the energy is always lost as heat. Much of the energy liberated during cell respiration and much of the energy released by the hydrolysis of ATP escapes as heat. Since thyroxine stimulates both ATP consumption and cell respiration, the actions of thyroxine result in the production of metabolic heat.

The heat-producing, or *calorigenic* (*calor* = heat), *effects* of thyroxine are required for cold adaptation. This does not mean that people who are cold-adapted have high levels of thyroxine secretion. Rather, thyroxine levels in the normal range coupled with the increased activity of the sympathoadrenal system and other responses previously discussed are responsible for cold adaptation. Thyroxine exerts a permissive effect on the ability of the sympathoadrenal system to increase heat production in response to cold stress.

***Thyroxine in Growth and Development.*** Through its stimulation of cell respiration, thyroxine stimulates the increased consumption of circulating energy substrates such as glucose, fatty acids, and other molecules. These effects, however, are mediated at least in part by the activation of genes; thyroxine thus stimulates both RNA and protein synthesis. As a result of its stimulation of protein synthesis throughout the body, thyroxine is considered to be an anabolic hormone like insulin and growth hormone.

**Figure 18.16.** Cretinism is a disease of infancy caused by an underactive thyroid gland.

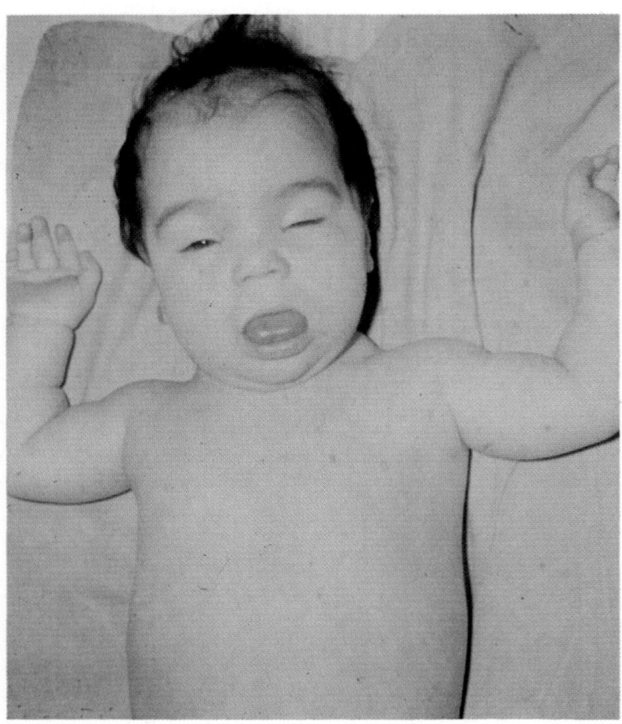

Because of its stimulation of protein synthesis, thyroxine is needed for growth of the skeleton and, most importantly, for the proper development of the central nervous system. Recent evidence has demonstrated the presence of receptor proteins for $T_3$ in the neurons and astrocytes of the brain. This need for thyroxine is particularly great during the time the brain is undergoing its greatest rate of development, from the end of the first trimester of prenatal life to six months after birth. Hypothyroidism during this time may result in **cretinism** (fig. 18.16). Unlike dwarfs, who have normal thyroxine secretion but a low secretion of growth hormone, cretins suffer from severe mental retardation. Treatment with thyroxine soon after birth, particularly before one month of age, has been found to completely or almost completely restore development of intelligence as measured by I.Q. tests administered five years later.

***Hypothyroidism and Hyperthyroidism.*** As might be predicted from the effects of thyroxine, people who are hypothyroid have an abnormally low basal metabolic rate (BMR) and experience weight gain and lethargy. There is also a decreased ability to adapt to cold stress when there is a thyroxine deficiency. Another symptom of hypothyroidism is **myxedema**—accumulation of mucoproteins in subcutaneous connective tissues. Hypothyroidism can be produced by a variety of causes, including insufficient thyrotropin-releasing hormone (TRH) secretion from the hypothalamus, insufficient TSH secretion from the pituitary, or insufficient iodine in the diet. Hypothyroidism due to lack of iodine is accompanied by excessive TSH secretion, which stimulates abnormal growth of the thyroid (goiter).

**Table 18.11** Comparison of hypothyroidism and hyperthyroidism

| | Hypothyroid | Hyperthyroid |
|---|---|---|
| Growth and Development | Impaired growth | Accelerated growth |
| Activity and Sleep | Decreased activity; increased sleep | Increased activity; decreased sleep |
| Temperature Tolerance | Intolerance to cold | Intolerance to heat |
| Skin Characteristics | Coarse, dry skin | Smooth skin |
| Perspiration | Absent | Excessive |
| Pulse | Slow | Rapid |
| Gastrointestinal Symptoms | Constipation; decreased appetite; increased weight | Frequent bowel movements; increased appetite; decreased weight |
| Reflexes | Slow | Rapid |
| Psychological Aspects | Depression and apathy | Nervous, "emotional" |
| Plasma $T_4$ Levels | Decreased | Increased |

A goiter can also be produced by another mechanism. In **Graves's disease,** autoantibodies are produced that have TSH-like effects on the thyroid. Since the production of these autoantibodies is not controlled by negative feedback, this results in the excessive stimulation of the thyroid. A goiter is thus produced that is associated with a hyperthyroid state. Hyperthyroidism produces a high BMR accompanied by weight loss, nervousness, irritability, and an intolerance to heat. The symptoms of hypothyroidism and hyperthyroidism are compared in table 18.11.

## Growth Hormone

The anterior pituitary secretes growth hormone, also called somatotropic hormone, in larger amounts than any other of its hormones. As its name implies, growth hormone stimulates growth in children and adolescents. The continued high secretion of growth hormone in adults, particularly under the conditions of fasting and other forms of stress, implies that this hormone can have important metabolic effects even after the growing years have ended.

**Figure 18.17.** The effects of growth hormone. The growth-promoting, or anabolic, effects of growth hormone are mediated indirectly via stimulation of somatomedin production by the liver.

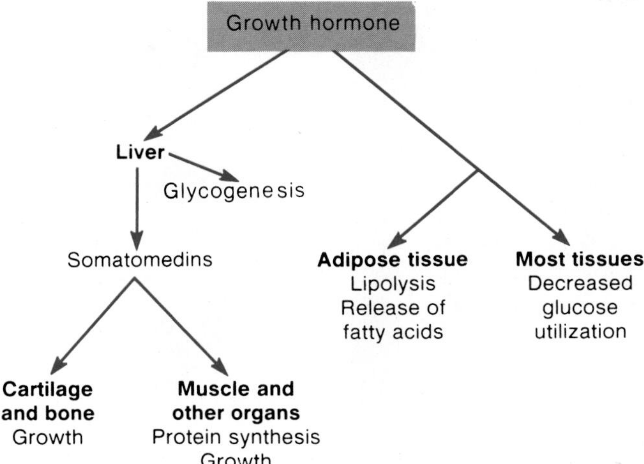


***Regulation of Growth Hormone Secretion.*** The secretion of growth hormone is inhibited by somatostatin, which is produced by the hypothalamus and secreted into the hypothalamo-hypophyseal portal system (chapter 11). In addition, a newly discovered hypothalamic-releasing hormone appears to stimulate growth hormone secretion. Growth hormone thus appears to be unique among the anterior pituitary hormones in that its secretion is controlled by both a releasing and an inhibiting hormone from the hypothalamus. The secretion of growth hormone follows a circadian ("about a day") pattern, increasing during sleep and decreasing during periods of wakefulness.

Growth hormone secretion is stimulated by an increase in the plasma concentrations of amino acids and by a decrease in the plasma glucose concentration. The secretion of growth hormone is, therefore, increased during absorption of a high protein meal, when amino acids are absorbed. The secretion of growth hormone is also increased during prolonged fasting when plasma glucose is low and plasma amino acid concentration is raised by the breakdown of muscle protein.

***Effects of Growth Hormone on Metabolism.*** The fact that growth hormone secretion is increased during fasting and also during absorption of a protein meal reflects the complex nature of this hormone's action. Growth hormone has both anabolic and catabolic effects; it promotes protein synthesis (anabolism) and it also stimulates the catabolism of fat and release of fatty acids from adipose tissue. Growth hormone is similar to insulin in its former effect and similar to glucagon in its latter effect.

In terms of its action on lipid and carbohydrate metabolism, growth hormone is said to have an anti-insulin effect. A rise in the plasma fatty acid concentration induced by growth hormone results in decreased rates of glycolysis in many organs. This inhibition of glycolysis by fatty acids, perhaps together with a more direct action of growth hormone, results in decreased glucose utilization by the tissues. Growth hormone thus acts to raise the blood glucose concentration.

Growth hormone stimulates the cellular uptake of amino acids and protein synthesis in many organs of the body. These actions are useful during a protein-rich meal; amino acids are removed from the blood and used to form proteins, and the plasma concentration of glucose and fatty acids is increased to provide alternate energy sources (fig. 18.17). The anabolic effect of growth hormone on protein synthesis is particularly important during the growing years, when it contributes to increases in bone length and in the mass of many soft tissues.

***Effects of Growth Hormone on Body Growth.*** The stimulatory effects of growth hormone on skeletal growth results from stimulation of mitosis in the epiphyseal discs of cartilage present in the long bones of growing children and adolescents. This effect, however, is believed to be indirect. Growth hormone stimulates the production of polypeptides called **somatomedins**, which in turn stimulate the cartilage cells to divide and secrete more cartilage matrix. Part of this growing cartilage is converted to bone, enabling the bone to grow in length. The liver is believed to be the primary source of the somatomedins found in plasma (fig. 18.18), but there is evidence that somatomedins may also be produced locally within the epiphyseal cartilages in response to growth hormone stimulation. In either case, the growth-promoting effects of growth hormone appear to be mediated by somatomedins. This skeletal growth stops when the epiphyseal discs are converted to bone after the growth spurt during puberty, despite the fact that growth hormone secretion continues throughout adulthood.

An excessive secretion of growth hormone in children can produce **gigantism** in which people may grow up to eight feet tall. An excessive growth hormone secretion that occurs after the epiphyseal discs have sealed, however, cannot produce increases in height. The oversecretion of growth hormone in adults results in an elongation of the jaw and deformities in the bones of the face, hands, and feet. This condition, called **acromegaly,** is accompanied by the growth of soft tissues and coarsening of the skin (fig. 18.19). It is interesting that athletes who (illegally) take growth hormone supplements to increase their muscle mass may also experience body changes similar to acromegaly.

**Figure 18.18.** The hepatic production of polypeptides called *somatomedins* is stimulated by growth hormone. These compounds in turn produce the anabolic effects that are characteristic of the actions of growth hormone in the body.

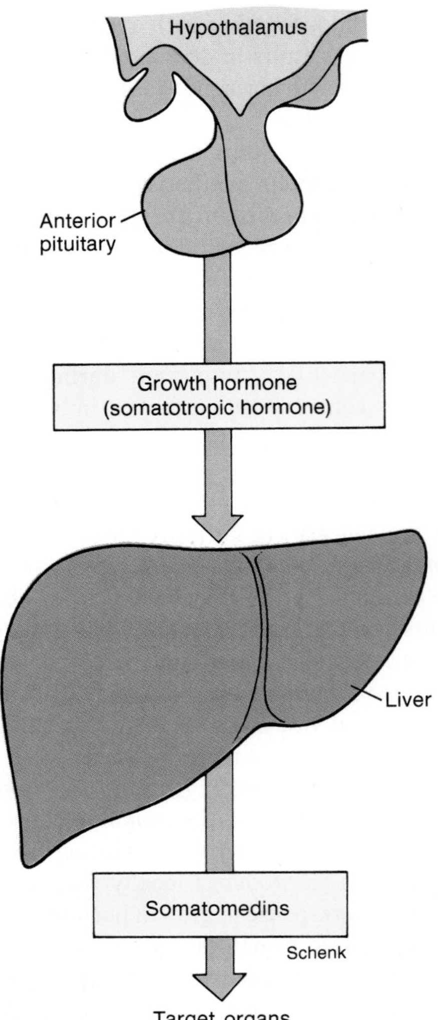


**Figure 18.19.** The progression of acromegaly in one individual, from age nine (*a*), sixteen (*b*), thirty-three (*c*), and fifty-two (*d*) years. The coarsening of features and disfigurement are evident by age thirty-three and severe at age fifty-two.

An inadequate secretion of growth hormone during the growing years results in **dwarfism.** An interesting variant of this is *Laron dwarfism,* in which there is a genetic insensitivity to the effects of growth hormone. This insensitivity is associated with, but may not be caused by, a reduction in the number of growth hormone receptors in the target cells.

It is also believed that the short stature of adult African pygmies may be due to a genetically low sensitivity to the effects of growth hormone, since these people have normal amounts of growth hormone secretion. In a recent study, it was observed that pygmy children grow normally and have blood levels of somatomedins that are similar to those of other children. During puberty, however, the pygmy adolescents had only one-third the normal blood levels of somatomedins and lacked the normal pubertal growth spurt, even though the blood concentrations of sex hormones was within the normal range.

These observations suggest that the primary cause of the pubertal growth spurt in normal children may be increased production of somatomedins in response to growth hormone stimulation. The increased secretion of sex hormones during puberty stimulates growth indirectly, according to this theory, by stimulating increased growth hormone secretion. (The sex hormones eventually also cause the conversion of the epiphyseal discs of cartilage to bone, thus stopping growth.) Further research is required to test this interesting theory and to learn the precise hormonal mechanisms responsible for the rapid growth that occurs during puberty.

**Table 18.12** The endocrine regulation of calcium and phosphate balance

| Hormone | Effect on Intestine | Effect on Kidneys | Effect on Bone | Associated Diseases |
|---|---|---|---|---|
| Parathyroid hormone (PTH) | No direct effect | Stimulates $Ca^{++}$ reabsorption<br>Inhibits $PO_4^{-3}$ reabsorption | Stimulates resorption | Osteitis fibrosa cystica with hypercalcemia due to excess PTH |
| 1,25-dihydroxyvitamin $D_3$ | Stimulates absorption of $Ca^{++}$ and $PO_4^{-3}$ | Stimulates reabsorption of $Ca^{++}$ and $PO_4^{-3}$ | Stimulates resorption | Osteomalacia (adults) and rickets (children) due to deficiency of 1,25-dihydroxyvitamin $D_3$ |
| Calcitonin | None | Inhibits resorption of $Ca^{++}$ and $PO_4^{-3}$ | Stimulates deposition | None |

An adequate diet, particularly of proteins, is required for the production of somatomedins. This helps to explain the common observation that many children are significantly taller than their parents, who may not have had an adequate diet in their youth. Children with protein malnutrition **(kwashiorkor)** have low growth rates and low somatomedin levels, despite the fact that their growth hormone levels are abnormally elevated. When these children eat an adequate diet, somatomedin levels and growth rates increase.

1. *Describe the effects of epinephrine and of the glucocorticoids on the metabolism of carbohydrates and lipids, and explain the significance of these effects as a response to stress.*
2. *Explain the actions of thyroxine on the basal metabolic rate, and explain why people who are hypothyroid have a tendency to gain weight and are less resistant to cold stress.*
3. *Describe the effects of growth hormone on the metabolism of lipids, glucose, and amino acids.*
4. *Explain how growth hormone stimulates skeletal growth.*

## Regulation of Calcium and Phosphate Balance

Parathyroid hormone, 1,25-dihydroxyvitamin $D_3$, and calcitonin help maintain the concentration of blood $Ca^{++}$ in the normal range, which is of critical importance for calcium-dependent physiological processes (such as contraction of skeletal and cardiac muscle). Parathyroid hormone promotes an elevation in blood $Ca^{++}$ by stimulating resorption of the calcium phosphate crystals from bone and renal excretion of phosphate. The primary action of 1,25-dihydroxyvitamin $D_3$ is to promote the intestinal absorption of calcium and phosphate. The secretion of parathyroid hormone, and the production of 1,25-dihydroxyvitamin $D_3$, are thus stimulated by a fall in blood $Ca^{++}$.

The calcium and phosphate concentrations of plasma are affected by bone formation and resorption, intestinal absorption, and urinary excretion of these ions. These processes are regulated by parathyroid hormone, 1,25-dihydroxyvitamin $D_3$, and calcitonin, as summarized in table 18.12.

The skeleton, in addition to providing support for the body, serves as a large store of calcium and phosphate in the form of *hydroxyapatite* crystals, which have the formula $Ca_{10}(PO_4)_6(OH)_2$. The calcium phosphate in these hydroxyapatite crystals is derived from the blood by the action of bone-forming cells, or **osteoblasts.** The osteoblasts secrete an organic matrix, composed largely of collagen protein, which becomes hardened by deposits of hydroxyapatite. This process is called **bone deposition. Bone resorption** (dissolution of hydroxyapatite), produced by the action of **osteoclasts** (fig. 18.20), results in the return of bone calcium and phosphate to the blood.

The formation and resorption of bone occur constantly at rates determined by the hormonal balance. Body growth during the first two decades of life occurs because bone formation proceeds at a faster rate than bone resorption. By age fifty or sixty, the rate of bone resorption often exceeds the rate of bone deposition. The constant activity of osteoblasts and osteoclasts allows bone to be remodeled throughout life. The position of the teeth, for example, can be changed by orthodontic appliances (braces), which cause bone resorption on the pressure-bearing side and bone formation on the opposite side of the alveolar sockets.

**Figure 18.20.** (*a*) The resorption of bone by osteoclasts and (*b*) the formation of new bone by osteoblasts. Both resorption and deposition (formation) occur simultaneously throughout the body.

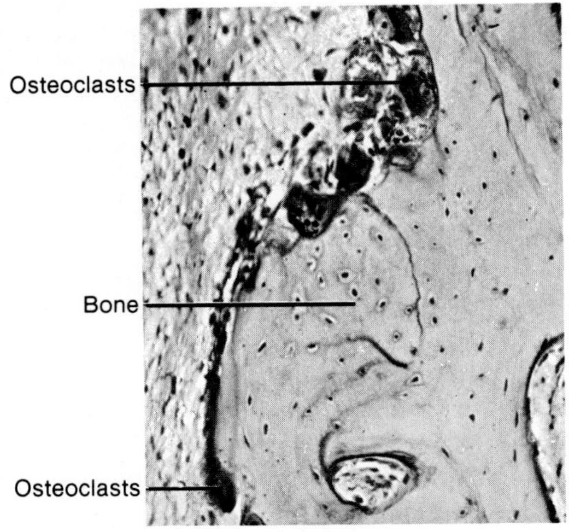


(a)

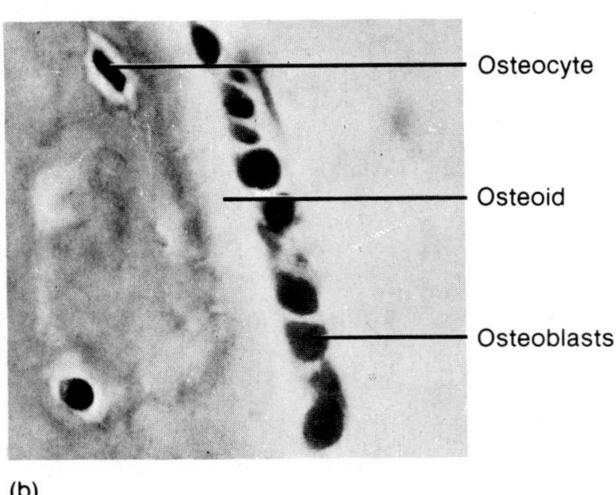


(b)

The plasma concentrations of calcium and phosphate are maintained, despite the changing rates of bone formation and resorption, by hormonal control of the intestinal absorption and urinary excretion of these ions. These hormonal control mechanisms are very effective at maintaining the plasma calcium and phosphate concentrations within narrow limits. Plasma calcium, for example, is normally maintained at about 2.5 millimolar, or 5 milliequivalents per liter (a milliequivalent equals a millimole times the valence of the ion, in this case, times two).

The maintenance of normal plasma calcium concentrations is important because of the wide variety of effects that calcium has in the body. In addition to its role in bone formation, excitation-contraction coupling in muscles, and as a second messenger in the action of some hormones, calcium is needed to maintain proper membrane permeability. An abnormally low plasma calcium concentration increases the permeability of the cell membranes to $Na^+$ and other ions. Hypocalcemia, therefore, enhances the excitability of nerves and muscles and can result in muscle spasm (tetany).

There are a variety of bone disorders associated with abnormal calcium and phosphate balance. In **osteomalacia** (in adults) and **rickets,** (in children) inadequate intake of vitamin D results in inadequate mineralization of the organic matrix of collagen. Excessive secretion of parathyroid hormone results in **osteitis fibrosa cystica**, in which excessive osteoclast activity causes resorption of both the mineral and organic components of bone, which is then replaced by fibrous tissue.

The most common bone disorder is **osteoporosis,** in which decreased bone mass per volume (fig. 18.21) due to inadequate activity of osteoblasts results in fractures in response to ordinary bone stress. Although the causes of osteoporosis are not well understood, age-related bone loss occurs more rapidly in women than men (osteoporosis is almost ten times more common in women after menopause than in men at comparable ages), suggesting that the fall in estrogen secretion at menopause contributes to this condition. Premenopausal women who have a very low percent body fat and amenorrhea can also have osteoporosis. It is interesting in this regard that estrogen receptors have recently been identified in osteoblasts, suggesting that estrogen may exert an unidentified effect on bone production. Women prior to menopause should have a daily calcium intake of at least 800 mg, and after menopause should increase this to 1500 mg per day in order to minimize the progression of osteoporosis. Many women also benefit from supplementary estrogen treatments after menopause.

## Parathyroid Hormone

Whenever the plasma concentration of $Ca^{++}$ begins to fall, the parathyroid glands are stimulated to secrete increased amounts of *parathyroid hormone (PTH)*, which acts to raise the blood $Ca^{++}$ back to normal levels. As might be predicted from this action of PTH, people who have their parathyroid glands removed (as may occur accidentally during surgical removal of the thyroid) will experience hypocalcemia. This can cause severe muscle tetany, for reasons previously discussed, and serves as a dramatic reminder of the importance of PTH.

**Figure 18.21.** Scanning electron micrographs of bone biopsy specimens from the iliac crest: (*a*) is normal and (*b*) is from a person with osteoporosis.

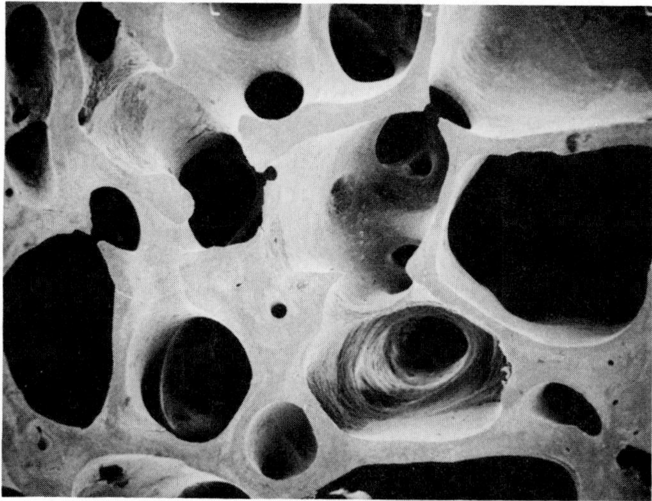

(a)

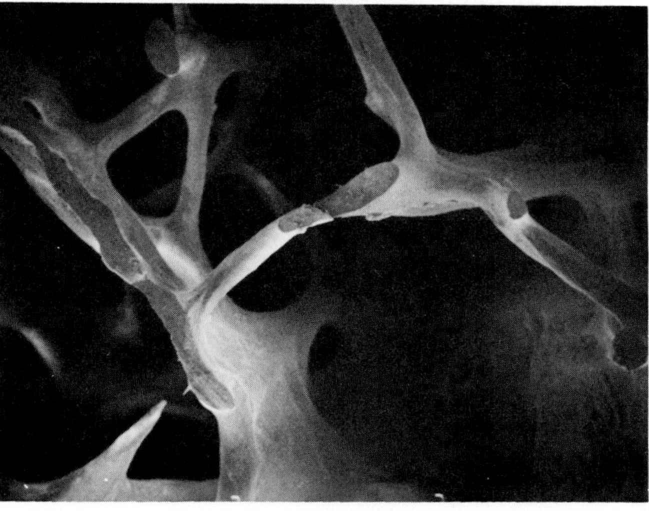

(b)

Parathyroid hormone helps to raise the blood $Ca^{++}$ concentration primarily by stimulating the activity of osteoclasts to resorb bone. In addition, PTH stimulates the kidneys to reabsorb $Ca^{++}$ from the glomerular filtrate while inhibiting the reabsorption of $PO_4^{-3}$. This raises blood $Ca^{++}$ levels without promoting the deposition of calcium phosphate crystals in bone. Finally, PTH promotes the formation of 1,25-dihydroxyvitamin $D_3$ (as described in the next section), and so it also helps to raise the blood calcium levels indirectly through the effects of this other hormone.

## 1,25-Dihydroxyvitamin $D_3$

The production of **1,25-dihydroxyvitamin $D_3$** begins in the skin, where vitamin $D_3$ is produced from its precursor molecule (7-dehydrocholesterol) under the influence of sunlight. When the skin does not make sufficient vitamin $D_3$ because of insufficient exposure to sunlight, this compound must be ingested in the diet—that is why it is called a vitamin. Whether this compound is secreted into the blood from the skin or enters the blood after being absorbed from the intestine, vitamin $D_3$ functions as a *prehormone,* which must be chemically changed in order to be biologically active (chapter 11).

An enzyme in the liver adds a hydroxyl group (OH) to carbon 25, which converts vitamin $D_3$ into 25-hydroxyvitamin $D_3$. In order to be active, however, another hydroxyl group must be added to the first carbon. Hydroxylation of the first carbon is accomplished by an enzyme in the kidneys, which converts the molecule to 1,25-dihydroxyvitamin $D_3$ (fig. 18.22). The activity of this enzyme in the kidneys is stimulated by parathyroid hormone (fig. 18.23). Increased secretion of PTH, stimulated by low blood $Ca^{++}$, is thus accompanied by the increased production of 1,25-dihydroxyvitamin $D_3$.

The hormone 1,25-dihydroxyvitamin $D_3$ helps to raise the plasma concentrations of calcium and phosphate by stimulating (1) the intestinal absorption of calcium and phosphate, (2) the resorption of bones, and (3) the renal reabsorption of calcium and phosphate so that less is excreted in the urine. Notice that 1,25-dihydroxyvitamin $D_3$, but not parathyroid hormone, directly stimulates intestinal absorption of calcium and phosphate and promotes the reabsorption of phosphate in the kidneys. The effect of simultaneously raising the blood concentrations of $Ca^{++}$ and $PO_4^{-3}$ results in the increased tendency of these two ions to precipitate as hydroxyapatite crystals in bone.

Since 1,25-dihydroxyvitamin $D_3$ directly stimulates bone resorption, it seems paradoxical that this hormone is needed for proper bone deposition and, in fact, that inadequate amounts of 1,25-dihydroxyvitamin $D_3$ result in the bone demineralization of osteomalacia and rickets. This apparent paradox may be explained logically by the fact that *the primary function of 1,25-dihydroxyvitamin $D_3$ is stimulation of intestinal $Ca^{++}$ and $PO_4^{-3}$ absorption.* When calcium intake is adequate, the major result of 1,25-dihydroxyvitamin $D_3$ action is the availability of $Ca^{++}$ and $PO_4^{-3}$ in sufficient amounts to promote bone deposition. Only when calcium intake is inadequate does the direct effect of 1,25-dihydroxyvitamin $D_3$ on bone resorption become significant, acting to assure proper blood $Ca^{++}$ levels.

**Figure 18.22.** The pathway for the production of the hormone 1,25-dihydroxyvitamin $D_3$. This hormone is produced in the kidneys from the inactive precursor, 25-hydroxyvitamin $D_3$ (formed in the liver). This latter molecule is produced from vitamin $D_3$ secreted by the skin.

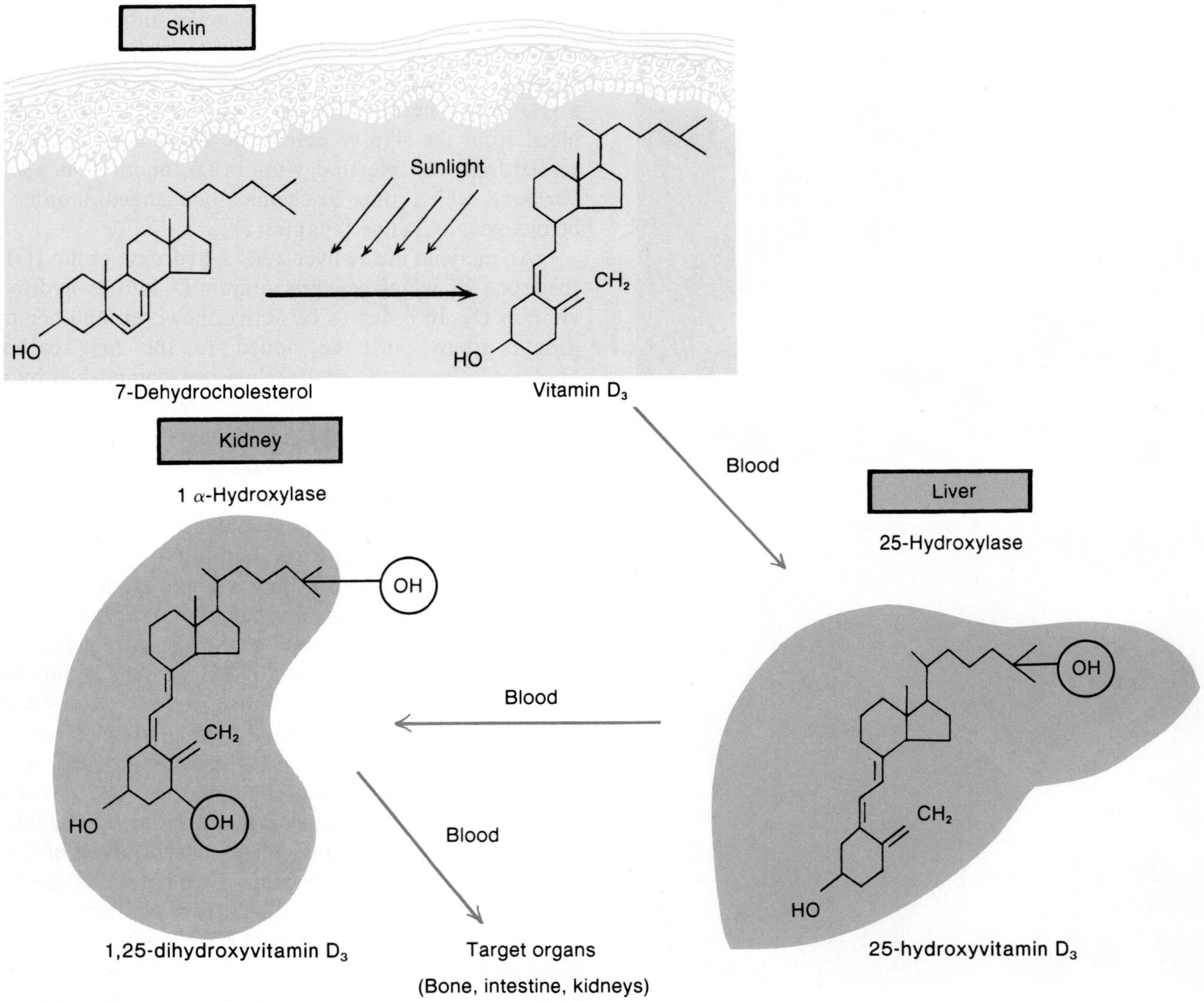

Skin
Sunlight
HO
7-Dehydrocholesterol
$CH_2$
HO
Vitamin $D_3$
Blood
Kidney
1 α-Hydroxylase
OH
$CH_2$
HO
OH
1,25-dihydroxyvitamin $D_3$
Blood
Blood
Target organs
(Bone, intestine, kidneys)
Liver
25-Hydroxylase
OH
$CH_2$
HO
25-hydroxyvitamin $D_3$

**Figure 18.23.** A decrease in plasma $Ca^{++}$ directly stimulates the secretion of parathyroid hormone (*PTH*). The production of 1,25-dihydroxyvitamin $D_3$ also rises when $Ca^{++}$ is low because PTH stimulates the final hydroxylation step in the formation of this compound in the kidneys.

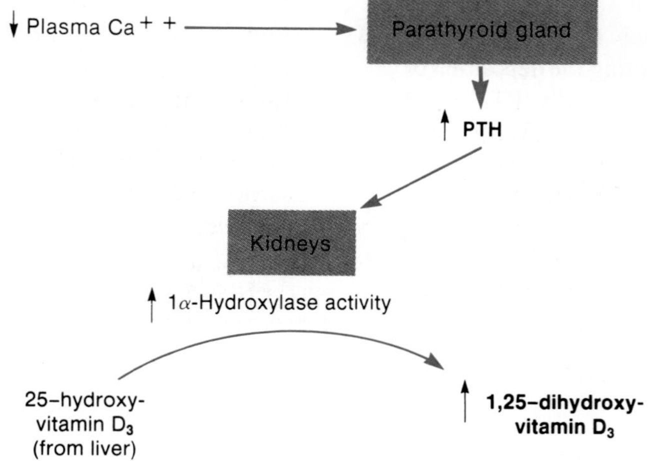

**Figure 18.24.** The negative feedback loop that returns low blood $Ca^{++}$ concentrations to normal without simultaneously raising blood phosphate levels above normal.

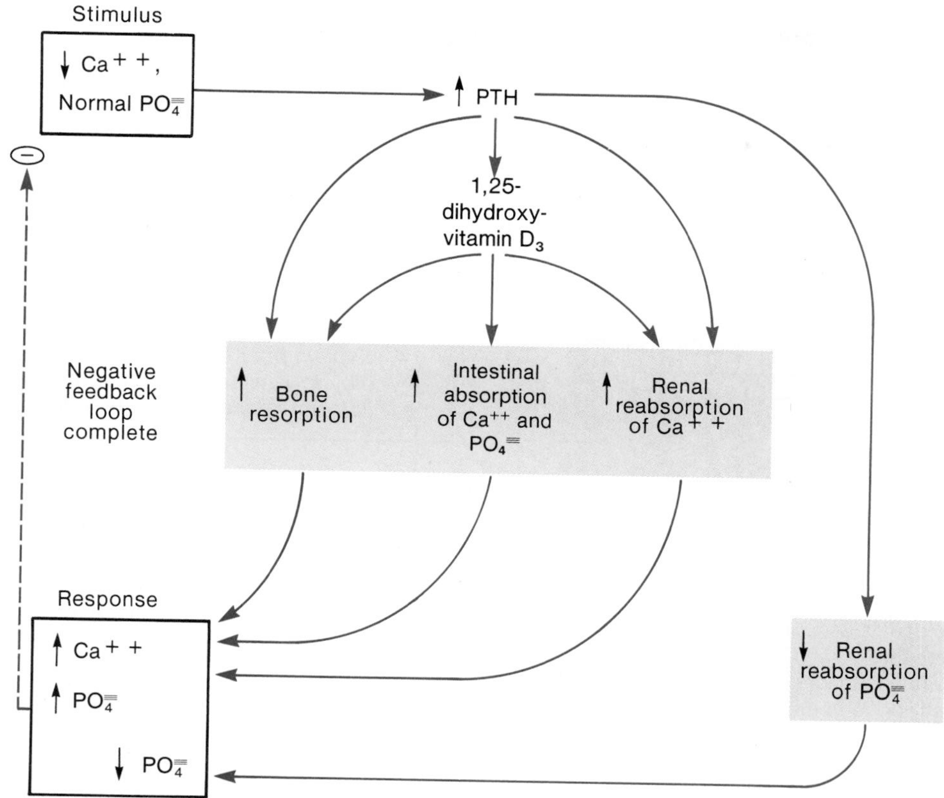


## Negative Feedback Control of Calcium and Phosphate Balance

The secretion of parathyroid hormone is controlled by the plasma calcium concentrations. Its secretion is stimulated by low-calcium and inhibited by high-calcium concentrations. Since parathyroid hormone stimulates the final hydroxylation step in the formation of 1,25-dihydroxyvitamin $D_3$, a rise in parathyroid hormone results in an increase in production of 1,25-dihydroxyvitamin $D_3$. Low blood calcium can thus be corrected by the effects of increased parathyroid hormone and 1,25-dihydroxyvitamin $D_3$ (fig. 18.24).

It is possible that plasma calcium levels might fall while phosphate levels remain normal. In this case, the increased secretion of parathyroid hormone and the production of 1,25-dihydroxyvitamin $D_3$ that result could abnormally raise phosphate levels while acting to restore normal calcium levels. This is prevented by the fact that parathyroid hormone inhibits phosphate reabsorption in the kidneys, so that more phosphate is excreted in the urine (fig. 18.23). In this way blood calcium levels can be raised to normal without excessively raising blood phosphate concentrations.

## Calcitonin

Experiments in the 1960s revealed that high blood calcium in dogs may be lowered by a hormone secreted from the thyroid gland. This hormone thus has an effect opposite to that of parathyroid hormone and 1,25-dihydroxyvitamin $D_3$. The calcium-lowering hormone, called **calcitonin,** was found to be a thirty-two-amino-acid polypeptide secreted by *parafollicular cells,* or *C cells,* in the thyroid, which are distinct from the follicular cells that secreted thyroxine.

The secretion of calcitonin is stimulated by high plasma calcium levels and acts to lower calcium levels by (1) inhibiting the activity of osteoclasts, thus reducing bone resorption and (2) stimulating the urinary excretion of calcium and phosphate by inhibiting their reabsorption in the kidneys (fig. 18.25).

Although it is attractive to think that calcium balance is regulated by the effects of antagonistic hormones, the significance of calcitonin in human physiology remains unclear. Patients with their thyroid gland surgically removed (as for thyroid cancer) are *not* hypercalcemic, as would have been expected if calcitonin

**Figure 18.25.** Negative feedback control of calcitonin secretion.

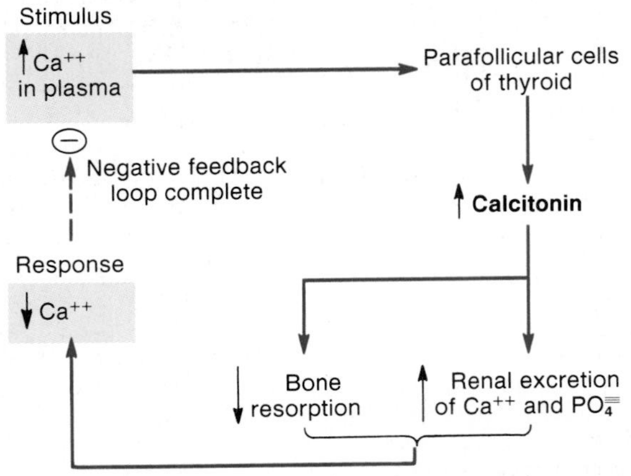


were needed to lower blood calcium levels. The ability of very large, pharmacological doses of calcitonin to inhibit osteoclast activity and bone resorption, however, is clinically useful in the treatment of *Paget's disease,* in which osteoclast activity causes softening of bone.

1. *Describe how the secretion of parathyroid hormone and of calcitonin is regulated.*
2. *List the steps involved in the formation of 1,25-dihydroxyvitamin $D_3$, and indicate how this formation is influenced by parathyroid hormone.*
3. *Describe the actions of parathyroid hormone, 1,25-dihydroxyvitamin $D_3$, and calcitonin on the intestine, skeletal system, and kidneys, and explain how these actions affect the blood levels of calcium.*
4. *Explain how the effects of 1,25-dihydroxyvitamin $D_3$ on bones differ when calcium intake is adequate or inadequate.*

## Summary

### Nutritional Requirements p. 564

I. Food provides molecules used in cell respiration for energy.
   A. The metabolic rate is influenced by physical activity, temperature and eating; the basal metabolic rate is measured as the rate of oxygen consumption when such influences are standardized and minimal.
   B. The energy consumed by the body when measures of metabolic rate are taken, and the energy provided in food, is measured in kilocalories.
   C. When the caloric intake is greater than the energy expenditure over a period of time, the excess calories are stored primarily as fat.

II. Vitamins and elements serve primarily as cofactors and coenzymes.
   A. Vitamins are divided into those that are fat-soluble (A, D, E, and K) and those that are water-soluble.
   B. Many water-soluble vitamins are needed for the activity of the enzymes involved in cell respiration.

### Regulation of Energy Metabolism p. 569

I. The body tissues can use circulating energy substrates, including glucose, fatty acids, ketone bodies, lactic acid, amino acid, and others, for cell respiration.
   A. Different organs have different preferred energy sources.
   B. Circulating energy substrates can be obtained from food or from the energy reserves of glycogen, fat, and protein in the body.

II. Eating behavior is regulated, at least in part, by the hypothalamus.
   A. Lesions of the ventromedial area of the hypothalamus produce hyperphagia, whereas lesions of the lateral hypothalamus produce hypophagia.
   B. A variety of neurotransmitters have been implicated in the control of eating behavior; these include the endorphins, norepinephrine, serotonin, and cholecystokinin.

III. The control of energy balance in the body is regulated by the anabolic and catabolic effects of a variety of hormones.

### Energy Regulation by the Islets of Langerhans p. 572

I. A rise in plasma glucose concentration stimulates insulin and inhibits glucagon secretion.
   A. Amino acids stimulate the secretion of both insulin and glucagon.
   B. Insulin secretion is also stimulated by parasympathetic innervation of the islets and by the action of gastric inhibitory peptide (GIP), secreted by the intestine.

II. During the intestinal absorption of a meal, insulin promotes the uptake of blood glucose into tissue cells.
   A. This lowers the blood glucose concentration and increases the energy reserves of glycogen, fat, and protein.
   B. Insulin is required for the production of fat by adipose cells.

III. During periods of fasting, insulin secretion decreases and glucagon secretion increases.
   A. Glucagon stimulates glycogenolysis in the liver, gluconeogenesis, lipolysis, and ketogenesis.
   B. These effects help maintain adequate levels of blood glucose for the brain and provide alternate energy sources for other organs.

## Diabetes Mellitus and Hypoglycemia p. 578

I. Diabetes mellitus and reactive hypoglycemia represent disorders of the islets of Langerhans.
   A. Type I, or juvenile-onset, diabetes occurs when the beta cells are destroyed; the resulting lack of insulin and excessive glucagon secretion produce the symptoms of this disease.
   B. Type II, or maturity-onset, diabetes occurs as a result of a relative tissue insensitivity to insulin and inadequate insulin secretion; this condition is aggravated by obesity.
   C. Reactive hypoglycemia occurs when the islets secrete excessive amounts of insulin in response to a rise in blood glucose concentration.

## Metabolic Regulation by Adrenal Hormones, Thyroxine, and Growth Hormone p. 580

I. The adrenal hormones involved in energy regulation include epinephrine from the adrenal medulla and glucocorticoids (mainly hydrocortisone) from the adrenal cortex.
   A. The effects of epinephrine are similar to those of glucagon.
   B. Glucocorticoids promote the breakdown of muscle protein and the conversion of amino acids to glucose in the liver.

II. Thyroxine stimulates the rate of cell respiration in almost all cells in the body.
   A. Thyroxine thus sets the basal metabolic rate (BMR), which is the rate at which energy is consumed by the body under resting conditions.
   B. Thyroxine also promotes protein synthesis and by this means is needed for proper body growth and development, particularly of the central nervous system.

III. The secretion of growth hormone is regulated by releasing and inhibiting hormones from the hypothalamus.
   A. The secretion of growth hormone is stimulated by a protein meal and by a fall in glucose, as occurs during fasting.
   B. Growth hormone stimulates catabolism of lipids and inhibits glucose utilization.
   C. Growth hormone also stimulates protein synthesis and thus promotes body growth.
   D. The anabolic effects of growth hormone, including the stimulation of bone growth in childhood, are believed to be produced indirectly via polypeptides called somatomedins.

## Regulation of Calcium and Phosphate Balance p. 587

I. Bone contains calcium and phosphate in the form of hydroxyapatite crystals; this serves as a reserve supply of calcium and phosphate for the blood.
   A. The formation and resorption of bone are produced by the action of osteoblasts and osteoclasts, respectively.
   B. The plasma concentrations of calcium and phosphate are also affected by absorption from the intestine and by the urinary excretion of these ions.

II. Parathyroid hormone stimulates bone resorption and calcium reabsorption in the kidneys; this hormone thus acts to raise the blood calcium concentration.
   A. The secretion of parathyroid hormone is stimulated by a fall in blood calcium levels.
   B. Parathyroid hormone also inhibits reabsorption of phosphate in the kidneys, so that more phosphate is excreted in the urine.

III. 1,25-dihydroxyvitamin $D_3$ is derived from vitamin D by hydroxylation reactions in the liver and kidneys.
   A. The last hydroxylation step is stimulated by parathyroid hormone.
   B. 1,25-dihydroxyvitamin $D_3$ stimulates the intestinal absorption of calcium and phosphate, resorption of bone, and renal reabsorption of phosphate.

IV. A rise in parathyroid hormone, accompanied by the increased production of 1,25-dihydroxyvitamin $D_3$, helps to maintain proper blood levels of calcium and phosphate in response to a fall in calcium levels.

V. Calcitonin is secreted by the parafollicular cells of the thyroid gland.
   A. Calcitonin secretion is stimulated by a rise in blood calcium levels.
   B. Calcitonin, at least at pharmacological levels, acts to lower blood calcium by inhibiting bone resorption and stimulating the urinary excretion of calcium and phosphate.

## Review Activities

### Objective Questions

Match the following:

1. Absorption of carbohydrate meal
2. Absorption of protein meal
3. Fasting

(a) rise in insulin, rise in glucagon
(b) fall in insulin, rise in glucagon
(c) rise in insulin, fall in glucagon
(d) fall in insulin, fall in glucagon

Match the following:

4. Growth hormone
5. Thyroxine
6. Hydrocortisone

(a) increased protein synthesis, increased cell respiration
(b) protein catabolism in muscles; gluconeogenesis in liver
(c) protein synthesis in muscles, decreased glucose utilization
(d) fall in blood glucose, increased fat synthesis

7. A lowering of blood glucose concentration promotes
   (a) decreased lipogenesis
   (b) increased lipolysis
   (c) increased glycogenolysis
   (d) all of the above
8. Glucose can be secreted into the blood by the
   (a) liver
   (b) muscles
   (c) liver and muscles
   (d) liver, muscles, and brain
9. The basal metabolic rate is determined primarily by
   (a) hydrocortisone
   (b) insulin
   (c) growth hormone
   (d) thyroxine
10. Somatomedins are required for the anabolic effects of
    (a) hydrocortisone
    (b) insulin
    (c) growth hormone
    (d) thyroxine
11. The increased intestinal absorption of calcium is stimulated directly by
    (a) parathyroid hormone
    (b) 1,25-dihydroxyvitamin $D_3$
    (c) calcitonin
    (d) all of the above
12. A rise in blood calcium levels directly stimulates
    (a) parathyroid hormone secretion
    (b) calcitonin secretion
    (c) 1,25-dihydroxyvitamin $D_3$ formation
    (d) all of the above
13. At rest, about 12% of the total calories consumed are used for
    (a) protein synthesis
    (b) cell transport
    (c) the $Na^+/K^+$ pumps
    (d) DNA replication
14. Which of the following hormones stimulates anabolism of proteins and catabolism of fat?
    (a) growth hormone
    (b) thyroxine
    (c) insulin
    (d) glucagon
    (e) epinephrine
15. If a person eats 600 kilocalories of protein in a meal, all of the following statements will be true *except*
    (a) insulin secretion will be increased
    (b) the metabolic rate will be increased over basal conditions
    (c) the tissue cells will use some of the amino acids for resynthesis of body proteins
    (d) the tissue cells will obtain 600 kilocalories worth of energy
    (e) body heat production and oxygen consumption will be increased over basal conditions
16. Ketoacidosis in untreated diabetes mellitus is due to
    (a) excessive fluid loss
    (b) hypoventilation
    (c) excessive eating and obesity
    (d) excessive fat catabolism

### Essay Questions

1. Compare the metabolic effects of fasting to the state of uncontrolled type I diabetes. Explain the hormonal similarities of these conditions.
2. Glucocorticoids stimulate the breakdown of protein in muscles but the synthesis of protein in the liver. Explain the significance of these differences.
3. Describe how thyroxine affects cell respiration, and explain why a hypothyroid person has a tendency to gain weight and has a reduced tolerance to cold.
4. Compare and contrast the metabolic effects of thyroxine and growth hormone.
5. Why is vitamin D considered to be both a vitamin and a prehormone? Explain why people with osteoporosis might be helped by taking controlled amounts of vitamin D.

## Selected Readings


Austin, L. A., and H. Heath, III. 1981. Calcitonin: Physiology and pathophysiology. *New England Journal of Medicine* 304:269.

Cahill, G. F., and H. O. McDevitt. 1981. Insulin-dependent diabetes mellitus: The initial lesion. *New England Journal of Medicine* 304:454.

Cheng, K., and J. Larner. 1985. Intracellular mediators of insulin action. *Annual Review of Physiology* 47:405.

DeLuca, H. F. 1980. The vitamin D hormonal system: Implications for bone disease. *Hospital Practice* 15:57.

Dussault, H., and J. Ruel. 1987. Thyroid hormones and brain development. *Annual Review of Physiology* 49:321.

Eisenbarth, G. S. 1986. Type I diabetes mellitus: A chronic autoimmune disease. *New England Journal of Medicine* 314:1360.

Gardner, L. I. July 1972. Deprivation dwarfism. *Scientific American.*

Goodman, D. S. 1984. Vitamin A and retinoids in health and disease. *New England Journal of Medicine* 310:1023.

Habener, J. F., and J. E. Mahaffey. 1978. Osteomalacia and disorders of vitamin D metabolism. *Annual Review of Medicine* 29:327.

Hahn, T. J. 1986. Physiology of bone: Mechanisms of osteopenic disorders. *Hospital Practice* 21:73.

Haussler, M. R., and T. A. McCain. 1977. Basic and clinical concepts related to vitamin D metabolism and action. *New England Journal of Medicine* 297: first part, p. 974; second part, p. 1041.

Hirsch, J. 1984. Hypothalamic control of appetite. *Hospital Practice* 19:131.

Isaksson, O. G. P., S. Edn, and J.-O. Jansson. 1985. Mode of action of pituitary growth hormone on target cells. *Annual Review of Physiology* 47:483.

Kitabchi, A. E., and R. C. Goodman. 1987. Hypoglycemia: Pathophysiology and diagnosis. *Hospital Practice* 22:45.

Levine, M. 1986. New concepts in the biology and biochemistry of ascorbic acid. *New England Journal of Medicine* 314:892.

Marcus, R. 1989. Understanding and preventing osteoporosis. *Hospital Practice* 24:189.

Nadel, E. R., and S. R. Bussolari. July–August 1988. The Daedalus project: Physiological problems and solutions. *American Scientist*, p. 351.

Notkins, A. L. November 1979. The cause of diabetes. *Scientific American.*

Oppenheimer, J. H. 1979. Thyroid hormone action at the cellular level. *Science* 203:971.

Orci, L., J. D. Vassalli, and A. Perrelet. September 1988. The insulin factory. *Scientific American.*

Phillips, L. S., and R. Vassilopoulou-Sellin. 1980. Somatomedins. *New England Journal of Medicine* 302: first part, p. 371; second part, p. 438.

Raisz, L. G., and B. E. Kream. 1981. Hormonal control of skeletal growth. *Annual Review of Physiology* 43:225.

Raisz, L. G. 1988. Local and systemic factors in the pathogenesis of osteoporosis. *New England Journal of Medicine* 318:818.

Reichel, H., H. P. Koeffler, and A. W. Norman. 1989. The role of the vitamin D endocrine system in health and disease. *New England Journal of Medicine* 320:980.

Siperstein, M. D. 1985. Type II diabetes: Some problems in diagnosis and treatment. *Hospital Practice* 20:55.

Sterling, S. 1979. Thyroid hormone action at the cellular level. *New England Journal of Medicine* 300: first part, p. 117; second part, p. 173.

Tepperman, J. 1980. *Metabolic and Endocrine Physiology.* 4th ed. Chicago: Year Book Medical Publishers.

Unger, R. H., R. E. Dobbs, and L. Orci. 1979. Insulin, glucagon, and somatostatin secretion in the regulation of metabolism. *Annual Review of Physiology* 40:307.

Unger, R. H., and L. Orci. 1981. Glucagon and the A cell: Physiology and pathophysiology. *New England Journal of Medicine* 304: first part, p. 1518; second part, p. 1575.

Van Wyk, J., and L. E. Underwood. 1978. Growth hormone, somatomedins, and growth failure. *Hospital Practice* 13:57.

# 19

# The Immune System

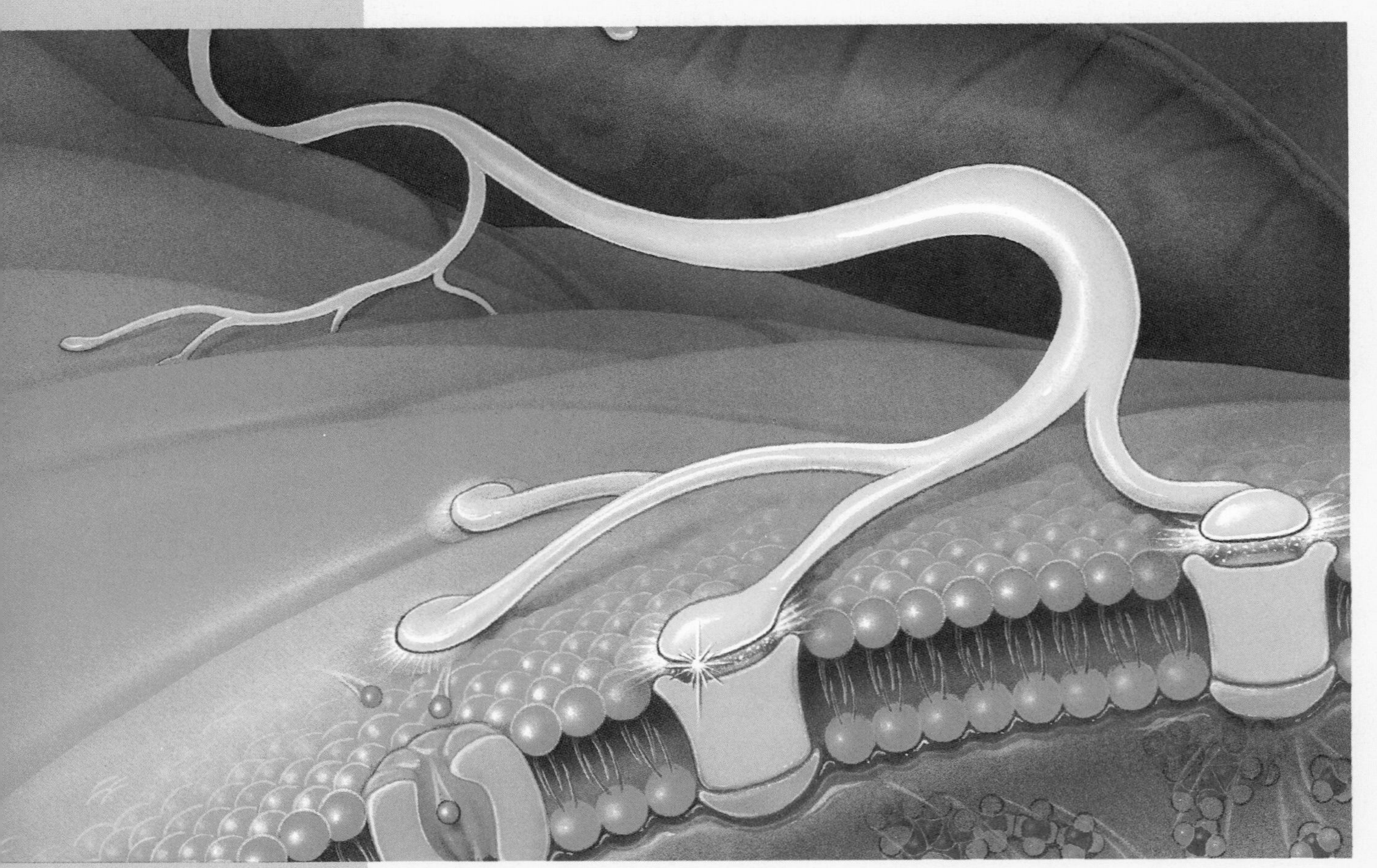

## Objectives

By studying this chapter, you should be able to

1. describe some of the mechanisms of nonspecific immunity and distinguish between nonspecific and specific immune defenses
2. describe how B lymphocytes respond to antigens and define the terms *memory cell* and *plasma cell*
3. describe the structure and classifications of antibodies and the nature of antigens
4. describe the complement system and explain how antigen-antibody reactions lead to the destruction of an invading pathogen
5. describe the events that occur during a local inflammation
6. describe the process of active immunity and explain how the clonal selection theory may account for this process
7. describe the mechanisms of passive immunity and give natural and clinical examples of this form of immunization
8. explain the meaning of the term *monoclonal antibodies* and describe some of their clinical uses
9. explain how T lymphocytes are classified and describe the function of the thymus gland
10. define the term *lymphokines* and list some of these molecules and their functions
11. describe the histocompatibility antigens and explain the importance of these antigens in the function of the T cell receptor proteins
12. describe the interaction between macrophages and helper T lymphocytes and explain how the helper T cells affect immunological defense by cytotoxic T cells and B cells
13. explain the possible role of suppressor T lymphocytes in the negative feedback control of the immune response
14. explain the possible mechanisms responsible for tolerance of self-antigens
15. describe some of the characteristics of cancer and explain how natural killer cells and cytotoxic T lymphocytes provide immunological surveillance against cancer
16. define the term *autoimmune disease,* provide examples of different autoimmune diseases, and explain some of the mechanisms that may be responsible for autoimmune diseases
17. explain how immune complex diseases may be produced and give examples of such diseases
18. distinguish between immediate hypersensitivity and delayed hypersensitivity and describe the mechanisms responsible for each form of allergy

## Outline

**Table 19.1** Structures and defense mechanisms of nonspecific immunity

| | Structure | Mechanisms |
|---|---|---|
| External | Skin | Anatomic barrier to penetration by pathogens; secretions have lysozyme (enzyme that destroys bacteria) |
| | Digestive tract | High acidity of stomach<br>Protection by normal bacterial population of colon |
| | Respiratory tract | Secretion of mucus; movement of mucus by cilia; alveolar macrophages |
| | Genitourinary tract | Acidity of urine<br>Vaginal lactic acid |
| Internal | Phagocytic cells | Ingest and destroy bacteria, cellular debris, denatured proteins, and toxins |
| | Interferons | Inhibit replication of viruses |
| | Complement proteins | Promote destruction of bacteria and other effects of inflammation |
| | Endogenous pyrogen | Secreted by leukocytes and other cells; produces fever |

## *DEFENSE MECHANISMS*

Nonspecific immune protection is provided by such mechanisms as phagocytosis, fever, and the release of interferons. Specific immunity involves the functions of lymphocytes and is directed at specific molecules, or parts of molecules, known as antigens.

The immune system includes all of the structures and processes that provide a defense against potential pathogens. These defenses can be grouped into *nonspecific* and *specific* categories.

Nonspecific defense mechanisms are inherited as part of the structure of each organism. Epithelial membranes that cover the body surfaces, for example, restrict infection by most pathogens. The strong acidity of gastric juice (pH 1–2) also helps to kill many microorganisms before they can invade the body. These external defenses are backed by internal defenses, such as phagocytosis, which function in both a specific and nonspecific manner (table 19.1).

Each individual can acquire the ability to defend against specific pathogens (disease-causing agents) by a prior exposure to those pathogens. This specific immune response is a function of lymphocytes. Internal specific and nonspecific defense mechanisms function together to combat infection, with lymphocytes interacting in a coordinated effort with phagocytic cells.

### Nonspecific Immunity

Invading pathogens, such as bacteria, that have crossed epithelial barriers enter connective tissues. These invaders—or chemicals, called *toxins,* secreted from them—may enter blood or lymphatic capillaries and be carried to other areas of the body. The invasion and spread of infection is fought in two stages: (1) nonspecific immunological defenses are employed; if these are sufficiently effective, the pathogens may be destroyed without progression to the next step; (2) lymphocytes may be recruited, and their specific actions used to reinforce the nonspecific immune defenses.

**Table 19.2** Phagocytic cells and their locations

| Phagocyte | Location |
|---|---|
| Neutrophils | Blood and all tissues |
| Monocytes | Blood and all tissues |
| Tissue macrophages (histiocytes) | All tissues (including spleen, lymph nodes, bone marrow) |
| Kupffer cells | Liver |
| Alveolar macrophages | Lungs |
| Microglia | Central nervous system |

***Phagocytosis.*** There are three major groups of phagocytic cells: (1) **neutrophils**; (2) the cells of the **mononuclear phagocyte system**; this includes *monocytes* in the blood and *macrophages* (derived from monocytes) in the connective tissues; and (3) **organ-specific phagocytes** in the liver, spleen, lymph nodes, lungs, and brain (table 19.2).

The *Kupffer cells* in the liver, together with phagocytic cells in the spleen and lymph nodes, are **fixed phagocytes.** This term refers to the fact that these cells are immobile ("fixed") in the channels within these organs. As blood flows through the liver and spleen and as lymph percolates through the lymph nodes, foreign chemicals and debris are removed by phagocytosis and chemically inactivated within the phagocytic cells. Invading pathogens are very effectively removed in this manner, so that blood is usually sterile after a few passes through the liver and spleen.

Connective tissues contain a resident population of all leukocyte types. Neutrophils and monocytes in particular can be highly mobile within connective tissues as they scavenge for invaders and cellular debris. These leukocytes are recruited to the site of an infection by a process

**Figure 19.1.** Diapedesis. White blood cells squeeze through openings between capillary endothelial cells to enter underlying connective tissues.

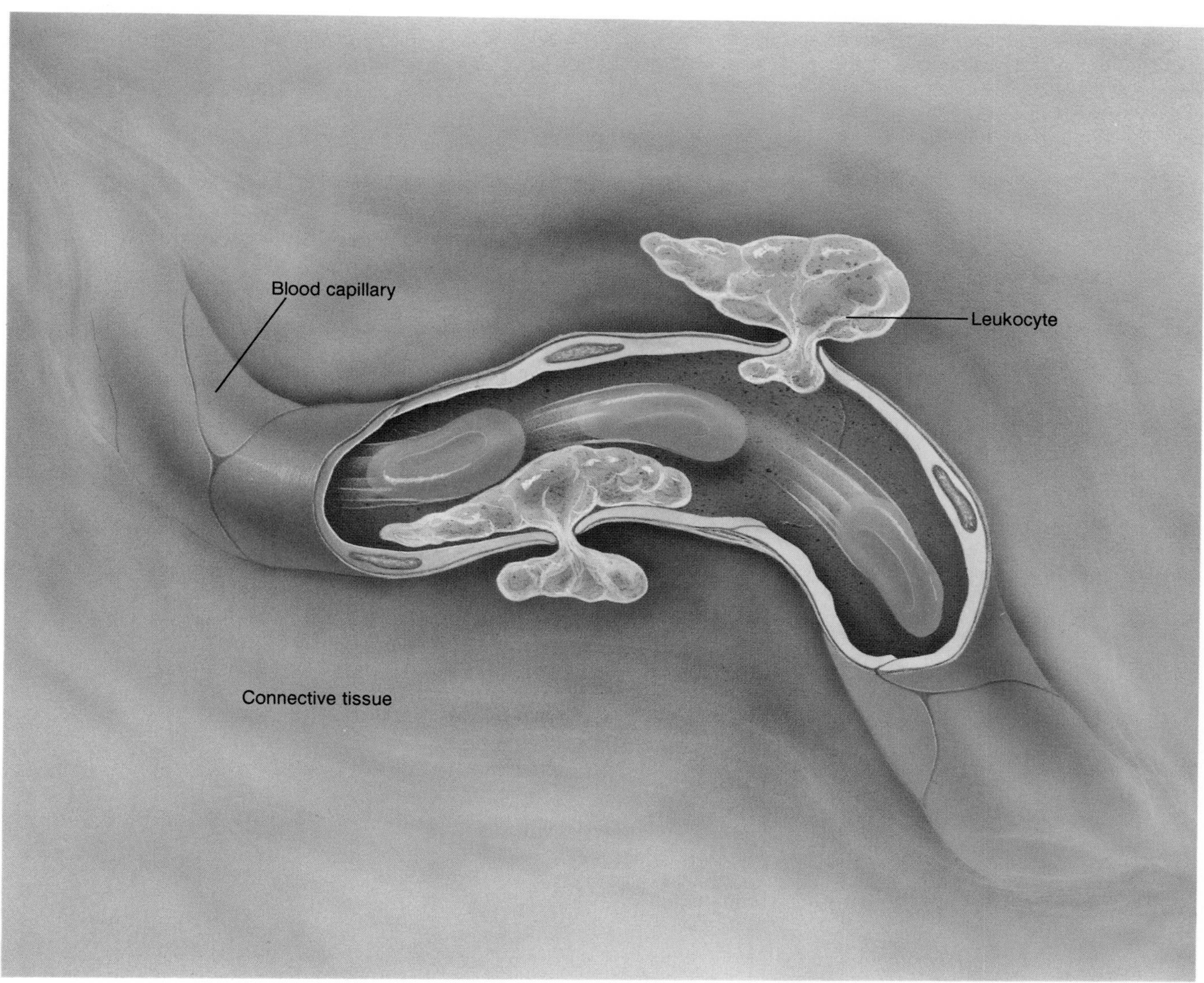


known as **chemotaxis**—movement toward chemical attractants. Neutrophils are the first to arrive at the site of an infection; monocytes arrive later and can be transformed into macrophages as the battle progresses.

If the infection is sufficiently large, new phagocytic cells from the blood may join those already in the connective tissue. These new neutrophils and monocytes are able to squeeze through the tiny gaps between adjacent endothelial cells in the capillary wall and enter the connective tissues. This process, called **diapedesis,** is illustrated in figure 19.1.

Phagocytic cells engulf particles in a manner similar to the way an amoeba eats. The particle becomes surrounded by cytoplasmic extensions called pseudopods, which ultimately fuse together. The particle thus becomes surrounded by a membrane derived from the plasma membrane (fig. 19.2) and contained within an organelle analogous to a food vacuole in an amoeba. This vacuole then fuses with lysosomes (organelles that contain digestive enzymes), so that the ingested particle and the digestive enzymes remain separated from the cytoplasm by a continuous membrane. Often, however, lysosomal enzymes are released before the food vacuole has completely formed. When this occurs, free lysosomal enzymes may be released into the infected area and contribute to inflammation.

**Figure 19.2.** Phagocytosis by a neutrophil or macrophage. A phagocytic cell extends its pseudopods around the object to be engulfed (such as a bacterium). Dots represent lysosomal enzymes (*L* = lysosomes). If the pseudopods fuse to form a complete food vacuole, lysosomal enzymes are restricted to the organelle formed by the lysosome and food vacuole. If the lysosome fuses with the vacuole before fusion of the pseudopods is complete, lysosomal enzymes are released into the infected area of tissue.

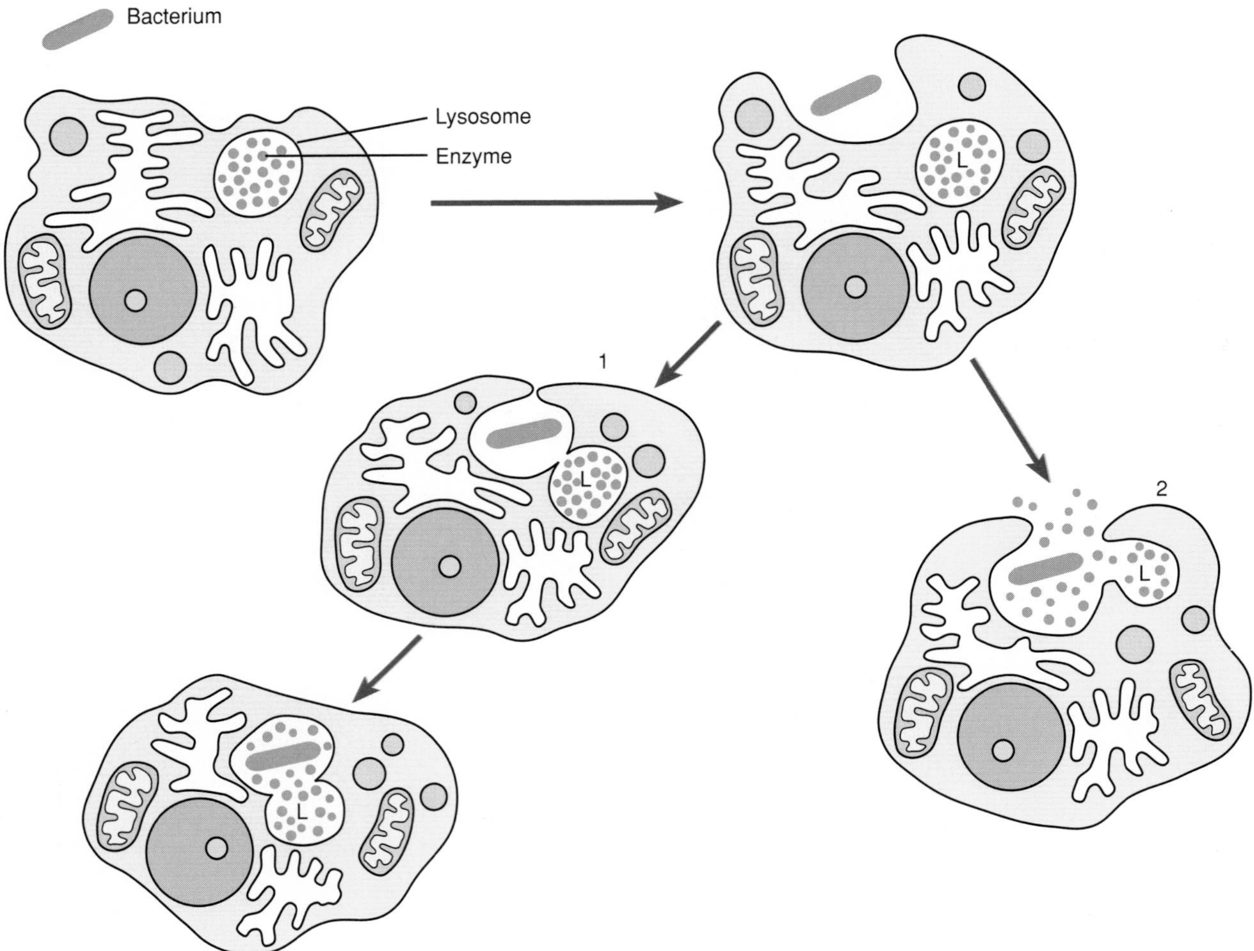


***Fever.*** Fever may be a component of the nonspecific defense system. Body temperature is regulated by the hypothalamus, which contains a thermoregulatory control center (a "thermostat") that coordinates skeletal muscle shivering and the activity of the sympathoadrenal system to maintain body temperature at about 37° C. This thermostat is reset upwards in response to a chemical called **endogenous pyrogen,** secreted by leukocytes. Endogenous pyrogen secretion is stimulated by a chemical called *endotoxin,* which is released by certain bacteria.

Although high fevers are definitely dangerous, many believe that a mild to moderate fever may be a beneficial response that aids recovery from bacterial infections. There is some evidence to support this view, but the mechanisms involved are not clearly understood. One theory is that elevated body temperature may interfere with the uptake of iron by some bacteria.

***Interferons.*** In 1957, researchers demonstrated that cells infected with a virus produced polypeptides that interfered with the ability of a second, unrelated strain of virus to infect other cells in the same culture. These **interferons,** as they were called, thus produced a nonspecific, short-acting resistance to viral infection. This discovery produced a great deal of excitement, but further research in this area was hindered by the fact that human interferons could only be obtained in very small quantities and that animal interferons had little effect in humans. In 1980, however, technological breakthroughs

**Table 19.3** Some of the proposed effects of interferons

| Stimulation | Inhibition |
|---|---|
| Macrophage phagocytosis | Cell division |
| Activity of cytotoxic ("killer") T cells | Tumor growth |
| Activity of natural killer cells | Maturation of adipose cells |
| Production of antibodies | Maturation of erythrocytes |

allowed researchers to introduce human interferon genes into bacteria—through a technique called *genetic recombination*—so that bacteria could act as interferon factories.

Leukocytes, fibroblasts, and probably many other cells make their own characteristic types of interferons. These polypeptides act as messengers that protect other cells in the vicinity from viral infection. The viruses are still able to penetrate these other cells, but the ability of the viruses to replicate and assemble new virus particles (fig. 19.3) is inhibited.

Lymphocytes called *T lymphocytes* (described in a later section) release interferons in response to viral infections and perhaps as part of their immunological surveillance against cancer. Interferons may destroy cancer cells directly and indirectly by the activation of T lymphocytes and *natural killer cells* (cells related to T lymphocytes). Some of the proposed effects of interferons are summarized in table 19.3.

Recent clinical trials have suggested that interferons may be effective in the treatment of such viral infections as viral hepatitis and herpes-induced "cold sores." Current evidence suggests that interferons may not prove to be a "magic bullet" against cancer but may, when combined with other methods, significantly augment the treatment of some forms of cancer. Certain cancers, particularly one known as hairy-cell leukemia, has been shown to respond very well to treatment with interferons produced by genetic engineering techniques. The emphasis is on treatment rather than prevention with interferons because the protective effects of interferons are of short duration.

## Specific Immunity

In 1890, a scientist named von Behring demonstrated that a guinea pig which had been previously injected with a sublethal dose of diphtheria toxin could survive subsequent injections of otherwise lethal doses of that toxin. Further, von Behring showed that this immunity could be transferred to a second, nonexposed animal by injections of serum from the immunized guinea pig. He concluded that the immunized animal had chemicals in its serum—which he called **antibodies**—that were responsible for the immunity. He also showed that these antibodies conferred immunity only to subsequent diphtheria infections; the

**Figure 19.3.** The sequence of events that can occur when human cells are infected with virus particles.

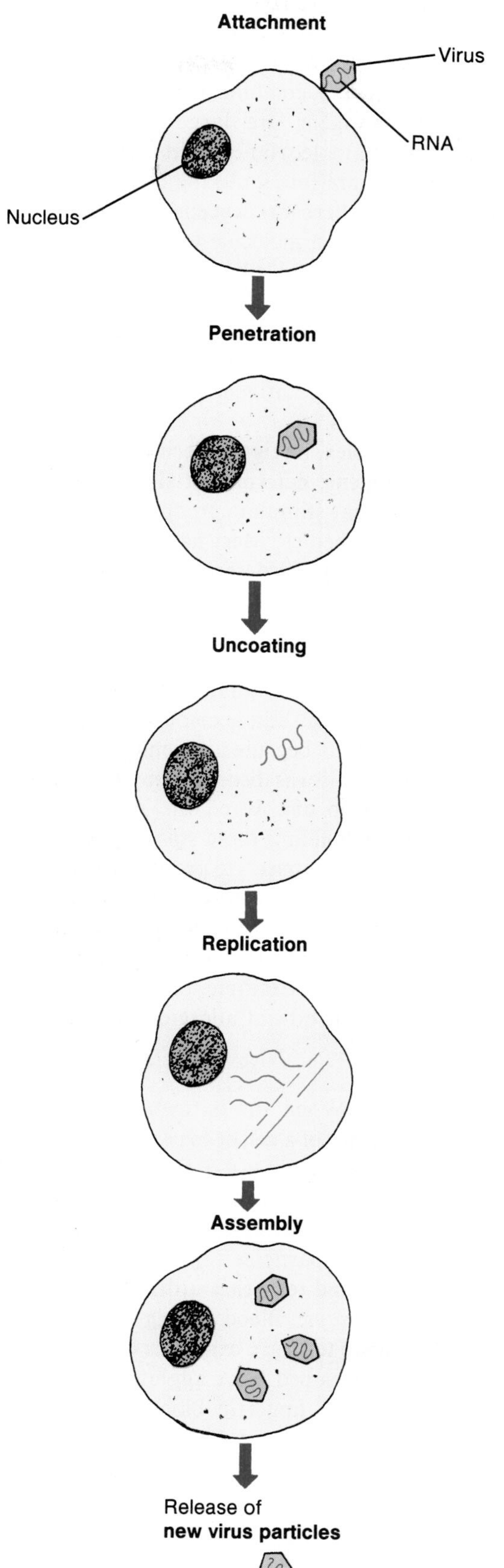

antibodies were *specific* in their actions. It was later learned that antibodies are proteins produced by a particular type of lymphocyte.

***Antigens.*** Antigens are molecules that stimulate antibody production and combine with these specific antibodies. Most antigens are large molecules (such as proteins) with a molecular weight greater than about 10,000, and they are foreign to the blood and other body fluids (although there are exceptions to both descriptions). The ability of a molecule to function as an antigen depends not only on its size but also on the complexity of its structure. Proteins are therefore more antigenic than polysaccharides, which have a simpler structure. Plastics used in artificial implants are composed of large molecules but are not very antigenic because of their simple, repeating structures.

A large, complex, foreign molecule can have a number of different **antigenic determinant sites,** which are areas of the molecule that stimulate production of and combine with different antibodies. Most naturally occurring antigens have many antigenic determinant sites and stimulate the production of different antibodies with specificities for these sites.

***Haptens.*** Many small organic molecules are not antigenic by themselves but can become antigens if they bond to proteins (and thus become antigenic determinant sites on the proteins). This was discovered by Karl Landsteiner, the same man who discovered the ABO blood groups (chapter 13). By bonding these small molecules—which Landsteiner called **haptens**—to proteins in the laboratory, new antigens could be created for research or diagnostic purposes. The bonding of foreign haptens to a person's own proteins can also occur in the body; by this means, derivatives of penicillin, for example, that would otherwise be harmless can produce fatal allergic reactions in susceptible people.

***Immunoassays.*** When the antigen or antibody is attached to the surface of a cell or to particles of latex rubber (in commercial diagnostic tests), the antigen-antibody reaction becomes visible because the particles *agglutinate* (clump) as a result of antigen-antibody bonding (fig. 19.4). These agglutinated particles can be used to assay a variety of antigens, and tests that utilize this procedure are called *immunoassays.* Blood typing (chapter 13) and modern pregnancy tests are examples of such immunoassays. A newly developed latex agglutination test for detecting AIDS using fingertip blood may also soon be available.

**Figure 19.4.** Immunoassay using the agglutination technique. Antibodies against a particular antigen are adsorbed to red blood cells or latex particles. When these are mixed with a solution that contains the appropriate antigen, the formation of the antigen-antibody complexes produces clumping (agglutination) that can be seen with the unaided eye.

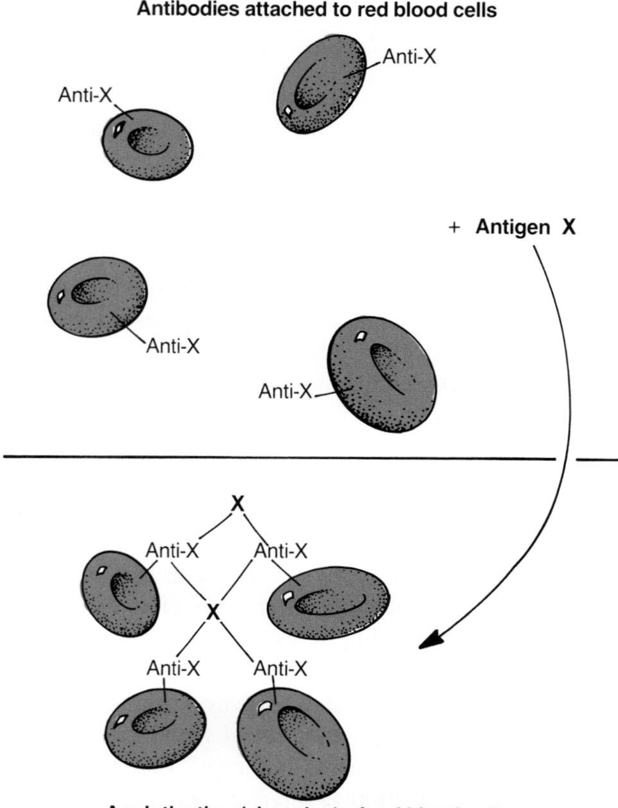


## Lymphocytes

Leukocytes, erythrocytes, and blood platelets are all ultimately derived from ("stem from") unspecialized cells in the bone marrow. These *stem cells* produce the specialized blood cells, and they replace themselves by cell division so that the stem cell population is not exhausted. Lymphocytes produced in this manner seed the thymus, spleen, and lymph nodes, producing self-replacing lymphocyte colonies in these organs.

The lymphocytes that become seeded in the thymus become **T lymphocytes.** These cells have surface characteristics and an immunological function that is different from other lymphocytes. The thymus, in turn, seeds other organs; about 65% to 85% of the lymphocytes in blood and most of the lymphocytes in lymph nodes are T lymphocytes. T lymphocytes, therefore, come from or had an ancestor that came from the thymus gland.

**Table 19.4** Comparison of B and T lymphocytes

| Characteristic | B Lymphocytes | T Lymphocytes |
|---|---|---|
| Site where processed | Bone marrow | Thymus |
| Type of immunity | Humoral (secretes antibodies) | Cell-mediated |
| Subpopulations | Memory cells and plasma cells | Cytotoxic (killer) T cells, helper cells, suppressor cells |
| Presence of surface antibodies | Yes—IgM or IgD | Not detectable |
| Receptors for antigens | Present—are surface antibodies | Present—are related to immunoglobulins |
| Life span | Short | Long |
| Tissue distribution | High in spleen, low in blood | High in blood and lymph |
| Percent of blood lymphocytes | 10%–15% | 75%–80% |
| Transformed by antigens to | Plasma cells | Small lymphocytes |
| Secretory product | Antibodies | Lymphokines |
| Immunity to viral infections | Enteroviruses, poliomyelitis | Most others |
| Immunity to bacterial infections | *Streptococcus, Staphylococcus,* many others | Tuberculosis, leprosy |
| Immunity to fungal infections | None known | Many |
| Immunity to parasitic infections | Trypanosomiasis, maybe to malaria | Most others |

Most of the lymphocytes that are not T lymphocytes are called **B lymphocytes.** The letter *B* is derived from immunological research performed in chickens. Chickens have an organ called the *bursa of Fabricius*, which processes B lymphocytes. Since mammals do not have a bursa, the *B* is often translated as the "bursa equivalent" for humans and other mammals. It is currently believed that the B lymphocytes in mammals are processed in the bone marrow, which conveniently also begins with the letter *B*.

Both B and T lymphocytes function in specific immunity. The B lymphocytes combat bacterial and some viral infections by secreting antibodies into the blood and lymph. They are therefore said to provide **humoral immunity.** T lymphocytes attack host cells that have become infected with viruses or fungi, transplanted human cells, and cancerous cells. The T lymphocytes do not secrete antibodies; they must come into close proximity or have actual physical contact with the victim cell in order to destroy it. T lymphocytes are therefore said to provide **cell-mediated immunity** (table 19.4).

1. *List the phagocytic cells in blood and lymph, and indicate which organs contain fixed phagocytes.*
2. *Describe the actions of interferons.*
3. *List four properties that antigens usually have. Explain why proteins are more antigenic than polysaccharides.*
4. *Describe the meaning of the term* hapten, *and give an example.*
5. *Distinguish between B and T lymphocytes in terms of their origins and immune functions.*

## Functions of B Lymphocytes

The specific antibody that a particular B lymphocyte produces will bind only to a specific antigen. When that antigen interacts with its appropriate antibody at the surface of that B cell it stimulates the B cell to divide many times, producing inactive memory cells and active, antibody-secreting cells called plasma cells. Binding of these secreted antibodies to antigens stimulates a cascade of reactions whereby a system of proteins in the plasma called complement is activated. Some of these activated complement proteins kill the cells containing the antigen; others promote phagocytosis and additional effects, which help the immune system to destroy the invading pathogens.

Exposure of a B lymphocyte to the appropriate antigen results in cell growth followed by many cell divisions. Some of the progeny become **memory cells,** which are indistinguishable from the original cell; others are transformed into **plasma cells** (fig. 19.5). Plasma cells are protein factories that produce about two thousand antibody proteins per second in their brief (five to seven day) life span.

The antibodies that are produced by plasma cells when B lymphocytes are exposed to a particular antigen react specifically with that antigen. Such antigens may be isolated molecules, or they may be molecules at the surface of an invading foreign cell. The specific bonding of antibodies to antigens serves to identify the enemy and to activate defense mechanisms that lead to the invader's destruction.

**Figure 19.5.** B lymphocytes have antibodies on their surface that function as receptors for specific antigens. The interaction of antigens and antibodies on the surface stimulates cell division and the maturation of the B cell progeny into memory cells and plasma cells. Plasma cells produce and secrete large amounts of the antibody (note the extensive rough endoplasmic reticulum in these cells).

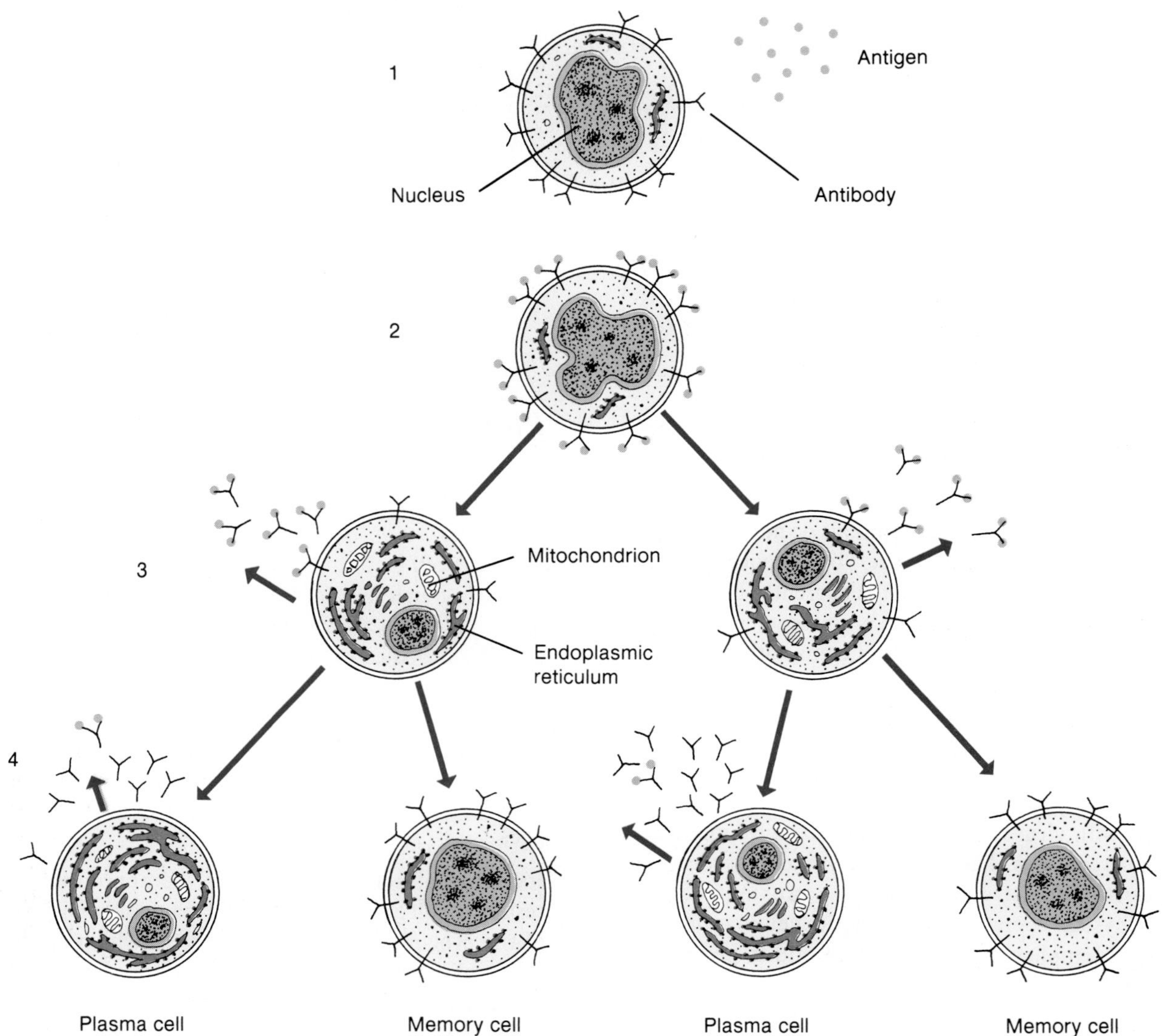


## Antibodies

Antibody proteins are also known as **immunoglobulins.** These are found in the gamma globulin class of plasma proteins, as identified by a technique called *electrophoresis,* in which classes of plasma proteins are separated by their movement in an electric field (fig. 19.6). With this technique, five distinct bands of proteins appear: albumin, alpha-1-globulin, alpha-2-globulin, beta globulin, and gamma globulin.

The gamma globulin band is wide and diffuse because it represents a heterogeneous class of molecules. Since antibodies are specific in their actions, it could be predicted that different types of antibodies have different structures. An antibody against smallpox, for example, does not confer immunity to poliomyelitis and, therefore, must have a slightly different structure than an antibody against polio. Despite these differences, antibodies are structurally related and form only a few subclasses.

**Figure 19.6.** The separation of serum protein by electrophoresis. *A* = albumin, alpha-1 ($\alpha_1$), alpha-2 ($\alpha_2$), beta ($\beta$), and gamma ($\gamma$) globulin.

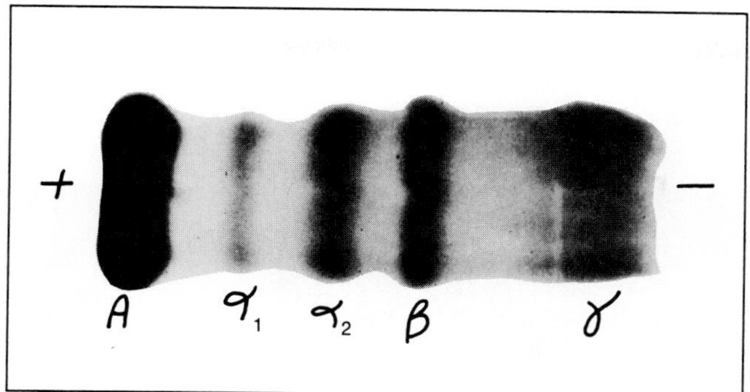


There are five subclasses of immunoglobulins (abbreviated Ig): *IgG, IgA, IgM, IgD,* and *IgE.* Most of the antibodies in serum are in the IgG subclass, whereas most of the antibodies in external secretions (saliva and milk) are IgA (table 19.5). Antibodies in the IgE subclass are involved in allergic reactions.

***Antibody Structure.*** All antibody molecules consist of four interconnected polypeptide chains. Two longer, higher molecular weight chains (the *H chains*) are joined to two shorter, lighter *L chains.* Research has shown that these four chains are arranged in the form of a Y. The stalk of the Y has been called the "crystallizable fragment" (abbreviated $F_c$), whereas the top of the Y is the "antigen-binding fragment" ($F_{ab}$). This is shown in figure 19.7.

The amino acid sequences of some antibodies have been determined by using antibodies derived from people with multiple myelomas. These lymphocyte tumors are derived from the division of a single B lymphocyte, forming a population of genetically identical cells (a clone) that secretes identical antibodies. These populations and the antibodies they secrete are different, however, in different patients. Analyses of these antibodies have shown that the $F_c$ regions of different antibodies are the same (are *constant*), whereas the $F_{ab}$ regions are *variable.* Variability of the antigen-binding regions is required for the specificity of antibodies for antigens. The $F_{ab}$ region of an antibody thus provides a specific site for bonding with a particular antigen (fig. 19.8).

B lymphocytes have antibodies on their cell membrane that serve as **receptors** for antigens. Combination of antigens with these antibody receptors stimulates the B cell to divide and produce more of these antibodies, which are secreted. Exposure to a given antigen thus results in increased amounts of the specific type of antibody that can attack that antigen. This provides active immunity, as described in the next major section.

**Table 19.5** The immunoglobulins

| Immunoglobulin | Examples of Functions |
|---|---|
| IgG | Main form of antibodies in circulation: production increased after immunization |
| IgA | Main antibody type in external secretions, such as saliva and mother's milk |
| IgE | Responsible for allergic symptoms in immediate hypersensitivity reactions |
| IgM | Function as antigen receptors on lymphocyte surface prior to immunization; secreted during primary response |
| IgD | Function as antigen receptors on lymphocyte surface prior to immunization; other functions unknown |

***Diversity of Antibodies.*** It is estimated that there are about 100 million trillion ($10^{20}$) antibody molecules in each individual, representing a few million different specificities for different antigens. Considering that antibodies to particular antigens can cross-react to some degree with closely related antigens, this tremendous antibody diversity usually ensures that there are some antibodies that can combine with almost any antigen the person may encounter. These observations evoke a question that has long fascinated scientists: How can a few million different antibodies be produced? A person cannot possibly inherit a correspondingly large number of genes devoted to antibody production.

Two mechanisms have been proposed to explain antibody diversity. First, since different combinations of heavy and light chains can produce different antibody specificities, a person does not have to inherit a million different genes to code for a million different antibodies. If a few hundred genes code for different H chains and a few hundred code for different L chains, different combinations of these polypeptide chains could produce millions of different antibodies. Second, the diversity of

**Figure 19.7.** Antibodies are composed of four polypeptide chains—two are heavy (*H*) and two are light (*L*). (*a*) A computer-generated model of antibody structure. (*b*) A simplified diagram showing the constant and variable regions. The variable regions are abbreviated V, and the constant regions are abbreviated C. Antigens combine with the variable regions. Each antibody molecule is divided into an $F_{ab}$ (antigen-binding) fragment and an $F_c$ (crystallizable) fragment.

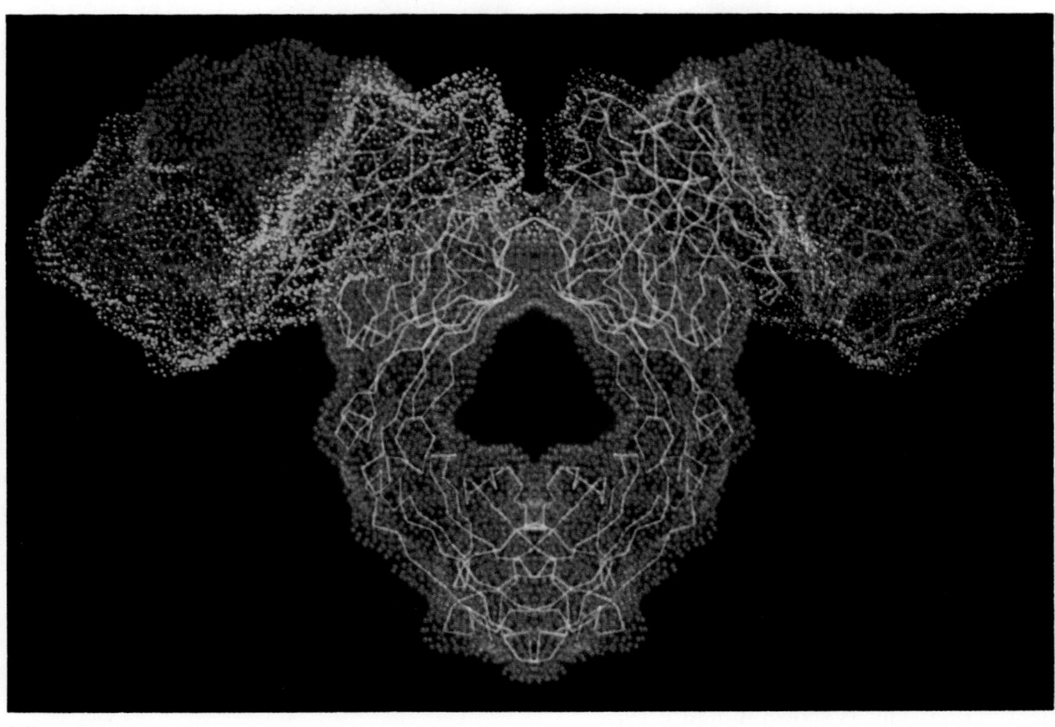

(a)

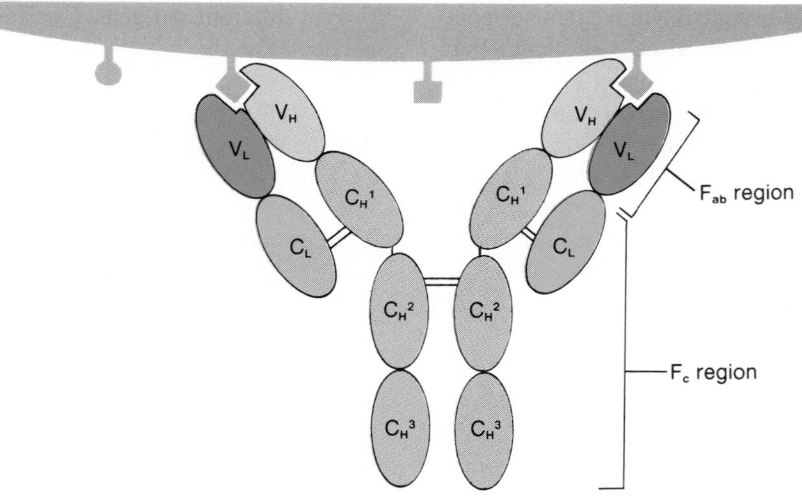


(b)

**Figure 19.8.** Structure of the $F_{ab}$ portion of an antibody molecule and the antigen with which it combines as determined by X-ray diffraction. The heavy and light chains of the antibody are shown in blue and yellow, respectively, and the antigen is shown in green. Note the complementary shape at the region where the two join together (*a* to *b*).

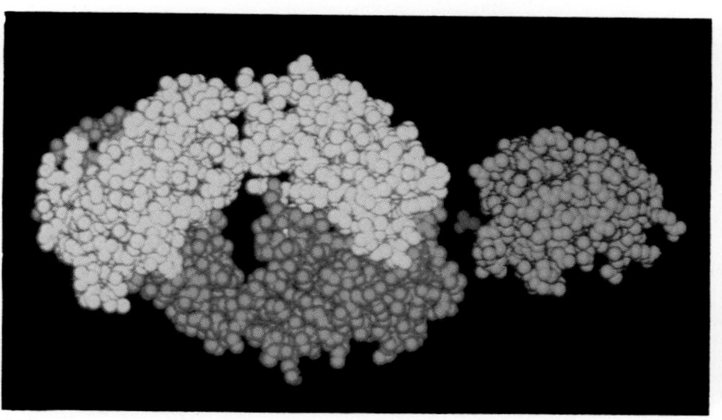

(a)

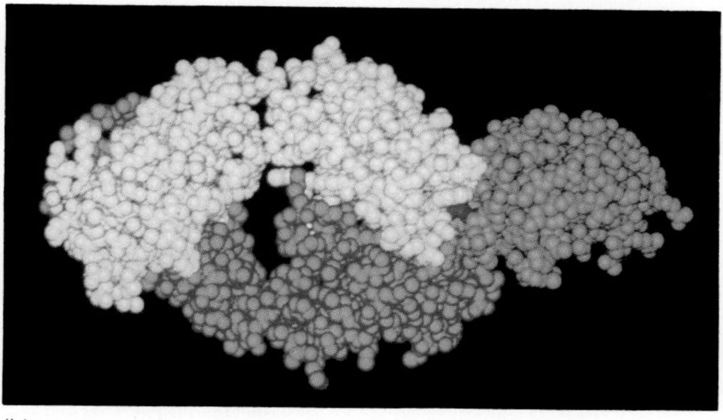

(b)

antibodies could increase during development if, when some lymphocytes divided, the progeny received antibody genes that have been slightly altered by mutations. This is called *somatic mutation* since the mutation occurs in a body cell rather than in a sperm or ovum. The diversity of antibodies would thus increase as the lymphocyte population increases.

***Opsonization.*** The combination of antibodies with antigens does not itself produce destruction of the antigens or of the pathogenic organisms that contain these antigens. Antibodies, rather, serve to identify the targets for immunological attack and to activate nonspecific immune processes that destroy the invader. Bacteria that are buttered with antibodies, for example, are better targets for phagocytosis by neutrophils and macrophages. The ability of antibodies to stimulate phagocytosis is termed **opsonization.** Immune destruction of bacteria is also promoted by antibody-induced activation of a system of serum proteins known as *complement.*

## The Complement System

It was learned in the early part of the twentieth century that rabbit antibodies to sheep red blood cell antigens could not lyse (destroy) these cells unless certain protein components of serum were present. These proteins, called **complement,** are a nonspecific defense system that is activated by the bonding of antibodies to antigens and by this means is directed against specific invaders that have been identified by antibodies.

There are eleven complement proteins, designated C1 (which has three protein components) through C9. These proteins are present in an inactive state within plasma and other body fluids and become activated by the attachment of antibodies to antigens. In terms of their functions, the complement proteins can be subdivided into three components: (1) recognition (C1); (2) activation (C4, C2, and C3, in that order); and (3) attack (C5–C9). The attack phase consists of **complement fixation,** in which complement proteins attach to the cell membrane and destroy the victim cell.

Antibodies of the IgG and IgM subclasses attach to antigens on the invading cell's membrane, bond to C1, and by this means activate its enzyme activity. Activated C1 catalyzes the hydrolysis of C4 into two fragments (fig. 19.9), designated C4a and C4b. The C4b fragment bonds to the cell membrane (is "fixed") and becomes an active enzyme that splits C2 into two fragments, C2a and C2b. The C2a becomes attached to C4b and cleaves C3 into C3a and C3b. Fragment C3b becomes attached to the growing complex of complement proteins on the cell membrane. The C3b converts C5 to C5a and C5b. The C5b and eventually C6 through C9 become fixed to the cell membrane.

Complement proteins C5 through C9 create large pores in the membrane (fig. 19.10). These pores allow the osmotic influx of water, so that the victim cell swells and bursts. Notice that the complement proteins, not the antibodies directly, kill the cell; antibodies only serve as activators of this process. Other molecules can also activate the complement system in an alternate nonspecific pathway that bypasses the early phases of the specific pathway described here.

Figure 19.9. The fixation of complement proteins. The formation of an antibody-antigen complex causes complement protein C4 to be split into two subunits—C4a and C4b. The C4b subunit attaches (is fixed) to the membrane of the cell to be destroyed (such as a bacterium). This event triggers the activation of other complement proteins, some of which attach (are fixed) to the C4b on the membrane surface.

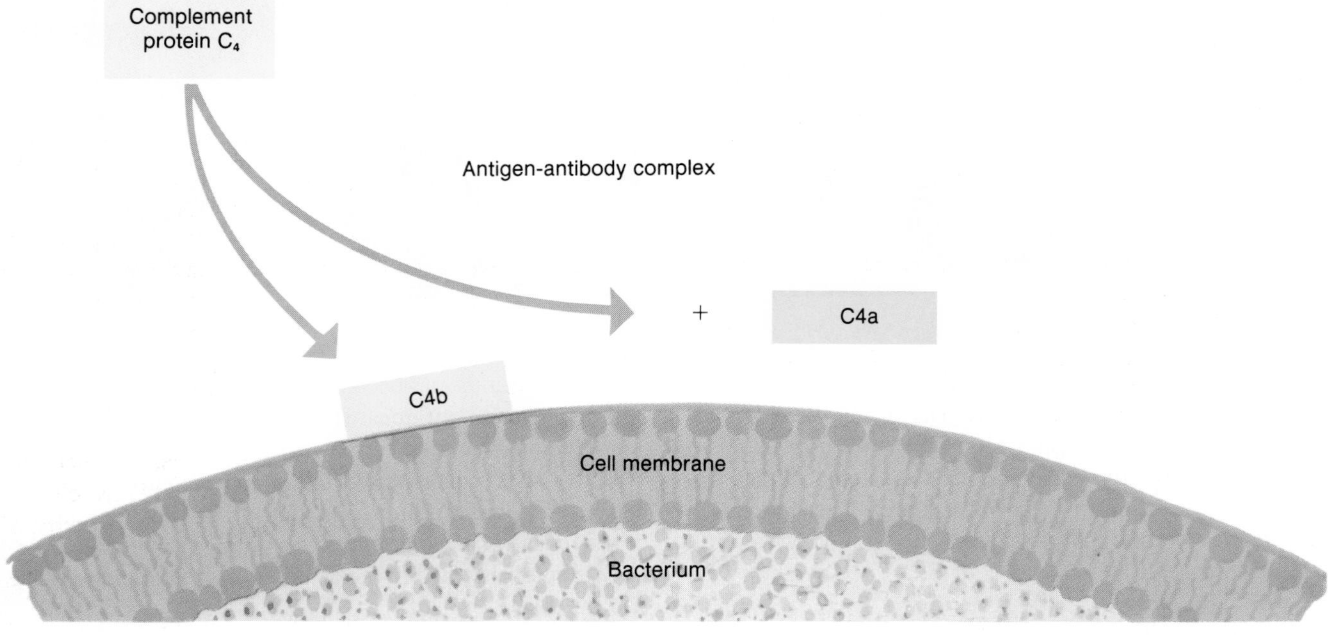


Figure 19.10. Complement proteins C5 through C9 (illustrated as a doughnut-shaped ring) puncture the membrane of the cell to which they are attached (fixed). This aids destruction of the cell.

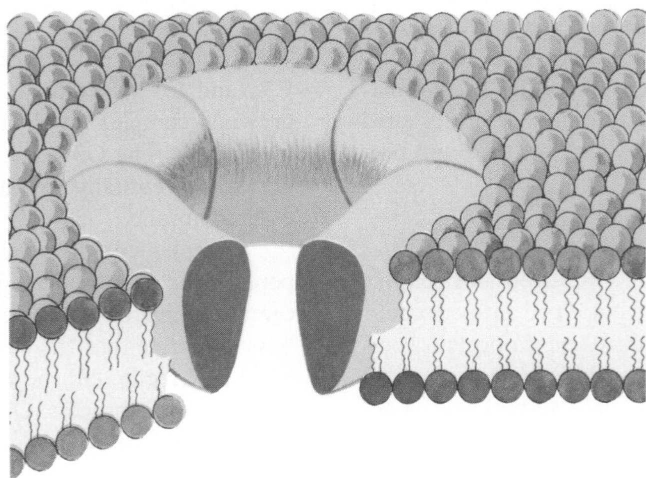

**Table 19.6** Summary of events that occur in a local inflammation when a break in the skin permits entry of bacteria

| Category | Events |
|---|---|
| Nonspecific immunity | Bacteria enter through break in anatomic barrier of skin.<br>Resident phagocytic cells—neutrophils and macrophages—engulf bacteria.<br>Nonspecific activation of complement proteins occurs. |
| Specific immunity | B cells are stimulated to produce specific antibodies.<br>Phagocytosis is enhanced by antibodies attached to bacterial surface antigens.<br>Specific activation of complement proteins occurs, which stimulates phagocytosis, chemotaxis of new phagocytes to the infected area, and secretion of histamine from tissue mast cells.<br>Diapedesis allows new phagocytic leukocytes (neutrophils and monocytes) to invade the infected area.<br>Vasodilation and increased capillary permeability (as a result of histamine secretion) produce redness and edema. |

Complement fragments that are not fixed but instead are liberated into the surrounding fluid have a number of effects. These effects include (1) *chemotaxis*—the liberated complement fragments attract phagocytic cells to the site of complement activation; (2) *opsonization*—phagocytic cells have receptors for C3b, so that this fragment may form bridges between the phagocyte and the victim cell that facilitates phagocytosis; and (3) fragments C3a and C5a *stimulate the release of histamine* from mast cells (a connective tissue cell type) and basophils. As a result of histamine release, there is increased blood flow to the infected area due to vasodilation and increased capillary permeability. The latter effect can result in the leakage of plasma proteins into the surrounding tissue fluid, producing local edema.

## Local Inflammation

Aspects of the nonspecific and specific immune responses and their interactions are well illustrated by the events that occur when bacteria enter a break in the skin and produce a local inflammation (table 19.6). The inflammatory reaction is initiated by the nonspecific mechanisms of phagocytosis and complement activation. Activated complement further increases this nonspecific response by attracting new phagocytes to the area and by increasing the activity of these phagocytic cells.

After some time, B lymphocytes are stimulated to produce antibodies against specific antigens that are part of the invading bacteria. Attachment of these antibodies to antigens in the bacteria greatly increases the previously nonspecific response. This occurs because of greater activation of complement, which directly destroys the bacteria and which—together with the antibodies themselves—promotes the phagocytic activity of neutrophils, macrophages, and monocytes (fig. 19.11).

As inflammation progresses, the release of lysosomal enzymes from macrophages causes the destruction of leukocytes and other tissue cells. These effects, together with those produced by histamine and other chemicals released from mast cells, produce the characteristic symptoms of a local inflammation: *redness and warmth* (due to vasodilation); *swelling* (edema); and *pus* (the accumulation of dead leukocytes). If the infection continues, the release of endogenous pyrogen from leukocytes and macrophages may produce a fever.

1. *Illustrate the structure of an antibody molecule. Label the constant and variable regions, the* $F_c$ *and* $F_{ab}$ *parts, and the heavy and light chains.*
2. *Define* opsonization, *and name two types of molecules that promote this process.*
3. *Describe complement fixation, and explain the roles of complement fragments that do not become fixed.*
4. *Explain how nonspecific and specific immune mechanisms cooperate during a local inflammation.*

**Figure 19.11.** The entry of bacteria through a cut in the skin produces a local inflammatory reaction. In this reaction, antigens on the bacterial surface are coated with antibodies and ingested by phagocytic cells. Symptoms of inflammation are produced by the release of lysosomal enzymes and by the secretion of histamine and other chemicals from tissue mast cells.

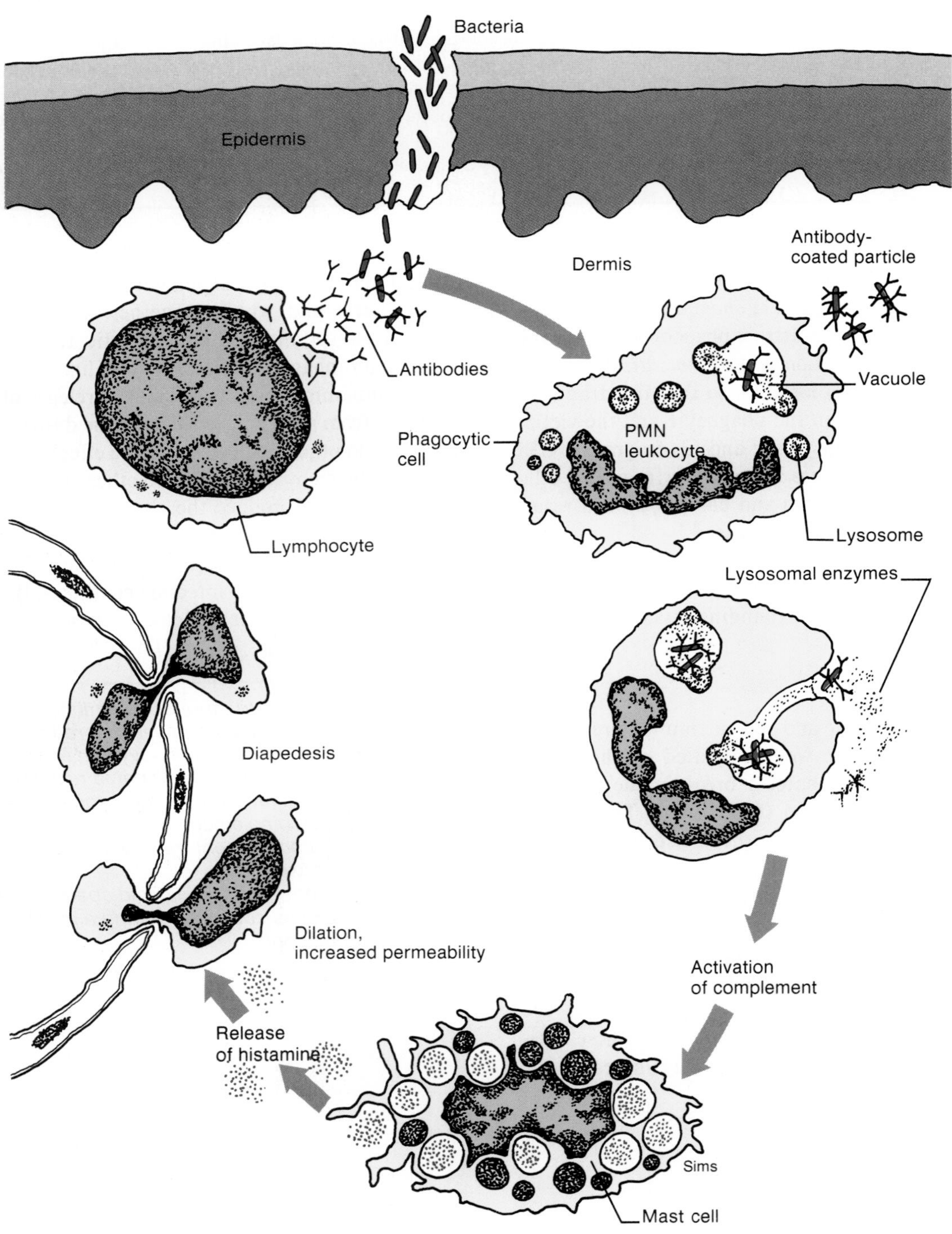

**Figure 19.12.** Active immunity to a pathogen can be gained by exposure to the fully virulent form or by inoculation with a pathogen whose virulence (ability to cause disease) has been attenuated (reduced) but whose antigens are the same as in the virulent form.

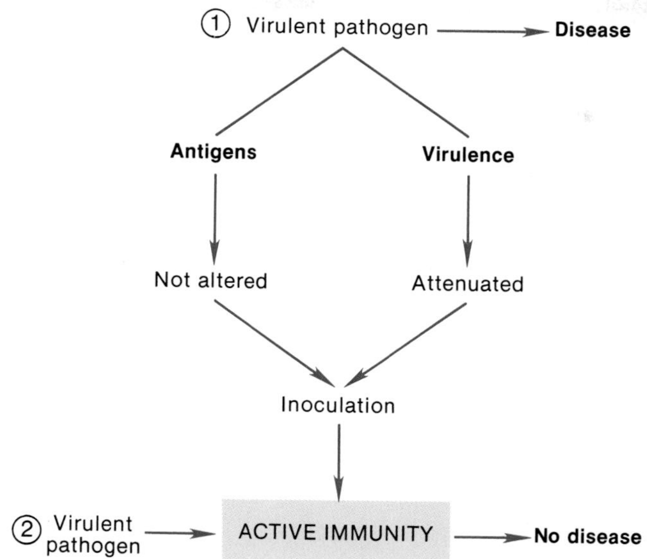


## *Active and Passive Immunity*

When a person is first exposed to antigens in a pathogen, too few antibodies may be produced to combat the disease. In the process, however, the lymphocytes that have specificity for those antigens are stimulated to divide many times and produce a clone. This is active immunity, and can protect the person from getting the disease upon subsequent exposures. A person may be passively immunized to a disease by obtaining antibodes produced by another organism.

It was first known in Western Europe in the mid-eighteenth century that the fatal effects of smallpox could be prevented by inducing mild cases of the disease. This was accomplished at that time by rubbing needles into the pustules of people who had mild forms of smallpox and injecting these needles into healthy people. Understandably, this method of immunization was not widely popular.

Acting on the observation that milkmaids who contracted cowpox—a disease similar to smallpox but less *virulent* (pathogenic)—were immune to smallpox, an English physician, named Edward Jenner, inoculated a healthy boy with cowpox. When the boy recovered, Jenner inoculated him with an otherwise deadly amount of smallpox, from which he also proved to be immune. (This was fortunate for both the boy—who was an orphan—and Jenner; Jenner's fame spread, and as the boy grew into manhood he proudly gave testimonials on Jenner's behalf.) This experiment, performed in 1796, began the first widespread immunization program.

A similar, but more sophisticated, demonstration of the effectiveness of immunizations was performed by Louis Pasteur almost a century later. Pasteur isolated the bacteria that cause anthrax and heated them until their ability to cause disease was greatly reduced (their virulence was *attenuated*), but the nature of their antigens was not significantly changed (fig. 19.12). He then injected these attenuated bacteria into twenty-five cows, leaving twenty-five unimmunized. Several weeks later, before a gathering of scientists, he injected all fifty cows with the completely active anthrax bacteria. All twenty-five of the unimmunized cows died—all twenty-five of the immunized animals survived.

### Active Immunity and the Clonal Selection Theory

When a person is exposed to a particular pathogen for the first time, there is a latent period of five to ten days before measurable amounts of specific antibodies appear in the blood. This sluggish **primary response** may not be sufficient to protect the individual against the disease caused by the pathogen. Antibody concentrations in the blood during this primary response reach a plateau in a few days and decline after a few weeks.

**Figure 19.13.** A comparison of antibody production in the primary response (upon first exposure to an antigen) to antibody production in the secondary response (upon subsequent exposure to the antigen). The greater secondary response is believed to be due to the development of lymphocyte clones produced during the primary response.

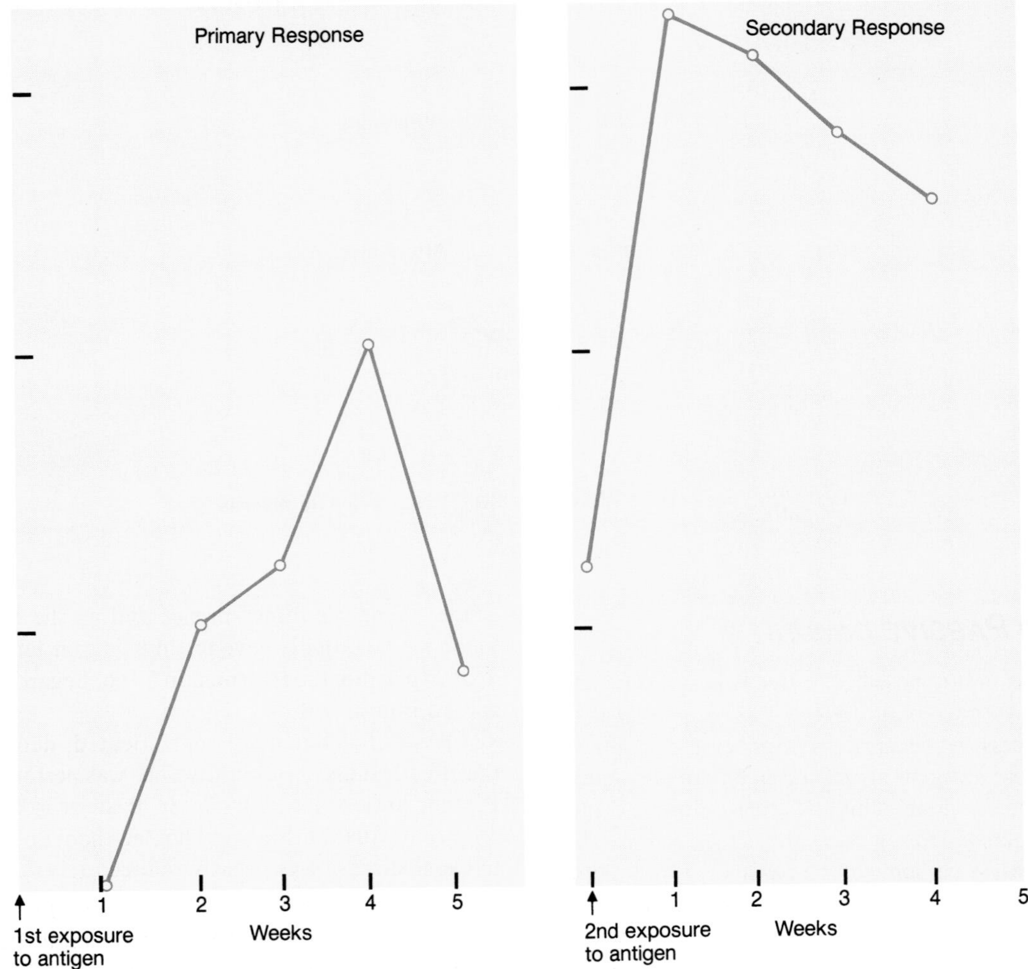


A subsequent exposure of the same individual to the same antigen results in a **secondary response** (fig. 19.13). Compared to the primary response, antibody production during the secondary response is much more rapid. Maximum antibody concentrations in the blood are reached in less than two hours and are maintained for a longer time than in the primary response. This rapid rise in antibody production is usually sufficient to prevent the disease.

Before stimulation by a particular antigen, the B cells that make the appropriate antibodies produce mainly IgM antibodies. These IgM antibodies therefore account for a high proportion of the antibodies made during the primary response. In contrast, most antibodies made during the secondary response are in the IgG subclass. This is shown in figure 19.14.

***Clonal Selection Theory.*** The immunization procedures of Jenner and Pasteur were effective because the people who were inoculated produced a secondary rather than a primary response when exposed to the virulent pathogens. This protection is not simply due to accumulations of antibodies in the blood, because secondary responses occur even after antibodies produced by the primary response have disappeared. Immunizations, therefore, seem to produce a type of "learning" in which the ability of the immune system to combat a particular pathogen is improved by prior exposure.

**Figure 19.14.** The greater antibody production in the secondary response compared to the primary response is due mainly to increased amounts of IgG antibodies. Notice that the amount of IgM antibodies is about the same in both primary and secondary responses.

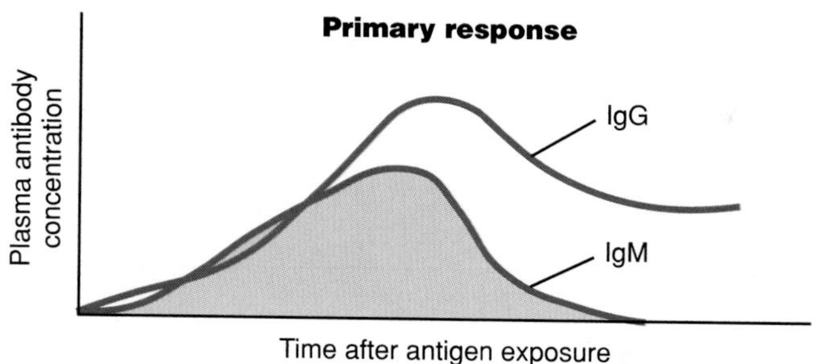


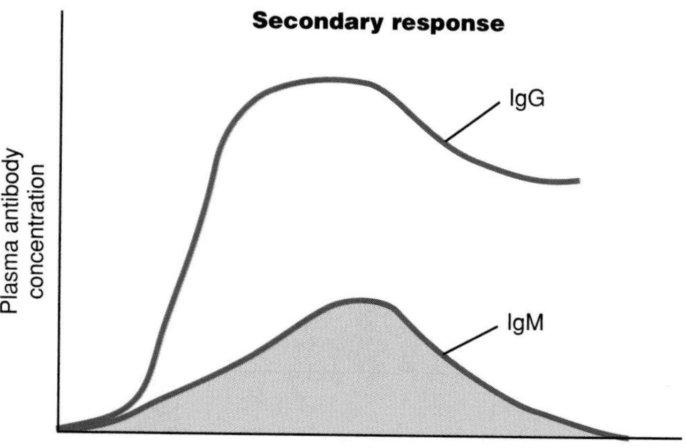


The mechanisms by which secondary responses are produced are not completely understood; the **clonal selection theory,** however, appears to account for most of the evidence. According to this theory, B lymphocytes *inherit* the ability to produce particular antibodies (and T lymphocytes inherit the ability to respond to particular antigens). One B lymphocyte can only produce one type of antibody, with specificity for one antigen. Since this ability is genetically inherited rather than acquired, some lymphocytes, for example, can respond to smallpox and produce antibodies against it even if the person has never been previously exposed to this disease.

The inherited specificity of each lymphocyte is reflected in the antigen receptor proteins on the surface of the lymphocyte's plasma membrane. Exposure to smallpox antigens thus stimulates these specific lymphocytes to divide many times until a large population of genetically identical cells—a clone—is produced. Some of these cells become plasma cells that secrete antibodies for the primary response; others become memory cells that can be stimulated to secrete antibodies during the secondary response (fig. 19.15).

Notice that according to the clonal selection theory (table 19.7), antigens do not induce lymphocytes to make the appropriate antibodies. Rather, antigens select lymphocytes (through interaction with surface receptors) that are already able to make antibodies against that antigen. This is analogous to evolution by natural selection. An environmental agent (in this case, antigens) acts on the genetic diversity already present in a population of organisms (lymphocytes) to cause increasing numbers of the individuals that are selected.

***Active Immunity.*** The development of a secondary response provides **active immunity** against the specific pathogens. The development of active immunity requires prior exposure to the specific antigens, during which time a primary response is produced and the person may get sick. Some parents, for example, deliberately expose their children to others who have measles, chicken pox, and mumps so that their children will be immune to these diseases in later life.

**Figure 19.15.** According to the clonal selection theory, exposure to an antigen stimulates the production of lymphocyte clones and the maturation of some members of B cell clones into antibody-secreting plasma cells.

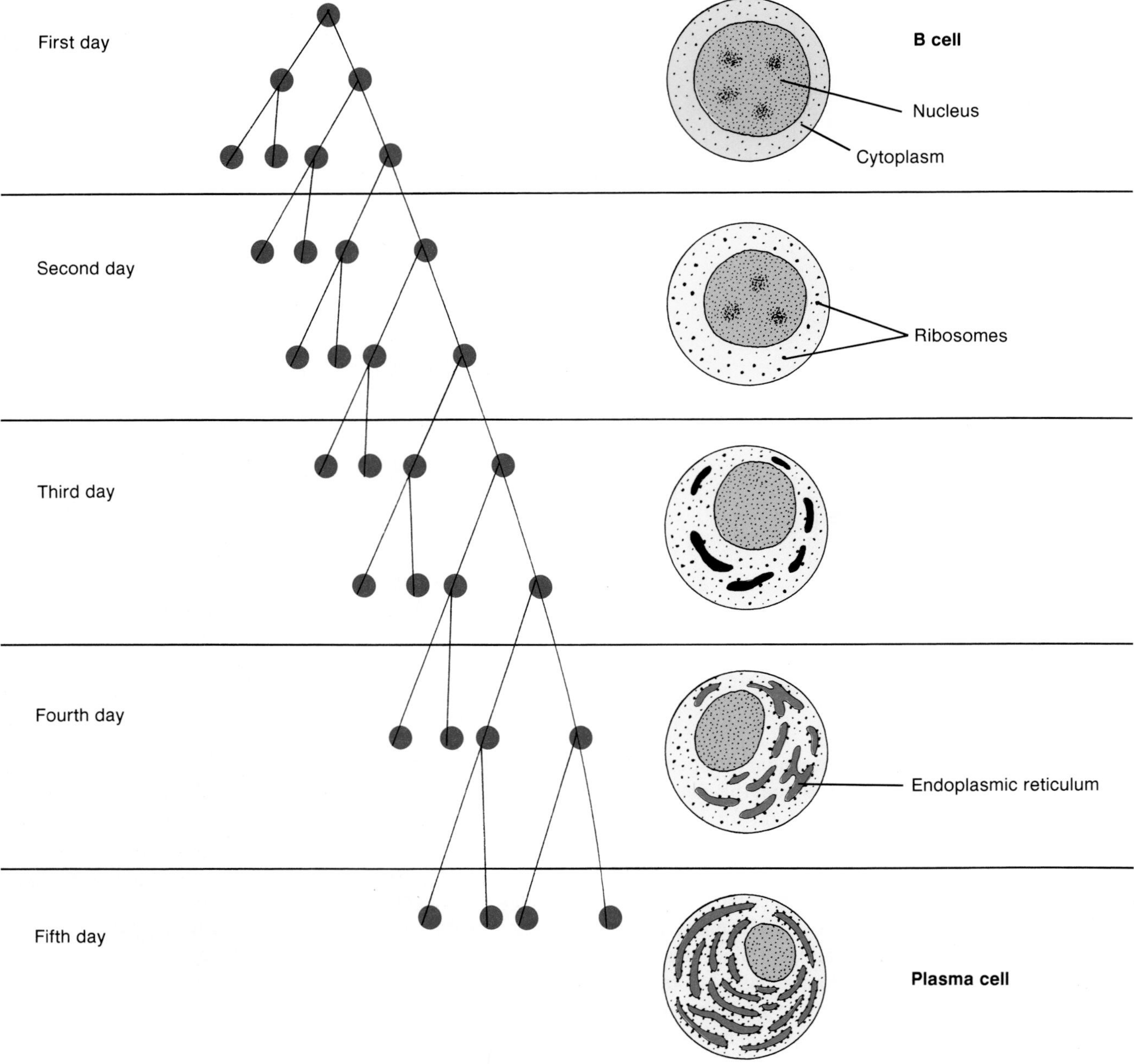


**Table 19.7** Summary of the clonal selection theory (with regard to B cells)

| Process | Results |
|---|---|
| Lymphocytes inherit the ability to produce specific antibodies. | Prior to antigen exposure, lymphocytes are already present in the body that can make the appropriate antibodies. |
| Antigens interact with antibody receptors on the lymphocyte surface. | Antigen-antibody interaction stimulates cell division and the development of lymphocyte clones containing memory cells and plasma cells that secrete antibodies. |
| Subsequent exposure to the specific antigens produce a more efficient response. | Exposure of lymphocyte clones to specific antigens results in greater and more rapid production of specific antibodies. |

Clinical immunization programs induce primary responses by inoculating people with pathogens whose virulence has been attenuated or destroyed (such as Pasteur's heat-inactivated anthrax bacteria) or by using closely related strains of microorganisms that are antigenically similar but less pathogenic (such as Jenner's cowpox inoculations). The name for these procedures—**vaccinations** (after the Latin word *vacca* = cow)—reflects the history of this technique. All of these procedures cause the development of lymphocyte clones that can combat the virulent pathogens by producing secondary responses.

The first successful polio vaccine (the Salk vaccine) was composed of viruses that had been inactivated by treatment with formaldehyde. These "killed" viruses were injected into the body, in contrast to the currently used oral (Sabin) vaccine. The oral vaccine contains "living" viruses that have attenuated virulence. These viruses invade the epithelial lining of the intestine and multiply but do not invade nerve tissue. The immune system can, therefore, become sensitized to polio antigens and produce a secondary response if polio viruses that attack the nervous system are later encountered.

There is always a danger when vaccines are prepared using attenuated viruses or toxins that some virulence may remain and cause disease in vaccinated people. One example is the commonly used vaccine for *pertussis* (whooping cough). This disease is responsible for about one million infant deaths annually worldwide, and is caused by a toxin released from a species of bacteria. Though the vaccine against pertussis prepared from the bacterial cells or their toxin is relatively effective, it occasionally produces severe side effects. Scientists have recently produced a protein subunit of pertussis toxin using genetic engineering techniques which appear to confer immunity without having the virulent actions of the toxin. In the future, production of specific proteins from cloned DNA may provide other vaccines that are safer and more effective than those prepared by traditional methods.

## Passive Immunity

The term **passive immunity** refers to the immune protection that can be produced by transferring antibodies produced by another person or by an animal. The person or animal that produced the antibodies to particular antigens was actively immunized as explained by the clonal selection theory. The person who receives these ready-made antibodies is thus passively immunized to the same antigens. Passive immunity occurs naturally in the transfer of immunity from a mother to her infant, and can be produced artificially by injections of antibodies to protect against certain diseases.

The ability to mount an immune response—called **immunological competence**—does not develop until about a month after birth. The fetus, therefore, cannot immunologically reject its mother. The immune system of the mother is fully competent but does not usually respond to fetal antigens for reasons that are not completely understood. Some IgG antibodies from the mother do cross the placenta and enter the fetal circulation, however, and these serve to confer passive immunity to the fetus.

The fetus and the newborn baby are, therefore, immune to the same antigens as the mother. Since the baby did not itself produce the lymphocyte clones needed to form these antibodies, such passive immunity disappears when the infant is about one month old. If the baby is breast-fed it can receive additional antibodies of the IgA subclass in its mother's first milk (the *colostrum*).

Passive immunizations are used clinically to protect people who have been exposed to extremely virulent infections or toxins, such as snake venom, tetanus, and others. In these cases the affected person is injected with *antiserum* (serum-containing antibodies), also called *antitoxin,* from an animal that has been previously exposed to the pathogen. The animal develops the lymphocyte clones and active immunity and thus has a high concentration of antibodies in its blood. Since the person who is injected with these antibodies does not develop active immunity, he or she must again be injected with antitoxin upon subsequent exposures. A comparison of active and passive immunity is shown in table 19.8.

**Table 19.8** Comparison of active and passive immunity

| Characteristic | Active Immunity | Passive Immunity |
|---|---|---|
| Injection of person with | Antigens | Antibodies |
| Source of antibodies | The person inoculated | Natural—the mother; artificial—injection with antibodies |
| Method | Injection with killed or attenuated pathogens or their toxins | Natural—transfer of antibodies across the placenta; artificial—injection with antibodies |
| Time to develop resistance | 5 to 14 days | Immediately after injection |
| Duration of resistance | Long (perhaps years) | Short (days to weeks) |
| When used | Before exposure to pathogen | Before or after exposure to pathogen |

### Monoclonal Antibodies

In addition to their use in passive immunity, antibodies are also commercially prepared for use in research and clinical laboratory tests. In the past, antibodies were obtained by chemically purifying a specific antigen and then injecting this antigen into animals. Since an antigen typically has many different antigenic determinant sites, however, the antibodies obtained by this method were polyclonal; they had different specificities. This decreased their sensitivity to a particular antigenic site and resulted in some degree of cross-reaction with closely related antigen molecules.

In the preparation of monoclonal antibodies, an animal (mice are commonly used) is injected with an antigen and then subsequently killed. B lymphocytes are then obtained from the spleen and placed in thousands of different *in vitro* incubation vessels. These cells soon die, however, unless they are hybridized with cancerous multiple myeloma cells. Cell fusion is promoted by a chemical, polyethylene glycol. The fusion of a B lymphocyte with a cancerous cell produces a hybrid that undergoes cell division and produces a clone, called a *hybridoma.* Each hybridoma secretes large amounts of identical, **monoclonal antibodies.** From among the thousands of hybridomas produced in this way, the one that produces the desired antibody is cultured while the rest are discarded.

The availability of large quantities of pure monoclonal antibodies has resulted in the development of much more sensitive clinical laboratory tests (of pregnancy, for example). These pure antibodies have also been used to pick one molecule (the specific antigen interferon, for example) out of a solution of many molecules and thus isolate and concentrate it. In the future, monoclonal antibodies against specific tumor antigens may aid the diagnosis of cancer. Even more exciting, cytotoxic drugs that can kill normal as well as cancerous cells might be aimed directly at a tumor by combining these drugs with monoclonal antibodies against specific tumor antigens.

---

1. *Describe three methods used to induce active immunity.*
2. *Explain the characteristics of the primary and secondary immune responses, and draw graphs to illustrate your discussion.*
3. *Explain the clonal selection theory and how this theory accounts for the secondary response.*
4. *Describe passive immunity, and give natural and clinical examples of this type of immunization.*

---

## Functions of T Lymphocytes

T lymphocytes are activated by specific antigens and respond by secreting chemicals called lymphokines. Through their secretion of lymphokines, T cells provide numerous functions that assist all aspects of the immune system. These functions include cell-mediated destruction by cytotoxic T cells and supporting roles by helper and suppressor T cells. T cells are only activated by foreign antigens when those antigens are presented to the T cells in conjunction with a group of self-antigens that are normally present on the surface of the host cells. Macrophages present foreign antigens in this way to T cells, and also secrete lymphokines of their own.

The thymus processes lymphocytes in such a way that their functions become quite distinct from those of B cells. Lymphocytes that reside in the thymus, or came from the thymus, or which are derived from cells that came from the thymus are all T lymphocytes. These cells can be distinguished from B cells by specialized techniques. Unlike B cells, the T lymphocytes provide specific immune protection without secreting antibodies. This is accomplished in different ways by the three subpopulations of T lymphocytes, which will be described shortly.

### Thymus Gland

The thymus gland extends from below the thyroid in the neck into the thoracic cavity. This organ grows during childhood but gradually regresses after puberty. Lymphocytes from the fetal liver and spleen and from the bone marrow postnatally seed the thymus and become transformed into T cells. These lymphocytes in turn enter the blood and seed lymph nodes and other organs, where they divide to produce new T cells when stimulated by antigens.

Small T lymphocytes that have not yet been stimulated by antigens have very long life spans—months or perhaps years. Still, new T cells must be continuously produced to provide efficient cell-mediated immunity. Since the thymus atrophies after puberty, this organ may not be able to provide new T cells in later life. Colonies of T cells in the lymph nodes and other organs are apparently able to produce new T cells under the stimulation of various **thymus hormones.**

Two hormones that are believed to be secreted by the thymus—*thymopoietin I* and *thymopoietin II*—may promote the transformation of lymphocytes into T cells. Another thymus hormone, called *thymosin,* may promote the maturation of T lymphocytes.

There is some experimental evidence supporting the notion that the administration of these thymus hormones may be able to restore cell-mediated immunity in some cases where T cell function has declined. This decline occurs in some congenital and acquired diseases as well as naturally in the course of aging in conjunction with an increased susceptibility to viral infections and cancer. Clinical manipulations of T cell function—through the administration of thymus hormones, interferon, interleukin-2, and other techniques—is an exciting possibility that may result from further experimental research.

### Cytotoxic, Helper, and Suppressor T Lymphocytes

The **cytotoxic**, or **killer**, **T lymphocytes** destroy specific victim cells that are identified by specific antigens on their surface. In order to effect this *cell-mediated* destruction, the T lymphocytes must be in actual contact with their victim cells (in contrast to B cells, which kill at a distance). Although the mechanisms by which the cytotoxic lymphocytes kill their victims are not completely understood, there is evidence that this is accomplished by the secretion of certain molecules at the region of contact. Among these molecules, specific polypeptides called *perforins* have been identified, which polymerize in the cell membrane of the victim cell and form cylindrical channels through the membrane. This is analogous to the channels formed by complement proteins previously discussed, and can result in osmotic destruction of the victim cell.

The cytotoxic lymphocytes defend against viral and fungal infections and are also responsible for transplant rejection reactions and for immunological surveillance against cancer. Although most bacterial infections are fought by B lymphocytes, some are the targets of cell-mediated attack by cytotoxic T lymphocytes. This is the case with the *tubercle bacilli* that cause tuberculosis. Injections of some of these bacteria under the skin produce inflammation after a latent period of forty-eight to seventy-two hours. This *delayed hypersensitivity* reaction is cell mediated rather than humoral, as shown by the fact that it can be induced in an unexposed guinea pig by an infusion of lymphocytes, but not of serum, from an exposed animal.

The **helper T lymphocytes** and **suppressor T lymphocytes** indirectly participate in the specific immune response by regulating the responses of the B cells (fig. 19.16) and the cytotoxic T cells. The activity of B cells and cytotoxic T cells is increased by helper T lymphocytes and decreased by suppressor T lymphocytes. The amount of antibodies secreted in response to antigens is thus affected by the relative numbers of helper to suppressor T cells that develop in response to a given antigen.

As a result of advances in recombinant DNA technology (genetic engineering) that allow the production of monoclonal antibodies, it is now possible for clinical laboratories to distinguish between the different subcategories of lymphocytes by means of antigen "markers" on their surfaces. Counting the lymphocytes in each of these subcategories provides far more information about diseases and their causes than was previously available. Tests of this sort have provided valuable information about the effects of the AIDS virus.

**Acquired immune deficiency syndrome (AIDS)** has increased overall adult male and female mortality in the United States by 0.7% and 0.07%, respectively. People at high risk for AIDS include homosexual and bisexual men (through anal intercourse), intravenous drug users (through sharing needles with infected individuals), hemophiliacs who received blood prior to 1985 (before plasma and clotting factor solutions were tested for AIDS), and women who have had sexual relations with men at high risk. Most children with AIDS were infected *in utero* by mothers who were intravenous drug users. In countries such as Haiti and those in central Africa, heterosexual contact is the primary route of transmission.

Most scientists now believe that AIDS is caused by a virus, known as the *human immunodeficiency virus (HIV),* which specifically destroys the helper T lymphocytes. Laboratory tests for AIDS include counting the relative number of helper to suppressor T cells. Helper cells have a particular antigen marker on their surface known as the T4 antigen, which can be distinguished from a different antigen (T8) on suppressor and cytotoxic cells. A normal person has a T4/T8 ratio greater than 1.0; a person with AIDS has a ration reduced to 0.5 or less, indicating a great reduction in helper lymphocytes. This results in decreased immunological function and greater susceptibility to opportunistic infections, including *Pneumocystis carinii pneumonia.* Many people with AIDS also develop a previously rare form of cancer known as *Kaposi's sarcoma.*

***Lymphokines.*** The T lymphocytes, as well as some other cells such as macrophages, secrete a number of polypeptides that serve to regulate many aspects of the immune system. These products are called **lymphokines**. When a lymphokine is first discovered, it is named according to its biological activity (e.g., *B cell–stimulating factor*). Since each lymphokine has many different actions (table 19.9), however, such names can be misleading. Scientists have thus agreed to use the name *interleukin*, followed by a number, to indicate a lymphokine once its amino acid sequence has been determined.

**Figure 19.16.** A given antigen can stimulate the production of both B and T cell clones. The ability to produce B cell clones, however, is also influenced by the relative effects of helper and suppressor T cells.

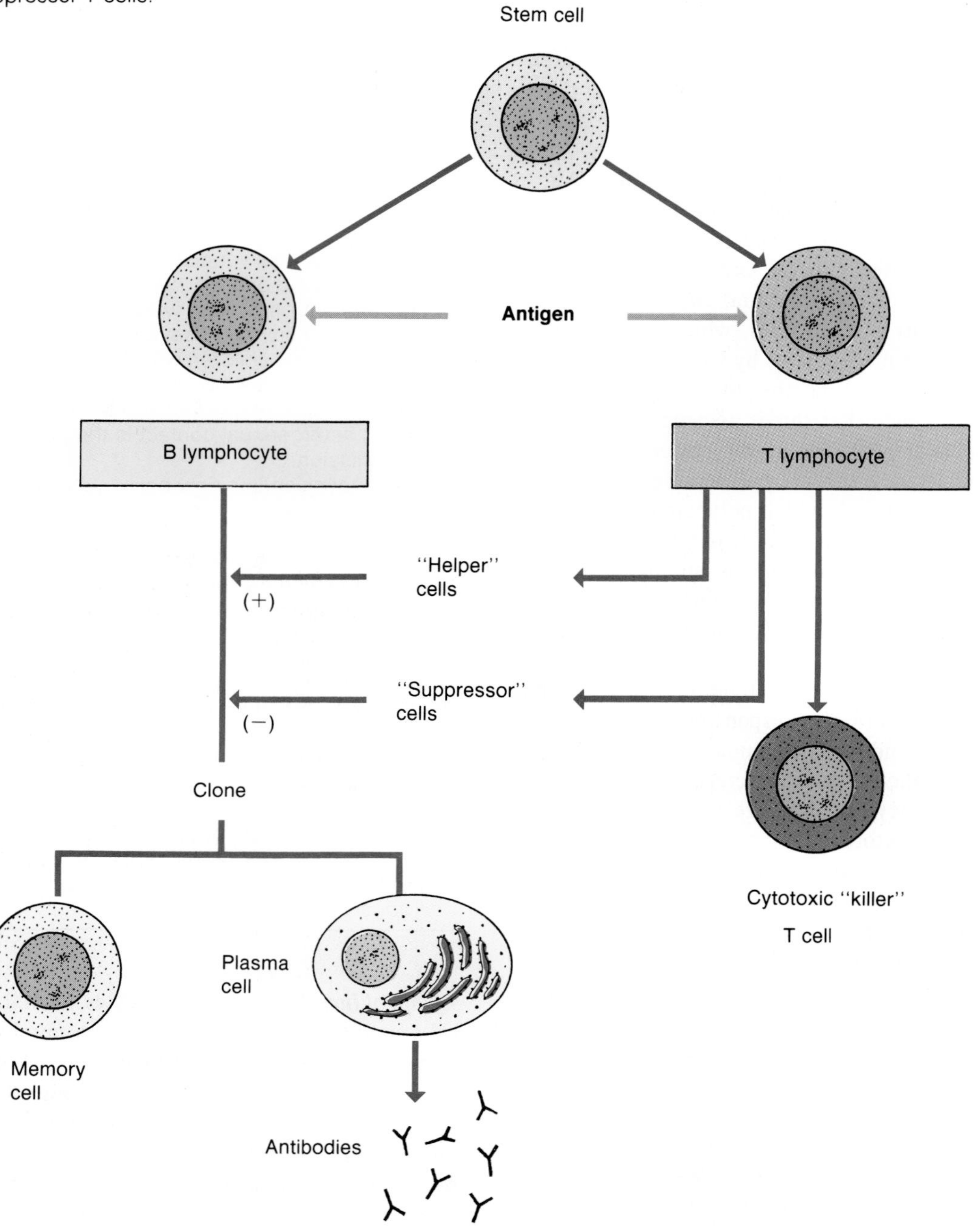

**Table 19.9** The major lymphokines

| Lymphokine | Biological Effects | Secreted By |
|---|---|---|
| Interleukin-1 | Activates resting T cells | Macrophages and others |
| Interleukin-2 | Growth factor for activated T cells; activates cytotoxic T lymphocytes | Helper T cells |
| Interleukin-3 | Promotes the growth of bone marrow stem cells; growth factor for mast cells | Helper T cells |
| Interleukin-4 (B cell–stimulating factor) | Promotes growth of activated B cells; promotes growth of resting T cells; enhances activity of cytotoxic T cells | Helper T cells |
| B cell–differentiating factor | Induces the conversion of activated B cells into antibody-secreting plasma cells | T cells and others |
| Colony-stimulating factors | Are different colony-stimulating factors for stimulating the proliferation of granulocytic leukocytes and of macrophages | T cells and others |
| Interferons | Activate macrophages; augment natural killer cell body activity; has antiviral activity | T cells and others |
| Tumor necrosis factors | Has direct cytotoxic effect on some tumor cells; stimulate production of other lymphokines | Macrophages and others |

Adapted from information appearing in *The New England Journal of Medicine* by C. A. Dinarello and J. W. Mier, Volume 317, page 940, 1987. 

*Interleukin-1,* for example, is secreted by macrophages and other cells and can activate the T cell system. B cell–stimulating factor, now called *interleukin-4,* is secreted by T lymphocytes and is required for the proliferation and clone development of B cells. *Interleukin-2* is released by helper T lymphocytes and, among its effects, is required for activation of cytotoxic T lymphocytes. *Macrophage colony-stimulating factor* is secreted by helper T lymphocytes and promotes the activity of macrophages. Three types of *interferons* (alpha, beta, and gamma) are also classified as lymphokines because of their diverse effects on the immune system.

Genetic engineering techniques have been used to generate recombinant bacteria that can produce large amounts of interleukin-2 for research and eventual clinical applications. This has provided the basis for investigation of possible immunological approaches to cancer therapy. In these experimental treatments, patients are given different combinations of interleukin-2, T cells (derived from the patient) which have been activated *in vitro* with interleukin-2, and interferons. Although there have been some cases of dramatic success with this immunological approach, it does not yet provide the hoped for "magic bullet" against cancer. More recently, clinical trials may proceed using specially prepared "tumor-infiltrating lymphocytes" in selected terminally ill cancer patients.

***T Cell Receptor Proteins.*** Unlike B cells, T cells do not make antibodies and thus do not have antibodies on their surface to serve as receptors for antigens. The T cells do, however, have specific receptors for antigens on their membrane surface, and these T cell receptors have recently been identified as molecules closely related to immunoglobulins. The T cell receptors differ from the antibody receptors on B cells in another, and very important, characteristic: they *cannot bond to free antigens.* In order for a T lymphocyte to respond to a foreign antigen, the antigen must be presented to the T lymphocyte on the membrane of an *antigen-presenting cell.* The chief antigen-presenting cells are macrophages, which present the foreign antigen together with other surface antigens, called *histocompatibility antigens,* to the T lymphocytes. Some knowledge of the histocompatibility antigens is thus required before T cell–macrophage interactions and T cell functions can be understood.

***Histocompatibility Antigens.*** Tissue that is transplanted from one person to another contains antigens that are foreign to the host. This is because all tissue cells, with the exception of mature red blood cells, are genetically marked with a characteristic combination of **histocompatibility antigens** on the membrane surface. The greater the difference in these antigens between the donor and the recipient in a transplant, the greater will be the chance of transplant rejection. Prior to organ transplantation, therefore, the "tissue type" of the recipient is matched to that of potential donors. Since the person's white blood cells are used for this purpose, an alternate name for histocompatibility antigens in humans is **human leukocyte antigens,** abbreviated **HLA.**

**Figure 19.17.** There are four human histocompatibility antigens (or human leukocyte antigens—HLA). Each of these antigens is coded by a gene located on chromosome number 6, as indicated in the expanded view on the right.

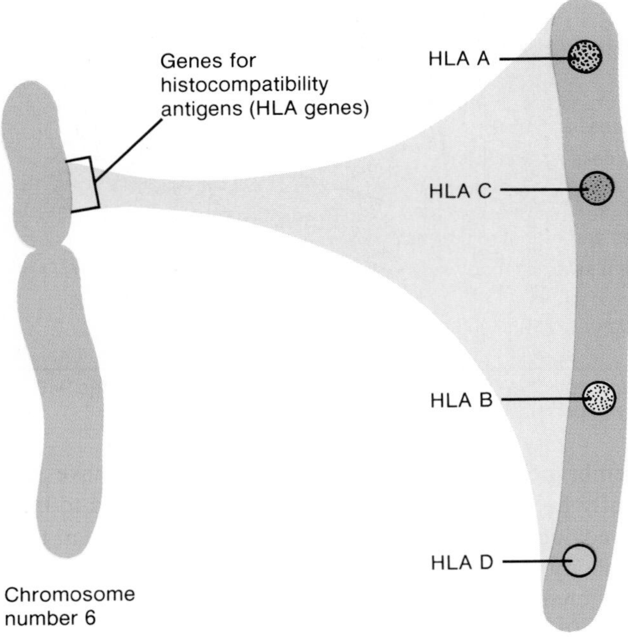


The histocompatibility antigens (HLA) are proteins that are coded by a group of genes, called the *major histocompatibility complex* (*MHC*), located on chromosome number 6 (fig. 19.17). These four genes are labeled A, B, C, and D. Each of these genes can code for only one protein in a given individual, but this protein can be different in different people. Two people, for example, may both have antigen A3, but one might have antigen B17 while the other has antigen B21. The closer two people are related, the more similar their histocompatibility antigens will be.

Clinical interest has been generated by the observation that certain diseases are much more common in people who have particular histocompatibility antigens (table 19.10). Ankylosing spondylitis (a type of rheumatoid arthritis), for example, is much more common in people who have antigen B27, and psoriasis (a skin disorder) is three times more common in people with antigen B17 than in the general population. Some other diseases that have a high correlation with particular histocompatibility antigens include Hodgkin's disease (a cancer of the lymph nodes), myasthenia gravis, Graves' disease, and type I diabetes mellitus.

**Table 19.10** Association between particular human histocompatibility antigens and diseases

| Disease | HLA Antigen | Frequency in Patients (%) | Frequency in Controls (%) |
|---|---|---|---|
| Ankylosing spondylitis | B27<br>B13 | 90<br>18 | 7<br>4 |
| Psoriasis | B17 | 29 | 8 |
| | B16 | 15 | 5 |
| Graves' disease | B8 | 47 | 21 |
| Coeliac disease | B8 | 78 | 24 |
| Dermatitis herpetiformis | B8 | 62 | 27 |
| Myasthenia gravis | B8 | 52 | 24 |
| SLE (systemic lupus erythematosus) | B15 | 33 | 8 |
| Multiple sclerosis | A3 | 36 | 25 |
| | B7 | 36 | 25 |
| Acute lymphatic leukemia | A2<br>B35 | 63<br>25 | 37<br>16 |
| Hodgkin's disease | A1 | 39 | 32 |
| | B8 | 26 | 22 |
| Chronic hepatitis | B8 | 68 | 18 |
| Ragweed hay fever | B7 | 50 | 19 |

From H. McDevitt and W. Bodmer, in *The Lancet 1,* p. 1269, 1974. Copyright © 1974 *The Lancet,* London, England. Reprinted by permission of the publisher and author.

## Interactions between Macrophages and T Lymphocytes

The major histocompatibility complex of genes produces two classes of HLA antigens, designated *class 1* and *class 2*. The class-1 molecules are made by all cells in the body except red blood cells. When these are combined with a foreign antigen, as will be discussed, they serve to identify the cell as a target for T cell destruction. Class-2 molecules comprise the D group of HLA antigens and, most importantly, are produced only by macrophages and B lymphocytes. The class-2 HLA antigens thus serve to promote the interactions between T cells and the other cells of the immune system.

When a foreign particle, such as a virus, infects the body, it is taken into macrophages by phagocytosis and partially digested. Within the macrophage, the partially digested virus particles provide foreign antigens that are moved to the surface of the cell membrane. At the membrane, these foreign antigens form a complex with the class-2 HLA antigens. This combination of HLA and foreign antigens is required for interaction with the receptors on the surface of helper T cells. The macrophages thus "present" the antigens to the helper T cells, and in this way stimulate activation of the T cells (fig. 19.18). It should be remembered that T cells are "blind" to free antigens; they can only respond to antigens presented to them by macrophages (and some other cells) in combination with class-2 HLA antigens.

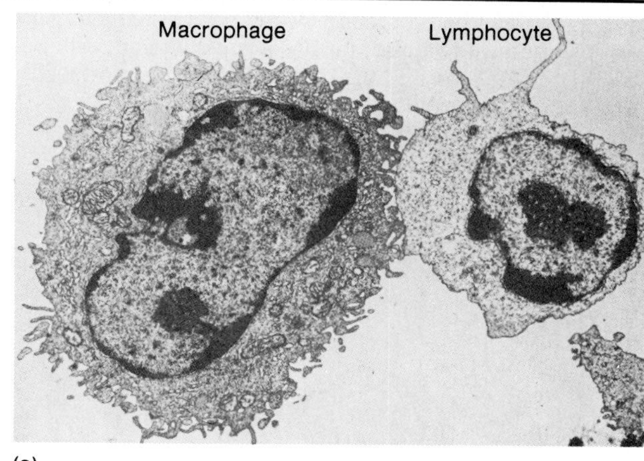


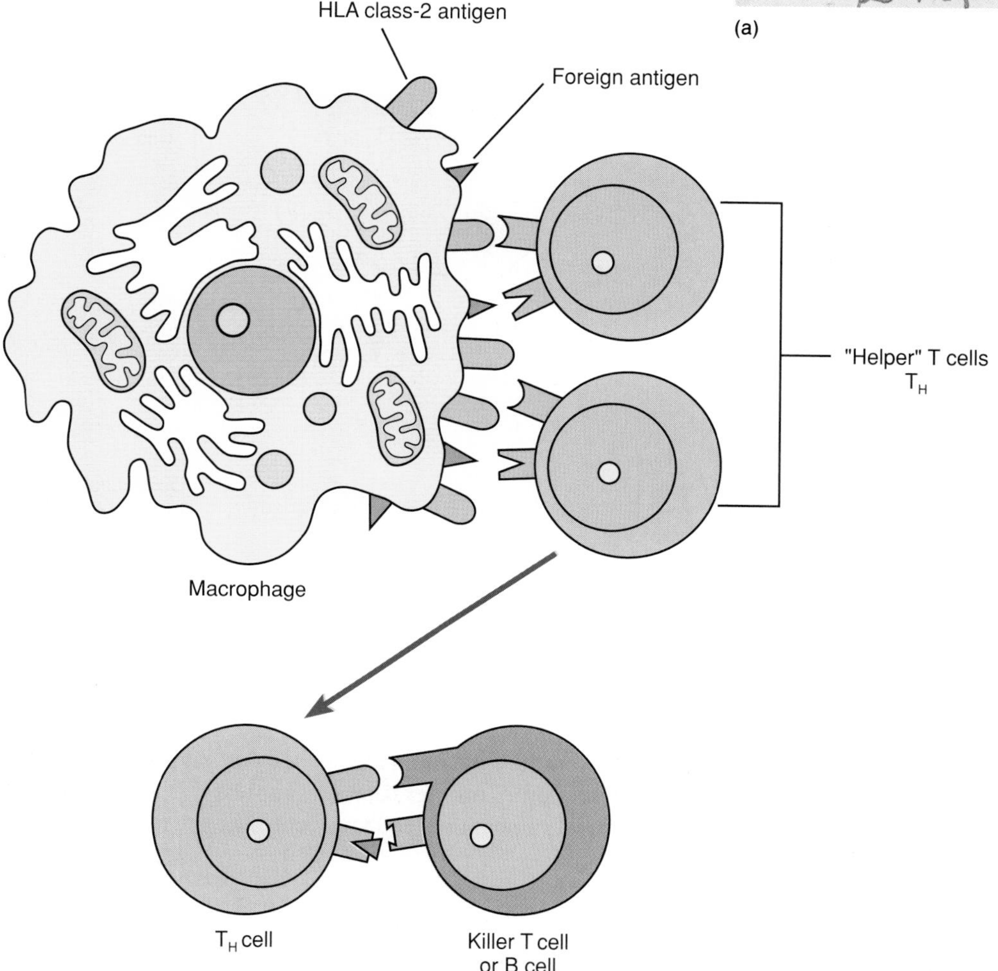


**Figure 19.18.** (*a*) An electron micrograph showing contact between a macrophage (*left*) and a lymphocyte (*right*). As illustrated in (*b*), such contact between a macrophage and a T cell requires that the helper T cell interact with both the foreign antigen and the HLA class-2 antigen on the surface of the macrophage.

The first phase of macrophage–T cell interaction then occurs: the macrophage is stimulated to secrete the lymphokine known as interleukin-1. As previously discussed, interleukin-1 stimulates cell division and proliferation of B and T lymphocytes. The activated helper T cells, in turn, secrete macrophage colony-stimulating factor and gamma interferon, which promote the activity of macrophages. In addition, interleukin-2 is secreted by the T lymphocytes and stimulates the macrophages to secrete *tumor necrosis factor*, which is particularly effective in killing cancer cells.

Cytotoxic T cells can only destroy infected cells if those cells display the foreign antigen together with their class-1 HLA antigens (fig. 19.19). Such interaction of cytotoxic T cells with the foreign antigen–HLA class-1 complex also stimulates proliferation of those cytotoxic T cells. This proliferation is supported by interleukin-2 secreted by the helper T cells, which were activated by macrophages as previously described (fig. 19.20).

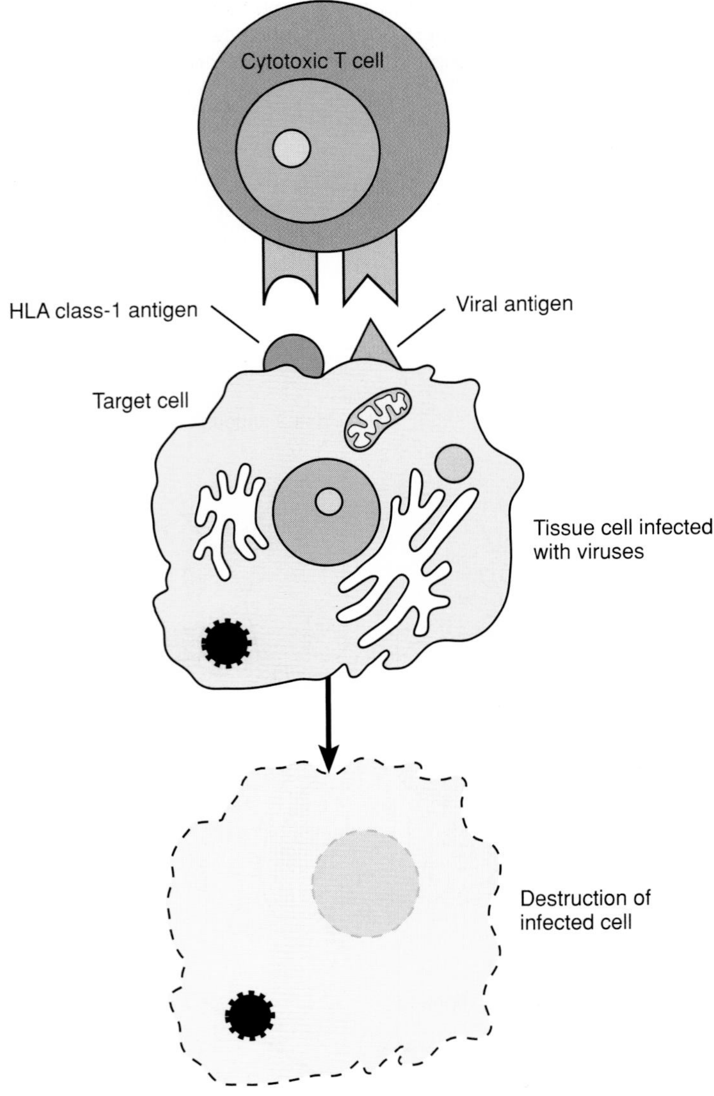


**Figure 19.19.** In order for a killer T cell to destroy a tissue cell infected with viruses, the T cell must interact with both the foreign antigen and the class-1 HLA antigen on the surface of the infected cell.

The network of interaction among different cell types of the immune system now spreads outwards. Helper T cells, activated to an antigen by macrophages, can also promote the humoral immune response of B cells. In order to do this, the membrane receptor proteins on the surface of the helper T lymphocytes must interact with molecules on the surface of the B cells. This occurs when the foreign antigen attaches to the immunoglobulin receptors on the B cells, so that the B cells can present this antigen together with its HLA class-2 antigens to the receptors on the helper T cells (fig. 19.21). This interaction stimulates proliferation of the B cells, their conversion to plasma cells, and their secretion of antibodies against the foreign antigens.

Suppressor T lymphocytes may also be activated by the interleukin-2 secreted by the helper T cells. Proliferation of the suppressor T cells is believed to occur at a slower rate than proliferation of cytotoxic T cells or B cells. This slower proliferation of suppressor T cells may help provide a negative feedback control of the immune system.

Glucocorticoids (such as hydrocortisone), secreted by the adrenal cortex, can act to suppress the activity of the immune system and inflammation. This is why **cortisone** and its analogues are used clinically in the treatment of inflammatory disorders and to inhibit the immune rejection of transplanted organs. The immunosuppressive effect of these hormones may result from the fact that they inhibit the secretion of the lymphokines. It is interesting in this regard that interleukin-1 has recently been shown to stimulate ACTH secretion. Rising ACTH, in turn, stimulates glucocorticoid secretion (chapter 11), which inhibits the immune system and suppresses interleukin-1 secretion. Immune control thus involves interaction not only among the cells of the immune system, but also between the immune and endocrine systems.

**Figure 19.20.** Interaction between macrophages, helper T lymphocytes, cytotoxic T lymphocytes, and infected cells in the immunological defense against viral infections.

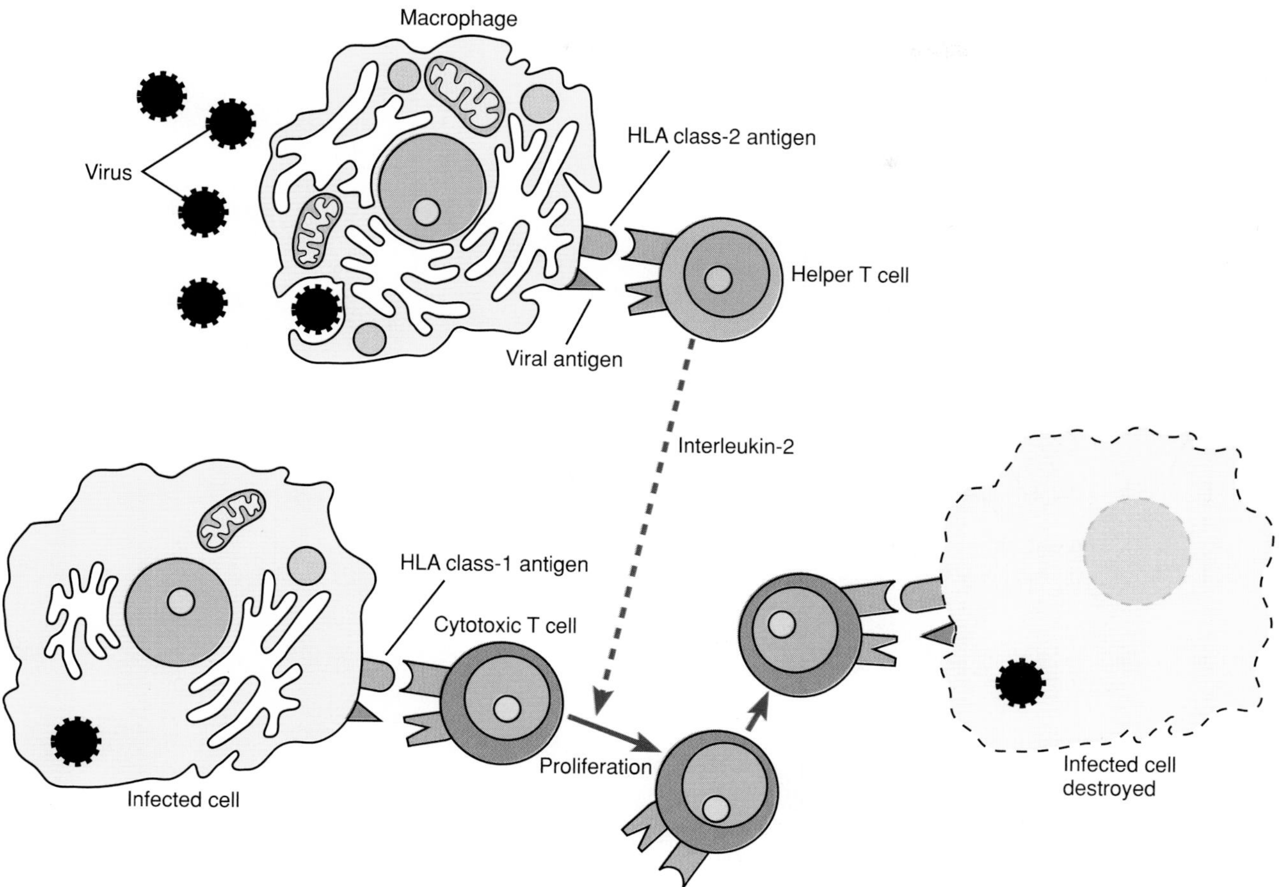


## Tolerance

The ability to produce antibodies against foreign, **non-self** antigens, while tolerating (not producing antibodies against) **self-**antigens occurs during the first month or so of postnatal life when immunological competence is established. If a fetal mouse of one strain receives transplanted antigens from a different strain, therefore, it will not recognize tissue transplanted later in life from the other strain as foreign and, as a result, will not immunologically reject the transplant.

The ability of an individual's immune system to recognize and tolerate self-antigens requires continuous exposure of the immune system to those antigens. If this exposure begins when the immune system is weak—such as in fetal and early postnatal life—tolerance is more complete and long lasting than when exposure occurs later in life. Some self-antigens, however, are normally hidden from the blood, such as thyroglobulin within the thyroid gland and lens protein in the eye. An exposure to these self-antigens results in antibody production just as if these proteins were foreign. Antibodies made against self-antigens are called **autoantibodies.**

The reasons why autoantibodies are not normally produced—that is, why tolerance to self-antigens occurs—, are not well understood. There are two general types of theories that have been proposed to account for immunological tolerance: (1) *clone deletion;* and (2) *immunological suppression.* According to the clone deletion theory, tolerance to self-antigens is achieved by destruction of the lymphocytes that inherit the ability to make autoantibodies. Presumably this occurs primarily during fetal life, when those lymphocytes that have receptors on their surface for self-antigens are recognized and destroyed. There is evidence that clonal deletion does in fact occur, but there is also evidence for the alternate theory.

According to the immunological suppression theory, the lymphocytes that make autoantibodies are present throughout life but are normally inhibited from attacking self-antigens. This can be due to the effects of suppressor T lymphocytes and/or antibodies that block the actions of autoantibodies. An alteration in the ratio of suppressor to helper T lymphocytes in later life, or a shift in the interactions among autoantibodies and their antibody blockers, therefore, might result in the production of autoantibodies.

**Figure 19.21.** Schematic diagram of the events that are believed to occur in the interactions of macrophages, helper T lymphocytes, and B lymphocytes.

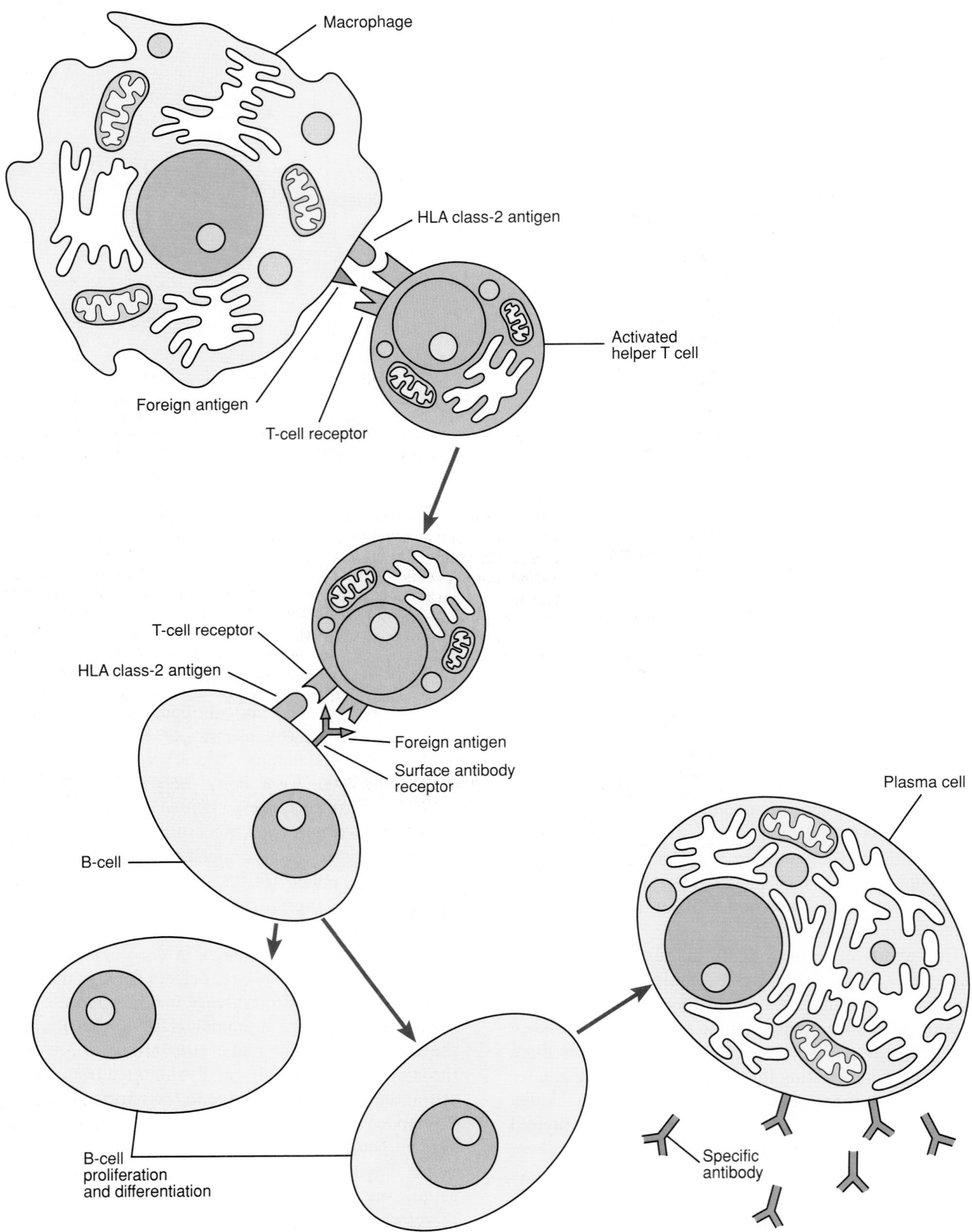

When foreign tissue is transplanted to an immunologically competent host, it is normally rejected. This may be due to the fact that the T lymphocytes of the host respond to the foreign HLA class-1 antigens as if they represented a combination of foreign antigens and host HLA antigens.

A spontaneous mutation in mice leads to the development of a strain that suffers from severe combined immunodeficiency (SCID). This condition is similar to a rare congenital condition in humans, and results in the absence of B and T lymphocytes. Grafts in SCID mice are therefore not rejected. This inability to reject transplants has recently been exploited by reconstituting a human immune system in the SCID mice, using lymphocytes from peripheral human blood or human fetal liver, thymus, and lymph node grafts. This technique may provide a means for studying the function of the immune system, and for studying diseases of the human immune system. Since the HIV virus, for example, only infects human (and chimpanzee) lymphocytes, the mouse-human chimera may provide an animal model for experimental investigation of AIDS.

1. *Describe the role of the thymus in cell-mediated immunity.*
2. *Define the term* lymphokines, *identify their origin, and list the different functions of these molecules.*
3. *Define the term* histocompatibility antigens, *and explain the importance of class-1 and class-2 HLA antigens in the function of T cells.*
4. *Describe the requirements for activation of helper T cells by macrophages, and explain how helper T cells promote the immunological defenses provided by cytotoxic T cells and by B cells.*
5. *Describe the mechanisms that have been proposed to explain tolerance of self-antigens.*

## *Tumor Immunology*

Tumor cells usually reveal antigens that may activate an immune reaction that destroys the tumor. When cancers develop, this immunological surveillance system—primarily the function of T cells and natural killer cells—has failed to prevent the growth and metastasis of the tumor.

Oncology (the study of tumors) has revealed that tumor biology is similar to and interrelated with the functions of the immune system. Most tumors appear to be clones of single cells that have become transformed. This is similar to the development of lymphocyte clones in response to specific antigens. Lymphocyte clones, however, are under complex inhibitory control systems—such as those exerted by suppressor T lymphocytes and negative feedback by antibodies. The division of tumor cells, in contrast, is not effectively controlled by normal inhibitory mechanisms. Tumor cells are also relatively unspecialized—they *dedifferentiate,* which means that they become similar to the less-specialized cells of an embryo.

Tumors are described as *benign* when they are relatively slow growing and limited to a specific location (warts, for example). *Malignant* tumors grow more rapidly and undergo **metastasis,** a term that refers to the dispersion of tumor cells and the resulting seeding of new tumors in different locations. The term **cancer,** as it is usually used, refers to malignant, life-threatening tumors.

As tumors dedifferentiate, they reveal surface antigens that can stimulate the immune destruction of the tumor cells. Consistent with the concept of dedifferentiation, some of these antigens are proteins produced in embryonic or fetal life that are not normally produced postnatally. Since they are absent at the time immunological competence is established they are treated as foreign and fit subjects for immunological attack when they are produced by cancerous cells. The release of two such antigens into the blood has provided the basis for a laboratory diagnosis of some cancers. *Carcinoembryonic antigen* tests are useful in the diagnosis of colon cancer, for example, and tests for *alpha-fetoprotein* (normally produced only by the fetal liver) help in the diagnosis of liver cancer.

Tumor antigens activate the immune system, initiating an attack primarily by cytotoxic T lymphocytes (fig. 19.22) and natural killer cells (described below). The concept of **immunological surveillance** against cancer was introduced in the early 1970s to describe the proposed role of the immune system in fighting cancer. According to this concept, tumor cells originate frequently in the body but are normally recognized and destroyed by the immune system before they can cause cancer. There is evidence that immunological surveillance does prevent some types of cancer; this explains why, for example, AIDS victims (with a depressed immune system) have a high incidence of Kaposi's sarcoma. It is not clear, however, why all types of cancers do not appear with high frequency in AIDS patients and others whose immune system is suppressed. For these reasons, the generality of the immunological surveillance system concept is currently controversial.

### Natural Killer Cells

Researchers observed that a strain of hairless mice, which genetically lacks a thymus and T lymphocytes, does not suffer from a particularly high incidence of tumor production. This surprising observation led to the discovery of **natural killer (NK) cells,** which are lymphocytes that are related to, but distinct from, T lymphocytes. Unlike killer T cells, NK cells destroy tumors in a nonspecific fashion and do not require prior exposure for sensitization to the tumor antigens. The NK cells thus provide a first line of cell-mediated defense, which is subsequently backed

**Figure 19.22.** A killer T cell (*a*) contacts a cancer cell (the larger cell), in a manner that requires specific interaction with antigens on the cancer cell. The killer T cell releases lymphokines, including toxins that cause the death of the cancer cell (*b*). (Scanning electron micrographs © Andrejs Liepens.)

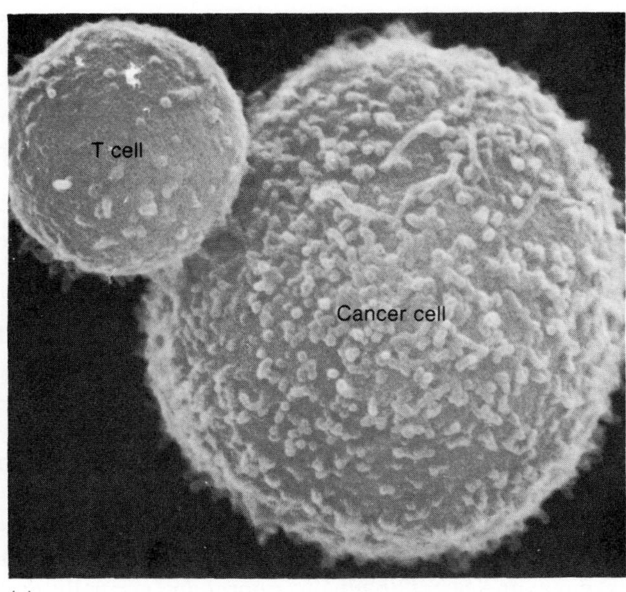


(a)

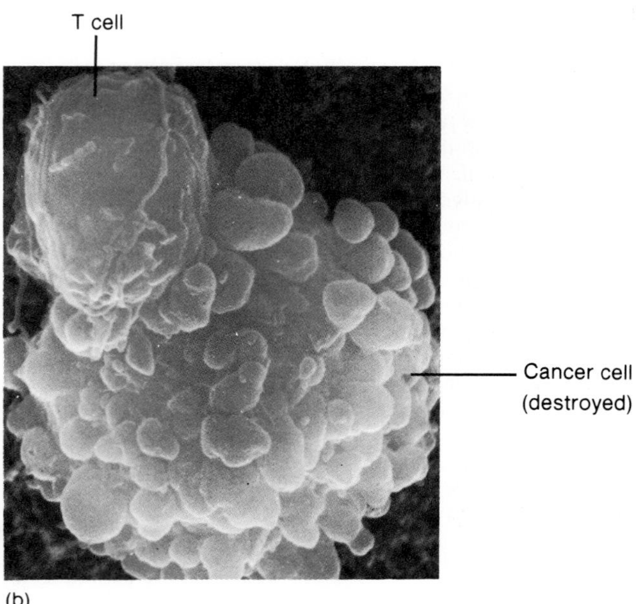


(b)

up by a specific response mediated by killer T cells. These two cell types interact, however; the activity of NK cells is stimulated by interferon, released as one of the lymphokines from T lymphocytes.

## Effects of Aging and Stress

Susceptibility to cancer varies greatly. The Epstein-Barr virus that causes Burkitt's lymphoma in a few individuals in Africa, for example, can also be found in healthy people throughout the world. In most cases the virus is harmless; in some cases mononucleosis (involving a limited proliferation of white blood cells) is produced. Only rarely does this virus cause the uncontrolled proliferation of leukocytes occurring in Burkitt's lymphoma. The reasons for these different responses to the Epstein-Barr virus and indeed for the different susceptibilities of people to other forms of cancer are not well understood.

It is known that cancer risk increases with age. According to one theory, this is due to the fact that aging lymphocytes accumulate genetic errors over the years that decrease their effectiveness. The secretion of thymus hormones also decreases with age in parallel with a decrease in cell-mediated immune competence. Both of these changes and perhaps others not yet discovered could increase susceptibility to cancer.

Numerous experiments have demonstrated that tumors grow faster in experimental animals subject to stress than in unstressed control animals. This is generally believed to result from the fact that stressed animals, including humans, have increased secretion of corticosteroid hormones, which act to suppress the immune system (this is why cortisone is given to people who receive organ transplants and to people with chronic inflammatory diseases). Some recent experiments, however, suggest that the stress-induced suppression of the immune system may also be due to other factors that do not involve the adrenal cortex. Future advances in cancer therapy may incorporate methods of strengthening the immune system together with methods that directly destroy tumors.

1. *Explain why cancer cells are believed to be dedifferentiated, and describe some of the clinical applications of this concept.*
2. *Describe what the term "immunological surveillance" against cancer refers to, and identify the cells involved in this function.*
3. *Explain the possible relationships between stress and cancer susceptibility.*

**Figure 19.23.** Autoimmune thyroiditis in a rabbit, induced experimentally by injection with thyroglobulin. Compare the picture of a normal thyroid (*left*) with that of the diseased thyroid (*right*). The grainy appearance of the diseased thyroid is due to the infiltration of large numbers of lymphocytes and macrophages.

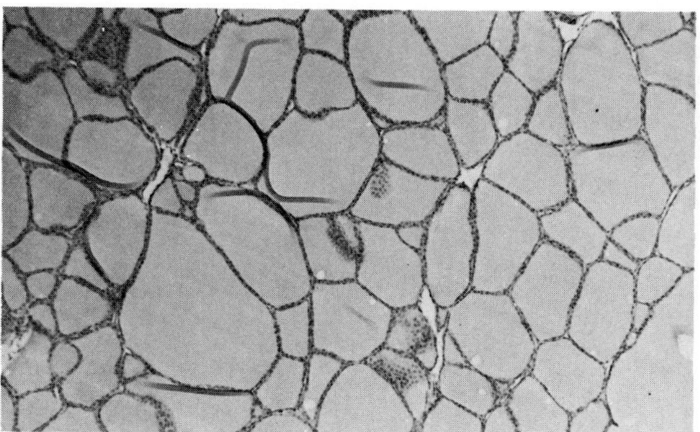

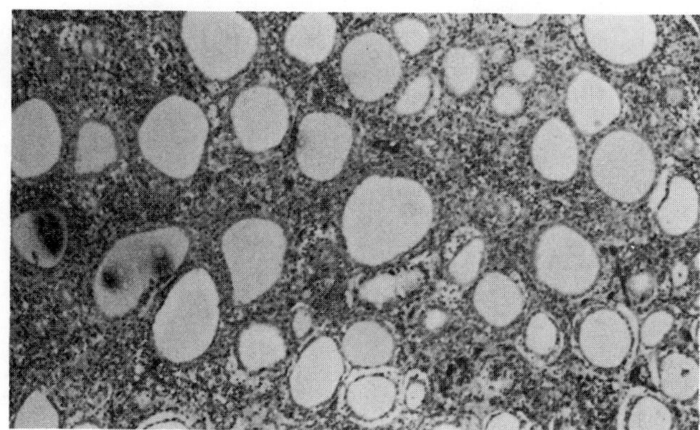

## Diseases Caused by the Immune System

Immune mechanisms that normally protect the body are very complex and subject to errors that can result in diseases. Autoimmune diseases and allergies are two categories of disease that are not caused by an invading pathogen, but rather by a derangement in the normal functions of the immune system.

The ability of the normal immune system to tolerate self-antigens while it identifies and attacks foreign antigens provides a specific defense against invading pathogens. In every individual, however, this system of defense against invaders at times commits domestic offenses. This can result in diseases that range in severity from the sniffles to sudden death.

Diseases caused by the immune system can be grouped into three interrelated categories: (1) *autoimmune diseases;* (2) *immune complex diseases;* and (3) *allergy,* or *hypersensitivity.* It is important to remember that these diseases are not caused by foreign pathogens but by abnormal responses of the immune system.

### Autoimmunity

Autoimmune diseases are those produced by failure of the immune system to recognize and tolerate self-antigens. This failure results in the production of autoantibodies that can cause inflammation and organ damage. Such autoimmune destruction may occur as a result of the following mechanisms.

**An antigen that does not normally circulate in the blood may become exposed to the immune system.** Thyroglobulin protein that is normally trapped within the thyroid follicles, for example, can stimulate the production of autoantibodies that cause the destruction of the thyroid (fig. 19.23); this occurs in *Hashimoto's thyroiditis.* Similarly, autoantibodies developed against lens protein in a damaged eye may cause the destruction of a healthy eye (in *sympathetic ophthalmia*).

**A self-antigen, which is otherwise tolerated, may be altered by combining with a foreign hapten.** The disease *thrombocytopenia* (low platelet count), for example, can be caused by the autoimmune destruction of thrombocytes (platelets). This occurs when drugs such as aspirin, sulfonamide, antihistamines, digoxin, and others combine with platelet proteins to produce new antigens. The symptoms of this disease usually stop when the person stops taking these drugs.

**Antibodies may be produced that are directed against other antibodies. Such interactions may be necessary for the prevention of autoimmunity, as previously suggested, but imbalances may actually cause autoimmune diseases.** *Rheumatoid arthritis,* for example, is an autoimmune disease associated with the abnormal production of one group of antibodies (of the IgM type) that attack other antibodies (of the IgG type). This results in an inflammation reaction of the joints characteristic of the disease.

**Table 19.11** Examples of autoimmune diseases

| Disease | Antigen | Ig and/or T Cell Response |
|---|---|---|
| Postvaccinal and postinfectious encephalomyelitis | Myelin, cross-reactive | T cell |
| Aspermatogenesis | Sperm | T cell |
| Sympathetic ophthalmia | Uvea | T cell |
| Hashimoto's disease | Thyroglobulin | IgG and T cell |
| Graves' disease | Receptor proteins for TSH | Thyroid-stimulating antibody (TSAb) |
| Autoimmune hemolytic disease | I, Rh, and others on surface of RBCs | IgM and IgG |
| Thrombocytopenic purpura | Hapten-platelet or hapten-adsorbed antigen complex | IgG |
| Myasthenia gravis | Acetylcholine receptors | IgG |
| Rheumatic fever | Streptococcal cross-reactive with heart | IgG and IgM |
| Glomerulonephritis | Streptococcal cross-reactive with kidney | IgG and IgM |
| Rheumatoid arthritis | IgG | IgM to Fc($\gamma$) |
| Systemic lupus erythematosus | DNA, nucleoprotein, RNA, etc. | IgG |

Barrett, James T.: Textbook of immunology, ed. 5, St. Louis, 1988, The C. V. Mosby Co.

**Antibodies produced against foreign antigens may cross-react with self-antigens.** Autoimmune diseases of this sort can occur, for example, as a result of *Streptococcus* bacterial infections. Antibodies produced in response to antigens in this bacterium may cross-react with self-antigens in the heart and kidneys. The inflammation induced by such autoantibodies can produce heart damage (including the valve defects characteristic of *rheumatic fever*) and damage to the glomerular capillaries in the kidneys (*glomerulonephritis*).

**Self-antigens, such as receptor proteins, may be presented to the helper T lymphocytes together with class-2 HLA antigens.** Normally, only macrophages and B lymphocytes produce class-2 HLA antigens, which are associated with foreign antigens and are recognized by helper T cells. Perhaps as a result of viral infection, however, cells that do not normally produce class-2 HLA antigens may start to do so and, in this way, present a self-antigen to the helper T cells. In *Graves' disease,* for example, the thyroid cells produce class-2 HLA antigens, and the immune system produces autoantibodies against the TSH receptor proteins in the thyroid cells. These autoantibodies, called *TSAb* for "thyroid-stimulating antibody," interact with the TSH receptors and overstimulate the thyroid gland. Similarly, in type I *diabetes mellitus*, the beta cells of the pancreatic islets abnormally produce class-2 HLA antigens, resulting in autoimmune destruction of the insulin-producing cells.

**Genes that code for antibodies against foreign antigens may mutate to produce autoantibodies.** This would not cause particular diseases but would lead to the increased frequency of autoimmune diseases in general. An increased frequency of autoimmune diseases does in fact occur with age, as predicted by this mutation theory. Table 19.11 provides some examples of autoimmune diseases.

## Immune Complex Diseases

The term *immune complexes* refers to combinations of antibodies with antigens that are free rather than attached to bacterial or other cells. The formation of such complexes activates complement proteins and promotes inflammation. This inflammation normally is self-limiting because the immune complexes are removed by phagocytic cells. When large numbers of immune complexes are continuously formed, however, the inflammation may be prolonged. Also, the dispersion of immune complexes to other sites can lead to widespread inflammations and organ damage. The damage produced by the inflammatory response to antigens is called **immune complex disease.**

Immune complex diseases can result from infections by bacteria, parasites, and viruses. In viral hepatitis B, for example, an immune complex that consists of viral antigens and antibodies can cause widespread inflammation of arteries *(periarteritis)*. Note that the arterial damage is not caused by the hepatitis virus itself but by the inflammatory process.

Immune complex diseases can also result from the formation of complexes between self-antigens and autoantibodies. This is the case in rheumatoid arthritis, where the inflammation is produced by complexes of altered IgG antibodies (the antigens in this case) and IgM antibodies. Another immune complex disease that has an autoimmune basis is *systemic lupus erythematosus (SLE)*. People with SLE produce antibodies against their own DNA and nuclear proteins. This can result in the formation of immune complexes throughout the body, including the glomerular capillaries where glomerulonephritis may be produced.

**Table 19.12** Allergy: comparison between immediate and delayed hypersensitivity reaction

| Characteristic | Immediate Reaction | Delayed Reaction |
|---|---|---|
| Time for onset of symptoms | Within several minutes | Within one to three days |
| Lymphocytes involved | B cells | T cells |
| Immune effector | IgE antibodies | Cell-mediated immunity |
| Allergies most commonly produced | Hay fever, asthma, and most other allergic conditions | Contact dermatitis (such as to poison ivy and poison oak) |
| Therapy | Antihistamines and adrenergic drugs | Corticosteroids (such as cortisone) |

## Allergy

The term *allergy,* usually used synonymously with *hypersensitivity,* refers to particular types of abnormal immune responses to antigens, which are called *allergens* in these cases. There are two major forms of allergy: (1) **immediate hypersensitivity,** which is due to an abnormal B lymphocyte response to an allergen that produces symptoms within seconds or minutes; and (2) **delayed hypersensitivity,** which is an abnormal T cell response that produces symptoms within about forty-eight hours after exposure to an allergen. A comparison between these two types of hypersensitivity is provided in table 19.12.

***Immediate Hypersensitivity.*** Immediate hypersensitivity can produce such symptoms as allergic rhinitis (chronic runny or stuffy nose), conjunctivitis (red eyes), allergic asthma, atopic dermatitis (urticaria, or hives), and others. These symptoms result from the production of antibodies of the IgE subclass, instead of the normal IgG antibodies.

Unlike IgG antibodies, IgE antibodies do not circulate in the blood but instead attach to tissue mast cells (which have membrane receptors for these antibodies). When the person is again exposed to the same allergen, the allergen bonds to the antibodies attached to the mast cells. This stimulates the mast cells to secrete various chemicals, including **histamine** (fig. 19.24). During this process, leukocytes may also secrete **prostaglandins** and related molecules called **leukotrienes.** These chemicals (table 19.13) produce the symptoms of the allergic reactions.

**Figure 19.24.** Allergy (immediate hypersensitivity) is produced when antibodies of the IgE subclass attach to tissue mast cells. The combination of these antibodies with allergens (antigens that provoke an allergic reaction) cause the mast cell to secrete histamine and other chemicals that produce the symptoms of allergy.

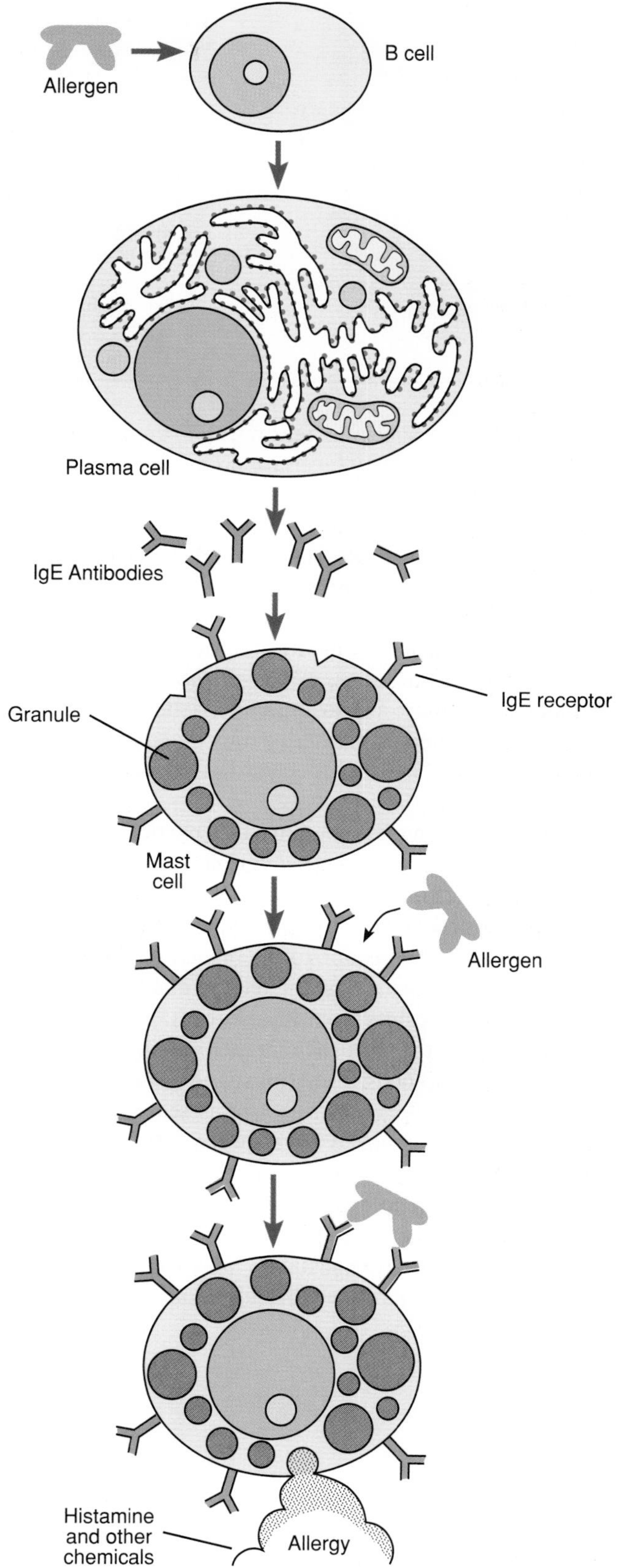

**Table 19.13** Some of the chemicals released by mast cells and leukocytes that are responsible for immediate hypersensitivity

| Chemical | Derivation | Action |
|---|---|---|
| Histamine | From histidine | Contracts smooth muscles in bronchioles; dilates blood vessels; increases capillary permeability |
| Serotonin | From tryptophane | Contracts smooth muscles |
| Prostaglandins and leukotrienes | Arachidonic acid | Prolonged contraction of smooth muscle; increased capillary permeability |
| Eosinophil chemotactic factor | Polypeptides | Attracts eosinophils |
| Bradykinin and related compounds | Polypeptides | Slow smooth muscle contraction |

The itching, sneezing, tearing, and runny nose of persons suffering from hay fever are produced largely by histamine and can be treated effectively by antihistamine drugs. Food allergies, causing diarrhea and colic, are mediated primarily by prostaglandins and can be treated with aspirin, which inhibits prostaglandin synthesis (these are the only allergies that respond positively to aspirin). Asthma, produced by smooth muscle constriction in the bronchioles in the lungs, is due to the release of leukotrienes. Since there are no antileukotriene drugs presently available, asthma is treated with epinephrinelike compounds (which cause bronchodilation) and corticosteroids.

Immediate hypersensitivity to a particular antigen is commonly tested by injecting various antigens under the skin (fig. 19.25). Within a short time a *flare-and-wheal reaction* is produced if the person is allergic to that antigen. This reaction is due to the release of histamine and other chemical mediators: the flare is due to vasodilation, and the wheal results from local edema.

Allergens that provoke immediate hypersensitivity include various foods, bee stings, and pollen grains. The most common allergy of this type is seasonal hay fever, which may be provoked by ragweed *(Ambrosia)* pollen grains. People with chronic allergic rhinitis and asthma due to an allergy to dust or feathers are usually allergic to a tiny mite (fig. 19.26) that lives in dust and eats the scales of skin that are constantly shed from the body. Actually, most of the antigens from the dust mite are not in its body but rather in its feces, which are tiny particles that can enter the nasal mucosa much like pollen grains.

**Figure 19.25.** Skin test for allergy. If an allergen is injected into the skin of a sensitive individual, a typical flare-and-wheal response occurs within several minutes.

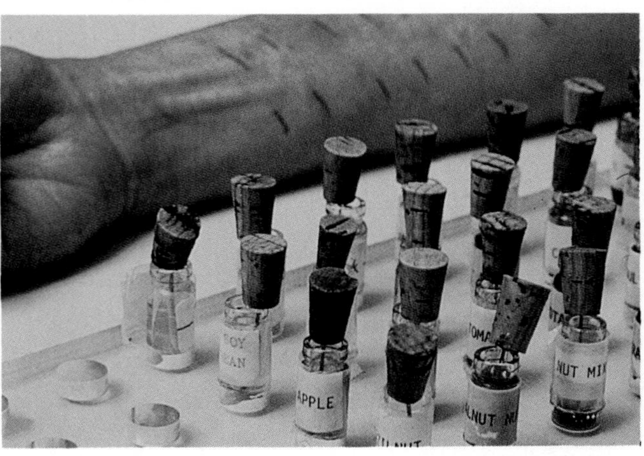


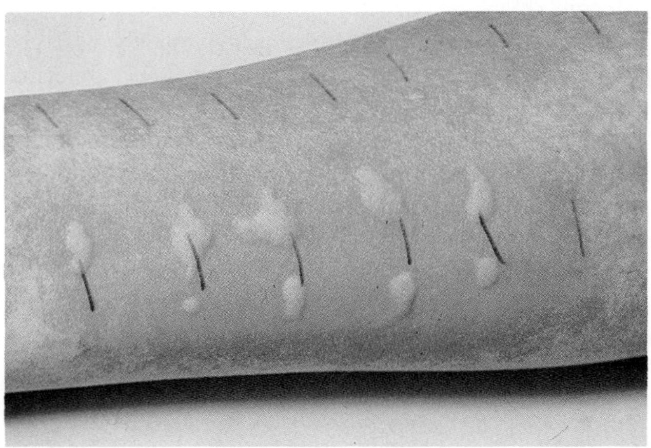

***Delayed Hypersensitivity.*** As the name implies, a longer time is required for the development of symptoms in delayed hypersensitivity (hours to days) than in immediate hypersensitivity. This may be due to the fact that immediate hypersensitivity is mediated by antibodies, whereas delayed hypersensitivity is a cell-mediated T lymphocyte response. Since the symptoms are caused by the secretion of lymphokines, rather than by the secretion of histamine, treatment with antihistamines provides little benefit. At present, corticosteroids are the only drugs that can effectively treat delayed hypersensitivity.

One of the best-known examples of delayed hypersensitivity is **contact dermatitis,** caused by poison ivy, poison oak, and poison sumac. The skin tests for tuberculosis—the tine test and the Mantoux test—also rely on

**Figure 19.26.** (*a*) A scanning electron micrograph of ragweed (*Ambrosia*), which is responsible for hay fever. (*b*) A scanning electron micrograph of the house dust mite (*Dermatophagoides farinae*), which lives in dust and is often responsible for yearlong allergic rhinitis and asthma.
([a] *From Scanning Electron Microscopy in Biology* by R. G. Kessel and C. Y. Shih. © Springer-Verlag, 1976.)

(a)

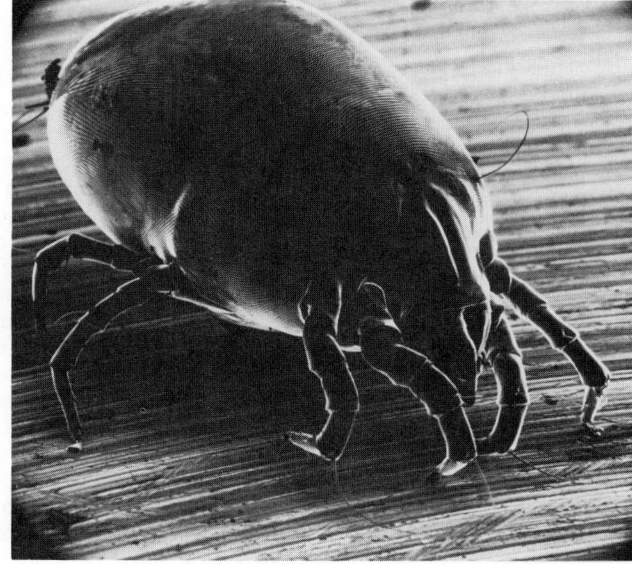
(b)

delayed hypersensitivity reactions. If a person has been exposed to the tubercle bacillus and has, as a result, developed T cell clones, skin reactions appear within a few days after the tubercle antigens are rubbed into the skin with small needles (tine test) or are injected under the skin (Mantoux test).

1. *Explain the possible mechanisms responsible for autoimmune diseases.*
2. *Distinguish between immediate and delayed hypersensitivity.*
3. *Describe the sequence of events by which allergens can produce symptoms of runny nose, skin rash, and asthma.*

## SUMMARY

### Defense Mechanisms p. 598

I. Nonspecific defense mechanisms include barriers to penetration of the body and internal defenses.
   A. Phagocytic cells engulf invading pathogens.
   B. Interferons are polypeptides secreted by cells infected with viruses that help protect other cells from viral infections.

II. Specific immune responses are directed against antigens.
   A. Antigens are molecules or parts of molecules that are usually large, complex, and foreign.
   B. A given molecule can have a number of antigenic determinant sites that stimulate the production of different antibodies.

III. Specific immunity is a function of lymphocytes.
   A. B lymphocytes secrete antibodies and provide humoral immunity.
   B. T lymphocytes provide cell-mediated immunity.

### Functions of B Lymphocytes p. 603

I. There are five subclasses of antibodies, or immunoglobulins.
   A. These subclasses differ in regard to the polypeptides in the constant region of the heavy chains.
   B. The five subclasses are IgG, IgA, IgM, IgD, and IgE.
   C. Each type of antibody has two variable regions that combine with specific antigens.
   D. The combination of antibodies with antigens promotes phagocytosis, a process called opsonization.

II. Antigen-antibody complexes activate a system of proteins called the complement system.
   A. This results in complement fixation, where complement proteins attach to a cell membrane and promote the destruction of the cell.
   B. Free complement proteins promote opsonization and chemotaxis and stimulate the release of histamine from tissue mast cells.

III. Specific and nonspecific immune mechanisms cooperate in the development of a local inflammation.

## Active and Passive Immunity p. 611

I. A primary response is produced when a person is first exposed to a pathogen; a subsequent exposure results in a secondary response.
   A. During the primary response IgM antibodies are produced slowly, and the person is likely to get sick.
   B. During the secondary response IgG antibodies are produced quickly, and the person has resistance to the pathogen.
   C. In active immunizations, the person is exposed to pathogens of attenuated virulence that have the same antigenicity as the virulent pathogen.
   D. The secondary response is believed to be due to the development of lymphocyte clones after the first exposure as a result of the antigen-stimulated proliferation of appropriate lymphocytes.

II. Passive immunity is provided by antibodies made by a different organism.
   A. This occurs naturally from mother to fetus.
   B. Injections of antiserum provide passive immunity to some pathogenic organisms and toxins.

III. Monoclonal antibodies are made by hybridomas, which are formed artificially by the fusion of B lymphocytes with multiple myeloma cells.

## Functions of T Lymphocytes p. 616

I. The thymus processes T lymphocytes and secretes hormones that are believed to be required for the proper function of the immune response of T lymphocytes throughout the body.

II. There are three subcategories of T lymphocytes.
   A. Cytotoxic, or killer, T lymphocytes kill victim cells by a mechanism that does not involve antibodies but that does require close contact between the cytotoxic T cell and the victim cell.
   B. Cytotoxic T lymphocytes are responsible for transplant rejection and for the immunological defense against fungal and viral infections, as well as for the defense against some bacterial infections.
   C. Helper T lymphocytes stimulate, and suppressor T lymphocytes suppress, the function of B lymphocytes and cytotoxic T lymphocytes.
   D. The T lymphocytes secrete a family of compounds called lymphokines, which promote the action of lymphocytes and macrophages; one of the more important lymphokines is interleukin-2, which stimulates proliferation of lymphocytes.
   E. Receptor proteins on the cell membrane of T lymphocytes must bond to both a foreign antigen and a histocompatibility antigen in order for the T cell to become activated.
   F. Histocompatibility antigens, or HLA antigens, are a family of molecules on the membranes of cells that are present in different combinations in different individuals.

III. Macrophages partially digest a foreign body, such as a virus, and present the antigens to the lymphocytes on the surface of the macrophage in combination with class-2 HLA antigens.
   A. Helper T lymphocytes require such interaction with macrophages in order to be activated by a foreign antigen; when activated in this way, the helper T cells secrete interleukin-2.
   B. Interleukin-2 stimulates proliferation of cytotoxic T lymphocytes that are specific for the foreign antigen.
   C. In order for the cytotoxic T lymphocytes to attack a victim cell, the victim cell must present the foreign antigen in combination with a class-1 HLA antigen.
   D. Interleukin-2, secreted by activated helper T lymphocytes, is also believed to stimulate proliferation of suppressor T lymphocytes, which proliferate more slowly than the cytotoxic T cells and provide a negative feedback control of the immune response.
   E. The interleukin-2 secreted by helper T cells also stimulates proliferation of B lymphocytes and so promotes the secretion of antibodies in response to the foreign antigen.

IV. Tolerance to self-antigens may be due to the destruction of lymphocytes that can recognize the self-antigens, or it may be due to suppression of the immune response by the action of specific suppressor T lymphocytes.

## Tumor Immunology p. 625

I. Immunological surveillance against cancer is provided mainly by cytotoxic T lymphocytes and natural killer cells.
   A. Cancerous cells dedifferentiate and may produce fetal antigens; these or other antigens may be presented to lymphocytes in association with abnormally produced class-2 HLA antigens.
   B. Natural killer cells are nonspecific; T lymphocytes are directed against specific antigens on the cancer cell surface.
   C. Immunological surveillance against cancer is weakened by stress.

## Diseases Caused by the Immune System p. 627

I. Autoimmune diseases are caused by the production of autoantibodies against self-antigens.
   A. The production of autoantibodies may result from lack of suppression by suppressor T cells or from the exposure of antigens that are normally hidden from the immune system or from the presentation of self-antigens to the immune system in association with class-2 HLA antigens, or from other mechanisms.
   B. Autoantibodies can cause inflammation and destruction of an organ, or—if the autoantibodies are directed against hormone receptor proteins—may cause overstimulation of an organ.

II. Immune complex diseases are those caused by the inflammation that results when free antigens are bonded to antibodies.

III. There are two types of allergic responses, which are characterized as immediate hypersensitivity and delayed hypersensitivity.
   A. Immediate hypersensitivity results when an allergen provokes the production of antibodies in the IgE class, which attach to tissue mast cells and stimulate the release of chemicals from the mast cells.

B. Mast cells secrete histamine, leukotrienes, and prostaglandins, which are believed to produce the symptoms of allergy.

C. Delayed hypersensitivity, as in contact dermatitis, is a cell-mediated response of T lymphocytes, which secrete the lymphokines that produce the allergic symptoms.

## Review Activities

### Objective Questions

1. Which of the following offers a nonspecific defense against viral infection?
   (a) antibodies
   (b) leukotrienes
   (c) interferon
   (d) histamine

Match the cell type with its secretion.

2. killer T cells
3. mast cells
4. plasma cells
5. macrophages

(a) antibodies
(b) perforins
(c) lysosomal enzymes
(d) histamine

6. Which of the following statements about the $F_{ab}$ portion of antibodies is *true?*
   (a) It bonds to antigens.
   (b) Its amino acid sequences are variable.
   (c) It consists of both H and L chains.
   (d) All of the above are true.
7. Which of the following statements about complement proteins C3a and C5a is *false*?
   (a) They are released during the complement fixation process.
   (b) They stimulate chemotaxis of phagocytic cells.
   (c) They promote the activity of phagocytic cells.
   (d) They produce pores in the victim cell membrane.
8. Mast cell secretion during an immediate hypersensitivity reaction is stimulated when antigens combine with
   (a) IgG antibodies
   (b) IgE antibodies
   (c) IgM antibodies
   (d) IgA antibodies
9. During a secondary immune response
   (a) antibodies are made quickly and in great amounts
   (b) antibody production lasts longer than in a primary response
   (c) antibodies of the IgG class are produced
   (d) lymphocyte clones are believed to develop
   (e) all of the above
10. Which of the following cells aids the activation of lymphocytes by antigens?
   (a) macrophages
   (b) neutrophils
   (c) mast cells
   (d) natural killer cells
11. Which of the following statements about T lymphocytes is *false?*
   (a) Some T cells promote the activity of B cells.
   (b) Some T cells suppress the activity of B cells.
   (c) Some T cells secrete interferon.
   (d) Some T cells produce antibodies.
12. Delayed hypersensitivity is mediated by
   (a) T cells
   (b) B cells
   (c) plasma cells
   (d) natural killer cells
13. Active immunity may be produced by
   (a) having a disease
   (b) receiving a vaccine
   (c) receiving gamma globulin injections
   (d) both a and b
   (e) both b and c
14. Which of the following statements about class-2 HLA antigens is *false?*
   (a) They are found on the surface of B lymphocytes.
   (b) They are found on the surface of macrophages.
   (c) They are required for B cell activation by a foreign antigen.
   (d) They are needed for interaction of helper and cytotoxic T cells.
   (e) They are presented together with foreign antigens by macrophages.

### Essay Questions

1. Explain how antibodies help destroy invading bacterial cells.
2. Explain how T lymphocytes interact with macrophages and the infected cells in fighting viral infections.
3. Explain the possible roles of helper and suppressor T lymphocytes in (a) defense against infections and (b) tolerance to self-antigens
4. Describe the clonal selection theory, and use this theory to explain how active immunity is produced.
5. Explain the physiological and clinical significance of histocompatibility antigens.

## Selected Readings


Acuto, O., and E. Reinherz. 1985. The human T cell receptor. Structure and function. *New England Journal of Medicine* 312:1100.

Ada, G. L., and G. Nossal. August 1987. The clonal selection theory. *Scientific American.*

Amit, A. G. et al. 1986. Three-dimensional structure of an antigen-antibody complex at 2.8 Å resolution. *Science* 233:747.

Baglioni, C., and T. W. Nilsen. 1981. The action of interferon at the molecular level. *American Scientist* 69:392.

Biusseret, P. D. August 1982. Allergy. *Scientific American.*

Capra, J. D., and A. B. Edmunson. January 1977. The antibody combining site. *Scientific American.*

Cohen, I. R. April 1988. The self, the world, and autoimmunity. *Scientific American.*

Croce, C. M. 1985. Chromosomal translocations, oncogenes, and B-cell tumors. *Hospital Practice* 20:41.

Cunningham, B. A. October 1977. The structure and function of histocompatibility antigens. *Scientific American.*

Dausset, J. 1981. The major histocompatibility complex in man: Past, present, and future concepts. *Science* 213:1469.

DiNome, M. A., and D. E. Young. August 1987. The clonal selection theory. *Scientific American.*

Hamburger, R. N. 1976. Allergy and the immune system. *American Scientist* 64:157.

Herberman, R. B., and J. R. Ortaldo. 1981. Natural killer cells: Their role in defense against disease. *Science* 214:24.

Kapp, J. A., C. W. Pierce, and C. M. Sorensen. August 1984. Antigen-specific suppressor T cell factors. *Hospital Practice* 19:85.

Koffler, D. July 1980. Systemic lupus erythematosus. *Scientific American.*

Laurence, J. December 1985. The immune system in AIDS. *Scientific American.*

Leder, P. November 1982. The genetics of antibody diversity. *Scientific American.*

McDevitt, H. O. 1985. The HLA system and its relation to disease. *Hospital Practice* 20:57.

Marrack, P., and J. Kappler. February 1986. The T cell and its receptor. *Scientific American.*

Matthews, T. J., and D. P. Bolognesi. October 1988. AIDS vaccines. *Scientific American.*

Milstein, C. October 1980. Monoclonal antibodies. *Scientific American.*

Milstein, C. 1986. From antibody structure to immunological diversification of immune response. *Science* 231:1261.

Nossal, G. J. V. 1987. The basic components of the immune system. *New England Journal of Medicine* 316:1320.

Oettgen, H. F. 1981. Immunological aspects of cancer. *Hospital Practice* 16:93.

Old, L. J. May 1977. Cancer immunology. *Scientific American.*

Old, L. J. May, 1988. Tumor necrosis factor. *Scientific American.*

Oldham, R. K. 1985. Biologicals for cancer treatment: Interferons. *Hospital Practice* 20:71.

Rose, N. R. February 1981. Autoimmune diseases. *Scientific American.*

Sachs, L. January 1986. Growth, differentiation, and the reversal of malignancy. *Scientific American.*

Samuelsson, B. 1983. Leukotrienes: Mediators of immediate hypersensitivity and inflammation. *Science* 220:568.

Steinmetz, M., and L. Hood. 1983. Genes of the major histocompatibility complex in mice and men. *Science* 222:727.

Tannock, I. F. 1983. Biology of tumor growth. *Hospital Practice* 18:81.

Tonegawa, S. October 1985. The molecules of the immune system. *Scientific American.*

Unanue, E. R., and P. M. Allen. 1987. The immunoregulatory role of the macrophage. *Hospital Practice* 22:87.

Vaughan, J. A. 1984. Rheumatoid arthritis: Evidence of a defect in T-cell function. *Hospital Practice* 19:101.

Weber, J. N., and R. A. Weiss. October 1988. HIV infection: The cellular picture. *Scientific American.*

Wolde-Mariam, W., and J. B. Peter. March 1984. Recent diagnostic advances in cellular immunology. *Diagnostic Medicine,* p. 25.

Yelton, D. E., and M. D. Scharff. 1980. Monoclonal antibodies. *American Scientist* 63:510.

Young, J. D., and Z. A. Cohen. January 1988. How killer cells kill. *Scientific American.*

# 20

# Reproduction

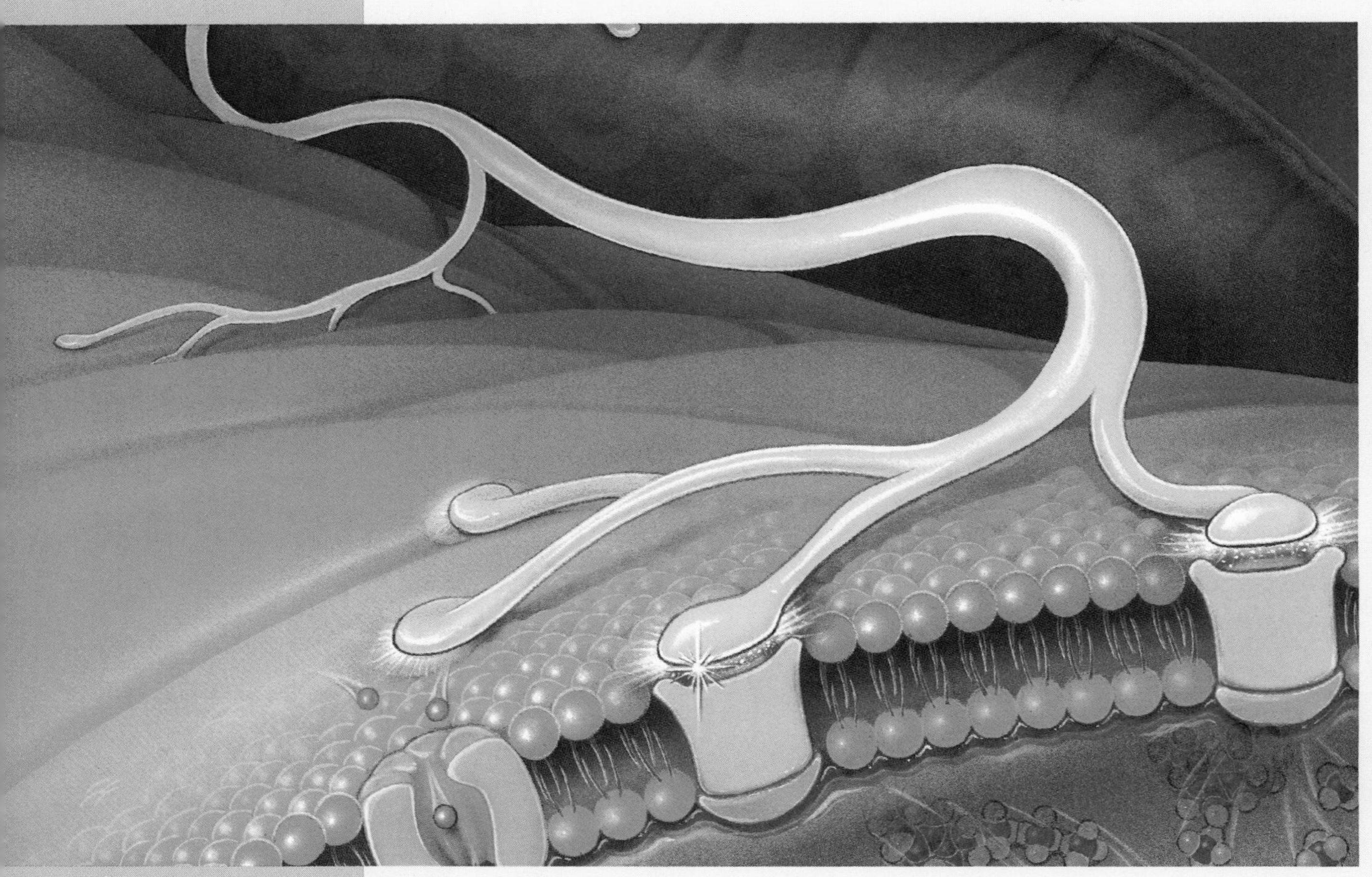

## Objectives

By studying this chapter, you should be able to

1. describe how the sex of an embryo is determined by the chromosomal content and how this relates to the development of testes or ovaries
2. explain how the development of sex accessory organs and external genitalia is affected by the presence or absence of testes in the embryo
3. describe the hormonal changes that occur during puberty, the mechanisms that might control the onset of puberty, and the changes in secondary sexual characteristics that result during puberty
4. explain how the secretions of pituitary gonadotropic hormones (FSH and LH) are regulated in the male and describe the actions of FSH and LH on the testis
5. describe the structure of the testis and how the interstitial Leydig cells and seminiferous tubules may interact
6. describe the stages of spermatogenesis and the roles of Sertoli cells in spermatogenesis
7. explain the hormonal control of spermatogenesis and the effects of androgens on the male sex accessory sexual organs
8. explain the physiology of erection and ejaculation and describe the composition of semen and the requirements for male fertility
9. describe oogenesis and the stages of ovarian follicle development through ovulation and the formation of a corpus luteum
10. explain the hormonal interactions involved in the control of ovulation
11. describe the changes that occur in ovarian sex steroid secretion during a nonfertile cycle and the function and fate of a corpus luteum during a nonfertile cycle
12. explain how the secretion of FSH and LH is controlled through negative and positive feedback mechanisms during a menstrual cycle
13. explain how contraceptive pills function to prevent ovulation
14. describe the cyclic changes that occur in the endometrium and the hormonal mechanisms that cause these changes
15. describe the acrosomal reaction and the events that occur at fertilization, blastocyst formation, and implantation
16. explain how menstruation and further ovulation are normally prevented during pregnancy
17. describe the structure and functions of the placenta
18. list the hormones secreted by the placenta and describe their actions
19. describe the factors that stimulate uterine contractions during labor and parturition and explain how the onset of labor may be regulated
20. explain the hormonal requirements for development of the mammary glands during pregnancy and how lactation is prevented during pregnancy
21. describe the milk-ejection reflex

## Outline

**Figure 20.1.** The human life cycle. Numbers in parentheses indicate haploid state (23 chromosomes) and diploid state (46 chromosomes).

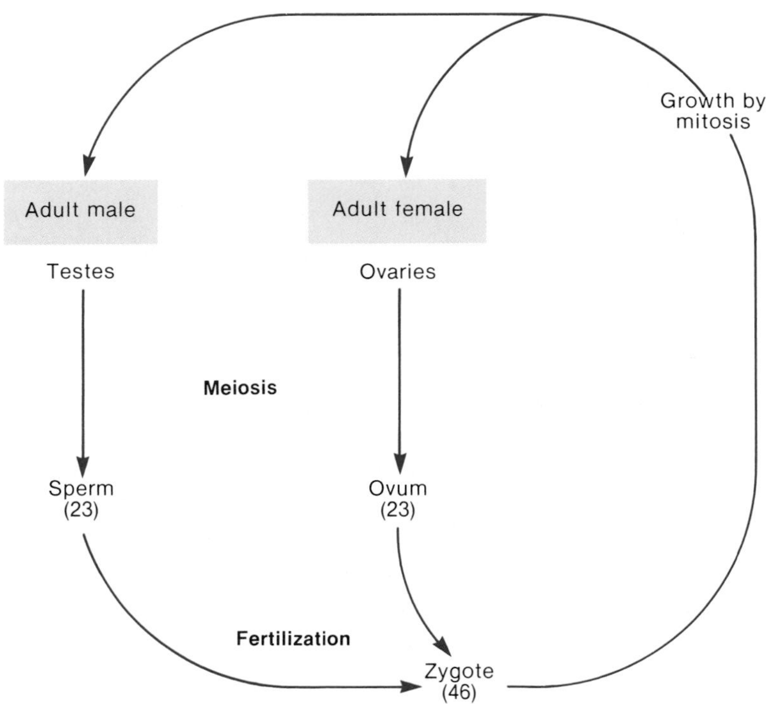


## Sexual Reproduction

The testes and ovaries develop from an embryonic organ which is indifferent until it reaches a critical stage of development. At that point, the genetic activity of the Y chromosome, if present, causes the gonads to become testes. Absence of the Y chromosome causes the development of ovaries. The embryonic testes secrete hormones that cause the development of male accessory sexual organs and external genitalia. The absence of testes (rather than the presence of ovaries) in a female embryo causes the development of the female accessory sexual organs.

"A chicken is an egg's way of making another egg." Phrased in more modern terms, genes are "selfish." Genes, according to this view, do not exist in order to make a well-functioning chicken (or other organism). The organism, rather, exists and functions so that the genes can survive beyond the mortal life of individual members of a species. Whether or not one accepts this rather cynical view, it is clear that reproduction is an essential function of life. The incredible complexity of structure and function in living organisms could not be produced in successive generations by chance; mechanisms must exist to transmit the blueprint (genetic code) from one generation to the next. *Sexual reproduction,* in which genes from two individuals are combined in random and novel ways with each new generation, offers the further advantage of introducing great variability into a population. This variability of genetic constitution helps to ensure that some members of a population will survive changes in the environment over evolutionary time.

In sexual reproduction, **germ cells,** or **gametes** (sperm and ova), are formed within the *gonads* (testes and ovaries) by a process of reduction division, or *meiosis* (chapter 3). During this type of cell division, the normal number of chromosomes in most human cells—forty-six—is halved, so that each gamete receives twenty-three chromosomes. Fusion of a sperm and egg cell (ovum) in the act of **fertilization** results in restoration of the original chromosome number of forty-six in the fertilized egg (the *zygote*). Growth of the zygote into an adult member of the next generation occurs by means of mitotic cell divisions, as described in chapter 3. When this individual reaches puberty, mature sperm or ova will be formed by meiosis within the gonads so that the life cycle can be continued (fig. 20.1).

### Sex Determination

Each zygote inherits twenty-three chromosomes from its mother and twenty-three chromosomes from its father. This does not produce forty-six different chromosomes, but rather twenty-three pairs of *homologous chromosomes.* Each member of a homologous pair, with the important exception of the sex chromosomes, looks like the other and contains similar genes (such as those coding for eye color,

**Figure 20.2.** Homologous pairs of chromosomes obtained from a human diploid cell. The first twenty-two pairs of chromosomes are called the autosomal chromosomes. The sex chromosomes are (*a*) XY for a male and (*b*) XX for a female.

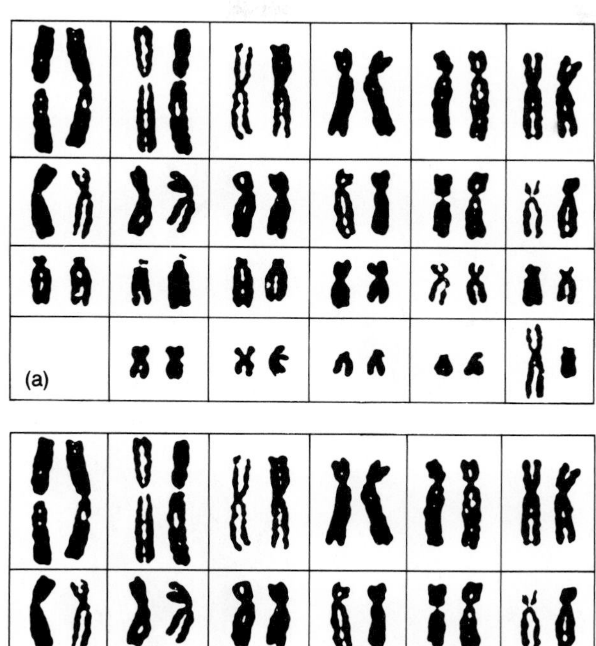


height, and so on). These homologous pairs of chromosomes can be photographed and numbered (as shown in fig. 20.2). Each cell that contains forty-six chromosomes (that is *diploid*) has two chromosomes number 1, two chromosomes number 2, and so on through chromosomes number 22. The first twenty-two pairs of chromosomes are called **autosomal chromosomes.**

The twenty-third pair of chromosomes are the **sex chromosomes.** In a female these consist of two X chromosomes, whereas in a male there is one X chromosome and one Y chromosome. The X and Y chromosomes look different and contain different genes. This is the exceptional pair of homologous chromosomes mentioned earlier.

When a diploid cell (with forty-six chromosomes) undergoes meiotic division, its daughter cells receive only one chromosome from each homologous pair of chromosomes. The gametes are therefore said to be *haploid* (they contain only half the number of chromosomes in the diploid parent cell). Each sperm, for example, will receive only one chromosome of homologous pair number 5—either the one originally contributed by the organism's mother, or the one originally contributed by the father. Which of the two chromosomes—maternal or paternal—ends up in a given sperm is completely random. This is also true for the sex chromosomes, so that approximately half of the sperm produced will contain an X and approximately half will contain a Y chromosome.

A similar random assortment of maternal and paternal chromosomes into ova will occur in a woman's ovary. Since females have two X chromosomes, however, all of the ova will normally contain one X chromosome. Because all ova contain one X chromosome, whereas some sperm are X bearing and others are Y bearing, the *chromosomal sex of the zygote is determined by the sperm.* If a Y-bearing sperm fertilizes the ovum, the zygote will be XY and male; if an X-bearing sperm fertilizes the ovum, the zygote will be XX and female.

Although each diploid cell in a woman's body inherits two X chromosomes, it appears that only one of each pair of X chromosomes remains active. The other X chromosome forms a clump of inactive "heterochromatin," which can often be seen as a dark spot, called a *Barr body,* at the edge of the nucleus of cheek cells (fig. 20.3). This provides a convenient test for chromosomal sex in cases in which there is suspicion that the chromosomal sex may

**Figure 20.3.** The nuclei of cheek cells from females (*a* and *b*) have Barr bodies (*arrows*). These are formed from one of the X chromosomes, which is inactive. No Barr body is present in the cell obtained from a male (*c*) because males have only one X chromosome, which remains active. Some white blood cells from females have a "drumsticklike" appendage that is not found in white blood cells from males (*d*).

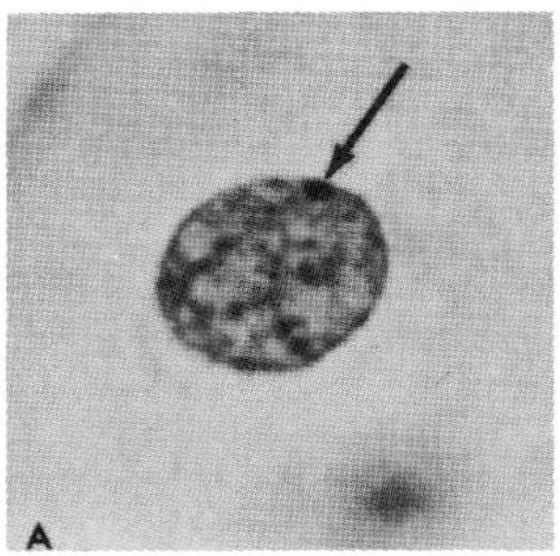


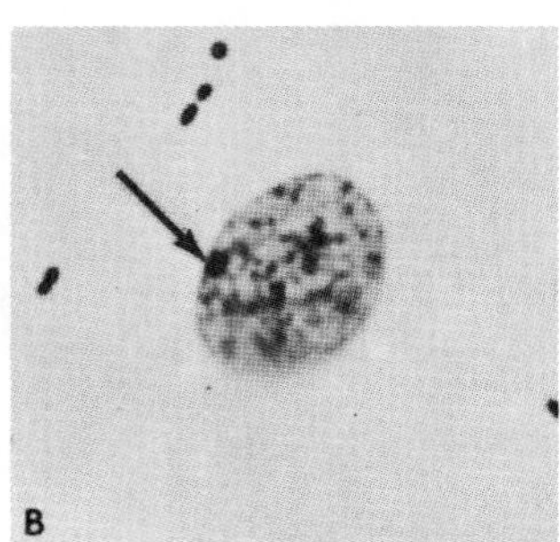


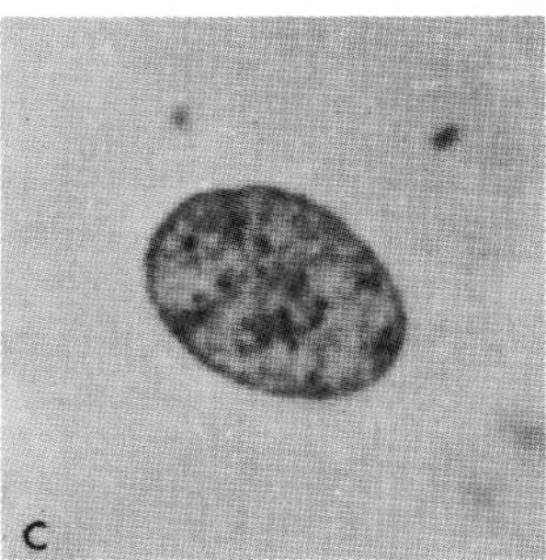


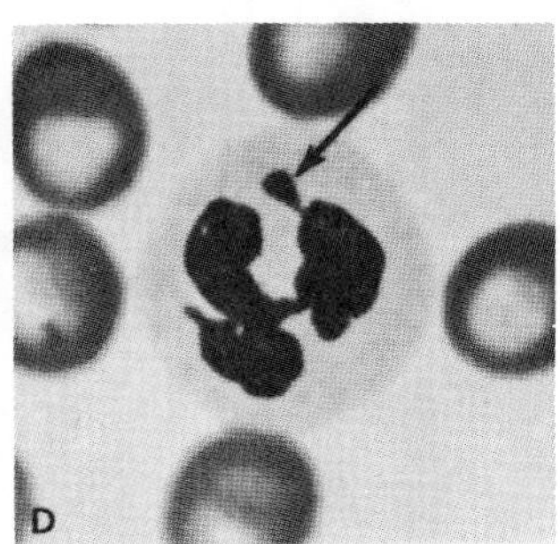


differ from the apparent ("phenotypic") sex of the individual. Also, some of the nuclei in polymorphonuclear leukocytes of females have a "drumstick" appendage not seen in neutrophils from males.

***Formation of Testes and Ovaries.*** The gonads of males and females appear similar for the first forty or so days of development following conception. During this time, cells that will give rise to sperm (called *spermatogonia*) and cells that will give rise to ova (called *oogonia*) migrate from the yolk sac to the embryonic gonads. At this stage in development the gonads have the potential to become either testes or ovaries.

Though it has long been recognized that male sex is determined by the presence of a Y chromosome and female sex by the absence of the Y chromosome, the genes involved have only recently been localized. Scientists have discovered that, in rare male babies with XX genotypes, one of the X chromosomes contains a segment of the Y chromosome. This error occurred during the meiotic cell division which formed the sperm. Similarly, rare female babies with XY genotypes were found to be missing the same portion of the Y chromosome that was erroneously inserted into the X chromosome of XX males. Through these and other observations, it has been shown that the gene for sex determination, called the **testis-determining factor (TDF)**, is located on the short arm of the Y chromosome (fig. 20.4).

Notice that it is normally the presence or absence of the Y chromosome that determines whether the embryo will have testes or ovaries. This point is well illustrated by two genetic abnormalities. In **Klinefelter's syndrome** the affected person has forty-seven instead of forty-six chromosomes because of the presence of an extra X chromosome (fig. 20.5). These people, with XXY genotypes, develop testes despite the presence of two X chromosomes. Patients with **Turner's syndrome,** who have the genotype XO (and therefore have only forty-five chromosomes) have poorly developed ("streak") gonads.

**Figure 20.4.** The formation of the chromosomal sex of the embryo and the development of the gonads. The very early embryo has "indifferent gonads" that can develop into either testes or ovaries. The testis-determining factor (TDF) is a gene located on the Y chromosome. In the absence of TDF, ovaries will develop.

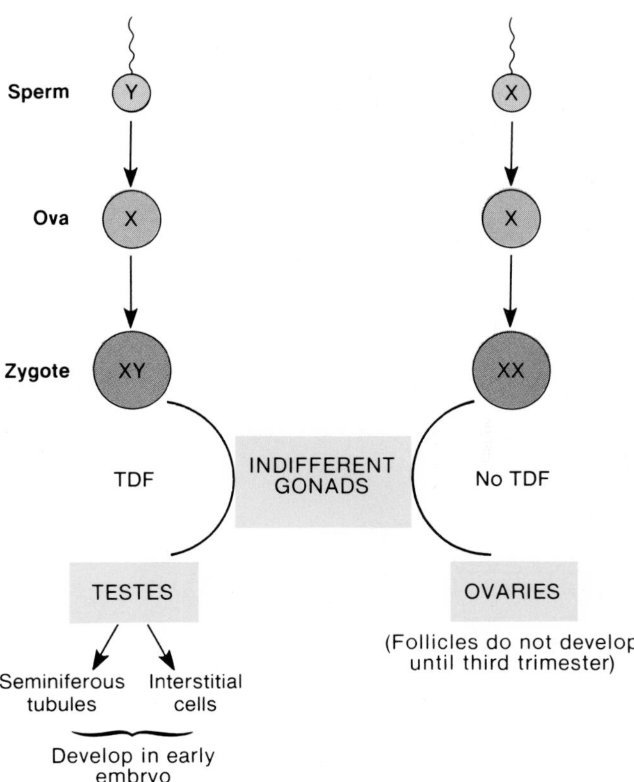


**Figure 20.5.** Chromosomes from a person with Turner's syndrome (*a*) and a variant of Klinefelter's syndrome (*b*).

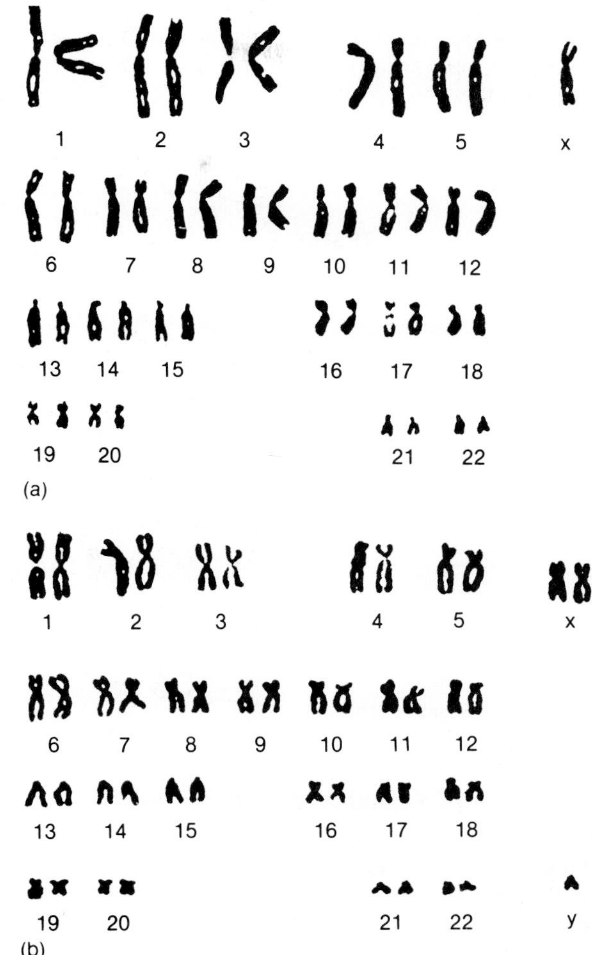


The structures that will eventually produce sperm within the testes—the **seminiferous tubules**—appear very early in embryonic development (between forty-three and fifty days following conception). Although spermatogenesis begins during embryonic life, it is arrested until the onset of puberty (this will be described in a later section). At about day sixty-five the **Leydig cells** appear in the embryonic testes. These cells are located in the *interstitial tissue* between adjacent convolutions of the seminiferous tubules and constitute the endocrine tissue of the testes. In contrast to the rapid development of the testes, the functional units of the ovaries—called the **ovarian follicles**—do not appear until the second trimester of pregnancy (at about day 105).

The early appearing Leydig cells in the embryonic testes secrete large amounts of male sex hormones, or *androgens,* (*andro* = man; *gen* = forming). The major androgen secreted by these cells is **testosterone.** Testosterone secretion begins as early as nine to ten weeks after conception, reaches a peak at twelve to fourteen weeks, and thereafter declines to very low levels by the end of the second trimester (at about twenty-one weeks). Testosterone secretion during embryonic development in the male serves a very important function (described in the next section); similarly high levels of testosterone will not appear again in the life of the individual until the time of puberty.

As the testes develop they move within the abdominal cavity and gradually descend into the *scrotum.* Descent of the testes is sometimes not complete until shortly after birth. The cooler temperature of the scrotum, which is usually three to four degrees Celsius lower than the temperature of the body cavity, is needed for spermatogenesis. This requirement is illustrated by the fact that spermatogenesis does not occur in males with undescended testes—a condition called *cryptorchidism* (*crypt* = hidden; *orchid* = testes).

**Figure 20.6.** The embryonic development of male and female sex accessory organs and external genitalia. In the presence of testosterone and müllerian inhibition factor (MIF) secreted by the testes, male structures develop. In the absence of these secretions, female structures develop.

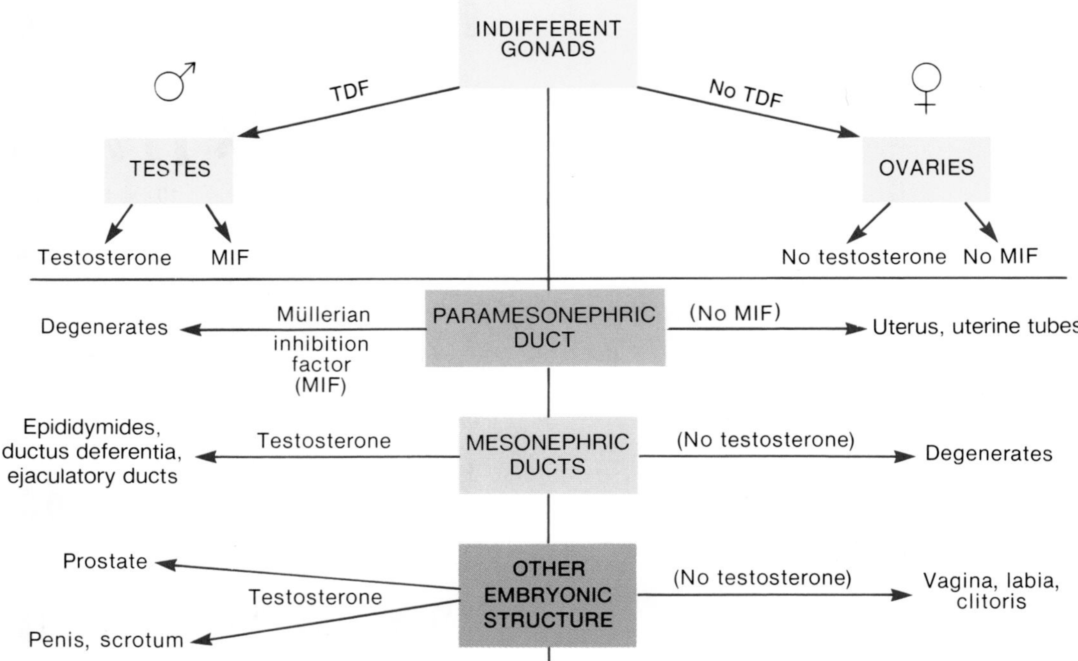


## Development of Sexual Accessory Organs and External Genitalia

In addition to testes and ovaries, various internal accessory sexual organs are needed for reproductive function. Most of these sex accessory organs are derived from two systems of embryonic ducts. Male accessory organs are derived from the **wolffian (mesonephric) ducts,** and female accessory organs are derived from the **müllerian (paramesonephric) ducts** (fig. 20.6). Interestingly, both male and female embryos between about day twenty-five and day fifty have both duct systems and, therefore, have the potential to form the accessory organs characteristic of either sex.

Experimental removal of the testes (castration) from male embryonic animals results in regression of the wolffian ducts and development of the müllerian ducts into female accessory organs: the **uterus** and **uterine (fallopian) tubes.** *Female sex accessory organs, therefore, develop as a result of the absence of testes rather than as a result of the presence of ovaries.*

The developing seminiferous tubules within the testes secrete a polypeptide called *müllerian inhibition factor,* which causes regression of the müllerian ducts beginning about day sixty. The secretion of testosterone by the Leydig cells of the testes subsequently causes growth and development of the wolffian ducts into *male sex accessory organs:* the **epididymis, vas deferens, seminal vesicles,** and **ejaculatory duct.** The structure and function of the sex accessory organs will be described in later sections.

There are other structures that male and female embryos share in common—the *urogenital sinus, genital tubercle, urogenital folds,* and *labiosacral folds.* The secretions of the testes masculinize these structures to form the **penis** and penile urethra, **prostate,** and **scrotum.** The genital tubercle that forms the penis in a male will, in the absence of testes, become the clitoris in a female. Penis and clitoris are thus said to be *homologous structures.* Similarly, the labiosacral folds form the scrotum in a male or the labia majora of a female; these structures are therefore also homologous (fig. 20.7).

**Figure 20.7.** Differentiation of the external genitalia in the male and female. (*a*) and ($a_1$, sagittal view) at six weeks, the genital tubercle, urogenital fold, and labioscrotal swelling have differentiated from the genital tubercle. (*b*) at eight weeks, a distinct phallus is present during the indifferent stage. By the twelfth week, the genitalia are distinctly male (*c*) or female (*d*), being derived from homologous structures. (*e* and *f*) at sixteen weeks, the genitalia are formed.

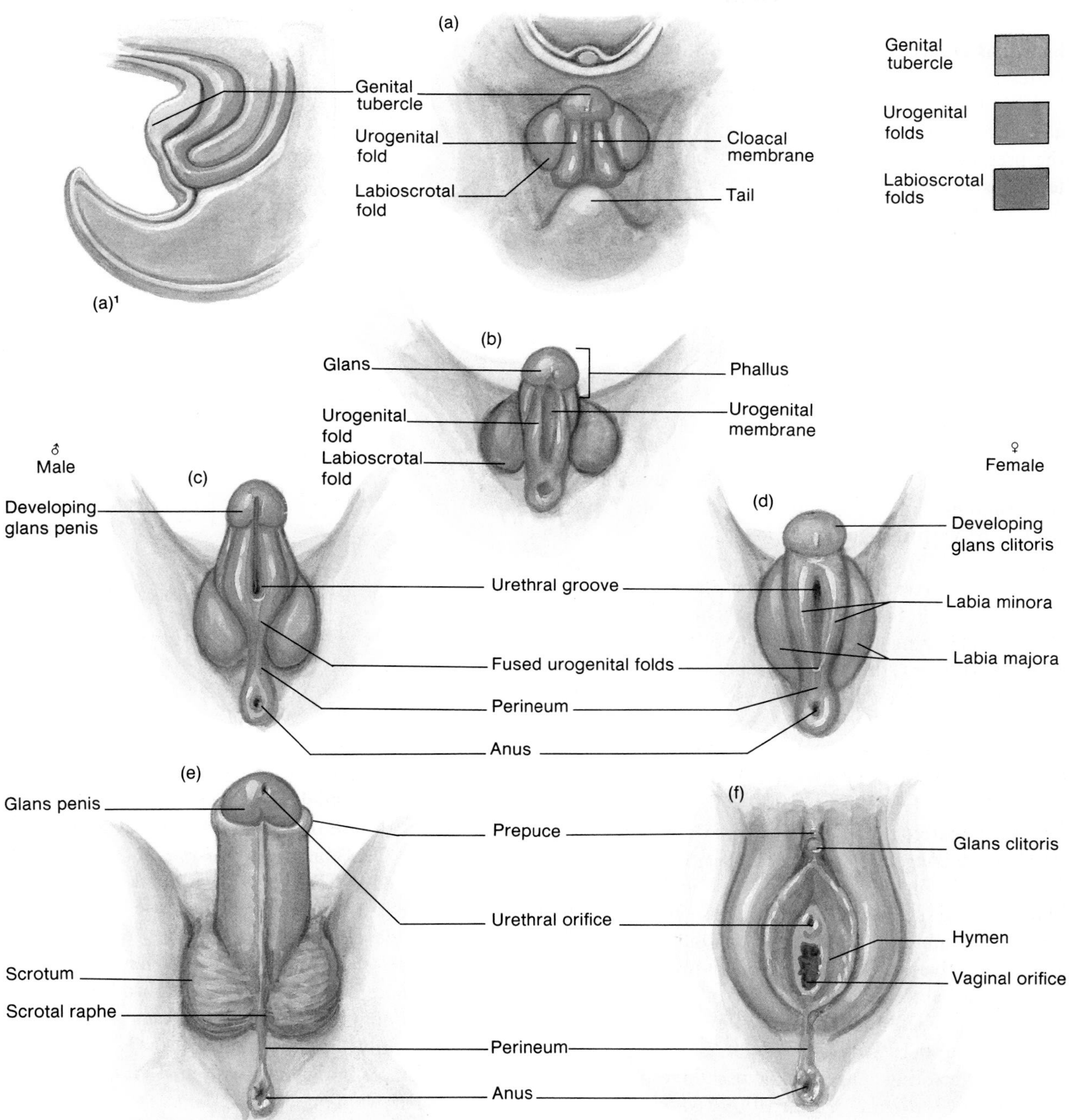

Masculinization of the embryonic structures described occurs as a result of testosterone, secreted by the embryonic testes. Testosterone itself, however, is not the active agent within these target organs. Once inside the target cells, testosterone is converted by means of an enzyme called *5α-reductase* into the active hormone known as **dihydrotestosterone (DHT),** which directly mediates the androgenic effect in these organs (fig. 20.8).

In summary, the genetic sex is determined by whether a Y-bearing or an X-bearing sperm fertilizes the ovum; the presence or absence of a Y chromosome in turn determines whether the gonads of the embryo will be testes or ovaries; the presence or absence of testes, finally, determines whether the sex accessory organs and external genitalia will be male or female (table 20.1). This regulatory pattern of sex determination makes sense in light of the fact that both male and female embryos develop within an environment high in estrogen, which is secreted by the mother's ovaries and the placenta. If the secretions of the ovaries determined the sex, all embryos would be female.

**Figure 20.8.** The conversion of testosterone, secreted by the interstitial cells of the testes, into dihydrotestosterone (DHT) within the target cells. This reaction involves the addition of a hydrogen (and the removal of the double carbon bond) in the first (*A*) ring of the steroid.

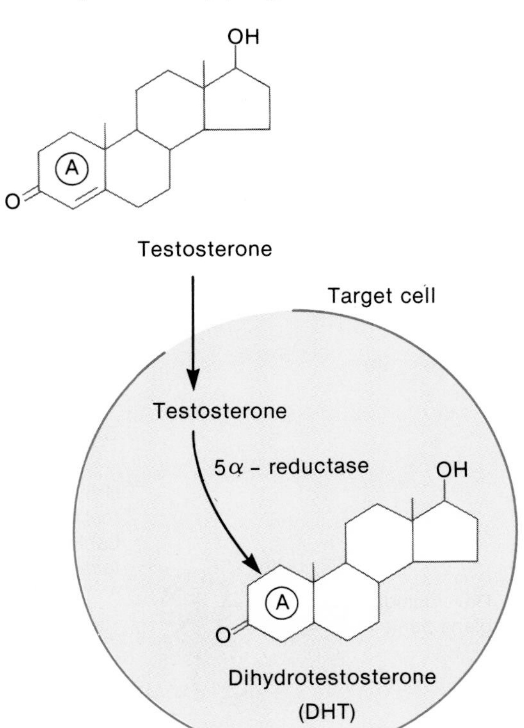


## Disorders of Embryonic Sexual Development

A *hermaphrodite* is an individual with both testes and ovaries or with testicular and ovarian tissue located together in the same gonad. This condition is extremely rare and appears to be caused by the fact that some embryonic cells receive the short arm of the Y chromosome, with its testis-determining factor, whereas others do not. More common (though still rare) disorders of sex determination involve individuals with either testes or ovaries, but not both, who have sex accessory organs and external genitalia that are incompletely developed or are inappropriate for their chromosomal sex. These individuals are called *pseudohermaphrodites* (*pseudo* = false).

The most common cause of female pseudohermaphroditism is *congenital adrenal hyperplasia.* This condition, which is inherited as a recessive trait, is caused by the excessive secretion of androgens from the adrenal cortex. A female with this condition would have müllerian duct derivatives (uterus and fallopian tubes), because the adrenal doesn't secrete müllerian inhibition factor, but would have partially masculinized external genitalia and wolffian duct derivatives.

An interesting cause of male pseudohermaphroditism is known as *testicular feminization syndrome.* These individuals have normally functioning testes but lack receptors for testosterone. Thus, although large amounts of testosterone are secreted, the embryonic tissues cannot respond to this hormone. Female genitalia therefore develop, but the vagina ends blindly (uterus and fallopian tubes don't develop because of the secretion of müllerian inhibition factor). Male sex accessory organs likewise cannot develop because the wolffian ducts lack testosterone receptors. A child with this condition appears externally to be a normal prepubertal girl, but she has testes in her body cavity and no sex accessory organs. These testes secrete an exceedingly large amount of testosterone at puberty because of the absence of negative feedback inhibition. This abnormally large amount of testosterone is converted by the liver and adrenal cortex into estrogens. As a result, the person with testicular feminization syndrome develops into a voluptuous female who never menstruates (and who, of course, can never become pregnant).

Some male pseudohermaphrodites have normally functioning testes and normal testosterone receptors, but genetically lack the ability to produce the enzyme 5α-reductase. Individuals with *5α-reductase deficiency* have normal wolffian duct derivatives (epididymis, vas deferens, seminal vesicles, and ejaculatory duct) because the development of these structures is stimulated directly by testosterone. These male accessory organs terminate in a vagina, however, because the development of male external genitalia requires the action of DHT, which cannot be produced from testosterone in the absence of 5α-reductase.

**Table 20.1** Timetable for developmental changes of the gonads and genital tracts

| Approximate Time after Fertilization | | Developmental Changes | | |
|---|---|---|---|---|
| Days | Trimester | Indifferent | Male | Female |
| 19 | First | Germ cells migrate from yolk sac. | | |
| 25–30 | | Wolffian ducts begin development. | | |
| 44–48 | | Müllerian ducts begin development. | | |
| 50–52 | | Urogenital sinus and tubercle develop. | | |
| 43–60 | | | Tubules and Sertoli cells appear. Müllerian ducts begin to regress. | |
| 60–75 | | | Leydig cells appear and begin testosterone production. Wolffian ducts grow. | Formation of vagina begins. Regression of wolffian ducts begins. |
| 105 | Second | | | Development of ovarian follicles begins. |
| 120 | | | | Uterus is formed. |
| 160–260 | Third | | Testes descend into scrotum. Growth of external genitalia occurs. | Formation of vagina complete. |



1. *Explain the meaning of the terms* diploid *and* haploid, *and describe how the chromosomal sex of an individual is determined.*
2. *Explain how the chromosomal sex determines whether testes or ovaries will be formed.*
3. *List the male and female sex accessory organs, and explain how the development of one or the other sets of these organs are determined.*
4. *Describe the abnormalities in testicular feminization syndrome and in 5α-reductase deficiency, and explain how these abnormalities are produced.*

## *Endocrine Regulation of Reproduction*

The functions of the testes and ovaries are regulated by gonadotropic hormones secreted by the anterior pituitary gland. The gonadotropic hormones stimulate the gonads to secrete their sex steroid hormones, and these steroid hormones, in turn, have an inhibitory effect on the secretion of the gonadotropic hormones. This interaction between the anterior pituitary and the gonads forms a negative feedback loop. Secretion of the gonadotropins is pulsatile rather than continuous. This pulsatile method of secretion is required to prevent desensitization of the gonads to the effects of the gonadotropins.

The embryonic testes during the first trimester of pregnancy are active endocrine glands, secreting the high amounts of testosterone needed to masculinize the male embryo's external genitalia and sex accessory organs. Ovaries, in contrast, don't mature until the third trimester of pregnancy. During the second trimester of pregnancy testosterone secretion in the male declines, so that, at the time the baby is born, the gonads of both sexes are relatively inactive.

There is no difference in the blood concentrations of *sex steroids*—androgens and estrogens—in prepubertal boys and girls. This does not appear to be due to deficiencies in the ability of the gonads to produce these hormones, but rather to lack of sufficient stimulation. During *puberty,* the gonads secrete increased amounts of sex steroid hormones as a result of increased stimulation by **gonadotropic hormones** from the anterior pituitary gland.

### Interactions between the Hypothalamus, Pituitary, and Gonads

The anterior pituitary gland produces and secretes two gonadotropic hormones—**FSH (follicle-stimulating hormone)** and **LH (luteinizing hormone).** Although these two hormones are named according to their actions in the female, the same hormones are secreted by the male's pituitary. The gonadotropic hormones of both sexes have three primary effects on the gonads: (1) stimulation of

**Figure 20.9.** A simplified biosynthetic pathway for the steroid hormones. Also indicated are the major sources of sex hormones in the blood.

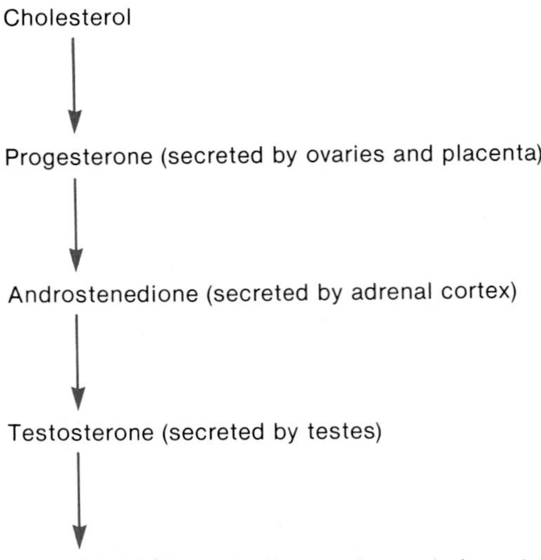


spermatogenesis or oogenesis (formation of sperm or ova); (2) stimulation of gonadal hormone secretion; and (3) maintenance of the structure of the gonads (the gonads atrophy if the pituitary is removed).

The secretion of both LH and FSH from the anterior pituitary is stimulated by a hormone produced by the hypothalamus and secreted into the hypothalamo-hypophyseal portal vessels (see chapter 11). This releasing hormone is sometimes called *LHRH (luteinizing hormone-releasing hormone)*. Since attempts to find a separate FSH-releasing hormone have thus far failed, and since LHRH stimulates FSH as well as LH secretion, LHRH is often referred to as **gonadotropin-releasing hormone** (and accordingly abbreviated **GnRH**).

If a male or female animal is castrated (has its gonads surgically removed), the secretion of FSH and LH increases to much higher levels than those measured in the intact animal. This demonstrates that the gonads secrete products that have a **negative feedback inhibition** of gonadotropin secretion. This negative feedback is exerted in large part by sex steroids: estrogen and progesterone in the female, and testosterone in the male. The biosynthetic pathways of these steroids are shown in figure 20.9.

The negative feedback effects of steroid hormones are believed to occur by means of two mechanisms: (1) inhibition of GnRH secretion from the hypothalamus, and (2) inhibition of the pituitary's response to a given amount of GnRH. In addition to steroid hormones, there is evidence that the testes (and probably the ovaries as well) secrete a polypeptide hormone called **inhibin.** This

**Figure 20.10.** Interactions between the hypothalamus, anterior pituitary, and gonads. Sex steroids secreted by the gonads have a negative feedback effect on the secretion of GnRH (gonadotropin-releasing hormone) and on the secretion of gonadotropins. The gonads may also secrete a polypeptide hormone called *inhibin* that exerts negative feedback control of FSH secretion.

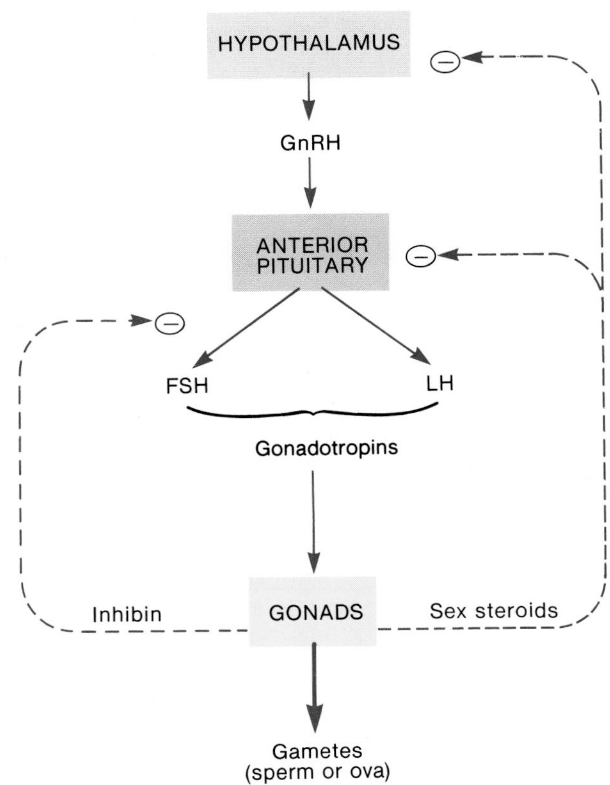


hormone specifically inhibits FSH secretion in response to GnRH, without affecting the secretion of LH. Inhibin will be discussed more fully in later sections.

Figure 20.10 illustrates the similarities in gonadal regulation of males and females. Important differences in hypothalamus-pituitary-gonad interactions exist, however, between males and females. Secretion of gonadotropins and sex steroids is more or less constant in adult males. Secretion of gonadotropins and sex steroids in adult females, in contrast, shows cyclic variations (during the menstrual cycle). Also, during one phase of the female cycle, estrogen exerts a positive feedback effect on LH secretion. This will be discussed in a later section.

Studies have shown that secretion of GnRH from the hypothalamus is pulsatile rather than continuous, and thus the secretion of FSH and LH follows this pulsatile pattern. These pulsatile patterns of secretion appear to be required to prevent desensitization and down-regulation of the target glands (discussed in chapter 11). It appears that the frequency of the pulses of secretion, as well as their amplitude (how much hormone is secreted per pulse), affects the target glands response to the hormone.

**Table 20.2** Development of secondary sexual characteristics and other changes that occur during puberty in girls

| Characteristic | Age of First Appearance | Hormonal Stimulation |
|---|---|---|
| Appearance of breast bud | 8–13 | Estrogen, progesterone, growth hormone, thyroxine, insulin, cortisol |
| Pubic hair | 8–14 | Adrenal androgens |
| Menarche (first menstrual flow) | 10–16 | Estrogen and progesterone |
| Axillary (underarm) hair | About two years after the appearance of pubic hair | Adrenal androgens |
| Eccrine sweat glands and sebaceous glands; acne (from blocked sebaceous glands) | About the same time as axillary hair growth | Adrenal androgens |

**Table 20.3** Development of secondary sexual characteristics and other changes that occur during puberty in boys

| Characteristic | Age of First Appearance | Hormonal Stimulation |
|---|---|---|
| Growth of testes | 10–14 | Testosterone, FSH, growth hormone |
| Pubic hair | 10–15 | Testosterone |
| Body growth | 11–16 | Testosterone, growth hormone |
| Growth of penis | 11–15 | Testosterone |
| Growth of larynx (voice lowers) | Same time as growth of penis | Testosterone |
| Facial and axillary (underarm) hair | About two years after the appearance of pubic hair | Testosterone |
| Eccrine sweat glands and sebaceous glands; acne (from blocked sebaceous glands) | About the same time as facial and axillary hair growth | Testosterone |

If a powerful synthetic analogue of GnRH (such as *nafarelin*) is administered, the anterior pituitary first increases and then decreases its secretion of FSH and LH. The decrease is due to a desensitization of the anterior pituitary, which is normally prevented by the fact that GnRH is secreted in a pulsatile fashion. The decrease in LH causes a fall in testosterone secretion from the testes, or of estradiol secretion from the ovaries. The decreased testosterone secretion is useful in the treatment of men who have *benign prostatic hyperplasia,* a condition common in older men in which testosterone supports abnormal growth of the prostate. The fall in estradiol secretion in women given synthetic GnRH analogues can be useful in the treatment of *endometriosis.* In this condition, ectopic endometrial tissue from the uterus (dependent on estradiol for growth) enters the peritoneal cavity. These treatments illustrate the reasons why GnRH and the gonadotropins are normally secreted in a pulsatile fashion, and are particularly beneficial clinically because they are reversible.

## The Onset of Puberty

Secretion of FSH and LH is high in the newborn, but falls to very low levels a few weeks after birth. Gonadotropin secretion remains low until the beginning of puberty, which is marked by rising levels of FSH followed by LH secretion. Experimental evidence suggests that this rise in gonadotropin secretion is a result of two processes: (1) maturational changes in the brain that result in increased GnRH secretion by the hypothalamus, and (2) decreased sensitivity of gonadotropin secretion to the negative feedback effects of sex steroid hormones.

The maturation of the hypothalamus or other regions of the brain that leads to increased GnRH secretion at the time of puberty appears to be programmed—children without gonads show increased FSH secretion at the normal time. Also during this period of time, a given amount of sex steroids has less of a suppressive effect on gonadotropin secretion than the same dose would have if administered prior to puberty. This suggests that the sensitivity of the hypothalamus and the pituitary to negative feedback effects decreases at puberty, which would also help account for rising gonadotropin secretion at this time.

During late puberty there is a pulsatile secretion of gonadotropins—FSH and LH secretion increase during periods of sleep, and decrease during periods of wakefulness. These pulses of increased gonadotropin secretion during puberty stimulate a rise in sex steroid secretion from the gonads. Increased secretion of testosterone from the testes and of **estradiol-17$\beta$** (estradiol is the major *estrogen,* or female sex steroid) from the ovaries during puberty, in turn, produces changes in body appearance characteristic of the two sexes. Such **secondary sexual characteristics** (tables 20.2 and 20.3) are the physical manifestations of the hormonal changes occurring during puberty.

**Figure 20.11.** A simplified biosynthetic pathway for the pineal gland hormone *melatonin.*

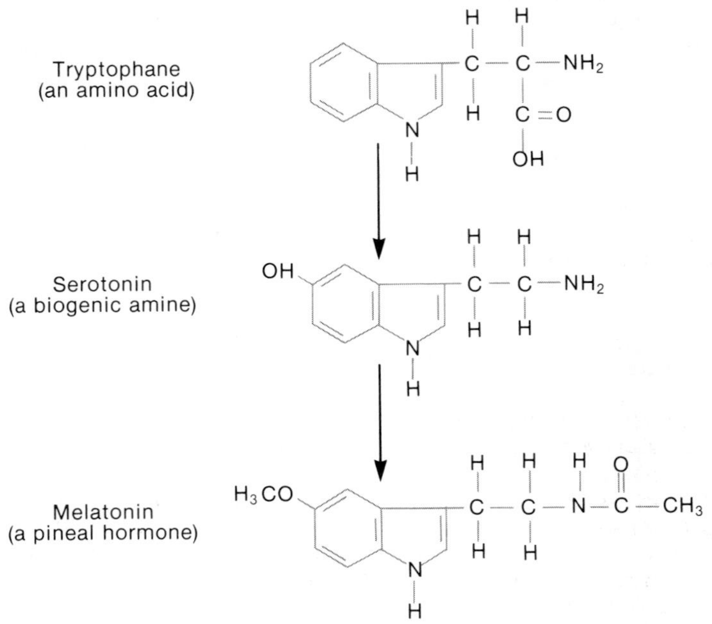

Tryptophane
(an amino acid)

Serotonin
(a biogenic amine)

Melatonin
(a pineal hormone)

It has been observed that the age of onset of puberty is related to the amount of body fat and level of physical activity of the child. Ballerinas, and other girls who are very physically active, reach *menarche*—the age when menstrual flow first occurs—at a later average age (15 years old) than in the general population (12.6 years old). This appears to be due to a requirement for a minimum percentage of body fat for menstruation to begin, and may represent a mechanism favored by natural selection for the ability to successfully complete a pregnancy and nurse the baby. Later in life, women who are very lean and physically active may have irregular cycles and *amenorrhea* (cessation of menstruation). This may also be related to the percentage of body fat. In addition, there is evidence that physical exercise may, through activation of neural pathways involving endorphin neurotransmitters (chapter 7), act to inhibit GnRH and gonadotropin secretion.

## The Pineal Gland

The role of the pineal gland in human physiology is poorly understood. It is known that the pineal, a gland located deep within the brain, secretes the hormone **melatonin** as a derivative of the amino acid tryptophane (fig. 20.11) and that production of this hormone is influenced by light-dark cycles.

The pineal glands of some lower vertebrates have photoreceptors that are directly sensitive to environmental light. Although no such photoreceptors are present in the pineal glands of mammals, the secretion of melatonin has been shown to increase at night and decrease during daylight. The inhibitory effect of light on melatonin secretion in mammals is indirect. Pineal secretion is stimulated by postganglionic sympathetic neurons that originate in the superior cervical ganglion; activity of these neurons, in turn, is inhibited by nerve tracts that are activated by light striking the retina.

Receptors for melatonin have been localized to the suprachiasmatic nucleus of the hypothalamus. This area of the brain is believed to be the site of the biological clock, which entrains (synchronizes) various processes of the body to a *circadian rhythm* (one that repeats every 24 hours). The rhythms of melatonin secretion may be entrained to cycles of light and dark through the action of the suprachiasmatic nucleus, and melatonin, in turn, may have some effect on this area of the brain. It is interesting in this regard that melatonin has been used clinically to relieve the symptoms of jet lag and help reestablish the disturbed circadian rhythms of a blind person.

There is abundant experimental evidence that, in rats and other lower animals, melatonin can inhibit gonadotropin secretion and thus have an "anti-gonad" effect. In other experimental situations, however, melatonin can stimulate the reproductive system. The role of melatonin in the regulation of human reproduction has long been suspected but, because of conflicting and inconclusive data, has not as yet been established.

**Figure 20.12.** A diagrammatic representation of seminiferous tubules: (*a*) A sagittal section of a testis; (*b*) A transverse section of a seminiferous tubule.

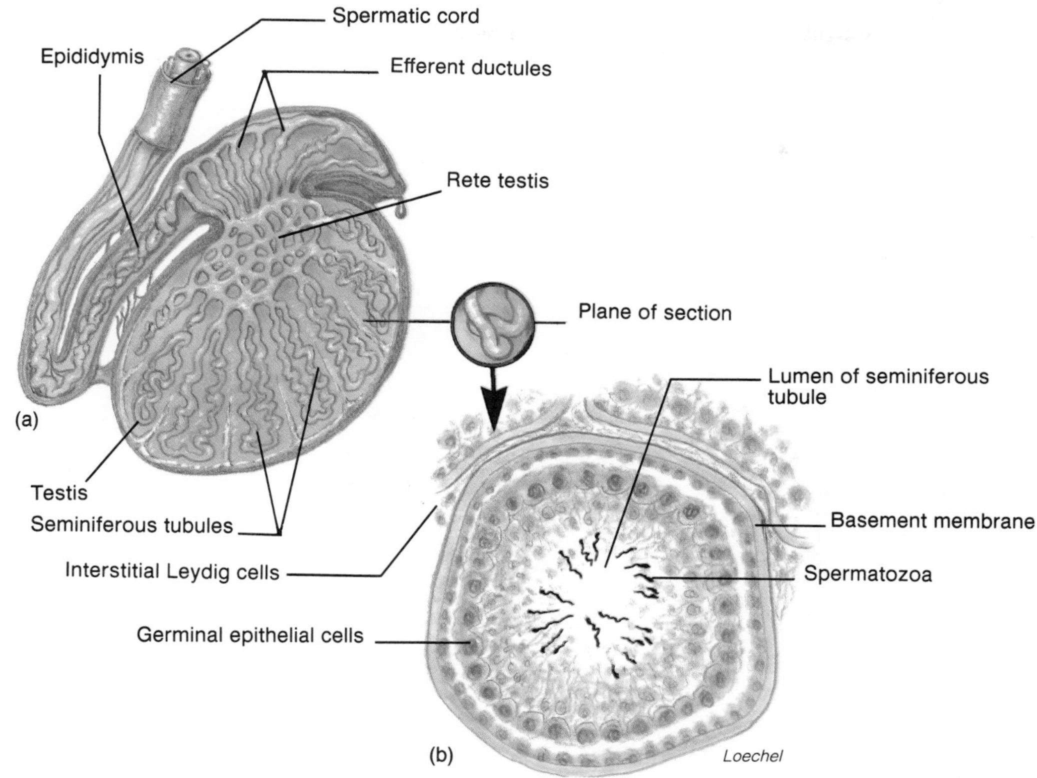


1. *Using a flow diagram, show the negative feedback control that the gonads exert on GnRH and gonadotropin secretion. Explain the effects of castration on FSH and LH secretion and the effects of removal of the pituitary on the structure of the gonads and accessory sexual organs.*
2. *Explain the significance of the pulsatile secretion of GnRH and the gonadotropic hormones.*
3. *Describe the two mechanisms that have been proposed to explain the rise in sex steroid secretion that occurs at puberty. Explain the possible effects of body fat and intense exercise on the timing of puberty.*
4. *Describe the effect of light on the pineal secretion of melatonin and the possible functions of this hormone.*

## Male Reproductive System

The Leydig cells in the interstitial tissue of the testes are stimulated by LH to secrete testosterone, a potent androgen that acts to maintain the structure and function of the male accessory sexual organs and to promote the development of male secondary sexual characteristics. In most cases, testosterone functions as a prehormone that is converted to other products within its target cells. These products include dihydrotestosterone and estradiol. The Sertoli cells in the seminiferous tubules of the testes are stimulated by FSH. This action is needed to initiate the production of sperm and stimulates the Sertoli cells to secrete various products, including inhibin. Spermatogenesis requires the cooperative actions of both FSH and testosterone.

The testes consist of two parts, or "compartments"—the seminiferous tubules, where spermatogenesis occurs, and the interstitial tissue, which contains androgen-secreting Leydig cells (fig. 20.12). About 90% of the weight of an adult testis (which averages 20 g) is comprised of seminiferous tubules. The interstitial tissue is a thin web of connective tissue (containing Leydig cells) between convolutions of the tubules.

There is a strict compartmentation in the testes with regard to gonadotropin action. Cellular receptor proteins for FSH are located exclusively in the seminiferous tubules, where they are confined to the **Sertoli cells** (discussed in a later section). LH receptor proteins are located exclusively in the interstitial Leydig cells. Secretion of testosterone by the Leydig cells is stimulated by LH but not by FSH. Spermatogenesis in the tubules is stimulated by FSH. The apparent simplicity of this compartmentation, however, is an illusion because the two compartments can interact with each other in complex ways.

Figure 20.13. Negative feedback relationships between the anterior pituitary and testes. The seminiferous tubules are the targets of FSH action; the interstitial Leydig cells are targets of LH action. Testosterone secreted by the Leydig cells inhibits LH secretion; inhibin secreted by the tubules may inhibit FSH secretion.

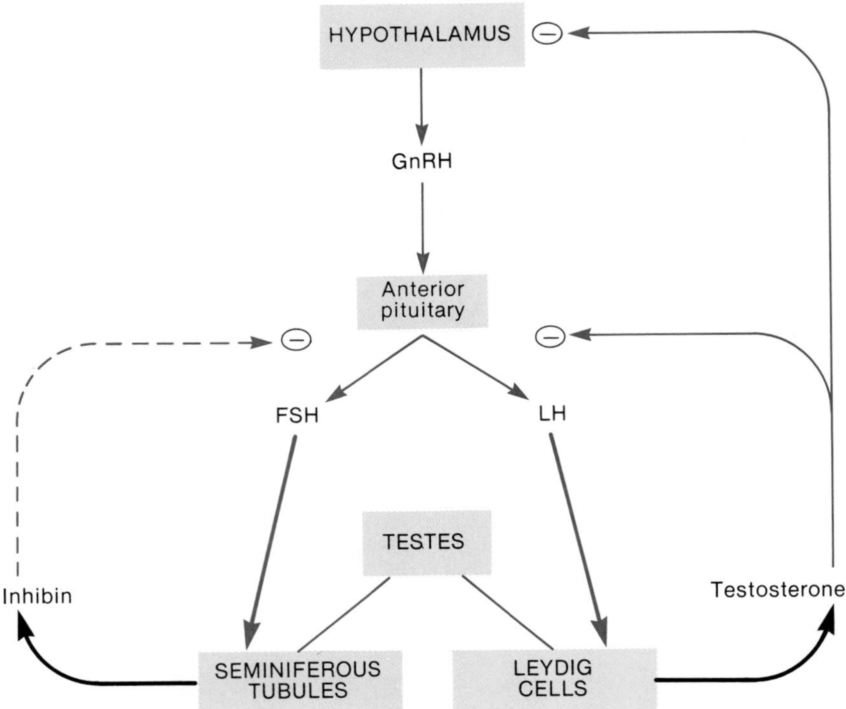


## Control of Gonadotropin Secretion

Castration of a male animal results in an immediate rise in FSH and LH secretion. This demonstrates that hormones secreted by the testes exert negative feedback inhibition of gonadotropin secretion. If testosterone is injected into the castrated animal, the secretion of LH can be returned to the previous (precastration) levels. This provides a classical example of negative feedback—LH stimulates testosterone secretion by the Leydig cells, and testosterone inhibits pituitary secretion of LH (fig. 20.13).

The amount of testosterone that is sufficient to suppress LH, however, is not sufficient to suppress the postcastration rise in FSH secretion in most experimental animals. In rams and bulls, a water-soluble (and, therefore, peptide rather than steroid) product of the seminiferous tubules specifically suppresses FSH secretion. This hormone, produced by the Sertoli cells, is called **inhibin.** There is now good evidence that the seminiferous tubules of the human testes also produce inhibin.

***Testosterone Derivatives in the Brain.*** The brain contains testosterone receptors and is a target organ for this hormone. The effects of testosterone on the brain, such as the suppression of LH secretion, are not mediated directly by testosterone, however, but rather by its derivatives that are produced within the brain cells. Testosterone may be converted by the enzyme 5$\alpha$-reductase to dihydrotestosterone (DHT), as previously described. The DHT, in turn, can be changed by other enzymes into other 5$\alpha$-reduced androgens—abbreviated 3$\alpha$-diol and 3$\beta$-diol (fig. 20.14). Alternatively, testosterone is also converted within the brain to estradiol-17$\beta$. Although usually regarded as a female sex steroid, estradiol is therefore an active compound in normal male physiology! Estradiol is formed from testosterone by an enzyme called *aromatase,* in a reaction known as *aromatization* (this term refers to the presence of an aromatic carbon ring—see chapter 2—not to an odor). The estradiol formed from testosterone in the brain is believed to be required for the negative feedback effects of testosterone on LH secretion.

***Testosterone Secretion and Age.*** The negative feedback effects of testosterone and inhibin help to maintain a constant secretion of gonadotropins in men, resulting in relatively constant levels of androgen secretion from the testes. This contrasts with the cyclic secretion of gonadotropins and ovarian steroids in women. Women experience an abrupt cessation in sex steroid secretion during menopause (as described in a later section). In contrast, the secretion of androgens declines only gradually and to

**Figure 20.14.** Testosterone secreted by the Leydig cells of the testes can be converted into active metabolites in the brain and other target organs. These active metabolites include DHT and other 5α-reduced androgens and estradiol.

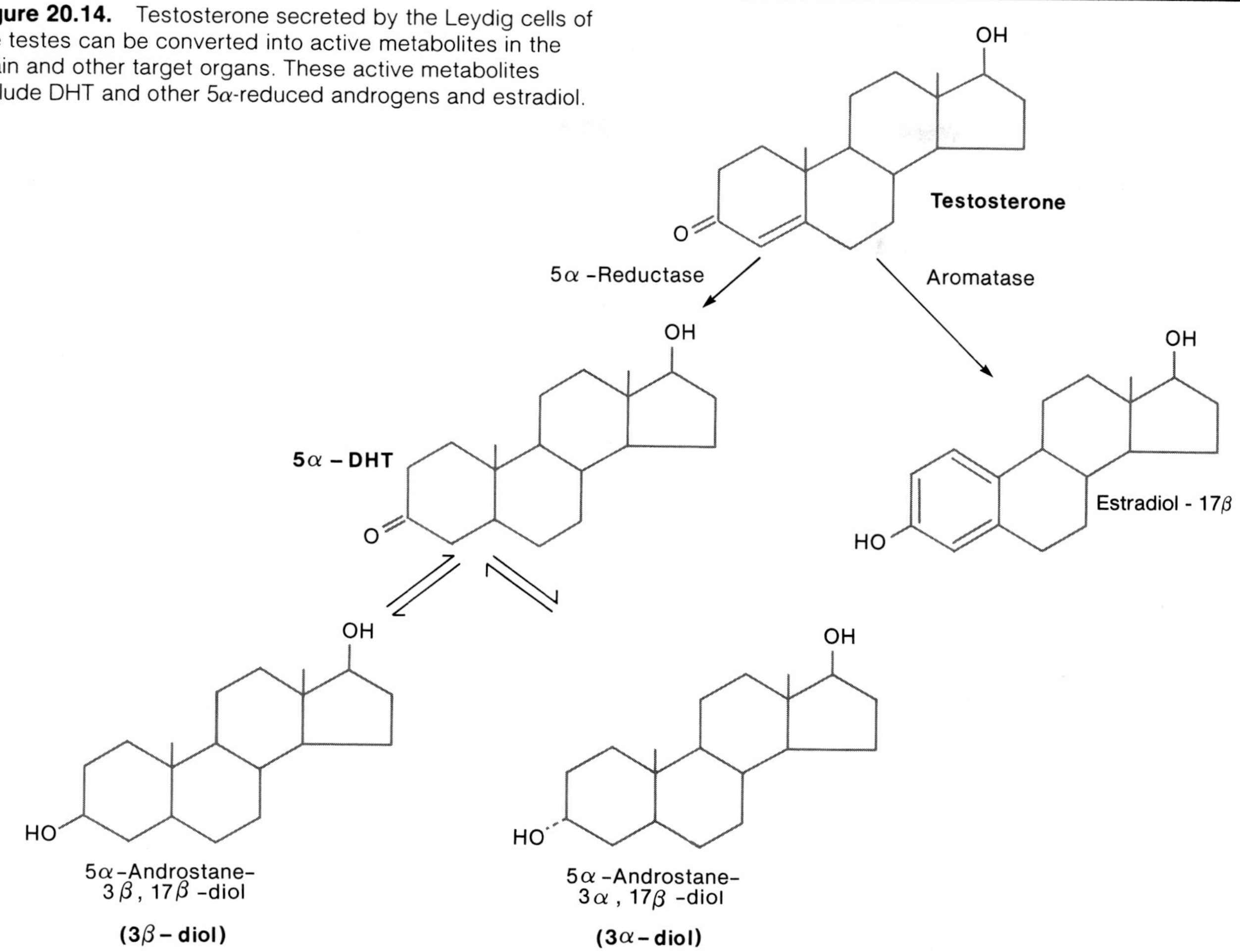

varying degrees in men over fifty years of age. The causes of this age-related change in testicular function are not currently known. The decline in testosterone secretion cannot be due to decreasing gonadotropin secretion, since gonadotropin levels in the blood are, in fact, elevated (due to less negative feedback) at the time that testosterone levels are declining.

## Endocrine Functions of the Testes

Testosterone is by far the major androgen secreted by the adult testis. This hormone, or derivatives of it (the 5α-reduced androgens), is responsible for initiation and maintenance of the body changes associated with puberty in men. Androgens are sometimes called *anabolic steroids* because they stimulate the growth of muscles and other structures (table 20.4). Increased testosterone secretion during puberty is also required for growth of the sex accessory organs—primarily the seminal vesicles and prostate. Removal of androgens by castration results in atrophy of these organs.

Androgens stimulate growth of the larynx (causing lowering of the voice), increased hemoglobin synthesis (males have higher hemoglobin levels than females), and bone growth. The effect of androgens on bone growth is self-limiting, however, because androgens ultimately cause conversion of cartilage to bone in the epiphyseal discs, thus "sealing" the discs and preventing further lengthening of the bones (as described in chapter 18).

It was once thought that the Leydig cells of the interstitial tissue were the only source of testicular androgens. Although the Leydig cells are the major source of testosterone, it is now known that the seminiferous tubules are also able to produce small amounts of testosterone if they are supplied with appropriate precursors (such as progesterone). Probably of more importance, the tubules in adult testes are able to convert testosterone, supplied to them by the Leydig cells, into 5α-reduced androgens (DHT and 3α-diol). The physiological significance of these products of the tubules is not currently understood.

Although androgens are by far the major secretory product of the testes, the testes do produce and secrete small amounts of estradiol. There is evidence that both the Sertoli cells of the tubules and the Leydig cells can produce estradiol, although estradiol receptors in the testes appear to be located only in the Leydig cells. Experiments suggest that, when LH is present in high amounts and not

**Table 20.4** Summary of some of the actions of androgens in the male

| Category | Action |
|---|---|
| Sex determination | Growth and development of wolffian ducts into epididymis, vas deferens, seminal vesicles, and ejaculatory ducts<br>Development of urogenital sinus and tubercle into prostate<br>Development of male external genitalia (penis and scrotum) |
| Spermatogenesis | At puberty: completion of meiotic division and early maturation of spermatids<br>After puberty: maintenance of spermatogenesis |
| Secondary sexual characteristics | Growth and maintenance of accessory sexual organs<br>Growth of penis<br>Growth of facial and axillary hair<br>Body growth |
| Anabolic effects | Protein synthesis and muscle growth<br>Growth of bones<br>Growth of other organs (including larynx)<br>Erythropoiesis (red blood cell formation) |

**Figure 20.15.** Interactions between the two compartments of the testes. Testosterone secreted by the interstitial Leydig cells stimulates spermatogenesis in the tubules. Leydig cells may also secrete ACTH, MSH, and $\beta$-endorphin. Secretion of inhibin of the tubules may affect the sensitivity of the interstitial cells to LH stimulation.

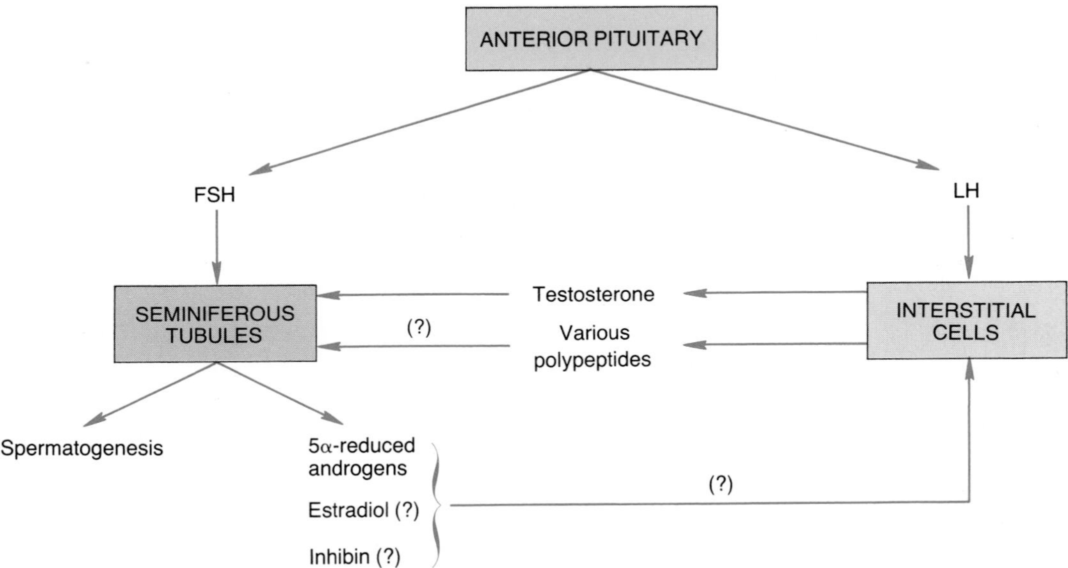


secreted in a pulsatile fashion, the desensitization and down-regulation of Leydig cell function that results may be partly mediated by estradiol.

The two compartments of the testes interact with each other in an autocrine fashion (fig. 20.15). Autocrine regulation, as described in chapter 11, refers to chemical regulation that occurs within an organ. Testosterone from the Leydig cells is metabolized by the tubules into other active androgens and is required for spermatogenesis (as described in the next section). The tubules also secrete products that could influence Leydig cell function. Such interactions are suggested by evidence that, in the pubertal male rat, exposure to FSH augments the responsiveness of the Leydig cells to LH. Since FSH can only directly stimulate the Sertoli cells of the tubules, the FSH-induced enhancement of LH responsiveness must be mediated by products secreted from the Sertoli cells.

Recent evidence suggests that inhibin secreted by the Sertoli cells in response to FSH can facilitate the Leydig cell's response to LH, as measured by the amount of testosterone secreted. Further, it has recently been shown that the Leydig cells are capable of producing a family of polypeptides previously associated only with the pituitary gland—ACTH, MSH, and $\beta$-endorphin. Experiments suggest that ACTH and MSH can stimulate Sertoli cell function, whereas $\beta$-endorphin can inhibit Sertoli function. The physiological significance of these fascinating autocrine interactions between the two compartments of the testes remains to be demonstrated.

## Spermatogenesis

The germ cells that migrate from the yolk sac to the testes during early embryonic development become "stem cells" called **spermatogonia** within the outer region of the seminiferous tubules. Spermatogonia are diploid cells (with forty-six chromosomes) that ultimately give rise to mature haploid gametes by a process of cell division called meiosis.

Meiosis, or reduction division, occurs in two parts (described in chapter 3). In the first part of this process, the DNA duplicates, and homologous chromosomes are separated into two daughter cells. Since each daughter cell contains only one of each homologous pair of chromosomes, the cells formed at the end of this *first meiotic division* contain twenty-three chromosomes each and are haploid. Each of the twenty-three chromosomes at this stage, however, consists of two strands (called chromatids) of identical DNA. During the second meiotic division, these duplicate chromatids are separated into daughter cells. Meiosis of one diploid spermatogonia cell therefore produces four haploid cells.

Actually, only about 1,000–2,000 stem cells migrate from the yolk sac into the embryonic testes. In order to produce many millions of sperm throughout adult life, these spermatogonia cells duplicate themselves by mitotic division, and only one of the two cells—now called a **primary spermatocyte**—undergoes meiotic division (fig. 20.16). In this way, spermatogenesis can occur continuously without exhausting the number of spermatogonia.

When a diploid primary spermatocyte completes the first meiotic division (at telophase I), the two haploid cells thus produced are called **secondary spermatocytes.** At the end of the second meiotic division, each of the two secondary spermatocytes produce two haploid **spermatids.** One primary spermatocyte therefore produces four spermatids.

The stages of spermatogenesis are arranged sequentially in the wall of the seminiferous tubule. The epithelial wall of the tubule—called the *germinal epithelium*—is indeed composed of germ cells in different stages of spermatogenesis. The spermatogonia and primary spermatocytes are located toward the outer side of the tubule, whereas spermatids and mature spermatozoa are located on the side of the tubule facing the lumen.

At the end of the second meiotic division, the four spermatids produced by meiosis of one primary spermatocyte are interconnected with each other—their cytoplasm does not completely pinch off at the end of each division. Development of these interconnected spermatids into separate, mature spermatozoa (a process called **spermiogenesis**) requires the participation of another type of cell in the tubules, the **Sertoli cells** (fig. 20.17).

**Figure 20.16.** Spermatogonia undergo mitotic division to replace themselves and produce a daughter cell that will undergo meiotic division. This cell is called a primary spermatocyte. Upon completion of the first meiotic division, the daughter cells are called secondary spermatocytes. Each of these completes a second meiotic division to form spermatids. Notice that the four spermatids produced by the meiosis of a primary spermatocyte are interconnected. Each spermatid forms a mature spermatozoan.

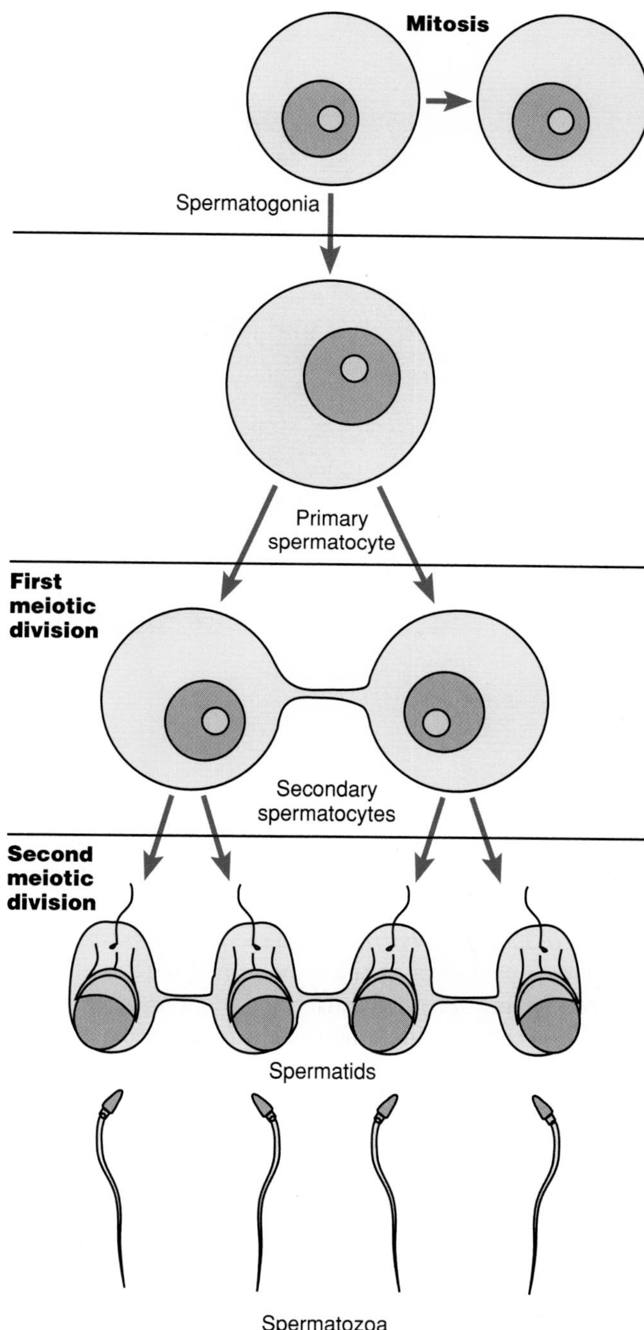

**Figure 20.17.** Seminiferous tubules. (*a*) A cross section with surrounding interstitial tissue. (*b*) The stages of spermatogenesis within the germinal epithelium of a seminiferous tubule, showing the relationship between nurse cells and developing spermatozoa. ([*a*] From: *Tissues and Organs: A Text Atlas of Scanning Electron Microscopy* by R. G. Kessel and R. Kardon. W. H. Freeman and Company. © 1979.)

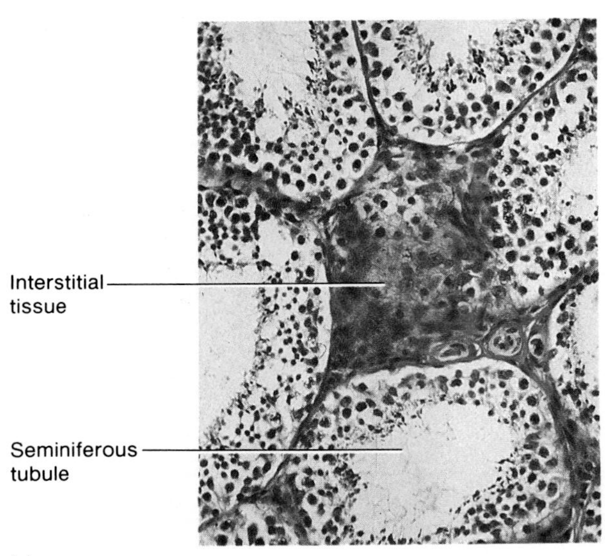


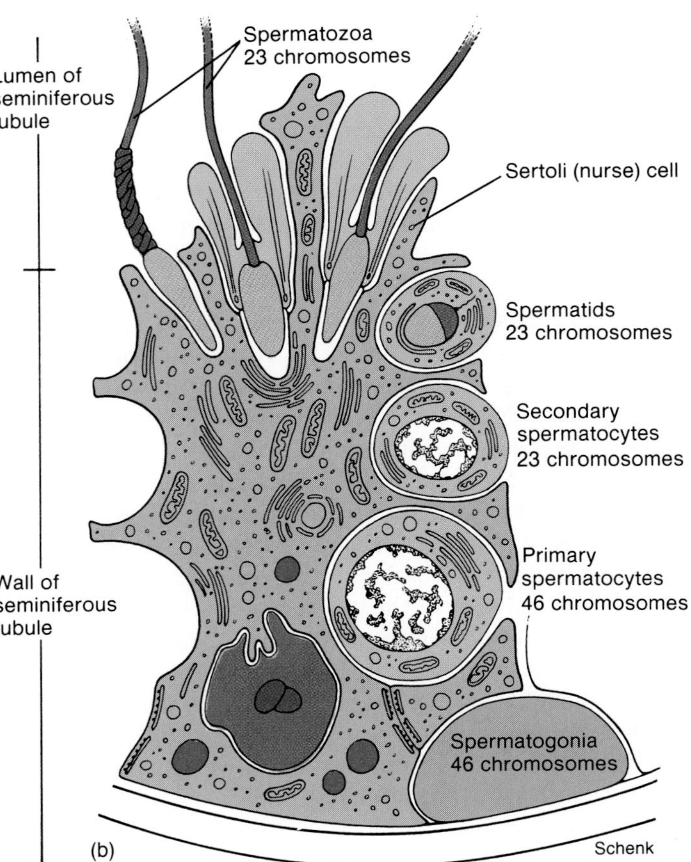


***Sertoli Cells.*** The Sertoli cells are the only nongerminal cell type in the tubules. They form a continuous layer, connected by tight junctions, around the circumference of each tubule. In this way the Sertoli cells comprise a *blood-testis barrier,* because molecules from the blood must pass through the cytoplasm of the Sertoli cells before entering germinal cells. Similarly, this barrier normally prevents the immune system from becoming sensitized to antigens in the developing sperm, and thus prevents autoimmune destruction of the sperm. The cytoplasm of the Sertoli cells extends through the width of the tubule and envelops the developing germ cells, so that it is often difficult to tell where the cytoplasm of the Sertoli cells and germ cells is separated.

In the process of *spermiogenesis* (conversion of spermatids to spermatozoa), most of the spermatid cytoplasm is eliminated. This occurs through phagocytosis by Sertoli cells of the "residual bodies" of cytoplasm from the spermatids (fig. 20.18). Many believe that phagocytosis of residual bodies may transmit informational molecules from germ cells to Sertoli cells. The Sertoli cells, in turn, may provide many molecules needed by the germ cells. It is known, for example, that the X chromosome of germ cells is inactive during meiosis. Since this chromosome contains genes needed to produce many essential molecules, it is believed that these molecules are provided by the Sertoli cells during this time.

**Figure 20.18.** Processing of spermatids into spermatozoa (spermiogenesis). As the spermatids develop into spermatozoa most of their cytoplasm is pinched off as residual bodies and ingested by the surrounding Sertoli cell cytoplasm.

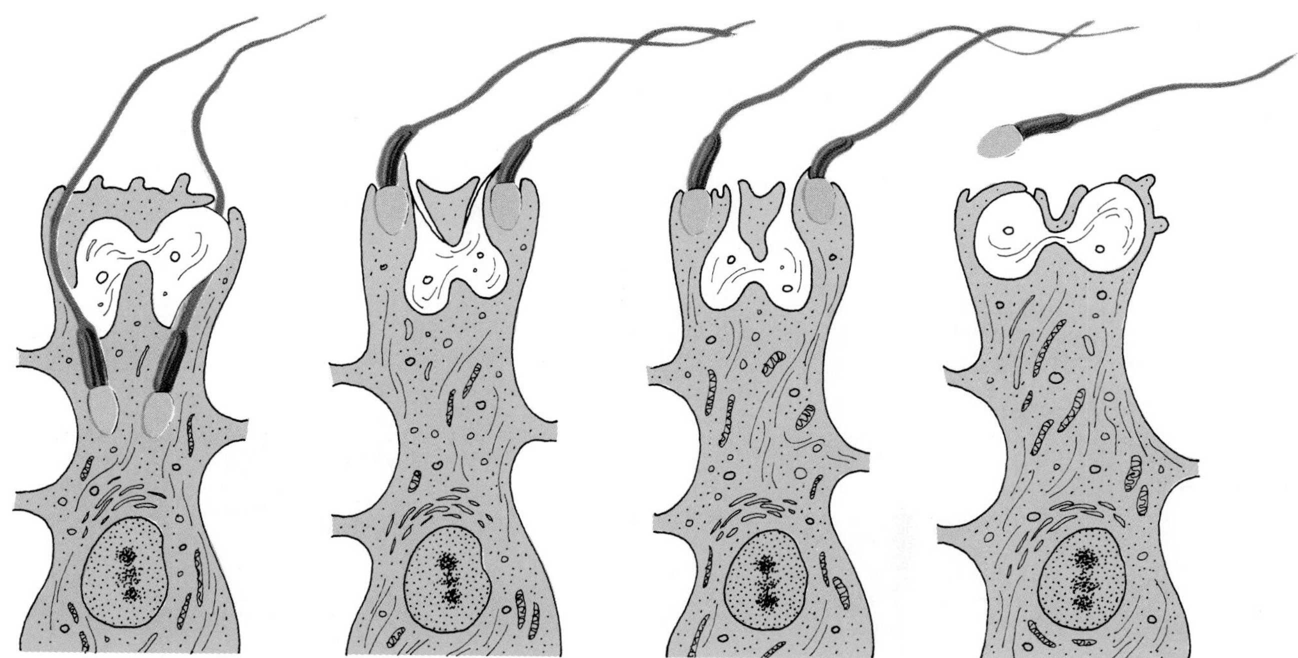

The importance of Sertoli cells in tubular function is further evidenced by the fact that FSH receptors are confined to the Sertoli cells. Any effect of FSH on the tubules, therefore, must be mediated through the action of Sertoli cells. These include the FSH-induced stimulation of spermiogenesis and the autocrine interactions between Sertoli cells and Leydig cells that have been previously described.

***Hormonal Control of Spermatogenesis.*** The very beginning of spermatogenesis—formation of primary spermatocytes and entry into early prophase I—is apparently somewhat independent of hormonal control and, in fact, starts during embryonic development. Spermatogenesis is arrested at this stage, however, until puberty, when testosterone secretion rises. Testosterone is required for completion of meiotic division and for the early stages of spermatid maturation. This effect is probably not produced by testosterone directly, but rather by some of the molecules derived from testosterone in the tubules.

The later stages of spermatid maturation during puberty appear to require stimulation by FSH (fig. 20.19). This FSH effect is mediated by the Sertoli cells which contain FSH receptors and surround and interact with the spermatids. During puberty, therefore, both FSH and androgens are needed for the initiation of spermatogenesis.

Experiments in rats, and more recently in humans, have revealed that spermatogenesis within the adult testis can be maintained by androgens alone, in the absence of FSH. It appears, in other words, that FSH is needed to initiate spermatogenesis at puberty, but that once spermatogenesis has begun, FSH may no longer be required for this function.

At the conclusion of spermiogenesis, spermatozoa are released into the lumen of the seminiferous tubules. The spermatozoa contain a *head* (with DNA inside), a *body*, and a *tail* (fig. 20.20). Although the tail will ultimately be capable of flagellar movement, the sperm at this stage are nonmotile. Motility and other maturational changes occur outside the testes in the epididymis.

**Figure 20.19.** The endocrine control of spermatogenesis. During puberty both testosterone and FSH are required to initiate spermatogenesis. In the adult, however, testosterone alone can maintain spermatogenesis.

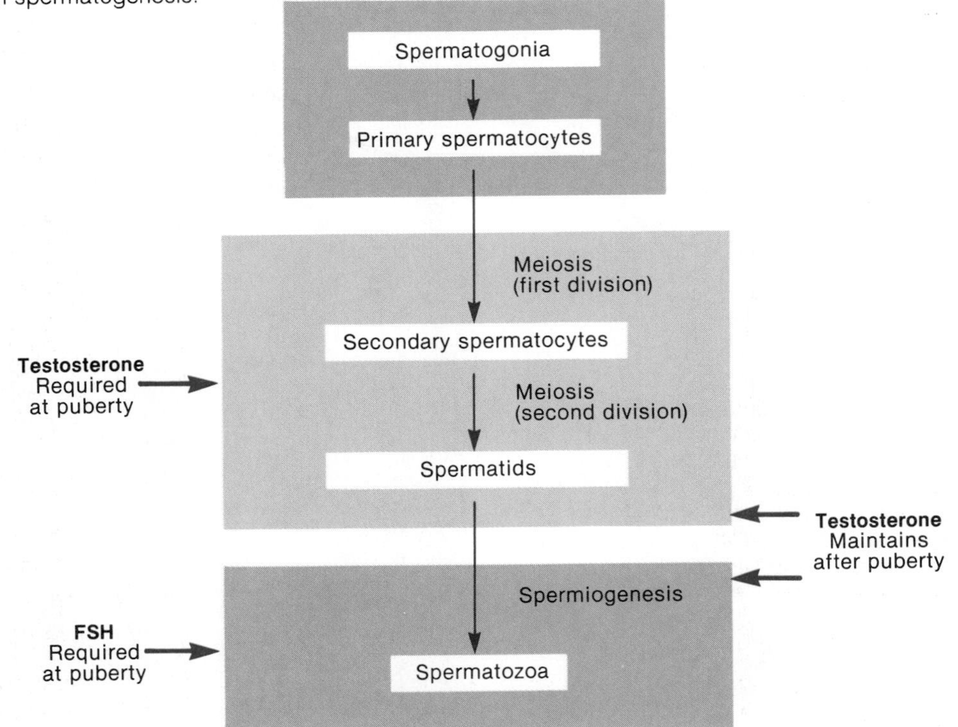


**Figure 20.20.** A human spermatozoan. (*a*) A diagrammatic representation; (*b*) a scanning electron micrograph. (From: *Tissues and Organs: A Text Atlas of Scanning Electron Microscopy* by R. G. Kessel and R. Kardon. W. H. Freeman and Company. © 1979.)

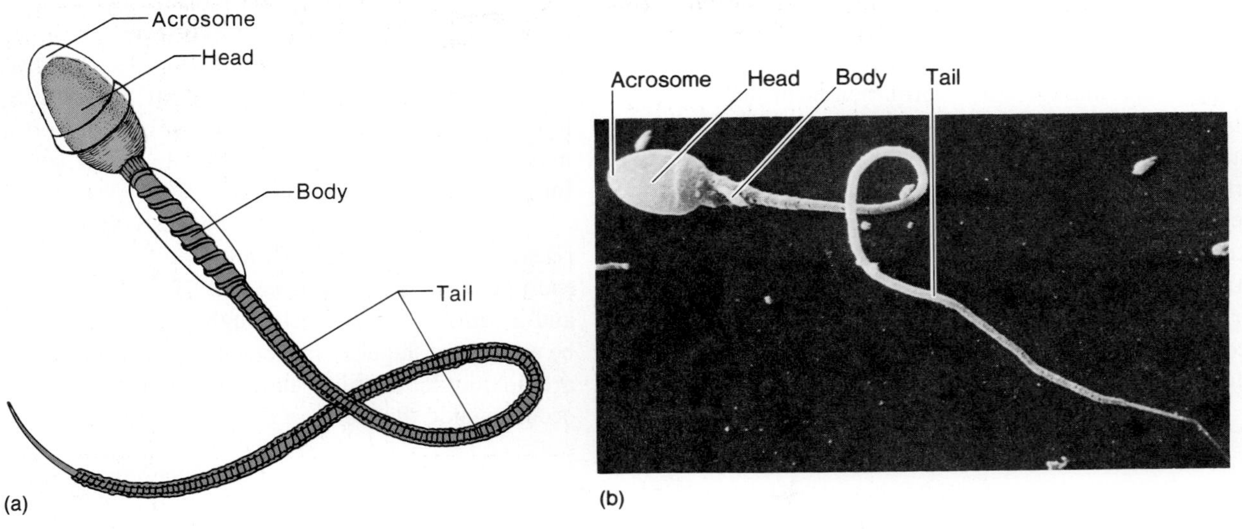

**Figure 20.21.** Organs of the male reproductive system in sagittal view.

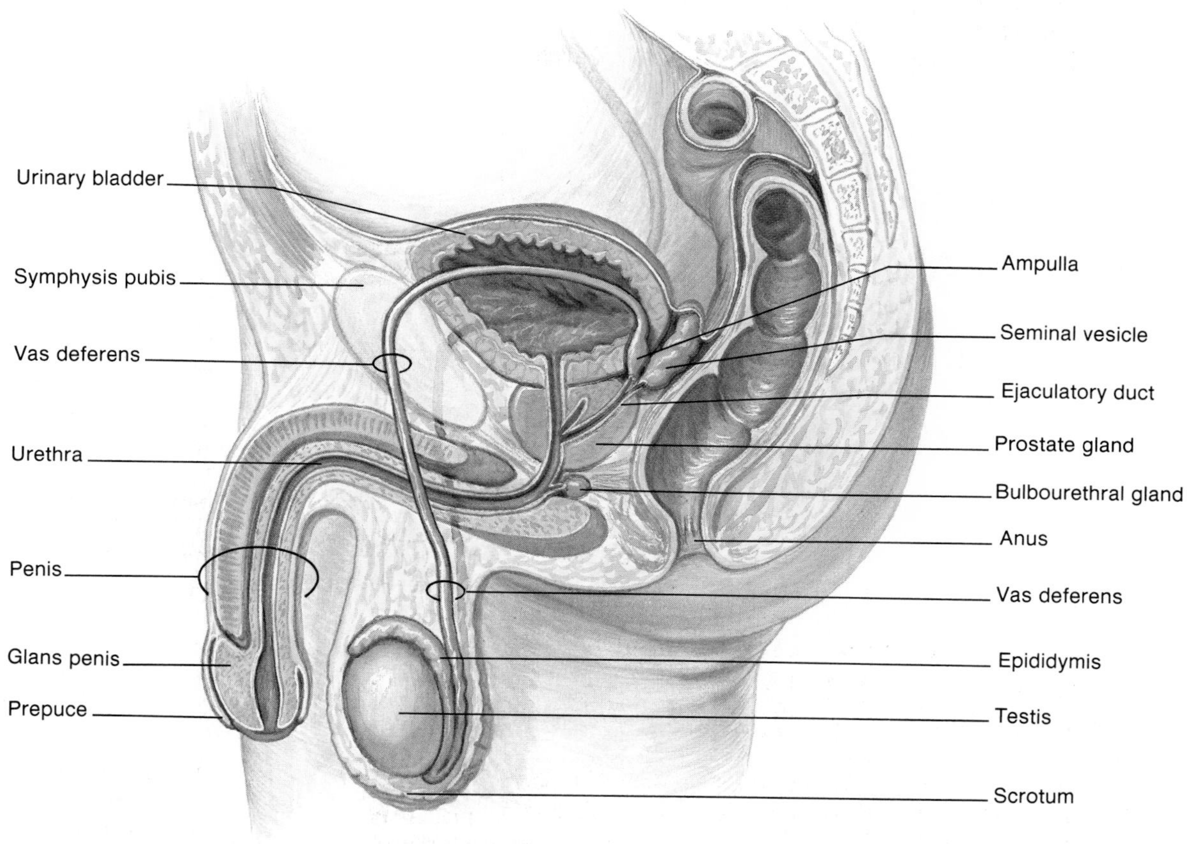


## Male Sex Accessory Organs

The seminiferous tubules are connected at both ends to the *rete testis* (see fig. 20.12). Spermatozoa and tubular secretions are moved to this area of the testis and are drained via the *ductuli efferentes* into the **epididymis.** The epididymis is a single-coiled tube, four to five meters long if stretched out, that receives the tubular products. Spermatozoa enter at the "head" of the epididymis and are drained from its "tail" by a single tube, the **vas deferens.**

Spermatozoa that enter the head of the epididymis are nonmotile. During their passage through the epididymis, spermatozoa gain motility and undergo other maturational changes—sperm that leave the epididymis are more resistant to changes in pH and temperature, are motile, and can become able to fertilize an ovum once they spend some time in the female reproductive tract. Sperm obtained from the seminiferous tubules, in contrast, cannot fertilize an ovum. The epididymis serves as a site for sperm maturation and for the storage of sperm between ejaculations.

The vas deferens carries sperm from the epididymis out of the scrotum and into the body cavity. In its passage, the vas deferens obtains fluid secretions of the **seminal vesicles** and **prostate** gland. This fluid, now called *semen,* is carried by the ejaculatory duct to the *urethra* (fig. 20.21). Sperm can be stored within a widened area of the vas deferens known as the *ampulla.*

The seminal vesicles and prostate are androgen-dependent accessory sexual organs—they atrophy if androgen is withdrawn by castration. The seminal vesicles secrete fluid containing fructose (which serves as an energy source for the spermatozoa), citric acid, coagulation proteins, and prostaglandins. The prostate secretes a liquefying agent and the enzyme *acid phosphatase,* which is often measured clinically to assess prostate function.

**Figure 20.22.** The structure of the penis showing the attachment, blood and nerve supply, and the arrangement of the erectile tissue.

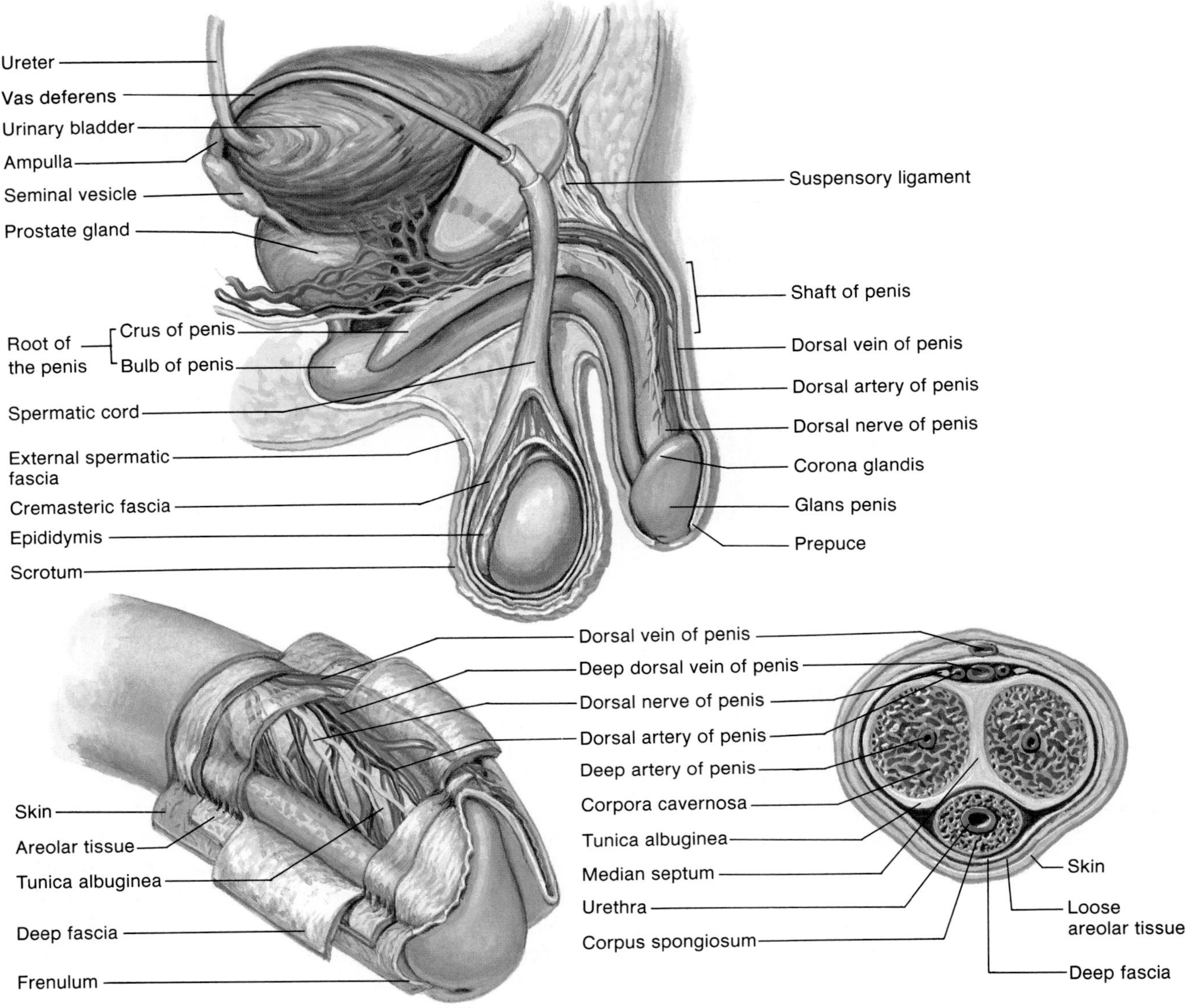


## Erection, Emission, and Ejaculation

Erection, accompanied by increases in the length and width of the penis, is achieved as a result of blood flow into the "erectile tissues" of the penis. These erectile tissues include two paired structures—the *corpora cavernosa*—located on the dorsal side of the penis, and one unpaired *corpus spongiosum* on the ventral side (fig. 20.22). The urethra runs through the center of the corpus spongiosum. The erectile tissue forms columns extending the length of the penis, although the corpora cavernosa do not extend all the way to the tip.

Erection is achieved as a result of parasympathetic nerve-induced vasodilation of arterioles that allows blood to flow into the penis (table 20.5). As the erectile tissues become engorged with blood and the penis becomes turgid, venous outflow of blood is partially occluded, thus aiding erection. The term *emission* refers to the movement of semen into the urethra, and *ejaculation* refers to the forcible expulsion of semen from the urethra out of the penis. Emission and ejaculation are stimulated by sympathetic nerves, which cause peristaltic contractions of the tubular system, contractions of the seminal vesicles and prostate, and contractions of muscles at the base of the penis. Sexual function in the male thus requires the synergistic action (rather than antagonistic action) of the parasympathetic and sympathetic systems.

Erection is controlled by two portions of the central nervous sytem—the hypothalamus in the brain and the sacral portion of the spinal cord. Conscious sexual thoughts orginating in the cerebral cortex act via the hypothalamus

**Table 20.5** Control of erection and ejaculation

| Regulation | Effect | Result |
|---|---|---|
| Parasympathetic nerves | Vasodilation produces increased blood flow into erectile tissues—corpora cavernosa and corpus spongiosum. As tissues become turgid, venous outflow is partially occluded, further increasing accumulation of blood. | Erection |
| Sympathetic nerves | Peristaltic waves of contraction in tubular system—primarily epididymis, vas deferens, and ejaculatory ducts, and contraction of prostate and seminal vesicles. | Ejaculation |

to control the sacral region, which in turn increases parasympathetic nerve activity to promote vasodilation and erection in the penis. Conscious thought is not required for erection however, because sensory stimulation of the penis can more directly activate the sacral region of the spinal cord and cause an erection.

## Male Fertility

The male ejaculates about 1.5–5.0 ml of semen. The bulk of this fluid (45% to 80%) is produced by the seminal vesicles, and about 15% to 30% is contributed by the prostate. The sperm content in human males averages 60–150 million per milliliter in the ejaculated semen. Some of the values of normal human semen are summarized in table 20.6.

A sperm concentration below about twenty million per milliliter is termed *oligospermia* (*oligo* = few), and is associated with decreased fertility. A total sperm count below about fifty million per ejaculation is clinically significant in male infertility. In addition to low sperm counts as a cause of infertility, some men and women have antibodies against sperm antigens (this is very common in men with vasectomies). Such antibodies do not appear to affect health, but do reduce fertility.

In order for fertilization to occur, specific protein receptors in the head of the sperm must interact with particular carbohydrate molecules in the *zona pellucida,* which is the outer translucent layer surrounding an ovum. Researchers have recently utilized this fact in an experimental testing of a new method of contraception. They isolated and purified a specific protein component of guinea pig sperm, and used this as an antigen to vaccinate both female and male guinea pigs. The vaccinated guinea pigs developed active immunity (chapter 19) to this protein which resulted in complete contraceptive success. Moreover, fertility was restored several months after the vaccination, showing that this method of contraception is reversible. Much further research, however, is required before a similar technique might be applied in human contraception.

**Vasectomy** (fig. 20.23) is commonly performed as a contraceptive method. In this procedure each vas deferens is cut and tied or, in some cases, a valve or similar device

**Table 20.6** Some characteristics and reference values used in the clinical examination of semen

| Characteristic | Reference Value |
|---|---|
| Volume of ejaculate | 1.5–5.0 ml |
| Sperm count | 40–250 million/ml |
| Sperm motility | |
| Percent of motile forms: | |
| 1 hour after ejaculation | 70% or more |
| 3 hours after ejaculation | 60% or more |
| Leukocyte count | 0–2000/ml |
| pH | 7.2–7.8 |
| Fructose concentration | 150–600 mg/100 ml |

Modified from Glasser, L. Seminal fluid and subfertility. *Diagnostic Medicine,* July/August 1981, p. 28. By permission.

is inserted. This procedure interferes with sperm transport but does not directly affect the secretion of androgens from Leydig cells in the interstitial tissue. Since spermatogenesis continues, the sperm produced cannot be drained from the testes and instead accumulates in "crypts" that form in the seminiferous tubules and vas deferens. These crypts present sites of inflammatory reactions in which spermatozoa are phagocytosed and destroyed by the immune system. It is thus not surprising that approximately 70% of men with vasectomies develop antisperm antibodies. These antibodies do not appear to cause autoimmune damage to the testes but they do significantly diminish the possibility of reversing the procedure and restoring fertility.

1. *Describe the effects of castration on FSH and LH secretion in the male, and explain the experimental evidence that suggests that the testes produce a polypeptide that specifically inhibits FSH secretion.*
2. *Describe the two compartments of the testes with respect to their (a) structure, (b) function, and (c) response to gonadotropin stimulation. Describe two ways that these compartments appear to interact.*
3. *Using a diagram, describe the stages of spermatogenesis. Explain why spermatogenesis can continue throughout life without using up all of the spermatogonia.*
4. *Describe the structure and proposed functions of the Sertoli cells in the seminiferous tubules.*
5. *Explain how FSH and androgens synergize to stimulate sperm production at puberty. Describe the hormonal requirements for spermatogenesis after puberty.*

**Figure 20.23.** A simplified illustration of a vasectomy, in which a segment of the ductus deferentia is removed through an incision in the scrotum.

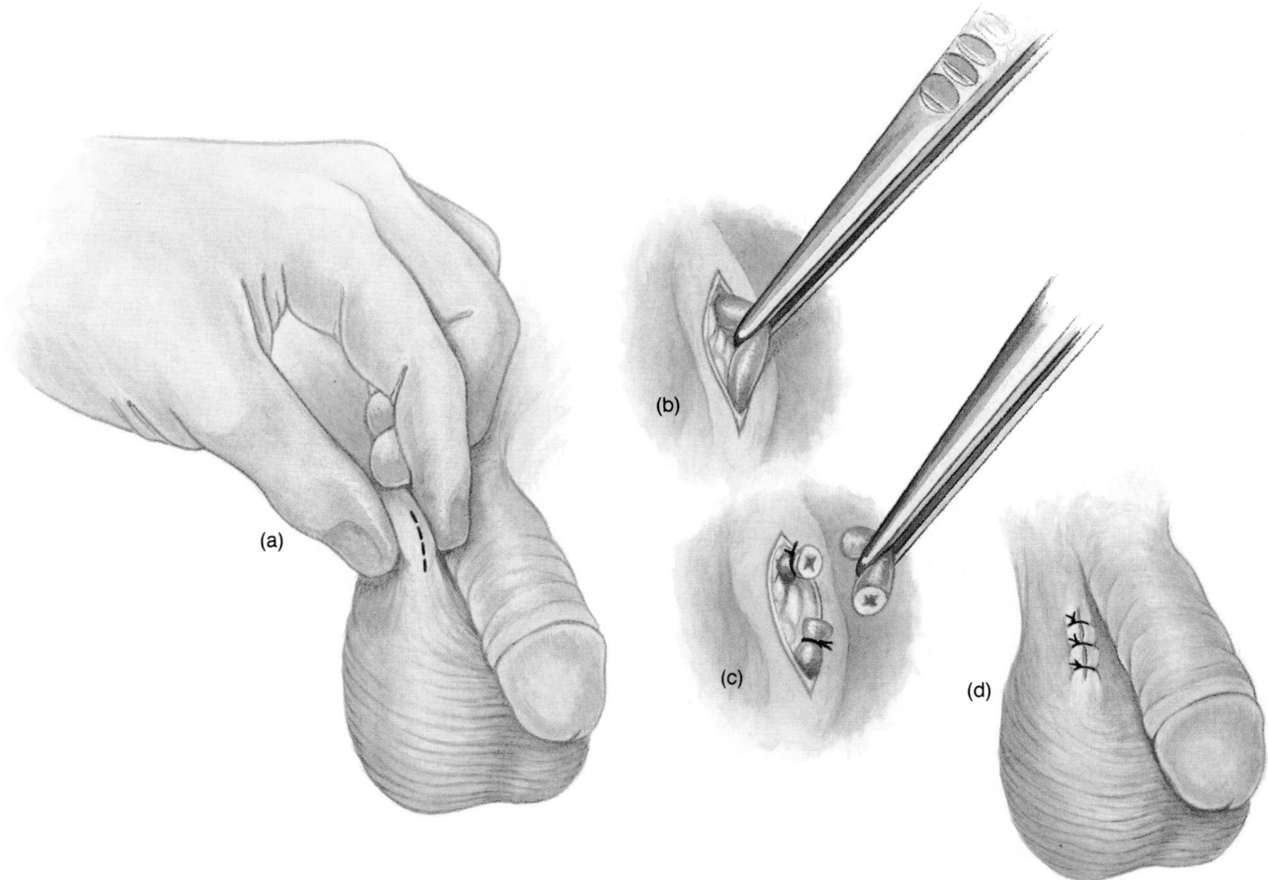


## *Female Reproductive System*

The ova are contained within the ovary in hollow structures known as follicles. As a result of gonadotropin stimulation, a primary follicle may grow and develop a cavity, called an antrum, within surrounding granulosa cells and become a secondary follicle. The ovum also undergoes changes during this time. A primary oocyte within a primary follicle progresses through to metaphase II of meiosis within a secondary follicle. In this process a large secondary oocyte is formed, together with a small polar body. The secondary oocyte contained within a fully mature, or graafian, follicle is expelled from the ovary in the process of ovulation. This secondary oocyte can then enter the uterine tube.

The two ovaries (fig. 20.24) are located within the body cavity, suspended by means of ligaments from the pelvic girdle. Extensions, or *fimbriae,* of the **uterine,** or **fallopian, tubes** partially cover each ovary. Ova that are released from the ovary—in a process called *ovulation*—are normally drawn into the fallopian tubes by the action of the ciliated epithelial lining of the tubes. The lumen of each fallopian tube is continuous with the **uterus** (or womb), a pear-shaped muscular organ also suspended within the pelvic girdle by ligaments.

The uterus narrows to form the *cervix* (*cervix* = neck), which opens to the **vagina.** The only physical barrier between the vagina and uterus is a plug of *cervical mucus.* These structures—the vagina, uterus, and fallopian tubes—constitute the sex accessory organs of the female (fig. 20.25). Like the sex accessory organs of the male, the female genital tract is affected by gonadal steroid hormones. Cyclic changes in ovarian secretion, as will be described in the next section, cause cyclic changes in the epithelial lining of the genital tract. The epithelial lining of the uterus, known as the *endometrium,* is most dramatically altered during the ovarian cycle.

**Figure 20.24.** A dorsal view of the female reproductive organs showing the relationship of the ovaries, uterine tubes, uterus, cervix, and vagina.

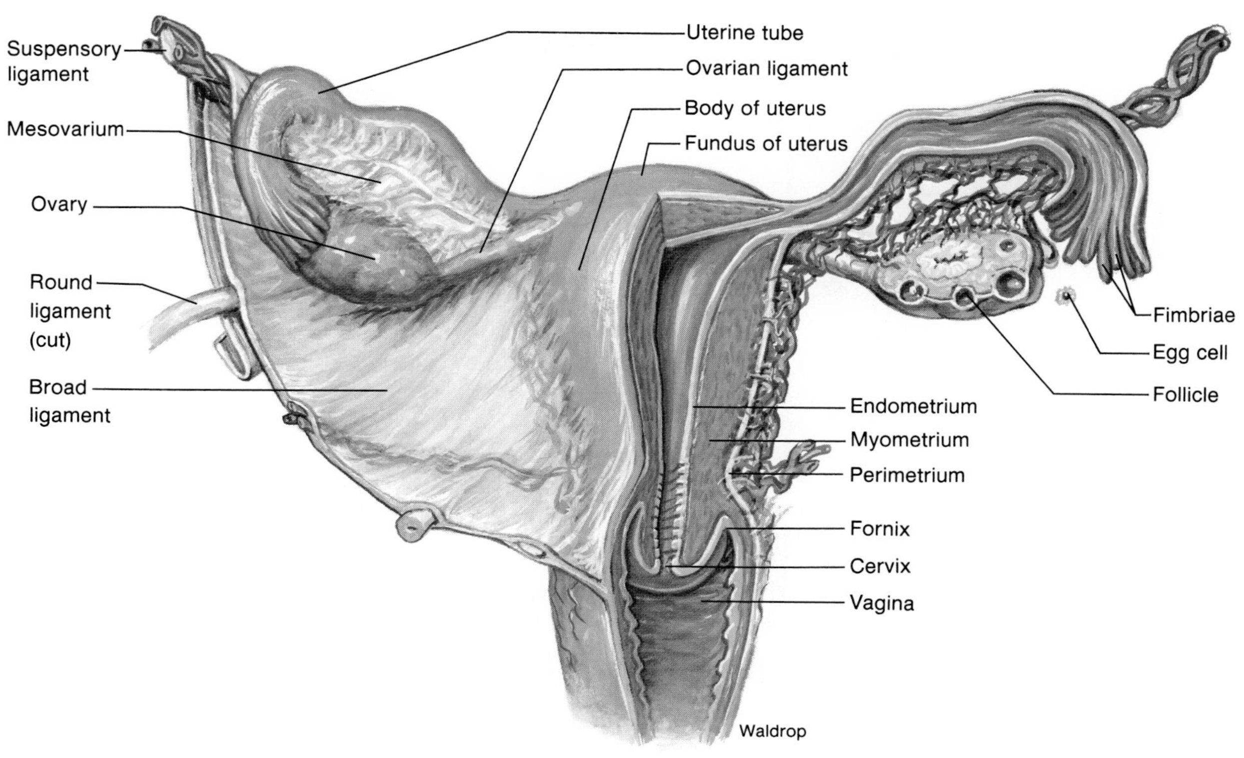


**Figure 20.25.** Organs of the female reproductive system seen in sagittal section.

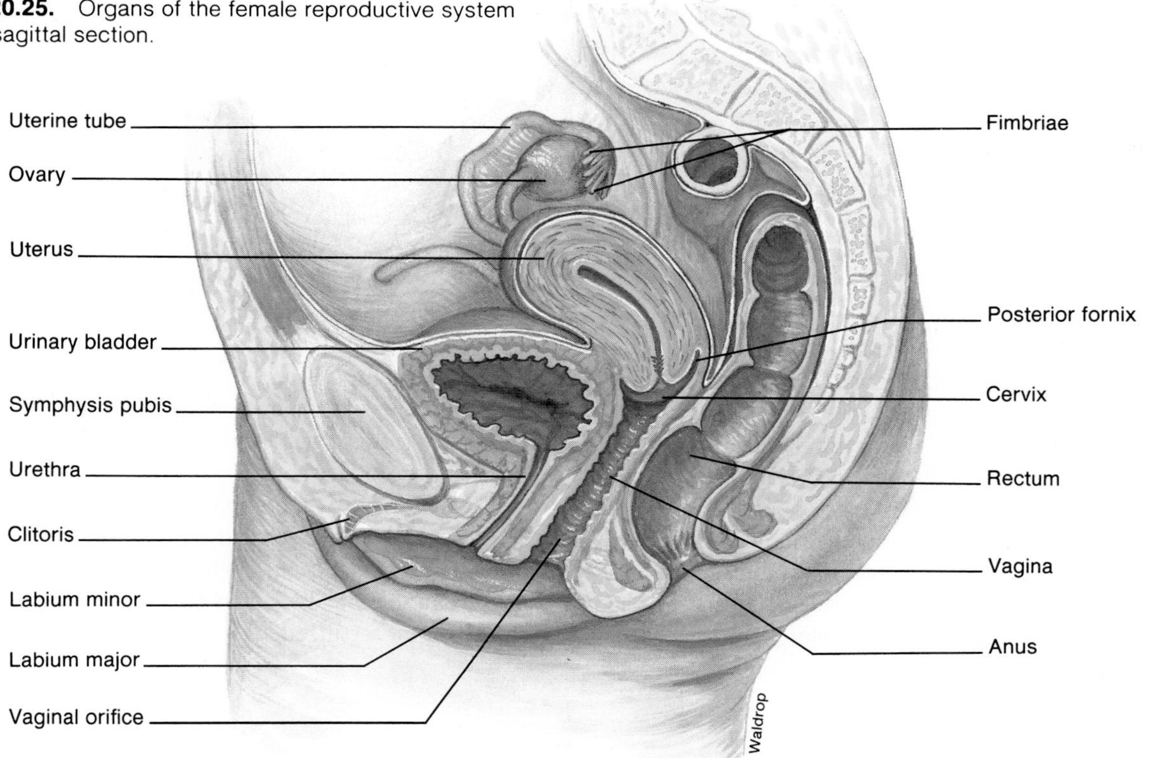

**Figure 20.26.** The external female genitalia.

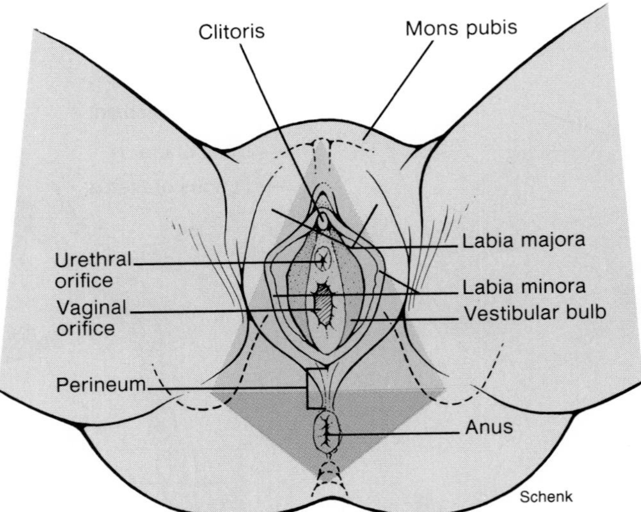


The vaginal opening is located immediately posterior to the opening of the urethra. Both openings are covered by inner **labia minora** and outer **labia majora** (fig. 20.26). The **clitoris**, which is homologous to the penis as previously described, is located at the anterior margin of the labia minora.

## Ovarian Cycle

The germ cells that migrate into the ovaries during early embryonic development multiply, so that by about five months of gestation (prenatal life) the ovaries contain approximately six to seven million oogonia. The production of new oogonia stops at this point and never resumes again. Toward the end of gestation the oogonia begin meiosis, at which time they are called **primary oocytes.** Like spermatogenesis in the prenatal male, oogenesis is arrested at prophase I of the first meiotic division. The number of primary oocytes decreases throughout a woman's reproductive years. The ovaries of a newborn girl contain about two million oocytes, but this number is reduced to about 300,000–400,000 by the time the girl enters puberty. Oogenesis ceases entirely at menopause (the time menstruation stops).

Primary oocytes that are not stimulated to complete the first meiotic division are contained within tiny follicles, called **primordial follicles.** In response to gonadotropin stimulation some of these oocytes and follicles get larger, and the follicular cells divide to produce numerous small **granulosa cells** that surround the oocyte and fill the follicle. A follicle at this stage in development is called a **primary follicle.**

Some primary follicles will be stimulated to grow still bigger and develop a fluid-filled cavity, called an *antrum,* at which time they are called **secondary follicles** (fig. 20.27). The granulosa cells of secondary follicles form a ring around the circumference of the follicle and form a mound that supports the ovum. This mound is called the *cumulus oophorus.* The ring of granulosa cells surrounding the ovum is the *corona radiata.* Between the oocyte and the corona radiata is a thin gellike layer of proteins and polysaccharides, called the *zona pellucida.* Under the stimulation of FSH from the anterior pituitary, the granulosa cells secrete increasing amounts of estrogen as the follicles grow. Interestingly, the granulosa cells produce estrogen from its precursor testosterone, which is supplied by cells of the *theca interna* layer immediately outside the follicle.

As the follicle develops, the primary oocyte completes its first meiotic division. This does not form two complete cells, however, because only one cell—the **secondary oocyte**—gets all the cytoplasm. The other cell formed at this time becomes a small *polar body* (fig. 20.28), which eventually fragments and disappears. This unequal division of cytoplasm ensures that the ovum will be large enough to become a viable embryo should fertilization later occur. The secondary oocyte then begins the second meiotic division but stops at metaphase II. Meiosis is arrested at metaphase II, before separation of chromosomes and division of cytoplasm occurs, while the ovum is in the ovary. The second meiotic division is only completed by an ovum that has been fertilized.

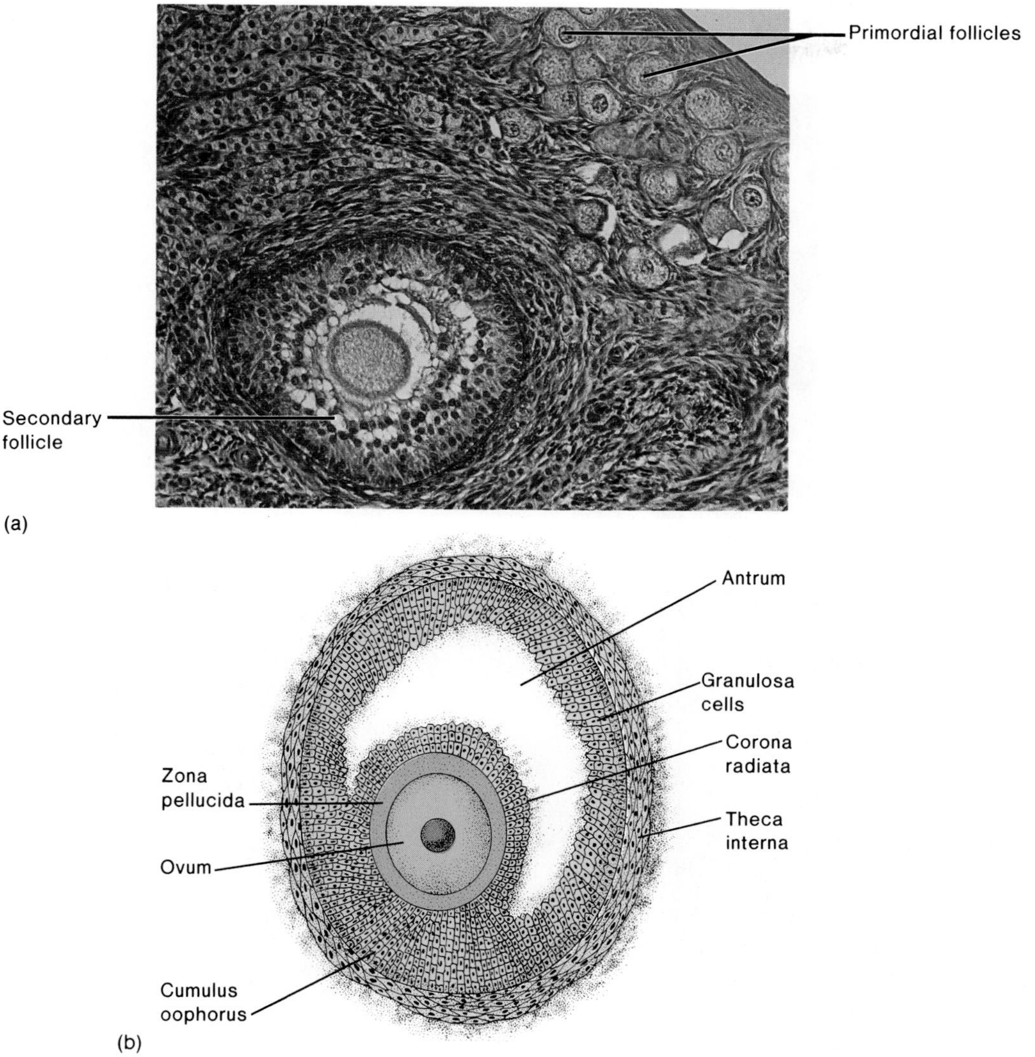


**Figure 20.27.** (*a*) A photomicrograph of primordial and secondary follicles. (*b*) A diagram of the parts of a secondary follicle.

## Ovulation

Usually, by about ten to fourteen days after the first day of menstruation, only one follicle has continued its growth to become a fully mature **graafian follicle** (fig. 20.29); other secondary follicles during that cycle regress and become *atretic.* The graafian follicle is so large that it forms a bulge on the surface of the ovary. Under proper hormonal stimulation this follicle will rupture—much like the popping of a blister—and extrude its oocyte into the uterine tube in the process of **ovulation** (fig. 20.30).

The released cell is a secondary oocyte, surrounded by the zona pellucida and corona radiata. If it is not fertilized it disintegrates in a couple of days. If a sperm passes through the corona radiata and zona pellucida and enters the cytoplasm of the secondary oocyte, the oocyte will then complete the second meiotic division. In this process the cytoplasm is again not divided equally; most of the cytoplasm remains in the zygote (fertilized egg), leaving another polar body which, like the first, disintegrates (fig. 20.31).

Changes continue in the ovary following ovulation. The empty follicle, under the influence of luteinizing hormone from the anterior pituitary, undergoes structural and biochemical changes to become a **corpus luteum** (*corpus luteum* = yellow body). Unlike the ovarian follicles, which secrete only estrogen, the corpus luteum secretes two sex steroid hormones: estrogen and progesterone. Toward the end of a nonfertile cycle the corpus luteum regresses and is changed into a nonfunctional *corpus albicans.* These cyclic changes in the ovary are summarized in figure 20.32.

**Figure 20.28.** (*a*) A primary oocyte at metaphase I of meiosis. Note the alignment of chromosomes (arrow). (*b*) A human secondary oocyte formed at the end of the first meiotic division and the first polar body (arrow).

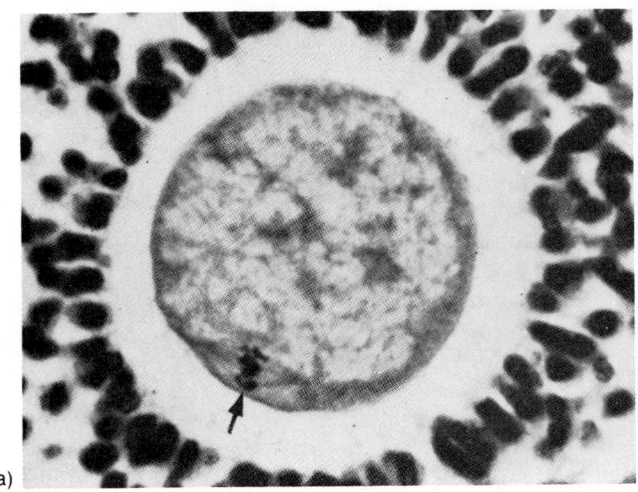

(a)

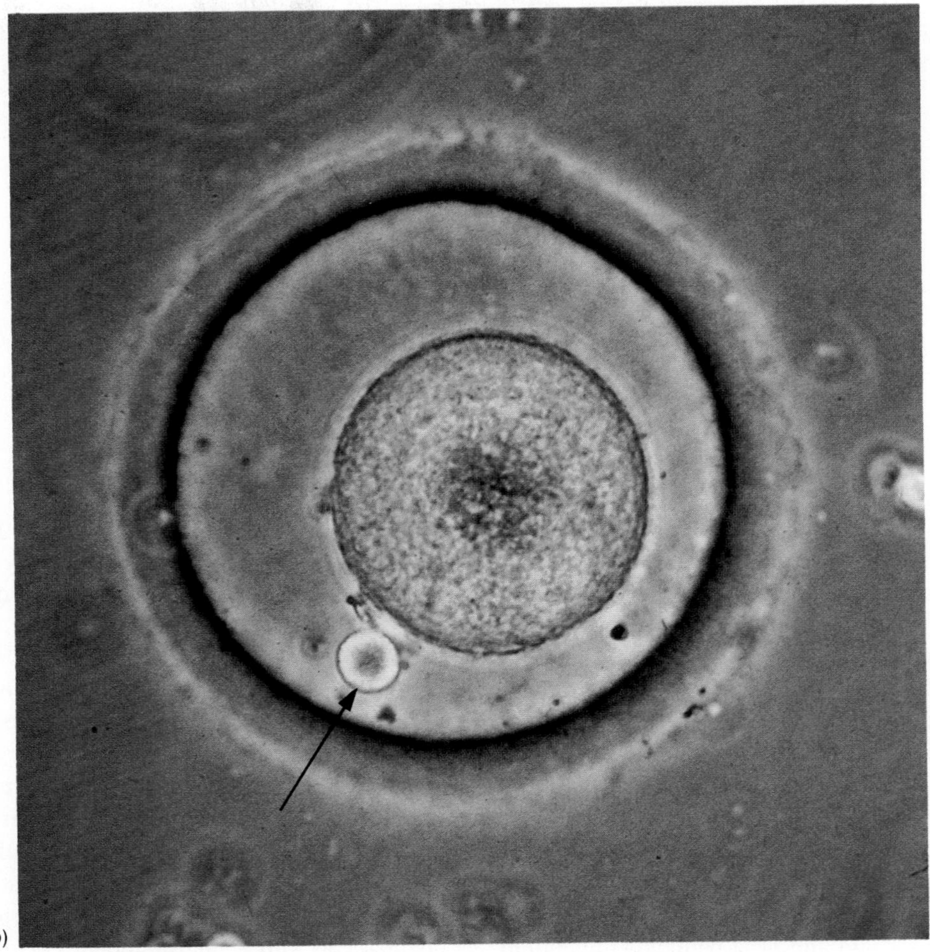

(b)

**Figure 20.29.** A graafian follicle within the ovary of a monkey.

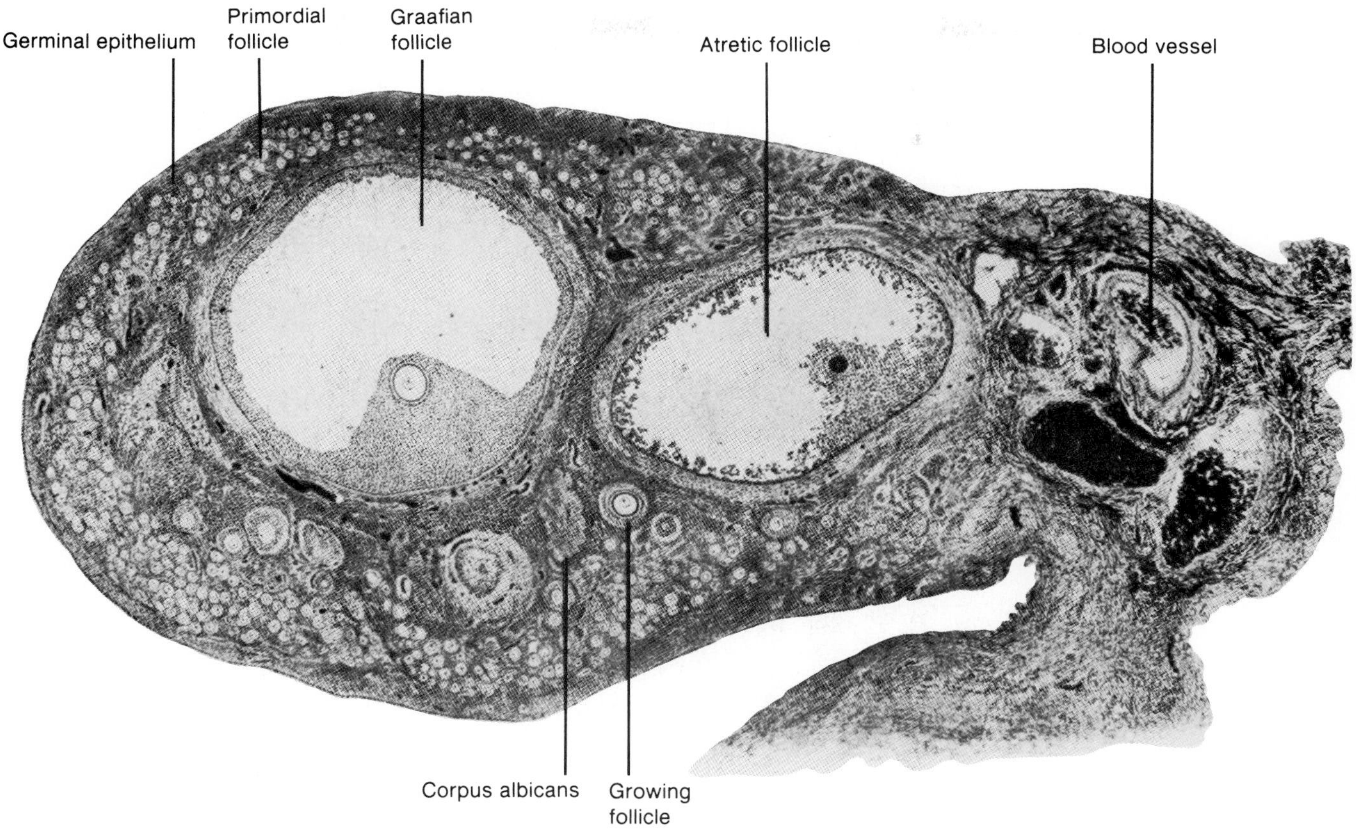

**Figure 20.30.** Ovulation from a human ovary.

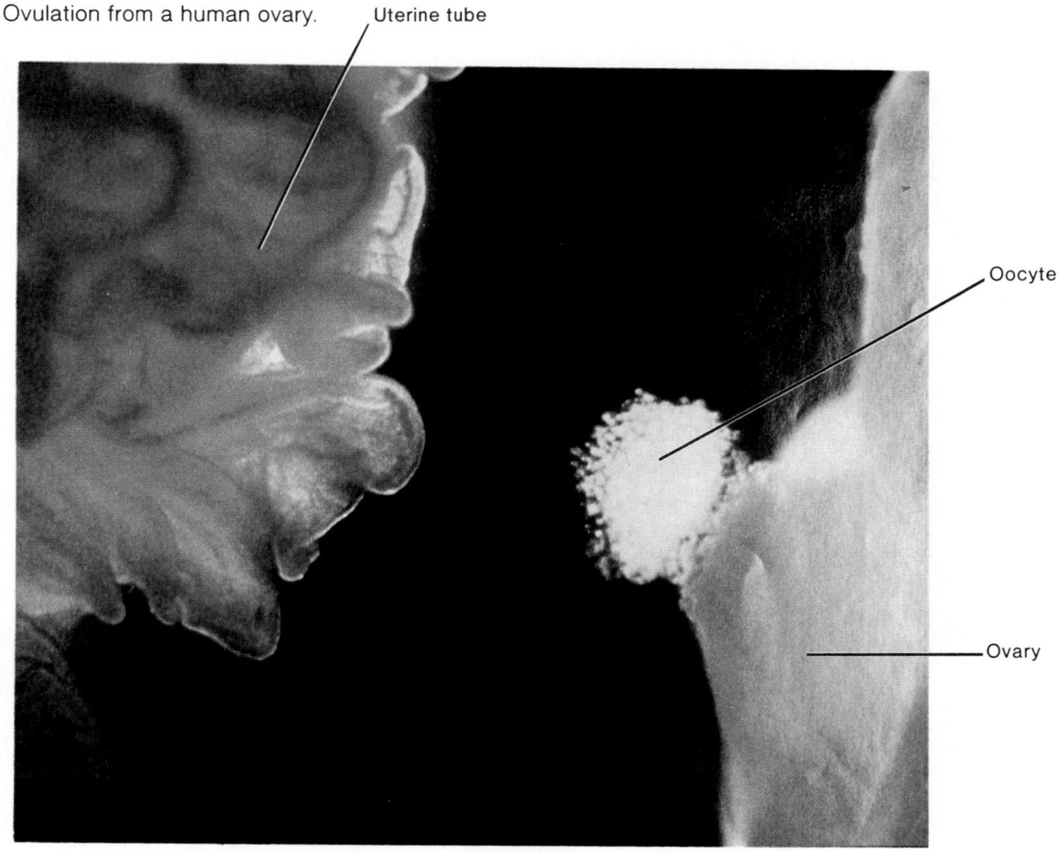

**Figure 20.31.** A schematic diagram of the process of oogenesis. During meiosis, each primary oocyte produces a single haploid gamete. If the secondary oocyte is fertilized, it forms a secondary polar body and becomes a zygote.

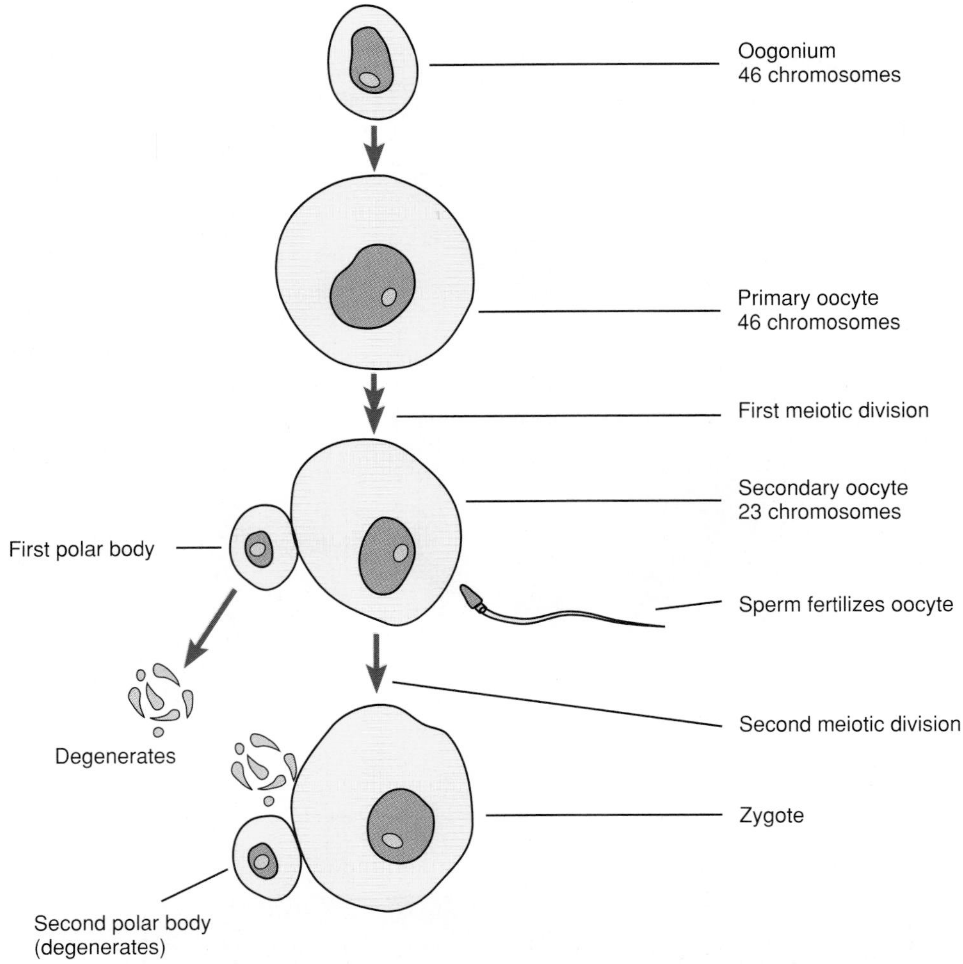


**Figure 20.32.** A schematic diagram of an ovary showing the various stages of ovum and follicle development.

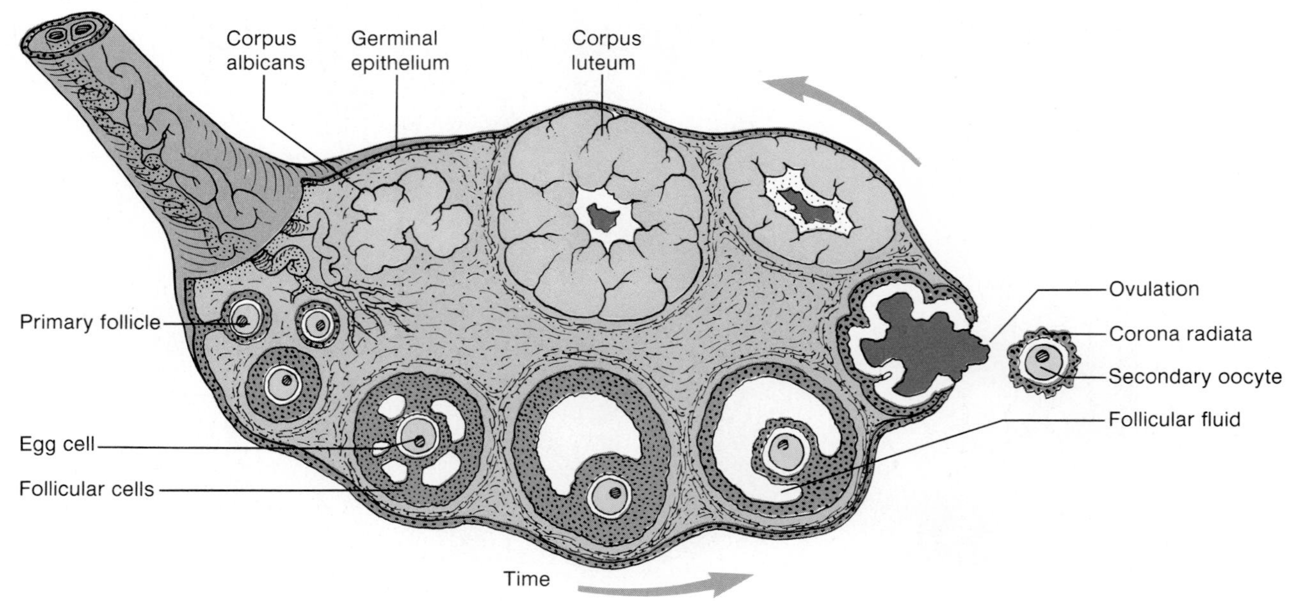

### Pituitary-Ovarian Axis

The term *pituitary-ovarian axis* refers to the hormonal interactions between the anterior pituitary and the ovaries. The anterior pituitary secretes two gonadotropic hormones—follicle-stimulating hormone (FSH) and luteinizing hormone (LH)—that promote cyclic changes in the structure and function of the ovaries. The secretion of both gonadotropic hormones, as previously discussed, is controlled by a single releasing hormone from the hypothalamus—called gonadotropin releasing hormone (GnRH)—and by feedback effects from hormones secreted from the ovaries. The nature of these interactions will be described in detail in the next section.

Since one releasing hormone can stimulate the secretion of both FSH and LH, one might expect to always see parallel changes in the secretion of these gonadotropins. This, however, is not the case. During an early phase of the menstrual cycle FSH secretion is slightly greater than LH secretion, and just prior to ovulation LH secretion greatly exceeds FSH secretion. These differences are believed to be a result of the feedback effects of ovarian sex steroids, which can change the amount of GnRH secreted, the pulse frequency of GnRH secretion, and the ability of the anterior pituitary cells to secrete FSH and/or LH. These complex interactions result in a pattern of hormone secretion that is responsible for, and characteristic of, each phase of the menstrual cycle.

---

1. *Compare the structure and contents of a primordial follicle, primary follicle, secondary follicle, and graafian follicle.*
2. *Define* ovulation, *and describe the changes that occur in the ovary following ovulation in a nonfertile cycle.*
3. *Describe oogenesis, and explain why only one mature ovum is produced by this process.*
4. *Compare the hormonal secretions of the ovarian follicles with those of a corpus luteum.*

---

## MENSTRUAL CYCLE

Cyclic changes in the secretion of gonadotropic hormones from the anterior pituitary cause the changes observed in the ovaries during a menstrual cycle. FSH stimulates development of the ovarian follicles, and LH promotes ovulation and the development of a corpus luteum from the empty follicle. Growth of the follicles and the development of the corpus luteum, in turn, produces cyclic changes in the secretion of the sex steroids estradiol and progesterone. These cyclic changes in ovarian hormone secretion interact with the hypothalamus and pituitary to regulate secretion of FSH and LH. The rise in secretion first of estradiol and then of progesterone promote growth and development of the endometrium, and the fall in the secretion of these hormones causes menstruation.

Humans, apes, and old-world monkeys have cycles of ovarian activity that repeat at approximately one-month intervals; hence the name **menstrual cycle** (*menstru* = monthly). The term *menstruation* is used to indicate the periodic shedding of the stratum functionale of the endometrium (epithelial lining of the uterus), which becomes thickened prior to menstruation under stimulation by ovarian steroid hormones. In primates (other than new-world monkeys) this shedding of the endometrium is accompanied by bleeding. There is no bleeding, in contrast, when other mammals shed the endometrium; their cycles, therefore, are not called menstrual cycles.

Human females and other primates that have menstrual cycles may permit copulation at any time of the cycle. Nonprimate female mammals, in contrast, are sexually receptive only at a particular time in their cycles (shortly before or shortly after ovulation). These animals are therefore said to have *estrous cycles*. Bleeding occurs in some animals (such as dogs and cats) that have estrous cycles shortly before they permit copulation. This bleeding is a result of high estrogen secretion and is not associated with shedding of the endometrium.

The bleeding that accompanies menstruation, in contrast, can be experimentally induced by removing the ovaries. This demonstrates that menstrual bleeding is caused by the withdrawal of ovarian hormones. During the normal menstrual cycle the secretion of ovarian hormones rises and falls in a regular fashion, causing cyclic changes in the endometrium and other sex steroid-dependent tissues.

### Phases of the Menstrual Cycle: Pituitary and Ovary

The average menstrual cycle has a duration of about twenty-eight days. Since it is a cycle, there is no beginning or end, and the changes that occur are generally gradual. It is convenient, however, to call the first day of menstruation "day one" of the cycle, because menstrual blood flow is the most apparent of the changes that occur. It is also convenient to divide the cycle into four phases based on changes that occur in the ovary and in the endometrium. The ovaries are in the **follicular phase** starting on the first day of menstruation and ending on the day of ovulation. After ovulation, the ovaries are in the **luteal phase** until the first day of menstruation. The cyclic changes that occur in the endometrium are called the *menstrual, proliferative,* and *secretory phases* and will be discussed separately.

***Follicular Phase.*** Menstruation lasts from day one to day four or five of the average cycle. During this time, the secretions of ovarian steroid hormones are at their lowest ebb, and the ovaries contain only primordial and primary follicles. During the *follicular phase of the ovaries,* which lasts from day one to about day thirteen of the cycle (this is highly variable), some of the primary follicles grow, form

**Figure 20.33.** Sample values for LH, FSH, progesterone, and estradiol during the menstrual cycle. The midcycle peak of LH is used as a reference day. (IU = international unit.)

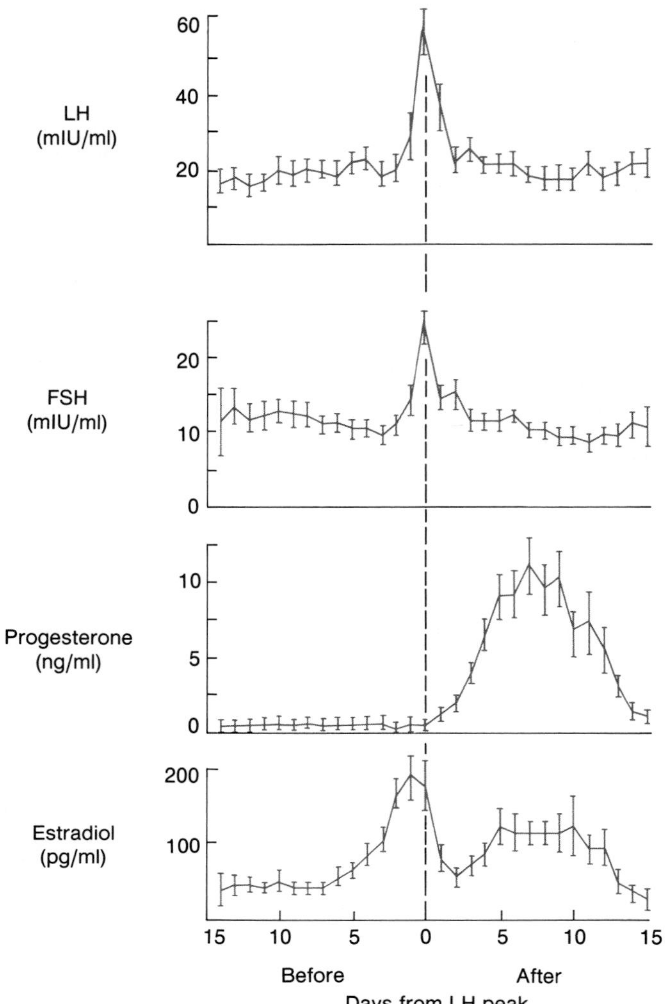


an antrum, and become secondary follicles. Toward the end of the follicular phase one follicle in one ovary reaches maturity and becomes a graafian follicle. As follicles grow, the granulosa cells secrete an increasing amount of **estradiol** (the principal estrogen), which reaches its highest concentration in the blood at about day twelve of the cycle (two days before ovulation).

The growth of the follicles and the secretion of estradiol are stimulated by, and dependent upon, FSH secreted from the anterior pituitary. The amount of FSH secreted during the early follicular phase is believed to be slightly greater than the amount secreted in the late follicular phase, although this can vary from cycle to cycle (a measure of variance is shown by vertical bars in fig. 20.33). FSH stimulates the production of FSH receptors in the granulosa cells, so that the follicles become increasingly sensitive to a given amount of FSH. This increased sensitivity is augmented by estradiol, which also stimulates the production of new FSH receptors in the follicles. As a result, the stimulatory effect of FSH on the follicles increases despite the fact that FSH levels in the blood do not increase throughout the follicular phase. Toward the end of the follicular phase, FSH and estradiol also stimulate the production of LH receptors in the graafian follicle. This prepares the graafian follicle for the next major event in the cycle.

The rapid rise in estradiol secretion from the granulosa cells during the follicular phase acts on the hypothalamus to increase the frequency of GnRH pulses. In addition, estradiol augments the ability of the pituitary to respond to GnRH with an increase in LH secretion. As a result of this stimulatory, or **positive feedback**, effect of estradiol on the pituitary, there is an increase in LH secretion in the late follicular phase that culminates in an **LH surge** (fig. 20.33).

**Figure 20.34.** The cycle of ovulation and menstruation.

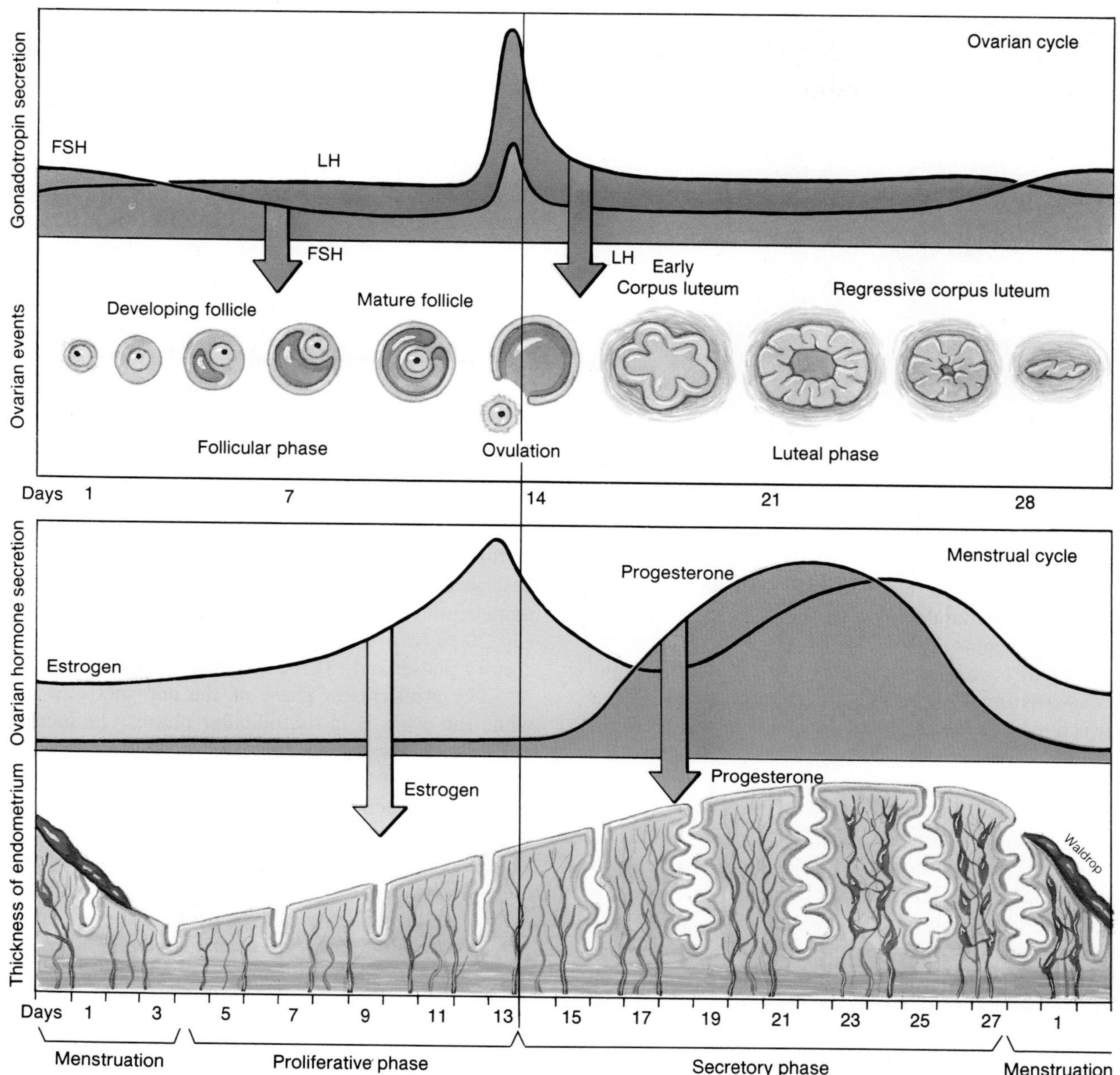


The LH surge begins about twenty-four hours before ovulation and reaches its peak about sixteen hours before ovulation. It is this surge that acts to trigger ovulation. Since GnRH stimulates the anterior pituitary to secrete both FSH and LH, there is a simultaneous, though smaller, surge in FSH secretion. Some investigators believe that this midcycle peak in FSH acts as a stimulus for the development of new follicles for the next month's cycle.

***Ovulation.*** Under the influence of FSH stimulation, the graafian follicle grows so large that it becomes a thin-walled "blister" on the surface of the ovary. The growth of the follicle is accompanied by a rapid rate of increase in estradiol secretion. This rapid increase in estradiol, in turn, triggers the LH surge at about day thirteen. Finally, the surge in LH secretion causes the wall of the graafian follicle to rupture at about day fourteen (fig. 20.34 *top*). Ovulation occurs, therefore, as a result of the sequential effects of FSH followed by LH on the ovarian follicles.

By means of the positive feedback effect of estradiol on LH secretion, the follicle in a sense sets the time for its own ovulation. This is because ovulation is triggered by an LH surge, and the LH surge is triggered by increased estradiol secretion, which occurs while the follicle grows. In this way the graafian follicle does not normally ovulate until it has reached the proper size and degree of maturation.

**Figure 20.35.** A corpus luteum in a human ovary.

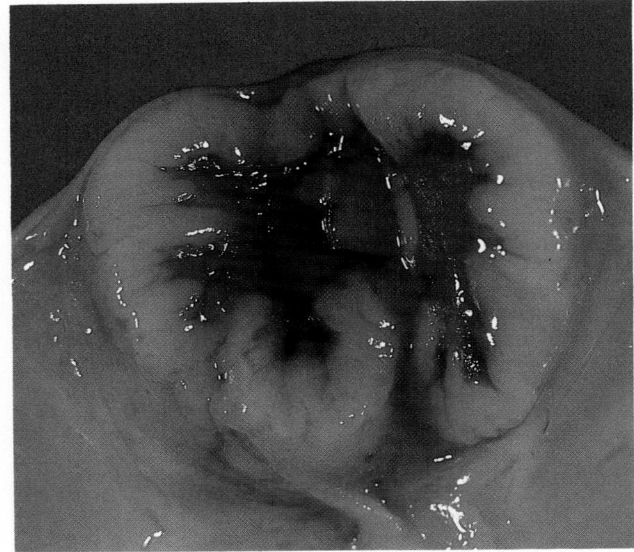

In ovulation, a secondary oocyte, arrested at metaphase II of meiosis, is released into a uterine tube. This oocyte is still surrounded by a zona pellucida and corona radiata as it begins its journey to the uterus. Normally only one ovary ovulates per cycle, with the left and right ovary alternating in successive cycles. Interestingly, if one ovary is removed the remaining ovary does not skip cycles, but ovulates every month. The mechanisms by which this regulation is achieved are not understood.

***Luteal Phase.*** After ovulation, the empty follicle is stimulated by LH to become a new structure, the **corpus luteum** (fig. 20.35). This change in structure is accompanied by a change in function. Whereas the developing follicles secrete only estradiol, the corpus luteum secretes both estradiol and **progesterone.** Progesterone levels in the blood are negligible before ovulation but rise rapidly to reach a peak during the luteal phase at approximately one week after ovulation (see figs. 20.33 and 20.34).

The combined high levels of estradiol and progesterone during the luteal phase exert a **negative feedback inhibition** of FSH and LH secretion. This serves to retard development of new follicles, so that further ovulation does not normally occur during that cycle. In this way multiple ovulations (and possible pregnancies) on succeeding days of the cycle are prevented.

High levels of estrogen and progesterone during the nonfertile cycle do not persist for very long, however, and new follicles do start to develop toward the end of one cycle, in preparation for the next cycle. Estrogen and progesterone levels fall during the late luteal phase (starting about day twenty-two), because the corpus luteum regresses and stops functioning. In lower mammals, the decline in corpus luteum function is caused by a hormone secreted by the uterus called *luteolysin.* A similar hormone has not yet been identified in humans, and the cause of corpus luteum regression in humans is not well understood. Luteolysis (breakdown of the corpus luteum) can be prevented by high LH secretion, but LH levels remain low during the luteal phase as a result of negative feedback inhibition by ovarian steroids. In a sense, therefore, the corpus causes its own demise.

With the declining function of the corpus luteum, estrogen and progesterone fall to very low levels by day twenty-eight of the cycle. The withdrawal of ovarian steroids causes menstruation and permits a new cycle of ovarian follicle development to progress.

### Cyclic Changes in the Endometrium

In addition to a description of the female cycle in terms of the phases of ovarian function, the cycle can also be described in terms of the changes that occur in the endometrium. Three phases can be identified on this basis (fig. 20.34 *bottom*): (1) the proliferative phase; (2) the secretory phase; and (3) the menstrual phase.

The **proliferative phase** of the endometrium occurs while the ovary is in its follicular phase. The increasing amounts of estradiol secreted by the developing follicles stimulates growth (proliferation) of the stratum functionale of the endometrium. In humans and other primates, spiral arteries develop in the endometrium during this phase. Estradiol may also stimulate the production of receptor proteins for progesterone at this time, in preparation for the next phase of the cycle.

The **secretory phase** of the endometrium occurs when the ovary is in its luteal phase. In this phase, increased progesterone secretion stimulates the development of mucous glands. As a result of the combined actions of estradiol and progesterone, the endometrium becomes thick, vascular, and "spongy" in appearance during the time of the cycle following ovulation. It is therefore well prepared to accept and nourish an embryo if fertilization occurs.

The **menstrual phase** occurs as a result of the fall in ovarian hormone secretion during the late luteal phase. Necrosis (cellular death) and sloughing of the stratum functionale of the endometrium may be produced by constriction of the spiral arteries. The spiral arteries appear to be responsible for bleeding during menstruation, because lower animals that lack spiral arteries don't bleed when they shed their endometrium. The phases of the menstrual cycle are summarized in figure 20.36 and in table 20.7.

**Figure 20.36.** The sequence of events in the endocrine control of the ovarian cycle and their correlation within the phases of the endometrium during the menstrual cycle.

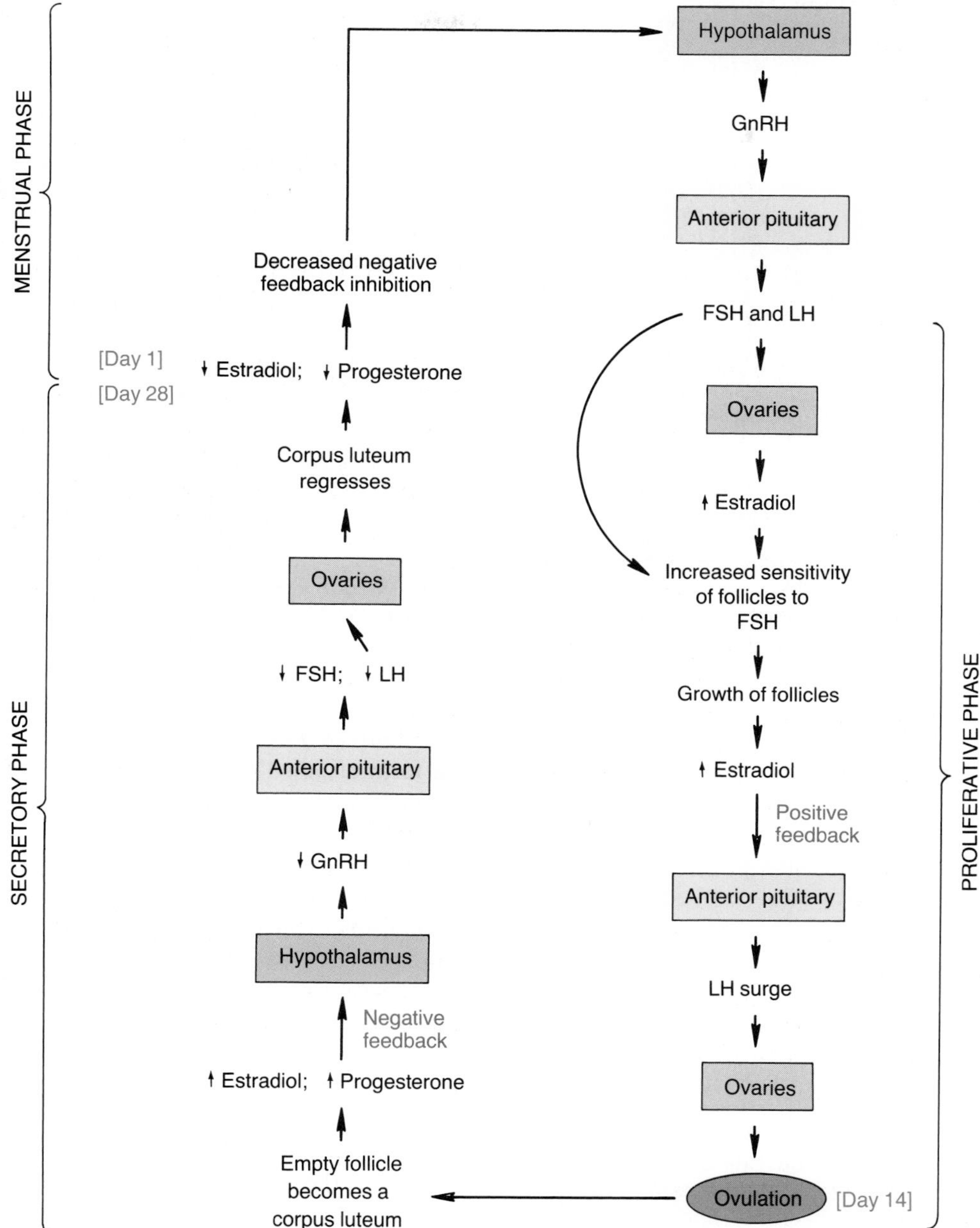

**Table 20.7** Phases of the menstrual cycle

| Phase of cycle | | Hormonal changes | | Tissue changes | |
|---|---|---|---|---|---|
| Ovarian | Endometrial | Pituitary | Ovary | Ovarian | Endometrial |
| Follicular (days 1–4) | Menstrual | FSH and LH secretion low | Estradiol and progesterone remain low | Primary follicles grow | Outer two-thirds of endometrium is shed with accompanying bleeding |
| Follicular (days 5–13) | Proliferative | FSH slightly higher than LH secretion in early follicular phase | Estradiol secretion rises (due to FSH stimulation of follicles) | Follicles grow; graffian follicle develops (due to FSH stimulation) | Mitotic division increases thickness of endometrium; spiral arteries develop (due to estradiol stimulation) |
| Ovulatory (day 14) | Proliferative | LH surge (and increased FSH) stimulated by positive feedback from estradiol | Fall in estradiol secretion | Graafian follicle is ruptured and ovum is extruded into fallopian tube | No change |
| Luteal (days 15–28) | Secretory | LH and FSH decrease (due to negative feedback of steroids) | Progesterone and estrogen secretion increase, then fall | Development of corpus luteum (due to LH stimulation); regression of corpus luteum | Glandular development in endometrium (due to progesterone stimulation) |

The cyclic changes in ovarian secretion cause other cyclic changes in the female genital ducts. High levels of estradiol secretion, for example, cause cornification of the vaginal epithelium (the upper cells die and become filled with keratin). High levels of estradiol also cause the production of a thin, watery cervical mucus, which can be easily penetrated by spermatozoa. During the luteal phase of the cycle, the high levels of progesterone cause the cervical mucus to become thick and sticky after ovulation has occurred.

Cyclic changes in ovarian hormone secretion also cause cyclic changes in *basal body temperature.* In the **rhythm method** of birth control, a woman measures her oral basal body temperature upon waking to determine when ovulation has occurred. On the day of the LH peak, when estradiol secretion begins to decline, there is a slight drop in basal body temperature. Starting about one day after the LH peak, the basal body temperature sharply rises as a result of progesterone secretion and remains elevated throughout the luteal phase of the cycle (fig. 20.37). The day of ovulation can be accurately determined by this method, making the method useful in increasing fertility if conception is desired. Since the day of the cycle in which ovulation occurs is quite variable in many women, however, the rhythm method is not very reliable for contraception by predicting when the next ovulation will occur. The contraceptive pill is a statistically more effective means of birth control.

## Contraceptive Pill

About ten million women in the United States and sixty million women in the world are currently using **oral contraceptives.** These contraceptives usually consist of a synthetic estrogen combined with a synthetic progesterone in the form of pills that are taken once each day for three weeks after the last day of a menstrual period. This procedure causes an immediate increase in blood levels of ovarian steroids (from the pill), which is maintained for the normal duration of a monthly cycle. As a result of *negative feedback inhibition* of gonadotropin secretion, *ovulation never occurs.* The entire cycle is like a false luteal phase, with high levels of progesterone and estrogen and low levels of gonadotropins.

Since the contraceptive pills contain ovarian steroid hormones, the endometrium proliferates and becomes secretory just as it does during a normal cycle. In order to prevent an abnormal growth of the endometrium, women stop taking the steroid pills after three weeks (placebo pills are taken during the fourth week). This causes estrogen and progesterone levels to fall, and permits menstruation to occur. The contraceptive pill is an extremely effective method of birth control, but it does have potentially serious side effects—including an increased incidence of thromboembolism, cardiovascular disorders, and endometrial and breast cancer. It has been pointed out, however, that the mortality risk of contraceptive pills is still much lower than the risk of death from the complications of pregnancy. Additionally, there is evidence that oral contraceptives may actually provide some protection against cancer of the uterus and ovary.

**Figure 20.37.** Changes in basal body temperature during the menstrual cycle.

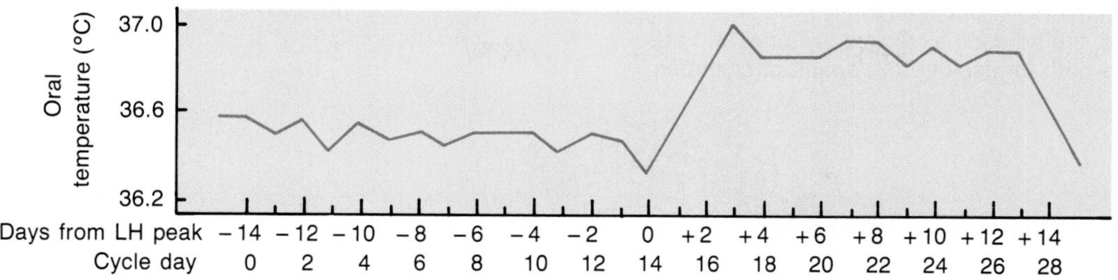


## The Menopause

The term *menopause* means literally "pause in the menses" and refers to the cessation of ovarian activity that occurs at about the age of fifty. During the postmenopausal years, which account for about a third of a woman's life span, no new ovarian follicles develop and the ovaries cease secreting estradiol. This fall in estradiol is due to changes in the ovaries, not in the pituitary; indeed, FSH and LH secretion by the pituitary is elevated due to the absence of negative feedback inhibition from estradiol. Like prepubertal boys and girls, the only estrogen found in the blood of postmenopausal women is that formed by aromatization of the weak androgen androstenedione, secreted principally by the adrenal cortex, into a weak estrogen called estrone.

It is the withdrawal of estradiol secretion from the ovaries that is most responsible for the many debilitating symptoms of menopause. These include vasomotor disturbances (which produce "hot flashes"), urogenital atrophy, and the increased development of osteoporosis (see chapter 18). Estrogen replacement therapy helps to alleviate these symptoms, although this treatment is given cautiously and often in combination with a progesterone derivative because of the increased risks associated with taking estrogen alone.

1. *Describe the changes that occur in the ovary and endometrium during the follicular phase, and explain the hormonal control of these changes.*
2. *Describe the hormonal regulation of ovulation.*
3. *Describe the formation, function, and fate of the corpus luteum, and describe the changes that occur in the endometrium during the luteal phase.*
4. *Explain the significance of negative feedback inhibition during the luteal phase, and explain the hormonal control of menstruation.*

## *Fertilization, Pregnancy, and Parturition*

Upon fertilization, the secondary oocyte completes meiotic division and then undergoes mitosis to form first a ball of cells and then an early embryonic structure called a blastocyst. Cells of the blastocyst secrete a hormone known as human chorionic gonadotropin, which maintains the mother's corpus luteum and its production of estradiol and progesterone during the first ten weeks of the pregnancy. This prevents menstruation, so that the embryo can implant into the endometrium, develop, and form a placenta. The placenta, composed of both fetal and maternal tissues, has numerous functions. Birth is dependent upon strong contractions of the uterus which are stimulated by oxytocin from the posterior pituitary. Lactation involves the cooperative effects of both parts of the pituitary gland and the nervous system.

During the act of sexual intercourse a man ejaculates an average of 300 million sperm into the vagina. This tremendous number is needed because of the high sperm fatality rate—only about 100 survive to enter each fallopian tube. During their passage through the female reproductive tract the sperm gain the ability to fertilize an ovum. This process is called **capacitation.** The changes that occur in capacitation are incompletely understood. Experiments have shown, however, that freshly ejaculated sperm are infertile; they must be present in the female tract for at least seven hours before they can fertilize an ovum.

A woman usually ovulates only one ovum a month, producing a total number of less than 450 during her reproductive years. Each ovulation releases a secondary oocyte arrested at metaphase of the second meiotic division. The secondary oocyte, as previously described, enters

**Figure 20.38.** The process of fertilization. (*a, b*) Diagrammatic representations. As the head of the sperm encounters the gelatinous corona radiata of the egg, which has progressed in meiotic development to the secondary oocyte stage (*2*), the acrosomal vesicle ruptures and the sperm digests a path for itself by the action of enzymes released from the acrosome (*3, 4*). When the cell membrane of the sperm contacts the cell membrane of the egg (*5*), they become continuous, and the sperm nucleus and other contents move into the egg cytoplasm. (*c*) A scanning electron micrograph of sperm bound to the egg surface.

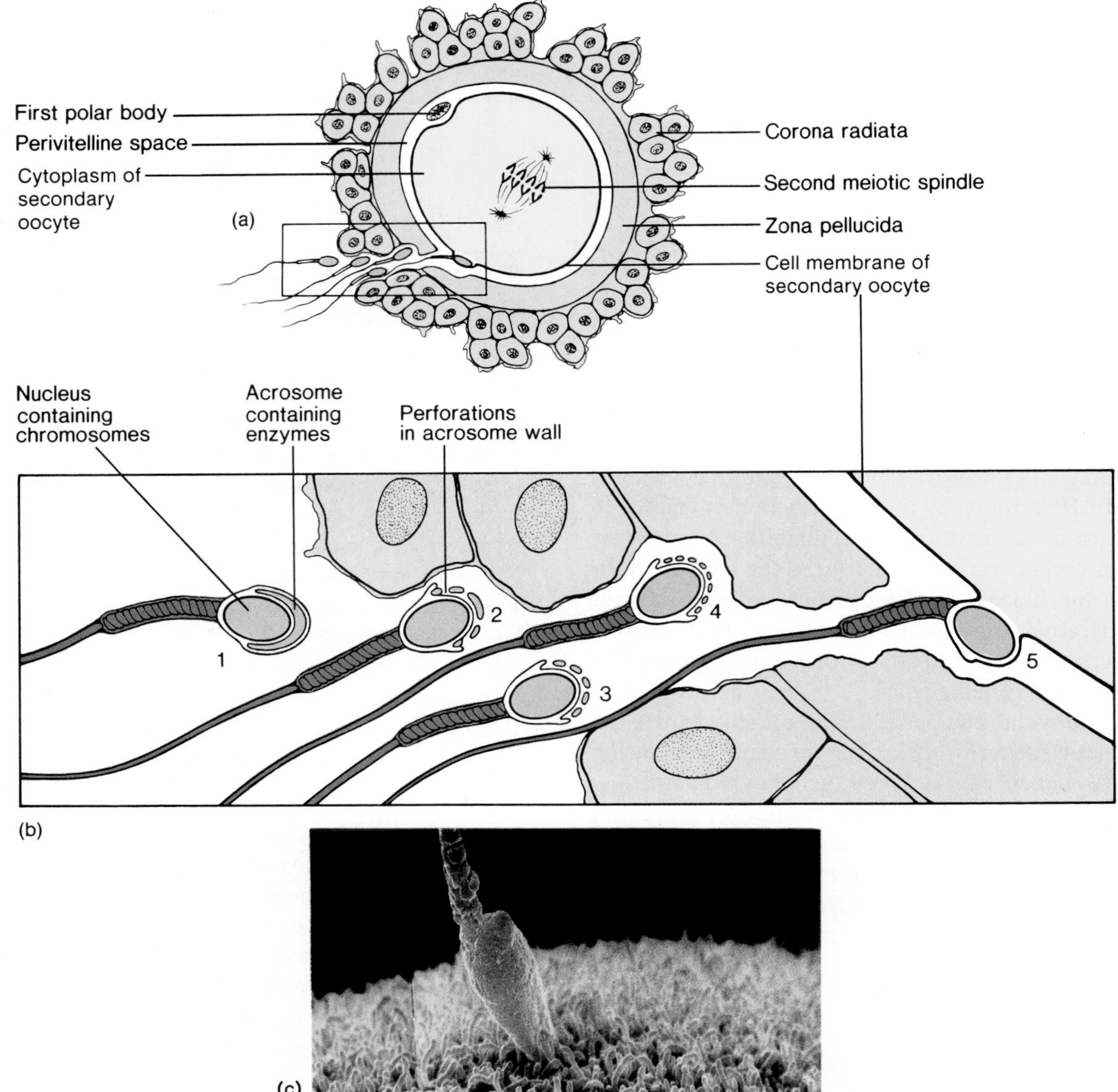


the fallopian tube surrounded by its zona pellucida (a thin transparent layer of protein and polysaccharides) and corona radiata of granulosa cells (fig. 20.38).

Fertilization normally occurs in the fallopian tubes. The head of each spermatozoan is capped by an organelle called an *acrosome* (fig. 20.39), which contains a trypsin-like protein-digesting enzyme and hyaluronidase (which digests hyaluronic acid, an important constituent of connective tissues). When sperm meets ovum, an **acrosomal reaction** occurs that exposes the acrosome's digestive enzymes and allows the sperm to penetrate the corona radiata and the zona pellucida. The acrosomal enzymes are not released in this process; rather, the sperm tunnels its way through these barriers by digestion reactions that are localized to its acrosomal cap.

As the first sperm tunnels its way through the zona pellucida, a chemical change in the zona occurs that prevents other sperm from entering. Only one sperm, therefore, is allowed to fertilize one ovum. As fertilization occurs, the secondary oocyte is stimulated to complete its second meiotic division (fig. 20.40). Like the first meiotic division, the second produces one cell that contains all of the cytoplasm—the mature ovum or egg cell—and one polar body. The second polar body, like the first, ultimately fragments and disintegrates.

**Figure 20.39.** An electron micrograph showing the head of a human sperm with its nucleus and acrosomal cap.

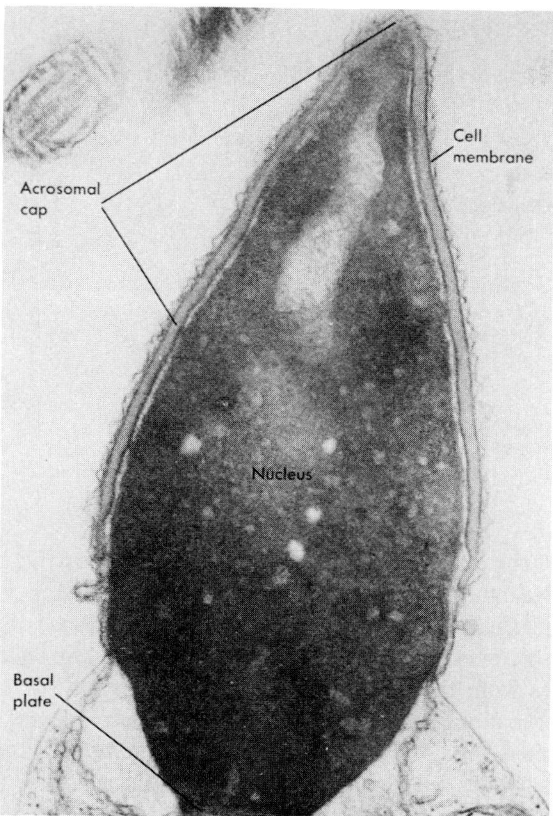


**Figure 20.40.** A secondary oocyte, arrested at metaphase II of meiosis, is released at ovulation. If this cell is fertilized, it completes its second meiotic division and produces a second polar body.

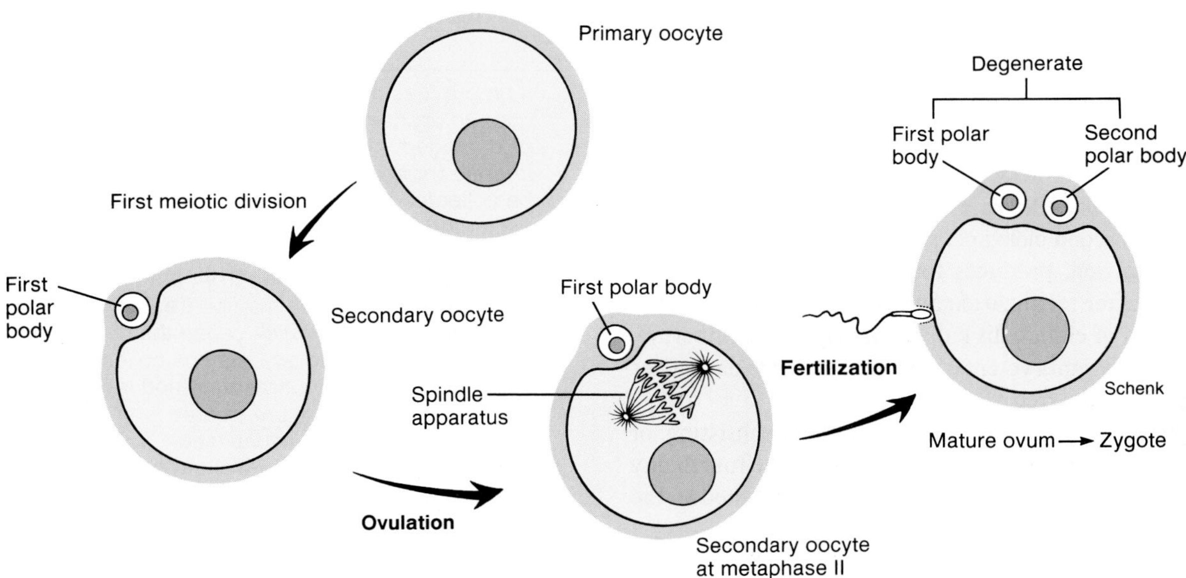

**Figure 20.41.** Fertilization and the union of chromosomes from the sperm and ovum to form the zygote.

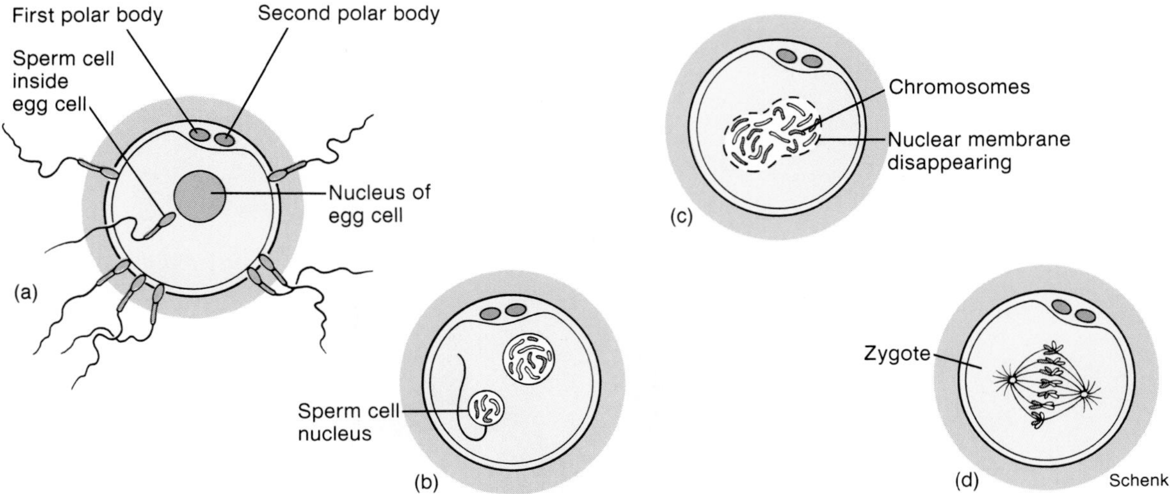


At fertilization, the sperm enters the cytoplasm of the much larger egg cell. Within twelve hours the nuclear membrane in the ovum disappears, and the haploid number of chromosomes (twenty-three) in the ovum is joined by the haploid number of chromosomes from the sperm. A fertilized egg, or *zygote,* containing the diploid number of chromosomes (forty-six) is thus formed (fig. 20.41).

A secondary oocyte that is ovulated but not fertilized does not complete its second meiotic division, but instead disintegrates twelve to twenty-four hours after ovulation. Fertilization therefore cannot occur if intercourse takes place beyond one day following ovulation. Sperm, in contrast, can survive up to three days in the female reproductive tract. Fertilization therefore can occur if intercourse is performed within three days prior to the day of ovulation.

## Cleavage and Formation of a Blastocyst

At about thirty to thirty-six hours after fertilization the zygote divides by mitosis—a process called *cleavage*—into two smaller cells. The rate of cleavage is thereafter accelerated. A second cleavage, performed about forty hours after fertilization, produces four cells. By about fifty to sixty hours after fertilization a third cleavage occurs, producing a ball of eight cells called a **morula** (= mulberry). This very early embryo enters the uterus three days after ovulation has occurred (fig. 20.42).

Cleavage continues so that a morula consisting of thirty-two to sixty-four cells is produced by the fourth day after fertilization. The embryo remains unattached to the uterine wall for the next two days, during which time it undergoes changes that convert it into a hollow structure called a **blastocyst** (fig. 20.43). The blastocyst consists of two parts: (1) an *inner cell mass,* which will become the fetus; and (2) a surrounding *chorion,* which will become part of the placenta. The cells that form the chorion are called *trophoblast cells.*

On the sixth day following fertilization, the blastocyst attaches to the uterine wall, with the side containing the inner cell mass against the endometrium. The trophoblast cells produce enzymes that allow the blastocyst to "eat its way" into the thick endometrium. This begins the process of **implantation,** or **nidation,** and by the seventh day the blastocyst is usually completely buried in the endometrium (fig. 20.44).

The process of **in vitro fertilization** is sometimes used to produce pregnancies in women with absent or damaged fallopian tubes, or in women who are infertile for a variety of other reasons. An ovum may be collected by aspiration following ovulation (as estimated by waiting 36 to 38 hours after the LH surge) and placed in a petri dish for two to three days with sperm collected from the husband. Alternatively, a woman may be treated with powerful FSH-like hormones which cause the development of multiple follicles, and ova may be collected by aspiration guided by ultrasound and laparoscopy. The husband's sperm are treated with procedures that duplicate normal capacitation. The embryos are usually transferred to the woman's uterus at their four-cell stage, 48 to 72 hours after fertilization. In some cases, the embryos may be transferred to the end of the fallopian tube. The chance of a successful implantation is only 10%–15% per procedure, but can approach 60% after six such attempted procedures.

**Figure 20.42.** A diagrammatic representation of the ovarian cycle, fertilization, and the morphogenic events of the first week. Implantation of the blastocyst begins between the fifth and seventh day and is generally completed by the tenth day.

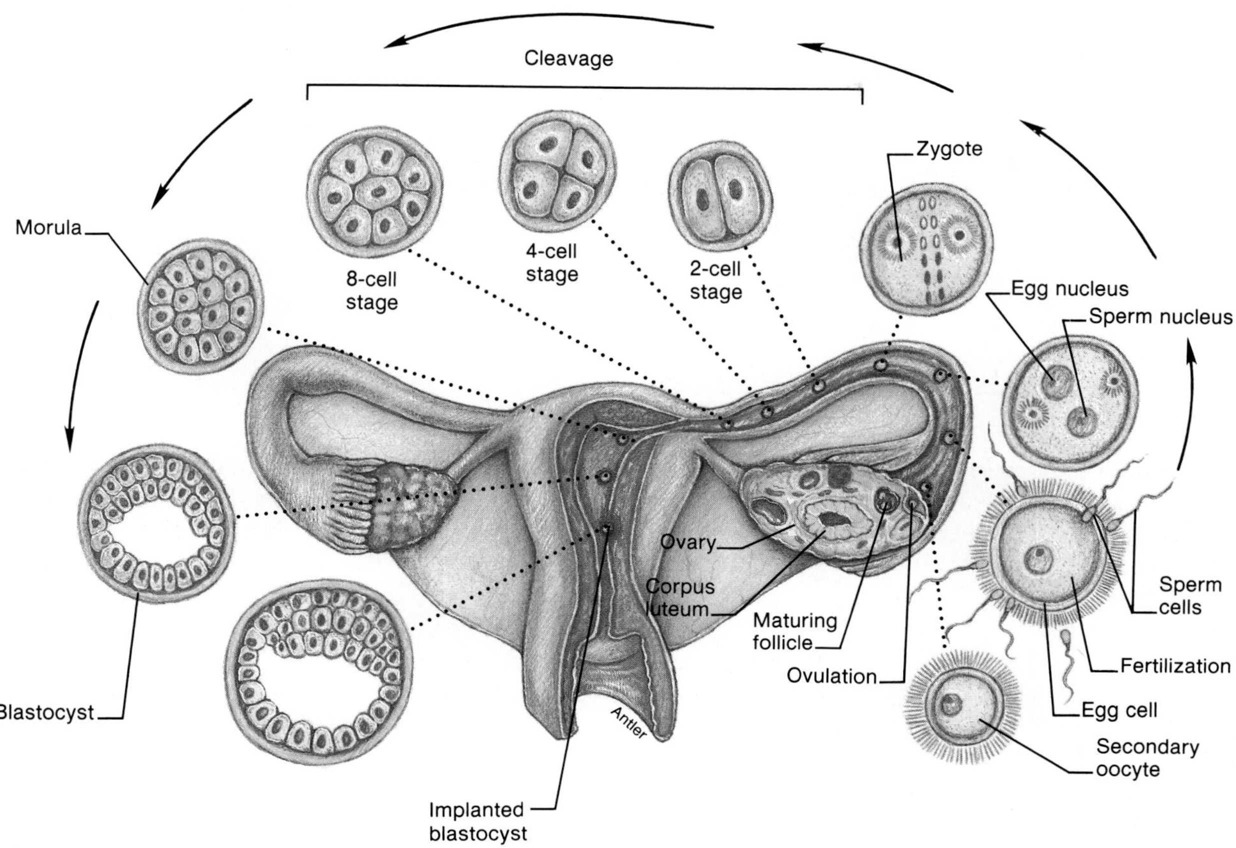


## Implantation and Formation of a Placenta

If fertilization does not take place, the corpus luteum begins to decrease its secretion of steroids about ten days after ovulation. This withdrawal of steroids, as previously described, causes necrosis and sloughing of the endometrium following day twenty-eight of the cycle. If fertilization and implantation have occurred, however, these events must obviously be prevented to maintain the pregnancy.

***Chorionic Gonadotropin.*** The blastocyst saves itself from being eliminated with the endometrium by secreting a hormone that indirectly prevents menstruation. Even before the sixth day when implantation occurs, the trophoblast cells of the chorion secrete **chorionic gonadotropin (hCG**—the *h* stands for *human*). This hormone is identical to LH in its effects and therefore is able to maintain the corpus luteum past the time when it would otherwise regress. The secretion of estradiol and progesterone is thus maintained and menstruation is normally prevented.

The secretion of hCG declines by the tenth week of pregnancy (fig. 20.45). Actually, this hormone is only required for the first five to six weeks of pregnancy, because the placenta itself becomes an active steroid hormone-secreting gland. At the fifth to sixth week the mother's corpus luteum begins to regress (even in the presence of hCG), but by this time the placenta is secreting more than sufficient amounts of steroids to maintain the endometrium and prevent menstruation.

All **pregnancy tests** assay for the presence of hCG in blood or urine, because this hormone is secreted by the blastocyst but not by the mother's endocrine glands. Modern pregnancy tests detect the presence of hCG by the use of antibodies against hCG or by use of cellular receptor proteins for hCG. Extremely sensitive techniques utilizing monoclonal antibodies against a subunit of hCG and radioimmunoassay (chapter 19) permit pregnancy to be detected in a clinical laboratory as early as seven to ten days after conception.

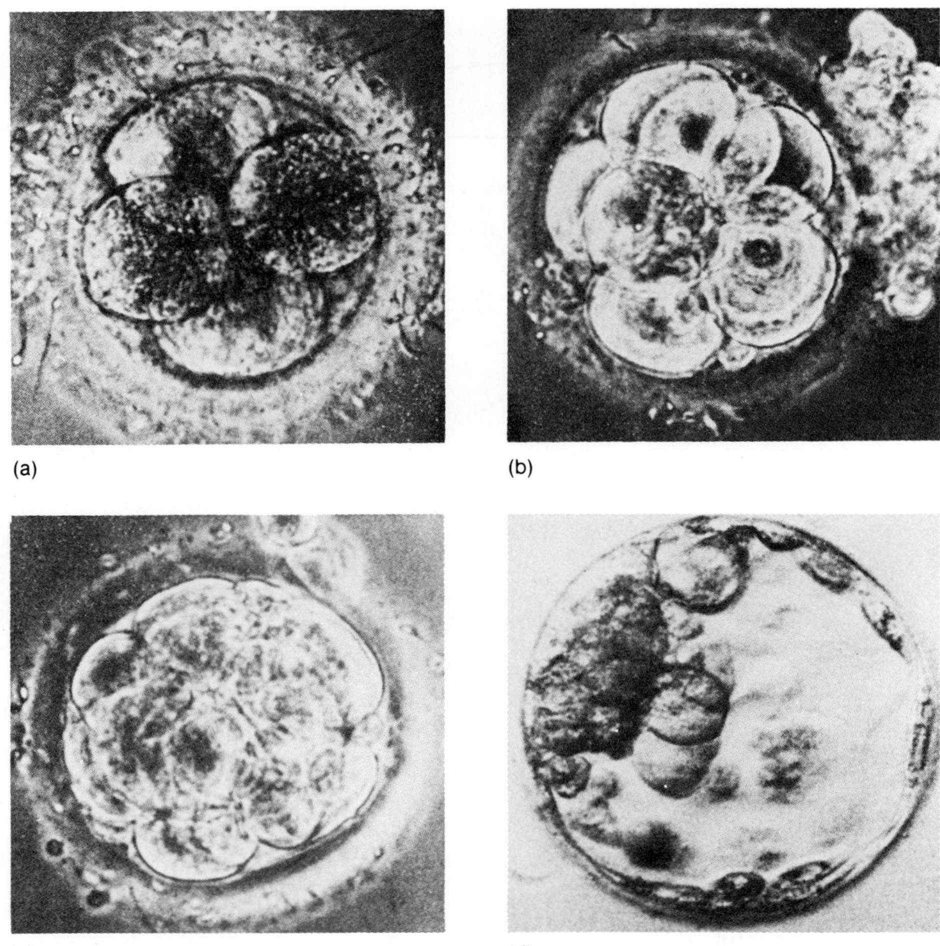


**Figure 20.43.** Stages of pre-embryonic development of a human ovum fertilized in a laboratory (*in vitro*) as seen in scanning electron micrographs. (*a*) 4-cell stage; (*b*) cleavage at the 16-cell stage; (*c*) a morula; and (*d*) a blastocyst.

***Chorionic Membranes.*** Between days seven and twelve, as the blastocyst becomes completely embedded in the endometrium, the chorion becomes a two-cell-thick structure, consisting of an inner *cytotrophoblast* layer and an outer *syncytiotrophoblast* layer. The inner cell mass (which will become the fetus), meanwhile, also develops two cell layers. These are the *ectoderm* (which will form such organs as the nervous system and skin) and the *endoderm* (which will eventually form the gut and its derivatives). A third, middle embryonic layer—the *mesoderm*—is not yet seen at this stage. The embryo at this stage is a two-layer-thick disc separated from the cytotrophoblast of the chorion by an *amniotic cavity.*

As the syncytiotrophoblast invades the endometrium, it secretes protein-digesting enzymes that create many small, blood-filled cavities in the maternal tissue. The cytotrophoblast then forms projections, or *villi,* (fig. 20.46) that grow into these pools of venous blood, producing a leafy-appearing structure called the *chorion frondosum* (*frond* = leaf). This occurs only on the side of the chorion that faces the uterine wall. As the embryonic structures grow, the other side of the chorion bulges into the cavity of the uterus, loses its villi, and becomes smooth in appearance.

**Figure 20.44.** The blastocyst adheres to the endometrium on about the sixth day as shown in diagram (*a*); (*b*) is a scanning electron micrograph showing the surface of the endometrium and implantation at twelve days following fertilization.

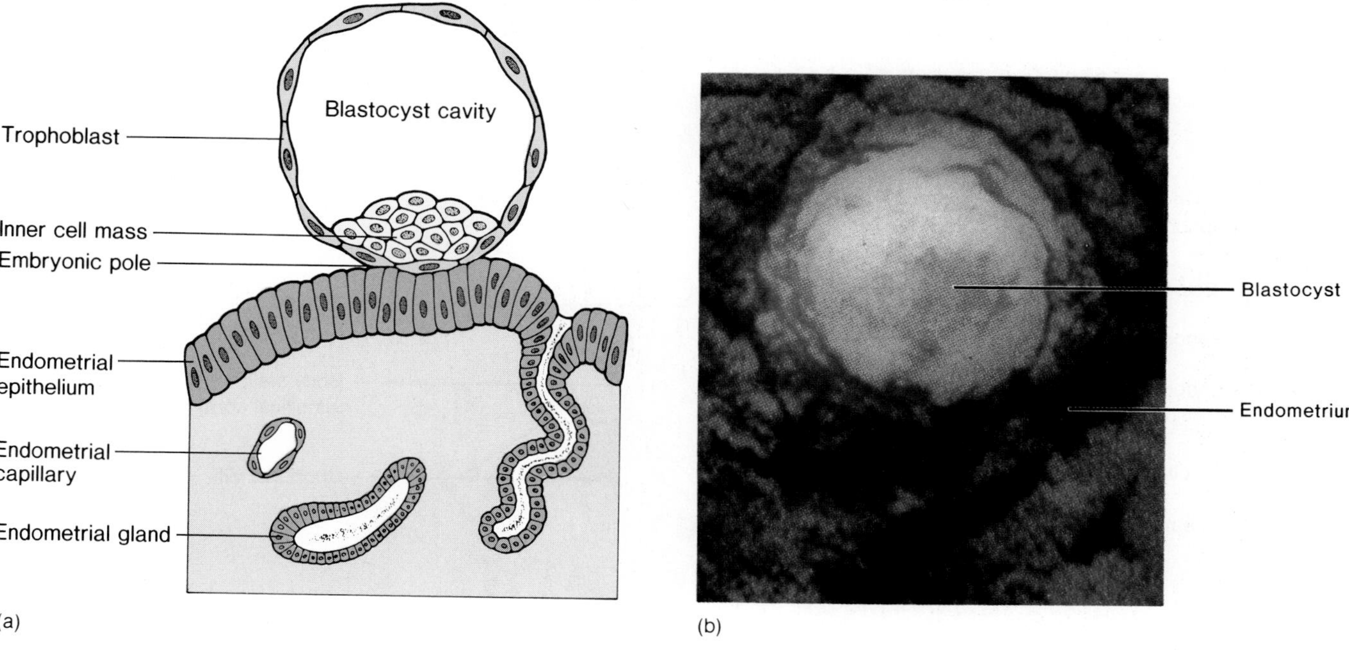


**Figure 20.45.** Human chorionic gonadotropin (hCG) is secreted by trophoblast cells during the first trimester of pregnancy. This hormone maintains the mother's corpus luteum for the first five and a half weeks. After that time the placenta becomes the major sex-hormone-producing gland, secreting increasing amounts of estrogen and progesterone throughout pregnancy.

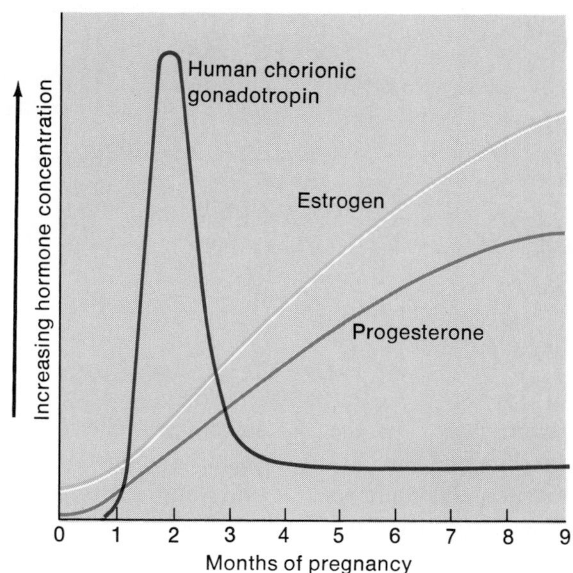

**Figure 20.46.** After the syncytiotrophoblast has created blood-filled cavities in the endometrium, these cavities are invaded by extensions of the cytotrophoblast (*a*). These extensions, called villi, branch extensively to produce the chorion frondosum (*b*). The developing embryo is surrounded by a membrane known as the amnion.

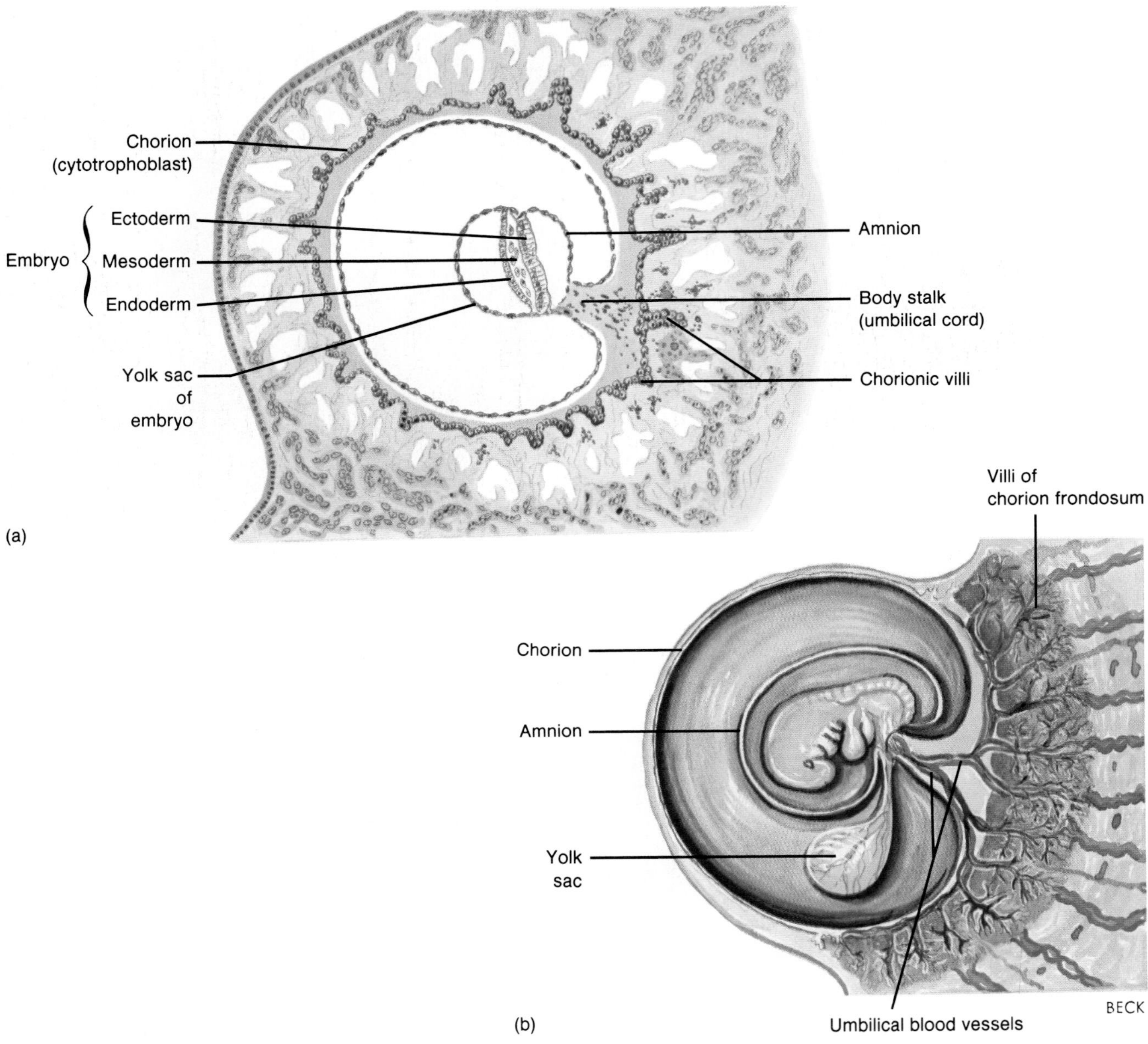


***Formation of the Placenta and Amniotic Sac.*** As the blastocyst is implanted in the endometrium and the chorion develops, the cells of the endometrium also undergo changes. These changes, including cellular growth and the accumulation of glycogen, are called the **decidual reaction.** The maternal tissue in contact with the chorion frondosum is called the *decidua basalis.* These two structures—chorion frondosum (fetal tissue) and decidua basalis (maternal tissue)—together form the functional unit known as the **placenta.**

The human placenta is a disc-shaped structure that is continuous at its outer surface with the smooth part of the chorion, which bulges into the uterine cavity. Immediately beneath the chorionic membrane is the amnion, which has grown to envelop the entire fetus (fig. 20.47). The fetus, together with its umbilical cord, is therefore located within the fluid-filled *amniotic sac.*

**Figure 20.47.** Blood from the fetus is carried to and from the chorion frondosum by umbilical arteries and veins. The maternal tissue between the chorionic villi is known as the decidua basalis, and this tissue, together with the villi, form the functioning placenta. The space between chorion and amnion is obliterated, and the fetus lies within the fluid-filled amniotic sac.

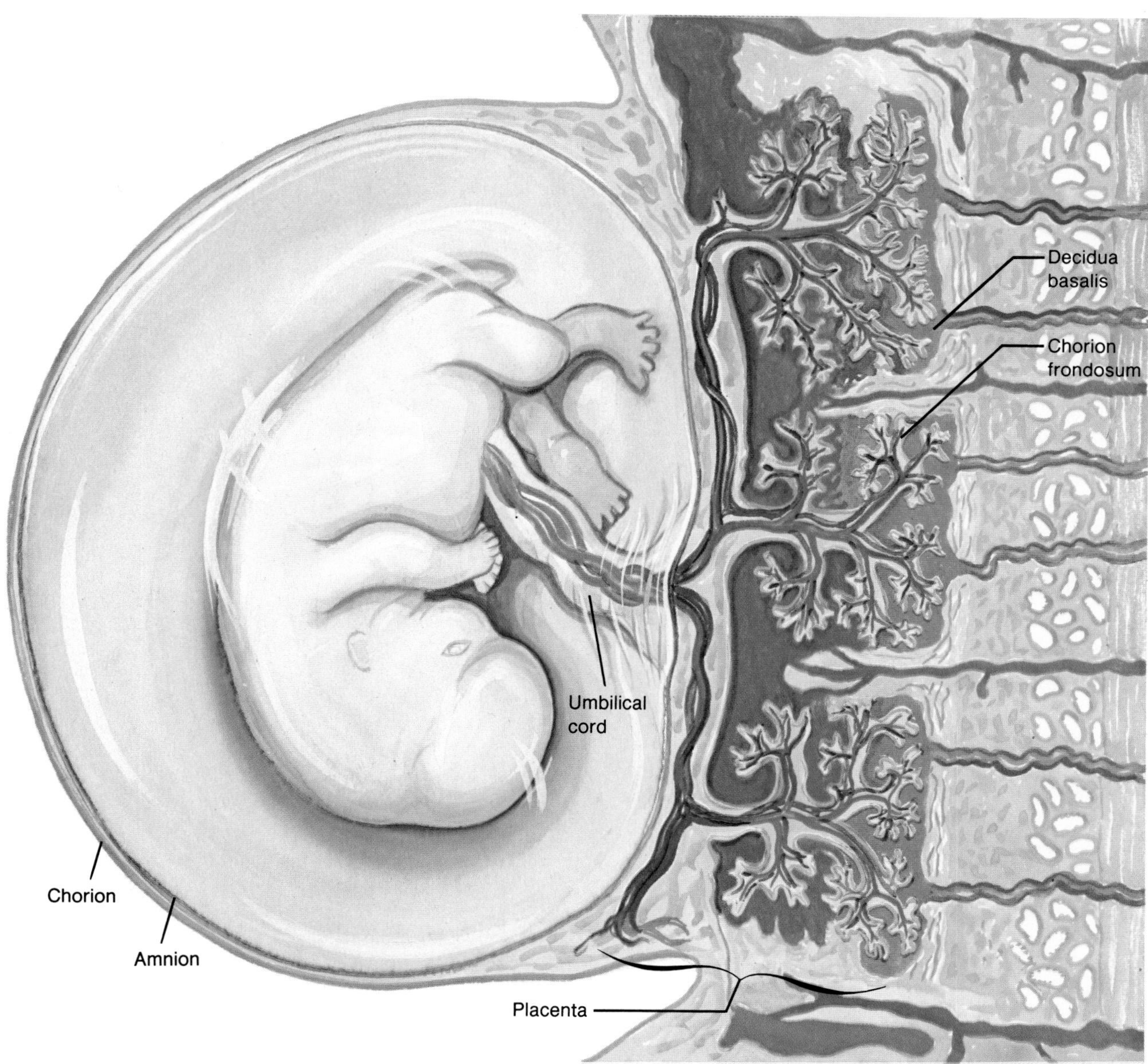


Amniotic fluid is formed initially as an isotonic secretion, which is later increased in volume and decreased in concentration by urine from the fetus. Amniotic fluid also contains cells that are sloughed off from the fetus, placenta, and amniotic sac. Since all of these cells are derived from the same fertilized ovum, all have the same genetic composition. Many genetic abnormalities can be detected by aspiration of this fluid and examination of the cells thus obtained. This procedure is called **amniocentesis** (fig. 20.48).

Amniocentesis is usually performed at the fourteenth or fifteenth week of pregnancy, when the amniotic sac contains 175–225 ml of fluid. Genetic diseases such as Down's syndrome (where there are three instead of two chromosomes number 21) can be detected by examining chromosomes; diseases such as Tay-Sachs disease, in which there is a defective enzyme involved in the formation of myelin sheaths, can be detected by biochemical techniques.

**Figure 20.48.** Amniocentesis. In this procedure amniotic fluid, together with suspended cells, is withdrawn for examination. Various genetic diseases can be detected prenatally by this means.

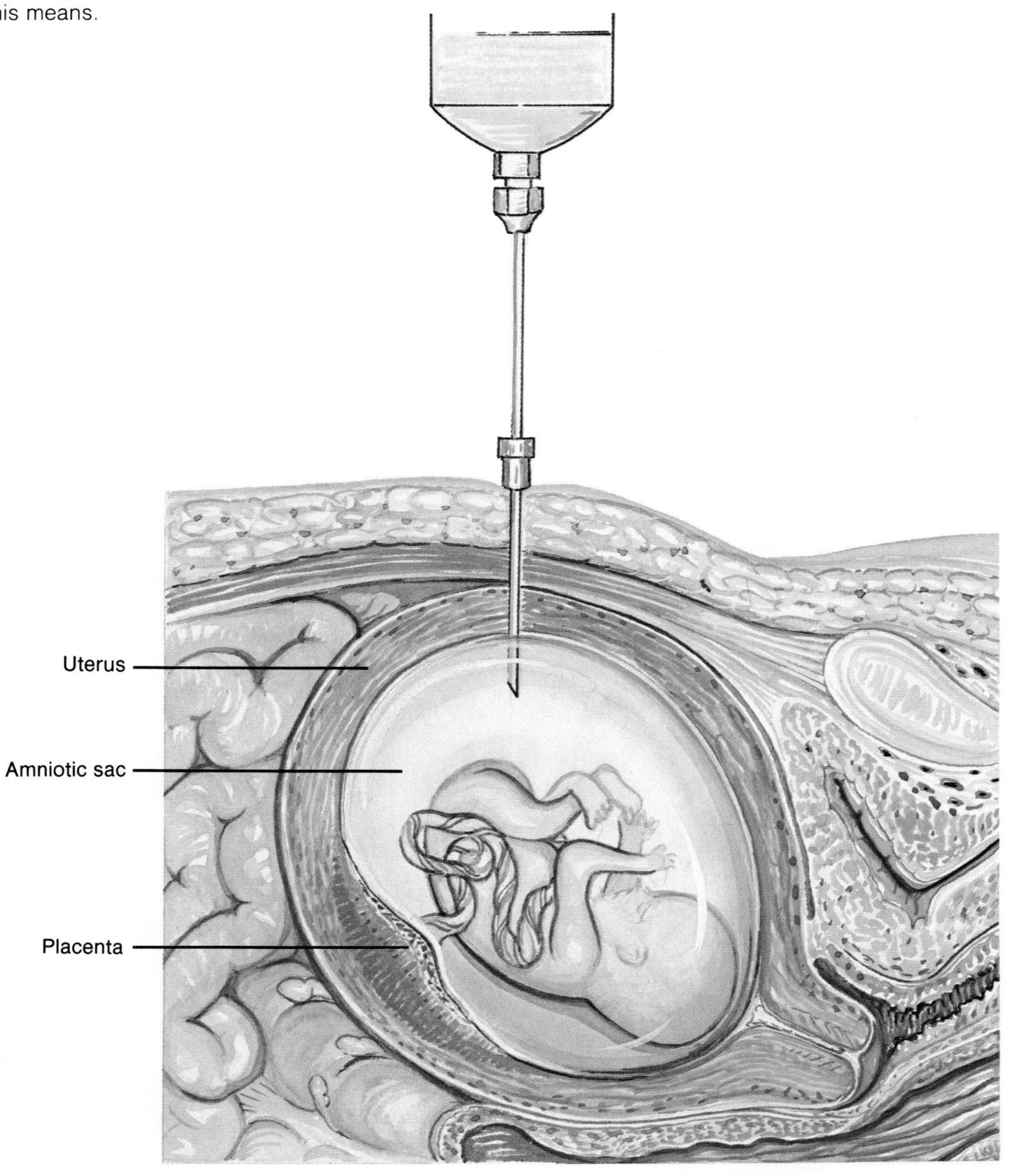

**Figure 20.49.** The circulation of blood within the placenta. Maternal blood is delivered to and drained from the spaces between the chorionic villi. Fetal blood is brought to blood vessels within the villi by branches of the umbilical artery and is drained by branches of the umbilical vein.

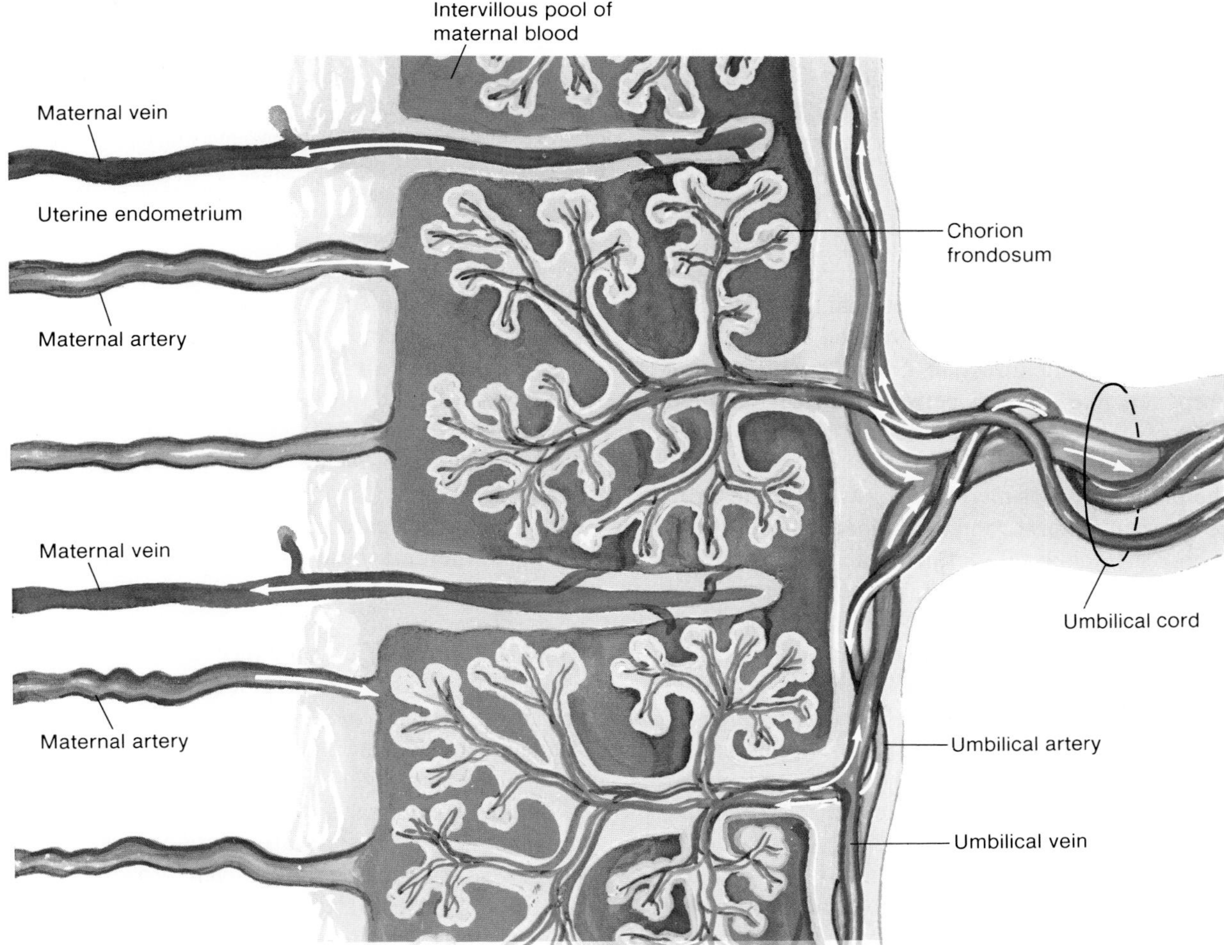


The amniotic fluid that is withdrawn contains fetal cells at a concentration that is too low to permit direct determination of genetic or chromosomal disorders. These cells must therefore be cultured *in vitro* for at least two weeks before they are present in sufficient numbers for the laboratory tests required. A newer method, called **chorionic villus biopsy,** is now available to detect genetic disorders much earlier than permitted by amniocentesis. In chorionic villus biopsy, a catheter is inserted through the cervix to the chorion, and a sample of a chorionic villus is obtained by suction or cutting. Genetic tests can be performed directly on the villus sample, since this sample contains much larger numbers of fetal cells than does a sample of amniotic fluid. Chorionic villus biopsy can provide genetic information at ten to twelve weeks' gestation; such information obtained by amniocentesis, in comparison, is not generally available before twenty weeks.

Major structural abnormalities, which may not be predictable from genetic analysis, can often be detected by *ultrasound.* Sound wave vibrations are reflected from the interface of tissues with different densities—such as the interface between the fetus and amniotic fluid—and used to produce an image. This technique is so sensitive that it can be used to detect a fetal heartbeat several weeks before it can be heard using a stethoscope.

## Exchange of Molecules across the Placenta

The *umbilical artery* delivers fetal blood to vessels within the villi of the chorion frondosum of the placenta. This blood circulates within the villi and returns to the fetus via the *umbilical vein.* Maternal blood is delivered to and drained from the cavities within the decidua basalis that are located between the chorionic villi (fig. 20.49). In this way, maternal and fetal blood are brought close together but never mix within the placenta.

**Table 20.8** Hormones secreted by placenta

| Hormones | Effects |
|---|---|
| **Pituitary-Like Hormones** | |
| Chorionic gonadotropin (hCG) | Similar to LH; maintains mother's corpus luteum for first 5½ weeks of pregnancy; may be involved in suppressing immunological rejection of embryo; also has TSH-like activity |
| Chorionic somatomammotropin (hCS) | Similar to prolactin and growth hormone; in the mother, hCS acts to promote increased fat breakdown and fatty acid release from adipose tissue and to promote the sparing of glucose use by maternal tissues ("diabeticlike" effects) |
| **Sex Steroids** | |
| Progesterone | Helps maintain endometrium during pregnancy; helps suppress gonadotropin secretion; promotes uterine sensitivity to oxytocin; helps stimulate mammary gland development |
| Estrogens | Help maintain endometrium during pregnancy; help suppress gonadotropin secretion; help stimulate mammary gland development; inhibit prolactin secretion |

The placenta serves as a site for the exchange of gases and other molecules between the maternal and fetal blood. Oxygen diffuses from mother to fetus, and carbon dioxide diffuses in the opposite direction. Nutrient molecules and waste products likewise pass between maternal and fetal blood; the placenta is, after all, the only link between the fetus and the outside world.

The placenta is not merely a passive conduit for exchange between maternal and fetal blood, however. It has a very high metabolic rate, utilizing about a third of all the oxygen and glucose supplied by the maternal blood. The rate of protein synthesis is, in fact, higher in the placenta than it is in the liver. Like the liver, the placenta produces a great variety of enzymes capable of converting biologically active molecules (such as hormones and drugs) into less active, more water-soluble forms. In this way, potentially dangerous molecules in the maternal blood are often prevented from harming the fetus.

### Endocrine Functions of the Placenta

The placenta secretes both steroid hormones and protein hormones that have actions similar to those of some anterior pituitary hormones. This latter category of hormones includes **chorionic gonadotropin (hCG)** and **chorionic somatomammotropin (hCS)** (table 20.8). Chorionic gonadotropin has LH-like effects, as previously described; it also has thyroid-stimulating ability, like pituitary TSH. Chorionic somatomammotropin likewise has actions that are similar to two pituitary hormones: growth hormone and prolactin. The placental hormones hCG and hCS thus duplicate the actions of four anterior pituitary hormones.

***Pituitary-Like Hormones from the Placenta.*** The importance of chorionic gonadotropin in maintaining the mother's corpus luteum for the first five and a half weeks of pregnancy has been previously discussed. There is also some evidence that hCG may in some way help to prevent immunological rejection of the implanting embryo. Chorionic somatomammotropin acts together (synergizes with) growth hormone from the mother's pituitary to produce a "diabeticlike" effect in the pregnant woman. The effects of these two hormones stimulate (1) lipolysis and increased plasma fatty acid concentration; (2) decreased maternal utilization of glucose and, therefore, increased blood glucose concentrations; and (3) polyuria (excretion of large volumes of urine), thereby producing a degree of dehydration and thirst. This diabeticlike effect in the mother helps to spare glucose for the placenta and fetus that, like the brain, use glucose as their primary energy source.

***Steroid Hormones from the Placenta.*** After the first five and a half weeks of pregnancy, when the corpus luteum regresses, the placenta becomes the major sex steroid–producing gland. The blood concentration of estrogens, as a result of placental secretion, rises to levels more than 100 times greater than those existing at the beginning of pregnancy. The placenta also secretes large amounts of progesterone, changing the estrogen/progesterone ratio in the blood from 100:1 at the beginning of pregnancy to a ratio of close to 1:1 toward full-term.

The placenta, however, is an "incomplete endocrine gland" because it cannot produce estrogen and progesterone without the aid of precursors supplied to it by both the mother and the fetus. The placenta, for example, cannot produce cholesterol from acetate and so must be supplied with cholesterol from the mother's circulation. Cholesterol, which is a steroid containing twenty-seven carbons, can then be converted by enzymes in the placenta into steroids that contain twenty-one carbons—such as progesterone. The placenta, however, lacks the enzymes needed to convert progesterone into androgens (which have nineteen carbons) and estrogens (which have eighteen carbons).

In order for the placenta to produce estrogens, it needs to cooperate with steroid-producing tissues in the fetus. Fetus and placenta, therefore, form a single functioning system in terms of steroid hormone production. This system has been called the **fetal-placental unit** (fig. 20.50).

Figure 20.50. The secretion of progesterone and estrogen from the placenta requires a supply of cholesterol from the mother's blood and the cooperation of fetal enzymes that convert progesterone to androgens.

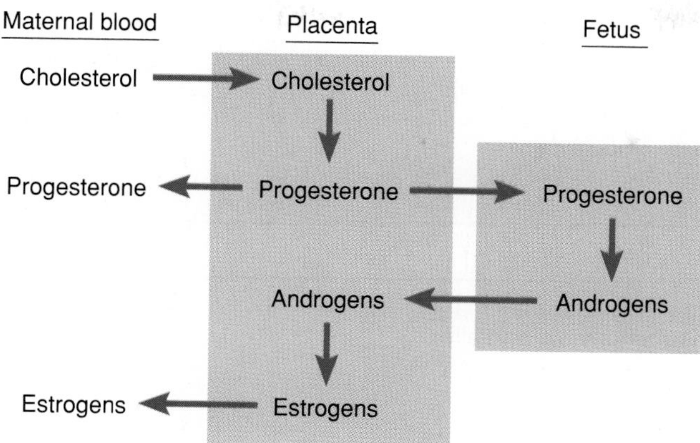


The ability of the placenta to convert androgens into estrogen helps to protect the female embryo from becoming masculinized by the androgens secreted from the mother's adrenal glands. In addition to producing estradiol, the placenta secretes large amounts of a weak estrogen called **estriol.** The production of estriol increases tenfold during pregnancy, so that by the third trimester estriol accounts for about 90% of the estrogens excreted in the mother's urine. Since almost all of this estriol comes from the placenta (rather than from maternal tissues), measurements of urinary estriol can be used clinically to assess the health of the placenta.

## Labor and Parturition

Powerful contractions of the uterus in *labor* are needed for childbirth *(parturition).* These uterine contractions are known to be stimulated by two agents: (1) *oxytocin,* a polypeptide hormone produced in the hypothalamus and secreted by the posterior pituitary; and (2) *prostaglandins,* a class of cyclic fatty acids with autocrine function produced within the uterus itself. Labor can indeed be induced artificially by injections of oxytocin or by insertion of prostaglandins into the vagina as a suppository.

Although the mechanisms operating during labor are well understood, the factors that lead to the initiation of parturition in humans remain unclear. In some lower mammals, the concentrations of estrogen and progesterone drop prior to parturition; this does not appear to be true in humans. In sheep and goats, labor seems to be initiated by corticosteroids from the fetal adrenal cortex, which stimulate production of prostaglandins in the uterus. The uterine contractions that result, in turn, stimulate increased secretion of oxytocin. The timing of parturition in humans, however, does not appear to be similarly dependent on the fetal adrenal cortex.

It appears that both oxytocin and prostaglandins are required in humans; oxytocin-induced contractions do not lead to dilation of the cervix and progressive labor in the absence of prostaglandins. Indeed, both alcohol—which inhibits oxytocin secretion—and indomethacin (which inhibits prostaglandin production) can be used to prevent preterm births. A mechanism has recently been proposed to explain how oxytocin and prostaglandins might cooperate to initiate labor in humans.

It has been demonstrated that the concentration of oxytocin receptors in the myometrium (smooth muscle layer) of the uterus increases dramatically during gestation. Although the blood concentrations of oxytocin remain constant, the sensitivity of the uterus to oxytocin increases as a result of the increased receptors. The threshold amount of oxytocin required to induce uterine contractions in a woman at term is almost one-hundredth the amount required in a nonpregnant woman. Increased oxytocin receptors and sensitivity may result from the effects of the rising levels of estrogens that occur during gestation.

In addition to finding increased oxytocin receptors in the myometrium, increased oxytocin receptors were also found in the nonmuscular decidua. It has been proposed that oxytocin (possibly coming from the fetus as well as from the mother) may stimulate prostaglandin production in the decidua. Prostaglandins diffusing into the myometrium from the decidua may then act together with the increased sensitivity of the myometrium to oxytocin and stimulate the onset of labor. These events are summarized in table 20.9.

**Table 20.9** Possible sequence of events leading to the onset of labor in humans

| Step | Event | Step | Event |
|---|---|---|---|
| 1 | High estrogen secretion from the placenta stimulates production of oxytocin receptors in the uterus | 4 | Prostaglandins may stimulate uterine contractions |
| 2 | Uterine muscle (myometrium) becomes increasingly sensitive to effects of oxytocin during pregnancy | 5 | Contractions of the uterus stimulate oxytocin |
| 3 | Oxytocin may stimulate production of prostaglandins in the uterus | 6 | Increased oxytocin secretion stimulates increased uterine contractions, creating a positive feedback loop and resulting in labor |

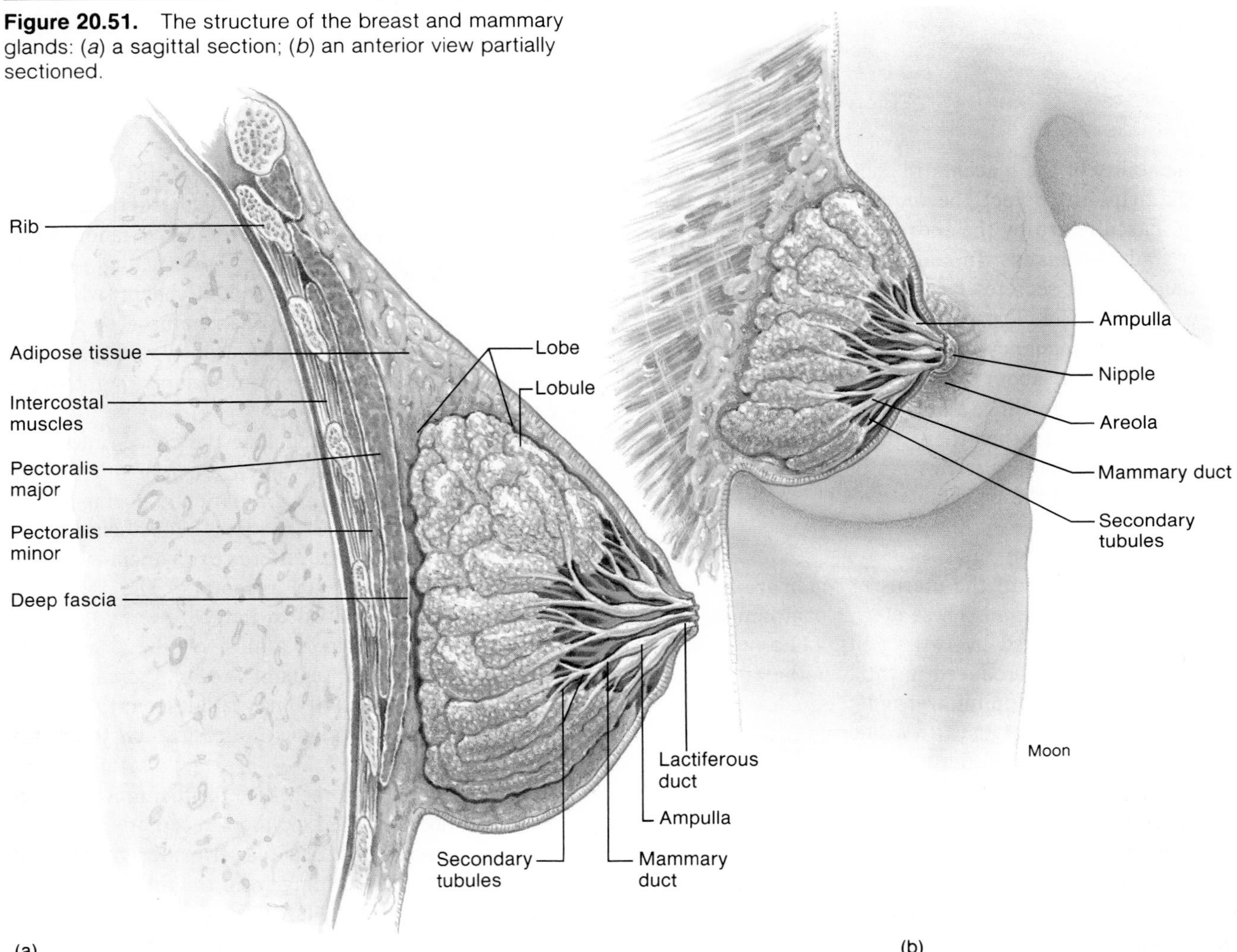


**Figure 20.51.** The structure of the breast and mammary glands: (*a*) a sagittal section; (*b*) an anterior view partially sectioned.

## Lactation

Each mammary gland is composed of fifteen to twenty *lobes,* divided by adipose tissue and each with its own drainage pathway to the outside. The amount of adipose tissue determines the size and shape of the breast but has nothing to do with the ability of a woman to nurse. Each lobe is subdivided into *lobules,* which contain the glandular **alveoli** (fig. 20.51) that secrete the milk of a lactating female. The clustered alveoli secrete milk into a series of *secondary tubules,* which converge to form a **lactiferous duct** that drains at the tip of the nipple.

The changes that occur in the mammary glands during pregnancy and the regulation of lactation provide excellent examples of hormonal interactions and neuroendocrine regulation (table 20.10). Growth and development of the mammary glands during pregnancy requires the permissive actions of insulin, cortisol, and thyroid hormones; in the presence of adequate amounts of these hormones, high levels of estrogen stimulate the development

**Table 20.10** Hormonal factors affecting lactation

| Hormones | Major Source | Effects |
|---|---|---|
| Insulin, cortisol, thyroid hormones | Pancreas, adrenal cortex, and thyroid | Permissive effects—adequate amounts of these must be present for other hormones to exert their effects on mammary glands |
| Estrogen and progesterone | Placenta | Growth and development of secretory units (alveoli) and ducts in mammary glands |
| Prolactin | Anterior pituitary | Production of milk proteins, including casein and lactalbumin |
| Oxytocin | Posterior pituitary | Stimulates milk-ejection |

**Figure 20.52.** The hormonal control of mammary gland development during pregnancy and lactation. Note that milk production is prevented during pregnancy by estrogen inhibition of prolactin secretion. This inhibition is accomplished by the stimulation of PIH (prolactin-inhibiting hormone) secretion from the hypothalamus.

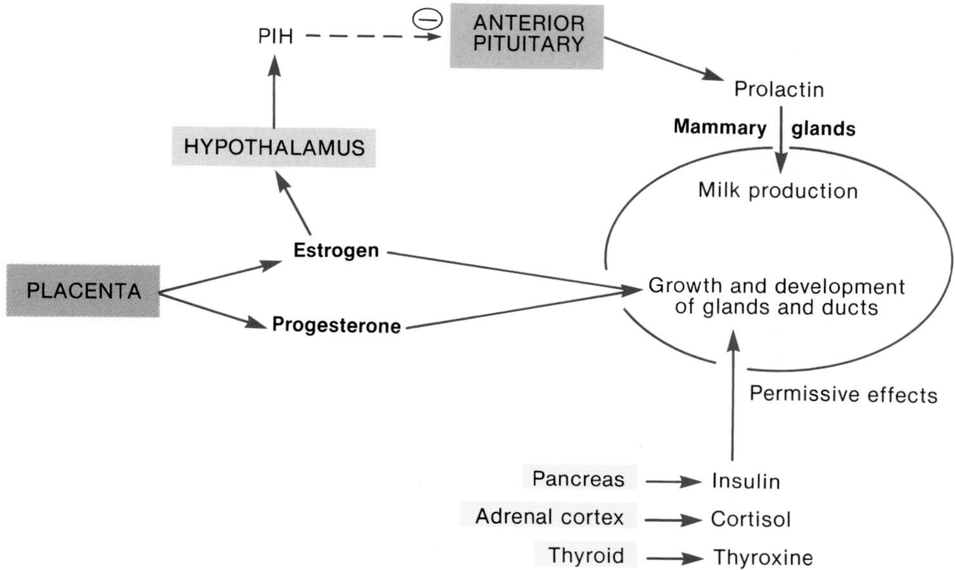


of the mammary alveoli and progesterone stimulates proliferation of the tubules and ducts (fig. 20.52).

The production of milk proteins, including casein and lactalbumin, is stimulated after parturition by **prolactin,** a hormone secreted by the anterior pituitary gland. The secretion of prolactin is controlled primarily by *prolactin-inhibiting hormone (PIH),* which is believed to be dopamine, produced by the hypothalamus and secreted into the portal blood vessels. Recent evidence also suggests that the posterior pituitary may pass dopamine through blood vessels to the anterior pituitary. The secretion of PIH is stimulated by high levels of estrogen. In addition, high levels of estrogen act directly on the mammary glands to block their stimulation by prolactin and hCS. During pregnancy, consequently, the high levels of estrogen prepare the breasts for lactation but prevent prolactin secretion and action.

After parturition, when the placenta is eliminated, declining levels of estrogen are accompanied by an increase in the secretion of prolactin. Lactation, therefore, commences. If a woman does not wish to breast-feed her baby, she may be injected with a powerful synthetic estrogen (diethylstilbestrol, or DES), which inhibits further prolactin secretion.

The act of nursing helps to maintain high levels of prolactin secretion via a *neuroendocrine reflex* (fig. 20.53). Sensory endings in the breast, activated by the stimulus of suckling, relay impulses to the hypothalamus and inhibit the secretion of PIH. There is also indirect evidence that the stimulus of suckling may cause the secretion of a *prolactin-releasing hormone* from the posterior pituitary, which travels through portal blood vessels to the anterior pituitary. If this is true, it would be the only example of a releasing hormone that is produced by the posterior pituitary rather than by the hypothalamus. Suckling thus results in the reflex secretion of high levels of prolactin, which promotes the secretion of milk from the alveoli into the ducts. In order for the baby to get the milk, however, the action of another hormone is needed.

**Figure 20.53.** Lactation occurs in two stages: milk production (stimulated by prolactin) and milk ejection (stimulated by oxytocin). The stimulus of suckling triggers a neuroendocrine reflex that results in increased secretion of oxytocin and prolactin.

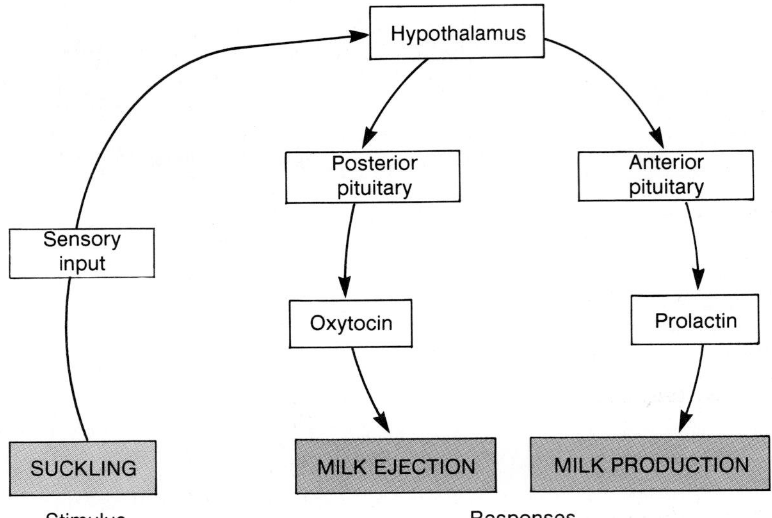


The stimulus of suckling also results in the reflex secretion of **oxytocin** from the posterior pituitary. This hormone is produced in the hypothalamus and stored in the posterior pituitary; its secretion results in the **milk-ejection reflex,** or milk let-down. Oxytocin stimulates contraction of the lactiferous ducts, as well as of the uterus (this is why women who breast-feed regain uterine muscle tone faster than women who bottle-feed).

Breast-feeding, acting through reflex inhibition of GnRH secretion, can inhibit the secretion of gonadotropins from the mother's anterior pituitary and thus inhibit ovulation. Breast-feeding is thus a natural contraceptive mechanism which helps to space births. This mechanism appears to be most effective in women with limited caloric intake and those who breast-feed their babies at frequent intervals throughout the day and night. In the traditional societies of the less-industrialized nations, therefore, breast-feeding is an effective contraceptive. Breast-feeding has much less of a contraceptive effect in women who are well-nourished, and who breast feed their babies at more widely spaced intervals.

## Concluding Remarks

It may seem strange to end a textbook of physiology with the topics of pregnancy and parturition. This is done in part for practical reasons; these topics are complex, and require much prior knowledge of subjects covered earlier in the text. Also, it seems appropriate to end at the beginning, at the start of a new life. Although generations of researchers have accumulated an impressive body of knowledge, the study of physiology is still young and rapidly growing. I hope that this introductory textbook will serve students' immediate practical needs as a foundation for understanding current applications, and will provide a good basis for a lifetime of further study.

---

1. *Describe the changes that occur in the sperm and ovum during fertilization.*
2. *Describe the source of hCG, and explain why it is needed to maintain pregnancy for the first ten weeks.*
3. *Name the fetal and maternal components of the placenta, and describe the circulation in these two components. Explain how fetal and maternal gas exchange occurs.*
4. *List the names and major functions of the protein hormones and sex steroids secreted by the placenta.*
5. *Name the two factors that stimulate uterine contraction during labor, and describe the proposed mechanisms that may initiate labor in humans.*
6. *Describe the hormonal interactions required for breast development during pregnancy and for lactation after delivery.*

---

## SUMMARY

### Sexual Reproduction p. 638

I. Sperm that bear X chromosomes produce XX zygotes, and sperm that carry Y chromosomes produce XY zygotes, when they fertilize X-containing ova.
  A. Embryos at first contain indeterminate structures that could develop into either male or female gonads and accessory organs.
  B. Embryos that have the XY genotype develop testes; those without a Y chromosome produce ovaries.
  C. The testes of a male embryo secrete testosterone and müllerian inhibition factor; these act to cause degeneration of female accessory organs and to promote the formation of male accessory organs.

II. The male sex accessory organs are the epididymis, vas deferens, seminal vesicles, prostate, and ejaculatory duct.
  A. The female sex accessory organs are the uterus and uterine (fallopian) tubes; these develop when testosterone and müllerian inhibition factor are absent.
  B. Testosterone indirectly (acting via conversion to dihydrotestosterone) promotes the formation of male external genitalia; female genitalia are formed when testosterone is absent.

III. There are a variety of disorders of embryonic sexual development which can be understood in terms of the normal physiology of the developmental processes.

### Endocrine Regulation of Reproduction p. 645

I. The gonads are stimulated by two anterior pituitary hormones: FSH (follicle-stimulating hormone) and LH (luteinizing hormone).
  A. The secretion of FSH and LH is stimulated by gonadotropin-releasing hormone (GnRH), which is secreted by the hypothalamus.
  B. The secretion of FSH and LH is also under the control of the gonads by means of negative feedback inhibition by gonadal steroid hormones and by a peptide called inhibin.

II. The rise in FSH and LH secretion that occurs at puberty may be due to maturational changes in the brain that result in increased GnRH secretion and to decreased sensitivity of the hypothalamus and pituitary to the negative feedback effects of sex steroid hormones.

III. The pineal gland secretes the hormone melatonin; this has an inhibitory effect on gonadal function in some species of mammals, but its role in human physiology is presently controversial.

### Male Reproductive System p. 649

I. In the male, the pituitary secretion of LH is controlled by negative feedback from testosterone, whereas the secretion of FSH is controlled by the secretion of inhibin from the testes.
  A. The negative feedback effect of testosterone is actually produced by the conversion of testosterone to 5$\alpha$-reduced androgens and to estradiol.
  B. The secretion of testosterone is relatively constant, rather than cyclic, and it does not decline sharply at a particular age.

II. Testosterone promotes the growth of soft tissue and bones before the epiphyseal discs have sealed; testosterone and related androgens are thus anabolic steroids.
  A. Testosterone is secreted by the interstitial cells of Leydig under stimulation by LH.
  B. The seminiferous tubules are able to convert testosterone to other derivatives, some of which may have regulatory functions in the Leydig cells.
  C. LH receptor proteins are located in the interstitial tissue; FSH receptors are located in the Sertoli cells within the seminiferous tubules.
  D. The Leydig cells of the interstitial compartment and the Sertoli cells of the tubular compartment of the testes secrete autocrine regulatory molecules which allow these compartments to interact.

III. Diploid spermatogonia in the seminiferous tubules undergo meiotic cell division to produce haploid sperm.
  A. At the end of meiosis, four spermatids are formed; these change into spermatozoa by a maturational process called spermiogenesis.
  B. Sertoli cells in the seminiferous tubules are required for spermatogenesis.
  C. At puberty, testosterone is required for the completion of meiosis and FSH is required for spermiogenesis.

IV. Spermatozoa in the seminiferous tubules are conducted to the epididymis and drained from the epididymis into the vas deferens; the prostate and seminal vesicles add fluid to the semen.

V. Penile erection is produced by parasympathetic-induced vasodilation; ejaculation is produced by sympathetic nerve stimulation of peristaltic contraction of the male sex accessory organs.
  A. The ejaculate has an average sperm count of about 300 million; at least 60 million are required for fertility.
  B. Vasectomy interferes with sperm transport in the vas deferens.

### Female Reproductive System p. 660

I. Primordial follicles in the ovary contain primary oocytes that have become arrested at prophase of the first meiotic division; the number of these is maximal at birth and declines thereafter.
  A. A small number of oocytes in each cycle are stimulated to complete their first meiotic division and become secondary oocytes.
  B. Secondary oocytes are contained within secondary follicles; these also contain granulosa cells, which form a corona radiata around the ovum.
  C. At the completion of the first meiotic division, the secondary oocyte is the only complete cell formed; the other product of this division is a tiny polar body, which disintegrates.

II. One of the secondary follicles grows very large, becomes a graafian follicle, and ovulates.
   A. Upon ovulation, the secondary oocyte is extruded from the ovary; it does not complete the second meiotic division unless it becomes fertilized.
   B. The empty follicle that has ovulated becomes a new endocrine gland called a corpus luteum.
   C. Whereas the ovarian follicles secrete only estradiol, the corpus luteum secretes both estradiol and progesterone.
III. The hypothalamus secretes GnRH in a pulsatile fashion, causing pulsatile secretion of gonadotropins; this is needed to prevent desensitization and down-regulation of the target glands.

### Menstrual Cycle p. 667

I. During the follicular phase of the cycle, the ovarian follicles are stimulated by FSH from the anterior pituitary.
   A. Under FSH stimulation, the follicles grow, mature, and secrete increasing concentrations of estradiol.
   B. At about day thirteen of an average cycle, the rapid rise in estradiol secretion stimulates a surge of LH secretion from the anterior pituitary; this represents positive feedback.
   C. FSH produces a follicle that is ready to ovulate; the LH surge stimulates ovulation at about day fourteen.
   D. After ovulation, the empty follicle is stimulated by LH to become a corpus luteum; the ovary is now in a luteal phase.
   E. The secretion of progesterone and estradiol rises during the first part of the luteal phase and exerts negative feedback inhibition of FSH and LH secretion.
   F. Without continued stimulation by LH, the corpus luteum regresses at the end of the luteal phase, and the secretion of estradiol and progesterone declines; this decline results in menstruation and the beginning of a new cycle.
II. The rising estradiol concentration during the follicular phase produces the proliferative phase of the endometrium; the secretion of progesterone during the luteal phase produces the secretory phase of the endometrium.
III. Oral contraceptive pills usually contain combinations of estrogen and progesterone, which produce negative feedback inhibition of FSH and LH secretion.
   A. This therefore mimics the endocrine events that occur in the luteal phase of the cycle; no new follicles develop and ovulation cannot occur.
   B. When the pill is stopped after three weeks, levels of estrogen and progesterone decline, resulting in menstruation.

### Fertilization, Pregnancy, and Parturition p. 673

I. The sperm undergoes an acrosomal reaction, which allows it to penetrate the corona radiata and zona pellucida.
   A. Upon fertilization, the secondary oocyte completes meiotic division and produces a second polar body, which degenerates.
   B. A diploid zygote is formed, which undergoes cleavage to form a morula and then a blastocyst, which implants in the endometrium.
II. The trophoblast cells of the blastocyst secrete human chorionic gonadotropin (hCG), which functions like LH and maintains the mother's corpus luteum for the first ten weeks of pregnancy.
   A. The trophoblast cells become the fetal contribution to the placenta; the placenta is also formed from adjacent maternal tissue in the endometrium.
   B. Oxygen, nutrients, and wastes are exchanged by diffusion between the fetal and maternal blood.
III. The placenta secretes chorionic somatomammotropin (hCS), chorionic gonadotropin (hCG), and steroid hormones.
   A. hCS acts like prolactin and growth hormone; hCG acts like LH and TSH.
   B. The major steroid hormone secreted by the placenta is estriol; the placenta and fetal glands cooperate in the production of steroid hormones.
IV. Contraction of the uterus in labor is stimulated by oxytocin from the posterior pituitary and by prostaglandins, produced within the uterus.
   A. The uterus contains increasing numbers of oxytocin receptors during gestation and thus becomes increasingly sensitive to oxytocin.
   B. Oxytocin may stimulate the production of prostaglandins, which cause uterine contractions.
   C. Uterine contractions stimulate a reflex secretion of oxytocin, thus causing a positive feedback loop to be produced that results in the contractions needed for labor and delivery.
V. The high levels of estrogen during pregnancy, acting synergistically with other hormones, stimulate growth and development of the mammary glands.
   A. Prolactin (and the prolactinlike effects of hCS) can stimulate the production of milk proteins; prolactin secretion and action, however, are blocked during pregnancy by the high levels of estrogen secreted by the placenta.
   B. After delivery, when the estrogen levels fall, prolactin stimulates milk production.
   C. The milk-ejection reflex is a neuroendocrine reflex; the stimulus of suckling causes reflex secretion of oxytocin, which causes contractions of the lactiferous ducts and the ejection of milk from the nipple.

## Review Activities

### Objective Questions

Match the following:

1. Menstrual phase
2. Follicular phase
3. Luteal phase
4. Ovulation

(a) high estrogen and progesterone; low FSH and LH
(b) low estrogen and progesterone
(c) LH surge
(d) increasing estrogen; low LH and low progesterone

5. A person with the genotype XO has
   (a) ovaries
   (b) testes
   (c) both ovaries and testes
   (d) neither ovaries nor testes
6. An embryo with the genotype XX develops female sex accessory organs because of
   (a) androgens
   (b) estrogens
   (c) absence of androgens
   (d) absence of estrogens
7. In the male
   (a) FSH is not secreted by the pituitary
   (b) FSH receptors are located in the Leydig cells
   (c) FSH receptors are located in the spermatogonia
   (d) FSH receptors are located in the Sertoli cells
8. The secretion of FSH in a male is inhibited by negative feedback effects of
   (a) inhibin secreted from the tubules
   (b) inhibin secreted from the Leydig cells
   (c) testosterone secreted from the tubules
   (d) testosterone secreted from the Leydig cells
9. Which of the following statements is *true?*
   (a) Sperm are not motile until they pass through the epididymis.
   (b) Sperm require capacitation in the female genital tract before they can fertilize an ovum.
   (c) An ovum does not complete meiotic division until it has been fertilized.
   (d) All of the above are true.
10. The corpus luteum is maintained for the first ten weeks of pregnancy by
    (a) hCG
    (b) LH
    (c) estrogen
    (d) progesterone
11. Fertilization normally occurs in
    (a) the ovaries
    (b) the fallopian tubes
    (c) the uterus
    (d) the vagina
12. The placenta is formed from
    (a) the fetal chorion frondosum
    (b) the maternal decidua basalis
    (c) both of the above
    (d) neither of the above
13. Uterine contractions are stimulated by
    (a) oxytocin
    (b) prostaglandins
    (c) prolactin
    (d) both *a* and *b*
    (e) both *b* and *c*
14. Contraction of the mammary glands and ducts during the milk-ejection reflex is stimulated by
    (a) prolactin
    (b) oxytocin
    (c) estrogen
    (d) progesterone
15. If GnRH were secreted in large amounts and at a constant rate rather than in a pulsatile fashion, the following would be *true.*
    (a) LH secretion would increase at first and then decrease.
    (b) LH secretion would increase indefinitely.
    (c) Testosterone secretion in a male would be continuously high.
    (d) Estradiol secretion in a woman would be continuously high.

### Essay Questions

1. Describe the conversion products of testosterone and their functions in the brain, prostate, and seminiferous tubules.
2. Explain why a testis is said to be composed of two separate compartments, and describe the interactions that may occur between these compartments.
3. Describe the role of the Sertoli cells in spermatogenesis, and describe the hormonal control of spermatogenesis.
4. Explain the hormonal interactions that control ovulation and make it occur at the proper time.
5. Compare menstrual bleeding and bleeding that occurs during the estrous cycle of a dog, in terms of hormonal control mechanisms and the ovarian cycle.
6. "The [contraceptive] pill tricks the brain into thinking you're pregnant." Interpret this popularized explanation in terms of physiological mechanisms.
7. Why does menstruation normally occur? Under what conditions does menstruation not occur? Explain.
8. Explain the proposed mechanisms whereby the act of a mother nursing her baby results in lactation.

## Selected Readings


Bardin, C. W. 1979. The neuroendocrinology of male reproduction. *Hospital Practice* 14:65.

Bartke, A., A. A. Hafiez, F. J. Bex, and S. Dalterio. 1978. Hormonal interaction in the regulation of androgen secretion. *Biology of Reproduction* 18:44.

Beaconsfield, P., G. Birdwood, and R. Beaconsfield. July 1980. The placenta. *Scientific American.*

Ben-Jonathon, N., J. F. Hyde, and I. Murai. 1988. Suckling-induced rise in prolactin: Mediation by a prolactin-releasing factor from the posterior pituitary. *News in Physiological Science* 3:172.

Boyar, R. M. 1978. Control of the onset of puberty. *Annual Review of Medicine* 31:329.

Chervenak, F. A., G. Isaacson, and M. J. Mahoney. 1986. Advances in the diagnosis of fetal defects. *New England Journal of Medicine* 315:305.

Dufau, M. L. 1988. Endocrine regulation and communicating functions of the Leydig cell. *Annual Review of Physiology* 50:483.

Fink, G. 1979. Feedback action of target hormones on hypothalamus and pituitary with special reference to gonadal steroids. *Annual Review of Physiology* 41:571.

Frantz, A. G. 1979. Prolactin. *New England Journal of Medicine* 298:112.

Frisch, R. E. March 1988. Fatness and fertility. *Scientific American.*

Grobstein, C. March 1979. External human fertilization. *Scientific American.*

Grumbach, M. M. 1979. The neuroendocrinology of puberty. *Hospital Practice* 14:65.

Jackson, L. G. 1985. First-trimester diagnosis of fetal genetic disorders. *Hospital Practice* 20:39.

Keys, P. L., and M. C. Wiltbank. 1988. Endocrine regulation of the corpus luteum. *Annual Review of Physiology* 50:465.

Lagerkrantz, H., and T. A. Slotkin. April 1986. The "stress" of being born. *Scientific American.*

Leong, D. A., L. S. Frawley, and J. D. Neill. 1983. Neuroendocrine control of prolactin secretion. *Annual Review of Physiology* 45:109.

Lipsett, M. B. 1980. Physiology and pathology of the Leydig cell. *New England Journal of Medicine* 303:682.

Mader, S. S. 1980. *Human Reproductive Biology.* Dubuque: Wm. C. Brown Company.

Masters, W. H. 1986. Sex and aging—expectations and reality. *Hospital Practice* 21:175.

Means, A. R., et al. 1980. Regulation of the testis Sertoli cell by follicle-stimulating hormone. *Annual Review of Physiology* 42:59.

Naftolin, F. March 1981. Understanding the basis of sex differences. *Science* 211 (No. 4488):1263.

Odell, W. D., and D. L. Moyer. 1971. *Physiology of Reproduction.* St. Louis: C. V. Mosby Company.

Reiter, E. O., and M. M. Grumbach. 1982. Neuroendocrine control mechanisms and the onset of puberty. *Annual Review of Physiology* 44:595.

Richards, J. S., and L. Hedin. 1988. Molecular aspects of hormone action in ovarian follicular development, ovulation, and luteinization. *Annual Review of Physiology* 50:441.

Segal, S. J. September 1974. The physiology of human reproduction. *Scientific American.*

Seibel, M. M. 1988. A new era in reproductive technology. *New England Journal of Medicine* 318:828.

Simpson, E. R., and P. C. MacDonald. 1981. Endocrine physiology of the placenta. *Annual Review of Physiology* 43:163.

Tamarkin, L., C. J. Baird, and O. F. X. Almeida. 1985. Melatonin: A coordinating signal for mammalian reproduction? *Science* 227:714.

Warkentin, D. L. April 1984. From A to hCG in pregnancy testing. *Diagnostic Medicine,* p. 35.

Wilson, J. D. 1978. Sexual differentiation. *Annual Review of Physiology* 40:279.

Wilson, J. D., F. W. George, and J. E. Griffin. March 1981. The hormonal control of sexual development. *Science* 211 (No. 4488):1278.

Yen, S. S. C. 1977. Regulation of the hypothalamic-pituitary-ovarian axis in women. *Journal of Reproduction and Fertility* 51:181.

Yen, S. S. C. 1979. Neuroendocrine regulation of the menstrual cycle. *Hospital Practice* 14:83.

# Appendix

## Answers to Objective Questions

### Chapter 1

1. d 5. d 9. a
2. d 6. c 10. c
3. b 7. b 11. c
4. b 8. b

### Chapter 2

1. c 5. c 9. d
2. b 6. b 10. b
3. a 7. c 11. d
4. d 8. d 12. b

### Chapter 3

1. d 6. b 11. d
2. b 7. a 12. e
3. a 8. c 13. b
4. c 9. a 14. a
5. d 10. b

### Chapter 4

1. b 5. d 8. d
2. d 6. e 9. d
3. d 7. e 10. d
4. a

### Chapter 5

1. b 6. c 10. d
2. a 7. a 11. b
3. c 8. c 12. d
4. e 9. a 13. b
5. d

### Chapter 6

1. c 5. b 9. b
2. b 6. d 10. d
3. a 7. a 11. b
4. c 8. a

### Chapter 7

1. c 6. d 11. b
2. d 7. d 12. d
3. a 8. a 13. d
4. a 9. c 14. b
5. c 10. c 15. a

### Chapter 8

1. d 6. e 11. a
2. b 7. c 12. b
3. e 8. d 13. d
4. a 9. b 14. a
5. b 10. c

### Chapter 9

1. d 7. c 12. b
2. a 8. c 13. b
3. c 9. d 14. c
4. d 10. b 15. b
5. c 11. d 16. c
6. a

### Chapter 10

1. c 5. c 8. e
2. c 6. b 9. c
3. c 7. b 10. c
4. a

### Chapter 11

1. d 5. d 9. d
2. d 6. a 10. a
3. e 7. b 11. d
4. e 8. e 12. c

### Chapter 12

1. b 6. b 11. d
2. d 7. a 12. a
3. c 8. c 13. e
4. b 9. b 14. c
5. e 10. b 15. b

### Chapter 13

1. c 7. c 12. d
2. b 8. a 13. d
3. e 9. d 14. c
4. a 10. b 15. c
5. b 11. c 16. b
6. c

### Chapter 14

1. a 7. a 12. d
2. d 8. c 13. b
3. c 9. d 14. d
4. e 10. b 15. c
5. b 11. c 16. e
6. c

### Chapter 15

1. c 7. b 13. a
2. d 8. a 14. d
3. c 9. e 15. b
4. a 10. c 16. a
5. c 11. a 17. c
6. c 12. c

### Chapter 16

1. b 7. d 12. c
2. a 8. c 13. a
3. c 9. d 14. d
4. b 10. a 15. b
5. e 11. e 16. e
6. d

### Chapter 17

1. e 5. d 9. d
2. d 6. b 10. e
3. b 7. d 11. e
4. c 8. a 12. d

### Chapter 18

1. c 7. d 12. b
2. a 8. a 13. c
3. b 9. d 14. a
4. c 10. c 15. d
5. a 11. b 16. d
6. b

### Chapter 19

1. c 6. d 11. d
2. b 7. d 12. a
3. d 8. b 13. d
4. a 9. e 14. c
5. c 10. a

### Chapter 20

1. b 6. c 11. b
2. d 7. d 12. c
3. a 8. a 13. d
4. c 9. d 14. b
5. a 10. a 15. a

# GLOSSARY

## A

**a-, an-** (Gr.) Not, without, lacking.
**ab-** (La.) Off, away from.
**abdomen** A body cavity between the thorax and pelvis.
**abductor** A muscle that moves the skeleton away from the median plane of the body or away from the axial line of a limb.
**ABO system** The most common system of classification for red blood cell antigens. On the basis of antigens on the red blood cell surface, individuals can be type A, type B, type AB, or type O.
**absorption** The transport of molecules across epithelial membranes into the body fluids.
**accommodation** To fit or adjust; specifically, the ability of the eyes to adjust their curvature so that an image of an object is focused on the retina at different distances.
**acetyl CoA** Acetyl coenzyme A; an intermediate molecule in aerobic cell respiration which, together with oxaloacetic acid, begins the Krebs cycle. Acetyl CoA is also an intermediate in the synthesis of fatty acids.
**acetylcholine (ACh)** A molecule—which is an acetic acid ester of choline—that functions as a neurotransmitter chemical in somatic motor nerve and parasympathetic nerve fibers.
**acetylcholinesterase** An enzyme in the membrane of postsynaptic cells that catalyzes the conversion of ACh into choline and acetic acid. This enzymatic reaction inactivates the neurotransmitter.
**acidosis** An abnormal increase in the $H^+$ concentration of the blood that lowers arterial pH below 7.35.
**acromegaly** A condition caused by the hypersecretion of growth hormone from the pituitary after maturity and characterized by enlargement of the extremities, such as the nose, jaws, fingers, and toes.
**ACTH** Abbreviation for adrenocorticotropic hormone. A hormone secreted by the anterior pituitary gland which stimulates the adrenal cortex.
**actin** A structural protein of muscle that, along with myosin, is responsible for muscle contraction.
**action potential** An all-or-none electrical event in an axon or muscle fiber, in which the polarity of the membrane potential is rapidly reversed and reestablished.
**active immunity** Immunity involving sensitization, in which antibody production is stimulated by prior exposure to an antigen.
**active transport** The movement of molecules or ions across the cell membranes of epithelial cells by membrane carriers; the expenditure of cellular energy (ATP) is required.
**ad-** (La.) Toward, next to.
**adductor** A muscle that moves the skeleton toward the midline of the body or toward the axial plane of a limb.
**adenylate cyclase** An enzyme, found in cell membranes, that catalyzes the conversion of ATP to cyclic AMP and pyrophosphate ($PP_i$). This enzyme is activated by an interaction between a specific hormone and its membrane receptor protein.
**ADH** Antidiuretic hormone, also known as *vasopressin.* A hormone produced by the hypothalamus and secreted by the posterior pituitary gland; it acts on the kidneys to promote water reabsorption, thus decreasing the urine volume.
**adipose tissue** Fatty tissue; a type of connective tissue consisting of fat cells in a loose connective tissue matrix.
**ADP** Adenosine diphosphate; a molecule that, together with inorganic phosphate, is used to make ATP (adenosine triphosphate).
**adrenal cortex** The outer part of the adrenal gland. Derived from embryonic mesoderm, the adrenal cortex secretes corticosteroid hormones (such as aldosterone and hydrocortisone).
**adrenal medulla** The inner part of the adrenal gland. Derived from embryonic postganglionic sympathetic neurons, the adrenal medulla secretes catecholamine hormones—epinephrine and (to a lesser degree) norepinephrine.
**adrenergic** An adjective describing the actions of epinephrine, norepinephrine, or other molecules with similar activity (as in *adrenergic receptor* and *adrenergic stimulation*).
**aerobic capacity** The ability of an organ to utilize oxygen and respire aerobically to meet its energy needs.
**afferent** The carrying of something toward a center. Afferent neurons, for example, conduct impulses toward the central nervous system; afferent arterioles carry blood toward the glomerulus.
**agglutinate** A clumping of cells (usually erythrocytes) due to specific chemical interaction between surface antigens and antibodies.
**agranular leukocytes** White blood cells (leukocytes) that do not contain cytoplasmic granules; specifically, lymphocytes and monocytes.
**albumin** A water-soluble protein, produced in the liver, that is the major component of the plasma proteins.
**aldosterone** The principal corticosteroid hormone involved in the regulation of electrolyte balance (mineralocorticoid).
**allergens** Antigens that evoke an allergic response rather than a normal immune response.
**allergy** A state of hypersensitivity caused by exposure to allergens; it results in the liberation of histamine and other molecules with histaminelike effects.
**all-or-none law** A given response will be produced to its maximum extent in response to any stimulus equal to or greater than a threshold value. Action potentials obey an all-or-none law.
**allosteric** Denotes the alteration of an enzyme's activity by its combination with a regulator molecule; allosteric inhibition by an end product represents negative feedback control of an enzyme's activity.
**alpha motoneurons** Somatic motor neurons that stimulate extrafusal skeletal muscle fibers.
**alveoli** Plural of *alveolus;* an anatomical term for small, saclike dilations (as in *lung alveoli*).
**amniocentesis** A procedure to obtain amniotic fluid and fetal cells in this fluid through transabdominal perforation of the uterus.
**amnion** The inner fetal membrane that contains the fetus in amniotic fluid; this is commonly called the "bag of waters."
**amphoteric** Pertaining to having opposite characteristics; denoting a molecule that can be positively or negatively charged, depending on the pH of its environment.
**an-** (Gr.) Without, not.

**anabolic steroids** Steroids with androgenlike stimulatory effects on protein synthesis.

**anabolism** Chemical reactions within cells that result in the production of larger molecules from smaller ones; specifically, the synthesis of protein, glycogen, and fat.

**anaerobic respiration** A form of cell respiration, involving the conversion of glucose to lactic acid, in which energy is obtained without the use of molecular oxygen.

**anaerobic threshold** The maximum exercise that can be performed before a significant amount of lactic acid is produced by the exercising skeletal muscles through anaerobic respiration. This generally occurs when about 60% of the total aerobic capacity of the person has been reached.

**anaphylaxis** An unusually severe allergic reaction, which can result in cardiovascular shock and death.

**androgens** Steroids, containing eighteen carbons, that have masculinizing effects; primarily those hormones (such as testosterone) secreted by the testes, although weaker androgens are also secreted by the adrenal cortex.

**anemia** An abnormal reduction in the red blood cell count, hemoglobin concentration, or hematocrit, or any combination of these measurements. This condition is associated with a decreased ability of the blood to carry oxygen.

**angina pectoris** A thoracic pain, often referred to the left pectoral and arm area, caused by myocardial ischemia.

**angiotensin II** An eight-amino-acid polypeptide formed from angiotensin I (a ten-amino-acid precursor), which in turn is formed from the cleavage of a protein (angiotensinogen) by the action of renin, an enzyme secreted by the kidneys. Angiotensin II is a powerful vasoconstrictor and a stimulator of aldosterone secretion from the adrenal cortex.

**anions** Ions that are negatively charged, such as chloride, bicarbonate, and phosphate.

**antagonistic effects** Actions of regulators such as hormones or nerves that counteract the effects of other regulators. The actions of sympathetic and parasympathetic neurons on the heart, for example, are antagonistic.

**anterior** An anatomical term referring to a forward position of a structure.

**anterior pituitary** Also called the *adenohypophysis;* this part of the pituitary secretes FSH (follicle-stimulating hormone), LH (luteinizing hormone), ACTH (adrenocorticotropic hormone), TSH (thyroid-stimulating hormone), GH (growth hormone), and prolactin. Secretions of the anterior pituitary are controlled by hormones secreted by the hypothalamus.

**antibodies** Immunoglobulin proteins secreted by B lymphocytes that have transformed into plasma cells. Antibodies are responsible for humoral immunity. Their synthesis is induced by specific antigens, and they combine with these specific antigens but not with unrelated antigens.

**anticoagulant** A substance that inhibits blood clotting.

**anticodon** A base triplet provided by three nucleotides within a loop of transfer RNA, which is complementary in its base pairing properties to a triplet (the codon) in mRNA; the matching of codon to anticodon provides the mechanism for translation of the genetic code into a specific sequence of amino acids.

**antigen** A molecule able to induce the production of antibodies and to react in a specific manner with antibodies.

**antigenic determinant site** The region of an antigen molecule that specifically reacts with particular antibodies. A large antigen molecule may have a number of such sites.

**antiserum** A serum that contains specific antibodies.

**aphasia** Speech and language disorders caused by damage to the brain. This can result from damage to Broca's area, Wernicke's area, the arcuate fasciculus, or the angular gyrus.

**apnea** The cessation of breathing.

**apneustic center** A collection of neurons in the brain stem that participates in the rhythmic control of breathing.

**aqueous humor** A fluid produced by the ciliary body that fills the anterior and posterior chambers of the eye.

**arteriosclerosis** A group of diseases characterized by thickening and hardening of the artery wall and narrowing of its lumen.

**arteriovenous anastomoses** Direct connections between arteries and veins that bypass capillary beds.

**artery** A vessel that carries blood away from the heart.

**astigmatism** Unequal curvature of the refractive surfaces of the eye (cornea and/or lens), so that light that enters the eye along certain meridians does not focus on the retina.

**atherosclerosis** A common type of arteriosclerosis found in medium and large arteries, in which raised areas, or plaque, within the tunica intima are formed from smooth muscle cells, cholesterol, and other lipids. These plaques occlude arteries and serve as sites for the formation of thrombi.

**atomic number** A whole number representing the number of positively charged protons in the nucleus of an atom.

**atopic dermatitis** An allergic skin reaction to agents such as poison ivy and poison oak; a type of delayed hypersensitivity.

**ATP** Adenosine triphosphate; the universal energy carrier of the cell.

**atretic** Without an opening; atretic ovarian follicles are those that fail to ovulate.

**atrial natriuretic factor** A chemical secreted by the atria which acts as a natriuretic hormone (a hormone that promotes the urinary excretion of sodium).

**atrioventricular node** The atrioventricular, or AV, node is the specialized mass of conducting tissue located in the right atrium near the junction of the interventricular septum. It transmits the impulse into the bundle of His.

**atrioventricular valves** The atrioventricular, or AV, valves are one-way valves located between the atria and ventricles. The AV valve on the right side of the heart is the tricuspid, and the AV valve on the left side is the bicuspid or mitral valve.

**atrophy** The decrease in mass and size of an organ; the opposite of hypertrophy.

**atropine** An alkaloid drug, obtained from a plant of the species *belladonna,* that acts as an anticholinergic agent. Used medically to inhibit parasympathetic nerve effects, dilate the pupils of the eye, increase the heart rate, and inhibit movements of the intestine.

**auto-** (Gr.) Self, same.

**autoantibodies** Antibodies that are formed in response to and that react with molecules that are part of one's own body.

**autocrine** A type of regulation in which one part of an organ releases chemicals that help regulate another part of the same organ.

**autonomic nervous system** The part of the nervous system that involves control of smooth muscle, cardiac muscle, and glands. The autonomic nervous system is subdivided into the sympathetic and parasympathetic divisions.

**autoregulation** The ability of an organ to intrinsically modify the degree of constriction or dilation of its small arteries and arterioles and to thus regulate the rate of its own blood flow. Autoregulation may occur through myogenic or metabolic mechanisms.

**autosomal chromosomes** The paired chromosomes; those other than the sex chromosomes.

**axon** The process of a nerve cell that conducts impulses away from the cell body.

**axonal transport** The transport of materials through the axon of a neuron. This usually occurs from the cell body to the end of the axon, but retrograde transport in the opposite direction can also occur.

## B

**baroreceptors** Receptors for arterial blood pressure located in the aortic arch and the carotid sinuses.

**Barr body** A microscopic structure in the cell nucleus produced from an inactive X chromosome in females.

**basal ganglia** Gray matter, or nuclei, within the cerebral hemispheres, comprising the corpus striatum, amygdaloid nucleus, and claustrum.

**basal metabolic rate (BMR)** The rate of metabolism (expressed as oxygen consumption or heat production) under resting or basal conditions fourteen to eighteen hours after eating.

**basophil** The rarest type of leukocyte; a granular leukocyte with an affinity for blue stain in the standard staining procedure.

**B cell lymphocytes** Lymphocytes that can be transformed by antigens into plasma cells that secrete antibodies (and are thus responsible for humoral immunity). The *B* stands for *bursa equivalent,* which is believed to be the bone marrow.

**benign** Not malignant or life threatening.

**bi-** (La.) Two, twice.

**bile** Fluid, produced by the liver and stored in the gallbladder, that contains bile salts, bile pigments, cholesterol, and other molecules. The bile is secreted into the small intestine.

**bile salts** Salts of derivatives of cholesterol in bile that are polar on one end and nonpolar on the other end of the molecule. Bile salts have detergent or surfactant effects and act to emulsify fat in the lumen of the small intestine.

**bilirubin** Bile pigment derived from the breakdown of the heme portion of hemoglobin.

**blocking antibodies** Antibodies that are specific for antigens attacked by other antibodies or by T cells and, therefore, may block this attack.

**blood-brain barrier** The structures and cells that selectively prevent particular molecules in the plasma from entering the central nervous system.

**Bohr effect** The effect of blood pH on the dissociation of oxyhemoglobin. Dissociation is promoted by a decrease in the pH.

**Boyle's Law** According to Boyle's law, the pressure of a given quantity of a gas is inversely proportional to its volume.

**bradycardia** A slow cardiac rate; less than sixty beats per minute.

**bradykinins** Short polypeptides that stimulate vasodilation and other cardiovascular changes.

**bronchioles** The smallest air passages in the lungs, which contain smooth muscle and cuboidal epithelial cells.

**brush border enzymes** Digestive enzymes in the small intestine that are located in the cell membrane of the microvilli of intestinal epithelial cells.

**buffer** A molecule that serves to prevent large changes in pH by either combining with $H^+$ or by releasing $H^+$ into solution.

**bundle of His** A band of rapidly conducting cardiac fibers that originate in the AV node and extend down the atrioventricular septum to the apex of the heart. This tissue conducts action potentials from the atria into the ventricles.

## C

**cable properties** The ability of neurons to conduct an electrical current. This occurs, for example, between nodes of Ranvier where action potentials are produced in a myelinated fiber.

**calcitonin** Also called *thyrocalcitonin.* A polypeptide hormone produced by the parafollicular cells of the thyroid and secreted in response to hypercalcemia. It acts to lower blood calcium and phosphate concentrations and may serve as an antagonist of parathyroid hormone.

**calmodulin** A receptor protein for $Ca^{++}$ located within the cytoplasm of target cells. It appears to mediate the effects of this ion on cellular activities.

**calorie** A unit of heat equal to the amount of heat needed to raise the temperature of one gram of water by 1° C.

**cAMP** Cyclic adenosine monophosphate; a second messenger in the action of many hormones, such as catecholamines, polypeptides, and glycoprotein hormones. It serves to mediate the effects of these hormones on their target cells.

**cancer** A tumor characterized by abnormally rapid cell division and the loss of specialized tissue characteristics. This term usually refers to malignant tumors.

**capacitation** Changes that occur within spermatozoa in the female reproductive tract that enable the sperm to fertilize ova; sperm that have not been capacitated in the female tract cannot fertilize ova.

**capillary** The smallest vessel in the vascular system. Capillary walls are only one cell thick, and all exchanges of molecules between the blood and tissue fluid occur across the capillary wall.

**carbohydrate** Organic molecules containing carbon, hydrogen, and oxygen in a ratio of 1:2:1. This class of molecules is subdivided into monosaccharides, disaccharides, and polysaccharides.

**carbonic anhydrase** An enzyme that catalyzes the formation or breakdown of carbonic acid. When carbon dioxide concentrations are relatively high, this enzyme catalyzes the formation of carbonic acid from $CO_2$ and $H_2O$. When carbon dioxide concentrations are low, the breakdown of carbonic acid to $CO_2$ and $H_2O$ is catalyzed. These reactions aid the transport of carbon dioxide from tissues to alveolar air.

**carboxyhemoglobin** An abnormal form of hemoglobin in which the heme is bonded to carbon monoxide.

**cardiac muscle** Muscle of the heart, consisting of striated muscle cells. These cells are interconnected into a mass called the myocardium.

**cardiac output** The volume of blood pumped by either the right or the left ventricle per minute.

**cardiogenic shock** Shock that results from low cardiac output in heart disease.

**carrier-mediated transport** The transport of molecules or ions across a cell membrane by means of specific protein carriers. It includes both facilitated diffusion and active transport.

**casts** An accumulation of proteins that produce molds of kidney tubules and that appear in urine sediment.

**catabolism** Chemical reactions in a cell whereby larger, more complex molecules are converted into smaller molecules.

**catalyst** A catalyst is a substance that increases the rate of a chemical reaction without changing the nature of the reaction or being changed by the reaction.

**catecholamines** A group of molecules, including epinephrine, norepinephrine, L-dopa, and related molecules, that have effects that are similar to those produced by activation of the sympathetic nervous system.

**cations** Positively charged ions, such as sodium, potassium, calcium, and magnesium.

**cell-mediated immunity** Immunological defense provided by T cell lymphocytes, which come into close proximity to their victim cells (as opposed to humoral immunity provided by the secretion of antibodies by plasma cells).

**cellular respiration** The energy-releasing metabolic pathways in a cell that oxidize organic molecules such as glucose, fatty acids, and others.

**centri-** (La.) Center.

**centrioles** Cell organelles that form the spindle apparatus during cell division.

**centromere** The central region of a chromosome to which the chromosomal arms are attached.

**cerebellum** A part of the metencephalon of the brain which serves as a major center of control in the extrapyramidal motor system.

**cerebral lateralization** The specialization of function of each cerebral hemisphere. Language ability, for example, is lateralized to the left hemisphere in most people.

**chemiosmotic theory** The theory that oxidative phosphorylation within mitochondria is driven by the development of a $H^+$ gradient across the inner mitochondrial membrane.

**chemoreceptors** Neural receptors sensitive to chemical changes in blood and other body fluids.

**chemotaxis** The movement of an organism or a cell, such as a leukocyte, toward a chemical stimulus.

**Cheyne-Stokes respiration** Breathing characterized by rhythmic waxing and waning of the depth of respiration, with regularly occurring periods of apnea (lack of breathing).

**chloride shift** The diffusion of $Cl^-$ into red blood cells as $HCO_3^-$ diffuses out of the cells. This occurs in tissue capillaries due to the production of carbonic acid from carbon dioxide.

**cholecystokinin** Abbreviated CCK, this hormone is secreted by the duodenum and acts to stimulate contraction of the gallbladder and to promote the secretion of pancreatic juice.
**cholesterol** A twenty-seven-carbon steroid that serves as the precursor of steroid hormones.
**cholinergic** Denoting nerve endings that liberate acetylcholine as a neurotransmitter, such as those of the parasympathetic system.
**chondrocytes** Cartilage-forming cells.
**chorea** The occurrence of a wide variety of rapid, complex, jerky movements that appear to be well coordinated but are performed involuntarily.
**chromatids** Duplicated chromosomes that are joined together at the centromere and separate during cell division.
**chromatin** Threadlike structures in the cell nucleus consisting primarily of DNA and protein. This represents the extended form of chromosomes during interphase.
**chromosomes** Structures in the cell nucleus, containing DNA and associated proteins, as well as RNA, that are made according to the genetic instructions in the DNA. The chromosomes are in a compact form during cell division and thus become visible as discrete structures in the light microscope during this time.
**chylomicrons** Particles of lipids and protein that are secreted by the intestinal epithelial cells into the lymph and are transported by the lymphatic system to the blood.
**chyme** A mixture of partially digested food and digestive juices within the stomach and small intestine.
**cilia** Plural of *cilium;* tiny hairlike processes that extend from the cell surface and beat in a coordinated fashion.
**circadian rhythms** Physiological changes that repeat at about a twenty-four-hour period, which are often synchronized to changes in the external environment, such as the day-night cycles.
**cirrhosis** Liver disease characterized by the loss of normal microscopic structure, which is replaced by fibrosis and nodular regeneration.
**clonal selection theory** The theory in immunology that active immunity is produced by the development of clones of lymphocytes able to respond to a particular antigen.
**clone** A group of cells derived from a single parent cell by mitotic cell division; since reproduction is asexual, the descendants of the parent cell are genetically identical. Term used when cells are separate individuals (as in white blood cells) rather than part of a growing organ.
**CNS** Central nervous system; that part of the nervous system consisting of the brain and spinal cord.
**cochlea** The organ of hearing in the inner ear where nerve impulses are generated in response to sound waves.
**codon** The sequence of three nucleotide bases in mRNA that specifies a given amino acid and determines the position of that amino acid in a polypeptide chain through complementary base pairing with an anticodon in transfer RNA.
**coenzyme** An organic molecule, usually derived from a water-soluble vitamin, that combines with and activates specific enzyme proteins.
**cofactor** A substance needed for the catalytic action of an enzyme; it usually refers to inorganic ions such as $Ca^{++}$ and $Mg^{++}$.
**colloid osmotic pressure** Osmotic pressure exerted by plasma proteins, which are present as a colloidal suspension. Also called oncotic pressure.
**com-, con-** (La.) With, together.
**compliance** A measure of the ease with which a structure, such as the lungs, expands under pressure; a measure of the change in volume as a function of pressure changes.
**conducting zone** The structures and airways that transmit inspired air into the respiratory zone of the lungs, where gas exchange occurs. The conducting zone includes such structures as the trachea, bronchi, and larger bronchioles.
**cones** Photoreceptors in the retina of the eye that provide color vision and high visual acuity.
**congestive heart failure** The inability of the heart to deliver an adequate blood flow, due to heart disease or hypertension. It is associated with breathlessness, salt and water retention, and edema.
**conjunctivitis** Inflammation of the conjunctiva of the eye, which is sometimes called "pink eye."
**connective tissue** One of the four primary tissues, characterized by an abundance of extracellular material.
**Conn's syndrome** Primary hyperaldosteronism; excessive secretion of aldosterone produces electrolyte imbalances.
**contralateral** Affecting the opposite side of the body.
**cornea** The transparent structure forming the anterior part of the connective tissue covering of the eye.
**corpora quadrigemina** A region of the mesencephalon consisting of the superior and inferior colliculi. The superior colliculi are centers for the control of visual reflexes, whereas the inferior colliculi are centers for the control of auditory reflexes.
**corpus callosum** A large, transverse tract of nerve fibers connecting the cerebral hemispheres.
**cortex** The outer covering or layer of an organ.
**corticosteroids** Steroid hormones of the adrenal cortex, consisting of glucocorticoids (such as hydrocortisone) and mineralocorticoids (such as aldosterone).
**co-transport** Also called coupled transport, this is a form of active transport in which a single carrier transports an ion (e.g., $Na^{+}$) down its concentration gradient while transporting a specific molecule (e.g., glucose) against its concentration gradient. The hydrolysis of ATP is indirectly required for co-transport because it is needed to maintain the steep concentration gradient of the ion.
**countercurrent exchange** The process that occurs in the vasa recta of the renal medulla in which blood flows in U-shaped loops. This allows sodium chloride to be trapped in the interstitial fluid while water is carried away from the kidneys.
**countercurrent multiplier system** The interaction that occurs between the descending limb and the ascending limb of the loop of Henle in the kidney. This interaction results in the multiplication of the solute concentration in the interstitial fluid of the renal medulla.
**creatine phosphate** An organic phosphate molecule in muscle cells that serves as a source of high-energy phosphate for the synthesis of ATP. Also called phosphocreatine.
**crenation** A notched or scalloped appearance of the red blood cell membrane caused by the osmotic loss of water from these cells.
**cretinism** A condition caused by insufficient thyroid secretion during prenatal development or the years of early childhood. It results in stunted growth and inadequate mental development.
**crypt-** (Gr.) Hidden, concealed.
**cryptorchidism** A developmental defect in which the testes fail to descend into the scrotum and, instead, remain in the body cavity.
**curare** A chemical derived from plant sources that causes flaccid paralysis by blocking ACh receptor proteins in muscle cell membranes.
**Cushing's syndrome** Symptoms caused by hypersecretion of adrenal steroid hormones, due to tumors of the adrenal cortex or to ACTH-secreting tumors of the anterior pituitary.
**cyanosis** A blue color given to the skin or mucous membranes by deoxyhemoglobin; it indicates inadequate oxygen concentration in the blood.
**cyto-** (Gr.) Cell.
**cytochrome** A pigment in mitochondria that transports electrons in the process of aerobic respiration.
**cytokinesis** A division of the cytoplasm that occurs in mitosis and meiosis when a parent cell divides to produce two daughter cells.
**cytoplasm** The semifluid part of the cell between the cell membrane and the nucleus, exclusive of membrane-bound organelles. It contains many enzymes and structural proteins.
**cytoskeleton** A latticework of structural proteins in the cytoplasm arranged in the form of microfilaments and microtubules.

## D

**Dalton's law** The physical law that describes the fact that the total pressure of a gas mixture is equal to the sum that each individual gas in the mixture would exert independently. The part contributed by each gas is known as the partial pressure of the gas.

**dark adaptation** The ability of the eyes to increase their sensitivity to low light levels over a period of time. Part of this adaptation involves increased amounts of visual pigment in the photoreceptors.

**dark current** The steady inward diffusion of $Na^+$ into the rods and cones when the photoreceptors are in the dark. Stimulation by light causes this dark current to be blocked and thus hyperpolarizes the photoreceptors.

**delayed hypersensitivity** An allergic response in which the onset of symptoms takes as long as two to three days after exposure to an antigen. Produced by T cells, it is a type of cell-mediated immunity.

**dendrite** A relatively short, highly branched neural process that carries electrical activity to the cell body.

**denervation hypersensitivity** The increased sensitivity of smooth muscles to neural stimulation after their innervation has been blocked or removed for a period of time.

**dentin** One of the hard tissues of the teeth; it covers the pulp cavity and is itself covered on its exposed surface by enamel and on its root surface by cementum.

**deoxyhemoglobin** The form of hemoglobin in which the heme groups are in the normal reduced form but are not bonded to a gas. Deoxyhemoglobin is produced when oxyhemoglobin releases oxygen.

**depolarization** The loss of membrane polarity in which the inside of the cell membrane becomes less negative in comparison to the outside of the membrane. The term is also used to indicate the reversal of membrane polarity that occurs during the production of action potentials in nerve and muscle cells.

**deposition, bone** The formation of the extracellular matrix of bone by osteoblasts. This process includes secretion of collagen and precipitation of calcium phosphate in the form of hydroxyapatite crystals.

**diabetes insipidus** A condition in which inadequate amounts of antidiuretic hormone (ADH) are secreted by the posterior pituitary. It results in inadequate reabsorption of water by the kidney tubules and, thus, in the excretion of a large volume of dilute urine.

**diabetes mellitus** The appearance of glucose in the urine due to the presence of high plasma glucose concentrations, even in the fasting state. This disease is caused by either a lack of sufficient insulin secretion or by inadequate responsiveness of the target tissues to the effects of insulin.

**diapedesis** The migration of white blood cells through the endothelial walls of blood capillaries into the surrounding connective tissues.

**diarrhea** Abnormal frequency of defecation accompanied by abnormal liquidity of the feces.

**diastole** The phase of relaxation in which the heart fills with blood. Unless accompanied by the modifier term *atrial,* diastole usually refers to the resting phase of the ventricles.

**diastolic blood pressure** The minimum pressure in the arteries that is produced during the phase of diastole of the heart. It is indicated by the last sound of Korotkoff when taking a blood pressure measurement.

**diffusion** The net movement of molecules or ions from regions of higher to regions of lower concentration.

**digestion** The process of converting food into molecules that can be absorbed through the intestine into the blood.

**1,25-dihydroxyvitamin $D_3$** The active form of vitamin D produced within the body by hydroxylation reactions in the liver and kidneys of vitamin D formed by the skin. This is a hormone which promotes the intestinal absorption of $Ca^{++}$.

**diploid** Denoting cells having two of each chromosome or twice the number of chromosomes that are present in sperm or ova.

**disaccharide** A class of double sugars; carbohydrates that yield two simple sugars, or monosaccharides, upon hydrolysis.

**diuretic** A substance that increases the rate of urine production, and thus decreases the blood volume.

**DNA** Deoxyribonucleic acid; composed of nucleotide bases and deoxyribose sugar, it contains the genetic code.

**dopa** A derivative of the amino acid tyrosine, L-dopa serves as the precursor for the neurotransmitter molecule dopamine. L-dopa is given to patients with Parkinson's disease to stimulate dopamine production.

**dopamine** Serves as a neurotransmitter in the central nervous system; also is the precursor of norepinephrine, another neurotransmitter molecule.

**2,3-DPG** Abbreviation for 2,3-diphosphoglyceric acid. A product of red blood cells, 2,3-DPG bonds with the protein component of hemoglobin and increases the ability of oxyhemoglobin to dissociate and release its oxygen.

**ductus arteriosus** A fetal blood vessel connecting the pulmonary artery directly to the aorta.

**dwarfism** A condition in which a person is undersized due to inadequate secretion of growth hormone.

**dyspnea** Subjective difficulty in breathing.

## E

**ECG** Electrocardiogram (also abbreviated EKG); a recording of electrical currents produced by the heart.

***E. coli*** A species of bacteria normally found in the human intestine; full name is *Escherichia coli.*

**ecto-** (Gr.) Outside, outer.

**-ectomy** (Gr.) Surgical removal of a structure.

**ectopic** Foreign, out of place.

**ectopic focus** An area of the heart other than the SA node that assumes pacemaker activity.

**ectopic pregnancy** Embryonic development that occurs anywhere other than in the uterus (as in the fallopian tubes or body cavity).

**edema** Swelling due to an increase in tissue fluid.

**EEG** Electroencephalogram; a recording of the electrical activity of the brain from electrodes placed on the scalp.

**effector organs** A collective term for muscles and glands that are activated by motor neurons.

**efferent** The transport of something away from a central location. Efferent nerve fibers conduct impulses away from the central nervous system, for example, and efferent arterioles transport blood away from the glomerulus.

**elasticity** The tendency of a structure to recoil to its initial dimensions after being distended (stretched).

**electrolytes** Ions and molecules that are able to ionize and thus carry an electric current. The most common electrolytes in the plasma are $Na^+$, $HCO_3^-$, and $K^+$.

**electrophoresis** A biochemical technique in which different molecules can be separated and identified by their rate of movement in an electric field.

**element, chemical** A substance that cannot be broken down by chemical means into simpler compounds. An element is composed of atoms that all have the same atomic number. An element can, however, contain different isotopes of that atom which have different numbers of neutrons and which thus have different atomic weights.

**elephantiasis** A disease caused by infection with a nematode worm, in which the larvae block lymphatic drainage and produce edema; the lower areas of the body can become enormously swollen as a result.

**EMG** Electromyogram; an electrical recording of the activity of skeletal muscles through the use of surface electrodes.

**emmetropia** A condition of normal vision in which the image of objects is focused on the retina, as opposed to nearsightedness (myopia) or farsightedness (hypermetropia).

**emphysema** A lung disease in which alveoli are destroyed and the remaining alveoli become larger. It results in decreased vital capacity and increased airways resistance.

**emulsification** The process of producing an emulsion or fine suspension; in the intestine, fat globules are emulsified by the detergent action of bile.

**end-diastolic volume** The volume of blood in each ventricle at the end of diastole, immediately before the ventricles contract at systole.

**endergonic** A chemical reaction that requires the input of energy from an external source in order to proceed.

**endo-** (Gr.) Within, inner.
**endocrine glands** Glands that secrete hormones into the circulation rather than into a duct; also called ductless glands.
**endocytosis** The cellular uptake of particles that are too large to cross the cell membrane; this occurs by invagination of the cell membrane until a membrane-enclosed vesicle is pinched off within the cytoplasm.
**endoderm** The innermost of the three primary germ layers of an embryo; it gives rise to the digestive tract and associated structures, respiratory tract, bladder, and urethra.
**endogenous** A product or process arising from within the body; as opposed to exogenous products or influences, which arise from external sources.
**endolymph** The fluid contained within the membranous labyrinth of the inner ear.
**endometrium** The mucous membrane of the uterus, the thickness and structure of which vary with the phase of the menstrual cycle.
**endoplasmic reticulum** An extensive system of membrane-enclosed cavities within the cytoplasm of the cell; those with ribosomes on their surface are called rough endoplasmic reticulum and participate in protein synthesis.
**endorphins** A group of endogenous opioid molecules that may act as a natural analgesic.
**endothelium** The simple squamous epithelium that lines blood vessels and the heart.
**endotoxin** A toxin contained within certain types of bacteria that is able to stimulate the release of endogenous pyrogen and produce a fever.
**end-plate potential** The graded depolarization produced by ACh at the neuromuscular junction. This is equivalent to the excitatory postsynaptic potential produced at neuron-neuron synapses.
**end-product inhibition** The inhibition of enzymatic steps of a metabolic pathway by products formed at the end of that pathway.
**enkephalins** Short polypeptides, containing five amino acids, that have analgesic effects; they may function as neurotransmitters in the brain. The two known enkephalins (which differ in only one amino acid) are endorphins.
**enteric** A term referring to the intestine.
**enterohepatic circulation** The recirculation of a compound secreted by the liver in the bile into the small intestine which is reabsorbed and returned to the liver via the hepatic portal vein.
**entropy** The energy of a system that is not available to perform work; a measure of the degree of disorder in a system, entropy increases whenever energy is transformed.
**enzyme** A protein catalyst that increases the rate of specific chemical reactions.
**epi-** (Gr.) Upon, over, outer.
**epidermis** The stratified squamous epithelium of the skin, the outer layer of which is dead and filled with keratin.
**epididymis** A tubelike structure outside the testes; sperm pass from the seminiferous tubules into the head of the epididymis and then pass from the tail of the epididymis to the vas deferens. The sperm mature, becoming motile, as they pass through the epididymis.
**epinephrine** Also known as adrenalin; a catecholamine hormone secreted by the adrenal medulla in response to sympathetic nerve stimulation; it acts together with norepinephrine released from sympathetic nerve endings to prepare the organism for "fight or flight."
**epithelium** One of the four primary tissues, forming membranes that cover and line the body surfaces and forming exocrine and endocrine glands.
**EPSP** Excitatory postsynaptic potential; a graded depolarization of a postsynaptic membrane in response to stimulation by a neurotransmitter chemical. EPSPs can be summated, but can only be transmitted short distances; EPSPs can stimulate the production of action potentials when a threshold level of depolarization is attained.
**equilibrium potential** The hypothetical membrane potential that would be created if only one ion were able to diffuse across a membrane and reach a stable, or equilibrium, state. In this stable state, the concentrations of the ion would remain constant inside and outside the membrane, and the membrane potential would be equal to a particular value. In a real cell, the equilibrium potential indicates the direction the membrane potential will approach as the permeability of the cell membrane to a particular ion is increased.
**erythroblastosis fetalis** Hemolytic anemia in a newborn Rh-positive baby caused by maternal antibodies against the Rh factor that have crossed the placenta.
**erythrocytes** Red blood cells; the formed elements of blood that contain hemoglobin and transport oxygen.
**erythropoietin** A hormone secreted by the kidneys which stimulates the bone marrow to produce red blood cells.
**essential amino acids** Those eight amino acids in adults or nine amino acids in children that cannot be made by the human body and therefore must be obtained in the diet.
**estradiol** The major estrogen (female sex steroid hormone) secreted by the ovaries.
**estrous cycle** Cyclic changes in the structure and function of the ovaries and female reproductive tract, accompanied by periods of "heat" (estrus), or sexual receptivity; the lower mammalian equivalent of the menstrual cycle, but differing from the menstrual cycle in that the endometrium is not shed with accompanying bleeding.
**ex-** (La.) Out, off, from.
**excitation-contraction coupling** The means by which electrical excitation of a muscle results in muscle contraction. This coupling is achieved by $Ca^{++}$, which enters the muscle cell cytoplasm in response to electrical excitation and which stimulates the events culminating in contraction.
**exergonic** Denoting chemical reactions that liberate energy.
**exo-** (Gr.) Outside or outward.
**exocrine gland** A gland that discharges its secretion through a duct to the outside of an epithelial membrane.
**exocytosis** The process of cellular secretion in which the secretory products are contained within a membrane-enclosed vesicle; the vesicle fuses with the cell membrane so that the lumen of the vesicle is open to the extracellular environment.
**exons** Exons are the regions of bases in DNA that code for the production of messenger RNA.
**extensor** A muscle that upon contraction increases the angle of a joint.
**exteroceptors** Sensory receptors that are sensitive to changes in the external environment (as opposed to interoceptors).
**extra-** (La.) Outside, beyond.
**extrafusal fibers** The ordinary muscle fibers within a skeletal muscle; not found within the muscle spindles.
**extraocular muscles** The muscles that insert into the sclera of the eye and act to change the position of the eye in its orbit (as opposed to the intraocular muscles such as those of the iris and ciliary body within the eye).
**extrapyramidal tracts** The major extrapyramidal tract is the reticulospinal tract, which originates in the reticular formation of the brain stem and receives excitatory and inhibitory input from both the cerebrum and the cerebellum. The extrapyramidal tracts are thus influenced by the activity in the brain that involves many synapses and appear to be required for fine control of voluntary movements.

## F

**facilitated diffusion** The carrier-mediated transport of molecules through the cell membrane along the direction of their concentration gradients; it does not require the expenditure of metabolic energy.
**FAD** Flavin adenine dinucleotide; a coenzyme derived from riboflavin that participates in electron transport within the mitochondria.
**feces** The excrement discharged from the large intestine.
**fertilization** Fusion of an ovum and sperm.
**fiber, muscle** A skeletal muscle cell.
**fiber, nerve** An axon of a motor neuron or the dendrite of a pseudounipolar sensory neuron in the PNS.

**fibrillation** A condition of cardiac muscle characterized electrically by random and continuously changing patterns of electrical activity and resulting in the inability of the myocardium to contract as a unit and pump blood. It can be fatal if it occurs in the ventricles.

**fibrin** The insoluble protein formed from fibrinogen by the enzymatic action of thrombin during the process of blood clot formation.

**fibrinogen** Also called factor I; a soluble plasma protein that serves as the precursor of fibrin.

**flaccid paralysis** The inability to contract muscles, resulting in a loss of muscle tone. This may be due to damage to lower motor neurons or to factors that block neuromuscular transmission.

**flagellum** A whiplike structure that provides motility for sperm.

**flare-and-wheal reaction** A cutaneous reaction to skin injury or the administration of antigens, produced by release of histamine and related molecules and characterized by local edema and a red flare.

**flavoprotein** A conjugated protein that contains a flavin pigment and is involved in electron transport within the mitochondria.

**flexor** A muscle that decreases the angle of a joint when it contracts.

**follicle** A microscopic, hollow structure within an organ. Follicles are the functional units of the thyroid gland and of the ovary.

**foramen ovale** An opening that is normally present in the atrial septum of a fetal heart and allows direct communication between the right and left atria.

**fovea centralis** A tiny pit in the macula lutea of the retina that contains slim, elongated cones and provides the highest visual acuity (clearest vision).

**Frank-Starling Law of the Heart** The Frank-Starling law describes the relationship between end-diastolic volume and stroke volume of the heart. A greater amount of blood in a ventricle prior to contraction results in greater stretch of the myocardium, and by this means produces a contraction of greater strength.

**FSH** Follicle-stimulating hormone; one of the two gonadotropic hormones secreted from the anterior pituitary. In females FSH stimulates the development of the ovarian follicles, whereas in males it stimulates the production of sperm in the seminiferous tubules.

## G

**GABA** Gamma-aminobutyric acid; it is believed to function as an inhibitory neurotransmitter in the central nervous system.

**gametes** A collective term for haploid germ cells: sperm and ova.

**gamma motoneuron** The type of somatic motor neuron which stimulates contraction of the intrafusal fibers within the muscle spindles.

**ganglion** A grouping of nerve cell bodies located outside the brain and spinal cord.

**gap junctions** Specialized regions of fusion between the cell membranes of two adjacent cells, which permit the diffusion of ions and small molecules from one cell to the next. These regions serve as electrical synapses in certain areas, such as in cardiac muscle.

**gas exchange** The diffusion of oxygen and carbon dioxide down their concentration gradients that occurs between pulmonary capillaries and alveoli, and between systemic capillaries and the surrounding tissue cells.

**gastric intrinsic factor** A glycoprotein secreted by the stomach and needed for the absorption of vitamin $B_{12}$.

**gastric juice** The secretions of the gastric mucosa. Gastric juice contains water, hydrochloric acid, and pepsinogen as the major components.

**gastrin** A hormone secreted by the stomach that stimulates the gastric secretion of hydrochloric acid and pepsin.

**gates** A term used to describe structures within the cell membrane that regulate the passage of ions through membrane channels. Such gates may be chemically regulated (by neurotransmitters) or voltage regulated (in which case they open in response to a threshold level of depolarization).

**gen-** (Gr.) Producing.

**generator potential** The graded depolarization produced by stimulation of a sensory receptor. The generator potential (also called the receptor potential) results in the production of action potentials by a sensory neuron.

**genetic recombination** The formation of new combinations of genes, as by crossing-over between homologous chromosomes.

**genetic transcription** The process by which RNA is produced with a sequence of nucleotide bases that is complementary to a region of DNA.

**genetic translation** The process by which proteins are produced with amino acid sequences specified by the sequence of codons in messenger RNA.

**gigantism** Abnormal body growth due to the excessive secretion of growth hormone.

**glomerular filtration rate** Abbreviated GFR, this is the volume of blood plasma filtered out of the glomeruli of both kidneys per minute. The GFR is measured by the renal plasma clearance of inulin.

**glomerular ultrafiltrate** Fluid filtered through the glomerular capillaries into Bowman's capsule of the kidney tubules.

**glomeruli** A general term for a tuft or cluster; it is most often used to describe the tufts of capillaries in the kidneys that filter fluid into the kidney tubules.

**glomerulonephritis** Inflammation of the renal glomeruli, associated with fluid retention, edema, hypertension, and the appearance of protein in the urine.

**glucagon** A polypeptide hormone that is secreted by the alpha cells of the islets of Langerhans in the pancreas and acts to promote glycogenolysis and raise the blood glucose levels.

**glucocorticoids** Steroid hormones secreted by the adrenal cortex (corticosteroids) that affect the metabolism of glucose, protein, and fat. These hormones also have anti-inflammatory and immunosuppressive effects; the major glucocorticoid in humans is hydrocortisone (cortisol).

**gluconeogenesis** The formation of glucose from noncarbohydrate molecules such as amino acids and lactic acid.

**glycogen** A polysaccharide of glucose—also called *animal starch*— produced primarily in the liver and skeletal muscles. Similar to plant starch in composition, glycogen contains more highly branched chains of glucose subunits than does plant starch.

**glycogenesis** The formation of glycogen from glucose.

**glycogenolysis** The hydrolysis of glycogen to glucose-1-phosphate, which can be converted to glucose-6-phosphate, which then may be oxidized via glycolysis or (in the liver) converted to free glucose.

**glycolysis** The metabolic pathway that converts glucose to pyruvic acid and yields a net production of two ATP molecules and two molecules of reduced NAD. In anaerobic respiration the reduced NAD is oxidized by the conversion of pyruvic acid to lactic acid. In aerobic respiration pyruvic acid enters the Krebs cycle in mitochondria, and reduced NAD is ultimately oxidized by oxygen to yield water.

**glycosuria** The excretion of an abnormal amount of glucose in the urine (urine normally only contains trace amounts of glucose).

**Golgi apparatus** Stacks of flattened membranous sacs within the cytoplasm of cells, which are believed to bud off vesicles containing secretory proteins.

**Golgi tendon organ** A tension receptor in the tendons of muscles that becomes activated by the pull exerted by a muscle on its tendons.

**gonadotropic hormones** Hormones of the anterior pituitary gland that stimulate gonadal function—the formation of gametes and secretion of sex steroids. The two gonadotropins are FSH (follicle-stimulating hormone) and LH (luteinizing hormone), which are essentially the same in males and females.

**gonads** A collective term for testes and ovaries.

**graafian follicle** A mature ovarian follicle, containing a single fluid-filled cavity, with the ovum located toward one side of the follicle, and perched on top of a hill of granulosa cells.

**granular leukocytes** Leukocytes with granules in the cytoplasm; on the basis of the staining properties of the granules, these cells are of three types: neutrophils, eosinophils, and basophils.

**Graves' disease** A hyperthyroid condition believed to be caused by excessive stimulation of the thyroid gland by autoantibodies; it is associated with exophthalmos (bulging eyes), high pulse rate, high metabolic rate, and other symptoms of hyperthyroidism.

**gray matter** The part of the central nervous system that contains neuron cell bodies and dendrites, but few myelinated axons. This forms the cortex of the cerebrum, cerebral nuclei, and the central region of the spinal cord.

**growth hormone** A hormone secreted by the anterior pituitary that stimulates growth of the skeleton and soft tissues during the growing years and that influences the metabolism of protein, carbohydrate, and fat throughout life.

**gyrus** A fold or convolution in the cerebrum.

## H

**haploid** A cell that has one of each chromosome type and therefore half the number of chromosomes present in most other body cells; only the gametes (sperm and ova) are haploid.

**haptens** Small molecules that are not antigenic by themselves, but which—when combined with proteins—become antigenic and thus able to stimulate the production of specific antibodies.

**haversian system** A haversian canal and its concentrically arranged layers, or lamellae, of bone; it constitutes the basic structural unit of compact bone.

**hay fever** A seasonal type of allergic rhinitis caused by pollen; it is characterized by itching and tearing of the eyes, swelling of the nasal mucosa, attacks of sneezing, and often by asthma.

**hCG** Abbreviation for human chorionic gonadotropin. This is a hormone secreted by the embryo which has LH-like actions, and is required for maintenance of the mother's corpus luteum for the first ten weeks of pregnancy.

**heart murmur** Abnormal heart sounds caused by an abnormal flow of blood in the heart due to structural defects, usually of the valves or septum.

**heart sounds** The sounds produced by closing of the AV valves of the heart during systole (the first sound) and by closing of the semilunar valves of the aorta and pulmonary trunk during diastole (the second sound).

**helper T cells** A subpopulation of T cells (lymphocytes) that help stimulate antibody production of B lymphocytes by antigens.

**hematocrit** The ratio of packed red blood cells to total blood volume in a centrifuged sample of blood, expressed as a percentage.

**heme** The iron-containing red pigment that, together with the protein globin, forms hemoglobin.

**hemoglobin** The combination of heme pigment and protein within red blood cells that acts to transport oxygen and (to a lesser degree) carbon dioxide. Hemoglobin also serves as a weak buffer within red blood cells.

**Henderson-Hasselbalch equation** This equation determines the blood pH that is produced by a given ratio of bicarbonate to carbon dioxide concentrations.

**Henry's law** The physical law that states that the concentration of gas dissolved in a fluid is directly proportional to the partial pressure of that gas.

**heparin** A mucopolysaccharide present in many tissues, but most abundant in the lungs and liver, that is used medically as an anticoagulant.

**hepatic** Pertaining to the liver.

**hepatitis** Inflammation of the liver.

**Hering-Breuer reflex** A reflex in which distension of the lungs stimulates stretch receptors, which in turn act to inhibit further distension of the lungs.

**hermaphrodite** An organism with both testes and ovaries.

**hetero-** (Gr.) Different, other.

**heterochromatin** A condensed, inactive form of chromatin.

**hiatal hernia** A protrusion of an abdominal structure through the esophageal hiatus of the diaphragm into the thoracic cavity.

**high-density lipoproteins** Combinations of lipids and proteins that migrate rapidly to the bottom of a test tube during centrifugation; carrier proteins that are believed to transport cholesterol from blood vessels to the liver, and thus to offer some protection from atherosclerosis.

**higher motor neurons** Neurons in the brain which, as part of the pyramidal or extrapyramidal system, influence the activity of the lower motor neurons in the spinal cord.

**histamine** A compound secreted by tissue mast cells and other connective tissue cells that stimulates vasodilation and increases capillary permeability; it is responsible for many of the symptoms of inflammation and allergy.

**histocompatibility antigens** A group of cell-surface antigens found on all cells of the body except mature red blood cells. These are important for the function of T lymphocytes, and differences between these antigens are responsible for the rejection of tissue transplants.

**histone** A basic protein associated with DNA that is believed to repress genetic expression.

**homeo** (Gr.) Same.

**homeostasis** The dynamic constancy of the internal environment, which serves as the principal function of physiological regulatory mechanisms. The concept of homeostasis provides a framework for the understanding of most physiological processes.

**homologous chromosomes** The matching pairs of chromosomes in a diploid cell.

**hormones** Regulatory chemicals secreted into the blood by endocrine cells and carried by the blood to target cells that respond to the hormones by an alteration in their metabolism.

**humoral immunity** The form of acquired immunity in which antibody molecules are secreted in response to antigenic stimulation (as opposed to cell-mediated immunity).

**hyaline membrane disease** A disease of some premature infants who lack pulmonary surfactant, it is characterized by collapse of the alveoli (atelectasis) and pulmonary edema. Also called respiratory distress syndrome.

**hydrocortisone** Also called cortisol; the principal corticosteroid hormone secreted by the adrenal cortex with glucocorticoid action.

**hydrophilic** A substance that readily absorbs water; literally, "water loving."

**hydrophobic** A substance that repels, and that is repelled by, water; literally, "water fearing."

**hyper-** (Gr.) Over, above, excessive.

**hyperbaric oxygen** Oxygen gas present at greater than atmospheric pressure.

**hypercapnia** Excessive concentration of carbon dioxide in the blood.

**hyperglycemia** Abnormally increased concentration of glucose in the blood.

**hyperkalemia** Abnormally high concentration of potassium in the blood.

**hyperopia** Also called farsightedness; a refractive disorder in which rays of light are brought to a focus behind the retina as a result of the eyeball being too short.

**hyperplasia** An increase in organ size due to an increase in cell numbers as a result of mitotic cell division.

**hyperpnea** Increased total minute volume during exercise. Unlike hyperventilation, the arterial blood carbon dioxide values are not changed during hyperpnea, because the increased ventilation is matched to an increased metabolic rate.

**hyperpolarization** An increase in the negativity of the inside of a cell membrane with respect to the resting membrane potential.

**hypersensitivity** Another name for allergy; an abnormal immune response that may be immediate (due to antibodies of the IgE class) or delayed (due to cell-mediated immunity).

**hypertension** High blood pressure. Divided into primary, or essential, hypertension of unknown cause and secondary hypertension that develops as a result of other, known disease processes.

**hypertonic** A solution with a greater solute concentration and thus a greater osmotic pressure than plasma.

**hypertrophy** Growth of an organ due to an increase in the size of its cells.

**hyperventilation** A high rate and depth of breathing that results in a decrease in the blood carbon dioxide concentration below normal.

**hypo-** (Gr.) Under, below, less.

**hypodermis** A layer of fat beneath the dermis of the skin.

**hypotension** Abnormally low blood pressure.

**hypothalamic hormones** Hormones produced by the hypothalamus; these include antidiuretic hormone and oxytocin, which are secreted by the posterior pituitary gland, and both releasing and inhibiting hormones that regulate the secretion of the anterior pituitary.

**hypothalamo-hypophyseal portal system** A vascular system that transports releasing and inhibiting hormones from the hypothalamus to the anterior pituitary.

**hypothalamo-hypophyseal tract** The tract of nerve fibers (axons) that transports antidiuretic hormone and oxytocin from the hypothalamus to the posterior pituitary gland.

**hypothalamus** An area of the brain below the thalamus and above the pituitary gland. The hypothalamus regulates the pituitary gland and contributes to the regulation of the autonomic nervous system, among many other functions.

**hypovolemic shock** A rapid fall in blood pressure due to diminished blood volume.

**hypoxemia** A low oxygen concentration of the arterial blood.

## I

**immediate hypersensitivity** Hypersensitivity (allergy) that is mediated by antibodies of the IgE class and that results in the release of histamine and related compounds from tissue cells.

**immunization** The process of increasing one's resistance to pathogens. In active immunity a person is injected with antigens that stimulate the development of clones of specific B or T lymphocytes; in passive immunity a person is injected with antibodies made by another organism.

**immunoassay** The detection or measurement of a molecule that acts as an antigen by its reaction with specific antibodies. The antigen-antibody reaction may be followed in a variety of ways, such as agglutination (if the antigen is attached to visible cells or particles), fluorescence, or radioactivity (a radioimmunoassay, or RIA).

**immunoglobulins** Subclasses of the gamma globulin fraction of plasma proteins that have antibody functions, providing humoral immunity.

**immunosurveillance** The function of the immune system to recognize and attack malignant cells that produce antigens not recognized as "self." This function is believed to be cell mediated rather than humoral.

**implantation** The process by which a blastocyst attaches itself to and penetrates into the endometrium of the uterus.

**inhibin** Believed to be a water-soluble hormone secreted by the seminiferous tubules of the testes that specifically exerts negative feedback inhibition of FSH secretion from the anterior pituitary gland.

**inositol triphosphate** A second messenger in hormone action which is produced by the cell membrane of a target cell in response to the action of a hormone. This compound is also believed to stimulate the release of $Ca^{++}$ from the sarcoplasmic reticulum in response to action potentials in a skeletal muscle fiber.

**insulin** A polypeptide hormone, secreted by the beta cells of the islets of Langerhans in the pancreas, that promotes the anabolism of carbohydrates, fat, and protein. Insulin acts to promote the cellular uptake of blood glucose and, therefore, to lower the blood glucose concentration; insulin deficiency produces hyperglycemia and diabetes mellitus.

**inter-** (La.) Between, among.

**interferons** A group of small proteins that inhibit the multiplication of viruses inside host cells and also have antitumor properties.

**interleukin-2** A lymphokine secreted by T lymphocytes that stimulates the proliferation of both B and T lymphocytes.

**interneurons** Also called association neurons; those neurons within the central nervous system that do not extend into the peripheral nervous system; they are interposed between sensory (afferent) and motor (efferent) neurons.

**interoceptors** Sensory receptors that respond to changes in the internal environment (as opposed to exteroceptors).

**interphase** The interval between successive cell divisions; during this time the chromosomes are in an extended state and are active in directing RNA synthesis.

**intra** (La.) Within, inside.

**intrafusal fibers** Modified muscle fibers that are encapsulated to form muscle spindle organs, which are muscle stretch receptors.

**intrapleural space** An actual or potential space between the visceral pleural membrane covering the lungs and the somatic pleural membrane lining the thoracic wall.

**intrapulmonary space** The space within the air sacs and airways of the lungs.

**introns** Noncoding regions of DNA bases that interrupt the coding regions (exons) for mRNA.

**inulin** A polysaccharide of fructose, produced by certain plants, that is filtered by the human kidneys, but neither reabsorbed nor secreted. The clearance rate of injected inulin is thus used to measure the glomerular filtration rate.

**in vitro** Occurring outside the body, in a test tube or other artificial environment.

**in vivo** Occurring within the body.

**ion** An atom or a group of atoms that has either lost or gained electrons and thus has a net positive or a net negative charge.

**ionization** The dissociation of a solute to form ions.

**ipsilateral** On the same side (as opposed to contralateral).

**IPSP** Inhibitory postsynaptic potential; hyperpolarization of the postsynaptic membrane in response to a particular neurotransmitter chemical, which makes it more difficult for the postsynaptic cell to attain a threshold level of depolarization required to produce action potentials; responsible for postsynaptic inhibition.

**ischemia** A rate of blood flow to an organ that is inadequate to supply sufficient oxygen and maintain aerobic respiration in that organ.

**islets of Langerhans** Encapsulated groupings of endocrine cells within the exocrine tissue of the pancreas. The islets contain alpha cells that secrete glucagon and beta cells that secrete insulin.

**iso-** (Gr.) Equal, same.

**isoenzymes** Enzymes, usually produced by different organs, that catalyze the same reaction but that differ from each other in amino acid composition.

**isometric contraction** Muscle contraction in which there is no appreciable shortening of the muscle.

**isotonic contraction** Muscle contraction in which the muscle shortens in length and maintains approximately the same amount of tension throughout the shortening process.

**isotonic solution** A solution having the same total solute concentration, osmolality, and osmotic pressure as the solution with which it is compared; a solution with the same solute concentration and osmotic pressure as plasma.

## J

**jaundice** A condition characterized by high blood bilirubin levels and staining of the tissues with bilirubin, which gives skin and mucous membranes a yellow color.

**junctional complexes** The structures that join adjacent epithelial cells together, including the zonula occludens, zonula adherens, and macula adherens (desmosome).

**juxta-** (La.) Near to, next to.

**juxtaglomerular apparatus** The region of the nephron tubule and afferent arteriole that are in contact with each other. Cells in the afferent arteriole of the juxtaglomerular apparatus secrete the enzyme renin into the blood, which activates the renin-angiotensin system.

## K

**keratin** A protein that forms the principal component of the outer layer of the epidermis and of hair and nails.

**ketoacidosis** A type of metabolic acidosis produced by the excessive production of ketone bodies, as in diabetes mellitus.

**ketogenesis** The production of ketone bodies.

**ketone bodies** These include acetone, acetoacetic acid, and $\beta$-hydroxybutyric acid, which are derived from fatty acids via acetyl coenzyme A in the liver. Ketone bodies are oxidized by skeletal muscles for energy.

**ketosis** An abnormal elevation in the blood concentration of ketone bodies that does not necessarily produce acidosis.

**kilocalorie** A unit of measurement equal to 1,000 calories, which are units of heat (a kilocalorie is the amount of heat required to raise the temperature of 1 kilogram of water by 1° C). In nutrition the kilocalorie is called a big calorie (Calorie).

**Klinefelter's syndrome** The syndrome produced in a male by the presence of an extra X chromosome (genotype XXY).

**Krebs cycle** A cyclic metabolic pathway in the matrix of mitochondria by which the acetic acid part of acetyl CoA is oxidized and substrates are provided for reactions that are coupled to the formation of ATP.

**Kupffer cells** Phagocytic cells that line the sinusoids of the liver and are part of the reticuloendothelial system.

## L

**lactose** Milk sugar; a disaccharide of glucose and galactose.

**lactose intolerance** The inability of many adults to digest lactose due to loss of the ability of the intestine to produce lactase enzyme.

**LaPlace, Law of** The law of LaPlace describes the fact that the pressure within an alveolus is directly proportional to its surface tension and inversely proportional to its radius.

**larynx** A structure, consisting of epithelial tissue, muscle, and cartilage, that serves as a sphincter guarding the entrance of the trachea and as the organ responsible for voice production.

**lateral inhibition** The sharpening of perception that occurs within the neural processing of sensory input. Input from those receptors that are most greatly stimulated is enhanced while input from other receptors is reduced. This results, for example, in improved pitch discrimination in hearing.

**lesion** A wounded or damaged area.

**leukocytes** White blood cells.

**Leydig cells** The interstitial cells of the testes that serve an endocrine function by secreting testosterone and other androgenic hormones.

**ligament** Dense regular connective tissue, containing many parallel arrangements of collagen fibers, that connects bones or cartilages and serves to strengthen joints.

**limbic system** A group of brain structures, including the hippocampus, cingulate gyrus, dentate gyrus, and amygdala. The limbic system appears to be important in memory, the control of autonomic function, and some aspects of emotion and behavior.

**lipids** Organic molecules that are nonpolar and thus insoluble in water. This class includes triglycerides, steroids, and phospholipids.

**lipogenesis** The formation of fat or triglycerides.

**lipolysis** The hydrolysis of triglycerides into free fatty acids and glycerol.

**long-term potentiation** A type of post-tetanic potentiation. High-frequency stimulation of a presynaptic axon may increase its ability to stimulate a postsynaptic neuron for a period of weeks or months. This may be a mechanism of neural learning.

**low-density lipoproteins** Plasma proteins that transport triglycerides and cholesterol and are believed to contribute to arteriosclerosis.

**lower motoneurons** The motor neurons that have their cell bodies in the gray matter of the spinal cord and contribute axons to peripheral nerves. These neurons innervate muscles and glands.

**lumen** The cavity of a tube or hollow organ.

**lung surfactant** A mixture of lipoproteins (containing phospholipids) secreted by type II alveolar cells into the alveoli of the lungs; it lowers surface tension and prevents collapse of the lungs as occurs in hyaline membrane disease, in which surfactant is absent.

**luteinizing hormone (LH)** A gonadotropic hormone secreted by the anterior pituitary that, in a female, stimulates ovulation and the development of a corpus luteum. In a male, LH stimulates the Leydig cells to secrete androgens.

**lymph** The fluid in lymphatic vessels that is derived from tissue fluid.

**lymphatic system** The lymphatic vessels and lymph nodes.

**lymphocytes** A type of mononuclear leukocyte; the cells responsible for humoral and cell-mediated immunity.

**lymphokines** A group of chemicals released from T cells that contribute to cell-mediated immunity.

**-lysis** (Gr.) Breakage, disintegration.

**lysosome** Organelle containing digestive enzymes and responsible for intracellular digestion.

## M

**macro-** (Gr.) Large.

**macromolecules** Large molecules; a term that usually refers to protein, RNA, and DNA.

**macrophage** A large phagocytic cell in connective tissue that contributes to both specific and nonspecific immunity.

**macula lutea** The region of the distal tubule of the renal nephron in contact with the afferent arteriole. This region functions as a sensory receptor for the amount of sodium excreted in the urine, and acts to inhibit the secretion of renin from the juxtaglomerular apparatus.

**malignant** A structure or process that is life threatening.

**mast cells** A type of connective tissue cells that produce and secrete histamine and heparin.

**maximal oxygen uptake** The maximum amount of oxygen that can be consumed by the body per unit time during heavy exercise.

**mean arterial pressure** The mean arterial pressure is an adjusted average of the systolic and diastolic blood pressures. It averages about 100 mm Hg in the systemic circulation and 10 mm Hg in the pulmonary circulation.

**mechanoreceptors** Sensory receptors that are stimulated by mechanical means. These include stretch receptors, hair cells in the inner ear, and pressure receptors.

**medulla oblongata** A part of the brain stem that contains neural centers for the control of breathing and for regulation of the cardiovascular system via autonomic nerves.

**mega-** Large, great.

**megakaryocyte** A bone marrow cell that gives rise to blood platelets.

**meiosis** A type of cell division in which a diploid parent cell gives rise to haploid daughter cells; it occurs in the process of gamete production in the gonads.

**melanin** A dark pigment found in the skin, hair, choroid layer of the eye, and substantia nigra of the brain; it may also be present in certain tumors (melanomas).

**melatonin** A hormone secreted by the pineal gland that produces lightening of the skin in lower animals and that may contribute to the regulation of gonadal function in mammals.

**membrane potential** The potential difference or voltage that exists between the inner and outer sides of a cell membrane; it exists in all cells, but is capable of being changed by excitable cells (neurons and muscle cells).

**membranous labyrinth** A system of communicating sacs and ducts within the bony labyrinth of the inner ear.

**menarche** The age at which menstruation begins.

**Meniere's disease** Deafness, tinnitus, and vertigo resulting from a disease of the labyrinth.

**menopause** The cessation of menstruation, usually occurring at about age forty-eight to fifty.

**menstrual cycle** The cyclic changes in the ovaries and endometrium of the uterus, which lasts about a month and is accompanied by shedding of the endometrium, with bleeding. This occurs only in humans and the higher primates.

**menstruation** Shedding of the outer two-thirds of the endometrium with accompanying bleeding, due to lowering of estrogen secretion by the ovaries at the end of the monthly cycle. The first day of menstruation is taken as day one of the menstrual cycle.

**meso-** (Gr.) Middle.

**mesoderm** The middle embryonic tissue layer that gives rise to connective tissue (including blood, bone, and cartilage), blood vessels, muscles, the adrenal cortex, and other organs.

**messenger RNA (mRNA)** A type of RNA that contains a base sequence complementary to a part of the DNA that specifies the synthesis of a particular protein.

**meta-** (Gr.) Change.

**metabolic acidosis and alkalosis** Abnormal changes in arterial blood pH due to changes in nonvolatile acid concentration (such as lactic acid or ketone bodies) or to changes in blood bicarbonate concentration.

**metabolism** All of the chemical reactions in the body; it includes those that result in energy storage (anabolism) and those that result in the liberation of energy (catabolism).

**metastasis** A process whereby cells of a malignant tumor can separate from the tumor, travel to a different site, and divide to produce a new tumor.
**methemoglobin** The abnormal form of hemoglobin in which the iron atoms in heme are oxidized to the ferrous form. Methemoglobin is incapable of bonding to oxygen.
**micelles** Colloidal particles formed by the aggregation of many molecules.
**micro-** (La.) Small; also, one-millionth.
**microvilli** Tiny fingerlike projections of a cell membrane; they occur on the apical (lumenal) surface of the cells of the small intestine and in the renal tubules.
**micturition** Urination.
**milliequivalents** The millimolar concentration of an ion multiplied by its number of charges.
**mineralocorticoids** Steroid hormones of the adrenal cortex (corticosteroids) that regulate electrolyte balance.
**mitosis** Cell division in which the two daughter cells receive the same number of chromosomes as the parent cell (both daughters and parent are diploid).
**molal** Pertaining to the number of moles of solute per kilogram of solvent.
**molar** Pertaining to the number of moles of solute per liter of solution.
**mole** The number of grams of a chemical that is equal to its formula weight (atomic weight for an element or molecular weight for a compound).
**mono-** (Gr.) One, single.
**monoamine oxidase** Abbreviated MAO, this enzyme degrades monoamine neurotransmitters within presynaptic axon endings. Drugs that inhibit the action of this enzyme thus potentiate the pathways that use monoamines as neurotransmitters.
**monoamines** A class of neurotransmitter molecules that includes serotonin, dopamine, and norepinephrine.
**monoclonal antibodies** Identical antibodies derived from a clone of genetically identical plasma cells.
**monocyte** A mononuclear, nongranular leukocyte that is phagocytic and able to be transformed into a macrophage.
**monomers** A single molecular unit of a longer, more complex molecule; monomers are joined together to form dimers, trimers, and polymers; the hydrolysis of polymers eventually yields separate monomers.
**mononuclear leukocytes** A category of white blood cell that includes the lympocytes and monocytes.
**mononuclear phagocyte system** A term for monocytes and tissue macrophages.
**monosaccharides** Also called simple sugars; the monomers of more complex carbohydrates. Examples include glucose, fructose, and galactose.
**-morph, morpho-** (Gr.) Form, shape.
**motile** Capable of self-propelled movement.
**motor cortex** The precentral gyrus of the frontal lobe of the cerebrum. Axons from this area form the descending pyramidal motor tracts.
**motor neuron** An efferent neuron that conducts action potentials away from the central nervous system and innervates effector organs (muscles and glands). It forms the ventral roots of spinal nerves.
**motor unit** A lower motor neuron and all of the skeletal muscle fibers stimulated by branches of its axon. Larger motor units (more muscle fibers per neuron) produce more power when the unit is activated, but smaller motor units afford a finer degree of neural control over muscle contraction.
**mucous membrane** The layers of visceral organs that include the lining epithelium, submucosal connective tissue, and (in some cases) a thin layer of smooth muscle (the muscularis mucosa).
**muscarinic receptors** Receptors for acetylcholine that are stimulated by postganglionic parasympathetic neurons. The name is derived from the fact that these receptors are also stimulated by the chemical muscarine, derived from a mushroom.
**muscle spindles** Sensory organs within skeletal muscles that are composed of intrafusal fibers and are sensitive to muscle stretch; they provide a length detector within muscles.
**myelin sheath** A sheath surrounding axons, which is formed by successive wrappings of a neuroglial cell membrane. Myelin sheaths are formed by Schwann cells in the peripheral nervous system and by oligodendrocytes within the central nervous system.
**myocardial infarction** An area of necrotic tissue in the myocardium that is filled in by scar (connective) tissue.
**myofibrils** Subunits of striated muscle fibers that consist of successive sarcomeres; myofibrils run parallel to the long axis of the muscle fiber, and the pattern of their filaments provides the striations characteristic of striated muscle cells.
**myogenic** Originating within muscle cells; this term is used to describe self-excitation by cardiac and smooth muscle cells.
**myoglobin** A molecule composed of globin protein and heme pigment, related to hemoglobin, but containing only one subunit (instead of the four in hemoglobin), and found in striated muscles. Myoglobin serves to store oxygen in skeletal and cardiac muscle cells.
**myoneural junction** Also called the neuromuscular junction; a synapse between a motor neuron and the muscle cell that it innervates.
**myopia** A condition of the eyes, also called nearsightedness, in which light is brought to a focus in front of the retina due to the eye being too long.
**myosin** The protein that forms the A bands of striated muscle cells; together with the protein actin, myosin provides the basis for muscle contraction.
**myxedema** A type of edema associated with hypothyroidism; it is characterized by accumulation of mucoproteins in tissue fluid.

## N

**NAD** Nicotinamide adenine dinucleotide; a coenzyme derived from niacin that functions to transport electrons in oxidation-reduction reactions; it helps to transport electrons to the electron transport chain within mitochondria.
**naloxone** A drug that antagonizes the effects of morphine and endorphins.
**natriuretic hormone** A hormone that increases the urinary excretion of sodium. Currently, this hormone is believed to be secreted by the atria of the heart.
**necrosis** Cellular death within tissues and organs.
**negative feedback** Mechanisms in the body that act to maintain a state of internal constancy, or homeostasis; effectors are activated by changes in the internal environment, and the actions of the effectors serve to counteract these changes and maintain a state of balance.
**neoplasm** A new, abnormal growth of tissue, as in a tumor.
**nephron** The functional unit of the kidneys, consisting of a system of renal tubules and a vascular component that includes capillaries of the glomerulus and the peritubular capillaries.
**Nernst equation** The equation that allows the equilibrium membrane potential to be calculated for given ions when the concentrations of those ions on each side of the membrane are known.
**nerve** A collection of motor axons and sensory dendrites in the peripheral nervous system.
**neurilemma** The basement membrane surrounding the sheath of Schwann of nerve fibers in the peripheral nervous system.
**neuroglia** The supporting tissue of the nervous system, consisting of neuroglial, or glial, cells. In addition to providing support, the neuroglial cells participate in the metabolic and bioelectrical processes of the nervous system.
**neurons** Nerve cells, consisting of a cell body that contains the nucleus, short branching processes, called dendrites, that carry electrical charges to the cell body, and a single fiber, or axon, that conducts nerve impulses away from the cell body.
**neurotransmitter** A chemical contained in synaptic vesicles in nerve endings, which is released into the synaptic cleft and stimulates the production of either excitatory or inhibitory postsynaptic potentials.
**neutrons** Electrically neutral particles that exist together with positively charged protons in the nucleus of atoms.
**nexus** A bond between members of a group; the type of bonds present in single-unit smooth muscles.
**nicotinic receptors** Receptors for acetylcholine located in the autonomic ganglia and in neuromuscular junctions. The name is derived from the fact that these receptors can also be stimulated by nicotine, derived from the tobacco plant.

**nidation** Implantation of the blastocyst into the endometrium of the uterus.

**Nissl bodies** Granular-appearing structures in the cell bodies of neurons that have an affinity for basic stain; they correspond to ribonucleoprotein.

**nocioceptors** Receptors for pain that are stimulated by tissue damage.

**nodes of Ranvier** Gaps in the myelin sheath of myelinated axons, located approximately 1 mm apart. Action potentials are produced only at the nodes of Ranvier in myelinated axons.

**norepinephrine** A catecholamine released as a neurotransmitter from postganglionic sympathetic nerve endings and as a hormone (together with epinephrine) by the adrenal medulla.

**nucleolus** A dark-staining area within a cell nucleus; the site where ribosomal RNA is produced.

**nucleoplasm** The protoplasm of a nucleus.

**nucleotide** The subunit of DNA and RNA macromolecules; each nucleotide is composed of a nitrogenous base (adenine, guanine, cytosine, and thymine or uracil), a sugar (deoxyribose or ribose), and a phosphate group.

**nucleus, brain** Aggregation of neuron cell bodies within the brain. Nuclei within the brain are surrounded by white matter and are deep to the cerebral cortex.

**nucleus, cell** The organelle, surrounded by a double saclike membrane, called the nuclear envelope, that contains the DNA and genetic information of the cell.

**nystagmus** Involuntary, oscillatory movements of the eye.

## O

**obese** Excessively fat.

**oligo-** (Gr.) Few, small.

**oligodendrocytes** A type of neuroglial cell; it forms myelin sheaths around axons in the central nervous system.

**oncology** The study of tumors.

**oncotic pressure** The colloid osmotic pressure of solutions produced by proteins; in plasma, it serves to counterbalance the outward filtration of fluid from capillaries due to hydrostatic pressure.

**oo-** (Gr.) Pertaining to an egg.

**oogenesis** The formation of ova in the ovaries.

**opsonization** The role of antibodies in enhancing the ability of phagocytic cells to attack bacteria.

**optic disc** The area of the retina where axons from ganglion cells gather to form the optic nerve and where blood vessels enter and leave the eye; it corresponds to the blind spot in the visual field due to the absence of photoreceptors.

**organ** A structure in the body composed of a number of primary tissues that perform particular functions.

**organelle** A structure within cells that performs specialized tasks. The term includes mitochondria, Golgi apparatus, endoplasmic reticulum, nuclei, and lysosomes; it is also used for some structures not enclosed by a membrane, such as ribosomes and centrioles.

**organ of Corti** The structure within the cochlea responsible for hearing. It consists of hair cells and supporting cells on the basilar membrane that help transduce sound waves into nerve impulses.

**osmolality** A measure of the total concentration of a solution; the number of moles of solute per kilogram of solvent.

**osmoreceptors** Sensory neurons that respond to changes in the osmotic pressure of the surrounding fluid.

**osmosis** The passage of solvent (water) from a more dilute to a more concentrated solution through a membrane that is more permeable to water than to the solute.

**osmotic pressure** A measure of the tendency for a solution to gain water by osmosis when separated by a membrane from pure water; directly related to the osmolality of the solution, it is the pressure required to just prevent osmosis.

**osteo-** (Gr.) Pertaining to bone.

**osteoblasts** Cells that produce bone.

**osteocytes** Bone-forming cells that have become entrapped within a matrix of bone; these cells remain alive due to nourishment supplied by canaliculi within the extracellular material of bone.

**osteomalacia** Softening of bones due to a deficiency of vitamin D and calcium.

**osteoporosis** Demineralization of bone, seen most commonly in the elderly. It may be accompanied by pain, loss of stature, and other deformities and fractures.

**ovaries** The gonads of a female that produce ova and secrete female sex steroids.

**ovi-** (La.) Pertaining to egg.

**oviduct** The part of the female reproductive tract that transports ova from the ovaries to the uterus. Also called the uterine or fallopian tube.

**ovulation** The extrusion of a secondary oocyte out of the ovary.

**oxidative phosphorylation** The formation of ATP by using energy derived from electron transport to oxygen; this occurs in the mitochondria.

**oxidizing agent** An atom that accepts electrons in an oxidation-reduction reaction.

**oxygen debt** The extra amount of oxygen required by the body after exercise to metabolize lactic acid and to supply the higher metabolic rate of muscles warmed during exercise.

**oxyhemoglobin** A compound formed by the bonding of molecular oxygen to hemoglobin.

**oxyhemoglobin saturation** The ratio, expressed as a percentage, of the amount of oxyhemoglobin compared to the total amount of hemoglobin in blood.

**oxytocin** One of the two hormones produced in the hypothalamus and secreted by the posterior pituitary (the other hormone is vasopressin); oxytocin stimulates the contraction of uterine smooth muscles and promotes milk ejection in females.

## P

**pacemaker** A group of cells that has the fastest spontaneous rate of depolarization and contraction in a mass of electrically coupled cells; in the heart, this is the sinoatrial, or SA, node.

**pacesetter potentials** Changes in membrane potential produced spontaneously by pacemaker cells of single-unit smooth muscles.

**Pacinian corpuscle** A cutaneous sensory receptor sensitive to pressure. Characterized by an onionlike layering of cells around a central sensory dendrite.

**PAH** Para-aminohippuric acid; a substance used to measure total renal plasma flow because its clearance rate is equal to the total rate of plasma flow to the kidneys; PAH is filtered and secreted but not reabsorbed by the renal nephrons.

**pancreatic juice** The secretions of the pancreas that are transported by the pancreatic duct to the duodenum. Pancreatic juice contains bicarbonate and such digestive enzymes as trypsin, lipase, and amylase.

**parathyroid hormone (PTH)** A polypeptide hormone secreted by the parathyroid glands, PTH acts to raise the blood $Ca^{++}$ levels primarily by stimulating reabsorption of bone.

**Parkinson's disease** A tremor of the resting muscles and other symptoms caused by inadequate dopamine-producing neurons in the basal ganglia of the cerebrum. Also called paralysis agitans.

**parturition** Birth.

**passive immunity** Specific immunity granted by the administration of antibodies made by another organism.

**Pasteur effect** A decrease in the rate of glucose utilization and lactic acid production in tissues or organisms by their exposure to oxygen.

**pathogen** Any disease-producing microorganism or substance.

**pepsin** The protein-digesting enzyme secreted in gastric juice.

**peptic ulcer** An injury to the mucosa of the esophagus, stomach, or small intestine caused by acidic gastric juice.

**perfusion** The flow of blood through an organ.

**peri-** (Gr.) Around, surrounding.

**perilymph** The fluid between the membranous and bony labyrinth of the inner ear.

**perimysium** The connective tissue surrounding a fascicle of skeletal muscle fibers.

**periosteum** Connective tissue covering bones; it contains osteoblasts and is therefore capable of forming new bone.

**peripheral resistance** The resistance to blood flow through the arterial system. Peripheral resistance is largely a function of the radius of small arteries and arterioles. The resistance to blood flow is proportional to the fourth power of the radius of the vessel.

**peristalsis** Waves of smooth muscle contraction in smooth muscles of the tubular digestive tract, involving circular and longitudinal muscle fibers at successive locations along the tract; it serves to propel the contents of the tract in one direction.

**permissive effect** A category of hormonal interaction in which one hormone acts to increase the effectiveness of another hormone. This may be due to promotion of the synthesis of the active form of the second hormone, or it may be due to an increase in the sensitivity of the target tissue to the effects of the second hormone.

**pH** The pH of a solution is equal to the logarithm of 1 over the hydrogen ion concentration. The pH scale goes from zero to 14; a pH of 7.0 is neutral, whereas solutions with lower pH are acidic and solutions with higher pH are basic.

**phagocytosis** Cellular eating; the ability of some cells (such as white bloodcells) to engulf large particles (such as bacteria) and digest these particles by merging the food vacuole containing these particles with a lysosome containing digestive enzymes.

**phonocardiogram** A visual display of the heart sounds.

**phosphodiesterase** An enzyme that cleaves cyclic AMP into inactive products, thus inhibiting the action of cyclic AMP as a second messenger.

**phospholipids** Lipids containing a phosphate group. These molecules (such as lecithin) are polar on one end and nonpolar on the other end. Phospholipids make up a large part of the cell membrane and function in the lung alveoli as surfactants.

**photoreceptors** Sensory cells (rods and cones) that respond electrically to light; they are located in the retinas of the eyes.

**pineal gland** A gland within the brain that secretes the hormone melatonin and is affected by sensory input from the photoreceptors of the eyes.

**pinocytosis** Cell drinking; invagination of the cell membrane to form narrow channels that pinch off into vacuoles; it provides cellular intake of extracellular fluid and dissolved molecules.

**plasma** The fluid portion of the blood. Unlike serum (which lacks fibrinogen), plasma is capable of forming insoluble fibrin threads when in contact with test tubes.

**plasma cells** Cells derived from B lymphocytes that produce and secrete large amounts of antibodies; they are responsible for humoral immunity.

**plasmalemma** Another term for the cell membrane.

**platelets** Disc-shaped structures, 2 to 4 micrometers in diameter, that are derived from bone marrow cells called megakaryocytes. Platelets circulate in the blood and participate (together with fibrin) in forming blood clots.

**pluripotent** A property of early embryonic cells, which are able to specialize in a number of ways to produce tissues characteristic of different organs.

**pneumotaxic center** A neural center in the pons that rhythmically inhibits inspiration in a manner independent of sensory input.

**pneumothorax** An abnormal condition in which air enters the intrapleural space, either through an open chest wound or from a tear in the lungs. This can lead to collapse of one lobe of the lungs.

**PNS** The peripheral nervous system, including nerves and ganglia.

**-pod, -podium** (Gr.) Foot, leg, extension.

**Poiseuille's law** According to the equation described by Poiseuille's law, the rate of blood flow through a vessel is directly proportional to the pressure difference between the two ends of the vessel and inversely proportional to the length of the vessel, the viscosity of the blood, and the fourth power of the radius of the vessel.

**polar body** A small daughter cell formed by meiosis that degenerates in the process of oocyte production.

**polar molecule** A molecule in which the shared electrons are not evenly distributed, so that one side of the molecule is relatively negatively (or positively) charged in comparison with the other side; polar molecules are soluble in polar solvents such as water.

**poly-** (Gr.) Many.

**polycythemia** An abnormally high red blood cell count.

**polydipsia** Excessive thirst.

**polymer** A large molecule formed by the combination of smaller subunits, or monomers.

**polymorphonuclear leukocyte** A granular leukocyte containing a nucleus with a number of lobes connected by thin, cytoplasmic strands; this term includes neutrophils, eosinophils, and basophils.

**polypeptide** A chain of amino acids connected by covalent bonds called peptide bonds. A very large polypeptide is called a protein.

**polyphagia** Excessive eating.

**polysaccharide** A carbohydrate formed by covalent bonding of numerous monosaccharides; examples include glycogen and starch.

**polyuria** Excretion of an excessively large volume of urine in a given period.

**portal system** Two capillary beds in series, where blood from the first is drained by veins into a second capillary bed, which in turn is drained by veins that return blood to the heart. The two major portal systems in the body are the hepatic portal system and the hypothalamo-hypophyseal portal system.

**positive feedback** Cause-and-effect relationships that result in the amplification of changes. Positive feedback results in avalanchelike effects, as occurs in the formation of a blood clot or in the production of the LH surge by the stimulatory effect of estrogen.

**posterior** Anatomical term denoting a backside position.

**posterior pituitary** The part of the pituitary gland that is derived from the brain; it secretes vasopressin (ADH) and oxytocin, produced in the hypothalamus. Also called the neurohypophysis.

**postsynaptic inhibition** The inhibition of a postsynaptic neuron by axon endings that release a neurotransmitter that induces hyperpolarization (inhibitory postsynaptic potentials).

**post-tetanic potentiation** High-frequency stimulation of a presynaptic axon improves its ability to release neurotransmitter when it is stimulated again at a later time.

**prehormone** An inactive form of a hormone which is secreted by an endocrine gland. The prehormone is converted within its target cells to the active form of the hormone.

**presynaptic inhibition** Neural inhibition in which axoaxonic synapses inhibit the release of neurotransmitter chemicals from the presynaptic axon.

**pro-** (Gr.) Before, in front of, forward.

**process, cell** Any thin cytoplasmic extension of a cell, such as the dendrites and axon of an neuron.

**progesterone** A steroid hormone secreted by the corpus luteum of the ovaries and by the placenta. Secretion of progesterone during the luteal phase of the menstrual cycle promotes the final maturation of the endometrium.

**prohormone** The precursor of a polypeptide hormone that is larger and less active than the hormone. This is produced within the cells of an endocrine gland and is normally converted into the shorter, active hormone prior to secretion.

**prolactin** A hormone secreted by the anterior pituitary that stimulates lactation (acting together with other hormones) in the postpartum female. It may also participate (along with the gonadotropins) in regulating gonadal function in some mammals.

**prophylaxis** Prevention or protection.

**proprioceptor** A sensory receptor that provides information about body position and movement; examples include receptors in muscles, tendons, and joints as well as the sense of equilibrium provided by the semicircular canals of the inner ear.

**prostaglandins** A family of fatty acids containing a cyclic ring which serves numerous autocrine regulatory functions.

**protein** The class of organic molecules composed of large polypeptides, in which over a hundred amino acids are bonded together by peptide bonds.

**protein kinase** The enzyme that is activated by cyclic AMP; catalyzes the phosphorylation of specific proteins (enzymes). Such phosphorylation may activate or inactivate enzymes.
**proto-** (Gr.) First, original.
**proton** A unit of positive charge in the nucleus of atoms.
**protoplasm** A general term that includes cytoplasm and nucleoplasm.
**pseudo-** (Gr.) False.
**pseudohermaphrodite** An individual who has some of the physical characteristics of both sexes, but lacks functioning gonads of both sexes; a true hermaphrodite has both testes and ovaries.
**pseudopods** Footlike extensions of the cytoplasm that enable some cells (with amoeboid motion) to move across a substrate; pseudopods also are used to surround food particles in the process of phagocytosis.
**ptyalin** Also called salivary amylase; an enzyme in saliva that catalyzes the hydrolysis of starch into smaller molecules.
**puberty** The period of time in an individual's life span when secondary sexual characteristics and fertility develop.
**pulmonary circulation** The part of the vascular system that includes the pulmonary artery and pulmonary veins; it transports blood from the right ventricle of the heart, through the lungs, and back to the left atrium of the heart.
**pupil** The opening at the center of the iris of the eye.
**purkinje fibers** Specialized conducting tissue in the ventricles of the heart that carry the impulse from the bundle of His to the myocardium of the ventricles.
**pyramidal tracts** Also called corticospinal tracts, these motor tracts descend without synaptic interruption from the cerebrum to the spinal cord. In the spinal cord, these fibers synapse either directly or indirectly (via spinal interneurons) with the lower motor neurons of the spinal cord.
**pyrogen** A fever-producing substance.

## Q

**QRS complex** The part of an electrocardiogram that is produced by depolarization of the ventricles.

## R

**reabsorption** The transport of a substance from the lumen of the renal nephron to the blood.
**receptive field** The area of the body that, when stimulated by a sensory stimulus, activates a particular sensory receptor.
**reciprocal innervation** The process whereby the motor neurons to an antagonistic muscle are inhibited when the motor neurons to an agonist muscle are stimulated. In this way, for example, the extensor muscle of the elbow joint is inhibited when the flexor muscles of this joint are stimulated to contract.
**recruitment** In terms of muscle contraction, the successive stimulation of more and larger motor units in order to produce increasing strengths of muscle contraction.
**reduced hemoglobin** Hemoglobin with iron in the reduced ferrous state, which is able to bond with oxygen but is not combined with oxygen. Also called deoxyhemoglobin.
**reducing agent** An electron donor in a coupled oxidation-reduction reaction.
**refraction** The bending of light rays when light passes from a medium of one density to a medium of another density. Refraction of light by the cornea and lens acts to focus the image on the retina of the eye.
**refractory period** The period of time during which a region of axon or muscle cell membrane cannot be stimulated to produce an action potential (absolute refractory period), or can only be stimulated by a very strong stimulus (relative refractory period).
**releasing hormones** Polypeptide hormones secreted by neurons in the hypothalamus that travel in the hypothalamo-hypophyseal portal system to the anterior pituitary gland and stimulate the anterior pituitary to secrete specific hormones.
**REM sleep** The stage of sleep in which dreaming occurs; it is associated with rapid eye movements (REM). REM sleep occurs three to four times each night and lasts from a few minutes to over an hour.
**renal** Pertaining to the kidneys.
**renal plasma clearance** The milliliters of plasma that are cleared of a particular solute per minute by the excretion of that solute in the urine. If there is no reabsorption or secretion of that solute by the nephron tubules, this is equal to the glomerular filtration rate.
**renal pyramids** The medulla of the human kidney.
**renin** An enzyme secreted into the blood by the juxtaglomerular apparatus of the kidneys. Renin catalyzes the conversion of angiotensinogen into angiotensin II.
**rennin** A digestive enzyme secreted in the gastric juice of infants which catalyzes the digestion of the milk protein casein.
**repolarization** The reestablishment of the resting membrane potential after depolarization has occurred.
**resorption, bone** The dissolution of the calcium phosphate crystals of bone by the action of osteoclasts.
**respiratory acidosis** A lowering of the blood pH below 7.35 due to the accumulation of $CO_2$ as a result of hypoventilation.
**respiratory alkalosis** A rise in blood pH above 7.45 due to the excessive elimination of blood $CO_2$ as a result of hyperventilation.
**respiratory distress syndrome** Also called hyaline membrane disease; most frequently occurring in premature infants, this syndrome is caused by abnormally high alveolar surface tension as a result of a deficiency in lung surfactant.
**respiratory zone** The region of the lungs including the respiratory bronchioles, in which individual alveoli are found, and terminal alveoli. Gas exchange between the inspired air and pulmonary blood occurs in the respiratory zone of the lungs.
**resting potential** The potential difference across a cell membrane when the cell is in an unstimulated state. The resting potential is always negatively charged on the inside of the membrane compared to the outside.
**reticular activating system (RAS)** A complex network of nuclei and fiber tracts within the brain stem that produces nonspecific arousal of the cerebrum to incoming sensory information. The RAS thus maintains a state of alert consciousness, and must be depressed during sleep.
**retina** The layer of the eye that contains neurons and photoreceptors (rods and cones).
**rhodopsin** Visual purple; a pigment in rod cells that undergoes a photochemical dissociation in response to light and, in so doing, stimulates electrical activity in the photoreceptors.
**ribosomes** Particles of protein and ribosomal RNA that form the organelles responsible for the translation of messenger RNA and protein synthesis.
**rickets** A condition caused by a deficiency of vitamin D and associated with interference of the normal ossification of bone.
**rigor mortis** The stiffening of a dead body, due to the depletion of ATP and the production of rigor complexes between actin and myosin in muscles.
**RNA** Ribonucleic acid; a nucleic acid consisting of the nitrogenous bases adenine, guanine, cytosine, and uracil; the sugar ribose; and phosphate groups. There are three types of RNA found in cytoplasm: messenger RNA (mRNA), transfer RNA (tRNA), and ribosomal RNA (rRNA).
**rods** One of the two categories of photoreceptors (along with cones) in the retina of the eye; rods are responsible for black-and-white vision under low illumination.

## S

**saccadic eye movements** Very rapid eye movements that occur constantly and that change the focus on the retina from one point to another.
**saltatory conduction** The rapid passage of action potentials from one node of Ranvier to another in myelinated axons.
**sarcomere** The structural subunit of a myofibril in a striated muscle, equal to the distance between two successive Z lines.
**sarcoplasm** The cytoplasm of striated muscle cells.
**sarcoplasmic reticulum** The smooth or agranular endoplasmic reticulum of skeletal muscle cells; it surrounds each myofibril and serves to store $Ca^{++}$ when the muscle is at rest.

**Schwann cell** A neuroglial cell of the peripheral nervous system that forms sheaths around peripheral nerve fibers. Schwann cells also direct regeneration of peripheral nerve fibers to their target cells.

**sclera** The tough white outer coat of the eyeball that is continuous anteriorly with the clear cornea.

**second messenger** A molecule or ion whose concentration within a target cell is increased by the action of a regulator compound (e.g., hormone or neurotransmitter) and which stimulates the metabolism of that target cell in a way characteristic of the actions of that regulator molecule—that is, in a way that mediates the intracellular effects of that regulatory compound.

**secretin** A polypeptide hormone secreted by the small intestine in response to acidity of the intestinal lumen; along with cholecystokinin, secretin stimulates the secretion of pancreatic juice into the small intestine.

**secretion, renal** The transport of a substance from the blood through the wall of the nephron tubule into the urine.

**semicircular canals** Three canals of the bony labyrinth that contain endolymph, which is continuous with the endolymph of the membranous labyrinth of the cochlea; the semicircular canals provide a sense of equilibrium.

**semilunar valves** The valve flaps of the aorta and pulmonary artery at their juncture with the ventricles.

**seminal vesicles** The paired organs located on the posterior border of the urinary bladder that empty their contents into the vas deferens and thus contribute to the semen.

**seminiferous tubules** The tubules within the testes that produce spermatozoa by meiotic division of their germinal epithelium.

**semipermeable membrane** A membrane with pores of a size that permits the passage of solvent and some solute molecules but restricts the passage of other solute molecules.

**sensory neuron** An afferent neuron that conducts impulses from peripheral sensory organs into the central nervous system.

**serosa** An outer epithelial membrane that covers the surface of a visceral organ.

**Sertoli cells** Nongerminal, supporting cells in the seminiferous tubules. Sertoli cells envelop spermatids and appear to participate in the transformation of spermatids into spermatozoa.

**serum** The fluid squeezed out of a clot as it retracts; the supernatant when a sample of blood clots in a test tube and is centrifuged; serum is plasma without fibrinogen (which has been converted to fibrin in clot formation).

**sex chromosomes** The X and Y chromosomes; the unequal pairs of chromosomes involved in sex determination (which is due to the presence or absence of a Y chromosome). Females lack a Y chromosome and normally have the genotype XX; males have a Y chromosome and normally have the genotype XY.

**shock** As it relates to the cardiovascular system, this refers to a rapid, uncontrolled fall in blood pressure, which in some cases becomes irreversible and leads to death.

**sickle-cell anemia** A hereditary, autosomal recessive trait that occurs primarily in people of African ancestry, in which it evolved apparently as a protection (in the carrier state) against malaria. In the homozygous state, hemoglobin S is made instead of hemoglobin A; this leads to the characteristic sickling of red blood cells, hemolytic anemia, and organ damage.

**sinoatrial node** The sinoatrial, or SA, node, is the normal pacemaker region of the heart and is located in the right atrium near the junction of the vena cavae.

**sinus** A cavity.

**sinusoids** Blood channels that appear as cavitylike in the surrounding tissue; they function as a type of capillary space that is relatively large and lined (in the liver sinusoids) by phagocytic cells of the reticuloendothelial system.

**skeletal muscle pump** The skeletal muscle pump is the effect skeletal muscle contraction has on the flow of blood in veins. As the muscles contract, they squeeze the veins and in this way help move the blood toward the heart.

**sleep apnea** A temporary cessation of breathing during sleep, usually lasting for several seconds.

**sliding filament theory** The theory that the thick and thin filaments of a myofibril slide past each other, while maintaining their initial length, during muscle contraction.

**smooth muscle** Nonstriated, spindle-shaped muscle cells with a single nucleus in the center; involuntary muscle in visceral organs that is innervated by autonomic nerve fibers.

**sodium/potassium pump** An active transport carrier, with ATPase enzymatic activity, that acts to accumulate $K^+$ within cells and extrude $Na^+$ from cells, thus maintaining gradients for these ions across the cell membrane.

**soma-, somato-, -some** (Gr.) Body, unit.

**somatesthetic sensations** Sensations arising from cutaneous, muscle, tendon, and joint receptors. These sensations project to the postcentral gyrus of the cerebrum.

**somatic motor neurons** Motor neurons in the spinal cord which innervate skeletal muscles. Somatic motor neurons are divided into alpha and gamma motoneurons.

**somatomammotropic hormone** A hormone secreted by the placenta which has actions similar to the pituitary hormones growth hormone and prolactin.

**somatomedins** A group of small polypeptides that are believed to be produced in the liver in response to growth hormone stimulation and to mediate the actions of growth hormone on the skeleton and other tissues.

**somatostatin** A polypeptide produced in the hypothalamus that acts to inhibit the secretion of growth hormone from the anterior pituitary; somatostatin is also produced in the islets of Langerhans of the pancreas, but its function there has not been established.

**somatotropic hormone** Growth hormone; an anabolic hormone secreted by the anterior pituitary that stimulates skeletal growth and protein synthesis in many organs.

**sounds of Korotkoff** The sounds heard when blood pressure measurements are taken. These sounds are produced by the turbulent flow of blood through an artery that has been partially constricted by a pressure cuff.

**spastic paralysis** Paralysis in which the muscles have such a high tone that they remain in a state of contracture. This may be caused by inability to degrade ACh released at the neuromuscular junction (as caused by certain drugs) or by damage to the spinal cord.

**spermatogenesis** The formation of spermatozoa, including meiosis and maturational processes in the seminiferous tubules.

**spermiogenesis** The maturational changes that transform spermatids into spermatozoa.

**sphygmo-** (Gr.) The pulse.

**sphygmomanometer** A manometer (pressure transducer) used to measure the blood pressure.

**spindle fibers** Filaments that extend from the poles of a cell to its equator and attach to chromosomes during the metaphase stage of cell division. Contraction of the spindle fibers pulls the chromosomes to opposite poles of the cell.

**spironolactones** Diuretic drugs that act as an aldosterone antagonist.

**Starling forces** The Starling forces are the hydrostatic pressures and the colloid osmotic pressures of the blood and tissue fluid. The balance of the Starling forces determine the net movement of fluid out of or into blood capillaries.

**steroid** A lipid, derived from cholesterol, that has three six-sided carbon rings and one five-sided carbon ring. These form the steroid hormones of the adrenal cortex and gonads.

**stretch reflex** The monosynaptic reflex whereby stretching a muscle results in a reflex contraction. The knee-jerk reflex is an example of a stretch reflex.

**striated muscle** Skeletal and cardiac muscle, the cells of which exhibit cross-banding, or striations, due to arrangement of thin and thick filaments into sarcomeres.

**stroke volume** The amount of blood ejected from each ventricle at each heartbeat.

**sub-** (La.) Under, below.

**substrate** In enzymatic reactions, the molecules that combine with the active sites of an enzyme and are converted to products by catalysis of the enzyme.
**sulcus** A groove or furrow; a depression in the cerebrum that separates folds, or gyri, of the cerebral cortex.
**summation** In neural physiology, summation refers to the additive effects of graded synaptic potentials. In muscle physiology, summation refers to the additive effects of contractions of different muscle fibers.
**super-, supra-** (La.) Above, over.
**suppressor T cells** A subpopulation of T lymphocytes that acts to inhibit the production of antibodies against specific antigens by B lymphocytes.
**surfactant** In the lungs, a mixture of phospholipids and proteins produced by alveolar cells that reduces the surface tension of the alveoli and contributes to the elastic properties of the lungs.
**sym-, syn-** (Gr.) With, together.
**synapse** A region where a nerve fiber comes into close or actual contact with another cell and across which nerve impulses are transmitted either directly or indirectly (via the release of chemical neurotransmitters).
**synaptic plasticity** The ability of synapses to change at a cellular or molecular level. At a cellular level, plasticity refers to the ability to form new synaptic associations. At a molecular level, plasticity refers to the ability of a presynaptic axon to release more than one type of neurotransmitter.
**syncitium** The organization of cells in a tissue into a single functional unit. Because the atria and ventricles of the heart have gap junctions between their cells, these myocardia behave as syncitia.
**synergistic** Pertaining to regulatory processes or molecules (such as hormones) that have complementary or additive effects.
**systemic circulation** The circulation that carries oxygenated blood from the left ventricle in arteries to the tissue cells and that carries blood depleted in oxygen via veins to the right atrium; the general circulation, as compared to the pulmonary circulation.
**systole** The phase of contraction in the cardiac cycle. When unmodified, this term refers to contraction of the ventricles; the term *atrial systole* refers to contraction of the atria.

## T

**tachycardia** Excessively rapid heart rate, usually applied to rates in excess of 100 beats per minute. In contrast to an excessively slow heart rate (below 60 beats per minute), which is termed *bradycardia.*
**target organ** The organ that is specifically affected by the action of a hormone or other regulatory process.
**T cell** A type of lymphocyte that provides cell-mediated immunity, in contrast to B lymphocytes that provide humoral immunity through the secretion of antibodies. There are three subpopulations of T cells: cytotoxic, helper, and suppressor.
**telo-** (Gr.) An end; complete; final.
**tendon** The dense regular connective tissue that attaches a muscle to the bones of its origin and insertion.
**testis-determining factor** The product of a gene located on the short arm of the Y chromosome which causes the indeterminate embryonic gonads to develop into testes.
**testosterone** The major androgenic steroid secreted by the Leydig cells of the testes after puberty.
**tetanus** A term used to mean either a smooth contraction of a muscle (as opposed to muscle twitching) or a state of maintained contracture of high tension.
**thalassemia** A group of hemolytic anemias caused by the hereditary inability to produce either the alpha or beta chain of hemoglobin. It is found primarily among Mediterranean people.
**thorax** The part of the body cavity above the diaphragm; the chest.
**threshold** The minimum stimulus that just produces a response.
**thrombin** A protein formed in blood plasma during clotting which enzymatically converts the soluble protein fibrinogen into insoluble fibrin.
**thrombocytes** Blood platelets; disc-shaped structures in blood that participate in clot formation.
**thrombus** A blood clot, produced by the formation of fibrin threads around a platelet plug.
**thyroxine** Also called tetraiodothyronine, or $T_4$. The major hormone secreted by the thyroid gland, which regulates the basal metabolic rate and stimulates protein synthesis in many organs; a deficiency of this hormone in early childhood produces cretinism.
**tinnitus** A ringing sound or other noise that is heard but is not related to external sounds.
**tolerance, immunological** The ability of the immune system to distinguish self from non-self, and to not attack those antigens that are part of one's own tissues.
**total minute volume** The product of tidal volume (ml per breath) and ventilation rate (breaths per minute).
**toxin** A poison.
**toxoid** A modified bacterial endotoxin that has lost toxicity but retains its ability to act as an antigen and stimulate antibody production.
**tracts** A collection of axons within the central nervous system, forming the white matter of the CNS.
**trans-** (La.) Across, through.
**transamination** The transfer of an amino group from an amino acid to an alpha-keto acid, forming a new keto acid and a new amino acid, without the appearance of free ammonia.
**transcription, genetic** The production of RNA that is a complementary copy of the base sequence in a region of DNA.
**translation, genetic** The production of proteins according to the sequence of codons in messenger RNA.
**transpulmonary pressure** The pressure difference across the wall of the lung, equal to the difference between intrapulmonary pressure and intrapleural pressure.
**triiodothyronine** Abbreviated $T_3$; a hormone secreted in small amounts by the thyroid; the active hormone in target cells formed from thyroxine.
**tropomyosin** A filamentous protein that attaches to actin in the thin filaments and that acts, together with another protein called troponin, to inhibit and regulate the attachment of myosin cross-bridges to actin.
**troponin** A protein found in the thin filaments of the sarcomeres of skeletal muscle. A subunit of troponin binds to $Ca^{++}$, and as a result, causes tropomyosin to change position in the thin filament.
**trypsin** A protein-digesting enzyme in pancreatic juice that is released into the small intestine.
**TSH** Abbreviation for thyroid-stimulating hormone, also called thyrotropin. This hormone is secreted by the anterior pituitary and stimulates the thyroid gland.
**twitch** A rapid contraction and relaxation of a muscle fiber or a group of muscle fibers.
**tympanic membrane** The eardrum; a membrane separating the external from the middle ear that transduces sound waves into movements of the middle ear ossicles.

## U

**universal donor** A person with blood type O, who is able to donate blood to people with other blood types in emergency blood transfusions.
**universal recipient** A person with blood type AB, who can receive blood of any type in emergency transfusions.
**urea** The chief nitrogenous waste product of protein catabolism in the urine, formed in the liver from amino acids.
**uremia** The retention of urea and other products of protein catabolism due to inadequate kidney function.
**urobilinogen** A compound formed from bilirubin in the intestine; some is excreted in the feces and some is absorbed and enters the enterohepatic circulation, where it may be excreted either in the bile or in the urine.

## V

**vaccination** The clinical induction of active immunity by introducing antigens to the body, so that the immune system becomes sensitized to those antigens. The immune system will mount a secondary response to those antigens upon subsequent exposures.

**vagus nerve** The tenth cranial nerve, composed of sensory dendrites from visceral organs and preganglionic parasympathetic nerve fibers. The vagus is the major parasympathetic nerve in the body.

**vasa-, vaso-** (La.) Pertaining to blood vessels.

**vasa vasora** Blood vessels that supply blood to the walls of large blood vessels.

**vasectomy** Surgical removal of a portion of the vas (ductus) deferens to induce infertility.

**vasoconstriction** Narrowing of the lumen of blood vessels due to contraction of the smooth muscles in their walls.

**vasodilation** Widening of the lumen of blood vessels due to relaxation of the smooth muscles in their walls.

**vein** A blood vessel that returns blood to the heart.

**ventilation** Breathing; the process of moving air into and out of the lungs.

**vertigo** A feeling of movement or loss of equilibrium.

**vestibular apparatus** The parts of the inner ear, including the semicircular canals, utricle, and saccule, which function to provide a sense of equilibirum.

**villi** Fingerlike folds of the mucosa of the small intestine.

**virulent** Pathogenic, or able to cause disease.

**vital capacity** The maximum amount of air that can be forcibly expired after a maximal inspiration.

**vitamin** A term for unrelated organic molecules present in foods which are required in small amounts by the body for normal metabolic function. Vitamins are classified as water-soluble and fat-soluble.

## W

**white matter** The portion of the central nervous system composed primarily of myelinated fiber tracts. This forms the region deep to the cerebral cortex in the brain and the outer portion of the spinal cord.

## Z

**zygote** A fertilized ovum.

**zymogens** Inactive enzymes that become active when part of their structure is removed by another enzyme or by some other means.

# CREDITS

## Illustrators

### J & R Art Services

Figs. 1.20, 3.18, 3.19, 3.23, 4.18, 5.2, 5.8, 5.10, 12.16, 14.1, 15.32, 16.16, 18.7, 20.15, 20.36

### Sam Collins

Figs. 1.15, 7.4, 7.7, 9.1, 12.9, 19.1

### Nancy Marshburn

Fig. 13.18

### Steve Moon

Figs. 8.23, 9.38, 12.8, 12.20, 12.22, 12.23, 12.24, 12.25

### Precision Graphics

Figs. 9.40, 15.31, 16.15, 16.17, 16.22, 16.28, 19.2, 19.14, 19.18 (*a* and *b*), 19.19, 19.20, 19.21, 19.24, 20.16, 20.31, 20.50

### Rolin Graphics

Fig. 19.13

### Tom Waldrop

Figs. 3.29 (*b*), 11.11, 12.1, 12.7, 12.10, 12.19, 12.21, 13.21, 15.24, 17.12, 20.34

## Photos

### Chapter 1

**Figures 1.1–1.13:** Edwin Reschke.

### Chapter 2

**Figure 2.26:** Edwin Reschke.

### Chapter 3

**3.3a, b:** Richard Chao; **3.4a, b:** Kwang W. Jeon; **3.5 (all):** R. H. Albertin; **3.6:** Sandra L. Wolin; **3.8:** Richard Chao; **3.9a, 3.10a:** Keith R. Porter; **3.11:** E. G. Pollack; **3.12:** Richard Chao; **3.17:** O. L. Miller, Jr., *Journal of Cell Physiology,* 74(1969); **3.20:** Alexander Rich; **3.25a:** © Daniel S. Friend; **3.28:** Wisniewski L. P. and Hirschhorn, K.: *"A Guide to Human Chromosome Defects,"* 2nd ed., White Plains: The March of Dimes Birth Defects Foundation, BD:OAS 16(6), 1980; **3.29a:** © David M. Phillips, 1986/Visuals Unlimited; **3.30a–e:** © Ed Reschke.

### Chapter 5

**Figure 5.9:** Fernandes-Moran V.M.D. P.H.D.

### Chapter 6

**Figure 6.4a:** Carolyn Chambers; **6.4b:** Kessel & Kardon/W. H. Freeman Co.; **6.11:** Richard Chao.

### Chapter 7

**Figure 7.2:** John D. Cunningham/ Visuals Unlimited; **7.6:** H. Webster, from Hubbard, John *The Vertebrate Peripheral Nervous System.* © Plenum Press, 1974; **7.9:** Andreas Karschin, Heinz Wassle, and Jutta Schnitzer/ Scientific American; **7.18a, b:** Gilula, Reeves, and Steinbach. *Nature,* 235:262–265. © Macmillan Journals Limited; **7.19:** © John Heuser, Washington University School of Medicine, St. Louis, MO.

### Chapter 8

**Figure 8.8a:** © Martin Rotker/ Taurus; **8.13b:** © Igaku-Shoin, Ltd.

### Chapter 9

**Figure 9.14:** © Dean E. Hillman; **9.35:** Thomas Sims; **9.38b:** P. N. Farnsworth, University of Medicine and Dentistry, New Jersey Medical School; **9.42b:** © F. Werblin; **9.49:** Hubel, David H.: *American Scientist* 67 (1959) 534.

### Chapter 11

**Figure 11.1b:** Ed Reschke; **11.17b:** © Journalism Services; **11.18:** © Martin Rotker/Taurus Photos: **11.20, 11.22:** © Lester Bergman & Associates; **11.26:** © Ed Reschke.

### Chapter 12

**Figure 12.3:** © John D. Cunningham/Visuals Unlimited; **12.6a:** International Biomedical, Inc.; **12.7b:** © John D. Cunningham/Visuals Unlimited; **12.11–12.12b, 12.13:** H. E. Huxley; **12.28, 12.30:** Hans Hoppler, *Respiratory Physiology* 44:94(1981); **12.31:** Ed Reschke.

### Chapter 13

**Figure 13.5:** © Manfred Kage/Peter Arnold, Inc.; **13.10b:** © Igaku-Shoin, Ltd.; **13.28:** © Don Fawcett; **13.31:** Historical Pictures Service, Chicago; **13.32a, b:** American Heart Association; **13.36:** © Igaku-Shoin, Ltd.

### Chapter 14

**Figure 14.8:** Markell, E. K., and Voge, M.: *Medical Parasitology,* 6th ed. © W. B. Saunders Company, 1986; **14.15a, b:** Donald S. Baim, from Hurst et al.: *The Heart,* 5th ed. © McGraw-Hill Book Company, 1982; **14.19a, b:** © Neils A. Lassen, Copenhagen, Denmark.

### Chapter 15

**Figure 15.2:** Murray, John F.: *The Normal Lung,* 2nd ed. © W. B. Saunders Company, 1986; **15.3a, b:** American Lung Association; **15.6:** Yokochi, C., and Rohen, J. W.: *Photographic Anatomy of the Human Body,* 2nd ed. © Igaku-Shoin, Ltd., 1978; **15.7:** American Lung Association; **15.9a–15.11:** Edward C. Vasquez, R. T., C.R.T., Department of Radiologic Technology, Los Angeles City College; **15.15a, b:** Comroe, J. H.: Physiology of Respiration © 1974 Yearbook Medical Publishers, Inc.; **15.18:** Courtesy of Warren E. Collins, Inc., Braintree, MA; **15.21a, b:** © Martin Rotker/Peter Arnold, Inc.; **15.38a–c:** McCurdy, P. R., Sickle-Cell Disease. © MedCom, Inc. 1973.

### Chapter 16

**Figure 16.3:** CNRI/Science Photo Library/Photo Researchers, Inc.; **16.8a:** Daniel Friend, from Bloom, William, and Fawcett, Don W. A. *Textbook of Histology* 10/e, W. B. Saunders Co.

### Chapter 17

**17.4b, 17.6:** Courtesy of Utah Valley Hospital, Department of Radiology; **17.7:** © Edwin Reschke; **17.11:** © Manfred Kage/Peter Arnold, Inc.; **17.14a:** © Keith Porter; **17.14b:** Alpers, D. H., and Seetharam, B. *New England Journal of Medicine* 296(1977)1047; **17.18:** Sodeman, W. A., and Watson, T. M. *Pathologic Physiology,* 6th ed. © W. B. Saunders Company, 1969; **17.19:** After Sobotta, from Bloom, W., and Fawcett, D. W.: *A Textbook of Histology,* 10th ed. © W. B. Saunders Company, 1975; **17.20a:** From Tissues and Organs: A TEXT ATLAS OF SCANNING ELECTRON MICROSCOPY by R. G. Kessel and R. Kardon. W. H. Freeman and Company © 1979; **17.26a:** © Carroll Weiss/RBP; **17.26b:** © Sheril D. Burton.

### Chapter 18

**18.16:** © Lester V. Bergman and Associates, Inc.; **18.19a–d:** *American Journal of Medicine* 20 (1956)133; **18.20a, b:** Bhasker, S. N., ed.: *Orban's Oral Histology and Embryology,* 9th ed. © C. V. Mosby Company, 1980; **18.21a, b:** Raisz, L. G., Dempster, D. W. et al, 1986. *J. Bone Miner Res* 1:15–21; **19.6:** © Stuart I. Fox.

## Chapter 19

**19.7a:** Computer graphic by Arthur J. Olson, Ph.D./© 1985 All rights reserved. Scripps Clinic and Research Foundation, LaJolla, CA 92037; **19.8a, b:** © S.E.V. Phillips, Biophysics, The University of Leeds, From *Science* 233(1986)749; **1918a:** Rosenthal, Alan S.: *New England Journal of Medicine,* 303(1980)1153; **19.22a, b:** © Andrejs Liepens, Department of Biology, Memorial University; St. John's Newfoundland; **19.23 (both):** © Noel R. Rose; **19.25 (both):** Custom Medical Stock Photo; **19.26b:** Catherin Ellis/ Science Photo Library/Photo Researchers, Inc.

## Chapter 20

**20.2a, b:** Wisniewski, L. P., and Hirschhorn, K.: *"A Guide to Human Chromosome Defects,"* 2nd ed., White Plains: The March of Dimes Birth Defects Foundation, BD: OAS 16(6), 1980; **20.3a–d:** Williams, R. H.: *Textbook of Endocrinology,* 6th ed. © W. B. Saunders Company, 1981; **20.5a, b:** From Redding, A., and Hirschhorn, K.: *"A Guide to Human Chromosome Defects."* D. Bergsma (ed). New York: The National Foundation–March of Dimes, BD:OAS IV (4), 1968; **20.27a:** © Edwin Reschke; **20.28a:** Blandau, R. J.: *A Textbook of Histology,* 10th ed. © W. B. Saunders Company, 1975; **20.28b:** © Landrum Shettles; **20.29:** Bloom, W., and Fawcett, D. W. *A Textbook of Histology,* 10th ed. © W. B. Saunders Company, 1975; **20.30:** © Landrum Shettles; **20.35:** © Martin Rotker/Taurus Photos; **20.38c:** © David Phillips/Visuals Unlimited; **20.39:** Lucian Zamboni, from Greep, Roy, and Weiss, Leon: *Histology,* 3rd ed. © McGraw-Hill Book Company, 1973; **20.43a–d:** R. G. Edwards; **20.44b:** © Landrum Shettles.

# Line Art

## Chapter 1

**Figures 1.4b and 1.9:** From Kent M. Van De Graaff, *Human Anatomy,* 2d ed. Copyright © 1988 Wm. C. Brown Publishers, Dubuque, Iowa. All Rights Reserved. Reprinted by permission. **Figure 1.14:** From John W. Hole, Jr., *Human Anatomy and Physiology,* 4th ed. Copyright © 1987 Wm. C. Brown Publishers, Dubuque, Iowa. All Rights Reserved. Reprinted by permission.

## Chapter 2

**Figure 2.5:** From John W. Hole, Jr., *Human Anatomy and Physiology,* 4th ed. Copyright © 1987 Wm. C. Brown Publishers, Dubuque, Iowa. All Rights Reserved. Reprinted by permission.

## Chapter 3

**Figure 3.9b:** From Leland G. Johnson, *Biology,* 2d ed. Copyright © 1987 Wm. C. Brown Publishers, Dubuque, Iowa. All Rights Reserved. Reprinted by permission. **Figure 3.29b:** From John W. Hole, Jr., *Human Anatomy and Physiology,* 4th ed. Copyright © 1987 Wm. C. Brown Publishers, Dubuque, Iowa. All Rights Reserved. Reprinted by permission.

## Chapter 7

**Figure 7.1:** From John W. Hole, Jr., *Human Anatomy and Physiology,* 3d ed. Copyright © 1984 Wm. C. Brown Publishers, Dubuque, Iowa. All Rights Reserved. Reprinted by permission. **Figure 7.3:** From Kent M. Van De Graaff, *Human Anatomy,* 2d ed. Copyright © 1988 Wm. C. Brown Publishers, Dubuque, Iowa. All Rights Reserved. Reprinted by permission. **Figure 7.8:** From John W. Hole, Jr., *Human Anatomy and Physiology,* 4th ed. Copyright © 1987 Wm. C. Brown Publishers, Dubuque, Iowa. All Rights Reserved. Reprinted by permission. **Figure 7.14:** From H. C. Kutchai, "Generation and Conduction of Action Potentials" in *Physiology,* 1983, edited by R. M. Bern and M. N. Levy. Copyright © 1983 The C.V. Mosby Company, St. Louis, MO. **Figure 7.18c:** From Leland G. Johnson, *Biology,* 2d ed. Copyright © 1987 Wm. C. Brown Publishers, Dubuque, Iowa. All Rights Reserved. Reprinted by permission.

## Chapter 8

**Figures 8.2, 8.3, 8.5b, 8.8b, 8.16:** From Kent M. Van De Graaff, *Human Anatomy,* 2d ed. Copyright © 1988 Wm. C. Brown Publishers, Dubuque, Iowa. All Rights Reserved. Reprinted by permission. **Figures 8.4, 8.6, and 8.17:** From John W. Hole, Jr., *Human Anatomy and Physiology,* 4th ed. Copyright © 1987 Wm. C. Brown Publishers, Dubuque, Iowa. All Rights Reserved. Reprinted by permission. **Figures 8.5a and 8.13a:** From Kent M. Van De Graaff, *Human Anatomy Laboratory Textbook,* 2d ed. Copyright © 1984 Wm. C. Brown Publishers, Dubuque, Iowa. All Rights Reserved. Reprinted by permission. **Figure 8.7:** From T. L. Peele, *Neuroanatomical Basis for Clinical Neurology,* 2d ed. Copyright © 1961 McGraw-Hill Publishing Company, New York, NY. **Figure 8.14:** From Charles R. Noback and Robert J. Demarest, *The Human Nervous System: Basic Principles of Neurobiology,* 2d ed. Copyright © 1975 McGraw-Hill Publishing Company, New York, NY. **Figures 8.18, 8.19, and 8.22:** From John W. Hole, Jr., *Human Anatomy and Physiology,* 3d ed. Copyright © 1984 Wm. C. Brown Publishers, Dubuque, Iowa. All Rights Reserved. Reprinted by permission.

## Chapter 9

**Figures 9.10, 9.11, 9.13, 9.16, 9.17a, 9.19, 9.42, and 9.46:** From John W. Hole, Jr., *Human Anatomy and Physiology,* 4th ed. Copyright © 1987 Wm. C. Brown Publishers, Dubuque, Iowa. All Rights Reserved. Reprinted by permission. **Figure 9.17b:** From John W. Hole, Jr., *Human Anatomy and Physiology,* 2d ed. Copyright © 1981 Wm. C. Brown Publishers, Dubuque, Iowa. All Rights Reserved. Reprinted by permission. **Figure 9.27a:** From J. R. McClintic, *Physiology of the Human Body,* 2d ed. Copyright © 1978 John Wiley & Sons, Inc., New York, NY. **Figures 9.31, 9.41, and 9.47:** From Kent M. Van De Graaff, *Human Anatomy,* 2d ed. Copyright © 1988 Wm. C. Brown Publishers, Dubuque, Iowa. All Rights Reserved. Reprinted by permission.

## Chapter 10

**Figure 10.3:** From Kent M. Van De Graaff, *Human Anatomy,* 2d ed. Copyright © 1988 Wm. C. Brown Publishers, Dubuque, Iowa. All Rights Reserved. Reprinted by permission. **Figure 10.6:** From Kent M. Van De Graaff, *Human Anatomy.* Copyright © 1984 Wm. C. Brown Publishers, Dubuque, Iowa. All Rights Reserved. Reprinted by permission.

## Chapter 11

**Figures 11.1a and 11.8:** From Kent M. Van De Graaff, *Human Anatomy,* 2d ed. Copyright © 1988 Wm. C. Brown Publishers, Dubuque, Iowa. All Rights Reserved. Reprinted by permission. **Figures 11.14, 11.23, and 11.27:** From John W. Hole, Jr., *Human Anatomy and Physiology,* 4th ed. Copyright © 1987 Wm. C. Brown Publishers, Dubuque, Iowa. All Rights Reserved. Reprinted by permission. **Figure 11.24:** From John W. Hole, Jr., *Essentials of Human Anatomy and Physiology,* 3d ed. Copyright © 1989 Wm. C. Brown Publishers, Dubuque, Iowa. All Rights Reserved. Reprinted by permission. **Figure 11.28:** Reprinted with permission of Medical Economics Co., Inc., Oradell, NJ.

## Chapter 12

**Figure 12.2:** From John W. Hole, Jr., *Human Anatomy and Physiology,* 4th ed. Copyright © 1987 Wm. C. Brown Publishers, Dubuque, Iowa. All Rights Reserved. Reprinted by permission. **Figure 12.15:** From Eugene Morkin, "Contractile Proteins of the Heart." *Hospital Practice* Volume 18, Issue 6. Copyright © June 1983, New York: H.P. Publishing Co., Inc.

## Chapter 13

**Figures 13.2, 13.9, 13.18, and 13.35:** From John W. Hole, Jr., *Human Anatomy and Physiology,* 4th ed. Copyright © 1987 Wm. C. Brown Publishers, Dubuque, Iowa. All Rights Reserved. Reprinted by permission. **Figure 13.7:** Reprinted with permission of Medical Economics Co., Inc., Oradell, NJ. **Figures 13.10a, 13.13, 13.26, and 13.34:** From Kent M. Van De Graaff, *Human Anatomy,* 2d ed. Copyright © 1988 Wm. C. Brown Publishers, Dubuque, Iowa. All Rights Reserved. Reprinted by permission. **Figure 13.22:** From EMERGENCY MEDICINE, Vol. II, Issue 2, February 15, 1979. Reprinted by permission.

## Chapter 14

**Figure 14.16:** Reproduced by permission of *Practical Cardiology.* **Figure 14.17:** From P. Astrand and K. Rodahl, *Textbook of Work Physiology.* Copyright © 1977 McGraw-Hill Book Company, New York, NY. Reprinted by permission. **Figure 14.22:** From E. O. Feigal, "Physics in the Cardiovascular System" in *Physiology and Biophysics,* Volume II, edited by T. C. Ruch and H. D. Pattor. Copyright © 1974 W.B. Saunders, Philadelphia, PA. Reprinted by permission. **Figure 14.29:** From Leslie A. Geddes, M.E., Ph.D., F.A.C.C., *The Direct and Indirect Measurement of Blood Pressure.* Copyright © 1970 Year Book Medical Publishers, Chicago, IL. Reproduced with permission.

## Chapter 15

**Figures 15.5a, 15.8, 15.16, 15.19, 15.40, and 15.41:** From John W. Hole, Jr., *Human Anatomy and Physiology,* 4th ed. Copyright © 1987 Wm. C. Brown Publishers, Dubuque, Iowa. All Rights Reserved. Reprinted by permission. **Figure 15.4:** From E. R. Weibel, *Morphometry of the Human Lung.* Copyright © 1963 Springer-Verlag, Heidelberg, West Germany. Reprinted by permission. **Figures 15.17 and 15.29:** From Kent M. Van De Graaff, *Human Anatomy,* 2d ed. Copyright © 1988 Wm. C. Brown Publishers,

Dubuque, Iowa. All Rights Reserved. Reprinted by permission. **Figure 15.32:** From R. D. Dripps and J. H. Comroe, *American Journal of Physiology,* 149:43 (1947). **Figure 15.43:** From K. Wasserman et al., *Journal of Applied Physiology,* 22: 71–85, 1967. Copyright © 1967 American Physiological Society, Bethesda, MD. Reprinted by permission.

## Chapter 16

**Figures 16.2, 16.6, and 16.7b:** From John W. Hole, Jr., *Human Anatomy and Physiology,* 4th ed. Copyright © 1987 Wm. C. Brown Publishers, Dubuque, Iowa. All Rights Reserved. Reprinted by permission. **Figure 16.28:** From Neil A. Kurtzman, "Renal tubular acidosis: A constellation of syndromes." *Hospital Practice* Copyright © November, 1987, New York: H.P. Publishing Co., Inc.

## Chapter 17

**Figures 17.1, 17.4a, and 17.8:** From John W. Hole, Jr., *Human Anatomy and Physiology,* 4th ed. Copyright © 1987 Wm. C. Brown Publishers, Dubuque, Iowa. All Rights Reserved. Reprinted by permission. **Figures 17.2, 17.5, and 17.10:** From Kent M. Van De Graaff, *Human Anatomy,* 2d ed. Copyright © 1988 Wm. C. Brown Publishers, Dubuque, Iowa. All Rights Reserved. Reprinted by permission.

## Chapter 18

**Figures 18.5 and 18.6:** Mary L. Parker, Rosita S. Pildes, Kuen-Lan Chao, Marvin Cornblath, David M. Kipnis: "Junvenile Diabetes Mellitus, A Deficiency in Insulin." DIABETES 1968; 17:27–32. Reproduced with permission from the American Diabetes Association Inc. **Figure 18.11:** From E. J. Barrett and R. A. De Fronao, "Diabetic Ketoacidosis: Diagnosis and Management" in *Hospital Practice,* 19(4):91 April 1984. New York: H.P. Publishing Co., Inc. Reprinted by permission. Illustration by Mr. Albert Miller.

## Chapter 19

**Figure 19.13:** From Ivan Roitt, *Essential Immunology.* Copyright © 1974 Blackwell Scientific Publications, Ltd., Oxford, England.

## Chapter 20

**Figures 20.7, 20.12, 20.26, 20.32, 20.38a&b, and 20.44a:** From Kent M. Van De Graaff, *Human Anatomy,* 2d ed. Copyright © 1988 Wm. C. Brown Publishers, Dubuque, Iowa. All Rights Reserved. Reprinted by permission. **Figure 20.18:** From D. W. Fawcett, "Endocrinology: Male Reproductive System" in *Handbook of Physiology,* Section 7, Chapter 2, page 43. Copyright © 1975 The American Physiological Society, Bethesda MD. Reprinted by permission of the publisher and author. **Figures 20.20a, 20.21, 20.25, 20.42, and 20.45:** From John W. Hole, Jr., *Human Anatomy and Physiology,* 4th ed. Copyright © 1987 Wm. C. Brown Publishers, Dubuque, Iowa. All Rights Reserved. Reprinted by permission. **Figure 20.33:** From Ian H. Thorneycroft, et al., *American Journal of Obstetrics and Gynecology,* Vol. III, page 947, 1971. Copyright © 1971 The C.V. Mosby Company, St. Louis, MO. Reprinted by permission of the publisher and author.

# INDEX

## D

## E

## F



## G



## H

## I

## N



## O

## T

## U



## V



## W



## Z